v	Specific volume, m^3/kg
v_{cr}	Critical specific volume, m^3/kg
v_r	Relative specific volume
v_R	Pseudoreduced specific volume
V	Total volume, m^3
$\dot{V}$	Volume flow rate, m^3/s
V	Voltage, V
V	Velocity, m/s
V_{avg}	Average velocity
w	Work per unit mass, kJ/kg
W	Total work, kJ
$\dot{W}$	Power, kW
W_{in}	Work input, kJ
W_{out}	Work output, kJ
W_{rev}	Reversible work, kJ
x	Quality
x	Specific exergy, kJ/kg
X	Total exergy, kJ
x_{dest}	Specific exergy destruction, kJ/kg
X_{dest}	Total exergy destruction, kJ
$\dot{X}_{dest}$	Rate of total exergy destruction, kW
y	Mole fraction
z	Elevation, m
Z	Compressibility factor
Z_h	Enthalpy departure factor
Z_s	Entropy departure factor

Greek Letters

α	Absorptivity
α	Isothermal compressibility, 1/kPa
β	Volume expansivity, 1/K
Δ	Finite change in quantity
ε	Emissivity; effectiveness
η_{th}	Thermal efficiency
η_{II}	Second-law efficiency
θ	Total energy of a flowing fluid, kJ/kg
μ_{JT}	Joule-Thomson coefficient, K/kPa
μ	Chemical potential, kJ/kg
ν	Stoichiometric coefficient
ρ	Density, kg/m^3
σ	Stefan–Boltzmann constant
σ_n	Normal stress, N/m^2
σ_s	Surface tension, N/m
ϕ	Relative humidity
ϕ	Specific closed system exergy, kJ/kg
Φ	Total closed system exergy, kJ
ψ	Stream exergy, kJ/kg
ω	Specific or absolute humidity, kg H_2O/kg dry air

Subscripts

a	Air
abs	Absolute
act	Actual
atm	Atmospheric
avg	Average
c	Combustion; cross-section
cr	Critical point
CV	Control volume
e	Exit conditions
f	Saturated liquid
fg	Difference in property between saturated liquid and saturated vapor
g	Saturated vapor
gen	Generation
H	High temperature (as in T_H and Q_H)
i	Inlet conditions
i	ith component
L	Low temperature (as in T_L and Q_L)
m	Mixture
r	Relative
R	Reduced
rev	Reversible
s	Isentropic
sat	Saturated
surr	Surroundings
sys	System
v	Water vapor
0	Dead state
1	Initial or inlet state
2	Final or exit state

Superscripts

˙ (over dot)	Quantity per unit time
¯ (over bar)	Quantity per unit mole
° (circle)	Standard reference state
* (asterisk)	Quantity at 1 atm pressure

Selected Materials from

Thermodynamics
An Engineering Approach

Seventh Edition

Yunus A. Çengel
University of Nevada, Reno

Michael A. Boles
North Carolina State University

Texas A&M University

Learning Solutions

Boston Burr Ridge, IL Dubuque, IA New York San Francisco
St. Louis Bangkok Bogotá Caracas Lisbon London Madrid
Mexico City Milan New Delhi Seoul Singapore Sydney Taipei Toronto

The McGraw-Hill Companies

Selected Materials from
Thermodynamics: An Engineering Approach, Seventh Edition
Texas A&M University

This book is a McGraw-Hill Learning Solutions textbook and contains select materials from *Thermodynamics: An Engineering Approach*, Seventh Edition by Yunus A. Çengel and Michael A. Boles. Copyright © 2011, 2008, 2006, 2002 by The McGraw-Hill Companies, Inc. Reprinted with permission of the publisher. Many custom published texts are modified versions or adaptations of our best-selling textbooks. Some adaptations are printed in black and white to keep prices at a minimum, while others are in color.

5 6 7 8 9 0 DIG DIG 12

ISBN-13: 978-0-07-746384-7
ISBN-10: 0-07-746384-6

Learning Solutions Representative: Katherine Kilburg
Production Editor: Tina Hermsen
Printer/Binder: Digital Impressions

The mind is like a parachute—it works only when it is open.
Unknown

Nature's laws are the invisible government of the earth.
Alfred A. Montapert

The true measure of a man is how he treats someone who can do him absolutely no good.
Samuel Johnson

Greatness lies not in being strong, but in the right use of strength.
Henry W. Beecher

The superior man is modest in his speech, but exceeds in his actions.
Confucius

Try not to be a man of success, but rather to be a man of value.
Albert Einstein

To ignore evil is to become an accomplice to it.
Martin Luther King, Jr

Character, in the long run, is the decisive factor in the life of an individual and of nations alike.
Theodore Roosevelt

A person who sees the good in things has good thoughts. And he who has good thoughts receives pleasure from life.
Said Nursi

To different minds, the same world is a hell, and a heaven.
Ralph W. Emerson

A leader is one who sees more than others see, who sees farther than others see, and who sees before others see.
Leroy Eims

Never mistake knowledge for wisdom. One helps you make a living, the other helps you make a life.
Sandra Carey

As one person I cannot change the world, but I can change the world of one person.
Paul S. Spear

About the Authors

Yunus A. Çengel is Professor Emeritus of Mechanical Engineering at the University of Nevada, Reno. He received his B.S. in mechanical engineering from Istanbul Technical University and his M.S. and Ph.D. in mechanical engineering from North Carolina State University. His areas of interest are renewable energy, energy efficiency, energy policies, heat transfer enhancement, and engineering education. He served as the director of the Industrial Assessment Center (IAC) at the University of Nevada, Reno, from 1996 to 2000. He has led teams of engineering students to numerous manufacturing facilities in Northern Nevada and California to perform industrial assessments, and has prepared energy conservation, waste minimization, and productivity enhancement reports for them. He has also served as an advisor for various government organizations and corporations.

Dr. Çengel is also the author or coauthor of the widely adopted textbooks *Fundamentals of Thermal-Fluid Sciences* (3rd ed., 2008), *Heat and Mass Transfer: Fundamentals and Applications* (4th ed., 2011), *Introduction to Thermodynamics and Heat Transfer* (2nd ed., 2008), *Fluid Mechanics: Fundamentals and Applications* (2nd ed., 2010), and *Essentials of Fluid Mechanics: Fundamentals and Applications* (1st ed., 2008), all published by McGraw-Hill. Some of his textbooks have been translated into Chinese, Japanese, Korean, Thai, Spanish, Portuguese, Turkish, Italian, Greek, and French.

Dr. Çengel is the recipient of several outstanding teacher awards, and he has received the ASEE Meriam/Wiley Distinguished Author Award for excellence in authorship in 1992 and again in 2000. Dr. Çengel is a registered Professional Engineer in the State of Nevada, and is a member of the American Society of Mechanical Engineers (ASME) and the American Society for Engineering Education (ASEE).

Michael A. Boles is Associate Professor of Mechanical and Aerospace Engineering at North Carolina State University, where he earned his Ph.D. in mechanical engineering and is an Alumni Distinguished Professor. Dr. Boles has received numerous awards and citations for excellence as an engineering educator. He is a past recipient of the SAE Ralph R. Teetor Education Award and has been twice elected to the NCSU Academy of Outstanding Teachers. The NCSU ASME student section has consistently recognized him as the outstanding teacher of the year and the faculty member having the most impact on mechanical engineering students.

Dr. Boles specializes in heat transfer and has been involved in the analytical and numerical solution of phase change and drying of porous media. He is a member of the American Society of Mechanical Engineers (ASME), the American Society for Engineering Education (ASEE), and Sigma Xi. Dr. Boles received the ASEE Meriam/Wiley Distinguished Author Award in 1992 for excellence in authorship.

Brief Contents

CONTENTS

CHAPTER THREE

PROPERTIES OF PURE SUBSTANCES 111

CHAPTER FOUR

ENERGY ANALYSIS OF CLOSED SYSTEMS 163

CHAPTER FIVE

MASS AND ENERGY ANALYSIS OF CONTROL VOLUMES 215

CHAPTER SIX

THE SECOND LAW OF THERMODYNAMICS 277

CHAPTER SEVEN

ENTROPY 331

CHAPTER EIGHT

EXERGY: A MEASURE OF WORK POTENTIAL 423

CHAPTER NINE

GAS POWER CYCLES 487

CHAPTER TEN

VAPOR AND COMBINED POWER CYCLES 555

CHAPTER ELEVEN
REFRIGERATION CYCLES 611

APPENDIX 1
PROPERTY TABLES AND CHARTS (SI UNITS) 907

APPENDIX 2

PROPERTY TABLES AND CHARTS (ENGLISH UNITS) 957

PREFACE

BACKGROUND

Thermodynamics is an exciting and fascinating subject that deals with energy, which is essential for sustenance of life, and thermodynamics has long been an essential part of engineering curricula all over the world. It has a broad application area ranging from microscopic organisms to common household appliances, transportation vehicles, power generation systems, and even philosophy. This introductory book contains sufficient material for two sequential courses in thermodynamics. Students are assumed to have an adequate background in calculus and physics.

OBJECTIVES

This book is intended for use as a textbook by undergraduate engineering students in their sophomore or junior year, and as a reference book for practicing engineers. The objectives of this text are

- To cover the *basic principles* of thermodynamics.
- To present a wealth of real-world *engineering examples* to give students a feel for how thermodynamics is applied in engineering practice.
- To develop an *intuitive understanding* of thermodynamics by emphasizing the physics and physical arguments.

It is our hope that this book, through its careful explanations of concepts and its use of numerous practical examples and figures, helps students develop the necessary skills to bridge the gap between knowledge and the confidence to properly apply knowledge.

PHILOSOPHY AND GOAL

The philosophy that contributed to the overwhelming popularity of the prior editions of this book has remained unchanged in this edition. Namely, our goal has been to offer an engineering textbook that

- Communicates directly to the minds of tomorrow's engineers in a *simple yet precise* manner.
- Leads students toward a clear understanding and firm grasp of the *basic principles* of thermodynamics.
- Encourages *creative thinking* and development of a *deeper understanding* and *intuitive feel* for thermodynamics.
- Is *read* by students with *interest* and *enthusiasm* rather than being used as an aid to solve problems.

Special effort has been made to appeal to students' natural curiosity and to help them explore the various facets of the exciting subject area of thermodynamics. The enthusiastic responses we have received from users of prior editions—from small colleges to large universities all over the world—and the continued translations into new languages indicate that our objectives have largely been achieved. It is our philosophy that the best way to learn is by practice. Therefore, special effort is made throughout the book to reinforce material that was presented earlier.

Yesterday's engineer spent a major portion of his or her time substituting values into the formulas and obtaining numerical results. However, formula manipulations and number crunching are now being left mainly to computers. Tomorrow's engineer will need a clear understanding and a firm grasp of the *basic principles* so that he or she can understand even the most complex problems, formulate them, and interpret the results. A conscious effort is made to emphasize these basic principles while also providing students with a perspective of how computational tools are used in engineering practice.

The traditional *classical,* or *macroscopic,* approach is used throughout the text, with microscopic arguments serving in a supporting role as appropriate. This approach is more in line with students' intuition and makes learning the subject matter much easier.

NEW IN THIS EDITION

The primary change in this seventh edition of the text is the upgrade of a large number of line artwork to realistic three-dimensional figures and the incorporation of about 400 new problems. All the popular features of the previous editions are retained, and the main body of all chapters and all the tables and charts in the Appendices remain mostly unchanged. Each chapter now contains at least one new solved example problem, and a significant part of existing problems are modified. In Chapter 1, the section on Dimensions and Units is updated, and a new subsection is added to Chapter 6 on the Performance of Refrigerators, Air-Conditioners, and Heat Pumps. In Chapter 8, the material on the second-law efficiency is updated, and some second-law efficiency definitions are revised for consistency. Also, the discussions in the section Second-Law Aspects of Daily Life have been extended. Chapter 11 now has a new section titled Second-Law Analysis of Vapor-Compression Refrigeration Cycle.

OVER 400 NEW PROBLEMS

This edition includes over 400 new problems with a variety of applications. Problems whose solutions require parametric investigations, and thus the use of a computer, are identified by a computer-EES icon, as before. Some existing problems from previous editions have been removed from the text.

LEARNING TOOLS

EARLY INTRODUCTION OF THE FIRST LAW OF THERMODYNAMICS

The first law of thermodynamics is introduced early in Chapter 2, "Energy, Energy Transfer, and General Energy Analysis." This introductory chapter sets the framework of establishing a general understanding of various forms of energy, mechanisms of energy transfer, the concept of energy balance, thermo-economics, energy conversion, and conversion efficiency using familiar settings that involve mostly electrical and mechanical forms of energy. It also exposes students to some exciting real-world applications of thermodynamics early in the course, and helps them establish a sense of the monetary value of energy. There is special emphasis on the utilization of renewable energy such as wind power and hydroulic energy, and the efficient use of existing resources.

EMPHASIS ON PHYSICS

A distinctive feature of this book is its emphasis on the physical aspects of the subject matter in addition to mathematical representations and manipulations. The authors believe that the emphasis in undergraduate education should remain on *developing a sense of underlying physical mechanisms* and a *mastery of solving practical problems* that an engineer is likely to face in the real world. Developing an intuitive understanding should also make the course a more motivating and worthwhile experience for students.

EFFECTIVE USE OF ASSOCIATION

An observant mind should have no difficulty understanding engineering sciences. After all, the principles of engineering sciences are based on our *everyday experiences* and *experimental observations.* Therefore, a physical, intuitive approach is used throughout this text. Frequently, *parallels are drawn* between the subject matter and students' everyday experiences so that they can relate the subject matter to what they already know. The process of cooking, for example, serves as an excellent vehicle to demonstrate the basic principles of thermodynamics.

SELF-INSTRUCTING

The material in the text is introduced at a level that an average student can follow comfortably. It speaks *to* students, not *over* students. In fact, it is *self-instructive.* The order of coverage is from *simple* to *general.* That is, it starts with the simplest case and adds complexities gradually. In this way, the basic principles are repeatedly applied to different systems, and students master how to apply the principles instead of how to simplify a general formula. Noting that the principles of sciences are based on experimental observations, all the derivations in this text are based on physical arguments, and thus they are easy to follow and understand.

EXTENSIVE USE OF ARTWORK

Figures are important learning tools that help students "get the picture," and the text makes very effective use of graphics. This edition of *Thermodynamics: An Engineering Approach,* Seventh Edition contains more figures and illustrations than any other book in this category. Further, a large number of figures have been upgraded to become three-dimensional and thus more real-life. Figures attract attention and stimulate curiosity and interest. Most of the figures in this text are intended to serve as a means of emphasizing some key concepts that would otherwise go unnoticed; some serve as page summaries. The popular cartoon feature "Blondie" is used to make some important points in a humorous way and also to break the ice and ease the nerves. Who says studying thermodynamics can't be fun?

LEARNING OBJECTIVES AND SUMMARIES

Each chapter begins with an *overview* of the material to be covered and chapter-specific *learning objectives.* A *summary* is included at the end of each chapter, providing a quick review of basic concepts and important relations, and pointing out the relevance of the material.

NUMEROUS WORKED-OUT EXAMPLES WITH A SYSTEMATIC SOLUTIONS PROCEDURE

Each chapter contains several worked-out *examples* that clarify the material and illustrate the use of the basic principles. An *intuitive* and *systematic* approach is used in the solution of the example problems, while maintaining an informal conversational style. The problem is first stated, and the objectives are identified. The assumptions are then stated, together with their justifications. The properties needed to solve the problem are listed separately if appropriate. Numerical values are used together with their units to emphasize that numbers without units are meaningless, and that unit manipulations are as important as manipulating the numerical values with a calculator. The significance of the findings is discussed following the solutions. This approach is also used consistently in the solutions presented in the instructor's solutions manual.

A WEALTH OF REAL-WORLD END-OF-CHAPTER PROBLEMS

The end-of-chapter problems are grouped under specific topics to make problem selection easier for both instructors and students. Within each group of problems are *Concept Questions,* indicated by "C," to check the students' level of understanding of basic concepts. The problems under *Review Problems* are more comprehensive in nature and are not directly tied to any specific section of a chapter—in some cases they require review of material learned in previous chapters. Problems designated as *Design and Essay* are intended to encourage students to make engineering judgments, to conduct independent exploration of topics of interest, and to communicate their findings in a professional manner. Problems designated by an "E" are in English units, and SI users can ignore them. Problems with the are solved using EES, and complete solutions together with parametric studies are included on the enclosed DVD. Problems with the are comprehensive in nature and are intended to be solved with a computer, preferably using the EES software that accompanies this text. Several economics- and safety-related problems are incorporated throughout to enhance cost and safety awareness among engineering students. Answers to selected problems are listed immediately following the problem for convenience to students. In addition, to prepare students for the Fundamentals of Engineering Exam (that is becoming more important for the outcome-based ABET 2000 criteria) and to facilitate multiple-choice tests, over 200 *multiple-choice problems* are included in the end-of-chapter problem sets. They are placed under the title *Fundamentals of Engineering (FE) Exam Problems* for easy recognition. These problems are intended to check the understanding of fundamentals and to help readers avoid common pitfalls.

RELAXED SIGN CONVENTION

The use of a formal sign convention for heat and work is abandoned as it often becomes counterproductive. A physically meaningful and engaging approach is adopted for interactions instead of a mechanical approach. Subscripts "in" and "out," rather than the plus and minus signs, are used to indicate the directions of interactions.

PHYSICALLY MEANINGFUL FORMULAS

The physically meaningful forms of the balance equations rather than formulas are used to foster deeper understanding and to avoid a cookbook approach.

The mass, energy, entropy, and exergy balances for *any system* undergoing *any process* are expressed as

Mass balance:

$$m_{\text{in}} - m_{\text{out}} = \Delta m_{\text{system}}$$

Energy balance:

$$\underbrace{E_{\text{in}} - E_{\text{out}}}_{\substack{\text{Net energy transfer}\\ \text{by heat, work, and mass}}} = \underbrace{\Delta E_{\text{system}}}_{\substack{\text{Change in internal, kinetic,}\\ \text{potential, etc., energies}}}$$

Entropy balance:

$$\underbrace{S_{\text{in}} - S_{\text{out}}}_{\substack{\text{Net entropy transfer}\\ \text{by heat and mass}}} + \underbrace{S_{\text{gen}}}_{\substack{\text{Entropy}\\ \text{generation}}} = \underbrace{\Delta S_{\text{system}}}_{\substack{\text{Change}\\ \text{in entropy}}}$$

Exergy balance:

$$\underbrace{X_{\text{in}} - X_{\text{out}}}_{\substack{\text{Net exergy transfer}\\ \text{by heat, work, and mass}}} - \underbrace{X_{\text{destroyed}}}_{\substack{\text{Exergy}\\ \text{destruction}}} = \underbrace{\Delta X_{\text{system}}}_{\substack{\text{Change}\\ \text{in exergy}}}$$

These relations reinforce the fundamental principles that during an actual process mass and energy are conserved, entropy is generated, and exergy is destroyed. Students are encouraged to use these forms of balances in early chapters after they specify the system, and to simplify them for the particular problem. A more relaxed approach is used in later chapters as students gain mastery.

A CHOICE OF SI ALONE OR SI/ENGLISH UNITS

In recognition of the fact that English units are still widely used in some industries, both SI and English units are used in this text, with an emphasis on SI. The material in this text can be covered using combined SI/English units or SI units alone, depending on the preference of the instructor. The property tables and charts in the appendices are presented in both units, except the ones that involve dimensionless quantities. Problems, tables, and charts in English units are designated by "E" after the number for easy recognition, and they can be ignored by SI users.

TOPICS OF SPECIAL INTEREST

Most chapters contain a section called "Topic of Special Interest" where interesting aspects of thermodynamics are discussed. Examples include *Thermodynamic Aspects of Biological Systems* in Chapter 4, *Household Refrigerators* in Chapter 6, *Second-Law Aspects of Daily Life* in Chapter 8, and *Saving Fuel and Money by Driving Sensibly* in Chapter 9. The topics selected for these sections provide intriguing extensions to thermodynamics, but they can be ignored if desired without a loss in continuity.

GLOSSARY OF THERMODYNAMIC TERMS

Throughout the chapters, when an important key term or concept is introduced and defined, it appears in **boldface** type. Fundamental thermodynamic terms and concepts also appear in a glossary located on our accompanying website (www.mhhe.com/cengel). This unique glossary helps to reinforce key terminology and is an excellent learning and review tool for students as they move forward in their study of thermodynamics. In addition, students can test their knowledge of these fundamental terms by using the flash cards and other interactive resources.

CONVERSION FACTORS

Frequently used conversion factors and physical constants are listed on the inner cover pages of the text for easy reference.

SUPPLEMENTS

The following supplements are available to users of the book.

STUDENT RESOURCE DVD

Engineering Equation Solver (EES)

Packaged free with every new text, the Student Resource DVD contains the Limited Academic Version of EES (Engineering Equation Solver) software with scripted solutions to selected text problems.

Developed by Sanford Klein and William Beckman from the University of Wisconsin—Madison, this software combines equation-solving capability and engineering property data. EES can do optimization, parametric analysis, and linear and nonlinear regression, and provides publication-quality plotting capabilities. Thermodynamics and transport properties for air, water, and many other fluids are built in, and EES allows the user to enter property data or functional relationships.

EES is a powerful equation solver with built-in functions and property tables for thermodynamic and transport properties as well as automatic unit checking capability. It requires less time than a calculator for data entry and allows more time for thinking critically about modeling and solving engineering problems. Look for the EES icons in the homework problems sections of the text.

PROPERTIES TABLE BOOKLET

(ISBN 0-07-735999-2)

This booklet provides students with an easy reference to the most important property tables and charts, many of which are found at the back of the textbook in both the SI and English units.

COSMOS

McGraw-Hill's COSMOS (Complete Online Solutions Manual Organization System) allows instructors to streamline the creation of assignments, quizzes, and tests by using problems and solutions from the textbook, as well as their own custom material. COSMOS is now available online at http://cosmos.mhhe.com/

HANDS-ON MECHANICS

Hands-on Mechanics is a website designed for instructors who are interested in incorporating three-dimensional, hands-on teaching aids into their lectures. Developed through a partnership between the McGraw-Hill Engineering Team and the Department of Civil and Mechanical Engineering at the United States Military Academy at West Point, this website not only provides detailed instructions for how to build 3-D teaching tools using materials found in any lab or local hardware store, but also provides a community where educators can share ideas, trade best practices, and submit their own original demonstrations for posting on the site. Visit www.handsonmechanics.com for more information.

ACKNOWLEDGMENTS

The authors would like to acknowledge with appreciation the numerous and valuable comments, suggestions, constructive criticisms, and praise from the following evaluators and reviewers:

Edward Anderson
Texas Tech University

John Biddle
Cal Poly Pomona University

Gianfranco DiGiuseppe
Kettering University

Shoeleh Di Julio
California State University-Northridge

Afshin Ghajar
Oklahoma State University

Harry Hardee
New Mexico State University

Kevin Lyons
North Carolina State University

Kevin Macfarlan
John Brown University

Saeed Manafzadeh
University of Illinois-Chicago

Alex Moutsoglou
South Dakota State University

Rishi Raj
The City College of New York

Maria Sanchez
California State University-Fresno

Kalyan Srinivasan
Mississippi State University

Robert Stiger
Gonzaga University

Their suggestions have greatly helped to improve the quality of this text. In particular we would like to express our gratitude to Mehmet Kanoglu of the University of Gaziantep, Turkey, for his valuable contributions, his critical review of the manuscript, and for his special attention to accuracy and detail.

We also would like to thank our students, who provided plenty of feedback from students' perspectives. Finally, we would like to express our appreciation to our wives, Zehra Çengel and Sylvia Boles, and to our children for their continued patience, understanding, and support throughout the preparation of this text.

Yunus A. Çengel
Michael A. Boles

Online Resources for the Student and Instructor

NEW TO THIS EDITION!

McGRAW-HILL CONNECT ENGINEERING

McGraw-Hill Connect Engineering is a web-based assignment and assessment platform that gives students the means to better connect with their coursework, with their instructors, and with the important concepts that they will need to know for success now and in the future. With Connect Engineering, instructors can deliver assignments, quizzes, and tests easily online. Students can practice important skills at their own pace and on their own schedule.

Connect Engineering for *Thermodynamics: An Engineering Approach,* Seventh Edition is available via the text website at www.mhhe.com/cengel

COSMOS

McGraw-Hill's COSMOS (Complete Online Solutions Manual Organization System) allows instructors to streamline the creation of assignments, quizzes, and tests by using problems and solutions from the textbook, as well as their own custom material. COSMOS is now available online at http://cosmos.mhhe.com/

WWW.MHHE.COM/CENGEL

This site offers resources for students and instructors.

The following resources are available for students:

- **Glossary of Key Terms in Thermodynamics**—Bolded terms in the text are defined in this accessible glossary. Organized at the chapter level or available as one large file.
- **Student Study Guide**—This resource outlines the fundamental concepts of the text and is a helpful guide that allows students to focus on the most important concepts. The guide can also serve as a lecture outline for instructors.
- **Learning Objectives**—The chapter learning objectives are outlined here. Organized by chapter and tied to ABET objectives.
- **Self-Quizzing**—Students can test their knowledge using multiple-choice quizzing. These self-tests provide immediate feedback and are an excellent learning tool.
- **Flashcards**—Interactive flashcards test student understanding of the text terms and their definitions. The program also allows students to flag terms that require further understanding.
- **Crossword Puzzles**—An interactive, timed puzzle that provides hints as well as a notes section.

- **Concentration**—An interactive matching game that enhances understanding of basic thermodynamic concepts.
- **Errata**—If errors should be found in the text, they will be reported here.

The following resources are available for instructors under password protection:

- **Instructor Testbank**—Additional problems prepared for instructors to assign to students. Solutions are given, and use of EES is recommended to verify accuracy.
- **Correlation Guide**—New users of this text will appreciate this resource. The guide provides a smooth transition for instructors not currently using the Çengel/Boles text.
- **Image Library**—The electronic version of the figures are supplied for easy integration into course presentations, exams, and assignments.
- **Instructor's Guide**—Provides instructors with helpful tools such as sample syllabi and exams, an ABET conversion guide, a thermodynamics glossary, and chapter objectives.
- **Errata**—If errors should be found in the solutions manual, they will be reported here.
- **Solutions Manual**—The detailed solutions to all text homework problems are provided in PDF form.
- **EES Solutions Manual**—The entire solutions manual is also available in EES. Any problem in the text can be modified and the solution of the modified problem can readily be obtained by copying and pasting the given EES solution on a blank EES screen and hitting the solve button.
- **PP slides**—Powerpoint presentation slides for all chapters in the text are available for use in lectures
- **Appendices**—These are provided in PDF form for ease of use.

CHAPTER

1

INTRODUCTION AND BASIC CONCEPTS

Every science has a unique vocabulary associated with it, and thermodynamics is no exception. Precise definition of basic concepts forms a sound foundation for the development of a science and prevents possible misunderstandings. We start this chapter with an overview of thermodynamics and the unit systems, and continue with a discussion of some basic concepts such as *system, state, state postulate, equilibrium, and process*. We also discuss *temperature* and *temperature scales* with particular emphasis on the International Temperature Scale of 1990. We then present *pressure,* which is the normal force exerted by a fluid per unit area and discuss *absolute* and *gage* pressures, the variation of pressure with depth, and pressure measurement devices, such as manometers and barometers. Careful study of these concepts is essential for a good understanding of the topics in the following chapters. Finally, we present an intuitive systematic *problem-solving technique* that can be used as a model in solving engineering problems.

Objectives

The objectives of Chapter 1 are to:

- Identify the unique vocabulary associated with thermodynamics through the precise definition of basic concepts to form a sound foundation for the development of the principles of thermodynamics.
- Review the metric SI and the English unit systems that will be used throughout the text.
- Explain the basic concepts of thermodynamics such as system, state, state postulate, equilibrium, process, and cycle.
- Review concepts of temperature, temperature scales, pressure, and absolute and gage pressure.
- Introduce an intuitive systematic problem-solving technique.

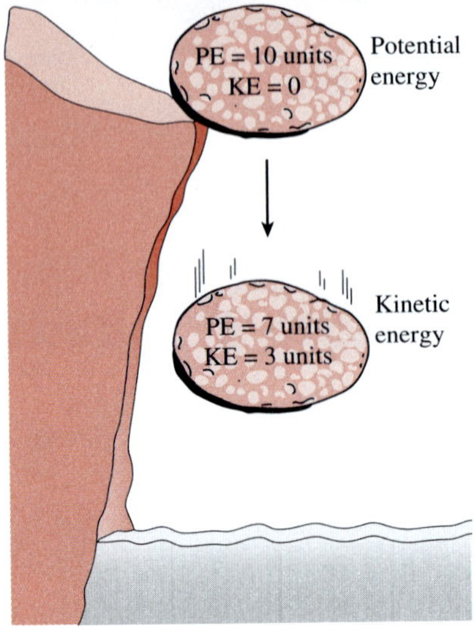

FIGURE 1–1
Energy cannot be created or destroyed; it can only change forms (the first law).

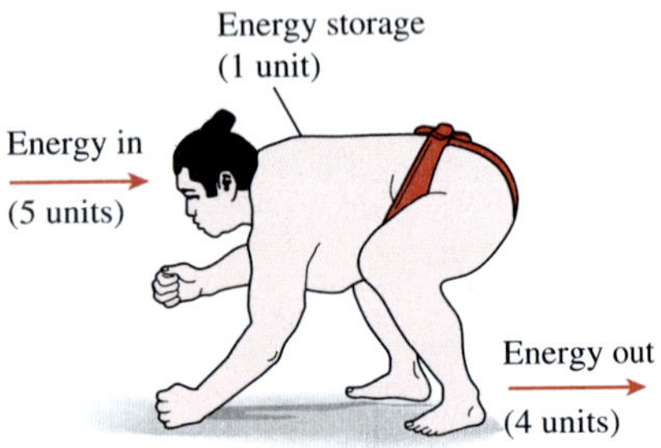

FIGURE 1–2
Conservation of energy principle for the human body.

FIGURE 1–3
Heat flows in the direction of decreasing temperature.

1–1 ■ THERMODYNAMICS AND ENERGY

Thermodynamics can be defined as the science of *energy.* Although everybody has a feeling of what energy is, it is difficult to give a precise definition for it. Energy can be viewed as the ability to cause changes.

The name *thermodynamics* stems from the Greek words *therme* (heat) and *dynamis* (power), which is most descriptive of the early efforts to convert heat into power. Today the same name is broadly interpreted to include all aspects of energy and energy transformations including power generation, refrigeration, and relationships among the properties of matter.

One of the most fundamental laws of nature is the **conservation of energy principle**. It simply states that during an interaction, energy can change from one form to another but the total amount of energy remains constant. That is, energy cannot be created or destroyed. A rock falling off a cliff, for example, picks up speed as a result of its potential energy being converted to kinetic energy (Fig. 1–1). The conservation of energy principle also forms the backbone of the diet industry: A person who has a greater energy input (food) than energy output (exercise) will gain weight (store energy in the form of fat), and a person who has a smaller energy input than output will lose weight (Fig. 1–2). The change in the energy content of a body or any other system is equal to the difference between the energy input and the energy output, and the energy balance is expressed as $E_{in} - E_{out} = \Delta E$.

The **first law of thermodynamics** is simply an expression of the conservation of energy principle, and it asserts that *energy* is a thermodynamic property. The **second law of thermodynamics** asserts that energy has *quality* as well as *quantity,* and actual processes occur in the direction of decreasing quality of energy. For example, a cup of hot coffee left on a table eventually cools, but a cup of cool coffee in the same room never gets hot by itself (Fig. 1–3). The high-temperature energy of the coffee is degraded (transformed into a less useful form at a lower temperature) once it is transferred to the surrounding air.

Although the principles of thermodynamics have been in existence since the creation of the universe, thermodynamics did not emerge as a science until the construction of the first successful atmospheric steam engines in England by Thomas Savery in 1697 and Thomas Newcomen in 1712. These engines were very slow and inefficient, but they opened the way for the development of a new science.

The first and second laws of thermodynamics emerged simultaneously in the 1850s, primarily out of the works of William Rankine, Rudolph Clausius, and Lord Kelvin (formerly William Thomson). The term *thermodynamics* was first used in a publication by Lord Kelvin in 1849. The first thermodynamic textbook was written in 1859 by William Rankine, a professor at the University of Glasgow.

It is well-known that a substance consists of a large number of particles called *molecules.* The properties of the substance naturally depend on the behavior of these particles. For example, the pressure of a gas in a container is the result of momentum transfer between the molecules and the walls of the container. However, one does not need to know the behavior of the gas particles to determine the pressure in the container. It would be sufficient to attach a pressure gage to the container. This macroscopic approach to the study of thermodynamics that does not require a knowledge of the behavior of individual particles is called **classical thermodynamics**. It provides a direct and easy way to the solution of engineering problems. A more elaborate approach, based on the average behavior of large groups of individual particles, is called **statistical thermodynamics**. This microscopic approach is rather involved and is used in this text only in the supporting role.

Application Areas of Thermodynamics

All activities in nature involve some interaction between energy and matter; thus, it is hard to imagine an area that does not relate to thermodynamics in some manner. Therefore, developing a good understanding of basic principles of thermodynamics has long been an essential part of engineering education.

Thermodynamics is commonly encountered in many engineering systems and other aspects of life, and one does not need to go very far to see some application areas of it. In fact, one does not need to go anywhere. The heart is constantly pumping blood to all parts of the human body, various energy conversions occur in trillions of body cells, and the body heat generated is constantly rejected to the environment. The human comfort is closely tied to the rate of this metabolic heat rejection. We try to control this heat transfer rate by adjusting our clothing to the environmental conditions.

Other applications of thermodynamics are right where one lives. An ordinary house is, in some respects, an exhibition hall filled with wonders of thermodynamics (Fig. 1–4). Many ordinary household utensils and appliances are designed, in whole or in part, by using the principles of thermodynamics. Some examples include the electric or gas range, the heating and air-conditioning systems, the refrigerator, the humidifier, the pressure cooker, the water heater, the shower, the iron, and even the computer and the TV. On a larger scale, thermodynamics plays a major part in the design and analysis of automotive engines, rockets, jet engines, and conventional or nuclear power plants, solar collectors, and the design of vehicles from ordinary cars to airplanes (Fig. 1–5). The energy-efficient home that you may be living in, for example, is designed on the basis of minimizing heat loss in winter and heat gain in summer. The size, location, and the power input of the fan of your computer is also selected after an analysis that involves thermodynamics.

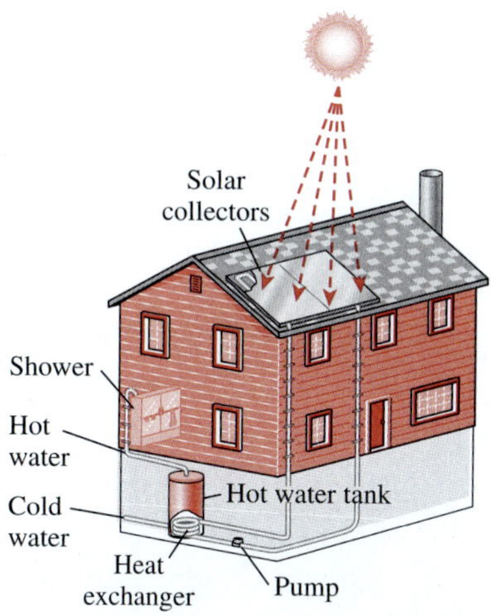

FIGURE 1–4
The design of many engineering systems, such as this solar hot water system, involves thermodynamics.

1–2 ▪ IMPORTANCE OF DIMENSIONS AND UNITS

Any physical quantity can be characterized by **dimensions**. The magnitudes assigned to the dimensions are called **units**. Some basic dimensions such as mass m, length L, time t, and temperature T are selected as **primary** or **fundamental dimensions**, while others such as velocity V, energy E, and volume V are expressed in terms of the primary dimensions and are called **secondary dimensions**, or **derived dimensions**.

A number of unit systems have been developed over the years. Despite strong efforts in the scientific and engineering community to unify the world with a single unit system, two sets of units are still in common use today: the **English system**, which is also known as the *United States Customary System*

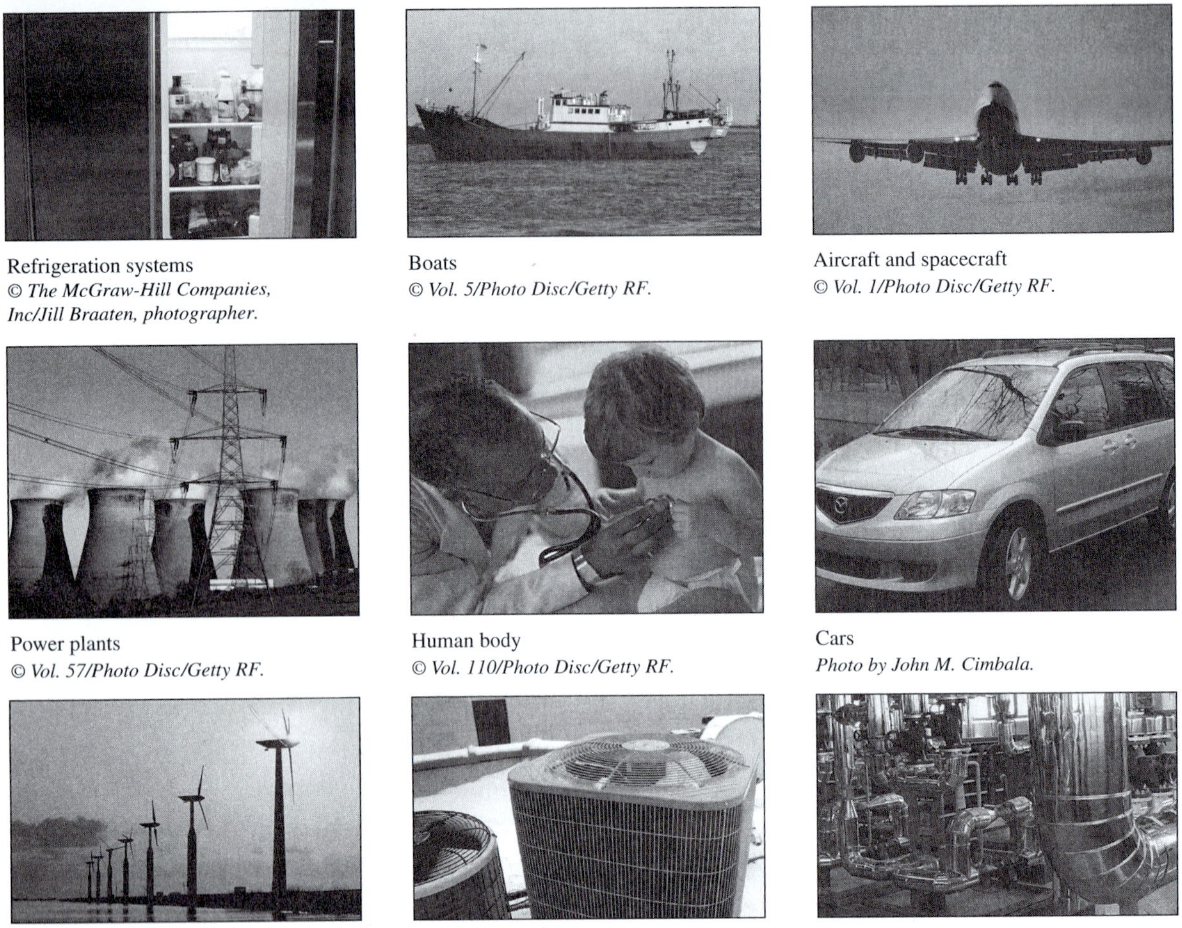

Refrigeration systems
© The McGraw-Hill Companies, Inc/Jill Braaten, photographer.

Boats
© Vol. 5/Photo Disc/Getty RF.

Aircraft and spacecraft
© Vol. 1/Photo Disc/Getty RF.

Power plants
© Vol. 57/Photo Disc/Getty RF.

Human body
© Vol. 110/Photo Disc/Getty RF.

Cars
Photo by John M. Cimbala.

Wind turbines
© Vol. 17/Photo Disc/Getty RF.

Air conditioning systems
© The McGraw-Hill Companies, Inc/Jill Braaten, photographer.

Industrial applications
Courtesy UMDE Engineering, Contracting, and Trading. Used by permission.

FIGURE 1–5
Some application areas of thermodynamics.

(USCS), and the metric **SI** (from *Le Système International d' Unités*), which is also known as the *International System.* The SI is a simple and logical system based on a decimal relationship between the various units, and it is being used for scientific and engineering work in most of the industrialized nations, including England. The English system, however, has no apparent systematic numerical base, and various units in this system are related to each other rather arbitrarily (12 in = 1 ft, 1 mile = 5280 ft, 4 qt = gal, etc.), which makes it confusing and difficult to learn. The United States is the only industrialized country that has not yet fully converted to the metric system.

The systematic efforts to develop a universally acceptable system of units dates back to 1790 when the French National Assembly charged the French Academy of Sciences to come up with such a unit system. An early version of the metric system was soon developed in France, but it did not find universal acceptance until 1875 when *The Metric Convention Treaty* was prepared and signed by 17 nations, including the United States. In this international treaty, meter and gram were established as the metric units for length and mass, respectively, and a *General Conference of Weights and Measures* (CGPM) was established that was to meet every six years. In 1960, the CGPM produced the SI, which was based on six fundamental quantities, and their units were adopted in 1954 at the Tenth General Conference of Weights and Measures: *meter* (m) for length, *kilogram* (kg) for mass, *second* (s) for time, *ampere* (A) for electric current, *degree Kelvin* (°K) for temperature, and *candela* (cd) for luminous intensity (amount of light). In 1971, the CGPM added a seventh fundamental quantity and unit: *mole* (mol) for the amount of matter.

Based on the notational scheme introduced in 1967, the degree symbol was officially dropped from the absolute temperature unit, and all unit names were to be written without capitalization even if they were derived from proper names (Table 1–1). However, the abbreviation of a unit was to be capitalized if the unit was derived from a proper name. For example, the SI unit of force, which is named after Sir Isaac Newton (1647–1723), is *newton* (not Newton), and it is abbreviated as N. Also, the full name of a unit may be pluralized, but its abbreviation cannot. For example, the length of an object can be 5 m or 5 meters, *not* 5 ms or 5 meter. Finally, no period is to be used in unit abbreviations unless they appear at the end of a sentence. For example, the proper abbreviation of meter is m (not m.).

The recent move toward the metric system in the United States seems to have started in 1968 when Congress, in response to what was happening in the rest of the world, passed a Metric Study Act. Congress continued to promote a voluntary switch to the metric system by passing the Metric Conversion Act in 1975. A trade bill passed by Congress in 1988 set a September 1992 deadline for all federal agencies to convert to the metric system. However, the deadlines were relaxed later with no clear plans for the future.

The industries that are heavily involved in international trade (such as the automotive, soft drink, and liquor industries) have been quick in converting to the metric system for economic reasons (having a single worldwide design, fewer sizes, smaller inventories, etc.). Today, nearly all the cars manufactured in the United States are metric. Most car owners probably do not realize this until they try an English socket wrench on a metric bolt. Most industries, however, resisted the change, thus slowing down the conversion process.

Presently the United States is a dual-system society, and it will stay that way until the transition to the metric system is completed. This puts an extra burden on today's engineering students, since they are expected to retain their understanding of the English system while learning, thinking, and working in terms of the SI. Given the position of the engineers in the transition period, both unit systems are used in this text, with particular emphasis on SI units.

As pointed out, the SI is based on a decimal relationship between units. The prefixes used to express the multiples of the various units are listed in Table 1–2. They are standard for all units, and the student is encouraged to memorize them because of their widespread use (Fig. 1–6).

TABLE 1–1

The seven fundamental (or primary) dimensions and their units in SI

Dimension	Unit
Length	meter (m)
Mass	kilogram (kg)
Time	second (s)
Temperature	kelvin (K)
Electric current	ampere (A)
Amount of light	candela (cd)
Amount of matter	mole (mol)

TABLE 1–2

Standard prefixes in SI units

Multiple	Prefix
10^{24}	yotta, Y
10^{21}	zetta, Z
10^{18}	exa, E
10^{15}	peta, P
10^{12}	tera, T
10^{9}	giga, G
10^{6}	mega, M
10^{3}	kilo, k
10^{2}	hecto, h
10^{1}	deka, da
10^{-1}	deci, d
10^{-2}	centi, c
10^{-3}	milli, m
10^{-6}	micro, μ
10^{-9}	nano, n
10^{-12}	pico, p
10^{-15}	femto, f
10^{-18}	atto, a
10^{-21}	zepto, z
10^{-24}	yocto, y

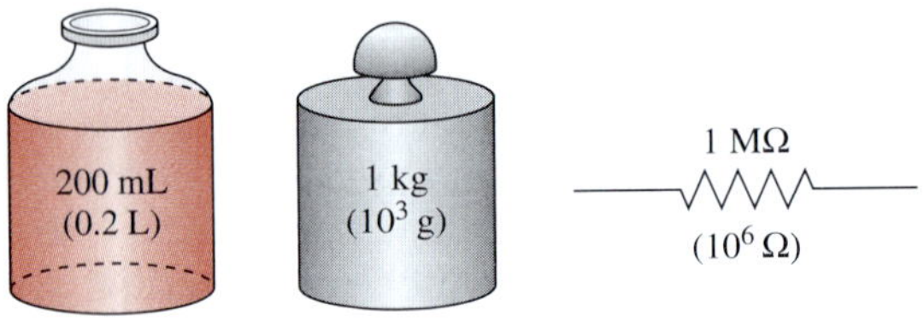

FIGURE 1–6
The SI unit prefixes are used in all branches of engineering.

Some SI and English Units

In SI, the units of mass, length, and time are the kilogram (kg), meter (m), and second (s), respectively. The respective units in the English system are the pound-mass (lbm), foot (ft), and second (s). The pound symbol *lb* is actually the abbreviation of *libra*, which was the ancient Roman unit of weight. The English retained this symbol even after the end of the Roman occupation of Britain in 410. The mass and length units in the two systems are related to each other by

$$1 \text{ lbm} = 0.45359 \text{ kg}$$

$$1 \text{ ft} = 0.3048 \text{ m}$$

In the English system, force is usually considered to be one of the primary dimensions and is assigned a nonderived unit. This is a source of confusion and error that necessitates the use of a dimensional constant (g_c) in many formulas. To avoid this nuisance, we consider force to be a secondary dimension whose unit is derived from Newton's second law, that is,

$$\text{Force} = (\text{Mass})(\text{Acceleration})$$

or

$$F = ma \tag{1–1}$$

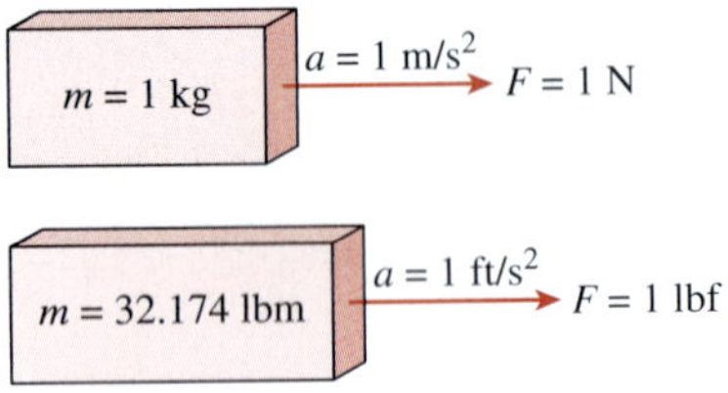

FIGURE 1–7
The definition of the force units.

In SI, the force unit is the newton (N), and it is defined as the *force required to accelerate a mass of 1 kg at a rate of 1 m/s²*. In the English system, the force unit is the **pound-force** (lbf) and is defined as the *force required to accelerate a mass of 32.174 lbm (1 slug) at a rate of 1 ft/s²* (Fig. 1–7). That is,

$$1 \text{ N} = 1 \text{ kg·m/s}^2$$

$$1 \text{ lbf} = 32.174 \text{ lbm·ft/s}^2$$

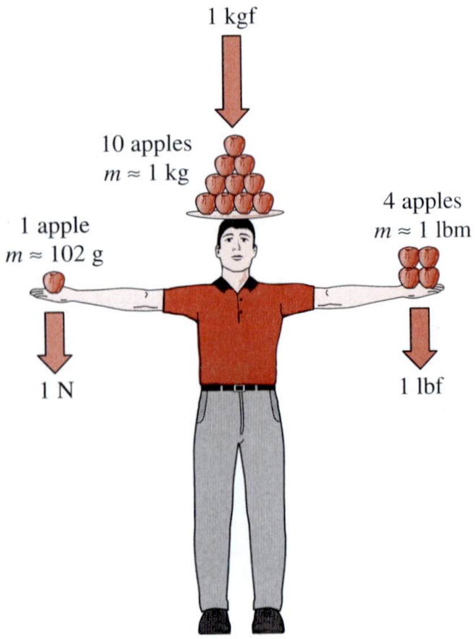

FIGURE 1–8
The relative magnitudes of the force units newton (N), kilogram-force (kgf), and pound-force (lbf).

A force of 1 N is roughly equivalent to the weight of a small apple (m = 102 g), whereas a force of 1 lbf is roughly equivalent to the weight of four medium apples (m_{total} = 454 g), as shown in Fig. 1–8. Another force unit in common use in many European countries is the *kilogram-force* (kgf), which is the weight of 1 kg mass at sea level (1 kgf = 9.807 N).

The term **weight** is often incorrectly used to express mass, particularly by the "weight watchers." Unlike mass, weight W is a *force*. It is the gravitational force applied to a body, and its magnitude is determined from Newton's second law,

$$W = mg \quad \text{(N)} \tag{1–2}$$

where m is the mass of the body, and g is the local gravitational acceleration (g is 9.807 m/s² or 32.174 ft/s² at sea level and 45° latitude). An ordinary bathroom scale measures the gravitational force acting on a body. The weight of a unit volume of a substance is called the **specific weight** γ and is determined from $\gamma = \rho g$, where ρ is density.

The mass of a body remains the same regardless of its location in the universe. Its weight, however, changes with a change in gravitational acceleration. A body weighs less on top of a mountain since g decreases with altitude. On the surface of the moon, an astronaut weighs about one-sixth of what she or he normally weighs on earth (Fig. 1–9).

At sea level a mass of 1 kg weighs 9.807 N, as illustrated in Fig. 1–10. A mass of 1 lbm, however, weighs 1 lbf, which misleads people to believe that pound-mass and pound-force can be used interchangeably as pound (lb), which is a major source of error in the English system.

It should be noted that the *gravity force* acting on a mass is due to the *attraction* between the masses, and thus it is proportional to the magnitudes of the masses and inversely proportional to the square of the distance between them. Therefore, the gravitational acceleration g at a location depends on the *local density* of the earth's crust, the *distance* to the center of the earth, and to a lesser extent, the positions of the moon and the sun. The value of g varies with location from 9.832 m/s^2 at the poles (9.789 at the equator) to 7.322 m/s^2 at 1000 km above sea level. However, at altitudes up to 30 km, the variation of g from the sea-level value of 9.807 m/s^2 is less than 1 percent. Therefore, for most practical purposes, the gravitational acceleration can be assumed to be *constant* at 9.807 m/s^2, often rounded to 9.81 m/s^2. It is interesting to note that at locations below sea level, the value of g increases with distance from the sea level, reaches a maximum at about 4500 m, and then starts decreasing. (What do you think the value of g is at the center of the earth?)

The primary cause of confusion between mass and weight is that mass is usually measured *indirectly* by measuring the *gravity force* it exerts. This approach also assumes that the forces exerted by other effects such as air buoyancy and fluid motion are negligible. This is like measuring the distance to a star by measuring its red shift, or measuring the altitude of an airplane by measuring barometric pressure. Both of these are also indirect measurements. The correct *direct* way of measuring mass is to compare it to a known mass. This is cumbersome, however, and it is mostly used for calibration and measuring precious metals.

Work, which is a form of energy, can simply be defined as force times distance; therefore, it has the unit "newton-meter (N·m)," which is called a **joule** (J). That is,

$$1 \text{ J} = 1 \text{ N·m} \tag{1-3}$$

A more common unit for energy in SI is the kilojoule (1 kJ = 10^3 J). In the English system, the energy unit is the **Btu** (British thermal unit), which is defined as the energy required to raise the temperature of 1 lbm of water at 68°F by 1°F. In the metric system, the amount of energy needed to raise the temperature of 1 g of water at 14.5°C by 1°C is defined as 1 **calorie** (cal), and 1 cal = 4.1868 J. The magnitudes of the kilojoule and Btu are almost identical (1 Btu = 1.0551 kJ). Here is a good way to get a feel for these units: If you light a typical match and let it burn itself out, it yields approximately one Btu (or one kJ) of energy (Fig. 1–11).

The unit for time rate of energy is joule per second (J/s), which is called a **watt** (W). In the case of work, the time rate of energy is called *power*.

FIGURE 1–9
A body weighing 150 lbf on earth will weigh only 25 lbf on the moon.

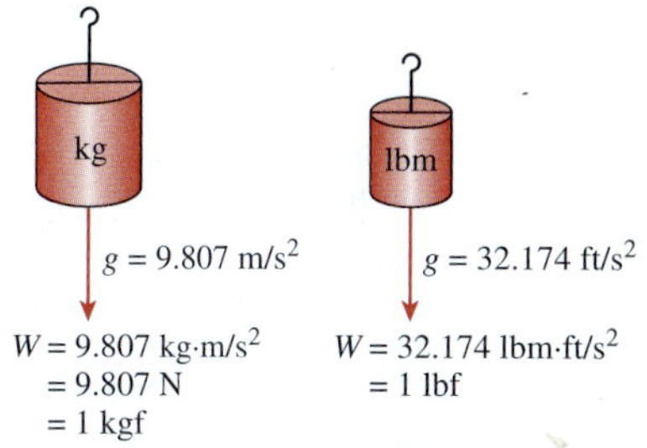

FIGURE 1–10
The weight of a unit mass at sea level.

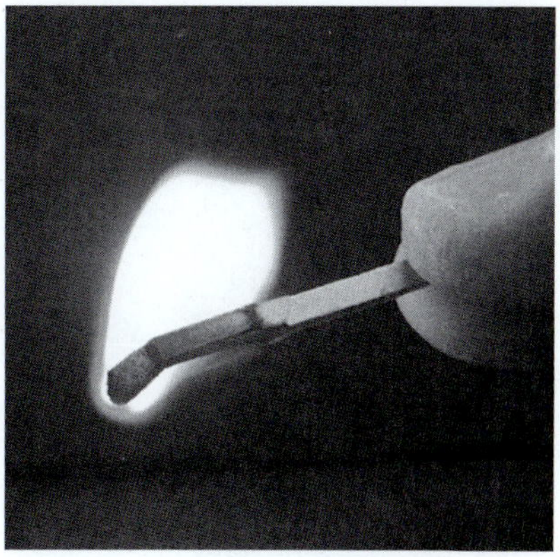

FIGURE 1–11
A typical match yields about one Btu (or one kJ) of energy if completely burned.
Photo by John M. Cimbala.

FIGURE 1–12
To be dimensionally homogeneous, all the terms in an equation must have the same dimensions.
BLONDIE © KING FEATURES SYNDICATE.

A commonly used unit of power is horsepower (hp), which is equivalent to 746 W. Electrical energy typically is expressed in the unit kilowatt-hour (kWh), which is equivalent to 3600 kJ. An electric appliance with a rated power of 1 kW consumes 1 kWh of electricity when running continuously for one hour. When dealing with electric power generation, the units kW and kWh are often confused. Note that kW or kJ/s is a unit of power, whereas kWh is a unit of energy. Therefore, statements like "the new wind turbine will generate 50 kW of electricity per year" are meaningless and incorrect. A correct statement should be something like "the new wind turbine with a rated power of 50 kW will generate 120,000 kWh of electricity per year."

Dimensional Homogeneity

We all know that apples and oranges do not add. But we somehow manage to do it (by mistake, of course). In engineering, all equations must be *dimensionally homogeneous*. That is, every term in an equation must have the same unit (Fig. 1–12). If, at some stage of an analysis, we find ourselves in a position to add two quantities that have different units, it is a clear indication that we have made an error at an earlier stage. So checking dimensions can serve as a valuable tool to spot errors.

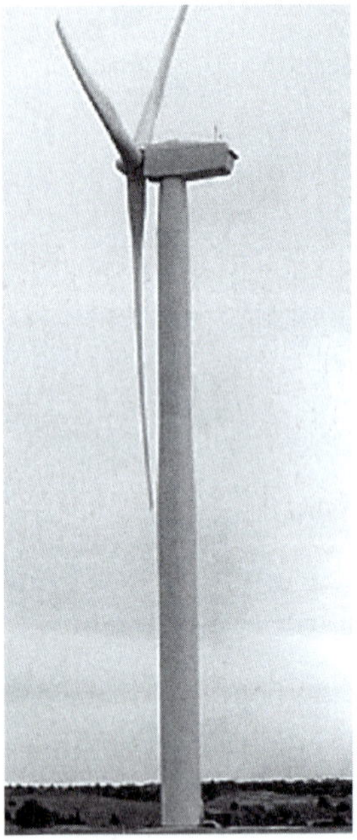

FIGURE 1–13
A wind turbine, as discussed in Example 1–1.
Courtesy of Steve Stadler, Oklahoma Wind Power Initiative.

EXAMPLE 1–1 **Electric Power Generation by a Wind Turbine**

A school is paying \$0.09/kWh for electric power. To reduce its power bill, the school installs a wind turbine (Fig. 1–13) with a rated power of 30 kW. If the turbine operates 2200 hours per year at the rated power, determine the amount of electric power generated by the wind turbine and the money saved by the school per year.

SOLUTION A wind turbine is installed to generate electricity. The amount of electric energy generated and the money saved per year are to be determined.

Analysis The wind turbine generates electric energy at a rate of 30 kW or 30 kJ/s. Then the total amount of electric energy generated per year becomes

$$\begin{aligned}\text{Total energy} &= (\text{Energy per unit time})(\text{Time interval})\\ &= (30\text{ kW})(2200\text{ h})\\ &= \mathbf{66{,}000\ kWh}\end{aligned}$$

The money saved per year is the monetary value of this energy determined as

$$\begin{aligned}\text{Money saved} &= (\text{Total energy})(\text{Unit cost of energy})\\ &= (66{,}000\text{ kWh})(\$0.09/\text{kWh})\\ &= \mathbf{\$5940}\end{aligned}$$

Discussion The annual electric energy production also could be determined in kJ by unit manipulations as

$$\text{Total energy} = (30\text{ kW})(2200\text{ h})\left(\frac{3600\text{ s}}{1\text{ h}}\right)\left(\frac{1\text{ kJ/s}}{1\text{ kW}}\right) = 2.38\times10^8\text{ kJ}$$

which is equivalent to 66,000 kWh (1 kWh = 3600 kJ).

We all know from experience that units can give terrible headaches if they are not used carefully in solving a problem. However, with some attention and skill, units can be used to our advantage. They can be used to check formulas; sometimes they can even be used to *derive* formulas, as explained in the following example.

EXAMPLE 1–2 Obtaining Formulas from Unit Considerations

A tank is filled with oil whose density is $\rho = 850$ kg/m^3. If the volume of the tank is $V = 2$ m^3, determine the amount of mass m in the tank.

SOLUTION The volume of an oil tank is given. The mass of oil is to be determined.

Assumptions Oil is a nearly incompressible substance and thus its density is constant.

Analysis A sketch of the system just described is given in Fig. 1–14. Suppose we forgot the formula that relates mass to density and volume. However, we know that mass has the unit of kilograms. That is, whatever calculations we do, we should end up with the unit of kilograms. Putting the given information into perspective, we have

$$\rho = 850 \text{ kg/m}^3 \qquad \text{and} \qquad V = 2 \text{ m}^3$$

It is obvious that we can eliminate m^3 and end up with kg by multiplying these two quantities. Therefore, the formula we are looking for should be

$$m = \rho V$$

Thus,

$$m = (850 \text{ kg/m}^3)(2 \text{ m}^3) = \mathbf{1700\ kg}$$

Discussion Note that this approach may not work for more complicated formulas. Nondimensional constants also may be present in the formulas, and these cannot be derived from unit considerations alone.

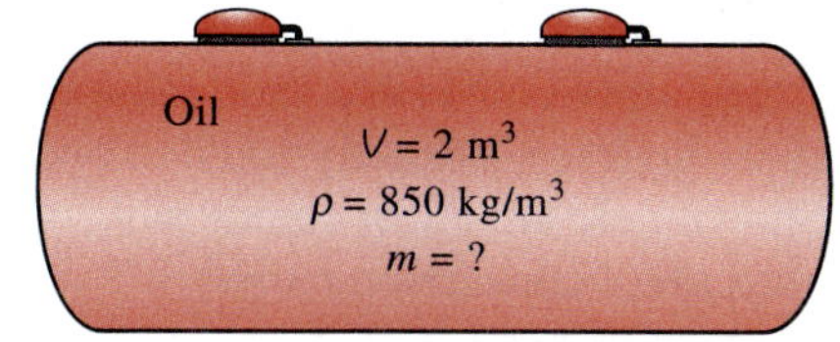

FIGURE 1–14
Schematic for Example 1–2.

You should keep in mind that a formula that is not dimensionally homogeneous is definitely wrong (Fig. 1–15), but a dimensionally homogeneous formula is not necessarily right.

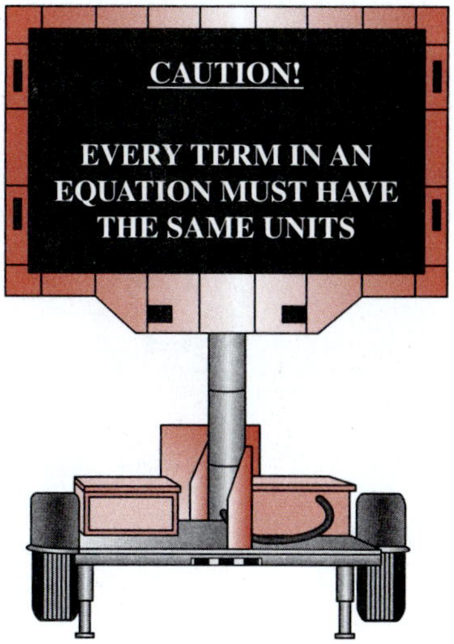

FIGURE 1–15
Always check the units in your calculations.

Unity Conversion Ratios

Just as all nonprimary dimensions can be formed by suitable combinations of primary dimensions, *all nonprimary units* **(secondary units)** *can be formed by combinations of primary units.* Force units, for example, can be expressed as

$$\text{N} = \text{kg}\,\frac{\text{m}}{\text{s}^2} \qquad \textit{and} \qquad \textit{lbf} = 32.174 \text{ lbm}\,\frac{\text{ft}}{\text{s}^2}$$

They can also be expressed more conveniently as **unity conversion ratios** as

$$\frac{\text{N}}{\text{kg·m/s}^2} = 1 \qquad \text{and} \qquad \frac{\text{lbf}}{32.174 \text{ lbm·ft/s}^2} = 1$$

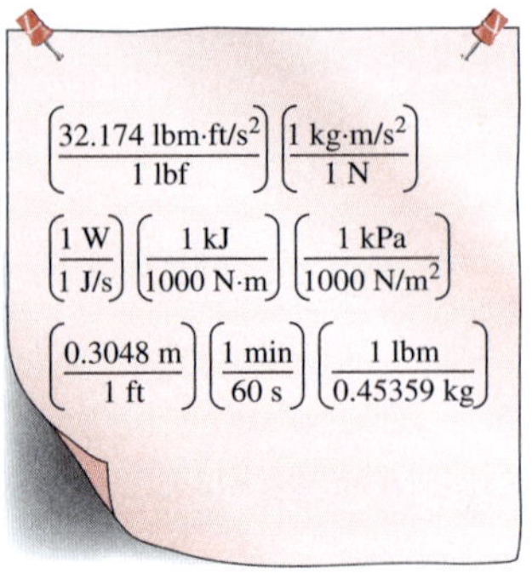

FIGURE 1–16
Every unity conversion ratio (as well as its inverse) is exactly equal to one. Shown here are a few commonly used unity conversion ratios.
Courtesy Steve Stadler, Oklahoma Wind Power Initiative. Used by permission.

Unity conversion ratios are identically equal to 1 and are unitless, and thus such ratios (or their inverses) can be inserted conveniently into any calculation to properly convert units (Fig. 1–16). You are encouraged to always use unity conversion ratios such as those given here when converting units. Some textbooks insert the archaic gravitational constant g_c defined as $g_c = 32.174 \text{ lbm·ft/lbf·s}^2 = \text{kg·m/N·s}^2 = 1$ into equations in order to force units to match. This practice leads to unnecessary confusion and is strongly discouraged by the present authors. We recommend that you instead use unity conversion ratios.

EXAMPLE 1–3 The Weight of One Pound-Mass

Using unity conversion ratios, show that 1.00 lbm weighs 1.00 lbf on earth (Fig. 1–17).

SOLUTION A mass of 1.00 lbm is subjected to standard earth gravity. Its weight in lbf is to be determined.
Assumptions Standard sea-level conditions are assumed.
Properties The gravitational constant is $g = 32.174 \text{ ft/s}^2$.
Analysis We apply Newton's second law to calculate the weight (force) that corresponds to the known mass and acceleration. The weight of any object is equal to its mass times the local value of gravitational acceleration. Thus,

$$W = mg = (1.00 \text{ lbm})(32.174 \text{ ft/s}^2)\left(\frac{1 \text{ lbf}}{32.174 \text{ lbm·ft/s}^2}\right) = \mathbf{1.00 \text{ lbf}}$$

Discussion The quantity in large parentheses in this equation is a unity conversion ratio. Mass is the same regardless of its location. However, on some other planet with a different value of gravitational acceleration, the weight of 1 lbm would differ from that calculated here.

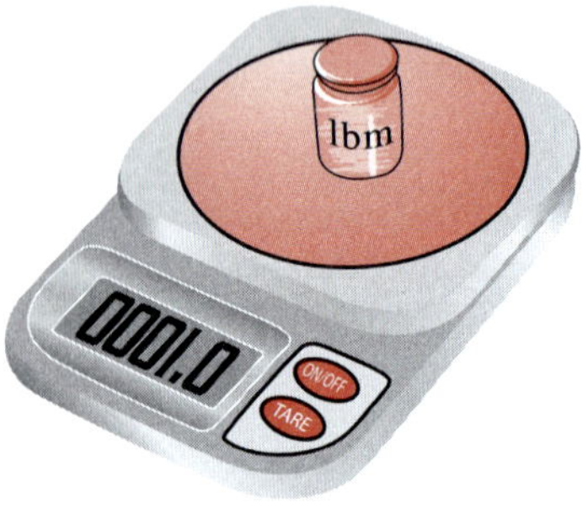

FIGURE 1–17
A mass of 1 lbm weighs 1 lbf on earth.

FIGURE 1–18
A quirk in the metric system of units.

When you buy a box of breakfast cereal, the printing may say "Net weight: One pound (454 grams)." (See Fig. 1–18.) Technically, this means that the cereal inside the box weighs 1.00 lbf on earth and has a *mass* of 453.6 g (0.4536 kg). Using Newton's second law, the actual weight of the cereal on earth is

$$W = mg = (453.6 \text{ g})(9.81 \text{ m/s}^2)\left(\frac{1 \text{ N}}{1 \text{ kg·m/s}^2}\right)\left(\frac{1 \text{ kg}}{1000 \text{ g}}\right) = 4.49 \text{ N}$$

1–3 ▪ SYSTEMS AND CONTROL VOLUMES

A system is defined as a *quantity of matter or a region in space chosen for study*. The mass or region outside the system is called the **surroundings**. The real or imaginary surface that separates the system from its surroundings is called the **boundary** (Fig. 1–19). The boundary of a system can be *fixed* or *movable*. Note that the boundary is the contact surface shared by both the system and the surroundings. Mathematically speaking, the boundary has zero thickness, and thus it can neither contain any mass nor occupy any volume in space.

Systems may be considered to be *closed* or *open,* depending on whether a fixed mass or a fixed volume in space is chosen for study. A **closed system** (also known as a **control mass** or just *system* when the context makes it clear) consists of a fixed amount of mass, and no mass can cross its boundary. That is, no mass can enter or leave a closed system, as shown in Fig. 1–20. But energy, in the form of heat or work, can cross the boundary; and the volume of a closed system does not have to be fixed. If, as a special case, even energy is not allowed to cross the boundary, that system is called an **isolated system**.

Consider the piston-cylinder device shown in Fig. 1–21. Let us say that we would like to find out what happens to the enclosed gas when it is heated. Since we are focusing our attention on the gas, it is our system. The inner surfaces of the piston and the cylinder form the boundary, and since no mass is crossing this boundary, it is a closed system. Notice that energy may cross the boundary, and part of the boundary (the inner surface of the piston, in this case) may move. Everything outside the gas, including the piston and the cylinder, is the surroundings.

An **open system**, or a **control volume**, as it is often called, is a properly selected region in space. It usually encloses a device that involves mass flow such as a compressor, turbine, or nozzle. Flow through these devices is best studied by selecting the region within the device as the control volume. Both mass and energy can cross the boundary of a control volume.

A large number of engineering problems involve mass flow in and out of a system and, therefore, are modeled as *control volumes.* A water heater, a car radiator, a turbine, and a compressor all involve mass flow and should be analyzed as control volumes (open systems) instead of as control masses (closed systems). In general, *any arbitrary region in space* can be selected as a control volume. There are no concrete rules for the selection of control volumes, but the proper choice certainly makes the analysis much easier. If we were to analyze the flow of air through a nozzle, for example, a good choice for the control volume would be the region within the nozzle.

The boundaries of a control volume are called a *control surface,* and they can be real or imaginary. In the case of a nozzle, the inner surface of the nozzle forms the real part of the boundary, and the entrance and exit areas form the imaginary part, since there are no physical surfaces there (Fig. 1–22*a*).

A control volume can be fixed in size and shape, as in the case of a nozzle, or it may involve a moving boundary, as shown in Fig. 1–22*b*. Most control volumes, however, have fixed boundaries and thus do not involve any moving boundaries. A control volume can also involve heat and work interactions just as a closed system, in addition to mass interaction.

As an example of an open system, consider the water heater shown in Fig. 1–23. Let us say that we would like to determine how much heat we must transfer to the water in the tank in order to supply a steady stream of hot water. Since hot water will leave the tank and be replaced by cold water, it is not convenient to choose a fixed mass as our system for the analysis. Instead, we can concentrate our attention on the volume formed by the interior surfaces of the tank and consider the hot and cold water streams as mass leaving and entering the control volume. The interior surfaces of the

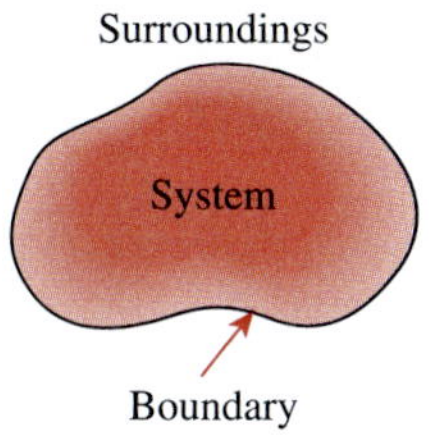

FIGURE 1–19
System, surroundings, and boundary.

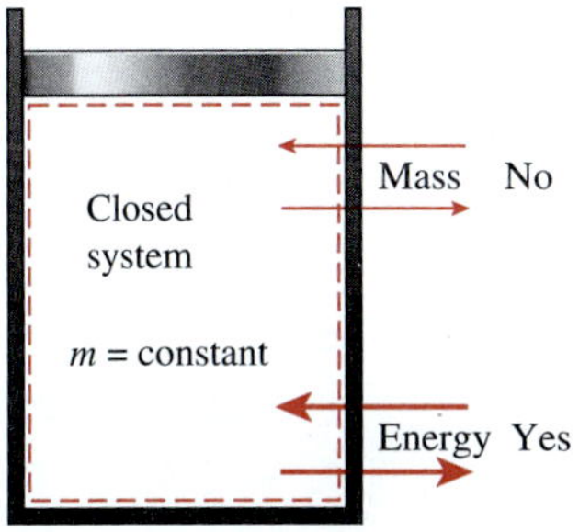

FIGURE 1–20
Mass cannot cross the boundaries of a closed system, but energy can.

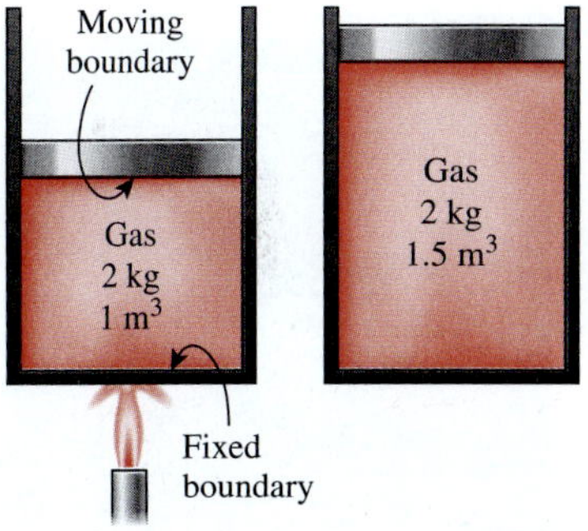

FIGURE 1–21
A closed system with a moving boundary.

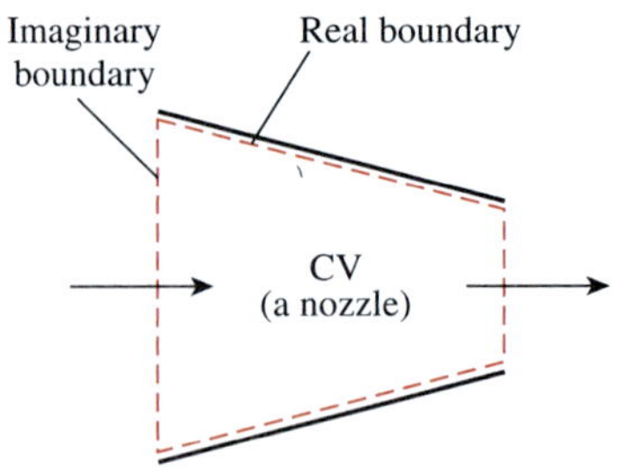

(*a*) A control volume with real and imaginary boundaries

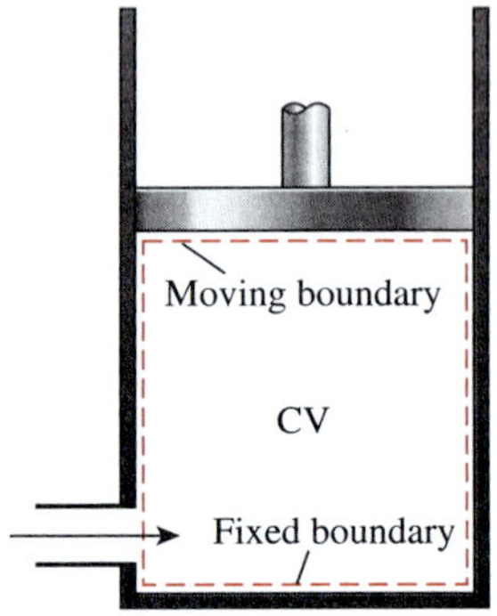

(*b*) A control volume with fixed and moving boundaries

FIGURE 1–22
A control volume can involve fixed, moving, real, and imaginary boundaries.

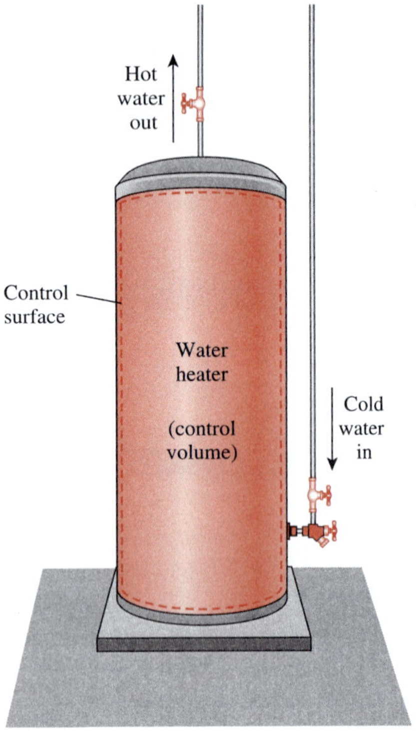

FIGURE 1–23
An open system (a control volume) with one inlet and one exit.

tank form the control surface for this case, and mass is crossing the control surface at two locations.

In an engineering analysis, the system under study *must* be defined carefully. In most cases, the system investigated is quite simple and obvious, and defining the system may seem like a tedious and unnecessary task. In other cases, however, the system under study may be rather involved, and a proper choice of the system may greatly simplify the analysis.

1–4 ▪ PROPERTIES OF A SYSTEM

Any characteristic of a system is called a **property**. Some familiar properties are pressure P, temperature T, volume V, and mass m. The list can be extended to include less familiar ones such as viscosity, thermal conductivity, modulus of elasticity, thermal expansion coefficient, electric resistivity, and even velocity and elevation.

Properties are considered to be either *intensive* or *extensive*. **Intensive properties** are those that are independent of the mass of a system, such as temperature, pressure, and density. **Extensive properties** are those whose values depend on the size—or extent—of the system. Total mass, total volume, and total momentum are some examples of extensive properties. An easy way to determine whether a property is intensive or extensive is to divide the system into two equal parts with an imaginary partition, as shown in Fig. 1–24. Each part will have the same value of intensive properties as the original system, but half the value of the extensive properties.

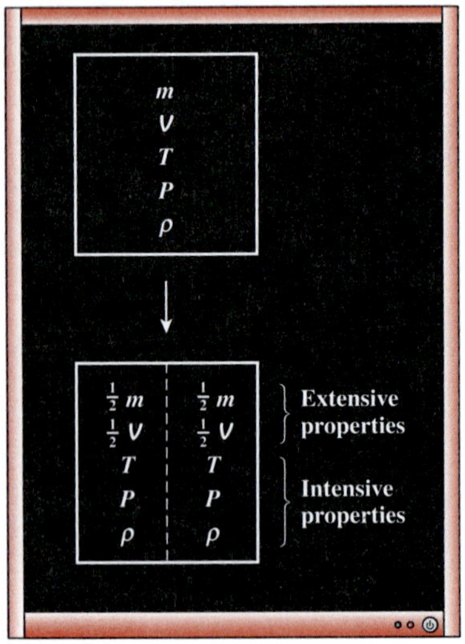

FIGURE 1–24
Criterion to differentiate intensive and extensive properties.

Generally, uppercase letters are used to denote extensive properties (with mass m being a major exception), and lowercase letters are used for intensive properties (with pressure P and temperature T being the obvious exceptions).

Extensive properties per unit mass are called **specific properties**. Some examples of specific properties are specific volume ($v = V/m$) and specific total energy ($e = E/m$).

Continuum

Matter is made up of atoms that are widely spaced in the gas phase. Yet it is very convenient to disregard the atomic nature of a substance and view it as a continuous, homogeneous matter with no holes, that is, a **continuum**. The continuum idealization allows us to treat properties as point functions and to assume the properties vary continually in space with no jump discontinuities. This idealization is valid as long as the size of the system we deal with is large relative to the space between the molecules. This is the case in practically all problems, except some specialized ones. The continuum idealization is implicit in many statements we make, such as "the density of water in a glass is the same at any point."

To have a sense of the distance involved at the molecular level, consider a container filled with oxygen at atmospheric conditions. The diameter of the oxygen molecule is about 3×10^{-10} m and its mass is 5.3×10^{-26} kg. Also, the *mean free path* of oxygen at 1 atm pressure and 20°C is 6.3×10^{-8} m. That is, an oxygen molecule travels, on average, a distance of 6.3×10^{-8} m (about 200 times of its diameter) before it collides with another molecule.

Also, there are about 3×10^{16} molecules of oxygen in the tiny volume of 1 mm^3 at 1 atm pressure and 20°C (Fig. 1–25). The continuum model is applicable as long as the characteristic length of the system (such as its diameter) is much larger than the mean free path of the molecules. At very high vacuums or very high elevations, the mean free path may become large (for example, it is about 0.1 m for atmospheric air at an elevation of 100 km). For such cases the **rarefied gas flow theory** should be used, and the impact of individual molecules should be considered. In this text we will limit our consideration to substances that can be modeled as a continuum.

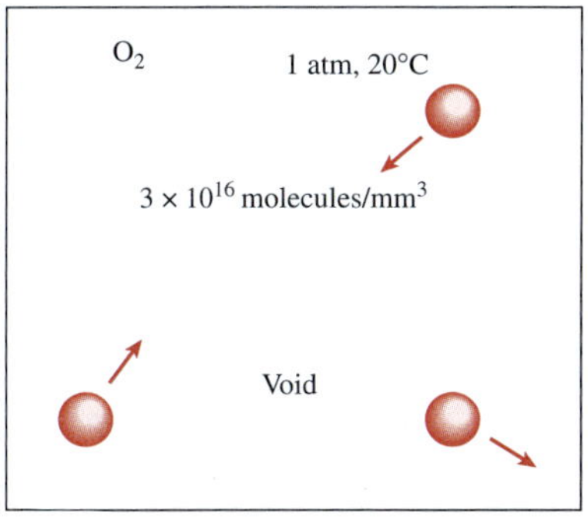

FIGURE 1–25
Despite the large gaps between molecules, a substance can be treated as a continuum because of the very large number of molecules even in an extremely small volume.

1–5 ▪ DENSITY AND SPECIFIC GRAVITY

Density is defined as *mass per unit volume* (Fig. 1–26).

Density:
$$\rho = \frac{m}{V} \quad (\text{kg/m}^3) \tag{1–4}$$

The reciprocal of density is the **specific volume** v, which is defined as *volume per unit mass*. That is,

$$v = \frac{V}{m} = \frac{1}{\rho} \tag{1–5}$$

For a differential volume element of mass δm and volume δV, density can be expressed as $\rho = \delta m/\delta V$.

The density of a substance, in general, depends on temperature and pressure. The density of most gases is proportional to pressure and inversely proportional to temperature. Liquids and solids, on the other hand, are essentially incompressible substances, and the variation of their density with pressure is usually negligible. At 20°C, for example, the density of water changes from 998 kg/m^3 at 1 atm to 1003 kg/m^3 at 100 atm, a change of just 0.5 percent. The density of liquids and solids depends more strongly on temperature than it does on pressure. At 1 atm, for example, the density of water changes from 998 kg/m^3 at 20°C to 975 kg/m^3 at 75°C, a change of 2.3 percent, which can still be neglected in many engineering analyses.

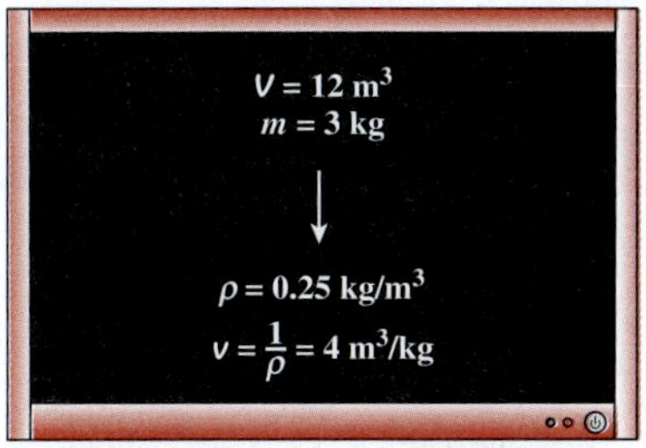

FIGURE 1–26
Density is mass per unit volume; specific volume is volume per unit mass.

TABLE 1–3

Specific gravities of some substances at 0°C

Substance	SG
Water	1.0
Blood	1.05
Seawater	1.025
Gasoline	0.7
Ethyl alcohol	0.79
Mercury	13.6
Wood	0.3–0.9
Gold	19.2
Bones	1.7–2.0
Ice	0.92
Air (at 1 atm)	0.0013

Sometimes the density of a substance is given relative to the density of a well-known substance. Then it is called **specific gravity**, or **relative density**, and is defined as *the ratio of the density of a substance to the density of some standard substance at a specified temperature* (usually water at 4°C, for which $\rho_{H_2O} = 1000$ kg/m³). That is,

Specific gravity:
$$SG = \frac{\rho}{\rho_{H_2O}} \quad \text{(1–6)}$$

Note that the specific gravity of a substance is a dimensionless quantity. However, in SI units, the numerical value of the specific gravity of a substance is exactly equal to its density in g/cm³ or kg/L (or 0.001 times the density in kg/m³) since the density of water at 4°C is 1 g/cm³ = 1 kg/L = 1000 kg/m³. The specific gravity of mercury at 0°C, for example, is 13.6. Therefore, its density at 0°C is 13.6 g/cm³ = 13.6 kg/L = 13,600 kg/m³. The specific gravities of some substances at 0°C are given in Table 1–3. Note that substances with specific gravities less than 1 are lighter than water, and thus they would float on water.

The weight of a unit volume of a substance is called **specific weight** and is expressed as

Specific weight:
$$\gamma_s = \rho g \qquad (\text{N/m}^3) \quad \text{(1–7)}$$

where g is the gravitational acceleration.

The densities of liquids are essentially constant, and thus they can often be approximated as being incompressible substances during most processes without sacrificing much in accuracy.

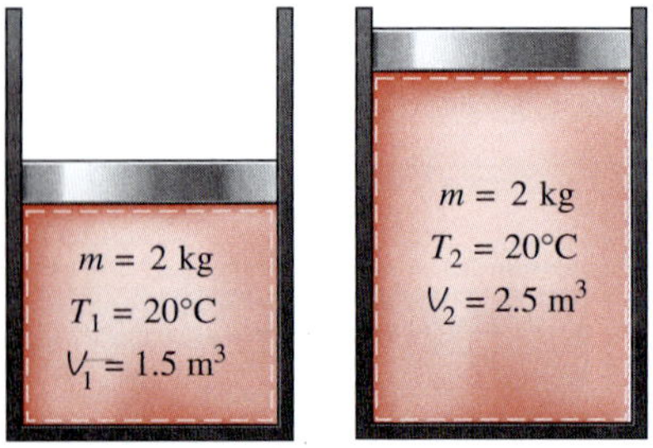

(*a*) State 1 (*b*) State 2

FIGURE 1–27
A system at two different states.

1–6 ▪ STATE AND EQUILIBRIUM

Consider a system not undergoing any change. At this point, all the properties can be measured or calculated throughout the entire system, which gives us a set of properties that completely describes the condition, or the **state**, of the system. At a given state, all the properties of a system have fixed values. If the value of even one property changes, the state will change to a different one. In Fig. 1–27 a system is shown at two different states.

Thermodynamics deals with *equilibrium* states. The word **equilibrium** implies a state of balance. In an equilibrium state there are no unbalanced potentials (or driving forces) within the system. A system in equilibrium experiences no changes when it is isolated from its surroundings.

There are many types of equilibrium, and a system is not in thermodynamic equilibrium unless the conditions of all the relevant types of equilibrium are satisfied. For example, a system is in **thermal equilibrium** if the temperature is the same throughout the entire system, as shown in Fig. 1–28. That is, the system involves no temperature differential, which is the driving force for heat flow. **Mechanical equilibrium** is related to pressure, and a system is in mechanical equilibrium if there is no change in pressure at any point of the system with time. However, the pressure may vary within the system with elevation as a result of gravitational effects. For example, the higher pressure at a bottom layer is balanced by the extra weight it must carry, and, therefore, there is no imbalance of forces. The variation of pressure as a result of gravity in most thermodynamic systems is relatively small and usually disregarded.

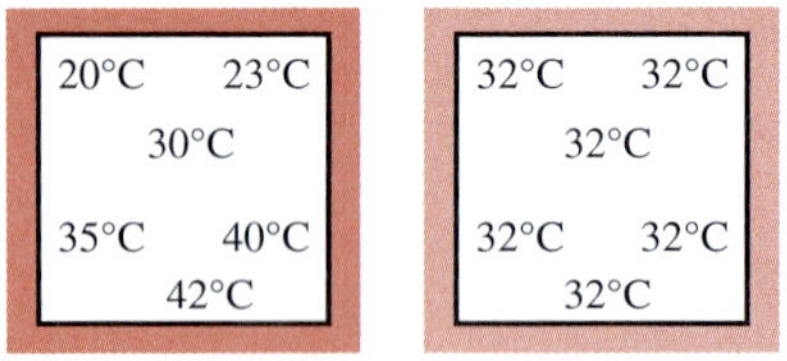

(*a*) Before (*b*) After

FIGURE 1–28
A closed system reaching thermal equilibrium.

If a system involves two phases, it is in **phase equilibrium** when the mass of each phase reaches an equilibrium level and stays there. Finally, a system is in **chemical equilibrium** if its chemical composition does not change with time, that is, no chemical reactions occur. A system will not be in equilibrium unless all the relevant equilibrium criteria are satisfied.

The State Postulate

As noted earlier, the state of a system is described by its properties. But we know from experience that we do not need to specify all the properties in order to fix a state. Once a sufficient number of properties are specified, the rest of the properties assume certain values automatically. That is, specifying a certain number of properties is sufficient to fix a state. The number of properties required to fix the state of a system is given by the **state postulate**:

> The state of a simple compressible system is completely specified by two independent, intensive properties.

A system is called a **simple compressible system** in the absence of electrical, magnetic, gravitational, motion, and surface tension effects. These effects are due to external force fields and are negligible for most engineering problems. Otherwise, an additional property needs to be specified for each effect that is significant. If the gravitational effects are to be considered, for example, the elevation z needs to be specified in addition to the two properties necessary to fix the state.

The state postulate requires that the two properties specified be independent to fix the state. Two properties are **independent** if one property can be varied while the other one is held constant. Temperature and specific volume, for example, are always independent properties, and together they can fix the state of a simple compressible system (Fig. 1–29). Temperature and pressure, however, are independent properties for single-phase systems, but are dependent properties for multiphase systems. At sea level (P = 1 atm), water boils at 100°C, but on a mountaintop where the pressure is lower, water boils at a lower temperature. That is, $T = f(P)$ during a phase-change process; thus, temperature and pressure are not sufficient to fix the state of a two-phase system. Phase-change processes are discussed in detail in Chap. 3.

FIGURE 1–29
The state of nitrogen is fixed by two independent, intensive properties.

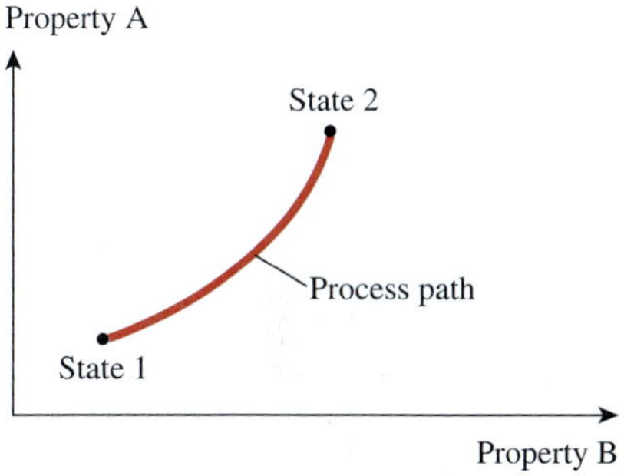

FIGURE 1–30
A process between states 1 and 2 and the process path.

1–7 ▪ PROCESSES AND CYCLES

Any change that a system undergoes from one equilibrium state to another is called a **process**, and the series of states through which a system passes during a process is called the **path** of the process (Fig. 1–30). To describe a process completely, one should specify the initial and final states of the process, as well as the path it follows, and the interactions with the surroundings.

When a process proceeds in such a manner that the system remains infinitesimally close to an equilibrium state at all times, it is called a **quasi-static**, or **quasi-equilibrium, process**. A quasi-equilibrium process can be viewed as a sufficiently slow process that allows the system to adjust itself internally so that properties in one part of the system do not change any faster than those at other parts.

This is illustrated in Fig. 1–31. When a gas in a piston-cylinder device is compressed suddenly, the molecules near the face of the piston will not

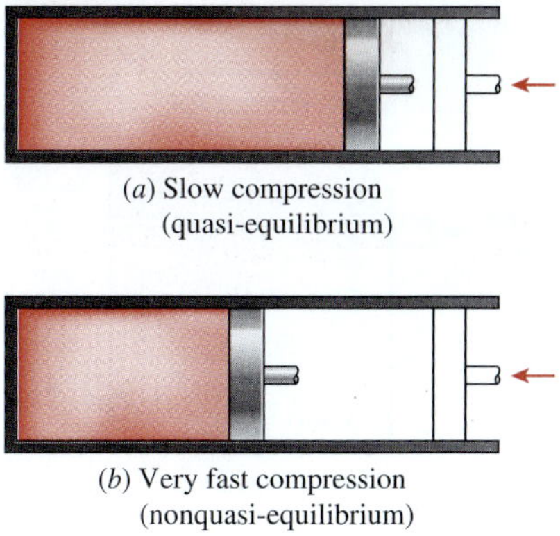

FIGURE 1–31
Quasi-equilibrium and nonquasi-equilibrium compression processes.

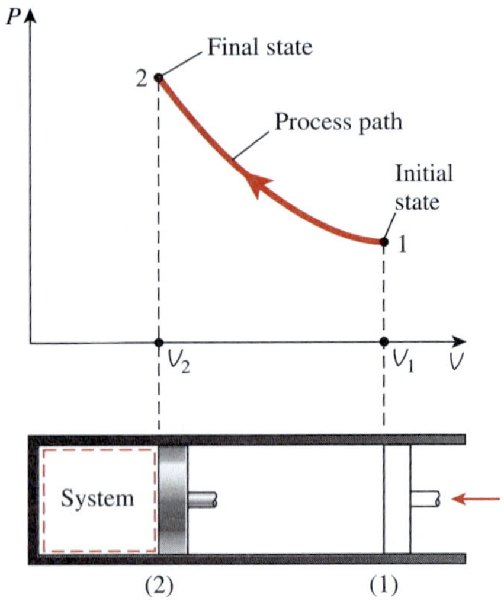

FIGURE 1–32
The P-V diagram of a compression process.

have enough time to escape and they will have to pile up in a small region in front of the piston, thus creating a high-pressure region there. Because of this pressure difference, the system can no longer be said to be in equilibrium, and this makes the entire process nonquasi-equilibrium. However, if the piston is moved slowly, the molecules will have sufficient time to redistribute and there will not be a molecule pileup in front of the piston. As a result, the pressure inside the cylinder will always be nearly uniform and will rise at the same rate at all locations. Since equilibrium is maintained at all times, this is a quasi-equilibrium process.

It should be pointed out that a quasi-equilibrium process is an idealized process and is not a true representation of an actual process. But many actual processes closely approximate it, and they can be modeled as quasi-equilibrium with negligible error. Engineers are interested in quasi-equilibrium processes for two reasons. First, they are easy to analyze; second, work-producing devices deliver the most work when they operate on quasi-equilibrium processes. Therefore, quasi-equilibrium processes serve as standards to which actual processes can be compared.

Process diagrams plotted by employing thermodynamic properties as coordinates are very useful in visualizing the processes. Some common properties that are used as coordinates are temperature T, pressure P, and volume V (or specific volume v). Figure 1–32 shows the P-V diagram of a compression process of a gas.

Note that the process path indicates a series of equilibrium states through which the system passes during a process and has significance for quasi-equilibrium processes only. For nonquasi-equilibrium processes, we are not able to characterize the entire system by a single state, and thus we cannot speak of a process path for a system as a whole. A nonquasi-equilibrium process is denoted by a dashed line between the initial and final states instead of a solid line.

The prefix *iso-* is often used to designate a process for which a particular property remains constant. An **isothermal process**, for example, is a process during which the temperature T remains constant; an **isobaric process** is a process during which the pressure P remains constant; and an **isochoric** (or **isometric**) **process** is a process during which the specific volume v remains constant.

A system is said to have undergone a **cycle** if it returns to its initial state at the end of the process. That is, for a cycle the initial and final states are identical.

The Steady-Flow Process

The terms *steady* and *uniform* are used frequently in engineering, and thus it is important to have a clear understanding of their meanings. The term *steady* implies *no change with time*. The opposite of steady is *unsteady,* or *transient.* The term *uniform,* however, implies *no change with location* over a specified region. These meanings are consistent with their everyday use (steady girlfriend, uniform properties, etc.).

A large number of engineering devices operate for long periods of time under the same conditions, and they are classified as *steady-flow devices*. Processes involving such devices can be represented reasonably well by a somewhat idealized process, called the **steady-flow process**, which can be defined as a *process during which a fluid flows through a control volume steadily* (Fig. 1–33). That is, the fluid properties can change from point to

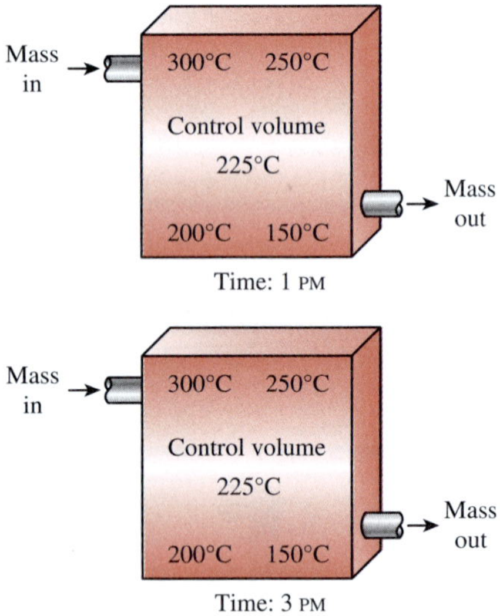

FIGURE 1–33
During a steady-flow process, fluid properties within the control volume may change with position but not with time.

point within the control volume, but at any fixed point they remain the same during the entire process. Therefore, the volume V, the mass m, and the total energy content E of the control volume remain constant during a steady-flow process (Fig. 1–34).

Steady-flow conditions can be closely approximated by devices that are intended for continuous operation such as turbines, pumps, boilers, condensers, and heat exchangers or power plants or refrigeration systems. Some cyclic devices, such as reciprocating engines or compressors, do not satisfy any of the conditions stated above since the flow at the inlets and the exits will be pulsating and not steady. However, the fluid properties vary with time in a periodic manner, and the flow through these devices can still be analyzed as a steady-flow process by using time-averaged values for the properties.

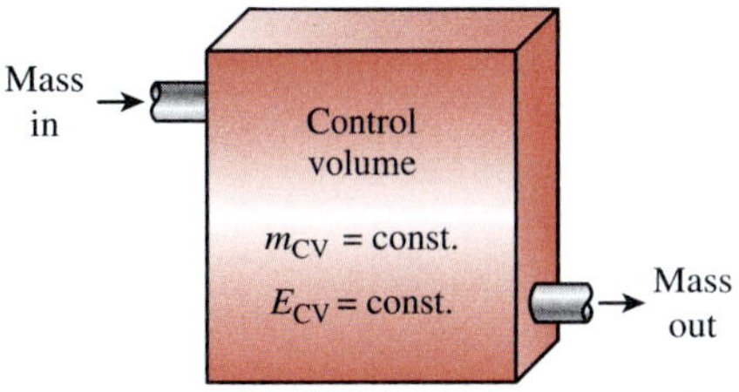

FIGURE 1–34
Under steady-flow conditions, the mass and energy contents of a control volume remain constant.

1–8 ▪ TEMPERATURE AND THE ZEROTH LAW OF THERMODYNAMICS

Although we are familiar with temperature as a measure of "hotness" or "coldness," it is not easy to give an exact definition for it. Based on our physiological sensations, we express the level of temperature qualitatively with words like *freezing cold, cold, warm, hot,* and *red-hot.* However, we cannot assign numerical values to temperatures based on our sensations alone. Furthermore, our senses may be misleading. A metal chair, for example, will feel much colder than a wooden one even when both are at the same temperature.

Fortunately, several properties of materials change with temperature in a *repeatable* and *predictable* way, and this forms the basis for accurate temperature measurement. The commonly used mercury-in-glass thermometer, for example, is based on the expansion of mercury with temperature. Temperature is also measured by using several other temperature-dependent properties.

It is a common experience that a cup of hot coffee left on the table eventually cools off and a cold drink eventually warms up. That is, when a body is brought into contact with another body that is at a different temperature, heat is transferred from the body at higher temperature to the one at lower temperature until both bodies attain the same temperature (Fig. 1–35). At that point, the heat transfer stops, and the two bodies are said to have reached **thermal equilibrium**. The equality of temperature is the only requirement for thermal equilibrium.

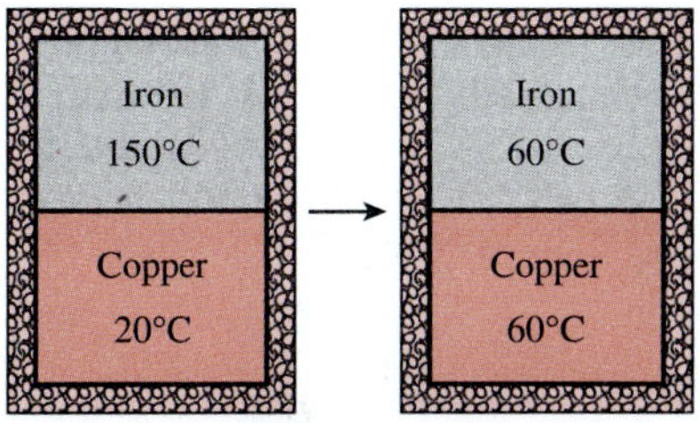

FIGURE 1–35
Two bodies reaching thermal equilibrium after being brought into contact in an isolated enclosure.

The **zeroth law of thermodynamics** states that if two bodies are in thermal equilibrium with a third body, they are also in thermal equilibrium with each other. It may seem silly that such an obvious fact is called one of the basic laws of thermodynamics. However, it cannot be concluded from the other laws of thermodynamics, and it serves as a basis for the validity of temperature measurement. By replacing the third body with a thermometer, the zeroth law can be restated as *two bodies are in thermal equilibrium if both have the same temperature reading even if they are not in contact.*

The zeroth law was first formulated and labeled by R. H. Fowler in 1931. As the name suggests, its value as a fundamental physical principle was recognized more than half a century after the formulation of the first and the second laws of thermodynamics. It was named the zeroth law since it should have preceded the first and the second laws of thermodynamics.

Temperature Scales

Temperature scales enable us to use a common basis for temperature measurements, and several have been introduced throughout history. All temperature scales are based on some easily reproducible states such as the freezing and boiling points of water, which are also called the *ice point* and the *steam point,* respectively. A mixture of ice and water that is in equilibrium with air saturated with vapor at 1 atm pressure is said to be at the ice point, and a mixture of liquid water and water vapor (with no air) in equilibrium at 1 atm pressure is said to be at the steam point.

The temperature scales used in the SI and in the English system today are the **Celsius scale** (formerly called the *centigrade scale;* in 1948 it was renamed after the Swedish astronomer A. Celsius, 1702–1744, who devised it) and the **Fahrenheit scale** (named after the German instrument maker G. Fahrenheit, 1686–1736), respectively. On the Celsius scale, the ice and steam points were originally assigned the values of 0 and 100°C, respectively. The corresponding values on the Fahrenheit scale are 32 and 212°F. These are often referred to as *two-point scales* since temperature values are assigned at two different points.

In thermodynamics, it is very desirable to have a temperature scale that is independent of the properties of any substance or substances. Such a temperature scale is called a **thermodynamic temperature scale**, which is developed later in conjunction with the second law of thermodynamics. The thermodynamic temperature scale in the SI is the **Kelvin scale**, named after Lord Kelvin (1824–1907). The temperature unit on this scale is the **kelvin**, which is designated by K (not °K; the degree symbol was officially dropped from kelvin in 1967). The lowest temperature on the Kelvin scale is absolute zero, or 0 K. Then it follows that only one nonzero reference point needs to be assigned to establish the slope of this linear scale. Using nonconventional refrigeration techniques, scientists have approached absolute zero kelvin (they achieved 0.000000002 K in 1989).

The thermodynamic temperature scale in the English system is the **Rankine scale**, named after William Rankine (1820–1872). The temperature unit on this scale is the **rankine**, which is designated by R.

A temperature scale that turns out to be nearly identical to the Kelvin scale is the **ideal-gas temperature scale**. The temperatures on this scale are measured using a **constant-volume gas thermometer**, which is basically a rigid vessel filled with a gas, usually hydrogen or helium, at low pressure. This thermometer is based on the principle that *at low pressures, the temperature of a gas is proportional to its pressure at constant volume*. That is, the temperature of a gas of fixed volume varies *linearly* with pressure at sufficiently low pressures. Then the relationship between the temperature and the pressure of the gas in the vessel can be expressed as

$$T = a + bP \qquad \textbf{(1–8)}$$

where the values of the constants a and b for a gas thermometer are determined experimentally. Once a and b are known, the temperature of a medium can be calculated from this relation by immersing the rigid vessel of the gas thermometer into the medium and measuring the gas pressure when thermal equilibrium is established between the medium and the gas in the vessel whose volume is held constant.

An ideal-gas temperature scale can be developed by measuring the pressures of the gas in the vessel at two reproducible points (such as the ice and the steam points) and assigning suitable values to temperatures at those two points. Considering that only one straight line passes through two fixed points on a plane, these two measurements are sufficient to determine the constants a and b in Eq. 1–8. Then the unknown temperature T of a medium corresponding to a pressure reading P can be determined from that equation by a simple calculation. The values of the constants will be different for each thermometer, depending on the type and the amount of the gas in the vessel, and the temperature values assigned at the two reference points. If the ice and steam points are assigned the values 0°C and 100°C, respectively, then the gas temperature scale will be identical to the Celsius scale. In this case the value of the constant a (which corresponds to an absolute pressure of zero) is determined to be -273.15°C regardless of the type and the amount of the gas in the vessel of the gas thermometer. That is, on a *P-T* diagram, all the straight lines passing through the data points in this case will intersect the temperature axis at -273.15°C when extrapolated, as shown in Fig. 1–36. This is the lowest temperature that can be obtained by a gas thermometer, and thus we can obtain an *absolute gas temperature scale* by assigning a value of zero to the constant a in Eq. 1–8. In that case, Eq. 1–8 reduces to $T = bP$, and thus we need to specify the temperature at only *one* point to define an absolute gas temperature scale.

It should be noted that the absolute gas temperature scale is not a thermodynamic temperature scale, since it cannot be used at very low temperatures (due to condensation) and at very high temperatures (due to dissociation and ionization). However, absolute gas temperature is identical to the thermodynamic temperature in the temperature range in which the gas thermometer can be used. Thus, we can view the thermodynamic temperature scale at this point as an absolute gas temperature scale that utilizes an "ideal" or "imaginary" gas that always acts as a low-pressure gas regardless of the temperature. If such a gas thermometer existed, it would read zero kelvin at absolute zero pressure, which corresponds to -273.15°C on the Celsius scale (Fig. 1–37).

The Kelvin scale is related to the Celsius scale by

$$T(\text{K}) = T(°\text{C}) + 273.15 \qquad \text{(1–9)}$$

The Rankine scale is related to the Fahrenheit scale by

$$T(\text{R}) = T(°\text{F}) + 459.67 \qquad \text{(1–10)}$$

It is common practice to round the constant in Eq. 1–9 to 273 and that in Eq. 1–10 to 460.

The temperature scales in the two unit systems are related by

$$T(\text{R}) = 1.8T(\text{K}) \qquad \text{(1–11)}$$

$$T(°\text{F}) = 1.8T(°\text{C}) + 32 \qquad \text{(1–12)}$$

A comparison of various temperature scales is given in Fig. 1–38.

The reference temperature chosen in the original Kelvin scale was 273.15 K (or 0°C), which is the temperature at which water freezes (or ice melts)

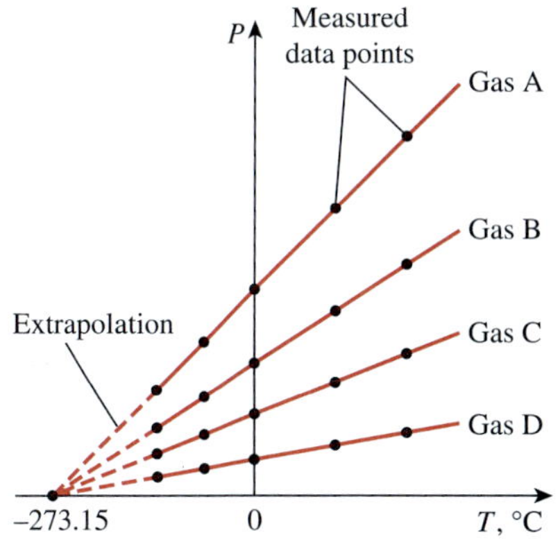

FIGURE 1–36
P versus *T* plots of the experimental data obtained from a constant-volume gas thermometer using four different gases at different (but low) pressures.

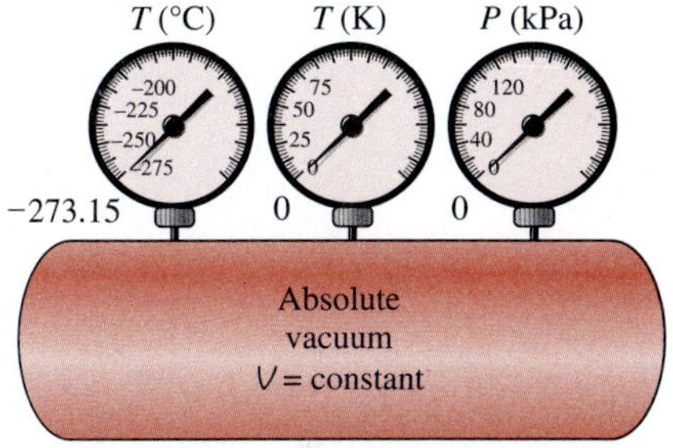

FIGURE 1–37
A constant-volume gas thermometer would read -273.15°C at absolute zero pressure.

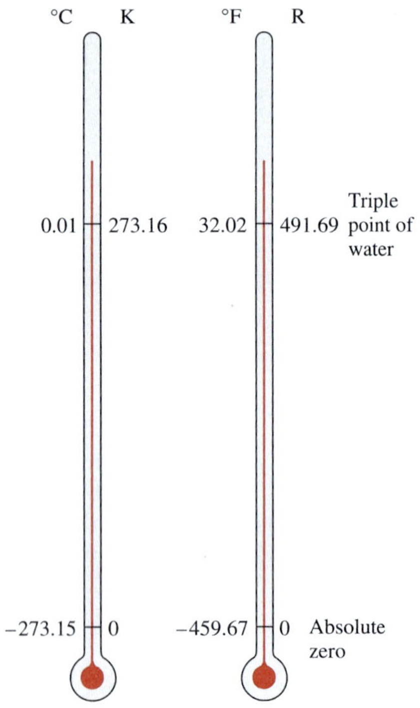

FIGURE 1–38
Comparison of temperature scales.

and water exists as a solid–liquid mixture in equilibrium under standard atmospheric pressure (the *ice point*). At the Tenth General Conference on Weights and Measures in 1954, the reference point was changed to a much more precisely reproducible point, the *triple point* of water (the state at which all three phases of water coexist in equilibrium), which is assigned the value 273.16 K. The Celsius scale was also redefined at this conference in terms of the ideal-gas temperature scale and a single fixed point, which is again the triple point of water with an assigned value of 0.01°C. The boiling temperature of water (the *steam point*) was experimentally determined to be again 100.00°C, and thus the new and old Celsius scales were in good agreement.

The International Temperature Scale of 1990 (ITS-90)

The *International Temperature Scale of 1990,* which supersedes the International Practical Temperature Scale of 1968 (IPTS-68), 1948 (ITPS-48), and 1927 (ITS-27), was adopted by the International Committee of Weights and Measures at its meeting in 1989 at the request of the Eighteenth General Conference on Weights and Measures. The ITS-90 is similar to its predecessors except that it is more refined with updated values of fixed temperatures, has an extended range, and conforms more closely to the thermodynamic temperature scale. On this scale, the unit of thermodynamic temperature T is again the kelvin (K), defined as the fraction 1/273.16 of the thermodynamic temperature of the triple point of water, which is sole defining fixed point of both the ITS-90 and the Kelvin scale and is the most important thermometric fixed point used in the calibration of thermometers to ITS-90.

The unit of Celsius temperature is the degree Celsius (°C), which is by definition equal in magnitude to the kelvin (K). A temperature difference may be expressed in kelvins or degrees Celsius. The ice point remains the same at 0°C (273.15 K) in both ITS-90 and ITPS-68, but the steam point is 99.975°C in ITS-90 (with an uncertainty of $\pm$0.005°C) whereas it was 100.000°C in IPTS-68. The change is due to precise measurements made by gas thermometry by paying particular attention to the effect of sorption (the impurities in a gas absorbed by the walls of the bulb at the reference temperature being desorbed at higher temperatures, causing the measured gas pressure to increase).

The ITS-90 extends upward from 0.65 K to the highest temperature practically measurable in terms of the Planck radiation law using monochromatic radiation. It is based on specifying definite temperature values on a number of fixed and easily reproducible points to serve as benchmarks and expressing the variation of temperature in a number of ranges and subranges in functional form.

In ITS-90, the temperature scale is considered in four ranges. In the range of 0.65 to 5 K, the temperature scale is defined in terms of the vapor pressure—temperature relations for ^3He and ^4He. Between 3 and 24.5561 K (the triple point of neon), it is defined by means of a properly calibrated helium gas thermometer. From 13.8033 K (the triple point of hydrogen) to 1234.93 K (the freezing point of silver), it is defined by means of platinum resistance thermometers calibrated at specified sets of defining fixed points.

Above 1234.93 K, it is defined in terms of the Planck radiation law and a suitable defining fixed point such as the freezing point of gold (1337.33 K).

We emphasize that the magnitudes of each division of 1 K and 1°C are identical (Fig. 1–39). Therefore, when we are dealing with temperature differences ΔT, the temperature interval on both scales is the same. Raising the temperature of a substance by 10°C is the same as raising it by 10 K. That is,

$$\Delta T(\text{K}) = \Delta T(°\text{C}) \quad \textbf{(1–13)}$$

$$\Delta T(\text{R}) = \Delta T(°\text{F}) \quad \textbf{(1–14)}$$

Some thermodynamic relations involve the temperature T and often the question arises of whether it is in K or °C. If the relation involves temperature differences (such as $a = b\Delta T$), it makes no difference and either can be used. However, if the relation involves temperatures only instead of temperature differences (such as $a = bT$) then K must be used. When in doubt, it is always safe to use K because there are virtually no situations in which the use of K is incorrect, but there are many thermodynamic relations that will yield an erroneous result if °C is used.

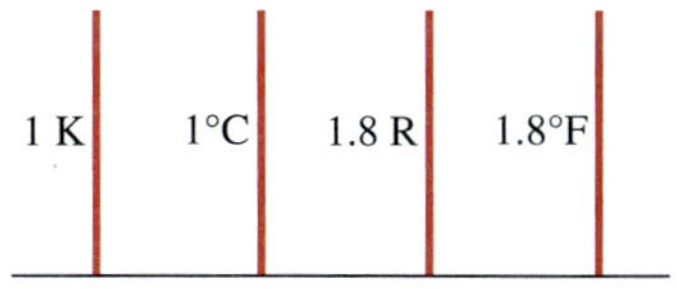

FIGURE 1–39
Comparison of magnitudes of various temperature units.

EXAMPLE 1–4 **Expressing Temperature Rise in Different Units**

During a heating process, the temperature of a system rises by 10°C. Express this rise in temperature in K, °F, and R.

SOLUTION The temperature rise of a system is to be expressed in different units.

Analysis This problem deals with temperature changes, which are identical in Kelvin and Celsius scales. Then,

$$\Delta T(\text{K}) = \Delta T(°\text{C}) = \mathbf{10\ K}$$

The temperature changes in Fahrenheit and Rankine scales are also identical and are related to the changes in Celsius and Kelvin scales through Eqs. 1–11 and 1–14:

$$\Delta T(\text{R}) = 1.8\ \Delta T(\text{K}) = (1.8)(10) = \mathbf{18\ R}$$

and

$$\Delta T(°\text{F}) = \Delta T(\text{R}) = \mathbf{18°F}$$

Discussion Note that the units °C and K are interchangeable when dealing with temperature differences.

1–9 ▪ PRESSURE

Pressure is defined as *a normal force exerted by a fluid per unit area.* We speak of pressure only when we deal with a gas or a liquid. The counterpart of pressure in solids is *normal stress.* Since pressure is defined as force per unit area, it has the unit of newtons per square meter (N/m^2), which is called a **pascal** (Pa). That is,

$$1\ \text{Pa} = 1\ \text{N/m}^2$$

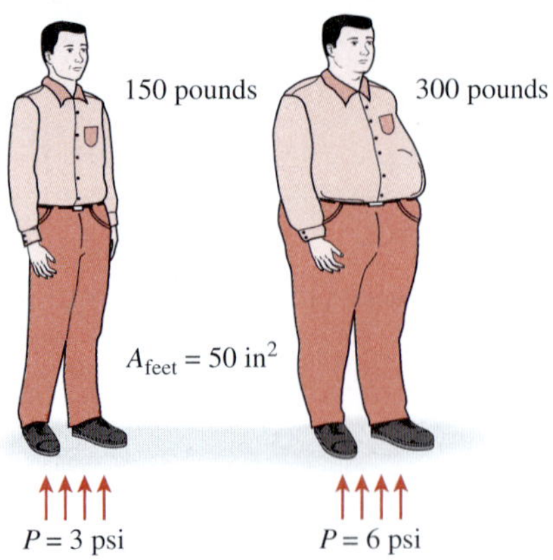

FIGURE 1–40
The normal stress (or "pressure") on the feet of a chubby person is much greater than on the feet of a slim person.

FIGURE 1–41
Some basic pressure gages.
Dresser Instruments, Dresser, Inc. Used by permission.

The pressure unit pascal is too small for pressures encountered in practice. Therefore, its multiples *kilopascal* (1 kPa = 10^3 Pa) and *megapascal* (1 MPa = 10^6 Pa) are commonly used. Three other pressure units commonly used in practice, especially in Europe, are *bar, standard atmosphere,* and *kilogram-force per square centimeter:*

$$1\ \text{bar} = 10^5\ \text{Pa} = 0.1\ \text{MPa} = 100\ \text{kPa}$$

$$1\ \text{atm} = 101{,}325\ \text{Pa} = 101.325\ \text{kPa} = 1.01325\ \text{bars}$$

$$1\ \text{kgf/cm}^2 = 9.807\ \text{N/cm}^2 = 9.807 \times 10^4\ \text{N/m}^2 = 9.807 \times 10^4\ \text{Pa}$$
$$= 0.9807\ \text{bar}$$
$$= 0.9679\ \text{atm}$$

Note that the pressure units bar, atm, and kgf/cm^2 are almost equivalent to each other. In the English system, the pressure unit is *pound-force per square inch* (lbf/in^2, or psi), and 1 atm = 14.696 psi. The pressure units kgf/cm^2 and lbf/in^2 are also denoted by kg/cm^2 and lb/in^2, respectively, and they are commonly used in tire gages. It can be shown that 1 kgf/cm^2 = 14.223 psi.

Pressure is also used for solids as synonymous to *normal stress,* which is force acting perpendicular to the surface per unit area. For example, a 150-pound person with a total foot imprint area of 50 in^2 exerts a pressure of 150 lbf/50 in^2 = 3.0 psi on the floor (Fig. 1–40). If the person stands on one foot, the pressure doubles. If the person gains excessive weight, he or she is likely to encounter foot discomfort because of the increased pressure on the foot (the size of the foot does not change with weight gain). This also explains how a person can walk on fresh snow without sinking by wearing large snowshoes, and how a person cuts with little effort when using a sharp knife.

The actual pressure at a given position is called the **absolute pressure**, and it is measured relative to absolute vacuum (i.e., absolute zero pressure). Most pressure-measuring devices, however, are calibrated to read zero in the atmosphere (Fig. 1–41), and so they indicate the difference between the absolute pressure and the local atmospheric pressure. This difference is called the **gage pressure**. Pressures below atmospheric pressure are called **vacuum pressures** and are measured by vacuum gages that indicate the difference between the atmospheric pressure and the absolute pressure. Absolute, gage, and vacuum pressures are all positive quantities and are related to each other by

$$P_{gage} = P_{abs} - P_{atm} \tag{1-15}$$

$$P_{vac} = P_{atm} - P_{abs} \tag{1-16}$$

This is illustrated in Fig. 1–42.

Like other pressure gages, the gage used to measure the air pressure in an automobile tire reads the gage pressure. Therefore, the common reading of 32 psi (2.25 kgf/cm^2) indicates a pressure of 32 psi above the atmospheric pressure. At a location where the atmospheric pressure is 14.3 psi, for example, the absolute pressure in the tire is 32 + 14.3 = 46.3 psi.

In thermodynamic relations and tables, absolute pressure is almost always used. Throughout this text, the pressure P will denote *absolute pressure*

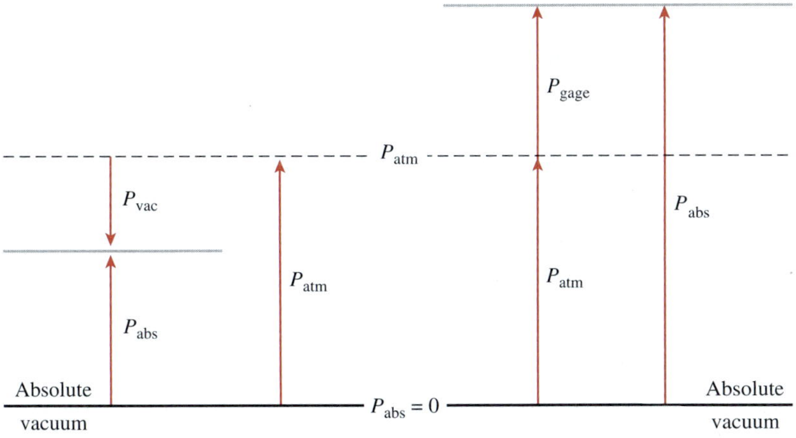

FIGURE 1–42
Absolute, gage, and vacuum pressures.

unless specified otherwise. Often the letters "a" (for absolute pressure) and "g" (for gage pressure) are added to pressure units (such as psia and psig) to clarify what is meant.

EXAMPLE 1–5 **Absolute Pressure of a Vacuum Chamber**

A vacuum gage connected to a chamber reads 5.8 psi at a location where the atmospheric pressure is 14.5 psi. Determine the absolute pressure in the chamber.

SOLUTION The gage pressure of a vacuum chamber is given. The absolute pressure in the chamber is to be determined.

Analysis The absolute pressure is easily determined from Eq. 1–16 to be

$$P_{abs} = P_{atm} - P_{vac} = 14.5 - 5.8 = \mathbf{8.7\ psi}$$

Discussion Note that the local value of the atmospheric pressure is used when determining the absolute pressure.

Pressure is the *compressive force* per unit area, and it gives the impression of being a vector. However, pressure at any point in a fluid is the same in all directions. That is, it has magnitude but not a specific direction, and thus it is a scalar quantity.

Variation of Pressure with Depth

It will come as no surprise to you that pressure in a fluid at rest does not change in the horizontal direction. This can be shown easily by considering a thin horizontal layer of fluid and doing a force balance in any horizontal direction. However, this is not the case in the vertical direction in a gravity field. Pressure in a fluid increases with depth because more fluid rests on deeper layers, and the effect of this "extra weight" on a deeper layer is balanced by an increase in pressure (Fig. 1–43).

To obtain a relation for the variation of pressure with depth, consider a rectangular fluid element of height Δz, length Δx, and unit depth (into the

FIGURE 1–43
The pressure of a fluid at rest increases with depth (as a result of added weight).

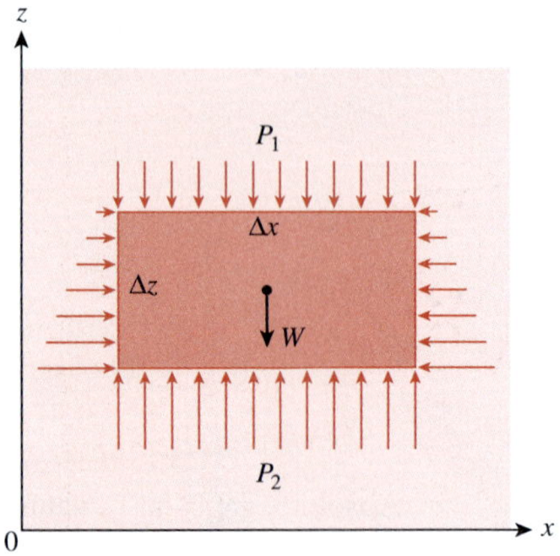

FIGURE 1–44
Free-body diagram of a rectangular fluid element in equilibrium.

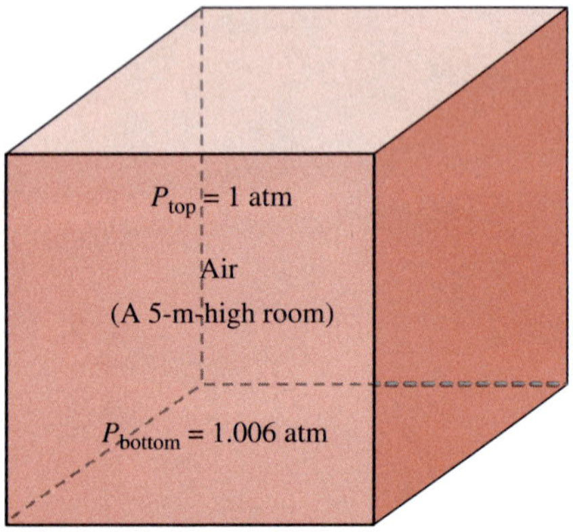

FIGURE 1–45
In a room filled with a gas, the variation of pressure with height is negligible.

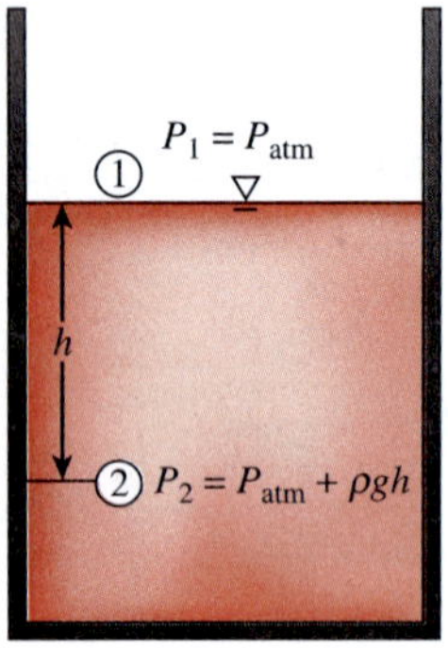

FIGURE 1–46
Pressure in a liquid at rest increases linearly with distance from the free surface.

page) in equilibrium, as shown in Fig. 1–44. Assuming the density of the fluid ρ to be constant, a force balance in the vertical z-direction gives

$$\sum F_z = ma_z = 0: \qquad P_2\,\Delta x - P_1\,\Delta x - \rho g\,\Delta x\,\Delta z = 0 \tag{1–17}$$

where $W = mg = \rho g\,\Delta x\,\Delta z$ is the weight of the fluid element. Dividing by Δx and rearranging gives

$$\Delta P = P_2 - P_1 = \rho g\,\Delta z = \gamma_s\,\Delta z \tag{1–18}$$

where $\gamma_s = \rho g$ is the *specific weight* of the fluid. Thus, we conclude that the pressure difference between two points in a constant density fluid is proportional to the vertical distance Δz between the points and the density ρ of the fluid. In other words, pressure in a fluid increases linearly with depth. This is what a diver experiences when diving deeper in a lake. For a given fluid, the vertical distance Δz is sometimes used as a measure of pressure, and it is called the *pressure head.*

We also conclude from Eq. 1–18 that for small to moderate distances, the variation of pressure with height is negligible for gases because of their low density. The pressure in a tank containing a gas, for example, can be considered to be uniform since the weight of the gas is too small to make a significant difference. Also, the pressure in a room filled with air can be assumed to be constant (Fig. 1–45).

If we take point 1 to be at the free surface of a liquid open to the atmosphere (Fig. 1–46), where the pressure is the atmospheric pressure P_{atm}, then the pressure at a depth h from the free surface becomes

$$P = P_{atm} + \rho g h \quad \text{or} \quad P_{gage} = \rho g h \tag{1–19}$$

Liquids are essentially incompressible substances, and thus the variation of density with depth is negligible. This is also the case for gases when the elevation change is not very large. The variation of density of liquids or gases with temperature can be significant, however, and may need to be considered when high accuracy is desired. Also, at great depths such as those encountered in oceans, the change in the density of a liquid can be significant because of the compression by the tremendous amount of liquid weight above.

The gravitational acceleration g varies from 9.807 m/s^2 at sea level to 9.764 m/s^2 at an elevation of 14,000 m where large passenger planes cruise. This is a change of just 0.4 percent in this extreme case. Therefore, g can be assumed to be constant with negligible error.

For fluids whose density changes significantly with elevation, a relation for the variation of pressure with elevation can be obtained by dividing Eq. 1–17 by $\Delta x\,\Delta z$, and taking the limit as $\Delta z \to 0$. It gives

$$\frac{dP}{dz} = -\rho g \tag{1–20}$$

The negative sign is due to our taking the positive z direction to be upward so that dP is negative when dz is positive since pressure decreases in an upward direction. When the variation of density with elevation is known, the pressure difference between points 1 and 2 can be determined by integration to be

$$\Delta P = P_2 - P_1 = -\int_1^2 \rho g\,dz \tag{1–21}$$

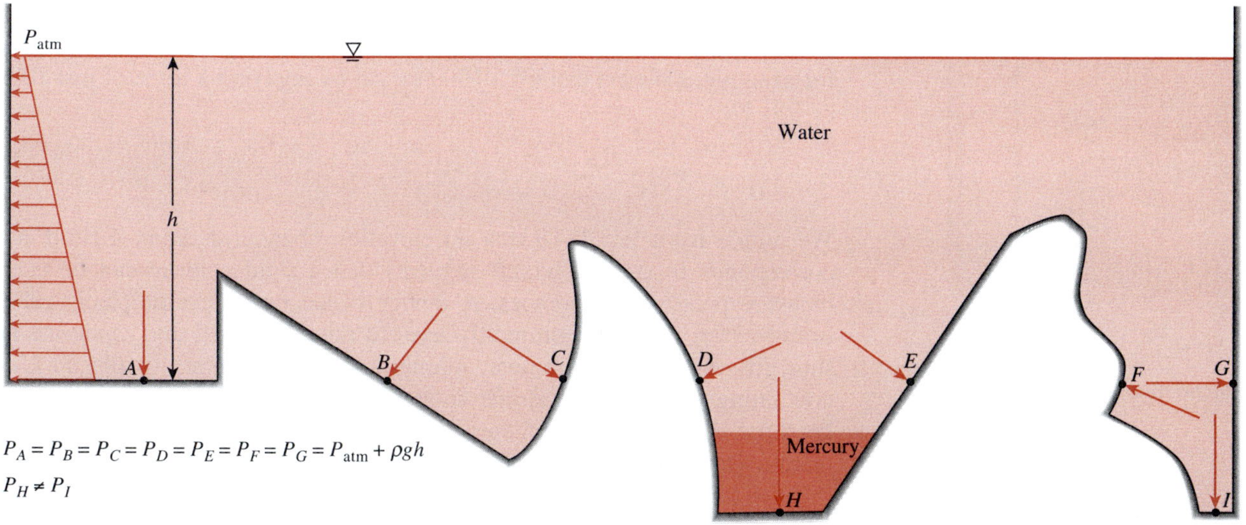

FIGURE 1–47
The pressure is the same at all points on a horizontal plane in a given fluid regardless of geometry, provided that the points are interconnected by the same fluid.

For constant density and constant gravitational acceleration, this relation reduces to Eq. 1–18, as expected.

Pressure in a fluid at rest is independent of the shape or cross section of the container. It changes with the vertical distance, but remains constant in other directions. Therefore, the pressure is the same at all points on a horizontal plane in a given fluid. The Dutch mathematician Simon Stevin (1548–1620) published in 1586 the principle illustrated in Fig. 1–47. Note that the pressures at points A, B, C, D, E, F, and G are the same since they are at the same depth, and they are interconnected by the same static fluid. However, the pressures at points H and I are not the same since these two points cannot be interconnected by the same fluid (i.e., we cannot draw a curve from point I to point H while remaining in the same fluid at all times), although they are at the same depth. (Can you tell at which point the pressure is higher?) Also, the pressure force exerted by the fluid is always normal to the surface at the specified points.

A consequence of the pressure in a fluid remaining constant in the horizontal direction is that *the pressure applied to a confined fluid increases the pressure throughout by the same amount*. This is called **Pascal's law**, after Blaise Pascal (1623–1662). Pascal also knew that the force applied by a fluid is proportional to the surface area. He realized that two hydraulic cylinders of different areas could be connected, and the larger could be used to exert a proportionally greater force than that applied to the smaller. "Pascal's machine" has been the source of many inventions that are a part of our daily lives such as hydraulic brakes and lifts. This is what enables us to lift a car easily by one arm, as shown in Fig. 1–48. Noting that $P_1 = P_2$ since both pistons are at the same level (the effect of small height differences is negligible, especially at high pressures), the ratio of output force to input force is determined to be

$$P_1 = P_2 \quad \rightarrow \quad \frac{F_1}{A_1} = \frac{F_2}{A_2} \quad \rightarrow \quad \frac{F_2}{F_1} = \frac{A_2}{A_1} \qquad \textbf{(1–22)}$$

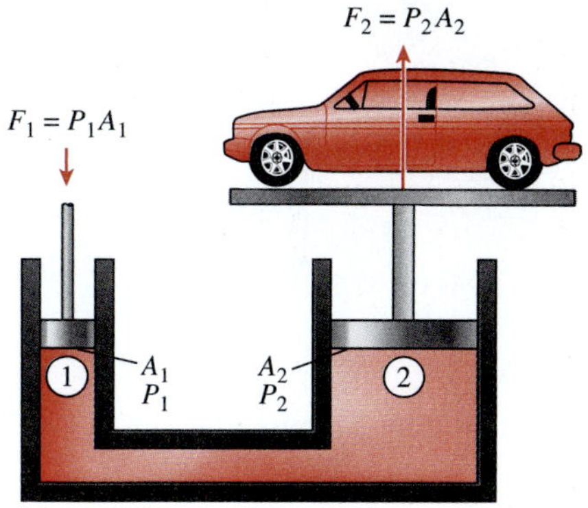

FIGURE 1–48
Lifting of a large weight by a small force by the application of Pascal's law.

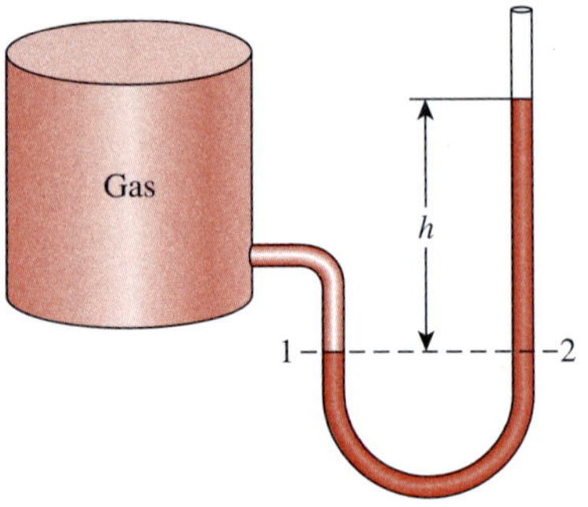

FIGURE 1–49
The basic manometer.

The area ratio A_2/A_1 is called the *ideal mechanical advantage* of the hydraulic lift. Using a hydraulic car jack with a piston area ratio of $A_2/A_1 = 10$, for example, a person can lift a 1000-kg car by applying a force of just 100 kgf (=981 N).

1–10 ▪ THE MANOMETER

We notice from Eq. 1–18 that an elevation change of Δz in a fluid at rest corresponds to $\Delta P/\rho g$, which suggests that a fluid column can be used to measure pressure differences. A device based on this principle is called a **manometer**, and it is commonly used to measure small and moderate pressure differences. A manometer mainly consists of a glass or plastic U-tube containing one or more fluids such as mercury, water, alcohol, or oil. To keep the size of the manometer to a manageable level, heavy fluids such as mercury are used if large pressure differences are anticipated.

Consider the manometer shown in Fig. 1–49 that is used to measure the pressure in the tank. Since the gravitational effects of gases are negligible, the pressure anywhere in the tank and at position 1 has the same value. Furthermore, since pressure in a fluid does not vary in the horizontal direction within a fluid, the pressure at point 2 is the same as the pressure at point 1, $P_2 = P_1$.

The differential fluid column of height h is in static equilibrium, and it is open to the atmosphere. Then the pressure at point 2 is determined directly from Eq. 1–19 to be

$$P_2 = P_{atm} + \rho gh \tag{1–23}$$

where ρ is the density of the fluid in the tube. Note that the cross-sectional area of the tube has no effect on the differential height h, and thus the pressure exerted by the fluid. However, the diameter of the tube should be large enough (more than a few millimeters) to ensure that the surface tension effect and thus the capillary rise is negligible.

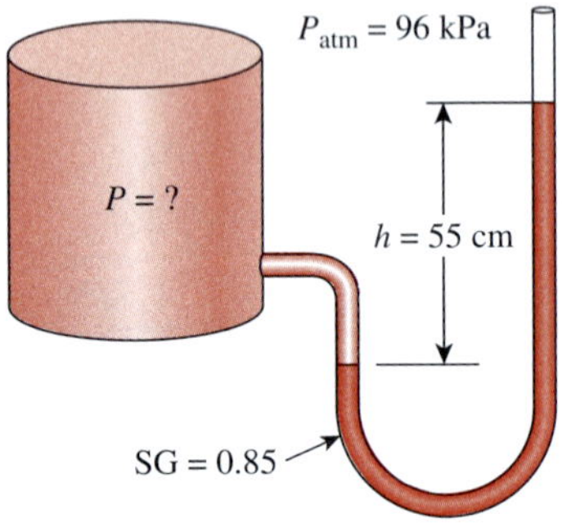

FIGURE 1–50
Schematic for Example 1–6.

EXAMPLE 1–6 Measuring Pressure with a Manometer

A manometer is used to measure the pressure in a tank. The fluid used has a specific gravity of 0.85, and the manometer column height is 55 cm, as shown in Fig. 1–50. If the local atmospheric pressure is 96 kPa, determine the absolute pressure within the tank.

SOLUTION The reading of a manometer attached to a tank and the atmospheric pressure are given. The absolute pressure in the tank is to be determined.
Assumptions The fluid in the tank is a gas whose density is much lower than the density of manometer fluid.
Properties The specific gravity of the manometer fluid is given to be 0.85. We take the standard density of water to be 1000 kg/m³.
Analysis The density of the fluid is obtained by multiplying its specific gravity by the density of water, which is taken to be 1000 kg/m³:

$$\rho = SG\,(\rho_{H_2O}) = (0.85)(1000\ kg/m^3) = 850\ kg/m^3$$

Then from Eq. 1–23,

$$P = P_{atm} + \rho gh$$

$$= 96 \text{ kPa} + (850 \text{ kg/m}^3)(9.81 \text{ m/s}^2)(0.55 \text{ m})\left(\frac{1 \text{ N}}{1 \text{ kg·m/s}^2}\right)\left(\frac{1 \text{ kPa}}{1000 \text{ N/m}^2}\right)$$

$$= \mathbf{100.6\ kPa}$$

Discussion Note that the gage pressure in the tank is 4.6 kPa.

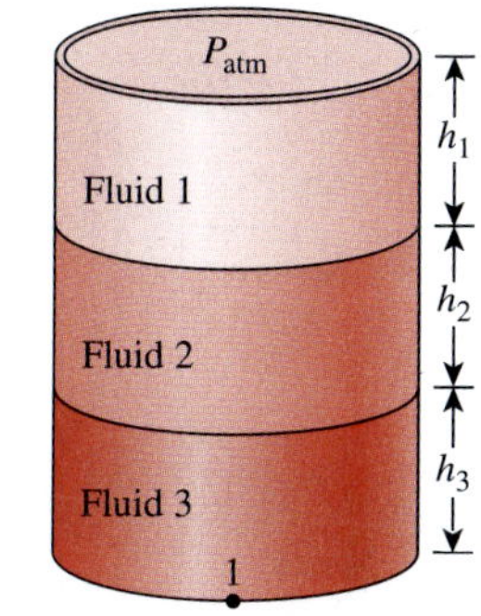

FIGURE 1–51
In stacked-up fluid layers, the pressure change across a fluid layer of density ρ and height h is ρgh.

Many engineering problems and some manometers involve multiple immiscible fluids of different densities stacked on top of each other. Such systems can be analyzed easily by remembering that (1) the pressure change across a fluid column of height h is $\Delta P = \rho gh$, (2) pressure increases downward in a given fluid and decreases upward (i.e., $P_{bottom} > P_{top}$), and (3) two points at the same elevation in a continuous fluid at rest are at the same pressure.

The last principle, which is a result of *Pascal's law,* allows us to "jump" from one fluid column to the next in manometers without worrying about pressure change as long as we don't jump over a different fluid, and the fluid is at rest. Then the pressure at any point can be determined by starting with a point of known pressure and adding or subtracting ρgh terms as we advance toward the point of interest. For example, the pressure at the bottom of the tank in Fig. 1–51 can be determined by starting at the free surface where the pressure is P_{atm}, moving downward until we reach point 1 at the bottom, and setting the result equal to P_1. It gives

$$P_{atm} + \rho_1 gh_1 + \rho_2 gh_2 + \rho_3 gh_3 = P_1$$

In the special case of all fluids having the same density, this relation reduces to Eq. 1–23, as expected.

Manometers are particularly well-suited to measure pressure drops across a horizontal flow section between two specified points due to the presence of a device such as a valve or heat exchanger or any resistance to flow. This is done by connecting the two legs of the manometer to these two points, as shown in Fig. 1–52. The working fluid can be either a gas or a liquid whose density is ρ_1. The density of the manometer fluid is ρ_2, and the differential fluid height is h.

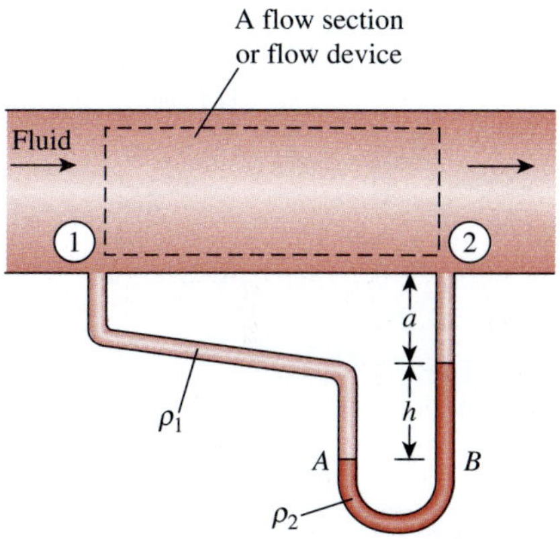

FIGURE 1–52
Measuring the pressure drop across a flow section or a flow device by a differential manometer.

A relation for the pressure difference $P_1 - P_2$ can be obtained by starting at point 1 with P_1, moving along the tube by adding or subtracting the ρgh terms until we reach point 2, and setting the result equal to P_2:

$$P_1 + \rho_1 g(a + h) - \rho_2 gh - \rho_1 ga = P_2 \qquad \textbf{(1–24)}$$

Note that we jumped from point A horizontally to point B and ignored the part underneath since the pressure at both points is the same. Simplifying,

$$P_1 - P_2 = (\rho_2 - \rho_1)gh \qquad \textbf{(1–25)}$$

Note that the distance a has no effect on the result, but must be included in the analysis. Also, when the fluid flowing in the pipe is a gas, then $\rho_1 \ll \rho_2$ and the relation in Eq. 1–25 simplifies to $P_1 - P_2 \cong \rho_2 gh$.

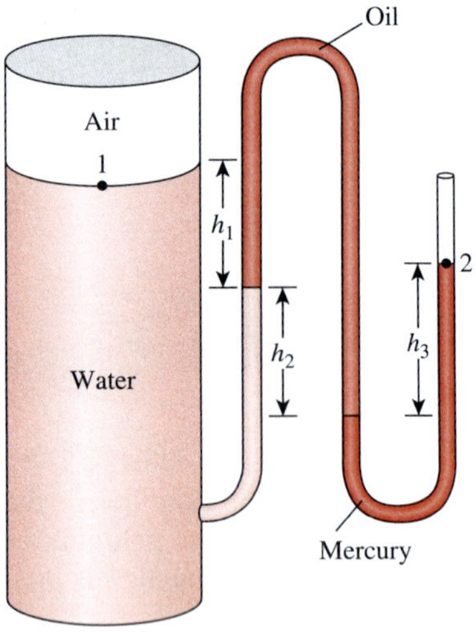

FIGURE 1–53
Schematic for Example 1–7. (Drawing not to scale.)

EXAMPLE 1–7 Measuring Pressure with a Multifluid Manometer

The water in a tank is pressurized by air, and the pressure is measured by a multifluid manometer as shown in Fig. 1–53. The tank is located on a mountain at an altitude of 1400 m where the atmospheric pressure is 85.6 kPa. Determine the air pressure in the tank if $h_1 = 0.1$ m, $h_2 = 0.2$ m, and $h_3 = 0.35$ m. Take the densities of water, oil, and mercury to be 1000 kg/m^3, 850 kg/m^3, and 13,600 kg/m^3, respectively.

SOLUTION The pressure in a pressurized water tank is measured by a multifluid manometer. The air pressure in the tank is to be determined.

Assumption The air pressure in the tank is uniform (i.e., its variation with elevation is negligible due to its low density), and thus we can determine the pressure at the air–water interface.

Properties The densities of water, oil, and mercury are given to be 1000 kg/m^3, 850 kg/m^3, and 13,600 kg/m^3, respectively.

Analysis Starting with the pressure at point 1 at the air–water interface, moving along the tube by adding or subtracting the ρgh terms until we reach point 2, and setting the result equal to P_{atm} since the tube is open to the atmosphere gives

$$P_1 + \rho_{water} g h_1 + \rho_{oil} g h_2 - \rho_{mercury} g h_3 = P_{atm}$$

Solving for P_1 and substituting,

$$\begin{aligned} P_1 &= P_{atm} - \rho_{water} g h_1 - \rho_{oil} g h_2 + \rho_{mercury} g h_3 \\ &= P_{atm} + g(\rho_{mercury} h_3 - \rho_{water} h_1 - \rho_{oil} h_2) \\ &= 85.6 \text{ kPa} + (9.81 \text{ m/s}^2)[(13{,}600 \text{ kg/m}^3)(0.35\text{m}) - 1000 \text{ kg/m}^3)(0.1 \text{ m}) \\ &\quad - (850 \text{ kg/m}^3)(0.2 \text{ m})]\left(\frac{1 \text{ N}}{1 \text{ kg·m/s}^2}\right)\left(\frac{1 \text{ kPa}}{1000 \text{ N/m}^2}\right) \\ &= \mathbf{130 \text{ kPa}} \end{aligned}$$

Discussion Note that jumping horizontally from one tube to the next and realizing that pressure remains the same in the same fluid simplifies the analysis considerably. Also note that mercury is a toxic fluid, and mercury manometers and thermometers are being replaced by ones with safer fluids because of the risk of exposure to mercury vapor during an accident.

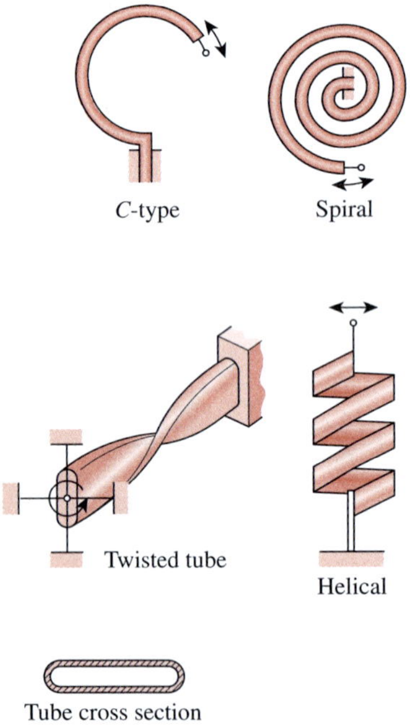

FIGURE 1–54
Various types of Bourdon tubes used to measure pressure.

Other Pressure Measurement Devices

Another type of commonly used mechanical pressure measurement device is the **Bourdon tube**, named after the French engineer and inventor Eugene Bourdon (1808–1884), which consists of a hollow metal tube bent like a hook whose end is closed and connected to a dial indicator needle (Fig. 1–54). When the tube is open to the atmosphere, the tube is undeflected, and the needle on the dial at this state is calibrated to read zero (gage pressure). When the fluid inside the tube is pressurized, the tube stretches and moves the needle in proportion to the pressure applied.

Electronics have made their way into every aspect of life, including pressure measurement devices. Modern pressure sensors, called **pressure transducers**, use various techniques to convert the pressure effect to an electrical effect such as a change in voltage, resistance, or capacitance. Pressure transducers are smaller and faster, and they can be more sensitive, reliable, and precise than their mechanical counterparts. They can measure pressures from less than a millionth of 1 atm to several thousands of atm.

A wide variety of pressure transducers is available to measure gage, absolute, and differential pressures in a wide range of applications. *Gage pressure transducers* use the atmospheric pressure as a reference by venting the back side of the pressure-sensing diaphragm to the atmosphere, and they give a zero signal output at atmospheric pressure regardless of altitude. The *absolute pressure transducers* are calibrated to have a zero signal output at full vacuum. *Differential pressure transducers* measure the pressure difference between two locations directly instead of using two pressure transducers and taking their difference.

Strain-gage pressure transducers work by having a diaphragm deflect between two chambers open to the pressure inputs. As the diaphragm stretches in response to a change in pressure difference across it, the strain gage stretches and a Wheatstone bridge circuit amplifies the output. A capacitance transducer works similarly, but capacitance change is measured instead of resistance change as the diaphragm stretches.

Piezoelectric transducers, also called solid-state pressure transducers, work on the principle that an electric potential is generated in a crystalline substance when it is subjected to mechanical pressure. This phenomenon, first discovered by brothers Pierre and Jacques Curie in 1880, is called the piezoelectric (or press-electric) effect. Piezoelectric pressure transducers have a much faster frequency response compared to the diaphragm units and are very suitable for high-pressure applications, but they are generally not as sensitive as the diaphragm-type transducers.

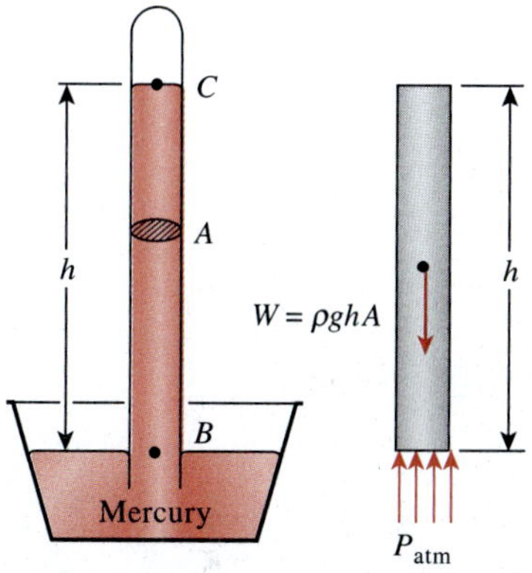

FIGURE 1–55
The basic barometer.

1–11 ▪ THE BAROMETER AND ATMOSPHERIC PRESSURE

Atmospheric pressure is measured by a device called a **barometer**; thus, the atmospheric pressure is often referred to as the *barometric pressure*.

The Italian Evangelista Torricelli (1608–1647) was the first to conclusively prove that the atmospheric pressure can be measured by inverting a mercury-filled tube into a mercury container that is open to the atmosphere, as shown in Fig. 1–55. The pressure at point B is equal to the atmospheric pressure, and the pressure at C can be taken to be zero since there is only mercury vapor above point C and the pressure is very low relative to P_{atm} and can be neglected to an excellent approximation. Writing a force balance in the vertical direction gives

$$P_{atm} = \rho gh \tag{1–26}$$

where ρ is the density of mercury, g is the local gravitational acceleration, and h is the height of the mercury column above the free surface. Note that the length and the cross-sectional area of the tube have no effect on the height of the fluid column of a barometer (Fig. 1–56).

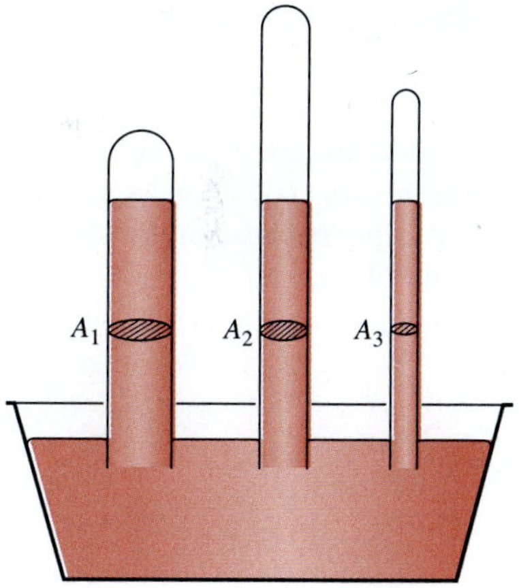

FIGURE 1–56
The length or the cross-sectional area of the tube has no effect on the height of the fluid column of a barometer, provided that the tube diameter is large enough to avoid surface tension (capillary) effects.

A frequently used pressure unit is the *standard atmosphere*, which is defined as the pressure produced by a column of mercury 760 mm in height at 0°C (ρ_{Hg} = 13,595 kg/m^3) under standard gravitational acceleration (g = 9.807 m/s^2). If water instead of mercury were used to measure the standard atmospheric pressure, a water column of about 10.3 m would be needed. Pressure is sometimes expressed (especially by weather forecasters) in terms of the height of the mercury column. The standard atmospheric pressure, for example, is 760 mmHg (29.92 inHg) at 0°C. The unit mmHg is also called the **torr** in honor of Torricelli. Therefore, 1 atm = 760 torr and 1 torr = 133.3 Pa.

The standard atmospheric pressure P_{atm} changes from 101.325 kPa at sea level to 89.88, 79.50, 54.05, 26.5, and 5.53 kPa at altitudes of 1000, 2000, 5000, 10,000, and 20,000 meters, respectively. The standard atmospheric pressure in Denver (elevation = 1610 m), for example, is 83.4 kPa.

Remember that the atmospheric pressure at a location is simply the weight of the air above that location per unit surface area. Therefore, it changes not only with elevation but also with weather conditions.

The decline of atmospheric pressure with elevation has far-reaching ramifications in daily life. For example, cooking takes longer at high altitudes since water boils at a lower temperature at lower atmospheric pressures. Nose bleeding is a common experience at high altitudes since the difference between the blood pressure and the atmospheric pressure is larger in this case, and the delicate walls of veins in the nose are often unable to withstand this extra stress.

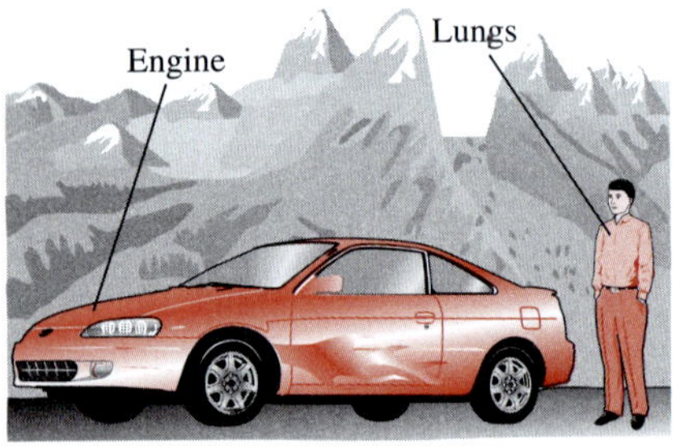

FIGURE 1–57
At high altitudes, a car engine generates less power and a person gets less oxygen because of the lower density of air.

For a given temperature, the density of air is lower at high altitudes, and thus a given volume contains less air and less oxygen. So it is no surprise that we tire more easily and experience breathing problems at high altitudes. To compensate for this effect, people living at higher altitudes develop more efficient lungs. Similarly, a 2.0-L car engine will act like a 1.7-L car engine at 1500 m altitude (unless it is turbocharged) because of the 15 percent drop in pressure and thus 15 percent drop in the density of air (Fig. 1–57). A fan or compressor will displace 15 percent less air at that altitude for the same volume displacement rate. Therefore, larger cooling fans may need to be selected for operation at high altitudes to ensure the specified mass flow rate. The lower pressure and thus lower density also affects lift and drag: airplanes need a longer runway at high altitudes to develop the required lift, and they climb to very high altitudes for cruising for reduced drag and thus better fuel efficiency.

EXAMPLE 1–8 Measuring Atmospheric Pressure with a Barometer

Determine the atmospheric pressure at a location where the barometric reading is 740 mm Hg and the gravitational acceleration is g = 9.81 m/s^2. Assume the temperature of mercury to be 10°C, at which its density is 13,570 kg/m^3.

SOLUTION The barometric reading at a location in height of mercury column is given. The atmospheric pressure is to be determined.
Assumptions The temperature of mercury is 10°C.
Properties The density of mercury is given to be 13,570 kg/m^3.

Analysis From Eq. 1–26, the atmospheric pressure is determined to be

$$P_{atm} = \rho g h$$
$$= (13{,}570\ \text{kg/m}^3)(9.81\ \text{m/s}^2)(0.74\ \text{m})\left(\frac{1\ \text{N}}{1\ \text{kg·m/s}^2}\right)\left(\frac{1\ \text{kPa}}{1000\ \text{N/m}^2}\right)$$
$$= \mathbf{98.5\ kPa}$$

Discussion Note that density changes with temperature, and thus this effect should be considered in calculations.

EXAMPLE 1–9 Effect of Piston Weight on Pressure in a Cylinder

The piston of a vertical piston–cylinder device containing a gas has a mass of 60 kg and a cross-sectional area of 0.04 m^2, as shown in Fig. 1–58. The local atmospheric pressure is 0.97 bar, and the gravitational acceleration is 9.81 m/s^2. (*a*) Determine the pressure inside the cylinder. (*b*) If some heat is transferred to the gas and its volume is doubled, do you expect the pressure inside the cylinder to change?

SOLUTION A gas is contained in a vertical cylinder with a heavy piston. The pressure inside the cylinder and the effect of volume change on pressure are to be determined.

Assumptions Friction between the piston and the cylinder is negligible.

Analysis (*a*) The gas pressure in the piston–cylinder device depends on the atmospheric pressure and the weight of the piston. Drawing the free-body diagram of the piston as shown in Fig. 1–58 and balancing the vertical forces yield

$$PA = P_{atm}A + W$$

Solving for P and substituting,

$$P = P_{atm} + \frac{mg}{A}$$
$$= 0.97\ \text{bar} + \frac{(60\ \text{kg})(9.81\ \text{m/s}^2)}{(0.04\ \text{m}^2)}\left(\frac{1\ \text{N}}{1\ \text{kg·m/s}^2}\right)\left(\frac{1\ \text{bar}}{10^5\ \text{N/m}^2}\right)$$
$$= \mathbf{1.12\ bar}$$

(*b*) The volume change will have no effect on the free-body diagram drawn in part (*a*), and therefore the pressure inside the cylinder will remain the same.

Discussion If the gas behaves as an ideal gas, the absolute temperature doubles when the volume is doubled at constant pressure.

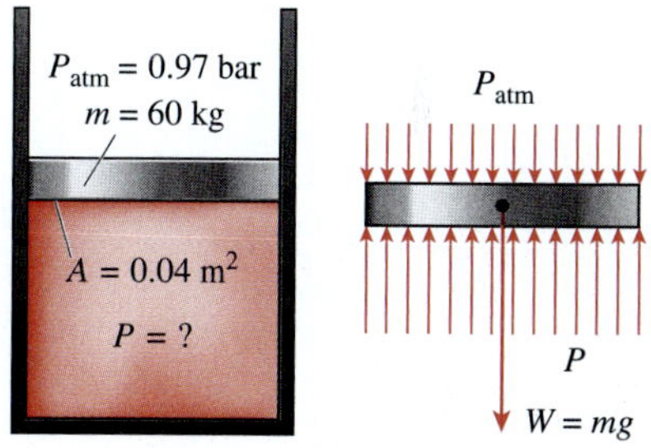

FIGURE 1–58
Schematic for Example 1–9; and the free-body diagram of the piston.

EXAMPLE 1–10 Hydrostatic Pressure in a Solar Pond with Variable Density

Solar ponds are small artificial lakes of a few meters deep that are used to store solar energy. The rise of heated (and thus less dense) water to the surface is prevented by adding salt at the pond bottom. In a typical salt gradient

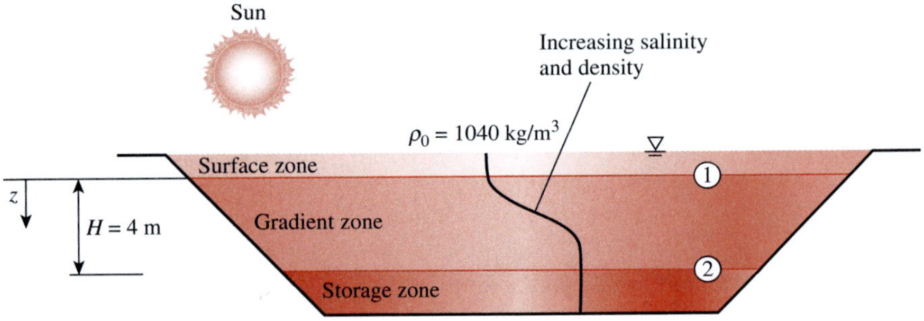

FIGURE 1–59
Schematic for Example 1–10.

solar pond, the density of water increases in the gradient zone, as shown in Fig. 1–59, and the density can be expressed as

$$\rho = \rho_0\sqrt{1 + \tan^2\left(\frac{\pi}{4}\frac{z}{H}\right)}$$

where ρ_0 is the density on the water surface, z is the vertical distance measured downward from the top of the gradient zone, and H is the thickness of the gradient zone. For $H = 4$ m, $\rho_0 = 1040$ kg/m^3, and a thickness of 0.8 m for the surface zone, calculate the gage pressure at the bottom of the gradient zone.

SOLUTION The variation of density of saline water in the gradient zone of a solar pond with depth is given. The gage pressure at the bottom of the gradient zone is to be determined.

Assumptions The density in the surface zone of the pond is constant.

Properties The density of brine on the surface is given to be 1040 kg/m^3.

Analysis We label the top and the bottom of the gradient zone as 1 and 2, respectively. Noting that the density of the surface zone is constant, the gage pressure at the bottom of the surface zone (which is the top of the gradient zone) is

$$P_1 = \rho g h_1 = (1040 \text{ kg/m}^3)(9.81 \text{ m/s}^2)(0.8 \text{ m})\left(\frac{1 \text{ kN}}{1000 \text{ kg·m/s}^2}\right) = 8.16 \text{ kPa}$$

since 1 kN/m^2 = 1 kPa. The differential change in hydrostatic pressure across a vertical distance of dz is given by

$$dP = \rho g\, dz$$

Integrating from the top of the gradient zone (point 1 where $z = 0$) to any location z in the gradient zone (no subscript) gives

$$P - P_1 = \int_0^z \rho g\, dz \quad \rightarrow \quad P = P_1 + \int_0^z \rho_0\sqrt{1 + \tan^2\left(\frac{\pi}{4}\frac{z}{H}\right)}\, g\, dz$$

Performing the integration gives the variation of gage pressure in the gradient zone to be

$$P = P_1 + \rho_0 g \frac{4H}{\pi} \sinh^{-1}\left(\tan \frac{\pi}{4}\frac{z}{H}\right)$$

Then the pressure at the bottom of the gradient zone ($z = H = 4$ m) becomes

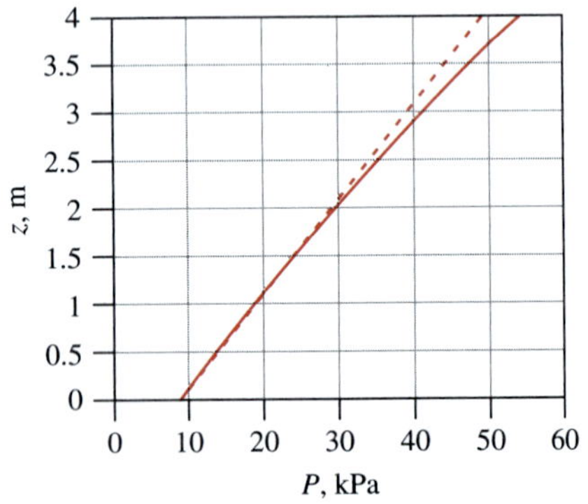

FIGURE 1–60
The variation of gage pressure with depth in the gradient zone of the solar pond.

$$P_2 = 8.16\text{ kPa} + (1040\text{ kg/m}^3)(9.81\text{ m/s}^2)\frac{4(4\text{ m})}{\pi}$$
$$\times \sinh^{-1}\left(\tan\frac{\pi}{4}\frac{4}{4}\right)\left(\frac{1\text{ kN}}{1000\text{ kg·m/s}^2}\right)$$
$$= \mathbf{54.0\text{ kPa (gage)}}$$

Discussion The variation of gage pressure in the gradient zone with depth is plotted in Fig. 1–60. The dashed line indicates the hydrostatic pressure for the case of constant density at 1040 kg/m^3 and is given for reference. Note that the variation of pressure with depth is not linear when density varies with depth.

1–12 ▪ PROBLEM-SOLVING TECHNIQUE

The first step in learning any science is to grasp the fundamentals and to gain a sound knowledge of it. The next step is to master the fundamentals by testing this knowledge. This is done by solving significant real-world problems. Solving such problems, especially complicated ones, requires a systematic approach. By using a step-by-step approach, an engineer can reduce the solution of a complicated problem into the solution of a series of simple problems (Fig. 1–61). When you are solving a problem, we recommend that you use the following steps zealously as applicable. This will help you avoid some of the common pitfalls associated with problem solving.

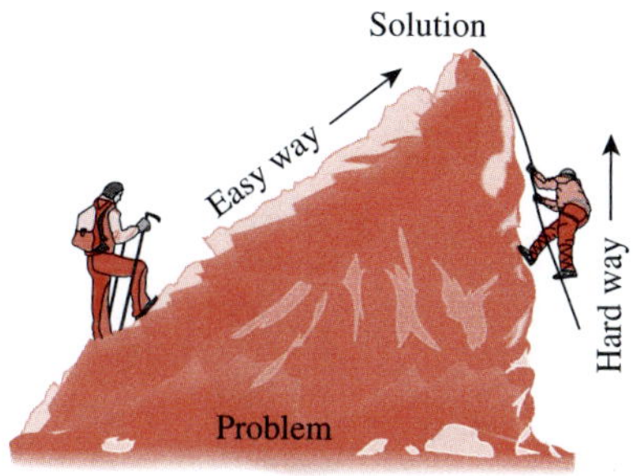

FIGURE 1–61
A step-by-step approach can greatly simplify problem solving.

Step 1: Problem Statement

In your own words, briefly state the problem, the key information given, and the quantities to be found. This is to make sure that you understand the problem and the objectives before you attempt to solve the problem.

Step 2: Schematic

Draw a realistic sketch of the physical system involved, and list the relevant information on the figure. The sketch does not have to be something elaborate, but it should resemble the actual system and show the key features. Indicate any energy and mass interactions with the surroundings. Listing the given information on the sketch helps one to see the entire problem at once. Also, check for properties that remain constant during a process (such as temperature during an isothermal process), and indicate them on the sketch.

Step 3: Assumptions and Approximations

State any appropriate assumptions and approximations made to simplify the problem to make it possible to obtain a solution. Justify the questionable assumptions. Assume reasonable values for missing quantities that are necessary. For example, in the absence of specific data for atmospheric pressure, it can be taken to be 1 atm. However, it should be noted in the analysis that the atmospheric pressure decreases with increasing elevation. For example, it drops to 0.83 atm in Denver (elevation 1610 m) (Fig. 1–62).

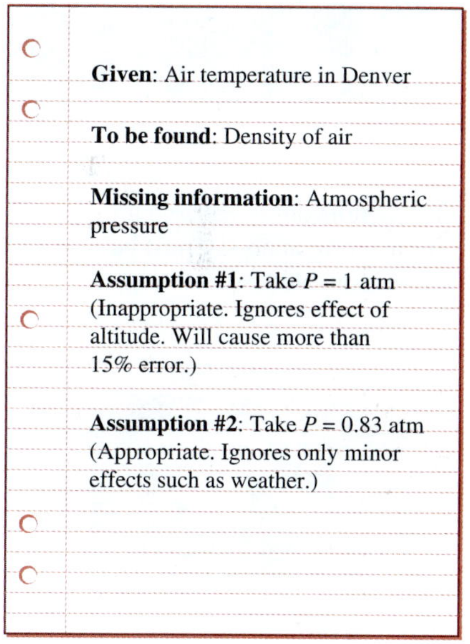

FIGURE 1–62
The assumptions made while solving an engineering problem must be reasonable and justifiable.

Step 4: Physical Laws

Apply all the relevant basic physical laws and principles (such as the conservation of mass), and reduce them to their simplest form by utilizing the assumptions made. However, the region to which a physical law is applied must be clearly identified first. For example, the increase in speed of water flowing through a nozzle is analyzed by applying conservation of mass between the inlet and outlet of the nozzle.

Step 5: Properties

Determine the unknown properties at known states necessary to solve the problem from property relations or tables. List the properties separately, and indicate their source, if applicable.

Step 6: Calculations

Substitute the known quantities into the simplified relations and perform the calculations to determine the unknowns. Pay particular attention to the units and unit cancellations, and remember that a dimensional quantity without a unit is meaningless. Also, don't give a false implication of high precision by copying all the digits from the screen of the calculator—round the results to an appropriate number of significant digits (see p. 37).

Step 7: Reasoning, Verification, and Discussion

Check to make sure that the results obtained are reasonable and intuitive, and verify the validity of the questionable assumptions. Repeat the calculations that resulted in unreasonable values. For example, insulating a water heater that uses $80 worth of natural gas a year cannot result in savings of $200 a year (Fig. 1–63).

FIGURE 1–63
The results obtained from an engineering analysis must be checked for reasonableness.

Also, point out the significance of the results, and discuss their implications. State the conclusions that can be drawn from the results, and any recommendations that can be made from them. Emphasize the limitations under which the results are applicable, and caution against any possible misunderstandings and using the results in situations where the underlying assumptions do not apply. For example, if you determined that wrapping a water heater with a $20 insulation jacket will reduce the energy cost by $30 a year, indicate that the insulation will pay for itself from the energy it saves in less than a year. However, also indicate that the analysis does not consider labor costs, and that this will be the case if you install the insulation yourself.

Keep in mind that the solutions you present to your instructors, and any engineering analysis presented to others, is a form of communication. Therefore neatness, organization, completeness, and visual appearance are of utmost importance for maximum effectiveness (Fig. 1–64). Besides, neatness also serves as a great checking tool since it is very easy to spot errors and inconsistencies in neat work. Carelessness and skipping steps to save time often end up costing more time and unnecessary anxiety.

FIGURE 1–64
Neatness and organization are highly valued by employers.

The approach described here is used in the solved example problems without explicitly stating each step, as well as in the Solutions Manual of this text. For some problems, some of the steps may not be applicable or necessary. For example, often it is not practical to list the properties separately.

However, we cannot overemphasize the importance of a logical and orderly approach to problem solving. Most difficulties encountered while solving a problem are not due to a lack of knowledge; rather, they are due to a lack of organization. You are strongly encouraged to follow these steps in problem solving until you develop your own approach that works best for you.

Engineering Software Packages

You may be wondering why we are about to undertake an in-depth study of the fundamentals of another engineering science. After all, almost all such problems we are likely to encounter in practice can be solved using one of several sophisticated software packages readily available in the market today. These software packages not only give the desired numerical results, but also supply the outputs in colorful graphical form for impressive presentations. It is unthinkable to practice engineering today without using some of these packages. This tremendous computing power available to us at the touch of a button is both a blessing and a curse. It certainly enables engineers to solve problems easily and quickly, but it also opens the door for abuses and misinformation. In the hands of poorly educated people, these software packages are as dangerous as sophisticated powerful weapons in the hands of poorly trained soldiers.

Thinking that a person who can use the engineering software packages without proper training on fundamentals can practice engineering is like thinking that a person who can use a wrench can work as a car mechanic. If it were true that the engineering students do not need all these fundamental courses they are taking because practically everything can be done by computers quickly and easily, then it would also be true that the employers would no longer need high-salaried engineers since any person who knows how to use a word-processing program can also learn how to use those software packages. However, the statistics show that the need for engineers is on the rise, not on the decline, despite the availability of these powerful packages.

We should always remember that all the computing power and the engineering software packages available today are just *tools*, and tools have meaning only in the hands of masters. Having the best word-processing program does not make a person a good writer, but it certainly makes the job of a good writer much easier and makes the writer more productive (Fig. 1–65). Hand calculators did not eliminate the need to teach our children how to add or subtract, and the sophisticated medical software packages did not take the place of medical school training. Neither will engineering software packages replace the traditional engineering education. They will simply cause a shift in emphasis in the courses from mathematics to physics. That is, more time will be spent in the classroom discussing the physical aspects of the problems in greater detail, and less time on the mechanics of solution procedures.

FIGURE 1–65
An excellent word-processing program does not make a person a good writer; it simply makes a good writer a more efficient writer.
© Vol. 80/PhotoDisc/Getty RF.

All these marvelous and powerful tools available today put an extra burden on today's engineers. They must still have a thorough understanding of the fundamentals, develop a "feel" of the physical phenomena, be able to put the data into proper perspective, and make sound engineering judgments, just like their predecessors. However, they must do it much better, and much faster, using more realistic models because of the powerful tools available today. The engineers in the past had to rely on hand calculations, slide rules, and later hand calculators and computers. Today they rely on software packages. The easy access to such power and the possibility of a simple misunderstanding or misinterpretation causing great damage make it more important today than ever to have solid training in the fundamentals of engineering. In this text we make an extra effort to put the emphasis on developing an intuitive and physical understanding of natural phenomena instead of on the mathematical details of solution procedures.

Engineering Equation Solver (EES)

EES is a program that solves systems of linear or nonlinear algebraic or differential equations numerically. It has a large library of built-in thermodynamic property functions as well as mathematical functions, and allows the user to supply additional property data. Unlike some software packages, EES does not solve engineering problems; it only solves the equations supplied by the user. Therefore, the user must understand the problem and formulate it by applying any relevant physical laws and relations. EES saves the user considerable time and effort by simply solving the resulting mathematical equations. This makes it possible to attempt significant engineering problems not suitable for hand calculations, and to conduct parametric studies quickly and conveniently. EES is a very powerful yet intuitive program that is very easy to use, as shown in Examples 1–11 and 1–12. The use and capabilities of EES are explained in Appendix 3 on the enclosed DVD.

EXAMPLE 1–11 Solving a System of Equations with EES

The difference of two numbers is 4, and the sum of the squares of these two numbers is equal to the sum of the numbers plus 20. Determine these two numbers.

SOLUTION Relations are given for the difference and the sum of the squares of two numbers. They are to be determined.

Analysis We start the EES program by double-clicking on its icon, open a new file, and type the following on the blank screen that appears:

```
x-y = 4
x^2 + y^2 = x + y + 20
```

which is an exact mathematical expression of the problem statement with x and y denoting the unknown numbers. The solution to this system of two nonlinear equations with two unknowns is obtained by a single click on the "calculator" icon on the taskbar. It gives

$$\mathbf{x = 5} \quad \text{and} \quad \mathbf{y = 1}$$

Discussion Note that all we did is formulate the problem as we would on paper; EES took care of all the mathematical details of solution. Also note that equations can be linear or nonlinear, and they can be entered in any order with unknowns on either side. Friendly equation solvers such as EES allow the user to concentrate on the physics of the problem without worrying about the mathematical complexities associated with the solution of the resulting system of equations.

EXAMPLE 1–12 Analyzing a Multifluid Manometer with EES

Reconsider the multifluid manometer discussed in Example 1–7 and replotted in Fig. 1–66. Determine the air pressure in the tank using EES. Also determine what the differential fluid height h_3 would be for the same air pressure if the mercury in the last column were replaced by seawater with a density of 1030 kg/m³.

SOLUTION The pressure in a water tank is measured by a multifluid manometer. The air pressure in the tank and the differential fluid height h_3 if mercury is replaced by seawater are to be determined using EES.

Analysis We start the EES program by double-clicking on its icon, open a new file, and type the following on the blank screen that appears (we express the atmospheric pressure in Pa for unit consistency):

```
g = 9.81
Patm = 85600
h1 = 0.1;     h2 = 0.2;     h3 = 0.35
rw = 1000;    roil = 850;   rm = 13600
P1 + rw*g*h1 + roil*g*h2-rm*g*h3 = Patm
```

Here P_1 is the only unknown, and it is determined by EES to be

$$P_1 = 129647 \text{ Pa} \cong \mathbf{130\ kPa}$$

which is identical to the result obtained before. The height of the fluid column h_3 when mercury is replaced by seawater is determined easily by replacing "h3 = 0.35" by "P1 = 129647" and "rm = 13600" by "rm = 1030," and clicking on the calculator symbol. It gives

$$h_3 = \mathbf{4.62\ m}$$

Discussion Note that we used the screen like a paper pad and wrote down the relevant information together with the applicable relations in an organized manner. EES did the rest. Equations can be written on separate lines or on the same line by separating them by semicolons, and blank or comment lines can be inserted for readability. EES makes it very easy to ask "what if" questions, and to perform parametric studies, as explained in Appendix 3 on the DVD.

EES also has the capability to check the equations for unit consistency if units are supplied together with numerical values. Units can be specified within brackets [ ] after the specified value. When this feature is utilized, the previous equations would take the following form:

```
g = 9.81 [m/s^2]
Patm = 85600 [Pa]
h1 = 0.1 [m];            h2 = 0.2 [m];        h3 = 0.35 [m]
rw = 1000 [kg/m^3];  roil = 850 [kg/m^3];  rm = 13600 [kg/m^3]
P1 + rw*g*h1 + roil*g*h2-rm*g*h3 = Patm
```

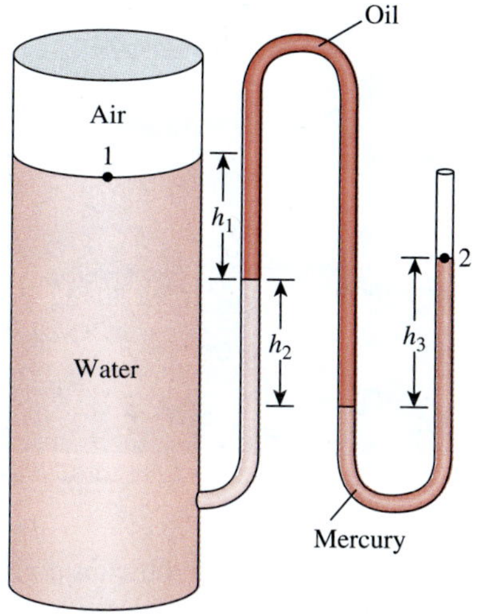

FIGURE 1–66
Schematic for Example 1–12.

A Remark on Significant Digits

In engineering calculations, the information given is not known to more than a certain number of significant digits, usually three digits. Consequently, the results obtained cannot possibly be accurate to more significant digits. Reporting results in more significant digits implies greater accuracy than exists, and it should be avoided.

For example, consider a 3.75-L container filled with gasoline whose density is 0.845 kg/L, and try to determine its mass. Probably the first thought that comes to your mind is to multiply the volume and density to obtain 3.16875 kg for the mass, which falsely implies that the mass determined is accurate to six significant digits. In reality, however, the mass cannot be more accurate than three significant digits since both the volume and the density are accurate to three significant digits only. Therefore, the result should be rounded to three significant digits, and the mass should be reported to be 3.17 kg instead of what appears in the screen of the calculator. The result 3.16875 kg would be correct only if the volume and density were given to be 3.75000 L and 0.845000 kg/L, respectively. The value 3.75 L implies that we are fairly confident that the volume is accurate within ±0.01 L, and it cannot be 3.74 or 3.76 L. However, the volume can be 3.746, 3.750, 3.753, etc., since they all round to 3.75 L (Fig. 1–67). It is more appropriate to retain all the digits during intermediate calculations, and to do the rounding in the final step since this is what a computer will normally do.

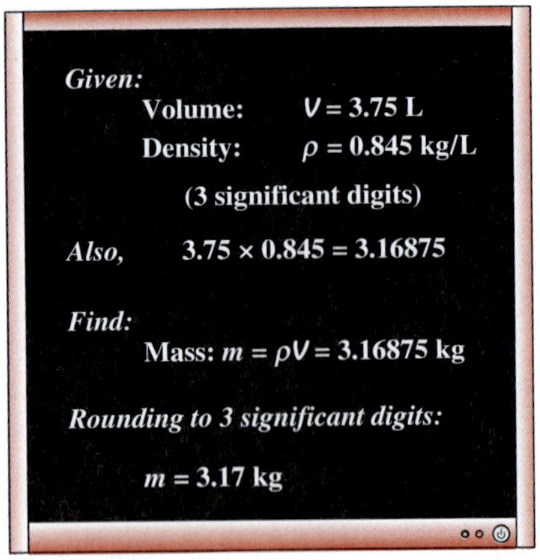

FIGURE 1–67
A result with more significant digits than that of given data falsely implies more accuracy.

When solving problems, we will assume the given information to be accurate to at least three significant digits. Therefore, if the length of a pipe is given to be 40 m, we will assume it to be 40.0 m in order to justify using three significant digits in the final results. You should also keep in mind that all experimentally determined values are subject to measurement errors and such errors will reflect in the results obtained. For example, if the density of a substance has an uncertainty of 2 percent, then the mass determined using this density value will also have an uncertainty of 2 percent.

You should also be aware that we sometimes knowingly introduce small errors in order to avoid the trouble of searching for more accurate data. For example, when dealing with liquid water, we just use the value of 1000 kg/m^3 for density, which is the density value of pure water at 0°C. Using this value at 75°C will result in an error of 2.5 percent since the density at this temperature is 975 kg/m^3. The minerals and impurities in the water will introduce additional error. This being the case, you should have no reservation in rounding the final results to a reasonable number of significant digits. Besides, having a few percent uncertainty in the results of engineering analysis is usually the norm, not the exception.

SUMMARY

In this chapter, the basic concepts of thermodynamics are introduced and discussed. *Thermodynamics* is the science that primarily deals with energy. The *first law of thermodynamics* is simply an expression of the conservation of energy principle, and it asserts that *energy* is a thermodynamic property. The *second law of thermodynamics* asserts that energy has *quality* as well as *quantity*, and actual processes occur in the direction of decreasing quality of energy.

A system of fixed mass is called a *closed system,* or *control mass,* and a system that involves mass transfer across its boundaries is called an *open system,* or *control volume*. The mass-dependent properties of a system are called *extensive properties* and the others *intensive properties*. *Density* is mass per unit volume, and *specific volume* is volume per unit mass.

A system is said to be in *thermodynamic equilibrium* if it maintains thermal, mechanical, phase, and chemical equilibrium. Any change from one state to another is called a *process*. A process with identical end states is called a *cycle.* During a *quasi-static* or *quasi-equilibrium process*, the system remains practically in equilibrium at all times. The state of a simple, compressible system is completely specified by two independent, intensive properties.

The *zeroth law of thermodynamics* states that two bodies are in thermal equilibrium if both have the same temperature reading even if they are not in contact.

The temperature scales used in the SI and the English system today are the *Celsius scale* and the *Fahrenheit scale*, respectively. They are related to absolute temperature scales by

$$T(\text{K}) = T(^\circ\text{C}) + 273.15$$

$$T(\text{R}) = T(^\circ\text{F}) + 459.67$$

The magnitudes of each division of 1 K and 1°C are identical, and so are the magnitudes of each division of 1 R and 1°F. Therefore,

$$\Delta T(\text{K}) = \Delta T(^\circ\text{C}) \quad \text{and} \quad \Delta T(\text{R}) = \Delta T(^\circ\text{F})$$

The normal force exerted by a fluid per unit area is called *pressure,* and its unit is the *pascal,* 1 Pa = 1 N/m^2. The pressure relative to absolute vacuum is called the *absolute pressure*, and the difference between the absolute pressure and the local atmospheric pressure is called the *gage pressure.* Pressures below atmospheric pressure are called *vacuum pressures.* The absolute, gage, and vacuum pressures are related by

$$P_{\text{gage}} = P_{\text{abs}} - P_{\text{atm}} \qquad \text{(for pressures above } P_{\text{atm}})$$

$$P_{\text{vac}} = P_{\text{atm}} - P_{\text{abs}} \qquad \text{(for pressures below } P_{\text{atm}})$$

The pressure at a point in a fluid has the same magnitude in all directions. The variation of pressure with elevation is given by

$$\frac{dP}{dz} = -\rho g$$

where the positive z direction is taken to be upward. When the density of the fluid is constant, the pressure difference across a fluid layer of thickness Δz is

$$\Delta P = P_2 - P_1 = \rho g \, \Delta z$$

The absolute and gage pressures in a liquid open to the atmosphere at a depth h from the free surface are

$$P = P_{\text{atm}} + \rho g h \quad \text{or} \quad P_{\text{gage}} = \rho g h$$

Small to moderate pressure differences are measured by a *manometer*. The pressure in a stationary fluid remains constant in the horizontal direction. *Pascal's principle* states that the pressure applied to a confined fluid increases the pressure throughout by the same amount. The atmospheric pressure is measured by a *barometer* and is given by

$$P_{\text{atm}} = \rho g h$$

where h is the height of the liquid column.

REFERENCES AND SUGGESTED READINGS

1. American Society for Testing and Materials. *Standards for Metric Practice.* ASTM E 380-79, January 1980.
2. A. Bejan. *Advanced Engineering Thermodynamics.* 2nd ed. New York: Wiley, 1997.
3. J. A. Schooley. *Thermometry*. Boca Raton, FL: CRC Press, 1986.

PROBLEMS*

Thermodynamics

1–1C Why does a bicyclist pick up speed on a downhill road even when he is not pedaling? Does this violate the conservation of energy principle?

1–2C One of the most amusing things a person can experience is that in certain parts of the world a still car on neutral going uphill when its brakes are released. Such occurrences are even broadcast on TV. Can this really happen or is it a bad eyesight? How can you verify if a road is really uphill or downhill?

1–3C An office worker claims that a cup of cold coffee on his table warmed up to 80°C by picking up energy from the surrounding air, which is at 25°C. Is there any truth to his claim? Does this process violate any thermodynamic laws?

Mass, Force, and Units

1–4C In a news article, it is stated that a recently developed geared turbofan engine produces 15,000 pounds of thrust to

* Problems designated by a "C" are concept questions, and students are encouraged to answer them all. Problems designated by an "E" are in English units, and the SI users can ignore them. Problems with the icon are solved using EES, and complete solutions together with parametric studies are included on the enclosed DVD. Problems with the icon are comprehensive in nature, and are intended to be solved with a computer, preferably using the EES software that accompanies this text.

propel the aircraft forward. Is "pound" mentioned here lbm or lbf? Explain.

1–5C Explain why the light-year has the dimension of length.

1–6C What is the net force acting on a car cruising at a constant velocity of 70 km/h (*a*) on a level road and (*b*) on an uphill road?

1–7E A man weighs 210 lbf at a location where $g = 32.10$ ft/s^2. Determine his weight on the moon, where $g = 5.47$ ft/s^2. *Answer:* 35.8 lbf

1–8 Determine the mass and the weight of the air contained in a room whose dimensions are 6 m $\times$ 6 m $\times$ 8 m. Assume the density of the air is 1.16 kg/m^3. *Answers:* 334.1 kg, 3277 N

1–9 At 45° latitude, the gravitational acceleration as a function of elevation z above sea level is given by $g = a - bz$, where $a = 9.807$ m/s^2 and $b = 3.32 \times 10^{-6}$ s^{-2}. Determine the height above sea level where the weight of an object will decrease by 0.5 percent. *Answer:* 14,770 m

1–10 What is the weight, in N, of an object with a mass of 200 kg at a location where $g = 9.6$ m/s^2?

1–11E The constant-pressure specific heat of air at 25°C is 1.005 kJ/kg·°C. Express this value in kJ/kg·K, J/g·°C, kcal/kg·°C, and Btu/lbm·°F.

1–12 A 3-kg rock is thrown upward with a force of 200 N at a location where the local gravitational acceleration is 9.79 m/s^2. Determine the acceleration of the rock, in m/s^2.

1–13 Solve Prob. 1–12 using EES (or other) software. Print out the entire solution, including the numerical results with proper units.

1–14 While solving a problem, a person ends up with the equation $E = 25 \text{ kJ} + 7 \text{ kJ/kg}$ at some stage. Here E is the total energy and has the unit of kilojoules. Determine how to correct the error and discuss what may have caused it.

1–15 A 4-kW resistance heater in a water heater runs for 2 hours to raise the water temperature to the desired level. Determine the amount of electric energy used in both kWh and kJ.

1–16 The gas tank of a car is filled with a nozzle that discharges gasoline at a constant flow rate. Based on unit considerations of quantities, obtain a relation for the filling time in terms of the volume V of the tank (in L) and the discharge rate of gasoline $\dot{V}$ (in L/s).

1–17 A pool of volume V (in m^3) is to be filled with water using a hose of diameter D (in m). If the average discharge velocity is V (in m/s) and the filling time is t (in s), obtain a relation for the volume of the pool based on until considerations of quantities involved.

1–18 Based on unit considerations alone show that the power needed to accelerate a car of mass m (in kg) from rest to velocity V (in m/s) in time interval t (in s) is proportional to mass and the square of the velocity of the car and inversely proportional to the time interval.

Systems, Properties, State, and Processes

1–19C You have been asked to do a metabolism (energy) analysis of a person. How would you define the system for this purpose? What type of system is this?

1–20C You are trying to understand how a reciprocating air compressor (a piston-cylinder device) works. What system would you use? What type of system is this?

1–21C How might you define a system to study the depletion of ozone in the upper layers of the earth's atmosphere?

1–22C What is the difference between intensive and extensive properties?

1–23C Is the weight of a system an extensive or intensive property?

1–24C The molar specific volume of a system $\bar{V}$ is defined as the ratio of the volume of the system to the number of moles of substance contained in the system. Is this an extensive or intensive property?

1–25C For a system to be in thermodynamic equilibrium, do the temperature and the pressure have to be the same everywhere?

1–26C What is a quasi-equilibrium process? What is its importance in engineering?

1–27C Define the isothermal, isobaric, and isochoric processes.

1–28C What is the state postulate?

1–29C How would you describe the state of the water in a bathtub? How would you describe the process that this water experiences as it cools?

1–30C When analyzing the acceleration of gases as they flow through a nozzle, what would you choose as your system? What type of system is this?

1–31C When is a steady-flow process?

1–32 The density of atmospheric air varies with elevation, decreasing with increasing altitude. (*a*) Using the data given in the table, obtain a relation for the variation of density with elevation, and calculate the density at an elevation of 7000 m. (*b*) Calculate the mass of the atmosphere using the correlation you obtained. Assume the earth to be a perfect sphere with a radius of 6377 km, and take the thickness of the atmosphere to be 25 km.

z, km	ρ, kg/m^3
6377	1.225
6378	1.112
6379	1.007
6380	0.9093
6381	0.8194
6382	0.7364
6383	0.6601
6385	0.5258
6387	0.4135
6392	0.1948
6397	0.08891
6402	0.04008

Temperature

1–33C What is the zeroth law of thermodynamics?

1–34C What are the ordinary and absolute temperature scales in the SI and the English system?

1–35C Consider an alcohol and a mercury thermometer that read exactly 0°C at the ice point and 100°C at the steam point. The distance between the two points is divided into 100 equal parts in both thermometers. Do you think these thermometers will give exactly the same reading at a temperature of, say, 60°C? Explain.

1–36 The deep body temperature of a healthy person is 37°C. What is it in kelvins?

1–37E What is the temperature of the heated air at 150°C in °F and R?

1–38 The temperature of a system rises by 45°C during a heating process. Express this rise in temperature in kelvins.

1–39E The flash point of an engine oil is 363°F. What is the absolute flash-point temperature in K and R?

1–40E The temperature of ambient air in a certain location is measured to be −40°C. Express this temperature in Fahrenheit (°F), Kelvin (K), and Rankine (R) units.

1–41E The temperature of water changes by 10°F during a process. Express this temperature change in Celcius (°C), Kelvin (K), and Rankine (R) units.

1–42E Humans are most comfortable when the temperature is between 65°F and 75°F. Express these temperature limits in °C. Convert the size of this temperature range (10°F) to K, °C, and R. Is there any difference in the size of this range as measured in relative or absolute units?

Pressure, Manometer, and Barometer

1–43C What is the difference between gage pressure and absolute pressure?

1–44C A health magazine reported that physicians measured 100 adults' blood pressure using two different arm positions: parallel to the body (along the side) and perpendicular to the body (straight out). Readings in the parallel position were up to 10 percent higher than those in the perpendicular position, regardless of whether the patient was standing, sitting, or lying down. Explain the possible cause for the difference.

1–45C Someone claims that the absolute pressure in a liquid of constant density doubles when the depth is doubled. Do you agree? Explain.

1–46C A tiny steel cube is suspended in water by a string. If the lengths of the sides of the cube are very small, how would you compare the magnitudes of the pressures on the top, bottom, and side surfaces of the cube?

1–47C Express Pascal's law, and give a real-world example of it.

1–48E The pressure at the exit of an air compressor is 150 psia. What is this pressure in kPa?

1–49 The pressure in a compressed air storage tank is 1500 kPa. What is the tank's pressure in (*a*) kN and m units; (*b*) kg, m, and s units; and (*c*) kg, km, and s units?

1–50E The pressure in a water line is 1500 kPa. What is the line pressure in (*a*) lb/ft^2 units and (*b*) lbf/in^2 (psi) units?

1–51E If the pressure inside a rubber balloon is 1500 mm Hg, what is this pressure in pounds-force per square inch (psi)? *Answer:* 29.0 psi

1–52 The pressure of the helium inside a toy balloon is 1250 mm Hg. What is this pressure in kPa?

1–53 The water in a tank is pressurized by air, and the pressure is measured by a multifluid manometer as shown in Fig. P1–53. Determine the gage pressure of air in the tank if h_1 = 0.2 m, h_2 = 0.3 m, and h_3 = 0.46 m. Take the densities of water, oil, and mercury to be 1000 kg/m^3, 850 kg/m^3, and 13,600 kg/m^3, respectively.

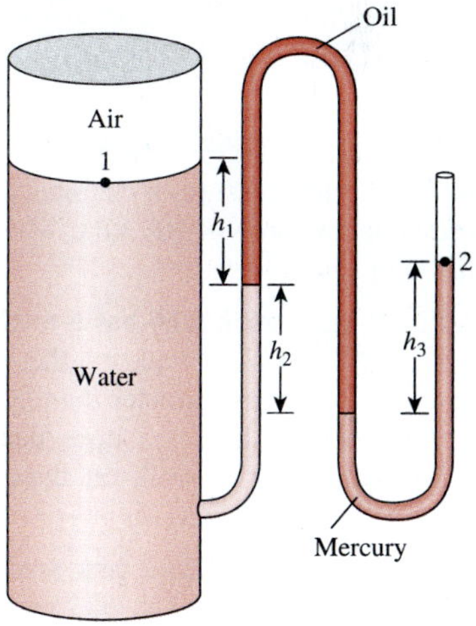

FIGURE P1–53

1–54 Determine the atmospheric pressure at a location where the barometric reading is 750 mm Hg. Take the density of mercury to be 13,600 kg/m^3.

1–55 The gage pressure in a liquid at a depth of 3 m is read to be 42 kPa. Determine the gage pressure in the same liquid at a depth of 9 m.

1–56 The absolute pressure in water at a depth of 5 m is read to be 145 kPa. Determine (*a*) the local atmospheric pressure, and (*b*) the absolute pressure at a depth of 5 m in a liquid whose specific gravity is 0.85 at the same location.

1–57E Show that 1 kgf/cm^2 = 14.223 psi.

1–58E The diameters of the pistons shown in Fig. P1–58E are D_1 = 3 in and D_2 = 1.5 in. Determine the pressure in chamber 3, in psia, when the other pressures are P_1 = 150 psia and P_2 = 250 psia.

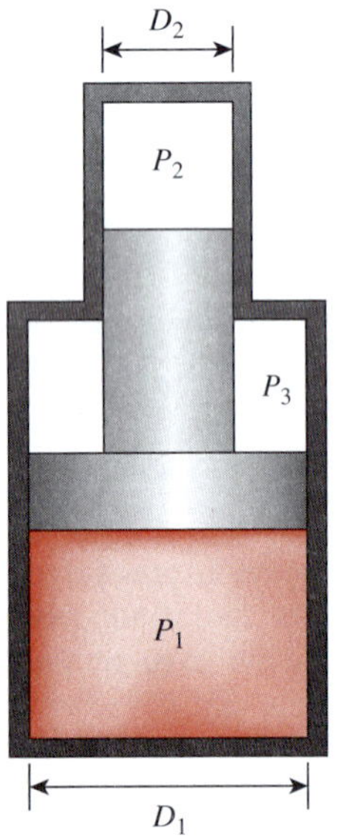

FIGURE P1–58E

1–59 The piston diameters in Fig. P1–58E are D_1 = 10 cm and D_2 = 4 cm. When the pressure in chamber 2 is 2000 kPa and the pressure in chamber 3 is 700 kPa, what is the pressure in chamber 1, in kPa? *Answer:* 908 kPa

1–60 Consider a 70-kg woman who has a total foot imprint area of 400 cm^2. She wishes to walk on the snow, but the snow cannot withstand pressures greater than 0.5 kPa. Determine the minimum size of the snowshoes needed (imprint area per shoe) to enable her to walk on the snow without sinking.

1–61 A vacuum gage connected to a tank reads 30 kPa at a location where the barometric reading is 750 mm Hg. Determine the absolute pressure in the tank. Take ρ_{Hg} = 13,590 kg/m^3. *Answer:* 70.0 kPa

1–62E The vacuum pressure of a condenser is given to be 80 kPa. If the atmospheric pressure is 98 kPa, what is the gage pressure and absolute pressure in kPa, kN/m^2, lbf/in^2, psi, and mm Hg.

1–63 The barometer of a mountain hiker reads 740 mbars at the beginning of a hiking trip and 630 mbars at the end. Neglecting the effect of altitude on local gravitational acceleration, determine the vertical distance climbed. Assume an average air density of 1.20 kg/m^3. *Answer:* 934 m

1–64 The basic barometer can be used to measure the height of a building. If the barometric readings at the top and at the bottom of a building are 675 and 695 mm Hg, respectively, determine the height of the building. Take the densities of air and mercury to be 1.18 kg/m^3 and 13,600 kg/m^3, respectively.

FIGURE P1–64

1–65 [EES] Solve Prob. 1–64 using EES (or other) software. Print out the entire solution, including the numerical results with proper units.

1–66 Consider a 1.75-m-tall man standing vertically in water and completely submerged in a pool. Determine the difference between the pressures acting at the head and at the toes of the man, in kPa.

1–67 A gas is contained in a vertical, frictionless piston–cylinder device. The piston has a mass of 3.2 kg and a cross-sectional area of 35 cm^2. A compressed spring above the piston exerts a force of 150 N on the piston. If the atmospheric pressure is 95 kPa, determine the pressure inside the cylinder. *Answer:* 147 kPa

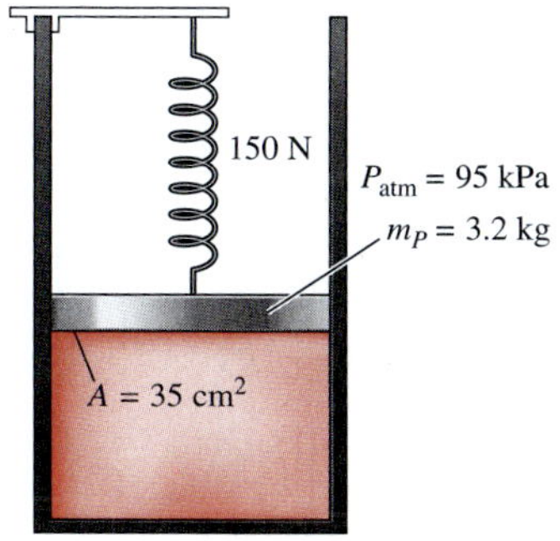

FIGURE P1–67

1–68 Reconsider Prob. 1–67. Using EES (or other) software, investigate the effect of the spring force in the range of 0 to 500 N on the pressure inside the cylinder. Plot the pressure against the spring force, and discuss the results.

1–69 Both a gage and a manometer are attached to a gas tank to measure its pressure. If the reading on the pressure gage is 80 kPa, determine the distance between the two fluid levels of the manometer if the fluid is (*a*) mercury (ρ = 13,600 kg/m^3) or (*b*) water (ρ = 1000 kg/m^3).

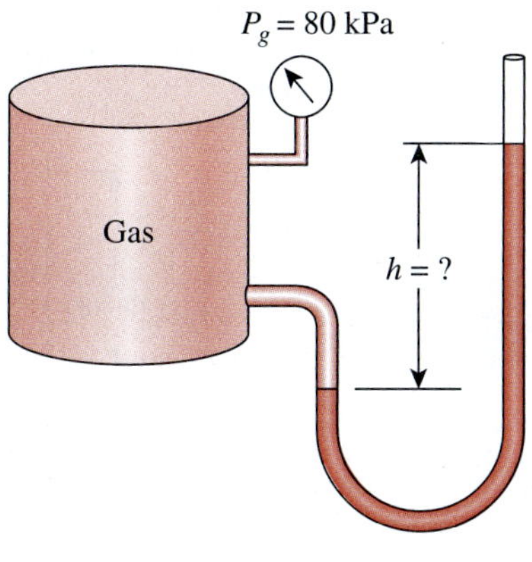

FIGURE P1–69

1–70 Reconsider Prob. 1–69. Using EES (or other) software, investigate the effect of the manometer fluid density in the range of 800 to 13,000 kg/m^3 on the differential fluid height of the manometer. Plot the differential fluid height against the density, and discuss the results.

1–71 A manometer containing oil (ρ = 850 kg/m^3) is attached to a tank filled with air. If the oil-level difference between the two columns is 36 cm and the atmospheric pressure is 98 kPa, determine the absolute pressure of the air in the tank. *Answer:* 101 kPa

1–72 A mercury manometer (ρ = 13,600 kg/m^3) is connected to an air duct to measure the pressure inside. The difference in the manometer levels is 15 mm, and the atmospheric pressure is 100 kPa. (*a*) Judging from Fig. P1–72, determine if the pressure in the duct is above or below the atmospheric pressure. (*b*) Determine the absolute pressure in the duct.

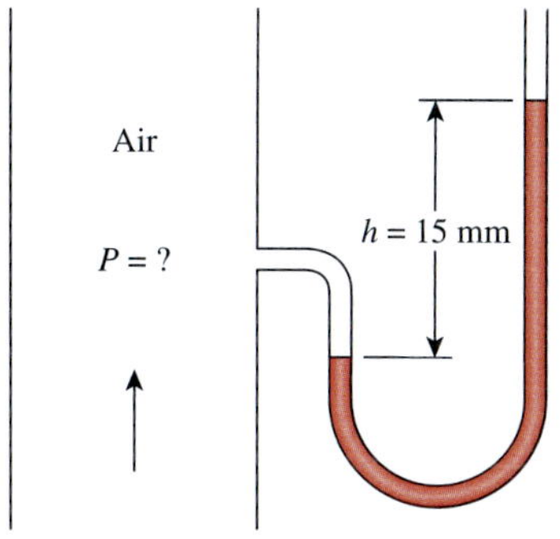

FIGURE P1–72

1–73 Repeat Prob. 1–72 for a differential mercury height of 45 mm.

1–74E Blood pressure is usually measured by wrapping a closed air-filled jacket equipped with a pressure gage around the upper arm of a person at the level of the heart. Using a mercury manometer and a stethoscope, the systolic pressure (the maximum pressure when the heart is pumping) and the diastolic pressure (the minimum pressure when the heart is resting) are measured in mm Hg. The systolic and diastolic pressures of a healthy person are about 120 mm Hg and 80 mm Hg, respectively, and are indicated as 120/80. Express both of these gage pressures in kPa, psi, and meter water column.

1–75 The maximum blood pressure in the upper arm of a healthy person is about 120 mm Hg. If a vertical tube open to the atmosphere is connected to the vein in the arm of the person, determine how high the blood will rise in the tube. Take the density of the blood to be 1050 kg/m^3.

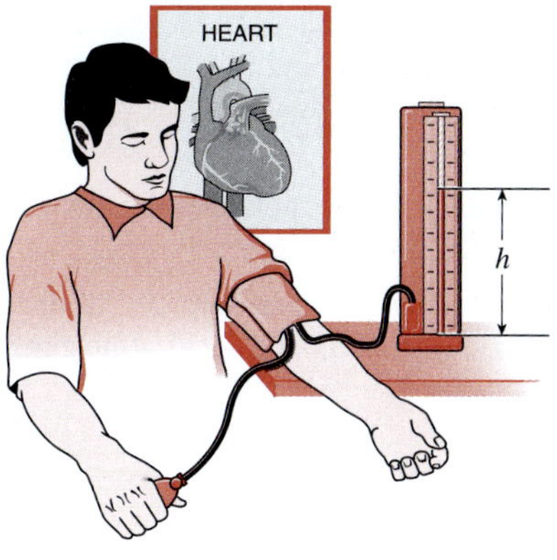

FIGURE P1–75

1–76 Determine the pressure exerted on a diver at 30 m below the free surface of the sea. Assume a barometric pressure of 101 kPa and a specific gravity of 1.03 for seawater. *Answer:* 404.0 kPa

1–77 Consider a U-tube whose arms are open to the atmosphere. Now water is poured into the U-tube from one arm, and light oil (ρ = 790 kg/m^3) from the other. One arm contains 70-cm-high water, while the other arm contains both fluids with an oil-to-water height ratio of 4. Determine the height of each fluid in that arm.

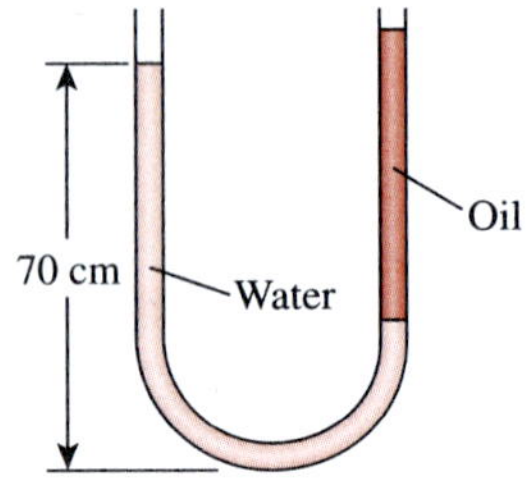

FIGURE P1–77

1–78 Freshwater and seawater flowing in parallel horizontal pipelines are connected to each other by a double U-tube manometer, as shown in Fig. P1–78. Determine the pressure difference between the two pipelines. Take the density of seawater at that location to be ρ = 1035 kg/m^3. Can the air column be ignored in the analysis?

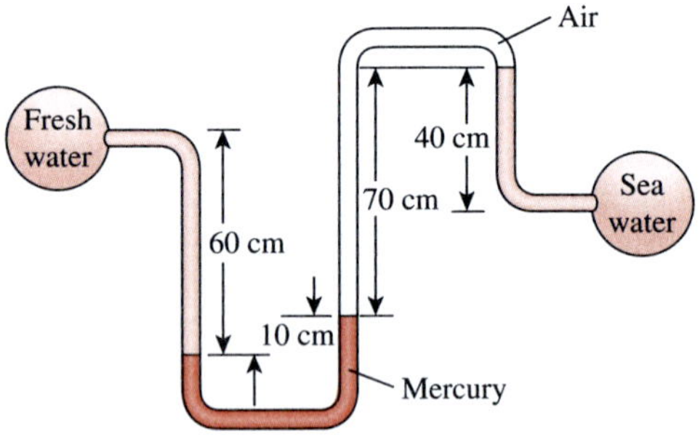

FIGURE P1–78

1–79 Repeat Prob. 1–78 by replacing the air with oil whose specific gravity is 0.72.

1–80 Calculate the absolute pressure, P_1, of the manometer shown in Fig. P1–80 in kPa. The local atmospheric pressure is 758 mm Hg.

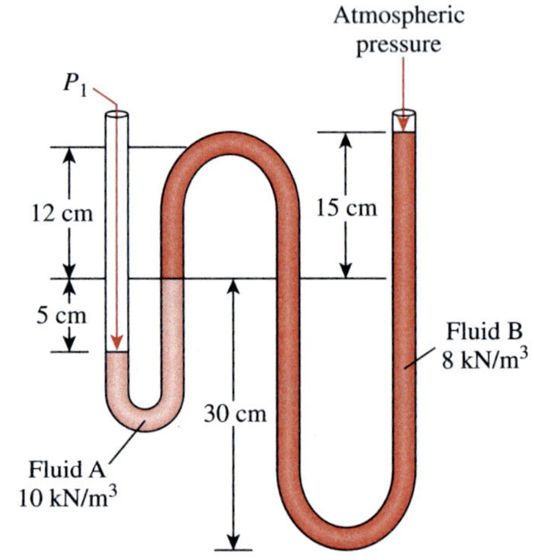

FIGURE P1–80

1–81 Consider the manometer in Fig. P1–80. If the specific weight of fluid A is 100 kN/m^3, what is the absolute pressure, in kPa, indicated by the manometer when the local atmospheric pressure is 90 kPa?

1–82 Consider the manometer in Fig. P1–80. If the specific weight of fluid B is 12 kN/m^3, what is the absolute pressure, in kPa, indicated by the manometer when the local atmospheric pressure is 720 mm Hg?

1–83 The gage pressure of the air in the tank shown in Fig. P1–83 is measured to be 80 kPa. Determine the differential height *h* of the mercury column.

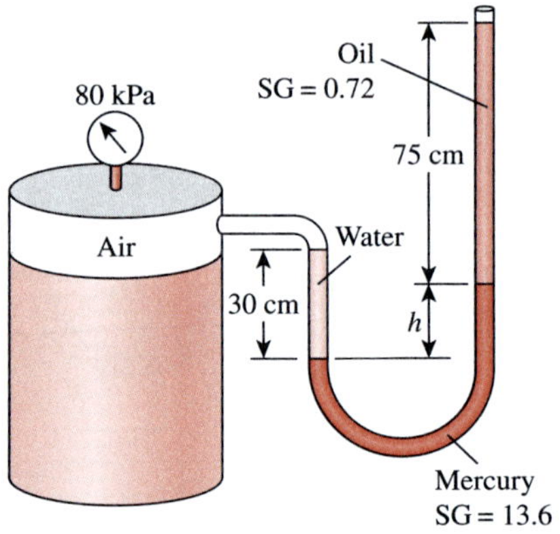

FIGURE P1–83

1–84 Repeat Prob. 1–83 for a gage pressure of 40 kPa.

1–85 The top part of a water tank is divided into two compartments, as shown in Fig. P1–85. Now a fluid with an unknown density is poured into one side, and the water level rises a certain amount on the other side to compensate for this effect. Based on the final fluid heights shown on the figure, determine the density of the fluid added. Assume the liquid does not mix with water.

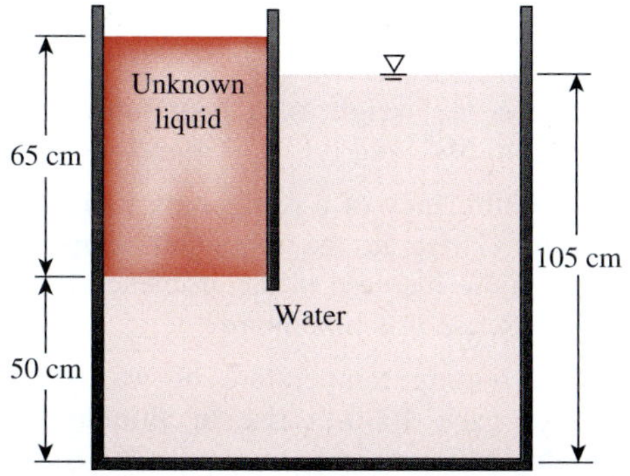

FIGURE P1–85

1–86 Consider the system shown in Fig. P1–86. If a change of 0.7 kPa in the pressure of air causes the brine–mercury interface in the right column to drop by 5 mm in the brine level in the right column while the pressure in the brine pipe remains constant, determine the ratio of A_2/A_1.

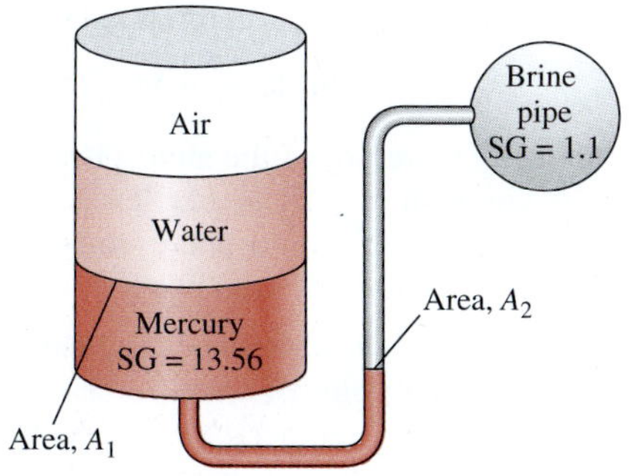

FIGURE P1–86

1–87 A multifluid container is connected to a U-tube, as shown in Fig. P1–87. For the given specific gravities and fluid column heights, determine the gage pressure at *A*. Also determine the height of a mercury column that would create the same pressure at *A*. *Answers:* 0.471 kPa, 0.353 cm

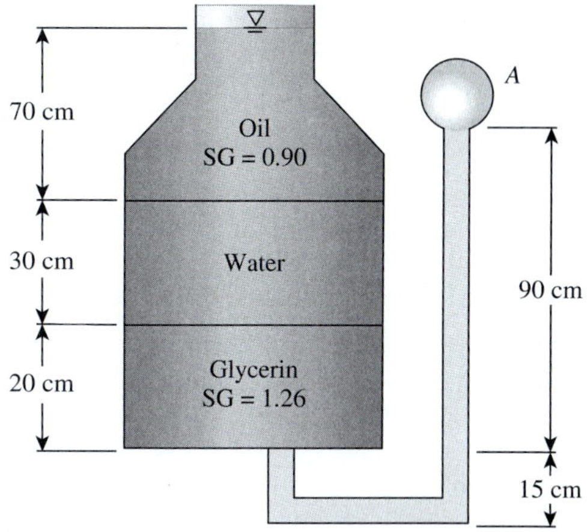

FIGURE P1–87

Solving Engineering Problems and EES

1–88C What is the value of the engineering software packages in (*a*) engineering education and (*b*) engineering practice?

1–89 Determine a positive real root of this equation using EES:

$$2x^3 - 10x^{0.5} - 3x = -3$$

1–90 Solve this system of two equations with two unknowns using EES:

$$x^3 - y^2 = 7.75$$

$$3xy + y = 3.5$$

1–91 Solve this system of three equations with three unknowns using EES:

$$2x - y + z = 7$$

$$3x^2 + 3y = z + 3$$

$$xy + 2z = 4$$

1–92 Solve this system of three equations with three unknowns using EES:

$$x^2y - z = 1$$

$$x - 3y^{0.5} + xz = -2$$

$$x + y - z = 2$$

1–93E Specific heat is defined as the amount of energy needed to increase the temperature of a unit mass of a substance by one degree. The specific heat of water at room temperature is 4.18 kJ/kg·°C in SI unit system. Using the unit conversion function capability of EES, express the specific heat of water in (*a*) kJ/kg·K, (*b*) Btu/lbm·°F, (*c*) Btu/lbm·R, and (*d*) kcal/kg·°C units. *Answers:* (*a*) 4.18, (*b*) (*c*) (*d*) 0.9984

Review Problems

1–94 A lunar exploration module weighs 2800 N at a location where $g = 9.8\ \text{m/s}^2$. Determine the weight of this module in newtons when it is on the moon where $g = 1.64\ \text{m/s}^2$.

1–95 The force generated by a spring is given by $F = kx$, where k is the spring constant and x is the deflection of the spring. The spring of Fig. P1–95 has a spring constant of 8 kN/cm. The pressures are $P_1 = 5000$ kPa, $P_2 = 10{,}000$ kPa, and $P_3 = 1000$ kPa. If the piston diameters are $D_1 = 8$ cm and $D_2 = 3$ cm, how far will the spring be deflected? *Answer:* 1.72 cm

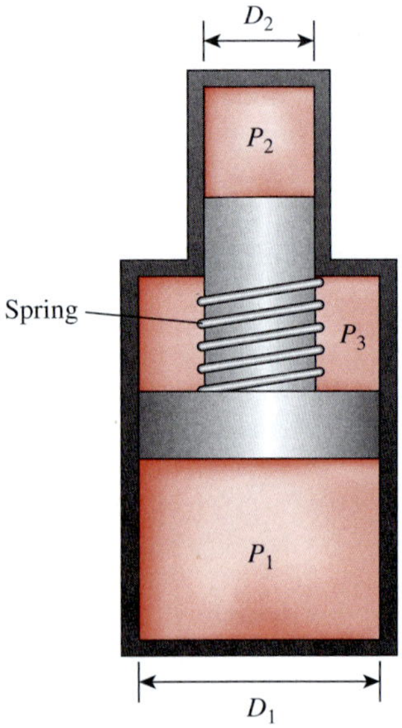

FIGURE P1–95

1–96 The pilot of an airplane reads the altitude 9000 m and the absolute pressure 25 kPa when flying over a city. Calculate the local atmospheric pressure in that city in kPa and in mm Hg. Take the densities of air and mercury to be 1.15 kg/m³ and 13,600 kg/m³, respectively.

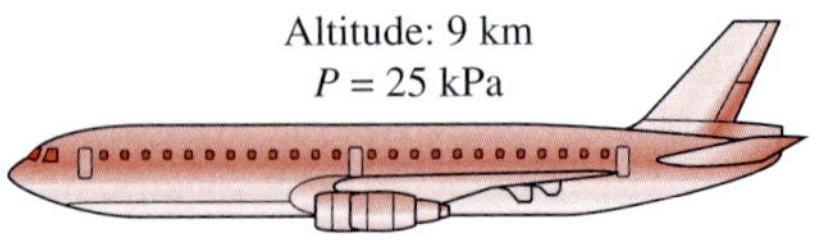

FIGURE P1–96

1–97 The weight of bodies may change somewhat from one location to another as a result of the variation of the gravitational acceleration g with elevation. Accounting for this variation using the relation in Prob. 1–9, determine the weight of an 80-kg person at sea level ($z = 0$), in Denver ($z = 1610$ m), and on the top of Mount Everest ($z = 8848$ m).

1–98E A man goes to a traditional market to buy a steak for dinner. He finds a 12-oz steak (1 lbm = 16 oz) for $3.15. He then goes to the adjacent international market and finds a 300-g steak of identical quality for $2.95. Which steak is the better buy?

1–99E What is the weight of a 1-kg substance in N, kN, kg·m/s², kgf, lbm·ft/s², and lbf?

1–100E The efficiency of a refrigerator increases by 3 percent for each °C rise in the minimum temperature in the device. What is the increase in the efficiency for each (*a*) K, (*b*) °F, and (*c*) R rise in temperature?

1–101E The boiling temperature of water decreases by about 3°C for each 1000-m rise in altitude. What is the decrease in the boiling temperature in (*a*) K, (*b*) °F, and (*c*) R for each 1000-m rise in altitude?

1–102E Hyperthermia of 5°C (i.e., 5°C rise above the normal body temperature) is considered fatal. Express this fatal level of hyperthermia in (*a*) K, (*b*) °F, and (*c*) R.

1–103E A house is losing heat at a rate of 2700 kJ/h per °C temperature difference between the indoor and the outdoor temperatures. Express the rate of heat loss from this house per (*a*) K, (*b*) °F, and (*c*) R difference between the indoor and the outdoor temperature.

1–104 The average temperature of the atmosphere in the world is approximated as a function of altitude by the relation

$$T_{\text{atm}} = 288.15 - 6.5z$$

where T_{atm} is the temperature of the atmosphere in K and z is the altitude in km with $z = 0$ at sea level. Determine the average temperature of the atmosphere outside an airplane that is cruising at an altitude of 12,000 m.

1–105 Joe Smith, an old-fashioned engineering student, believes that the boiling point of water is best suited for use as the reference point on temperature scales. Unhappy that the boiling point corresponds to some odd number in the current absolute temperature scales, he has proposed a new absolute temperature scale that he calls the Smith scale. The temperature unit on this scale is *smith*, denoted by S, and the boiling point of water on this scale is assigned to be 1000 S. From a thermodynamic point of view, discuss if it is an acceptable temperature scale. Also, determine the ice point of water on the Smith scale and obtain a relation between the Smith and Celsius scales.

1–106E It is well-known that cold air feels much colder in windy weather than what the thermometer reading indicates because of the "chilling effect" of the wind. This effect is due to the increase in the convection heat transfer coefficient with increasing air velocities. The *equivalent wind chill temperature* in °F is given by [ASHRAE, *Handbook of Fundamentals* (Atlanta, GA, 1993), p. 8.15]

$$T_{\text{equiv}} = 91.4 - (91.4 - T_{\text{ambient}}) \times (0.475 - 0.0203V + 0.304\sqrt{V})$$

where V is the wind velocity in mi/h and T_{ambient} is the ambient air temperature in °F in calm air, which is taken to be air with light winds at speeds up to 4 mi/h. The constant 91.4°F in the given equation is the mean skin temperature of a resting person in a comfortable environment. Windy air at temperature T_{ambient} and velocity V will feel as cold as the calm air at temperature T_{equiv}. Using proper conversion factors, obtain an equivalent relation in SI units where V is the wind velocity in km/h and T_{ambient} is the ambient air temperature in °C.

Answer: $T_{equiv} = 33.0 - (33.0 - T_{ambient}) \times (0.475 - 0.0126V + 0.240\sqrt{V})$

1–107E Reconsider Prob. 1–106E. Using EES (or other) software, plot the equivalent wind chill temperatures in °F as a function of wind velocity in the range of 4 to 40 mph for the ambient temperatures of 20, 40, and 60°F. Discuss the results.

1–108 An air-conditioning system requires a 20-m-long section of 15-cm diameter duct work to be laid underwater. Determine the upward force the water will exert on the duct. Take the densities of air and water to be 1.3 kg/m³ and 1000 kg/m³, respectively.

1–109 Balloons are often filled with helium gas because it weighs only about one-seventh of what air weighs under identical conditions. The buoyancy force, which can be expressed as $F_b = \rho_{\text{air}} g V_{\text{balloon}}$, will push the balloon upward. If the balloon has a diameter of 12 m and carries two people, 85 kg each, determine the acceleration of the balloon when it is first released. Assume the density of air is $\rho = 1.16$ kg/m³, and neglect the weight of the ropes and the cage. *Answer:* 22.4 m/s²

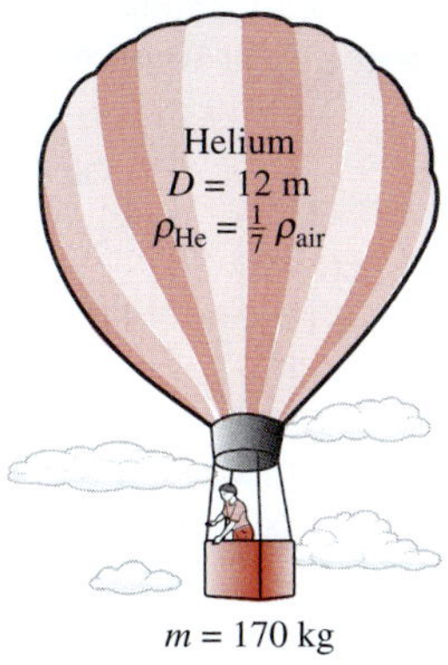

FIGURE P1–109

1–110 Reconsider Prob. 1–109. Using EES (or other) software, investigate the effect of the number of people carried in the balloon on acceleration. Plot the acceleration against the number of people, and discuss the results.

1–111 Determine the maximum amount of load, in kg, the balloon described in Prob. 1–109 can carry. *Answer:* 900 kg

1–112 The lower half of a 10-m-high cylindrical container is filled with water ($\rho = 1000$ kg/m³) and the upper half with oil that has a specific gravity of 0.85. Determine the pressure difference between the top and bottom of the cylinder. *Answer:* 90.7 kPa

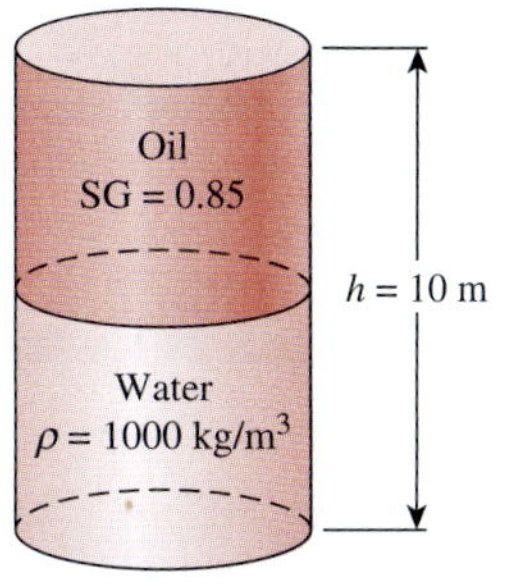

FIGURE P1–112

1–113 A vertical, frictionless piston–cylinder device contains a gas at 180 kPa absolute pressure. The atmospheric pressure outside is 100 kPa, and the piston area is 25 cm². Determine the mass of the piston.

1–114 A pressure cooker cooks a lot faster than an ordinary pan by maintaining a higher pressure and temperature inside. The lid of a pressure cooker is well sealed, and steam can escape only through an opening in the middle of the lid. A separate metal piece, the petcock, sits on top of this opening and prevents steam from escaping until the pressure force overcomes the weight of the petcock. The periodic escape of the steam in this manner prevents any potentially dangerous pressure buildup and keeps the pressure inside at a constant value. Determine the mass of the petcock of a pressure cooker whose operation pressure is 100 kPa gage and has an opening cross-sectional area of 4 mm². Assume an atmospheric pressure of 101 kPa, and draw the free-body diagram of the petcock. *Answer:* 40.8 g

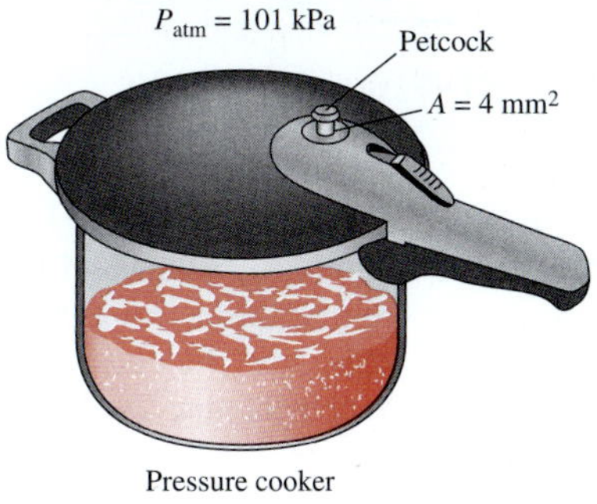

FIGURE P1–114

1–115 A glass tube is attached to a water pipe, as shown in Fig. P1–115. If the water pressure at the bottom of the tube is 120 kPa and the local atmospheric pressure is 99 kPa, determine how high the water will rise in the tube, in m. Take the density of water to be 1000 kg/m^3.

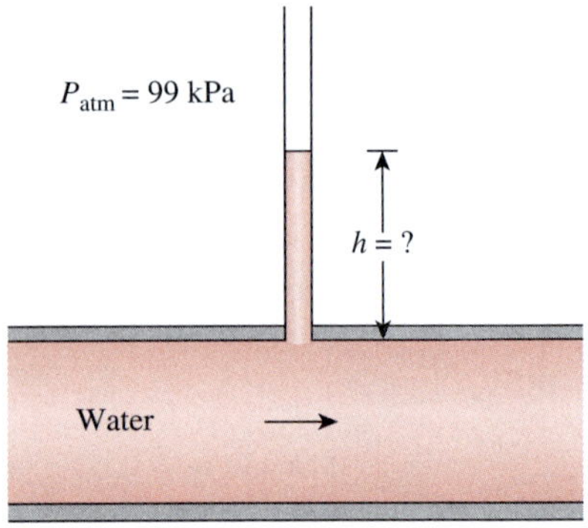

FIGURE P1–115

1–116 When measuring small pressure differences with a manometer, often one arm of the manometer is inclined to improve the accuracy of reading. (The pressure difference is still proportional to the *vertical* distance and not the actual length of the fluid along the tube.) The air pressure in a circular duct is to be measured using a manometer whose open arm is inclined 45° from the horizontal, as shown in Fig. P1–116 The density of the liquid in the manometer is 0.81 kg/L, and the vertical distance between the fluid levels in the two arms of the manometer is 12 cm. Determine the gage pressure of air in the duct and the length of the fluid column in the inclined arm above the fluid level in the vertical arm.

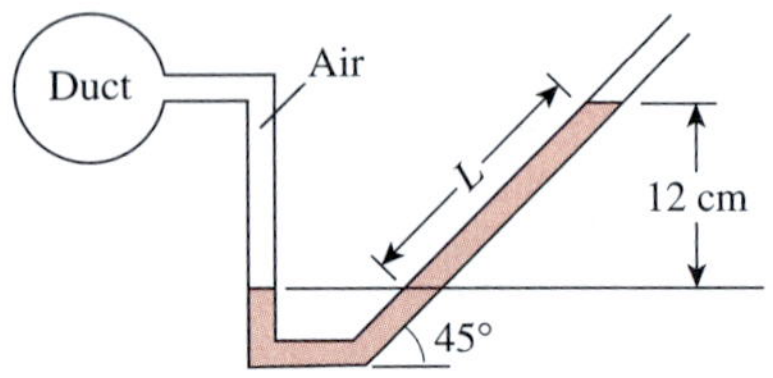

FIGURE P1–116

1–117E Consider a U-tube whose arms are open to the atmosphere. Now equal volumes of water and light oil (ρ = 49.3 lbm/ft^3) are poured from different arms. A person blows from the oil side of the U-tube until the contact surface of the two fluids moves to the bottom of the U-tube, and thus the liquid levels in the two arms are the same. If the fluid height in each arm is 30 in, determine the gage pressure the person exerts on the oil by blowing.

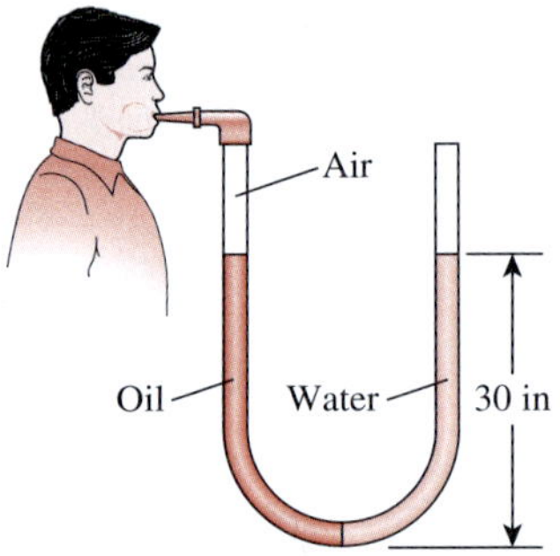

FIGURE P1–117E

1–118 Intravenous infusions are usually driven by gravity by hanging the fluid bottle at sufficient height to counteract the blood pressure in the vein and to force the fluid into the body. The higher the bottle is raised, the higher the flow rate of the fluid will be. (*a*) If it is observed that the fluid and the blood pressures balance each other when the bottle is 80 cm above the arm level, determine the gage pressure of the blood. (*b*) If the gage pressure of the fluid at the arm level needs to be 15 kPa for sufficient flow rate, determine how high the bottle must be placed. Take the density of the fluid to be 1020 kg/m^3.

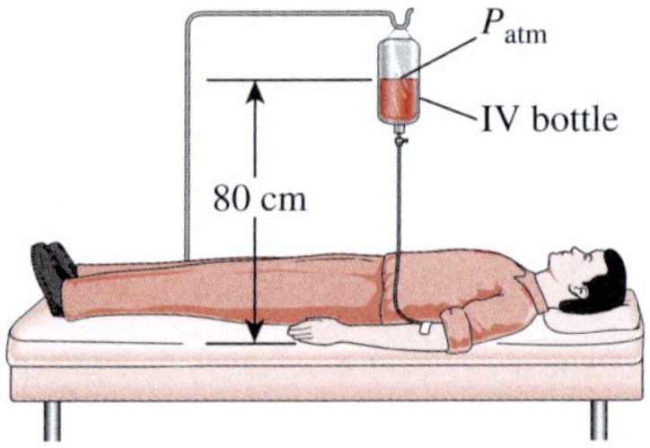

FIGURE P1–118

1–119E A water pipe is connected to a double-U manometer as shown in Fig. P1–119E at a location where the local atmospheric pressure is 14.2 psia. Determine the absolute pressure at the center of the pipe.

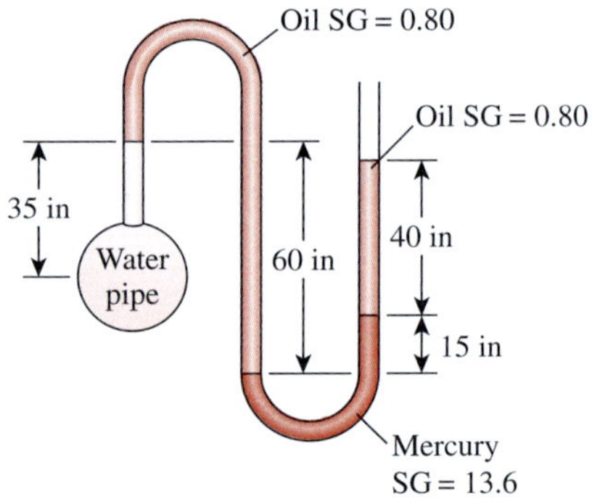

FIGURE P1–119E

1–120 The average atmospheric pressure on earth is approximated as a function of altitude by the relation $P_{atm} = 101.325\ (1 - 0.02256z)^{5.256}$, where P_{atm} is the atmospheric pressure in kPa and z is the altitude in km with $z = 0$ at sea level. Determine the approximate atmospheric pressures at Atlanta (z = 306 m), Denver (z = 1610 m), Mexico City (z = 2309 m), and the top of Mount Everest (z = 8848 m).

1–121 It is well-known that the temperature of the atmosphere varies with altitude. In the troposphere, which extends to an altitude of 11 km, for example, the variation of temperature can be approximated by $T = T_0 - \beta z$, where T_0 is the temperature at sea level, which can be taken to be 288.15 K, and β = 0.0065 K/m. The gravitational acceleration also changes with altitude as $g(z) = g_0/(1 + z/6{,}370{,}320)^2$ where g_0 = 9.807 m/s^2 and z is the elevation from sea level in m. Obtain a relation for the variation of pressure in the troposphere (*a*) by ignoring and (*b*) by considering the variation of g with altitude.

1–122 The variation of pressure with density in a thick gas layer is given by $P = C\rho^n$, where C and n are constants. Noting that the pressure change across a differential fluid layer of thickness dz in the vertical z-direction is given as $dP = -\rho g\, dz$, obtain a relation for pressure as a function of elevation z. Take the pressure and density at $z = 0$ to be P_0 and ρ_0, respectively.

1–123 Pressure transducers are commonly used to measure pressure by generating analog signals usually in the range of 4 mA to 20 mA or 0 V-dc to 10 V-dc in response to applied pressure. The system whose schematic is shown in Fig. P1–123 can be used to calibrate pressure transducers. A rigid container is filled with pressurized air, and pressure is measured by the manometer attached. A valve is used to regulate the pressure in the container. Both the pressure and the electric signal are measured simultaneously for various settings, and the results are tabulated. For the given set of measurements, obtain the calibration curve in the form of $P = aI + b$, where a and b are constants, and calculate the pressure that corresponds to a signal of 10 mA.

Δh, mm	28.0	181.5	297.8	413.1	765.9
I, mA	4.21	5.78	6.97	8.15	11.76

Δh, mm	1027	1149	1362	1458	1536
I, mA	14.43	15.68	17.86	18.84	19.64

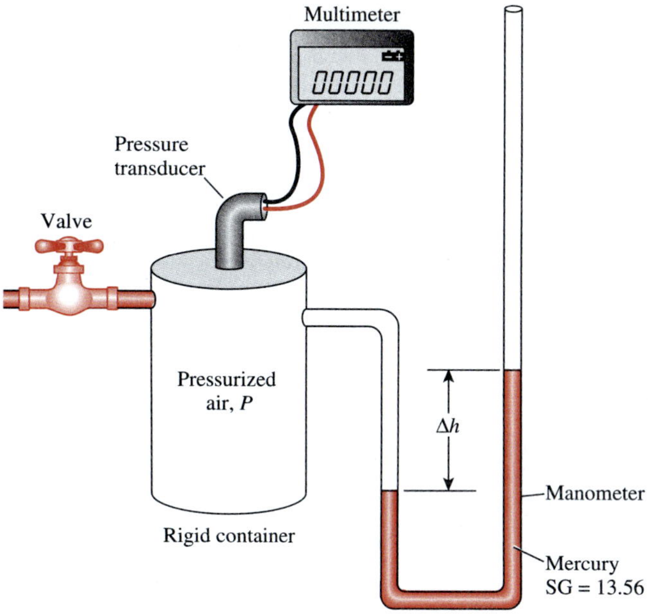

FIGURE P1–123

1–124 Consider the flow of air through a wind turbine whose blades sweep an area of diameter D (in m). The average air velocity through the swept area is V (in m/s). On the bases of the units of the quantities involved, show that the mass flow rate of air (in kg/s) through the swept area is proportional to air density, the wind velocity, and the square of the diameter of the swept area.

1–125 The drag force exerted on a car by air depends on a dimensionless drag coefficient, the density of air, the car velocity, and the frontal area of the car. That is, F_D = function (C_{Drag} $A_{front,}$ ρ, V). Based on unit considerations alone, obtain a relation for the drag force.

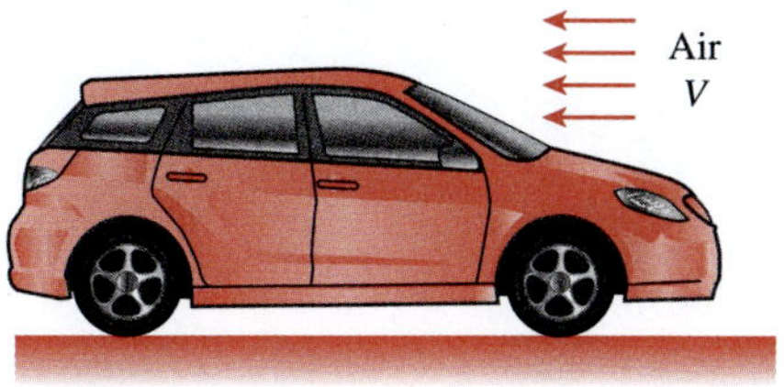

FIGURE P1–125

Fundamentals of Engineering (FE) Exam Problems

1–126 Consider a fish swimming 5 m below the free surface of water. The increase in the pressure exerted on the fish when it dives to a depth of 25 m below the free surface is

(*a*) 196 Pa (*b*) 5400 Pa (*c*) 30,000 Pa
(*d*) 196,000 Pa (*e*) 294,000 Pa

1–127 The atmospheric pressures at the top and the bottom of a building are read by a barometer to be 96.0 and 98.0 kPa. If the density of air is 1.0 kg/m^3, the height of the building is

(*a*) 17 m (*b*) 20 m (*c*) 170 m
(*d*) 204 m (*e*) 252 m

1–128 An apple loses 4.5 kJ of heat as it cools per °C drop in its temperature. The amount of heat loss from the apple per °F drop in its temperature is

(*a*) 1.25 kJ (*b*) 2.50 kJ (*c*) 5.0 kJ
(*d*) 8.1 kJ (*e*) 4.1 kJ

1–129 Consider a 2-m deep swimming pool. The pressure difference between the top and bottom of the pool is

(*a*) 12.0 kPa (*b*) 19.6 kPa (*c*) 38.1 kPa
(*d*) 50.8 kPa (*e*) 200 kPa

1–130 At sea level, the weight of 1 kg mass in SI units is 9.81 N. The weight of 1 lbm mass in English units is

(*a*) 1 lbf (*b*) 9.81 lbf (*c*) 32.2 lbf
(*d*) 0.1 lbf (*e*) 0.031 lbf

1–131 During a heating process, the temperature of an object rises by 10°C. This temperature rise is equivalent to a temperature rise of

(*a*) 10°F (*b*) 42°F (*c*) 18 K
(*d*) 18 R (*e*) 283 K

Design and Essay Problems

1–132 Write an essay on different temperature measurement devices. Explain the operational principle of each device, its advantages and disadvantages, its cost, and its range of applicability. Which device would you recommend for use in the following cases: taking the temperatures of patients in a doctor's office, monitoring the variations of temperature of a car engine block at several locations, and monitoring the temperatures in the furnace of a power plant?

1–133 Write an essay on the various mass- and volume-measurement devices used throughout history. Also, explain the development of the modern units for mass and volume.

CHAPTER

2

ENERGY, ENERGY TRANSFER, AND GENERAL ENERGY ANALYSIS

Whether we realize it or not, energy is an important part of most aspects of daily life. The quality of life, and even its sustenance, depends on the availability of energy. Therefore, it is important to have a good understanding of the sources of energy, the conversion of energy from one form to another, and the ramifications of these conversions.

Energy exists in numerous forms such as thermal, mechanical, electric, chemical, and nuclear. Even mass can be considered a form of energy. Energy can be transferred to or from a closed system (a fixed mass) in two distinct forms: *heat* and *work*. For control volumes, energy can also be transferred by mass flow. An energy transfer to or from a closed system is *heat* if it is caused by a temperature difference. Otherwise it is *work*, and it is caused by a force acting through a distance.

We start this chapter with a discussion of various forms of energy and energy transfer by heat. We then introduce various forms of work and discuss energy transfer by work. We continue with developing a general intuitive expression for the *first law of thermodynamics*, also known as the *conservation of energy principle*, which is one of the most fundamental principles in nature, and we then demonstrate its use. Finally, we discuss the efficiencies of some familiar energy conversion processes, and examine the impact on energy conversion on the environment. Detailed treatments of the first law of thermodynamics for closed systems and control volumes are given in Chaps. 4 and 5, respectively.

Objectives

The objectives of Chapter 2 are to:

- Introduce the concept of energy and define its various forms.
- Discuss the nature of internal energy.
- Define the concept of heat and the terminology associated with energy transfer by heat.
- Discuss the three mechanisms of heat transfer: conduction, convection, and radiation.
- Define the concept of work, including electrical work and several forms of mechanical work.
- Introduce the first law of thermodynamics, energy balances, and mechanisms of energy transfer to or from a system.
- Determine that a fluid flowing across a control surface of a control volume carries energy across the control surface in addition to any energy transfer across the control surface that may be in the form of heat and/or work.
- Define energy conversion efficiencies.
- Discuss the implications of energy conversion on the environment.

2–1 ▪ INTRODUCTION

We are familiar with the conservation of energy principle, which is an expression of the first law of thermodynamics, back from our high school years. We are told repeatedly that energy cannot be created or destroyed during a process; it can only change from one form to another. This seems simple enough, but let's test ourselves to see how well we understand and truly believe in this principle.

Consider a room whose door and windows are tightly closed, and whose walls are well-insulated so that heat loss or gain through the walls is negligible. Now let's place a refrigerator in the middle of the room with its door open, and plug it into a wall outlet (Fig. 2–1). You may even use a small fan to circulate the air in order to maintain temperature uniformity in the room. Now, what do you think will happen to the average temperature of air in the room? Will it be increasing or decreasing? Or will it remain constant?

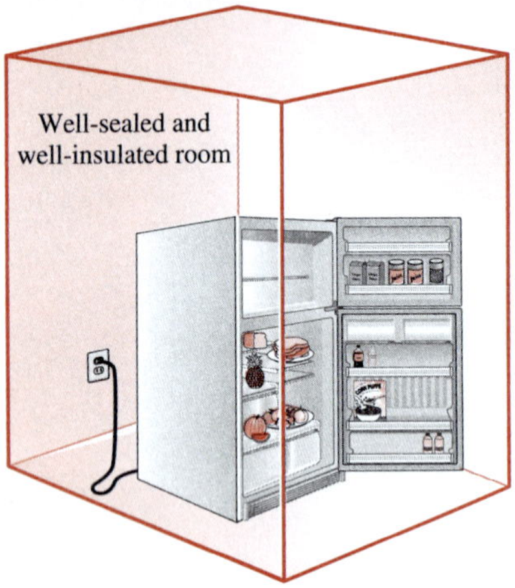

FIGURE 2–1
A refrigerator operating with its door open in a well-sealed and well-insulated room.

Probably the first thought that comes to mind is that the average air temperature in the room will decrease as the warmer room air mixes with the air cooled by the refrigerator. Some may draw our attention to the heat generated by the motor of the refrigerator, and may argue that the average air temperature may rise if this heating effect is greater than the cooling effect. But they will get confused if it is stated that the motor is made of superconducting materials, and thus there is hardly any heat generation in the motor.

Heated discussion may continue with no end in sight until we remember the conservation of energy principle that we take for granted: If we take the entire room—including the air and the refrigerator—as the system, which is an adiabatic closed system since the room is well-sealed and well-insulated, the only energy interaction involved is the electrical energy crossing the system boundary and entering the room. The conservation of energy requires the energy content of the room to increase by an amount equal to the amount of the electrical energy drawn by the refrigerator, which can be measured by an ordinary electric meter. The refrigerator or its motor does not store this energy. Therefore, this energy must now be in the room air, and it will manifest itself as a rise in the air temperature. The temperature rise of air can be calculated on the basis of the conservation of energy principle using the properties of air and the amount of electrical energy consumed. What do you think would happen if we had a window air conditioning unit instead of a refrigerator placed in the middle of this room? What if we operated a fan in this room instead (Fig. 2–2)?

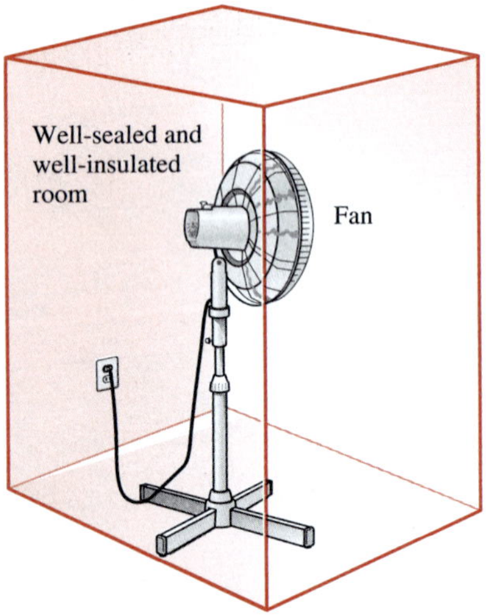

FIGURE 2–2
A fan running in a well-sealed and well-insulated room will raise the temperature of air in the room.

Note that energy is conserved during the process of operating the refrigerator placed in a room—the electrical energy is converted into an equivalent amount of thermal energy stored in the room air. If energy is already conserved, then what are all those speeches on energy conservation and the measures taken to conserve energy? Actually, by "energy conservation" what is meant is the conservation of the *quality* of energy, not the quantity. Electricity, which is of the highest quality of energy, for example, can always be converted to an equal amount of thermal energy (also called *heat*). But only a small fraction of thermal energy, which is the lowest quality of energy, can be converted back to electricity, as we discuss in Chap. 6. Think about the things that you can do with the electrical energy that the refrigerator has consumed, and the air in the room that is now at a higher temperature.

Now if asked to name the energy transformations associated with the operation of a refrigerator, we may still have a hard time answering because all we see is electrical energy entering the refrigerator and heat dissipated from the refrigerator to the room air. Obviously there is need to study the various forms of energy first, and this is exactly what we do next, followed by a study of the mechanisms of energy transfer.

2–2 ▪ FORMS OF ENERGY

Energy can exist in numerous forms such as thermal, mechanical, kinetic, potential, electric, magnetic, chemical, and nuclear, and their sum constitutes the **total energy** E of a system. The total energy of a system on a *unit mass* basis is denoted by e and is expressed as

$$e = \frac{E}{m} \qquad \text{(kJ/kg)} \tag{2–1}$$

Thermodynamics provides no information about the absolute value of the total energy. It deals only with the *change* of the total energy, which is what matters in engineering problems. Thus the total energy of a system can be assigned a value of zero ($E = 0$) at some convenient reference point. The change in total energy of a system is independent of the reference point selected. The decrease in the potential energy of a falling rock, for example, depends on only the elevation difference and not the reference level selected.

In thermodynamic analysis, it is often helpful to consider the various forms of energy that make up the total energy of a system in two groups: *macroscopic* and *microscopic*. The **macroscopic** forms of energy are those a system possesses as a whole with respect to some outside reference frame, such as kinetic and potential energies (Fig. 2–3). The **microscopic** forms of energy are those related to the molecular structure of a system and the degree of the molecular activity, and they are independent of outside reference frames. The sum of all the microscopic forms of energy is called the **internal energy** of a system and is denoted by U.

FIGURE 2–3
The macroscopic energy of an object changes with velocity and elevation.

The term *energy* was coined in 1807 by Thomas Young, and its use in thermodynamics was proposed in 1852 by Lord Kelvin. The term *internal energy* and its symbol U first appeared in the works of Rudolph Clausius and William Rankine in the second half of the nineteenth century, and it eventually replaced the alternative terms *inner work, internal work,* and *intrinsic energy* commonly used at the time.

The macroscopic energy of a system is related to motion and the influence of some external effects such as gravity, magnetism, electricity, and surface tension. The energy that a system possesses as a result of its motion relative to some reference frame is called **kinetic energy** (KE). When all parts of a system move with the same velocity, the kinetic energy is expressed as

$$\text{KE} = m\frac{V^2}{2} \qquad \text{(kJ)} \tag{2–2}$$

or, on a unit mass basis,

$$\text{ke} = \frac{V^2}{2} \qquad \text{(kJ/kg)} \tag{2–3}$$

where V denotes the velocity of the system relative to some fixed reference frame. The kinetic energy of a rotating solid body is given by $\frac{1}{2}I\omega^2$ where I is the moment of inertia of the body and ω is the angular velocity.

The energy that a system possesses as a result of its elevation in a gravitational field is called **potential energy** (PE) and is expressed as

$$\text{PE} = mgz \qquad \text{(kJ)} \tag{2–4}$$

or, on a unit mass basis,

$$\text{pe} = gz \qquad \text{(kJ/kg)} \tag{2–5}$$

where g is the gravitational acceleration and z is the elevation of the center of gravity of a system relative to some arbitrarily selected reference level.

The magnetic, electric, and surface tension effects are significant in some specialized cases only and are usually ignored. In the absence of such effects, the total energy of a system consists of the kinetic, potential, and internal energies and is expressed as

$$E = U + \text{KE} + \text{PE} = U + m\frac{V^2}{2} + mgz \qquad \text{(kJ)} \tag{2–6}$$

or, on a unit mass basis,

$$e = u + \text{ke} + \text{pe} = u + \frac{V^2}{2} + gz \qquad \text{(kJ/kg)} \tag{2–7}$$

Most closed systems remain stationary during a process and thus experience no change in their kinetic and potential energies. Closed systems whose velocity and elevation of the center of gravity remain constant during a process are frequently referred to as **stationary systems**. The change in the total energy ΔE of a stationary system is identical to the change in its internal energy ΔU. In this text, a closed system is assumed to be stationary unless stated otherwise.

Control volumes typically involve fluid flow for long periods of time, and it is convenient to express the energy flow associated with a fluid stream in the rate form. This is done by incorporating the **mass flow rate** $\dot{m}$, which is *the amount of mass flowing through a cross section per unit time*. It is related to the **volume flow rate** $\dot{V}$, which is the volume of a fluid flowing through a cross section per unit time, by

Mass flow rate: $$\dot{m} = \rho\dot{V} = \rho A_c V_{avg} \qquad \text{(kg/s)} \tag{2–8}$$

which is analogous to $m = \rho V$. Here ρ is the fluid density, A_c is the cross-sectional area of flow, and V_{avg} is the average flow velocity normal to A_c. The dot over a symbol is used to indicate *time rate* throughout the book. Then the energy flow rate associated with a fluid flowing at a rate of $\dot{m}$ is (Fig. 2–4)

Energy flow rate: $$\dot{E} = \dot{m}e \qquad \text{(kJ/s or kW)} \tag{2–9}$$

which is analogous to $E = me$.

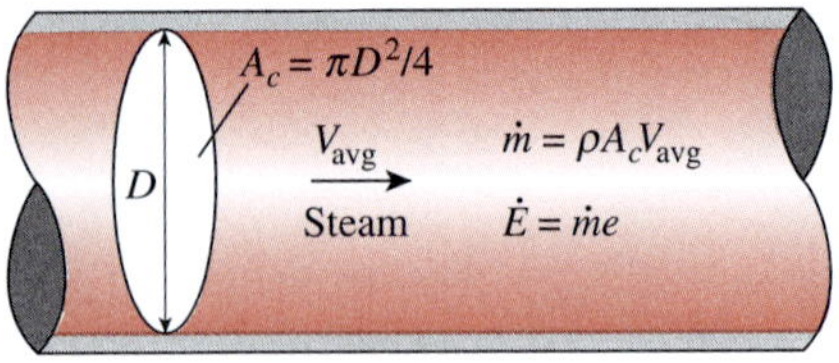

FIGURE 2–4
Mass and energy flow rates associated with the flow of steam in a pipe of inner diameter D with an average velocity of V_{avg}.

Some Physical Insight to Internal Energy

Internal energy is defined earlier as the sum of all the *microscopic* forms of energy of a system. It is related to the *molecular structure* and the degree of *molecular activity,* and can be viewed as the sum of the *kinetic* and *potential* energies of the molecules.

To have a better understanding of internal energy, let us examine a system at the molecular level. The molecules of a gas move through space with some velocity, and thus possess some kinetic energy. This is known as the *translational energy*. The atoms of polyatomic molecules rotate about an axis, and the energy associated with this rotation is the *rotational kinetic energy*. The atoms of a polyatomic molecule may also vibrate about their common center of mass, and the energy associated with this back-and-forth motion is the *vibrational kinetic energy*. For gases, the kinetic energy is mostly due to translational and rotational motions, with vibrational motion becoming significant at higher temperatures. The electrons in an atom rotate about the nucleus, and thus possess *rotational kinetic energy*. Electrons at outer orbits have larger kinetic energies. Electrons also spin about their axes, and the energy associated with this motion is the *spin energy*. Other particles in the nucleus of an atom also possess spin energy. The portion of the internal energy of a system associated with the kinetic energies of the molecules is called the **sensible energy** (Fig. 2–5). The average velocity and the degree of activity of the molecules are proportional to the temperature of the gas. Therefore, at higher temperatures, the molecules possess higher kinetic energies, and as a result the system has a higher internal energy.

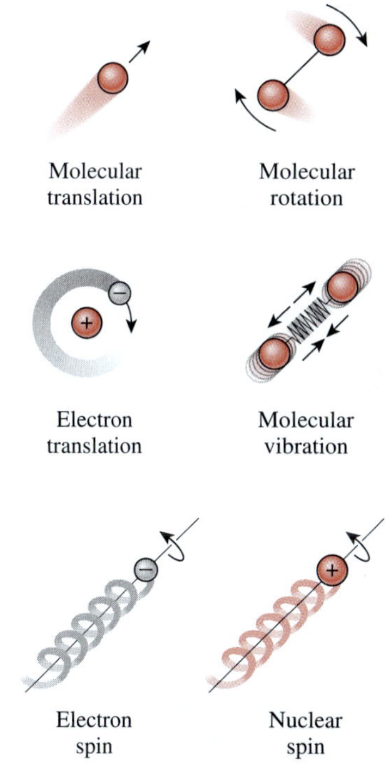

FIGURE 2–5
The various forms of microscopic energies that make up *sensible* energy.

The internal energy is also associated with various *binding forces* between the molecules of a substance, between the atoms within a molecule, and between the particles within an atom and its nucleus. The forces that bind the *molecules* to each other are, as one would expect, strongest in solids and weakest in gases. If sufficient energy is added to the molecules of a solid or liquid, the molecules overcome these molecular forces and break away, turning the substance into a gas. This is a phase-change process. Because of this added energy, a system in the gas phase is at a higher internal energy level than it is in the solid or the liquid phase. The internal energy associated with the phase of a system is called the **latent energy**. The phase-change process can occur without a change in the chemical composition of a system. Most practical problems fall into this category, and one does not need to pay any attention to the forces binding the atoms in a molecule to each other.

An atom consists of neutrons and positively charged protons bound together by very strong nuclear forces in the nucleus, and negatively charged electrons orbiting around it. The internal energy associated with the atomic bonds in a molecule is called **chemical energy**. During a chemical reaction, such as a combustion process, some chemical bonds are destroyed while others are formed. As a result, the internal energy changes. The nuclear forces are much larger than the forces that bind the electrons to the nucleus. The tremendous amount of energy associated with the strong bonds within the nucleus of the atom itself is called **nuclear energy** (Fig. 2–6). Obviously, we need not be concerned with nuclear energy in thermodynamics unless, of course, we deal with fusion or fission reactions. A chemical reaction involves changes in the structure of the electrons of the atoms, but a nuclear reaction involves changes in the core or nucleus. Therefore, an

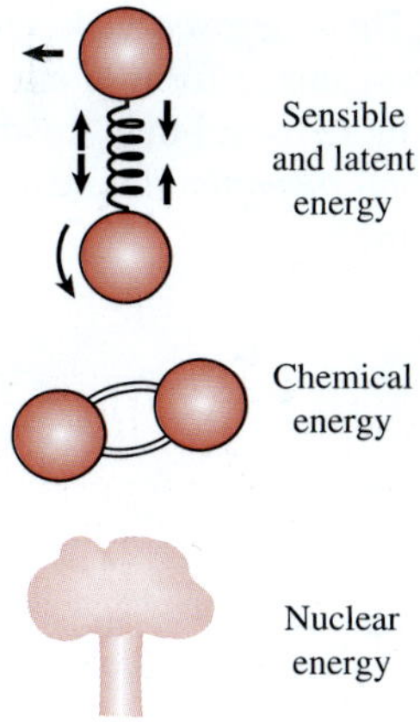

FIGURE 2–6
The internal energy of a system is the sum of all forms of the microscopic energies.

atom preserves its identity during a chemical reaction but loses it during a nuclear reaction. Atoms may also possess *electric* and *magnetic dipole-moment energies* when subjected to external electric and magnetic fields due to the twisting of the magnetic dipoles produced by the small electric currents associated with the orbiting electrons.

The forms of energy already discussed, which constitute the total energy of a system, can be *contained* or *stored* in a system, and thus can be viewed as the *static* forms of energy. The forms of energy not stored in a system can be viewed as the *dynamic* forms of energy or as *energy interactions*. The dynamic forms of energy are recognized at the system boundary as they cross it, and they represent the energy gained or lost by a system during a process. The only two forms of energy interactions associated with a closed system are **heat transfer** and **work**. An energy interaction is heat transfer if its driving force is a temperature difference. Otherwise it is work, as explained in the next section. A control volume can also exchange energy via mass transfer since any time mass is transferred into or out of a system, the energy content of the mass is also transferred with it.

In daily life, we frequently refer to the sensible and latent forms of internal energy as *heat*, and we talk about heat content of bodies. In thermodynamics, however, we usually refer to those forms of energy as **thermal energy** to prevent any confusion with *heat transfer*.

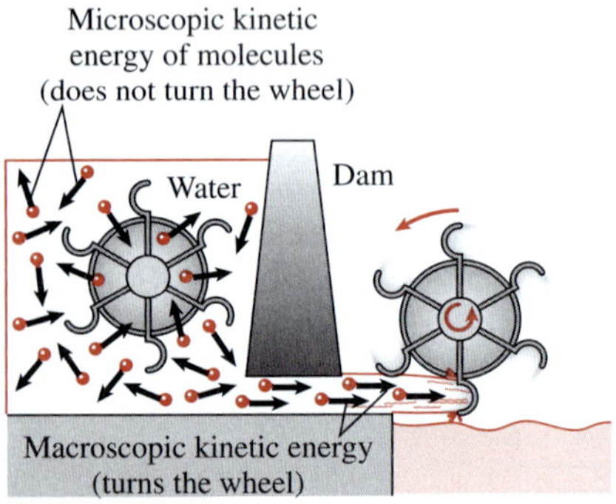

FIGURE 2–7
The *macroscopic* kinetic energy is an organized form of energy and is much more useful than the disorganized *microscopic* kinetic energies of the molecules.

Distinction should be made between the macroscopic kinetic energy of an object as a whole and the microscopic kinetic energies of its molecules that constitute the sensible internal energy of the object (Fig. 2–7). The kinetic energy of an object is an *organized* form of energy associated with the orderly motion of all molecules in one direction in a straight path or around an axis. In contrast, the kinetic energies of the molecules are completely *random* and highly *disorganized*. As you will see in later chapters, the organized energy is much more valuable than the disorganized energy, and a major application area of thermodynamics is the conversion of disorganized energy (heat) into organized energy (work). You will also see that the organized energy can be converted to disorganized energy completely, but only a fraction of disorganized energy can be converted to organized energy by specially built devices called *heat engines* (like car engines and power plants). A similar argument can be given for the macroscopic potential energy of an object as a whole and the microscopic potential energies of the molecules.

More on Nuclear Energy

The best known fission reaction involves the split of the uranium atom (the U-235 isotope) into other elements and is commonly used to generate electricity in nuclear power plants (440 of them in 2004, generating 363,000 MW worldwide), to power nuclear submarines and aircraft carriers, and even to power spacecraft as well as building nuclear bombs.

The percentage of electricity produced by nuclear power is 78 percent in France, 25 percent in Japan, 28 percent in Germany, and 20 percent in the United States. The first nuclear chain reaction was achieved by Enrico Fermi in 1942, and the first large-scale nuclear reactors were built in 1944 for the purpose of producing material for nuclear weapons. When a uranium-235 atom absorbs a neutron and splits during a fission process, it produces a cesium-140 atom, a rubidium-93 atom, 3 neutrons, and 3.2×10^{-11} J of energy. In practical terms, the complete fission of 1 kg of uranium-235 releases 6.73×10^{10} kJ of heat, which is more than the heat released when 3000 tons of coal are burned. Therefore, for the same amount of fuel, a nuclear fission reaction releases several million times more energy than a chemical reaction. The safe disposal of used nuclear fuel, however, remains a concern.

Nuclear energy by fusion is released when two small nuclei combine into a larger one. The huge amount of energy radiated by the sun and the other stars originates from such a fusion process that involves the combination of two hydrogen atoms into a helium atom. When two heavy hydrogen (deuterium) nuclei combine during a fusion process, they produce a helium-3 atom, a free neutron, and 5.1×10^{-13} J of energy (Fig. 2–8).

Fusion reactions are much more difficult to achieve in practice because of the strong repulsion between the positively charged nuclei, called the *Coulomb repulsion*. To overcome this repulsive force and to enable the two nuclei to fuse together, the energy level of the nuclei must be raised by heating them to about 100 million °C. But such high temperatures are found only in the stars or in exploding atomic bombs (the A-bomb). In fact, the uncontrolled fusion reaction in a hydrogen bomb (the H-bomb) is initiated by a small atomic bomb. The uncontrolled fusion reaction was achieved in the early 1950s, but all the efforts since then to achieve controlled fusion by massive lasers, powerful magnetic fields, and electric currents to generate power have failed.

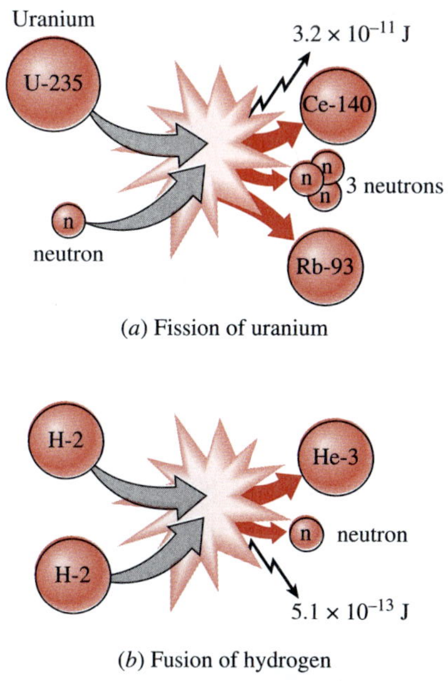

FIGURE 2–8
The fission of uranium and the fusion of hydrogen during nuclear reactions, and the release of nuclear energy.

EXAMPLE 2–1 A Car Powered by Nuclear Fuel

An average car consumes about 5 L of gasoline a day, and the capacity of the fuel tank of a car is about 50 L. Therefore, a car needs to be refueled once every 10 days. Also, the density of gasoline ranges from 0.68 to 0.78 kg/L, and its lower heating value is about 44,000 kJ/kg (that is, 44,000 kJ of heat is released when 1 kg of gasoline is completely burned). Suppose all the problems associated with the radioactivity and waste disposal of nuclear fuels are resolved, and a car is to be powered by U-235. If a new car comes equipped with 0.1-kg of the nuclear fuel U-235, determine if this car will ever need refueling under average driving conditions (Fig. 2–9).

Solution A car powered by nuclear energy comes equipped with nuclear fuel. It is to be determined if this car will ever need refueling.

Assumptions **1** Gasoline is an incompressible substance with an average density of 0.75 kg/L. **2** Nuclear fuel is completely converted to thermal energy.

Analysis The mass of gasoline used per day by the car is

$$m_{\text{gasoline}} = (\rho V)_{\text{gasoline}} = (0.75\ \text{kg/L})(5\ \text{L/day}) = 3.75\ \text{kg/day}$$

Noting that the heating value of gasoline is 44,000 kJ/kg, the energy supplied to the car per day is

$$\begin{aligned} E &= (m_{\text{gasoline}})(\text{Heating value}) \\ &= (3.75\ \text{kg/day})(44{,}000\ \text{kJ/kg}) = 165{,}000\ \text{kJ/day} \end{aligned}$$

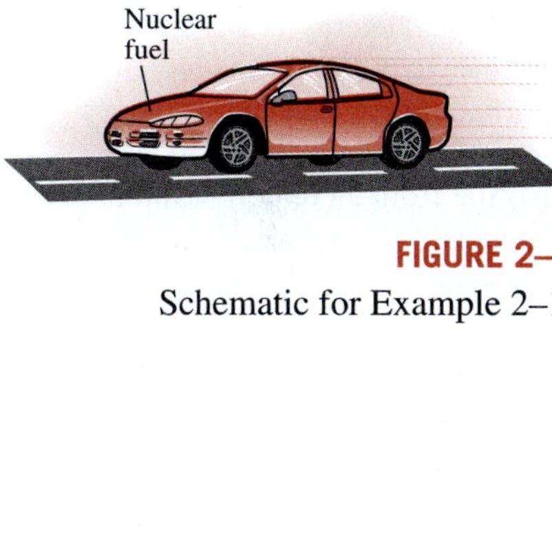

FIGURE 2–9
Schematic for Example 2–1.

The complete fission of 0.1 kg of uranium-235 releases

$$(6.73 \times 10^{10}\ \text{kJ/kg})(0.1\ \text{kg}) = 6.73 \times 10^{9}\ \text{kJ}$$

of heat, which is sufficient to meet the energy needs of the car for

$$\text{No. of days} = \frac{\text{Energy content of fuel}}{\text{Daily energy use}} = \frac{6.73 \times 10^{9}\ \text{kJ}}{165{,}000\ \text{kJ/day}} = \mathbf{40{,}790\ days}$$

which is equivalent to about 112 years. Considering that no car will last more than 100 years, this car will never need refueling. It appears that nuclear fuel of the size of a cherry is sufficient to power a car during its lifetime.

Discussion Note that this problem is not quite realistic since the necessary critical mass cannot be achieved with such a small amount of fuel. Further, all of the uranium cannot be converted in fission, again because of the critical mass problems after partial conversion.

Mechanical Energy

Many engineering systems are designed to transport a fluid from one location to another at a specified flow rate, velocity, and elevation difference, and the system may generate mechanical work in a turbine or it may consume mechanical work in a pump or fan during this process (Fig. 2–10). These systems do not involve the conversion of nuclear, chemical, or thermal energy to mechanical energy. Also, they do not involve any heat transfer in any significant amount, and they operate essentially at constant temperature. Such systems can be analyzed conveniently by considering the *mechanical forms of energy* only and the frictional effects that cause the mechanical energy to be lost (i.e., to be converted to thermal energy that usually cannot be used for any useful purpose).

FIGURE 2–10
Mechanical energy is a useful concept for flows that do not involve significant heat transfer or energy conversion, such as the flow of gasoline from an underground tank into a car.
© *Corbis RF.*

The **mechanical energy** can be defined as *the form of energy that can be converted to mechanical work completely and directly by an ideal mechanical device such as an ideal turbine*. Kinetic and potential energies are the familiar forms of mechanical energy. Thermal energy is not mechanical energy, however, since it cannot be converted to work directly and completely (the second law of thermodynamics).

A pump transfers mechanical energy to a fluid by raising its pressure, and a turbine extracts mechanical energy from a fluid by dropping its pressure. Therefore, the pressure of a flowing fluid is also associated with its mechanical energy. In fact, the pressure unit Pa is equivalent to Pa = N/m^2 = N·m/m^3 = J/m^3, which is energy per unit volume, and the product $P\text{v}$ or its equivalent P/ρ has the unit J/kg, which is energy per unit mass. Note that pressure itself is not a form of energy but a pressure force acting on a fluid through a distance produces work, called *flow work*, in the amount of P/ρ per unit mass. Flow work is expressed in terms of fluid properties, and it is convenient to view it as part of the energy of a flowing fluid and call it *flow energy*. Therefore, the mechanical energy of a flowing fluid can be expressed on a unit mass basis as

$$e_{\text{mech}} = \frac{P}{\rho} + \frac{V^2}{2} + gz \tag{2–10}$$

where P/ρ is the *flow energy*, $V^2/2$ is the *kinetic energy*, and gz is the *potential energy* of the fluid, all per unit mass. It can also be expressed in rate form as

$$\dot{E}_{\text{mech}} = \dot{m}e_{\text{mech}} = \dot{m}\left(\frac{P}{\rho} + \frac{V^2}{2} + gz\right) \quad \textbf{(2–11)}$$

where $\dot{m}$ is the mass flow rate of the fluid. Then the mechanical energy change of a fluid during incompressible (ρ = constant) flow becomes

$$\Delta e_{\text{mech}} = \frac{P_2 - P_1}{\rho} + \frac{V_2^2 - V_1^2}{2} + g(z_2 - z_1) \quad (\text{kJ/kg}) \quad \textbf{(2–12)}$$

and

$$\Delta \dot{E}_{\text{mech}} = \dot{m}\Delta e_{\text{mech}} = \dot{m}\left(\frac{P_2 - P_1}{\rho} + \frac{V_2^2 - V_1^2}{2} + g(z_2 - z_1)\right) \quad (\text{kW}) \quad \textbf{(2–13)}$$

Therefore, the mechanical energy of a fluid does not change during flow if its pressure, density, velocity, and elevation remain constant. In the absence of any irreversible losses, the mechanical energy change represents the mechanical work supplied to the fluid (if $\Delta e_{\text{mech}} > 0$) or extracted from the fluid (if $\Delta e_{\text{mech}} < 0$). The maximum (ideal) power generated by a turbine, for example, is $\dot{W}_{\text{max}} = \dot{m}\Delta e_{\text{mech}}$, as shown in Fig. 2–11.

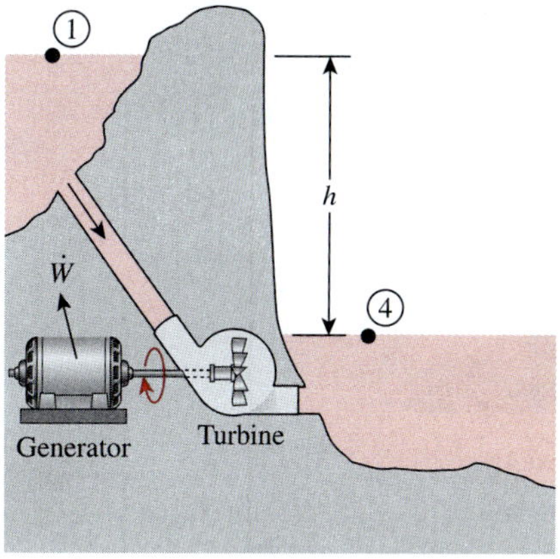

$\dot{W}_{\text{max}} = \dot{m}\Delta e_{\text{mech}} = \dot{m}g(z_1 - z_4) = \dot{m}gh$
since $P_1 \approx P_4 = P_{\text{atm}}$ and $V_1 = V_4 \approx 0$
(*a*)

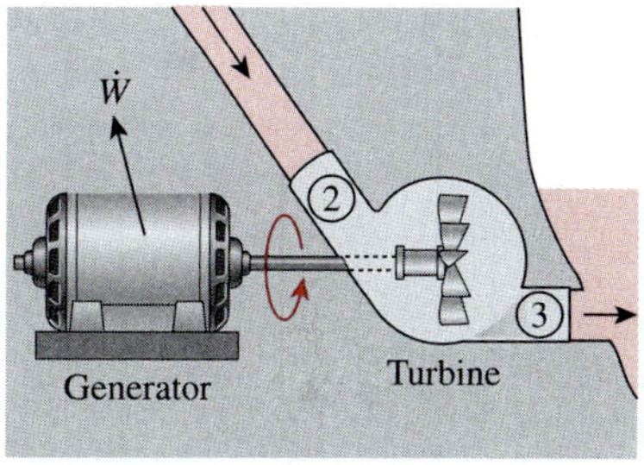

$\dot{W}_{\text{max}} = \dot{m}\Delta e_{\text{mech}} = \dot{m}\dfrac{(P_2 - P_3)}{\rho} = \dot{m}\dfrac{\Delta P}{\rho}$
since $V_2 \approx V_3$ and $z_2 = z_3$
(*b*)

FIGURE 2–11
Mechanical energy is illustrated by an ideal hydraulic turbine coupled with an ideal generator. In the absence of irreversible losses, the maximum produced power is proportional to (*a*) the change in water surface elevation from the upstream to the downstream reservoir or (*b*) (close-up view) the drop in water pressure from just upstream to just downstream of the turbine.

EXAMPLE 2–2 Wind Energy

A site evaluated for a wind farm is observed to have steady winds at a speed of 8.5 m/s (Fig. 2–12). Determine the wind energy (*a*) per unit mass, (*b*) for a mass of 10 kg, and (*c*) for a flow rate of 1154 kg/s for air.

Solution A site with a specified wind speed is considered. Wind energy per unit mass, for a specified mass, and for a given mass flow rate of air are to be determined.
Assumptions Wind flows steadily at the specified speed.
Analysis The only harvestable form of energy of atmospheric air is the kinetic energy, which is captured by a wind turbine.
(*a*) Wind energy per unit mass of air is

$$e = \text{ke} = \frac{V^2}{2} = \frac{(8.5\ \text{m/s})^2}{2}\left(\frac{1\ \text{J/kg}}{1\ \text{m}^2/\text{s}^2}\right) = \mathbf{36.1\ J/kg}$$

(*b*) Wind energy for an air mass of 10 kg is

$$E = me = (10\ \text{kg})(36.1\ \text{J/kg}) = \mathbf{361\ J}$$

(*c*) Wind energy for a mass flow rate of 1154 kg/s is

$$\dot{E} = \dot{m}e = (1154\ \text{kg/s})(36.1\ \text{J/kg})\left(\frac{1\ \text{kW}}{1000\ \text{J/s}}\right) = \mathbf{41.7\ kW}$$

Discussion It can be shown that the specified mass flow rate corresponds to a 12-m diameter flow section when the air density is 1.2 kg/m^3. Therefore, a wind turbine with a wind span diameter of 12 m has a power generation potential of 41.7 kW. Real wind turbines convert about one-third of this potential to electric power.

FIGURE 2–12
Potential site for a wind farm as discussed in Example 2–2.
© Vol. 36/PhotoDisc/Getty RF.

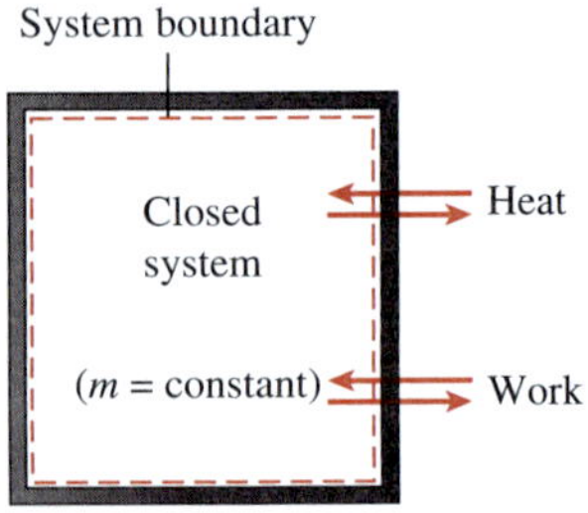

FIGURE 2–13
Energy can cross the boundaries of a closed system in the form of heat and work.

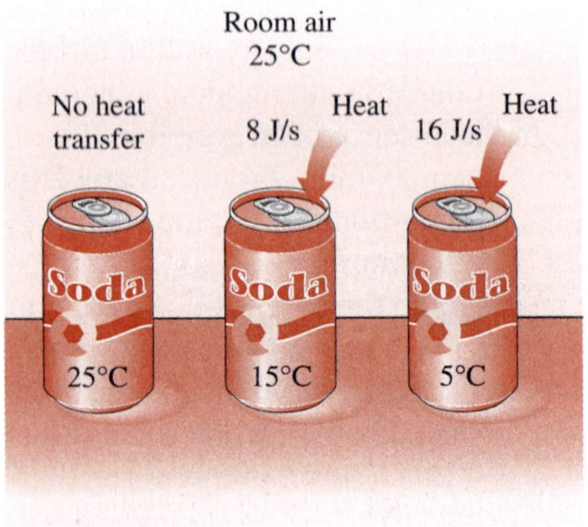

FIGURE 2–14
Temperature difference is the driving force for heat transfer. The larger the temperature difference, the higher is the rate of heat transfer.

2–3 ▪ ENERGY TRANSFER BY HEAT

Energy can cross the boundary of a closed system in two distinct forms: *heat* and *work* (Fig. 2–13). It is important to distinguish between these two forms of energy. Therefore, they will be discussed first, to form a sound basis for the development of the laws of thermodynamics.

We know from experience that a can of cold soda left on a table eventually warms up and that a hot baked potato on the same table cools down. When a body is left in a medium that is at a different temperature, energy transfer takes place between the body and the surrounding medium until thermal equilibrium is established, that is, the body and the medium reach the same temperature. The direction of energy transfer is always from the higher temperature body to the lower temperature one. Once the temperature equality is established, energy transfer stops. In the processes described above, energy is said to be transferred in the form of heat.

Heat is defined as *the form of energy that is transferred between two systems (or a system and its surroundings) by virtue of a temperature difference* (Fig. 2–14). That is, an energy interaction is heat only if it takes place because of a temperature difference. Then it follows that there cannot be any heat transfer between two systems that are at the same temperature.

Several phrases in common use today—such as heat flow, heat addition, heat rejection, heat absorption, heat removal, heat gain, heat loss, heat storage, heat generation, electrical heating, resistance heating, frictional heating, gas heating, heat of reaction, liberation of heat, specific heat, sensible heat, latent heat, waste heat, body heat, process heat, heat sink, and heat source—are not consistent with the strict thermodynamic meaning of the term *heat*, which limits its use to the *transfer* of thermal energy during a process. However, these phrases are deeply rooted in our vocabulary, and they are used by both ordinary people and scientists without causing any misunderstanding since they are usually interpreted properly instead of being taken literally. (Besides, no acceptable alternatives exist for some of these phrases.) For example, the phrase *body heat* is understood to mean *the thermal energy content* of a body. Likewise, *heat flow* is understood to mean *the transfer of thermal energy,* not the flow of a fluidlike substance called heat, although the latter incorrect interpretation, which is based on the caloric theory, is the origin of this phrase. Also, the transfer of heat into a system is frequently referred to as *heat addition* and the transfer of heat out of a system as *heat rejection*. Perhaps there are thermodynamic reasons for being so reluctant to replace *heat* by *thermal energy*: It takes less time and energy to say, write, and comprehend *heat* than it does *thermal energy*.

Heat is energy in transition. It is recognized only as it crosses the boundary of a system. Consider the hot baked potato one more time. The potato contains energy, but this energy is heat transfer only as it passes through the skin of the potato (the system boundary) to reach the air, as shown in Fig. 2–15. Once in the surroundings, the transferred heat becomes part of the internal energy of the surroundings. Thus, in thermodynamics, the term *heat* simply means *heat transfer*.

A process during which there is no heat transfer is called an **adiabatic process** (Fig. 2–16). The word *adiabatic* comes from the Greek word *adiabatos,* which means *not to be passed.* There are two ways a process can be adiabatic: Either the system is well insulated so that only a negligible amount of heat can pass through the boundary, or both the system and the surroundings are at the same temperature and therefore there is no driving force (temperature difference) for heat transfer. An adiabatic process should not be confused with an isothermal process. Even though there is no heat transfer during an adiabatic process, the energy content and thus the temperature of a system can still be changed by other means such as work.

As a form of energy, heat has energy units, kJ (or Btu) being the most common one. The amount of heat transferred during the process between two states (states 1 and 2) is denoted by Q_{12}, or just Q. Heat transfer *per unit mass* of a system is denoted q and is determined from

$$q = \frac{Q}{m} \qquad \text{(kJ/kg)} \tag{2–14}$$

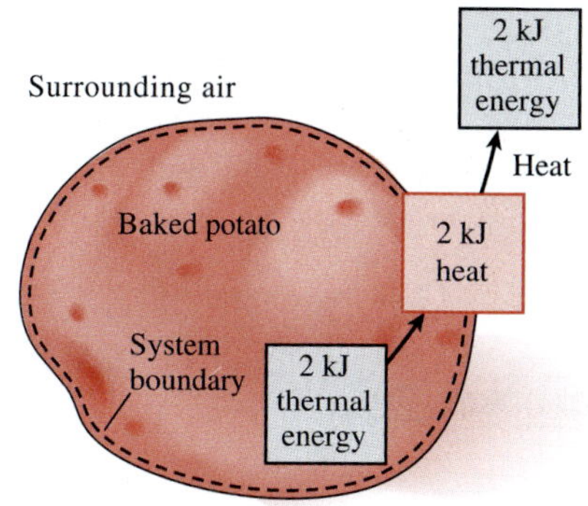

FIGURE 2–15
Energy is recognized as heat transfer only as it crosses the system boundary.

Sometimes it is desirable to know the *rate of heat transfer* (the amount of heat transferred per unit time) instead of the total heat transferred over some time interval (Fig. 2–17). The heat transfer rate is denoted $\dot{Q}$, where the overdot stands for the time derivative, or "per unit time." The heat transfer rate $\dot{Q}$ has the unit kJ/s, which is equivalent to kW. When $\dot{Q}$ varies with time, the amount of heat transfer during a process is determined by integrating $\dot{Q}$ over the time interval of the process:

$$Q = \int_{t_1}^{t_2} \dot{Q}\, dt \qquad \text{(kJ)} \tag{2–15}$$

When $\dot{Q}$ remains constant during a process, this relation reduces to

$$Q = \dot{Q}\, \Delta t \qquad \text{(kJ)} \tag{2–16}$$

where $\Delta t = t_2 - t_1$ is the time interval during which the process takes place.

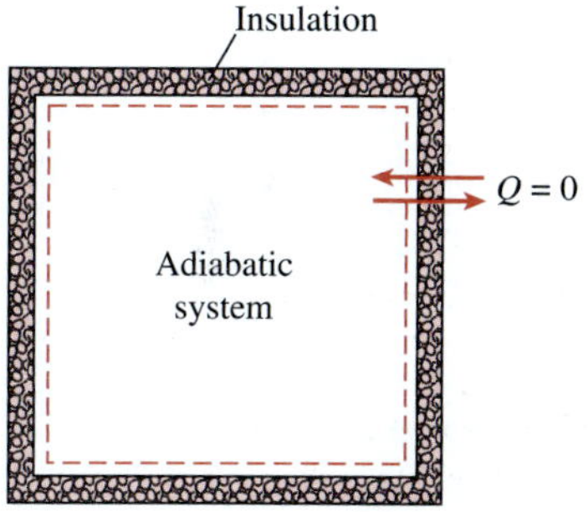

FIGURE 2–16
During an adiabatic process, a system exchanges no heat with its surroundings.

Historical Background on Heat

Heat has always been perceived to be something that produces in us a sensation of warmth, and one would think that the nature of heat is one of the first things understood by mankind. However, it was only in the middle of the nineteenth century that we had a true physical understanding of the nature of heat, thanks to the development at that time of the **kinetic theory**, which treats molecules as tiny balls that are in motion and thus possess kinetic energy. Heat is then defined as the energy associated with the random motion of atoms and molecules. Although it was suggested in the eighteenth and early nineteenth centuries that heat is the manifestation of motion at the molecular level (called the *live force*), the prevailing view of heat until the middle of the nineteenth century was based on the caloric theory proposed by the French chemist Antoine Lavoisier (1744–1794) in 1789. The caloric theory asserts that heat is a fluidlike substance called the **caloric** that is a massless, colorless, odorless, and tasteless substance that can be poured from one body into another (Fig. 2–18). When caloric was added to

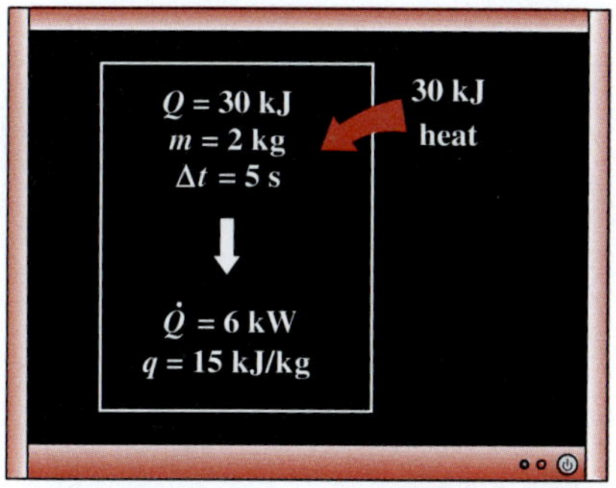

FIGURE 2–17
The relationships among q, Q, and $\dot{Q}$.

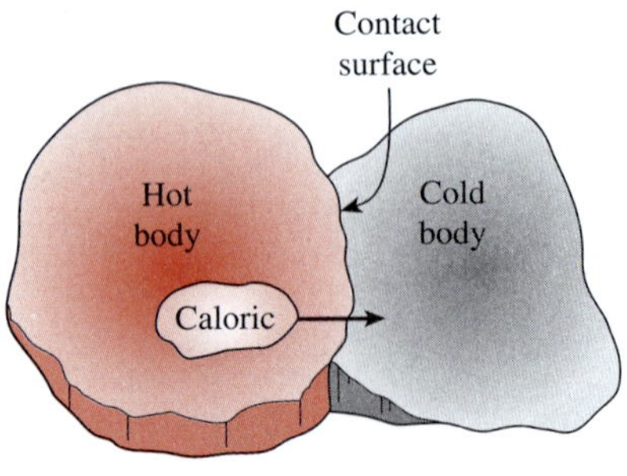

FIGURE 2–18
In the early nineteenth century, heat was thought to be an invisible fluid called the *caloric* that flowed from warmer bodies to the cooler ones.

a body, its temperature increased; and when caloric was removed from a body, its temperature decreased. When a body could not contain any more caloric, much the same way as when a glass of water could not dissolve any more salt or sugar, the body was said to be saturated with caloric. This interpretation gave rise to the terms *saturated liquid* and *saturated vapor* that are still in use today.

The caloric theory came under attack soon after its introduction. It maintained that heat is a substance that could not be created or destroyed. Yet it was known that heat can be generated indefinitely by rubbing one's hands together or rubbing two pieces of wood together. In 1798, the American Benjamin Thompson (Count Rumford) (1754–1814) showed in his papers that heat can be generated continuously through friction. The validity of the caloric theory was also challenged by several others. But it was the careful experiments of the Englishman James P. Joule (1818–1889) published in 1843 that finally convinced the skeptics that heat was not a substance after all, and thus put the caloric theory to rest. Although the caloric theory was totally abandoned in the middle of the nineteenth century, it contributed greatly to the development of thermodynamics and heat transfer.

Heat is transferred by three mechanisms: conduction, convection, and radiation. **Conduction** is the transfer of energy from the more energetic particles of a substance to the adjacent less energetic ones as a result of interaction between particles. **Convection** is the transfer of energy between a solid surface and the adjacent fluid that is in motion, and it involves the combined effects of conduction and fluid motion. **Radiation** is the transfer of energy due to the emission of electromagnetic waves (or photons). An overview of the three mechanisms of heat transfer is given at the end of this chapter as a Topic of Special Interest.

2–4 ▪ ENERGY TRANSFER BY WORK

Work, like heat, is an energy interaction between a system and its surroundings. As mentioned earlier, energy can cross the boundary of a closed system in the form of heat or work. Therefore, *if the energy crossing the boundary of a closed system is not heat, it must be work.* Heat is easy to recognize: Its driving force is a temperature difference between the system and its surroundings. Then we can simply say that an energy interaction that is not caused by a temperature difference between a system and its surroundings is work. More specifically, *work is the energy transfer associated with a force acting through a distance.* A rising piston, a rotating shaft, and an electric wire crossing the system boundaries are all associated with work interactions.

Work is also a form of energy transferred like heat and, therefore, has energy units such as kJ. The work done during a process between states 1 and 2 is denoted by W_{12}, or simply W. The work done *per unit mass* of a system is denoted by w and is expressed as

$$w = \frac{W}{m} \qquad \text{(kJ/kg)} \tag{2–17}$$

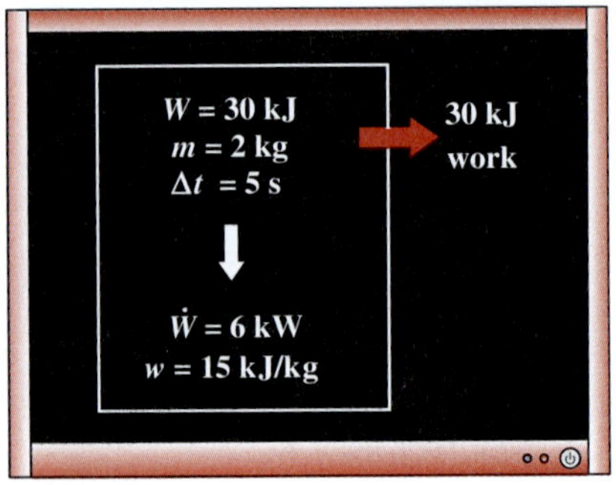

FIGURE 2–19
The relationships among w, W, and $\dot{W}$.

The work done *per unit time* is called **power** and is denoted $\dot{W}$ (Fig. 2–19). The unit of power is kJ/s, or kW.

Heat and work are *directional quantities*, and thus the complete description of a heat or work interaction requires the specification of both the *magnitude* and *direction*. One way of doing that is to adopt a sign convention. The generally accepted **formal sign convention** for heat and work interactions is as follows: *heat transfer to a system and work done by a system are positive; heat transfer from a system and work done on a system are negative*. Another way is to use the subscripts *in* and *out* to indicate direction (Fig. 2–20). For example, a work input of 5 kJ can be expressed as $W_{in} = 5$ kJ, while a heat loss of 3 kJ can be expressed as $Q_{out} = 3$ kJ. When the direction of a heat or work interaction is not known, we can simply *assume* a direction for the interaction (using the subscript *in* or *out*) and solve for it. A positive result indicates the assumed direction is right. A negative result, on the other hand, indicates that the direction of the interaction is the opposite of the assumed direction. This is just like assuming a direction for an unknown force when solving a statics problem, and reversing the direction when a negative result is obtained for the force. We will use this *intuitive approach* in this book as it eliminates the need to adopt a formal sign convention and the need to carefully assign negative values to some interactions.

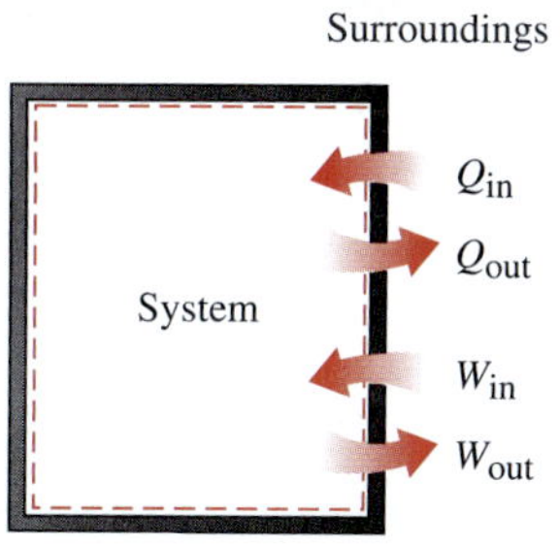

FIGURE 2–20
Specifying the directions of heat and work.

Note that a quantity that is transferred to or from a system during an interaction is not a property since the amount of such a quantity depends on more than just the state of the system. Heat and work are *energy transfer mechanisms* between a system and its surroundings, and there are many similarities between them:

1. Both are recognized at the boundaries of a system as they cross the boundaries. That is, both heat and work are *boundary* phenomena.
2. Systems possess energy, but not heat or work.
3. Both are associated with a *process,* not a state. Unlike properties, heat or work has no meaning at a state.
4. Both are *path functions* (i.e., their magnitudes depend on the path followed during a process as well as the end states).

Path functions have **inexact differentials** designated by the symbol δ. Therefore, a differential amount of heat or work is represented by δQ or δW, respectively, instead of dQ or dW. Properties, however, are **point functions** (i.e., they depend on the state only, and not on how a system reaches that state), and they have **exact differentials** designated by the symbol *d.* A small change in volume, for example, is represented by dV, and the total volume change during a process between states 1 and 2 is

$$\int_1^2 dV = V_2 - V_1 = \Delta V$$

That is, the volume change during process 1–2 is always the volume at state 2 minus the volume at state 1, regardless of the path followed (Fig. 2–21). The total work done during process 1–2, however, is

$$\int_1^2 \delta W = W_{12} \qquad (\textit{not } \Delta W)$$

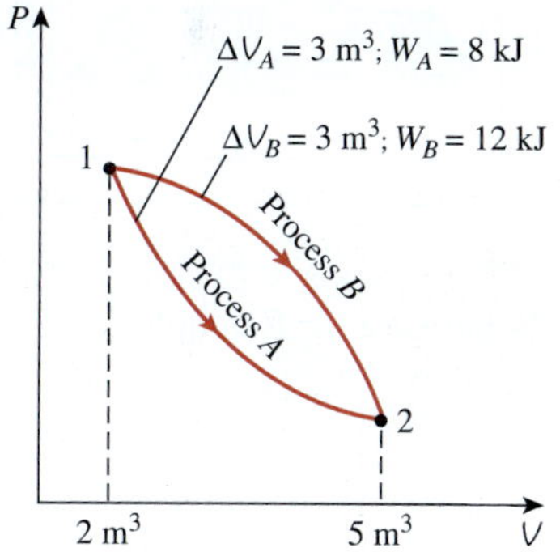

FIGURE 2–21
Properties are point functions; but heat and work are path functions (their magnitudes depend on the path followed).

That is, the total work is obtained by following the process path and adding the differential amounts of work (δW) done along the way. The integral of δW *is not* $W_2 - W_1$ (i.e., the work at state 2 minus work at state 1), which is meaningless since work is not a property and systems do not possess work at a state.

EXAMPLE 2–3 Burning of a Candle in an Insulated Room

A candle is burning in a well-insulated room. Taking the room (the air plus the candle) as the system, determine (*a*) if there is any heat transfer during this burning process and (*b*) if there is any change in the internal energy of the system.

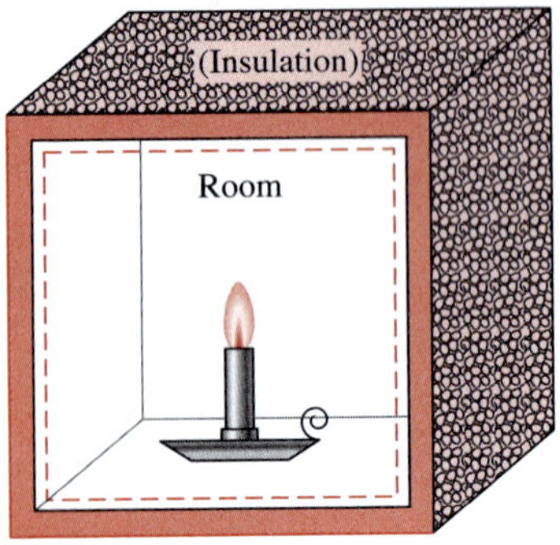

FIGURE 2–22
Schematic for Example 2–3.

Solution A candle burning in a well-insulated room is considered. It is to be determined whether there is any heat transfer and any change in internal energy.

Analysis (*a*) The interior surfaces of the room form the system boundary, as indicated by the dashed lines in Fig. 2–22. As pointed out earlier, heat is recognized as it crosses the boundaries. Since the room is well insulated, we have an adiabatic system and no heat will pass through the boundaries. Therefore, $Q = 0$ for this process.

(*b*) The internal energy involves energies that exist in various forms (sensible, latent, chemical, nuclear). During the process just described, part of the chemical energy is converted to sensible energy. Since there is no increase or decrease in the total internal energy of the system, $\Delta U = 0$ for this process.

EXAMPLE 2–4 Heating of a Potato in an Oven

A potato initially at room temperature (25°C) is being baked in an oven that is maintained at 200°C, as shown in Fig. 2–23. Is there any heat transfer during this baking process?

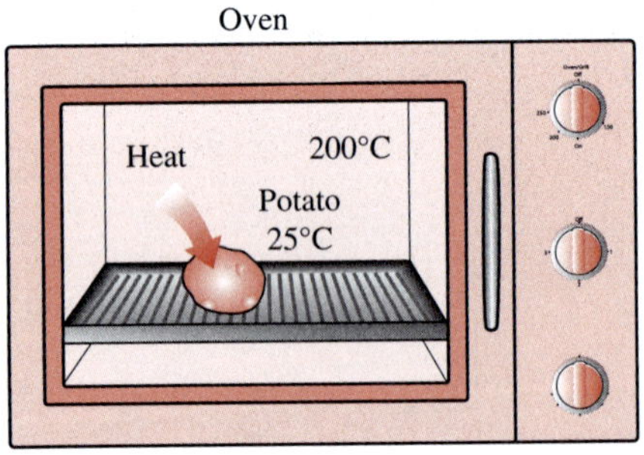

FIGURE 2–23
Schematic for Example 2–4.

Solution A potato is being baked in an oven. It is to be determined whether there is any heat transfer during this process.

Analysis This is not a well-defined problem since the system is not specified. Let us assume that we are observing the potato, which will be our system. Then the outer surface of the skin of the potato can be viewed as the system boundary. Part of the energy in the oven will pass through the skin to the potato. Since the driving force for this energy transfer is a temperature difference, this is a heat transfer process.

EXAMPLE 2–5 Heating of an Oven by Work Transfer

A well-insulated electric oven is being heated through its heating element. If the entire oven, including the heating element, is taken to be the system, determine whether this is a heat or work interaction.

Solution A well-insulated electric oven is being heated by its heating element. It is to be determined whether this is a heat or work interaction.

Analysis For this problem, the interior surfaces of the oven form the system boundary, as shown in Fig. 2–24. The energy content of the oven obviously increases during this process, as evidenced by a rise in temperature. This energy transfer to the oven is not caused by a temperature difference between the oven and the surrounding air. Instead, it is caused by *electrons* crossing the system boundary and thus doing work. Therefore, this is a work interaction.

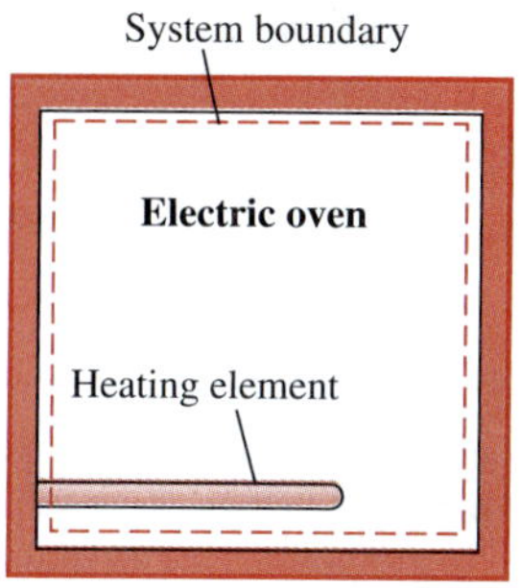

FIGURE 2–24
Schematic for Example 2–5.

EXAMPLE 2–6 Heating of an Oven by Heat Transfer

Answer the question in Example 2–5 if the system is taken as only the air in the oven without the heating element.

Solution The question in Example 2–5 is to be reconsidered by taking the system to be only the air in the oven.

Analysis This time, the system boundary will include the outer surface of the heating element and will not cut through it, as shown in Fig. 2–25. Therefore, no electrons will be crossing the system boundary at any point. Instead, the energy generated in the interior of the heating element will be transferred to the air around it as a result of the temperature difference between the heating element and the air in the oven. Therefore, this is a heat transfer process.

Discussion For both cases, the amount of energy transfer to the air is the same. These two examples show that an energy transfer can be heat or work, depending on how the system is selected.

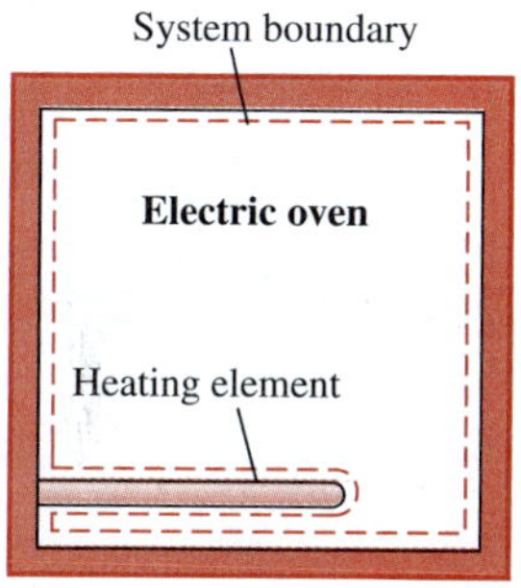

FIGURE 2–25
Schematic for Example 2–6.

Electrical Work

It was pointed out in Example 2–5 that electrons crossing the system boundary do electrical work on the system. In an electric field, electrons in a wire move under the effect of electromotive forces, doing work. When N coulombs of electrical charge move through a potential difference $\mathbf{V}$, the electrical work done is

$$W_e = \mathbf{V}N$$

which can also be expressed in the rate form as

$$\dot{W}_e = \mathbf{V}I \qquad \text{(W)} \tag{2-18}$$

where $\dot{W}_e$ is the **electrical power** and I is the number of electrical charges flowing per unit time, that is, the *current* (Fig. 2–26). In general, both $\mathbf{V}$ and I vary with time, and the electrical work done during a time interval Δt is expressed as

$$W_e = \int_1^2 \mathbf{V}I\,dt \qquad \text{(kJ)} \tag{2-19}$$

When both $\mathbf{V}$ and I remain constant during the time interval Δt, it reduces to

$$W_e = \mathbf{V}I\,\Delta t \qquad \text{(kJ)} \tag{2-20}$$

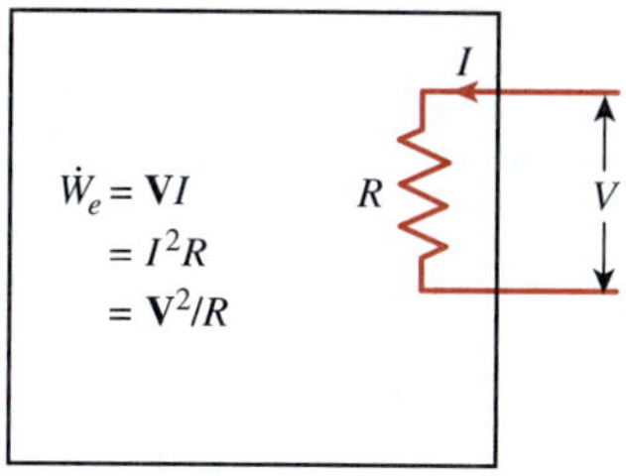

FIGURE 2–26
Electrical power in terms of resistance R, current I, and potential difference $\mathbf{V}$.

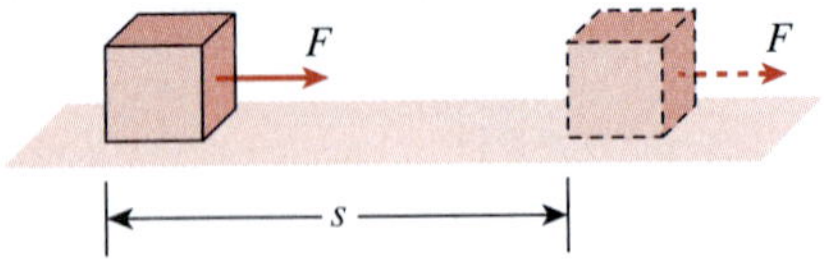

FIGURE 2–27
The work done is proportional to the force applied (F) and the distance traveled (s).

FIGURE 2–28
If there is no movement, no work is done.
BLONDIE © KING FEATURES SYNDICATE.

FIGURE 2–29
Energy transmission through rotating shafts is commonly encountered in practice.

2–5 ▪ MECHANICAL FORMS OF WORK

There are several different ways of doing work, each in some way related to a force acting through a distance (Fig. 2–27). In elementary mechanics, the work done by a constant force F on a body displaced a distance s in the direction of the force is given by

$$W = Fs \qquad \text{(kJ)} \tag{2–21}$$

If the force F is not constant, the work done is obtained by adding (i.e., integrating) the differential amounts of work,

$$W = \int_1^2 F\, ds \qquad \text{(kJ)} \tag{2–22}$$

Obviously, one needs to know how the force varies with displacement to perform this integration. Equations 2–21 and 2–22 give only the magnitude of the work. The sign is easily determined from physical considerations: The work done on a system by an external force acting in the direction of motion is negative, and work done by a system against an external force acting in the opposite direction to motion is positive.

There are two requirements for a work interaction between a system and its surroundings to exist: (1) there must be a *force* acting on the boundary, and (2) the boundary must *move.* Therefore, the presence of forces on the boundary without any displacement of the boundary does not constitute a work interaction. Likewise, the displacement of the boundary without any force to oppose or drive this motion (such as the expansion of a gas into an evacuated space) is not a work interaction since no energy is transferred.

In many thermodynamic problems, mechanical work is the only form of work involved. It is associated with the movement of the boundary of a system or with the movement of the entire system as a whole (Fig. 2–28). Some common forms of mechanical work are discussed next.

Shaft Work

Energy transmission with a rotating shaft is very common in engineering practice (Fig. 2–29). Often the torque T applied to the shaft is constant, which means that the force F applied is also constant. For a specified constant torque, the work done during n revolutions is determined as follows: A force F acting through a moment arm r generates a torque T of (Fig. 2–30)

$$\text{T} = Fr \quad \rightarrow \quad F = \frac{\text{T}}{r} \tag{2–23}$$

This force acts through a distance s, which is related to the radius r by

$$s = (2\pi r)n \tag{2–24}$$

Then the shaft work is determined from

$$W_{\text{sh}} = Fs = \left(\frac{\text{T}}{r}\right)(2\pi rn) = 2\pi n\text{T} \qquad \text{(kJ)} \tag{2–25}$$

The power transmitted through the shaft is the shaft work done per unit time, which can be expressed as

$$\dot{W}_{\text{sh}} = 2\pi\dot{n}\text{T} \qquad \text{(kW)} \tag{2–26}$$

where $\dot{n}$ is the number of revolutions per unit time.

EXAMPLE 2–7 Power Transmission by the Shaft of a Car

Determine the power transmitted through the shaft of a car when the torque applied is 200 N·m and the shaft rotates at a rate of 4000 revolutions per minute (rpm).

Solution The torque and the rpm for a car engine are given. The power transmitted is to be determined.

Analysis A sketch of the car is given in Fig. 2–31. The shaft power is determined directly from

$$\dot{W}_{sh} = 2\pi\dot{n}\text{T} = (2\pi)\left(4000\,\frac{1}{\text{min}}\right)(200\ \text{N}\cdot\text{m})\left(\frac{1\ \text{min}}{60\ \text{s}}\right)\left(\frac{1\ \text{kJ}}{1000\ \text{N}\cdot\text{m}}\right)$$

$$= \mathbf{83.8\ kW} \qquad (\text{or } 112\ \text{hp})$$

Discussion Note that power transmitted by a shaft is proportional to torque and the rotational speed.

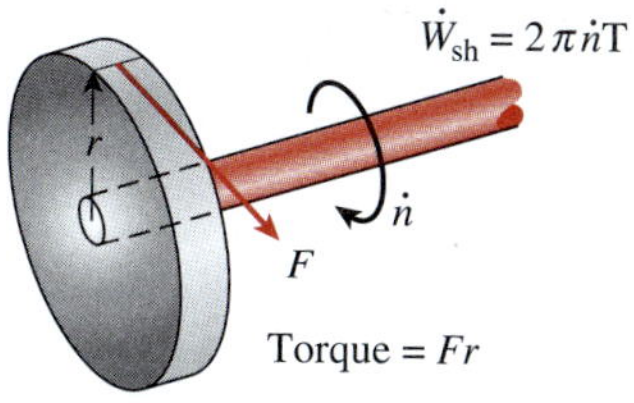

FIGURE 2–30
Shaft work is proportional to the torque applied and the number of revolutions of the shaft.

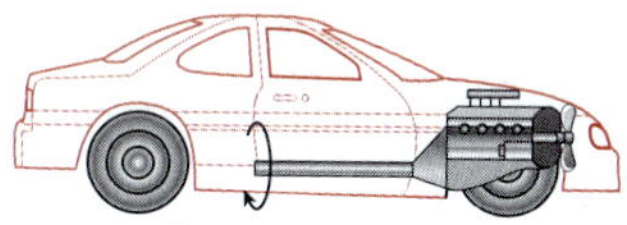

FIGURE 2–31
Schematic for Example 2–7.

Spring Work

It is common knowledge that when a force is applied on a spring, the length of the spring changes (Fig. 2–32). When the length of the spring changes by a differential amount dx under the influence of a force F, the work done is

$$\delta W_{spring} = F\,dx \tag{2–27}$$

To determine the total spring work, we need to know a functional relationship between F and x. For linear elastic springs, the displacement x is proportional to the force applied (Fig. 2–33). That is,

$$F = kx \qquad (\text{kN}) \tag{2–28}$$

where k is the spring constant and has the unit kN/m. The displacement x is measured from the undisturbed position of the spring (that is, $x = 0$ when $F = 0$). Substituting Eq. 2–28 into Eq. 2–27 and integrating yield

$$W_{spring} = \tfrac{1}{2}k(x_2^2 - x_1^2) \qquad (\text{kJ}) \tag{2–29}$$

where x_1 and x_2 are the initial and the final displacements of the spring, respectively, measured from the undisturbed position of the spring.

There are many other forms of mechanical work. Next we introduce some of them briefly.

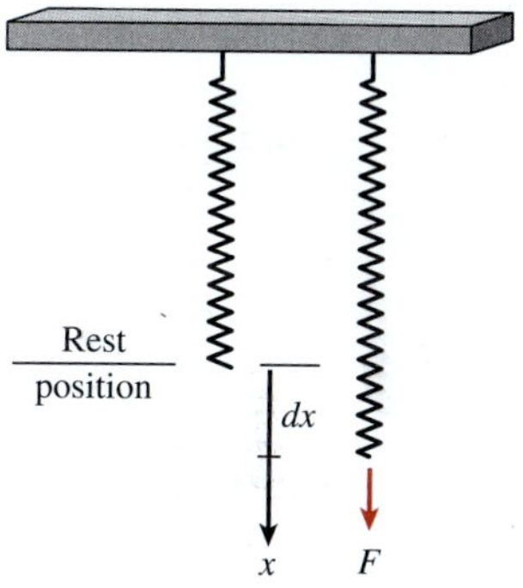

FIGURE 2–32
Elongation of a spring under the influence of a force.

Work Done on Elastic Solid Bars

Solids are often modeled as linear springs because under the action of a force they contract or elongate, as shown in Fig. 2–34, and when the force is lifted, they return to their original lengths, like a spring. This is true as long as the force is in the elastic range, that is, not large enough to cause permanent (plastic) deformations. Therefore, the equations given for a linear spring can also be used for elastic solid bars. Alternately, we can determine

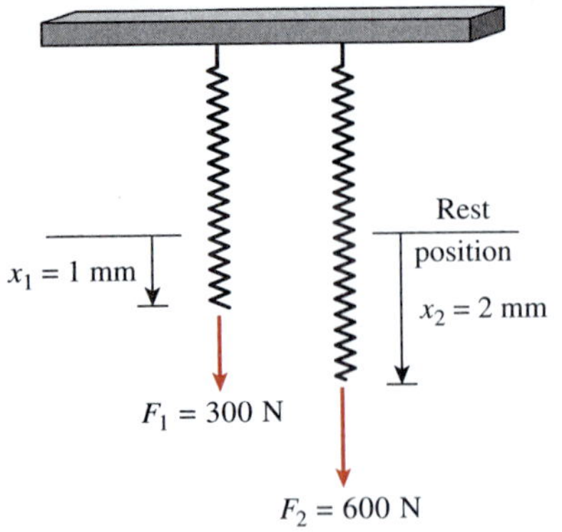

FIGURE 2–33
The displacement of a linear spring doubles when the force is doubled.

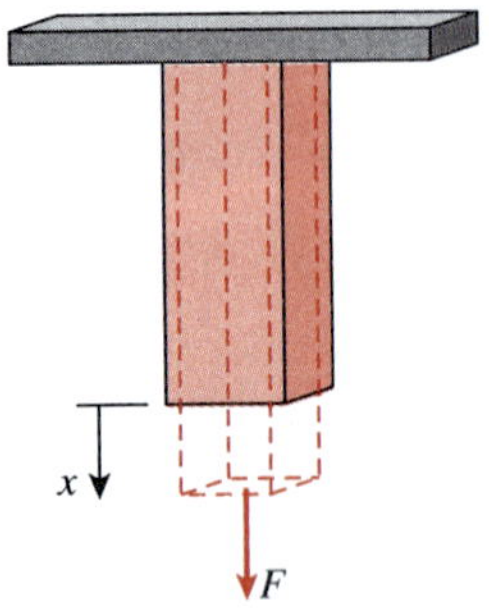

FIGURE 2–34
Solid bars behave as springs under the influence of a force.

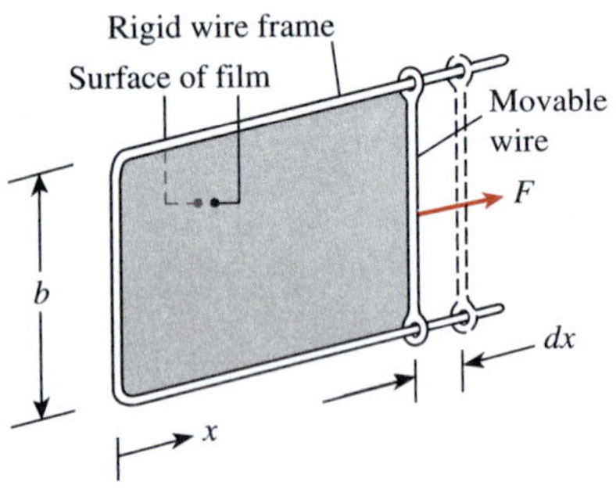

FIGURE 2–35
Stretching a liquid film with a movable wire.

the work associated with the expansion or contraction of an elastic solid bar by replacing pressure P by its counterpart in solids, *normal stress* $\sigma_n = F/A$, in the work expression:

$$W_{\text{elastic}} = \int_1^2 F\,dx = \int_1^2 \sigma_n A\,dx \qquad \text{(kJ)} \tag{2–30}$$

where A is the cross-sectional area of the bar. Note that the normal stress has pressure units.

Work Associated with the Stretching of a Liquid Film

Consider a liquid film such as soap film suspended on a wire frame (Fig. 2–35). We know from experience that it will take some force to stretch this film by the movable portion of the wire frame. This force is used to overcome the microscopic forces between molecules at the liquid–air interfaces. These microscopic forces are perpendicular to any line in the surface, and the force generated by these forces per unit length is called the **surface tension** σ_s, whose unit is N/m. Therefore, the work associated with the stretching of a film is also called *surface tension work.* It is determined from

$$W_{\text{surface}} = \int_1^2 \sigma_s\,dA \qquad \text{(kJ)} \tag{2–31}$$

where $dA = 2b\,dx$ is the change in the surface area of the film. The factor 2 is due to the fact that the film has two surfaces in contact with air. The force acting on the movable wire as a result of surface tension effects is $F = 2b\sigma_s$ where σ_s is the surface tension force per unit length.

Work Done to Raise or to Accelerate a Body

When a body is raised in a gravitational field, its potential energy increases. Likewise, when a body is accelerated, its kinetic energy increases. The conservation of energy principle requires that an equivalent amount of energy must be transferred to the body being raised or accelerated. Remember that energy can be transferred to a given mass by heat and work, and the energy transferred in this case obviously is not heat since it is not driven by a temperature difference. Therefore, it must be work. Then we conclude that (1) the work transfer needed to raise a body is equal to the change in the potential energy of the body, and (2) the work transfer needed to accelerate a body is equal to the change in the kinetic energy of the body (Fig. 2–36). Similarly, the potential or kinetic energy of a body represents the work that can be obtained from the body as it is lowered to the reference level or decelerated to zero velocity.

This discussion together with the consideration for friction and other losses form the basis for determining the required power rating of motors used to drive devices such as elevators, escalators, conveyor belts, and ski lifts. It also plays a primary role in the design of automotive and aircraft engines, and in the determination of the amount of hydroelectric power that can be produced from a given water reservoir, which is simply the potential energy of the water relative to the location of the hydraulic turbine.

EXAMPLE 2–8 Power Needs of a Car to Climb a Hill

Consider a 1200-kg car cruising steadily on a level road at 90 km/h. Now the car starts climbing a hill that is sloped 30° from the horizontal (Fig. 2–37). If the velocity of the car is to remain constant during climbing, determine the additional power that must be delivered by the engine.

Solution A car is to climb a hill while maintaining a constant velocity. The additional power needed is to be determined.

Analysis The additional power required is simply the work that needs to be done per unit time to raise the elevation of the car, which is equal to the change in the potential energy of the car per unit time:

$$\dot{W}_g = mg\ \Delta z/\Delta t = mgV_{\text{vertical}}$$

$$= (1200\text{ kg})(9.81\text{ m/s}^2)(90\text{ km/h})(\sin 30°)\left(\frac{1\text{ m/s}}{3.6\text{ km/h}}\right)\left(\frac{1\text{ kJ/kg}}{1000\text{ m}^2/\text{s}^2}\right)$$

$$= 147\text{ kJ/s} = \mathbf{147\ kW} \qquad (\text{or } 197\text{ hp})$$

Discussion Note that the car engine will have to produce almost 200 hp of additional power while climbing the hill if the car is to maintain its velocity.

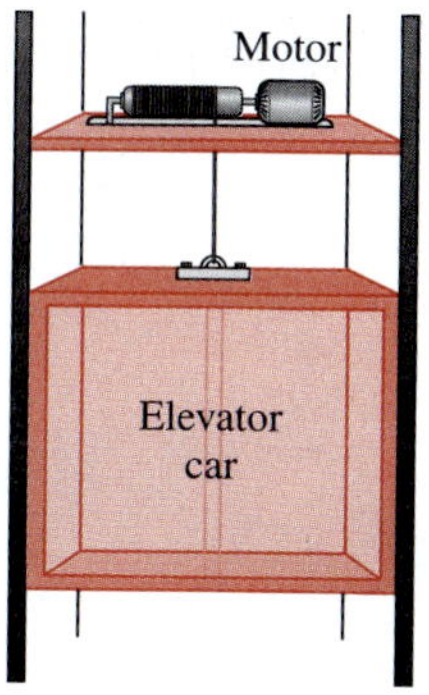

FIGURE 2–36
The energy transferred to a body while being raised is equal to the change in its potential energy.

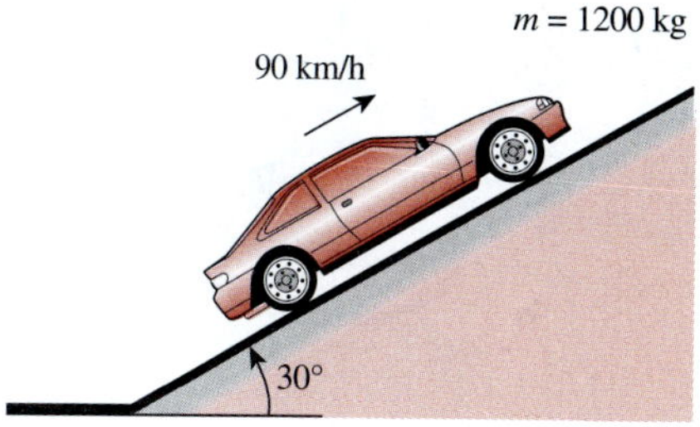

FIGURE 2–37
Schematic for Example 2–8.

EXAMPLE 2–9 Power Needs of a Car to Accelerate

Determine the power required to accelerate a 900-kg car shown in Fig. 2–38 from rest to a velocity of 80 km/h in 20 s on a level road.

Solution The power required to accelerate a car to a specified velocity is to be determined.

Analysis The work needed to accelerate a body is simply the change in the kinetic energy of the body,

$$W_a = \tfrac{1}{2}m(V_2^2 - V_1^2) = \tfrac{1}{2}(900\text{ kg})\left[\left(\frac{80{,}000\text{ m}}{3600\text{ s}}\right)^2 - 0^2\right]\left(\frac{1\text{ kJ/kg}}{1000\text{ m}^2/\text{s}^2}\right)$$

$$= 222\text{ kJ}$$

The average power is determined from

$$\dot{W}_a = \frac{W_a}{\Delta t} = \frac{222\text{ kJ}}{20\text{ s}} = \mathbf{11.1\ kW} \qquad (\text{or } 14.9\text{ hp})$$

Discussion This is in addition to the power required to overcome friction, rolling resistance, and other imperfections.

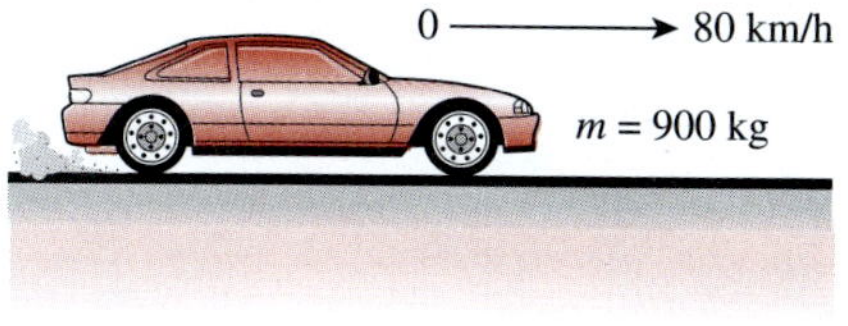

FIGURE 2–38
Schematic for Example 2–9.

Nonmechanical Forms of Work

The treatment in Section 2–5 represents a fairly comprehensive coverage of mechanical forms of work except the *moving boundary work* that is covered in Chap. 4. Some work modes encountered in practice are not mechanical in nature. However, these nonmechanical work modes can be treated in a similar manner by identifying a *generalized force F* acting in the direction of a *generalized displacement* x. Then the work associated with the differential displacement under the influence of this force is determined from $\delta W = F dx$.

Some examples of nonmechanical work modes are **electrical work**, where the generalized force is the *voltage* (the electrical potential) and the generalized displacement is the *electrical charge,* as discussed earlier; **magnetic work**, where the generalized force is the *magnetic field strength* and the generalized displacement is the total *magnetic dipole moment;* and **electrical polarization work**, where the generalized force is the *electric field strength* and the generalized displacement is the *polarization of the medium* (the sum of the electric dipole rotation moments of the molecules). Detailed consideration of these and other nonmechanical work modes can be found in specialized books on these topics.

2–6 ▪ THE FIRST LAW OF THERMODYNAMICS

So far, we have considered various forms of energy such as heat Q, work W, and total energy E individually, and no attempt is made to relate them to each other during a process. The *first law of thermodynamics,* also known as *the conservation of energy principle,* provides a sound basis for studying the relationships among the various forms of energy and energy interactions. Based on experimental observations, the first law of thermodynamics states that *energy can be neither created nor destroyed during a process; it can only change forms.* Therefore, every bit of energy should be accounted for during a process.

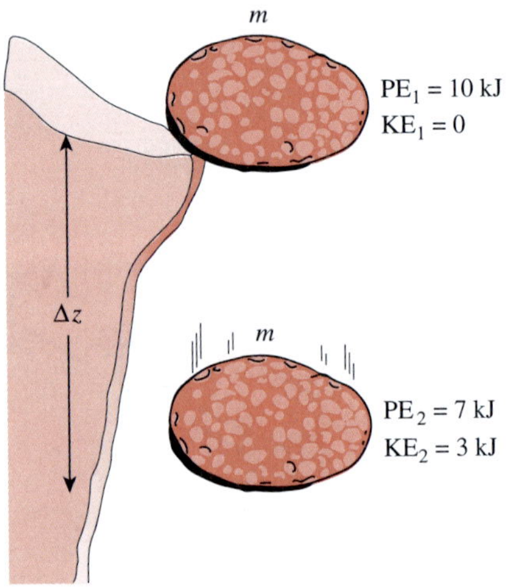

FIGURE 2–39
Energy cannot be created or destroyed; it can only change forms.

We all know that a rock at some elevation possesses some potential energy, and part of this potential energy is converted to kinetic energy as the rock falls (Fig. 2–39). Experimental data show that the decrease in potential energy ($mg\ \Delta z$) exactly equals the increase in kinetic energy $[m(V_2^2 - V_1^2)/2]$ when the air resistance is negligible, thus confirming the conservation of energy principle for mechanical energy.

Consider a system undergoing a series of *adiabatic* processes from a specified state 1 to another specified state 2. Being adiabatic, these processes obviously cannot involve any heat transfer, but they may involve several kinds of work interactions. Careful measurements during these experiments indicate the following: *For all adiabatic processes between two specified states of a closed system, the net work done is the same regardless of the nature of the closed system and the details of the process.* Considering that there are an infinite number of ways to perform work interactions under adiabatic conditions, this statement appears to be very powerful, with a potential for far-reaching implications. This statement, which is largely based on the experiments of Joule in the first half of the nineteenth century, cannot be drawn from any other known physical principle and is recognized as a fundamental principle. This principle is called the **first law of thermodynamics** or just the **first law**.

A major consequence of the first law is the existence and the definition of the property *total energy E.* Considering that the net work is the same for all adiabatic processes of a closed system between two specified states, the value of the net work must depend on the end states of the system only, and thus it must correspond to a change in a property of the system. This property is the *total energy.* Note that the first law makes no reference to the value of thc total energy of a closed system at a state. It simply states that the *change* in the total energy during an adiabatic process must be equal to the net work done. Therefore, any convenient arbitrary value can be assigned to total energy at a specified state to serve as a reference point.

Implicit in the first law statement is the conservation of energy. Although the essence of the first law is the existence of the property *total energy,* the first law is often viewed as a statement of the *conservation of energy* principle. Next, we develop the first law or the conservation of energy relation with the help of some familiar examples using intuitive arguments.

First, we consider some processes that involve heat transfer but no work interactions. The potato baked in the oven is a good example for this case (Fig. 2–40). As a result of heat transfer to the potato, the energy of the potato will increase. If we disregard any mass transfer (moisture loss from the potato), the increase in the total energy of the potato becomes equal to the amount of heat transfer. That is, if 5 kJ of heat is transferred to the potato, the energy increase of the potato will also be 5 kJ.

As another example, consider the heating of water in a pan on top of a range (Fig. 2–41). If 15 kJ of heat is transferred to the water from the heating element and 3 kJ of it is lost from the water to the surrounding air, the increase in energy of the water will be equal to the net heat transfer to water, which is 12 kJ.

Now consider a well-insulated (i.e., adiabatic) room heated by an electric heater as our system (Fig. 2–42). As a result of electrical work done, the energy of the system will increase. Since the system is adiabatic and cannot have any heat transfer to or from the surroundings ($Q = 0$), the conservation of energy principle dictates that the electrical work done on the system must equal the increase in energy of the system.

Next, let us replace the electric heater with a paddle wheel (Fig. 2–43). As a result of the stirring process, the energy of the system will increase. Again, since there is no heat interaction between the system and its surroundings ($Q = 0$), the shaft work done on the system must show up as an increase in the energy of the system.

Many of you have probably noticed that the temperature of air rises when it is compressed (Fig. 2–44). This is because energy is transferred to the air in the form of boundary work. In the absence of any heat transfer ($Q = 0$), the entire boundary work will be stored in the air as part of its total energy. The conservation of energy principle again requires that the increase in the energy of the system be equal to the boundary work done on the system.

We can extend these discussions to systems that involve various heat and work interactions simultaneously. For example, if a system gains 12 kJ of heat during a process while 6 kJ of work is done on it, the increase in the energy of the system during that process is 18 kJ (Fig. 2–45). That is, the change in the energy of a system during a process is simply equal to the net energy transfer to (or from) the system.

Energy Balance

In the light of the preceding discussions, the conservation of energy principle can be expressed as follows: *The net change (increase or decrease) in the total energy of the system during a process is equal to the difference*

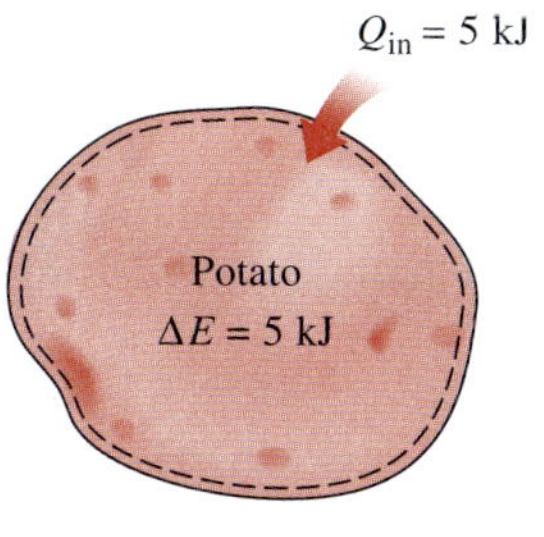

FIGURE 2–40
The increase in the energy of a potato in an oven is equal to the amount of heat transferred to it.

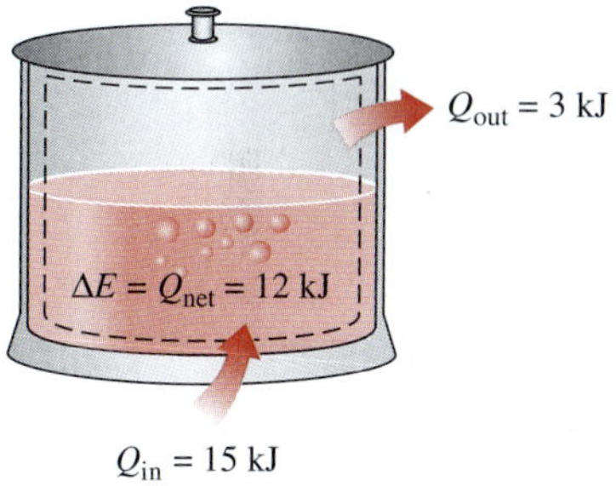

FIGURE 2–41
In the absence of any work interactions, the energy change of a system is equal to the net heat transfer.

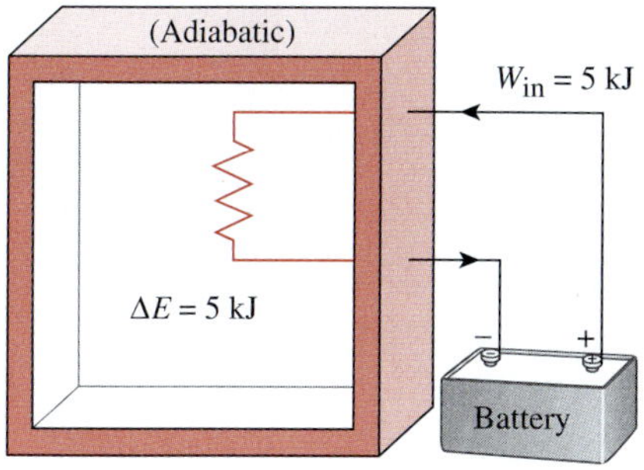

FIGURE 2–42
The work (electrical) done on an adiabatic system is equal to the increase in the energy of the system.

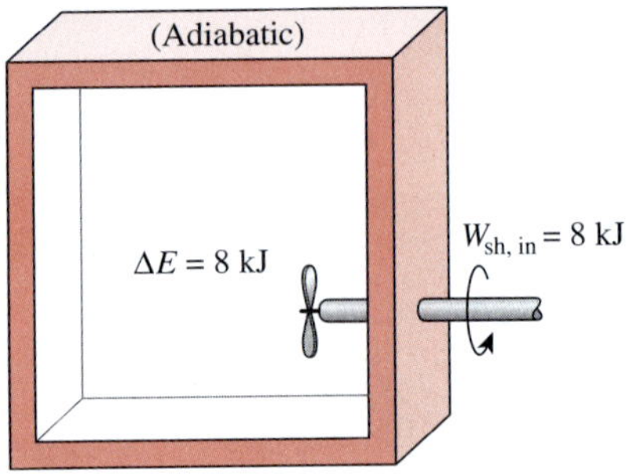

FIGURE 2–43
The work (shaft) done on an adiabatic system is equal to the increase in the energy of the system.

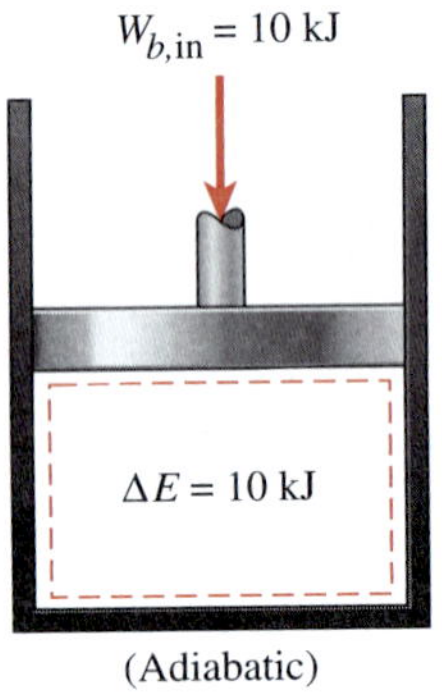

FIGURE 2–44
The work (boundary) done on an adiabatic system is equal to the increase in the energy of the system.

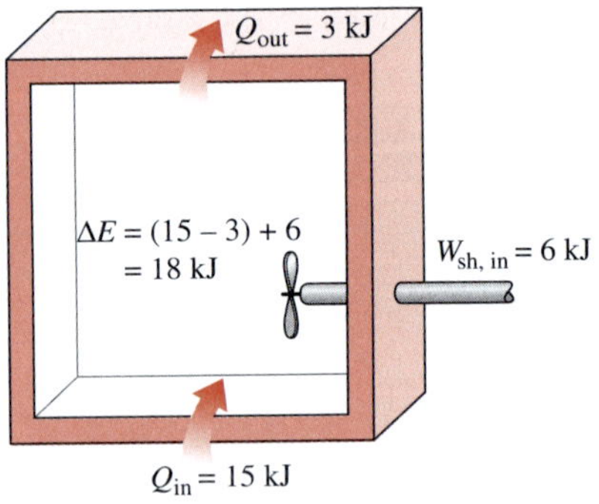

FIGURE 2–45
The energy change of a system during a process is equal to the *net* work and heat transfer between the system and its surroundings.

between the total energy entering and the total energy leaving the system during that process. That is,

$$\begin{pmatrix}\text{Total energy}\\ \text{entering the system}\end{pmatrix} - \begin{pmatrix}\text{Total energy}\\ \text{leaving the system}\end{pmatrix} = \begin{pmatrix}\text{Change in the total}\\ \text{energy of the system}\end{pmatrix}$$

or

$$E_{in} - E_{out} = \Delta E_{system}$$

This relation is often referred to as the **energy balance** and is applicable to any kind of system undergoing any kind of process. The successful use of this relation to solve engineering problems depends on understanding the various forms of energy and recognizing the forms of energy transfer.

Energy Change of a System, ΔE_{system}

The determination of the energy change of a system during a process involves the evaluation of the energy of the system at the beginning and at the end of the process, and taking their difference. That is,

$$\text{Energy change} = \text{Energy at final state} - \text{Energy at initial state}$$

or

$$\Delta E_{system} = E_{final} - E_{initial} = E_2 - E_1 \quad \textbf{(2–32)}$$

Note that energy is a property, and the value of a property does not change unless the state of the system changes. Therefore, the energy change of a system is zero if the state of the system does not change during the process. Also, energy can exist in numerous forms such as internal (sensible, latent, chemical, and nuclear), kinetic, potential, electric, and magnetic, and their sum constitutes the *total energy E* of a system. In the absence of electric, magnetic, and surface tension effects (i.e., for simple compressible systems), the change in the total energy of a system during a process is the sum of the changes in its internal, kinetic, and potential energies and can be expressed as

$$\Delta E = \Delta U + \Delta \text{KE} + \Delta \text{PE} \quad \textbf{(2–33)}$$

where

$$\Delta U = m(u_2 - u_1)$$

$$\Delta \text{KE} = \tfrac{1}{2}m(V_2^2 - V_1^2)$$

$$\Delta \text{PE} = mg(z_2 - z_1)$$

When the initial and final states are specified, the values of the specific internal energies u_1 and u_2 can be determined directly from the property tables or thermodynamic property relations.

Most systems encountered in practice are stationary, that is, they do not involve any changes in their velocity or elevation during a process (Fig. 2–46). Thus, for **stationary systems**, the changes in kinetic and potential energies are zero (that is, $\Delta \text{KE} = \Delta \text{PE} = 0$), and the total energy change relation in Eq. 2–33 reduces to $\Delta E = \Delta U$ for such systems. Also, the energy of a system during a process will change even if only one form of its energy changes while the other forms of energy remain unchanged.

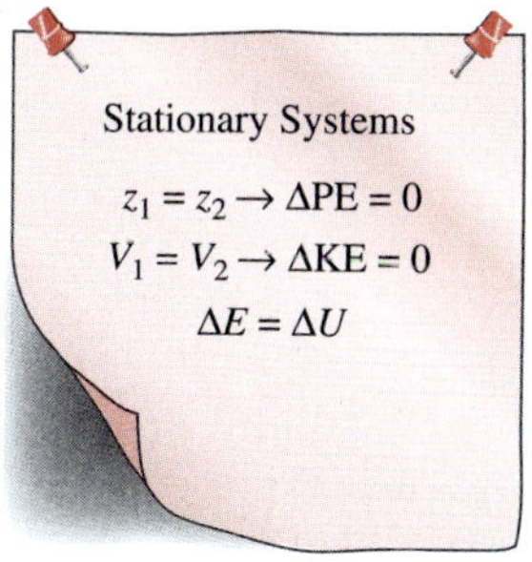

FIGURE 2–46
For stationary systems, $\Delta KE = \Delta PE = 0$; thus $\Delta E = \Delta U$.

Mechanisms of Energy Transfer, E_{in} and E_{out}

Energy can be transferred to or from a system in three forms: *heat, work,* and *mass flow.* Energy interactions are recognized at the system boundary as they cross it, and they represent the energy gained or lost by a system during a process. The only two forms of energy interactions associated with a fixed mass or closed system are *heat transfer* and *work.*

1. **Heat Transfer, Q** Heat transfer to a system (heat gain) increases the energy of the molecules and thus the internal energy of the system, and heat transfer from a system (heat loss) decreases it since the energy transferred out as heat comes from the energy of the molecules of the system.
2. **Work Transfer, W** An energy interaction that is not caused by a temperature difference between a system and its surroundings is work. A rising piston, a rotating shaft, and an electrical wire crossing the system boundaries are all associated with work interactions. Work transfer to a system (i.e., work done on a system) increases the energy of the system, and work transfer from a system (i.e., work done by the system) decreases it since the energy transferred out as work comes from the energy contained in the system. Car engines and hydraulic, steam, or gas turbines produce work while compressors, pumps, and mixers consume work.
3. **Mass Flow, m** Mass flow in and out of the system serves as an additional mechanism of energy transfer. When mass enters a system, the energy of the system increases because mass carries energy with it (in fact, mass is energy). Likewise, when some mass leaves the system, the energy contained within the system decreases because the leaving mass takes out some energy with it. For example, when some hot water is taken out of a water heater and is replaced by the same amount of cold water, the energy content of the hot-water tank (the control volume) decreases as a result of this mass interaction (Fig. 2–47).

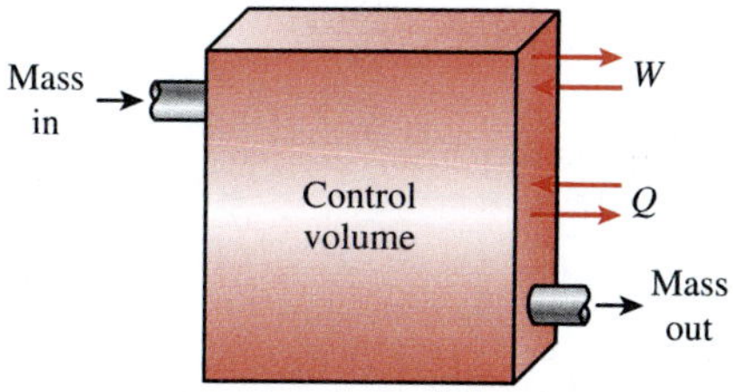

FIGURE 2–47
The energy content of a control volume can be changed by mass flow as well as heat and work interactions.

Noting that energy can be transferred in the forms of heat, work, and mass, and that the net transfer of a quantity is equal to the difference between the amounts transferred in and out, the energy balance can be written more explicitly as

$$E_{in} - E_{out} = (Q_{in} - Q_{out}) + (W_{in} - W_{out}) + (E_{mass,in} - E_{mass,out}) = \Delta E_{system} \quad \textbf{(2–34)}$$

where the subscripts "in" and "out" denote quantities that enter and leave the system, respectively. All six quantities on the right side of the equation represent "amounts," and thus they are *positive* quantities. The direction of any energy transfer is described by the subscripts "in" and "out."

The heat transfer Q is zero for adiabatic systems, the work transfer W is zero for systems that involve no work interactions, and the energy transport with mass E_{mass} is zero for systems that involve no mass flow across their boundaries (i.e., closed systems).

Energy balance for any system undergoing any kind of process can be expressed more compactly as

$$\underbrace{E_{in} - E_{out}}_{\substack{\text{Net energy transfer}\\ \text{by heat, work, and mass}}} = \underbrace{\Delta E_{system}}_{\substack{\text{Change in internal, kinetic,}\\ \text{potential, etc., energies}}} \quad \text{(kJ)} \tag{2-35}$$

or, in the **rate form**, as

$$\underbrace{\dot{E}_{in} - \dot{E}_{out}}_{\substack{\text{Rate of net energy transfer}\\ \text{by heat, work, and mass}}} = \underbrace{dE_{system}/dt}_{\substack{\text{Rate of change in internal,}\\ \text{kinetic, potential, etc., energies}}} \quad \text{(kW)} \tag{2-36}$$

For constant rates, the total quantities during a time interval Δt are related to the quantities per unit time as

$$Q = \dot{Q}\,\Delta t, \quad W = \dot{W}\,\Delta t, \quad \text{and} \quad \Delta E = (dE/dt)\,\Delta t \quad \text{(kJ)} \tag{2-37}$$

The energy balance can be expressed on a **per unit mass** basis as

$$e_{in} - e_{out} = \Delta e_{system} \quad \text{(kJ/kg)} \tag{2-38}$$

which is obtained by dividing all the quantities in Eq. 2–35 by the mass m of the system. Energy balance can also be expressed in the differential form as

$$\delta E_{in} - \delta E_{out} = dE_{system} \quad \text{or} \quad \delta e_{in} - \delta e_{out} = de_{system} \tag{2-39}$$

For a closed system undergoing a **cycle**, the initial and final states are identical, and thus $\Delta E_{system} = E_2 - E_1 = 0$. Then the energy balance for a cycle simplifies to $E_{in} - E_{out} = 0$ or $E_{in} = E_{out}$. Noting that a closed system does not involve any mass flow across its boundaries, the energy balance for a cycle can be expressed in terms of heat and work interactions as

$$W_{net,out} = Q_{net,in} \quad \text{or} \quad \dot{W}_{net,out} = \dot{Q}_{net,in} \quad \text{(for a cycle)} \tag{2-40}$$

That is, the net work output during a cycle is equal to net heat input (Fig. 2–48).

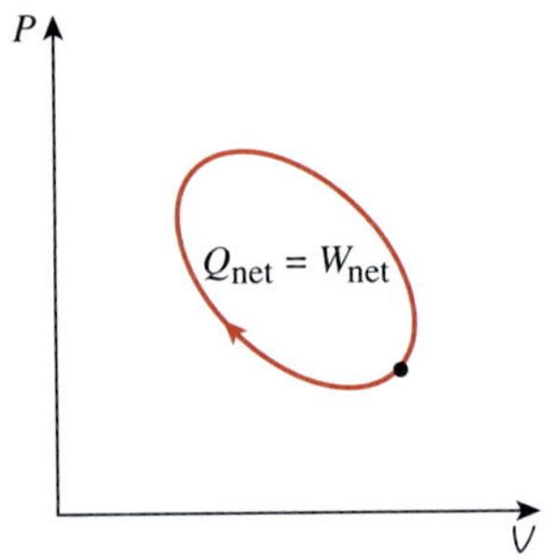

FIGURE 2–48
For a cycle $\Delta E = 0$, thus $Q = W$.

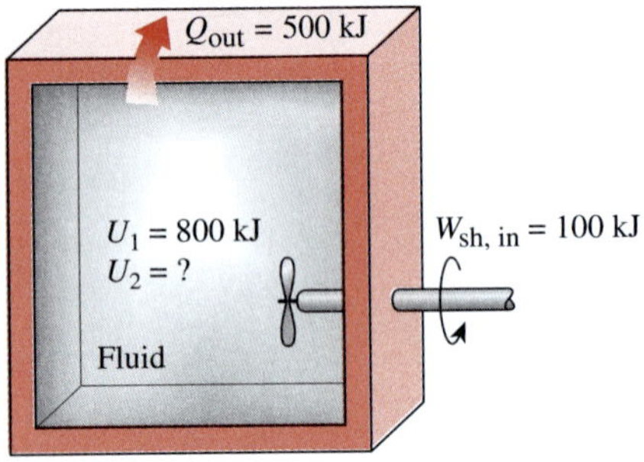

FIGURE 2–49
Schematic for Example 2–10.

EXAMPLE 2–10 Cooling of a Hot Fluid in a Tank

A rigid tank contains a hot fluid that is cooled while being stirred by a paddle wheel. Initially, the internal energy of the fluid is 800 kJ. During the cooling process, the fluid loses 500 kJ of heat, and the paddle wheel does 100 kJ of work on the fluid. Determine the final internal energy of the fluid. Neglect the energy stored in the paddle wheel.

Solution A fluid in a rigid tank looses heat while being stirred. The final internal energy of the fluid is to be determined.
Assumptions **1** The tank is stationary and thus the kinetic and potential energy changes are zero, $\Delta KE = \Delta PE = 0$. Therefore, $\Delta E = \Delta U$ and internal energy is the only form of the system's energy that may change during this process. **2** Energy stored in the paddle wheel is negligible.
Analysis Take the contents of the tank as the *system* (Fig. 2–49). This is a *closed system* since no mass crosses the boundary during the process. We observe that the volume of a rigid tank is constant, and thus there is no moving boundary work. Also, heat is lost from the system and shaft work is done on the system. Applying the energy balance on the system gives

$$\underbrace{E_{in} - E_{out}}_{\substack{\text{Net energy transfer}\\ \text{by heat, work, and mass}}} = \underbrace{\Delta E_{system}}_{\substack{\text{Change in internal, kinetic,}\\ \text{potential, etc., energies}}}$$

$$W_{sh,in} - Q_{out} = \Delta U = U_2 - U_1$$

$$100 \text{ kJ} - 500 \text{ kJ} = U_2 - 800 \text{ kJ}$$

$$U_2 = \mathbf{400 \text{ kJ}}$$

Therefore, the final internal energy of the system is 400 kJ.

EXAMPLE 2–11 Acceleration of Air by a Fan

A fan that consumes 20 W of electric power when operating is claimed to discharge air from a ventilated room at a rate of 1.0 kg/s at a discharge velocity of 8 m/s (Fig. 2–50). Determine if this claim is reasonable.

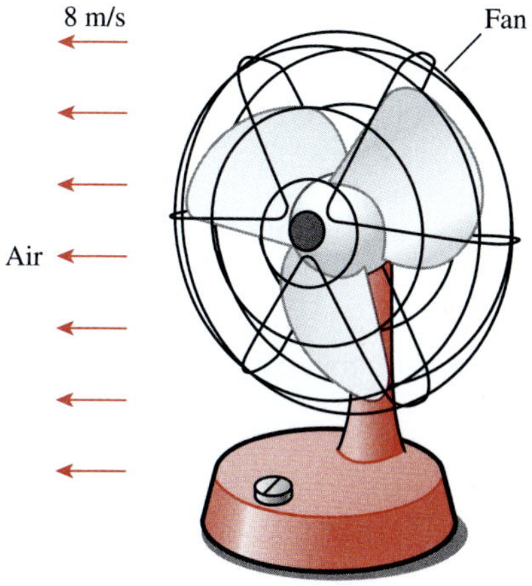

FIGURE 2–50
Schematic for Example 2–11.

Solution A fan is claimed to increase the velocity of air to a specified value while consuming electric power at a specified rate. The validity of this claim is to be investigated.

Assumptions The ventilating room is relatively calm, and air velocity in it is negligible.

Analysis First, let's examine the energy conversions involved: The motor of the fan converts part of the electrical power it consumes to mechanical (shaft) power, which is used to rotate the fan blades in air. The blades are shaped such that they impart a large fraction of the mechanical power of the shaft to air by mobilizing it. In the limiting ideal case of no losses (no conversion of electrical and mechanical energy to thermal energy) in steady operation, the electric power input will be equal to the rate of increase of the kinetic energy of air. Therefore, for a control volume that encloses the fan-motor unit, the energy balance can be written as

$$\underbrace{\dot{E}_{in} - \dot{E}_{out}}_{\substack{\text{Rate of net energy transfer}\\ \text{by heat, work, and mass}}} = \underbrace{dE_{system} / dt^{\nearrow 0 \text{ (steady)}}}_{\substack{\text{Rate of change in internal, kinetic,}\\ \text{potential, etc., energies}}} = 0 \quad \rightarrow \quad \dot{E}_{in} = \dot{E}_{out}$$

$$\dot{W}_{elect, in} = \dot{m}_{air} \, \text{ke}_{out} = \dot{m}_{air} \frac{V_{out}^2}{2}$$

Solving for V_{out} and substituting gives the maximum air outlet velocity to be

$$V_{out} = \sqrt{\frac{2\dot{W}_{elect,in}}{\dot{m}_{air}}} = \sqrt{\frac{2(20 \text{ J/s})}{1.0 \text{ kg/s}}\left(\frac{1 \text{ m}^2/\text{s}^2}{1 \text{ J/kg}}\right)} = 6.3 \text{ m/s}$$

which is less than 8 m/s. Therefore, the claim is **false**.

Discussion The conservation of energy principle requires the energy to be preserved as it is converted from one form to another, and it does not allow any energy to be created or destroyed during a process. From the first law point of view, there is nothing wrong with the conversion of the entire electrical energy into kinetic energy. Therefore, the first law has no objection to air velocity reaching 6.3 m/s—but this is the upper limit. Any claim of higher velocity is in violation of the first law, and thus impossible. In reality, the air velocity will be considerably lower than 6.3 m/s because of the losses associated with the conversion of electrical energy to mechanical shaft energy, and the conversion of mechanical shaft energy to kinetic energy or air.

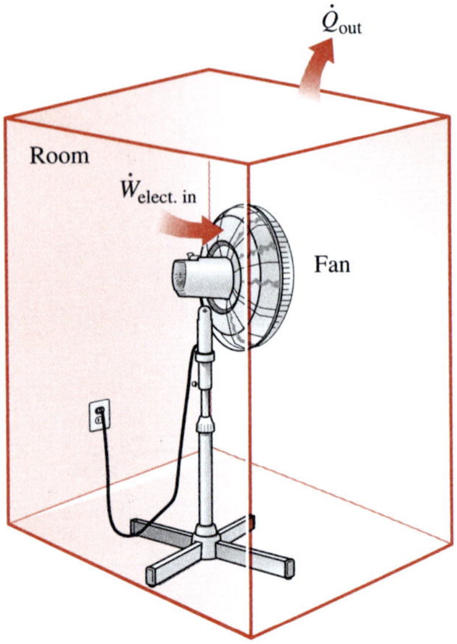

FIGURE 2–51
Schematic for Example 2–12.

EXAMPLE 2–12 Heating Effect of a Fan

A room is initially at the outdoor temperature of 25°C. Now a large fan that consumes 200 W of electricity when running is turned on (Fig. 2–51). The heat transfer rate between the room and the outdoor air is given as $\dot{Q} = UA(T_i - T_o)$ where $U = 6$ W/m²·°C is the overall heat transfer coefficient, $A = 30$ m² is the exposed surface area of the room, and T_i and T_o are the indoor and outdoor air temperatures, respectively. Determine the indoor air temperature when steady operating conditions are established.

Solution A large fan is turned on and kept on in a room that looses heat to the outdoors. The indoor air temperature is to be determined when steady operation is reached.

Assumptions **1** Heat transfer through the floor is negligible. **2** There are no other energy interactions involved.

Analysis The electricity consumed by the fan is energy input for the room, and thus the room gains energy at a rate of 200 W. As a result, the room air temperature tends to rise. But as the room air temperature rises, the rate of heat loss from the room increases until the rate of heat loss equals the electric power consumption. At that point, the temperature of the room air, and thus the energy content of the room, remains constant, and the conservation of energy for the room becomes

$$\underbrace{\dot{E}_{in} - \dot{E}_{out}}_{\text{Rate of net energy transfer by heat, work, and mass}} = \underbrace{dE_{system}/dt^{\nearrow 0\,(steady)}}_{\text{Rate of change in internal, kinetic, potential, etc., energies}} = 0 \rightarrow \dot{E}_{in} = \dot{E}_{out}$$

$$\dot{W}_{elect,in} = \dot{Q}_{out} = UA(T_i - T_o)$$

Substituting,

$$200\text{ W} = (6\text{ W/m}^2\cdot{}^\circ\text{C})(30\text{ m}^2)(T_i - 25^\circ\text{C})$$

It gives

$$T_i = \mathbf{26.1^\circ C}$$

Therefore, the room air temperature will remain constant after it reaches 26.1°C.

Discussion Note that a 200-W fan heats a room just like a 200-W resistance heater. In the case of a fan, the motor converts part of the electric energy it draws to mechanical energy in the form of a rotating shaft while the remaining part is dissipated as heat to the room air because of the motor inefficiency (no motor converts 100 percent of the electric energy it receives to mechanical energy, although some large motors come close with a conversion efficiency of over 97 percent). Part of the mechanical energy of the shaft is converted to kinetic energy of air through the blades, which is then converted to thermal energy as air molecules slow down because of friction. At the end, the entire electric energy drawn by the fan motor is converted to thermal energy of air, which manifests itself as a rise in temperature.

FIGURE 2–52
Fluorescent lamps lighting a classroom as discussed in Example 2–13.
© Vol. 24/PhotoDisc/Getty RF.

EXAMPLE 2–13 Annual Lighting Cost of a Classroom

The lighting needs of a classroom are met by 30 fluorescent lamps, each consuming 80 W of electricity (Fig. 2–52). The lights in the classroom are kept on for 12 hours a day and 250 days a year. For a unit electricity cost of

7 cents per kWh, determine annual energy cost of lighting for this classroom. Also, discuss the effect of lighting on the heating and air-conditioning requirements of the room.

Solution The lighting of a classroom by fluorescent lamps is considered. The annual electricity cost of lighting for this classroom is to be determined, and the lighting's effect on the heating and air-conditioning requirements is to be discussed.
Assumptions The effect of voltage fluctuations is negligible so that each fluorescent lamp consumes its rated power.
Analysis The electric power consumed by the lamps when all are on and the number of hours they are kept on per year are

$$\begin{aligned}\text{Lighting power} &= (\text{Power consumed per lamp}) \times (\text{No. of lamps})\\ &= (80\ \text{W/lamp})(30\ \text{lamps})\\ &= 2400\ \text{W} = 2.4\ \text{kW}\end{aligned}$$

$$\text{Operating hours} = (12\ \text{h/day})(250\ \text{days/year}) = 3000\ \text{h/year}$$

Then the amount and cost of electricity used per year become

$$\begin{aligned}\text{Lighting energy} &= (\text{Lighting power})(\text{Operating hours})\\ &= (2.4\ \text{kW})(3000\ \text{h/year}) = 7200\ \text{kWh/year}\end{aligned}$$

$$\begin{aligned}\text{Lighting cost} &= (\text{Lighting energy})(\text{Unit cost})\\ &= (7200\ \text{kWh/year})(\$0.07/\text{kWh}) = \mathbf{\$504/year}\end{aligned}$$

Light is absorbed by the surfaces it strikes and is converted to thermal energy. Disregarding the light that escapes through the windows, the entire 2.4 kW of electric power consumed by the lamps eventually becomes part of thermal energy of the classroom. Therefore, the lighting system in this room reduces the heating requirements by 2.4 kW, but increases the air-conditioning load by 2.4 kW.
Discussion Note that the annual lighting cost of this classroom alone is over $500. This shows the importance of energy conservation measures. If incandescent light bulbs were used instead of fluorescent tubes, the lighting costs would be four times as much since incandescent lamps use four times as much power for the same amount of light produced.

EXAMPLE 2–14 Conservation of Energy for an Oscillating Steel Ball

The motion of a steel ball in a hemispherical bowl of radius *h* shown in Fig. 2–53 is to be analyzed. The ball is initially held at the highest location at point *A*, and then it is released. Obtain relations for the conservation of energy of the ball for the cases of frictionless and actual motions.

Solution A steel ball is released in a bowl. Relations for the energy balance are to be obtained.
Assumptions The motion is frictionless, and thus friction between the ball, the bowl, and the air is negligible.
Analysis When the ball is released, it accelerates under the influence of gravity, reaches a maximum velocity (and minimum elevation) at point *B* at

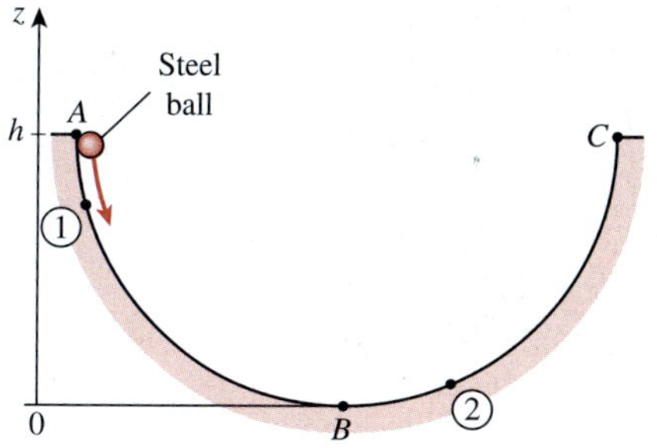

FIGURE 2–53
Schematic for Example 2–14.

FIGURE 2–54
The definition of performance is not limited to thermodynamics only.
BLONDIE © KING FEATURES SYNDICATE.

the bottom of the bowl, and moves up toward point C on the opposite side. In the ideal case of frictionless motion, the ball will oscillate between points A and C. The actual motion involves the conversion of the kinetic and potential energies of the ball to each other, together with overcoming resistance to motion due to friction (doing frictional work). The general energy balance for any system undergoing any process is

$$\underbrace{E_{\text{in}} - E_{\text{out}}}_{\substack{\text{Net energy transfer}\\ \text{by heat, work, and mass}}} = \underbrace{\Delta E_{\text{system}}}_{\substack{\text{Change in internal, kinetic,}\\ \text{potential, etc., energies}}}$$

Then the energy balance for the ball for a process from point 1 to point 2 becomes

$$-w_{\text{friction}} = (\text{ke}_2 + \text{pe}_2) - (\text{ke}_1 + \text{pe}_1)$$

or

$$\frac{V_1^2}{2} + gz_1 = \frac{V_2^2}{2} + gz_2 + w_{\text{friction}}$$

since there is no energy transfer by heat or mass and no change in the internal energy of the ball (the heat generated by frictional heating is dissipated to the surrounding air). The frictional work term w_{friction} is often expressed as e_{loss} to represent the loss (conversion) of mechanical energy into thermal energy.

For the idealized case of frictionless motion, the last relation reduces to

$$\frac{V_1^2}{2} + gz_1 = \frac{V_2^2}{2} + gz_2 \quad \text{or} \quad \frac{V^2}{2} + gz = C = \text{constant}$$

where the value of the constant is $C = gh$. That is, *when the frictional effects are negligible, the sum of the kinetic and potential energies of the ball remains constant.*

Discussion This is certainly a more intuitive and convenient form of the conservation of energy equation for this and other similar processes such as the swinging motion of the pendulum of a wall clock.

2–7 ■ ENERGY CONVERSION EFFICIENCIES

Efficiency is one of the most frequently used terms in thermodynamics, and it indicates how well an energy conversion or transfer process is accomplished. Efficiency is also one of the most frequently misused terms in thermodynamics and a source of misunderstandings. This is because efficiency is often used without being properly defined first. Next, we will clarify this further and define some efficiencies commonly used in practice.

Performance or efficiency, in general, can be expressed in terms of the desired output and the required input as (Fig. 2–54)

$$\text{Performance} = \frac{\text{Desired output}}{\text{Required input}} \tag{2–41}$$

If you are shopping for a water heater, a knowledgeable salesperson will tell you that the efficiency of a conventional electric water heater is about 90 percent (Fig. 2–55). You may find this confusing, since the heating elements of electric water heaters are resistance heaters, and the efficiency

Water heater

Type	Efficiency
Gas, conventional	55%
Gas, high-efficiency	62%
Electric, conventional	90%
Electric, high-efficiency	94%

FIGURE 2–55
Typical efficiencies of conventional and high-efficiency electric and natural gas water heaters.
© The McGraw-Hill Companies, Inc./Jill Braaten, photographer.

of all resistance heaters is 100 percent as they convert all the electrical energy they consume into thermal energy. A knowledgeable salesperson will clarify this by explaining that the heat losses from the hot-water tank to the surrounding air amount to 10 percent of the electrical energy consumed, and the **efficiency of a water heater** is defined as the ratio of the *energy delivered to the house by hot water* to the *energy supplied to the water heater.* A clever salesperson may even talk you into buying a more expensive water heater with thicker insulation that has an efficiency of 94 percent. If you are a knowledgeable consumer and have access to natural gas, you will probably purchase a gas water heater whose efficiency is only 55 percent since a gas unit costs about the same as an electric unit to purchase and install, but the annual energy cost of a gas unit will be much less than that of an electric unit.

Perhaps you are wondering how the efficiency for a gas water heater is defined, and why it is much lower than the efficiency of an electric heater. As a general rule, the efficiency of equipment that involves the combustion of a fuel is based on the **heating value of the fuel**, which is *the amount of heat released when a unit amount of fuel at room temperature is completely burned and the combustion products are cooled to the room temperature* (Fig. 2–56). Then the performance of combustion equipment can be characterized by **combustion efficiency**, defined as

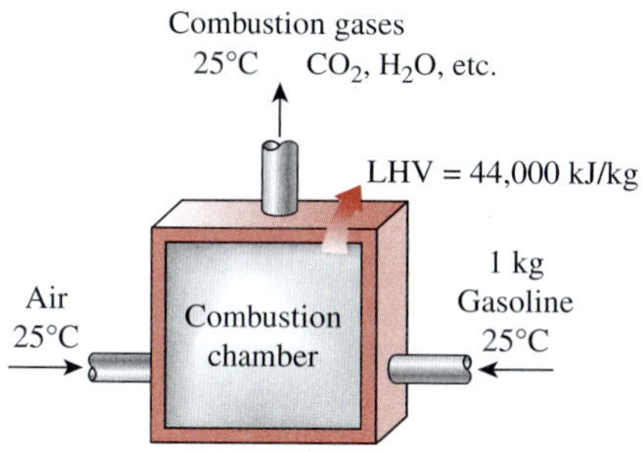

FIGURE 2–56
The definition of the heating value of gasoline.

$$\eta_{\text{combustion}} = \frac{Q}{\text{HV}} = \frac{\text{Amount of heat released during combustion}}{\text{Heating value of the fuel burned}} \quad \textbf{(2–42)}$$

A combustion efficiency of 100 percent indicates that the fuel is burned completely and the stack gases leave the combustion chamber at room temperature, and thus the amount of heat released during a combustion process is equal to the heating value of the fuel.

Most fuels contain hydrogen, which forms water when burned, and the heating value of a fuel will be different, depending on whether the water in combustion products is in the liquid or vapor form. The heating value is called the *lower heating value,* or LHV, when the water leaves as a vapor, and the *higher heating value,* or HHV, when the water in the combustion gases is completely condensed and thus the heat of vaporization is also recovered. The difference between these two heating values is equal to the product of the amount of water and the enthalpy of vaporization of water at room temperature. For example, the lower and higher heating values of gasoline are 44,000 kJ/kg and 47,300 kJ/kg, respectively. An efficiency definition should make it clear whether it is based on the higher or lower heating value of the fuel. Efficiencies of cars and jet engines are normally based on *lower heating values* since water normally leaves as a vapor in the exhaust gases, and it is not practical to try to recuperate the heat of vaporization. Efficiencies of furnaces, on the other hand, are based on *higher heating values.*

The efficiency of space heating systems of residential and commercial buildings is usually expressed in terms of the **annual fuel utilization efficiency**, or **AFUE**, which accounts for the combustion efficiency as well as other losses such as heat losses to unheated areas and start-up and cool-down losses. The AFUE of most new heating systems is about 85 percent,

TABLE 2–1

The efficacy of different lighting systems

Type of lighting	Efficacy, lumens/W
Combustion	
Candle	0.3
Kerosene lamp	1–2
Incandescent	
Ordinary	6–20
Halogen	15–35
Fluorescent	
Compact	40–87
Tube	60–120
High-intensity discharge	
Mercury vapor	40–60
Metal halide	65–118
High-pressure sodium	85–140
Low-pressure sodium	70–200
Solid-State	
LED	20–160
OLED	15–60
Theoretical limit	300*

*This value depends on the spectral distribution of the assumed ideal light source. For white light sources, the upper limit is about 300 lm/W for metal halide, 350 lm/W for fluorescents, and 400 lm/W for LEDs. Spectral maximum occurs at a wavelength of 555 nm (green) with a light output of 683 lm/W.

FIGURE 2–57
A 15-W compact fluorescent lamp provides as much light as a 60-W incandescent lamp.

although the AFUE of some old heating systems is under 60 percent. The AFUE of some new high-efficiency furnaces exceeds 96 percent, but the high cost of such furnaces cannot be justified for locations with mild to moderate winters. Such high efficiencies are achieved by reclaiming most of the heat in the flue gases, condensing the water vapor, and discharging the flue gases at temperatures as low as 38°C (or 100°F) instead of about 200°C (or 400°F) for the conventional models.

For *car engines,* the work output is understood to be the power delivered by the crankshaft. But for power plants, the work output can be the mechanical power at the turbine exit, or the electrical power output of the generator.

A generator is a device that converts mechanical energy to electrical energy, and the effectiveness of a generator is characterized by the **generator efficiency**, which is the ratio of the *electrical power output* to the *mechanical power input.* The *thermal efficiency* of a power plant, which is of primary interest in thermodynamics, is usually defined as the ratio of the net shaft work output of the turbine to the heat input to the working fluid. The effects of other factors are incorporated by defining an **overall efficiency** for the power plant as the ratio of the *net electrical power output* to the *rate of fuel energy input.* That is,

$$\eta_{\text{overall}} = \eta_{\text{combustion}}\eta_{\text{thermal}}\eta_{\text{generator}} = \frac{\dot{W}_{\text{net,electric}}}{\text{HHV} \times \dot{m}_{\text{net}}} \tag{2–43}$$

The overall efficiencies are about 26–30 percent for gasoline automotive engines, 34–40 percent for diesel engines, and up to 60 percent for large power plants.

We are all familiar with the conversion of electrical energy to *light* by incandescent lightbulbs, fluorescent tubes, and high-intensity discharge lamps. The efficiency for the conversion of electricity to light can be defined as the ratio of the energy converted to light to the electrical energy consumed. For example, common incandescent lightbulbs convert about 5 percent of the electrical energy they consume to light; the rest of the energy consumed is dissipated as heat, which adds to the cooling load of the air conditioner in summer. However, it is more common to express the effectiveness of this conversion process by **lighting efficacy**, which is defined as the *amount of light output in lumens per W of electricity consumed.*

The efficacy of different lighting systems is given in Table 2–1. Note that a compact fluorescent lightbulb produces about four times as much light as an incandescent lightbulb per W, and thus a 15-W fluorescent bulb can replace a 60-W incandescent lightbulb (Fig. 2–57). Also, a compact fluorescent bulb lasts about 10,000 h, which is 10 times as long as an incandescent bulb, and it plugs directly into the socket of an incandescent lamp. Therefore, despite their higher initial cost, compact fluorescents reduce the lighting costs considerably through reduced electricity consumption. Sodium-filled high-intensity discharge lamps provide the most efficient lighting, but their use is limited to outdoor use because of their yellowish light.

We can also define efficiency for cooking appliances since they convert electrical or chemical energy to heat for cooking. The **efficiency of a cooking appliance** can be defined as the ratio of the *useful energy transferred to the* *food to the energy consumed by the appliance* (Fig. 2–58). Electric ranges are more efficient than gas ranges, but it is much cheaper to cook with natural gas than with electricity because of the lower unit cost of natural gas (Table 2–2).

The cooking efficiency depends on user habits as well as the individual appliances. Convection and microwave ovens are inherently more efficient than conventional ovens. On average, convection ovens save about *one-third* and microwave ovens save about *two-thirds* of the energy used by conventional ovens. The cooking efficiency can be increased by using the smallest oven for baking, using a pressure cooker, using an electric slow cooker for stews and soups, using the smallest pan that will do the job, using the smaller heating element for small pans on electric ranges, using flat-bottomed pans on electric burners to assure good contact, keeping burner drip pans clean and shiny, defrosting frozen foods in the refrigerator before cooking, avoiding preheating unless it is necessary, keeping the pans covered during cooking, using timers and thermometers to avoid overcooking, using the self-cleaning feature of ovens right after cooking, and keeping inside surfaces of microwave ovens clean.

Using energy-efficient appliances and practicing energy conservation measures help our pocketbooks by reducing our utility bills. It also helps the **environment** by reducing the amount of pollutants emitted to the atmosphere during the combustion of fuel at home or at the power plants where electricity is generated. The combustion of *each therm of natural gas* produces 6.4 kg of carbon dioxide, which causes global climate change; 4.7 g of nitrogen oxides and 0.54 g of hydrocarbons, which cause smog; 2.0 g of carbon monoxide, which is toxic; and 0.030 g of sulfur dioxide, which causes acid rain. Each therm of natural gas saved eliminates the emission of these pollutants while saving $0.60 for the average consumer in the United States. Each kWh of electricity conserved saves 0.4 kg of coal and 1.0 kg of CO_2 and 15 g of SO_2 from a coal power plant.

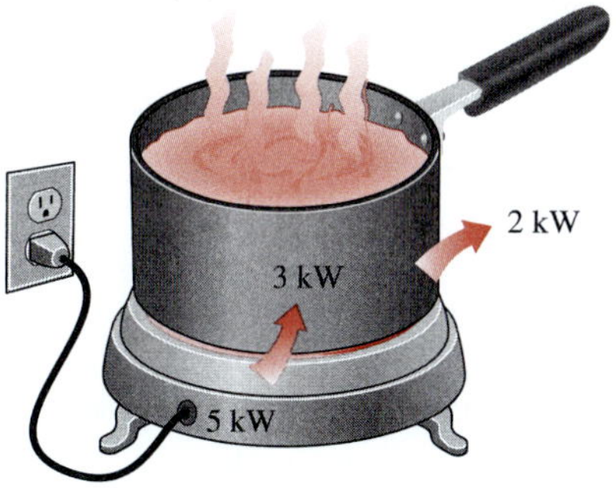

$$\text{Efficiency} = \frac{\text{Energy utilized}}{\text{Energy supplied to appliance}} = \frac{3\text{ kWh}}{5\text{ kWh}} = 0.60$$

FIGURE 2–58
The efficiency of a cooking appliance represents the fraction of the energy supplied to the appliance that is transferred to the food.

TABLE 2–2

Energy costs of cooking a casserole with different appliances*

[From J. T. Amann, A. Wilson, and K. Ackerly, *Consumer Guide to Home Energy Savings*, 9th ed., American Council for an Energy-Efficient Economy, Washington, D.C., 2007, p. 163.]

Cooking appliance	Cooking temperature	Cooking time	Energy used	Cost of energy
Electric oven	350°F (177°C)	1 h	2.0 kWh	$0.19
Convection oven (elect.)	325°F (163°C)	45 min	1.39 kWh	$0.13
Gas oven	350°F (177°C)	1 h	0.112 therm	$0.13
Frying pan	420°F (216°C)	1 h	0.9 kWh	$0.09
Toaster oven	425°F (218°C)	50 min	0.95 kWh	$0.09
Crockpot	200°F (93°C)	7 h	0.7 kWh	$0.07
Microwave oven	"High"	15 min	0.36 kWh	$0.03

*Assumes a unit cost of $0.095/kWh for electricity and $1.20/therm for gas.

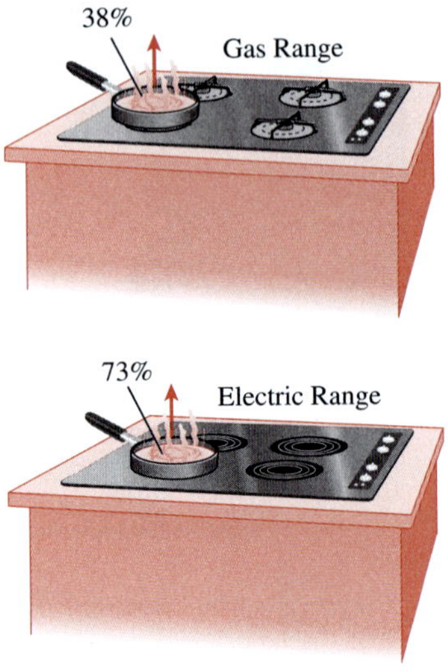

FIGURE 2–59
Schematic of the 73 percent efficient electric heating unit and 38 percent efficient gas burner discussed in Example 2–15.

EXAMPLE 2–15 Cost of Cooking with Electric and Gas Ranges

The efficiency of cooking appliances affects the internal heat gain from them since an inefficient appliance consumes a greater amount of energy for the same task, and the excess energy consumed shows up as heat in the living space. The efficiency of open burners is determined to be 73 percent for electric units and 38 percent for gas units (Fig. 2–59). Consider a 2-kW electric burner at a location where the unit costs of electricity and natural gas are $0.09/kWh and $1.20/therm, respectively. Determine the rate of energy consumption by the burner and the unit cost of utilized energy for both electric and gas burners.

Solution The operation of electric and gas ranges is considered. The rate of energy consumption and the unit cost of utilized energy are to be determined.

Analysis The efficiency of the electric heater is given to be 73 percent. Therefore, a burner that consumes 2 kW of electrical energy will supply

$$\dot{Q}_{\text{utilized}} = (\text{Energy input}) \times (\text{Efficiency}) = (2\text{ kW})(0.73) = \mathbf{1.46\ kW}$$

of useful energy. The unit cost of utilized energy is inversely proportional to the efficiency, and is determined from

$$\text{Cost of utilized energy} = \frac{\text{Cost of energy input}}{\text{Efficiency}} = \frac{\$0.09/\text{kWh}}{0.73} = \mathbf{\$0.123/kWh}$$

Noting that the efficiency of a gas burner is 38 percent, the energy input to a gas burner that supplies utilized energy at the same rate (1.46 kW) is

$$\dot{Q}_{\text{input, gas}} = \frac{\dot{Q}_{\text{utilized}}}{\text{Efficiency}} = \frac{1.46\text{ kW}}{0.38} = \mathbf{3.84\ kW} \qquad (=13{,}100\text{ Btu/h})$$

since 1 kW = 3412 Btu/h. Therefore, a gas burner should have a rating of at least 13,100 Btu/h to perform as well as the electric unit.

Noting that 1 therm = 29.3 kWh, the unit cost of utilized energy in the case of a gas burner is determined to be

$$\text{Cost of utilized energy} = \frac{\text{Cost of energy input}}{\text{Efficiency}} = \frac{\$1.20/29.3\text{ kWh}}{0.38}$$
$$= \mathbf{\$0.108/kWh}$$

Discussion The cost of utilized gas is less than that of utilized electricity. Therefore, despite its higher efficiency, cooking with an electric burner will cost about 14 percent more compared to a gas burner in this case. This explains why cost-conscious consumers always ask for gas appliances, and it is not wise to use electricity for heating purposes.

Efficiencies of Mechanical and Electrical Devices

The transfer of mechanical energy is usually accomplished by a rotating shaft, and thus mechanical work is often referred to as *shaft work*. A pump or a fan receives shaft work (usually from an electric motor) and transfers it to the fluid as mechanical energy (less frictional losses). A turbine, on the other hand, converts the mechanical energy of a fluid to shaft work. In the absence of any irreversibilities such as friction, mechanical energy can be converted entirely from one mechanical form to another, and the **mechanical efficiency** of a device or process can be defined as (Fig. 2–60)

$$\eta_{\text{mech}} = \frac{\text{Mechanical energy output}}{\text{Mechanical energy input}} = \frac{E_{\text{mech,out}}}{E_{\text{mech,in}}} = 1 - \frac{E_{\text{mech,loss}}}{E_{\text{mech,in}}} \quad \text{(2–44)}$$

A conversion efficiency of less than 100 percent indicates that conversion is less than perfect and some losses have occurred during conversion. A mechanical efficiency of 97 percent indicates that 3 percent of the mechanical energy input is converted to thermal energy as a result of frictional heating, and this will manifest itself as a slight rise in the temperature of the fluid.

In fluid systems, we are usually interested in increasing the pressure, velocity, and/or elevation of a fluid. This is done by *supplying mechanical energy* to the fluid by a pump, a fan, or a compressor (we will refer to all of them as pumps). Or we are interested in the reverse process of *extracting mechanical energy* from a fluid by a turbine and producing mechanical power in the form of a rotating shaft that can drive a generator or any other rotary device. The degree of perfection of the conversion process between the mechanical work supplied or extracted and the mechanical energy of the fluid is expressed by the **pump efficiency** and **turbine efficiency**, defined as

$$\eta_{\text{pump}} = \frac{\text{Mechanical energy increase of the fluid}}{\text{Mechanical energy input}} = \frac{\Delta \dot{E}_{\text{mech,fluid}}}{\dot{W}_{\text{shaft,in}}} = \frac{\dot{W}_{\text{pump},u}}{\dot{W}_{\text{pump}}} \quad \text{(2–45)}$$

where $\Delta \dot{E}_{\text{mech,fluid}} = \dot{E}_{\text{mech,out}} - \dot{E}_{\text{mech,in}}$ is the rate of increase in the mechanical energy of the fluid, which is equivalent to the **useful pumping power** $\dot{W}_{\text{pump},u}$ supplied to the fluid, and

$$\eta_{\text{turbine}} = \frac{\text{Mechanical energy output}}{\text{Mechanical energy decrease of the fluid}} = \frac{\dot{W}_{\text{shaft,out}}}{|\Delta \dot{E}_{\text{mech,fluid}}|} = \frac{\dot{W}_{\text{turbine}}}{\dot{W}_{\text{turbine},e}} \quad \text{(2–46)}$$

where $|\Delta \dot{E}_{\text{mech,fluid}}| = \dot{E}_{\text{mech,in}} - \dot{E}_{\text{mech,out}}$ is the rate of decrease in the mechanical energy of the fluid, which is equivalent to the mechanical power extracted from the fluid by the turbine $\dot{W}_{\text{turbine},e}$, and we use the absolute value sign to avoid negative values for efficiencies. A pump or turbine efficiency of 100 percent indicates perfect conversion between the shaft work and the mechanical energy of the fluid, and this value can be approached (but never attained) as the frictional effects are minimized.

Electrical energy is commonly converted to *rotating mechanical energy* by electric motors to drive fans, compressors, robot arms, car starters, and so forth. The effectiveness of this conversion process is characterized by the *motor efficiency* η_{motor}, which is the ratio of the *mechanical energy output* of the motor to the *electrical energy input*. The full-load motor efficiencies range from about 35 percent for small motors to over 97 percent for large high-efficiency motors. The difference between the electrical energy consumed and the mechanical energy delivered is dissipated as waste heat.

The mechanical efficiency should not be confused with the **motor efficiency** and the **generator efficiency**, which are defined as

Motor: $$\eta_{\text{motor}} = \frac{\text{Mechanical power output}}{\text{Electric power input}} = \frac{\dot{W}_{\text{shaft,out}}}{\dot{W}_{\text{elect,in}}} \quad \text{(2–47)}$$

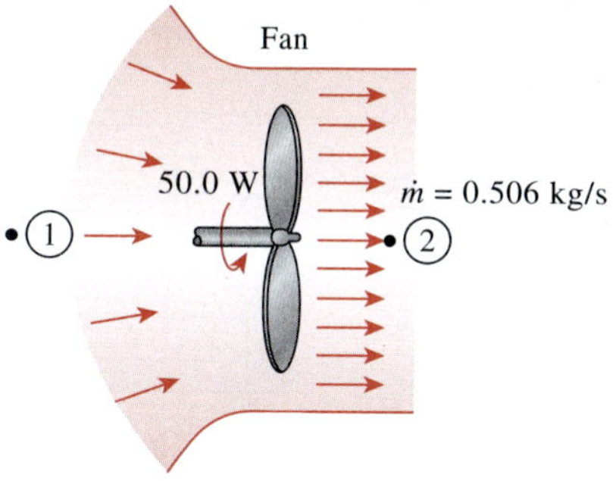

$V_1 \times 0$, $V_2 = 12.1$ m/s
$z_1 = z_2$
$P_1 \times P_{\text{atm}}$ and $P_2 \times P_{\text{atm}}$

$$\eta_{\text{mech, fan}} = \frac{\Delta \dot{E}_{\text{mech, fluid}}}{\dot{W}_{\text{shaft, in}}} = \frac{\dot{m}V_2^2/2}{\dot{W}_{\text{shaft, in}}} = \frac{(0.506 \text{ kg/s})(12.1 \text{ m/s})^2/2}{50.0 \text{ W}} = 0.741$$

FIGURE 2–60
The mechanical efficiency of a fan is the ratio of the rate of increase of the mechanical energy of air to the mechanical power input.

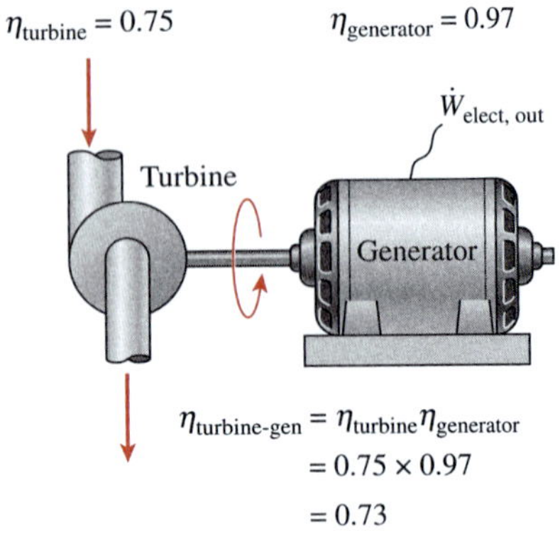

FIGURE 2–61
The overall efficiency of a turbine–generator is the product of the efficiency of the turbine and the efficiency of the generator, and represents the fraction of the mechanical power of the fluid converted to electrical power.

and

$$\textit{Generator:} \qquad \eta_{generator} = \frac{\text{Electric power output}}{\text{Mechanical power input}} = \frac{\dot{W}_{elect,out}}{\dot{W}_{shaft,in}} \tag{2–48}$$

A pump is usually packaged together with its motor, and a turbine with its generator. Therefore, we are usually interested in the **combined** or **overall efficiency** of pump–motor and turbine–generator combinations (Fig. 2–61), which are defined as

$$\eta_{pump-motor} = \eta_{pump}\eta_{motor} = \frac{\dot{W}_{pump,u}}{\dot{W}_{elect,in}} = \frac{\Delta\dot{E}_{mech,fluid}}{\dot{W}_{elect,in}} \tag{2–49}$$

and

$$\eta_{turbine-gen} = \eta_{turbine}\eta_{generator} = \frac{\dot{W}_{elect,out}}{\dot{W}_{turbine,e}} = \frac{\dot{W}_{elect,out}}{|\Delta\dot{E}_{mech,fluid}|} \tag{2–50}$$

All the efficiencies just defined range between 0 and 100 percent. The lower limit of 0 percent corresponds to the conversion of the entire mechanical or electric energy input to thermal energy, and the device in this case functions like a resistance heater. The upper limit of 100 percent corresponds to the case of perfect conversion with no friction or other irreversibilities, and thus no conversion of mechanical or electric energy to thermal energy.

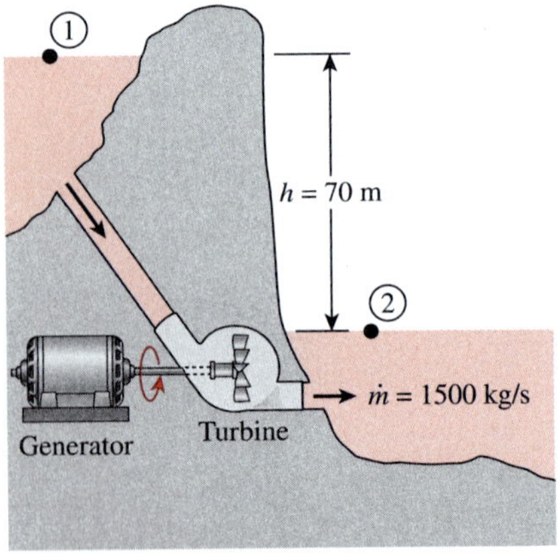

FIGURE 2–62
Schematic for Example 2–16.

EXAMPLE 2–16 Power Generation from a Hydroelectric Plant

Electric power is to be generated by installing a hydraulic turbine–generator at a site 70 m below the free surface of a large water reservoir that can supply water at a rate of 1500 kg/s steadily (Fig. 2–62). If the mechanical power output of the turbine is 800 kW and the electric power generation is 750 kW, determine the turbine efficiency and the combined turbine–generator efficiency of this plant. Neglect losses in the pipes.

Solution A hydraulic turbine-generator installed at a large reservoir is to generate electricity. The combined turbine–generator efficiency and the turbine efficiency are to be determined.

Assumptions **1** The water elevation in the reservoir remains constant. **2** The mechanical energy of water at the turbine exit is negligible.

Analysis We take the free surface of water in the reservoir to be point 1 and the turbine exit to be point 2. We also take the turbine exit as the reference level ($z_2 = 0$) so that the potential energies at 1 and 2 are $pe_1 = gz_1$ and $pe_2 = 0$. The flow energy P/ρ at both points is zero since both 1 and 2 are open to the atmosphere ($P_1 = P_2 = P_{atm}$). Further, the kinetic energy at both points is zero ($ke_1 = ke_2 = 0$) since the water at point 1 is essentially motionless, and the kinetic energy of water at turbine exit is assumed to be negligible. The potential energy of water at point 1 is

$$pe_1 = gz_1 = (9.81 \text{ m/s}^2)(70 \text{ m})\left(\frac{1 \text{ kJ/kg}}{1000 \text{ m}^2/\text{s}^2}\right) = 0.687 \text{ kJ/kg}$$

Then the rate at which the mechanical energy of water is supplied to the turbine becomes

$$|\Delta \dot{E}_{\text{mech,fluid}}| = \dot{m}(e_{\text{mech,in}} - e_{\text{mech,out}}) = \dot{m}(\text{pe}_1 - 0) = \dot{m}\text{pe}_1$$

$$= (1500 \text{ kg/s})(0.687 \text{ kJ/kg})$$

$$= 1031 \text{ kW}$$

The combined turbine–generator and the turbine efficiency are determined from their definitions to be

$$\eta_{\text{turbine-gen}} = \frac{\dot{W}_{\text{elect,out}}}{|\Delta \dot{E}_{\text{mech,fluid}}|} = \frac{750 \text{ kW}}{1031 \text{ kW}} = 0.727 \quad \text{or} \quad \mathbf{72.7\%}$$

$$\eta_{\text{turbine}} = \frac{\dot{W}_{\text{elect,out}}}{|\Delta \dot{E}_{\text{mech,fluid}}|} = \frac{800 \text{ kW}}{1031 \text{ kW}} = 0.776 \quad \text{or} \quad \mathbf{77.6\%}$$

Therefore, the reservoir supplies 1031 kW of mechanical energy to the turbine, which converts 800 kW of it to shaft work that drives the generator, which then generates 750 kW of electric power.

Discussion This problem can also be solved by taking point 1 to be at the turbine inlet, and using flow energy instead of potential energy. It would give the same result since the flow energy at the turbine inlet is equal to the potential energy at the free surface of the reservoir.

EXAMPLE 2–17 Cost Savings Associated with High-Efficiency Motors

A 60-hp electric motor (a motor that delivers 60 hp of shaft power at full load) that has an efficiency of 89.0 percent is worn out and is to be replaced by a 93.2 percent efficient high-efficiency motor (Fig. 2–63). The motor operates 3500 hours a year at full load. Taking the unit cost of electricity to be \$0.08/kWh, determine the amount of energy and money saved as a result of installing the high-efficiency motor instead of the standard motor. Also, determine the simple payback period if the purchase prices of the standard and high-efficiency motors are \$4520 and \$5160, respectively.

Solution A worn-out standard motor is to be replaced by a high-efficiency one. The amount of electrical energy and money saved as well as the simple payback period are to be determined.

Assumptions The load factor of the motor remains constant at 1 (full load) when operating.

Analysis The electric power drawn by each motor and their difference can be expressed as

$$\dot{W}_{\text{electric in,standard}} = \dot{W}_{\text{shaft}}/\eta_{\text{st}} = (\text{Rated power})(\text{Load factor})/\eta_{\text{st}}$$

$$\dot{W}_{\text{electric in,efficient}} = \dot{W}_{\text{shaft}}/\eta_{\text{eff}} = (\text{Rated power})(\text{Load factor})/\eta_{\text{eff}}$$

$$\text{Power savings} = \dot{W}_{\text{electric in,standard}} - \dot{W}_{\text{electric in,efficient}}$$

$$= (\text{Rated power})(\text{Load factor})(1/\eta_{\text{st}} - 1/\eta_{\text{eff}})$$

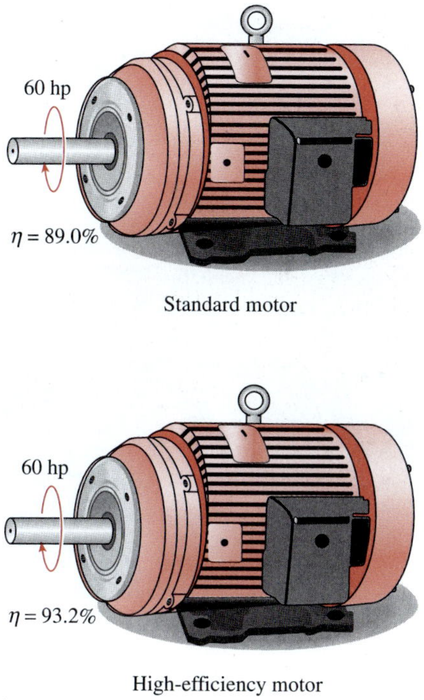

FIGURE 2–63
Schematic for Example 2–17.

where η_{st} is the efficiency of the standard motor, and η_{eff} is the efficiency of the comparable high-efficiency motor. Then the annual energy and cost savings associated with the installation of the high-efficiency motor become

$$\begin{aligned}\text{Energy savings} &= (\text{Power savings})(\text{Operating hours})\\ &= (\text{Rated power})(\text{Operating hours})(\text{Load factor})(1/\eta_{st} - 1/\eta_{eff})\\ &= (60\text{ hp})(0.7457\text{ kW/hp})(3500\text{ h/year})(1)(1/0.89 - 1/0.93.2)\\ &= \mathbf{7929\ kWh/year}\end{aligned}$$

$$\begin{aligned}\text{Cost savings} &= (\text{Energy savings})(\text{Unit cost of energy})\\ &= (7929\text{ kWh/year})(\$0.08/\text{ kWh})\\ &= \mathbf{\$634/year}\end{aligned}$$

Also,

$$\text{Excess initial cost} = \text{Purchase price differential} = \$5160 - \$4520 = \$640$$

This gives a simple payback period of

$$\text{Simple payback period} = \frac{\text{Excess initial cost}}{\text{Annual cost savings}} = \frac{\$640}{\$634/\text{year}} = \mathbf{1.01\ year}$$

Discussion Note that the high-efficiency motor pays for its price differential within about one year from the electrical energy it saves. Considering that the service life of electric motors is several years, the purchase of the higher efficiency motor is definitely indicated in this case.

2–8 ▪ ENERGY AND ENVIRONMENT

The conversion of energy from one form to another often affects the environment and the air we breathe in many ways, and thus the study of energy is not complete without considering its impact on the environment (Fig. 2–64). Fossil fuels such as coal, oil, and natural gas have been powering the industrial development and the amenities of modern life that we enjoy since the 1700s, but this has not been without any undesirable side effects. From the soil we farm and the water we drink to the air we breathe, the environment has been paying a heavy toll for it. Pollutants emitted during the combustion of fossil fuels are responsible for smog, acid rain, global warming, and climate change. The environmental pollution has reached such high levels that it became a serious threat to vegetation, wild life, and human health. Air pollution has been the cause of numerous health problems including asthma and cancer. It is estimated that over 60,000 people in the United States alone die each year due to heart and lung diseases related to air pollution.

FIGURE 2–64
Energy conversion processes are often accompanied by environmental pollution.
© Corbis Royalty Free.

Hundreds of elements and compounds such as benzene and formaldehyde are known to be emitted during the combustion of coal, oil, natural gas, and wood in electric power plants, engines of vehicles, furnaces, and even fireplaces. Some compounds are added to liquid fuels for various reasons (such as MTBE to raise the octane number of the fuel and also to oxygenate the fuel in winter months to reduce urban smog). The largest source of air pollution is the motor vehicles, and the pollutants released by the vehicles are usually grouped as hydrocarbons (HC), nitrogen oxides (NO_x), and carbon monoxide (CO) (Fig. 2–65). The HC emissions are a large component of volatile organic compounds (VOCs) emissions, and the two terms are generally used interchangeably for motor vehicle emissions. A significant portion of the VOC or HC emissions are caused by the evaporation of fuels during refueling or spillage during spitback or by evaporation from gas tanks with faulty caps that do not close tightly. The solvents, propellants, and household cleaning products that contain benzene, butane, or other HC products are also significant sources of HC emissions.

FIGURE 2–65
Motor vehicles are the largest source of air pollution.

The increase of environmental pollution at alarming rates and the rising awareness of its dangers made it necessary to control it by legislation and international treaties. In the United States, the Clean Air Act of 1970 (whose passage was aided by the 14-day smog alert in Washington that year) set limits on pollutants emitted by large plants and vehicles. These early standards focused on emissions of hydrocarbons, nitrogen oxides, and carbon monoxide. The new cars were required to have catalytic converters in their exhaust systems to reduce HC and CO emissions. As a side benefit, the removal of lead from gasoline to permit the use of catalytic converters led to a significant reduction in toxic lead emissions.

Emission limits for HC, NO_x, and CO from cars have been declining steadily since 1970. The Clean Air Act of 1990 made the requirements on emissions even tougher, primarily for ozone, CO, nitrogen dioxide, and particulate matter (PM). As a result, today's industrial facilities and vehicles emit a fraction of the pollutants they used to emit a few decades ago. The HC emissions of cars, for example, decreased from about 8 gpm (grams per mile) in 1970 to 0.4 gpm in 1980 and about 0.1 gpm in 1999. This is a significant reduction since many of the gaseous toxics from motor vehicles and liquid fuels are hydrocarbons.

Children are most susceptible to the damages caused by air pollutants since their organs are still developing. They are also exposed to more pollution since they are more active, and thus they breathe faster. People with heart and lung problems, especially those with asthma, are most affected by air pollutants. This becomes apparent when the air pollution levels in their neighborhoods rise to high levels.

Ozone and Smog

If you live in a metropolitan area such as Los Angeles, you are probably familiar with urban smog—the dark yellow or brown haze that builds up in a large stagnant air mass and hangs over populated areas on calm hot summer days. *Smog* is made up mostly of ground-level ozone (O_3), but it also contains numerous other chemicals, including carbon monoxide (CO), particulate matter such as soot and dust, volatile organic compounds (VOCs) such as benzene, butane, and other hydrocarbons. The harmful ground-level ozone should not be confused with the useful ozone layer high in the stratosphere that protects the earth from the sun's harmful ultraviolet rays. Ozone at ground level is a pollutant with several adverse health effects.

The primary source of both nitrogen oxides and hydrocarbons is the motor vehicles. Hydrocarbons and nitrogen oxides react in the presence of sunlight on hot calm days to form ground-level ozone, which is the primary

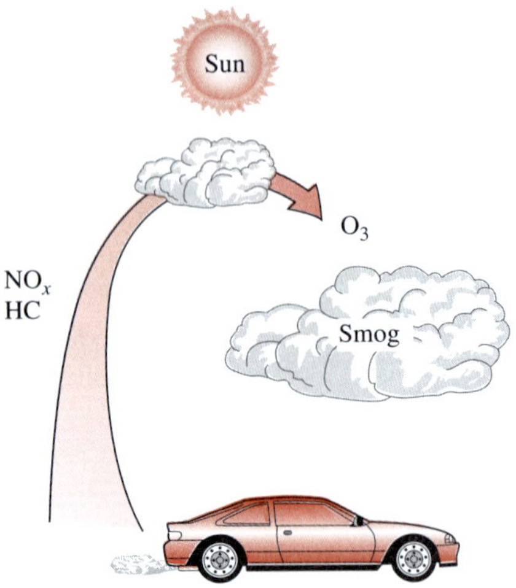

FIGURE 2–66
Ground-level ozone, which is the primary component of smog, forms when HC and NO_x react in the presence of sunlight in hot calm days.

component of smog (Fig. 2–66). The smog formation usually peaks in late afternoons when the temperatures are highest and there is plenty of sunlight. Although ground-level smog and ozone form in urban areas with heavy traffic or industry, the prevailing winds can transport them several hundred miles to other cities. This shows that pollution knows of no boundaries, and it is a global problem.

Ozone irritates eyes and damages the air sacs in the lungs where oxygen and carbon dioxide are exchanged, causing eventual hardening of this soft and spongy tissue. It also causes shortness of breath, wheezing, fatigue, headaches, and nausea, and aggravates respiratory problems such as asthma. Every exposure to ozone does a little damage to the lungs, just like cigarette smoke, eventually reducing the individual's lung capacity. Staying indoors and minimizing physical activity during heavy smog minimizes damage. Ozone also harms vegetation by damaging leaf tissues. To improve the air quality in areas with the worst ozone problems, reformulated gasoline (RFG) that contains at least 2 percent oxygen was introduced. The use of RFG has resulted in significant reduction in the emission of ozone and other pollutants, and its use is mandatory in many smog-prone areas.

The other serious pollutant in smog is *carbon monoxide,* which is a colorless, odorless, poisonous gas. It is mostly emitted by motor vehicles, and it can build to dangerous levels in areas with heavy congested traffic. It deprives the body's organs from getting enough oxygen by binding with the red blood cells that would otherwise carry oxygen. At low levels, carbon monoxide decreases the amount of oxygen supplied to the brain and other organs and muscles, slows body reactions and reflexes, and impairs judgment. It poses a serious threat to people with heart disease because of the fragile condition of the circulatory system and to fetuses because of the oxygen needs of the developing brain. At high levels, it can be fatal, as evidenced by numerous deaths caused by cars that are warmed up in closed garages or by exhaust gases leaking into the cars.

Smog also contains suspended particulate matter such as dust and soot emitted by vehicles and industrial facilities. Such particles irritate the eyes and the lungs since they may carry compounds such as acids and metals.

Acid Rain

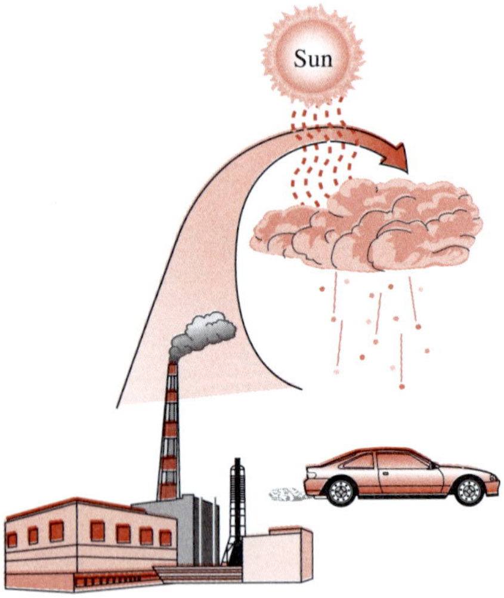

FIGURE 2–67
Sulfuric acid and nitric acid are formed when sulfur oxides and nitric oxides react with water vapor and other chemicals high in the atmosphere in the presence of sunlight.

Fossil fuels are mixtures of various chemicals, including small amounts of sulfur. The sulfur in the fuel reacts with oxygen to form sulfur dioxide (SO_2), which is an air pollutant. The main source of SO_2 is the electric power plants that burn high-sulfur coal. The Clean Air Act of 1970 has limited the SO_2 emissions severely, which forced the plants to install SO_2 scrubbers, to switch to low-sulfur coal, or to gasify the coal and recover the sulfur. Motor vehicles also contribute to SO_2 emissions since gasoline and diesel fuel also contain small amounts of sulfur. Volcanic eruptions and hot springs also release sulfur oxides (the cause of the rotten egg smell).

The sulfur oxides and nitric oxides react with water vapor and other chemicals high in the atmosphere in the presence of sunlight to form sulfuric and nitric acids (Fig. 2–67). The acids formed usually dissolve in the suspended water droplets in clouds or fog. These acid-laden droplets, which can be as acidic as lemon juice, are washed from the air on to the soil by rain or snow. This is known as **acid rain**. The soil is capable of neutralizing a certain amount of acid, but the amounts produced by the power plants using inexpensive high-sulfur coal has exceeded this capability, and as a result many lakes and rivers in industrial areas such as New York, Pennsylvania, and Michigan have become too acidic for fish to grow. Forests in those areas also experience a slow death due to absorbing the acids through their leaves, needles, and roots. Even marble structures deteriorate due to acid rain. The magnitude of the problem was not recognized until the early 1970s, and serious measures have been taken since then to reduce the sulfur dioxide emissions drastically by installing scrubbers in plants and by desulfurizing coal before combustion.

The Greenhouse Effect: Global Warming and Climate Change

You have probably noticed that when you leave your car under direct sunlight on a sunny day, the interior of the car gets much warmer than the air outside, and you may have wondered why the car acts like a heat trap. This is because glass at thicknesses encountered in practice transmits over 90 percent of radiation in the visible range and is practically opaque (nontransparent) to radiation in the longer wavelength infrared regions. Therefore, glass allows the solar radiation to enter freely but blocks the infrared radiation emitted by the interior surfaces. This causes a rise in the interior temperature as a result of the thermal energy buildup in the car. This heating effect is known as the **greenhouse effect**, since it is utilized primarily in greenhouses.

The greenhouse effect is also experienced on a larger scale on earth. The surface of the earth, which warms up during the day as a result of the absorption of solar energy, cools down at night by radiating part of its energy into deep space as infrared radiation. Carbon dioxide (CO_2), water vapor, and trace amounts of some other gases such as methane and nitrogen oxides act like a blanket and keep the earth warm at night by blocking the heat radiated from the earth (Fig. 2–68). Therefore, they are called "greenhouse gases," with CO_2 being the primary component. Water vapor is usually taken out of this list since it comes down as rain or snow as part of the water cycle and human activities in producing water (such as the burning of fossil fuels) do not make much difference on its concentration in the atmosphere (which is mostly due to evaporation from rivers, lakes, oceans, etc.). CO_2 is different, however, in that people's activities do make a difference in CO_2 concentration in the atmosphere.

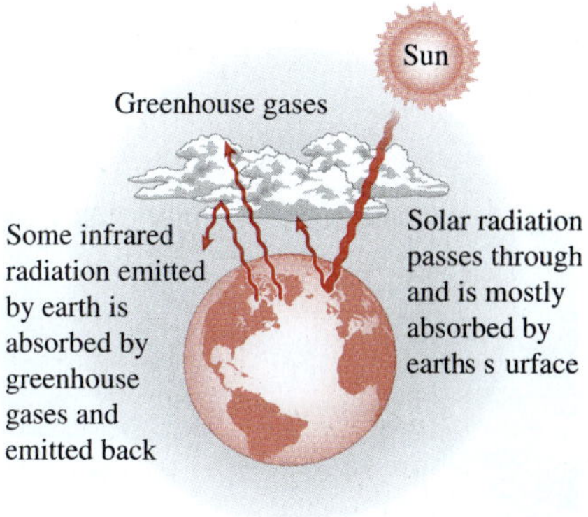

FIGURE 2–68
The greenhouse effect on earth.

The greenhouse effect makes life on earth possible by keeping the earth warm (about 30°C warmer). However, excessive amounts of these gases disturb the delicate balance by trapping too much energy, which causes the average temperature of the earth to rise and the climate at some localities to change. These undesirable consequences of the greenhouse effect are referred to as **global warming** or **global climate change**.

The global climate change is due to the excessive use of fossil fuels such as coal, petroleum products, and natural gas in electric power generation, transportation, buildings, and manufacturing, and it has been a concern in recent decades. In 1995, a total of 6.5 billion tons of carbon was released to the atmosphere as CO_2. The current concentration of CO_2 in the atmosphere is about 360 ppm (or 0.36 percent). This is 20 percent higher than the level a century ago, and it is projected to increase to over 700 ppm by the year 2100. Under normal conditions, vegetation consumes CO_2 and releases O_2 during the photosynthesis process, and thus keeps the CO_2 concentration in the atmosphere in check. A mature, growing tree consumes about 12 kg of CO_2 a year and exhales enough oxygen to support a family of four. However, deforestation and the huge increase in the CO_2 production in recent decades disturbed this balance.

In a 1995 report, the world's leading climate scientists concluded that the earth has already warmed about 0.5°C during the last century, and they estimate that the earth's temperature will rise another 2°C by the year 2100. A rise of this magnitude is feared to cause severe changes in weather patterns with storms and heavy rains and flooding at some parts and drought in others, major floods due to the melting of ice at the poles, loss of wetlands and coastal areas due to rising sea levels, variations in water supply, changes in the ecosystem due to the inability of some animal and plant species to adjust to the changes, increases in epidemic diseases due to the warmer temperatures, and adverse side effects on human health and socioeconomic conditions in some areas.

The seriousness of these threats has moved the United Nations to establish a committee on climate change. A world summit in 1992 in Rio de Janeiro, Brazil, attracted world attention to the problem. The agreement prepared by the committee in 1992 to control greenhouse gas emissions was signed by 162 nations. In the 1997 meeting in Kyoto (Japan), the world's industrialized countries adopted the Kyoto protocol and committed to reduce their CO_2 and other greenhouse gas emissions by 5 percent below the 1990 levels by 2008 to 2012. This can be done by increasing conservation efforts and improving conversion efficiencies, while meeting new energy demands by the use of renewable energy (such as hydroelectric, solar, wind, and geothermal energy) rather than by fossil fuels.

FIGURE 2–69

The average car produces several times its weight in CO_2 every year (it is driven 12,000 miles a year, consumes 600 gallons of gasoline, and produces 20 lbm of CO_2 per gallon).

© Vol. 39/PhotoDisc/Getty RF.

The United States is the largest contributor of greenhouse gases, with over 5 tons of carbon emissions per person per year. A major source of greenhouse gas emissions is transportation. Each liter of gasoline burned by a vehicle produces about 2.5 kg of CO_2 (or, each gallon of gasoline burned produces about 20 lbm of CO_2). An average car in the United States is driven about 12,000 miles a year, and it consumes about 600 gallons of gasoline. Therefore, a car emits about 12,000 lbm of CO_2 to the atmosphere a year, which is about four times the weight of a typical car (Fig. 2–69). This and other emissions can be reduced significantly by buying an energy-efficient car that burns less fuel over the same distance, and by driving sensibly. Saving fuel also saves money and the environment. For example, choosing a vehicle that gets 30 rather than 20 miles per gallon will prevent 2 tons of CO_2 from being released to the atmosphere every year while reducing the fuel cost by $400 per year (under average driving conditions of 12,000 miles a year and at a fuel cost of $2.00/gal).

It is clear from these discussions that considerable amounts of pollutants are emitted as the chemical energy in fossil fuels is converted to thermal, mechanical, or electrical energy via combustion, and thus power plants, motor vehicles, and even stoves take the blame for air pollution. In contrast, no pollution is emitted as electricity is converted to thermal, chemical, or mechanical energy, and thus electric cars are often touted as "zero emission" vehicles and their widespread use is seen by some as the ultimate solution to the air pollution problem. It should be remembered, however, that the electricity used by the electric cars is generated somewhere else mostly by burning fuel and thus emitting pollution. Therefore, each time an electric car consumes 1 kWh of electricity, it bears the responsibility for the pollutions emitted as 1 kWh of electricity (plus the conversion and transmission losses) is generated elsewhere. The electric cars can be claimed to be zero emission vehicles only when the electricity they consume is generated by emission-free renewable resources such as hydroelectric, solar, wind, and geothermal energy (Fig. 2–70). Therefore, the use of renewable energy should be encouraged worldwide, with incentives, as necessary, to make the earth a better place to live in. The advancements in thermodynamics have contributed greatly in recent decades to improve conversion efficiencies (in some cases doubling them) and thus to reduce pollution. As individuals, we can also help by practicing energy conservation measures and by making energy efficiency a high priority in our purchases.

FIGURE 2–70
Renewable energies such as wind are called "green energy" since they emit no pollutants or greenhouse gases.
© Corbis Royalty Free.

EXAMPLE 2–18 Reducing Air Pollution by Geothermal Heating

A geothermal power plant in Nevada is generating electricity using geothermal water extracted at 180°C, and reinjected back to the ground at 85°C. It is proposed to utilize the reinjected brine for heating the residential and commercial buildings in the area, and calculations show that the geothermal heating system can save 18 million therms of natural gas a year. Determine the amount of NO_x and CO_2 emissions the geothermal system will save a year. Take the average NO_x and CO_2 emissions of gas furnaces to be 0.0047 kg/therm and 6.4 kg/therm, respectively.

Solution The gas heating systems in an area are being replaced by a geothermal district heating system. The amounts of NO_x and CO_2 emissions saved per year are to be determined.

Analysis The amounts of emissions saved per year are equivalent to the amounts emitted by furnaces when 18 million therms of natural gas are burned,

$$\begin{aligned} NO_x \text{ savings} &= (NO_x \text{ emission per therm})(\text{No. of therms per year}) \\ &= (0.0047 \text{ kg/therm})(18 \times 10^6 \text{ therm/year}) \\ &= \mathbf{8.5 \times 10^4 \text{ kg/year}} \end{aligned}$$

$$\begin{aligned} CO_2 \text{ savings} &= (CO_2 \text{ emission per therm})(\text{No. of therms per year}) \\ &= (6.4 \text{ kg/therm})(18 \times 10^6 \text{ therm/year}) \\ &= \mathbf{1.2 \times 10^8 \text{ kg/year}} \end{aligned}$$

Discussion A typical car on the road generates about 8.5 kg of NO_x and 6000 kg of CO_2 a year. Therefore the environmental impact of replacing the gas heating systems in the area by the geothermal heating system is equivalent to taking 10,000 cars off the road for NO_x emission and taking 20,000 cars off the road for CO_2 emission. The proposed system should have a significant effect on reducing smog in the area.

TOPIC OF SPECIAL INTEREST* Mechanisms of Heat Transfer

Heat can be transferred in three different ways: *conduction, convection,* and *radiation.* We will give a brief description of each mode to familiarize the reader with the basic mechanisms of heat transfer. All modes of heat transfer require the existence of a temperature difference, and all modes of heat transfer are from the high-temperature medium to a lower temperature one.

Conduction is the transfer of energy from the more energetic particles of a substance to the adjacent less energetic ones as a result of interactions between the particles. Conduction can take place in solids, liquids, or gases. In gases and liquids, conduction is due to the collisions of the molecules during their random motion. In solids, it is due to the combination of vibrations of molecules in a lattice and the energy transport by free electrons. A cold canned drink in a warm room, for example, eventually warms up to the room temperature as a result of heat transfer from the room to the drink through the aluminum can by conduction (Fig. 2–71).

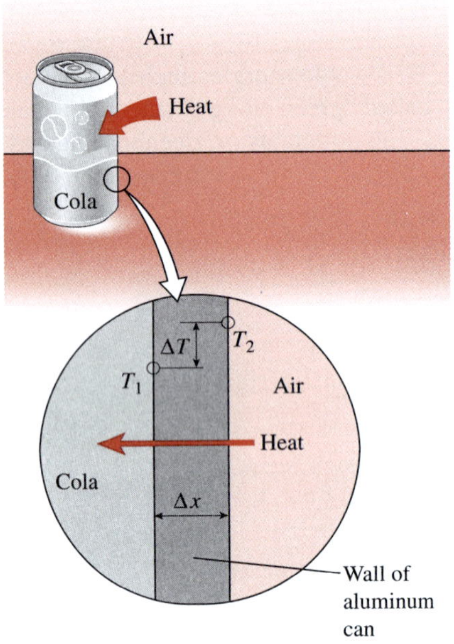

FIGURE 2–71
Heat conduction from warm air to a cold canned drink through the wall of the aluminum can.

It is observed that the rate of heat conduction $\dot{Q}_{cond}$ through a layer of constant thickness Δx is proportional to the temperature difference ΔT across the layer and the area A normal to the direction of heat transfer, and is inversely proportional to the thickness of the layer. Therefore,

$$\dot{Q}_{cond} = k_t A \frac{\Delta T}{\Delta x} \qquad \text{(W)} \tag{2–51}$$

where the constant of proportionality k_t is the **thermal conductivity** of the material, which is a measure of the ability of a material to conduct heat (Table 2–3). Materials such as copper and silver, which are good electric conductors, are also good heat conductors, and therefore have high k_t values. Materials such as rubber, wood, and styrofoam are poor conductors of heat, and therefore have low k_t values.

In the limiting case of $\Delta x \to 0$, the equation above reduces to the differential form

$$\dot{Q}_{cond} = -k_t A \frac{dT}{dx} \qquad \text{(W)} \tag{2–52}$$

which is known as **Fourier's law** of heat conduction. It indicates that the rate of heat conduction in a direction is proportional to the *temperature gradient* in that direction. Heat is conducted in the direction of decreasing temperature, and the temperature gradient becomes negative when temperature decreases with increasing x. Therefore, a negative sign is added in Eq. 2–52 to make heat transfer in the positive x direction a positive quantity.

Temperature is a measure of the kinetic energies of the molecules. In a liquid or gas, the kinetic energy of the molecules is due to the random motion of the molecules as well as the vibrational and rotational motions. When two molecules possessing different kinetic energies collide, part of the kinetic energy of the more energetic (higher temperature) molecule is transferred to the less energetic (lower temperature) particle, in much the same way as when two elastic balls of the same mass at different velocities collide, part of the kinetic energy of the faster ball is transferred to the slower one.

*This section can be skipped without a loss in continuity.

In solids, heat conduction is due to two effects: the lattice vibrational waves induced by the vibrational motions of the molecules positioned at relatively fixed position in a periodic manner called a *lattice*, and the energy transported via the free flow of electrons in the solid. The thermal conductivity of a solid is obtained by adding the lattice and the electronic components. The thermal conductivity of pure metals is primarily due to the electronic component, whereas the thermal conductivity of nonmetals is primarily due to the lattice component. The lattice component of thermal conductivity strongly depends on the way the molecules are arranged. For example, the thermal conductivity of diamond, which is a highly ordered crystalline solid, is much higher than the thermal conductivities of pure metals, as can be seen from Table 2–3.

TABLE 2–3

Thermal conductivities of some materials at room conditions

Material	Thermal conductivity, W/m·K
Diamond	2300
Silver	429
Copper	401
Gold	317
Aluminium	237
Iron	80.2
Mercury (ℓ)	8.54
Glass	1.4
Brick	0.72
Water (ℓ)	0.613
Human skin	0.37
Wood (oak)	0.17
Helium (*g*)	0.152
Soft rubber	0.13
Glass fiber	0.043
Air (*g*)	0.026
Urethane, rigid foam	0.026

Convection is the mode of energy transfer between a solid surface and the adjacent liquid or gas that is in motion, and it involves the combined effects of *conduction* and *fluid motion*. The faster the fluid motion, the greater the convection heat transfer. In the absence of any bulk fluid motion, heat transfer between a solid surface and the adjacent fluid is by pure conduction. The presence of bulk motion of the fluid enhances the heat transfer between the solid surface and the fluid, but it also complicates the determination of heat transfer rates.

Consider the cooling of a hot block by blowing of cool air over its top surface (Fig. 2–72). Energy is first transferred to the air layer adjacent to the surface of the block by conduction. This energy is then carried away from the surface by convection; that is, by the combined effects of conduction within the air, which is due to random motion of air molecules, and the bulk or macroscopic motion of the air, which removes the heated air near the surface and replaces it by the cooler air.

Convection is called **forced convection** if the fluid is *forced* to flow in a tube or over a surface by external means such as a fan, pump, or the wind. In contrast, convection is called **free** (or **natural**) **convection** if the fluid motion is caused by buoyancy forces induced by density differences due to the variation of temperature in the fluid (Fig. 2–73). For example, in the absence of a fan, heat transfer from the surface of the hot block in Fig. 2–72 will be by natural convection since any motion in the air in this case will be due to the rise of the warmer (and thus lighter) air near the surface and the fall of the cooler (and thus heavier) air to fill its place. Heat transfer between the block and surrounding air will be by conduction if the temperature difference between the air and the block is not large enough to overcome the resistance of air to move and thus to initiate natural convection currents.

Heat transfer processes that involve *change of phase* of a fluid are also considered to be convection because of the fluid motion induced during the process such as the rise of the vapor bubbles during *boiling* or the fall of the liquid droplets during *condensation*.

The rate of heat transfer by convection $\dot{Q}_{conv}$ is determined from **Newton's law of cooling**, expressed as

$$\dot{Q}_{conv} = hA(T_s - T_f) \qquad \text{(W)} \tag{2-53}$$

where h is the **convection heat transfer coefficient**, A is the surface area through which heat transfer takes place, T_s is the surface temperature, and T_f is bulk fluid temperature away from the surface. (At the surface, the fluid temperature equals the surface temperature of the solid.)

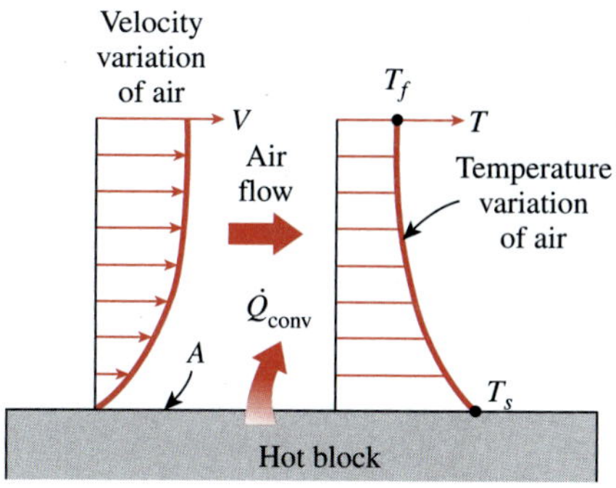

FIGURE 2–72
Heat transfer from a hot surface to air by convection.

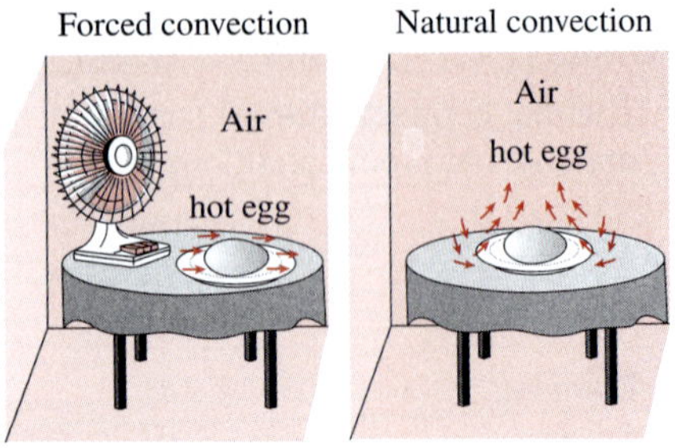

FIGURE 2–73
The cooling of a boiled egg by forced and natural convection.

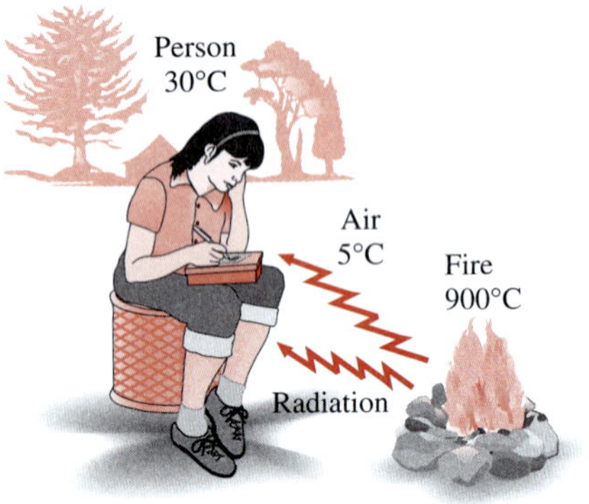

FIGURE 2–74
Unlike conduction and convection, heat transfer by radiation can occur between two bodies, even when they are separated by a medium colder than both of them.

The convection heat transfer coefficient h is not a property of the fluid. It is an experimentally determined parameter whose value depends on all the variables that influence convection such as the surface geometry, the nature of fluid motion, the properties of the fluid, and the bulk fluid velocity. Typical values of h, in W/m²·K, are in the range of 2–25 for the free convection of gases, 50–1000 for the free convection of liquids, 25–250 for the forced convection of gases, 50–20,000 for the forced convection of liquids, and 2500–100,000 for convection in boiling and condensation processes.

Radiation is the energy emitted by matter in the form of electromagnetic waves (or photons) as a result of the changes in the electronic configurations of the atoms or molecules. Unlike conduction and convection, the transfer of energy by radiation does not require the presence of an intervening medium (Fig. 2–74). In fact, energy transfer by radiation is fastest (at the speed of light) and it suffers no attenuation in a vacuum. This is exactly how the energy of the sun reaches the earth.

In heat transfer studies, we are interested in *thermal radiation,* which is the form of radiation emitted by bodies because of their temperature. It differs from other forms of electromagnetic radiation such as X-rays, gamma rays, microwaves, radio waves, and television waves that are not related to temperature. All bodies at a temperature above absolute zero emit thermal radiation.

Radiation is a *volumetric phenomenon,* and all solids, liquids, and gases emit, absorb, or transmit radiation of varying degrees. However, radiation is usually considered to be a *surface phenomenon* for solids that are opaque to thermal radiation such as metals, wood, and rocks since the radiation emitted by the interior regions of such material can never reach the surface, and the radiation incident on such bodies is usually absorbed within a few microns from the surface.

The maximum rate of radiation that can be emitted from a surface at an *absolute* temperature T_s is given by the *Stefan–Boltzmann law* as

$$\dot{Q}_{\text{emit,max}} = \sigma A T_s^4 \qquad \text{(W)} \tag{2–54}$$

where A is the surface area and $\sigma = 5.67 \times 10^{-8}$ W/m²·K⁴ is the **Stefan–Boltzmann constant**. The idealized surface that emits radiation at this maximum rate is called a **blackbody**, and the radiation emitted by a blackbody is called **blackbody radiation**. The radiation emitted by all *real* surfaces is less than the radiation emitted by a blackbody at the same temperatures and is expressed as

$$\dot{Q}_{\text{emit}} = \varepsilon \sigma A T_s^4 \qquad \text{(W)} \tag{2–55}$$

where ε is the **emissivity** of the surface. The property emissivity, whose value is in the range $0 \leq \varepsilon \leq 1$, is a measure of how closely a surface approximates a blackbody for which $\varepsilon = 1$. The emissivities of some surfaces are given in Table 2–4.

Another important radiation property of a surface is its **absorptivity**, α, which is the fraction of the radiation energy incident on a surface that is absorbed by the surface. Like emissivity, its value is in the range $0 \leq \alpha \leq 1$. A blackbody absorbs the entire radiation incident on it. That is, a blackbody is a perfect absorber ($\alpha = 1$) as well as a perfect emitter.

In general, both ε and α of a surface depend on the temperature and the wavelength of the radiation. **Kirchhoff's law** of radiation states that the emissivity and the absorptivity of a surface are equal at the same temperature and wavelength. In most practical applications, the dependence of ε and α on the temperature and wavelength is ignored, and the average absorptivity of a surface is taken to be equal to its average emissivity. The rate at which a surface absorbs radiation is determined from (Fig. 2–75)

$$\dot{Q}_{abs} = \alpha \dot{Q}_{incident} \qquad \text{(W)} \tag{2–56}$$

where $\dot{Q}_{incident}$ is the rate at which radiation is incident on the surface and α is the absorptivity of the surface. For opaque (nontransparent) surfaces, the portion of incident radiation that is not absorbed by the surface is reflected back.

The difference between the rates of radiation emitted by the surface and the radiation absorbed is the *net* radiation heat transfer. If the rate of radiation absorption is greater than the rate of radiation emission, the surface is said to be *gaining* energy by radiation. Otherwise, the surface is said to be *losing* energy by radiation. In general, the determination of the net rate of heat transfer by radiation between two surfaces is a complicated matter since it depends on the properties of the surfaces, their orientation relative to each other, and the interaction of the medium between the surfaces with radiation. However, in the special case of a relatively small surface of emissivity ε and surface area A at *absolute* temperature T_s that is completely enclosed by a much larger surface at *absolute* temperature T_{surr} separated by a gas (such as air) that does not intervene with radiation (i.e., the amount of radiation emitted, absorbed, or scattered by the medium is negligible), the net rate of radiation heat transfer between these two surfaces is determined from (Fig. 2–76)

$$\dot{Q}_{rad} = \varepsilon \sigma A (T_s^4 - T_{surr}^4) \qquad \text{(W)} \tag{2–57}$$

In this special case, the emissivity and the surface area of the surrounding surface do not have any effect on the net radiation heat transfer.

TABLE 2–4

Emissivity of some materials at 300 K

Material	Emissivity
Aluminium foil	0.07
Anodized aluminum	0.82
Polished copper	0.03
Polished gold	0.03
Polished silver	0.02
Polished stainless steel	0.17
Black paint	0.98
White paint	0.90
White paper	0.92–0.97
Asphalt pavement	0.85–0.93
Red brick	0.93–0.96
Human skin	0.95
Wood	0.82–0.92
Soil	0.93–0.96
Water	0.96
Vegetation	0.92–0.96

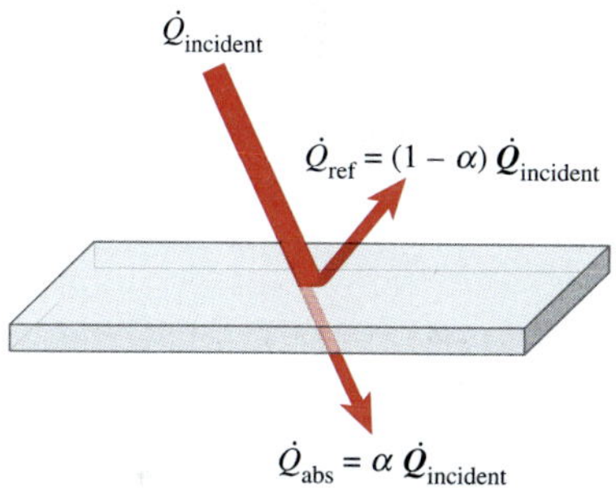

FIGURE 2–75
The absorption of radiation incident on an opaque surface of absorptivity α.

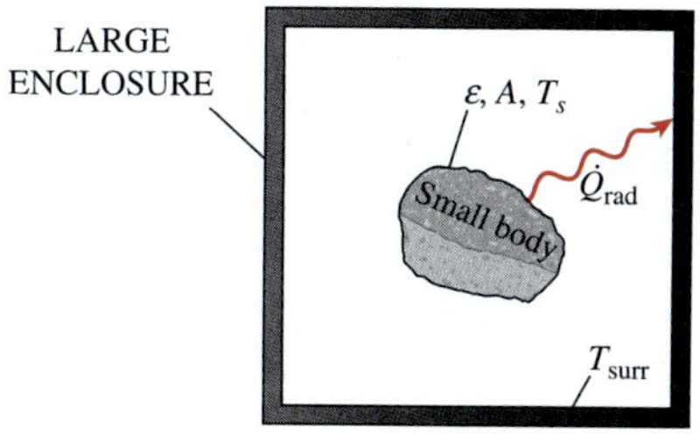

FIGURE 2–76
Radiation heat transfer between a body and the inner surfaces of a much larger enclosure that completely surrounds it.

EXAMPLE 2–19 **Heat Transfer from a Person**

Consider a person standing in a breezy room at 20°C. Determine the total rate of heat transfer from this person if the exposed surface area and the average outer surface temperature of the person are 1.6 m^2 and 29°C, respectively, and the convection heat transfer coefficient is 6 W/m^2·°C (Fig. 2–77).

Solution A person is standing in a breezy room. The total rate of heat loss from the person is to be determined.
Assumptions **1** The emissivity and heat transfer coefficient are constant and uniform. **2** Heat conduction through the feet is negligible. **3** Heat loss by evaporation is disregarded.
Analysis The heat transfer between the person and the air in the room will be by convection (instead of conduction) since it is conceivable that the air in the vicinity of the skin or clothing will warm up and rise as a result of heat transfer from the body, initiating natural convection currents. It appears

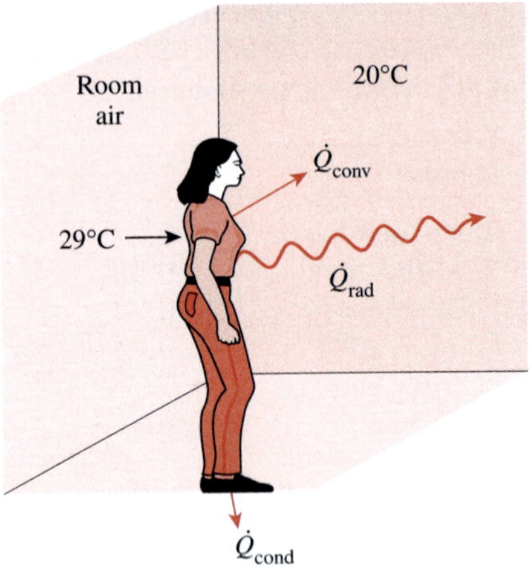

FIGURE 2–77
Heat transfer from the person described in Example 2–19.

that the experimentally determined value for the rate of convection heat transfer in this case is 6 W per unit surface area (m^2) per unit temperature difference (in K or °C) between the person and the air away from the person. Thus, the rate of convection heat transfer from the person to the air in the room is, from Eq. 2–53,

$$\begin{aligned}\dot{Q}_{conv} &= hA(T_s - T_f)\\ &= (6\ \text{W/m}^2\cdot°\text{C})(1.6\ \text{m}^2)(29 - 20)\ °\text{C}\\ &= 86.4\ \text{W}\end{aligned}$$

The person will also lose heat by radiation to the surrounding wall surfaces. We take the temperature of the surfaces of the walls, ceiling, and the floor to be equal to the air temperature in this case for simplicity, but we recognize that this does not need to be the case. These surfaces may be at a higher or lower temperature than the average temperature of the room air, depending on the outdoor conditions and the structure of the walls. Considering that air does not intervene with radiation and the person is completely enclosed by the surrounding surfaces, the net rate of radiation heat transfer from the person to the surrounding walls, ceiling, and the floor is, from Eq. 2–57,

$$\begin{aligned}\dot{Q}_{rad} &= \varepsilon\sigma A(T_s^4 - T_{surr}^4)\\ &= (0.95)(5.67 \times 10^{-8}\ \text{W/m}^2\cdot\text{K}^4)(1.6\ \text{m}^2) \times [(29 + 273)^4 - (20 + 273)^4]\text{K}^4\\ &= 81.7\ \text{W}\end{aligned}$$

Note that we must use *absolute* temperatures in radiation calculations. Also note that we used the emissivity value for the skin and clothing at room temperature since the emissivity is not expected to change significantly at a slightly higher temperature.

Then the rate of total heat transfer from the body is determined by adding these two quantities to be

$$\dot{Q}_{total} = \dot{Q}_{conv} + \dot{Q}_{rad} = 86.4 + 81.7 = \mathbf{168.1\ W}$$

The heat transfer would be much higher if the person were not dressed since the exposed surface temperature would be higher. Thus, an important function of the clothes is to serve as a barrier against heat transfer.

Discussion In the above calculations, heat transfer through the feet to the floor by conduction, which is usually very small, is neglected. Heat transfer from the skin by perspiration, which is the dominant mode of heat transfer in hot environments, is not considered here.

SUMMARY

The sum of all forms of energy of a system is called *total energy,* which consists of internal, kinetic, and potential energy for simple compressible systems. *Internal energy* represents the molecular energy of a system and may exist in sensible, latent, chemical, and nuclear forms.

Mass flow rate $\dot{m}$ is defined as the amount of mass flowing through a cross section per unit time. It is related to the *volume flow rate* $\dot{V}$, which is the volume of a fluid flowing through a cross section per unit time, by

$$\dot{m} = \rho\dot{V} = \rho A_c V_{avg}$$

The energy flow rate associated with a fluid flowing at a rate of $\dot{m}$ is

$$\dot{E} = \dot{m}e$$

which is analogous to $E = me$.

The *mechanical energy* is defined as *the form of energy that can be converted to mechanical work completely and directly by a mechanical device such as an ideal turbine*. It is expressed on a unit mass basis and rate form as

$$e_{\text{mech}} = \frac{P}{\rho} + \frac{V^2}{2} + gz$$

and

$$\dot{E}_{\text{mech}} = \dot{m}e_{\text{mech}} = \dot{m}\left(\frac{P}{\rho} + \frac{V^2}{2} + gz\right)$$

where P/ρ is the *flow energy*, $V^2/2$ is the *kinetic energy*, and gz is the *potential energy* of the fluid per unit mass.

Energy can cross the boundaries of a closed system in the form of heat or work. For control volumes, energy can also be transported by mass. If the energy transfer is due to a temperature difference between a closed system and its surroundings, it is *heat;* otherwise, it is *work.*

Work is the energy transferred as a force acts on a system through a distance. Various forms of work are expressed as follows:

Electrical work: $W_e = \mathbf{V}I\,\Delta t$

Shaft work: $W_{\text{sh}} = 2\pi n \mathrm{T}$

Spring work: $W_{\text{spring}} = \frac{1}{2}k(x_2^2 - x_1^2)$

The *first law of thermodynamics* is essentially an expression of the conservation of energy principle, also called the *energy balance*. The general mass and energy balances for *any system* undergoing *any process* can be expressed as

$$\underbrace{E_{\text{in}} - E_{\text{out}}}_{\substack{\text{Net energy transfer}\\ \text{by heat, work, and mass}}} = \underbrace{\Delta E_{\text{system}}}_{\substack{\text{Change in internal, kinetic,}\\ \text{potential, etc., energies}}} \quad (\text{kJ})$$

It can also be expressed in the *rate form* as

$$\underbrace{\dot{E}_{\text{in}} - \dot{E}_{\text{out}}}_{\substack{\text{Rate of net energy transfer}\\ \text{by heat, work, and mass}}} = \underbrace{dE_{\text{system}}/dt}_{\substack{\text{Rate of change in internal,}\\ \text{kinetic, potential, etc., energies}}} \quad (\text{kW})$$

The efficiencies of various devices are defined as

$$\eta_{\text{pump}} = \frac{\Delta\dot{E}_{\text{mech,fluid}}}{\dot{W}_{\text{shaft,in}}} = \frac{\dot{W}_{\text{pump},u}}{\dot{W}_{\text{pump}}}$$

$$\eta_{\text{turbine}} = \frac{\dot{W}_{\text{shaft,out}}}{|\Delta\dot{E}_{\text{mech,fluid}}|} = \frac{\dot{W}_{\text{turbine}}}{\dot{W}_{\text{turbine},e}}$$

$$\eta_{\text{motor}} = \frac{\text{Mechanical power output}}{\text{Electric power input}} = \frac{\dot{W}_{\text{shaft,out}}}{\dot{W}_{\text{elect,in}}}$$

$$\eta_{\text{generator}} = \frac{\text{Electric power output}}{\text{Mechanical power input}} = \frac{\dot{W}_{\text{elect,out}}}{\dot{W}_{\text{shaft,in}}}$$

$$\eta_{\text{pump-motor}} = \eta_{\text{pump}}\eta_{\text{motor}} = \frac{\Delta\dot{E}_{\text{mech,fluid}}}{\dot{W}_{\text{elect,in}}}$$

$$\eta_{\text{turbine-gen}} = \eta_{\text{turbine}}\eta_{\text{generator}} = \frac{\dot{W}_{\text{elect,out}}}{|\Delta\dot{E}_{\text{mech,fluid}}|}$$

The conversion of energy from one form to another is often associated with adverse effects on the environment, and environmental impact should be an important consideration in the conversion and utilization of energy.

REFERENCES AND SUGGESTED READINGS

1. ASHRAE *Handbook of Fundamentals*. SI version. Atlanta, GA: American Society of Heating, Refrigerating, and Air-Conditioning Engineers, Inc., 1993.
2. Y. A. Çengel. "An Intuitive and Unified Approach to Teaching Thermodynamics." ASME International Mechanical Engineering Congress and Exposition, Atlanta, Georgia, AES-Vol. 36, pp. 251–260, November 17–22, 1996.

PROBLEMS*

Forms of Energy

2–1C What is the difference between the macroscopic and microscopic forms of energy?

2–2C What is total energy? Identify the different forms of energy that constitute the total energy.

2–3C How are heat, internal energy, and thermal energy related to each other?

2–4C What is mechanical energy? How does it differ from thermal energy? What are the forms of mechanical energy of a fluid stream?

2–5C Natural gas, which is mostly methane CH_4, is a fuel and a major energy source. Can we say the same about hydrogen gas, H_2?

2–6E Calculate the total kinetic energy, in Btu, of an object with a mass of 15 lbm when its velocity is 100 ft/s. *Answer:* 3.0 Btu

2–7 Calculate the total kinetic energy, in kJ, of an object whose mass is 100 kg and whose velocity is 20 m/s.

2–8E The specific potential energy of an object with respect to some datum level is given by gz where g is the local gravitational acceleration and z is the elevation of the object above the datum. Determine the specific potential energy, in Btu/lbm, of an object elevated 100 ft above a datum at a location where $g = 32.1$ ft/s^2.

2–9E Calculate the total potential energy, in Btu, of an object with a mass of 200 lbm when it is 10 ft above a datum level at a location where standard gravitational acceleration exists.

2–10 Calculate the total potential energy, in kJ, of an object whose mass is 20 kg when it is located 20 m below a datum level in a location where $g = 9.5$ m/s^2.

2–11 A person gets into an elevator at the lobby level of a hotel together with his 30-kg suitcase, and gets out at the 10th floor 35 m above. Determine the amount of energy consumed by the motor of the elevator that is now stored in the suitcase.

2–12 Electric power is to be generated by installing a hydraulic turbine–generator at a site 160 m below the free surface of a large water reservoir that can supply water at a rate of 3500 kg/s steadily. Determine the power generation potential.

2–13 At a certain location, wind is blowing steadily at 10 m/s. Determine the mechanical energy of air per unit mass and the power generation potential of a wind turbine with 60-m-diameter blades at that location. Take the air density to be 1.25 kg/m^3.

2–14 A water jet that leaves a nozzle at 60 m/s at a flow rate of 120 kg/s is to be used to generate power by striking the buckets located on the perimeter of a wheel. Determine the power generation potential of this water jet.

2–15 Two sites are being considered for wind power generation. In the first site, the wind blows steadily at 7 m/s for 3000 hours per year, whereas in the second site the wind blows at 10 m/s for 2000 hours per year. Assuming the wind velocity is negligible at other times for simplicity, determine which is a better site for wind power generation. *Hint:* Note that the mass flow rate of air is proportional to wind velocity.

2–16 A river flowing steadily at a rate of 175 m^3/s is considered for hydroelectric power generation. It is determined that a dam can be built to collect water and release it from an elevation difference of 80 m to generate power. Determine how much power can be generated from this river water after the dam is filled.

2–17 Consider a river flowing toward a lake at an average velocity of 3 m/s at a rate of 500 m^3/s at a location 90 m above the lake surface. Determine the total mechanical energy of the river water per unit mass and the power generation potential of the entire river at that location.

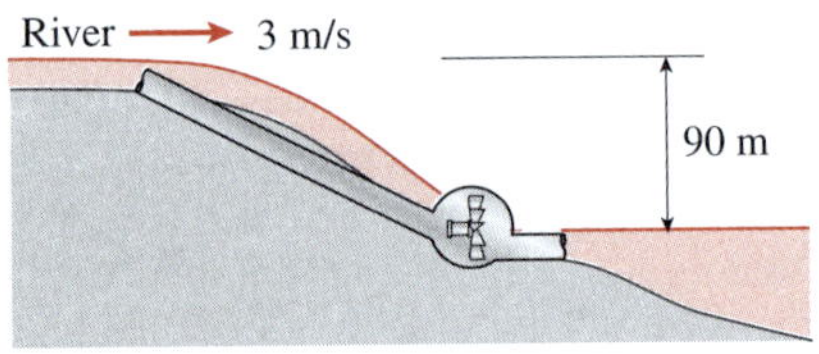

FIGURE P2–17

Energy Transfer by Heat and Work

2–18C In what forms can energy cross the boundaries of a closed system?

2–19C When is the energy crossing the boundaries of a closed system heat and when is it work?

2–20C What is an adiabatic process? What is an adiabatic system?

2–21C What are point and path functions? Give some examples.

* Problems designated by a "C" are concept questions, and students are encouraged to answer them all. Problems designated by an "E" are in English units, and the SI users can ignore them. Problems with the icon are solved using EES, and complete solutions together with parametric studies are included on the enclosed DVD. Problems with the icon are comprehensive in nature, and are intended to be solved with a computer, preferably using the EES software that accompanies this text.

2–22C Consider an automobile traveling at a constant speed along a road. Determine the direction of the heat and work interactions, taking the following as the system: (*a*) the car radiator, (*b*) the car engine, (*c*) the car wheels, (*d*) the road, and (*e*) the air surrounding the car.

2–23C The length of a spring can be changed by (*a*) applying a force to it or (*b*) changing its temperature (i.e., thermal expansion). What type of energy interaction between the system (spring) and surroundings is required to change the length of the spring in these two ways?

2–24C Consider an electric refrigerator located in a room. Determine the direction of the work and heat interactions (in or out) when the following are taken as the system: (*a*) the contents of the refrigerator, (*b*) all parts of the refrigerator including the contents, and (*c*) everything contained within the room during a winter day.

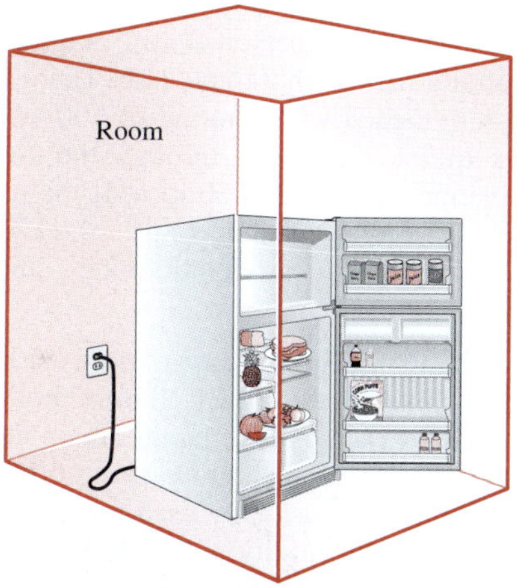

FIGURE P2–24C

2–25C A personal computer is to be examined from a thermodynamic perspective. Determine the direction of the work and heat transfers (in or out) when the (*a*) keyboard, (*b*) monitor, (*c*) processing unit, and (*d*) all of these are taken as the system.

2–26 A small electrical motor produces 5 W of mechanical power. What is this power in (*a*) N, m, and s units; and (*b*) kg, m, and s units? *Answers:* (*a*) 5 N·m/s, (*b*) 5 kg·m^2/s^3

2–27E A model aircraft internal-combustion engine produces 10 W of power. How much power is this in (*a*) lbf·ft/s and (*b*) hp?

Mechanical Forms of Work

2–28C A car is accelerated from rest to 85 km/h in 10 s. Would the energy transferred to the car be different if it were accelerated to the same speed in 5 s?

2–29 Determine the energy required to accelerate an 800-kg car from rest to 100 km/h on a level road. *Answer:* 309 kJ

2–30E A construction crane lifts a prestressed concrete beam weighing 3 tons from the ground to the top of piers that are 24 ft above the ground. Determine the amount of work done considering (*a*) the beam and (*b*) the crane as the system. Express your answers in both lbf·ft and Btu.

2–31E A man weighing 180 lbf is pushing a cart that weighs 100 lbf with its contents up a ramp that is inclined at an angle of 10° from the horizontal. Determine the work needed to move along this ramp a distance of 100 ft considering (*a*) the man and (*b*) the cart and its contents as the system. Express your answers in both lbf·ft and Btu.

FIGURE P2–31E

2–32E The force F required to compress a spring a distance x is given by $F - F_0 = kx$ where k is the spring constant and F_0 is the preload. Determine the work required to compress a spring whose spring constant is $k = 200$ lbf/in a distance of one inch starting from its free length where $F_0 = 0$ lbf. Express your answer in both lbf·ft and Btu.

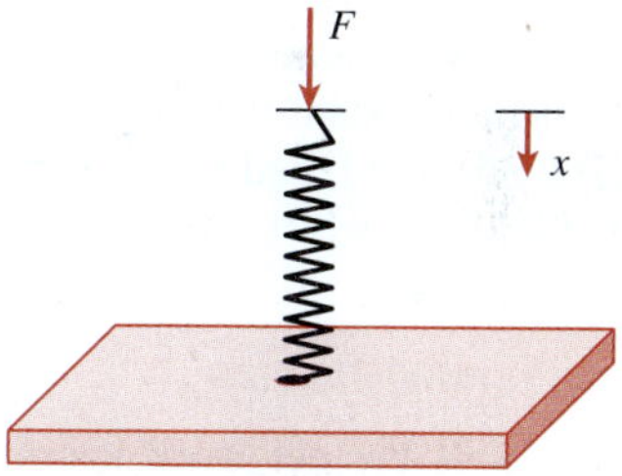

FIGURE P2–32E

2–33E A spherical soap bubble with a surface-tension of 0.005 lbf/ft is expanded from a diameter of 0.5 in to 2.0 in. How much work, in Btu, is required to expand this bubble? *Answer:* 1.05×10^{-6} Btu

2–34E A 0.5-in diameter, 12-in-long steel rod with a Young's modulus of 30,000 lbf/in^2 is stretched 0.125 in. How much work does this require, in Btu? The strain work is given by $W = \frac{V_0 E}{2}(\varepsilon_2^2 - \varepsilon_1^2)$ where V_0 is the original volume of the solid, E is Young's modulus and ε is the strain at the beginning and ending of the process.

2–35E A spring whose spring constant is 200 lbf/in has an initial force of 100 lbf acting on it. Determine the work, in Btu, required to compress it another 1 inch.

2–36 How much work, in kJ, can a spring whose spring constant is 3 kN/cm produce after it has been compressed 3 cm from its unloaded length?

2–37 A ski lift has a one-way length of 1 km and a vertical rise of 200 m. The chairs are spaced 20 m apart, and each chair can seat three people. The lift is operating at a steady speed of 10 km/h. Neglecting friction and air drag and assuming that the average mass of each loaded chair is 250 kg, determine the power required to operate this ski lift. Also estimate the power required to accelerate this ski lift in 5 s to its operating speed when it is first turned on.

2–38 Determine the power required for a 1150-kg car to climb a 100-m-long uphill road with a slope of 30° (from horizontal) in 12 s (*a*) at a constant velocity, (*b*) from rest to a final velocity of 30 m/s, and (*c*) from 35 m/s to a final velocity of 5 m/s. Disregard friction, air drag, and rolling resistance. *Answers:* (*a*) 47.0 kW, (*b*) 90.1 kW, (*c*) −10.5 kW

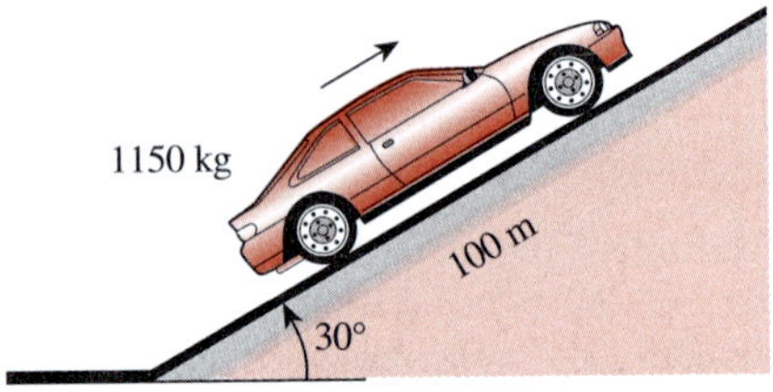

FIGURE P2–38

2–39 A damaged 1200-kg car is being towed by a truck. Neglecting the friction, air drag, and rolling resistance, determine the extra power required (*a*) for constant velocity on a level road, (*b*) for constant velocity of 50 km/h on a 30° (from horizontal) uphill road, and (*c*) to accelerate on a level road from stop to 90 km/h in 12 s. *Answers:* (*a*) 0, (*b*) 81.7 kW, (*c*) 31.3 kW

The First Law of Thermodynamics

2–40C For a cycle, is the net work necessarily zero? For what kind of systems will this be the case?

2–41C What are the different mechanisms for transferring energy to or from a control volume?

2–42C On a hot summer day, a student turns his fan on when he leaves his room in the morning. When he returns in the evening, will the room be warmer or cooler than the neighboring rooms? Why? Assume all the doors and windows are kept closed.

2–43E One way to improve the fuel efficiency of a car is to use tires that have a lower rolling resistance—tires that roll with less resistance, and highway tests at 65 mph showed that tires with the lowest rolling resistance can improve the fuel efficiency by nearly 2 mpg (miles per gallon). Consider a car that gets 35 mpg on high rolling resistance tires and is driven 15,000 miles per year. For a fuel cost of $2.20/gal, determine how much money will be saved per year by switching to low rolling resistance tires.

2–44 An adiabatic closed system is accelerated from 0 m/s to 30 m/s. Determine the specific energy change of this system, in kJ/kg.

2–45 An adiabatic closed system is raised 100 m at a location where the gravitational acceleration is 9.8 m/s^2. Determine the energy change of this system, in kJ/kg.

2–46E A water pump increases the water pressure from 10 psia to 50 psia. Determine the power input required, in hp, to pump 1.2 ft^3/s of water. Does the water temperature at the inlet have any significant effect on the required flow power? *Answer:* 12.6 hp

2–47 A classroom that normally contains 40 people is to be air-conditioned with window air-conditioning units of 5-kW cooling capacity. A person at rest may be assumed to dissipate heat at a rate of about 360 kJ/h. There are 10 lightbulbs in the room, each with a rating of 100 W. The rate of heat transfer to the classroom through the walls and the windows is estimated to be 15,000 kJ/h. If the room air is to be maintained at a constant temperature of 21°C, determine the number of window air-conditioning units required. *Answer:* 2 units

2–48 The lighting needs of a storage room are being met by 6 fluorescent light fixtures, each fixture containing four lamps rated at 60 W each. All the lamps are on during operating hours of the facility, which are 6 AM to 6 PM 365 days a year. The storage room is actually used for an average of 3 h a day. If the price of electricity is $0.08/kWh, determine the amount of energy and money that will be saved as a result of installing motion sensors. Also, determine the simple payback period if the purchase price of the sensor is $32 and it takes 1 hour to install it at a cost of $40.

2–49 A university campus has 200 classrooms and 400 faculty offices. The classrooms are equipped with 12 fluorescent tubes, each consuming 110 W, including the electricity used by the ballasts. The faculty offices, on average, have half as many tubes. The campus is open 240 days a year. The classrooms and faculty offices are not occupied an average of 4 h a day, but the lights are kept on. If the unit cost of electricity is $0.082/kWh, determine how much the campus will save a year if the lights in the classrooms and faculty offices are turned off during unoccupied periods.

2–50 Consider a room that is initially at the outdoor temperature of 20°C. The room contains a 100-W lightbulb, a 110-W TV set, a 200-W refrigerator, and a 1000-W iron. Assuming no heat transfer through the walls, determine the rate of increase of the energy content of the room when all of these electric devices are on.

2–51 A fan is to accelerate quiescent air to a velocity of 8 m/s at a rate of 9 m^3/s. Determine the minimum power that must be supplied to the fan. Take the density of air to be 1.18 kg/m^3. *Answer: 340 W*

2–52E Consider a fan located in a 3 ft × 3 ft square duct. Velocities at various points at the outlet are measured, and the average flow velocity is determined to be 22 ft/s. Taking the air density to 0.075 lbm/ft^3, estimate the minimum electric power consumption of the fan motor.

2–53 The driving force for fluid flow is the pressure difference, and a pump operates by raising the pressure of a fluid (by converting the mechanical shaft work to flow energy). A gasoline pump is measured to consume 3.8 kW of electric power when operating. If the pressure differential between the outlet and inlet of the pump is measured to be 7 kPa and the changes in velocity and elevation are negligible, determine the maximum possible volume flow rate of gasoline.

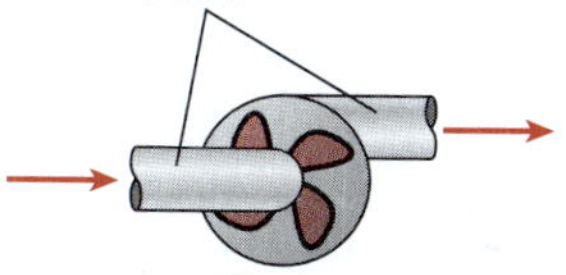

FIGURE P2–53

2–54 An escalator in a shopping center is designed to move 30 people, 75 kg each, at a constant speed of 0.8 m/s at 45° slope. Determine the minimum power input needed to drive this escalator. What would your answer be if the escalator velocity were to be doubled?

2–55 An automobile moving through the air causes the air velocity (measured with respect to the car) to decrease and fill a larger flow channel. An automobile has an effective flow channel area of 3 m^2. The car is traveling at 90 km/h on a day when the barometric pressure is 70 cm of mercury and the temperature is 20°C. Behind the car, the air velocity (with respect to the car) is measured to be 82 km/h, and the temperature is 20°C. Determine the power required to move this car through the air and the area of the effective flow channel behind the car.

FIGURE P2–55

Energy Conversion Efficiencies

2–56C What is mechanical efficiency? What does a mechanical efficiency of 100 percent mean for a hydraulic turbine?

2–57C How is the combined pump–motor efficiency of a pump and motor system defined? Can the combined pump–motor efficiency be greater than either the pump or the motor efficiency?

2–58C Define turbine efficiency, generator efficiency, and combined turbine–generator efficiency.

2–59C Can the combined turbine-generator efficiency be greater than either the turbine efficiency or the generator efficiency? Explain.

2–60 Consider a 24-kW hooded electric open burner in an area where the unit costs of electricity and natural gas are \$0.10/kWh and \$1.20/therm (1 therm = 105,500 kJ), respectively. The efficiency of open burners can be taken to be 73 percent for electric burners and 38 percent for gas burners. Determine the rate of energy consumption and the unit cost of utilized energy for both electric and gas burners.

2–61 A 75-hp (shaft output) motor that has an efficiency of 91.0 percent is worn out and is replaced by a high-efficiency 75-hp motor that has an efficiency of 95.4 percent. Determine the reduction in the heat gain of the room due to higher efficiency under full-load conditions.

2–62 A 90-hp (shaft output) electric car is powered by an electric motor mounted in the engine compartment. If the motor has an average efficiency of 91 percent, determine the rate of heat supply by the motor to the engine compartment at full load.

2–63 A 75-hp (shaft output) motor that has an efficiency of 91.0 percent is worn out and is to be replaced by a high-efficiency motor that has an efficiency of 95.4 percent. The motor operates 4368 hours a year at a load factor of 0.75. Taking the cost of electricity to be \$0.08/kWh, determine the amount of energy and money saved as a result of installing the high-efficiency motor instead of the standard motor. Also, determine the simple payback period if the purchase prices of the standard and high-efficiency motors are \$5449 and \$5520, respectively.

2–64E The steam requirements of a manufacturing facility are being met by a boiler whose rated heat input is 5.5×10^6 Btu/h. The combustion efficiency of the boiler is measured to be 0.7 by a hand-held flue gas analyzer. After tuning up the boiler, the combustion efficiency rises to 0.8. The boiler operates 4200 hours a year intermittently. Taking the unit cost of energy to be \$4.35/10^6 Btu, determine the annual energy and cost savings as a result of tuning up the boiler.

2–65E Reconsider Prob. 2–64E. Using EES (or other) software, study the effects of the unit cost of energy, the new combustion efficiency on the annual energy, and cost savings. Let the efficiency vary from 0.7 to 0.9, and the unit cost to vary from \$4 to \$6 per million Btu. Plot the annual energy and cost savings against the efficiency for unit costs of \$4, \$5, and \$6 per million Btu, and discuss the results.

2–66 An exercise room has eight weight-lifting machines that have no motors and four treadmills each equipped with a 2.5-hp (shaft output) motor. The motors operate at an average load factor of 0.7, at which their efficiency is 0.77. During peak evening hours, all 12 pieces of exercising equipment are used continuously, and there are also two people doing light exercises while waiting in line for one piece of the equipment. Assuming the average rate of heat dissipation from people in an exercise room is 525 W, determine the rate of heat gain of the exercise room from people and the equipment at peak load conditions.

2–67 A room is cooled by circulating chilled water through a heat exchanger located in a room. The air is circulated through the heat exchanger by a 0.25-hp (shaft output) fan. Typical efficiency of small electric motors driving 0.25-hp equipment is 54 percent. Determine the rate of heat supply by the fan–motor assembly to the room.

2–68 The water in a large lake is to be used to generate electricity by the installation of a hydraulic turbine-generator at a location where the depth of the water is 50 m (Fig. 2–62). Water is to be supplied at a rate of 5000 kg/s. If the electric power generated is measured to be 1862 kW and the generator efficiency is 95 percent, determine (*a*) the overall efficiency of the turbine—generator, (*b*) the mechanical efficiency of the turbine, and (*c*) the shaft power supplied by he turbine to the generator.

2–69 At a certain location, wind is blowing steadily at 7 m/s. Determine the mechanical energy of air per unit mass and the power generation potential of a wind turbine with 80-m-diameter blades at that location. Also determine the actual electric power generation assuming an overall efficiency of 30 percent. Take the air density to be 1.25 kg/m^3.

2–70 [EES] Reconsider Prob. 2–69. Using EES (or other) software, investigate the effect of wind velocity and the blade span diameter on wind power generation. Let the velocity vary from 5 to 20 m/s in increments of 5 m/s, and the diameter vary from 20 to 120 m in increments of 20 m. Tabulate the results, and discuss their significance.

2–71 Water is pumped from a lake to a storage tank 20 m above at a rate of 70 L/s while consuming 20.4 kW of electric power. Disregarding any frictional losses in the pipes and any changes in kinetic energy, determine (*a*) the overall efficiency of the pump–motor unit and (*b*) the pressure difference between the inlet and the exit of the pump.

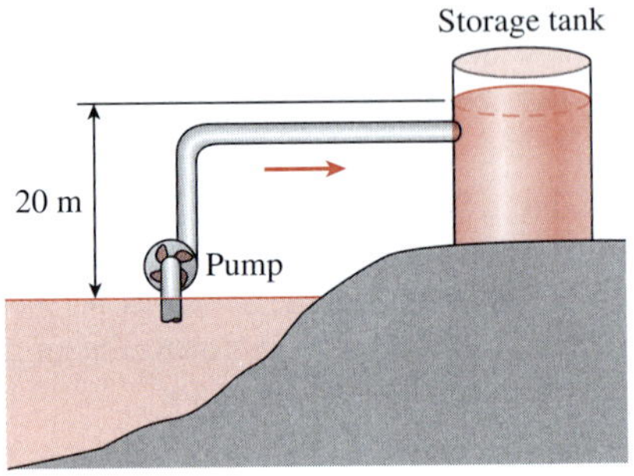

FIGURE P2–71

2–72 Large wind turbines with blade span diameters of over 100 m are available for electric power generation. Consider a wind turbine with a blade span diameter of 100 m installed at a site subjected to steady winds at 8 m/s. Taking the overall efficiency of the wind turbine to be 32 percent and the air density to be 1.25 kg/m^3, determine the electric power generated by this wind turbine. Also, assuming steady winds of 8 m/s during a 24-hour period, determine the amount of electric energy and the revenue generated per day for a unit price of \$0.06/kWh for electricity.

2–73E A water pump delivers 6 hp of shaft power when operating. If the pressure differential between the outlet and the inlet of the pump is measured to be 1.2 psi when the flow rate is 15 ft^3/s and the changes in velocity and elevation are negligible, determine the mechanical efficiency of this pump.

2–74 Water is pumped from a lower reservoir to a higher reservoir by a pump that provides 20 kW of shaft power. The free surface of the upper reservoir is 45 m higher than that of the lower reservoir. If the flow rate of water is measured to be 0.03 m^3/s, determine mechanical power that is converted to thermal energy during this process due to frictional effects.

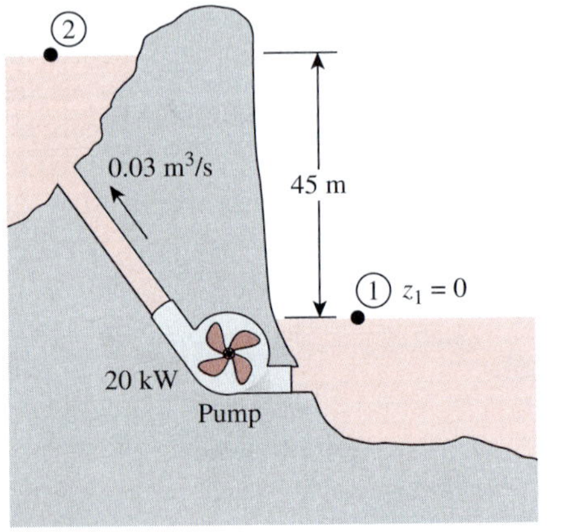

FIGURE P2–74

2–75 The water behind Hoover Dam in Nevada is 206 m higher than the Colorado River below it. At what rate must water pass through the hydraulic turbines of this dam to produce 100 MW of power if the turbines are 100 percent efficient?

FIGURE P2–75

Photo by Lynn Betts, USDA Natural Resources Conservation Service.

2–76 An oil pump is drawing 35 kW of electric power while pumping oil with ρ = 860 kg/m^3 at a rate of 0.1 m^3/s. The inlet and outlet diameters of the pipe are 8 cm and 12 cm, respectively. If the pressure rise of oil in the pump is measured to be 400 kPa and the motor efficiency is 90 percent, determine the mechanical efficiency of the pump.

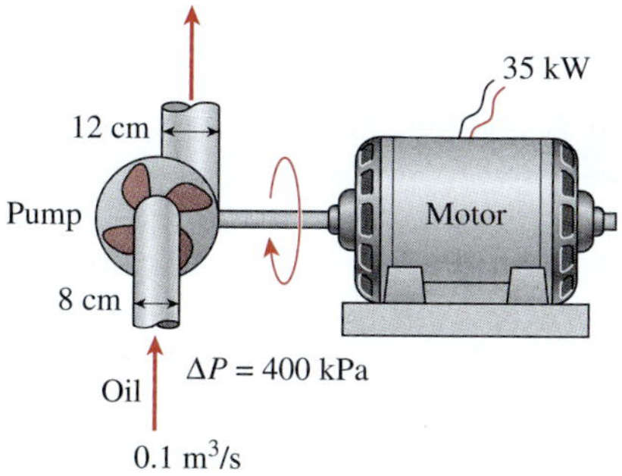

FIGURE P2–76

2–77E An 80-percent efficient pump with a power input of 20 hp is pumping water from a lake to a nearby pool at a rate of 1.5 ft^3/s through a constant-diameter pipe. The free surface of the pool is 80 ft above that of the lake. Determine the mechanical power used to overcome frictional effects in piping. *Answer:* 2.37 hp

2–78 A wind turbine is rotating at 15 rpm under steady winds flowing through the turbine at a rate of 42,000 kg/s. The tip velocity of the turbine blade is measured to be 250 km/h. If 180 kW power is produced by the turbine, determine (*a*) the average velocity of the air and (*b*) the conversion efficiency of the turbine. Take the density of air to be 1.31 kg/m^3.

Energy and Environment

2–79C How does energy conversion affect the environment? What are the primary chemicals that pollute the air? What is the primary source of these pollutants?

2–80C What is smog? What does it consist of? How does ground-level ozone form? What are the adverse effects of ozone on human health?

2–81C What is acid rain? Why is it called a "rain"? How do the acids form in the atmosphere? What are the adverse effects of acid rain on the environment?

2–82C Why is carbon monoxide a dangerous air pollutant? How does it affect human health at low and at high levels?

2–83C What is the greenhouse effect? How does the excess CO_2 gas in the atmosphere cause the greenhouse effect? What are the potential long-term consequences of greenhouse effect? How can we combat this problem?

2–84E A Ford Taurus driven 15,000 miles a year will use about 715 gallons of gasoline compared to a Ford Explorer that would use 940 gallons. About 19.7 lbm of CO_2, which causes global warming, is released to the atmosphere when a gallon of gasoline is burned. Determine the extra amount of CO_2 production a man is responsible for during a 5-year period if he trades his Taurus for an Explorer.

2–85 When a hydrocarbon fuel is burned, almost all of the carbon in the fuel burns completely to form CO_2 (carbon dioxide), which is the principal gas causing the greenhouse effect and thus global climate change. On average, 0.59 kg of CO_2 is produced for each kWh of electricity generated from a power plant that burns natural gas. A typical new household refrigerator uses about 700 kWh of electricity per year. Determine the amount of CO_2 production that is due to the refrigerators in a city with 300,000 households.

2–86 Repeat Prob. 2–85 assuming the electricity is produced by a power plant that burns coal. The average production of CO_2 in this case is 1.1 kg per kWh.

2–87E Consider a household that uses 11,000 kWh of electricity per year and 1500 gallons of fuel oil during a heating season. The average amount of CO_2 produced is 26.4 lbm/gallon of fuel oil and 1.54 lbm/kWh of electricity. If this household reduces its oil and electricity usage by 15 percent as a result of implementing some energy conservation measures, determine the reduction in the amount of CO_2 emissions by that household per year.

2–88 A typical car driven 20,000 km a year emits to the atmosphere about 11 kg per year of NO_x (nitrogen oxides), which cause smog in major population areas. Natural gas burned in the furnace emits about 4.3 g of NO_x per therm (1 therm = 105,500 kJ), and the electric power plants emit about 7.1 g of NO_x per kWh of electricity produced. Consider a household that has two cars and consumes 9000 kWh of electricity and 1200 therms of natural gas. Determine the amount of NO_x emission to the atmosphere per year for which this household is responsible.

FIGURE P2–88

Special Topic: Mechanisms of Heat Transfer

2–89C What are the mechanisms of heat transfer?

2–90C Which is a better heat conductor, diamond or silver?

2–91C Does any of the energy of the sun reach the earth by conduction or convection?

2–92C How does forced convection differ from natural convection?

2–93C What is blackbody? How do real bodies differ from a blackbody?

2–94C Define emissivity and absorptivity. What is Kirchhoff's law of radiation?

2–95 The inner and outer surfaces of a 5-m × 6-m brick wall of thickness 30 cm and thermal conductivity 0.69 W/m·°C are maintained at temperatures of 20°C and 5°C, respectively. Determine the rate of heat transfer through the wall, in W.

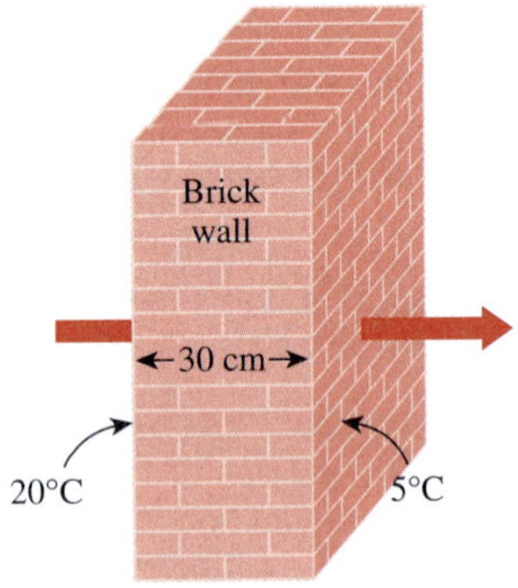

FIGURE P2–95

2–96 The inner and outer surfaces of a 0.5-cm-thick 2-m × 2-m window glass in winter are 15°C and 6°C, respectively. If the thermal conductivity of the glass is 0.78 W/m·°C, determine the amount of heat loss, in kJ, through the glass over a period of 10 h. What would your answer be if the glass were 1-cm thick?

2–97 [EES] Reconsider Prob. 2–96. Using EES (or other) software, investigate the effect of glass thickness on heat loss for the specified glass surface temperatures. Let the glass thickness vary from 0.2 to 2 cm. Plot the heat loss versus the glass thickness, and discuss the results.

2–98 An aluminum pan whose thermal conductivity is 237 W/m·°C has a flat bottom whose diameter is 20 cm and thickness 0.4 cm. Heat is transferred steadily to boiling water in the pan through its bottom at a rate of 500 W. If the inner surface of the bottom of the pan is 105°C, determine the temperature of the outer surface of the bottom of the pan.

2–99 The inner and outer glasses of a 2-m × 2-m double pane window are at 18°C and 6°C, respectively. If the 1-cm space between the two glasses is filled with still air, determine the rate of heat transfer through the air layer by conduction, in kW.

2–100 Two surfaces of a 2-cm-thick plate are maintained at 0°C and 100°C, respectively. If it is determined that heat is transferred through the plate at a rate of 500 W/m^2, determine its thermal conductivity.

2–101 For heat transfer purposes, a standing man can be modeled as a 30-cm diameter, 170-cm long vertical cylinder with both the top and bottom surfaces insulated and with the side surface at an average temperature of 34°C. For a convection heat transfer coefficient of 15 W/m^2·°C, determine the rate of heat loss from this man by convection in an environment at 20°C. *Answer:* 336 W

2–102 A 9-cm-diameter spherical ball whose surface is maintained at a temperature of 110°C is suspended in the middle of a room at 20°C. If the convection heat transfer coefficient is 15 W/m^2·C and the emissivity of the surface is 0.8, determine the total rate of heat transfer from the ball.

2–103 [EES] Reconsider Prob. 2–102. Using EES (or other) software, investigate the effect of the convection heat transfer coefficient and surface emissivity on the heat transfer rate from the ball. Let the heat transfer coefficient vary from 5 to 30 W/m^2·°C. Plot the rate of heat transfer against the convection heat transfer coefficient for the surface emissivities of 0.1, 0.5, 0.8, and 1, and discuss the results.

2–104 Hot air at 80°C is blown over a 2-m × 4-m flat surface at 30°C. If the convection heat transfer coefficient is 55 W/m^2·°C, determine the rate of heat transfer from the air to the plate, in kW.

2–105 A 1000-W iron is left on the ironing board with its base exposed to the air at 20°C. The convection heat transfer coefficient between the base surface and the surrounding air is 35 W/m^2·°C. If the base has an emissivity of 0.6 and a surface area of 0.02 m^2, determine the temperature of the base of the iron.

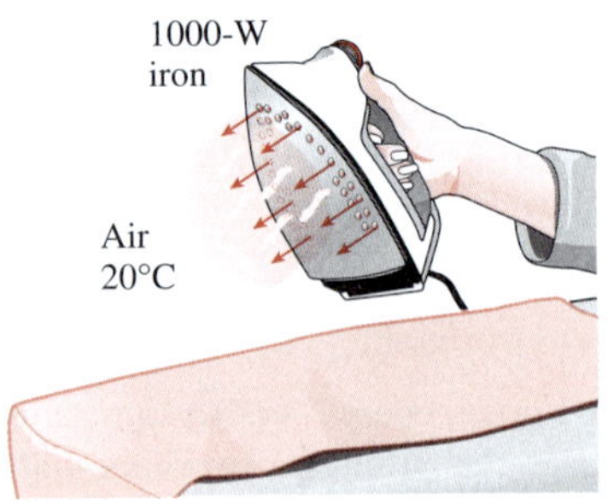

FIGURE P2–105

2–106 A thin metal plate is insulated on the back and exposed to solar radiation on the front surface. The exposed surface of the plate has an absorptivity of 0.8 for solar radiation. If solar radiation is incident on the plate at a rate of 450 W/m^2 and the surrounding air temperature is 25°C, determine the surface temperature of the plate when the heat loss by convection equals the solar energy absorbed by the plate. Assume the convection heat transfer coefficient to be 50 W/m^2·°C, and disregard heat loss by radiation.

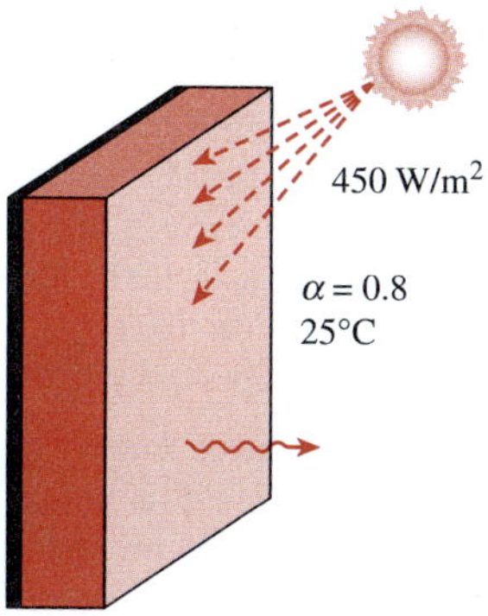

FIGURE P2–106

2–107 Reconsider Prob. 2–106. Using EES (or other) software, investigate the effect of the convection heat transfer coefficient on the surface temperature of the plate. Let the heat transfer coefficient vary from 10 to 90 W/m^2·°C. Plot the surface temperature against the convection heat transfer coefficient, and discuss the results.

2–108 A 5-cm-external-diameter, 10-m-long hot-water pipe at 80°C is losing heat to the surrounding air at 5°C by natural convection with a heat transfer coefficient of 25 W/m^2·°C. Determine the rate of heat loss from the pipe by natural convection, in kW.

2–109 The outer surface of a spacecraft in space has an emissivity of 0.6 and an absorptivity of 0.2 for solar radiation. If solar radiation is incident on the spacecraft at a rate of 1000 W/m^2, determine the surface temperature of the spacecraft when the radiation emitted equals the solar energy absorbed.

2–110 Reconsider Prob. 2–109. Using EES (or other) software, investigate the effect of the surface emissivity and absorptivity of the spacecraft on the equilibrium surface temperature. Plot the surface temperature against emissivity for solar absorptivities of 0.1, 0.5, 0.8, and 1, and discuss the results.

2–111 A hollow spherical iron container whose outer diameter is 20 cm and thickness is 0.4 cm is filled with iced water at 0°C. If the outer surface temperature is 5°C, determine the approximate rate of heat loss from the sphere, and the rate at which ice melts in the container.

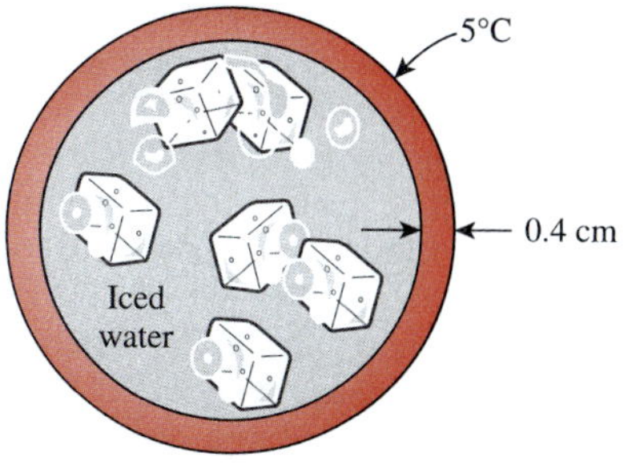

FIGURE P2–111

Review Problems

2–112 Consider a classroom for 55 students and one instructor, each generating heat at a rate of 100 W. Lighting is provided by 18 fluorescent lightbulbs, 40 W each, and the ballasts consume an additional 10 percent. Determine the rate of internal heat generation in this classroom when it is fully occupied.

2–113 Consider a homeowner who is replacing his 25-year-old natural gas furnace that has an efficiency of 55 percent. The homeowner is considering a conventional furnace that has an efficiency of 82 percent and costs $1600 and a high-efficiency furnace that has an efficiency of 95 percent and costs $2700. The homeowner would like to buy the high-efficiency furnace if the savings from the natural gas pay for the additional cost in less than 8 years. If the homeowner presently pays $1200 a year for heating, determine if he should buy the conventional or high-efficiency model.

2–114 Wind energy has been used since 4000 BC to power sailboats, grind grain, pump water for farms, and, more recently, generate electricity. In the United States alone, more than 6 million small windmills, most of them under 5 hp, have been used since the 1850s to pump water. Small windmills have been used to generate electricity since 1900, but the development of modern wind turbines occurred only recently in response to the energy crises in the early 1970s. The cost of wind power has dropped an order of magnitude from about $0.50/kWh in the early 1980s to about $0.05/kWh in the mid-1990s, which is about the price of electricity generated at coal-fired power plants. Areas with an average wind speed of 6 m/s (or 14 mph) are potential sites for economical wind power generation. Commercial wind turbines generate from 100 kW to 3.2 MW of electric power each at peak design conditions. The blade span (or rotor) diameter of the 3.2 MW wind turbine built by Boeing Engineering is 320 ft (97.5 m). The rotation speed of rotors of wind turbines is usually under 40 rpm (under 20 rpm for large turbines). Altamont Pass in California is the world's largest wind farm with 15,000 modern wind turbines. This farm and two others in California produced 2.8 billion kWh of electricity in 1991, which is enough power to meet the electricity needs of San Francisco.

In 2008, 27,260 MW of new wind energy generating capacity were installed worldwide, bringing the world's total wind energy capacity to 121,200 MW. The United States, Germany, Denmark, and Spain account for over 75 percent of current wind energy generating capacity worldwide. Denmark uses wind turbines to supply 10 percent of its national electricity.

Many wind turbines currently in operation have just two blades. This is because at tip speeds of 100 to 200 mph, the efficiency of the two-bladed turbine approaches the theoretical maximum, and the increase in the efficiency by adding a third or fourth blade is so little that they do not justify the added cost and weight.

Consider a wind turbine with an 80-m-diameter rotor that is rotating at 20 rpm under steady winds at an average velocity of 30 km/h. Assuming the turbine has an efficiency of 35 percent (i.e., it converts 35 percent of the kinetic energy of the wind to electricity), determine (*a*) the power produced, in kW; (*b*) the tip speed of the blade, in km/h; and (*c*) the revenue generated by the wind turbine per year if the electric power produced is sold to the utility at \$0.06/kWh. Take the density of air to be 1.20 kg/m^3.

FIGURE P2–114

2–115 Repeat Prob. 2–114 for an average wind velocity of 20 km/h.

2–116E The energy contents, unit costs, and typical conversion efficiencies of various energy sources for use in water heaters are given as follows: 1025 Btu/ft^3, \$0.012/ft^3, and 55 percent for natural gas; 138,700 Btu/gal, \$1.15/gal, and 55 percent for heating oil; and 1 kWh/kWh, \$0.084/kWh, and 90 percent for electric heaters, respectively. Determine the lowest-cost energy source for water heaters.

2–117 A homeowner is considering these heating systems for heating his house: Electric resistance heating with \$0.09/kWh and 1 kWh = 3600 kJ, gas heating with \$1.24/therm and 1 therm = 105,500 kJ, and oil heating with \$1.25/gal and 1 gal of oil = 138,500 kJ. Assuming efficiencies of 100 percent for the electric furnace and 87 percent for the gas and oil furnaces, determine the heating system with the lowest energy cost.

2–118 A typical household pays about \$1200 a year on energy bills, and the U.S. Department of Energy estimates that 46 percent of this energy is used for heating and cooling, 15 percent for heating water, 15 percent for refrigerating and freezing, and the remaining 24 percent for lighting, cooking, and running other appliances. The heating and cooling costs of a poorly insulated house can be reduced by up to 30 percent by adding adequate insulation. If the cost of insulation is \$200, determine how long it will take for the insulation to pay for itself from the energy it saves.

2–119 The U.S. Department of Energy estimates that up to 10 percent of the energy use of a house can be saved by caulking and weatherstripping doors and windows to reduce air leaks at a cost of about \$60 for materials for an average home with 12 windows and 2 doors. Caulking and weatherstripping every gas-heated home properly would save enough energy to heat about 4 million homes. The savings can be increased by installing storm windows. Determine how long it will take for the caulking and weatherstripping to pay for itself from the energy they save for a house whose annual energy use is \$1500.

2–120E The energy stored in the spring of a railroad car is 5000 lbf·ft. What is this energy in (*a*) lbm, ft, and s units; (*b*) lbf and yd units; and (*c*) lbm, mile, and hour units?

2–121E The force required to expand the gas in a gas spring a distance x is given by

$$F = \frac{\text{Constant}}{x^k}$$

where the constant is determined by the geometry of this device and k is determined by the gas used in the device. One such device has a constant of 200 lbf·in$^{1.4}$ and $k = 1.4$. Determine the work, in Btu, required to compress this device from 1 in to 4 in. *Answer:* 0.0228 Btu

2–122E A man weighing 180 lbf pushes a block weighing 100 lbf along a horizontal plane. The dynamic coefficient of friction between the block and plane is 0.2. Assuming that the block is moving at constant speed, calculate the work required to move the block a distance of 100 ft considering (*a*) the man and (*b*) the block as the system. Express your answers in both lbf·ft and Btu.

2–123 A diesel engine with an engine volume of 4.0 L and an engine speed of 2500 rpm operates on an air–fuel ratio of

18 kg air/kg fuel. The engine uses light diesel fuel that contains 750 ppm (parts per million) of sulfur by mass. All of this sulfur is exhausted to the environment where the sulfur is converted to sulfurous acid (H_2SO_3). If the rate of the air entering the engine is 336 kg/h, determine the mass flow rate of sulfur in the exhaust. Also, determine the mass flow rate of sulfurous acid added to the environment if for each kmol of sulfur in the exhaust, one kmol sulfurous acid will be added to the environment.

2–124 Leaded gasoline contains lead that ends up in the engine exhaust. Lead is a very toxic engine emission. The use of leaded gasoline in the United States has been unlawful for most vehicles since the 1980s. However, leaded gasoline is still used in some parts of the world. Consider a city with 5,000 cars using leaded gasoline. The gasoline contains 0.15 g/L of lead and 35 percent of lead is exhausted to the environment. Assuming that an average car travels 15,000 km per year with a gasoline consumption of 8.5 L/100 km, determine the amount of lead put into the atmosphere per year in that city. *Answer:* 335 kg

2–125E Water is pumped from a 200-ft-deep well into a 100-ft-high storage tank. Determine the power, in kW, that would be required to pump 200 gallons per minute.

2–126 A grist mill of the 1800s employed a water wheel that was 14 m high; 320 liters per minute of water flowed on to the wheel near the top. How much power, in kW, could this water wheel have produced? *Answer:* 0.732 kW

2–127 Windmills slow the air and cause it to fill a larger channel as it passes through the blades. Consider a circular windmill with a 7-m-diameter rotor in a 10 m/s wind on a day when the atmospheric pressure is 100 kPa and the temperature is 20°C. The wind speed behind the windmill is measured at 9 m/s. Determine the diameter of the wind channel downstream from the rotor and the power produced by this windmill, presuming that the air is incompressible.

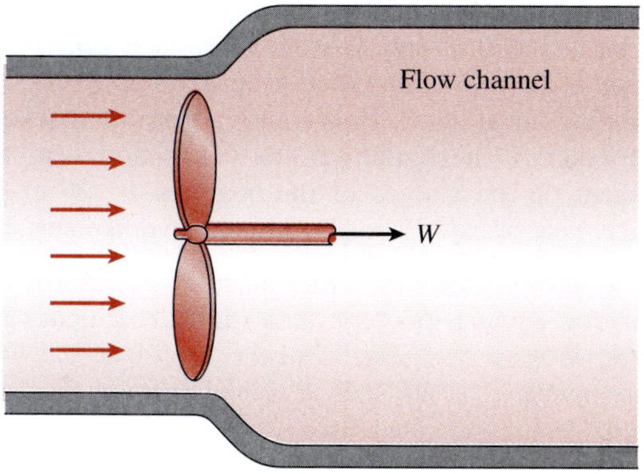

FIGURE P2–127

2–128 In a hydroelectric power plant, 65 m^3/s of water flows from an elevation of 90 m to a turbine, where electric power is generated. The overall efficiency of the turbine–generator is 84 percent. Disregarding frictional losses in piping, estimate the electric power output of this plant. *Answer:* 48.2 MW

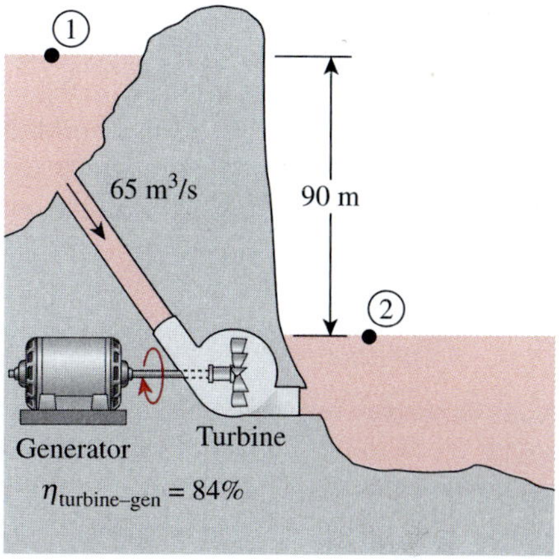

FIGURE P2–128

2–129 The demand for electric power is usually much higher during the day than it is at night, and utility companies often sell power at night at much lower prices to encourage consumers to use the available power generation capacity and to avoid building new expensive power plants that will be used only a short time during peak periods. Utilities are also willing to purchase power produced during the day from private parties at a high price.

Suppose a utility company is selling electric power for \$0.03/kWh at night and is willing to pay \$0.08/kWh for power produced during the day. To take advantage of this opportunity, an entrepreneur is considering building a large reservoir 40 m above the lake level, pumping water from the lake to the reservoir at night using cheap power, and letting the water flow from the reservoir back to the lake during the day, producing power as the pump–motor operates as a turbine–generator during reverse flow. Preliminary analysis shows that a water flow rate of 2 m^3/s can be used in either direction. The combined pump–motor and turbine–generator efficiencies are expected to be 75 percent each. Disregarding the frictional losses in piping and assuming the system operates for 10 h each in the pump and turbine modes during a typical day, determine the potential revenue this pump–turbine system can generate per year.

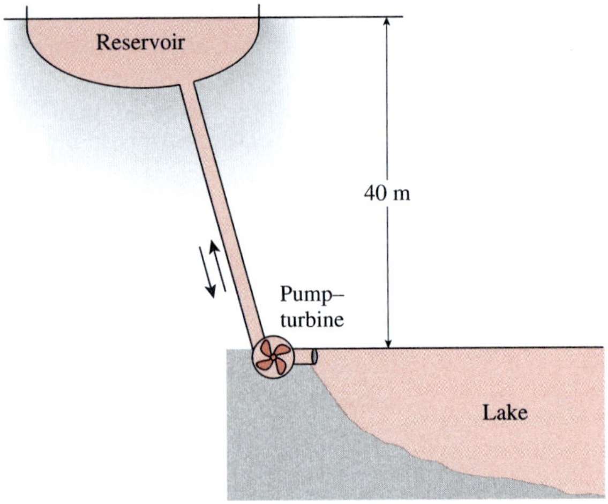

FIGURE P2–129

Fundamentals of Engineering (FE) Exam Problems

2–130 A 2-kW electric resistance heater in a room is turned on and kept on for 50 min. The amount of energy transferred to the room by the heater is

(*a*) 2 kJ (*b*) 100 kJ (*c*) 3000 kJ (*d*) 6000 kJ
(*e*) 12,000 kJ

2–131 On a hot summer day, the air in a well-sealed room is circulated by a 0.50-hp fan driven by a 65 percent efficient motor. (Note that the motor delivers 0.50 hp of net shaft power to the fan.) The rate of energy supply from the fan-motor assembly to the room is

(*a*) 0.769 kJ/s (*b*) 0.325 kJ/s (*c*) 0.574 kJ/s
(*d*) 0.373 kJ/s (*e*) 0.242 kJ/s

2–132 A fan is to accelerate quiescent air to a velocity to 12 m/s at a rate of 3 m^3/s. If the density of air is 1.15 kg/m^3, the minimum power that must be supplied to the fan is

(*a*) 248 W (*b*) 72 W (*c*) 497 W (*d*) 216 W
(*e*) 162 W

2–133 A 900-kg car cruising at a constant speed of 60 km/s is to accelerate to 100 km/h in 4 s. The additional power needed to achieve this acceleration is

(*a*) 56 kW (*b*) 222 kW (*c*) 2.5 kW (*d*) 62 kW
(*e*) 90 kW

2–134 The elevator of a large building is to raise a net mass of 400 kg at a constant speed of 12 m/s using an electric motor. Minimum power rating of the motor should be

(*a*) 0 kW (*b*) 4.8 kW (*c*) 47 kW (*d*) 12 kW
(*e*) 36 kW

2–135 Electric power is to be generated in a hydroelectric power plant that receives water at a rate of 70 m^3/s from an elevation of 65 m using a turbine–generator with an efficiency of 85 percent. When frictional losses in piping are disregarded, the electric power output of this plant is

(*a*) 3.9 MW (*b*) 38 MW (*c*) 45 MW (*d*) 53 MW
(*e*) 65 MW

2–136 A 75-hp compressor in a facility that operates at full load for 2500 h a year is powered by an electric motor that has an efficiency of 93 percent. If the unit cost of electricity is \$0.06/kWh, the annual electricity cost of this compressor is

(*a*) \$7802 (*b*) \$9021 (*c*) \$12,100 (*d*) \$8389
(*e*) \$10,460

2–137 Consider a refrigerator that consumes 320 W of electric power when it is running. If the refrigerator runs only one quarter of the time and the unit cost of electricity is \$0.09/kWh, the electricity cost of this refrigerator per month (30 days) is

(*a*) \$3.56 (*b*) \$5.18 (*c*) \$8.54 (*d*) \$9.28
(*e*) \$20.74

2–138 A 2-kW pump is used to pump kerosene (ρ = 0.820 kg/L) from a tank on the ground to a tank at a higher elevation. Both tanks are open to the atmosphere, and the elevation difference between the free surfaces of the tanks is 30 m. The maximum volume flow rate of kerosene is

(*a*) 8.3 L/s (*b*) 7.2 L/s (*c*) 6.8 L/s
(*d*) 12.1 L/s (*e*) 17.8 L/s

2–139 A glycerin pump is powered by a 5-kW electric motor. The pressure differential between the outlet and the inlet of the pump at full load is measured to be 211 kPa. If the flow rate through the pump is 18 L/s and the changes in elevation and the flow velocity across the pump are negligible, the overall efficiency of the pump is

(*a*) 69 percent (*b*) 72 percent (*c*) 76 percent
(*d*) 79 percent (*e*) 82 percent

The Following Problems Are Based on the Optional Special Topic of Heat Transfer

2–140 A 10-cm high and 20-cm wide circuit board houses on its surface 100 closely spaced chips, each generating heat at a rate of 0.08 W and transferring it by convection to the surrounding air at 25°C. Heat transfer from the back surface of the board is negligible. If the convection heat transfer coefficient on the surface of the board is 10 W/m^2·°C and radiation heat transfer is negligible, the average surface temperature of the chips is

(*a*) 26°C (*b*) 45°C (*c*) 15°C (*d*) 80°C
(*e*) 65°C

2–141 A 50-cm-long, 0.2-cm-diameter electric resistance wire submerged in water is used to determine the boiling heat transfer coefficient in water at 1 atm experimentally. The surface temperature of the wire is measured to be 130°C when a wattmeter indicates the electric power consumption to be 4.1 kW. Then the heat transfer coefficient is

(*a*) 43,500 W/m²·°C (*b*) 137 W/m²·°C
(*c*) 68,330 W/m²·°C (*d*) 10,038 W/m²·°C
(*e*) 37,540 W/m²·°C

2–142 A 3-m² hot black surface at 80°C is losing heat to the surrounding air at 25°C by convection with a convection heat transfer coefficient of 12 W/m²·°C, and by radiation to the surrounding surfaces at 15°C. The total rate of heat loss from the surface is

(*a*) 1987 W (*b*) 2239 W (*c*) 2348 W (*d*) 3451 W
(*e*) 3811 W

2–143 Heat is transferred steadily through a 0.2-m thick 8 m × 4 m wall at a rate of 2.4 kW. The inner and outer surface temperatures of the wall are measured to be 15°C and 5°C. The average thermal conductivity of the wall is

(*a*) 0.002 W/m·°C (*b*) 0.75 W/m·°C (*c*) 1.0 W/m·°C
(*d*) 1.5 W/m·°C (*e*) 3.0 W/m·°C

2–144 The roof of an electrically heated house is 7-m long, 10-m wide, and 0.25-m thick. It is made of a flat layer of concrete whose thermal conductivity is 0.92 W/m·°C. During a certain winter night, the temperatures of the inner and outer surfaces of the roof are measured to be 15°C and 4°C, respectively. The average rate of heat loss through the roof that night was

(*a*) 41 W (*b*) 177 W (*c*) 4894 W (*d*) 5567 W
(*e*) 2834 W

Design and Essay Problems

2–145 Conduct a literature survey that reviews that concepts of thermal pollution and its current state of the art.

2–146 An average vehicle puts out nearly 20 lbm of carbon dioxide into the atmosphere for every gallon of gasoline it burns, and thus one thing we can do to reduce global warming is to buy a vehicle with higher fuel economy. A U.S. government publication states that a vehicle that gets 25 rather than 20 miles per gallon will prevent 10 tons of carbon dioxide from being released over the lifetime of the vehicle. Making reasonable assumptions, evaluate if this is a reasonable claim or a gross exaggeration.

2–147 Solar energy reaching the earth is about 1350 W/m² outside the earth's atmosphere, and 950 W/m² on earth's surface normal to the sun on a clear day. Someone is marketing 2 m × 3 m photovoltaic cell panels with the claim that a single panel can meet the electricity needs of a house. How do you evaluate this claim? Photovoltaic cells have a conversion efficiency of about 15 percent.

2–148 Find out the prices of heating oil, natural gas, and electricity in your area, and determine the cost of each per kWh of energy supplied to the house as heat. Go through your utility bills and determine how much money you spent for heating last January. Also determine how much your January heating bill would be for each of the heating systems if you had the latest and most efficient system installed.

2–149 Prepare a report on the heating systems available in your area for residential buildings. Discuss the advantages and disadvantages of each system and compare their initial and operating costs. What are the important factors in the selection of a heating system? Give some guidelines. Identify the conditions under which each heating system would be the best choice in your area.

2–150 The performance of a device is defined as the ratio of the desired output to the required input, and this definition can be extended to nontechnical fields. For example, your performance in this course can be viewed as the grade you earn relative to the effort you put in. If you have been investing a lot of time in this course and your grades do not reflect it, you are performing poorly. In that case, perhaps you should try to find out the underlying cause and how to correct the problem. Give three other definitions of performance from nontechnical fields and discuss them.

2–151 An electrical-generation utility sometimes pumps liquid water into an elevated reservoir during periods of low electrical consumption. This water is used to generate electricity during periods when the demand for electricity exceeds the utility's ability to produce electricity. Discuss this energy-storage scheme from a conversion efficiency perspective as compared to storing a compressed phase-changing substance.

2–152 Some engineers have suggested that air compressed into tanks can be used to propel personal transportation vehicles. Current compressed-air tank technology permits us to compress and safely hold air at up to 4000 psia. Tanks made of composite materials require about 10 lbm of construction materials for each 1 ft³ of stored gas. Approximately 0.01 hp is required per pound of vehicle weight to move a vehicle at a speed of 30 miles per hour. What is the maximum range that this vehicle can have? Account for the weight of the tanks only and assume perfect conversion of the energy in the compressed air.

CHAPTER

3

PROPERTIES OF PURE SUBSTANCES

We start this chapter with the introduction of the concept of a *pure substance* and a discussion of the physics of phase-change processes. We then illustrate the various property diagrams and *P-v-T* surfaces of pure substances. After demonstrating the use of the property tables, the hypothetical substance *ideal gas* and the *ideal-gas equation of state* are discussed. The *compressibility factor*, which accounts for the deviation of real gases from ideal-gas behavior, is introduced, and some of the best-known equations of state such as the van der Waals, Beattie-Bridgeman, and Benedict-Webb-Rubin equations are presented.

Objectives

The objectives of Chapter 3 are to:

- Introduce the concept of a pure substance.
- Discuss the physics of phase-change processes.
- Illustrate the *P-v*, *T-v*, and *P-T* property diagrams and *P-v-T* surfaces of pure substances.
- Demonstrate the procedures for determining thermodynamic properties of pure substances from tables of property data.
- Describe the hypothetical substance "ideal gas" and the ideal-gas equation of state.
- Apply the ideal-gas equation of state in the solution of typical problems.
- Introduce the compressibility factor, which accounts for the deviation of real gases from ideal-gas behavior.
- Present some of the best-known equations of state.

FIGURE 3–1
Nitrogen and gaseous air are pure substances.

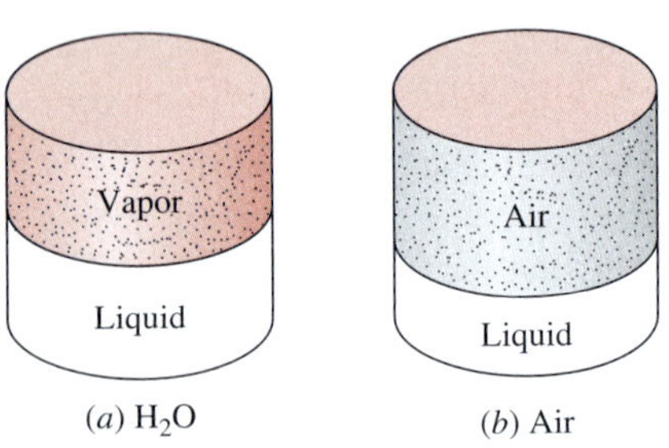

FIGURE 3–2
A mixture of liquid and gaseous water is a pure substance, but a mixture of liquid and gaseous air is not.

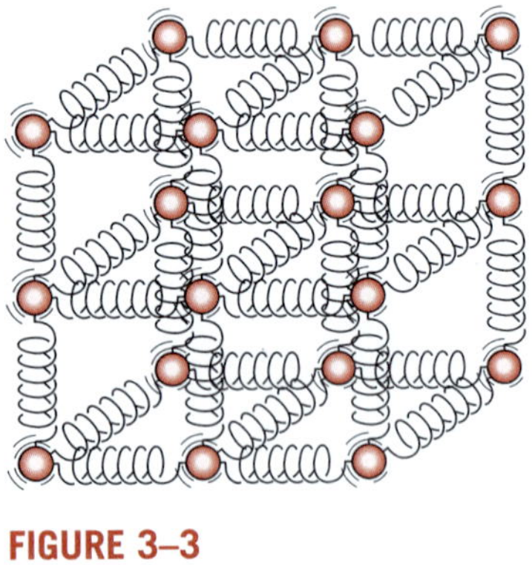

FIGURE 3–3
The molecules in a solid are kept at their positions by the large springlike intermolecular forces.

3–1 ▪ PURE SUBSTANCE

A substance that has a fixed chemical composition throughout is called a **pure substance**. Water, nitrogen, helium, and carbon dioxide, for example, are all pure substances.

A pure substance does not have to be of a single chemical element or compound, however. A mixture of various chemical elements or compounds also qualifies as a pure substance as long as the mixture is homogeneous. Air, for example, is a mixture of several gases, but it is often considered to be a pure substance because it has a uniform chemical composition (Fig. 3–1). However, a mixture of oil and water is not a pure substance. Since oil is not soluble in water, it will collect on top of the water, forming two chemically dissimilar regions.

A mixture of two or more phases of a pure substance is still a pure substance as long as the chemical composition of all phases is the same (Fig. 3–2). A mixture of ice and liquid water, for example, is a pure substance because both phases have the same chemical composition. A mixture of liquid air and gaseous air, however, is not a pure substance since the composition of liquid air is different from the composition of gaseous air, and thus the mixture is no longer chemically homogeneous. This is due to different components in air condensing at different temperatures at a specified pressure.

3–2 ▪ PHASES OF A PURE SUBSTANCE

We all know from experience that substances exist in different phases. At room temperature and pressure, copper is a solid, mercury is a liquid, and nitrogen is a gas. Under different conditions, each may appear in a different phase. Even though there are three principal phases—solid, liquid, and gas—a substance may have several phases within a principal phase, each with a different molecular structure. Carbon, for example, may exist as graphite or diamond in the solid phase. Helium has two liquid phases; iron has three solid phases. Ice may exist at seven different phases at high pressures. A phase is identified as having a distinct molecular arrangement that is homogeneous throughout and separated from the others by easily identifiable boundary surfaces. The two phases of H_2O in iced water represent a good example of this.

When studying phases or phase changes in thermodynamics, one does not need to be concerned with the molecular structure and behavior of different phases. However, it is very helpful to have some understanding of the molecular phenomena involved in each phase, and a brief discussion of phase transformations follows.

Intermolecular bonds are strongest in solids and weakest in gases. One reason is that molecules in solids are closely packed together, whereas in gases they are separated by relatively large distances.

The molecules in a **solid** are arranged in a three-dimensional pattern (lattice) that is repeated throughout (Fig. 3–3). Because of the small distances between molecules in a solid, the attractive forces of molecules on each other are large and keep the molecules at fixed positions (Fig. 3–4). Note that the attractive forces between molecules turn to repulsive forces as the distance between the molecules approaches zero, thus preventing the molecules from piling up on top of each other. Even though the molecules in a solid cannot move relative to each other, they continually oscillate about their equilibrium positions. The velocity of the molecules during these oscillations depends on the temperature. At sufficiently high temperatures, the velocity (and thus the momentum) of the molecules may reach a point where the intermolecular forces are partially overcome and groups of molecules break away (Fig. 3–5). This is the beginning of the melting process.

The molecular spacing in the **liquid** phase is not much different from that of the solid phase, except the molecules are no longer at fixed positions relative to each other and they can rotate and translate freely. In a liquid, the intermolecular forces are weaker relative to solids, but still relatively strong compared with gases. The distances between molecules generally experience a slight increase as a solid turns liquid, with water being a notable exception.

In the **gas** phase, the molecules are far apart from each other, and a molecular order is nonexistent. Gas molecules move about at random, continually colliding with each other and the walls of the container they are in. Particularly at low densities, the intermolecular forces are very small, and collisions are the only mode of interaction between the molecules. Molecules in the gas phase are at a considerably higher energy level than they are in the liquid or solid phases. Therefore, the gas must release a large amount of its energy before it can condense or freeze.

FIGURE 3–4
In a solid, the attractive and repulsive forces between the molecules tend to maintain them at relatively constant distances from each other.

BLONDIE © KING FEATURES SYNDICATE.

3–3 ▪ PHASE-CHANGE PROCESSES OF PURE SUBSTANCES

There are many practical situations where two phases of a pure substance coexist in equilibrium. Water exists as a mixture of liquid and vapor in the boiler and the condenser of a steam power plant. The refrigerant turns from liquid to vapor in the freezer of a refrigerator. Even though many home

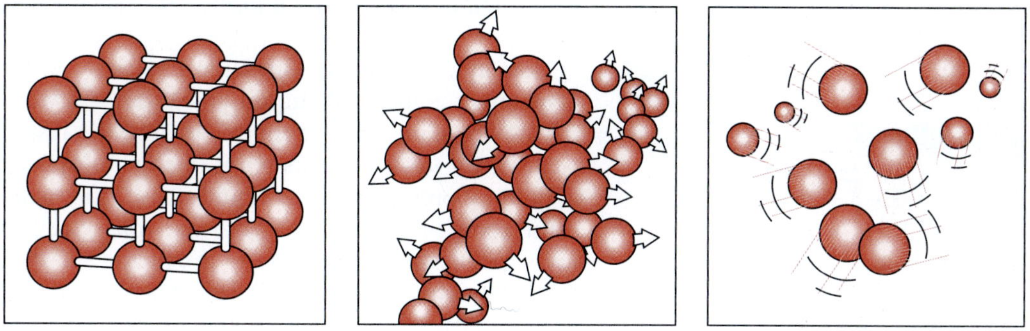

(*a*) (*b*) (*c*)

FIGURE 3–5
The arrangement of atoms in different phases: (*a*) molecules are at relatively fixed positions in a solid, (*b*) groups of molecules move about each other in the liquid phase, and (*c*) molecules move about at random in the gas phase.

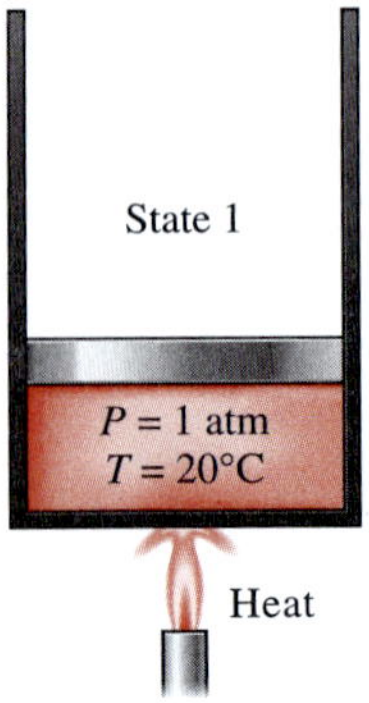

FIGURE 3–6

At 1 atm and 20°C, water exists in the liquid phase (*compressed liquid*).

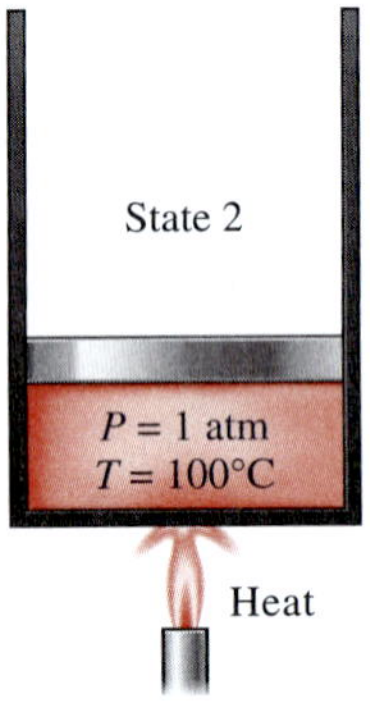

FIGURE 3–7

At 1 atm pressure and 100°C, water exists as a liquid that is ready to vaporize (*saturated liquid*).

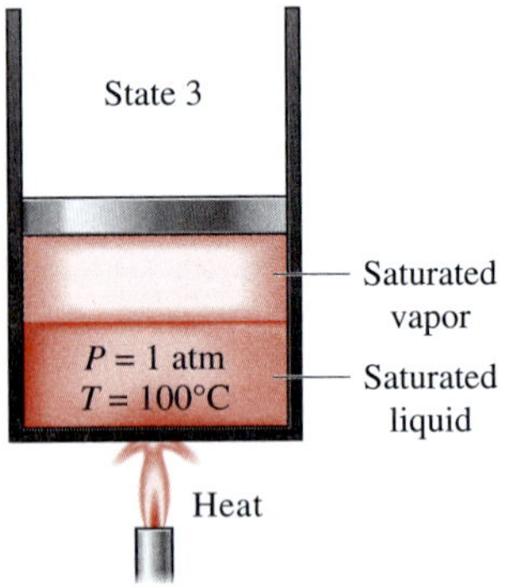

FIGURE 3–8

As more heat is transferred, part of the saturated liquid vaporizes (*saturated liquid–vapor mixture*).

owners consider the freezing of water in underground pipes as the most important phase-change process, attention in this section is focused on the liquid and vapor phases and their mixture. As a familiar substance, water is used to demonstrate the basic principles involved. Remember, however, that all pure substances exhibit the same general behavior.

Compressed Liquid and Saturated Liquid

Consider a piston–cylinder device containing liquid water at 20°C and 1 atm pressure (state 1, Fig. 3–6). Under these conditions, water exists in the liquid phase, and it is called a **compressed liquid**, or a **subcooled liquid**, meaning that it is *not about to vaporize.* Heat is now transferred to the water until its temperature rises to, say, 40°C. As the temperature rises, the liquid water expands slightly, and so its specific volume increases. To accommodate this expansion, the piston moves up slightly. The pressure in the cylinder remains constant at 1 atm during this process since it depends on the outside barometric pressure and the weight of the piston, both of which are constant. Water is still a compressed liquid at this state since it has not started to vaporize.

As more heat is transferred, the temperature keeps rising until it reaches 100°C (state 2, Fig. 3–7). At this point water is still a liquid, but any heat addition will cause some of the liquid to vaporize. That is, a phase-change process from liquid to vapor is about to take place. A liquid that is *about to vaporize* is called a **saturated liquid**. Therefore, state 2 is a saturated liquid state.

Saturated Vapor and Superheated Vapor

Once boiling starts, the temperature stops rising until the liquid is completely vaporized. That is, the temperature will remain constant during the entire phase-change process if the pressure is held constant. This can easily be verified by placing a thermometer into boiling pure water on top of a stove. At sea level (P = 1 atm), the thermometer will always read 100°C if the pan is uncovered or covered with a light lid. During a boiling process, the only change we will observe is a large increase in the volume and a steady decline in the liquid level as a result of more liquid turning to vapor.

Midway about the vaporization line (state 3, Fig. 3–8), the cylinder contains equal amounts of liquid and vapor. As we continue transferring heat, the vaporization process continues until the last drop of liquid is vaporized (state 4, Fig. 3–9). At this point, the entire cylinder is filled with vapor that is on the borderline of the liquid phase. Any heat loss from this vapor will cause some of the vapor to condense (phase change from vapor to liquid). A vapor that is *about to condense* is called a **saturated vapor**. Therefore, state 4 is a saturated vapor state. A substance at states between 2 and 4 is referred to as a **saturated liquid–vapor mixture** since the *liquid and vapor phases coexist* in equilibrium at these states.

Once the phase-change process is completed, we are back to a single-phase region again (this time vapor), and further transfer of heat results in an increase in both the temperature and the specific volume (Fig. 3–10). At state 5, the temperature of the vapor is, let us say, 300°C; and if we transfer some heat from the vapor, the temperature may drop somewhat but no condensation will take place as long as the temperature remains above 100°C (for $P = 1$ atm). A vapor that is *not about to condense* (i.e., not a saturated vapor) is called a **superheated vapor**. Therefore, water at state 5 is a superheated vapor. This constant-pressure phase-change process is illustrated on a *T-v* diagram in Fig. 3–11.

If the entire process described here is reversed by cooling the water while maintaining the pressure at the same value, the water will go back to state 1, retracing the same path, and in so doing, the amount of heat released will exactly match the amount of heat added during the heating process.

In our daily life, water implies liquid water and steam implies water vapor. In thermodynamics, however, both water and steam usually mean only one thing: H_2O.

FIGURE 3–9
At 1 atm pressure, the temperature remains constant at 100°C until the last drop of liquid is vaporized (*saturated vapor*).

Saturation Temperature and Saturation Pressure

It probably came as no surprise to you that water started to boil at 100°C. Strictly speaking, the statement "water boils at 100°C" is incorrect. The correct statement is "water boils at 100°C at 1 atm pressure." The only reason water started boiling at 100°C was because we held the pressure constant at 1 atm (101.325 kPa). If the pressure inside the cylinder were raised to 500 kPa by adding weights on top of the piston, water would start boiling at 151.8°C. That is, *the temperature at which water starts boiling depends on the pressure; therefore, if the pressure is fixed, so is the boiling temperature.*

At a given pressure, the temperature at which a pure substance changes phase is called the **saturation temperature** T_{sat}. Likewise, at a given temperature, the pressure at which a pure substance changes phase is called the **saturation pressure** P_{sat}. At a pressure of 101.325 kPa, T_{sat} is 99.97°C. Conversely, at a temperature of 99.97°C, P_{sat} is 101.325 kPa. (At 100.00°C, P_{sat} is 101.42 kPa in the ITS-90 discussed in Chap. 1.)

Saturation tables that list the saturation pressure against the temperature (or the saturation temperature against the pressure) are available for

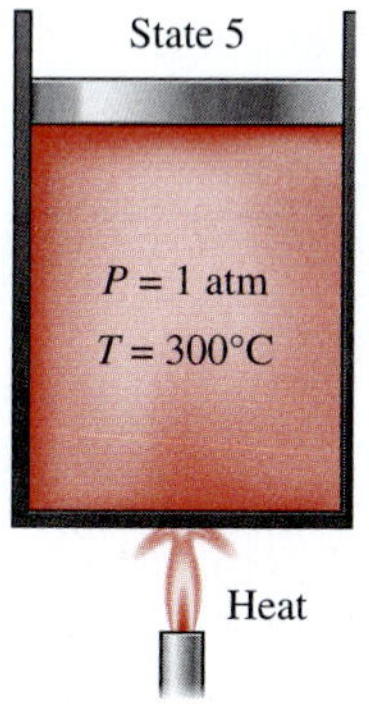

FIGURE 3–10
As more heat is transferred, the temperature of the vapor starts to rise (*superheated vapor*).

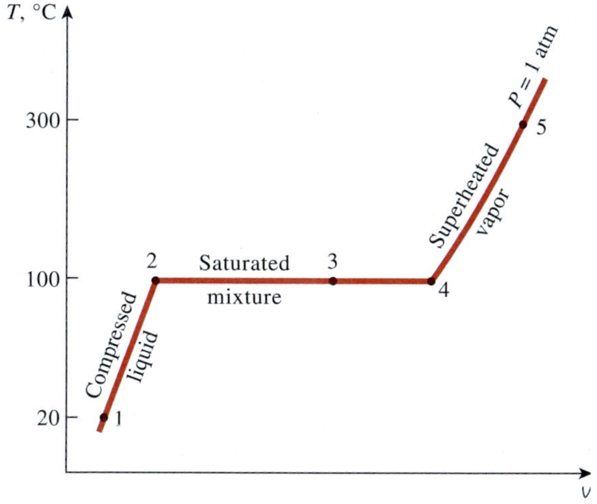

FIGURE 3–11
T-v diagram for the heating process of water at constant pressure.

TABLE 3–1

Saturation (boiling) pressure of water at various temperatures

Temperature, T, °C	Saturation pressure, P_{sat}, kPa
−10	0.26
−5	0.40
0	0.61
5	0.87
10	1.23
15	1.71
20	2.34
25	3.17
30	4.25
40	7.39
50	12.35
100	101.4
150	476.2
200	1555
250	3976
300	8588

practically all substances. A partial listing of such a table is given in Table 3–1 for water. This table indicates that the pressure of water changing phase (boiling or condensing) at 25°C must be 3.17 kPa, and the pressure of water must be maintained at 3976 kPa (about 40 atm) to have it boil at 250°C. Also, water can be frozen by dropping its pressure below 0.61 kPa.

It takes a large amount of energy to melt a solid or vaporize a liquid. The amount of energy absorbed or released during a phase-change process is called the **latent heat**. More specifically, the amount of energy absorbed during melting is called the **latent heat of fusion** and is equivalent to the amount of energy released during freezing. Similarly, the amount of energy absorbed during vaporization is called the **latent heat of vaporization** and is equivalent to the energy released during condensation. The magnitudes of the latent heats depend on the temperature or pressure at which the phase change occurs. At 1 atm pressure, the latent heat of fusion of water is 333.7 kJ/kg and the latent heat of vaporization is 2256.5 kJ/kg.

During a phase-change process, pressure and temperature are obviously dependent properties, and there is a definite relation between them, that is, $T_{sat} = f(P_{sat})$. A plot of T_{sat} versus P_{sat}, such as the one given for water in Fig. 3–12, is called a **liquid–vapor saturation curve**. A curve of this kind is characteristic of all pure substances.

It is clear from Fig. 3–12 that T_{sat} increases with P_{sat}. Thus, a substance at higher pressures boils at higher temperatures. In the kitchen, higher boiling temperatures mean shorter cooking times and energy savings. A beef stew, for example, may take 1 to 2 h to cook in a regular pan that operates at 1 atm pressure, but only 20 min in a pressure cooker operating at 3 atm absolute pressure (corresponding boiling temperature: 134°C).

The atmospheric pressure, and thus the boiling temperature of water, decreases with elevation. Therefore, it takes longer to cook at higher altitudes than it does at sea level (unless a pressure cooker is used). For example, the standard atmospheric pressure at an elevation of 2000 m is 79.50 kPa, which corresponds to a boiling temperature of 93.3°C as opposed to 100°C at sea level (zero elevation). The variation of the boiling temperature of water with altitude at standard atmospheric conditions is given in Table 3–2. For each 1000 m increase in elevation, the boiling temperature drops by a little over 3°C. Note that the atmospheric pressure at a location, and thus the boiling temperature, changes slightly with the weather conditions. But the corresponding change in the boiling temperature is no more than about 1°C.

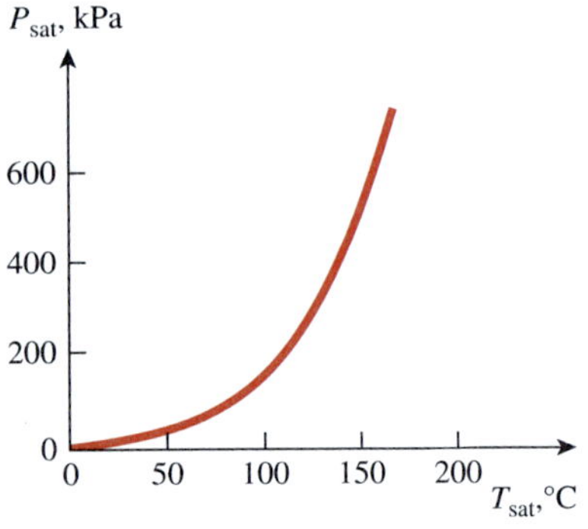

FIGURE 3–12

The liquid–vapor saturation curve of a pure substance (numerical values are for water).

Some Consequences of T_{sat} and P_{sat} Dependence

We mentioned earlier that a substance at a specified pressure boils at the saturation temperature corresponding to that pressure. This phenomenon allows us to control the boiling temperature of a substance by simply controlling the pressure, and it has numerous applications in practice. Below we give some examples. The natural drive to achieve phase equilibrium by allowing some liquid to evaporate is at work behind the scenes.

Consider a sealed can of *liquid refrigerant-134a* in a room at 25°C. If the can has been in the room long enough, the temperature of the refrigerant in the can is also 25°C. Now, if the lid is opened slowly and some refrigerant is allowed to escape, the pressure in the can will start dropping until it reaches the atmospheric pressure. If you are holding the can, you will notice its temperature dropping rapidly, and even ice forming outside the can if the air is humid. A thermometer inserted in the can will register −26°C when the pressure drops to 1 atm, which is the saturation temperature of refrigerant-134a at that pressure. The temperature of the liquid refrigerant will remain at −26°C until the last drop of it vaporizes.

Another aspect of this interesting physical phenomenon is that a liquid cannot vaporize unless it absorbs energy in the amount of the latent heat of vaporization, which is 217 kJ/kg for refrigerant-134a at 1 atm. Therefore, the rate of vaporization of the refrigerant depends on the rate of heat transfer to the can: the larger the rate of heat transfer, the higher the rate of vaporization. The rate of heat transfer to the can and thus the rate of vaporization of the refrigerant can be minimized by insulating the can heavily. In the limiting case of no heat transfer, the refrigerant will remain in the can as a liquid at −26°C indefinitely.

The boiling temperature of *nitrogen* at atmospheric pressure is −196°C (see Table A–3*a*). This means the temperature of liquid nitrogen exposed to the atmosphere must be −196°C since some nitrogen will be evaporating. The temperature of liquid nitrogen remains constant at −196°C until it is depleted. For this reason, nitrogen is commonly used in low-temperature scientific studies (such as superconductivity) and cryogenic applications to maintain a test chamber at a constant temperature of −196°C. This is done by placing the test chamber into a liquid nitrogen bath that is open to the atmosphere. Any heat transfer from the environment to the test section is absorbed by the nitrogen, which evaporates isothermally and keeps the test chamber temperature constant at −196°C (Fig. 3–13). The entire test section must be insulated heavily to minimize heat transfer and thus liquid nitrogen consumption. Liquid nitrogen is also used for medical purposes to burn off unsightly spots on the skin. This is done by soaking a cotton swap in liquid nitrogen and wetting the target area with it. As the nitrogen evaporates, it freezes the affected skin by rapidly absorbing heat from it.

A practical way of cooling leafy vegetables is **vacuum cooling**, which is based on *reducing the pressure* of the sealed cooling chamber to the saturation pressure at the desired low temperature, and evaporating some water from the products to be cooled. The heat of vaporization during evaporation is absorbed from the products, which lowers the product temperature. The saturation pressure of water at 0°C is 0.61 kPa, and the products can be cooled to 0°C by lowering the pressure to this level. The cooling rate can be increased by lowering the pressure below 0.61 kPa, but this is not desirable because of the danger of freezing and the added cost.

In vacuum cooling, there are two distinct stages. In the first stage, the products at ambient temperature, say at 25°C, are loaded into the chamber, and the operation begins. The temperature in the chamber remains constant until the *saturation pressure* is reached, which is 3.17 kPa at 25°C. In the second stage that follows, saturation conditions are maintained inside at progressively *lower pressures* and the corresponding *lower temperatures* until the desired temperature is reached (Fig. 3–14).

TABLE 3–2

Variation of the standard atmospheric pressure and the boiling (saturation) temperature of water with altitude

Elevation m	Atmospheric pressure, kPa	Boiling temperature, °C
0	101.33	100.0
1,000	89.55	96.5
2,000	79.50	93.3
5,000	54.05	83.3
10,000	26.50	66.3
20,000	5.53	34.7

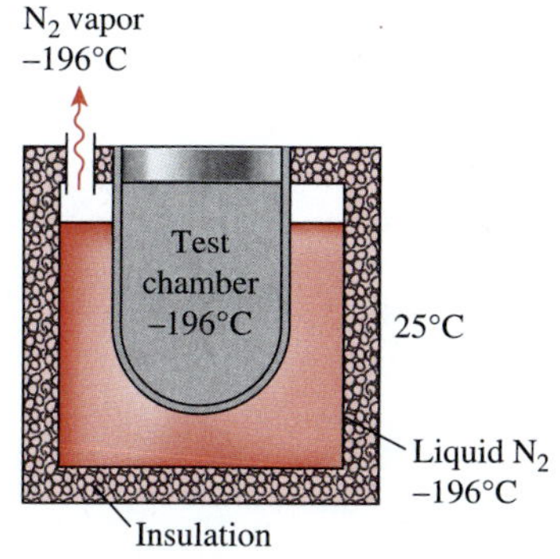

FIGURE 3–13
The temperature of liquid nitrogen exposed to the atmosphere remains constant at −196°C, and thus it maintains the test chamber at −196°C.

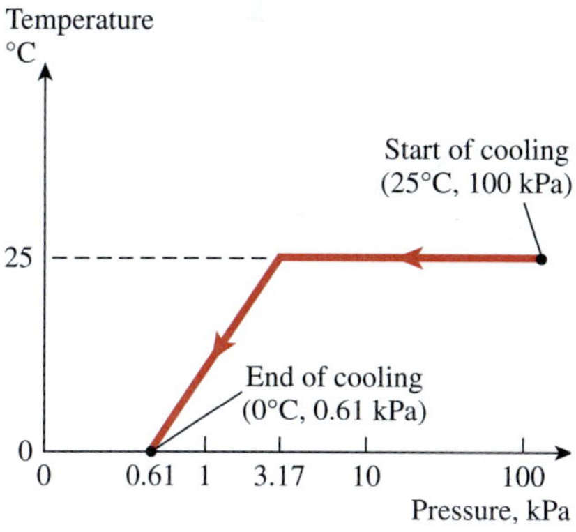

FIGURE 3–14
The variation of the temperature of fruits and vegetables with pressure during vacuum cooling from 25°C to 0°C.

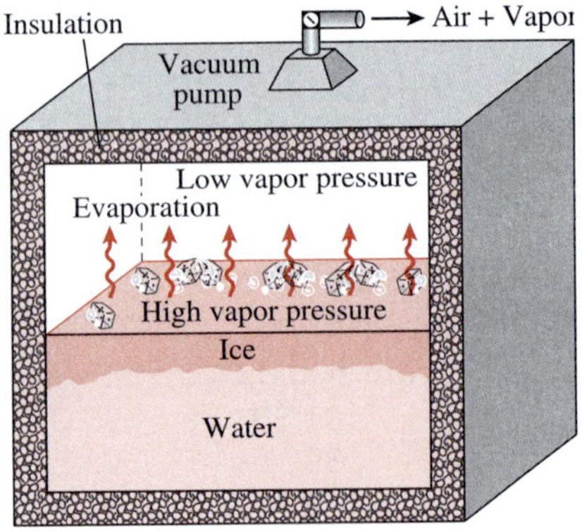

FIGURE 3–15
In 1775, ice was made by evacuating the air space in a water tank.

Vacuum cooling is usually more expensive than the conventional refrigerated cooling, and its use is limited to applications that result in much faster cooling. Products with large surface area per unit mass and a high tendency to release moisture such as lettuce and spinach are well-suited for vacuum cooling. Products with low surface area to mass ratio are not suitable, especially those that have relatively impervious peels such as tomatoes and cucumbers. Some products such as mushrooms and green peas can be vacuum cooled successfully by wetting them first.

The vacuum cooling just described becomes **vacuum freezing** if the vapor pressure in the vacuum chamber is dropped below 0.61 kPa, the saturation pressure of water at 0°C. The idea of making ice by using a vacuum pump is nothing new. Dr. William Cullen actually made ice in Scotland in 1775 by evacuating the air in a water tank (Fig. 3–15).

Package icing is commonly used in small-scale cooling applications to remove heat and keep the products cool during transit by taking advantage of the large latent heat of fusion of water, but its use is limited to products that are not harmed by contact with ice. Also, ice provides *moisture* as well as *refrigeration*.

3–4 ▪ PROPERTY DIAGRAMS FOR PHASE-CHANGE PROCESSES

The variations of properties during phase-change processes are best studied and understood with the help of property diagrams. Next, we develop and discuss the T-v, P-v, and P-T diagrams for pure substances.

1 The T-v Diagram

The phase-change process of water at 1 atm pressure was described in detail in the last section and plotted on a T-v diagram in Fig. 3–11. Now we repeat this process at different pressures to develop the T-v diagram.

Let us add weights on top of the piston until the pressure inside the cylinder reaches 1 MPa. At this pressure, water has a somewhat smaller specific volume than it does at 1 atm pressure. As heat is transferred to the water at this new pressure, the process follows a path that looks very much like the process path at 1 atm pressure, as shown in Fig. 3–16, but there are some noticeable differences. First, water starts boiling at a much higher temperature (179.9°C) at this pressure. Second, the specific volume of the saturated liquid is larger and the specific volume of the saturated vapor is smaller than the corresponding values at 1 atm pressure. That is, the horizontal line that connects the saturated liquid and saturated vapor states is much shorter.

As the pressure is increased further, this saturation line continues to shrink, as shown in Fig. 3–16, and it becomes a point when the pressure reaches 22.06 MPa for the case of water. This point is called the **critical point**, and it is defined as *the point at which the saturated liquid and saturated vapor states are identical.*

The temperature, pressure, and specific volume of a substance at the critical point are called, respectively, the *critical temperature* T_{cr}, *critical pressure* P_{cr}, and *critical specific volume* v_{cr}. The critical-point properties of

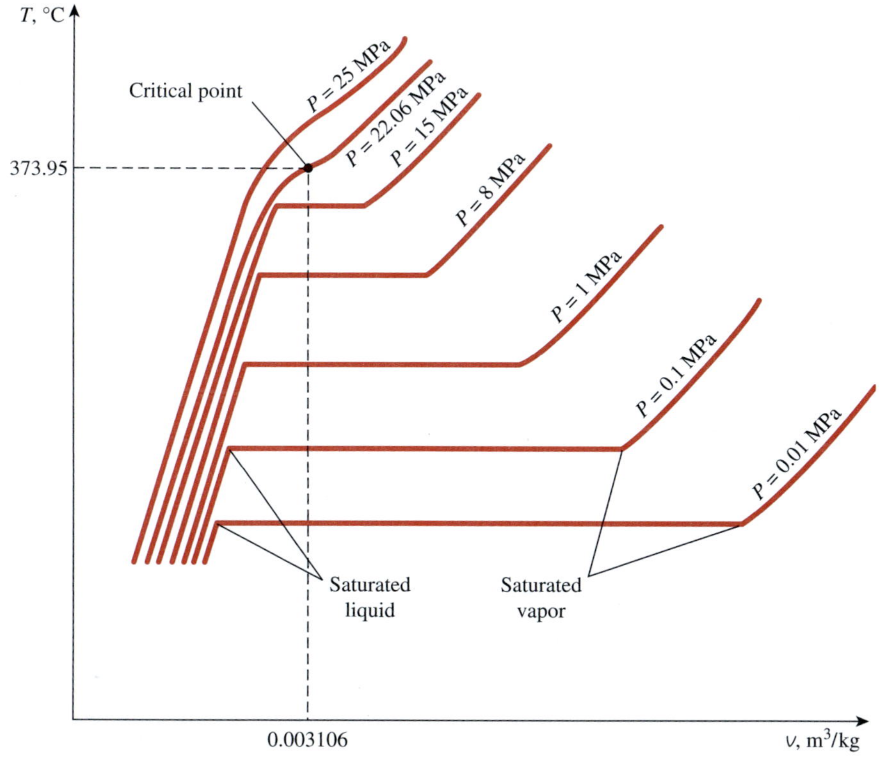

FIGURE 3–16
T-*v* diagram of constant-pressure phase-change processes of a pure substance at various pressures (numerical values are for water).

water are $P_{cr} = 22.06$ MPa, $T_{cr} = 373.95°C$, and $v_{cr} = 0.003106$ m^3/kg. For helium, they are 0.23 MPa, −267.85°C, and 0.01444 m^3/kg. The critical properties for various substances are given in Table A–1 in the appendix.

At pressures above the critical pressure, there is not a distinct phase-change process (Fig. 3–17). Instead, the specific volume of the substance continually increases, and at all times there is only one phase present. Eventually, it resembles a vapor, but we can never tell when the change has occurred. Above the critical state, there is no line that separates the compressed liquid region and the superheated vapor region. However, it is customary to refer to the substance as superheated vapor at temperatures above the critical temperature and as compressed liquid at temperatures below the critical temperature.

The saturated liquid states in Fig. 3–16 can be connected by a line called the **saturated liquid line**, and saturated vapor states in the same figure can be connected by another line, called the **saturated vapor line**. These two lines meet at the critical point, forming a dome as shown in Fig. 3–18*a*. All the compressed liquid states are located in the region to the left of the saturated liquid line, called the **compressed liquid region**. All the superheated vapor states are located to the right of the saturated vapor line, called the **superheated vapor region**. In these two regions, the substance exists in a single phase, a liquid or a vapor. All the states that involve both phases in equilibrium are located under the dome, called the **saturated liquid–vapor mixture region**, or the **wet region**.

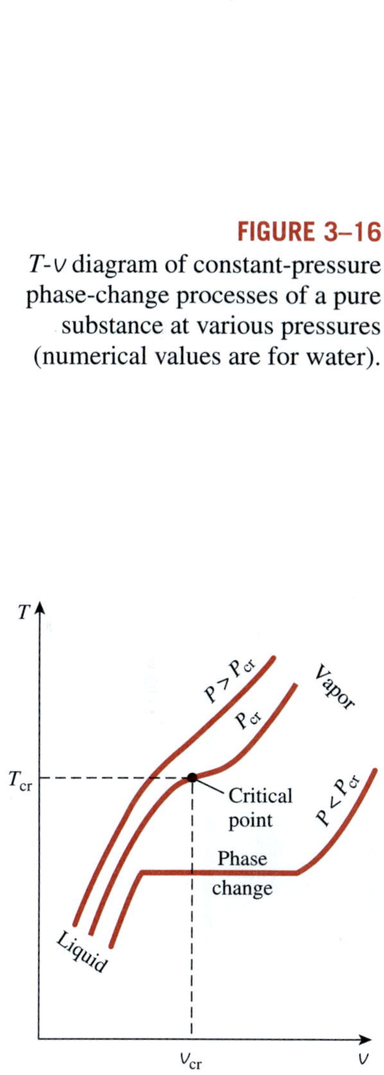

FIGURE 3–17
At supercritical pressures ($P > P_{cr}$), there is no distinct phase-change (boiling) process.

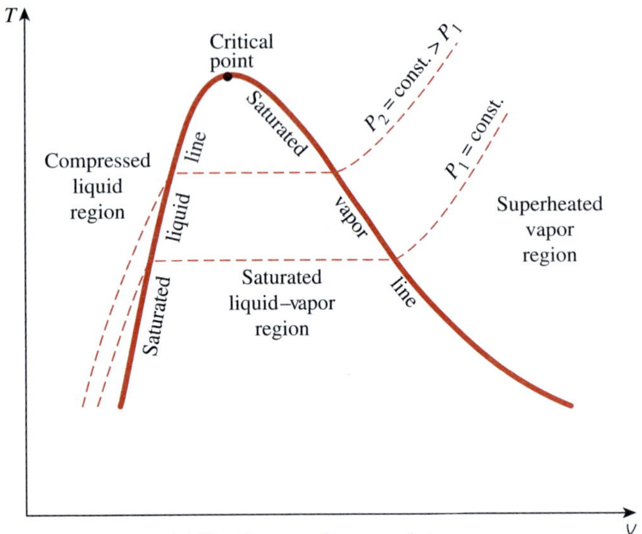

(*a*) *T*-v diagram of a pure substance

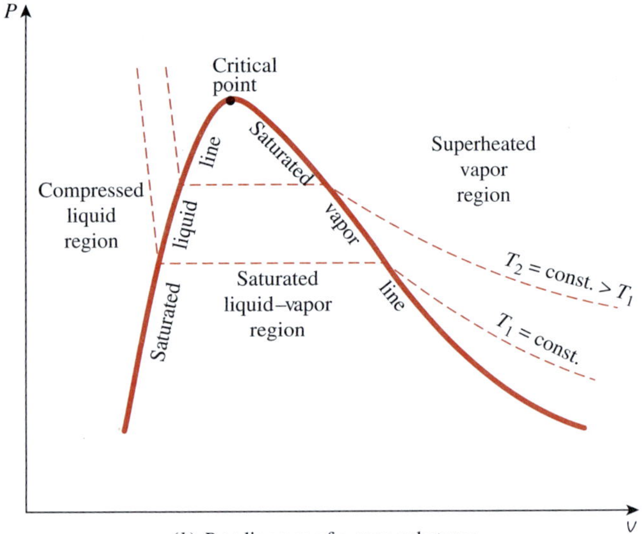

(*b*) *P*-v diagram of a pure substance

FIGURE 3–18
Property diagrams of a pure substance.

2 The *P*-v Diagram

The general shape of the *P*-v diagram of a pure substance is very much like the *T*-v diagram, but the *T* = constant lines on this diagram have a downward trend, as shown in Fig. 3–18*b*.

Consider again a piston–cylinder device that contains liquid water at 1 MPa and 150°C. Water at this state exists as a compressed liquid. Now the weights on top of the piston are removed one by one so that the pressure inside the cylinder decreases gradually (Fig. 3–19). The water is allowed to exchange heat with the surroundings so its temperature remains constant. As the pressure decreases, the volume of the water increases slightly. When the pressure reaches the saturation-pressure value at the specified temperature (0.4762 MPa), the water starts to boil. During this vaporization process, both the temperature and the pressure remain constant, but the specific volume increases. Once the last drop of liquid is vaporized, further reduction in pressure results in a further increase in specific volume. Notice that during the phase-change process, we did not remove any weights. Doing so would cause the pressure and therefore the temperature to drop [since $T_{sat} = f(P_{sat})$], and the process would no longer be isothermal.

When the process is repeated for other temperatures, similar paths are obtained for the phase-change processes. Connecting the saturated liquid and the saturated vapor states by a curve, we obtain the *P*-v diagram of a pure substance, as shown in Fig. 3–18*b*.

FIGURE 3–19
The pressure in a piston–cylinder device can be reduced by reducing the weight of the piston.

Extending the Diagrams to Include the Solid Phase

The two equilibrium diagrams developed so far represent the equilibrium states involving the liquid and the vapor phases only. However, these diagrams can easily be extended to include the solid phase as well as the

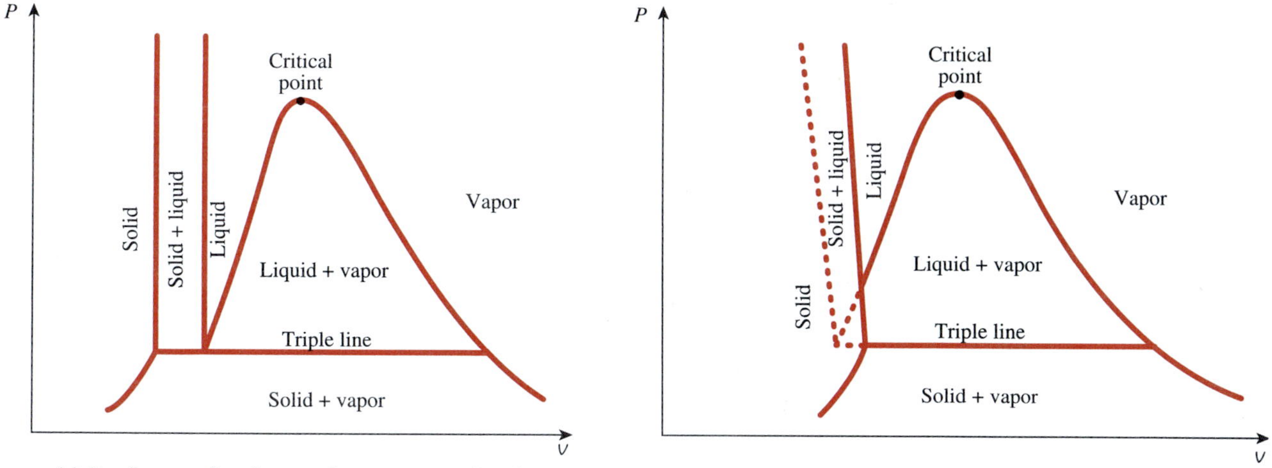

(*a*) P-v diagram of a substance that contracts on freezing

(*b*) P-v diagram of a substance that expands on freezing (such as water)

FIGURE 3–20
P-v diagrams of different substances.

solid–liquid and the solid–vapor saturation regions. The basic principles discussed in conjunction with the liquid–vapor phase-change process apply equally to the solid–liquid and solid–vapor phase-change processes. Most substances contract during a solidification (i.e., freezing) process. Others, like water, expand as they freeze. The P-v diagrams for both groups of substances are given in Figs. 3–20*a* and 3–20*b*. These two diagrams differ only in the solid–liquid saturation region. The T-v diagrams look very much like the P-v diagrams, especially for substances that contract on freezing.

The fact that water expands upon freezing has vital consequences in nature. If water contracted on freezing as most other substances do, the ice formed would be heavier than the liquid water, and it would settle to the bottom of rivers, lakes, and oceans instead of floating at the top. The sun's rays would never reach these ice layers, and the bottoms of many rivers, lakes, and oceans would be covered with ice at times, seriously disrupting marine life.

We are all familiar with two phases being in equilibrium, but under some conditions all three phases of a pure substance coexist in equilibrium (Fig. 3–21). On P-v or T-v diagrams, these triple-phase states form a line called the **triple line**. The states on the triple line of a substance have the same pressure and temperature but different specific volumes. The triple line appears as a point on the P-T diagrams and, therefore, is often called the **triple point**. The triple-point temperatures and pressures of various substances are given in Table 3–3. For water, the triple-point temperature and pressure are 0.01°C and 0.6117 kPa, respectively. That is, all three phases of water coexist in equilibrium only if the temperature and pressure have precisely these values. No substance can exist in the liquid phase in stable equilibrium at pressures below the triple-point pressure. The same can be said for temperature for substances that contract on freezing. However, substances at high pressures can exist in the liquid phase at temperatures below the triple-point temperature. For example,

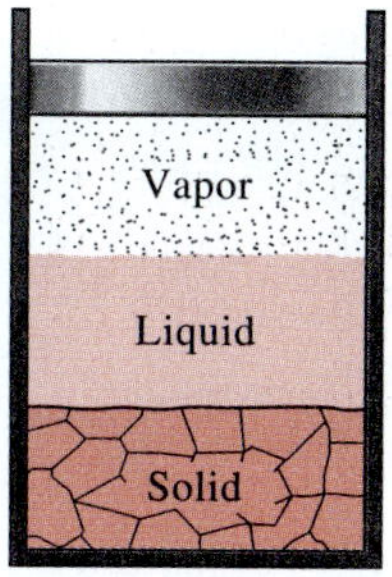

FIGURE 3–21
At triple-point pressure and temperature, a substance exists in three phases in equilibrium.

TABLE 3–3

Triple-point temperatures and pressures of various substances

Substance	Formula	T_{tp}, K	P_{tp}, kPa
Acetylene	C_2H_2	192.4	120
Ammonia	NH_3	195.40	6.076
Argon	A	83.81	68.9
Carbon (graphite)	C	3900	10,100
Carbon dioxide	CO_2	216.55	517
Carbon monoxide	CO	68.10	15.37
Deuterium	D_2	18.63	17.1
Ethane	C_2H_6	89.89	8×10^{-4}
Ethylene	C_2H_4	104.0	0.12
Helium 4 (λ point)	He	2.19	5.1
Hydrogen	H_2	13.84	7.04
Hydrogen chloride	HCl	158.96	13.9
Mercury	Hg	234.2	1.65×10^{-7}
Methane	CH_4	90.68	11.7
Neon	Ne	24.57	43.2
Nitric oxide	NO	109.50	21.92
Nitrogen	N_2	63.18	12.6
Nitrous oxide	N_2O	182.34	87.85
Oxygen	O_2	54.36	0.152
Palladium	Pd	1825	3.5×10^{-3}
Platinum	Pt	2045	2.0×10^{-4}
Sulfur dioxide	SO_2	197.69	1.67
Titanium	Ti	1941	5.3×10^{-3}
Uranium hexafluoride	UF_6	337.17	151.7
Water	H_2O	273.16	0.61
Xenon	Xe	161.3	81.5
Zinc	Zn	692.65	0.065

Source: Data from National Bureau of Standards (U.S.) Circ., 500 (1952).

water cannot exist in liquid form in equilibrium at atmospheric pressure at temperatures below 0°C, but it can exist as a liquid at −20°C at 200 MPa pressure. Also, ice exists at seven different solid phases at pressures above 100 MPa.

There are two ways a substance can pass from the solid to vapor phase: either it melts first into a liquid and subsequently evaporates, or it evaporates directly without melting first. The latter occurs at pressures below the triple-point value, since a pure substance cannot exist in the liquid phase at those pressures (Fig. 3–22). Passing from the solid phase directly into the vapor phase is called **sublimation**. For substances that have a triple-point pressure above the atmospheric pressure such as solid CO_2 (dry ice), sublimation is the only way to change from the solid to vapor phase at atmospheric conditions.

FIGURE 3–22
At low pressures (below the triple-point value), solids evaporate without melting first (*sublimation*).

3 The *P-T* Diagram

Figure 3–23 shows the *P-T* diagram of a pure substance. This diagram is often called the **phase diagram** since all three phases are separated from each other by three lines. The sublimation line separates the solid and vapor

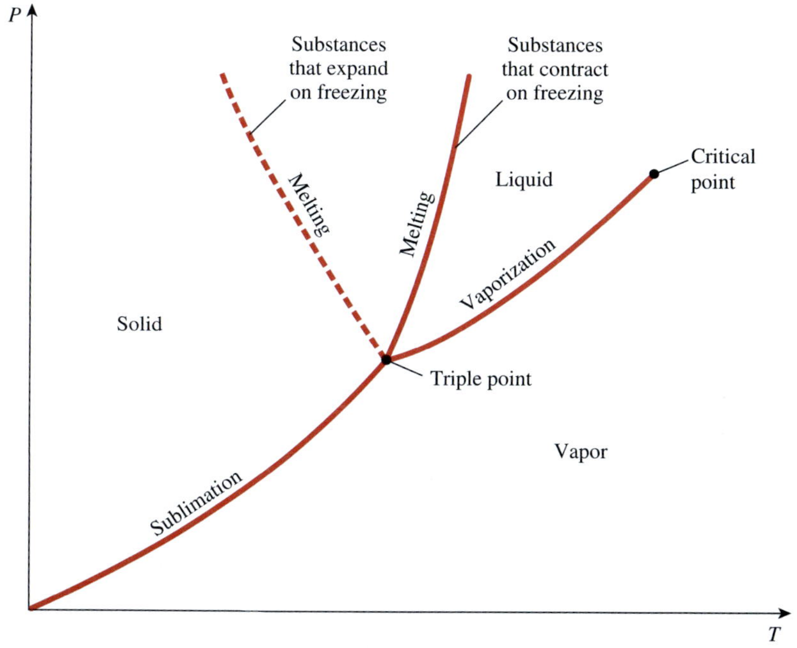

FIGURE 3–23
P-T diagram of pure substances.

regions, the vaporization line separates the liquid and vapor regions, and the melting (or fusion) line separates the solid and liquid regions. These three lines meet at the triple point, where all three phases coexist in equilibrium. The vaporization line ends at the critical point because no distinction can be made between liquid and vapor phases above the critical point. Substances that expand and contract on freezing differ only in the melting line on the *P-T* diagram.

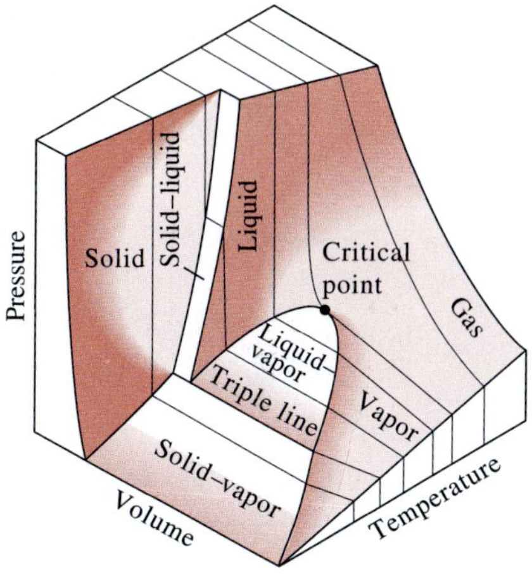

FIGURE 3–24
P-v-T surface of a substance that *contracts* on freezing.

The *P-v-T* Surface

The state of a simple compressible substance is fixed by any two independent, intensive properties. Once the two appropriate properties are fixed, all the other properties become dependent properties. Remembering that any equation with two independent variables in the form $z = z(x, y)$ represents a surface in space, we can represent the *P-v-T* behavior of a substance as a surface in space, as shown in Figs. 3–24 and 3–25. Here T and v may be viewed as the independent variables (the base) and P as the dependent variable (the height).

All the points on the surface represent equilibrium states. All states along the path of a quasi-equilibrium process lie on the *P-v-T* surface since such a process must pass through equilibrium states. The single-phase regions appear as curved surfaces on the *P-v-T* surface, and the two-phase regions as surfaces perpendicular to the *P-T* plane. This is expected since the projections of two-phase regions on the *P-T* plane are lines.

All the two-dimensional diagrams we have discussed so far are merely projections of this three-dimensional surface onto the appropriate planes. A *P-v* diagram is just a projection of the *P-v-T* surface on the *P-v* plane, and a *T-v* diagram is nothing more than the bird's-eye view of this surface.

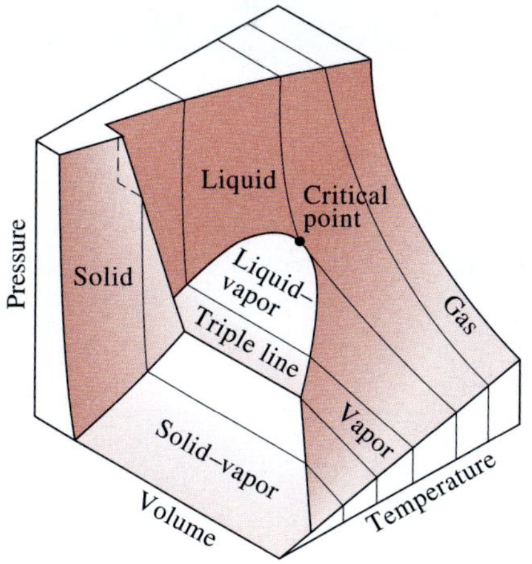

FIGURE 3–25
P-v-T surface of a substance that *expands* on freezing (like water).

The P-v-T surfaces present a great deal of information at once, but in a thermodynamic analysis it is more convenient to work with two-dimensional diagrams, such as the P-v and T-v diagrams.

3–5 ▪ PROPERTY TABLES

For most substances, the relationships among thermodynamic properties are too complex to be expressed by simple equations. Therefore, properties are frequently presented in the form of tables. Some thermodynamic properties can be measured easily, but others cannot and are calculated by using the relations between them and measurable properties. The results of these measurements and calculations are presented in tables in a convenient format. In the following discussion, the steam tables are used to demonstrate the use of thermodynamic property tables. Property tables of other substances are used in the same manner.

For each substance, the thermodynamic properties are listed in more than one table. In fact, a separate table is prepared for each region of interest such as the superheated vapor, compressed liquid, and saturated (mixture) regions. Property tables are given in the appendix in both SI and English units. The tables in English units carry the same number as the corresponding tables in SI, followed by an identifier E. Tables A–6 and A–6E, for example, list properties of superheated water vapor, the former in SI and the latter in English units. Before we get into the discussion of property tables, we define a new property called *enthalpy*.

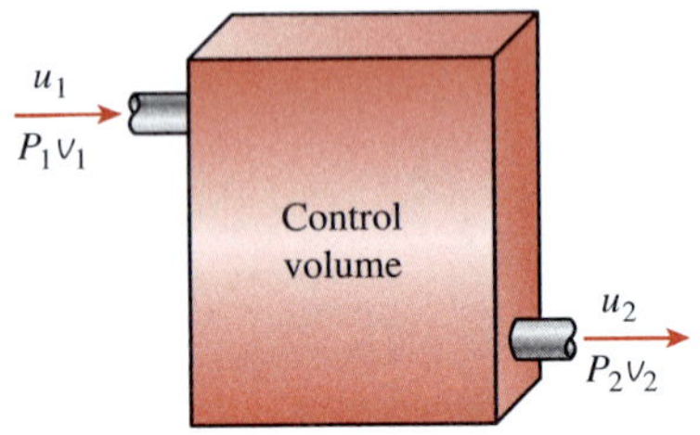

FIGURE 3–26
The combination $u + Pv$ is frequently encountered in the analysis of control volumes.

Enthalpy—A Combination Property

A person looking at the tables will notice two new properties: enthalpy h and entropy s. Entropy is a property associated with the second law of thermodynamics, and we will not use it until it is properly defined in Chap. 7. However, it is appropriate to introduce enthalpy at this point.

In the analysis of certain types of processes, particularly in power generation and refrigeration (Fig. 3–26), we frequently encounter the combination of properties $u + Pv$. For the sake of simplicity and convenience, this combination is defined as a new property, **enthalpy**, and given the symbol h:

$$h = u + Pv \qquad \text{(kJ/kg)} \tag{3–1}$$

or,

$$H = U + PV \qquad \text{(kJ)} \tag{3–2}$$

Both the total enthalpy H and specific enthalpy h are simply referred to as enthalpy since the context clarifies which one is meant. Notice that the equations given above are dimensionally homogeneous. That is, the unit of the pressure–volume product may differ from the unit of the internal energy by only a factor (Fig. 3–27). For example, it can be easily shown that 1 kPa·m^3 = 1 kJ. In some tables encountered in practice, the internal energy u is frequently not listed, but it can always be determined from $u = h - Pv$.

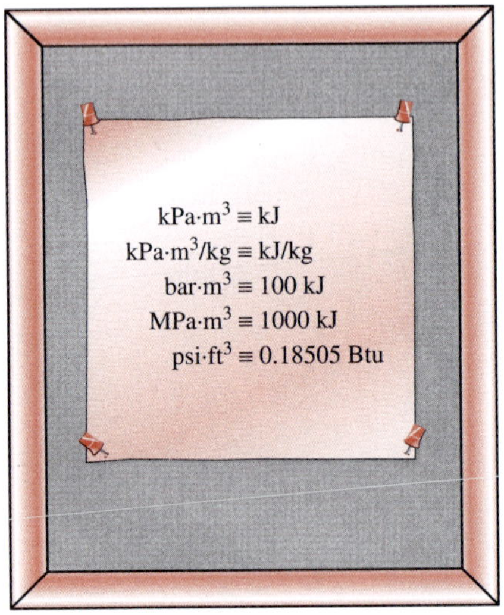

FIGURE 3–27
The product *pressure* × *volume* has energy units.

The widespread use of the property enthalpy is due to Professor Richard Mollier, who recognized the importance of the group $u + Pv$ in the analysis of steam turbines and in the representation of the properties of steam in tabular and graphical form (as in the famous Mollier chart). Mollier referred to the group $u + Pv$ as *heat content* and *total heat*. These terms were not quite consistent with the modern thermodynamic terminology and were replaced in the 1930s by the term *enthalpy* (from the Greek word *enthalpien*, which means *to heat*).

1a Saturated Liquid and Saturated Vapor States

The properties of saturated liquid and saturated vapor for water are listed in Tables A–4 and A–5. Both tables give the same information. The only difference is that in Table A–4 properties are listed under temperature and in Table A–5 under pressure. Therefore, it is more convenient to use Table A–4 when *temperature* is given and Table A–5 when *pressure* is given. The use of Table A–4 is illustrated in Fig. 3–28.

The subscript f is used to denote properties of a saturated liquid, and the subscript g to denote the properties of saturated vapor. These symbols are commonly used in thermodynamics and originated from German. Another subscript commonly used is fg, which denotes the difference between the saturated vapor and saturated liquid values of the same property. For example,

$$v_f = \text{specific volume of saturated liquid}$$

$$v_g = \text{specific volume of saturated vapor}$$

$$v_{fg} = \text{difference between } v_g \text{ and } v_f \text{ (that is, } v_{fg} = v_g - v_f)$$

The quantity h_{fg} is called the **enthalpy of vaporization** (or latent heat of vaporization). It represents the amount of energy needed to vaporize a unit mass of saturated liquid at a given temperature or pressure. It decreases as the temperature or pressure increases and becomes zero at the critical point.

Temp. °C T	Sat. press. kPa P_{sat}	Specific volume m^3/kg: Sat. liquid v_f	Sat. vapor v_g
85	57.868	0.001032	2.8261
90	70.183	0.001036	2.3593
95	84.609	0.001040	1.9808

Specific temperature

Corresponding saturation pressure

Specific volume of saturated liquid

Specific volume of saturated vapor

FIGURE 3–28
A partial list of Table A–4.

EXAMPLE 3–1 Pressure of Saturated Liquid in a Tank

A rigid tank contains 50 kg of saturated liquid water at 90°C. Determine the pressure in the tank and the volume of the tank.

SOLUTION A rigid tank contains saturated liquid water. The pressure and volume of the tank are to be determined.

Analysis The state of the saturated liquid water is shown on a T-v diagram in Fig. 3–29. Since saturation conditions exist in the tank, the pressure must be the saturation pressure at 90°C:

$$P = P_{sat @ 90°C} = \mathbf{70.183\ kPa} \qquad \text{(Table A–4)}$$

The specific volume of the saturated liquid at 90°C is

$$v = v_{f @ 90°C} = 0.001036\ \text{m}^3/\text{kg} \qquad \text{(Table A–4)}$$

Then the total volume of the tank becomes

$$V = mv = (50\ \text{kg})(0.001036\ \text{m}^3/\text{kg}) = \mathbf{0.0518\ m^3}$$

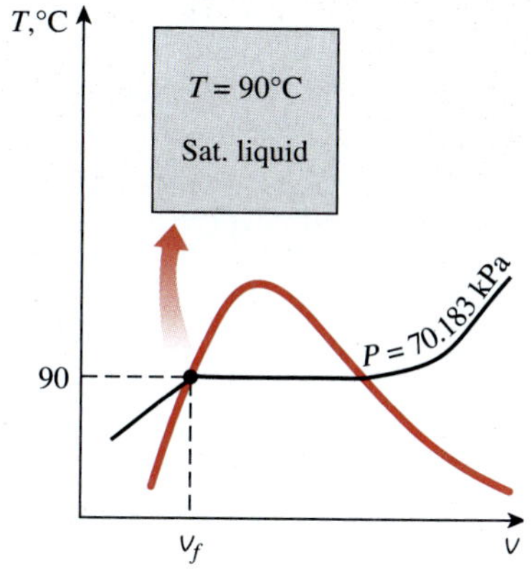

FIGURE 3–29
Schematic and T-v diagram for Example 3–1.

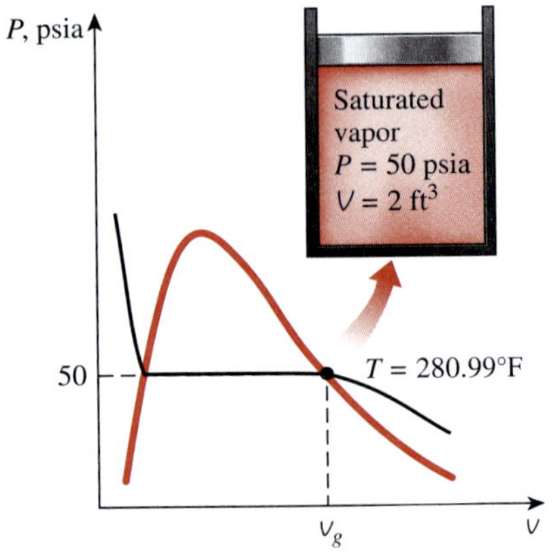

FIGURE 3–30
Schematic and P-v diagram for Example 3–2.

EXAMPLE 3–2 Temperature of Saturated Vapor in a Cylinder

A piston–cylinder device contains 2 ft^3 of saturated water vapor at 50-psia pressure. Determine the temperature and the mass of the vapor inside the cylinder.

SOLUTION A cylinder contains saturated water vapor. The temperature and the mass of vapor are to be determined.

Analysis The state of the saturated water vapor is shown on a P-v diagram in Fig. 3–30. Since the cylinder contains saturated vapor at 50 psia, the temperature inside must be the saturation temperature at this pressure:

$$T = T_{\text{sat @ 50 psia}} = \mathbf{280.99°F} \qquad \text{(Table A–5E)}$$

The specific volume of the saturated vapor at 50 psia is

$$v = v_{g\text{ @ 50 psia}} = 8.5175 \text{ ft}^3/\text{lbm} \qquad \text{(Table A–5E)}$$

Then the mass of water vapor inside the cylinder becomes

$$m = \frac{V}{v} = \frac{2 \text{ ft}^3}{8.5175 \text{ ft}^3/\text{lbm}} = \mathbf{0.235\ lbm}$$

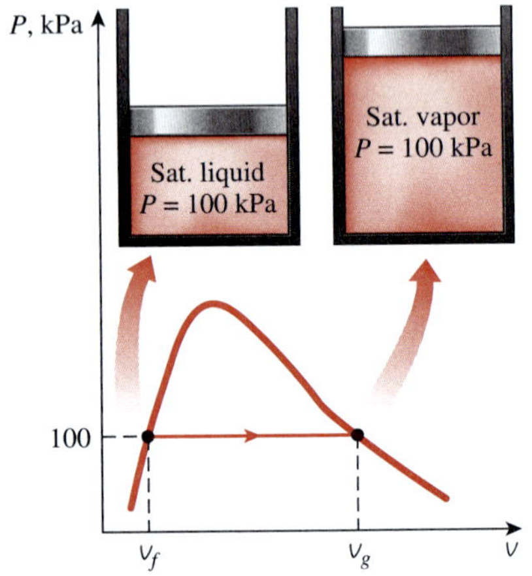

FIGURE 3–31
Schematic and P-v diagram for Example 3–3.

EXAMPLE 3–3 Volume and Energy Change during Evaporation

A mass of 200 g of saturated liquid water is completely vaporized at a constant pressure of 100 kPa. Determine (*a*) the volume change and (*b*) the amount of energy transferred to the water.

SOLUTION Saturated liquid water is vaporized at constant pressure. The volume change and the energy transferred are to be determined.

Analysis (*a*) The process described is illustrated on a P-v diagram in Fig. 3–31. The volume change per unit mass during a vaporization process is v_{fg}, which is the difference between v_g and v_f. Reading these values from Table A–5 at 100 kPa and substituting yield

$$v_{fg} = v_g - v_f = 1.6941 - 0.001043 = 1.6931 \text{ m}^3/\text{kg}$$

Thus,

$$\Delta V = m v_{fg} = (0.2 \text{ kg})(1.6931 \text{ m}^3/\text{kg}) = \mathbf{0.3386\ m^3}$$

(*b*) The amount of energy needed to vaporize a unit mass of a substance at a given pressure is the enthalpy of vaporization at that pressure, which is h_{fg} = 2257.5 kJ/kg for water at 100 kPa. Thus, the amount of energy transferred is

$$m h_{fg} = (0.2 \text{ kg})(2257.5 \text{ kJ/kg}) = \mathbf{451.5\ kJ}$$

Discussion Note that we have considered the first four decimal digits of v_{fg} and disregarded the rest. This is because v_g has significant numbers to the first four decimal places only, and we do not know the numbers in the other decimal places. Copying all the digits from the calculator would mean that we are assuming v_g = 1.694100, which is not necessarily the case. It could very well be that v_g = 1.694138 since this number, too, would truncate to 1.6941. All the digits in our result (1.6931) are significant. But if we did not truncate the result, we would obtain v_{fg} = 1.693057, which falsely implies that our result is accurate to the sixth decimal place.

1b Saturated Liquid–Vapor Mixture

During a vaporization process, a substance exists as part liquid and part vapor. That is, it is a mixture of saturated liquid and saturated vapor (Fig. 3–32). To analyze this mixture properly, we need to know the proportions of the liquid and vapor phases in the mixture. This is done by defining a new property called the **quality** x as the ratio of the mass of vapor to the total mass of the mixture:

$$x = \frac{m_{\text{vapor}}}{m_{\text{total}}} \quad \textbf{(3–3)}$$

where

$$m_{\text{total}} = m_{\text{liquid}} + m_{\text{vapor}} = m_f + m_g$$

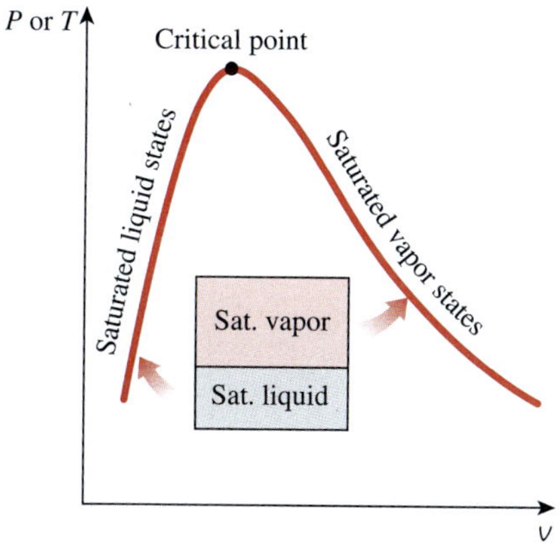

FIGURE 3–32
The relative amounts of liquid and vapor phases in a saturated mixture are specified by the *quality x*.

Quality has significance for *saturated mixtures* only. It has no meaning in the compressed liquid or superheated vapor regions. Its value is between 0 and 1. The quality of a system that consists of *saturated liquid* is 0 (or 0 percent), and the quality of a system consisting of *saturated vapor* is 1 (or 100 percent). In saturated mixtures, quality can serve as one of the two independent intensive properties needed to describe a state. Note that *the properties of the saturated liquid are the same whether it exists alone or in a mixture with saturated vapor.* During the vaporization process, only the amount of saturated liquid changes, not its properties. The same can be said about a saturated vapor.

A saturated mixture can be treated as a combination of two subsystems: the saturated liquid and the saturated vapor. However, the amount of mass for each phase is usually not known. Therefore, it is often more convenient to imagine that the two phases are mixed well, forming a homogeneous mixture (Fig. 3–33). Then the properties of this "mixture" will simply be the average properties of the saturated liquid–vapor mixture under consideration. Here is how it is done.

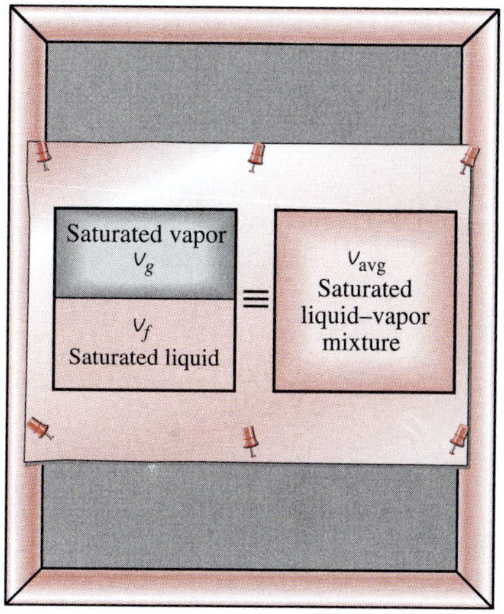

FIGURE 3–33
A two-phase system can be treated as a homogeneous mixture for convenience.

Consider a tank that contains a saturated liquid–vapor mixture. The volume occupied by saturated liquid is V_f, and the volume occupied by saturated vapor is V_g. The total volume V is the sum of the two:

$$V = V_f + V_g$$

$$V = mv \longrightarrow m_t v_{\text{avg}} = m_f v_f + m_g v_g$$

$$m_f = m_t - m_g \longrightarrow m_t v_{\text{avg}} = (m_t - m_g)v_f + m_g v_g$$

Dividing by m_t yields

$$v_{\text{avg}} = (1 - x)v_f + xv_g$$

since $x = m_g/m_t$. This relation can also be expressed as

$$v_{\text{avg}} = v_f + xv_{fg} \qquad (\text{m}^3/\text{kg}) \quad \textbf{(3–4)}$$

where $v_{fg} = v_g - v_f$. Solving for quality, we obtain

$$x = \frac{v_{\text{avg}} - v_f}{v_{fg}} \quad \textbf{(3–5)}$$

Based on this equation, quality can be related to the horizontal distances on a *P*-*v* or *T*-*v* diagram (Fig. 3–34). At a given temperature or pressure, the numerator of Eq. 3–5 is the distance between the actual state and the saturated liquid state, and the denominator is the length of the entire horizontal line that connects the saturated liquid and saturated vapor states. A state of 50 percent quality lies in the middle of this horizontal line.

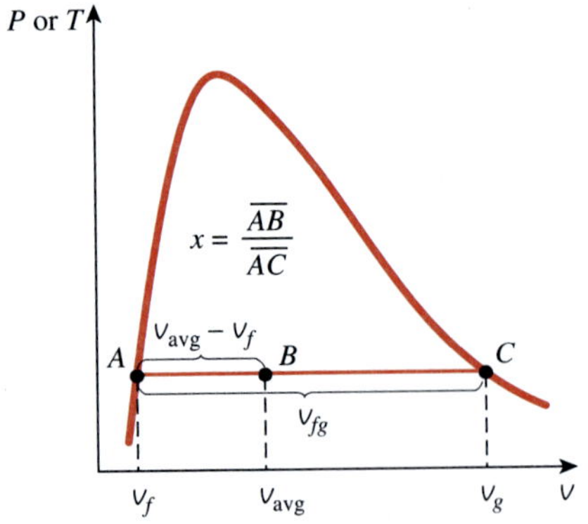

FIGURE 3–34
Quality is related to the horizontal distances on P-v and T-v diagrams.

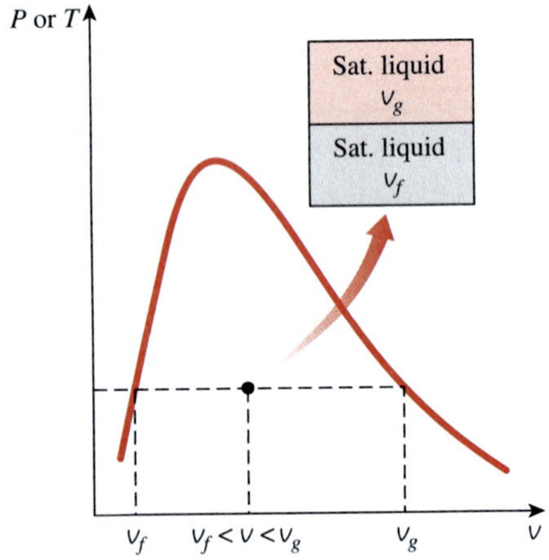

FIGURE 3–35
The v value of a saturated liquid–vapor mixture lies between the v_f and v_g values at the specified T or P.

The analysis given above can be repeated for internal energy and enthalpy with the following results:

$$u_{avg} = u_f + xu_{fg} \qquad \text{(kJ/kg)} \tag{3-6}$$

$$h_{avg} = h_f + xh_{fg} \qquad \text{(kJ/kg)} \tag{3-7}$$

All the results are of the same format, and they can be summarized in a single equation as

$$y_{avg} = y_f + xy_{fg}$$

where y is v, u, or h. The subscript "avg" (for "average") is usually dropped for simplicity. The values of the average properties of the mixtures are always *between* the values of the saturated liquid and the saturated vapor properties (Fig. 3–35). That is,

$$y_f \leq y_{avg} \leq y_g$$

Finally, all the saturated-mixture states are located under the saturation curve, and to analyze saturated mixtures, all we need are saturated liquid and saturated vapor data (Tables A–4 and A–5 in the case of water).

EXAMPLE 3–4 Pressure and Volume of a Saturated Mixture

A rigid tank contains 10 kg of water at 90°C. If 8 kg of the water is in the liquid form and the rest is in the vapor form, determine (*a*) the pressure in the tank and (*b*) the volume of the tank.

SOLUTION A rigid tank contains saturated mixture. The pressure and the volume of the tank are to be determined.

Analysis (*a*) The state of the saturated liquid–vapor mixture is shown in Fig. 3–36. Since the two phases coexist in equilibrium, we have a saturated mixture, and the pressure must be the saturation pressure at the given temperature:

$$P = P_{sat\,@\,90°C} = \mathbf{70.183\ kPa} \qquad \text{(Table A–4)}$$

(*b*) At 90°C, we have v_f = 0.001036 m³/kg and v_g = 2.3593 m³/kg (Table A-4). One way of finding the volume of the tank is to determine the volume occupied by each phase and then add them:

$$\begin{aligned} V &= V_f + V_g = m_f v_f + m_g v_g \\ &= (8\text{ kg})(0.001036\text{ m}^3/\text{kg}) + (2\text{ kg})(2.3593\text{ m}^3/\text{kg}) \\ &= \mathbf{4.73\ m^3} \end{aligned}$$

Another way is to first determine the quality x, then the average specific volume v, and finally the total volume:

$$x = \frac{m_g}{m_t} = \frac{2\text{ kg}}{10\text{ kg}} = 0.2$$

$$\begin{aligned} v &= v_f + xv_{fg} \\ &= 0.001036\text{ m}^3/\text{kg} + (0.2)[(2.3593 - 0.001036)\text{ m}^3/\text{kg}] \\ &= 0.473\text{ m}^3/\text{kg} \end{aligned}$$

and

$$V = mv = (10 \text{ kg})(0.473 \text{ m}^3/\text{kg}) = 4.73 \text{ m}^3$$

Discussion The first method appears to be easier in this case since the masses of each phase are given. In most cases, however, the masses of each phase are not available, and the second method becomes more convenient.

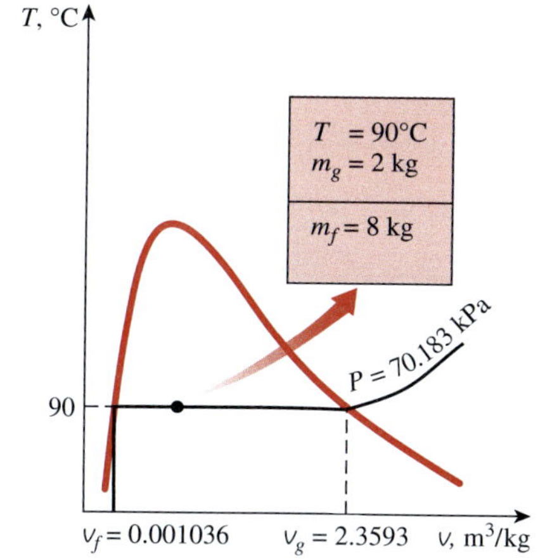

FIGURE 3–36
Schematic and T-v diagram for Example 3–4.

EXAMPLE 3–5 Properties of Saturated Liquid–Vapor Mixture

An 80-L vessel contains 4 kg of refrigerant-134a at a pressure of 160 kPa. Determine (*a*) the temperature, (*b*) the quality, (*c*) the enthalpy of the refrigerant, and (*d*) the volume occupied by the vapor phase.

SOLUTION A vessel is filled with refrigerant-134a. Some properties of the refrigerant are to be determined.

Analysis (*a*) The state of the saturated liquid–vapor mixture is shown in Fig. 3–37. At this point we do not know whether the refrigerant is in the compressed liquid, superheated vapor, or saturated mixture region. This can be determined by comparing a suitable property to the saturated liquid and saturated vapor values. From the information given, we can determine the specific volume:

$$v = \frac{V}{m} = \frac{0.080 \text{ m}^3}{4 \text{ kg}} = 0.02 \text{ m}^3/\text{kg}$$

At 160 kPa, we read

$$v_f = 0.0007437 \text{ m}^3/\text{kg} \qquad \text{(Table A–12)}$$
$$v_g = 0.12348 \text{ m}^3/\text{kg}$$

Obviously, $v_f < v < v_g$, and, the refrigerant is in the saturated mixture region. Thus, the temperature must be the saturation temperature at the specified pressure:

$$T = T_{\text{sat @ 160 kPa}} = \mathbf{-15.60°C}$$

(*b*) Quality can be determined from

$$x = \frac{v - v_f}{v_{fg}} = \frac{0.02 - 0.0007437}{0.12348 - 0.0007437} = \mathbf{0.157}$$

(*c*) At 160 kPa, we also read from Table A–12 that h_f = 31.21 kJ/kg and h_{fg} = 209.90 kJ/kg. Then,

$$\begin{aligned} h &= h_f + xh_{fg} \\ &= 31.21 \text{ kJ/kg} + (0.157)(209.90 \text{ kJ/kg}) \\ &= \mathbf{64.2 \text{ kJ/kg}} \end{aligned}$$

(*d*) The mass of the vapor is

$$m_g = xm_t = (0.157)(4 \text{ kg}) = 0.628 \text{ kg}$$

and the volume occupied by the vapor phase is

$$V_g = m_g v_g = (0.628 \text{ kg})(0.12348 \text{ m}^3/\text{kg}) = \mathbf{0.0775 \text{ m}^3} \text{ (or 77.5 L)}$$

The rest of the volume (2.5 L) is occupied by the liquid.

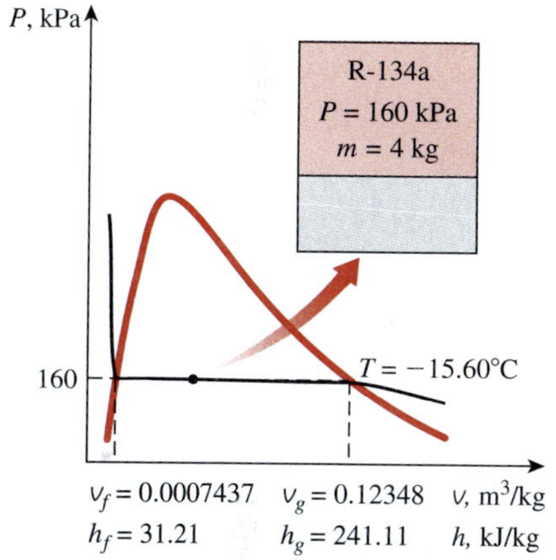

FIGURE 3–37
Schematic and P-v diagram for Example 3–5.

T, °C	v m³/kg	u kJ/kg	h kJ/kg
	P = 0.1 MPa (99.61°C)		
Sat.	1.6941	2505.6	2675.0
100	1.6959	2506.2	2675.8
150	1.9367	2582.9	2776.6
⋮	⋮	⋮	⋮
1300	7.2605	4687.2	5413.3
	P = 0.5 MPa (151.83°C)		
Sat.	0.37483	2560.7	2748.1
200	0.42503	2643.3	2855.8
250	0.47443	2723.8	2961.0

FIGURE 3–38
A partial listing of Table A–6.

Property tables are also available for saturated solid–vapor mixtures. Properties of saturated ice–water vapor mixtures, for example, are listed in Table A–8. Saturated solid–vapor mixtures can be handled just as saturated liquid–vapor mixtures.

2 Superheated Vapor

In the region to the right of the saturated vapor line and at temperatures above the critical point temperature, a substance exists as superheated vapor. Since the superheated region is a single-phase region (vapor phase only), temperature and pressure are no longer dependent properties and they can conveniently be used as the two independent properties in the tables. The format of the superheated vapor tables is illustrated in Fig. 3–38.

In these tables, the properties are listed against temperature for selected pressures starting with the saturated vapor data. The saturation temperature is given in parentheses following the pressure value.

Compared to saturated vapor, superheated vapor is characterized by

Lower pressures ($P < P_{sat}$ at a given T)
Higher tempreatures ($T > T_{sat}$ at a given P)
Higher specific volumes ($v > v_g$ at a given P or T)
Higher internal energies ($u > u_g$ at a given P or T)
Higher enthalpies ($h > h_g$ at a given P or T)

EXAMPLE 3–6 Internal Energy of Superheated Vapor

Determine the internal energy of water at 20 psia and 400°F.

SOLUTION The internal energy of water at a specified state is to be determined.
Analysis At 20 psia, the saturation temperature is 227.92°F. Since $T > T_{sat}$, the water is in the superheated vapor region. Then the internal energy at the given temperature and pressure is determined from the superheated vapor table (Table A–6E) to be

$$u = \mathbf{1145.1\ Btu/lbm}$$

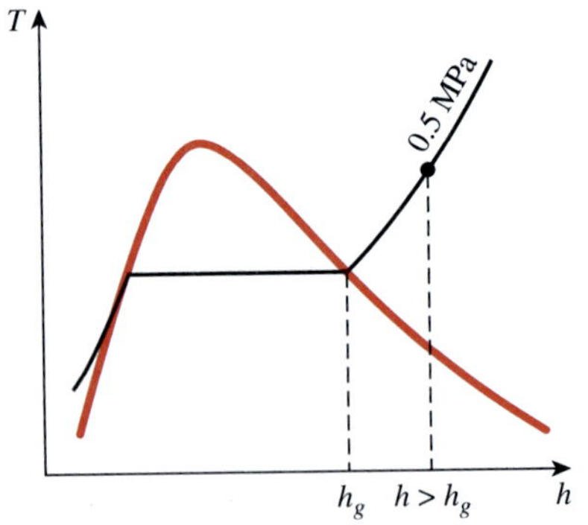

FIGURE 3–39
At a specified P, superheated vapor exists at a higher h than the saturated vapor (Example 3–7).

EXAMPLE 3–7 Temperature of Superheated Vapor

Determine the temperature of water at a state of P = 0.5 MPa and h = 2890 kJ/kg.

SOLUTION The temperature of water at a specified state is to be determined.
Analysis At 0.5 MPa, the enthalpy of saturated water vapor is h_g = 2748.1 kJ/kg. Since $h > h_g$, as shown in Fig. 3–39, we again have superheated vapor. Under 0.5 MPa in Table A–6 we read

T, °C	h, kJ/kg
200	2855.8
250	2961.0

Obviously, the temperature is between 200 and 250°C. By linear interpolation it is determined to be

$$T = \mathbf{216.3°C}$$

3 Compressed Liquid

Compressed liquid tables are not as commonly available, and Table A–7 is the only compressed liquid table in this text. The format of Table A–7 is very much like the format of the superheated vapor tables. One reason for the lack of compressed liquid data is the relative independence of compressed liquid properties from pressure. Variation of properties of compressed liquid with pressure is very mild. Increasing the pressure 100 times often causes properties to change less than 1 percent.

In the absence of compressed liquid data, a general approximation is *to treat compressed liquid as saturated liquid at the given temperature* (Fig. 3–40). This is because the compressed liquid properties depend on temperature much more strongly than they do on pressure. Thus,

$$y \cong y_{f\,@\,T} \tag{3–8}$$

for compressed liquids, where y is v, u, or h. Of these three properties, the property whose value is most sensitive to variations in the pressure is the enthalpy h. Although the above approximation results in negligible error in v and u, the error in h may reach undesirable levels. However, the error in h at low to moderate pressures and temperatures can be reduced significantly by evaluating it from

$$h \cong h_{f\,@\,T} + v_{f\,@\,T}(P - P_{\text{sat}\,@\,T}) \tag{3–9}$$

instead of taking it to be just h_f. Note, however, that the approximation in Eq. 3–9 does not yield any significant improvement at moderate to high temperatures and pressures, and it may even backfire and result in greater error due to overcorrection at very high temperatures and pressures (*see* Kostic, Ref. 4).

In general, a compressed liquid is characterized by

Higher pressures ($P > P_{\text{sat}}$ at a given T)
Lower tempreatures ($T < T_{\text{sat}}$ at a given P)
Lower specific volumes ($v < v_f$ at a given P or T)
Lower internal energies ($u < u_f$ at a given P or T)
Lower enthalpies ($h < h_f$ at a given P or T)

But unlike superheated vapor, the compressed liquid properties are not much different from the corresponding saturated liquid values.

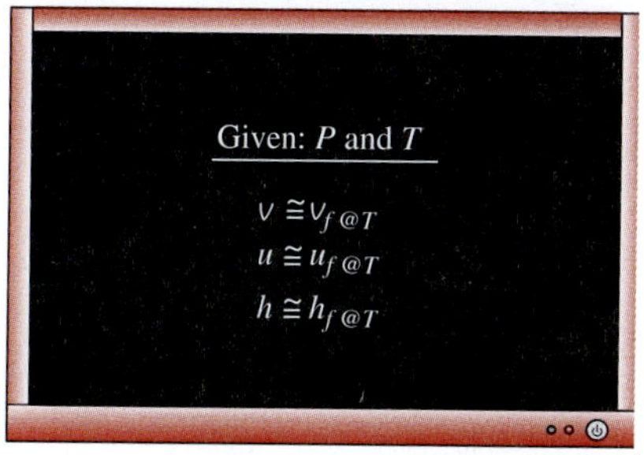

FIGURE 3–40
A compressed liquid may be approximated as a saturated liquid at the given temperature.

EXAMPLE 3–8 Approximating Compressed Liquid as Saturated Liquid

Determine the internal energy of compressed liquid water at 80°C and 5 MPa, using (*a*) data from the compressed liquid table and (*b*) saturated liquid data. What is the error involved in the second case?

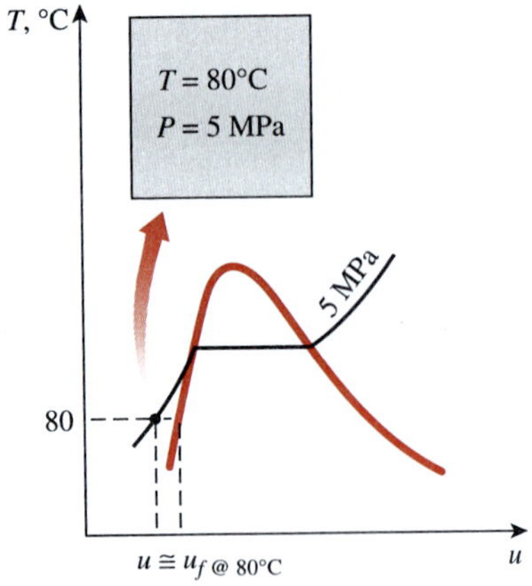

FIGURE 3–41
Schematic and T-u diagram for Example 3–8.

SOLUTION The exact and approximate values of the internal energy of liquid water are to be determined.

Analysis At 80°C, the saturation pressure of water is 47.416 kPa, and since 5 MPa $> P_{sat}$, we obviously have compressed liquid, as shown in Fig. 3–41.

(*a*) From the compressed liquid table (Table A–7)

$$\left.\begin{array}{l} P = 5 \text{ MPa} \\ T = 80°\text{C} \end{array}\right\} \quad u = \mathbf{333.82 \ kJ/kg}$$

(*b*) From the saturation table (Table A–4), we read

$$u \cong u_{f\,@\,80°\text{C}} = \mathbf{334.97 \ kJ/kg}$$

The error involved is

$$\frac{334.97 - 333.82}{333.82} \times 100 = \mathbf{0.34\%}$$

which is less than 1 percent.

Reference State and Reference Values

The values of u, h, and s cannot be measured directly, and they are calculated from measurable properties using the relations between thermodynamic properties. However, those relations give the *changes* in properties, not the values of properties at specified states. Therefore, we need to choose a convenient *reference state* and assign a value of *zero* for a convenient property or properties at that state. For water, the state of saturated liquid at 0.01°C is taken as the reference state, and the internal energy and entropy are assigned zero values at that state. For refrigerant-134a, the state of saturated liquid at −40°C is taken as the reference state, and the enthalpy and entropy are assigned zero values at that state. Note that some properties may have negative values as a result of the reference state chosen.

It should be mentioned that sometimes different tables list different values for some properties at the same state as a result of using a different reference state. However, in thermodynamics we are concerned with the *changes* in properties, and the reference state chosen is of no consequence in calculations as long as we use values from a single consistent set of tables or charts.

EXAMPLE 3–9 **The Use of Steam Tables to Determine Properties**

Determine the missing properties and the phase descriptions in the following table for water:

	T, °C	P, kPa	u, kJ/kg	x	Phase description
(*a*)		200		0.6	
(*b*)	125		1600		
(*c*)		1000	2950		
(*d*)	75	500			
(*e*)		850		0.0	

SOLUTION Properties and phase descriptions of water are to be determined at various states.

Analysis (*a*) The quality is given to be $x = 0.6$, which implies that 60 percent of the mass is in the vapor phase and the remaining 40 percent is in the liquid phase. Therefore, we have saturated liquid–vapor mixture at a pressure of 200 kPa. Then the temperature must be the saturation temperature at the given pressure:

$$T = T_{\text{sat @ 200 kPa}} = \mathbf{120.21°C} \qquad \text{(Table A–5)}$$

At 200 kPa, we also read from Table A–5 that $u_f = 504.50$ kJ/kg and $u_{fg} = 2024.6$ kJ/kg. Then the average internal energy of the mixture is

$$\begin{aligned} u &= u_f + xu_{fg} \\ &= 504.50 \text{ kJ/kg} + (0.6)(2024.6 \text{ kJ/kg}) \\ &= \mathbf{1719.26 \text{ kJ/kg}} \end{aligned}$$

(*b*) This time the temperature and the internal energy are given, but we do not know which table to use to determine the missing properties because we have no clue as to whether we have saturated mixture, compressed liquid, or superheated vapor. To determine the region we are in, we first go to the saturation table (Table A–4) and determine the u_f and u_g values at the given temperature. At 125°C, we read $u_f = 524.83$ kJ/kg and $u_g = 2534.3$ kJ/kg. Next we compare the given u value to these u_f and u_g values, keeping in mind that

if	$u < u_f$	we have *compressed liquid*
if	$u_f \le u \le u_g$	we have *saturated mixture*
if	$u > u_g$	we have *superheated vapor*

In our case the given u value is 1600, which falls between the u_f and u_g values at 125°C. Therefore, we have saturated liquid–vapor mixture. Then the pressure must be the saturation pressure at the given temperature:

$$P = P_{\text{sat @ 125°C}} = \mathbf{232.23 \text{ kPa}} \qquad \text{(Table A–4)}$$

The quality is determined from

$$x = \frac{u - u_f}{u_{fg}} = \frac{1600 - 524.83}{2009.5} = \mathbf{0.535}$$

The criteria above for determining whether we have compressed liquid, saturated mixture, or superheated vapor can also be used when enthalpy h or specific volume v is given instead of internal energy u, or when pressure is given instead of temperature.

(*c*) This is similar to case (*b*), except pressure is given instead of temperature. Following the argument given above, we read the u_f and u_g values at the specified pressure. At 1 MPa, we have $u_f = 761.39$ kJ/kg and $u_g = 2582.8$ kJ/kg. The specified u value is 2950 kJ/kg, which is greater than the u_g value at 1 MPa. Therefore, we have superheated vapor, and the temperature at this state is determined from the superheated vapor table by interpolation to be

$$T = \mathbf{395.2°C} \qquad \text{(Table A–6)}$$

We would leave the quality column blank in this case since quality has no meaning for a superheated vapor.

(*d*) In this case the temperature and pressure are given, but again we cannot tell which table to use to determine the missing properties because we do

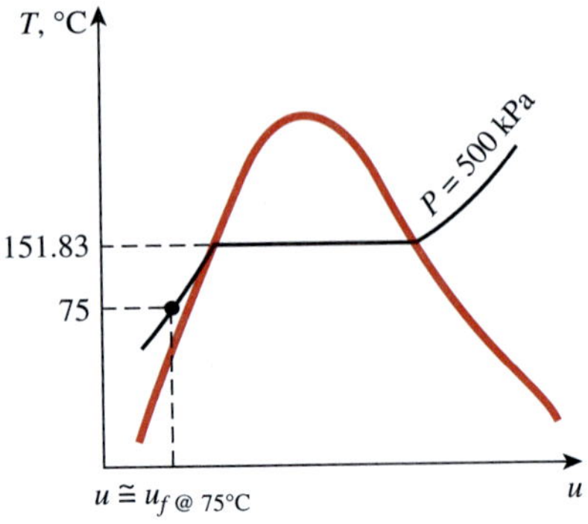

FIGURE 3–42
At a given P and T, a pure substance will exist as a compressed liquid if $T < T_{sat\ @\ P}$.

not know whether we have saturated mixture, compressed liquid, or superheated vapor. To determine the region we are in, we go to the saturation table (Table A–5) and determine the saturation temperature value at the given pressure. At 500 kPa, we have $T_{sat} = 151.83°C$. We then compare the given T value to this T_{sat} value, keeping in mind that

if	$T < T_{sat\ @\ given\ P}$	we have *compressed liquid*
if	$T = T_{sat\ @\ given\ P}$	we have *saturated mixture*
if	$T > T_{sat\ @\ given\ P}$	we have *superheated vapor*

In our case, the given T value is 75°C, which is less than the T_{sat} value at the specified pressure. Therefore, we have compressed liquid (Fig. 3–42), and normally we would determine the internal energy value from the compressed liquid table. But in this case the given pressure is much lower than the lowest pressure value in the compressed liquid table (which is 5 MPa), and therefore we are justified to treat the compressed liquid as saturated liquid at the given temperature (*not* pressure):

$$u \cong u_{f\ @\ 75°C} = \mathbf{313.99\ kJ/kg} \qquad \text{(Table A–4)}$$

We would leave the quality column blank in this case since quality has no meaning in the compressed liquid region.

(*e*) The quality is given to be $x = 0$, and thus we have saturated liquid at the specified pressure of 850 kPa. Then the temperature must be the saturation temperature at the given pressure, and the internal energy must have the saturated liquid value:

$$T = T_{sat\ @\ 850\ kPa} = \mathbf{172.94°C}$$

$$u = u_{f\ @\ 850\ kPa} = \mathbf{731.00\ kJ/kg} \qquad \text{(Table A–5)}$$

3–6 ▪ THE IDEAL-GAS EQUATION OF STATE

Property tables provide very accurate information about the properties, but they are bulky and vulnerable to typographical errors. A more practical and desirable approach would be to have some simple relations among the properties that are sufficiently general and accurate.

Any equation that relates the pressure, temperature, and specific volume of a substance is called an **equation of state**. Property relations that involve other properties of a substance at equilibrium states are also referred to as equations of state. There are several equations of state, some simple and others very complex. The simplest and best-known equation of state for substances in the gas phase is the ideal-gas equation of state. This equation predicts the P-v-T behavior of a gas quite accurately within some properly selected region.

Gas and *vapor* are often used as synonymous words. The vapor phase of a substance is customarily called a *gas* when it is above the critical temperature. *Vapor* usually implies a gas that is not far from a state of condensation.

In 1662, Robert Boyle, an Englishman, observed during his experiments with a vacuum chamber that the pressure of gases is inversely proportional to their volume. In 1802, J. Charles and J. Gay-Lussac,

Frenchmen, experimentally determined that at low pressures the volume of a gas is proportional to its temperature. That is,

$$P = R\left(\frac{T}{v}\right)$$

or

$$Pv = RT \quad \textbf{(3–10)}$$

where the constant of proportionality R is called the **gas constant**. Equation 3–10 is called the **ideal-gas equation of state**, or simply the **ideal-gas relation**, and a gas that obeys this relation is called an **ideal gas**. In this equation, P is the absolute pressure, T is the absolute temperature, and v is the specific volume.

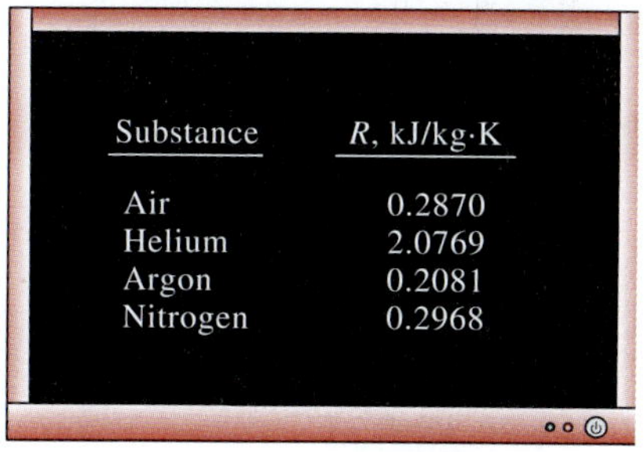

FIGURE 3–43
Different substances have different gas constants.

The gas constant R is different for each gas (Fig. 3–43) and is determined from

$$R = \frac{R_u}{M} \qquad \text{(kJ/kg·K or kPa·m}^3\text{/kg·K)}$$

where R_u is the **universal gas constant** and M is the molar mass (also called *molecular weight*) of the gas. The constant R_u is the same for all substances, and its value is

$$R_u = \begin{cases} 8.31447 \text{ kJ/kmol·K} \\ 8.31447 \text{ kPa·m}^3\text{/kmol·K} \\ 0.0831447 \text{ bar·m}^3\text{/kmol·K} \\ 1.98588 \text{ Btu/lbmol·R} \\ 10.7316 \text{ psia·ft}^3\text{/lbmol·R} \\ 1545.37 \text{ ft·lbf/lbmol·R} \end{cases} \quad \textbf{(3–11)}$$

The **molar mass** M can simply be defined as *the mass of one mole* (also called a *gram-mole*, abbreviated gmol) *of a substance in grams*, or *the mass of one kmol* (also called a *kilogram-mole*, abbreviated kgmol) *in kilograms*. In English units, it is the mass of 1 lbmol in lbm. Notice that the molar mass of a substance has the same numerical value in both unit systems because of the way it is defined. When we say the molar mass of nitrogen is 28, it simply means the mass of 1 kmol of nitrogen is 28 kg, or the mass of 1 lbmol of nitrogen is 28 lbm. That is, M = 28 kg/kmol = 28 lbm/lbmol. The mass of a system is equal to the product of its molar mass M and the mole number N:

$$m = MN \qquad \text{(kg)} \quad \textbf{(3–12)}$$

The values of R and M for several substances are given in Table A–1.

The ideal-gas equation of state can be written in several different forms:

$$V = mv \longrightarrow PV = mRT \quad \textbf{(3–13)}$$

$$mR = (MN)R = NR_u \longrightarrow PV = NR_uT \quad \textbf{(3–14)}$$

$$V = N\bar{v} \longrightarrow P\bar{v} = R_uT \quad \textbf{(3–15)}$$

where $\bar{v}$ is the molar specific volume, that is, the volume per unit mole (in m^3/kmol or ft^3/lbmol). A bar above a property denotes values on a *unit-mole basis* throughout this text (Fig. 3–44).

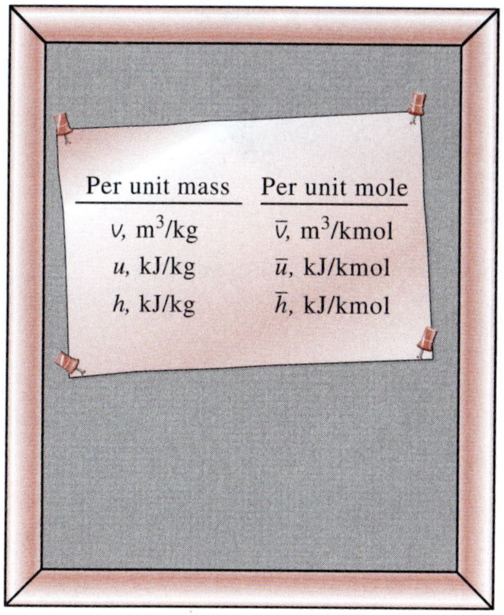

FIGURE 3–44
Properties per unit mole are denoted with a bar on the top.

FIGURE 3–45
The ideal-gas relation often is not applicable to real gases; thus, care should be exercised when using it.
BLONDIE © KING FEATURES SYNDICATE.

By writing Eq. 3–13 twice for a fixed mass and simplifying, the properties of an ideal gas at two different states are related to each other by

$$\frac{P_1 V_1}{T_1} = \frac{P_2 V_2}{T_2} \tag{3-16}$$

An ideal gas is an *imaginary* substance that obeys the relation $Pv = RT$ (Fig. 3–45). It has been experimentally observed that the ideal-gas relation given closely approximates the P-v-T behavior of real gases at low densities. At low pressures and high temperatures, the density of a gas decreases, and the gas behaves as an ideal gas under these conditions. What constitutes low pressure and high temperature is explained later.

In the range of practical interest, many familiar gases such as air, nitrogen, oxygen, hydrogen, helium, argon, neon, krypton, and even heavier gases such as carbon dioxide can be treated as ideal gases with negligible error (often less than 1 percent). Dense gases such as water vapor in steam power plants and refrigerant vapor in refrigerators, however, should not be treated as ideal gases. Instead, the property tables should be used for these substances.

EXAMPLE 3-10 Temperature Rise of Air in a Tire During a Trip

The gage pressure of an automobile tire is measured to be 210 kPa before a trip and 220 kPa after the trip at a location where the atmospheric pressure is 95 kPa (Fig. 3–46). Assuming the volume of the tire remains constant and the air temperature before the trip is 25°C, determine air temperature in the tire after the trip.

FIGURE 3–46
Schematic for Example 3-10.

SOLUTION The pressure in an automobile tire is measured before and after a trip. The temperature of air in the tire after the trip is to be determined.

Assumptions **1** The volume of the tire remains constant. **2** Air is an ideal gas.

Properties The local atmospheric pressure is 95 kPa.

Analysis The absolute pressures in the tire before and after the trip are

$$P_1 = P_{\text{gage.1}} + P_{\text{atm}} = 210 + 95 = 305 \text{ kPa}$$
$$P_2 = P_{\text{gage.2}} + P_{\text{atm}} = 220 + 90 = 315 \text{ kPa}$$

Note that air is an ideal gas and the volume is constant, the air temperatures after the trip is determined to be

$$\frac{P_1 V_1}{T_1} = \frac{P_2 V_2}{T_2} \longrightarrow T_2 = \frac{P_2}{P_1} T_1 = \frac{315 \text{ kPa}}{305 \text{ kPa}} (25 + 273 \text{ K}) = 307.8 \text{ K} = \mathbf{34.8°C}$$

Therefore, the absolute temperature of air in the tire will increase by 6.9% during this trip.

Discussion Note that the air temperature has risen nearly 10°C during this trip. This shows the importance of measuring the tire pressures before long trips to avoid errors due to temperature rise of air in tire. Also note that the unit Kelvin is used for temperature in the ideal gas relation.

Is Water Vapor an Ideal Gas?

This question cannot be answered with a simple yes or no. The error involved in treating water vapor as an ideal gas is calculated and plotted in Fig. 3–47. It is clear from this figure that at pressures below 10 kPa, water vapor can be treated as an ideal gas, regardless of its temperature, with negligible error (less than 0.1 percent). At higher pressures, however, the ideal-gas assumption yields unacceptable errors, particularly in the vicinity of the critical point and the saturated vapor line (over 100 percent). Therefore, in air-conditioning applications, the water vapor in the air can be treated as an ideal gas with essentially no error since the pressure of the water vapor is very low. In steam power plant applications, however, the pressures involved are usually very high; therefore, ideal-gas relations should not be used.

3–7 ▪ COMPRESSIBILITY FACTOR—A MEASURE OF DEVIATION FROM IDEAL-GAS BEHAVIOR

The ideal-gas equation is very simple and thus very convenient to use. However, as illustrated in Fig. 3–47, gases deviate from ideal-gas behavior significantly at states near the saturation region and the critical point.

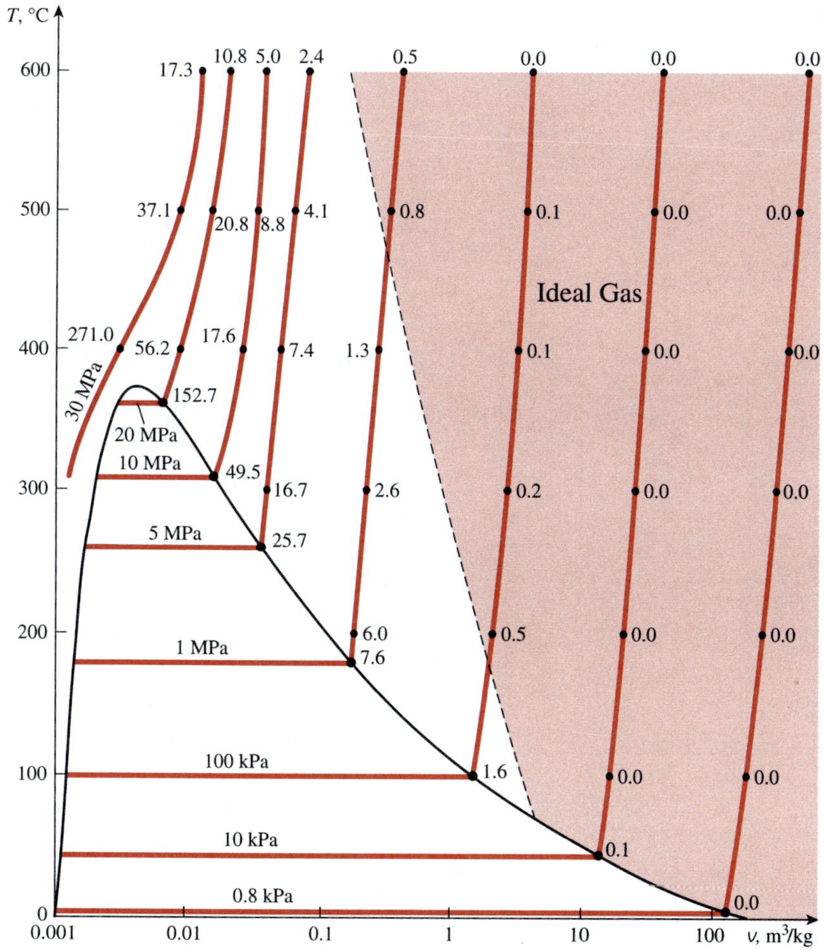

FIGURE 3–47

Percentage of error ($[|v_{table} - v_{ideal}|/v_{table}] \times 100$) involved in assuming steam to be an ideal gas, and the region where steam can be treated as an ideal gas with less than 1 percent error.

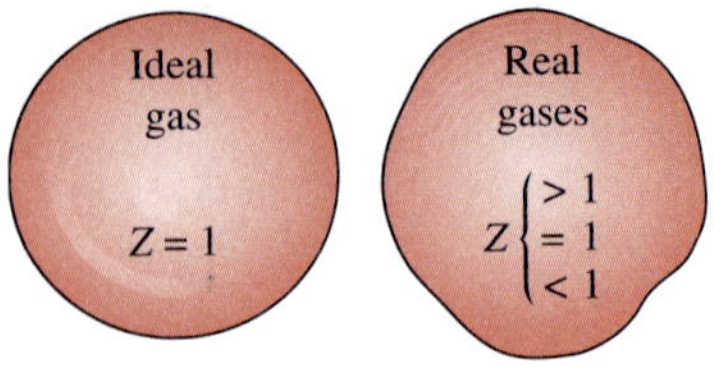

FIGURE 3–48
The compressibility factor is unity for ideal gases.

This deviation from ideal-gas behavior at a given temperature and pressure can accurately be accounted for by the introduction of a correction factor called the **compressibility factor** Z defined as

$$Z = \frac{Pv}{RT} \tag{3–17}$$

or

$$Pv = ZRT \tag{3–18}$$

It can also be expressed as

$$Z = \frac{v_{\text{actual}}}{v_{\text{ideal}}} \tag{3–19}$$

where $v_{\text{ideal}} = RT/P$. Obviously, $Z = 1$ for ideal gases. For real gases Z can be greater than or less than unity (Fig. 3–48). The farther away Z is from unity, the more the gas deviates from ideal-gas behavior.

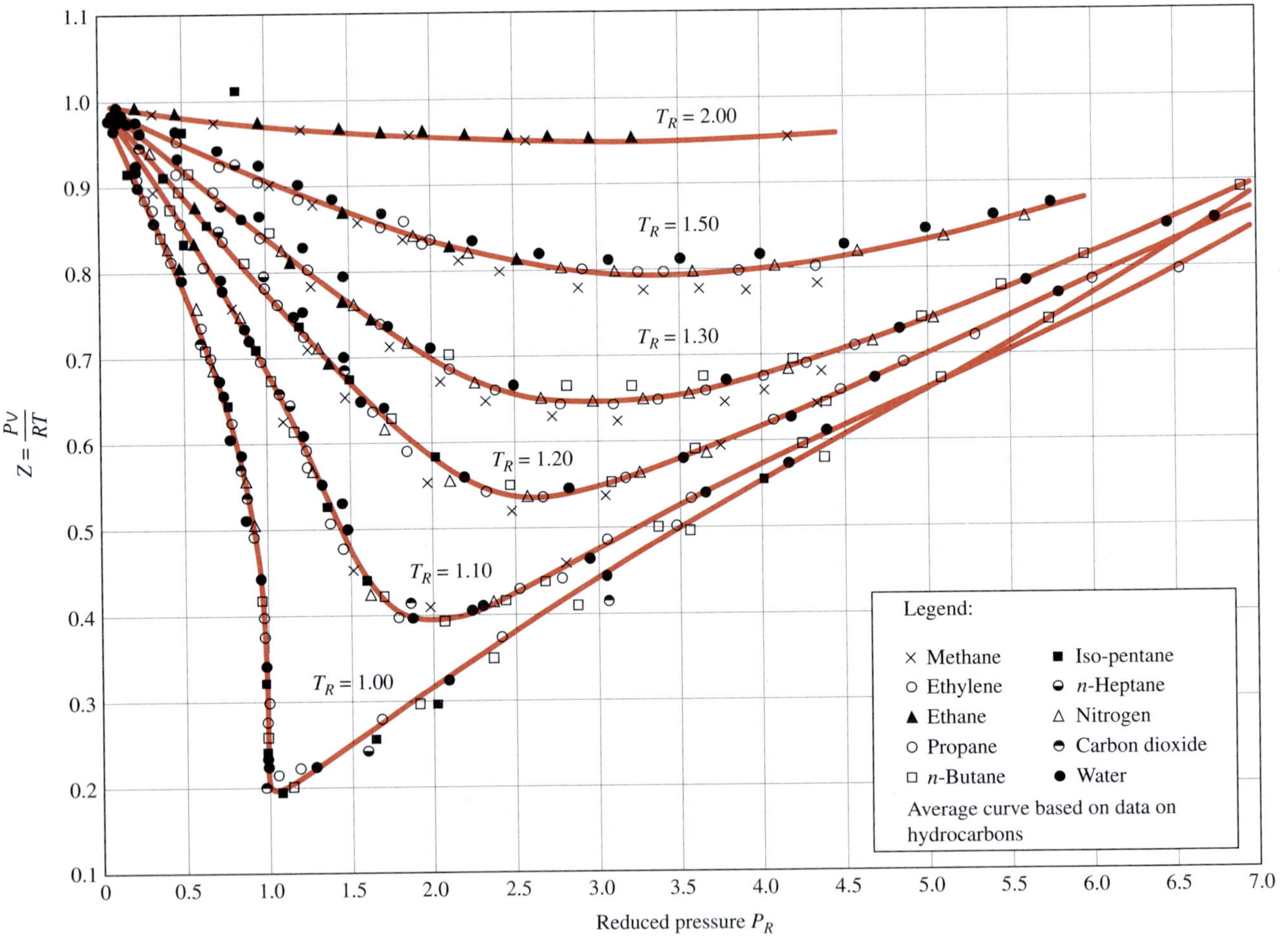

FIGURE 3–49
Comparison of Z factors for various gases.

Source: *Gour-Jen Su, "Modified Law of Corresponding States,"* Ind. Eng. Chem. *(international ed.) 38 (1946), p. 803.*

We have said that gases follow the ideal-gas equation closely at low pressures and high temperatures. But what exactly constitutes low pressure or high temperature? Is $-100°C$ a low temperature? It definitely is for most substances but not for air. Air (or nitrogen) can be treated as an ideal gas at this temperature and atmospheric pressure with an error under 1 percent. This is because nitrogen is well over its critical temperature ($-147°C$) and away from the saturation region. At this temperature and pressure, however, most substances would exist in the solid phase. Therefore, the pressure or temperature of a substance is high or low relative to its critical temperature or pressure.

Gases behave differently at a given temperature and pressure, but they behave very much the same at temperatures and pressures normalized with respect to their critical temperatures and pressures. The normalization is done as

$$P_R = \frac{P}{P_{cr}} \quad \text{and} \quad T_R = \frac{T}{T_{cr}} \tag{3-20}$$

Here P_R is called the **reduced pressure** and T_R the **reduced temperature**. The Z factor for all gases is approximately the same at the same reduced pressure and temperature. This is called the **principle of corresponding states**. In Fig. 3–49, on the previous page, the experimentally determined Z values are plotted against P_R and T_R for several gases. The gases seem to obey the principle of corresponding states reasonably well. By curve-fitting all the data, we obtain the **generalized compressibility chart** that can be used for all gases (Fig. A–15).

The following observations can be made from the generalized compressibility chart:

1. At very low pressures ($P_R \ll 1$), gases behave as an ideal gas regardless of temperature (Fig. 3–50),
2. At high temperatures ($T_R > 2$), ideal-gas behavior can be assumed with good accuracy regardless of pressure (except when $P_R \gg 1$).
3. The deviation of a gas from ideal-gas behavior is greatest in the vicinity of the critical point (Fig. 3–51).

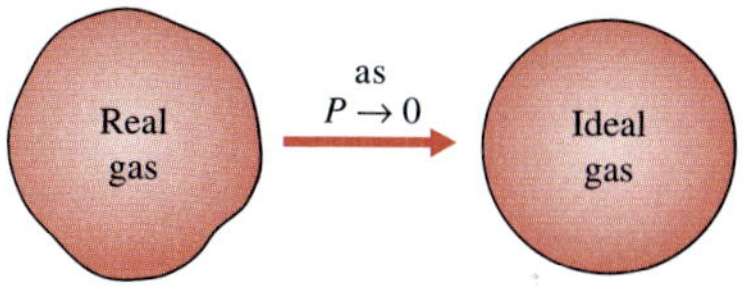

FIGURE 3–50
At very low pressures, all gases approach ideal-gas behavior (regardless of their temperature).

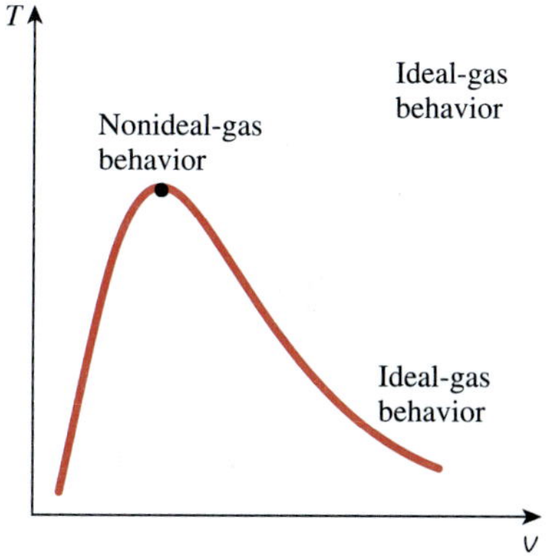

FIGURE 3–51
Gases deviate from the ideal-gas behavior the most in the neighborhood of the critical point.

EXAMPLE 3–11 The Use of Generalized Charts

Determine the specific volume of refrigerant-134a at 1 MPa and 50°C, using (*a*) the ideal-gas equation of state and (*b*) the generalized compressibility chart. Compare the values obtained to the actual value of 0.021796 m^3/kg and determine the error involved in each case.

SOLUTION The specific volume of refrigerant-134a is to be determined assuming ideal- and nonideal-gas behavior.

Analysis The gas constant, the critical pressure, and the critical temperature of refrigerant-134a are determined from Table A–1 to be

$$R = 0.0815 \text{ kPa·m}^3/\text{kg·K}$$
$$P_{cr} = 4.059 \text{ MPa}$$
$$T_{cr} = 374.2 \text{ K}$$

(*a*) The specific volume of refrigerant-134a under the ideal-gas assumption is

$$v = \frac{RT}{P} = \frac{(0.0815 \text{ kPa·m}^3/\text{kg·K})(323 \text{ K})}{1000 \text{ kPa}} = \mathbf{0.026325\ m^3/kg}$$

Therefore, treating the refrigerant-134a vapor as an ideal gas would result in an error of (0.026325 − 0.021796)/0.021796 = **0.208**, or 20.8 percent in this case.

(*b*) To determine the correction factor *Z* from the compressibility chart, we first need to calculate the reduced pressure and temperature:

$$\left.\begin{aligned} P_R &= \frac{P}{P_{cr}} = \frac{1 \text{ MPa}}{4.059 \text{ MPa}} = 0.246 \\ T_R &= \frac{T}{T_{cr}} = \frac{323 \text{ K}}{374.2 \text{ K}} = 0.863 \end{aligned}\right\} \quad Z = 0.84$$

Thus

$$v = Zv_{ideal} = (0.84)(0.026325 \text{ m}^3/\text{kg}) = \mathbf{0.022113\ m^3/kg}$$

Discussion The error in this result is less than **2 percent**. Therefore, in the absence of tabulated data, the generalized compressibility chart can be used with confidence.

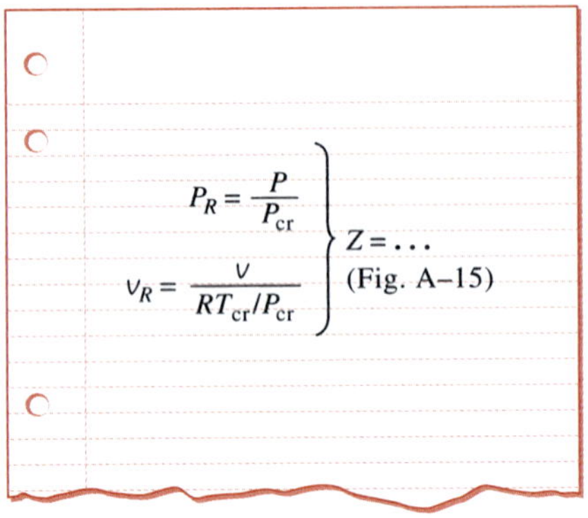

FIGURE 3–52
The compressibility factor can also be determined from a knowledge of P_R and v_R.

When *P* and v, or *T* and v, are given instead of *P* and *T*, the generalized compressibility chart can still be used to determine the third property, but it would involve tedious trial and error. Therefore, it is necessary to define one more reduced property called the **pseudo-reduced specific volume** v_R as

$$v_R = \frac{v_{actual}}{RT_{cr}/P_{cr}} \qquad \textbf{(3–21)}$$

Note that v_R is defined differently from P_R and T_R. It is related to T_{cr} and P_{cr} instead of v_{cr}. Lines of constant v_R are also added to the compressibility charts, and this enables one to determine *T* or *P* without having to resort to time-consuming iterations (Fig. 3–52).

EXAMPLE 3–12 Using Generalized Charts to Determine Pressure

Determine the pressure of water vapor at 600°F and 0.51431 ft³/lbm, using (*a*) the steam tables, (*b*) the ideal-gas equation, and (*c*) the generalized compressibility chart.

SOLUTION The pressure of water vapor is to be determined in three different ways.

Analysis A sketch of the system is given in Fig. 3–53. The gas constant, the critical pressure, and the critical temperature of steam are determined from Table A–1E to be

$$R = 0.5956 \text{ psia·ft}^3/\text{lbm·R}$$
$$P_{cr} = 3200 \text{ psia}$$
$$T_{cr} = 1164.8 \text{ R}$$

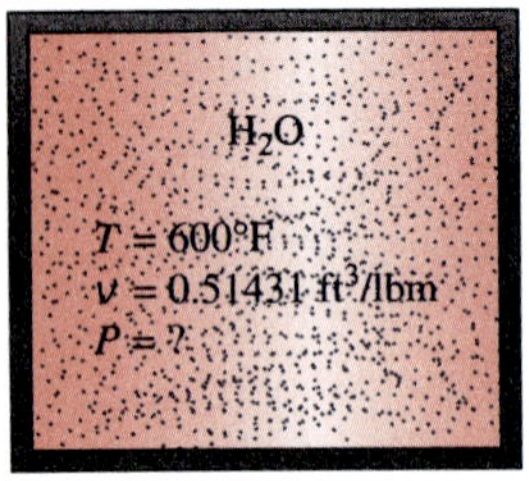

FIGURE 3–53
Schematic for Example 3–12.

(*a*) The pressure at the specified state is determined from Table A–6E to be

$$\left.\begin{array}{l} v = 0.51431 \text{ ft}^3/\text{lbm} \\ T = 600°\text{F} \end{array}\right\} \quad P = \mathbf{1000\ psia}$$

This is the experimentally determined value, and thus it is the most accurate.
(*b*) The pressure of steam under the ideal-gas assumption is determined from the ideal-gas relation to be

$$P = \frac{RT}{v} = \frac{(0.5956 \text{ psia·ft}^3/\text{lbm·R})(1060 \text{ R})}{0.51431 \text{ ft}^3/\text{lbm}} = \mathbf{1228\ psia}$$

Therefore, treating the steam as an ideal gas would result in an error of (1228 − 1000)/1000 = 0.228, or 22.8 percent in this case.
(*c*) To determine the correction factor Z from the compressibility chart (Fig. A–15), we first need to calculate the pseudo-reduced specific volume and the reduced temperature:

$$\left.\begin{array}{l} v_R = \dfrac{v_{\text{actual}}}{RT_{\text{cr}}/P_{\text{cr}}} = \dfrac{(0.51431 \text{ ft}^3/\text{lbm})(3200 \text{ psia})}{(0.5956 \text{ psia·ft}^3/\text{lbm·R})(1164.8 \text{ R})} = 2.372 \\ T_R = \dfrac{T}{T_{\text{cr}}} = \dfrac{1060 \text{ R}}{1164.8 \text{ R}} = 0.91 \end{array}\right\} \quad P_R = 0.33$$

Thus,

$$P = P_R P_{\text{cr}} = (0.33)(3200 \text{ psia}) = \mathbf{1056\ psia}$$

Discussion Using the compressibility chart reduced the error from 22.8 to 5.6 percent, which is acceptable for most engineering purposes (Fig. 3–54). A bigger chart, of course, would give better resolution and reduce the reading errors. Notice that we did not have to determine Z in this problem since we could read P_R directly from the chart.

	P, psia
Exact	1000
Z chart	1056
Ideal gas	1228

(from Example 3-12)

FIGURE 3–54
Results obtained by using the compressibility chart are usually within a few percent of actual values.

3–8 ▪ OTHER EQUATIONS OF STATE

The ideal-gas equation of state is very simple, but its range of applicability is limited. It is desirable to have equations of state that represent the *P-v-T* behavior of substances accurately over a larger region with no limitations. Such equations are naturally more complicated. Several equations have been proposed for this purpose (Fig. 3–55), but we shall discuss only three: the *van der Waals* equation because it is one of the earliest, the *Beattie-Bridgeman* equation of state because it is one of the best known and is reasonably accurate, and the *Benedict-Webb-Rubin* equation because it is one of the more recent and is very accurate.

van der Waals
Berthelet
Redlich-Kwang
Beattie-Bridgeman
Benedict-Webb-Rubin
Strobridge
Virial

FIGURE 3–55
Several equations of state have been proposed throughout history.

van der Waals Equation of State

The van der Waals equation of state was proposed in 1873, and it has two constants that are determined from the behavior of a substance at the critical point. It is given by

$$\left(P + \frac{a}{v^2}\right)(v - b) = RT \qquad \textbf{(3–22)}$$

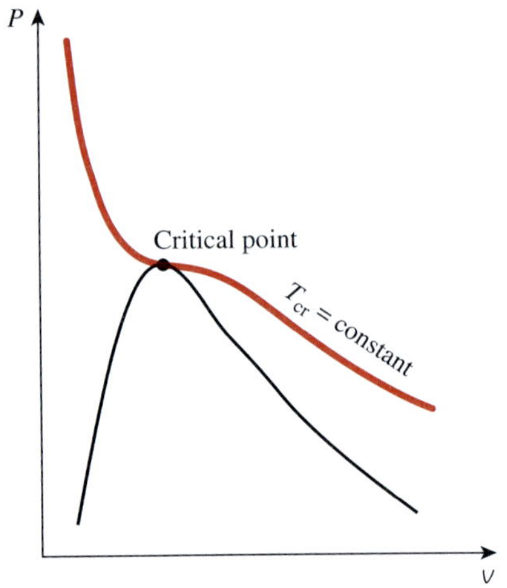

FIGURE 3–56
Critical isotherm of a pure substance has an inflection point at the critical state.

Van der Waals intended to improve the ideal-gas equation of state by including two of the effects not considered in the ideal-gas model: the *intermolecular attraction forces* and the *volume occupied by the molecules themselves*. The term a/v^2 accounts for the intermolecular forces, and b accounts for the volume occupied by the gas molecules. In a room at atmospheric pressure and temperature, the volume actually occupied by molecules is only about one-thousandth of the volume of the room. As the pressure increases, the volume occupied by the molecules becomes an increasingly significant part of the total volume. Van der Waals proposed to correct this by replacing v in the ideal-gas relation with the quantity $v - b$, where b represents the volume occupied by the gas molecules per unit mass.

The determination of the two constants appearing in this equation is based on the observation that the critical isotherm on a P-v diagram has a horizontal inflection point at the critical point (Fig. 3–56). Thus, the first and the second derivatives of P with respect to v at the critical point must be zero. That is,

$$\left(\frac{\partial P}{\partial v}\right)_{T=T_{cr}=\text{const}} = 0 \quad \text{and} \quad \left(\frac{\partial^2 P}{\partial v^2}\right)_{T=T_{cr}=\text{const}} = 0$$

By performing the differentiations and eliminating v_{cr}, the constants a and b are determined to be

$$a = \frac{27R^2T_{cr}^2}{64P_{cr}} \quad \text{and} \quad b = \frac{RT_{cr}}{8P_{cr}} \tag{3–23}$$

The constants a and b can be determined for any substance from the critical-point data alone (Table A–1).

The accuracy of the van der Waals equation of state is often inadequate, but it can be improved by using values of a and b that are based on the actual behavior of the gas over a wider range instead of a single point. Despite its limitations, the van der Waals equation of state has a historical value in that it was one of the first attempts to model the behavior of real gases. The van der Waals equation of state can also be expressed on a unit-mole basis by replacing the v in Eq. 3–22 by $\overline{v}$ and the R in Eqs. 3–22 and 3–23 by R_u.

Beattie-Bridgeman Equation of State

The Beattie-Bridgeman equation, proposed in 1928, is an equation of state based on five experimentally determined constants. It is expressed as

$$P = \frac{R_uT}{\overline{v}^2}\left(1 - \frac{c}{\overline{v}T^3}\right)(\overline{v} + B) - \frac{A}{\overline{v}^2} \tag{3–24}$$

where

$$A = A_0\left(1 - \frac{a}{\overline{v}}\right) \quad \text{and} \quad B = B_0\left(1 - \frac{b}{\overline{v}}\right) \tag{3–25}$$

The constants appearing in the above equation are given in Table 3–4 for various substances. The Beattie-Bridgeman equation is known to be reasonably accurate for densities up to about $0.8\rho_{cr}$, where ρ_{cr} is the density of the substance at the critical point.

TABLE 3–4

Constants that appear in the Beattie-Bridgeman and the Benedict-Webb-Rubin equations of state

(*a*) When P is in kPa, $\bar{v}$ is in m^3/kmol, T is in K, and R_u = 8.314 kPa·m^3/kmol·K, the five constants in the Beattie-Bridgeman equation are as follows:

Gas	A_0	a	B_0	b	c
Air	131.8441	0.01931	0.04611	−0.001101	4.34×10^4
Argon, Ar	130.7802	0.02328	0.03931	0.0	5.99×10^4
Carbon dioxide, CO_2	507.2836	0.07132	0.10476	0.07235	6.60×10^5
Helium, He	2.1886	0.05984	0.01400	0.0	40
Hydrogen, H_2	20.0117	−0.00506	0.02096	−0.04359	504
Nitrogen, N_2	136.2315	0.02617	0.05046	−0.00691	4.20×10^4
Oxygen, O_2	151.0857	0.02562	0.04624	0.004208	4.80×10^4

Source: Gordon J. Van Wylen and Richard E. Sonntag, *Fundamentals of Classical Thermodynamics,* English/SI Version, 3rd ed. (New York: John Wiley & Sons, 1986), p. 46, table 3.3.

(*b*) When P is in kPa, $\bar{v}$ is in m^3/kmol, T is in K, and R_u = 8.314 kPa·m^3/kmol·K, the eight constants in the Benedict-Webb-Rubin equation are as follows:

Gas	a	A_0	b	B_0	c	C_0	α	γ
n-Butane, C_4H_{10}	190.68	1021.6	0.039998	0.12436	3.205×10^7	1.006×10^8	1.101×10^{-3}	0.0340
Carbon dioxide, CO_2	13.86	277.30	0.007210	0.04991	1.511×10^6	1.404×10^7	8.470×10^{-5}	0.00539
Carbon monoxide, CO	3.71	135.87	0.002632	0.05454	1.054×10^5	8.673×10^5	1.350×10^{-4}	0.0060
Methane, CH_4	5.00	187.91	0.003380	0.04260	2.578×10^5	2.286×10^6	1.244×10^{-4}	0.0060
Nitrogen, N_2	2.54	106.73	0.002328	0.04074	7.379×10^4	8.164×10^5	1.272×10^{-4}	0.0053

Source: Kenneth Wark, *Thermodynamics,* 4th ed. (New York: McGraw-Hill, 1983), p. 815, table A-21M. Originally published in H. W. Cooper and J. C. Goldfrank, *Hydrocarbon Processing* 46, no. 12 (1967), p. 141.

Benedict-Webb-Rubin Equation of State

Benedict, Webb, and Rubin extended the Beattie-Bridgeman equation in 1940 by raising the number of constants to eight. It is expressed as

$$P = \frac{R_uT}{\bar{v}} + \left(B_0R_uT - A_0 - \frac{C_0}{T^2}\right)\frac{1}{\bar{v}^2} + \frac{bR_uT - a}{\bar{v}^3} + \frac{a\alpha}{\bar{v}^6} + \frac{c}{\bar{v}^3T^2}\left(1 + \frac{\gamma}{\bar{v}^2}\right)e^{-\gamma/\bar{v}^2} \quad \text{(3–26)}$$

The values of the constants appearing in this equation are given in Table 3–4. This equation can handle substances at densities up to about $2.5\rho_{cr}$. In 1962, Strobridge further extended this equation by raising the number of constants to 16 (Fig. 3–57).

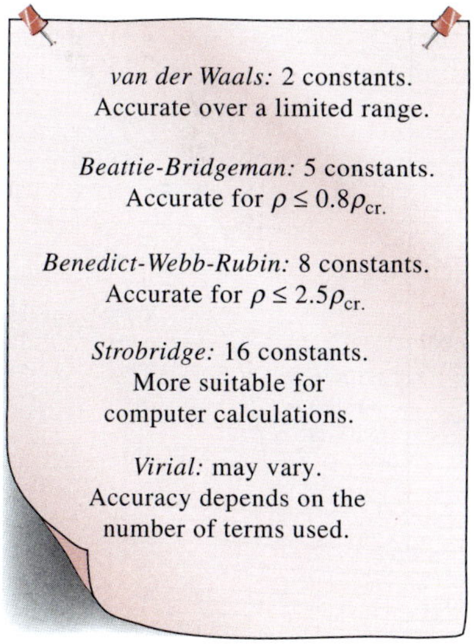

FIGURE 3–57
Complex equations of state represent the *P-v-T* behavior of gases more accurately over a wider range.

Virial Equation of State

The equation of state of a substance can also be expressed in a series form as

$$P = \frac{RT}{v} + \frac{a(T)}{v^2} + \frac{b(T)}{v^3} + \frac{c(T)}{v^4} + \frac{d(T)}{v^5} + \ldots \quad \text{(3–27)}$$

This and similar equations are called the *virial equations of state*, and the coefficients $a(T)$, $b(T)$, $c(T)$, and so on, that are functions of temperature alone are called *virial coefficients*. These coefficients can be determined experimentally or theoretically from statistical mechanics. Obviously, as the pressure approaches zero, all the virial coefficients will vanish and the equation will reduce to the ideal-gas equation of state. The *P*-v-*T* behavior of a substance can be represented accurately with the virial equation of state over a wider range by including a sufficient number of terms. The equations of state discussed here are applicable to the gas phase of the substances only, and thus should not be used for liquids or liquid–vapor mixtures.

Complex equations represent the *P*-v-*T* behavior of substances reasonably well and are very suitable for digital computer applications. For hand calculations, however, it is suggested that the reader use the property tables or the simpler equations of state for convenience. This is particularly true for specific-volume calculations since all the earlier equations are implicit in v and require a trial-and-error approach. The accuracy of the van der Waals, Beattie-Bridgeman, and Benedict-Webb-Rubin equations of state is illustrated in Fig. 3–58. It is apparent from this figure that the Benedict-Webb-Rubin equation of state is usually the most accurate.

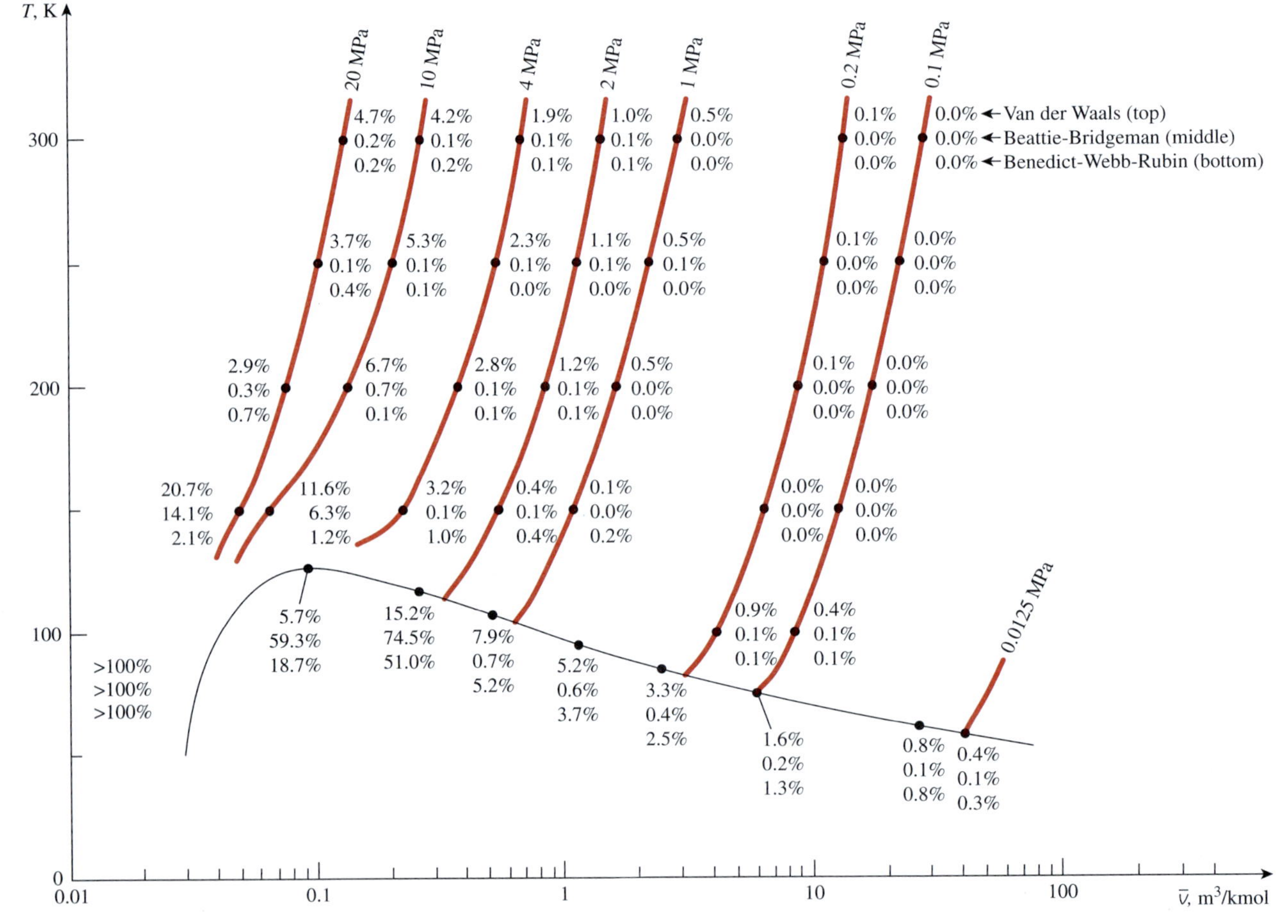

FIGURE 3–58

Percentage of error involved in various equations of state for nitrogen (% error $= [(|v_{\text{table}} - v_{\text{equation}}|)/v_{\text{table}}] \times 100$).

EXAMPLE 3–13 Different Methods of Evaluating Gas Pressure

Predict the pressure of nitrogen gas at T = 175 K and v = 0.00375 m³/kg on the basis of (*a*) the ideal-gas equation of state, (*b*) the van der Waals equation of state, (*c*) the Beattie-Bridgeman equation of state, and (*d*) the Benedict-Webb-Rubin equation of state. Compare the values obtained to the experimentally determined value of 10,000 kPa.

SOLUTION The pressure of nitrogen gas is to be determined using four different equations of state.

Properties The gas constant of nitrogen gas is 0.2968 kPa·m³/kg·K (Table A–1).

Analysis (*a*) Using the ideal-gas equation of state, the pressure is found to be

$$P = \frac{RT}{v} = \frac{(0.2968\ \text{kPa}\cdot\text{m}^3/\text{kg}\cdot\text{K})(175\ \text{K})}{0.00375\ \text{m}^3/\text{kg}} = \mathbf{13{,}851\ kPa}$$

which is in error by 38.5 percent.

(*b*) The van der Waals constants for nitrogen are determined from Eq. 3–23 to be

$$a = 0.175\ \text{m}^6\cdot\text{kPa/kg}^2$$

$$b = 0.00138\ \text{m}^3/\text{kg}$$

From Eq. 3–22,

$$P = \frac{RT}{v-b} - \frac{a}{v^2} = \mathbf{9471\ kPa}$$

which is in error by 5.3 percent.

(*c*) The constants in the Beattie-Bridgeman equation are determined from Table 3–4 to be

$$A = 102.29$$

$$B = 0.05378$$

$$c = 4.2 \times 10^4$$

Also, $\bar{v} = Mv$ = (28.013 kg/mol)(0.00375 m³/kg) = 0.10505 m³/kmol. Substituting these values into Eq. 3–24, we obtain

$$P = \frac{R_uT}{\bar{v}^2}\left(1 - \frac{c}{\bar{v}T^3}\right)(\bar{v} + B) - \frac{A}{\bar{v}^2} = \mathbf{10{,}110\ kPa}$$

which is in error by 1.1 percent.

(*d*) The constants in the Benedict-Webb-Rubin equation are determined from Table 3–4 to be

$$a = 2.54 \qquad A_0 = 106.73$$

$$b = 0.002328 \qquad B_0 = 0.04074$$

$$c = 7.379 \times 10^4 \qquad C_0 = 8.164 \times 10^5$$

$$\alpha = 1.272 \times 10^{-4} \qquad \gamma = 0.0053$$

Substituting these values into Eq. 3–26 gives

$$P = \frac{R_u T}{\overline{v}} + \left(B_0 R_u T - A_0 - \frac{C_0}{T^2}\right)\frac{1}{\overline{v}^2} + \frac{bR_u T - a}{\overline{v}^3}$$

$$+ \frac{a\alpha}{\overline{v}^6} + \frac{c}{\overline{v}^3 T^2}\left(1 + \frac{\gamma}{\overline{v}^2}\right)e^{-\gamma/\overline{v}^2}$$

$$= \mathbf{10{,}009\ kPa}$$

which is in error by only 0.09 percent. Thus, the accuracy of the Benedict-Webb-Rubin equation of state is rather impressive in this case.

TOPIC OF SPECIAL INTEREST* Vapor Pressure and Phase Equilibrium

The pressure in a gas container is due to the individual molecules striking the wall of the container and exerting a force on it. This force is proportional to the average velocity of the molecules and the number of molecules per unit volume of the container (i.e., molar density). Therefore, the pressure exerted by a gas is a strong function of the density and the temperature of the gas. For a gas mixture, the pressure measured by a sensor such as a transducer is the sum of the pressures exerted by the individual gas species, called the *partial pressure*. It can be shown (see Chap. 13) that the partial pressure of a gas in a mixture is proportional to the number of moles (or the mole fraction) of that gas.

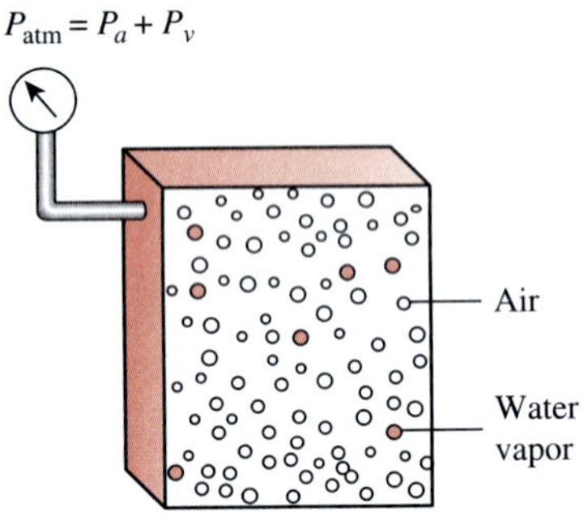

FIGURE 3–59
Atmospheric pressure is the sum of the dry air pressure P_a and the vapor pressure P_v.

Atmospheric air can be viewed as a mixture of dry air (air with zero moisture content) and water vapor (also referred to as moisture), and the atmospheric pressure is the sum of the pressure of dry air P_a and the pressure of water vapor, called the **vapor pressure** P_v (Fig. 3–59). That is,

$$P_{atm} = P_a + P_v \qquad (3\text{–}28)$$

(Note that in some applications, the phrase "vapor pressure" is used to indicate saturation pressure.) The vapor pressure constitutes a small fraction (usually under 3 percent) of the atmospheric pressure since air is mostly nitrogen and oxygen, and the water molecules constitute a small fraction (usually under 3 percent) of the total molecules in the air. However, the amount of water vapor in the air has a major impact on thermal comfort and many processes such as drying.

Air can hold a certain amount of moisture only, and the ratio of the actual amount of moisture in the air at a given temperature to the maximum amount air can hold at that temperature is called the **relative humidity** ϕ. The relative humidity ranges from 0 for dry air to 100 percent for **saturated air** (air that cannot hold any more moisture). The vapor pressure of saturated air at a given temperature is equal to the saturation pressure of water at that temperature. For example, the vapor pressure of saturated air at 25°C is 3.17 kPa.

*This section can be skipped without a loss in continuity.

The amount of moisture in the air is completely specified by the temperature and the relative humidity, and the vapor pressure is related to relative humidity ϕ by

$$P_v = \phi P_{\text{sat @ } T} \quad \text{(3–29)}$$

where $P_{\text{sat @ } T}$ is the saturation pressure of water at the specified temperature. For example, the vapor pressure of air at 25°C and 60 percent relative humidity is

$$P_v = \phi P_{\text{sat @ 25°C}} = 0.6 \times (3.17 \text{ kPa}) = 1.90 \text{ kPa}$$

The desirable range of relative humidity for thermal comfort is 40 to 60 percent.

Note that the amount of moisture air can hold is proportional to the saturation pressure, which increases with temperature. Therefore, air can hold more moisture at higher temperatures. Dropping the temperature of moist air reduces its moisture capacity and may result in the condensation of some of the moisture in the air as suspended water droplets (fog) or as a liquid film on cold surfaces (dew). So it is no surprise that fog and dew are common occurrences at humid locations especially in the early morning hours when the temperatures are the lowest. Both fog and dew disappear (evaporate) as the air temperature rises shortly after sunrise. You also may have noticed that electronic devices such as camcorders come with warnings against bringing them into moist indoors when the devices are cold to avoid moisture condensation on the sensitive electronics of the devices.

It is a common observation that whenever there is an imbalance of a commodity in a medium, nature tends to redistribute it until a "balance" or "equality" is established. This tendency is often referred to as the *driving force,* which is the mechanism behind many naturally occurring transport phenomena such as heat transfer, fluid flow, electric current, and mass transfer. If we define the amount of a commodity per unit volume as the *concentration* of that commodity, we can say that the flow of a commodity is always in the direction of decreasing concentration, that is, from the region of high concentration to the region of low concentration (Fig. 3–60). The commodity simply creeps away during redistribution, and thus the flow is a *diffusion process.*

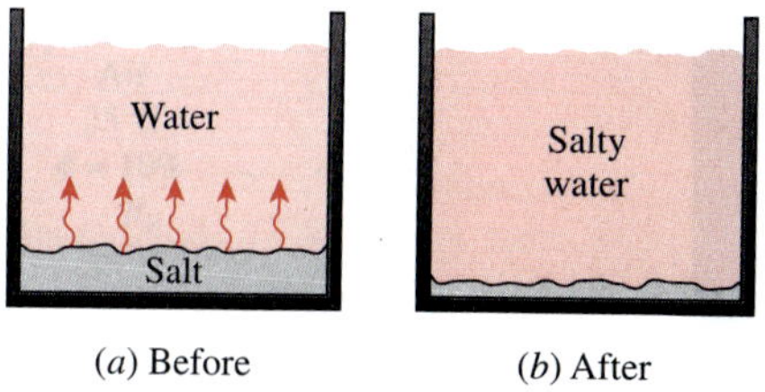

FIGURE 3–60
Whenever there is a concentration difference of a physical quantity in a medium, nature tends to equalize things by forcing a flow from the high to the low concentration region.

We know from experience that a wet T-shirt hanging in an open area eventually dries, a small amount of water left in a glass evaporates, and the aftershave in an open bottle quickly disappears. These and many other similar examples suggest that there is a driving force between the two phases of a substance that forces the mass to transform from one phase to another. The magnitude of this force depends on the relative concentrations of the two phases. A wet T-shirt dries much faster in dry air than it would in humid air. In fact, it does not dry at all if the relative humidity of the environment is 100 percent and thus the air is saturated. In this case, there is no transformation from the liquid phase to the vapor phase, and the two phases are in **phase equilibrium**. For liquid water that is open to the atmosphere, the criterion for phase equilibrium can be expressed as follows: *The vapor pressure in the air must be equal to the saturation pressure of water at the water temperature.* That is (Fig. 3–61),

$$\text{Phase equilibrium criterion for water exposed to air:} \quad P_v = P_{\text{sat @ } T} \quad \text{(3–30)}$$

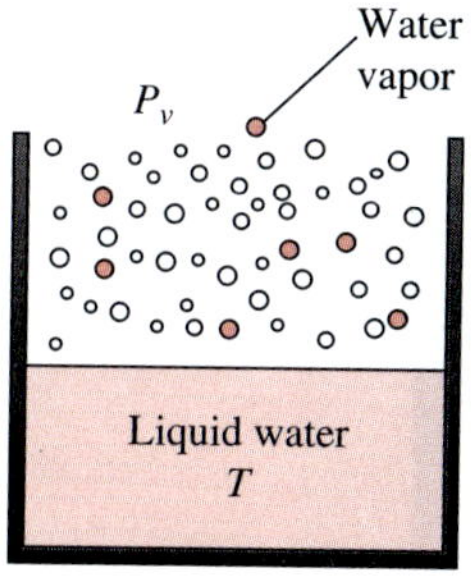

FIGURE 3–61
When open to the atmosphere, water is in phase equilibrium with the vapor in the air if the vapor pressure is equal to the saturation pressure of water.

3–41 Three kilograms of water in a container have a pressure of 100 kPa and temperature of 250°C. What is the volume of this container?

3–42 10-kg of R-134a at 300 kPa fills a rigid container whose volume is 14 L. Determine the temperature and total enthalpy in the container. The container is now heated until the pressure is 600 kPa. Determine the temperature and total enthalpy when the heating is completed.

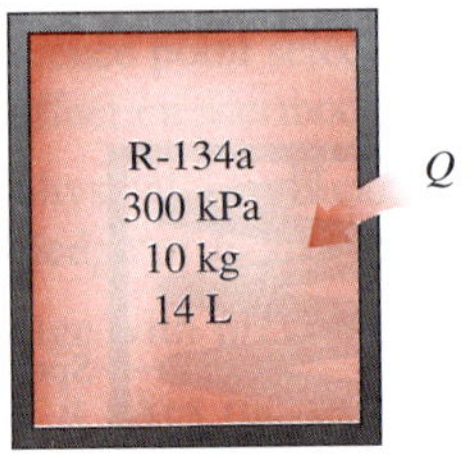

FIGURE P3–42

3–43 100-kg of R-134a at 200 kPa are contained in a piston-cylinder device whose volume is 12.322 m^3. The piston is now moved until the volume is one-half its original size. This is done such that the pressure of the R-134a does not change. Determine the final temperature and the change in the total internal energy of the R-134a.

3–44 Water initially at 300 kPa and 250°C is contained in a piston-cylinder device fitted with stops. The water is allowed to cool at constant pressure until it exists as a saturated vapor and the piston rests on the stops. Then the water continues to cool until the pressure is 100 kPa. On the *T*-*v* diagrams sketch, with respect to the saturation lines, the process curves passing through both the initial, intermediate, and final states of the water. Label the *T*, *P* and *v* values for end states on the process curves. Find the overall change in internal energy between the initial and final states per unit mass of water.

FIGURE P3–44

3–45E The atmospheric pressure at a location is usually specified at standard conditions, but it changes with the weather conditions. As the weather forecasters frequently state, the atmospheric pressure drops during stormy weather and it rises during clear and sunny days. If the pressure difference between the two extreme conditions is given to be 0.2 in of mercury, determine how much the boiling temperatures of water will vary as the weather changes from one extreme to the other.

3–46 A person cooks a meal in a 30-cm-diameter pot that is covered with a well-fitting lid and lets the food cool to the room temperature of 20°C. The total mass of the food and the pot is 8 kg. Now the person tries to open the pan by lifting the lid up. Assuming no air has leaked into the pan during cooling, determine if the lid will open or the pan will move up together with the lid.

3–47 Water is boiled at 1 atm pressure in a 25-cm-internal-diameter stainless steel pan on an electric range. If it is observed that the water level in the pan drops by 10 cm in 45 min, determine the rate of heat transfer to the pan.

3–48 Repeat Prob. 3–47 for a location at 2000-m elevation where the standard atmospheric pressure is 79.5 kPa.

3–49 Saturated steam coming off the turbine of a steam power plant at 40°C condenses on the outside of a 3-cm-outer-diameter, 35-m-long tube at a rate of 130 kg/h. Determine the rate of heat transfer from the steam to the cooling water flowing through the pipe.

3–50 Water in a 5-cm-deep pan is observed to boil at 98°C. At what temperature will the water in a 40-cm-deep pan boil? Assume both pans are full of water.

3–51 Water is being heated in a vertical piston–cylinder device. The piston has a mass of 20 kg and a cross-sectional area of 100 cm^2. If the local atmospheric pressure is 100 kPa, determine the temperature at which the water starts boiling.

3–52 A rigid tank with a volume of 1.8 m^3 contains 15 kg of saturated liquid–vapor mixture of water at 90°C. Now the water is slowly heated. Determine the temperature at which the liquid in the tank is completely vaporized. Also, show the process on a *T*-*v* diagram with respect to saturation lines. *Answer:* 202.9°C

3–53 A piston–cylinder device contains 0.005 m^3 of liquid water and 0.9 m^3 of water vapor in equilibrium at 600 kPa. Heat is transferred at constant pressure until the temperature reaches 200°C.

(*a*) What is the initial temperature of the water?
(*b*) Determine the total mass of the water.
(*c*) Calculate the final volume.
(*d*) Show the process on a *P*-*v* diagram with respect to saturation lines.

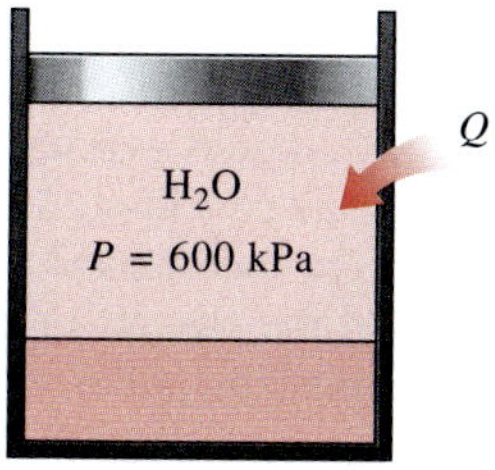

FIGURE P3–53

3–54 Reconsider Prob. 3–53. Using EES (or other) software, investigate the effect of pressure on the total mass of water in the tank. Let the pressure vary from 0.1 MPa to 1 MPa. Plot the total mass of water against pressure, and discuss the results. Also, show the process in Prob. 3–53 on a P-v diagram using the property plot feature of EES.

3–55E Superheated water vapor at 180 psia and 500°F is allowed to cool at constant volume until the temperature drops to 250°F. At the final state, determine (*a*) the pressure, (*b*) the quality, and (*c*) the enthalpy. Also, show the process on a T-v diagram with respect to saturation lines. *Answers:* (*a*) 29.84 psia, (*b*) 0.219, (*c*) 426.0 Btu/lbm

3–56E Reconsider Prob. 3–55E. Using EES (or other) software, investigate the effect of initial pressure on the quality of water at the final state. Let the pressure vary from 100 psi to 300 psi. Plot the quality against initial pressure, and discuss the results. Also, show the process in Prob. 3–55E on a T-v diagram using the property plot feature of EES.

3–57 Determine the specific volume, internal energy, and enthalpy of compressed liquid water at 80°C and 20 MPa using the saturated liquid approximation. Compare these values to the ones obtained from the compressed liquid tables.

3–58 Reconsider Prob. 3–57. Using EES (or other) software, determine the indicated properties of compressed liquid, and compare them to those obtained using the saturated liquid approximation.

3–59 A piston–cylinder device contains 0.8 kg of steam at 300°C and 1 MPa. Steam is cooled at constant pressure until one-half of the mass condenses.

(*a*) Show the process on a T-v diagram.
(*b*) Find the final temperature.
(*c*) Determine the volume change.

3–60 A rigid tank contains water vapor at 250°C and an unknown pressure. When the tank is cooled to 124°C, the vapor starts condensing. Estimate the initial pressure in the tank. *Answer:* 0.30 MPa

3–61 A piston-cylinder device initially contains 1.4-kg saturated liquid water at 200°C. Now heat is transferred to the water until the volume quadruples and the cylinder contains saturated vapor only. Determine (*a*) the volume of the tank, (*b*) the final temperature and pressure, and (*c*) the internal energy change of the water.

FIGURE P3–61

3–62E How much error would one expect in determining the specific enthalpy by applying the incompressible-liquid approximation to water at 1500 psia and 400°F?

3–63 How much error would result in calculating the specific volume and enthalpy of water at 20 MPa and 140°C by using the incompressible-liquid approximation?

3–64 100 grams of R-134a initially fill a weighted piston-cylinder device at 60 kPa and −20°C. The device is then heated until the temperature is 100°C. Determine the change in the device's volume as a result of the heating. *Answer:* 0.0168 m^3

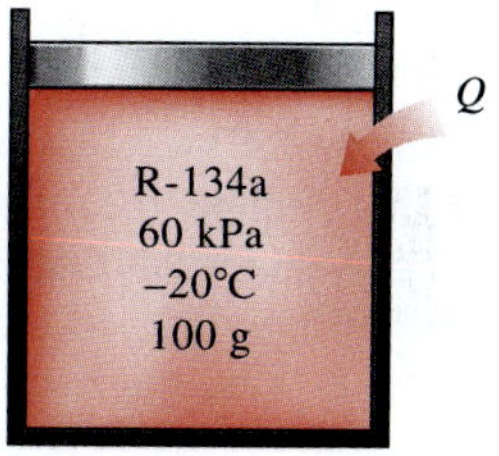

FIGURE P3–64

3–65 A rigid vessel contains 2 kg of refrigerant-134a at 800 kPa and 120°C. Determine the volume of the vessel and the total internal energy. *Answers:* 0.0753 m^3, 655.7 kJ

3–66 A 0.5-m^3 vessel contains 10 kg of refrigerant-134a at −20°C. Determine (*a*) the pressure, (*b*) the total internal energy, and (*c*) the volume occupied by the liquid phase. *Answers:* (*a*) 132.82 kPa, (*b*) 904.2 kJ, (*c*) 0.00489 m^3

3–67 Pressure-enthalpy state diagrams are useful for studying refrigeration and like systems. Using EES (or other) software and actual property data, plot a pressure-enthalpy diagram for R-134a that includes the saturation lines. Also, sketch isothermal and constant-entropy processes on this diagram.

3–68 A 0.3-m^3 rigid vessel initially contains saturated liquid-vapor mixture of water at 150°C. The water is now heated until it reaches the critical state. Determine the mass of the liquid water and the volume occupied by the liquid at the initial state. *Answers:* 96.1 kg, 0.105 m^3

Ideal Gas

3–69C What is the difference between mass and molar mass? How are these two related?

3–70C Under what conditions is the ideal-gas assumption suitable for real gases?

3–71C What is the difference between R and R_u? How are these two related?

3–72C Propane and methane are commonly used for heating in winter, and the leakage of these fuels, even for short periods, poses a fire danger for homes. Which gas leakage do you think poses a greater risk for fire? Explain.

3–73 What is the specific volume of nitrogen at 300 kPa and 227°C?

3–74E A 3-ft^3 container is filled with 2-lbm of oxygen at a pressure of 80 psia. What is the temperature of the oxygen?

3–75 A 2-kg mass of helium is maintained at 300 kPa and 27°C in a rigid container. How large is the container, in m^3?

3–76 A spherical balloon with a diameter of 9 m is filled with helium at 27°C and 200 kPa. Determine the mole number and the mass of the helium in the balloon. *Answers:* 30.6 kmol, 123 kg

3–77 EES Reconsider Prob. 3–76. Using EES (or other) software, investigate the effect of the balloon diameter on the mass of helium contained in the balloon for the pressures of (*a*) 100 kPa and (*b*) 200 kPa. Let the diameter vary from 5 m to 15 m. Plot the mass of helium against the diameter for both cases.

3–78 A 1-m^3 tank containing air at 25°C and 500 kPa is connected through a valve to another tank containing 5 kg of air at 35°C and 200 kPa. Now the valve is opened, and the entire system is allowed to reach thermal equilibrium with the surroundings, which are at 20°C. Determine the volume of the second tank and the final equilibrium pressure of air. *Answers:* 2.21 m^3, 284.1 kPa

3–79E 2.3 lbmol of air at 65°F and 32 psia is contained in an elastic tank. What is the volume of the tank? If the volume is doubled at the same pressure, determine the final temperature.

3–80 A rigid tank contains an ideal gas at 1227°C and 200 kPa gage. The gas is cooled until the gage pressure reads 50 kPa. If the atmospheric pressure is 100 kPa, determine the final temperature of the gas.

3–81 A rigid tank whose volume is unknown is divided into two parts by a partition. One side of the tank contains an ideal gas at 927°C. The other side is evacuated and has a volume twice the size of the part containing the gas. The partition is now removed and the gas expands to fill the entire tank. Heat is now applied to the gas until the pressure equals the initial pressure. Determine the final temperature of the gas. *Answer:* 3327°C

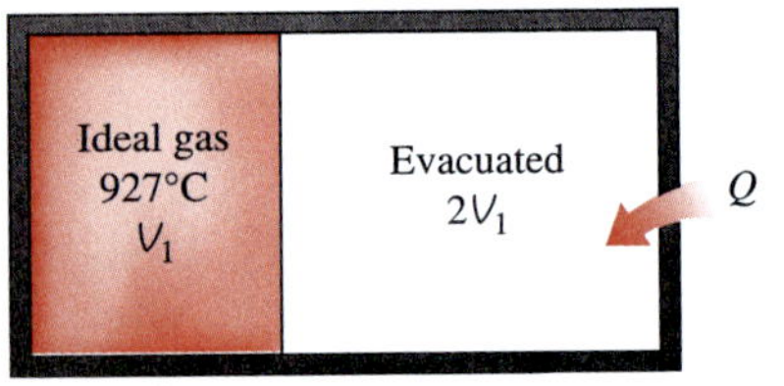

FIGURE P3–81

3–82 Argon in the amount of 0.6 kg fills a 0.05-m^3 piston-cylinder device at 550 kPa. The piston is now moved by changing the weights until the volume is twice its original size. During this process, argon's temperature is maintained constant. Determine the final pressure in the device.

3–83 The pressure in an automobile tire depends on the temperature of the air in the tire. When the air temperature is 25°C, the pressure gage reads 210 kPa. If the volume of the tire is 0.025 m^3, determine the pressure rise in the tire when the air temperature in the tire rises to 50°C. Also, determine the amount of air that must be bled off to restore pressure to its original value at this temperature. Assume the atmospheric pressure is 100 kPa.

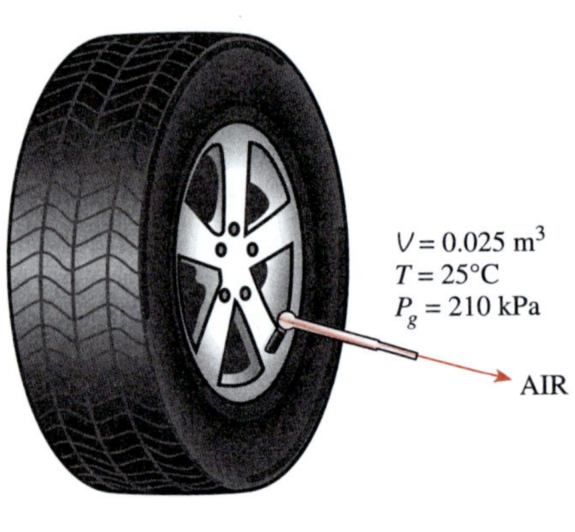

FIGURE P3–83

Compressibility Factor

3–84C What is the physical significance of the compressibility factor Z?

3–85C What is the principle of corresponding states?

3–86C How are the reduced pressure and reduced temperature defined?

3–87 Determine the specific volume of superheated water vapor at 15 MPa and 350°C, using (*a*) the ideal-gas equation, (*b*) the generalized compressibility chart, and (*c*) the steam tables. Also determine the error involved in the first two cases. *Answers:* (*a*) 0.01917 m^3/kg, 67.0 percent, (*b*) 0.01246 m^3/kg, 8.5 percent, (*c*) 0.01148 m^3/kg

3–88 Reconsider Prob. 3–87. Solve the problem using the generalized compressibility factor feature of the EES software. Again using EES, compare the specific volume of water for the three cases at 15 MPa over the temperature range of 350 to 600°C in 25°C intervals. Plot the percent error involved in the ideal-gas approximation against temperature, and discuss the results.

3–89 Determine the specific volume of refrigerant-134a vapor at 0.9 MPa and 70°C based on (*a*) the ideal-gas equation, (*b*) the generalized compressibility chart, and (*c*) data from tables. Also, determine the error involved in the first two cases.

3–90 Determine the specific volume of superheated water vapor at 3.5 MPa and 450°C based on (*a*) the ideal-gas equation, (*b*) the generalized compressibility chart, and (*c*) the steam tables. Determine the error involved in the first two cases.

3–91E Ethane in a rigid vessel is to be heated from 80 psia and 100°F until its temperature is 540°F. What is the final pressure of the ethane as predicted by the compressibility chart?

3–92 Ethylene is heated at constant pressure from 5 MPa and 20°C to 200°C. Using the compressibility chart, determine the change in the ethylene's specific volume as a result of this heating. *Answer:* 0.0172 m^3/kg

3–93 Saturated water vapor at 350°C is heated at constant pressure until its volume has doubled. Determine the final temperature using the ideal gas equation of state, the compressibility charts, and the steam tables.

3–94E Saturated water vapor at 400°F is heated at constant pressure until its volume has doubled. Determine the final temperature using the ideal gas equation of state, the compressibility charts, and the steam tables.

3–95 Methane at 8 MPa and 300 K is heated at constant pressure until its volume has increased by 50 percent. Determine the final temperature using the ideal gas equation of state and the compressibility factor. Which of these two results is more accurate?

3–96 What is the percentage of error involved in treating carbon dioxide at 5 MPa and 25°C as an ideal gas? *Answer:* 45 percent

3–97 Carbon dioxide gas enters a pipe at 3 MPa and 500 K at a rate of 2 kg/s. CO_2 is cooled at constant pressure as it flows in the pipe and the temperature of CO_2 drops to 450 K at the exit. Determine the volume flow rate and the density of carbon dioxide at the inlet and the volume flow rate at the exit of the pipe using (*a*) the ideal-gas equation and (*b*) the generalized compressibility chart. Also, determine (*c*) the error involved in the first case.

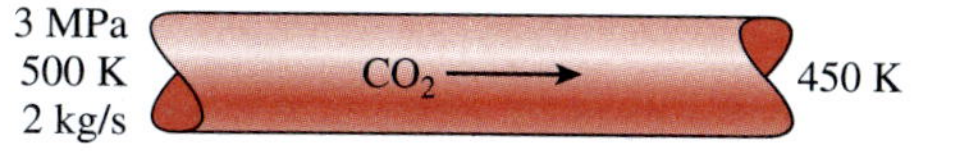

FIGURE P3–97

3–98 Determine the specific volume of nitrogen gas at 10 MPa and 150 K based on (*a*) the idea-gas equation and (*b*) the generalized compressibility chart. Compare these results with the experimental value of 0.002388 m^3/kg, and determine the error involved in each case. *Answers:* (*a*) 0.004452 m^3/kg, 86.4 percent; (*b*) 0.002404 m^3/kg, 0.7 percent

Other Equations of State

3–99C What is the physical significance of the two constants that appear in the van der Waals equation of state? On what basis are they determined?

3-100 100 grams of carbon monoxide are contained in a weighted piston-cylinder device. Initially the carbon monoxide is at 1000 kPa and 200°C. It is then heated until its temperature is 500°C. Determine the final volume of the carbon monoxide treating it as (*a*) an ideal gas and (*b*) a Benedict-Webb-Rubin gas.

3–101 Methane is heated in a rigid container from 80 kPa and 20°C to 300°C. Determine the final pressure of the methane treating it as (*a*) an ideal gas and (*b*) a Benedict-Webb-Rubin gas.

3–102E Carbon monoxide is heated in a rigid container from 14.7 psia and 70°F to 800°F. Determine the final pressure of the carbon monoxide treating it as (*a*) an ideal gas and (*b*) a Benedict-Webb-Rubin gas.

3–103E Refrigerant-134a at 160 psia has a specific volume of 0.3479 ft^3/lbm. Determine the temperature of the refrigerant based on (*a*) the ideal-gas equation, (*b*) the van der Waals equation, and (*c*) the refrigerant tables.

3–104 Nitrogen at 150 K has a specific volume of 0.041884 m^3/kg. Determine the pressure of the nitrogen, using (*a*) the ideal-gas equation and (*b*) the Beattie-Bridgeman equation. Compare your results to the experimental value of 1000 kPa. *Answers:* (*a*) 1063 kPa, (*b*) 1000.4 kPa

3–105 Reconsider Prob. 3–104. Using EES (or other) software, compare the pressure results of the ideal-gas and Beattie-Bridgeman equations with nitrogen data supplied by EES. Plot temperature versus specific volume for a pressure of 1000 kPa with respect to the saturated liquid and saturated vapor lines of nitrogen over the range of 110 K $< T <$ 150 K.

3–106 1-kg of carbon dioxide is compressed from 1 MPa and 200°C to 3 MPa in a piston-cylinder device arranged to execute a polytropic process for which $PV^{1.2}$ = constant. Determine the final temperature treating the carbon dioxide as (*a*) an ideal gas and (*b*) a van der Waals gas.

Special Topic: Vapor Pressure and Phase Equilibrium

3–107 Consider a glass of water in a room that is at 20°C and 40 percent relative humidity. If the water temperature is 15°C, determine the vapor pressure (*a*) at the free surface of the water and (*b*) at a location in the room far from the glass.

3–108 During a hot summer day at the beach when the air temperature is 30°C, someone claims the vapor pressure in the air to be 5.2 kPa. Is this claim reasonable?

3–109 On a certain day, the temperature and relative humidity of air over a large swimming pool are measured to be 20°C and 40 percent, respectively. Determine the water temperature of the pool when phase equilibrium conditions are established between the water in the pool and the vapor in the air.

3–110 During a hot summer day when the air temperature is 35°C and the relative humidity is 70 percent, you buy a supposedly "cold" canned drink from a store. The store owner claims that the temperature of the drink is below 10°C. Yet the drink does not feel so cold and you are skeptical since you notice no condensation forming outside the can. Can the store owner be telling the truth?

3–111E A thermos bottle is half-filled with water and is left open to the atmospheric air at 70°F and 35 percent relative humidity. If heat transfer to the water through the thermos walls and the free surface is negligible, determine the temperature of water when phase equilibrium is established.

3–112 Consider two rooms that are identical except that one is maintained at 25°C and 40 percent relative humidity while the other is maintained at 20°C and 55 percent relative humidity. Noting that the amount of moisture is proportional to the vapor pressure, determine which room contains more moisture.

Review Problems

3-113 A rigid tank contains nitrogen gas at 227°C and 100 kPa gage. The gas is heated until the gage pressure reads 250 kPa. If the atmospheric pressure is 100 kPa, determine the final temperature of the gas, in °C.

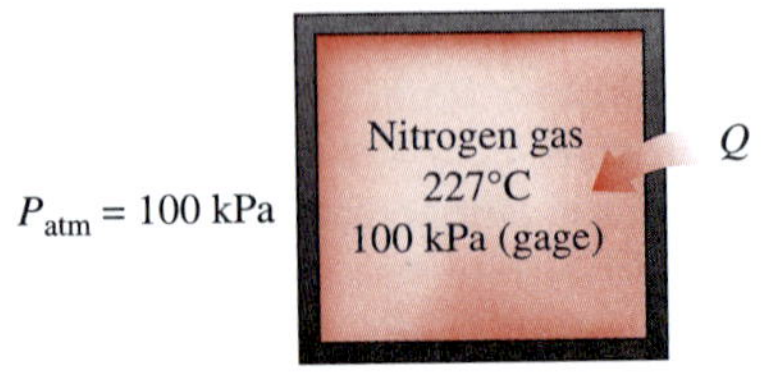

FIGURE P3–113

3–114 Carbon-dioxide gas at 3 MPa and 500 K flows steadily in a pipe at a rate of 0.4 kmol/s. Determine (*a*) the volume and mass flow rates and the density of carbon dioxide at this state. If CO_2 is cooled at constant pressure as it flows in the pipe so that the temperature of CO_2 drops to 450 K at the exit of the pipe, determine (*b*) the volume flow rate at the exit of the pipe.

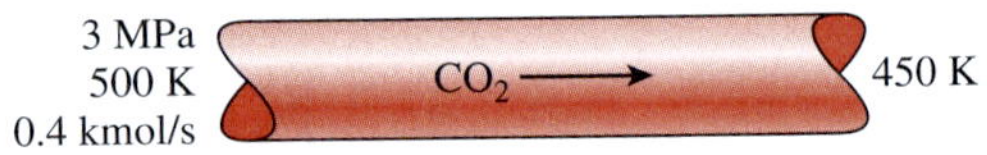

FIGURE P3–114

3–115 The combustion in a gasoline engine may be approximated by a constant volume heat addition process. There exists the air–fuel mixture in the cylinder before the combustion and the combustion gases after it, and both may be approximated as air, an ideal gas. In a gasoline engine, the cylinder conditions are 1.4 MPa and 450°C before the combustion and 1600°C after it. Determine the pressure at the end of the combustion process. *Answer:* 3.63 MPa

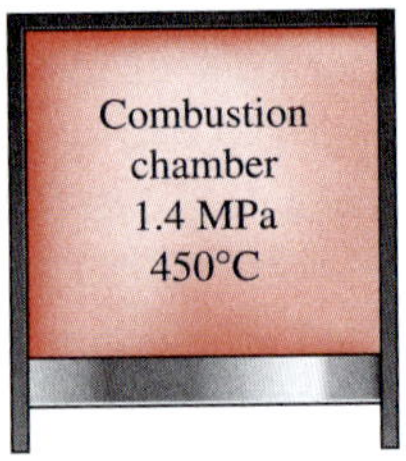

FIGURE P3–115

3–116 Combustion in a diesel engine may be modeled as a constant-pressure heat addition process with air in the cylinder before and after combustion. Consider a diesel engine with cylinder conditions of 950 K and 75 cm³ before combustion, and 150 cm³ after it. The engine operates with an air–fuel ratio of 22 kg air/kg fuel (the mass of the air divided by the mass of the fuel). Determine the temperature after the combustion process.

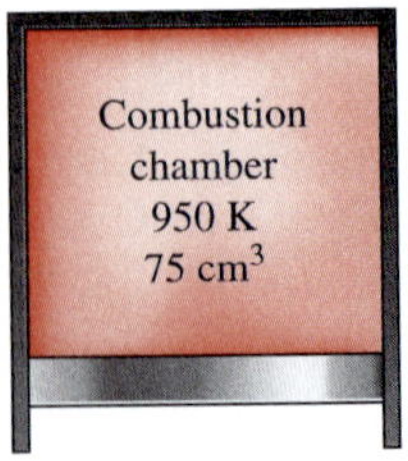

FIGURE P3–116

3–117 One kilogram of R-134a fills a 0.090 m^3 rigid container at an initial temperature of -40°C. The container is then heated until the pressure is 280 kPa. Determine the initial pressure and final temperature. *Answers:* 51.25 kPa, 50°C

3–118E One pound-mass of water fills a 2.649 ft^3 weighted piston-cylinder device at a temperature of 400°F. The piston-cylinder device is now cooled until its temperature is 100°F. Determine the final pressure and volume of the water.

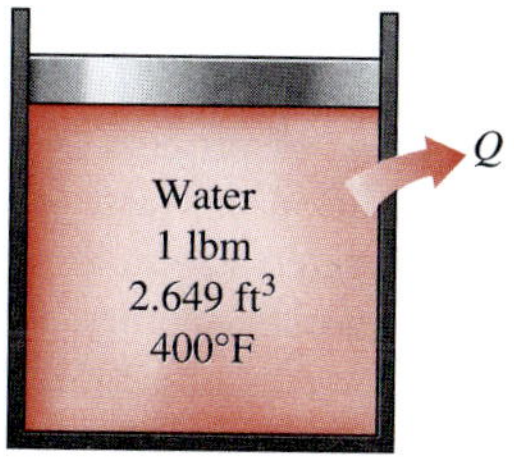

FIGURE P3–118E

3–119 The piston diameters in Fig. P3–119 are D_1 = 10 cm and D_2 = 4 cm. Chamber 1 contains 1 kg of helium, chamber 2 is filled with condensing water vapor, and chamber 3 is evacuated. The entire assembly is placed in an environment whose temperature is 200°C. Determine the volume of chamber 1 when thermodynamic equilibrium has been established. *Answer:* 3.95 m^3

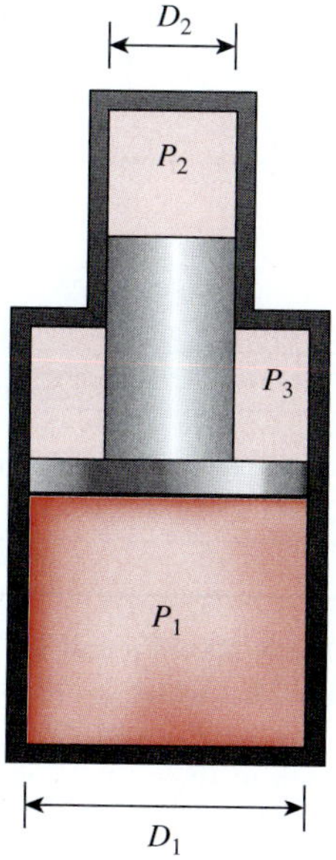

FIGURE P3–119

3–120E The piston diameters in Fig. P3–119 are D_1 = 3 in and D_2 = 2 in. The pressure in chamber 3 is 30 psia. Chamber 1 contains 0.5 lbm of air, and chamber 2 is filled with condensing R-134a. The entire assembly is maintained at 120°F. Determine the volume of chamber 1.

3–121E Reconsider Prob. 3–120E. How much will the volume of chamber 1 change when the pressure in chamber 3 is increased to 60 psia?

3–122 Ethane at 10 MPa and 100°C is heated at constant pressure until its volume has increased by 60 percent. Determine the final temperature using (*a*) the ideal gas equation of state and (*b*) the compressibility factor. Which of these two results is the more accurate?

3–123 A 20-m^3 tank contains nitrogen at 23°C and 600 kPa. Some nitrogen is allowed to escape until the pressure in the tank drops to 400 kPa. If the temperature at this point is 20°C, determine the amount of nitrogen that has escaped. *Answer:* 44.6 kg

3–124 A 4-L rigid tank contains 2 kg of saturated liquid–vapor mixture of water at 50°C. The water is now slowly heated until it exists in a single phase. At the final state, will the water be in the liquid phase or the vapor phase? What would your answer be if the volume of the tank were 400 L instead of 4 L?

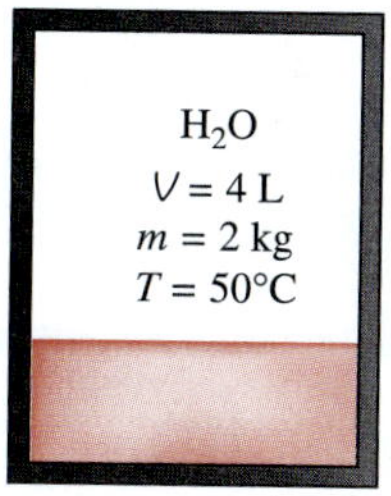

FIGURE P3–124

3–125 A 0.5-m^3 rigid tank containing hydrogen at 20°C and 400 kPa is connected by a valve to another 0.5-m^3 rigid tank that holds hydrogen at 50°C and 150 kPa. Now the valve is opened and the system is allowed to reach thermal equilibrium with the surroundings, which are at 15°C. Determine the final pressure in the tank.

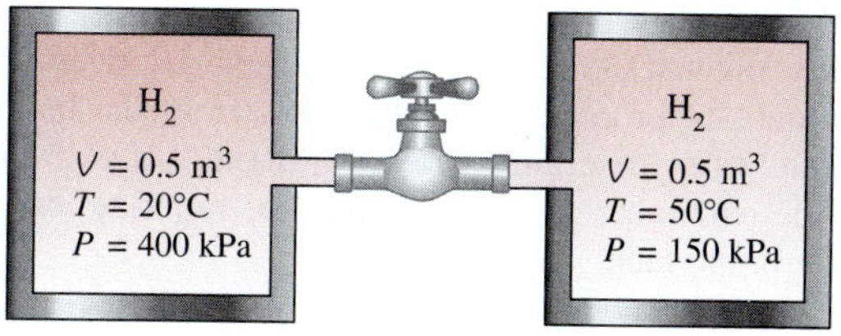

FIGURE P3–125

3–126 EES Reconsider Prob. 3–125. Using EES (or other) software, investigate the effect of the surroundings temperature on the final equilibrium pressure in the tanks. Assume the surroundings temperature to vary from -10 to 30°C. Plot the final pressure in the tanks versus the surroundings temperature, and discuss the results.

3–127 The gage pressure of an automobile tire is measured to be 200 kPa before a trip and 220 kPa after the trip at a location where the atmospheric pressure is 90 kPa. Assuming the volume of the tire remains constant at 0.035 m^3, determine the percent increase in the absolute temperature of the air in the tire.

3–128 Steam at 400°C has a specific volume of 0.02 m^3/kg. Determine the pressure of the steam based on (*a*) the ideal-gas equation, (*b*) the generalized compressibility chart, and (*c*) the steam tables. *Answers:* (*a*) 15,529 kPa, (*b*) 12,574 kPa, (*c*) 12,515 kPa

3–129 A tank whose volume is unknown is divided into two parts by a partition. One side of the tank contains 0.03 m^3 of refrigerant-134a that is a saturated liquid at 1.2 MPa, while the other side is evacuated. The partition is now removed, and the refrigerant fills the entire tank. If the final state of the refrigerant is 30°C and 400 kPa, determine the volume of the tank.

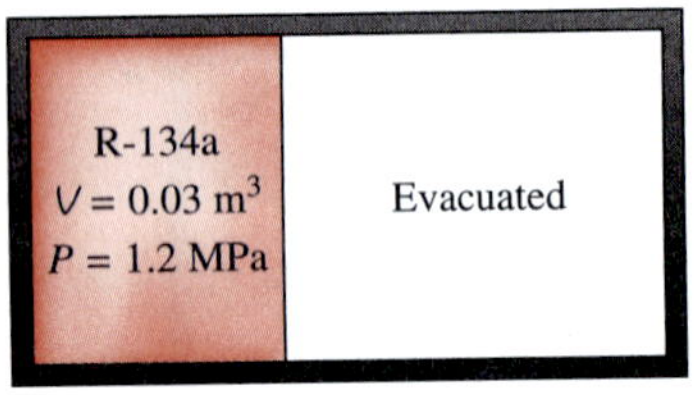

FIGURE P3–129

3–130 Reconsider Prob. 3–129. Using EES (or other) software, investigate the effect of the initial pressure of refrigerant-134a on the volume of the tank. Let the initial pressure vary from 0.5 to 1.5 MPa. Plot the volume of the tank versus the initial pressure, and discuss the results.

3–131 Liquid propane is commonly used as a fuel for heating homes, powering vehicles such as forklifts, and filling portable picnic tanks. Consider a propane tank that initially contains 5 L of liquid propane at the environment temperature of 20°C. If a hole develops in the connecting tube of a propane tank and the propane starts to leak out, determine the temperature of propane when the pressure in the tank drops to 1 atm. Also, determine the total amount of heat transfer from the environment to the tank to vaporize the entire propane in the tank.

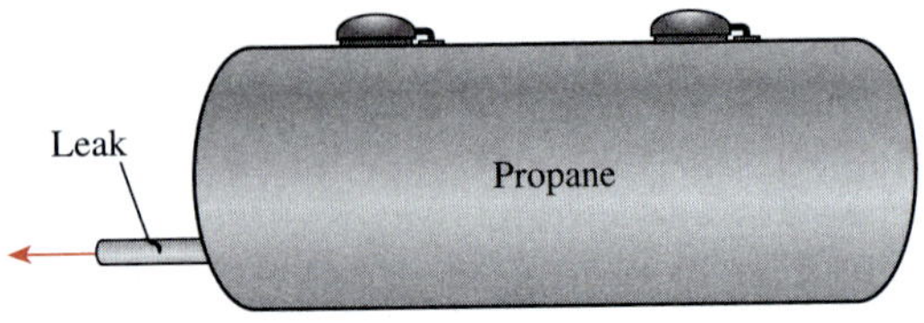

FIGURE P3–131

3–132 Repeat Prob. 3–131 for isobutane.

3–133 A tank contains helium at 77°C and 110 kPa gage. The helium is heated in a process by heat transfer from the surroundings such that the helium reaches a final equilibrium state at 300°C. Determine the final gage pressure of the helium. Assume atmospheric pressure is 100 kPa.

3–134 What are the first eight virial coefficients of a Benedict-Webb-Rubin gas? *Hint:* Expand the exponential term in a power series expansion.

3–135 If sufficient data are provided, complete the blank cells in the following table of properties of water. In the last column describe the condition of water as compressed liquid, saturated mixture, superheated vapor, or insufficient information; and, if applicable, give the quality.

P, kPa	*T*, °C	*v*, m³/kg	*u*, kJ/kg	Condition description and quality (if applicable)
	250		2728.9	
300			1560.0	
101.42	100			
3000	180			

3–136 If sufficient data are provided, complete the blank cells in the following table of properties of water. In the last column describe the condition of water as compressed liquid, saturated mixture, superheated vapor, or insufficient information; and, if applicable, give the quality.

P, kPa	*T*, °C	*v*, m³/kg	*u*, kJ/kg	Condition description and quality (if applicable)
200			2529.1	
	125			$x = 0.650$
	400		2967.2	
1000	30			
120.90	150			

3–137 Water initially at 300 kPa and 250°C is contained in a constant volume tank. The water is allowed to cool until its pressure is 150 kPa. On the *P*-*v* and *T*-*v* diagrams sketch, with respect to the saturation lines, the process curve passing through both the initial and final states of the water. Label the end states on the process curve. Also, on both the *P*-*v* and *T*-*v* diagrams, sketch the isotherms passing through both states and show their values, in °C, on the isotherms.

3-138 Water initially at 300 kPa and 0.5 m^3/kg is contained in a piston-cylinder device fitted with stops so that the water supports the weight of the piston and the force of the atmosphere. The water is heated until it reaches the saturated vapor state and the piston rests against the stops. With the piston against the stops, the water is further heated until the pressure is 600 kPa. On the *P*-*v* and *T*-*v* diagrams sketch, with respect to the saturation lines, the process curves passing through both the initial and final states of the water. Label the states on the process as 1, 2, and 3. On both the *P*-*v* and *T*-*v* diagrams, sketch the isotherms passing through the states and show their values, in °C, on the isotherms.

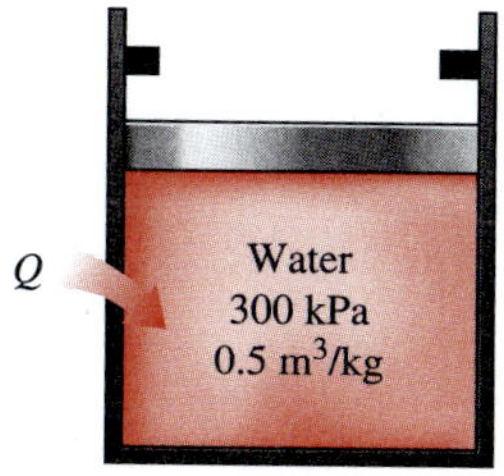

FIGURE P3–138

3–139E 0.5-lbm of argon is compressed from 1000 psia and 300°F to 2000 psia in a piston-cylinder device which executes a polytropic process for which $PV^{1.6}$ = constant. Determine the final temperature treating the argon as (*a*) an ideal gas and (*b*) a Beattie-Bridgeman gas.

3–140E Nitrogen is maintained at 400 psia and −100°F. Compare the specific volume of this nitrogen as predicted by (*a*) the ideal gas equation of state, (*b*) the Benedict-Webb-Rubin equation of state, and (*c*) with the compressibility factor.

Fundamentals of Engineering (FE) Exam Problems

3–141 A rigid tank contains 2 kg of an ideal gas at 4 atm and 40°C. Now a valve is opened, and half of mass of the gas is allowed to escape. If the final pressure in the tank is 2.2 atm, the final temperature in the tank is

(*a*) 71°C (*b*) 44°C (*c*) −100°C (*d*) 20°C (*e*) 172°C

3–142 The pressure of an automobile tire is measured to be 190 kPa (gage) before a trip and 215 kPa (gage) after the trip at a location where the atmospheric pressure is 95 kPa. If the temperature of air in the tire before the trip is 25°C, the air temperature after the trip is

(*a*) 51.1°C (*b*) 64.2°C (*c*) 27.2°C (*d*) 28.3°C (*e*) 25.0°C

3–143 A 300-m³ rigid tank is filled with saturated liquid–vapor mixture of water at 200 kPa. If 25 percent of the mass is liquid and 75 percent of the mass is vapor, the total mass in the tank is

(*a*) 451 kg (*b*) 556 kg (*c*) 300 kg (*d*) 331 kg (*e*) 195 kg

3–144 Water is boiled at 1 atm pressure in a coffee maker equipped with an immersion-type electric heating element. The coffee maker initially contains 1 kg of water. Once boiling started, it is observed that half of the water in the coffee maker evaporated in 10 minutes. If the heat loss from the coffee maker is negligible, the power rating of the heating element is

(*a*) 3.8 kW (*b*) 2.2 kW (*c*) 1.9 kW
(*d*) 1.6 kW (*e*) 0.8 kW

3–145 A 1-m³ rigid tank contains 10 kg of water (in any phase or phases) at 160°C. The pressure in the tank is

(*a*) 738 kPa (*b*) 618 kPa (*c*) 370 kPa
(*d*) 2000 MPa (*e*) 1618 kPa

3–146 Water is boiling at 1 atm pressure in a stainless steel pan on an electric range. It is observed that 2 kg of liquid water evaporates in 30 min. The rate of heat transfer to the water is

(*a*) 2.51 kW (*b*) 2.32 kW (*c*) 2.97 kW
(*d*) 0.47 kW (*e*) 3.12 kW

3–147 Water is boiled in a pan on a stove at sea level. During 10 min of boiling, it is observed that 200 g of water has evaporated. Then the rate of heat transfer to the water is

(*a*) 0.84 kJ/min (*b*) 45.1 kJ/min
(*c*) 41.8 kJ/min (*d*) 53.5 kJ/min
(*e*) 225.7 kJ/min

3–148 A 3-m³ rigid vessel contains steam at 4 MPa and 500°C. The mass of the steam is

(*a*) 3 kg (*b*) 9 kg (*c*) 26 kg (*d*) 35 kg (*e*) 52 kg

3–149 Consider a sealed can that is filled with refrigerant-134a. The contents of the can are at the room temperature of 25°C. Now a leak develops, and the pressure in the can drops to the local atmospheric pressure of 90 kPa. The temperature of the refrigerant in the can is expected to drop to (rounded to the nearest integer)

(*a*) 0°C (*b*) −29°C (*c*) −16°C (*d*) 5°C (*e*) 25°C

Design and Essay Problems

3–150 In an article on tire maintenance, it is stated that tires lose air over time, and pressure losses as high as 90 kPa (13 psi) per year are measured. The article recommends checking tire pressure at least once a month to avoid low tire pressure that hurts fuel efficiency and causes uneven thread wear on tires. Taking the beginning tire pressure to be 220 kPa (gage) and the atmospheric pressure to be 100 kPa, determine the fraction of air that can be lost from a tire per year.

3–151 It is well known that water freezes at 0°C at atmospheric pressure. The mixture of liquid water and ice at 0°C is said to be at stable equilibrium since it cannot undergo any changes when it is isolated from its surroundings. However, when water is free of impurities and the inner surfaces of the container are smooth, the temperature of water can be lowered to −2°C or even lower without any formation of ice at atmospheric pressure. But at that state even a small disturbance can initiate the formation of ice abruptly, and the water temperature stabilizes at 0°C following this sudden change. The water at −2°C is said to be in a *metastable state*. Write an essay on metastable states and discuss how they differ from stable equilibrium states.

CHAPTER

4

ENERGY ANALYSIS OF CLOSED SYSTEMS

In Chap. 2, we considered various forms of energy and energy transfer, and we developed a general relation for the conservation of energy principle or energy balance. Then in Chap. 3, we learned how to determine the thermodynamics properties of substances. In this chapter, we apply the energy balance relation to systems that do not involve any mass flow across their boundaries; that is, closed systems.

We start this chapter with a discussion of the *moving boundary work* or *P dV work* commonly encountered in reciprocating devices such as automotive engines and compressors. We continue by applying the *general energy balance* relation, which is simply expressed as $E_{in} - E_{out} = \Delta E_{system}$, to systems that involve pure substance. Then we define *specific heats*, obtain relations for the internal energy and enthalpy of *ideal gases* in terms of specific heats and temperature changes, and perform energy balances on various systems that involve ideal gases. We repeat this for systems that involve solids and liquids, which are approximated as *incompressible substances*.

Objectives

The objectives of Chapter 4 are to:

- Examine the moving boundary work or *P dV* work commonly encountered in reciprocating devices such as automotive engines and compressors.
- Identify the first law of thermodynamics as simply a statement of the conservation of energy principle for closed (fixed mass) systems.
- Develop the general energy balance applied to closed systems.
- Define the specific heat at constant volume and the specific heat at constant pressure.
- Relate the specific heats to the calculation of the changes in internal energy and enthalpy of ideal gases.
- Describe incompressible substances and determine the changes in their internal energy and enthalpy.
- Solve energy balance problems for closed (fixed mass) systems that involve heat and work interactions for general pure substances, ideal gases, and incompressible substances.

4–1 ▪ MOVING BOUNDARY WORK

One form of mechanical work frequently encountered in practice is associated with the expansion or compression of a gas in a piston–cylinder device. During this process, part of the boundary (the inner face of the piston) moves back and forth. Therefore, the expansion and compression work is often called **moving boundary work**, or simply **boundary work** (Fig. 4–1). Some call it the $P\,dV$ work for reasons explained later. Moving boundary work is the primary form of work involved in *automobile engines*. During their expansion, the combustion gases force the piston to move, which in turn forces the crankshaft to rotate.

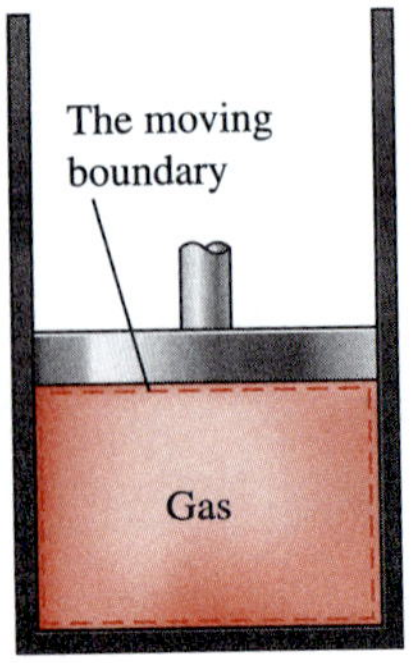

FIGURE 4–1
The work associated with a moving boundary is called *boundary work*.

The moving boundary work associated with real engines or compressors cannot be determined exactly from a thermodynamic analysis alone because the piston usually moves at very high speeds, making it difficult for the gas inside to maintain equilibrium. Then the states through which the system passes during the process cannot be specified, and no process path can be drawn. Work, being a path function, cannot be determined analytically without a knowledge of the path. Therefore, the boundary work in real engines or compressors is determined by direct measurements.

In this section, we analyze the moving boundary work for a *quasi-equilibrium process*, a process during which the system remains nearly in equilibrium at all times. A quasi-equilibrium process, also called a *quasi-static process*, is closely approximated by real engines, especially when the piston moves at low velocities. Under identical conditions, the work output of the engines is found to be a maximum, and the work input to the compressors to be a minimum when quasi-equilibrium processes are used in place of nonquasi-equilibrium processes. Below, the work associated with a moving boundary is evaluated for a quasi-equilibrium process.

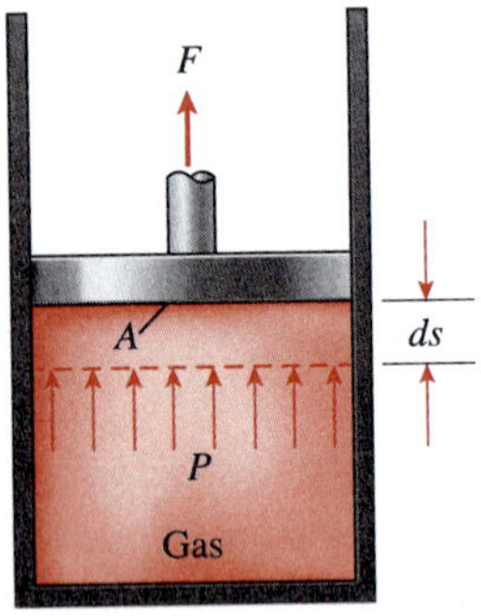

FIGURE 4–2
A gas does a differential amount of work δW_b as it forces the piston to move by a differential amount ds.

Consider the gas enclosed in the piston–cylinder device shown in Fig. 4–2. The initial pressure of the gas is P, the total volume is V, and the cross-sectional area of the piston is A. If the piston is allowed to move a distance ds in a quasi-equilibrium manner, the differential work done during this process is

$$\delta W_b = F\,ds = PA\,ds = P\,dV \qquad \textbf{(4–1)}$$

That is, the boundary work in the differential form is equal to the product of the absolute pressure P and the differential change in the volume dV of the system. This expression also explains why the moving boundary work is sometimes called the $P\,dV$ work.

Note in Eq. 4–1 that P is the absolute pressure, which is always positive. However, the volume change dV is positive during an expansion process (volume increasing) and negative during a compression process (volume decreasing). Thus, the boundary work is positive during an expansion process and negative during a compression process. Therefore, Eq. 4–1 can be viewed as an expression for boundary work output, $W_{b,\text{out}}$. A negative result indicates boundary work input (compression).

The total boundary work done during the entire process as the piston moves is obtained by adding all the differential works from the initial state to the final state:

$$W_b = \int_1^2 P\,dV \qquad \text{(kJ)} \qquad \textbf{(4–2)}$$

This integral can be evaluated only if we know the functional relationship between P and V during the process. That is, $P = f(V)$ should be available. Note that $P = f(V)$ is simply the equation of the process path on a P-V diagram.

The quasi-equilibrium expansion process described is shown on a P-V diagram in Fig. 4–3. On this diagram, the differential area dA is equal to $P\,dV$, which is the differential work. The total area A under the process curve 1–2 is obtained by adding these differential areas:

$$\text{Area} = A = \int_1^2 dA = \int_1^2 P\,dV \tag{4–3}$$

A comparison of this equation with Eq. 4–2 reveals that *the area under the process curve on a P-V diagram is equal, in magnitude, to the work done during a quasi-equilibrium expansion or compression process of a closed system.* (On the P-v diagram, it represents the boundary work done per unit mass.)

A gas can follow several different paths as it expands from state 1 to state 2. In general, each path will have a different area underneath it, and since this area represents the magnitude of the work, the work done will be different for each process (Fig. 4–4). This is expected, since work is a path function (i.e., it depends on the path followed as well as the end states). If work were not a path function, no cyclic devices (car engines, power plants) could operate as work-producing devices. The work produced by these devices during one part of the cycle would have to be consumed during another part, and there would be no net work output. The cycle shown in Fig. 4–5 produces a net work output because the work done by the system during the expansion process (area under path A) is greater than the work done on the system during the compression part of the cycle (area under path B), and the difference between these two is the net work done during the cycle (the colored area).

If the relationship between P and V during an expansion or a compression process is given in terms of experimental data instead of in a functional form, obviously we cannot perform the integration analytically. We can, however, plot the P-V diagram of the process using these data points and calculate the area underneath graphically to determine the work done.

Strictly speaking, the pressure P in Eq. 4–2 is the pressure at the inner surface of the piston. It becomes equal to the pressure of the gas in the cylinder only if the process is quasi-equilibrium and thus the entire gas in the cylinder is at the same pressure at any given time. Equation 4–2 can also be used for nonquasi-equilibrium processes provided that the pressure *at the inner face of the piston* is used for P. (Besides, we cannot speak of the pressure of a *system* during a nonquasi-equilibrium process since properties are defined for equilibrium states only.) Therefore, we can generalize the boundary work relation by expressing it as

$$W_b = \int_1^2 P_i\,dV \tag{4–4}$$

where P_i is the pressure at the inner face of the piston.

Note that work is a mechanism for energy interaction between a system and its surroundings, and W_b represents the amount of energy transferred

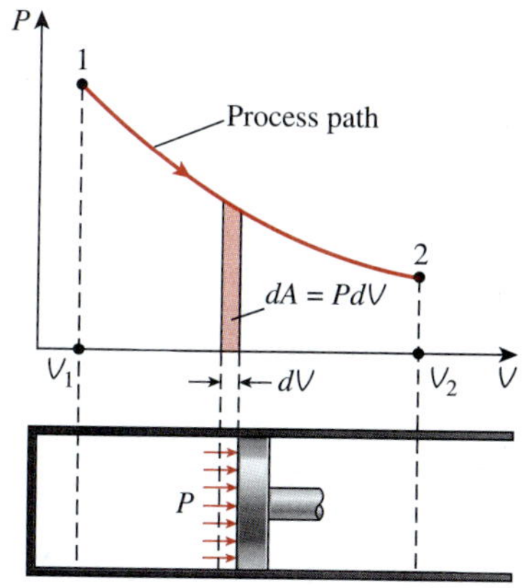

FIGURE 4–3
The area under the process curve on a P-V diagram represents the boundary work.

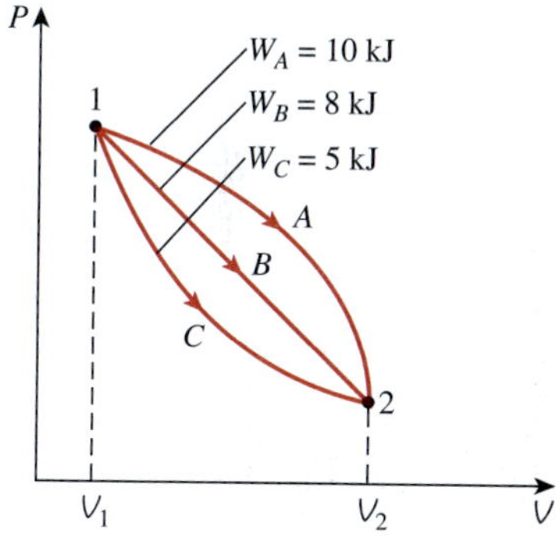

FIGURE 4–4
The boundary work done during a process depends on the path followed as well as the end states.

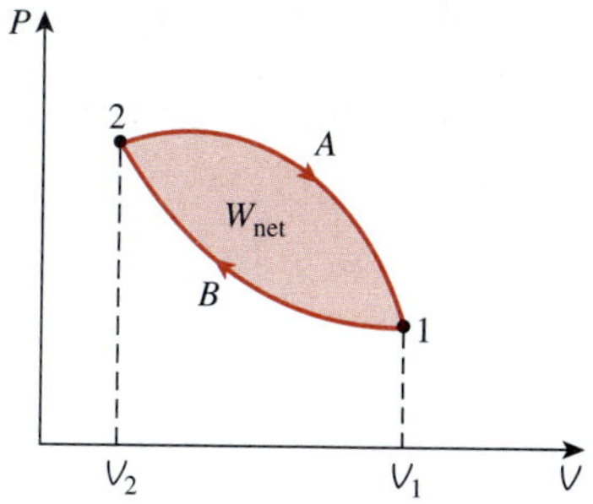

FIGURE 4–5
The net work done during a cycle is the difference between the work done by the system and the work done on the system.

from the system during an expansion process (or to the system during a compression process). Therefore, it has to appear somewhere else and we must be able to account for it since energy is conserved. In a car engine, for example, the boundary work done by the expanding hot gases is used to overcome friction between the piston and the cylinder, to push atmospheric air out of the way, and to rotate the crankshaft. Therefore,

$$W_b = W_{\text{friction}} + W_{\text{atm}} + W_{\text{crank}} = \int_1^2 (F_{\text{friction}} + P_{\text{atm}}A + F_{\text{crank}})\, dx \qquad \textbf{(4–5)}$$

Of course the work used to overcome friction appears as frictional heat and the energy transmitted through the crankshaft is transmitted to other components (such as the wheels) to perform certain functions. But note that the energy transferred by the system as work must equal the energy received by the crankshaft, the atmosphere, and the energy used to overcome friction. The use of the boundary work relation is not limited to the quasi-equilibrium processes of gases only. It can also be used for solids and liquids.

EXAMPLE 4–1 Boundary Work for a Constant-Volume Process

A rigid tank contains air at 500 kPa and 150°C. As a result of heat transfer to the surroundings, the temperature and pressure inside the tank drop to 65°C and 400 kPa, respectively. Determine the boundary work done during this process.

SOLUTION Air in a rigid tank is cooled, and both the pressure and temperature drop. The boundary work done is to be determined.

Analysis A sketch of the system and the *P*-*V* diagram of the process are shown in Fig. 4–6. The boundary work can be determined from Eq. 4–2 to be

$$W_b = \int_1^2 P\, \overset{0}{\cancel{dV}} = \mathbf{0}$$

Discussion This is expected since a rigid tank has a constant volume and $dV = 0$ in this equation. Therefore, there is no boundary work done during this process. That is, the boundary work done during a constant-volume process is always zero. This is also evident from the *P*-*V* diagram of the process (the area under the process curve is zero).

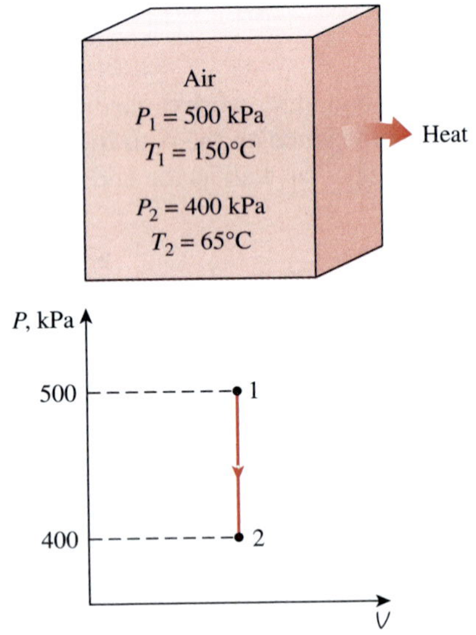

FIGURE 4–6
Schematic and *P*-*V* diagram for Example 4–1.

EXAMPLE 4–2 Boundary Work for a Constant-Pressure Process

A frictionless piston–cylinder device contains 10 lbm of steam at 60 psia and 320°F. Heat is now transferred to the steam until the temperature reaches 400°F. If the piston is not attached to a shaft and its mass is constant, determine the work done by the steam during this process.

SOLUTION Steam in a piston cylinder device is heated and the temperature rises at constant pressure. The boundary work done is to be determined.

Analysis A sketch of the system and the *P*-*v* diagram of the process are shown in Fig. 4–7.

Assumption The expansion process is quasi-equilibrium.
Analysis Even though it is not explicitly stated, the pressure of the steam within the cylinder remains constant during this process since both the atmospheric pressure and the weight of the piston remain constant. Therefore, this is a constant-pressure process, and, from Eq. 4–2

$$W_b = \int_1^2 P\,dV = P_0 \int_1^2 dV = P_0(V_2 - V_1) \qquad \textbf{(4–6)}$$

or

$$W_b = mP_0(v_2 - v_1)$$

since $V = mv$. From the superheated vapor table (Table A–6E), the specific volumes are determined to be $v_1 = 7.4863$ ft³/lbm at state 1 (60 psia, 320°F) and $v_2 = 8.3548$ ft³/lbm at state 2 (60 psia, 400°F). Substituting these values yields

$$W_b = (10\text{ lbm})(60\text{ psia})[(8.3548 - 7.4863)\text{ ft}^3/\text{lbm}]\left(\frac{1\text{ Btu}}{5.404\text{ psia}\cdot\text{ft}^3}\right)$$

$$= \mathbf{96.4\ Btu}$$

Discussion The positive sign indicates that the work is done by the system. That is, the steam used 96.4 Btu of its energy to do this work. The magnitude of this work could also be determined by calculating the area under the process curve on the *P-V* diagram, which is simply $P_0\,\Delta V$ for this case.

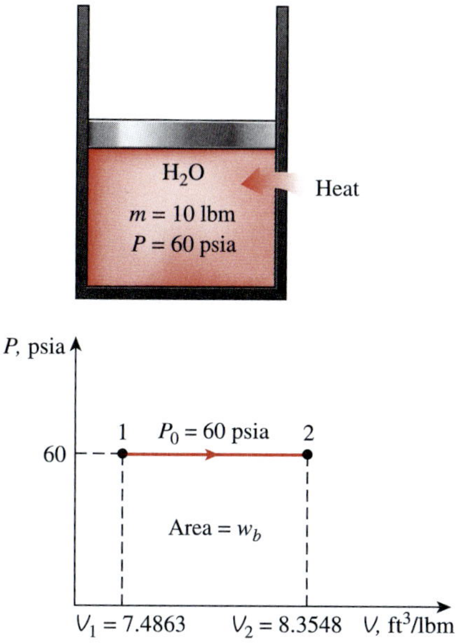

FIGURE 4–7
Schematic and *P-v* diagram for Example 4–2.

EXAMPLE 4–3 Isothermal Compression of an Ideal Gas

A piston–cylinder device initially contains 0.4 m³ of air at 100 kPa and 80°C. The air is now compressed to 0.1 m³ in such a way that the temperature inside the cylinder remains constant. Determine the work done during this process.

SOLUTION Air in a piston–cylinder device is compressed isothermally. The boundary work done is to be determined.
Analysis A sketch of the system and the *P-V* diagram of the process are shown in Fig. 4–8.
Assumptions **1** The compression process is quasi-equilibrium. **2** At specified conditions, air can be considered to be an ideal gas since it is at a high temperature and low pressure relative to its critical-point values.
Analysis For an ideal gas at constant temperature T_0,

$$PV = mRT_0 = C \quad \text{or} \quad P = \frac{C}{V}$$

where C is a constant. Substituting this into Eq. 4–2, we have

$$W_b = \int_1^2 P\,dV = \int_1^2 \frac{C}{V}\,dV = C\int_1^2 \frac{dV}{V} = C \ln\frac{V_2}{V_1} = P_1V_1 \ln\frac{V_2}{V_1} \qquad \textbf{(4–7)}$$

In Eq. 4–7, P_1V_1 can be replaced by P_2V_2 or mRT_0. Also, V_2/V_1 can be replaced by P_1/P_2 for this case since $P_1V_1 = P_2V_2$.

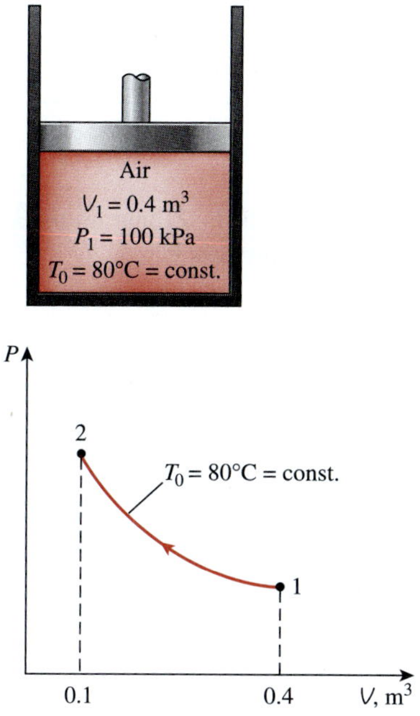

FIGURE 4–8
Schematic and *P-V* diagram for Example 4–3.

Substituting the numerical values into Eq. 4–7 yields

$$W_b = (100 \text{ kPa})(0.4 \text{ m}^3)\left(\ln \frac{0.1}{0.4}\right)\left(\frac{1 \text{ kJ}}{1 \text{ kPa}\cdot\text{m}^3}\right)$$

$$= \mathbf{-55.5 \text{ kJ}}$$

Discussion The negative sign indicates that this work is done on the system (a work input), which is always the case for compression processes.

Polytropic Process

During actual expansion and compression processes of gases, pressure and volume are often related by $PV^n = C$, where n and C are constants. A process of this kind is called a **polytropic process** (Fig. 4–9). Below we develop a general expression for the work done during a polytropic process. The pressure for a polytropic process can be expressed as

$$P = CV^{-n} \tag{4–8}$$

Substituting this relation into Eq. 4–2, we obtain

$$W_b = \int_1^2 P\,dV = \int_1^2 CV^{-n}\,dV = C\frac{V_2^{-n+1} - V_1^{-n+1}}{-n+1} = \frac{P_2V_2 - P_1V_1}{1-n} \tag{4–9}$$

since $C = P_1V_1^n = P_2V_2^n$. For an ideal gas ($PV = mRT$), this equation can also be written as

$$W_b = \frac{mR(T_2 - T_1)}{1-n} \qquad n \neq 1 \qquad \text{(kJ)} \tag{4–10}$$

For the special case of $n = 1$ the boundary work becomes

$$W_b = \int_1^2 P\,dV = \int_1^2 CV^{-1}\,dV = PV\ln\left(\frac{V_2}{V_1}\right)$$

For an ideal gas this result is equivalent to the isothermal process discussed in the previous example.

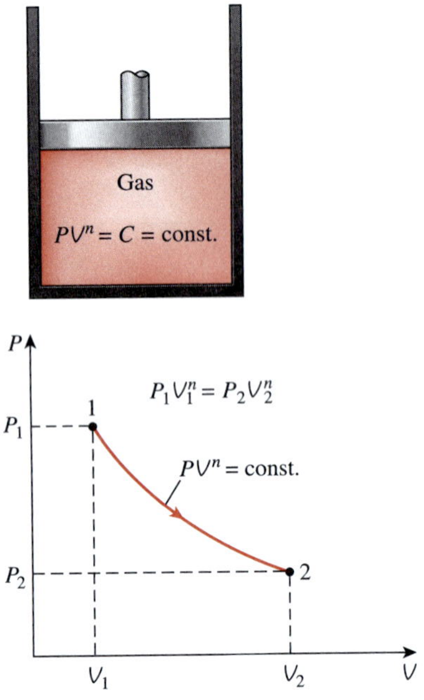

FIGURE 4–9
Schematic and P-V diagram for a polytropic process.

EXAMPLE 4–4 Expansion of a Gas against a Spring

A piston–cylinder device contains 0.05 m^3 of a gas initially at 200 kPa. At this state, a linear spring that has a spring constant of 150 kN/m is touching the piston but exerting no force on it. Now heat is transferred to the gas, causing the piston to rise and to compress the spring until the volume inside the cylinder doubles. If the cross-sectional area of the piston is 0.25 m^2, determine (*a*) the final pressure inside the cylinder, (*b*) the total work done by the gas, and (*c*) the fraction of this work done against the spring to compress it.

SOLUTION A gas in a piston–cylinder device equipped with a linear spring expands as a result of heating. The final gas pressure, the total work done, and the fraction of the work done to compress the spring are to be determined.

Assumptions **1** The expansion process is quasi-equilibrium. **2** The spring is linear in the range of interest.

Analysis A sketch of the system and the *P-V* diagram of the process are shown in Fig. 4–10.

(*a*) The enclosed volume at the final state is

$$V_2 = 2V_1 = (2)(0.05\ \text{m}^3) = 0.1\ \text{m}^3$$

Then the displacement of the piston (and of the spring) becomes

$$x = \frac{\Delta V}{A} = \frac{(0.1 - 0.05)\ \text{m}^3}{0.25\ \text{m}^2} = 0.2\ \text{m}$$

The force applied by the linear spring at the final state is

$$F = kx = (150\ \text{kN/m})(0.2\ \text{m}) = 30\ \text{kN}$$

The additional pressure applied by the spring on the gas at this state is

$$P = \frac{F}{A} = \frac{30\ \text{kN}}{0.25\ \text{m}^2} = 120\ \text{kPa}$$

Without the spring, the pressure of the gas would remain constant at 200 kPa while the piston is rising. But under the effect of the spring, the pressure rises linearly from 200 kPa to

$$200 + 120 = \mathbf{320\ kPa}$$

at the final state.

(*b*) An easy way of finding the work done is to plot the process on a *P-V* diagram and find the area under the process curve. From Fig. 4–10 the area under the process curve (a trapezoid) is determined to be

$$W = \text{area} = \frac{(200 + 320)\ \text{kPa}}{2}[(0.1 - 0.05)\ \text{m}^3]\left(\frac{1\ \text{kJ}}{1\ \text{kPa}\cdot\text{m}^3}\right) = \mathbf{13\ kJ}$$

Note that the work is done by the system.

(*c*) The work represented by the rectangular area (region I) is done against the piston and the atmosphere, and the work represented by the triangular area (region II) is done against the spring. Thus,

$$W_{\text{spring}} = \tfrac{1}{2}[(320 - 200)\ \text{kPa}](0.05\ \text{m}^3)\left(\frac{1\ \text{kJ}}{1\ \text{kPa}\cdot\text{m}^3}\right) = \mathbf{3\ kJ}$$

Discussion This result could also be obtained from

$$W_{\text{spring}} = \tfrac{1}{2}k(x_2^2 - x_1^2) = \tfrac{1}{2}(150\ \text{kN/m})[(0.2\ \text{m})^2 - 0^2]\left(\frac{1\ \text{kJ}}{1\ \text{kN}\cdot\text{m}}\right) = 3\ \text{kJ}$$

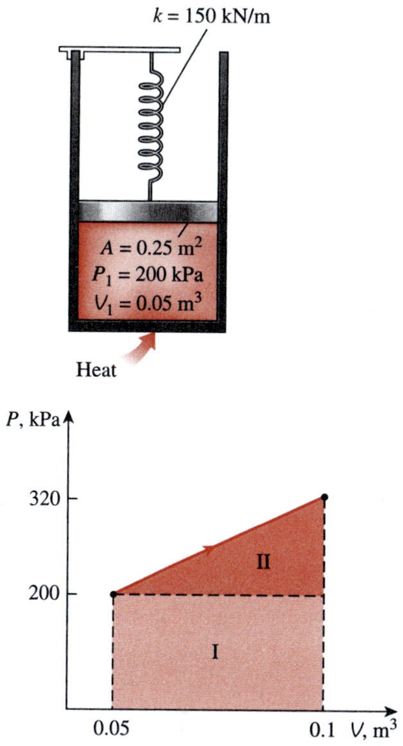

FIGURE 4–10
Schematic and *P-V* diagram for Example 4–4.

4–2 ▪ ENERGY BALANCE FOR CLOSED SYSTEMS

Energy balance for any system undergoing any kind of process was expressed as (see Chap. 2)

$$\underbrace{E_{\text{in}} - E_{\text{out}}}_{\substack{\text{Net energy transfer}\\ \text{by heat, work, and mass}}} = \underbrace{\Delta E_{\text{system}}}_{\substack{\text{Change in internal, kinetic,}\\ \text{potential, etc., energies}}} \quad (\text{kJ}) \tag{4–11}$$

or, in the **rate form**, as

$$\underbrace{\dot{E}_{in} - \dot{E}_{out}}_{\substack{\text{Rate of net energy transfer} \\ \text{by heat, work, and mass}}} = \underbrace{dE_{system}/dt}_{\substack{\text{Rate of change in internal,} \\ \text{kinetic, potential, etc., energies}}} \quad \text{(kW)} \tag{4–12}$$

For constant rates, the total quantities during a time interval Δt are related to the quantities per unit time as

$$Q = \dot{Q}\,\Delta t, \quad W = \dot{W}\,\Delta t, \quad \text{and} \quad \Delta E = (dE/dt)\Delta t \qquad \text{(kJ)} \tag{4–13}$$

The energy balance can be expressed on a **per unit mass** basis as

$$e_{in} - e_{out} = \Delta e_{system} \qquad \text{(kJ/kg)} \tag{4–14}$$

which is obtained by dividing all the quantities in Eq. 4–11 by the mass m of the system. Energy balance can also be expressed in the differential form as

$$\delta E_{in} - \delta E_{out} = dE_{system} \quad \text{or} \quad \delta e_{in} - \delta e_{out} = de_{system} \tag{4–15}$$

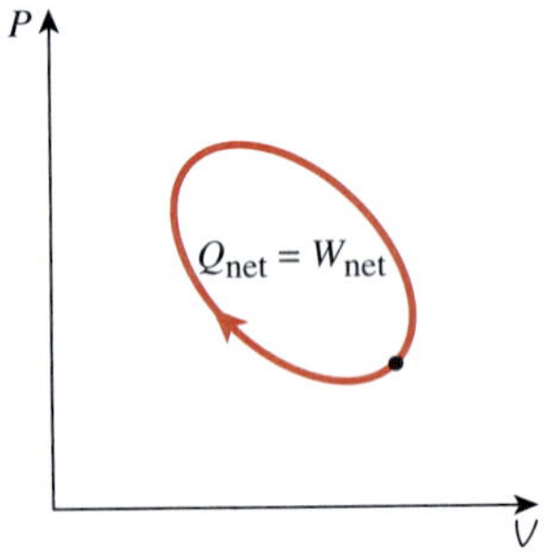

FIGURE 4–11
For a cycle $\Delta E = 0$, thus $Q = W$.

For a closed system undergoing a **cycle**, the initial and final states are identical, and thus $\Delta E_{system} = E_2 - E_1 = 0$. Then, the energy balance for a cycle simplifies to $E_{in} - E_{out} = 0$ or $E_{in} = E_{out}$. Noting that a closed system does not involve any mass flow across its boundaries, the energy balance for a cycle can be expressed in terms of heat and work interactions as

$$W_{net,out} = Q_{net,in} \quad \text{or} \quad \dot{W}_{net,out} = \dot{Q}_{net,in} \qquad \text{(for a cycle)} \tag{4–16}$$

That is, the net work output during a cycle is equal to net heat input (Fig. 4–11).

The energy balance (or the first-law) relations already given are intuitive in nature and are easy to use when the magnitudes and directions of heat and work transfers are known. However, when performing a general analytical study or solving a problem that involves an unknown heat or work interaction, we need to assume a direction for the heat or work interactions. In such cases, it is common practice to use the classical thermodynamics sign convention and to assume heat to be transferred *into the system* (heat input) in the amount of Q and work to be done *by the system* (work output) in the amount of W, and then to solve the problem. The energy balance relation in that case for a closed system becomes

$$Q_{net,in} - W_{net,out} = \Delta E_{system} \quad \text{or} \quad Q - W = \Delta E \tag{4–17}$$

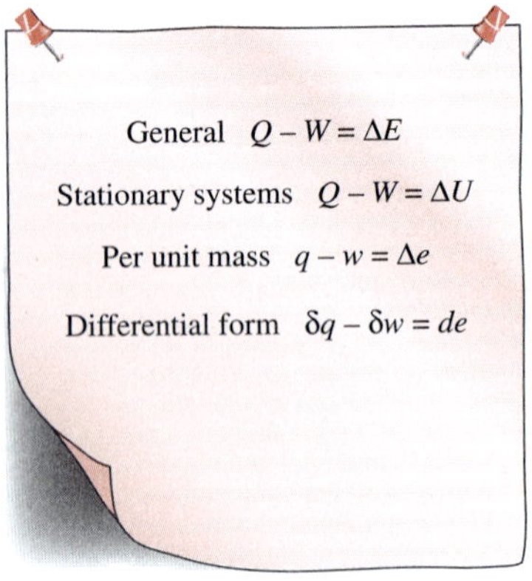

FIGURE 4–12
Various forms of the first-law relation for closed systems.

where $Q = Q_{net,in} = Q_{in} - Q_{out}$ is the *net heat input* and $W = W_{net,out} = W_{out} - W_{in}$ is the *net work output.* Obtaining a negative quantity for Q or W simply means that the assumed direction for that quantity is wrong and should be reversed. Various forms of this "traditional" first-law relation for closed systems are given in Fig. 4–12.

The first law cannot be proven mathematically, but no process in nature is known to have violated the first law, and this should be taken as sufficient proof. Note that if it were possible to prove the first law on the basis of other physical principles, the first law then would be a consequence of those principles instead of being a fundamental physical law itself.

As energy quantities, heat and work are not that different, and you probably wonder why we keep distinguishing them. After all, the change in the energy content of a system is equal to the amount of energy that crosses the system boundaries, and it makes no difference whether the energy crosses the boundary as heat or work. It seems as if the first-law relations would be much simpler if we had just one quantity that we could call *energy interaction* to represent both heat and work. Well, from the first-law point of view, heat and work are not different at all. From the second-law point of view, however, heat and work are very different, as is discussed in later chapters.

EXAMPLE 4–5 Electric Heating of a Gas at Constant Pressure

A piston–cylinder device contains 25 g of saturated water vapor that is maintained at a constant pressure of 300 kPa. A resistance heater within the cylinder is turned on and passes a current of 0.2 A for 5 min from a 120-V source. At the same time, a heat loss of 3.7 kJ occurs. (*a*) Show that for a closed system the boundary work W_b and the change in internal energy ΔU in the first-law relation can be combined into one term, ΔH, for a constant-pressure process. (*b*) Determine the final temperature of the steam.

SOLUTION Saturated water vapor in a piston–cylinder device expands at constant pressure as a result of heating. It is to be shown that $\Delta U + W_b = \Delta H$, and the final temperature is to be determined.

Assumptions **1** The tank is stationary and thus the kinetic and potential energy changes are zero, $\Delta KE = \Delta PE = 0$. Therefore, $\Delta E = \Delta U$ and internal energy is the only form of energy of the system that may change during this process. **2** Electrical wires constitute a very small part of the system, and thus the energy change of the wires can be neglected.

Analysis We take the contents of the cylinder, including the resistance wires, as the *system* (Fig. 4–13). This is a *closed system* since no mass crosses the system boundary during the process. We observe that a piston–cylinder device typically involves a moving boundary and thus boundary work W_b. The pressure remains constant during the process and thus $P_2 = P_1$. Also, heat is lost from the system and electrical work W_e is done on the system.

(*a*) This part of the solution involves a general analysis for a closed system undergoing a quasi-equilibrium constant-pressure process, and thus we consider a general closed system. We take the direction of heat transfer Q to be to the system and the work W to be done by the system. We also express the work as the sum of boundary and other forms of work (such as electrical and shaft). Then, the energy balance can be expressed as

$$\underbrace{E_{in} - E_{out}}_{\substack{\text{Net energy transfer}\\ \text{by heat, work, and mass}}} = \underbrace{\Delta E_{system}}_{\substack{\text{Change in internal, kinetic,}\\ \text{potential, etc., energies}}}$$

$$Q - W = \Delta U + \cancelto{0}{\Delta KE} + \cancelto{0}{\Delta PE}$$

$$Q - W_{other} - W_b = U_2 - U_1$$

For a constant-pressure process, the boundary work is given as $W_b = P_0(V_2 - V_1)$. Substituting this into the preceding relation gives

$$Q - W_{other} - P_0(V_2 - V_1) = U_2 - U_1$$

However,

$$P_0 = P_2 = P_1 \;\rightarrow\; Q - W_{other} = (U_2 + P_2V_2) - (U_1 + P_1V_1)$$

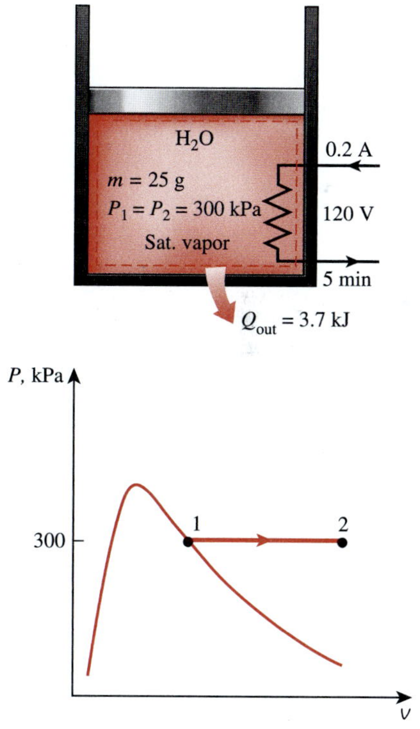

FIGURE 4–13
Schematic and P-v diagram for Example 4–5.

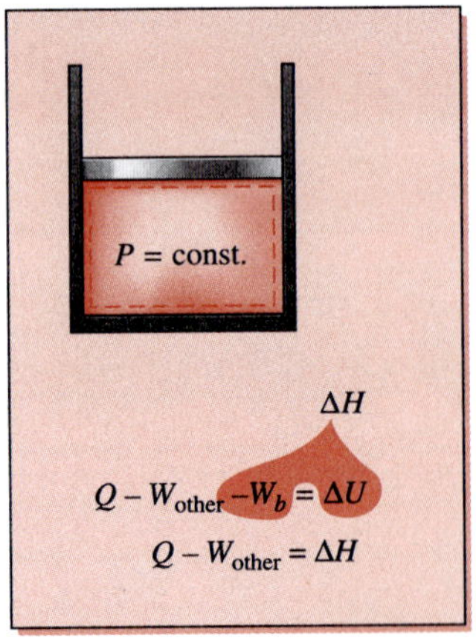

FIGURE 4–14
For a closed system undergoing a quasi-equilibrium, P = constant process, $\Delta U + W_b = \Delta H$. Note that this relation is NOT valid for closed systems processes during which pressure DOES NOT remain constant.

Also $H = U + PV$, and thus

$$Q - W_{\text{other}} = H_2 - H_1 \qquad \text{(kJ)} \tag{4–18}$$

which is the desired relation (Fig. 4–14). This equation is very convenient to use in the analysis of closed systems undergoing a ***constant-pressure quasi-equilibrium process*** since the boundary work is automatically taken care of by the enthalpy terms, and one no longer needs to determine it separately.

(*b*) The only other form of work in this case is the electrical work, which can be determined from

$$W_e = \mathbf{V}I\,\Delta t = (120\text{ V})(0.2\text{ A})(300\text{ s})\left(\frac{1\text{ kJ/s}}{1000\text{ VA}}\right) = 7.2\text{ kJ}$$

$$\textit{State 1:}\quad \left.\begin{array}{l} P_1 = 300\text{ kPa} \\ \text{sat. vapor} \end{array}\right\} \quad h_1 = h_{g\,@\,300\text{ kPa}} = 2724.9\text{ kJ/kg} \qquad \text{(Table A–5)}$$

The enthalpy at the final state can be determined directly from Eq. 4–18 by expressing heat transfer from the system and work done on the system as negative quantities (since their directions are opposite to the assumed directions). Alternately, we can use the general energy balance relation with the simplification that the boundary work is considered automatically by replacing ΔU by ΔH for a constant-pressure expansion or compression process:

$$\underbrace{E_{\text{in}} - E_{\text{out}}}_{\substack{\text{Net energy transfer} \\ \text{by heat, work, and mass}}} = \underbrace{\Delta E_{\text{system}}}_{\substack{\text{Change in internal, kinetic,} \\ \text{potential, etc., energies}}}$$

$$W_{e,\text{in}} - Q_{\text{out}} - W_b = \Delta U$$

$$W_{e,\text{in}} - Q_{\text{out}} = \Delta H = m(h_2 - h_1) \qquad \text{(since } P = \text{constant)}$$

$$7.2\text{ kJ} - 3.7\text{ kJ} = (0.025\text{ kg})(h_2 - 2724.9)\text{ kJ/kg}$$

$$h_2 = 2864.9\text{ kJ/kg}$$

Now the final state is completely specified since we know both the pressure and the enthalpy. The temperature at this state is

$$\textit{State 2:}\quad \left.\begin{array}{l} P_2 = 300\text{ kPa} \\ h_2 = 2864.9\text{ kJ/kg} \end{array}\right\} \quad T_2 = \mathbf{200^\circ C} \qquad \text{(Table A–6)}$$

Therefore, the steam will be at 200°C at the end of this process.

Discussion Strictly speaking, the potential energy change of the steam is not zero for this process since the center of gravity of the steam rose somewhat. Assuming an elevation change of 1 m (which is rather unlikely), the change in the potential energy of the steam would be 0.0002 kJ, which is very small compared to the other terms in the first-law relation. Therefore, in problems of this kind, the potential energy term is always neglected.

EXAMPLE 4–6 Unrestrained Expansion of Water

A rigid tank is divided into two equal parts by a partition. Initially, one side of the tank contains 5 kg of water at 200 kPa and 25°C, and the other side is evacuated. The partition is then removed, and the water expands into the entire tank. The water is allowed to exchange heat with its surroundings until the temperature in the tank returns to the initial value of 25°C. Determine (*a*) the volume of the tank, (*b*) the final pressure, and (*c*) the heat transfer for this process.

SOLUTION One half of a rigid tank is filled with liquid water while the other side is evacuated. The partition between the two parts is removed and water is allowed to expand and fill the entire tank while the temperature is maintained constant. The volume of the tank, the final pressure, and the heat transfer are to be to determined.

Assumptions **1** The system is stationary and thus the kinetic and potential energy changes are zero, $\Delta KE = \Delta PE = 0$ and $\Delta E = \Delta U$. **2** The direction of heat transfer is to the system (heat gain, Q_{in}). A negative result for Q_{in} indicates the assumed direction is wrong and thus it is a heat loss. **3** The volume of the rigid tank is constant, and thus there is no energy transfer as boundary work. **4** There is no electrical, shaft, or any other kind of work involved.

Analysis We take the contents of the tank, including the evacuated space, as the *system* (Fig. 4–15). This is a *closed system* since no mass crosses the system boundary during the process. We observe that the water fills the entire tank when the partition is removed (possibly as a liquid–vapor mixture).

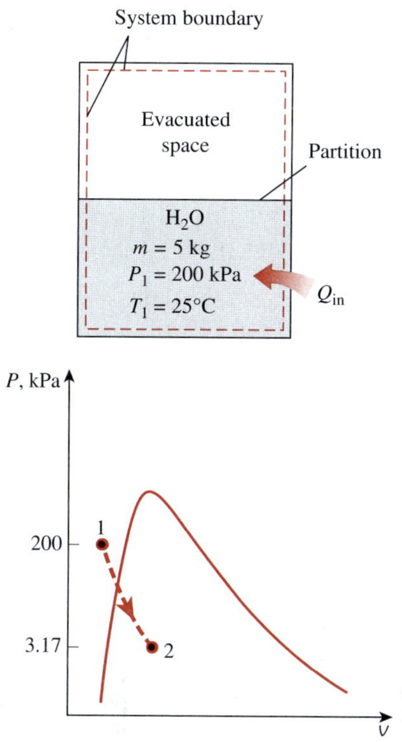

FIGURE 4–15
Schematic and P-v diagram for Example 4–6.

(*a*) Initially the water in the tank exists as a compressed liquid since its pressure (200 kPa) is greater than the saturation pressure at 25°C (3.1698 kPa). Approximating the compressed liquid as a saturated liquid at the given temperature, we find

$$v_1 \cong v_{f\,@\,25°C} = 0.001003 \text{ m}^3/\text{kg} \cong 0.001 \text{ m}^3/\text{kg} \qquad \text{(Table A–4)}$$

Then the initial volume of the water is

$$V_1 = mv_1 = (5 \text{ kg})(0.001 \text{ m}^3/\text{kg}) = 0.005 \text{ m}^3$$

The total volume of the tank is twice this amount:

$$V_{tank} = (2)(0.005 \text{ m}^3) = \mathbf{0.01\ m^3}$$

(*b*) At the final state, the specific volume of the water is

$$v_2 = \frac{V_2}{m} = \frac{0.01 \text{ m}^3}{5 \text{ kg}} = 0.002 \text{ m}^3/\text{kg}$$

which is twice the initial value of the specific volume. This result is expected since the volume doubles while the amount of mass remains constant.

$$\text{At 25°C:} \quad v_f = 0.001003 \text{ m}^3/\text{kg} \quad \text{and} \quad v_g = 43.340 \text{ m}^3/\text{kg} \quad \text{(Table A–4)}$$

Since $v_f < v_2 < v_g$, the water is a saturated liquid–vapor mixture at the final state, and thus the pressure is the saturation pressure at 25°C:

$$P_2 = P_{sat\,@\,25°C} = \mathbf{3.1698\ kPa} \qquad \text{(Table A–4)}$$

(*c*) Under stated assumptions and observations, the energy balance on the system can be expressed as

$$\underbrace{E_{in} - E_{out}}_{\substack{\text{Net energy transfer}\\ \text{by heat, work, and mass}}} = \underbrace{\Delta E_{system}}_{\substack{\text{Change in internal, kinetic,}\\ \text{potential, etc., energies}}}$$

$$Q_{in} = \Delta U = m(u_2 - u_1)$$

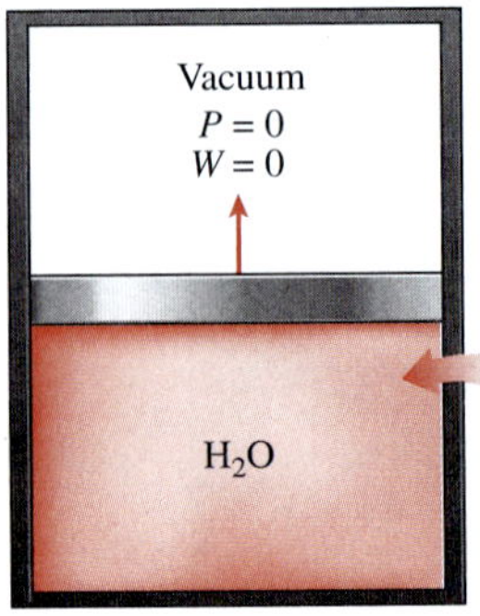

FIGURE 4–16
Expansion against a vacuum involves no work and thus no energy transfer.

Notice that even though the water is expanding during this process, the system chosen involves fixed boundaries only (the dashed lines) and therefore the moving boundary work is zero (Fig. 4–16). Then $W = 0$ since the system does not involve any other forms of work. (Can you reach the same conclusion by choosing the water as our system?) Initially,

$$u_1 \cong u_{f\,@\,25°C} = 104.83 \text{ kJ/kg}$$

The quality at the final state is determined from the specific volume information:

$$x_2 = \frac{v_2 - v_f}{v_{fg}} = \frac{0.002 - 0.001}{43.34 - 0.001} = 2.3 \times 10^{-5}$$

Then

$$\begin{aligned} u_2 &= u_f + x_2 u_{fg} \\ &= 104.83 \text{ kJ/kg} + (2.3 \times 10^{-5})(2304.3 \text{ kJ/kg}) \\ &= 104.88 \text{ kJ/kg} \end{aligned}$$

Substituting yields

$$Q_{in} = (5 \text{ kg})[(104.88 - 104.83) \text{ kJkg}] = \mathbf{0.25 \text{ kJ}}$$

Discussion The positive sign indicates that the assumed direction is correct, and heat is transferred to the water.

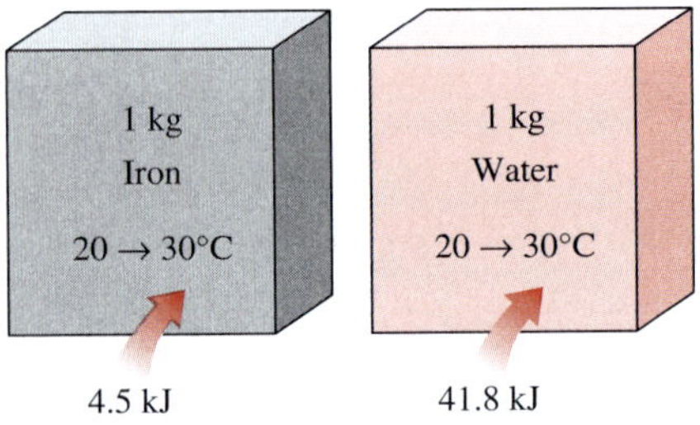

FIGURE 4–17
It takes different amounts of energy to raise the temperature of different substances by the same amount.

4–3 ▪ SPECIFIC HEATS

We know from experience that it takes different amounts of energy to raise the temperature of identical masses of different substances by one degree. For example, we need about 4.5 kJ of energy to raise the temperature of 1 kg of iron from 20 to 30°C, whereas it takes about 9 times this energy (41.8 kJ to be exact) to raise the temperature of 1 kg of liquid water by the same amount (Fig. 4–17). Therefore, it is desirable to have a property that will enable us to compare the energy storage capabilities of various substances. This property is the specific heat.

The **specific heat** is defined as *the energy required to raise the temperature of a unit mass of a substance by one degree* (Fig. 4–18). In general, this energy depends on how the process is executed. In thermodynamics, we are interested in two kinds of specific heats: **specific heat at constant volume** c_v and **specific heat at constant pressure** c_p.

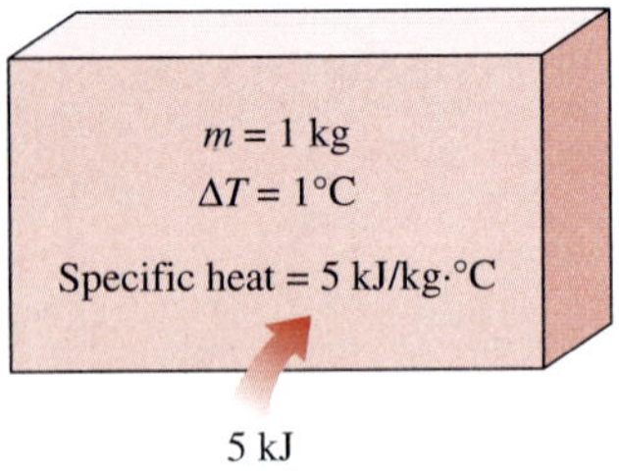

FIGURE 4–18
Specific heat is the energy required to raise the temperature of a unit mass of a substance by one degree in a specified way.

Physically, the specific heat at constant volume c_v can be viewed as *the energy required to raise the temperature of the unit mass of a substance by one degree as the volume is maintained constant*. The energy required to do the same as the pressure is maintained constant is the specific heat at constant pressure c_p. This is illustrated in Fig. 4–19. The specific heat at constant pressure c_p is always greater than c_v because at constant pressure the system is allowed to expand and the energy for this expansion work must also be supplied to the system.

Now we attempt to express the specific heats in terms of other thermodynamic properties. First, consider a fixed mass in a stationary closed system undergoing a constant-volume process (and thus no expansion or compression work is involved). The conservation of energy principle $e_{in} - e_{out} = \Delta e_{system}$ for this process can be expressed in the differential form as

$$\delta e_{in} - \delta e_{out} = du$$

The left-hand side of this equation represents the net amount of energy transferred to the system. From the definition of c_v, this energy must be equal to $c_v\, dT$, where dT is the differential change in temperature. Thus,

$$c_v{}^\circ dT = du \;^\circ\;^\circ \text{ at constant volume}$$

or

$$c_v = \left(\frac{\partial u}{\partial T}\right)_v \qquad \textbf{(4–19)}$$

Similarly, an expression for the specific heat at constant pressure c_p can be obtained by considering a constant-pressure expansion or compression process. It yields

$$c_p = \left(\frac{\partial h}{\partial T}\right)_p \qquad \textbf{(4–20)}$$

Equations 4–19 and 4–20 are the defining equations for c_v and c_p, and their interpretation is given in Fig. 4–20.

Note that c_v and c_p are expressed in terms of other properties; thus, they must be properties themselves. Like any other property, the specific heats of a substance depend on the state that, in general, is specified by two independent, intensive properties. That is, the energy required to raise the temperature of a substance by one degree is different at different temperatures and pressures (Fig. 4–21). But this difference is usually not very large.

A few observations can be made from Eqs. 4–19 and 4–20. First, these equations are *property relations* and as such *are independent of the type of processes*. They are valid for *any* substance undergoing *any* process. The only relevance c_v has to a constant-volume process is that c_v happens to be the energy transferred to a system during a constant-volume process per unit mass, per unit degree rise in temperature. This is how the values of c_v are determined. This is also how the name *specific heat at constant volume* originated. Likewise, the energy transferred to a system per unit mass per unit temperature rise during a constant-pressure process happens to be equal to c_p. This is how the values of c_p can be determined and also explains the origin of the name *specific heat at constant pressure*.

Another observation that can be made from Eqs. 4–19 and 4–20 is that c_v is related to the changes in *internal energy* and c_p to the changes in *enthalpy*. In fact, it would be more proper to define c_v as *the change in the internal energy of a substance per unit change in temperature at constant volume*. Likewise, c_p can be defined as *the change in the enthalpy of a substance per unit change in temperature at constant pressure*. In other words, c_v is a measure of the variation of internal energy of a substance with temperature, and c_p is a measure of the variation of enthalpy of a substance with temperature.

Both the internal energy and enthalpy of a substance can be changed by the transfer of *energy* in any form, with heat being only one of them. Therefore,

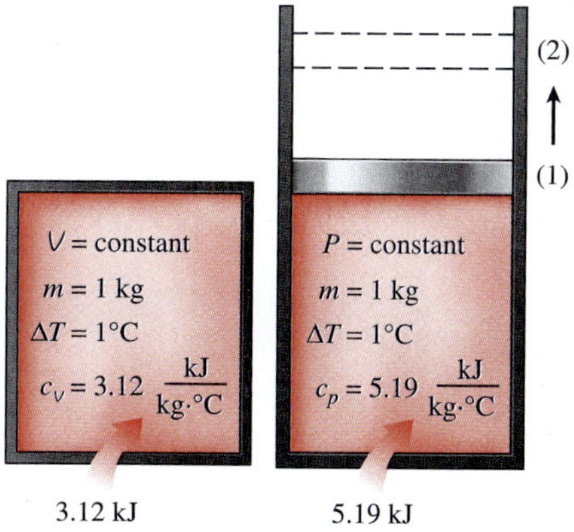

FIGURE 4–19
Constant-volume and constant-pressure specific heats c_v and c_p (values given are for helium gas).

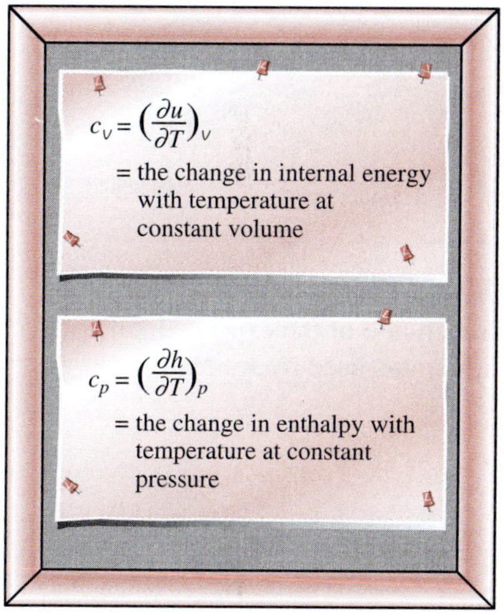

FIGURE 4–20
Formal definitions of c_v and c_p.

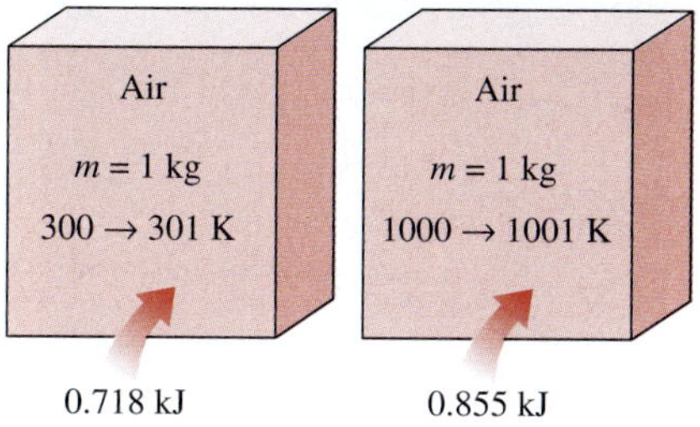

FIGURE 4–21
The specific heat of a substance changes with temperature.

the term *specific energy* is probably more appropriate than the term *specific heat*, which implies that energy is transferred (and stored) in the form of heat.

A common unit for specific heats is kJ/kg·°C or kJ/kg·K. Notice that these two units are *identical* since $\Delta T(°\text{C}) = \Delta T(\text{K})$, and 1°C change in temperature is equivalent to a change of 1 K. The specific heats are sometimes given on a *molar basis*. They are then denoted by $\bar{c}_v$ and $\bar{c}_p$ and have the unit kJ/kmol·°C or kJ/kmol·K.

4–4 ▪ INTERNAL ENERGY, ENTHALPY, AND SPECIFIC HEATS OF IDEAL GASES

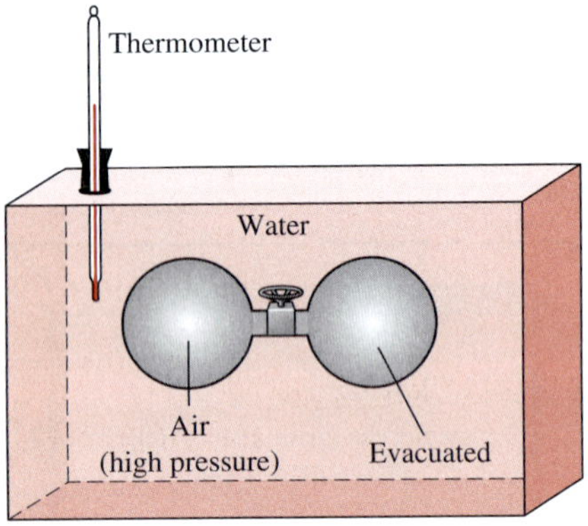

FIGURE 4–22
Schematic of the experimental apparatus used by Joule.

We defined an ideal gas as a gas whose temperature, pressure, and specific volume are related by

$$Pv = RT$$

It has been demonstrated mathematically (Chap. 12) and experimentally (Joule, 1843) that for an ideal gas the internal energy is a function of the temperature only. That is,

$$u = u(T) \tag{4–21}$$

In his classical experiment, Joule submerged two tanks connected with a pipe and a valve in a water bath, as shown in Fig. 4–22. Initially, one tank contained air at a high pressure and the other tank was evacuated. When thermal equilibrium was attained, he opened the valve to let air pass from one tank to the other until the pressures equalized. Joule observed no change in the temperature of the water bath and assumed that no heat was transferred to or from the air. Since there was also no work done, he concluded that the internal energy of the air did not change even though the volume and the pressure changed. Therefore, he reasoned, the internal energy is a function of temperature only and not a function of pressure or specific volume. (Joule later showed that for gases that deviate significantly from ideal-gas behavior, the internal energy is not a function of temperature alone.)

Using the definition of enthalpy and the equation of state of an ideal gas, we have

$$\left.\begin{aligned} h &= u + Pv \\ Pv &= RT \end{aligned}\right\} \quad h = u + RT$$

Since R is constant and $u = u(T)$, it follows that the enthalpy of an ideal gas is also a function of temperature only:

$$h = h(T) \tag{4–22}$$

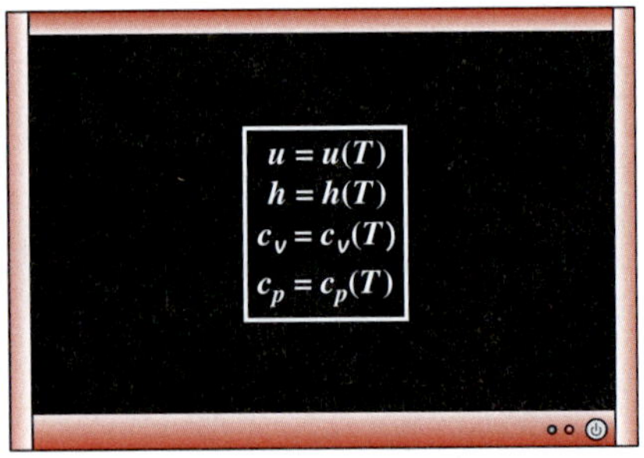

FIGURE 4–23
For ideal gases, u, h, c_v, and c_p vary with temperature only.

Since u and h depend only on temperature for an ideal gas, the specific heats c_v and c_p also depend, at most, on temperature only. Therefore, at a given temperature, u, h, c_v, and c_p of an ideal gas have fixed values regardless of the specific volume or pressure (Fig. 4–23). Thus, for ideal gases, the partial derivatives in Eqs. 4–19 and 4–20 can be replaced by ordinary derivatives. Then, the differential changes in the internal energy and enthalpy of an ideal gas can be expressed as

$$du = c_v(T)\,dT \tag{4–23}$$

and

$$dh = c_p(T)\,dT \tag{4–24}$$

The change in internal energy or enthalpy for an ideal gas during a process from state 1 to state 2 is determined by integrating these equations:

$$\Delta u = u_2 - u_1 = \int_1^2 c_v(T)\,dT \qquad \text{(kJ/kg)} \tag{4–25}$$

and

$$\Delta h = h_2 - h_1 = \int_1^2 c_p(T)\,dT \qquad \text{(kJ/kg)} \tag{4–26}$$

To carry out these integrations, we need to have relations for c_v and c_p as functions of temperature.

At low pressures, all real gases approach ideal-gas behavior, and therefore their specific heats depend on temperature only. The specific heats of real gases at low pressures are called *ideal-gas specific heats*, or *zero-pressure specific heats*, and are often denoted c_{p0} and c_{v0}. Accurate analytical expressions for ideal-gas specific heats, based on direct measurements or calculations from statistical behavior of molecules, are available and are given as third-degree polynomials in the appendix (Table A–2*c*) for several gases. A plot of $\bar{c}_{p0}(T)$ data for some common gases is given in Fig. 4–24.

The use of ideal-gas specific heat data is limited to low pressures, but these data can also be used at moderately high pressures with reasonable accuracy as long as the gas does not deviate from ideal-gas behavior significantly.

The integrations in Eqs. 4–25 and 4–26 are straightforward but rather time-consuming and thus impractical. To avoid these laborious calculations, *u* and *h* data for a number of gases have been tabulated over small temperature intervals. These tables are obtained by choosing an arbitrary reference point and performing the integrations in Eqs. 4–25 and 4–26 by treating state 1 as the reference state. In the ideal-gas tables given in the appendix, zero kelvin is chosen as the reference state, and both the enthalpy and the internal energy are assigned zero values at that state (Fig. 4–25). The choice of the reference state has no effect on Δu or Δh calculations. The *u* and *h* data are given in kJ/kg for air (Table A–17) and usually in kJ/kmol for other gases. The unit kJ/kmol is very convenient in the thermodynamic analysis of chemical reactions.

Some observations can be made from Fig. 4–24. First, the specific heats of gases with complex molecules (molecules with two or more atoms) are higher and increase with temperature. Also, the variation of specific heats with temperature is smooth and may be approximated as linear over small temperature intervals (a few hundred degrees or less). Therefore, the specific heat functions in Eqs. 4–25 and 4–26 can be replaced by the constant average specific heat values. Then, the integrations in these equations can be performed, yielding

$$u_2 - u_1 = c_{v,\text{avg}}(T_2 - T_1) \qquad \text{(kJ/kg)} \tag{4–27}$$

and

$$h_2 - h_1 = c_{p,\text{avg}}(T_2 - T_1) \qquad \text{(kJ/kg)} \tag{4–28}$$

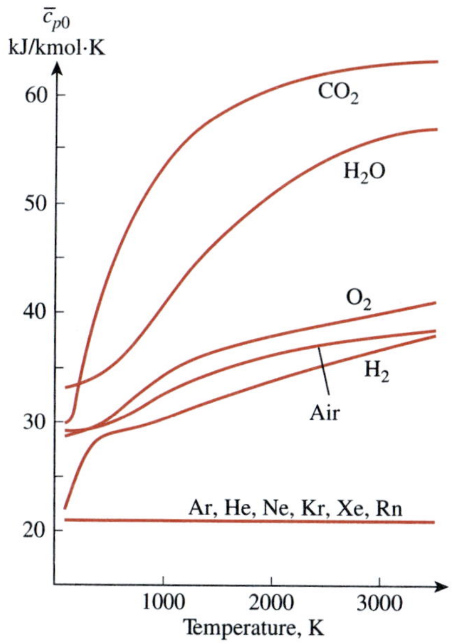

FIGURE 4–24
Ideal-gas constant-pressure specific heats for some gases (see Table A–2*c* for c_p equations).

Air

T, K	*u*, kJ/kg	*h*, kJ/kg
0	0	0
.	.	.
.	.	.
300	214.07	300.19
310	221.25	310.24
.	.	.
.	.	.

FIGURE 4–25
In the preparation of ideal-gas tables, 0 K is chosen as the reference temperature.

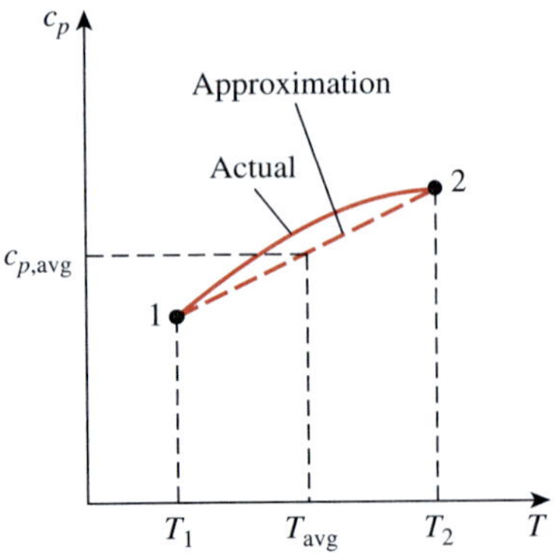

FIGURE 4–26
For small temperature intervals, the specific heats may be assumed to vary linearly with temperature.

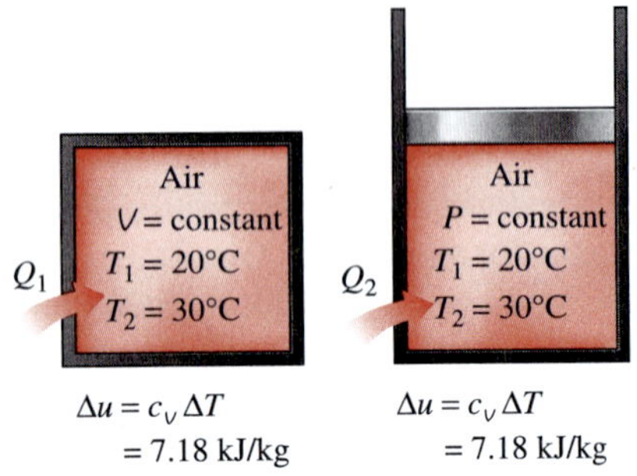

FIGURE 4–27
The relation $\Delta u = c_v \Delta T$ is valid for *any* kind of process, constant-volume or not.

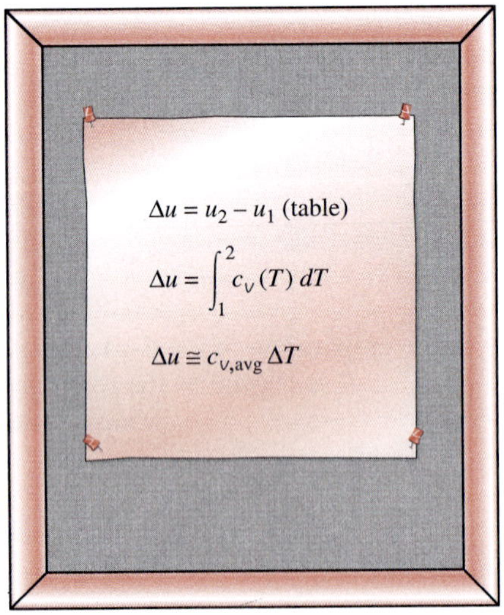

FIGURE 4–28
ways of calculating Δu.

The specific heat values for some common gases are listed as a function of temperature in Table A–2*b*. The average specific heats $c_{p,\text{avg}}$ and $c_{v,\text{avg}}$ are evaluated from this table at the average temperature $(T_1 + T_2)/2$, as shown in Fig. 4–26. If the final temperature T_2 is not known, the specific heats may be evaluated at T_1 or at the anticipated average temperature. Then T_2 can be determined by using these specific heat values. The value of T_2 can be refined, if necessary, by evaluating the specific heats at the new average temperature.

Another way of determining the average specific heats is to evaluate them at T_1 and T_2 and then take their average. Usually both methods give reasonably good results, and one is not necessarily better than the other.

Another observation that can be made from Fig. 4–24 is that the ideal-gas specific heats of *monatomic gases* such as argon, neon, and helium remain constant over the entire temperature range. Thus, Δu and Δh of monatomic gases can easily be evaluated from Eqs. 4–27 and 4–28.

Note that the Δu and Δh relations given previously are not restricted to any kind of process. They are valid for all processes. The presence of the constant-volume specific heat c_v in an equation should not lead one to believe that this equation is valid for a constant-volume process only. On the contrary, the relation $\Delta u = c_{v,\text{avg}}\, \Delta T$ is valid for *any* ideal gas undergoing *any* process (Fig. 4–27). A similar argument can be given for c_p and Δh.

To summarize, there are three ways to determine the internal energy and enthalpy changes of ideal gases (Fig. 4–28):

1. By using the tabulated u and h data. This is the easiest and most accurate way when tables are readily available.
2. By using the c_v or c_p relations as a function of temperature and performing the integrations. This is very inconvenient for hand calculations but quite desirable for computerized calculations. The results obtained are very accurate.
3. By using average specific heats. This is very simple and certainly very convenient when property tables are not available. The results obtained are reasonably accurate if the temperature interval is not very large.

Specific Heat Relations of Ideal Gases

A special relationship between c_p and c_v for ideal gases can be obtained by differentiating the relation $h = u + RT$, which yields

$$dh = du + R\,dT$$

Replacing dh by $c_p dT$ and du by $c_v dT$ and dividing the resulting expression by dT, we obtain

$$c_p = c_v + R \qquad \text{(kJ/kg·K)} \tag{4–29}$$

This is an important relationship for ideal gases since it enables us to determine c_v from a knowledge of c_p and the gas constant R.

When the specific heats are given on a molar basis, R in the above equation should be replaced by the universal gas constant R_u (Fig. 4–29).

$$\bar{c}_p = \bar{c}_v + R_u \qquad \text{(kJ/ kmol·K)} \tag{4–30}$$

At this point, we introduce another ideal-gas property called the **specific heat ratio** k, defined as

$$k = \frac{c_p}{c_v} \qquad \text{(4–31)}$$

The specific ratio also varies with temperature, but this variation is very mild. For monatomic gases, its value is essentially constant at 1.667. Many diatomic gases, including air, have a specific heat ratio of about 1.4 at room temperature.

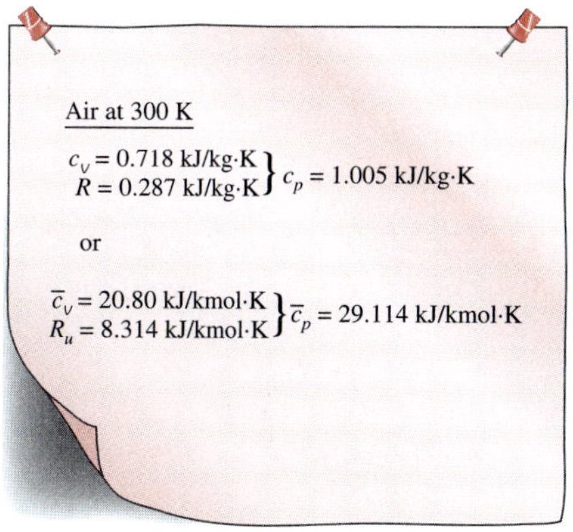

FIGURE 4–29
The c_p of an ideal gas can be determined from a knowledge of c_v and R.

EXAMPLE 4–7 Evaluation of the Δu of an Ideal Gas

Air at 300 K and 200 kPa is heated at constant pressure to 600 K. Determine the change in internal energy of air per unit mass, using (*a*) data from the air table (Table A–17), (*b*) the functional form of the specific heat (Table A–2*c*), and (*c*) the average specific heat value (Table A–2*b*).

SOLUTION The internal energy change of air is to be determined in three different ways.

Assumptions At specified conditions, air can be considered to be an ideal gas since it is at a high temperature and low pressure relative to its critical-point values.

Analysis The internal energy change Δu of ideal gases depends on the initial and final temperatures only, and not on the type of process. Thus, the following solution is valid for any kind of process.

(*a*) One way of determining the change in internal energy of air is to read the u values at T_1 and T_2 from Table A–17 and take the difference:

$$u_1 = u_{@\,300\text{ K}} = 214.07 \text{ kJ/kg}$$

$$u_2 = u_{@\,600\text{ K}} = 434.78 \text{ kJ/kg}$$

Thus,

$$\Delta u = u_2 - u_1 = (434.78 - 214.07) \text{ kJ/kg} = \mathbf{220.71 \text{ kJ/kg}}$$

(*b*) The $\bar{c}_p(T)$ of air is given in Table A–2*c* in the form of a third-degree polynomial expressed as

$$\bar{c}_p(T) = a + bT + cT^2 + dT^3$$

where $a = 28.11$, $b = 0.1967 \times 10^{-2}$, $c = 0.4802 \times 10^{-5}$, and $d = -1.966 \times 10^{-9}$. From Eq. 4-30,

$$\bar{c}_v(T) = \bar{c}_p - R_u = (a - R_u) + bT + cT^2 + dT^3$$

From Eq. 4–25,

$$\Delta \bar{u} = \int_1^2 \bar{c}_v(T)\, dT = \int_{T_1}^{T_2} [(a - R_u) + bT + cT^2 + dT^3]\, dT$$

Performing the integration and substituting the values, we obtain

$$\Delta \bar{u} = 6447 \text{ kJ/kmol}$$

The change in the internal energy on a unit-mass basis is determined by dividing this value by the molar mass of air (Table A–1):

$$\Delta u = \frac{\Delta \bar{u}}{M} = \frac{6447 \text{ kJ/kmol}}{28.97 \text{ kg/kmol}} = \mathbf{222.5 \text{ kJ/kg}}$$

which differs from the tabulated value by 0.8 percent.

(*c*) The average value of the constant-volume specific heat $c_{v,\text{avg}}$ is determined from Table A–2*b* at the average temperature of $(T_1 + T_2)/2 = 450$ K to be

$$c_{v,\text{avg}} = c_{v\,@\,450\text{ K}} = 0.733 \text{ kJ/kg·K}$$

Thus,

$$\Delta u = c_{v,\text{avg}}(T_2 - T_1) = (0.733 \text{ kJ/kg·K})[(600 - 300)\text{K}]$$
$$= 220 \text{ kJ/kg}$$

Discussion This answer differs from the tabulated value (220.71 kJ/kg) by only 0.4 percent. This close agreement is not surprising since the assumption that c_v varies linearly with temperature is a reasonable one at temperature intervals of only a few hundred degrees. If we had used the c_v value at $T_1 = 300$ K instead of at T_{avg}, the result would be 215.4 kJ/kg, which is in error by about 2 percent. Errors of this magnitude are acceptable for most engineering purposes.

EXAMPLE 4–8 Heating of a Gas in a Tank by Stirring

An insulated rigid tank initially contains 1.5 lbm of helium at 80°F and 50 psia. A paddle wheel with a power rating of 0.02 hp is operated within the tank for 30 min. Determine (*a*) the final temperature and (*b*) the final pressure of the helium gas.

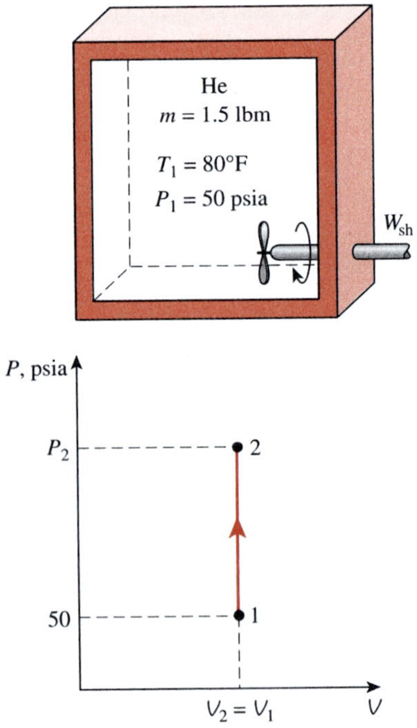

FIGURE 4–30
Schematic and *P-V* diagram for Example 4–8.

SOLUTION Helium gas in an insulated rigid tank is stirred by a paddle wheel. The final temperature and pressure of helium are to be determined.
Assumptions **1** Helium is an ideal gas since it is at a very high temperature relative to its critical-point value of −451°F. **2** Constant specific heats can be used for helium. **3** The system is stationary and thus the kinetic and potential energy changes are zero, $\Delta KE = \Delta PE = 0$ and $\Delta E = \Delta U$. **4** The volume of the tank is constant, and thus there is no boundary work. **5** The system is adiabatic and thus there is no heat transfer.
Analysis We take the contents of the tank as the *system* (Fig. 4–30). This is a *closed system* since no mass crosses the system boundary during the process. We observe that there is shaft work done on the system.

(*a*) The amount of paddle-wheel work done on the system is

$$W_{\text{sh}} = \dot{W}_{\text{sh}}\,\Delta t = (0.02 \text{ hp})(0.5 \text{ h})\left(\frac{2545 \text{ Btu/h}}{1 \text{ hp}}\right) = 25.45 \text{ Btu}$$

Under the stated assumptions and observations, the energy balance on the system can be expressed as

$$\underbrace{E_{\text{in}} - E_{\text{out}}}_{\substack{\text{Net energy transfer}\\ \text{by heat, work, and mass}}} = \underbrace{\Delta E_{\text{system}}}_{\substack{\text{Change in internal, kinetic,}\\ \text{potential, etc., energies}}}$$

$$W_{\text{sh,in}} = \Delta U = m(u_2 - u_1) = mc_{v,\text{avg}}(T_2 - T_1)$$

As we pointed out earlier, the ideal-gas specific heats of monatomic gases (helium being one of them) are constant. The c_v value of helium is determined from Table A–2E*a* to be c_v = 0.753 Btu/lbm·°F. Substituting this and other known quantities into the above equation, we obtain

$$25.45 \text{ Btu} = (1.5 \text{ lbm})(0.753 \text{ Btu/lbm·°F})(T_2 - 80\text{°F})$$

$$T_2 = \mathbf{102.5°F}$$

(*b*) The final pressure is determined from the ideal-gas relation

$$\frac{P_1V_1}{T_1} = \frac{P_2V_2}{T_2}$$

where V_1 and V_2 are identical and cancel out. Then the final pressure becomes

$$\frac{50 \text{ psia}}{(80 + 460) \text{ R}} = \frac{P_2}{(102.5 + 460)\text{R}}$$

$$P_2 = \mathbf{52.1 \text{ psia}}$$

Discussion Note that the pressure in the ideal-gas relation is always the absolute pressure.

EXAMPLE 4–9 Heating of a Gas by a Resistance Heater

A piston–cylinder device initially contains 0.5 m³ of nitrogen gas at 400 kPa and 27°C. An electric heater within the device is turned on and is allowed to pass a current of 2 A for 5 min from a 120-V source. Nitrogen expands at constant pressure, and a heat loss of 2800 J occurs during the process. Determine the final temperature of nitrogen.

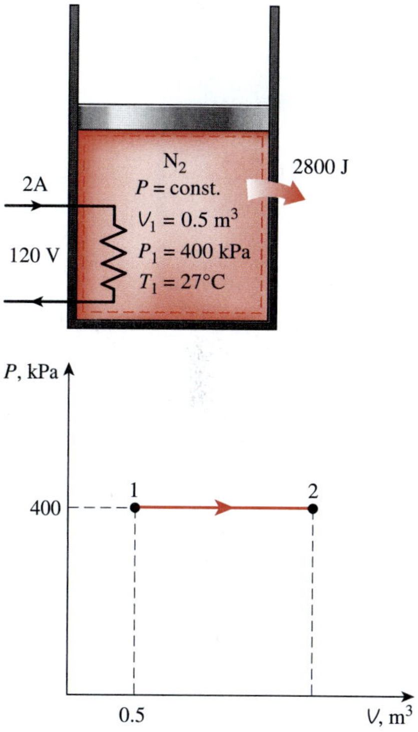

FIGURE 4–31
Schematic and *P*-*V* diagram for Example 4–9.

SOLUTION Nitrogen gas in a piston–cylinder device is heated by an electric resistance heater. Nitrogen expands at constant pressure while some heat is lost. The final temperature of nitrogen is to be determined.

Assumptions **1** Nitrogen is an ideal gas since it is at a high temperature and low pressure relative to its critical-point values of −147°C, and 3.39 MPa. **2** The system is stationary and thus the kinetic and potential energy changes are zero, ΔKE = ΔPE = 0 and $\Delta E = \Delta U$. **3** The pressure remains constant during the process and thus $P_2 = P_1$. **4** Nitrogen has constant specific heats at room temperature.

Analysis We take the contents of the cylinder as the *system* (Fig. 4–31). This is a *closed system* since no mass crosses the system boundary during the process. We observe that a piston–cylinder device typically involves a moving boundary and thus boundary work, W_b. Also, heat is lost from the system and electrical work W_e is done on the system.

First, let us determine the electrical work done on the nitrogen:

$$W_e = \mathbf{V}I\,\Delta t = (120 \text{ V})\ (2 \text{ A})\ (5 \times 60 \text{ s}) \left(\frac{1 \text{ kJ/s}}{1000 \text{ VA}}\right) = 72 \text{ kJ}$$

The mass of nitrogen is determined from the ideal-gas relation:

$$m = \frac{P_1V_1}{RT_1} = \frac{(400 \text{ kPa})\ (0.5 \text{ m}^3)}{(0.297 \text{ kPa·m}^3/\text{kg·K})\ (300 \text{ K})} = 2.245 \text{ kg}$$

Under the stated assumptions and observations, the energy balance on the system can be expressed as

$$\underbrace{E_{\text{in}} - E_{\text{out}}}_{\substack{\text{Net energy transfer}\\ \text{by heat, work, and mass}}} = \underbrace{\Delta E_{\text{system}}}_{\substack{\text{Change in internal, kinetic,}\\ \text{potential, etc., energies}}}$$

$$W_{e,\text{in}} - Q_{\text{out}} - W_{b,\text{out}} = \Delta U$$

$$W_{e,\text{in}} - Q_{\text{out}} = \Delta H = m(h_2 - h_1) = mc_p(T_2 - T_1)$$

since $\Delta U + W_b \equiv \Delta H$ for a closed system undergoing a quasi-equilibrium expansion or compression process at constant pressure. From Table A–2a, c_p = 1.039 kJ/kg·K for nitrogen at room temperature. The only unknown quantity in the previous equation is T_2, and it is found to be

$$72 \text{ kJ} - 2.8 \text{ kJ} = (2.245 \text{ kg})(1.039 \text{ kJ/kg·K})(T_2 - 27°\text{C})$$

$$T_2 = \mathbf{56.7°C}$$

Discussion Note that we could also solve this problem by determining the boundary work and the internal energy change rather than the enthalpy change.

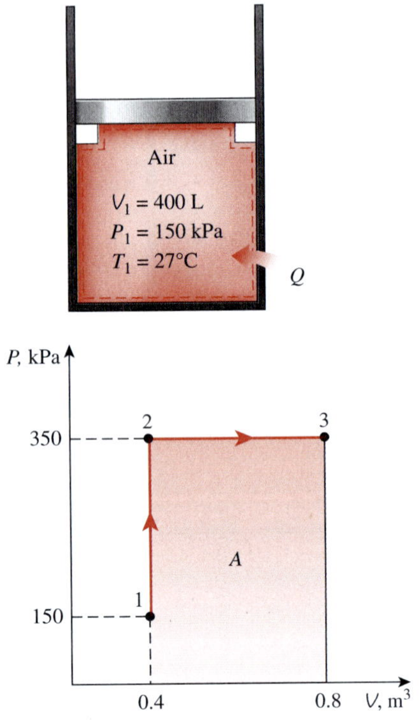

FIGURE 4–32
Schematic and P-V diagram for Example 4–10.

EXAMPLE 4–10 Heating of a Gas at Constant Pressure

A piston–cylinder device initially contains air at 150 kPa and 27°C. At this state, the piston is resting on a pair of stops, as shown in Fig. 4–32, and the enclosed volume is 400 L. The mass of the piston is such that a 350-kPa pressure is required to move it. The air is now heated until its volume has doubled. Determine (*a*) the final temperature, (*b*) the work done by the air, and (*c*) the total heat transferred to the air.

SOLUTION Air in a piston–cylinder device with a set of stops is heated until its volume is doubled. The final temperature, work done, and the total heat transfer are to be determined.

Assumptions **1** Air is an ideal gas since it is at a high temperature and low pressure relative to its critical-point values. **2** The system is stationary and thus the kinetic and potential energy changes are zero, $\Delta KE = \Delta PE = 0$ and $\Delta E = \Delta U$. **3** The volume remains constant until the piston starts moving, and the pressure remains constant afterwards. **4** There are no electrical, shaft, or other forms of work involved.

Analysis We take the contents of the cylinder as the *system* (Fig. 4–32). This is a *closed system* since no mass crosses the system boundary during the process. We observe that a piston-cylinder device typically involves a moving boundary and thus boundary work, W_b. Also, the boundary work is done by the system, and heat is transferred to the system.

(*a*) The final temperature can be determined easily by using the ideal-gas relation between states 1 and 3 in the following form:

$$\frac{P_1V_1}{T_1} = \frac{P_3V_3}{T_3} \longrightarrow \frac{(150 \text{ kPa})(V_1)}{300 \text{ K}} = \frac{(350 \text{ kPa})(2V_1)}{T_3}$$

$$T_3 = \mathbf{1400\ K}$$

(*b*) The work done could be determined by integration, but for this case it is much easier to find it from the area under the process curve on a *P-V* diagram, shown in Fig. 4–32:

$$A = (V_2 - V_1)P_2 = (0.4\ \text{m}^3)(350\ \text{kPa}) = 140\ \text{m}^3\cdot\text{kPa}$$

Therefore,

$$W_{13} = \mathbf{140\ kJ}$$

The work is done by the system (to raise the piston and to push the atmospheric air out of the way), and thus it is work output.

(*c*) Under the stated assumptions and observations, the energy balance on the system between the initial and final states (process 1–3) can be expressed as

$$\underbrace{E_{\text{in}} - E_{\text{out}}}_{\substack{\text{Net energy transfer}\\ \text{by heat, work, and mass}}} = \underbrace{\Delta E_{\text{system}}}_{\substack{\text{Change in internal, kinetic,}\\ \text{potential, etc., energies}}}$$

$$Q_{\text{in}} - W_{b,\text{out}} = \Delta U = m(u_3 - u_1)$$

The mass of the system can be determined from the ideal-gas relation:

$$m = \frac{P_1V_1}{RT_1} = \frac{(150\ \text{kPa})\ (0.4\ \text{m}^3)}{(0.287\ \text{kPa}\cdot\text{m}^3/\ \text{kg}\cdot\text{K})\ (300\ \text{K})} = 0.697\ \text{kg}$$

The internal energies are determined from the air table (Table A–17) to be

$$u_1 = u_{@\ 300\ \text{K}} = 214.07\ \text{kJ/kg}$$

$$u_3 = u_{@\ 1400\ \text{K}} = 1113.52\ \text{kJ/kg}$$

Thus,

$$Q_{\text{in}} - 140\ \text{kJ} = (0.697\ \text{kg})[(1113.52 - 214.07)\ \text{kJ/kg}]$$

$$Q_{\text{in}} = \mathbf{767\ kJ}$$

Discussion The positive sign verifies that heat is transferred to the system.

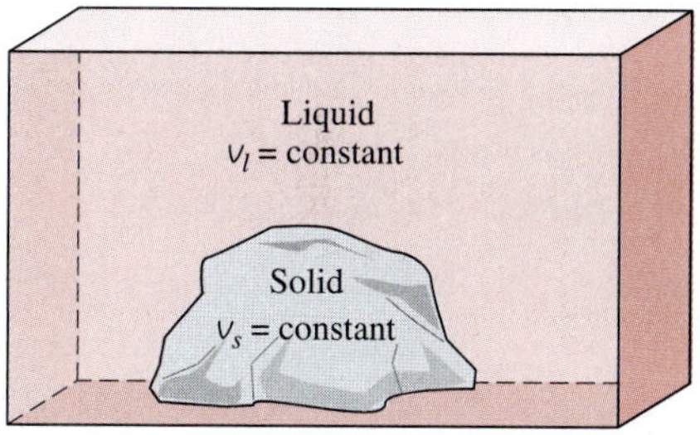

FIGURE 4–33
The specific volumes of incompressible substances remain constant during a process.

4–5 ▪ INTERNAL ENERGY, ENTHALPY, AND SPECIFIC HEATS OF SOLIDS AND LIQUIDS

A substance whose specific volume (or density) is constant is called an **incompressible substance**. The specific volumes of solids and liquids essentially remain constant during a process (Fig. 4–33). Therefore, liquids and solids can be approximated as incompressible substances without sacrificing much in accuracy. The constant-volume assumption should be taken to imply that the energy associated with the volume change is negligible compared with other forms of energy. Otherwise, this assumption would be ridiculous for studying the thermal stresses in solids (caused by volume change with temperature) or analyzing liquid-in-glass thermometers.

It can be mathematically shown that (see Chap. 12) the constant-volume and constant-pressure specific heats are identical for incompressible substances (Fig. 4–34). Therefore, for solids and liquids, the subscripts on c_p

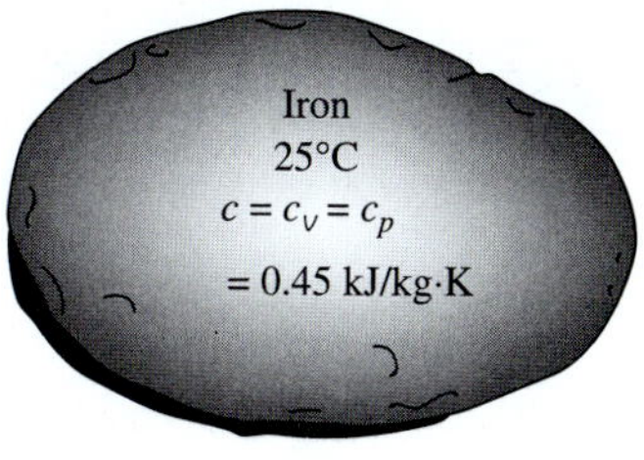

FIGURE 4–34
The c_v and c_p values of incompressible substances are identical and are denoted by c.

and c_v can be dropped, and both specific heats can be represented by a single symbol c. That is,

$$c_p = c_v = c \tag{4–32}$$

This result could also be deduced from the physical definitions of constant-volume and constant-pressure specific heats. Specific heat values for several common liquids and solids are given in Table A–3.

Internal Energy Changes

Like those of ideal gases, the specific heats of incompressible substances depend on temperature only. Thus, the partial differentials in the defining equation of c_v can be replaced by ordinary differentials, which yield

$$du = c_v\, dT = c(T)dT \tag{4–33}$$

The change in internal energy between states 1 and 2 is then obtained by integration:

$$\Delta u = u_2 - u_1 = \int_1^2 c(T)\,dT \qquad \text{(kJ/kg)} \tag{4–34}$$

The variation of specific heat c with temperature should be known before this integration can be carried out. For small temperature intervals, a c value at the average temperature can be used and treated as a constant, yielding

$$\Delta u \cong c_{\text{avg}}(T_2 - T_1) \qquad \text{(kJ/kg)} \tag{4–35}$$

Enthalpy Changes

Using the definition of enthalpy $h = u + Pv$ and noting that $v =$ constant, the differential form of the enthalpy change of incompressible substances can be determined by differentiation to be

$$dh = du + v\, dP + P\, \overset{\nearrow 0}{dv} = du + v\, dP \tag{4–36}$$

Integrating,

$$\Delta h = \Delta u + v\, \Delta P \cong c_{\text{avg}}\, \Delta T + v\, \Delta P \qquad \text{(kJ/ kg)} \tag{4–37}$$

For *solids*, the term $v\, \Delta P$ is insignificant and thus $\Delta h = \Delta u \cong c_{\text{avg}}\Delta T$. For *liquids*, two special cases are commonly encountered:

1. *Constant-pressure processes,* as in heaters ($\Delta P = 0$): $\Delta h = \Delta u \cong c_{\text{avg}}\Delta T$
2. *Constant-temperature processes,* as in pumps ($\Delta T = 0$): $\Delta h = v\, \Delta P$

For a process between states 1 and 2, the last relation can be expressed as $h_2 - h_1 = v(P_2 - P_1)$. By taking state 2 to be the compressed liquid state at a given T and P and state 1 to be the saturated liquid state at the same temperature, the enthalpy of the compressed liquid can be expressed as

$$h_{@P,T} \cong h_{f\,@\,T} + v_{f\,@\,T}(P - P_{\text{sat}\,@\,T}) \tag{4–38}$$

as discussed in Chap. 3. This is an improvement over the assumption that the enthalpy of the compressed liquid could be taken as h_f at the given temperature (that is, $h_{@\,P,T} \cong h_{f\,@\,T}$). However, the contribution of the last term is often very small, and is neglected. (Note that at high temperature and pressures, Eq. 4–38 may overcorrect the enthalpy and result in a larger error than the approximation $h \cong h_{f\,@\,T}$.)

EXAMPLE 4–11 Enthalpy of Compressed Liquid

Determine the enthalpy of liquid water at 100°C and 15 MPa (*a*) by using compressed liquid tables, (*b*) by approximating it as a saturated liquid, and (*c*) by using the correction given by Eq. 4–38.

SOLUTION The enthalpy of liquid water is to be determined exactly and approximately.

Analysis At 100°C, the saturation pressure of water is 101.42 kPa, and since $P > P_{sat}$, the water exists as a compressed liquid at the specified state.

(*a*) From compressed liquid tables, we read

$$\left.\begin{array}{l} P = 15 \text{ MPa} \\ T = 100°\text{C} \end{array}\right\} \quad h = \mathbf{430.39\ kJ/kg} \qquad \text{(Table A–7)}$$

This is the exact value.

(*b*) Approximating the compressed liquid as a saturated liquid at 100°C, as is commonly done, we obtain

$$h \cong h_{f\,@\,100°\text{C}} = \mathbf{419.17\ kJ/kg}$$

This value is in error by about 2.6 percent.

(*c*) From Eq. 4–38,

$$h_{@P,T} \cong h_{f\,@\,T} + v_{f\,@\,T}(P - P_{\text{sat}\,@\,T})$$

$$= (419.17 \text{ kJ/kg}) + (0.001 \text{ m}^3 \text{ kg})[(15{,}000 - 101.42) \text{ kPa}]\left(\frac{1 \text{ kJ}}{1 \text{ kPa}\cdot\text{m}^3}\right)$$

$$= \mathbf{434.07\ kJ/kg}$$

Discussion Note that the correction term reduced the error from 2.6 to about 1 percent in this case. However, this improvement in accuracy is often not worth the extra effort involved.

EXAMPLE 4–12 Cooling of an Iron Block by Water

A 50-kg iron block at 80°C is dropped into an insulated tank that contains 0.5 m³ of liquid water at 25°C. Determine the temperature when thermal equilibrium is reached.

SOLUTION An iron block is dropped into water in an insulated tank. The final temperature when thermal equilibrium is reached is to be determined.

Assumptions **1** Both water and the iron block are incompressible substances. **2** Constant specific heats at room temperature can be used for water and the iron. **3** The system is stationary and thus the kinetic and potential energy changes are zero, $\Delta KE = \Delta PE = 0$ and $\Delta E = \Delta U$. **4** There are no electrical, shaft, or other forms of work involved. **5** The system is well-insulated and thus there is no heat transfer.

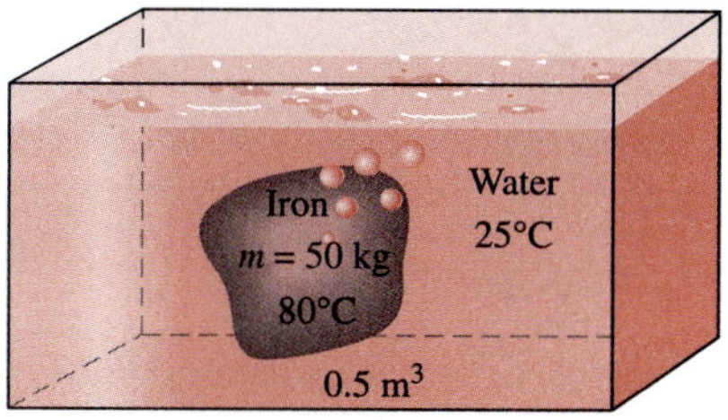

FIGURE 4–35
Schematic for Example 4–12.

Analysis We take the entire contents of the tank as the *system* (Fig. 4–35). This is a *closed system* since no mass crosses the system boundary during the process. We observe that the volume of a rigid tank is constant, and thus there is no boundary work. The energy balance on the system can be expressed as

$$\underbrace{E_{in} - E_{out}}_{\substack{\text{Net energy transfer}\\ \text{by heat, work, and mass}}} = \underbrace{\Delta E_{system}}_{\substack{\text{Change in internal, kinetic,}\\ \text{potential, etc., energies}}}$$

$$0 = \Delta U$$

The total internal energy U is an extensive property, and therefore it can be expressed as the sum of the internal energies of the parts of the system. Then the total internal energy change of the system becomes

$$\Delta U_{sys} = \Delta U_{iron} + \Delta U_{water} = 0$$

$$[mc(T_2 - T_1)]_{iron} + [mc(T_2 - T_1)]_{water} = 0$$

The specific volume of liquid water at or about room temperature can be taken to be 0.001 m³/kg. Then the mass of the water is

$$m_{water} = \frac{V}{v} = \frac{0.5 \text{ m}^3}{0.001 \text{ m}_3/\text{kg}} = 500 \text{ kg}$$

The specific heats of iron and liquid water are determined from Table A–3 to be c_{iron} = 0.45 kJ/kg·°C and c_{water} = 4.18 kJ/kg·°C. Substituting these values into the energy equation, we obtain

$$(50 \text{ kg})(0.45 \text{ kJ/kg·°C})(T_2 - 80°\text{C}) + (500 \text{ kg})(4.18 \text{ kJ/kg·°C})(T_2 - 25°\text{C}) = 0$$

$$T_2 = \mathbf{25.6°C}$$

Therefore, when thermal equilibrium is established, both the water and iron will be at 25.6°C.

Discussion The small rise in water temperature is due to its large mass and large specific heat.

EXAMPLE 4–13 Heating of Aluminum Rods in a Furnace

Long cylindrical aluminum rods (ρ = 2700 kg/m³ and c_p = 0.973 kJ/kg·K) of 5-cm diameter are heat treated from 20°C to an average temperature of 400°C by drawing them at a velocity of 8 m/min through a long oven. Determine the rate of heat transfer to the rods in the oven.

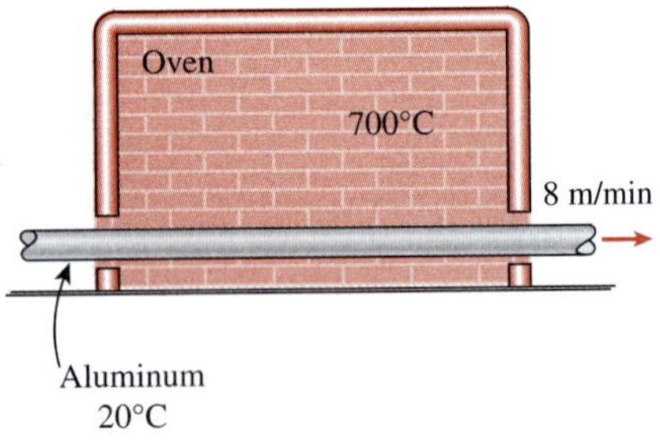

FIGURE 4-36
Schematic for Example 4-13

SOLUTION Aluminum rods are to be heated in an oven to a specified average temperature. The rate of heat transfer to the rods is to be determined.

Assumptions **1** The thermal properties of the rods are constant. **2** There are no changes in kinetic and potential energies. **3** The balls are at a uniform temperature when they leave the oven.

Analysis Aluminum rods pass through the oven at a constant speed of 8 m/min. That is, an external observer will see that an 8-m long section of cold rods enters and an 8-m long section of hot rods leaves the oven every minute. We take the 8-m long section of the rod as the system. The energy balance for this closed system can be expressed as

$$\underbrace{E_{in} - E_{out}}_{\substack{\text{Net energy transfer}\\ \text{by heat, work, and mass}}} = \underbrace{\Delta E_{system}}_{\substack{\text{Change in internal, kinetic,}\\ \text{potential, etc., energies}}}$$

$$Q_{in} = \Delta U_{rod} = m(u_2 - u_1)$$
$$Q_{in} = mc(T_2 - T_1)$$

The density and specific heat of the rods are given to be $\rho = 2700$ kg/m^3 and $c = 0.973$ kJ/kg·K $= 0.973$ kJ/kg·°C. The amount of heat transfer to an 8-m long section of the rod as it is heated to the specified temperature is determined to be

$$m = \rho V = \rho \frac{\pi D^2}{4} L = (2700 \text{ kg/m}^3) \frac{\pi (0.05 \text{ m})^2}{4} (8 \text{ m}) = 42.41 \text{ kg}$$

$$Q_{in} = mc(T_2 - T_1) = (42.41 \text{ kg})(0.973 \text{ kJ/kg·°C})(400 - 20)\text{°C}$$
$$= 15{,}680 \text{ kJ (per 8-m section)}$$

Considering that an 8-m long section of the rods is heated every minute, the rate of heat transfer to the rods in the oven becomes

$$\dot{Q}_{in} = Q_{in}/\Delta t = 15{,}680 \text{ kJ/min} = \mathbf{261\ kJ/s}$$

Discussion This problem can also be solved by working with the *rate* form of the equations as

$$\dot{m} = \rho \dot{V} = \rho \frac{\pi D^2}{4} L/\Delta t = \rho \frac{\pi D^2}{4} V = (2700 \text{ kg/m}^3) \frac{\pi (0.05 \text{ m})^2}{4} (8 \text{ m/min})$$
$$= 42.41 \text{ kg/min}$$

$$\dot{Q}_{in} = \dot{m}c(T_2 - T_1) = (42.41 \text{ kg/min})(0.973 \text{ kJ/kg·°C})(400 - 20)\text{°C}$$
$$= 15{,}680 \text{ kJ/min}$$

which is identical to the result obtained before.

TOPIC OF SPECIAL INTEREST* Thermodynamic Aspects of Biological Systems

An important and exciting application area of thermodynamics is biological systems, which are the sites of rather complex and intriguing energy transfer and transformation processes. Biological systems are not in thermodynamic equilibrium, and thus they are not easy to analyze. Despite their complexity, biological systems are primarily made up of four simple elements: hydrogen, oxygen, carbon, and nitrogen. In the human body, hydrogen accounts for 63 percent, oxygen 25.5 percent, carbon 9.5 percent, and nitrogen 1.4 percent of all the atoms. The remaining 0.6 percent of the atoms comes from 20 other elements essential for life. By mass, about 72 percent of the human body is water.

The building blocks of living organisms are *cells,* which resemble miniature factories performing functions that are vital for the survival of organisms.

*This section can be skipped without a loss in continuity.

FIGURE 4–37
An average person dissipates energy to the surroundings at a rate of 84 W when resting.

A biological system can be as simple as a single cell. The human body contains about 100 trillion cells with an average diameter of 0.01 mm. The membrane of the cell is a semipermeable wall that allows some substances to pass through it while excluding others.

In a typical cell, thousands of chemical reactions occur every second during which some molecules are broken down and energy is released and some new molecules are formed. This high level of chemical activity in the cells, which maintains the human body at a temperature of 37°C while performing the necessary bodily tasks, is called **metabolism**. In simple terms, metabolism refers to the burning of foods such as carbohydrates, fat, and protein. The rate of metabolism in the resting state is called the *basal metabolic rate*, which is the rate of metabolism required to keep a body performing the necessary functions (such as breathing and blood circulation) at zero external activity level. The metabolic rate can also be interpreted as the energy consumption rate for a body. For an average male (30 years old, 70 kg, 1.8-m^2 body surface area), the basal metabolic rate is 84 W. That is, the body dissipates energy to the environment at a rate of 84 W, which means that the body is converting chemical energy of the food (or of the body fat if the person has not eaten) into thermal energy at a rate of 84 W (Fig. 4–37). The metabolic rate increases with the level of activity, and it may exceed 10 times the basal metabolic rate when a body is doing strenuous exercise. That is, two people doing heavy exercising in a room may be supplying more energy to the room than a 1-kW electrical resistance heater (Fig. 4–38). The fraction of sensible heat varies from about 40 percent in the case of heavy work to about 70 percent in the case of light work. The rest of the energy is rejected from the body by perspiration in the form of latent heat.

The basal metabolic rate varies with sex, body size, general health conditions, and so forth, and decreases considerably with age. This is one of the reasons people tend to put on weight in their late twenties and thirties even though they do not increase their food intake. The brain and the liver are the major sites of metabolic activity. These two organs are responsible for almost 50 percent of the basal metabolic rate of an adult human body although they constitute only about 4 percent of the body mass. In small children, it is remarkable that about half of the basal metabolic activity occurs in the brain alone.

The biological reactions in cells occur essentially at constant temperature, pressure, and volume. The temperature of the cell tends to rise when some chemical energy is converted to heat, but this energy is quickly transferred to the circulatory system, which transports it to outer parts of the body and eventually to the environment through the skin.

The muscle cells function very much like an engine, converting the chemical energy into mechanical energy (work) with a conversion efficiency of close to 20 percent. When the body does no net work on the environment (such as moving some furniture upstairs), the entire work is also converted to heat. In that case, the entire chemical energy in the food released during metabolism in the body is eventually transferred to the environment. A TV set that consumes electricity at a rate of 300 W must reject heat to its environment at a rate of 300 W in steady operation regardless of what goes on in the set. That is, turning on a 300-W TV set or three 100-W light bulbs will produce the same heating effect in a room as a 300-W resistance heater

FIGURE 4–38
Two fast-dancing people supply more energy to a room than a 1-kW electric resistance heater.

(Fig. 4–39). This is a consequence of the conservation of energy principle, which requires that the energy input into a system must equal the energy output when the total energy content of a system remains constant during a process.

Food and Exercise

The energy requirements of a body are met by the food we eat. The nutrients in the food are considered in three major groups: carbohydrates, proteins, and fats. *Carbohydrates* are characterized by having hydrogen and oxygen atoms in a 2:1 ratio in their molecules. The molecules of carbohydrates range from very simple (as in plain sugar) to very complex or large (as in starch). Bread and plain sugar are the major sources of carbohydrates. *Proteins* are very large molecules that contain carbon, hydrogen, oxygen, and nitrogen, and they are essential for the building and repairing of the body tissues. Proteins are made up of smaller building blocks called *amino acids*. Complete proteins such as meat, milk, and eggs have all the amino acids needed to build body tissues. Plant source proteins such as those in fruits, vegetables, and grains lack one or more amino acids, and are called incomplete proteins. *Fats* are relatively small molecules that consist of carbon, hydrogen, and oxygen. Vegetable oils and animal fats are major sources of fats. Most foods we eat contain all three nutrition groups at varying amounts. The typical average American diet consists of 45 percent carbohydrate, 40 percent fat, and 15 percent protein, although it is recommended that in a healthy diet less than 30 percent of the calories should come from fat.

The energy content of a given food is determined by burning a small sample of the food in a device called a *bomb calorimeter,* which is basically a well-insulated rigid tank (Fig. 4–40). The tank contains a small combustion chamber surrounded by water. The food is ignited and burned in the combustion chamber in the presence of excess oxygen, and the energy released is transferred to the surrounding water. The energy content of the food is calculated on the basis of the conservation of energy principle by measuring the temperature rise of the water. The carbon in the food is converted into CO_2 and hydrogen into H_2O as the food burns. The same chemical reactions occur in the body, and thus the same amount of energy is released.

Using dry (free of water) samples, the average energy contents of the three basic food groups are determined by bomb calorimeter measurements to be 18.0 MJ/kg for carbohydrates, 22.2 MJ/kg for proteins, and 39.8 MJ/kg for fats. These food groups are not entirely metabolized in the human body, however. The fraction of metabolizable energy contents are 95.5 percent for carbohydrates, 77.5 percent for proteins, and 97.7 percent for fats. That is, the fats we eat are almost entirely metabolized in the body, but close to one quarter of the protein we eat is discarded from the body unburned. This corresponds to 4.1 Calories/g for proteins and carbohydrates and 9.3 Calories/g for fats (Fig. 4–41) commonly seen in nutrition books and on food labels. The energy contents of the foods we normally eat are much lower than the values above because of the large water content (water adds bulk to the food but it cannot be metabolized or burned, and thus it has no energy value). Most vegetables, fruits, and meats, for example, are mostly water. The average metabolizable energy contents of the three basic food groups are 4.2 MJ/kg for carbohydrates, 8.4 MJ/kg for proteins, and 33.1 MJ/kg for fats.

FIGURE 4–39
Some arrangements that supply a room the same amount of energy as a 300-W electric resistance heater.

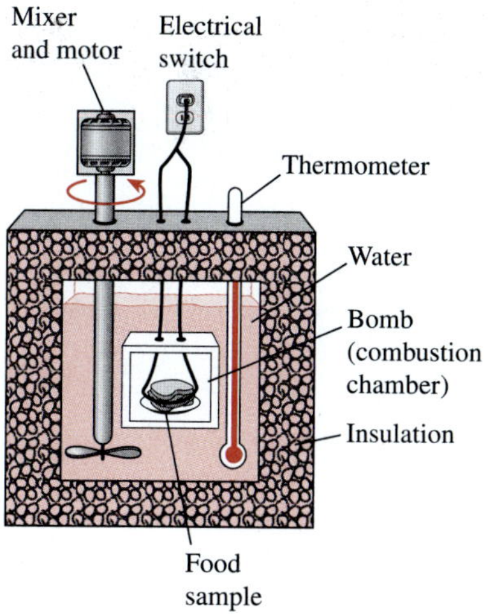

FIGURE 4–40
Schematic of a bomb calorimeter used to determine the energy content of food samples.

3 cookies (32 g)

Fat: (8 g)(9.3 Cal/g) = 74.4 Cal
Protein: (2 g)(4.1 Cal/g) = 8.2 Cal
Carbohydrates: (21 g)(4.1 Cal/g) = 86.1 Cal
Other: (1 g)(0 Cal/g) = 0

TOTAL (for 32 g): 169 Cal

FIGURE 4–41
Evaluating the calorie content of one serving of chocolate chip cookies (values are for Chips Ahoy cookies made by Nabisco).
©Vol. 30/PhotoDisc/Getty RF.

Note that 1 kg of natural fat contains almost 8 times the metabolizable energy of 1 kg of natural carbohydrates. Thus, a person who fills his stomach with fatty foods is consuming much more energy than a person who fills his stomach with carbohydrates such as bread or rice.

The metabolizable energy content of foods is usually expressed by nutritionists in terms of the capitalized *Calories*. One Calorie is equivalent to one *kilocalorie* (1000 calories), which is equivalent to 4.1868 kJ. That is,

$$1 \text{ Cal (Calorie)} = 1000 \text{ calories} = 1 \text{ kcal (kilocalorie)} = 4.1868 \text{ kJ}$$

The calorie notation often causes confusion since it is not always followed in the tables or articles on nutrition. When the topic is food or fitness, a calorie normally means a kilocalorie whether it is capitalized or not.

The **daily calorie needs** of people vary greatly with age, gender, the state of health, the activity level, the body weight, and the composition of the body as well as other factors. A small person needs fewer calories than a larger person of the same sex and age. An average man needs about 2400 to 2700 Calories a day. The daily need of an average woman varies from 1800 to 2200 Calories. The daily calorie needs are about 1600 for sedentary women and some older adults; 2000 for sedentary men and most older adults; 2200 for most children, teenage girls, and active women; 2800 for teenage boys, active men, and some very active women; and above 3000 for very active men. The *averag*e value of calorie intake is usually taken to be 2000 Calories per day. The daily calorie needs of a person can be determined by multiplying the body weight in pounds (which is 2.205 times the body weight in kg) by 11 for a sedentary person, 13 for a moderately active person, 15 for a moderate exerciser or physical laborer, and 18 for an extremely active exerciser or physical laborer. The extra calories a body consumes are usually stored as fat, which serves as the spare energy of the body for use when the energy intake of the body is less than the needed amount.

Like other natural fat, 1 kg of human body fat contains about 33.1 MJ of metabolizable energy. Therefore, a starving person (zero energy intake) who uses up 2200 Calories (9211 kJ) a day can meet his daily energy intake requirements by burning only 9211/33,100 = 0.28 kg of body fat. So it is no surprise that people are known to survive over 100 days without eating. (They still need to drink water, however, to replenish the water lost through the lungs and the skin to avoid the dehydration that may occur in just a few days.) Although the desire to get rid of the excess fat in a thin world may be overwhelming at times, starvation diets are not recommended because the body soon starts to consume its own muscle tissue in addition to fat. A healthy diet should involve regular exercise while allowing a reasonable amount of calorie intake.

The average metabolizable energy contents of various foods and the energy consumption during various activities are given in Tables 4–1 and 4–2. Considering that no two hamburgers are alike, and that no two people walk exactly the same way, there is some uncertainty in these values, as you would expect. Therefore, you may encounter somewhat different values in other books or magazines for the same items.

The rates of energy consumption listed in Table 4–2 during some activities are for a 68-kg adult. The energy consumed for smaller or larger adults can be determined using the proportionality of the metabolism rate and the

TABLE 4–1

Approximate metabolizable energy content of some common foods (1 Calorie = 4.1868 kJ = 3.968 Btu)

Food	Calories	Food	Calories	Food	Calories
Apple (one, medium)	70	Fish sandwich	450	Milk (skim, 200 ml)	76
Baked potato (plain)	250	French fries (regular)	250	Milk (whole, 200 ml)	136
Baked potato with cheese	550	Hamburger	275	Peach (one, medium)	65
Bread (white, one slice)	70	Hot dog	300	Pie (one $\frac{1}{8}$ slice, 23 cm diameter)	300
Butter (one teaspoon)	35	Ice cream (100 ml, 10% fat)	110	Pizza (large, cheese, one $\frac{1}{8}$ slice)	350
Cheeseburger	325	Lettuce salad with French dressing	150		
Chocolate candy bar (20 g)	105				
Cola (200 ml)	87				
Egg (one)	80				

body size. For example, the rate of energy consumption by a 68-kg bicyclist is listed in Table 4–2 to be 639 Calories/h. Then the rate of energy consumption by a 50-kg bicyclist is

$$(50 \text{ kg})\frac{639 \text{ Cal/h}}{68 \text{ kg}} = 470 \text{ Cal/h}$$

For a 100-kg person, it would be 940 Cal/h.

The thermodynamic analysis of the human body is rather complicated since it involves mass transfer (during breathing, perspiring, etc.) as well as energy transfer. As such, it should be treated as an open system. However, the energy transfer with mass is difficult to quantify. Therefore, the human body is often modeled as a closed system for simplicity by treating energy transported with mass as just energy transfer. For example, eating is modeled as the transfer of energy into the human body in the amount of the metabolizable energy content of the food.

Dieting

Most diets are based on *calorie counting*; that is, the conservation of energy principle: a person who consumes more calories than his or her body burns will gain weight whereas a person who consumes less calories than his or her body burns will lose weight. Yet, people who eat whatever they want whenever they want without gaining any weight are living proof that the calorie-counting technique alone does not work in dieting. Obviously there is more to dieting than keeping track of calories. It should be noted that the phrases *weight gain* and *weight loss* are misnomers. The correct phrases should be *mass gain* and *mass loss*. A man who goes to space loses practically all of his weight but none of his mass. When the topic is food and fitness, *weight* is understood to mean *mass,* and weight is expressed in mass units.

Researchers on nutrition proposed several theories on dieting. One theory suggests that some people have very "food efficient" bodies. These people need fewer calories than other people do for the same activity, just like a fuel-efficient car needing less fuel for traveling a given distance. It is interesting that we want our cars to be fuel efficient but we do not want the

TABLE 4–2

Approximate energy consumption of a 68-kg adult during some activities (1 Calorie = 4.1868 kJ = 3.968 Btu)

Activity	Calories/h
Basal metabolism	72
Basketball	550
Bicycling (21 km/h)	639
Cross-country skiing (13 km/h)	936
Driving a car	180
Eating	99
Fast dancing	600
Fast running (13 km/h)	936
Jogging (8 km/h)	540
Swimming (fast)	860
Swimming (slow)	288
Tennis (advanced)	480
Tennis (beginner)	288
Walking (7.2 km/h)	432
Watching TV	72

same high efficiency for our bodies. One thing that frustrates the dieters is that the body interprets dieting as *starvation* and starts using the energy reserves of the body more stringently. Shifting from a normal 2000-Calorie daily diet to an 800-Calorie diet without exercise is observed to lower the basal metabolic rate by 10 to 20 percent. Although the metabolic rate returns to normal once the dieting stops, extended periods of low-calorie dieting without adequate exercise may result in the loss of considerable muscle tissue together with fat. With less muscle tissue to burn calories, the metabolic rate of the body declines and stays below normal even after a person starts eating normally. As a result, the person regains the weight he or she has lost in the form of fat, plus more. The basal metabolic rate remains about the same in people who exercise while dieting.

Regular moderate exercise is part of any healthy dieting program for good reason: it builds or preserves muscle tissue that burns calories much faster than the fat tissue does. It is interesting that aerobic exercise continues burning calories for several hours after the workout, raising the overall metabolic rate considerably.

Another theory suggests that people with *too many fat cells* developed during childhood or adolescence are much more likely to gain weight. Some people believe that the fat content of the bodies is controlled by the setting of a "fat control" mechanism, much like the temperature of a house is controlled by the thermostat setting.

Some people put the blame for weight problems simply on the *genes*. Considering that 80 percent of the children of overweight parents are also overweight, heredity may indeed play an important role in the way a body stores fat. Researchers from the University of Washington and the Rockefeller University have identified a gene, called the RIIbeta, that seems to control the rate of metabolism. The body tries to keep the body fat at a particular level, called the **set point**, that differs from person to person (Fig. 4–42). This is done by *speeding up* the metabolism and thus burning extra calories much faster when a person tends to gain weight and by *slowing down* the metabolism and thus burning calories at a slower rate when a person tends to lose weight. Therefore, a person who just became slim burns fewer calories than does a person of the same size who has always been slim. Even exercise does not seem to change that. Then to keep the weight off, the newly slim person should consume no more calories than he or she can burn. Note that in people with high metabolic rates, the body dissipates the extra calories as body heat instead of storing them as fat, and thus there is no violation of the conservation of energy principle.

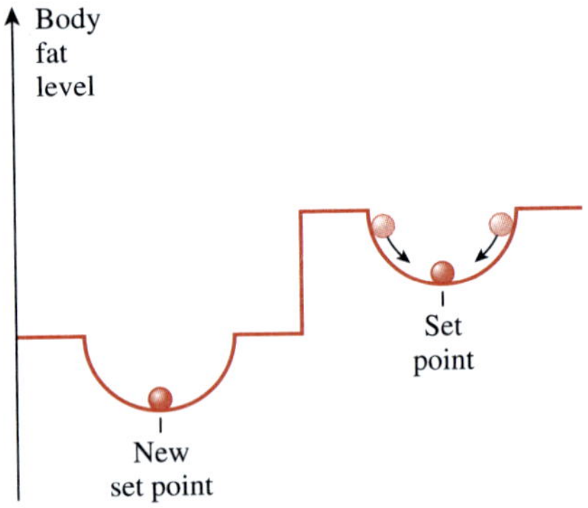

FIGURE 4–42
The body tends to keep the body fat level at a *set point* by speeding up metabolism when a person splurges and by slowing it down when the person starves.

In some people, a *genetic flaw* is believed to be responsible for the extremely low rates of metabolism. Several studies concluded that losing weight for such people is nearly impossible. That is, obesity is a biological phenomenon. However, even such people will not gain weight unless they eat more than their body can burn. They just must learn to be content with little food to remain slim, and forget about ever having a normal "eating" life. For most people, genetics determine the range of normal weights. A person may end up at the high or low end of that range, depending on eating and exercise habits. This also explains why some genetically identical twins are not so identical when it comes to body weight. *Hormone imbalance* is also believed to cause excessive weight gain or loss.

Based on his experience, the first author of this book has also developed a diet called the "*sensible diet*." It consists of two simple rules: eat *whatever* you want *whenever* you want *as much as* you want provided that (1) you do not eat unless you are hungry and (2) you stop eating before you get stuffed. In other words, *listen to your body and don't impose on it*. Don't expect to see this unscientific diet advertised anywhere since there is nothing to be sold and thus no money to be made. Also, it is not as easy as it sounds since food is at the center stage of most leisure activities in social life, and eating and drinking have become synonymous with having a good time. However, it is comforting to know that the human body is quite forgiving of occasional impositions.

Being *overweight* is associated with a long list of health risks from high blood pressure to some forms of cancer, especially for people who have a weight-related medical condition such as diabetes, hypertension, and heart disease. Therefore, people often wonder if their weight is in the proper range. Well, the answer to this question is not written in stone, but if you cannot see your toes or you can pinch your love handles more than an inch, you don't need an expert to tell you that you went over your range. On the other hand, some people who are obsessed with the weight issue try to lose more weight even though they are actually underweight. Therefore, it is useful to have a scientific criterion to determine physical fitness. The range of healthy weight for adults is usually expressed in terms of the **body mass index** (BMI), defined, in SI units, as

$$\text{BMI} = \frac{W\,(\text{kg})}{H^2(\text{m}^2)} \quad \text{with} \quad \begin{array}{ll} \text{BMI} < 19 & \text{underweight} \\ 19 \leq \text{BMI} \leq 25 & \text{healthy weight} \\ \text{BMI} > 25 & \text{overweight} \end{array} \tag{4-39}$$

where W is the weight (actually, the mass) of the person in kg and H is the height in m. Therefore, a BMI of 25 is the upper limit for the healthy weight and a person with a BMI of 27 is 8 percent overweight. It can be shown that the formula above is equivalent in English units to BMI = 705 W/H^2 where W is in pounds and H is in inches. The proper range of weight for adults of various heights is given in Table 4–3 in both SI and English units.

TABLE 4–3

The range of healthy weight for adults of various heights (Source: National Institute of Health)

English Units		*SI Units*	
Height in.	Healthy Weight, lbm*	Height, m	Healthy weight, kg*
58	91–119	1.45	40–53
60	97–127	1.50	43–56
62	103–136	1.55	46–60
64	111–146	1.60	49–64
66	118–156	1.65	52–68
68	125–165	1.70	55–72
70	133–175	1.75	58–77
72	140–185	1.80	62–81
74	148–195	1.85	65–86
76	156–205	1.90	69–90

*The upper and lower limits of healthy range correspond to mass body indexes of 19 and 25, respectively.

EXAMPLE 4–14 Burning Off Lunch Calories

A 90-kg man had two hamburgers, a regular serving of french fries, and a 200-ml Coke for lunch (Fig. 4–43). Determine how long it will take for him to burn the lunch calories off (*a*) by watching TV and (*b*) by fast swimming. What would your answers be for a 45-kg man?

SOLUTION A man had lunch at a restaurant. The time it will take for him to burn the lunch calories by watching TV and by fast swimming are to be determined.

Assumptions The values in Tables 4–1 and 4–2 are applicable for food and exercise.

Analysis (*a*) We take the human body as our *system* and treat it as *a closed system* whose energy content remains unchanged during the process. Then the conservation of energy principle requires that the energy input into the

FIGURE 4–43
A typical lunch discussed in Example 4–14.

body must be equal to the energy output. The net energy input in this case is the metabolizable energy content of the food eaten. It is determined from Table 4–1 to be

$$\begin{aligned} E_{\text{in}} &= 2 \times E_{\text{hamburger}} + E_{\text{fries}} + E_{\text{cola}} \\ &= 2 \times 275 + 250 + 87 \\ &= 887 \text{ Cal} \end{aligned}$$

The rate of energy output for a 68-kg man watching TV is given in Table 4–2 to be 72 Calories/h. For a 90-kg man it becomes

$$E_{\text{out}} = (90 \text{ kg}) \frac{72 \text{ Cal/h}}{68 \text{ kg}} = 95.3 \text{ Cal/h}$$

Therefore, it will take

$$\Delta t = \frac{887 \text{ Cal}}{95.3 \text{ Cal/h}} = \mathbf{9.3\ h}$$

to burn the lunch calories off by watching TV.

(*b*) It can be shown in a similar manner that it takes only **47 min** to burn the lunch calories off by fast swimming.

Discussion The 45-kg man is half as large as the 90-kg man. Therefore, expending the same amount of energy takes twice as long in each case: **18.6 h** by watching TV and **94 min** by fast swimming.

EXAMPLE 4–15 Losing Weight by Switching to Fat-Free Chips

The fake fat olestra passes through the body undigested, and thus adds zero calorie to the diet. Although foods cooked with olestra taste pretty good, they may cause abdominal discomfort and the long-term effects are unknown. A 1-oz (28.3-g) serving of regular potato chips has 10 g of fat and 150 Calories, whereas 1 oz of the so-called fat-free chips fried in olestra has only 75 Calories. Consider a person who eats 1 oz of regular potato chips every day at lunch without gaining or losing any weight. Determine how much weight this person will lose in one year if he or she switches to fat-free chips (Fig. 4–44).

FIGURE 4–44
Schematic for Example 4–15.

SOLUTION A person switches from regular potato chips to fat-free ones. The weight the person loses in one year is to be determined.

Assumptions Exercising and other eating habits remain the same.

Analysis The person who switches to the fat-free chips consumes 75 fewer Calories a day. Then the annual reduction in calories consumed becomes

$$E_{\text{reduced}} = (75 \text{ Cal/day})(365 \text{ day/year}) = 27{,}375 \text{ Cal/year}$$

The metabolizable energy content of 1 kg of body fat is 33,100 kJ. Therefore, assuming the deficit in the calorie intake is made up by burning body fat, the person who switches to fat-free chips will lose

$$m_{\text{fat}} = \frac{E_{\text{reduced}}}{\text{Energy content of fat}} = \frac{27{,}375 \text{ Cal}}{33{,}100 \text{ kJ/kg}} \left(\frac{4.1868 \text{ kJ}}{1 \text{ Cal}} \right) = \mathbf{3.46\ kg}$$

(about 7.6 pounds) of body fat that year.

SUMMARY

Work is the energy transferred as a force acts on a system through a distance. The most common form of mechanical work is the *boundary work,* which is the work associated with the expansion and compression of substances. On a P-V diagram, the area under the process curve represents the boundary work for a quasi-equilibrium process. Various forms of boundary work are expressed as follows:

(1) General $\qquad W_b = \int_1^2 P\,dV$

(2) Isobaric process

$$W_b = P_0(V_2 - V_1) \qquad (P_1 = P_2 = P_0 = \text{constant})$$

(3) Polytropic process

$$W_b = \frac{P_2V_2 - P_1V_1}{1 - n} \quad (n \neq 1) \qquad (PV^n = \text{constant})$$

(4) Isothernal process of an ideal gas

$$W_b = P_1V_1 \ln \frac{V_2}{V_1}$$

$$= mRT_0 \ln \frac{V_2}{V_1} \qquad (PV = mRT_0 = \text{constant})$$

The first law of thermodynamics is essentially an expression of the conservation of energy principle, also called the energy balance. The general energy balances for *any system* undergoing *any process* can be expressed as

$$\underbrace{E_{\text{in}} - E_{\text{out}}}_{\substack{\text{Net energy transfer}\\ \text{by heat, work, and mass}}} = \underbrace{\Delta E_{\text{system}}}_{\substack{\text{Change in internal, kinetic,}\\ \text{potential, etc., energies}}}$$

It can also be expressed in the *rate form* as

$$\underbrace{\dot{E}_{\text{in}} - \dot{E}_{\text{out}}}_{\substack{\text{Rate of net energy transfer}\\ \text{by heat, work, and mass}}} = \underbrace{dE_{\text{system}}/dt}_{\substack{\text{Rate of change in internal,}\\ \text{kinetic, potential, etc., energies}}}$$

Taking heat transfer *to* the system and work done *by* the system to be positive quantities, the energy balance for a closed system can also be expressed as

$$Q - W = \Delta U + \Delta \text{KE} + \Delta \text{PE}$$

where

$$W = W_{\text{other}} + W_b$$
$$\Delta U = m(u_2 - u_1)$$
$$\Delta \text{KE} = \tfrac{1}{2}m(V_2^2 - V_1^2)$$
$$\Delta \text{PE} = mg(z_2 - z_1)$$

For a *constant-pressure process,* $W_b + \Delta U = \Delta H$. Thus,

$$Q - W_{\text{other}} = \Delta H + \Delta \text{KE} + \Delta \text{PE}$$

Note that the relation above is limited to constant pressure processes of closed system, and is NOT valid for processes during which pressure varies.

The amount of energy needed to raise the temperature of a unit mass of a substance by one degree is called the *specific heat at constant volume* c_v for a constant-volume process and the *specific heat at constant pressure* c_p for a constant-pressure process. They are defined as

$$c_v = \left(\frac{\partial u}{\partial T}\right)_v \quad \text{and} \quad c_p = \left(\frac{\partial h}{\partial T}\right)_p$$

For ideal gases u, h, c_v, and c_p are functions of temperature alone. The Δu and Δh of ideal gases are expressed as

$$\Delta u = u_2 - u_1 = \int_1^2 c_v(T)\,dT \cong c_{v,\text{avg}}(T_2 - T_1)$$

$$\Delta h = h_2 - h_1 = \int_1^2 c_p(T)\,dT \cong c_{p,\text{avg}}(T_2 - T_1)$$

For ideal gases, c_v and c_p are related by

$$c_p = c_v + R$$

where R is the gas constant. The *specific heat ratio* k is defined as

$$k = \frac{c_p}{c_v}$$

For *incompressible substances* (liquids and solids), both the constant-pressure and constant-volume specific heats are identical and denoted by c:

$$c_p = c_v = c$$

The Δu and Δh of imcompressible substances are given by

$$\Delta u = \int_1^2 c(T)\,dT \cong c_{\text{avg}}(T_2 - T_1)$$

$$\Delta h = \Delta u + v\Delta P$$

1. ASHRAE *Handbook of Fundamentals.* SI version. Atlanta, GA: American Society of Heating, Refrigerating, and Air-Conditioning Engineers, Inc., 1993.

2. ASHRAE *Handbook of Refrigeration.* SI version. Atlanta, GA: American Society of Heating, Refrigerating, and Air-Conditioning Engineers, Inc., 1994.

PROBLEMS*

Moving Boundary Work

4–1C Is the boundary work associated with constant-volume systems always zero?

4–2C An ideal gas at a given state expands to a fixed final volume first at constant pressure and then at constant temperature. For which case is the work done greater?

4–3 Show that 1 kPa·m^3 = 1 kJ.

4–4 The volume of 1 kg of helium in a piston-cylinder device is initially 7 m^3. Now helium is compressed to 3 m^3 while its pressure is maintained constant at 150 kPa. Determine the initial and final temperatures of helium as well as the work required to compress it, in kJ.

4–5E Calculate the total work, in Btu, for process 1–3 shown in Fig. P4–5E.

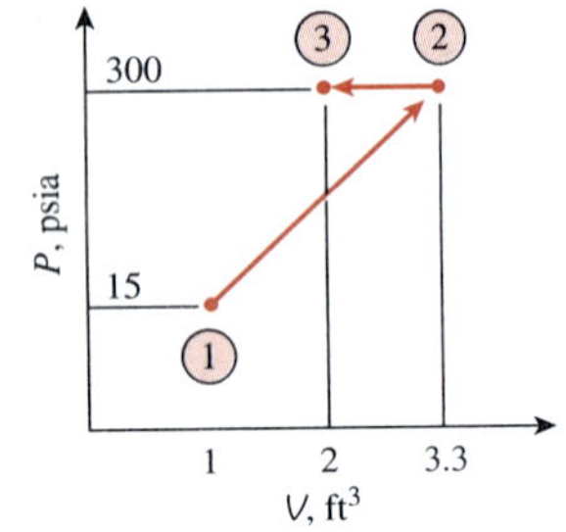

FIGURE P4–5E

4–6 Calculate the total work, in kJ, produced by the isothermal process of Fig. P4–6 when the system consists of 3 kg of oxygen.

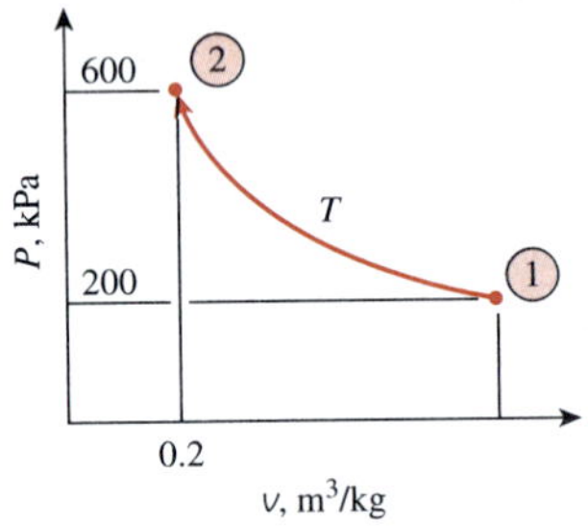

FIGURE P4–6

4–7 A piston–cylinder device initially contains 0.07 m^3 of nitrogen gas at 130 kPa and 120°C. The nitrogen is now expanded polytropically to a state of 100 kPa and 100°C. Determine the boundary work done during this process.

4–8 A piston–cylinder device with a set of stops initially contains 0.3 kg of steam at 1.0 MPa and 400°C. The location of the stops corresponds to 60 percent of the initial volume. Now the steam is cooled. Determine the compression work if the final state is (*a*) 1.0 MPa and 250°C and (*b*) 500 kPa. (*c*) Also determine the temperature at the final state in part (*b*).

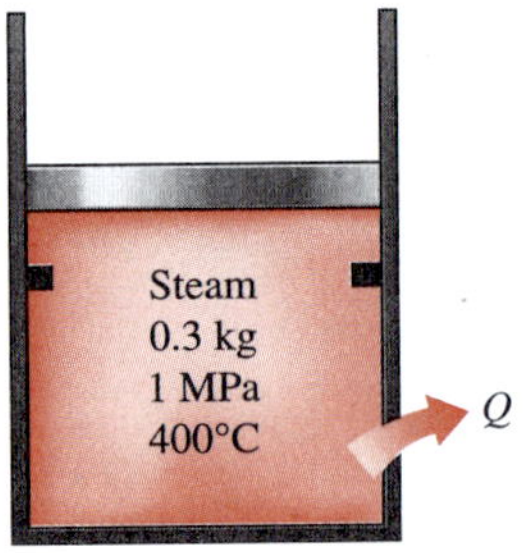

FIGURE P4–8

4–9 A piston–cylinder device initially contains 0.07 m^3 of nitrogen gas at 130 kPa and 180°C. The nitrogen is now expanded to a pressure of 80 kPa polytropically with a polytropic exponent whose value is equal to the specific heat ratio (called *isentropic expansion*). Determine the final temperature and the boundary work done during this process.

4–10 A mass of 5 kg of saturated water vapor at 300 kPa is heated at constant pressure until the temperature reaches 200°C. Calculate the work done by the steam during this process. *Answer:* 165.9 kJ

4–11 A frictionless piston–cylinder device initially contains 50 L of saturated liquid refrigerant-134a. The piston is free to move, and its mass is such that it maintains a pressure of 500 kPa on the refrigerant. The refrigerant is now heated until its temperature rises to 70°C. Calculate the work done during this process. *Answer:* 1600 kJ

FIGURE P4–11

* Problems designated by a "C" are concept questions, and students are encouraged to answer them all. Problems designated by an "E" are in English units, and the SI users can ignore them. Problems with the icon are solved using EES, and complete solutions together with parametric studies are included on the enclosed DVD. Problems with the icon are comprehensive in nature, and are intended to be solved with a computer, preferably using the EES software that accompanies this text.

4–12 Reconsider Prob. 4–11. Using EES (or other) software, investigate the effect of pressure on the work done. Let the pressure vary from 200 to 1200 kPa. Plot the work done versus the pressure, and discuss the results. Explain why the plot is not linear. Also plot the process described in Prob. 4–11 on the *P*-v diagram.

4–13 1-m^3 of saturated liquid water at 200°C is expanded isothermally in a closed system until its quality is 80 percent. Determine the total work produced by this expansion, in kJ.

4–14 A mass of 2.4 kg of air at 150 kPa and 12°C is contained in a gas-tight, frictionless piston–cylinder device. The air is now compressed to a final pressure of 600 kPa. During the process, heat is transferred from the air such that the temperature inside the cylinder remains constant. Calculate the work input during this process. *Answer:* 272 kJ

4–15 Determine the boundary work done by a gas during an expansion process if the pressure and volume values at various states are measured to be 300 kPa, 1 L; 290 kPa, 1.1 L; 270 kPa, 1.2 L; 250 kPa, 1.4 L; 220 kPa, 1.7 L; and 200 kPa, 2 L.

4–16 During some actual expansion and compression processes in piston–cylinder devices, the gases have been observed to satisfy the relationship $PV^n = C$, where *n* and *C* are constants. Calculate the work done when a gas expands from 350 kPa and 0.03 m^3 to a final volume of 0.2 m^3 for the case of $n = 1.5$.

4–17 Reconsider Prob. 4–16. Using the EES (or other) software, plot the process described in the problem on a *P*-*V* diagram, and investigate the effect of the polytropic exponent *n* on the boundary work. Let the polytropic exponent vary from 1.1 to 1.6. Plot the boundary work versus the polytropic exponent, and discuss the results.

4–18 A frictionless piston–cylinder device contains 2 kg of nitrogen at 100 kPa and 300 K. Nitrogen is now compressed slowly according to the relation $PV^{1.4}$ = constant until it reaches a final temperature of 360 K. Calculate the work input during this process. *Answer:* 89 kJ

FIGURE P4–18

4–19 The equation of state of a gas is given as $\overline{v}(P + 10/\overline{v}^2) = R_uT$, where the units of $\overline{v}$ and P are m^3/kmol and kPa, respectively. Now 0.2 kmol of this gas is expanded in a quasi-equilibrium manner from 2 to 4 m^3 at a constant temperature of 350 K. Determine (*a*) the unit of the quantity 10 in the equation and (*b*) the work done during this isothermal expansion process.

4–20 Reconsider Prob. 4–19. Using the integration feature of the EES software, calculate the work done, and compare your result with the "hand-calculated" result obtained in Prob. 4–19. Plot the process described in the problem on a *P*-v diagram.

4–21 Carbon dioxide contained in a piston–cylinder device is compressed from 0.3 to 0.1 m^3. During the process, the pressure and volume are related by $P = aV^{-2}$, where a = 8 kPa·m^6. Calculate the work done on the carbon dioxide during this process. *Answer:* 53.3 kJ

4–22E During an expansion process, the pressure of a gas changes from 15 to 100 psia according to the relation $P = aV + b$, where a = 5 psia/ft^3 and b is a constant. If the initial volume of the gas is 7 ft^3, calculate the work done during the process. *Answer:* 181 Btu

4–23 A piston–cylinder device initially contains 0.25 kg of nitrogen gas at 130 kPa and 180°C. The nitrogen is now expanded isothermally to a pressure of 80 kPa. Determine the boundary work done during this process. *Answer:* 16.3 kJ

FIGURE P4–23

4–24 A piston–cylinder device contains 0.15 kg of air initially at 2 MPa and 350°C. The air is first expanded isothermally to 500 kPa, then compressed polytropically with a polytropic exponent of 1.2 to the initial pressure, and finally compressed at the constant pressure to the initial state. Determine the boundary work for each process and the net work of the cycle.

4–25 1-kg of water that is initially at 90°C with a quality of 10 percent occupies a spring-loaded piston–cylinder device, such as that in Fig. P4–25. This device is now heated until the pressure rises to 800 kPa and the temperature is 250°C.

Determine the total work produced during this process, in kJ. *Answer:* 24.5 kJ

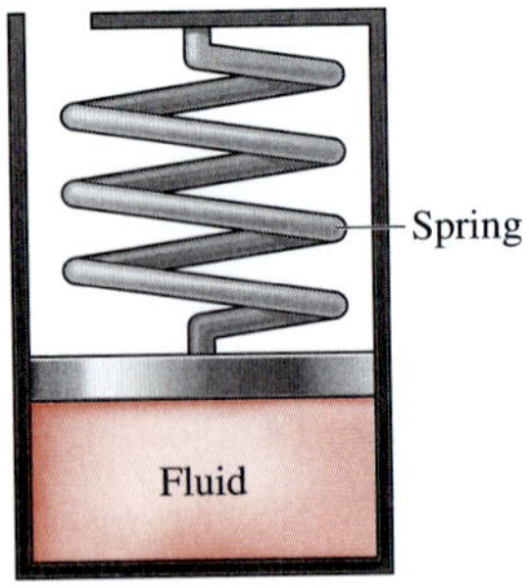

FIGURE P4–25

4–26 1.5-kg water that is initially at 1 MPa and 30 percent quality occupies a spring-loaded piston–cylinder device. This device is now cooled until the water is a saturated liquid at 100°C. Calculate the total work produced during this process, in kJ.

4–27 An ideal gas undergoes two processes in a piston-cylinder device as follows:

1-2 Polytropic compression from T_1 and P_1 with a polytropic exponent n and a compression ratio of $r = V_1/V_2$.

2-3 Constant pressure expansion at $P_3 = P_2$ until $V_3 = V_1$.

(*a*) Sketch the processes on a single P-V diagram.

(*b*) Obtain an expression for the ratio of the compression-to-expansion work as a function of n and r.

(*c*) Find the value of this ratio for values of $n = 1.4$ and $r = 6$. *Answers:* (*b*) $\frac{1}{n-1}\left(\frac{1-r^{1-n}}{r-1}\right)$, (*c*) 0.256

Closed System Energy Analysis

4–28E Complete the table below on the basis of the conservation of energy principle for a closed system.

Q_{in} Btu	W_{out} Btu	E_1 Btu	E_2 Btu	m lbm	$e_2 - e_1$ Btu/lbm
350	—	1020	860	3	—
350	130	550	—	5	—
—	260	600	—	2	150
-500	—	1400	900	7	—
—	-50	1000	—	3	-200

4–29E A closed system like that shown in Fig. P4–29E is operated in an adiabatic manner. First, 15,000 lbf·ft of work are done by this system. Then, work is applied to the stirring device to raise the internal energy of the fluid by 10.28 Btu. Calculate the net increase in the internal energy of this system.

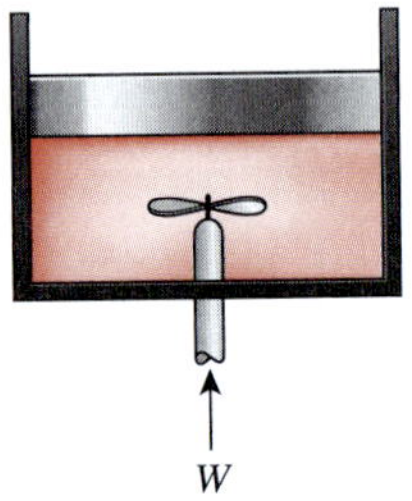

FIGURE P4–29E

4–30E A closed system undergoes a process in which there is no internal energy change. During this process, the system produces 1.1×10^6 lbf·ft of work. Calculate the heat transfer for this process, in Btu.

4–31 A rigid container equipped with a stirring device contains 1.5 kg of motor oil. Determine the rate of specific energy increase when heat is transferred to the oil at a rate of 1 W, and 1.5 W of power is applied to the stirring device.

4–32 A well-insulated rigid tank contains 2 kg of a saturated liquid–vapor mixture of water at 150 kPa. Initially, three-quarters of the mass is in the liquid phase. An electric resistor placed in the tank is connected to a 110-V source, and a current of 8 A flows through the resistor when the switch is turned on. Determine how long it will take to vaporize all the liquid in the tank. Also, show the process on a T-v diagram with respect to saturation lines.

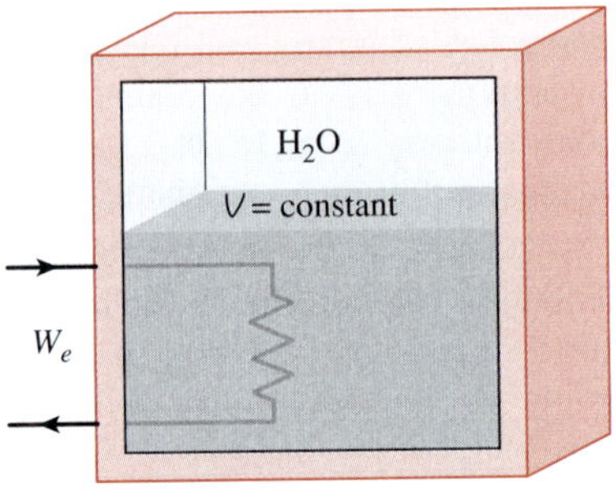

FIGURE P4–32

4–33 Reconsider Prob. 4–32. Using EES (or other) software, investigate the effect of the initial mass of water on the length of time required to completely vaporize the liquid. Let the initial mass vary from 1 to 10 kg. Plot the vaporization time against the initial mass, and discuss the results.

4–34 Saturated water vapor at 200°C is isothermally condensed to a saturated liquid in a piston–cylinder device. Calculate the heat transfer and the work done during this process, in kJ/kg. *Answers:* 1940 kJ/kg, 196 kJ/kg

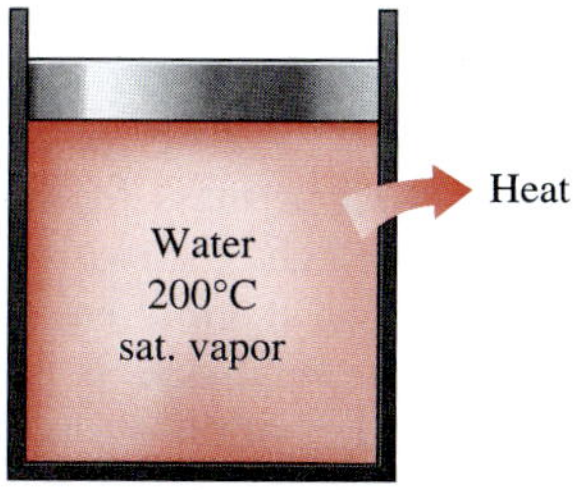

FIGURE P4–34

4–35 A rigid 10-L vessel initially contains a mixture of liquid water and vapor at 100°C with 12.3 percent quality. The mixture is then heated until its temperature is 150°C. Calculate the heat transfer required for this process. *Answer:* 46.9 kJ

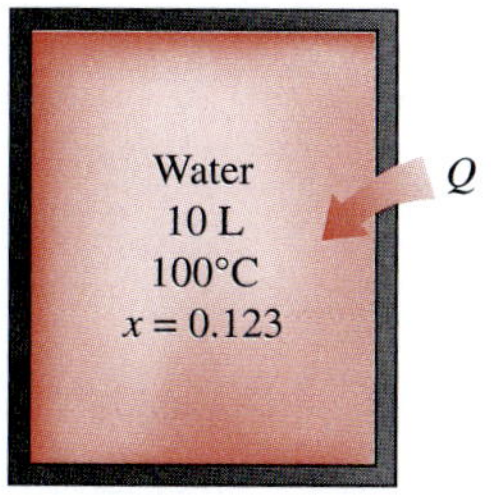

FIGURE P4–35

4–36 A fixed mass of saturated water vapor at 300 kPa is isothermally cooled until it is a saturated liquid. Calculate the amount of heat rejected during this process, in kJ/kg.

4–37 Saturated water vapor in a closed system is condensed by cooling it at constant pressure to a saturated liquid at 40 kPa. Determine the heat transferred and the work done during this process, in kJ/kg.

4–38 An insulated piston–cylinder device contains 5 L of saturated liquid water at a constant pressure of 175 kPa. Water is stirred by a paddle wheel while a current of 8 A flows for 45 min through a resistor placed in the water. If one-half of the liquid is evaporated during this constant-pressure process and the paddle-wheel work amounts to 400 kJ, determine the voltage of the source. Also, show the process on a *P-v* diagram with respect to saturation lines. *Answer:* 224 V

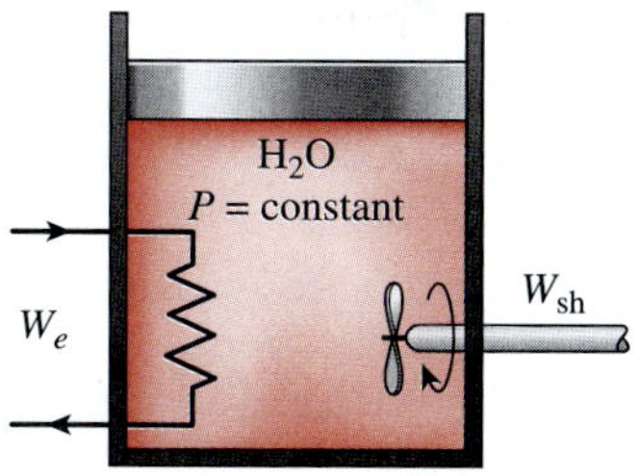

FIGURE P4–38

4–39 A piston–cylinder device initially contains steam at 200 kPa, 200°C, and 0.4 m^3. At this state, a linear spring ($F \propto x$) is touching the piston but exerts no force on it. Heat is now slowly transferred to the steam, causing the pressure and the volume to rise to 250 kPa and 0.6 m^3, respectively. Show the process on a *P-v* diagram with respect to saturation lines and determine (*a*) the final temperature, (*b*) the work done by the steam, and (*c*) the total heat transferred. *Answers:* (*a*) 606°C, (*b*) 45 kJ, (*c*) 288 kJ

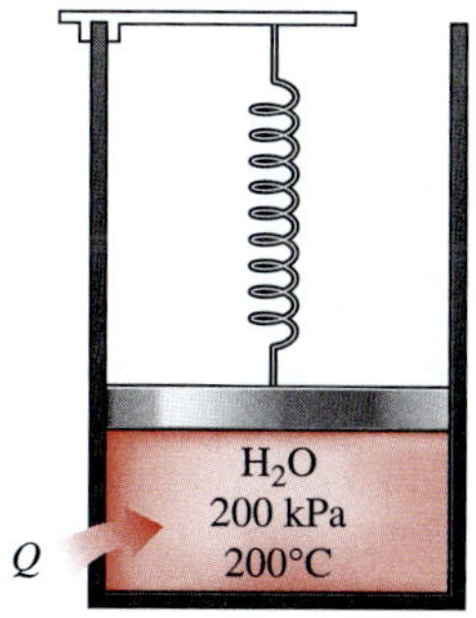

FIGURE P4–39

4–40 Reconsider Prob. 4–39. Using EES (or other) software, investigate the effect of the initial temperature of steam on the final temperature, the work done, and the total heat transfer. Let the initial temperature vary from 150 to 250°C. Plot the final results against the initial temperature, and discuss the results.

4–41 A 30-L electrical radiator containing heating oil is placed in a 50-m^3 room. Both the room and the oil in the radiator are initially at 10°C. The radiator with a rating of 1.8 kW is now turned on. At the same time, heat is lost from the room at an average rate of 0.35 kJ/s. After some time, the average temperature is measured to be 20°C for the air in the room, and 50°C for the oil in the radiator. Taking the density and the specific heat of the oil to be 950 kg/m^3 and 2.2 kJ/kg·°C, respectively, determine how long the heater is kept on. Assume the room is well-sealed so that there are no air leaks.

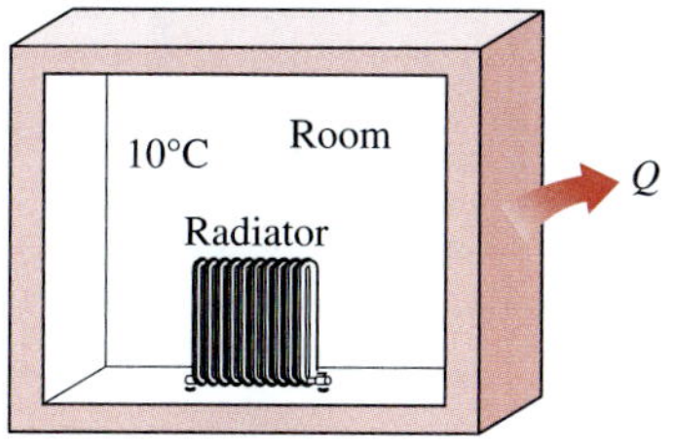

FIGURE P4–41

4–42 Steam at 75 kPa and 8 percent quality is contained in a spring-loaded piston–cylinder device, as shown in Fig. P4–42, with an initial volume of 2 m^3. Steam is now heated until its volume is 5 m^3 and its pressure is 225 kPa. Determine the heat transferred to and the work produced by the steam during this process.

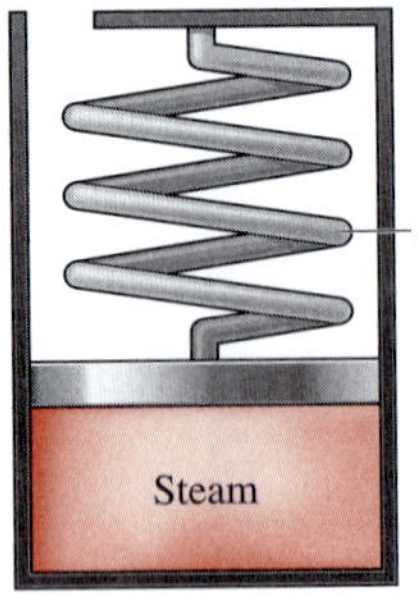

FIGURE P4–42

4–43 Refrigerant-134a at 600 kPa and 150°C is contained in a spring-loaded piston–cylinder device with an initial volume of 0.3 m^3. The refrigerant is now cooled until its temperature is −30°C and its volume is 0.1 m3. Determine the heat transferred to and the work produced by the refrigerant during this process. *Answers:* 1849 kJ heat out, 68.4 kJ work in

4–44E Saturated R-134a vapor at 100°F is condensed at constant pressure to a saturated liquid in a closed piston–cylinder system. Calculate the heat transfer and work done during this process, in Btu/lbm.

4–45 0.8-kg of saturated liquid R-134a with an initial temperature of −5°C is contained in a well-insulated, weighted piston–cylinder device. This device contains an electrical resistor, as shown in Fig. P4–45, to which 10 volts are applied causing a current of 2 amperes to flow through the resistor. Determine the time required for the refrigerant to be converted to a saturated vapor, and the final temperature.

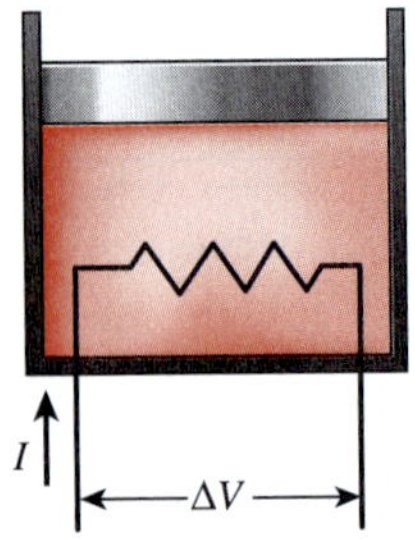

FIGURE P4–45

4–46 Two tanks (Tank A and Tank B) are separated by a partition. Initially Tank A contains 2-kg steam at 1 MPa and 300°C while Tank B contains 3-kg saturated liquid–vapor mixture at 150°C with a vapor mass fraction of 50 percent. The partition is now removed and the two sides are allowed to mix until the mechanical and thermal equilibrium are established. If the pressure at the final state is 300 kPa, determine (*a*) the temperature and quality of the steam (if mixture) at the final state and (*b*) the amount of heat lost from the tanks.

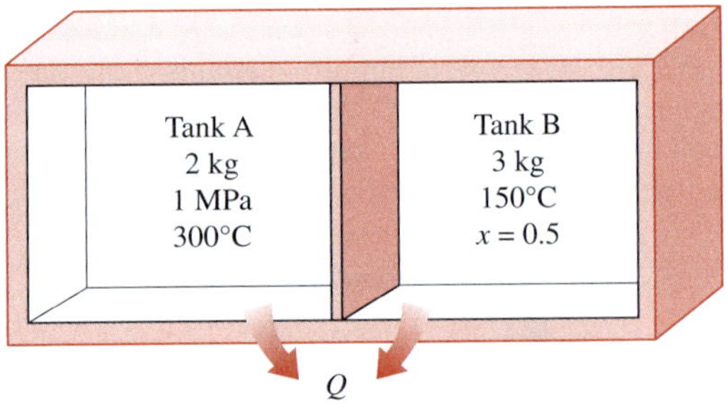

FIGURE P4–46

Specific Heats, Δu, and Δh of Ideal Gases

4–47C In the relation $\Delta u = mc_v \, \Delta T$, what is the correct unit of c_v — kJ/kg·°C or kJ/kg·K?

4–48C Is the relation $\Delta u = mc_{v,\text{avg}}\Delta T$ restricted to constant-volume processes only, or can it be used for any kind of process of an ideal gas?

4–49C Is the relation $\Delta h = mc_{p,\text{avg}}\Delta T$ restricted to constant-pressure processes only, or can it be used for any kind of process of an ideal gas?

4–50C Is the energy required to heat air from 295 to 305 K the same as the energy required to heat it from 345 to 355 K? Assume the pressure remains constant in both cases.

4–51C A fixed mass of an ideal gas is heated from 50 to 80°C at a constant pressure of (*a*) 1 atm and (*b*) 3 atm. For which case do you think the energy required will be greater? Why?

4–52C A fixed mass of an ideal gas is heated from 50 to 80°C at a constant volume of (*a*) 1 m^3 and (*b*) 3 m^3. For which case do you think the energy required will be greater? Why?

4–53C When undergoing a given temperature change, which of the two gases—air or oxygen—experiences the largest change in its (*a*) enthalpy, *h* and (*b*) internal energy, *u*?

4–54 Show that for an ideal gas $\bar{c}_p = \bar{c}_v + R_u$.

4–55 What is the change in the enthalpy, in kJ/kg, of oxygen as its temperature changes from 150 to 200°C? Is there any difference if the temperature change were from 0 to 50°C? Does the pressure at the beginning and end of this process have any effect on the enthalpy change?

4–56E Air is compressed from 20 psia and 70°F to 150 psia in a compressor. The compressor is operated such that the air temperature remains constant. Calculate the change in the specific volume of air as it passes through this compressor.

4–57 The temperature of 2 kg of neon is increased from 20 to 180°C. Calculate the change in the total internal energy of the neon, in kJ. Would the internal energy change be any different if the neon were replaced with argon?

4–58 Calculate the change in the enthalpy of argon, in kJ/kg, when it is cooled from 100 to 25°C. If neon had undergone this same change of temperature, would its enthalpy change have been any different?

4–59E Determine the enthalpy change Δh of oxygen, in Btu/lbm, as it is heated from 800 to 1500 R, using (*a*) the empirical specific heat equation as a function of temperature (Table A–2E*c*), (*b*) the c_p value at the average temperature (Table A–2E*b*), and (*c*) the c_p value at room temperature (Table A–2E*a*). *Answers:* (*a*) 170.1 Btu/lbm, (*b*) 178.5 Btu/lbm, (*c*) 153.3 Btu/lbm

4–60 Determine the internal energy change Δu of hydrogen, in kJ/kg, as it is heated from 200 to 800 K, using (*a*) the empirical specific heat equation as a function of temperature (Table A–2*c*), (*b*) the c_v value at the average temperature (Table A–2*b*), and (*c*) the c_v value at room temperature (Table A–2*a*).

4–61E 1-ft^3 of air is contained in the spring-loaded piston-cylinder device shown in Fig. P4–61E. The spring constant is 5 lbf/in, and the piston diameter is 10 in. When no force is exerted by the spring on the piston, the state of the air is 250 psia and 460°F. This device is now cooled until the volume is one-half its original size. Determine the change in the specific internal energy and enthalpy of the air. *Answers:* 78.9 Btu/lbm, 111 Btu/lbm

FIGURE P4–61E

Closed-System Energy Analysis: Ideal Gases

4–62C Is it possible to compress an ideal gas isothermally in an adiabatic piston–cylinder device? Explain.

4–63 1-kg of oxygen is heated from 20 to 120°C. Determine the amount of heat transfer required when this is done during a (*a*) constant-volume process and (*b*) isobaric process.

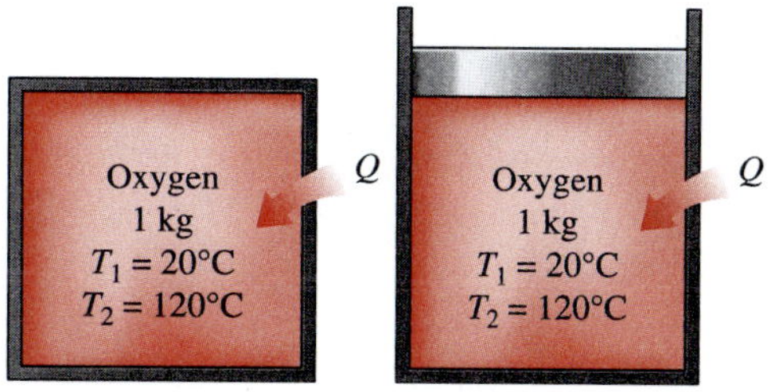

FIGURE P4–63

4–64E A rigid tank contains 20 lbm of air at 50 psia and 80°F. The air is now heated until its pressure doubles. Determine (*a*) the volume of the tank and (*b*) the amount of heat transfer. *Answers:* (*a*) 80 ft^3, (*b*) 1898 Btu

4–65E Determine the internal energy change of air in a rigid container as its temperature is increased by transferring 50 Btu/lbm of heat.

4–66E Nitrogen gas to 20 psia and 100°F initially occupies a volume of 1 ft^3 in a rigid container equipped with a stirring paddle wheel. After 5000 lbf·ft of paddle wheel work is done on nitrogen, what is its final temperature? *Answer:* 489°F

4–67 Consider a piston-cylinder device that contains nitrogen gas as the system. Initially, the system is at 1 MPa and 427°C. It now undergoes an isobaric process until its temperature is 27°C. Determine the final pressure and the heat transfer, in kJ/kg, associated with this process. *Answer:* 416 kJ/kg

4–68 A 4-m × 5-m × 6-m room is to be heated by a baseboard resistance heater. It is desired that the resistance heater be able to raise the air temperature in the room from 5 to 25°C within 11 min. Assuming no heat losses from the room and an atmospheric pressure of 100 kPa, determine the required power of the resistance heater. Assume constant specific heats at room temperature. *Answer:* 3.28 kW

4–69 A student living in a 4-m × 6-m × 6-m dormitory room turns on her 150-W fan before she leaves the room on a summer day, hoping that the room will be cooler when she comes back in the evening. Assuming all the doors and windows are tightly closed and disregarding any heat transfer through the walls and the windows, determine the temperature in the room when she comes back 10 h later. Use specific heat values at room temperature, and assume the room to be at 100 kPa and 15°C in the morning when she leaves. *Answer:* 58.2°C

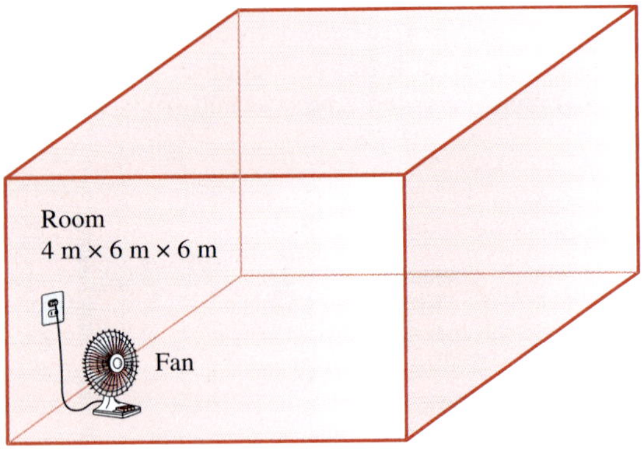

FIGURE P4–69

4–70E A 3-ft^3 adiabatic rigid container is divided into two equal volumes by a thin membrane, as shown in Fig. P4–70E. Initially, one of these chambers is filled with air at 100 psia and 100°F while the other chamber is evacuated. Determine the internal energy change of the air when the membrane is ruptured. Also determine the final air pressure in the container.

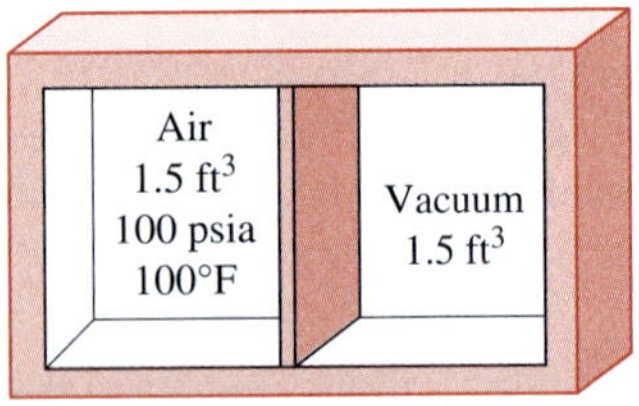

FIGURE P4–70E

4–71 A closed system containing 2-kg of air undergoes an isothermal process from 600 kPa and 200°C to 80 kPa. Determine the initial volume of this system, the work done, and the heat transfer during this process. *Answers:* 0.453 m^3, 547 kJ, 547 kJ

4–72 A piston–cylinder device containing argon gas as the system undergoes an isothermal process from 200 kPa and 100°C to 50 kPa. During the process, 1500 kJ of heat is transferred to the system. Determine the mass of this system and the amount of work produced.

4–73 Argon is compressed in a polytropic process with n = 1.2 from 120 kPa and 10°C to 800 kPa in a piston–cylinder device. Determine the work produced and heat transferred during this compression process, in kJ/kg.

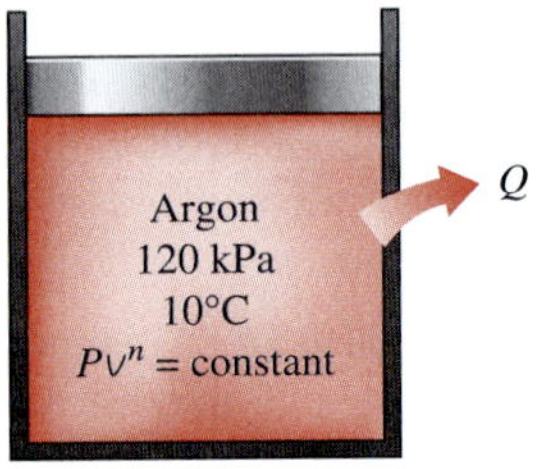

FIGURE P4–73

4–74 A spring-loaded piston-cylinder device contains 1 kg of carbon dioxide. This system is heated from 100 kPa and 25°C to 1000 kPa and 300°C. Determine the total heat transfer to and work produced by this system.

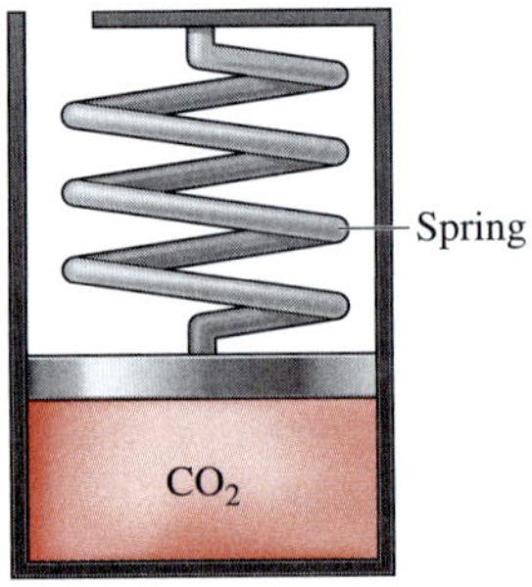

FIGURE P4–74

4–75 Air is contained in a variable-load piston-cylinder device equipped with a paddle wheel. Initially, air is at 500 kPa and 27°C. The paddle wheel is now turned by an external electric motor until 50 kJ/kg of work has been transferred to air. During this process, heat is transferred to maintain a constant air temperature while allowing the gas volume to triple. Calculate the required amount of heat transfer, in kJ/kg. *Answer:* 44.6 kJ/kg

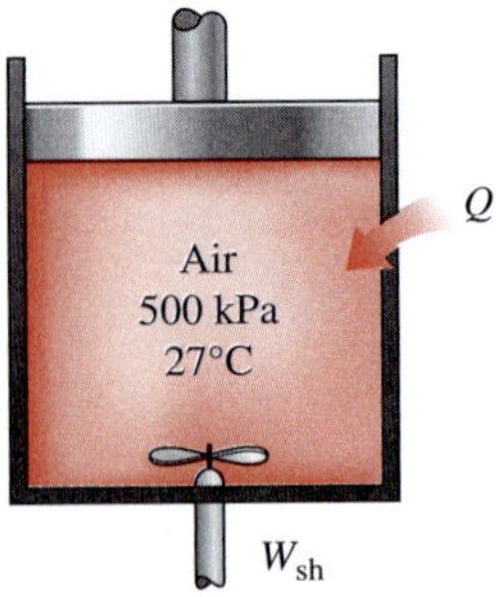

FIGURE P4–75

4–76 A mass of 15 kg of air in a piston–cylinder device is heated from 25 to 77°C by passing current through a resistance heater inside the cylinder. The pressure inside the cylinder is held constant at 300 kPa during the process, and a heat loss of 60 kJ occurs. Determine the electric energy supplied, in kWh. *Answer:* 0.235 kWh

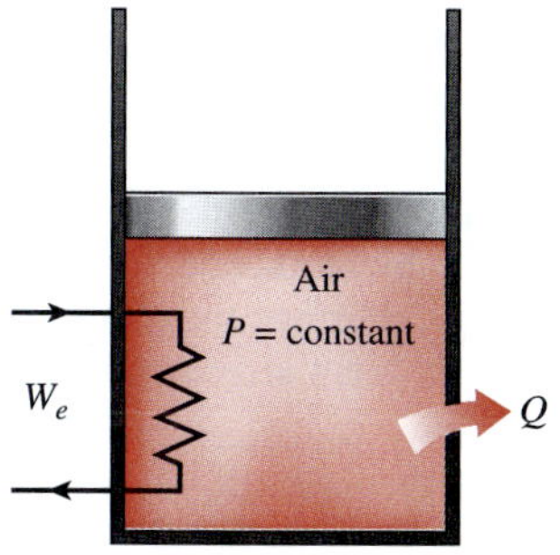

FIGURE P4–76

4–77 A piston–cylinder device contains 1.5 kg of nitrogen initially at 100 kPa and 17°C. The nitrogen is now compressed slowly in a polytropic process during which $PV^{1.3}$ = constant until the volume is reduced by one-half. Determine the work done and the heat transfer for this process.

4–78 Reconsider Prob. 4–77. Using EES (or other) software, plot the process described in the problem on a *P*-*V* diagram, and investigate the effect of the polytropic exponent *n* on the boundary work and heat transfer. Let the polytropic exponent vary from 1.1 to 1.6. Plot the boundary work and the heat transfer versus the polytropic exponent, and discuss the results.

4–79 A room is heated by a baseboard resistance heater. When the heat losses from the room on a winter day amount to 6500 kJ/h, the air temperature in the room remains constant even though the heater operates continuously. Determine the power rating of the heater, in kW.

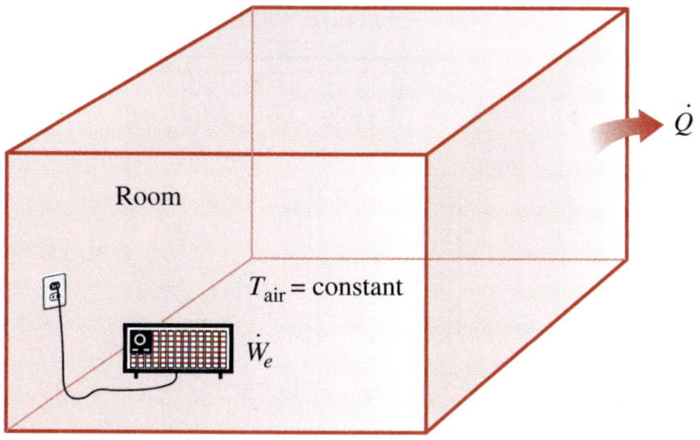

FIGURE P4–79

4–80 A piston–cylinder device, with a set of stops on the top, initially contains 3 kg of air at 200 kPa and 27°C. Heat is now transferred to the air, and the piston rises until it hits the stops, at which point the volume is twice the initial volume. More heat is transferred until the pressure inside the cylinder also doubles. Determine the work done and the amount of heat transfer for this process. Also, show the process on a *P*-*V* diagram.

4–81 Air is contained in a cylinder device fitted with a piston-cylinder. The piston initially rests on a set of stops, and a pressure of 300 kPa is required to move the piston. Initially, the air is at 100 kPa and 27°C and occupies a volume of 0.4 m^3. Determine the amount of heat transferred to the air, in kJ, while increasing the temperature to 1200 K. Assume air has constant specific heats evaluated at 300 K. *Answer:* 340 kJ

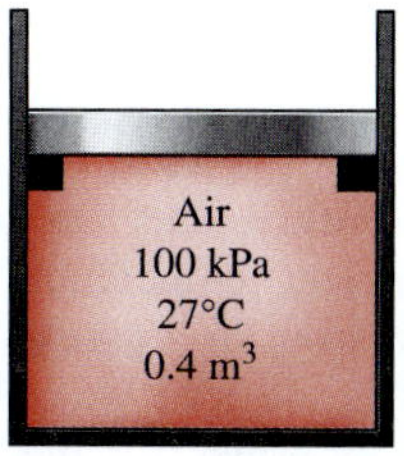

FIGURE P4–81

4–82 Air is contained in a piston-cylinder device at 600 kPa and 927°C, and occupies a volume of 0.8 m^3. The air undergoes and isothermal (constant temperature) process until the pressure in reduced to 300 kPa. The piston is now fixed in place and not allowed to move while a heat transfer process takes place until the air reaches 27°C.

(*a*) Sketch the system showing the energies crossing the boundary and the *P*-*V* diagram for the combined processes.

(*b*) For the combined processes determine the net amount of heat transfer, in kJ, and its direction.

Assume air has constant specific heats evaluated at 300 K.

4–83 Argon is contained in a cylinder device fitted with a piston. Initially, the argon is at 100 kPa and 27°C and occupies a volume of 0.4 m^3. The argon is first compressed while the temperature is held constant until the volume is 0.2 m^3. Then the argon expands while the pressure is held constant until the volume is 0.6 m^3.

(*a*) Sketch both processes on a single *P*-*V* diagram showing the direction of the processes and label the end states as 1, 2, and 3.

(*b*) Determine the total amount of net heat transferred to the argon, in kJ, during the combined processes. *Answer:* (*b*) 172 kJ

4–84 An ideal gas contained in a piston-cylinder device undergoes an isothermal compression process which begins with an initial pressure and volume of 100 kPa and 0.6 m^3, respectively. During the process there is a heat transfer of 60 kJ from the ideal gas to the surroundings. Determine the volume and pressure at the end of the process. *Answers:* 0.221 m^3, 272 kPa

Closed-System Energy Analysis: Solids and Liquids

4–85 A 1-kg block of iron is heated from 25 to 75°C. What is the change in the iron's total internal energy and enthalpy?

4–86E The state of liquid water is changed from 50 psia and 50°F to 2000 psia and 100°F. Determine the change in the internal energy and enthalpy of water on the basis of the (*a*) compressed liquid tables, (*b*) incompressible substance approximation and property tables, and (*c*) specific-heat model.

4–87E During a picnic on a hot summer day, all the cold drinks disappeared quickly, and the only available drinks were those at the ambient temperature of 75°F. In an effort to cool a 12-fluid-oz drink in a can, a person grabs the can and starts shaking it in the iced water of the chest at 32°F. Using the properties of water for the drink, determine the mass of ice that will melt by the time the canned drink cools to 45°F.

4–88 Consider a 1000-W iron whose base plate is made of 0.5-cm-thick aluminum alloy 2024-T6 (ρ = 2770 kg/m^3 and c_p = 875 J/kg·°C). The base plate has a surface area of 0.03 m^2. Initially, the iron is in thermal equilibrium with the ambient air at 22°C. Assuming 90 percent of the heat generated in the resistance wires is transferred to the plate, determine the minimum time needed for the plate temperature to reach 200°C.

FIGURE P4–88

4–89 Stainless steel ball bearings (ρ = 8085 kg/m^3 and c_p = 0.480 kJ/kg·°C) having a diameter of 1.2 cm are to be quenched in water at a rate of 800 per minute. The balls leave the oven at a uniform temperature of 900°C and are exposed to air at 25°C for a while before they are dropped into the water. If the temperature of the balls drops to 850°C prior to quenching, determine the rate of heat transfer from the balls to the air.

4–90 Carbon steel balls (ρ = 7833 kg/m^3 and c_p = 0.465 kJ/kg·°C) 8 mm in diameter are annealed by heating them first to 900°C in a furnace, and then allowing them to cool slowly to 100°C in ambient air at 35°C. If 2500 balls are to be annealed per hour, determine the total rate of heat transfer from the balls to the ambient air. *Answer:* 542 W

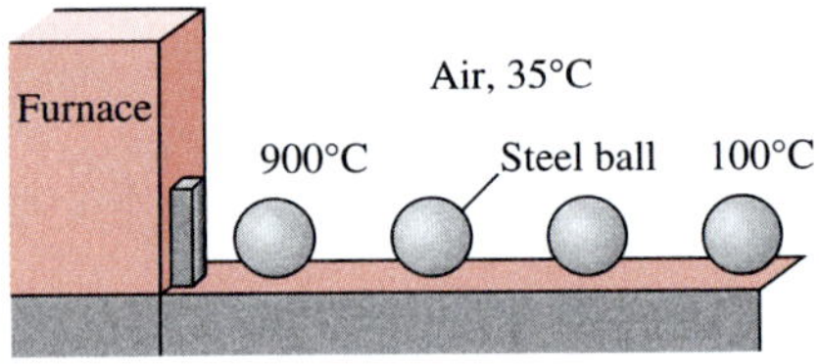

FIGURE P4–90

4–91 An ordinary egg can be approximated as a 5.5-cm-diameter sphere. The egg is initially at a uniform temperature of 8°C and is dropped into boiling water at 97°C. Taking the properties of the egg to be ρ = 1020 kg/m^3 and c_p = 3.32 kJ/kg·°C, determine how much heat is transferred to the egg by the time the average temperature of the egg rises to 80°C.

4–92E In a production facility, 1.2-in-thick 2-ft × 2-ft square brass plates (ρ = 532.5 lbm/ft^3 and c_p = 0.091 Btu/lbm·°F) that are initially at a uniform temperature of 75°F are heated by passing them through an oven at 1300°F at a rate of 300 per minute. If the plates remain in the oven until their average temperature rises to 1000°F, determine the rate of heat transfer to the plates in the furnace.

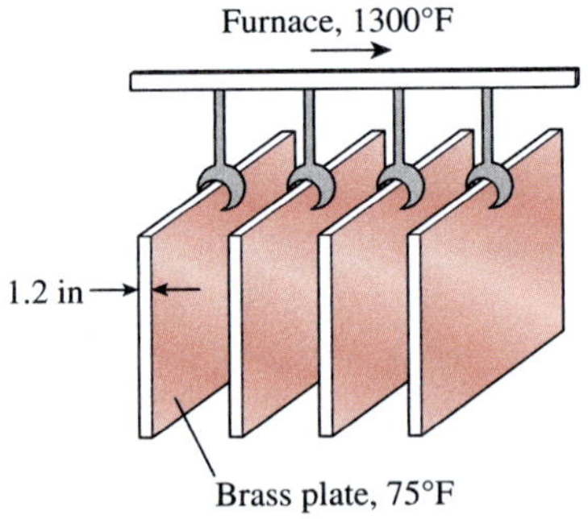

FIGURE P4–92E

4–93 Long cylindrical steel rods (ρ = 7833 kg/m^3 and c_p = 0.465 kJ/kg·°C) of 8-cm diameter are heat-treated by drawing them at a velocity of 2 m/min through an oven maintained at 900°C. If the rods enter the oven at 30°C and leave at a mean temperature of 700°C, determine the rate of heat transfer to the rods in the oven.

4–94 An electronic device dissipating 25 W has a mass of 20 g and a specific heat of 850 J/kg·°C. The device is lightly used, and it is on for 5 min and then off for several hours, during which it cools to the ambient temperature of 25°C. Determine the highest possible temperature of the device at the end of the 5-min operating period. What would your answer be if the device were attached to a 0.5-kg aluminum heat sink? Assume the device and the heat sink to be nearly isothermal.

4–95 Reconsider Prob. 4–94. Using EES (or other) software, investigate the effect of the mass of the heat sink on the maximum device temperature. Let the mass of heat sink vary from 0 to 1 kg. Plot the maximum temperature against the mass of heat sink, and discuss the results.

4–96 If you ever slapped someone or got slapped yourself, you probably remember the burning sensation. Imagine you had the unfortunate occasion of being slapped by an angry person, which caused the temperature of the affected area of your face to rise by 1.8°C (ouch!). Assuming the slapping hand has a mass of 1.2 kg and about 0.150 kg of the tissue on the face and the hand is affected by the incident, estimate the velocity of the hand just before impact. Take the specific heat of the tissue to be 3.8 kJ/kg·K.

Special Topic: Biological Systems

4–97C What is metabolism? What is basal metabolic rate? What is the value of basal metabolic rate for an average man?

4–98C Is the metabolizable energy content of a food the same as the energy released when it is burned in a bomb calorimeter? If not, how does it differ?

4–99C Is the number of prospective occupants an important consideration in the design of heating and cooling systems of classrooms? Explain.

4–100C What do you think of a diet program that allows for generous amounts of bread and rice provided that no butter or margarine is added?

4–101 Consider two identical rooms, one with a 2-kW electric resistance heater and the other with three couples fast dancing. In which room will the air temperature rise faster?

4–102 Consider two identical 80-kg men who are eating identical meals and doing identical things except that one of them jogs for 30 min every day while the other watches TV. Determine the weight difference between the two in a month. *Answer:* 1.045 kg

4–103 A 68-kg woman is planning to bicycle for an hour. If she is to meet her entire energy needs while bicycling by eating 30-g chocolate candy bars, determine how many candy bars she needs to take with her.

4–104 A 75-kg man gives in to temptation and eats an entire 1-L box of ice cream. How long does this man need to jog to burn off the calories he consumed from the ice cream? *Answer:* 1.85 h

4–105 Consider a man who has 20 kg of body fat when he goes on a hunger strike. Determine how long he can survive on his body fat alone.

4–106 Consider two identical 50-kg women, Candy and Wendy, who are doing identical things and eating identical food except that Candy eats her baked potato with four teaspoons of butter while Wendy eats hers plain every evening. Determine the difference in the weights of Candy and Wendy after one year. *Answer:* 6.5 kg

4–107 A woman who used to drink about one liter of regular cola every day switches to diet cola (zero calorie) and starts eating two slices of apple pie every day. Is she now consuming fewer or more calories?

4–108 A 12-oz serving of a regular beer contains 13 g of alcohol and 13 g of carbohydrates, and thus 150 Calories. A 12-oz serving of a light beer contains 11 g of alcohol and 5 g of carbohydrates, and thus 100 Calories. An average person burns 700 Calories per hour while exercising on a treadmill. Determine how long it will take to burn the calories from a 12-oz can of (*a*) regular beer and (*b*) light beer on a treadmill.

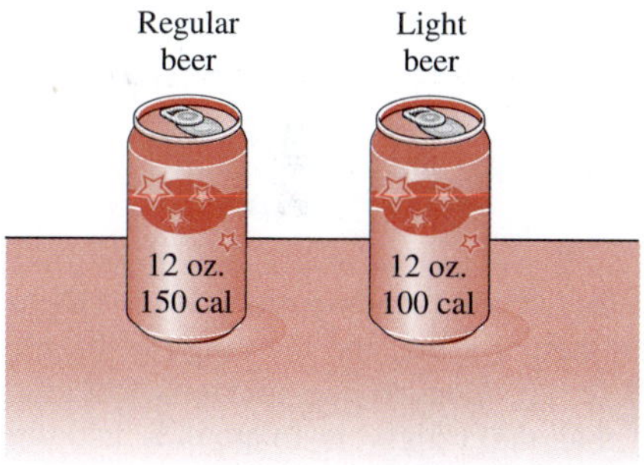

FIGURE P4–108

4–109 A 190-pound man and a 110-pound woman went to Burger King for lunch. The man had a BK Big Fish sandwich (720 Cal), medium french fries (400 Cal), and a large Coke (225 Cal). The woman had a basic hamburger (330 Cal), medium french fries (400 Cal), and a diet Coke (0 Cal). After lunch, they start shoveling snow and burn calories at a rate of 360 Cal/h for the woman and 480 Cal/h for the man. Determine how long each one of them needs to shovel snow to burn off the lunch calories.

4–110 Consider two friends who go to Burger King every day for lunch. One of them orders a Double Whopper sandwich, large fries, and a large Coke (total Calories = 1600) while the other orders a Whopper Junior, small fries, and a small Coke (total Calories = 800) every day. If these two friends are very much alike otherwise and they have the same metabolic rate, determine the weight difference between these two friends in a year.

4–111E A 150-pound person goes to Hardee's for dinner and orders a regular roast beef (270 Cal) and a big roast beef (410 Cal) sandwich together with a 12-oz can of Pepsi (150 Cal). A 150-pound person burns 400 Calories per hour while climbing stairs. Determine how long this person needs to climb stairs to burn off the dinner calories.

4–112 A person eats a McDonald's Big Mac sandwich (530 Cal), a second person eats a Burger King Whopper sandwich (640 Cal), and a third person eats 50 olives with regular french fries (350 Cal) for lunch. Determine who consumes the most calories. An olive contains about 5 Calories.

4–113 A 100-kg man decides to lose 10 kg without cutting down his intake of 4000 Calories a day. Instead, he starts fast swimming, fast dancing, jogging, and biking each for an hour every day. He sleeps or relaxes the rest of the day. Determine how long it will take him to lose 10 kg.

4–114 Many fast food restaurants offer supersizing your burger, fries, and soda at a bargain price, but supersizing may cost you up to 1000 extra calories. Determine how much extra weight supersizing once a day can cause a person to gain per month.

4–115E The range of healthy weight for adults is usually expressed in terms of the *body mass index* (BMI), defined, in SI units, as

$$\text{BMI} = \frac{W(\text{kg})}{H^2(\text{m}^2)}$$

where W is the weight (actually, the mass) of the person in kg and H is the height in m, and the range of healthy weight is $19 \leq \text{BMI} \geq 25$. Convert the previous formula to English units such that the weight is in pounds and the height in inches. Also, calculate your own BMI, and if it is not in the healthy range, determine how many pounds (or kg) you need to gain or lose to be fit.

4–116 The body mass index (BMI) of a 1.6-m tall woman who normally has 3 large slices of cheese pizza and a 400-ml Coke for lunch is 30. She now decides to change her lunch to 2 slices of pizza and a 200-ml Coke. Assuming that the deficit in the calorie intake is made up by burning body fat, determine how long it will take for the BMI of this person to drop to 20. Use the data in the text for calories and take the metabolizable energy content of 1 kg of body fat to be 33,100 kJ. *Answer:* 463 days

Review Problems

4–117 Does the concept of a specific heat property (either at constant volume or constant pressure) have a meaning for substances that are undergoing a phase change? Why or why not?

4–118 10-kg of nitrogen is heated from 20°C to 250°C. Determine the total amount of heat required when this is done in (*a*) a constant-volume process and (*b*) an isobaric process.

4–119 Consider a classroom that is losing heat to the outdoors at a rate of 12,000 kJ/h. If there are 40 students in class, each dissipating sensible heat at a rate of 84 W, determine if it is necessary to turn the heater in the classroom on to prevent the room temperature from dropping.

4–120E Air in the amount of 2 lbm is contained in a well-insulated, rigid vessel equipped with a stirring paddle wheel. The initial state of this air is 30 psia and 60°F. How much work, in Btu, must be transferred to the air with the paddle wheel to raise the air pressure to 40 psia? Also, what is the final temperature of air?

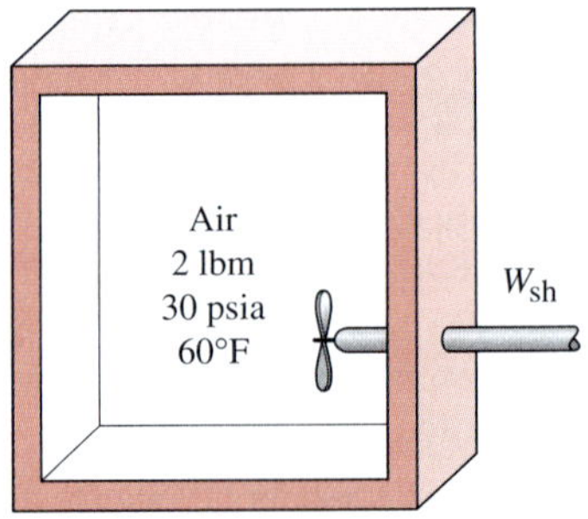

FIGURE P4–120E

4–121 Air is expanded in a polytropic process with $n = 1.5$ from 2 MPa and 300°C to 80 kPa in a piston-cylinder device. Determine the final temperature of the air.

4–122 Nitrogen at 100 kPa and 25°C in a rigid vessel is heated until its pressure is 300 kPa. Calculate the work done and the heat transferred during this process, in kJ/kg.

4–123 A well-insulated rigid vessel contains 3 kg of saturated liquid water at 40°C. The vessel also contains an electrical resistor that draws 10 amperes when 50 volts are applied. Determine the final temperature in the vessel after the resistor has been operating for 30 minutes. *Answer:* 119°C

4–124 Derive a general expression for the work produced by an ideal gas as it undergoes a polytropic process in a closed system from initial state 1 to final state 2. Your result should be in terms of the initial pressure and temperature and the final pressure as well as the gas constant R and the polytropic exponent n.

4–125 A frictionless piston–cylinder device initially contains air at 100 kPa and 0.15 m^3. At this state, a linear spring ($F \propto x$) is touching the piston but exerts no force on it. The air is now heated to a final state of 0.45 m^3 and 800 kPa. Determine (*a*) the total work done by the air and (*b*) the work done against the spring. Also, show the process on a *P*-V diagram. *Answers:* (*a*) 135 kJ, (*b*) 105 kJ

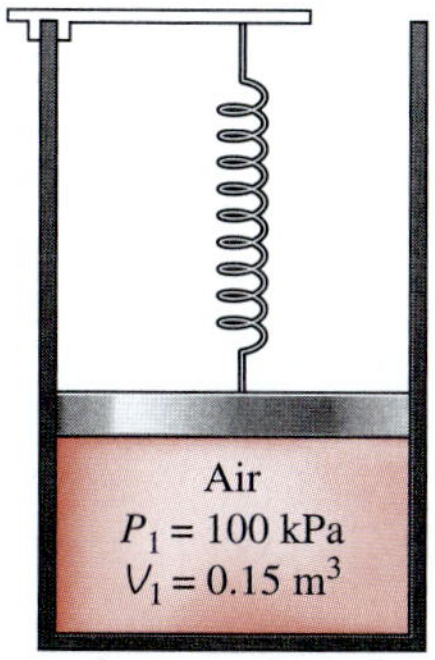

FIGURE P4–125

4–126 A mass of 5 kg of saturated liquid–vapor mixture of water is contained in a piston–cylinder device at 125 kPa. Initially, 2 kg of the water is in the liquid phase and the rest is in the vapor phase. Heat is now transferred to the water, and the piston, which is resting on a set of stops, starts moving when the pressure inside reaches 300 kPa. Heat transfer continues until the total volume increases by 20 percent. Determine (*a*) the initial and final temperatures, (*b*) the mass of liquid water when the piston first starts moving, and (*c*) the work done during this process. Also, show the process on a *P*-V diagram.

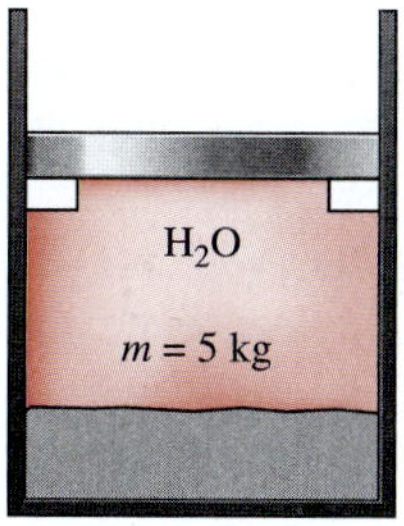

FIGURE P4–126

4–127E A spherical balloon contains 7 lbm of air at 30 psia and 600 R. The balloon material is such that the pressure inside is always proportional to the square of the diameter. Determine the work done when the volume of the balloon doubles as a result of heat transfer. *Answer:* 376 Btu

4–128E Reconsider Prob. 4–127E. Using the integration feature of the EES software, determine the work done. Compare the result with your "hand-calculated" result.

4–129 A mass of 12 kg of saturated refrigerant-134a vapor is contained in a piston–cylinder device at 240 kPa. Now 300 kJ of heat is transferred to the refrigerant at constant pressure while a 110-V source supplies current to a resistor within the cylinder for 6 min. Determine the current supplied if the final temperature is 70°C. Also, show the process on a *T*-V diagram with respect to the saturation lines. *Answer:* 12.8 A

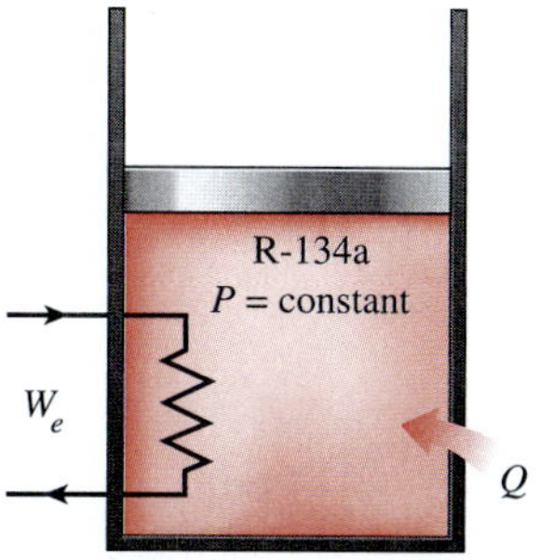

FIGURE P4–129

4–130 Saturated water vapor at 200°C is condensed to a saturated liquid at 50°C in a spring-loaded piston-cylinder device. Determine the heat transfer for this process, in kJ/kg.

4–131 A piston–cylinder device contains helium gas initially at 150 kPa, 20°C, and 0.5 m^3. The helium is now compressed in a polytropic process (PV^n = constant) to 400 kPa and 140°C. Determine the heat loss or gain during this process. *Answer:* 11.2 kJ loss

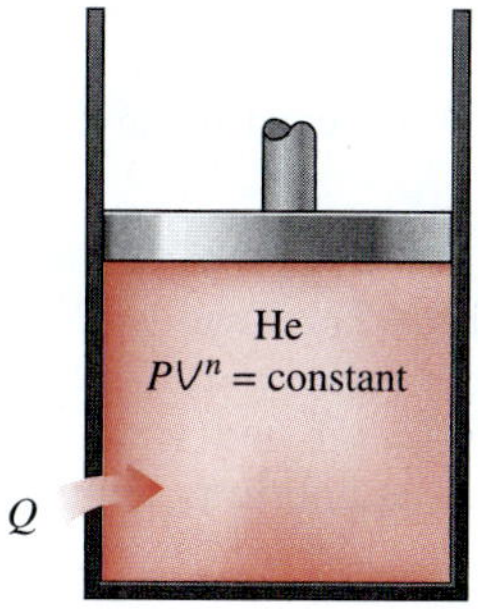

FIGURE P4–131

4–132 Nitrogen gas is expanded in a polytropic process with n = 1.45 from 2 MPa and 1200 K to 200 kPa in a piston–cylinder device. How much work is produced and heat is transferred during this expansion process, in kJ/kg?

4–133 A frictionless piston–cylinder device and a rigid tank initially contain 12 kg of an ideal gas each at the same temperature, pressure, and volume. It is desired to raise the temperatures of both systems by 15°C. Determine the amount of extra heat that must be supplied to the gas in the cylinder which is maintained at constant pressure to achieve this result. Assume the molar mass of the gas is 25.

4–134 A passive solar house that is losing heat to the outdoors at an average rate of 50,000 kJ/h is maintained at 22°C at all times during a winter night for 10 h. The house is to be heated by 50 glass containers each containing 20 L of water that is heated to 80°C during the day by absorbing solar energy. A thermostat-controlled 15-kW back-up electric resistance heater turns on whenever necessary to keep the house at 22°C. (*a*) How long did the electric heating system run that night? (*b*) How long would the electric heater run that night if the house incorporated no solar heating? *Answers:* (*a*) 4.77 h, (*b*) 9.26 h

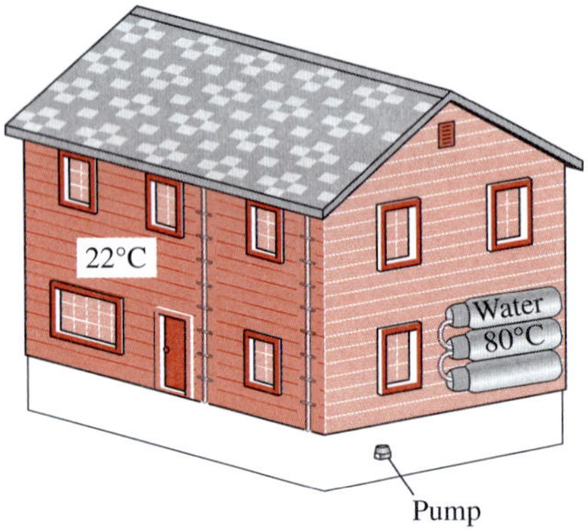

FIGURE P4–134

4–135 An 1800-W electric resistance heating element is immersed in 40 kg of water initially at 20°C. Determine how long it will take for this heater to raise the water temperature to 80°C.

4–136 One ton (1000 kg) of liquid water at 50°C is brought into a well-insulated and well-sealed 4-m × 5-m × 6-m room initially at 15°C and 95 kPa. Assuming constant specific heats for both air and water at room temperature, determine the final equilibrium temperature in the room. *Answer:* 49.2°C

4–137 A 4-m × 5-m × 6-m room is to be heated by one ton (1000 kg) of liquid water contained in a tank that is placed in the room. The room is losing heat to the outside at an average rate of 8000 kJ/h. The room is initially at 20°C and 100 kPa and is maintained at an average temperature of 20°C at all times. If the hot water is to meet the heating requirements of this room for a 24-h period, determine the minimum temperature of the water when it is first brought into the room. Assume constant specific heats for both air and water at room temperature.

4–138 The energy content of a certain food is to be determined in a bomb calorimeter that contains 3 kg of water by burning a 2-g sample of it in the presence of 100 g of air in the reaction chamber. If the water temperature rises by 3.2°C when equilibrium is established, determine the energy content of the food, in kJ/kg, by neglecting the thermal energy stored in the reaction chamber and the energy supplied by the mixer. What is a rough estimate of the error involved in neglecting the thermal energy stored in the reaction chamber? *Answer:* 20,060 kJ/kg

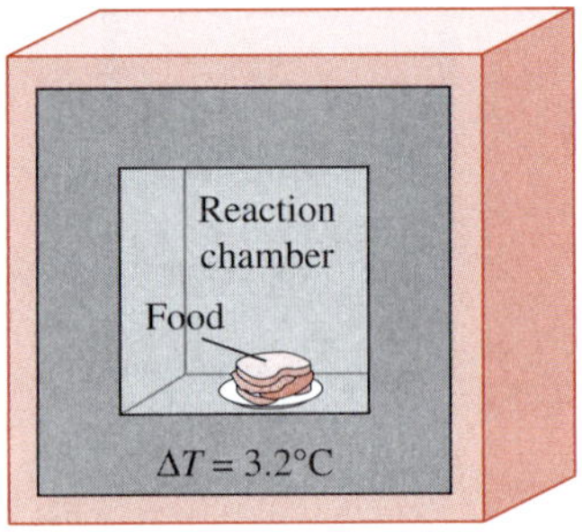

FIGURE P4–138

4–139 A 68-kg man whose average body temperature is 39°C drinks 1 L of cold water at 3°C in an effort to cool down. Taking the average specific heat of the human body to be 3.6 kJ/kg·°C, determine the drop in the average body temperature of this person under the influence of this cold water.

4–140 A 0.3-L glass of water at 20°C is to be cooled with ice to 5°C. Determine how much ice needs to be added to the water, in grams, if the ice is at (*a*) 0°C and (*b*) −20°C. Also determine how much water would be needed if the cooling is to be done with cold water at 0°C. The melting temperature and the heat of fusion of ice at atmospheric pressure are 0°C and 333.7 kJ/kg, respectively, and the density of water is 1 kg/L.

4–141 Reconsider Prob. 4–140. Using EES (or other) software, investigate the effect of the initial temperature of the ice on the final mass required. Let the ice temperature vary from –26 to 0°C. Plot the mass of ice against the initial temperature of ice, and discuss the results.

4–142 A rigid tank containing 0.4 m^3 of air at 400 kPa and 30°C is connected by a valve to a piston–cylinder device with zero clearance. The mass of the piston is such that a pressure of 200 kPa is required to raise the piston. The valve is now opened slightly, and air is allowed to flow into the cylinder until the pressure in the tank drops to 200 kPa. During this process, heat is exchanged with the surroundings such that the entire air remains at 30°C at all times. Determine the heat transfer for this process.

FIGURE P4–142

4–143 A well-insulated 4-m × 4-m × 5-m room initially at 10°C is heated by the radiator of a steam heating system. The radiator has a volume of 15 L and is filled with superheated vapor at 200 kPa and 200°C. At this moment both the inlet and the exit valves to the radiator are closed. A 120-W fan is used to distribute the air in the room. The pressure of the steam is observed to drop to 100 kPa after 30 min as a result of heat transfer to the room. Assuming constant specific heats for air at room temperature, determine the average temperature of air in 30 min. Assume the air pressure in the room remains constant at 100 kPa.

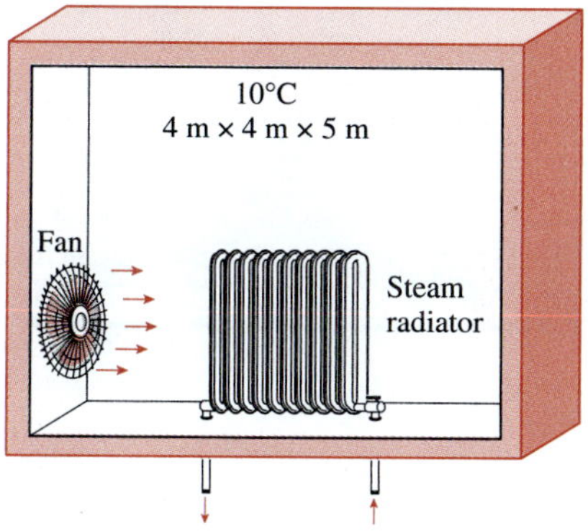

FIGURE P4–143

4–144 Consider a well-insulated horizontal rigid cylinder that is divided into two compartments by a piston that is free to move but does not allow either gas to leak into the other side. Initially, one side of the piston contains 1 m³ of N_2 gas at 500 kPa and 120°C while the other side contains 1 m³ of He gas at 500 kPa and 40°C. Now thermal equilibrium is established in the cylinder as a result of heat transfer through the piston. Using constant specific heats at room temperature, determine the final equilibrium temperature in the cylinder. What would your answer be if the piston were not free to move?

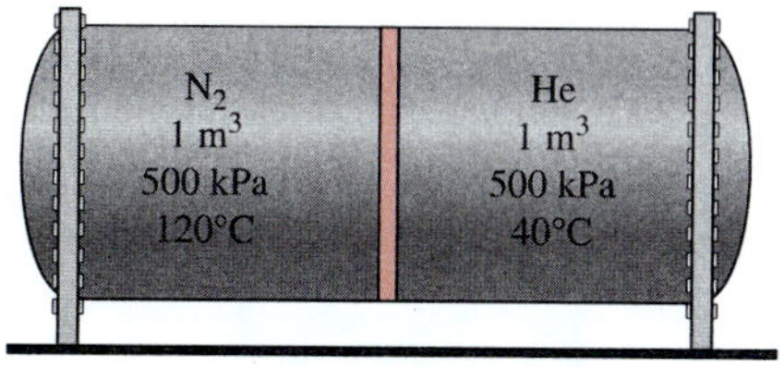

FIGURE P4–144

4–145 Repeat Prob. 4–144 by assuming the piston is made of 8 kg of copper initially at the average temperature of the two gases on both sides. *Answer:* 83.7°C

4–146 Reconsider Prob. 4–145. Using EES (or other) software, investigate the effect of the mass of the copper piston on the final equilibrium temperature. Let the mass of piston vary from 1 to 10 kg. Plot the final temperature against the mass of piston, and discuss the results.

4–147 An insulated piston-cylinder device initially contains 1.4-kg saturated liquid water at 200°C. Now an electric resistor placed in the tank is turned on for 20 min until the volume quadruples. Determine (*a*) the volume of the tank, (*b*) the final temperature, and (*c*) the electrical power rating of the resistor. *Answers:* (*a*) 0.00648 m³, (*b*) 200°C, (*c*) 0.0623 kW

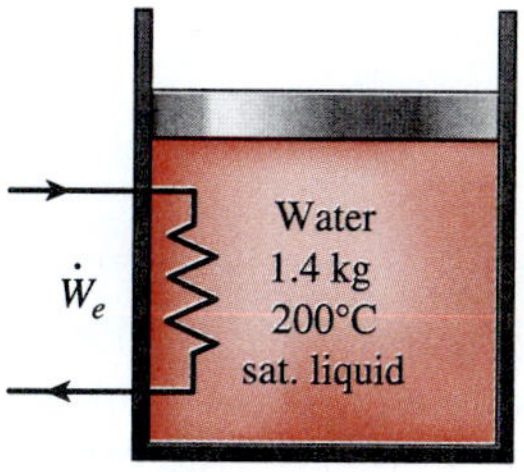

FIGURE P4–147

4–148 A vertical 12-cm diameter piston–cylinder device contains an ideal gas at the ambient conditons of 1 bar and 24°C. Initially, the inner face of the piston is 20 cm from the base of the cylinder. Now an external shaft connected to the piston exerts a force corresponding to a boundary work input of 0.1 kJ. The temperature of the gas remains constant during the process. Determine (*a*) the amount of heat transfer, (*b*) the final pressure in the cylinder, and (*c*) the distance that the piston is displaced.

4–149 A piston–cylinder device initially contains 0.35-kg steam at 3.5 MPa, superheated by 7.4°C. Now the steam loses heat to the surroundings and the piston moves down, hitting a set of stops at which point the cylinder contains saturated liquid water. The cooling continues until the cylinder contains water at 200°C. Determine (*a*) the final pressure and the quality (if mixture), (*b*) the boundary work, (*c*) the amount of heat transfer when the piston first hits the stops, (*d*) and the total heat transfer.

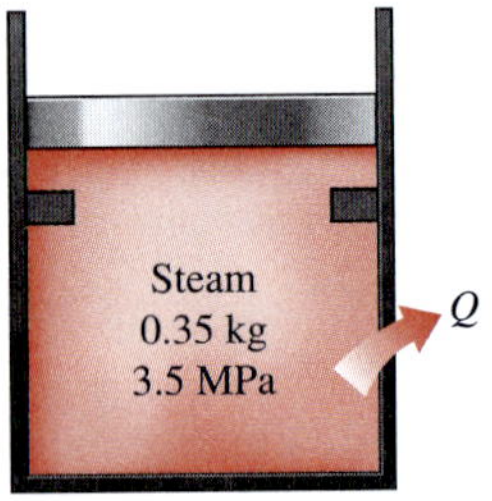

FIGURE P4–149

4–150 An insulated rigid tank is divided into two compartments of different volumes. Initially, each compartment contains the same ideal gas at identical pressure but at different temperatures and masses. The wall separating the two compartments is removed and the two gases are allowed to mix. Assuming constant specific heats, find the simplest expression for the mixture temperature written in the form

$$T_3 = f\left(\frac{m_1}{m_3}, \frac{m_2}{m_3}, T_1, T_2\right)$$

where m_3 and T_3 are the mass and temperature of the final mixture, respectively.

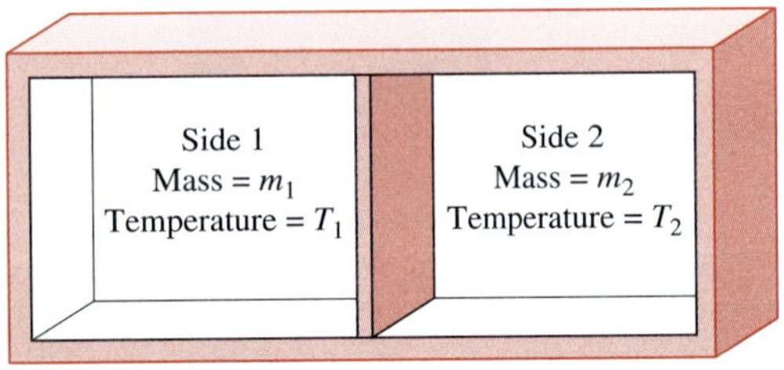

FIGURE P4–150

4–151 Catastrophic explosions of steam boilers in the 1800s and early 1900s resulted in hundreds of deaths, which prompted the development of the ASME Boiler and Pressure Vessel Code in 1915. Considering that the pressurized fluid in a vessel eventually reaches equilibrium with its surroundings shortly after the explosion, the work that a pressurized fluid would do if allowed to expand adiabatically to the state of the surroundings can be viewed as the *explosive energy* of the pressurized fluid. Because of the very short time period of the explosion and the apparent stability afterward, the explosion process can be considered to be adiabatic with no changes in kinetic and potential energies. The closed-system conservation of energy relation in this case reduces to $W_{out} = m(u_1 - u_2)$. Then the explosive energy E_{exp} becomes

$$E_{exp} = m(u_1 - u_2)$$

where the subscripts 1 and 2 refer to the state of the fluid before and after the explosion, respectively. The specific explosion energy e_{exp} is usually expressed *per unit volume,* and it is obtained by dividing the quantity above by the total V of the vessel:

$$e_{exp} = \frac{u_1 - u_2}{v_1}$$

where v_1 is the specific volume of the fluid before the explosion.

Show that the specific explosion energy of an ideal gas with constant specific heat is

$$e_{exp} = \frac{P_1}{k-1}\left(1 - \frac{T_2}{T_1}\right)$$

Also, determine the total explosion energy of 20 m^3 of air at 5 MPa and 100°C when the surroundings are at 20°C.

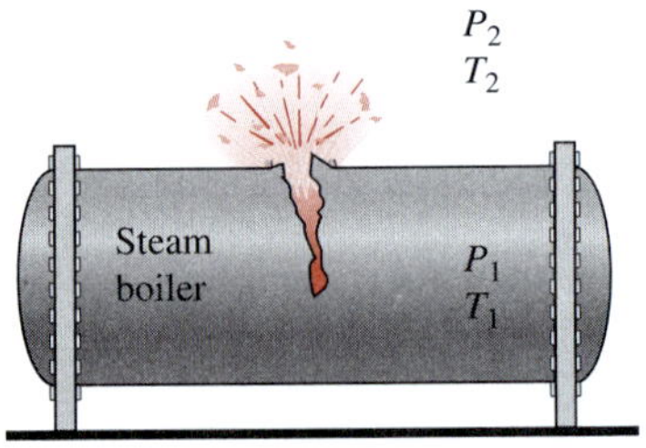

FIGURE P4–151

4–152 Using the relations in Prob. 4–151, determine the explosive energy of 20 m^3 of steam at 10 MPa and 500°C assuming the steam condenses and becomes a liquid at 25°C after the explosion. To how many kilograms of TNT is this explosive energy equivalent? The explosive energy of TNT is about 3250 kJ/kg.

4–153 One kilogram of carbon dioxide is compressed from 1 MPa and 200°C to 3 MPa in a piston-cylinder device arranged to execute to polytropic process $PV^{1.5}$ = constant. Determine the final temperature treating the carbon dioxide as (*a*) an ideal gas and (*b*) a van der Waals gas. *Answers:* (*a*) 682.1 k. (*b*) 680.9 k

FIGURE P4–153

4–154 One kilogram of carbon dioxide is compressed from 1 MPa and 200°C to 3 MPa in a piston-cylinder device arranged to execute a polytropic process with $n = 1.2$. Use the compressibility factor to determine the final temperature.

4–155E Two 10-ft^3 adiabatic tanks are connected by a valve. Initially, one tank contains water at 450 psia with 10 percent quality, while the second contains water at 15 psia with 75 percent quality. The valve is now opened, allowing the water vapor from the high-pressure tank to move to the low-pressure tank until the pressure in the two becomes equal. Determine the final pressure and the final mass in each tank. *Answers:* 313 psia, 41.6 lbm

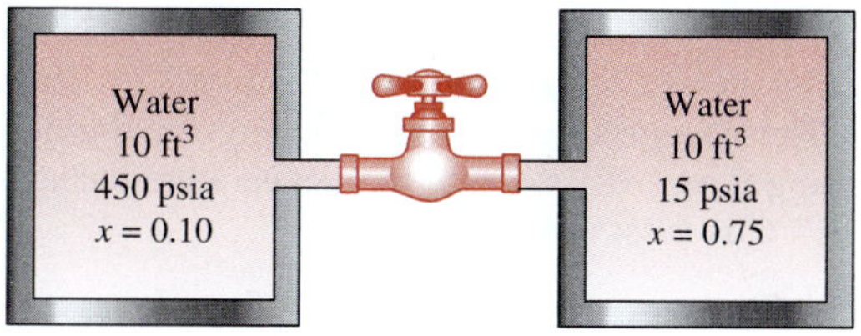

FIGURE P4–155E

Fundamentals of Engineering (FE) Exam Problems

4–156 A frictionless piston–cylinder device and a rigid tank contain 3 kmol of an ideal gas at the same temperature, pressure, and volume. Now heat is transferred, and the temperature of both systems is raised by 10°C. The amount of extra heat that must be supplied to the gas in the cylinder that is maintained at constant pressure is

(*a*) 0 kJ
(*b*) 27 kJ
(*c*) 83 kJ
(*d*) 249 kJ
(*e*) 300 kJ

4–157 The specific heat of a material is given in a strange unit to be c = 3.60 kJ/kg·°F. The specific heat of this material in the SI units of kJ/kg·°C is

(*a*) 2.00 kJ/kg·°C
(*b*) 3.20 kJ/kg·°C
(*c*) 3.60 kJ/kg·°C
(*d*) 4.80 kJ/kg·°C
(*e*) 6.48 kJ/kg·°C

4–158 A 3-m^3 rigid tank contains nitrogen gas at 500 kPa and 300 K. Now heat is transferred to the nitrogen in the tank and the pressure of nitrogen rises to 800 kPa. The work done during this process is

(*a*) 500 kJ
(*b*) 1500 kJ
(*c*) 0 kJ
(*d*) 900 kJ
(*e*) 2400 kJ

4–159 A 0.5-m^3 rigid tank contains nitrogen gas at 600 kPa and 300 K. Now the gas is compressed isothermally to a volume of 0.1 m^3. The work done on the gas during this compression process is

(*a*) 720 kJ
(*b*) 483 kJ
(*c*) 240 kJ
(*d*) 175 kJ
(*e*) 143 kJ

4–160 A well-sealed room contains 60 kg of air at 200 kPa and 25°C. Now solar energy enters the room at an average rate of 0.8 kJ/s while a 120-W fan is turned on to circulate the air in the room. If heat transfer through the walls is negligible, the air temperature in the room in 30 min will be

(*a*) 25.6°C
(*b*) 49.8°C
(*c*) 53.4°C
(*d*) 52.5°C
(*e*) 63.4°C

4–161 A 2-kW baseboard electric resistance heater in a vacant room is turned on and kept on for 15 min. The mass of the air in the room is 75 kg, and the room is tightly sealed so that no air can leak in or out. The temperature rise of air at the end of 15 min is

(*a*) 8.5°C
(*b*) 12.4°C
(*c*) 24.0°C
(*d*) 33.4°C
(*e*) 54.8°C

4–162 A room contains 75 kg of air at 100 kPa and 15°C. The room has a 250-W refrigerator (the refrigerator consumes 250 W of electricity when running), a 120-W TV, a 1.8-kW electric resistance heater, and a 50-W fan. During a cold winter day, it is observed that the refrigerator, the TV, the fan, and the electric resistance heater are running continuously but the air temperature in the room remains constant. The rate of heat loss from the room that day is

(*a*) 5832 kJ/h
(*b*) 6192 kJ/h
(*c*) 7560 kJ/h
(*d*) 7632 kJ/h
(*e*) 7992 kJ/h

4–163 A piston–cylinder device contains 5 kg of air at 400 kPa and 30°C. During a quasi-equilibium isothermal expansion process, 15 kJ of boundary work is done by the system, and 3 kJ of paddle-wheel work is done on the system. The heat transfer during this process is

(*a*) 12 kJ (*b*) 18 kJ
(*c*) 2.4 kJ (*d*) 3.5 kJ
(*e*) 60 kJ

4–164 A container equipped with a resistance heater and a mixer is initially filled with 3.6 kg of saturated water vapor at 120°C. Now the heater and the mixer are turned on; the steam is compressed; and there is heat loss to the surrounding air. At the end of the process, the temperature and pressure of steam in the container are measured to be 300°C and 0.5 MPa. The net energy transfer to the steam during this process is

(*a*) 274 kJ (*b*) 914 kJ
(*c*) 1213 kJ (*d*) 988 kJ
(*e*) 1291 kJ

4–165 A 6-pack canned drink is to be cooled from 18°C to 3°C. The mass of each canned drink is 0.355 kg. The drinks can be treated as water, and the energy stored in the aluminum can itself is negligible. The amount of heat transfer from the 6 canned drinks is

(*a*) 22 kJ (*b*) 32 kJ
(*c*) 134 kJ (*d*) 187 kJ
(*e*) 223 kJ

4–166 A glass of water with a mass of 0.45 kg at 20°C is to be cooled to 0°C by dropping ice cubes at 0°C into it. The latent heat of fusion of ice is 334 kJ/kg, and the specific heat of water is 4.18 kJ/kg·°C. The amount of ice that needs to be added is

(*a*) 56 g (*b*) 113 g
(*c*) 124 g (*d*) 224 g
(*e*) 450 g

4–167 A 2-kW electric resistance heater submerged in 5-kg water is turned on and kept on for 10 min. During the process, 300 kJ of heat is lost from the water. The temperature rise of water is

(*a*) 0.4°C (*b*) 43.1°C
(*c*) 57.4°C (*d*) 71.8°C
(*e*) 180°C

4–168 1.5 kg of liquid water initially at 12°C is to be heated at 95°C in a teapot equipped with a 800-W electric heating element inside. The specific heat of water can be taken to be 4.18 kJ/kg·°C, and the heat loss from the water during heating can be neglected. The time it takes to heat water to the desired temperature is

(*a*) 5.9 min (*b*) 7.3 min
(*c*) 10.8 min (*d*) 14.0 min
(*e*) 17.0 min

4–169 An ordinary egg with a mass of 0.1 kg and a specific heat of 3.32 kJ/kg·°C is dropped into boiling water at 95°C. If the initial temperature of the egg is 5°C, the maximum amount of heat transfer to the egg is

(*a*) 12 kJ (*b*) 30 kJ
(*c*) 24 kJ (*d*) 18 kJ
(*e*) infinity

4–170 An apple with an average mass of 0.18 kg and average specific heat of 3.65 kJ/kg·°C is cooled from 22°C to 5°C. The amount of heat transferred from the apple is

(*a*) 0.85 kJ (*b*) 62.1 kJ
(*c*) 17.7 kJ (*d*) 11.2 kJ
(*e*) 7.1 kJ

4–171 The specific heat at constant pressure for an ideal gas is given by $c_p = 0.9 + (2.7 \times 10^{-4})T$ (kJ/kg·K) where T is in kelvin. The change in the enthalpy for this ideal gas under-going a process in which the temperature changes from 27 to 147°C is most nearly

(*a*) 19.7 kJ/kg (*b*) 22.0 kJ/kg
(*c*) 25.5 kJ/kg (*d*) 29.7 kJ/kg
(*e*) 32.1 kJ/kg

4–172 The specific heat at constant volume for an ideal gas is given by $c_v = 0.7 + (2.7 \times 10^{-4})T$ (kJ/kg·K) where T is in kelvin. The change in the internal energy for this ideal gas undergoing a process in which the temperature changes from 27 to 127°C is most nearly

(*a*) 70 kJ/kg (*b*) 72.1 kJ/kg
(*c*) 79.5 kJ/kg (*d*) 82.1 kJ/kg
(*e*) 84.0 kJ/kg

4–173 An ideal gas has a gas constant $R = 0.3$ kJ/kg·K and a constant-volume specific heat $c_v = 0.7$ kJ/kg·K. If the gas has a temperature change of 100°C, choose the correct answer for each of the following:

1. The change in enthalpy is, in kJ/kg
   (*a*) 30 (*b*) 70 (*c*) 100
   (*d*) insufficient information to determine
2. The change in internal energy is, in kJ/kg
   (*a*) 30 (*b*) 70 (*c*) 100
   (*d*) insufficient information to determine
3. The work done is, in kJ/kg
   (*a*) 30 (*b*) 70 (*c*) 100
   (*d*) insufficient information to determine
4. The heat transfer is, in kJ/kg
   (*a*) 30 (*b*) 70 (*c*) 100
   (*d*) insufficient information to determine

5. The change in the pressure-volume product is, in kJ/kg
   (*a*) 30 (*b*) 70 (*c*) 100
   (*d*) insufficient information to determine

4–174 An ideal gas undergoes a constant temperature (isothermal) process in a closed system. The heat transfer and work are, respectively

(*a*) 0, $-c_v\Delta T$ (*b*) $c_v\Delta T$, 0 (*c*) $c_p\Delta T$, $R\Delta T$
(*d*) $R\ln(T_2/T_1)$, $R\ln(T_2/T_1)$

4–175 An ideal gas undergoes a constant volume (isochoric) process in a closed system. The heat transfer and work are, respectively

(*a*) 0, $-c_v\Delta T$ (*b*) $c_v\Delta T$, 0 (*c*) $c_p\Delta T$, $R\Delta T$
(*d*) $R\ln(T_2/T_1)$, $R\ln(T_2/T_1)$

4–176 An ideal gas undergoes a constant pressure (isobaric) process in a closed system. The heat transfer and work are, respectively

(*a*) 0, $-c_v\Delta T$ (*b*) $c_v\Delta T$, 0 (*c*) $c_p\Delta T$, $R\Delta T$
(*d*) $R\ln(T_2/T_1)$, $R\ln(T_2/T_1)$

4–177 An ideal gas undergoes a constant entropy (isentropic) process in a closed system. The heat transfer and work are, respectively

(*a*) 0, $-c_v\Delta T$ (*b*) $c_v\Delta T$, 0 (*c*) $c_p\Delta T$, $R\Delta T$
(*d*) $R\ln(T_2/T_1)$, $R\ln(T_2/T_1)$

Design and Essay Problems

4–178 Compressed gases and phase-changing liquids are used to store energy in rigid containers. What are the advantages and disadvantages of each substance as a means of storing energy?

4–179 Someone has suggested that the device shown in Fig. P4–179 be used to move the maximum force *F* against the spring, which has a spring constant of *k*. This is accomplished by changing the temperature of the liquid–vapor mixture in the container. You are to design such a device to close sun-blocking window shutters that require a maximum force of 0.5 lbf. The piston must move 6 inches to close these shutters completely. You elect to use R-134a as the working fluid and arrange the liquid–vapor mixture container such that the temperature changes from 70°F when shaded from the sun to 100°F when exposed to the full sun. Select the sizes of the various components in this system to do this task. Also select the necessary spring constant and the amount of R-134a to be used.

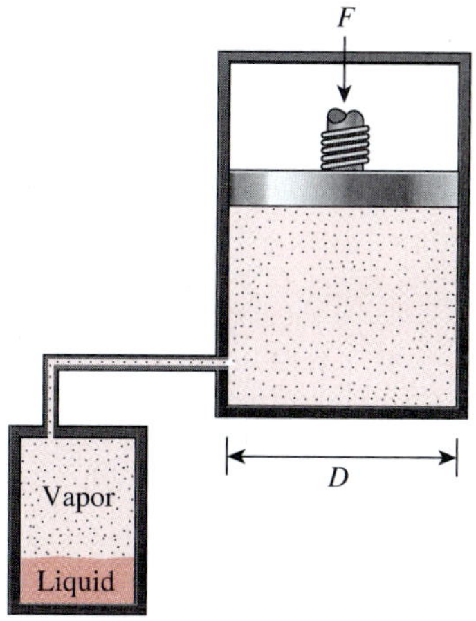

FIGURE P4–179

4–180 Design an experiment complete with instrumentation to determine the specific heats of a liquid using a resistance heater. Discuss how the experiment will be conducted, what measurements need to be taken, and how the specific heats will be determined. What are the sources of error in your system? How can you minimize the experimental error? How would you modify this system to determine the specific heat of a solid?

4–181 You are asked to design a heating system for a swimming pool that is 2 m deep, 25 m long, and 25 m wide. Your client desires that the heating system be large enough to raise the water temperature from 20 to 30°C in 3 h. The rate of heat loss from the water to the air at the outdoor design conditions is determined to be 960 W/m^2, and the heater must also be able to maintain the pool at 30°C at those conditions. Heat losses to the ground are expected to be small and can be disregarded. The heater considered is a natural gas furnace whose efficiency is 80 percent. What heater size (in kW input) would you recommend to your client?

4–182 It is claimed that fruits and vegetables are cooled by 6°C for each percentage point of weight loss as moisture during vacuum cooling. Using calculations, demonstrate if this claim is reasonable.

4–183 A 1982 U.S. Department of Energy article (FS #204) states that a leak of one drip of hot water per second can cost $1.00 per month. Making reasonable assumptions about the drop size and the unit cost of energy, determine if this claim is reasonable.

CHAPTER

5

MASS AND ENERGY ANALYSIS OF CONTROL VOLUMES

In Chap. 4, we applied the general energy balance relation expressed as $E_{in} - E_{out} = \Delta E_{system}$ to closed systems. In this chapter, we extend the energy analysis to systems that involve mass flow across their boundaries i.e., control volumes, with particular emphasis to steady-flow systems.

We start this chapter with the development of the general *conservation of mass* relation for control volumes, and we continue with a discussion of flow work and the energy of fluid streams. We then apply the energy balance to systems that involve *steady-flow processes* and analyze the common steady-flow devices such as nozzles, diffusers, compressors, turbines, throttling devices, mixing chambers, and heat exchangers. Finally, we apply the energy balance to general *unsteady-flow processes* such as the charging and discharging of vessels.

Objectives

The objectives of Chapter 5 are to:

- Develop the conservation of mass principle.
- Apply the conservation of mass principle to various systems including steady- and unsteady-flow control volumes.
- Apply the first law of thermodynamics as the statement of the conservation of energy principle to control volumes.
- Identify the energy carried by a fluid stream crossing a control surface as the sum of internal energy, flow work, kinetic energy, and potential energy of the fluid and to relate the combination of the internal energy and the flow work to the property enthalpy.
- Solve energy balance problems for common steady-flow devices such as nozzles, compressors, turbines, throttling valves, mixers, heaters, and heat exchangers.
- Apply the energy balance to general unsteady-flow processes with particular emphasis on the uniform-flow process as the model for commonly encountered charging and discharging processes.

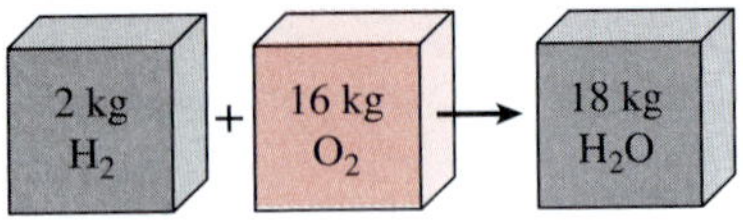

FIGURE 5–1
Mass is conserved even during chemical reactions.

5–1 ▪ CONSERVATION OF MASS

Conservation of mass is one of the most fundamental principles in nature. We are all familiar with this principle, and it is not difficult to understand. A person does not have to be a rocket scientist to figure out how much vinegar-and-oil dressing is obtained by mixing 100 g of oil with 25 g of vinegar. Even chemical equations are balanced on the basis of the conservation of mass principle. When 16 kg of oxygen reacts with 2 kg of hydrogen, 18 kg of water is formed (Fig. 5–1). In an electrolysis process, the water separates back to 2 kg of hydrogen and 16 kg of oxygen.

Mass, like energy, is a conserved property, and it cannot be created or destroyed during a process. However, mass m and energy E can be converted to each other according to the well-known formula proposed by Albert Einstein (1879–1955):

$$E = mc^2 \tag{5–1}$$

where c is the speed of light in a vacuum, which is $c = 2.9979 \times 10^8$ m/s. The equation suggests that there is equivalence between mass and energy. All physical and chemical systems exhibit energy interactions with their surroundings. But the amount of energy interacted is equivalent to an extremely small mass compared to the system's total mass. For example, when 1 kg of liquid water is formed from oxygen and hydrogen at normal atmospheric conditions, the amount of energy released is 15.8 MJ, which corresponds to a mass of 1.76×10^{-10} kg. However, in nuclear reactions, the mass equivalence of the amount of energy interacted is a significant fraction of the total mass involved. Therefore, in most engineering analysis, we consider both mass and energy as conserved quantities.

For *closed systems*, the conservation of mass principle is implicitly used by requiring that the mass of the system remain constant during a process. For *control volumes*, however, mass can cross the boundaries, and so we must keep track of the amount of mass entering and leaving the control volume.

Mass and Volume Flow Rates

The amount of mass flowing through a cross section per unit time is called the **mass flow rate** and is denoted by $\dot{m}$. The dot over a symbol is used to indicate *time rate of change*, as explained in Chap. 2.

A fluid usually flows into or out of a control volume through pipes or ducts. The differential mass flow rate of fluid flowing across a small area element dA_c on a flow cross section is proportional to dA_c itself, the fluid density ρ, and the component of the flow velocity normal to dA_c, which we denote as V_n, and is expressed as (Fig. 5–2)

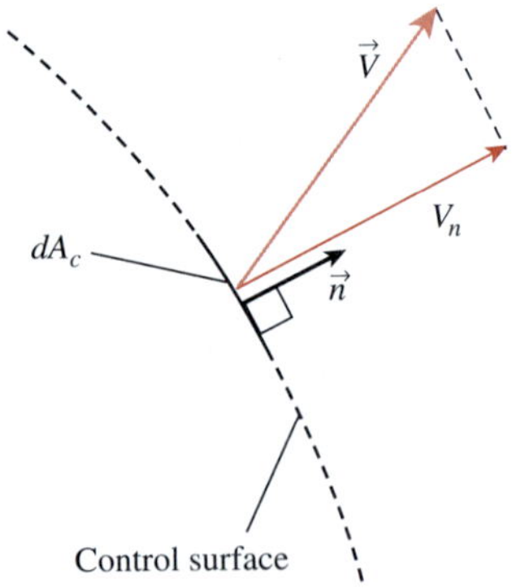

FIGURE 5–2
The normal velocity V_n for a surface is the component of velocity perpendicular to the surface.

$$\delta\dot{m} = \rho V_n \, dA_c \tag{5–2}$$

Note that both δ and d are used to indicate differential quantities, but δ is typically used for quantities (such as heat, work, and mass transfer) that are *path functions* and have *inexact differentials*, while d is used for quantities (such as properties) that are *point functions* and have *exact differentials*. For flow through an annulus of inner radius r_1 and outer radius r_2, for example, $\int_1^2 dA_c = A_{c2} - A_{c1} = \pi(r_2^2 - r_1^2)$ but $\int_1^2 \delta\dot{m} = \dot{m}_{\text{total}}$ (total mass flow rate through the annulus), not $\dot{m}_2 - \dot{m}_1$. For specified values of r_1 and r_2, the value of the integral of dA_c is fixed (thus the names point function and exact differential), but this is not the case for the integral of dm· (thus the names path function and inexact differential).

The mass flow rate through the entire cross-sectional area of a pipe or duct is obtained by integration:

$$\dot{m} = \int_{A_c} \delta\dot{m} = \int_{A_c} \rho V_n \, dA_c \qquad \text{(kg/s)} \tag{5–3}$$

While Eq. 5–3 is always valid (in fact it is *exact*), it is not always practical for engineering analyses because of the integral. We would like instead to express mass flow rate in terms of average values over a cross section of the pipe. In a general compressible flow, both ρ and V_n vary across the pipe. In many practical applications, however, the density is essentially uniform over the pipe cross section, and we can take ρ outside the integral of Eq. 5–3. Velocity, however, is *never* uniform over a cross section of a pipe because of the fluid sticking to the surface and thus having zero velocity at the wall (the no-slip condition). Rather, the velocity varies from zero at the walls to some maximum value at or near the centerline of the pipe. We define the **average velocity** V_{avg} as the average value of V_n across the entire cross section (Fig. 5–3),

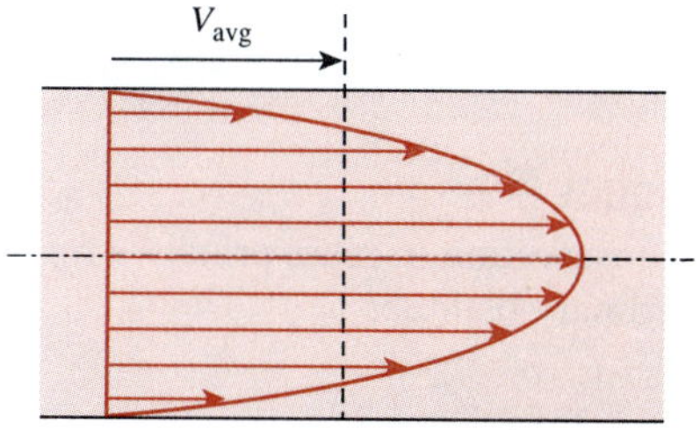

FIGURE 5–3
The average velocity V_{avg} is defined as the average speed through a cross section.

Average velocity:
$$V_{avg} = \frac{1}{A_c} \int_{A_c} V_n \, dA_c \tag{5–4}$$

where A_c is the area of the cross section normal to the flow direction. Note that if the velocity were V_{avg} all through the cross section, the mass flow rate would be identical to that obtained by integrating the actual velocity profile. Thus for incompressible flow or even for compressible flow where ρ is uniform across A_c, Eq. 5–3 becomes

$$\dot{m} = \rho V_{avg} A_c \qquad \text{(kg/s)} \tag{5–5}$$

For compressible flow, we can think of ρ as the bulk average density over the cross section, and then Eq. 5–5 can still be used as a reasonable approximation.

For simplicity, we drop the subscript on the average velocity. Unless otherwise stated, V denotes the average velocity in the flow direction. Also, A_c denotes the cross-sectional area normal to the flow direction.

The volume of the fluid flowing through a cross section per unit time is called the **volume flow rate** $\dot{V}$ (Fig. 5–4) and is given by

$$\dot{V} = \int_{A_c} V_n \, dA_c = V_{avg} A_c = V A_c \qquad \text{(m}^3\text{/s)} \tag{5–6}$$

An early form of Eq. 5–6 was published in 1628 by the Italian monk Benedetto Castelli (circa 1577–1644). Note that most fluid mechanics textbooks use Q instead of $\dot{V}$ for volume flow rate. We use $\dot{V}$ to avoid confusion with heat transfer.

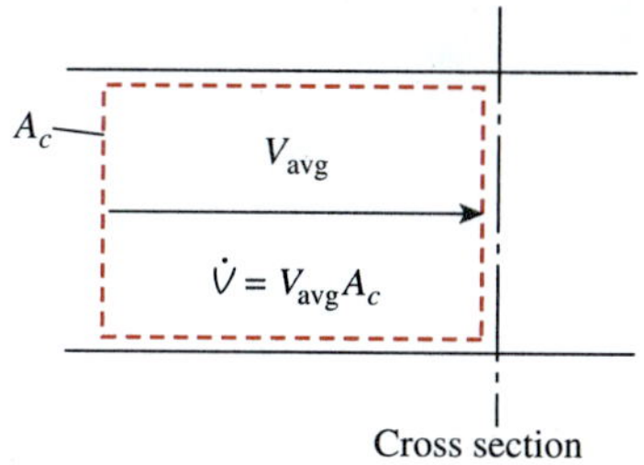

FIGURE 5–4
The volume flow rate is the volume of fluid flowing through a cross section per unit time.

The mass and volume flow rates are related by

$$\dot{m} = \rho \dot{V} = \frac{\dot{V}}{v} \tag{5–7}$$

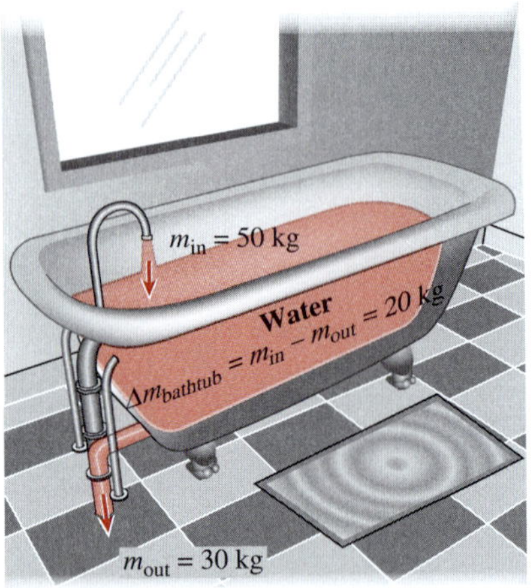

FIGURE 5–5
Conservation of mass principle for an ordinary bathtub.

where v is the specific volume. This relation is analogous to $m = \rho V = V/v$, which is the relation between the mass and the volume of a fluid in a container.

Conservation of Mass Principle

The **conservation of mass principle** for a control volume can be expressed as: *The net mass transfer to or from a control volume during a time interval* Δt *is equal to the net change (increase or decrease) in the total mass within the control volume during* Δt. That is,

$$\begin{pmatrix}\text{Total mass entering}\\ \text{the CV during } \Delta t\end{pmatrix} - \begin{pmatrix}\text{Total mass leaving}\\ \text{the CV during } \Delta t\end{pmatrix} = \begin{pmatrix}\text{Net change of mass}\\ \text{within the CV during } \Delta t\end{pmatrix}$$

or

$$m_{in} - m_{out} = \Delta m_{CV} \qquad \text{(kg)} \tag{5–8}$$

where $\Delta m_{CV} = m_{final} - m_{initial}$ is the change in the mass of the control volume during the process (Fig. 5–5). It can also be expressed in *rate form* as

$$\dot{m}_{in} - \dot{m}_{out} = dm_{CV}/dt \qquad \text{(kg/s)} \tag{5–9}$$

where $\dot{m}_{in}$ and $\dot{m}_{out}$ are the total rates of mass flow into and out of the control volume, and dm_{CV}/dt is the time rate of change of mass within the control volume boundaries. Equations 5–8 and 5–9 are often referred to as the **mass balance** and are applicable to any control volume undergoing any kind of process.

Consider a control volume of arbitrary shape, as shown in Fig. 5–6. The mass of a differential volume dV within the control volume is $dm = \rho\, dV$. The total mass within the control volume at any instant in time t is determined by integration to be

Total mass within the CV:
$$m_{CV} = \int_{CV} \rho\, dV \tag{5–10}$$

Then the time rate of change of the amount of mass within the control volume can be expressed as

Rate of change of mass within the CV:
$$\frac{dm_{CV}}{dt} = \frac{d}{dt}\int_{CV} \rho\, dV \tag{5–11}$$

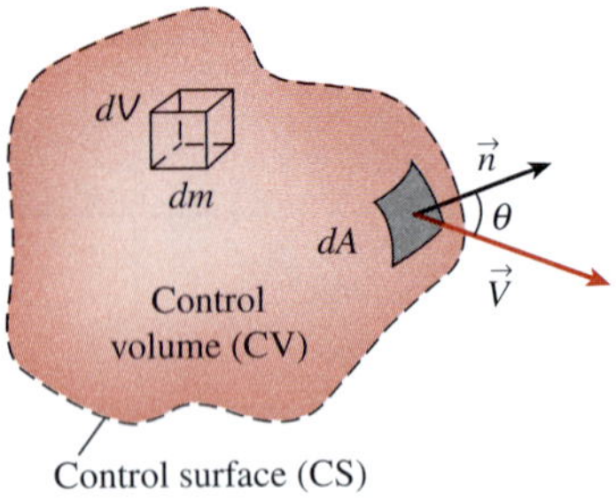

FIGURE 5–6
The differential control volume dV and the differential control surface dA used in the derivation of the conservation of mass relation.

For the special case of no mass crossing the control surface (i.e., the control volume is a closed system), the conservation of mass principle reduces to $dm_{CV}/dt = 0$. This relation is valid whether the control volume is fixed, moving, or deforming.

Now consider mass flow into or out of the control volume through a differential area dA on the control surface of a fixed control volume. Let $\vec{n}$ be the outward unit vector of dA normal to dA and $\vec{V}$ be the flow velocity at dA relative to a fixed coordinate system, as shown in Fig. 5–6. In general, the velocity may cross dA at an angle θ off the normal of dA, and the mass flow rate is proportional to the normal component of velocity $\vec{V}_n = \vec{V}\cos\theta$ ranging from a maximum outflow of $\vec{V}$ for $\theta = 0$ (flow is normal to dA) to a minimum of zero for $\theta = 90°$ (flow is tangent to dA) to a maximum *inflow* of $\vec{V}$ for $\theta = 180°$ (flow is normal to dA but in the opposite direction). Making use of the concept of dot product of two vectors, the magnitude of the normal component of velocity can be expressed as

Normal component of velocity: $$V_n = V\cos\theta = \vec{V}\cdot\vec{n} \quad \textbf{(5–12)}$$

The mass flow rate through dA is proportional to the fluid density ρ, normal velocity V_n, and the flow area dA, and can be expressed as

Differential mass flow rate: $$\delta\dot{m} = \rho V_n\,dA = \rho(V\cos\theta)\,dA = \rho(\vec{V}\cdot\vec{n})\,dA \quad \textbf{(5–13)}$$

The net flow rate into or out of the control volume through the entire control surface is obtained by integrating $\delta\dot{m}$ over the entire control surface,

Net mass flow rate: $$\dot{m}_{\text{net}} = \int_{\text{CS}} \delta\dot{m} = \int_{\text{CS}} \rho V_n\,dA = \int_{\text{CS}} \rho(\vec{V}\cdot\vec{n})\,dA \quad \textbf{(5–14)}$$

Note that $\vec{V}\cdot\vec{n} = V\cos\theta$ is positive for $\theta < 90°$ (outflow) and negative for $\theta > 90°$ (inflow). Therefore, the direction of flow is automatically accounted for, and the surface integral in Eq. 5–14 directly gives the *net* mass flow rate. A positive value for $\dot{m}_{\text{net}}$ indicates net outflow, and a negative value indicates a net inflow of mass.

Rearranging Eq. 5–9 as $dm_{\text{CV}}/dt + \dot{m}_{\text{out}} - \dot{m}_{\text{in}} = 0$, the conservation of mass relation for a fixed control volume can then be expressed as

General conservation of mass: $$\frac{d}{dt}\int_{\text{CV}} \rho\,dV + \int_{\text{CS}} \rho(\vec{V}\cdot\vec{n})\,dA = 0 \quad \textbf{(5–15)}$$

It states that *the time rate of change of mass within the control volume plus the net mass flow rate through the control surface is equal to zero.*

Splitting the surface integral in Eq. 5–15 into two parts—one for the outgoing flow streams (positive) and one for the incoming streams (negative)—the general conservation of mass relation can also be expressed as

$$\frac{d}{dt}\int_{\text{CV}} \rho\,dV + \sum_{\text{out}} \rho|V_n|\,dA - +\sum_{\text{in}} \rho|V_n|\,dA = 0 \quad \textbf{(5–16)}$$

where A represents the area for an inlet or outlet, and the summation signs are used to emphasize that *all* the inlets and outlets are to be considered. Using the definition of mass flow rate, Eq. 5–16 can also be expressed as

$$\frac{d}{dt}\int_{\text{CV}} \rho\,dV = \sum_{\text{in}} \dot{m} - \sum_{\text{out}} \dot{m} \quad \text{or} \quad \frac{dm_{\text{CV}}}{dt} = \sum_{\text{in}} \dot{m} - \sum_{\text{out}} \dot{m} \quad \textbf{(5–17)}$$

Equations 5–15 and 5–16 are also valid for moving or deforming control volumes provided that the *absolute velocity* $\vec{V}$ is replaced by the *relative velocity* $\vec{V}_r$, which is the fluid velocity relative to the control surface.

Mass Balance for Steady-Flow Processes

During a steady-flow process, the total amount of mass contained within a control volume does not change with time (m_{CV} = constant). Then the conservation of mass principle requires that the total amount of mass

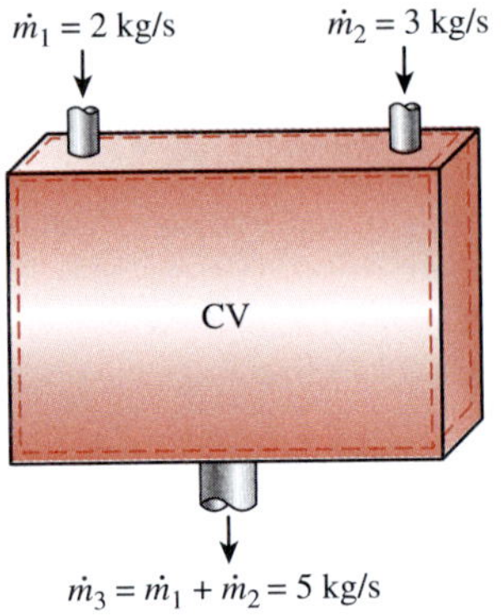

FIGURE 5–7
Conservation of mass principle for a two-inlet–one-outlet steady-flow system.

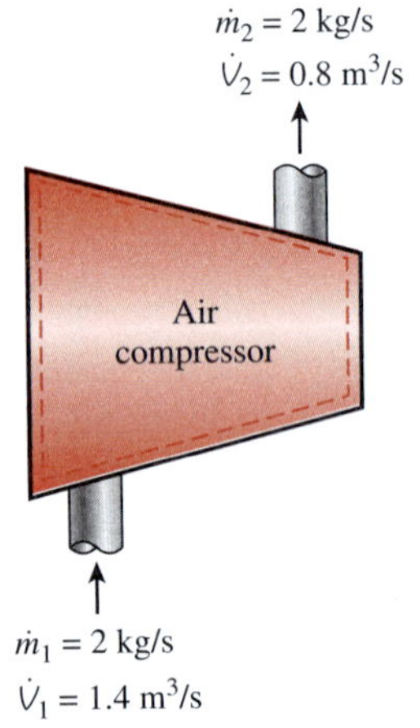

FIGURE 5–8
During a steady-flow process, volume flow rates are not necessarily conserved although mass flow rates are.

entering a control volume equal the total amount of mass leaving it. For a garden hose nozzle in steady operation, for example, the amount of water entering the nozzle per unit time is equal to the amount of water leaving it per unit time.

When dealing with steady-flow processes, we are not interested in the amount of mass that flows in or out of a device over time; instead, we are interested in the amount of mass flowing per unit time, that is, *the mass flow rate* $\dot{m}$. *The conservation of mass principle* for a general steady-flow system with multiple inlets and outlets can be expressed in rate form as (Fig. 5–7)

Steady flow: $$\sum_{\text{in}} \dot{m} = \sum_{\text{out}} \dot{m} \qquad \text{(kg/s)} \tag{5–18}$$

It states that *the total rate of mass entering a control volume is equal to the total rate of mass leaving it.*

Many engineering devices such as nozzles, diffusers, turbines, compressors, and pumps involve a single stream (only one inlet and one outlet). For these cases, we denote the inlet state by the subscript 1 and the outlet state by the subscript 2, and drop the summation signs. Then Eq. 5–18 reduces, for *single-stream steady-flow systems*, to

Steady flow (single stream): $$\dot{m}_1 = \dot{m}_2 \quad \rightarrow \quad \rho_1 V_1 A_1 = \rho_2 V_2 A_2 \tag{5–19}$$

Special Case: Incompressible Flow

The conservation of mass relations is simplified even further when the fluid is incompressible, which is usually the case for liquids. Canceling the density from both sides of the general steady-flow relation gives

Steady, incompressible flow: $$\sum_{\text{in}} \dot{V} = \sum_{\text{out}} \dot{V} \qquad (\text{m}^3/\text{s}) \tag{5–20}$$

For single-stream steady-flow systems it becomes

Steady, incompressible flow (single stream): $$\dot{V}_1 = \dot{V}_2 \longrightarrow V_1 A_1 = V_2 A_2 \tag{5–21}$$

It should always be kept in mind that there is no such thing as a "conservation of volume" principle. Therefore, the volume flow rates into and out of a steady-flow device may be different. The volume flow rate at the outlet of an air compressor is much less than that at the inlet even though the mass flow rate of air through the compressor is constant (Fig. 5–8). This is due to the higher density of air at the compressor exit. For steady flow of liquids, however, the volume flow rates, as well as the mass flow rates, remain nearly constant since liquids are essentially incompressible (constant-density) substances. Water flow through the nozzle of a garden hose is an example of the latter case.

The conservation of mass principle requires every bit of mass to be accounted for during a process. If you can balance your checkbook (by keeping track of deposits and withdrawals, or by simply observing the "conservation of money" principle), you should have no difficulty applying the conservation of mass principle to engineering systems.

EXAMPLE 5–1 Water Flow through a Garden Hose Nozzle

A garden hose attached with a nozzle is used to fill a 10-gal bucket. The inner diameter of the hose is 2 cm, and it reduces to 0.8 cm at the nozzle exit (Fig. 5–9). If it takes 50 s to fill the bucket with water, determine (*a*) the volume and mass flow rates of water through the hose, and (*b*) the average velocity of water at the nozzle exit.

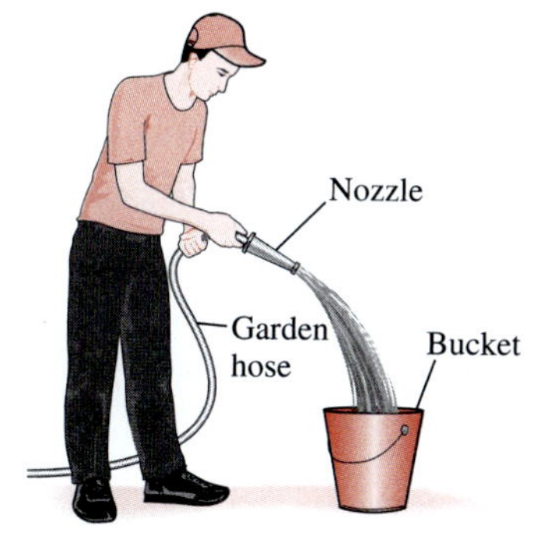

FIGURE 5–9
Schematic for Example 5–1.

SOLUTION A garden hose is used to fill a water bucket. The volume and mass flow rates of water and the exit velocity are to be determined.
Assumptions **1** Water is an incompressible substance. **2** Flow through the hose is steady. **3** There is no waste of water by splashing.
Properties We take the density of water to be 1000 kg/m³ = 1 kg/L.
Analysis (*a*) Noting that 10 gal of water are discharged in 50 s, the volume and mass flow rates of water are

$$\dot{V} = \frac{V}{\Delta t} = \frac{10\text{ gal}}{50\text{ s}}\left(\frac{3.7854\text{ L}}{1\text{ gal}}\right) = \mathbf{0.757\ L/s}$$

$$\dot{m} = \rho\dot{V} = (1\text{ kg/L})(0.757\text{ L/s}) = \mathbf{0.757\ kg/s}$$

(*b*) The cross-sectional area of the nozzle exit is

$$A_e = \pi r_e^2 = \pi(0.4\text{ cm})^2 = 0.5027\text{ cm}^2 = 0.5027 \times 10^{-4}\text{ m}^2$$

The volume flow rate through the hose and the nozzle is constant. Then the average velocity of water at the nozzle exit becomes

$$V_e = \frac{\dot{V}}{A_e} = \frac{0.757\text{ L/s}}{0.5027 \times 10^{-4}\text{ m}^2}\left(\frac{1\text{ m}^3}{1000\text{ L}}\right) = \mathbf{15.1\ m/s}$$

Discussion It can be shown that the average velocity in the hose is 2.4 m/s. Therefore, the nozzle increases the water velocity by over six times.

EXAMPLE 5–2 Discharge of Water from a Tank

A 4-ft-high, 3-ft-diameter cylindrical water tank whose top is open to the atmosphere is initially filled with water. Now the discharge plug near the bottom of the tank is pulled out, and a water jet whose diameter is 0.5 in streams out (Fig. 5–10). The average velocity of the jet is approximated as $V = \sqrt{2gh}$, where h is the height of water in the tank measured from the center of the hole (a variable) and g is the gravitational acceleration. Determine how long it takes for the water level in the tank to drop to 2 ft from the bottom.

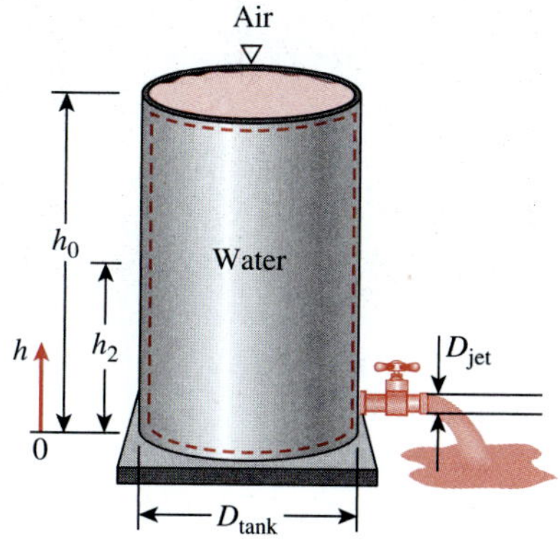

FIGURE 5–10
Schematic for Example 5–2.

SOLUTION The plug near the bottom of a water tank is pulled out. The time it takes for half of the water in the tank to empty is to be determined.
Assumptions **1** Water is an incompressible substance. **2** The distance between the bottom of the tank and the center of the hole is negligible compared to the total water height. **3** The gravitational acceleration is 32.2 ft/s².
Analysis We take the volume occupied by water as the control volume. The size of the control volume decreases in this case as the water level drops,

and thus this is a variable control volume. (We could also treat this as a fixed control volume that consists of the interior volume of the tank by disregarding the air that replaces the space vacated by the water.) This is obviously an unsteady-flow problem since the properties (such as the amount of mass) within the control volume change with time.

The conservation of mass relation for a control volume undergoing any process is given in rate form as

$$\dot{m}_{\text{in}} - \dot{m}_{\text{out}} = \frac{dm_{\text{CV}}}{dt} \tag{1}$$

During this process no mass enters the control volume ($\dot{m}_{\text{in}} = 0$), and the mass flow rate of discharged water can be expressed as

$$\dot{m}_{\text{out}} = (\rho VA)_{\text{out}} = \rho\sqrt{2gh}A_{\text{jet}} \tag{2}$$

where $A_{\text{jet}} = \pi D^2{}_{\text{jet}}/4$ is the cross-sectional area of the jet, which is constant. Noting that the density of water is constant, the mass of water in the tank at any time is

$$m_{\text{CV}} = \rho = \rho A_{\text{tank}}h \tag{3}$$

where $A_{\text{tank}} = \pi D^2{}_{\text{tank}}/4$ is the base area of the cylindrical tank. Substituting Eqs. 2 and 3 into the mass balance relation (Eq. 1) gives

$$-\rho\sqrt{2gh}A_{\text{jet}} = \frac{d(\rho A_{\text{tank}}h)}{dt} \rightarrow -\rho\sqrt{2gh}(\pi D_{\text{jet}}^2/4) = \frac{\rho(\pi D_{\text{tank}}^2/4)\,dh}{dt}$$

Canceling the densities and other common terms and separating the variables give

$$dt = \frac{D_{\text{tank}}^2}{D_{\text{jet}}^2}\frac{dh}{\sqrt{2gh}}$$

Integrating from $t = 0$ at which $h = h_0$ to $t = t$ at which $h = h_2$ gives

$$\int_0^t dt = -\frac{D_{\text{tank}}^2}{D_{\text{jet}}^2\sqrt{2g}}\int_{h_0}^{h_2}\frac{dh}{\sqrt{h}} \rightarrow t = \frac{\sqrt{h_0} - \sqrt{h_2}}{\sqrt{g/2}}\left(\frac{D_{\text{tank}}}{D_{\text{jet}}}\right)^2$$

Substituting, the time of discharge is

$$t = \frac{\sqrt{4\text{ ft}} - \sqrt{2\text{ ft}}}{\sqrt{32.2/2\text{ ft/s}^2}}\left(\frac{3 \times 12\text{ in}}{0.5\text{ in}}\right)^2 = 757\text{ s} = \mathbf{12.6\ min}$$

Therefore, half of the tank is emptied in 12.6 min after the discharge hole is unplugged.

Discussion Using the same relation with $h_2 = 0$ gives $t = 43.1$ min for the discharge of the entire amount of water in the tank. Therefore, emptying the bottom half of the tank takes much longer than emptying the top half. This is due to the decrease in the average discharge velocity of water with decreasing h.

5–2 ▪ FLOW WORK AND THE ENERGY OF A FLOWING FLUID

Unlike closed systems, control volumes involve mass flow across their boundaries, and some work is required to push the mass into or out of the control volume. This work is known as the **flow work**, or **flow energy**, and is necessary for maintaining a continuous flow through a control volume.

To obtain a relation for flow work, consider a fluid element of volume V as shown in Fig. 5–11. The fluid immediately upstream forces this fluid element to enter the control volume; thus, it can be regarded as an imaginary piston. The fluid element can be chosen to be sufficiently small so that it has uniform properties throughout.

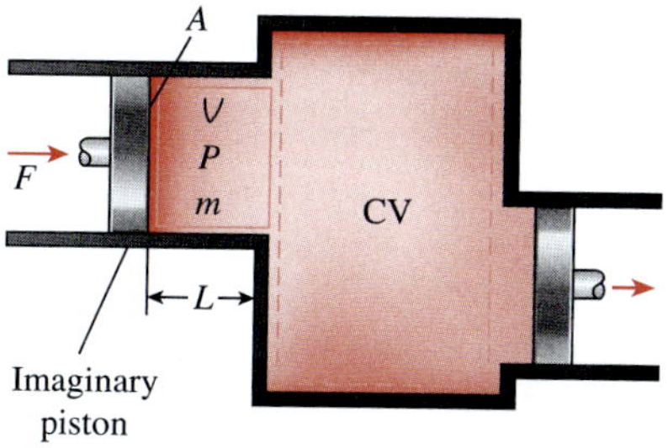

FIGURE 5–11
Schematic for flow work.

If the fluid pressure is P and the cross-sectional area of the fluid element is A (Fig. 5–12), the force applied on the fluid element by the imaginary piston is

$$F = PA \qquad \text{(5–22)}$$

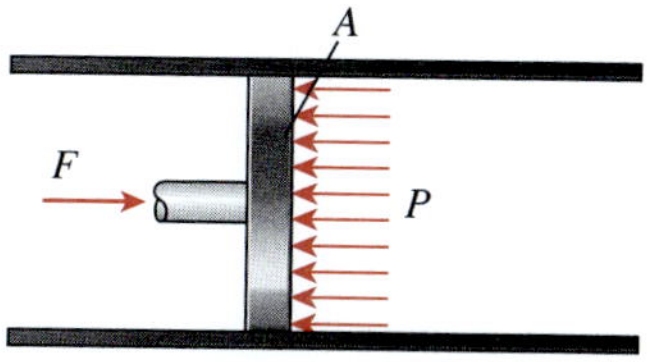

FIGURE 5–12
In the absence of acceleration, the force applied on a fluid by a piston is equal to the force applied on the piston by the fluid.

To push the entire fluid element into the control volume, this force must act through a distance L. Thus, the work done in pushing the fluid element across the boundary (i.e., the flow work) is

$$W_{\text{flow}} = FL = PAL = PV \qquad \text{(kJ)} \qquad \text{(5–23)}$$

The flow work per unit mass is obtained by dividing both sides of this equation by the mass of the fluid element:

$$w_{\text{flow}} = Pv \qquad \text{(kJ/kg)} \qquad \text{(5–24)}$$

The flow work relation is the same whether the fluid is pushed into or out of the control volume (Fig. 5–13).

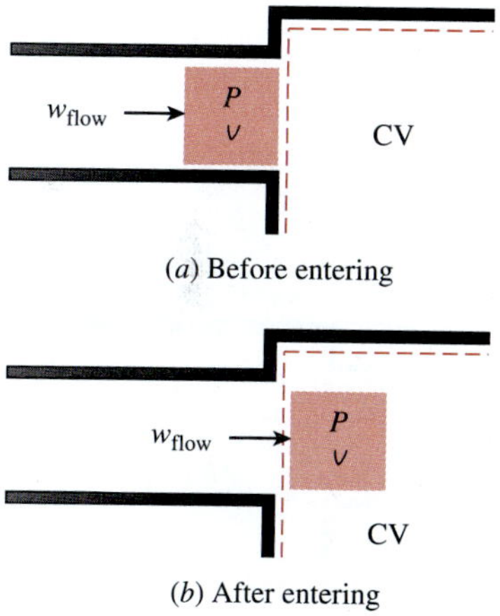

FIGURE 5–13
Flow work is the energy needed to push a fluid into or out of a control volume, and it is equal to Pv.

It is interesting that unlike other work quantities, flow work is expressed in terms of properties. In fact, it is the product of two properties of the fluid. For that reason, some people view it as a *combination property* (like enthalpy) and refer to it as *flow energy*, *convected energy*, or *transport energy* instead of flow work. Others, however, argue rightfully that the product PV represents energy for flowing fluids only and does not represent any form of energy for nonflow (closed) systems. Therefore, it should be treated as work. This controversy is not likely to end, but it is comforting to know that both arguments yield the same result for the energy balance equation. In the discussions that follow, we consider the flow energy to be part of the energy of a flowing fluid, since this greatly simplifies the energy analysis of control volumes.

Total Energy of a Flowing Fluid

As we discussed in Chap. 2, the total energy of a simple compressible system consists of three parts: internal, kinetic, and potential energies (Fig. 5–14). On a unit-mass basis, it is expressed as

$$e = u + \text{ke} + \text{pe} = u + \frac{V^2}{2} + gz \qquad \text{(kJ/kg)} \qquad \text{(5–25)}$$

where V is the velocity and z is the elevation of the system relative to some external reference point.

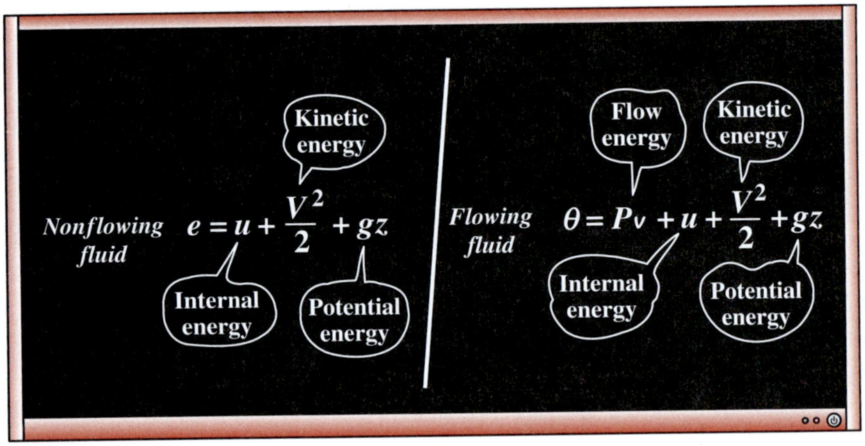

FIGURE 5–14
The total energy consists of three parts for a nonflowing fluid and four parts for a flowing fluid.

The fluid entering or leaving a control volume possesses an additional form of energy—the *flow energy* Pv, as already discussed. Then the total energy of a **flowing fluid** on a unit-mass basis (denoted by θ) becomes

$$\theta = Pv + e = Pv + (u + \text{ke} + \text{pe}) \tag{5–26}$$

But the combination $Pv + u$ has been previously defined as the enthalpy h. So the relation in Eq. 5–26 reduces to

$$\theta = h + \text{ke} + \text{pe} = h + \frac{V^2}{2} + gz \qquad (\text{kJ/kg}) \tag{5–27}$$

By using the enthalpy instead of the internal energy to represent the energy of a flowing fluid, one does not need to be concerned about the flow work. The energy associated with pushing the fluid into or out of the control volume is automatically taken care of by enthalpy. In fact, this is the main reason for defining the property enthalpy. From now on, the energy of a fluid stream flowing into or out of a control volume is represented by Eq. 5–27, and no reference will be made to flow work or flow energy.

Energy Transport by Mass

Noting that θ is total energy per unit mass, the total energy of a flowing fluid of mass m is simply $m\theta$, provided that the properties of the mass m are uniform. Also, when a fluid stream with uniform properties is flowing at a mass flow rate of $\dot{m}$, the rate of energy flow with that stream is $\dot{m}\theta$ (Fig. 5–15). That is,

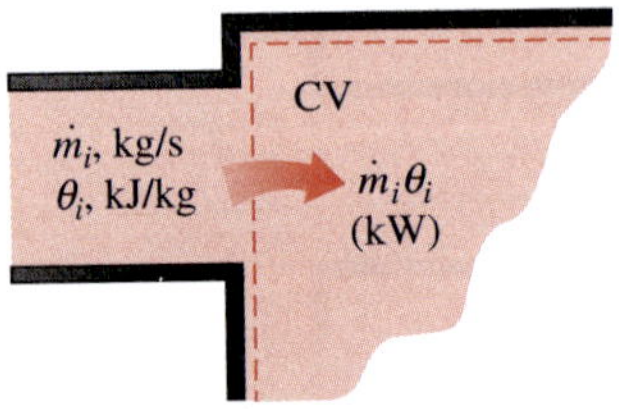

FIGURE 5–15
The product $\dot{m}_i\theta_i$ is the energy transported into control volume by mass per unit time.

Amount of energy transport: $$E_{\text{mass}} = m\theta = m\left(h + \frac{V^2}{2} + gz\right) \qquad (\text{kJ}) \tag{5–28}$$

Rate of energy transport: $$\dot{E}_{\text{mass}} = \dot{m}\theta = \dot{m}\left(h + \frac{V^2}{2} + gz\right) \qquad (\text{kW}) \tag{5–29}$$

When the kinetic and potential energies of a fluid stream are negligible, as is often the case, these relations simplify to $E_{\text{mass}} = mh$ and $\dot{E}_{\text{mass}} = \dot{m}h$.

In general, the total energy transported by mass into or out of the control volume is not easy to determine since the properties of the mass at each inlet or exit may be changing with time as well as over the cross section.

Thus, the only way to determine the energy transport through an opening as a result of mass flow is to consider sufficiently small differential masses δm that have uniform properties and to add their total energies during flow.

Again noting that θ is total energy per unit mass, the total energy of a flowing fluid of mass δm is $\theta\, \delta m$. Then the total energy transported by mass through an inlet or exit ($m_i\theta_i$ and $m_e\theta_e$) is obtained by integration. At an inlet, for example, it becomes

$$E_{\text{in,mass}} = \int_{m_i} \theta_i\, \delta m_i = \int_{m_i} \left(h_i + \frac{V_i^2}{2} + gz_i \right) \delta m_i \qquad \text{(5–30)}$$

Most flows encountered in practice can be approximated as being steady and one-dimensional, and thus the simple relations in Eqs. 5–28 and 5–29 can be used to represent the energy transported by a fluid stream.

EXAMPLE 5–3 Energy Transport by Mass

Steam is leaving a 4-L pressure cooker whose operating pressure is 150 kPa (Fig. 5–16). It is observed that the amount of liquid in the cooker has decreased by 0.6 L in 40 min after the steady operating conditions are established, and the cross-sectional area of the exit opening is 8 mm². Determine (*a*) the mass flow rate of the steam and the exit velocity, (*b*) the total and flow energies of the steam per unit mass, and (*c*) the rate at which energy leaves the cooker by steam.

SOLUTION Steam leaves a pressure cooker at a specified pressure. The velocity, flow rate, the total and flow energies, and the rate of energy transfer by mass are to be determined.

Assumptions **1** The flow is steady, and the initial start-up period is disregarded. **2** The kinetic and potential energies are negligible, and thus they are not considered. **3** Saturation conditions exist within the cooker at all times so that steam leaves the cooker as a saturated vapor at the cooker pressure.

Properties The properties of saturated liquid water and water vapor at 150 kPa are v_f = 0.001053 m³/kg, v_g = 1.1594 m³/kg, u_g = 2519.2 kJ/kg, and h_g = 2693.1 kJ/kg (Table A–5).

Analysis (*a*) Saturation conditions exist in a pressure cooker at all times after the steady operating conditions are established. Therefore, the liquid has the properties of saturated liquid and the exiting steam has the properties of saturated vapor at the operating pressure. The amount of liquid that has evaporated, the mass flow rate of the exiting steam, and the exit velocity are

$$m = \frac{\Delta V_{\text{liquid}}}{v_f} = \frac{0.6\text{ L}}{0.001053\text{ m}^3/\text{kg}} \left(\frac{1\text{ m}^3}{1000\text{ L}} \right) = 0.570\text{ kg}$$

$$\dot{m} = \frac{m}{\Delta t} = \frac{0.570\text{ kg}}{40\text{ min}} = 0.0142\text{ kg/min} = \mathbf{2.37 \times 10^{-4}\ kg/s}$$

$$V = \frac{\dot{m}}{\rho_g A_c} = \frac{\dot{m} v_g}{A_c} = \frac{(2.37 \times 10^{-4}\text{ kg/s})(1.1594\text{ m}^3/\text{kg})}{8 \times 10^{-6}\text{ m}^2} = \mathbf{34.3\ m/s}$$

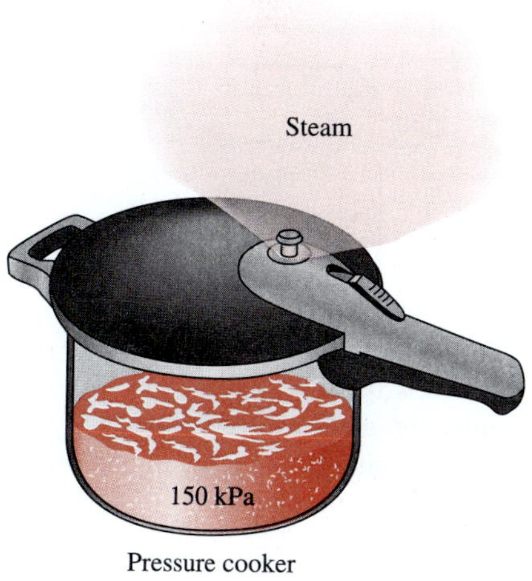

FIGURE 5–16
Schematic for Example 5–3.

(*b*) Noting that $h = u + Pv$ and that the kinetic and potential energies are disregarded, the flow and total energies of the exiting steam are

$$e_{\text{flow}} = Pv = h - u = 2693.1 - 2519.2 = \mathbf{173.9\ kJ/kg}$$

$$\theta = h + \text{ke} + \text{pe} \cong h = \mathbf{2693.1\ kJ/kg}$$

Note that the kinetic energy in this case is $\text{ke} = V^2/2 = (34.3\ \text{m/s})^2/2 = 588\ \text{m}^2/\text{s}^2 = 0.588\ \text{kJ/kg}$, which is small compared to enthalpy.

(*c*) The rate at which energy is leaving the cooker by mass is simply the product of the mass flow rate and the total energy of the exiting steam per unit mass,

$$\dot{E}_{\text{mass}} = \dot{m}\theta = (2.37 \times 10^{-4}\ \text{kg/s})(2693.1\ \text{kJ/kg}) = 0.638\ \text{kJ/s} = \mathbf{0.638\ kW}$$

Discussion The numerical value of the energy leaving the cooker with steam alone does not mean much since this value depends on the reference point selected for enthalpy (it could even be negative). The significant quantity is the difference between the enthalpies of the exiting vapor and the liquid inside (which is h_{fg}) since it relates directly to the amount of energy supplied to the cooker.

FIGURE 5–17
Many engineering systems such as power plants operate under steady conditions.
© Vol. 57/PhotoDisc/Getty RF.

5–3 ▪ ENERGY ANALYSIS OF STEADY-FLOW SYSTEMS

A large number of engineering devices such as turbines, compressors, and nozzles operate for long periods of time under the same conditions once the transient start-up period is completed and steady operation is established, and they are classified as *steady-flow devices* (Fig. 5–17). Processes involving such devices can be represented reasonably well by a somewhat idealized process, called the **steady-flow process**, which was defined in Chap. 1 as *a process during which a fluid flows through a control volume steadily.* That is, the fluid properties can change from point to point within the control volume, but at any point, they remain constant during the entire process. (Remember, *steady* means *no change with time.*)

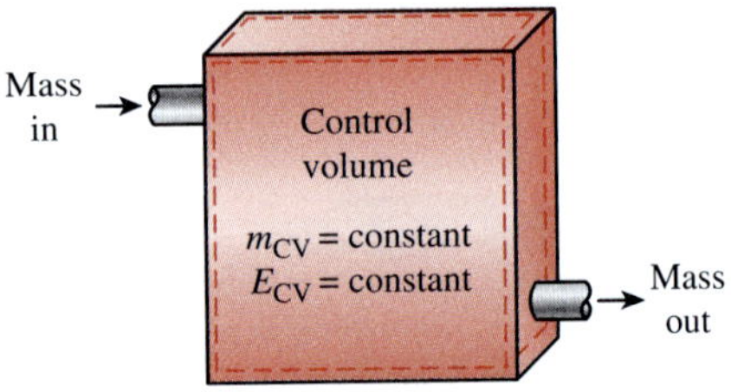

FIGURE 5–18
Under steady-flow conditions, the mass and energy contents of a control volume remain constant.

During a steady-flow process, no intensive or extensive properties *within the control volume* change with time. Thus, the volume V, the mass m, and the total energy content E of the control volume remain constant (Fig. 5–18). As a result, the boundary work is zero for steady-flow systems (since V_{CV} = constant), and the total mass or energy entering the control volume must be equal to the total mass or energy leaving it (since m_{CV} = constant and E_{CV} = constant). These observations greatly simplify the analysis.

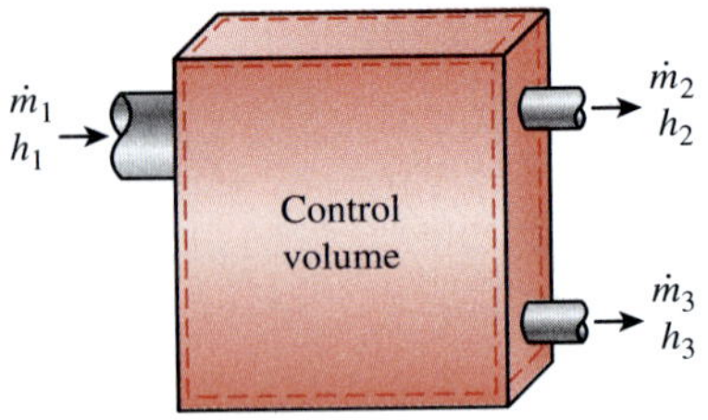

FIGURE 5–19
Under steady-flow conditions, the fluid properties at an inlet or exit remain constant (do not change with time).

The fluid properties at an inlet or exit remain constant during a steady-flow process. The properties may, however, be different at different inlets and exits. They may even vary over the cross section of an inlet or an exit. However, all properties, including the velocity and elevation, must remain constant with time at a fixed point at an inlet or exit. It follows that the mass flow rate of the fluid at an opening must remain constant during a steady-flow process (Fig. 5–19). As an added simplification, the fluid properties at an opening are usually considered to be uniform (at some average value) over the cross section. Thus, the fluid properties at an inlet or exit may be specified by the average single values. Also, the *heat* and *work* interactions between a steady-flow system and its surroundings do not change with time. Thus, the power delivered by a system and the rate of heat transfer to or from a system remain constant during a steady-flow process.

The *mass balance* for a general steady-flow system was given in Sec. 5–1 as

$$\sum_{\text{in}} \dot{m} = \sum_{\text{out}} \dot{m} \qquad \text{(kg/s)} \tag{5–31}$$

The mass balance for a single-stream (one-inlet and one-outlet) steady-flow system was given as

$$\dot{m}_1 = \dot{m}_2 \quad \longrightarrow \quad \rho_1 V_1 A_1 = \rho_2 V_2 A_2 \tag{5–32}$$

where the subscripts 1 and 2 denote the inlet and the exit states, respectively, ρ is density, V is the average flow velocity in the flow direction, and A is the cross-sectional area normal to flow direction.

During a steady-flow process, the total energy content of a control volume remains constant (E_{CV} = constant), and thus the change in the total energy of the control volume is zero ($\Delta E_{CV} = 0$). Therefore, the amount of energy entering a control volume in all forms (by heat, work, and mass) must be equal to the amount of energy leaving it. Then the rate form of the general energy balance reduces for a steady-flow process to

$$\underbrace{\dot{E}_{\text{in}} - \dot{E}_{\text{out}}}_{\substack{\text{Rate of net energy transfer} \\ \text{by heat, work, and mass}}} = \underbrace{dE_{\text{system}}/dt}_{\substack{\text{Rate of change in internal, kinetic,} \\ \text{potential, etc., energies}}} \overset{\nearrow 0 \text{ (steady)}}{} = 0 \tag{5–33}$$

or

Energy balance:

$$\underbrace{\dot{E}_{\text{in}}}_{\substack{\text{Rate of net energy transfer in} \\ \text{by heat, work, and mass}}} = \underbrace{\dot{E}_{\text{out}}}_{\substack{\text{Rate of net energy transfer out} \\ \text{by heat, work, and mass}}} \qquad \text{(kW)} \tag{5–34}$$

Noting that energy can be transferred by heat, work, and mass only, the energy balance in Eq. 5–34 for a general steady-flow system can also be written more explicitly as

$$\dot{Q}_{\text{in}} + \dot{W}_{\text{in}} + \sum_{\text{in}} \dot{m}\theta = \dot{Q}_{\text{out}} + \dot{W}_{\text{out}} + \sum_{\text{out}} \dot{m}\theta \tag{5–35}$$

or

$$\dot{Q}_{\text{in}} + \dot{W}_{\text{in}} + \underbrace{\sum_{\text{in}} \dot{m}\left(h + \frac{V^2}{2} + gz\right)}_{\text{for each inlet}} = \dot{Q}_{\text{out}} + \dot{W}_{\text{out}} + \underbrace{\sum_{\text{out}} \dot{m}\left(h + \frac{V^2}{2} + gz\right)}_{\text{for each exit}} \tag{5–36}$$

since the energy of a flowing fluid per unit mass is $\theta = h + \text{ke} + \text{pe} = h + V^2/2 + gz$. The energy balance relation for steady-flow systems first appeared in 1859 in a German thermodynamics book written by Gustav Zeuner.

Consider, for example, an ordinary electric hot-water heater under steady operation, as shown in Fig. 5–20. A cold-water stream with a mass flow rate $\dot{m}$ is continuously flowing into the water heater, and a hot-water stream of the same mass flow rate is continuously flowing out of it. The water heater (the control volume) is losing heat to the surrounding air at a rate of $\dot{Q}_{\text{out}}$,

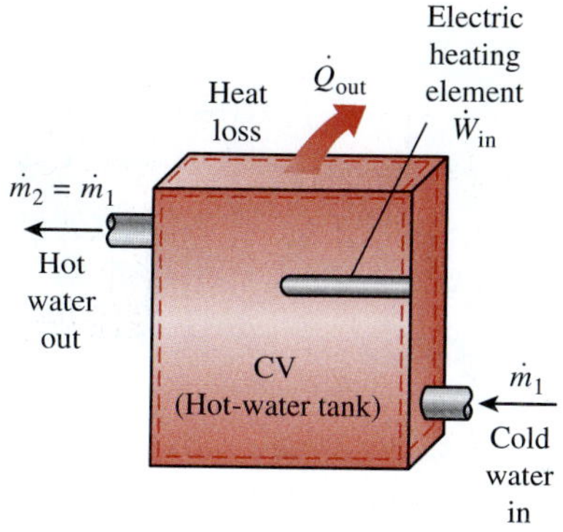

FIGURE 5–20
A water heater in steady operation.

and the electric heating element is supplying electrical work (heating) to the water at a rate of $\dot{W}_{\text{in}}$. On the basis of the conservation of energy principle, we can say that the water stream experiences an increase in its total energy as it flows through the water heater that is equal to the electric energy supplied to the water minus the heat losses.

The energy balance relation just given is intuitive in nature and is easy to use when the magnitudes and directions of heat and work transfers are known. When performing a general analytical study or solving a problem that involves an unknown heat or work interaction, however, we need to assume a direction for the heat or work interactions. In such cases, it is common practice to assume heat to be transferred *into the system* (heat input) at a rate of $\dot{Q}$, and work produced *by the system* (work output) at a rate of $\dot{W}$, and then solve the problem. The first-law or energy balance relation in that case for a general steady-flow system becomes

$$\dot{Q} - \dot{W} = \underbrace{\sum_{\text{out}} \dot{m}\left(h + \frac{V^2}{2} + gz\right)}_{\text{for each exit}} - \underbrace{\sum_{\text{in}} \dot{m}\left(h + \frac{V^2}{2} + gz\right)}_{\text{for each inlet}} \tag{5–37}$$

Obtaining a negative quantity for $\dot{Q}$ or $\dot{W}$ simply means that the assumed direction is wrong and should be reversed. For single-stream devices, the steady-flow energy balance equation becomes

$$\dot{Q} - \dot{W} = \dot{m}\left[h_2 - h_1 + \frac{V_2^2 - V_1^2}{2} + g(z_2 - z_1)\right] \tag{5–38}$$

Dividing Eq. 5–38 by $\dot{m}$ gives the energy balance on a unit-mass basis as

$$q - w = h_2 - h_1 + \frac{V_2^2 - V_1^2}{2} + g(z_2 - z_1) \tag{5–39}$$

where $q = \dot{Q}/\dot{m}$ and $w = \dot{W}/\dot{m}$ are the heat transfer and work done per unit mass of the working fluid, respectively. When the fluid experiences negligible changes in its kinetic and potential energies (that is, $\Delta\text{ke} \cong 0$, $\Delta\text{pe} \cong 0$), the energy balance equation is reduced further to

$$q - w = h_2 - h_1 \tag{5–40}$$

The various terms appearing in the above equations are as follows:

$\dot{Q}$ = **rate of heat transfer between the control volume and its surroundings**. When the control volume is losing heat (as in the case of the water heater), $\dot{Q}$ is negative. If the control volume is well insulated (i.e., adiabatic), then $\dot{Q} = 0$.

$\dot{W}$ = **power**. For steady-flow devices, the control volume is constant; thus, there is no boundary work involved. The work required to push mass into and out of the control volume is also taken care of by using enthalpies for the energy of fluid streams instead of internal energies. Then $\dot{W}$ represents the remaining forms of work done per unit time (Fig. 5–21). Many steady-flow devices, such as turbines, compressors, and pumps, transmit power through a shaft, and $\dot{W}$ simply becomes the shaft power for those devices. If the control surface is crossed by electric wires (as in the case of an electric water heater), $\dot{W}$ represents the electrical work done per unit time. If neither is present, then $\dot{W} = 0$.

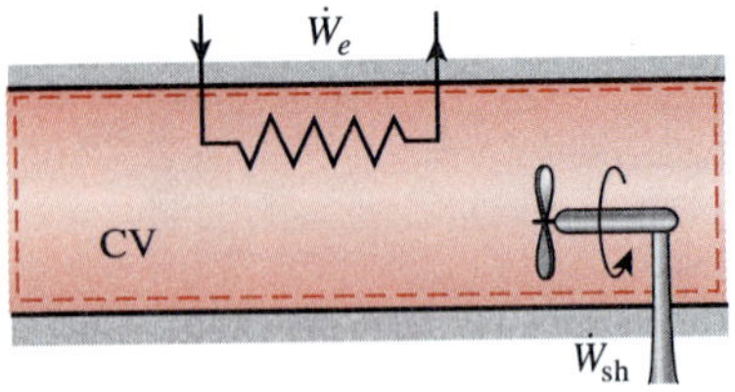

FIGURE 5–21

Under steady operation, shaft work and electrical work are the only forms of work a simple compressible system may involve.

$\Delta h = h_2 - h_1$. The enthalpy change of a fluid can easily be determined by reading the enthalpy values at the exit and inlet states from the tables. For ideal gases, it can be approximated by $\Delta h = c_{p,\text{avg}}(T_2 - T_1)$. Note that $(\text{kg/s})(\text{kJ/kg}) \equiv \text{kW}$.

$\Delta \text{ke} = (V_2^2 - V_1^2)/2$. The unit of kinetic energy is m^2/s^2, which is equivalent to J/kg (Fig. 5–22). The enthalpy is usually given in kJ/kg. To add these two quantities, the kinetic energy should be expressed in kJ/kg. This is easily accomplished by dividing it by 1000. A velocity of 45 m/s corresponds to a kinetic energy of only 1 kJ/kg, which is a very small value compared with the enthalpy values encountered in practice. Thus, the kinetic energy term at low velocities can be neglected. When a fluid stream enters and leaves a steady-flow device at about the same velocity ($V_1 \cong V_2$), the change in the kinetic energy is close to zero regardless of the velocity. Caution should be exercised at high velocities, however, since small changes in velocities may cause significant changes in kinetic energy (Fig. 5–23).

$\Delta \text{pe} = g(z_2 - z_1)$. A similar argument can be given for the potential energy term. A potential energy change of 1 kJ/kg corresponds to an elevation difference of 102 m. The elevation difference between the inlet and exit of most industrial devices such as turbines and compressors is well below this value, and the potential energy term is always neglected for these devices. The only time the potential energy term is significant is when a process involves pumping a fluid to high elevations and we are interested in the required pumping power.

$$\frac{\text{J}}{\text{kg}} \equiv \frac{\text{N}\cdot\text{m}}{\text{kg}} \equiv \left(\text{kg}\frac{\text{m}}{\text{s}^2}\right)\frac{\text{m}}{\text{kg}} \equiv \frac{\text{m}^2}{\text{s}^2}$$

$$\left(\text{Also, } \frac{\text{Btu}}{\text{lbm}} \equiv 25{,}037\ \frac{\text{ft}^2}{\text{s}^2}\right)$$

FIGURE 5–22
The units m^2/s^2 and J/kg are equivalent.

V_1 m/s	V_2 m/s	Δke kJ/kg
0	45	1
50	67	1
100	110	1
200	205	1
500	502	1

FIGURE 5–23
At very high velocities, even small changes in velocities can cause significant changes in the kinetic energy of the fluid.

5–4 ▪ SOME STEADY-FLOW ENGINEERING DEVICES

Many engineering devices operate essentially under the same conditions for long periods of time. The components of a steam power plant (turbines, compressors, heat exchangers, and pumps), for example, operate nonstop for months before the system is shut down for maintenance (Fig. 5–24). Therefore, these devices can be conveniently analyzed as steady-flow devices.

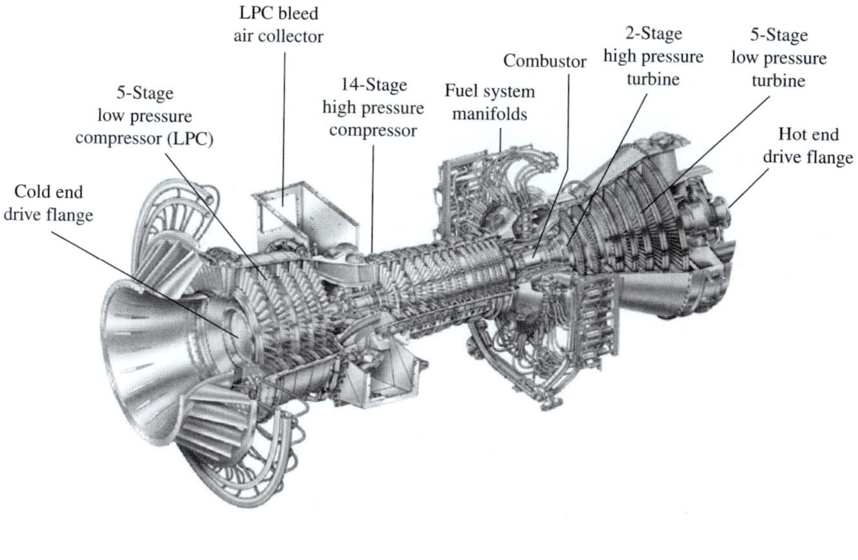

FIGURE 5–24
A modern land-based gas turbine used for electric power production. This is a General Electric LM5000 turbine. It has a length of 6.2 m, it weighs 12.5 tons, and produces 55.2 MW at 3600 rpm with steam injection.
Courtesy of GE Power Systems.

In this section, some common steady-flow devices are described, and the thermodynamic aspects of the flow through them are analyzed. The conservation of mass and the conservation of energy principles for these devices are illustrated with examples.

1 Nozzles and Diffusers

Nozzles and diffusers are commonly utilized in jet engines, rockets, spacecraft, and even garden hoses. A **nozzle** is a device that *increases the velocity of a fluid* at the expense of pressure. A **diffuser** is a device that *increases the pressure of a fluid* by slowing it down. That is, nozzles and diffusers perform opposite tasks. The cross-sectional area of a nozzle decreases in the flow direction for subsonic flows and increases for supersonic flows. The reverse is true for diffusers.

The rate of heat transfer between the fluid flowing through a nozzle or a diffuser and the surroundings is usually very small ($\dot{Q} \approx 0$) since the fluid has high velocities, and thus it does not spend enough time in the device for any significant heat transfer to take place. Nozzles and diffusers typically involve no work ($\dot{W} = 0$) and any change in potential energy is negligible ($\Delta\text{pe} \cong 0$). But nozzles and diffusers usually involve very high velocities, and as a fluid passes through a nozzle or diffuser, it experiences large changes in its velocity (Fig. 5–25). Therefore, the kinetic energy changes must be accounted for in analyzing the flow through these devices ($\Delta\text{ke} \neq 0$).

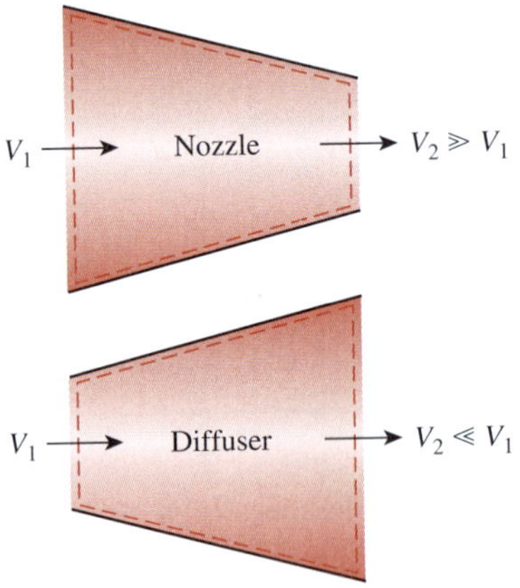

FIGURE 5–25
Nozzles and diffusers are shaped so that they cause large changes in fluid velocities and thus kinetic energies.

EXAMPLE 5–4 Deceleration of Air in a Diffuser

Air at 10°C and 80 kPa enters the diffuser of a jet engine steadily with a velocity of 200 m/s. The inlet area of the diffuser is 0.4 m². The air leaves the diffuser with a velocity that is very small compared with the inlet velocity. Determine (*a*) the mass flow rate of the air and (*b*) the temperature of the air leaving the diffuser.

SOLUTION Air enters the diffuser of a jet engine steadily at a specified velocity. The mass flow rate of air and the temperature at the diffuser exit are to be determined.

Assumptions **1** This is a steady-flow process since there is no change with time at any point and thus $\Delta m_{CV} = 0$ and $\Delta E_{CV} = 0$. **2** Air is an ideal gas since it is at a high temperature and low pressure relative to its critical-point values. **3** The potential energy change is zero, $\Delta\text{pe} = 0$. **4** Heat transfer is negligible. **5** Kinetic energy at the diffuser exit is negligible. **6** There are no work interactions.

Analysis We take the *diffuser* as the system (Fig. 5–26). This is a *control volume* since mass crosses the system boundary during the process. We observe that there is only one inlet and one exit and thus $\dot{m}_1 = \dot{m}_2 = \dot{m}$.

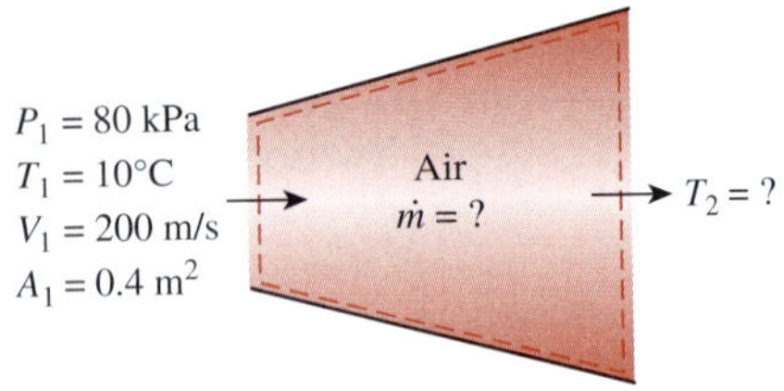

FIGURE 5–26
Schematic for Example 5–4.

(*a*) To determine the mass flow rate, we need to find the specific volume of the air first. This is determined from the ideal-gas relation at the inlet conditions:

$$v_1 = \frac{RT_1}{P_1} = \frac{0.287\ \text{kPa}\cdot\text{m}^3/\text{kg}\cdot\text{K})(283\ \text{K})}{80\ \text{kPa}} = 1.015\ \text{m}^3/\text{kg}$$

Then,

$$\dot{m} = \frac{1}{v_1} V_1 A_1 = \frac{1}{1.015 \text{ m}^3/\text{kg}} (200 \text{ m/s})(0.4 \text{ m}^2) = \mathbf{78.8 \text{ kg/s}}$$

Since the flow is steady, the mass flow rate through the entire diffuser remains constant at this value.

(*b*) Under stated assumptions and observations, the energy balance for this steady-flow system can be expressed in the rate form as

$$\underbrace{\dot{E}_{in} - \dot{E}_{out}}_{\substack{\text{Rate of net energy transfer} \\ \text{by heat, work, and mass}}} = \underbrace{dE_{system}/dt}_{\substack{\text{Rate of change in internal, kinetic,} \\ \text{potential, etc., energies}}} \overset{\nearrow 0 \text{ (steady)}}{} = 0$$

$$\dot{E}_{in} = \dot{E}_{out}$$

$$\dot{m}\left(h_1 + \frac{V_1^2}{2}\right) = \dot{m}\left(h_2 + \frac{V_2^2}{2}\right) \qquad (\text{since } \dot{Q} \cong 0, \dot{W} = 0, \text{ and } \Delta\text{pe} \cong 0)$$

$$h_2 = h_1 - \frac{V_2^2 - V_1^2}{2}$$

The exit velocity of a diffuser is usually small compared with the inlet velocity ($V_2 \ll V_1$); thus, the kinetic energy at the exit can be neglected. The enthalpy of air at the diffuser inlet is determined from the air table (Table A–17) to be

$$h_1 = h_{@ 283 \text{ K}} = 283.14 \text{ kJ/kg}$$

Substituting, we get

$$h_2 = 283.14 \text{ kJ/kg} - \frac{0 - (200 \text{ m/s})^2}{2}\left(\frac{1 \text{ kJ/kg}}{1000 \text{ m}^2/\text{s}^2}\right)$$

$$= 303.14 \text{ kJ/kg}$$

From Table A–17, the temperature corresponding to this enthalpy value is

$$T_2 = \mathbf{303 \text{ K}}$$

Discussion This result shows that the temperature of the air increases by about 20°C as it is slowed down in the diffuser. The temperature rise of the air is mainly due to the conversion of kinetic energy to internal energy.

EXAMPLE 5–5 Acceleration of Steam in a Nozzle

Steam at 250 psia and 700°F steadily enters a nozzle whose inlet area is 0.2 ft². The mass flow rate of steam through the nozzle is 10 lbm/s. Steam leaves the nozzle at 200 psia with a velocity of 900 ft/s. Heat losses from the nozzle per unit mass of the steam are estimated to be 1.2 Btu/lbm. Determine (*a*) the inlet velocity and (*b*) the exit temperature of the steam.

SOLUTION Steam enters a nozzle steadily at a specified flow rate and velocity. The inlet velocity of steam and the exit temperature are to be determined.

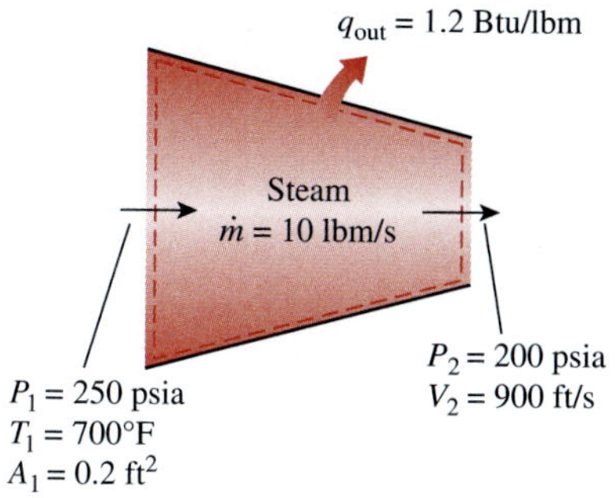

FIGURE 5–26
Schematic for Example 5–5.

Assumptions **1** This is a steady-flow process since there is no change with time at any point and thus $\Delta m_{CV} = 0$ and $\Delta E_{CV} = 0$. **2** There are no work interactions. **3** The potential energy change is zero, $\Delta pe = 0$.

Analysis We take the *nozzle* as the system (Fig. 5–26). This is a *control volume* since mass crosses the system boundary during the process. We observe that there is only one inlet and one exit and thus $\dot{m}_1 = \dot{m}_2 = \dot{m}$.

(*a*) The specific volume and enthalpy of steam at the nozzle inlet are

$$\left.\begin{aligned} P_1 &= 250 \text{ psia} \\ T_1 &= 700°\text{F} \end{aligned}\right\} \quad \begin{aligned} v_1 &= 2.6883 \text{ ft}^3/\text{lbm} \\ h_1 &= 1371.4 \text{ Btu/lbm} \end{aligned} \qquad \text{(Table A–6E)}$$

Then,

$$\dot{m} = \frac{1}{v_1} V_1 A_1$$

$$10 \text{ lbm/s} = \frac{1}{2.6883 \text{ ft}^3/\text{lbm}} (V_1)(0.2 \text{ ft}^2)$$

$$V_1 = \mathbf{134.4 \text{ ft/s}}$$

(*b*) Under stated assumptions and observations, the energy balance for this steady-flow system can be expressed in the rate form as

$$\underbrace{\dot{E}_{in} - \dot{E}_{out}}_{\substack{\text{Rate of net energy transfer} \\ \text{by heat, work, and mass}}} = \underbrace{dE_{system}/dt}_{\substack{\text{Rate of change in internal, kinetic,} \\ \text{potential, etc., energies}}} \overset{\nearrow 0 \text{ (steady)}}{} = 0$$

$$\dot{E}_{in} = \dot{E}_{out}$$

$$\dot{m}\left(h_1 + \frac{V_1^2}{2}\right) = \dot{Q}_{out} + \dot{m}\left(h_2 + \frac{V_2^2}{2}\right) \qquad (\text{since } \dot{W} = 0, \text{ and } \Delta pe \cong 0)$$

Dividing by the mass flow rate $\dot{m}$ and substituting, h_2 is determined to be

$$h_2 = h_1 - q_{out} - \frac{V_2^2 - V_1^2}{2}$$

$$= (1371.4 - 1.2) \text{ Btu/lbm} - \frac{(900 \text{ ft/s})^2 - (134.4 \text{ ft/s})^2}{2}\left(\frac{1 \text{ Btu/lbm}}{25{,}037 \text{ ft}^2/\text{s}^2}\right)$$

$$= 1354.4 \text{ Btu/lbm}$$

Then,

$$\left.\begin{aligned} P_2 &= 200 \text{ psia} \\ h_2 &= 1354.4 \text{ Btu/lbm} \end{aligned}\right\} \quad T_2 = \mathbf{662.0°F} \qquad \text{(Table A–6E)}$$

Discussion Note that the temperature of steam drops by 38.0°F as it flows through the nozzle. This drop in temperature is mainly due to the conversion of internal energy to kinetic energy. (The heat loss is too small to cause any significant effect in this case.)

2 Turbines and Compressors

In steam, gas, or hydroelectric power plants, the device that drives the electric generator is the turbine. As the fluid passes through the turbine, work is done against the blades, which are attached to the shaft. As a result, the shaft rotates, and the turbine produces work.

Compressors, as well as pumps and fans, are devices used to increase the pressure of a fluid. Work is supplied to these devices from an external source through a rotating shaft. Therefore, compressors involve work inputs. Even though these three devices function similarly, they do differ in the tasks they perform. A *fan* increases the pressure of a gas slightly and is mainly used to mobilize a gas. A *compressor* is capable of compressing the gas to very high pressures. *Pumps* work very much like compressors except that they handle liquids instead of gases.

Note that turbines produce power output whereas compressors, pumps, and fans require power input. Heat transfer from turbines is usually negligible ($\dot{Q} \approx 0$) since they are typically well insulated. Heat transfer is also negligible for compressors unless there is intentional cooling. Potential energy changes are negligible for all of these devices ($\Delta \text{pe} \cong 0$). The velocities involved in these devices, with the exception of turbines and fans, are usually too low to cause any significant change in the kinetic energy ($\Delta \text{ke} \cong 0$). The fluid velocities encountered in most turbines are very high, and the fluid experiences a significant change in its kinetic energy. However, this change is usually very small relative to the change in enthalpy, and thus it is often disregarded.

EXAMPLE 5–6 Compressing Air by a Compressor

Air at 100 kPa and 280 K is compressed steadily to 600 kPa and 400 K. The mass flow rate of the air is 0.02 kg/s, and a heat loss of 16 kJ/kg occurs during the process. Assuming the changes in kinetic and potential energies are negligible, determine the necessary power input to the compressor.

SOLUTION Air is compressed steadily by a compressor to a specified temperature and pressure. The power input to the compressor is to be determined.

Assumptions **1** This is a steady-flow process since there is no change with time at any point and thus $\Delta m_{CV} = 0$ and $\Delta E_{CV} = 0$. **2** Air is an ideal gas since it is at a high temperature and low pressure relative to its critical-point values. **3** The kinetic and potential energy changes are zero, $\Delta \text{ke} = \Delta \text{pe} = 0$.

Analysis We take the *compressor* as the system (Fig. 5–27). This is a *control volume* since mass crosses the system boundary during the process. We observe that there is only one inlet and one exit and thus $\dot{m}_1 = \dot{m}_2 = \dot{m}$. Also, heat is lost from the system and work is supplied to the system.

Under stated assumptions and observations, the energy balance for this steady-flow system can be expressed in the rate form as

$$\underbrace{\dot{E}_{in} - \dot{E}_{out}}_{\substack{\text{Rate of net energy transfer}\\ \text{by heat, work, and mass}}} = \underbrace{dE_{system}/dt}_{\substack{\text{Rate of change in internal, kinetic,}\\ \text{potential, etc., energies}}} \nearrow^{0\ \text{(steady)}} = 0$$

$$\dot{E}_{in} = \dot{E}_{out}$$

$$\dot{W}_{in} + \dot{m}h_1 = \dot{Q}_{out} + \dot{m}h_2 \qquad (\text{since } \Delta \text{ke} = \Delta \text{pe} \cong 0)$$

$$\dot{W}_{in} = \dot{m}q_{out} + \dot{m}(h_2 - h_1)$$

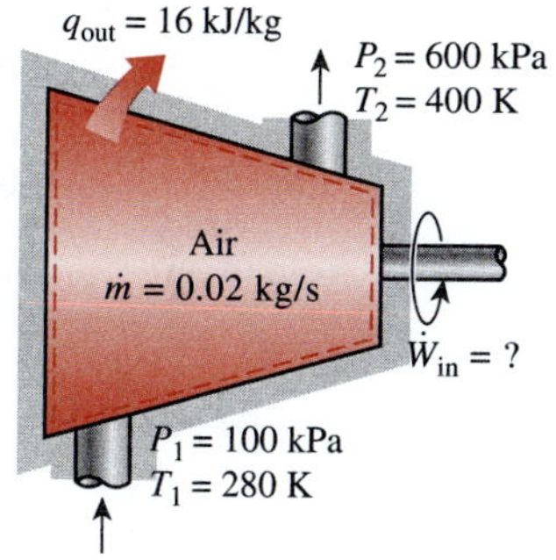

FIGURE 5–27
Schematic for Example 5–6.

The enthalpy of an ideal gas depends on temperature only, and the enthalpies of the air at the specified temperatures are determined from the air table (Table A–17) to be

$$h_1 = h_{@\ 280\ \text{K}} = 280.13\ \text{kJ/kg}$$

$$h_2 = h_{@\ 400\ \text{K}} = 400.98\ \text{kJ/kg}$$

Substituting, the power input to the compressor is determined to be

$$\dot{W}_{\text{in}} = (0.02\ \text{kg/s})(16\ \text{kJ/kg}) + (0.02\ \text{kg/s})(400.98 - 280.13)\ \text{kJ/kg}$$
$$= \mathbf{2.74\ kW}$$

Discussion Note that the mechanical energy input to the compressor manifests itself as a rise in enthalpy of air and heat loss from the compressor.

EXAMPLE 5–7 Power Generation by a Steam Turbine

The power output of an adiabatic steam turbine is 5 MW, and the inlet and the exit conditions of the steam are as indicated in Fig. 5–28.

(*a*) Compare the magnitudes of Δh, Δke, and Δpe.
(*b*) Determine the work done per unit mass of the steam flowing through the turbine.
(*c*) Calculate the mass flow rate of the steam.

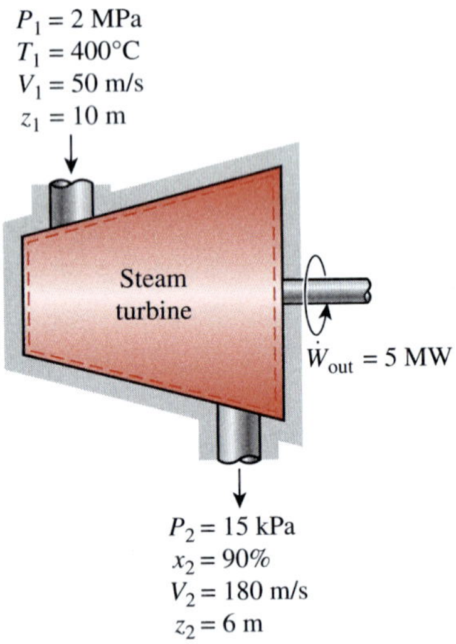

FIGURE 5–28
Schematic for Example 5–7.

SOLUTION The inlet and exit conditions of a steam turbine and its power output are given. The changes in kinetic energy, potential energy, and enthalpy of steam, as well as the work done per unit mass and the mass flow rate of steam are to be determined.

Assumptions **1** This is a steady-flow process since there is no change with time at any point and thus $\Delta m_{\text{CV}} = 0$ and $\Delta E_{\text{CV}} = 0$. **2** The system is adiabatic and thus there is no heat transfer.

Analysis We take the *turbine* as the system. This is a *control volume* since mass crosses the system boundary during the process. We observe that there is only one inlet and one exit and thus $\dot{m}_1 = \dot{m}_2 = \dot{m}$. Also, work is done by the system. The inlet and exit velocities and elevations are given, and thus the kinetic and potential energies are to be considered.

(*a*) At the inlet, steam is in a superheated vapor state, and its enthalpy is

$$\left.\begin{array}{l} P_1 = 2\ \text{MPa} \\ T_1 = 400^\circ\text{C} \end{array}\right\} \quad h_1 = 3248.4\ \text{kJ/kg} \qquad \text{(Table A–6)}$$

At the turbine exit, we obviously have a saturated liquid–vapor mixture at 15-kPa pressure. The enthalpy at this state is

$$h_2 = h_f + x_2 h_{fg} = [225.94 + (0.9)(2372.3)]\ \text{kJ/kg} = 2361.01\ \text{kJ/kg}$$

Then

$$\Delta h = h_2 - h_1 = (2361.01 - 3248.4)\ \text{kJ/kg} = \mathbf{-887.39\ kJ/kg}$$

$$\Delta\text{ke} = \frac{V_2^2 - V_1^2}{2} = \frac{(180\ \text{m/s})^2 - (50\ \text{m/s})^2}{2}\left(\frac{1\ \text{kJ/kg}}{1000\ \text{m}^2/\text{s}^2}\right) = \mathbf{14.95\ kJ/kg}$$

$$\Delta\text{pe} = g(z_2 - z_1) = (9.81\ \text{m/s}^2)[(6 - 10)\ \text{m}]\left(\frac{1\ \text{kJ/kg}}{1000\ \text{m}^2/\text{s}^2}\right) = \mathbf{-0.04\ kJ/kg}$$

(*b*) The energy balance for this steady-flow system can be expressed in the rate form as

$$\underbrace{\dot{E}_{\text{in}} - \dot{E}_{\text{out}}}_{\substack{\text{Rate of net energy transfer} \\ \text{by heat, work, and mass}}} = \underbrace{dE_{\text{system}}/dt}_{\substack{\text{Rate of change in internal, kinetic,} \\ \text{potential, etc., energies}}} \overset{\nearrow 0 \text{ (steady)}}{} = 0$$

$$\dot{E}_{\text{in}} = \dot{E}_{\text{out}}$$

$$\dot{m}\left(h_1 + \frac{V_1^2}{2} + gz_1\right) = \dot{W}_{\text{out}} + \dot{m}\left(h_2 + \frac{V_2^2}{2} + gz_2\right) \qquad (\text{since } \dot{Q} = 0)$$

Dividing by the mass flow rate $\dot{m}$ and substituting, the work done by the turbine per unit mass of the steam is determined to be

$$w_{\text{out}} = -\left[(h_2 - h_1) + \frac{V_2^2 - V_1^2}{2} + g(z_2 - z_1)\right] = -(\Delta h + \Delta \text{ke} + \Delta \text{pe})$$

$$= -[-887.39 + 14.95 - 0.04] \text{ kJ/kg} = \mathbf{872.48 \text{ kJ/kg}}$$

(*c*) The required mass flow rate for a 5-MW power output is

$$\dot{m} = \frac{\dot{W}_{\text{out}}}{w_{\text{out}}} = \frac{5000 \text{ kJ/s}}{872.48 \text{ kJ/kg}} = \mathbf{5.73 \text{ kg/s}}$$

Discussion Two observations can be made from these results. First, the change in potential energy is insignificant in comparison to the changes in enthalpy and kinetic energy. This is typical for most engineering devices. Second, as a result of low pressure and thus high specific volume, the steam velocity at the turbine exit can be very high. Yet the change in kinetic energy is a small fraction of the change in enthalpy (less than 2 percent in our case) and is therefore often neglected.

3 Throttling Valves

Throttling valves are *any kind of flow-restricting devices* that cause a significant pressure drop in the fluid. Some familiar examples are ordinary adjustable valves, capillary tubes, and porous plugs (Fig. 5–29). Unlike turbines, they produce a pressure drop without involving any work. The pressure drop in the fluid is often accompanied by a *large drop in temperature*, and for that reason throttling devices are commonly used in refrigeration and air-conditioning applications. The magnitude of the temperature drop (or, sometimes, the temperature rise) during a throttling process is governed by a property called the *Joule-Thomson coefficient*, discussed in Chap. 12.

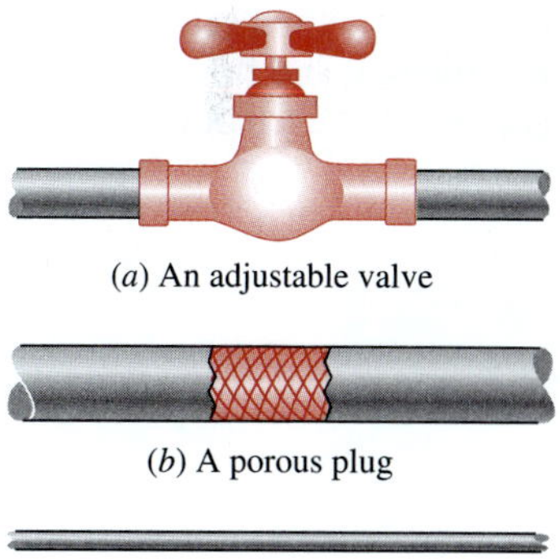

(*a*) An adjustable valve

(*b*) A porous plug

(*c*) A capillary tube

FIGURE 5–29
Throttling valves are devices that cause large pressure drops in the fluid.

Throttling valves are usually small devices, and the flow through them may be assumed to be adiabatic ($q \cong 0$) since there is neither sufficient time nor large enough area for any effective heat transfer to take place. Also, there is no work done ($w = 0$), and the change in potential energy, if any, is very small ($\Delta\text{pe} \cong 0$). Even though the exit velocity is often considerably higher than the inlet velocity, in many cases, the increase in kinetic energy is insignificant ($\Delta\text{ke} \cong 0$). Then the conservation of energy equation for this single-stream steady-flow device reduces to

$$h_2 \cong h_1 \qquad (\text{kJ/kg}) \tag{5–41}$$

That is, enthalpy values at the inlet and exit of a throttling valve are the same. For this reason, a throttling valve is sometimes called an *isenthalpic device*. Note, however, that for throttling devices with large exposed surface areas such as capillary tubes, heat transfer may be significant.

To gain some insight into how throttling affects fluid properties, let us express Eq. 5–41 as follows:

$$u_1 + P_1 v_1 = u_2 + P_2 v_2$$

or

$$\text{Internal energy} + \text{Flow energy} = \text{Constant}$$

Thus the final outcome of a throttling process depends on which of the two quantities increases during the process. If the flow energy increases during the process ($P_2 v_2 > P_1 v_1$), it can do so at the expense of the internal energy. As a result, internal energy decreases, which is usually accompanied by a drop in temperature. If the product Pv decreases, the internal energy and the temperature of a fluid will increase during a throttling process. In the case of an ideal gas, $h = h(T)$, and thus the temperature has to remain constant during a throttling process (Fig. 5–30).

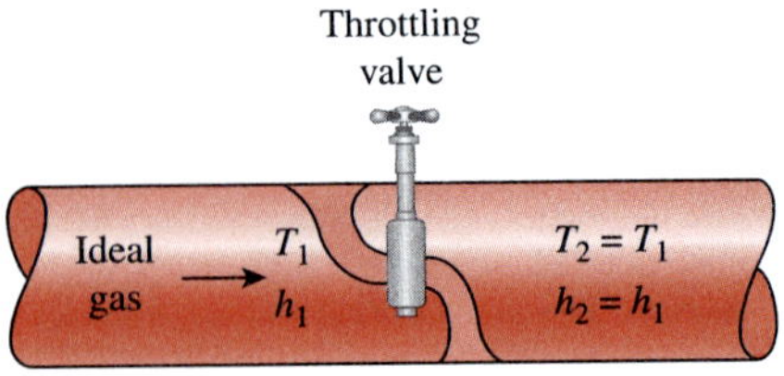

FIGURE 5–30
The temperature of an ideal gas does not change during a throttling (h = constant) process since $h = h(T)$.

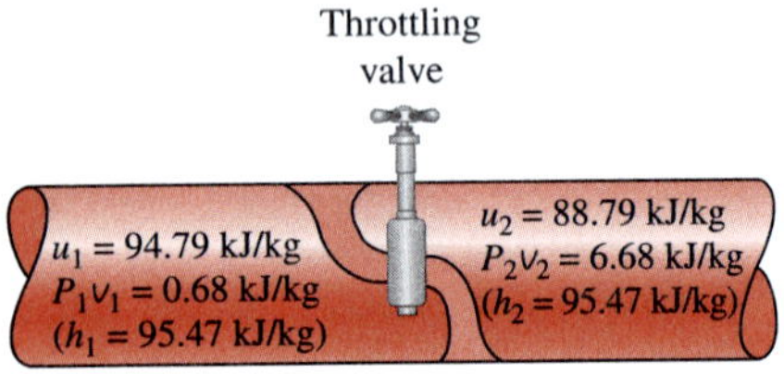

FIGURE 5–31
During a throttling process, the enthalpy (flow energy + internal energy) of a fluid remains constant. But internal and flow energies may be converted to each other.

EXAMPLE 5–8 Expansion of Refrigerant-134a in a Refrigerator

Refrigerant-134a enters the capillary tube of a refrigerator as saturated liquid at 0.8 MPa and is throttled to a pressure of 0.12 MPa. Determine the quality of the refrigerant at the final state and the temperature drop during this process.

SOLUTION Refrigerant-134a that enters a capillary tube as saturated liquid is throttled to a specified pressure. The exit quality of the refrigerant and the temperature drop are to be determined.

Assumptions **1** Heat transfer from the tube is negligible. **2** Kinetic energy change of the refrigerant is negligible.

Analysis A capillary tube is a simple flow-restricting device that is commonly used in refrigeration applications to cause a large pressure drop in the refrigerant. Flow through a capillary tube is a throttling process; thus, the enthalpy of the refrigerant remains constant (Fig. 5–31).

$$\textit{At inlet:}\quad \left.\begin{array}{l} P_1 = 0.8\text{ MPa} \\ \text{sat. liquid} \end{array}\right\} \quad \begin{array}{l} T_1 = T_{\text{sat @ 0.8 MPa}} = 31.31^\circ\text{C} \\ h_1 = h_{f\text{ @ 0.8 MPa}} = 95.47\text{ kJ/kg} \end{array} \qquad \text{(Table A–12)}$$

$$\textit{At exit:}\quad \begin{array}{l} P_2 = 0.12\text{ MPa} \\ (h_2 = h_1) \end{array} \quad \longrightarrow \quad \begin{array}{l} h_f = 22.49\text{ kJ/kg} \\ h_g = 236.97\text{ kJ/kg} \end{array} \qquad T_{\text{sat}} = -22.32^\circ\text{C}$$

Obviously $h_f < h_2 < h_g$; thus, the refrigerant exists as a saturated mixture at the exit state. The quality at this state is

$$x_2 = \frac{h_2 - h_f}{h_{fg}} = \frac{95.47 - 22.49}{236.97 - 22.49} = \mathbf{0.340}$$

Since the exit state is a saturated mixture at 0.12 MPa, the exit temperature must be the saturation temperature at this pressure, which is −22.32°C. Then the temperature change for this process becomes

$$\Delta T = T_2 - T_1 = (-22.32 - 31.31)^\circ\text{C} = \mathbf{-53.63^\circ C}$$

Discussion Note that the temperature of the refrigerant drops by 53.63°C during this throttling process. Also note that 34.0 percent of the refrigerant vaporizes during this throttling process, and the energy needed to vaporize this refrigerant is absorbed from the refrigerant itself.

4a Mixing Chambers

In engineering applications, mixing two streams of fluids is not a rare occurrence. The section where the mixing process takes place is commonly referred to as a **mixing chamber**. The mixing chamber does not have to be a distinct "chamber." An ordinary T-elbow or a Y-elbow in a shower, for example, serves as the mixing chamber for the cold- and hot-water streams (Fig. 5–32).

The conservation of mass principle for a mixing chamber requires that the sum of the incoming mass flow rates equal the mass flow rate of the outgoing mixture.

Mixing chambers are usually well insulated ($q \cong 0$) and usually do not involve any kind of work ($w = 0$). Also, the kinetic and potential energies of the fluid streams are usually negligible (ke $\cong$ 0, pe $\cong$ 0). Then all there is left in the energy equation is the total energies of the incoming streams and the outgoing mixture. The conservation of energy principle requires that these two equal each other. Therefore, the conservation of energy equation becomes analogous to the conservation of mass equation for this case.

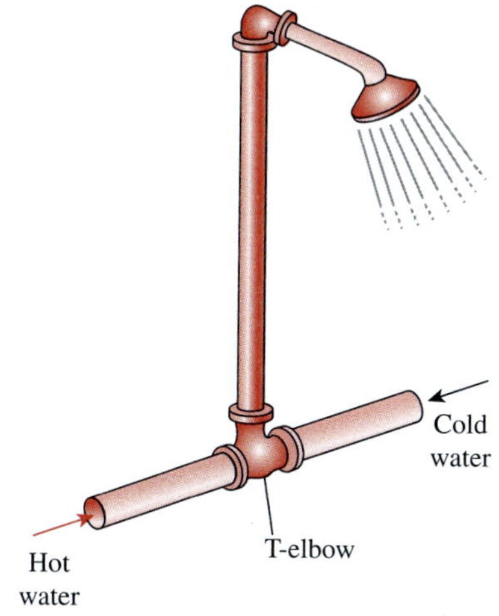

FIGURE 5–32
The T-elbow of an ordinary shower serves as the mixing chamber for the hot- and the cold-water streams.

EXAMPLE 5–9 Mixing of Hot and Cold Waters in a Shower

Consider an ordinary shower where hot water at 140°F is mixed with cold water at 50°F. If it is desired that a steady stream of warm water at 110°F be supplied, determine the ratio of the mass flow rates of the hot to cold water. Assume the heat losses from the mixing chamber to be negligible and the mixing to take place at a pressure of 20 psia.

SOLUTION In a shower, cold water is mixed with hot water at a specified temperature. For a specified mixture temperature, the ratio of the mass flow rates of the hot to cold water is to be determined.

Assumptions **1** This is a steady-flow process since there is no change with time at any point and thus $\Delta m_{CV} = 0$ and $\Delta E_{CV} = 0$. **2** The kinetic and potential energies are negligible, ke $\cong$ pe $\cong$ 0. **3** Heat losses from the system are negligible and thus $\dot{Q} \cong 0$. **4** There is no work interaction involved.

Analysis We take the *mixing chamber* as the system (Fig. 5–33). This is a *control volume* since mass crosses the system boundary during the process. We observe that there are two inlets and one exit.

Under the stated assumptions and observations, the mass and energy balances for this steady-flow system can be expressed in the rate form as follows:

Mass balance:
$$\dot{m}_{in} - \dot{m}_{out} = dm_{system}/dt^{\nearrow 0\ (steady)} = 0$$
$$\dot{m}_{in} = \dot{m}_{out} \rightarrow \dot{m}_1 + \dot{m}_2 = \dot{m}_3$$

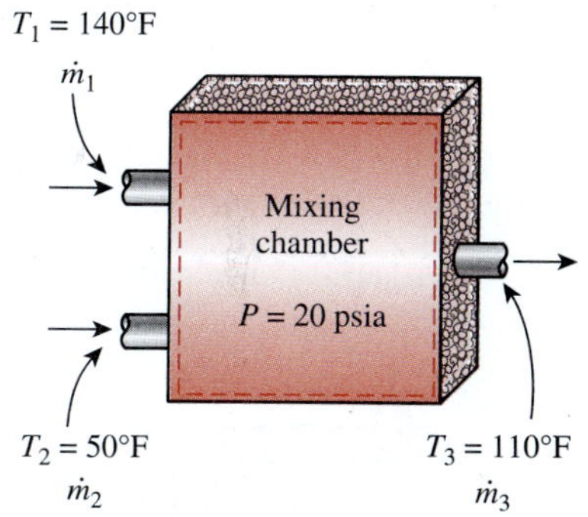

FIGURE 5–33
Schematic for Example 5–9.

$$\text{Energy balance:} \quad \underbrace{\dot{E}_{\text{in}} - \dot{E}_{\text{out}}}_{\substack{\text{Rate of net energy transfer}\\ \text{by heat, work, and mass}}} = \underbrace{dE_{\text{system}}/dt}_{\substack{\text{Rate of change in internal, kinetic,}\\ \text{potential, etc., energies}}}^{\nearrow 0 \text{ (steady)}} = 0$$

$$\dot{E}_{\text{in}} = \dot{E}_{\text{out}}$$

$$\dot{m}_1 h_1 + \dot{m}_2 h_2 = \dot{m}_3 h_3 \quad (\text{since } \dot{Q} \cong 0, \dot{W} = 0, \text{ke} \cong \text{pe} \cong 0)$$

Combining the mass and energy balances,

$$\dot{m}_1 h_1 + \dot{m}_2 h_2 = (\dot{m}_1 + \dot{m}_2) h_3$$

Dividing this equation by $\dot{m}_2$ yields

$$y h_1 + h_2 = (y + 1) h_3$$

where $y = \dot{m}_1 / \dot{m}_2$ is the desired mass flow rate ratio.

The saturation temperature of water at 20 psia is 227.92°F. Since the temperatures of all three streams are below this value ($T < T_{\text{sat}}$), the water in all three streams exists as a compressed liquid (Fig. 5–34). A compressed liquid can be approximated as a saturated liquid at the given temperature. Thus,

$$h_1 \cong h_{f\,@\,140°\text{F}} = 107.99 \text{ Btu/lbm}$$

$$h_2 \cong h_{f\,@\,50°\text{F}} = 18.07 \text{ Btu/lbm}$$

$$h_3 \cong h_{f\,@\,110°\text{F}} = 78.02 \text{ Btu/lbm}$$

Solving for y and substituting yields

$$y = \frac{h_3 - h_2}{h_1 - h_3} = \frac{78.02 - 18.07}{107.99 - 78.02} = \mathbf{2.0}$$

Discussion Note that the mass flow rate of the hot water must be twice the mass flow rate of the cold water for the mixture to leave at 110°F.

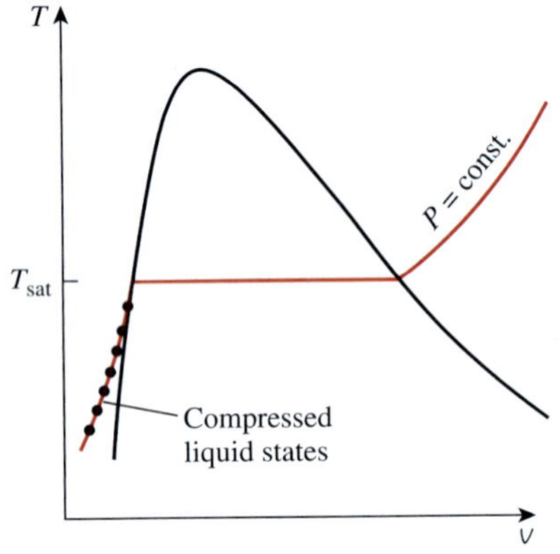

FIGURE 5–34
A substance exists as a compressed liquid at temperatures below the saturation temperatures at the given pressure.

4b Heat Exchangers

As the name implies, **heat exchangers** are devices where two moving fluid streams exchange heat without mixing. Heat exchangers are widely used in various industries, and they come in various designs.

The simplest form of a heat exchanger is a *double-tube* (also called *tube-and-shell*) *heat exchanger*, shown in Fig. 5–35. It is composed of two concentric pipes of different diameters. One fluid flows in the inner pipe, and the other in the annular space between the two pipes. Heat is transferred from the hot fluid to the cold one through the wall separating them. Sometimes the inner tube makes a couple of turns inside the shell to increase the heat transfer area, and thus the rate of heat transfer. The mixing chambers discussed earlier are sometimes classified as *direct-contact* heat exchangers.

The conservation of mass principle for a heat exchanger in steady operation requires that the sum of the inbound mass flow rates equal the sum of the outbound mass flow rates. This principle can also be expressed as follows: *Under steady operation, the mass flow rate of each fluid stream flowing through a heat exchanger remains constant.*

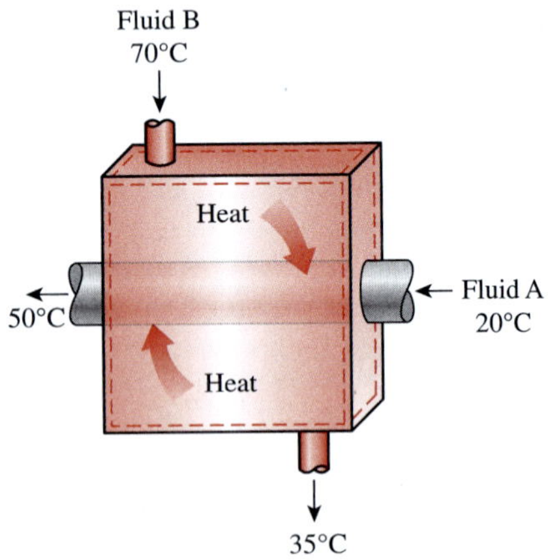

FIGURE 5–35
A heat exchanger can be as simple as two concentric pipes.

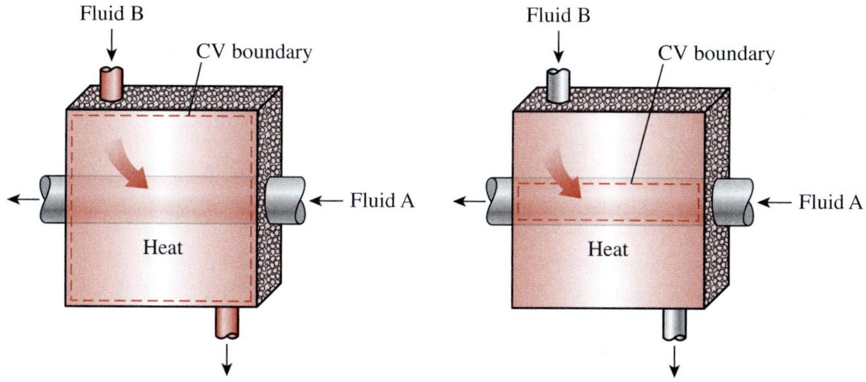

(*a*) System: Entire heat exchanger ($Q_{CV} = 0$)

(*b*) System: Fluid A ($Q_{CV} \neq 0$)

FIGURE 5–36
The heat transfer associated with a heat exchanger may be zero or nonzero depending on how the control volume is selected.

Heat exchangers typically involve no work interactions ($w = 0$) and negligible kinetic and potential energy changes ($\Delta ke \cong 0$, $\Delta pe \cong 0$) for each fluid stream. The heat transfer rate associated with heat exchangers depends on how the control volume is selected. Heat exchangers are intended for heat transfer between two fluids *within* the device, and the outer shell is usually well insulated to prevent any heat loss to the surrounding medium.

When the entire heat exchanger is selected as the control volume, $\dot{Q}$ becomes zero, since the boundary for this case lies just beneath the insulation and little or no heat crosses the boundary (Fig. 5–36). If, however, only one of the fluids is selected as the control volume, then heat will cross this boundary as it flows from one fluid to the other and $\dot{Q}$ will not be zero. In fact, $\dot{Q}$ in this case will be the rate of heat transfer between the two fluids.

EXAMPLE 5–10 Cooling of Refrigerant-134a by Water

Refrigerant-134a is to be cooled by water in a condenser. The refrigerant enters the condenser with a mass flow rate of 6 kg/min at 1 MPa and 70°C and leaves at 35°C. The cooling water enters at 300 kPa and 15°C and leaves at 25°C. Neglecting any pressure drops, determine (*a*) the mass flow rate of the cooling water required and (*b*) the heat transfer rate from the refrigerant to water.

SOLUTION Refrigerant-134a is cooled by water in a condenser. The mass flow rate of the cooling water and the rate of heat transfer from the refrigerant to the water are to be determined.

Assumptions **1** This is a steady-flow process since there is no change with time at any point and thus $\Delta m_{CV} = 0$ and $\Delta E_{CV} = 0$. **2** The kinetic and potential energies are negligible, ke ≅ pe ≅ 0. **3** Heat losses from the system are negligible and thus $\dot{Q} \cong 0$. **4** There is no work interaction.

Analysis We take the *entire heat exchanger* as the system (Fig. 5–37). This is a *control volume* since mass crosses the system boundary during the process. In general, there are several possibilities for selecting the control volume for multiple-stream steady-flow devices, and the proper choice depends on the situation at hand. We observe that there are two fluid streams (and thus two inlets and two exits) but no mixing.

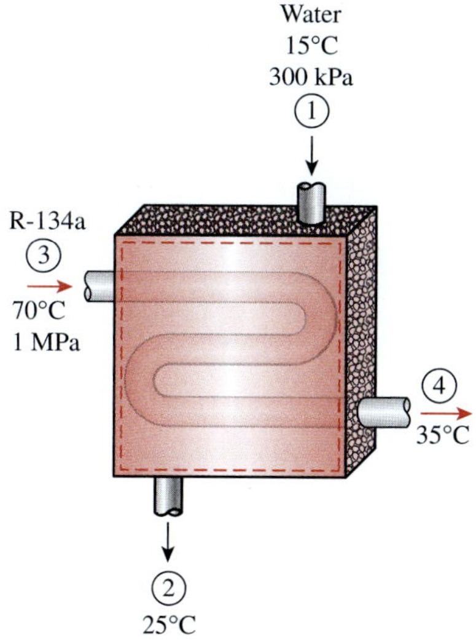

FIGURE 5–37
Schematic for Example 5–10.

(*a*) Under the stated assumptions and observations, the mass and energy balances for this steady-flow system can be expressed in the rate form as follows:

Mass balance: $$\dot{m}_{\text{in}} = \dot{m}_{\text{out}}$$

for each fluid stream since there is no mixing. Thus,

$$\dot{m}_1 = \dot{m}_2 = \dot{m}_w$$

$$\dot{m}_3 = \dot{m}_4 = \dot{m}_R$$

Energy balance:

$$\underbrace{\dot{E}_{\text{in}} - \dot{E}_{\text{out}}}_{\substack{\text{Rate of net energy transfer}\\ \text{by heat, work, and mass}}} = \underbrace{dE_{\text{system}}/dt^{\nearrow 0\text{ (steady)}}}_{\substack{\text{Rate of change in internal, kinetic,}\\ \text{potential, etc., energies}}} = 0$$

$$\dot{E}_{\text{in}} = \dot{E}_{\text{out}}$$

$$\dot{m}_1 h_1 + \dot{m}_3 h_3 = \dot{m}_2 h_2 + \dot{m}_4 h_4 \qquad (\text{since } \dot{Q} \cong 0,\ \dot{W} = 0,\ \text{ke} \cong \text{pe} \cong 0)$$

Combining the mass and energy balances and rearranging give

$$\dot{m}_w(h_1 - h_2) = \dot{m}_R(h_4 - h_3)$$

Now we need to determine the enthalpies at all four states. Water exists as a compressed liquid at both the inlet and the exit since the temperatures at both locations are below the saturation temperature of water at 300 kPa (133.52°C). Approximating the compressed liquid as a saturated liquid at the given temperatures, we have

$$\left.\begin{aligned} h_1 &\cong h_{f\,@\,15^\circ\text{C}} = 62.982 \text{ kJ/kg} \\ h_2 &\cong h_{f\,@\,25^\circ\text{C}} = 104.83 \text{ kJ/kg} \end{aligned}\right. \qquad \text{(Table A–4)}$$

The refrigerant enters the condenser as a superheated vapor and leaves as a compressed liquid at 35°C. From refrigerant-134a tables,

$$\left.\begin{aligned} P_3 &= 1 \text{ MPa} \\ T_3 &= 70^\circ\text{C} \end{aligned}\right\} \quad h_3 = 303.85 \text{ kJ/kg} \qquad \text{(Table A–13)}$$

$$\left.\begin{aligned} P_4 &= 1 \text{ MPa} \\ T_4 &= 35^\circ\text{C} \end{aligned}\right\} \quad h_4 \cong h_{f\,@\,35^\circ\text{C}} = 100.87 \text{ kJ/kg} \qquad \text{(Table A–11)}$$

Substituting, we find

$$\dot{m}_w(62.982 - 104.83) \text{ kJ/kg} = (6 \text{ kg/min})[(100.87 - 303.85) \text{ kJ/kg}]$$

$$\dot{m}_w = \mathbf{29.1\ kg/min}$$

(*b*) To determine the heat transfer from the refrigerant to the water, we have to choose a control volume whose boundary lies on the path of heat transfer. We can choose the volume occupied by either fluid as our control volume. For no particular reason, we choose the volume occupied by the water. All the assumptions stated earlier apply, except that the heat transfer is no longer zero. Then assuming heat to be transferred to water, the energy balance for this single-stream steady-flow system reduces to

$$\underbrace{\dot{E}_{\text{in}} - \dot{E}_{\text{out}}}_{\substack{\text{Rate of net energy transfer}\\ \text{by heat, work, and mass}}} = \underbrace{dE_{\text{system}}/dt^{\nearrow 0\text{ (steady)}}}_{\substack{\text{Rate of change in internal, kinetic,}\\ \text{potential, etc., energies}}} = 0$$

$$\dot{E}_{\text{in}} = \dot{E}_{\text{out}}$$

$$\dot{Q}_{w,\text{ in}} + \dot{m}_w h_1 = \dot{m}_w h_2$$

Rearranging and substituting,

$$\dot{Q}_{w,\,\text{in}} = \dot{m}_w(h_2 - h_1) = (29.1\text{ kg/min})[(104.83 - 62.982)\text{ kJ/kg}]$$
$$= \mathbf{1218\ kJ/min}$$

Discussion Had we chosen the volume occupied by the refrigerant as the control volume (Fig. 5–38), we would have obtained the same result for $\dot{Q}_{R,\text{out}}$ since the heat gained by the water is equal to the heat lost by the refrigerant.

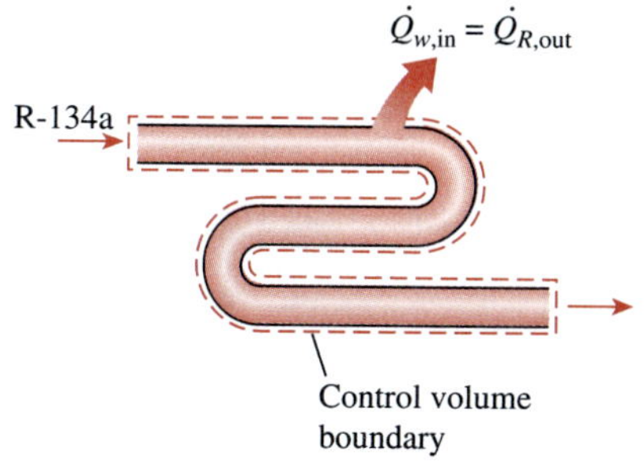

FIGURE 5–38
In a heat exchanger, the heat transfer depends on the choice of the control volume.

5 Pipe and Duct Flow

The transport of liquids or gases in pipes and ducts is of great importance in many engineering applications. Flow through a pipe or a duct usually satisfies the steady-flow conditions and thus can be analyzed as a steady-flow process. This, of course, excludes the transient start-up and shut-down periods. The control volume can be selected to coincide with the interior surfaces of the portion of the pipe or the duct that we are interested in analyzing.

Under normal operating conditions, the amount of heat gained or lost by the fluid may be very significant, particularly if the pipe or duct is long (Fig. 5–39). Sometimes heat transfer is desirable and is the sole purpose of the flow. Water flow through the pipes in the furnace of a power plant, the flow of refrigerant in a freezer, and the flow in heat exchangers are some examples of this case. At other times, heat transfer is undesirable, and the pipes or ducts are insulated to prevent any heat loss or gain, particularly when the temperature difference between the flowing fluid and the surroundings is large. Heat transfer in this case is negligible.

If the control volume involves a heating section (electric wires), a fan, or a pump (shaft), the work interactions should be considered (Fig. 5–40). Of these, fan work is usually small and often neglected in energy analysis.

The velocities involved in pipe and duct flow are relatively low, and the kinetic energy changes are usually insignificant. This is particularly true when the pipe or duct diameter is constant and the heating effects are negligible. Kinetic energy changes may be significant, however, for gas flow in ducts with variable cross-sectional areas especially when the compressibility effects are significant. The potential energy term may also be significant when the fluid undergoes a considerable elevation change as it flows in a pipe or duct.

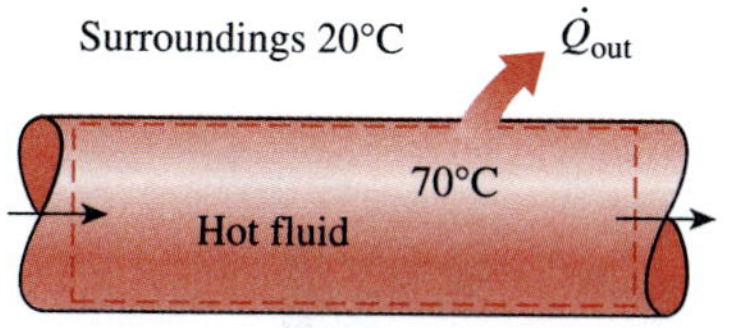

FIGURE 5–39
Heat losses from a hot fluid flowing through an uninsulated pipe or duct to the cooler environment may be very significant.

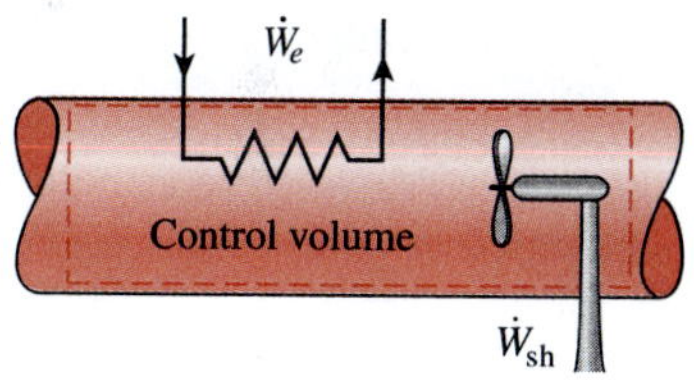

FIGURE 5–40
Pipe or duct flow may involve more than one form of work at the same time.

EXAMPLE 5–11 **Electric Heating of Air in a House**

The electric heating systems used in many houses consist of a simple duct with resistance heaters. Air is heated as it flows over resistance wires. Consider a 15-kW electric heating system. Air enters the heating section at 100 kPa and 17°C with a volume flow rate of 150 m^3/min. If heat is lost from the air in the duct to the surroundings at a rate of 200 W, determine the exit temperature of air.

SOLUTION The electric heating system of a house is considered. For specified electric power consumption and air flow rate, the air exit temperature is to be determined.

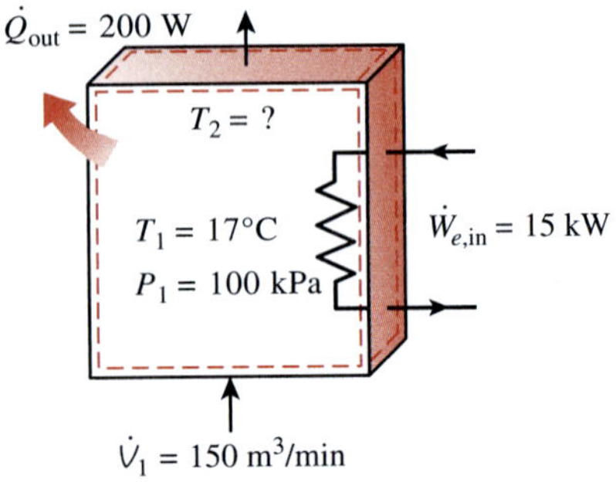

FIGURE 5–41
Schematic for Example 5–11.

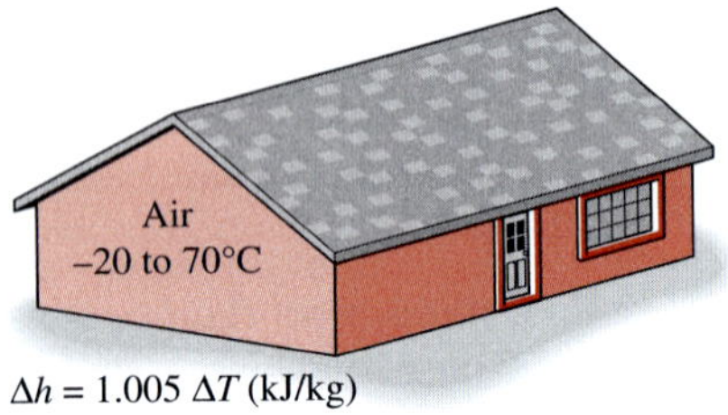

FIGURE 5–42
The error involved in $\Delta h = c_p\,\Delta T$, where c_p = 1.005 kJ/kg·°C, is less than 0.5 percent for air in the temperature range −20 to 70°C.

Assumptions **1** This is a steady-flow process since there is no change with time at any point and thus $\Delta m_{CV} = 0$ and $\Delta E_{CV} = 0$. **2** Air is an ideal gas since it is at a high temperature and low pressure relative to its critical-point values. **3** The kinetic and potential energy changes are negligible, $\Delta ke \cong \Delta pe \cong 0$. **4** Constant specific heats at room temperature can be used for air.

Analysis We take the *heating section portion of the duct* as the system (Fig. 5–41). This is a *control volume* since mass crosses the system boundary during the process. We observe that there is only one inlet and one exit and thus $\dot{m}_1 = \dot{m}_2 = \dot{m}$. Also, heat is lost from the system and electrical work is supplied to the system.

At temperatures encountered in heating and air-conditioning applications, Δh can be replaced by $c_p\,\Delta T$ where c_p = 1.005 kJ/kg·°C—the value at room temperature—with negligible error (Fig. 5–42). Then the energy balance for this steady-flow system can be expressed in the rate form as

$$\underbrace{\dot{E}_{in} - \dot{E}_{out}}_{\substack{\text{Rate of net energy transfer}\\ \text{by heat, work, and mass}}} = \underbrace{dE_{system}/dt}_{\substack{\text{Rate of change in internal, kinetic,}\\ \text{potential, etc., energies}}}\overset{\nearrow 0\ \text{(steady)}}{} = 0$$

$$\dot{E}_{in} = \dot{E}_{out}$$

$$\dot{W}_{e,in} + \dot{m}h_1 = \dot{Q}_{out} + \dot{m}h_2 \quad (\text{since } \Delta ke \cong \Delta pe \cong 0)$$

$$\dot{W}_{e,in} - \dot{Q}_{out} = \dot{m}c_p(T_2 - T_1)$$

From the ideal-gas relation, the specific volume of air at the inlet of the duct is

$$v_1 = \frac{RT_1}{P_1} = \frac{(0.287\ \text{kPa·m}^3/\text{kg·K})(290\ \text{K})}{100\ \text{kPa}} = 0.832\ \text{m}^3/\text{kg}$$

The mass flow rate of the air through the duct is determined from

$$\dot{m} = \frac{\dot{V}_1}{v_1} = \frac{150\ \text{m}^3/\text{min}}{0.832\ \text{m}^3/\text{kg}}\left(\frac{1\ \text{min}}{60\ \text{s}}\right) = 3.0\ \text{kg/s}$$

Substituting the known quantities, the exit temperature of the air is determined to be

$$(15\ \text{kJ/s}) - (0.2\ \text{kJ/s}) = (3\ \text{kg/s})(1.005\ \text{kJ/kg·°C})(T_2 - 17)\text{°C}$$

$$T_2 = \mathbf{21.9°C}$$

Discussion Note that heat loss from the duct reduces the exit temperature of air.

5–5 ▪ ENERGY ANALYSIS OF UNSTEADY-FLOW PROCESSES

During a steady-flow process, no changes occur within the control volume; thus, one does not need to be concerned about what is going on within the boundaries. Not having to worry about any changes within the control volume with time greatly simplifies the analysis.

Many processes of interest, however, involve *changes* within the control volume with time. Such processes are called *unsteady-flow,* or *transient-flow,* processes. The steady-flow relations developed earlier are obviously not applicable to these processes. When an unsteady-flow process is analyzed, it is important to keep track of the mass and energy contents of the control volume as well as the energy interactions across the boundary.

Some familiar unsteady-flow processes are the charging of rigid vessels from supply lines (Fig. 5–43), discharging a fluid from a pressurized vessel, driving a gas turbine with pressurized air stored in a large container, inflating tires or balloons, and even cooking with an ordinary pressure cooker.

Unlike steady-flow processes, unsteady-flow processes start and end over some finite time period instead of continuing indefinitely. Therefore in this section, we deal with changes that occur over some time interval Δt instead of with the rate of changes (changes per unit time). An unsteady-flow system, in some respects, is similar to a closed system, except that the mass within the system boundaries does not remain constant during a process.

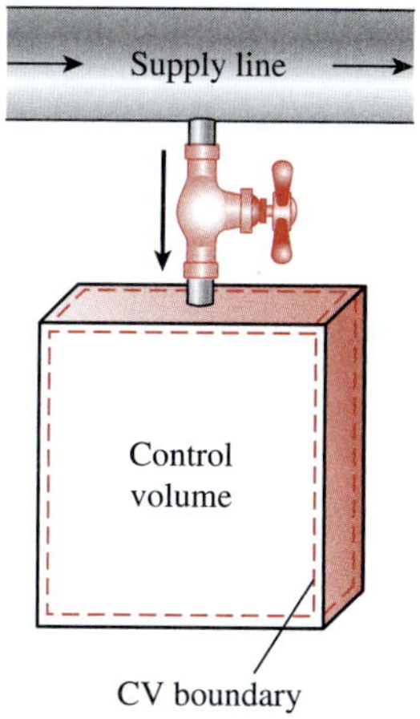

FIGURE 5–43
Charging of a rigid tank from a supply line is an unsteady-flow process since it involves changes within the control volume.

Another difference between steady- and unsteady-flow systems is that steady-flow systems are fixed in space, size, and shape. Unsteady-flow systems, however, are not (Fig. 5–44). They are usually stationary; that is, they are fixed in space, but they may involve moving boundaries and thus boundary work.

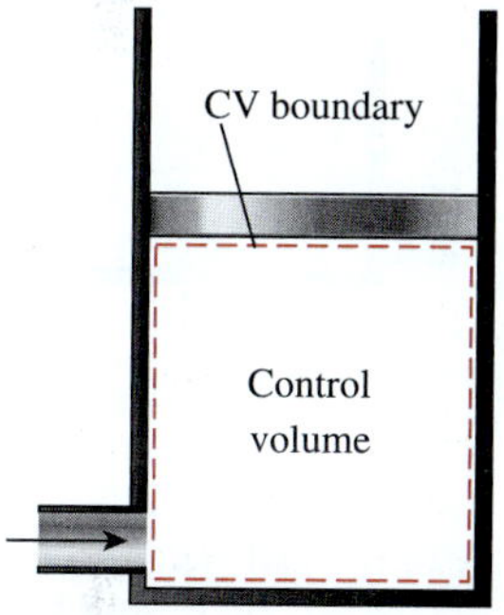

FIGURE 5–44
The shape and size of a control volume may change during an unsteady-flow process.

The *mass balance* for any system undergoing any process can be expressed as (*see* Sec. 5–1)

$$m_{\text{in}} - m_{\text{out}} = \Delta m_{\text{system}} \qquad (\text{kg}) \tag{5–42}$$

where $\Delta m_{\text{system}} = m_{\text{final}} - m_{\text{initial}}$ is the change in the mass of the system. For control volumes, it can also be expressed more explicitly as

$$m_i - m_e = (m_2 - m_1)_{\text{CV}} \tag{5–43}$$

where i = inlet, e = exit, 1 = initial state, and 2 = final state of the control volume. Often one or more terms in the equation above are zero. For example, $m_i = 0$ if no mass enters the control volume during the process, $m_e = 0$ if no mass leaves, and $m_1 = 0$ if the control volume is initially evacuated.

The energy content of a control volume changes with time during an unsteady-flow process. The magnitude of change depends on the amount of energy transfer across the system boundaries as heat and work as well as on the amount of energy transported into and out of the control volume by mass during the process. When analyzing an unsteady-flow process, we must keep track of the energy content of the control volume as well as the energies of the incoming and outgoing flow streams.

The general energy balance was given earlier as

Energy balance:
$$\underbrace{E_{\text{in}} - E_{\text{out}}}_{\substack{\text{Net energy transfer}\\ \text{by heat, work, and mass}}} = \underbrace{\Delta E_{\text{system}}}_{\substack{\text{Change in internal, kinetic,}\\ \text{potential, etc., energies}}} \qquad (\text{kJ}) \tag{5–44}$$

The general unsteady-flow process, in general, is difficult to analyze because the properties of the mass at the inlets and exits may change during a process. Most unsteady-flow processes, however, can be represented reasonably well by the **uniform-flow process**, which involves the following idealization: *The fluid flow at any inlet or exit is uniform and steady, and thus the fluid properties do not change with time or position over the cross*

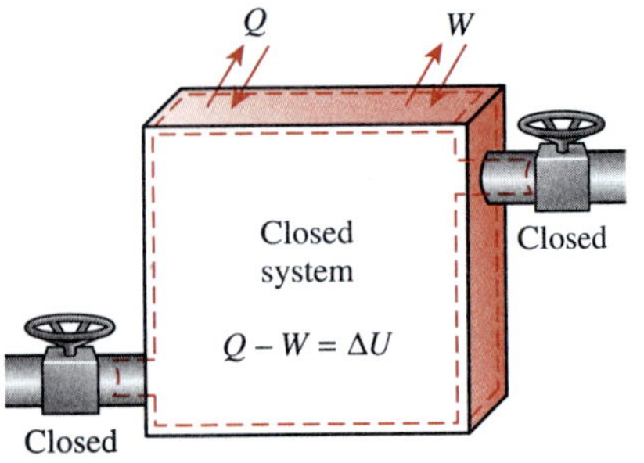

FIGURE 5–45
The energy equation of a uniform-flow system reduces to that of a closed system when all the inlets and exits are closed.

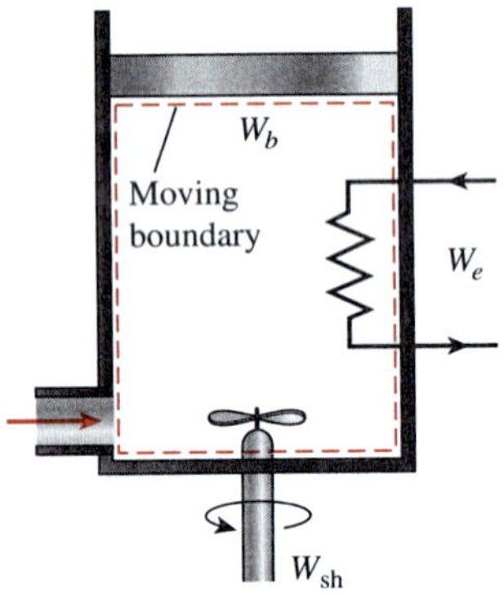

FIGURE 5–46
A uniform-flow system may involve electrical, shaft, and boundary work all at once.

section of an inlet or exit. If they do, they are averaged and treated as constants for the entire process.

Note that unlike the steady-flow systems, the state of an unsteady-flow system may change with time, and that the state of the mass leaving the control volume at any instant is the same as the state of the mass in the control volume at that instant. The initial and final properties of the control volume can be determined from the knowledge of the initial and final states, which are completely specified by two independent intensive properties for simple compressible systems.

Then the energy balance for a uniform-flow system can be expressed explicitly as

$$\left(Q_{\text{in}} + W_{\text{in}} + \sum_{\text{in}} m\theta\right) - \left(Q_{\text{out}} + W_{\text{out}} + \sum_{\text{out}} m\theta\right) = (m_2e_2 - m_1e_1)_{\text{system}} \quad \text{(5–45)}$$

where $\theta = h + \text{ke} + \text{pe}$ is the energy of a fluid stream at any inlet or exit per unit mass, and $e = u + \text{ke} + \text{pe}$ is the energy of the nonflowing fluid within the control volume per unit mass. When the kinetic and potential energy changes associated with the control volume and fluid streams are negligible, as is usually the case, the energy balance above simplifies to

$$Q - W = \sum_{\text{out}} mh - \sum_{\text{in}} mh + (m_2u_2 - m_1u_1)_{\text{system}} \quad \text{(5–46)}$$

where $Q = Q_{\text{net,in}} = Q_{\text{in}} - Q_{\text{out}}$ is the net heat input and $W = W_{\text{net,out}} = W_{\text{out}} - W_{\text{in}}$ is the net work output. Note that if no mass enters or leaves the control volume during a process ($m_i = m_e = 0$, and $m_1 = m_2 = m$), this equation reduces to the energy balance relation for closed systems (Fig. 5–45). Also note that an unsteady-flow system may involve boundary work as well as electrical and shaft work (Fig. 5–46).

Although both the steady-flow and uniform-flow processes are somewhat idealized, many actual processes can be approximated reasonably well by one of these with satisfactory results. The degree of satisfaction depends on the desired accuracy and the degree of validity of the assumptions made.

EXAMPLE 5–12 Charging of a Rigid Tank by Steam

A rigid, insulated tank that is initially evacuated is connected through a valve to a supply line that carries steam at 1 MPa and 300°C. Now the valve is opened, and steam is allowed to flow slowly into the tank until the pressure reaches 1 MPa, at which point the valve is closed. Determine the final temperature of the steam in the tank.

SOLUTION A valve connecting an initially evacuated tank to a steam line is opened, and steam flows in until the pressure inside rises to the line level. The final temperature in the tank is to be determined.

Assumptions **1** This process can be analyzed as a *uniform-flow process* since the properties of the steam entering the control volume remain constant during the entire process. **2** The kinetic and potential energies of the streams are negligible, ke ≅ pe ≅ 0. **3** The tank is stationary and thus its kinetic and potential energy changes are zero; that is, $\Delta KE = \Delta PE = 0$ and $\Delta E_{\text{system}} = \Delta U_{\text{system}}$.

4 There are no boundary, electrical, or shaft work interactions involved. **5** The tank is well insulated and thus there is no heat transfer.

Analysis We take the *tank* as the system (Fig. 5–47). This is a *control volume* since mass crosses the system boundary during the process. We observe that this is an unsteady-flow process since changes occur within the control volume. The control volume is initially evacuated and thus $m_1 = 0$ and $m_1 u_1 = 0$. Also, there is one inlet and no exits for mass flow.

Noting that microscopic energies of flowing and nonflowing fluids are represented by enthalpy h and internal energy u, respectively, the mass and energy balances for this uniform-flow system can be expressed as

$$\textit{Mass balance:} \quad m_{\text{in}} - m_{\text{out}} = \Delta m_{\text{system}} \quad \rightarrow \quad m_i = m_2 - \overset{\nearrow 0}{m_1} = m_2$$

$$\textit{Energy balance:} \quad \underbrace{E_{\text{in}} - E_{\text{out}}}_{\substack{\text{Net energy transfer} \\ \text{by heat, work, and mass}}} = \underbrace{\Delta E_{\text{system}}}_{\substack{\text{Change in internal, kinetic,} \\ \text{potential, etc., energies}}}$$

$$m_i h_i = m_2 u_2 \qquad (\text{since } W = Q = 0,\ \text{ke} \cong \text{pe} \cong 0,\ m_1 = 0)$$

Combining the mass and energy balances gives

$$u_2 = h_i$$

That is, the final internal energy of the steam in the tank is equal to the enthalpy of the steam entering the tank. The enthalpy of the steam at the inlet state is

$$\left.\begin{array}{l} P_i = 1\ \text{MPa} \\ T_i = 300^\circ\text{C} \end{array}\right\} \quad h_i = 3051.6\ \text{kJ/kg} \qquad (\text{Table A–6})$$

which is equal to u_2. Since we now know two properties at the final state, it is fixed and the temperature at this state is determined from the same table to be

$$\left.\begin{array}{l} P_2 = 1\ \text{MPa} \\ u_2 = 3051.6\ \text{kJ/kg} \end{array}\right\} \quad T_2 = \mathbf{456.1^\circ C}$$

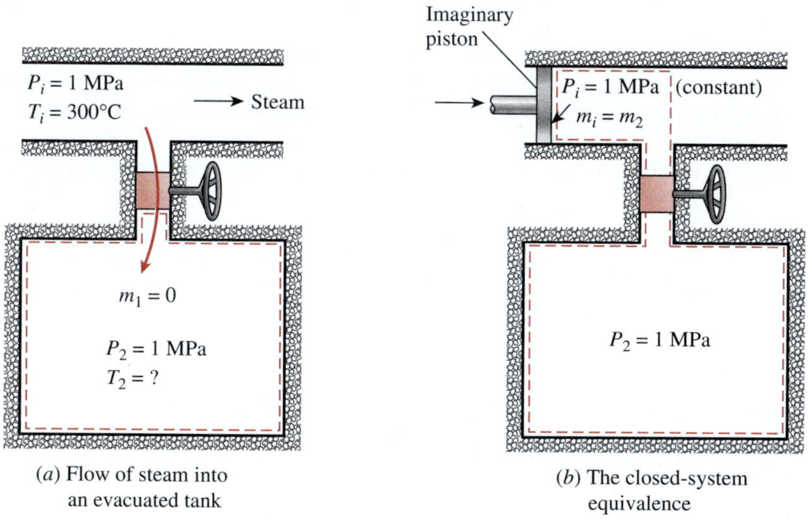

FIGURE 5–47
Schematic for Example 5–12.

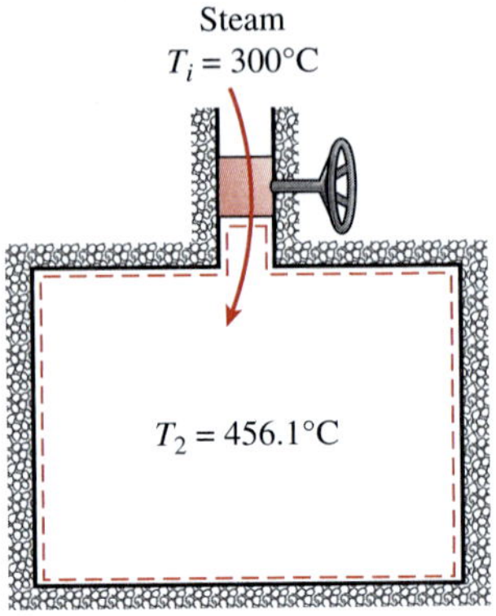

FIGURE 5–48
The temperature of steam rises from 300 to 456.1°C as it enters a tank as a result of flow energy being converted to internal energy.

Discussion Note that the temperature of the steam in the tank has increased by 156.1°C. This result may be surprising at first, and you may be wondering where the energy to raise the temperature of the steam came from. The answer lies in the enthalpy term $h = u + Pv$. Part of the energy represented by enthalpy is the flow energy Pv, and this flow energy is converted to sensible internal energy once the flow ceases to exist in the control volume, and it shows up as an increase in temperature (Fig. 5–48).

Alternative solution This problem can also be solved by considering the region within the tank and the mass that is destined to enter the tank as a closed system, as shown in Fig. 5–47*b*. Since no mass crosses the boundaries, viewing this as a closed system is appropriate.

During the process, the steam upstream (the imaginary piston) will push the enclosed steam in the supply line into the tank at a constant pressure of 1 MPa. Then the boundary work done during this process is

$$W_{b,\text{in}} = -\int_1^2 P_i\,dV = -P_i(V_2 - V_1) = -P_i[V_{\text{tank}} - (V_{\text{tank}} + V_i)] = P_iV_i$$

where V_i is the volume occupied by the steam before it enters the tank and P_i is the pressure at the moving boundary (the imaginary piston face). The energy balance for the closed system gives

$$\underbrace{E_{\text{in}} - E_{\text{out}}}_{\substack{\text{Net energy transfer}\\ \text{by heat, work, and mass}}} = \underbrace{\Delta E_{\text{system}}}_{\substack{\text{Change in internal, kinetic,}\\ \text{potential, etc., energies}}}$$

$$W_{b,\text{in}} = \Delta U$$

$$m_iP_iV_i = m_2u_2 - m_iu_i$$

$$u_2 = u_i + P_iV_i = h_i$$

since the initial state of the system is simply the line conditions of the steam. This result is identical to the one obtained with the uniform-flow analysis. Once again, the temperature rise is caused by the so-called flow energy or flow work, which is the energy required to move the fluid during flow.

EXAMPLE 5–13 Discharge of Heated Air at Constant Temperature

An insulated 8-m^3 rigid tank contains air at 600 kPa and 400 K. A valve connected to the tank is now opened, and air is allowed to escape until the pressure inside drops to 200 kPa. The air temperature during the process is maintained constant by an electric resistance heater placed in the tank. Determine the electrical energy supplied to air during this process.

SOLUTION Pressurized air in an insulated rigid tank equipped with an electric heater is allowed to escape at constant temperature until the pressure inside drops to a specified value. The amount of electrical energy supplied to air is to be determined.

Assumptions **1** This is an unsteady process since the conditions within the device are changing during the process, but it can be analyzed as a uniform-flow process since the exit conditions remain constant. **2** Kinetic and potential energies are negligible. **3** The tank is insulated and thus heat transfer is negligible. **4** Air is an ideal gas with variable specific heats.

Analysis We take the contents of the tank as the system, which is a control volume since mass crosses the boundary (Fig. 5–49). Noting that the microscopic energies of flowing and nonflowing fluids are represented by enthalpy h and internal energy u, respectively, the mass and energy balances for this uniform-flow system can be expressed as

Mass balance: $$m_{\text{in}} - m_{\text{out}} = \Delta m_{\text{system}} \rightarrow m_e = m_1 - m_2$$

Energy balance: $$\underbrace{E_{\text{in}} - E_{\text{out}}}_{\substack{\text{Net energy transfer}\\ \text{by heat, work, and mass}}} = \underbrace{\Delta E_{\text{system}}}_{\substack{\text{Change in internal, kinetic,}\\ \text{potential, etc., energies}}}$$

$$W_{e,\text{in}} - m_e h_e = m_2 u_2 - m_1 u_1 \qquad (\text{since } Q \cong \text{ke} \cong \text{pe} \cong 0)$$

The gas constant of air is $R = 0.287$ kPa·m³/kg·K (Table A-1). The initial and final masses of air in the tank and the discharged amount are determined from the ideal gas relation to be

$$m_1 = \frac{P_1 V_1}{RT_1} = \frac{(600 \text{ kPa})(8 \text{ m}^3)}{(0.287 \text{ kPa}\cdot\text{m}^3/\text{kg}\cdot\text{K})(400 \text{ K})} = 41.81 \text{ kg}$$

$$m_2 = \frac{P_2 V_2}{RT_2} = \frac{(200 \text{ kPa})(8 \text{ m}^3)}{(0.287 \text{ kPa}\cdot\text{m}^3/\text{kg}\cdot\text{K})(400 \text{ K})} = 13.94 \text{ kg}$$

$$m_e = m_1 - m_2 = 41.81 - 13.94 = 27.87 \text{ kg}$$

The enthalpy and internal energy of air at 400 K are $h_e = 400.98$ kJ/kg and $u_1 = u_2 = 286.16$ kJ/kg (Table A-17). The the electrical energy supplied to air is determined from the energy balance to be

$$\begin{aligned} W_{e,\text{in}} &= m_e h_e + m_2 u_2 - m_1 u_1 \\ &= (27.87 \text{ kg})(400.98 \text{ kJ/kg}) + (13.94 \text{ kg})(286.16 \text{ kJ/kg}) \\ &\quad - (41.81 \text{ kg})(286.16 \text{ kJ/kg}) \\ &= 3200 \text{ kJ} = \mathbf{0.889 \text{ kWh}} \end{aligned}$$

since 1 kWh = 3600 kJ.

Discussion If the temperature of discharged air changes during the process, the problem can be solved with reasonable accuracy by evaluating h_e at the average discharge temperature $T_e = (T_2 + T_1)/2$, and treating it as constant.

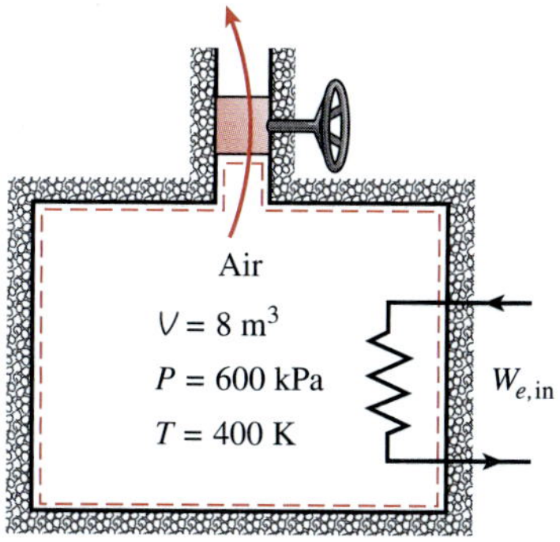

FIGURE 5–49
Schematic for Example 5–13.

TOPIC OF SPECIAL INTEREST* General Energy Equation

One of the most fundamental laws in nature is the **first law of thermodynamics**, also known as the **conservation of energy principle**, which provides a sound basis for studying the relationships among the various forms of energy and energy interactions. It states that *energy can be neither created nor destroyed during a process; it can only change forms.*

*This section can be skipped without a loss in continuity.

The energy content of a fixed quantity of mass (a closed system) can be changed by two mechanisms: *heat transfer Q* and *work transfer W*. Then the conservation of energy for a fixed quantity of mass can be expressed in rate form as

$$\dot{Q} - \dot{W} = \frac{dE_{sys}}{dt} \quad \text{or} \quad \dot{Q} - \dot{W} = \frac{d}{dt}\int_{sys} \rho e\, dV \qquad \textbf{(5–47)}$$

where $\dot{Q} = \dot{Q}_{net,in} = \dot{Q}_{in} - \dot{Q}_{out}$ is the net rate of heat transfer to the system (negative, if from the system), $\dot{W} = \dot{W}_{net,out} = \dot{W}_{out} - \dot{W}_{in}$ is the net power output from the system in all forms (negative, if power input) and dE_{sys}/dt is the rate of change of the total energy content of the system. The overdot stands for time rate. For simple compressible systems, total energy consists of internal, kinetic, and potential energies, and it is expressed on a unit-mass basis as

$$e = u + \text{ke} + \text{pe} = u + \frac{V^2}{2} + gz \qquad \textbf{(5–48)}$$

Note that total energy is a property, and its value does not change unless the state of the system changes.

An energy interaction is *heat* if its driving force is a temperature difference, and it is *work* if it is associated with a force acting through a distance, as explained in Chap. 2. A system may involve numerous forms of work, and the total work can be expressed as

$$W_{total} = W_{shaft} + W_{pressure} + W_{viscous} + W_{other} \qquad \textbf{(5–49)}$$

where W_{shaft} is the work transmitted by a rotating shaft, $W_{pressure}$ is the work done by the pressure forces on the control surface, $W_{viscous}$ is the work done by the normal and shear components of viscous forces on the control surface, and W_{other} is the work done by other forces such as electric, magnetic, and surface tension, which are insignificant for simple compressible systems and are not considered in this text. We do not consider $W_{viscous}$ either since it is usually small relative to other terms in control volume analysis. But it should be kept in mind that the work done by shear forces as the blades shear through the fluid may need to be considered in a refined analysis of turbomachinery.

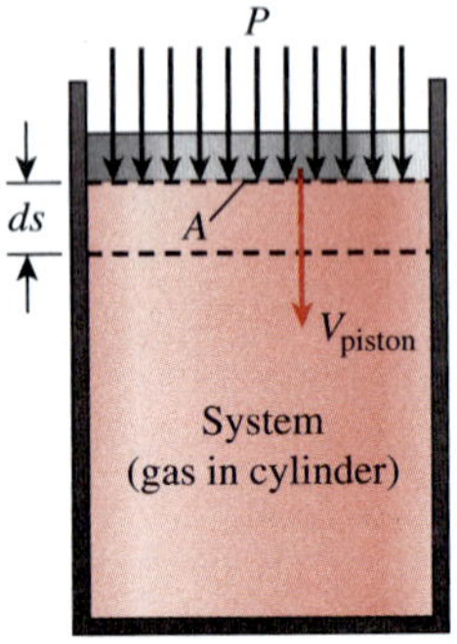

(*a*)

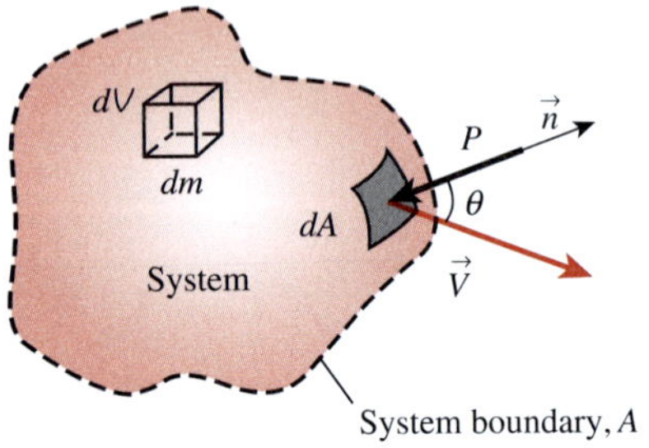

(*b*)

FIGURE 5–50
The pressure force acting on (*a*) the moving boundary of a system in a piston–cylinder device, and (*b*) the differential surface area of a system of arbitrary shape.

Work Done by Pressure Forces

Consider a gas being compressed in the piston–cylinder device shown in Fig. 5–50*a*. When the piston moves down a differential distance *ds* under the influence of the pressure force *PA*, where *A* is the cross-sectional area of the piston, the boundary work done *on* the system is $\delta W_{boundary} = PA\,ds$. Dividing both sides of this relation by the differential time interval *dt* gives the time rate of boundary work (i.e., *power*),

$$\delta\dot{W}_{pressure} = \delta\dot{W}_{boundary} = PAV_{piston}$$

where $V_{piston} = ds/dt$ is the piston velocity, which is the velocity of the moving boundary at the piston face.

Now consider a material chunk of fluid (a system) of arbitrary shape, which moves with the flow and is free to deform under the influence of pressure, as shown in Fig. 5–50*b*. Pressure always acts inward and normal to the surface, and the pressure force acting on a differential area dA is $P\,dA$. Again noting that work is force times distance and distance traveled per unit time is velocity, the time rate at which work is done by pressure forces on this differential part of the system is

$$\delta \dot{W}_{\text{pressure}} = P\,dA\,V_n = P\,dA(\vec{V}\cdot\vec{n}) \tag{5–50}$$

since the normal component of velocity through the differential area dA is $V_n = V\cos\theta = \vec{V}\cdot\vec{n}$. Note that $\vec{n}$ is the outer normal of dA, and thus the quantity $\vec{V}\cdot\vec{n}$ is positive for expansion and negative for compression. The total rate of work done by pressure forces is obtained by integrating $\delta\dot{W}_{\text{pressure}}$ over the entire surface A,

$$\dot{W}_{\text{pressure,net out}} = \int_A P(\vec{V}\cdot\vec{n})\,dA = \int_A \frac{P}{\rho}\,\rho(\vec{V}\cdot\vec{n})\,dA \tag{5–51}$$

In light of these discussions, the net power transfer can be expressed as

$$\dot{W}_{\text{net,out}} = \dot{W}_{\text{shaft,net out}} + \dot{W}_{\text{pressure,net out}} = \dot{W}_{\text{shaft,net out}} + \int_A P(\vec{V}\cdot\vec{n})\,dA \tag{5–52}$$

Then the rate form of the conservation of energy relation for a closed system becomes

$$\dot{Q}_{\text{net,in}} - \dot{W}_{\text{shaft,net out}} - \dot{W}_{\text{pressure,net out}} = \frac{dE_{\text{sys}}}{dt} \tag{5–53}$$

To obtain a relation for the conservation of energy for a *control volume*, we apply the Reynolds transport theorem by replacing the extensive property B with total energy E, and its associated intensive property b with total energy per unit mass e, which is $e = u + \text{ke} + \text{pe} = u + V^2/2 + gz$ (Fig. 5–51). This yields

$$\frac{dE_{\text{sys}}}{dt} = \frac{d}{dt}\int_{\text{CV}} e\rho\,dV + \int_{\text{CS}} e\rho(\vec{V}_r\cdot\vec{n})\,A \tag{5–54}$$

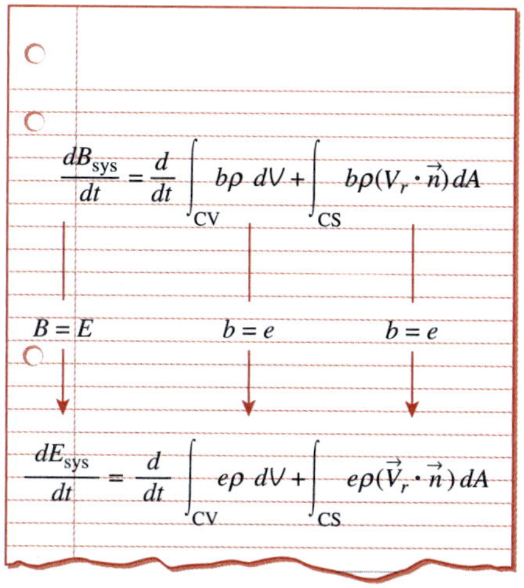

FIGURE 5–51
The conservation of energy equation is obtained by replacing an extensive property B in the Reynolds transport theorem by energy E and its associated intensive property b by e (Ref. 3).

Substituting the left-hand side of Eq. 5–53 into Eq. 5–54, the general form of the energy equation that applies to fixed, moving, or deforming control volumes becomes

$$\dot{Q}_{\text{net,in}} - \dot{W}_{\text{shaft,net out}} - \dot{W}_{\text{pressure,net out}} = \frac{d}{dt}\int_{\text{CV}} e\rho\,dV + \int_{\text{CS}} e\rho(\vec{V}_r\cdot\vec{n})\,dA \tag{5–55}$$

which can be stated as

$$\begin{pmatrix}\text{The net rate of energy}\\ \text{transfer into a CV by}\\ \text{heat and work transfer}\end{pmatrix} = \begin{pmatrix}\text{The time rate of}\\ \text{change of the energy}\\ \text{content of the CV}\end{pmatrix} + \begin{pmatrix}\text{The net flow rate of}\\ \text{energy out of the control}\\ \text{surface by mass flow}\end{pmatrix}$$

Here $\vec{V}_r = \vec{V} - \vec{V}_{\text{CS}}$ is the fluid velocity relative to the control surface, and the product $\rho(\vec{V}_r\cdot\vec{n})\,dA$ represents the mass flow rate through area element dA into or out of the control volume. Again noting that $\vec{n}$ is the outer normal of dA, the quantity $\vec{V}_r\cdot\vec{n}$ and thus mass flow is positive for outflow and negative for inflow.

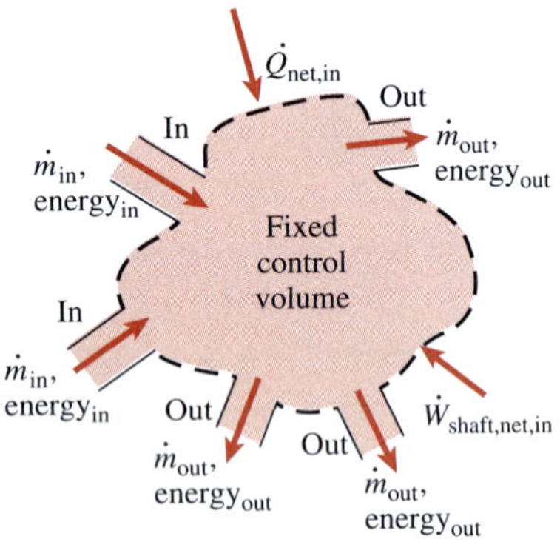

FIGURE 5–52

In a typical engineering problem, the control volume may contain many inlets and outlets; energy flows in at each inlet, and energy flows out at each outlet. Energy also enters the control volume through net heat transfer and net shaft work.

Substituting the surface integral for the rate of pressure work from Eq. 5–51 into Eq. 5–55 and combining it with the surface integral on the right give

$$\dot{Q}_{net,in} - \dot{W}_{shaft,net\ out} = \frac{d}{dt}\int_{CV} e\rho\, dV + \int_{CS}\left(\frac{P}{\rho} + e\right)\rho(\vec{V}_r \cdot \vec{n})\, dA \quad \textbf{(5–56)}$$

This is a very convenient form for the energy equation since pressure work is now combined with the energy of the fluid crossing the control surface and we no longer have to deal with pressure work.

The term $P/\rho = Pv = w_{flow}$ is the *flow work,* which is the work associated with pushing a fluid into or out of a control volume per unit mass. Note that the fluid velocity at a solid surface is equal to the velocity of the solid surface because of the no-slip condition and is zero for nonmoving surfaces. As a result, the pressure work along the portions of the control surface that coincide with nonmoving solid surfaces is zero. Therefore, pressure work for fixed control volumes can exist only along the imaginary part of the control surface where the fluid enters and leaves the control volume (i.e., inlets and outlets).

This equation is not in a convenient form for solving practical engineering problems because of the integrals, and thus it is desirable to rewrite it in terms of average velocities and mass flow rates through inlets and outlets. If $P/\rho + e$ is nearly uniform across an inlet or outlet, we can simply take it outside the integral. Noting that $\dot{m} = \int_{A_c} \rho(\vec{V}_r \cdot \vec{n})\, dA_c$ is the mass flow rate across an inlet or outlet, the rate of inflow or outflow of energy through the inlet or outlet can be approximated as $\dot{m}(P/\rho + e)$. Then the energy equation becomes (Fig. 5–52)

$$\dot{Q}_{net,in} - \dot{W}_{shaft,net\ out} = \frac{d}{dt}\int_{CV} e\rho\, dV + \sum_{out}\dot{m}\left(\frac{P}{\rho} + e\right) - \sum_{in}\dot{m}\left(\frac{P}{\rho} + e\right) \quad \textbf{(5–57)}$$

where $e = u + V^2/2 + gz$ is the total energy per unit mass for both the control volume and flow streams. Then,

$$\dot{Q}_{net,in} - \dot{W}_{shaft,net\ out} = \frac{d}{dt}\int_{CV} e\rho\, dV + \sum_{out}\dot{m}\left(\frac{P}{\rho} + u + \frac{V^2}{2} + gz\right) - \sum_{in}\dot{m}\left(\frac{P}{\rho} + u + \frac{V^2}{2} + gz\right) \quad \textbf{(5–58)}$$

or

$$\dot{Q}_{net,in} - \dot{W}_{shaft,net\ out} = \frac{d}{dt}\int_{CV} e\rho\, dV + \sum_{out}\dot{m}\left(h + \frac{V^2}{2} + gz\right) - \sum_{in}\dot{m}\left(h + \frac{V^2}{2} + gz\right) \quad \textbf{(5–59)}$$

where we used the definition of enthalpy $h = u + Pv = u + P/\rho$. The last two equations are fairly general expressions of conservation of energy, but their use is still limited to uniform flow at inlets and outlets and negligible work due to viscous forces and other effects. Also, the subscript "net,in" stands for "net input," and thus any heat or work transfer is positive if *to* the system and negative if *from* the system.

SUMMARY

The *conservation of mass principle* states that the net mass transfer to or from a system during a process is equal to the net change (increase or decrease) in the total mass of the system during that process, and is expressed as

$$m_{in} - m_{out} = \Delta m_{system} \quad \text{and} \quad \dot{m}_{in} - \dot{m}_{out} = dm_{system}/dt$$

where $\Delta m_{system} = m_{final} - m_{initial}$ is the change in the mass of the system during the process, $\dot{m}_{in}$ and $\dot{m}_{out}$ are the total rates of mass flow into and out of the system, and dm_{system}/dt is the rate of change of mass within the system boundaries. The relations above are also referred to as the mass balance and are applicable to any system undergoing any kind of process.

The amount of mass flowing through a cross section per unit time is called the *mass flow rate*, and is expressed as

$$\dot{m} = \rho VA$$

where ρ = density of fluid, V = average fluid velocity normal to A, and A = cross-sectional area normal to flow direction. The volume of the fluid flowing through a cross section per unit time is called the *volume flow rate* and is expressed as

$$\dot{V} = VA = \dot{m}/\rho$$

The work required to push a unit mass of fluid into or out of a control volume is called *flow work* or *flow energy*, and is expressed as $w_{flow} = Pv$. In the analysis of control volumes, it is convenient to combine the flow energy and internal energy into *enthalpy*. Then the total energy of a flowing fluid is expressed as

$$\theta = h + \text{ke} + \text{pe} = h + \frac{V^2}{2} + gz$$

The total energy transported by a flowing fluid of mass m with uniform properties is $m\theta$. The rate of energy transport by a fluid with a mass flow rate of $\dot{m}$ is $\dot{m}\theta$. When the kinetic and potential energies of a fluid stream are negligible, the amount and rate of energy transport become $E_{mass} = mh$ and $\dot{E}_{mass} = \dot{m}h$, respectively.

The *first law of thermodynamics* is essentially an expression of the conservation of energy principle, also called the *energy balance*. The general mass and energy balances for *any system* undergoing *any process* can be expressed as

$$\underbrace{E_{in} - E_{out}}_{\substack{\text{Net energy transfer}\\ \text{by heat, work, and mass}}} = \underbrace{\Delta E_{system}}_{\substack{\text{Changes in internal, kinetic,}\\ \text{potential, etc., energies}}}$$

It can also be expressed in the *rate form* as

$$\underbrace{\dot{E}_{in} - \dot{E}_{out}}_{\substack{\text{Rate of net energy transfer}\\ \text{by heat, work, and mass}}} = \underbrace{dE_{system}/dt}_{\substack{\text{Rate of change in internal, kinetic,}\\ \text{potential, etc., energies}}}$$

Thermodynamic processes involving control volumes can be considered in two groups: steady-flow processes and unsteady-flow processes. During a *steady-flow process*, the fluid flows through the control volume steadily, experiencing no change with time at a fixed position. The mass and energy content of the control volume remain constant during a steady-flow process. Taking heat transfer *to* the system and work done *by* the system to be positive quantities, the conservation of mass and energy equations for steady-flow processes are expressed as

$$\sum_{in} \dot{m} = \sum_{out} \dot{m}$$

$$\dot{Q} - \dot{W} = \sum_{out} \underbrace{\dot{m}\left(h + \frac{V^2}{2} + gz\right)}_{\text{for each exit}} - \sum_{in} \underbrace{\dot{m}\left(h + \frac{V^2}{2} + gz\right)}_{\text{for each inlet}}$$

These are the most general forms of the equations for steady-flow processes. For single-stream (one-inlet–one-exit) systems such as nozzles, diffusers, turbines, compressors, and pumps, they simplify to

$$\dot{m}_1 = \dot{m}_2 \longrightarrow \frac{1}{v_1} V_1 A_1 = \frac{1}{v_2} V_2 A_2$$

$$\dot{Q} - \dot{W} = \dot{m}\left[h_2 - h_1 + \frac{V_2^2 - V_1^2}{2} + g(z_2 - z_1)\right]$$

In these relations, subscripts 1 and 2 denote the inlet and exit states, respectively.

Most unsteady-flow processes can be modeled as a *uniform-flow process*, which requires that the fluid flow at any inlet or exit is uniform and steady, and thus the fluid properties do not change with time or position over the cross section of an inlet or exit. If they do, they are averaged and treated as constants for the entire process. When kinetic and potential energy changes associated with the control volume and the fluid streams are negligible, the mass and energy balance relations for a uniform-flow system are expressed as

$$m_{in} - m_{out} = \Delta m_{system}$$

$$Q - W = \sum_{out} mh - \sum_{in} mh + (m_2u_2 - m_1u_1)_{system}$$

where $Q = Q_{net,in} = Q_{in} - Q_{out}$ is the net heat input and $W = W_{net,out} = W_{out} - W_{in}$ is the net work output.

When solving thermodynamic problems, it is recommended that the general form of the energy balance $E_{in} - E_{out} = \Delta E_{system}$ be used for all problems, and simplify it for the particular problem instead of using the specific relations given here for different processes.

REFERENCES AND SUGGESTED READINGS

1. ASHRAE Handbook of Fundamentals. SI version. Atlanta, GA: American Society of Heating, Refrigerating, and Air-Conditioning Engineers, Inc., 1993.
2. ASHRAE Handbook of Refrigeration. SI version. Atlanta, GA: American Society of Heating, Refrigerating, and Air-Conditioning Engineers, Inc., 1994.
3. Y. A. Çengel and J. M. Cimbala, *Fluid Mechanics: Fundamentals and Applications, Second Edition.* New York: McGraw-Hill, 2009.

PROBLEMS*

Conservation of Mass

5–1C Define mass and volume flow rates. How are they related to each other?

5–2C When is the flow through a control volume steady?

5–3C Does the amount of mass entering a control volume have to be equal to the amount of mass leaving during an unsteady-flow process?

5–4C Consider a device with one inlet and one outlet. If the volume flow rates at the inlet and at the outlet are the same, is the flow through this device necessarily steady? Why?

5–5E A pneumatic accumulator arranged to maintain a constant pressure as air enters or leaves it is set for 200 psia. Initially, the volume is 0.2 ft³ and the temperature is 80°F. Air is now added to the accumulator until its volume is 1 ft³ and its temperature is 80°F. How much air has been added to the accumulator?

5–6E A steady-flow compressor is used to compress helium from 15 psia and 70°F at the inlet to 200 psia and 600°F at the outlet. The outlet area and velocity are 0.01 ft² and 100 ft/s, respectively, and the inlet velocity is 50 ft/s. Determine the mass flow rate and the inlet area. *Answers:* 0.0704 lbm/s, 0.133 ft²

5–7 Air enters a nozzle steadily at 2.21 kg/m³ and 40 m/s and leaves at 0.762 kg/m³ and 180 m/s. If the inlet area of the nozzle is 90 cm², determine (*a*) the mass flow rate through the nozzle, and (*b*) the exit area of the nozzle. *Answers:* (*a*) 0.796 kg/s, (*b*) 58 cm²

* Problems designated by a "C" are concept questions, and students are encouraged to answer them all. Problems designated by an "E" are in English units, and the SI users can ignore them. Problems with the EES icon are solved using EES, and complete solutions together with parametric studies are included on the enclosed DVD. Problems with the computer icon are comprehensive in nature, and are intended to be solved with a computer, preferably using the EES software that accompanies this text.

5–8 Water enters the constant 130-mm inside-diameter tubes of a boiler at 7 MPa and 65°C, and leaves the tubes at 6 MPa and 450°C with a velocity of 80 m/s. Calculate the velocity of the water at the tube inlet and the inlet volume flow rate.

5–9 A hair dryer is basically a duct of constant diameter in which a few layers of electric resistors are placed. A small fan pulls the air in and forces it through the resistors where it is heated. If the density of air is 1.20 kg/m³ at the inlet and 0.95 kg/m³ at the exit, determine the percent increase in the velocity of air as it flows through the dryer.

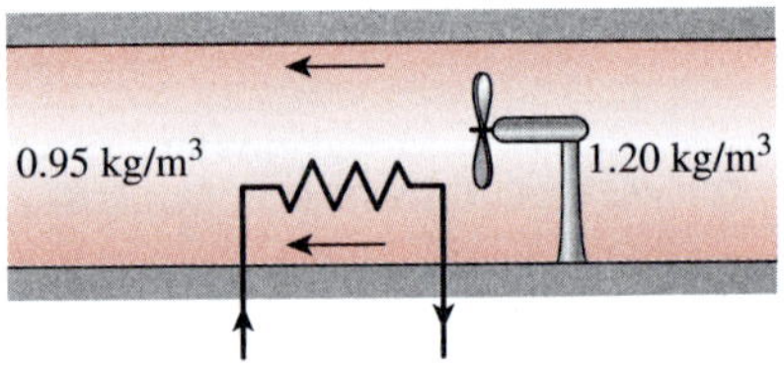

FIGURE P5–9

5–10 A 1-m³ rigid tank initially contains air whose density is 1.18 kg/m³. The tank is connected to a high-pressure supply line through a valve. The valve is opened, and air is allowed to enter the tank until the density in the tank rises to 7.20 kg/m³. Determine the mass of air that has entered the tank. *Answer:* 6.02 kg

5–11 A cyclone separator like that in Fig. P5–11 is used to remove fine solid particles, such as fly ash, that are suspended in a gas stream. In the flue-gas system of an electrical power plant, the weight fraction of fly ash in the exhaust gases is approximately 0.001. Determine the mass flow rates at the two outlets (flue gas and fly ash) when 10 kg/s of flue gas and ash mixture enters this unit. Also determine the amount of fly ash collected per year.

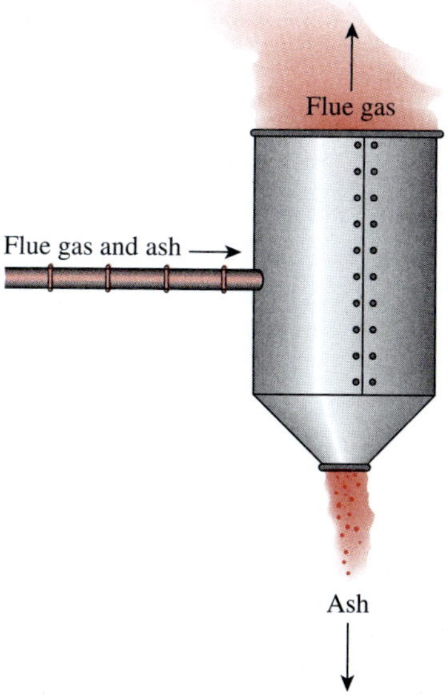

FIGURE P5–11

5–12 Air enters the 1-m^2 inlet of an aircraft engine at 100 kPa and 20°C with a velocity of 180 m/s. Determine the volume flow rate, in m^3/s, at the engine's inlet and the mass flow rate, in kg/s, at the engine's exit.

5–13 A spherical hot-air balloon is initially filled with air at 120 kPa and 20°C with an initial diameter of 5 m. Air enters this balloon at 120 kPa and 20°C with a velocity of 3 m/s through a 1-m diameter opening. How many minutes will it take to inflate this balloon to a 15-m diameter when the pressure and temperature of the air in the balloon remain the same as the air entering the balloon? *Answer:* 12.0 min

FIGURE P5–13

©VOl. 27/PhotoDisc/Getty RF.

5–14 A pump increases the water pressure from 70 kPa at the inlet to 700 kPa at the outlet. Water enters this pump at 15°C through a 1-cm-diameter opening and exits through a 1.5-cm-diameter opening. Determine the velocity of the water at the inlet and outlet when the mass flow rate through the pump is 0.5 kg/s. Will these velocities change significantly if the inlet temperature is raised to 40°C?

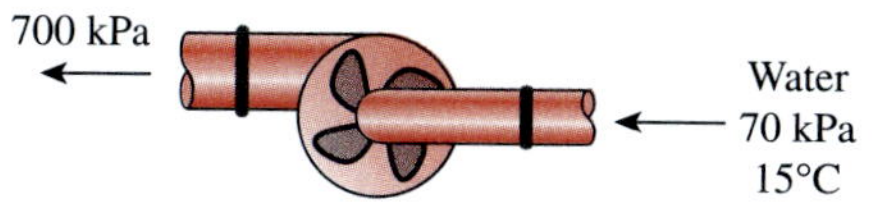

FIGURE P5–14

5–15 Refrigerant-134a enters a 28-cm diameter pipe steadily at 200 kPa and 20°C with a velocity of 5 m/s. The refrigerant gains heat as it flows and leaves the pipe at 180 kPa and 40°C. Determine (*a*) the volume flow rate of the refrigerant at the inlet, (*b*) the mass flow rate of the refrigerant, and (*c*) the velocity and volume flow rate at the exit.

5–16 A smoking lounge is to accommodate 15 heavy smokers. The minimum fresh air requirement for smoking lounges is specified to be 30 L/s per person (ASHRAE, Standard 62, 1989). Determine the minimum required flow rate of fresh air that needs to be supplied to the lounge, and the diameter of the duct if the air velocity is not to exceed 8 m/s.

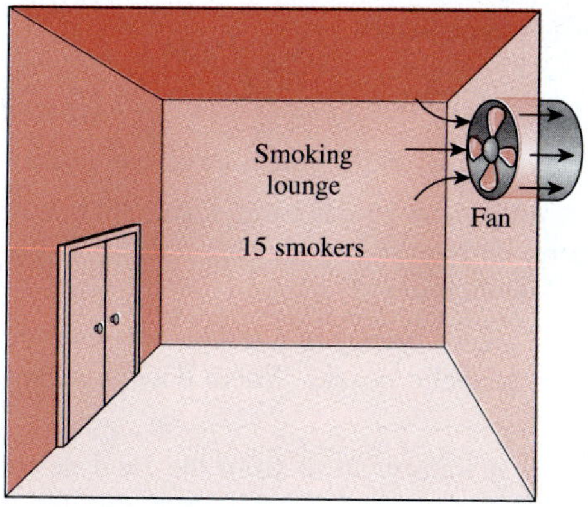

FIGURE P5–16

5–17 The minimum fresh air requirement of a residential building is specified to be 0.35 air change per hour (ASHRAE, Standard 62, 1989). That is, 35 percent of the entire air contained in a residence should be replaced by fresh outdoor air every hour. If the ventilation requirement of a 3.0-m-high, 200-m^2 residence is to be met entirely by a fan, determine the flow capacity in L/min of the fan that needs to be installed. Also determine the diameter of the duct if the air velocity is not to exceed 4 m/s.

Flow Work and Energy Transfer by Mass

5–18C What are the different mechanisms for transferring energy to or from a control volume?

5–19C What is flow energy? Do fluids at rest possess any flow energy?

5–20C How do the energies of a flowing fluid and a fluid at rest compare? Name the specific forms of energy associated with each case.

5–21E A water pump increases the water pressure from 10 psia to 50 psia. Determine the flow work, in Btu/lbm, required by the pump.

5–22 An air compressor compresses 6 L of air at 120 kPa and 20°C to 1000 kPa and 400°C. Determine the flow work, in kJ/kg, required by the compressor. *Answer:* 109 kJ/kg

5–23E Steam is leaving a pressure cooker whose operating pressure is 20 psia. It is observed that the amount of liquid in the cooker has decreased by 0.6 gal in 45 minutes after the steady operating conditions are established, and the cross-sectional area of the exit opening is 0.15 in^2. Determine (*a*) the mass flow rate of the steam and the exit velocity, (*b*) the total and flow energies of the steam per unit mass, and (*c*) the rate at which energy is leaving the cooker by steam.

5–24 Air flows steadily in a pipe at 300 kPa, 77°C, and 25 m/s at a rate of 18 kg/min. Determine (*a*) the diameter of the pipe, (*b*) the rate of flow energy, (*c*) the rate of energy transport by mass, and (*d*) the error involved in part (*c*) if the kinetic energy is neglected.

Steady-Flow Energy Balance: Nozzles and Diffusers

5–25C Can a steady-flow system involve boundary work?

5–26C A diffuser is an adiabatic device that decreases the kinetic energy of the fluid by slowing it down. What happens to this *lost* kinetic energy?

5–27C The kinetic energy of a fluid increases as it is accelerated in an adiabatic nozzle. Where does this energy come from?

5–28C Is heat transfer to or from the fluid desirable as it flows through a nozzle? How will heat transfer affect the fluid velocity at the nozzle exit?

5–29 Air at 80 kPa and 127°C enters an adiabatic diffuser steadily at a rate of 6000 kg/h and leaves at 100 kPa. The velocity of the air stream is decreased from 230 to 30 m/s as it passes through the diffuser. Find (*a*) the exit temperature of the air and (*b*) the exit area of the diffuser.

5–30 Air enters an adiabatic nozzle steadily at 300 kPa, 200°C, and 45 m/s and leaves at 100 kPa and 180 m/s. The inlet area of the nozzle is 110 cm^2. Determine (*a*) the mass flow rate through the nozzle, (*b*) the exit temperature of the air, and (*c*) the exit area of the nozzle. *Answers:* (*a*) 1.09 kg/s, (*b*) 185°C, (*c*) 79.9 cm^2

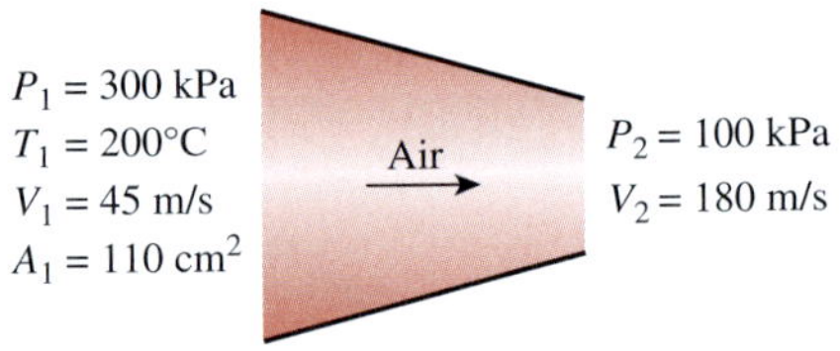

FIGURE P5–30

5–31 Reconsider Prob. 5–30. Using EES (or other) software, investigate the effect of the inlet area on the mass flow rate, exit temperature, and the exit area. Let the inlet area vary from 50 cm^2 to 150 cm^2. Plot the final results against the inlet area, and discuss the results.

5–32E The stators in a gas turbine are designed to increase the kinetic energy of the gas passing through them adiabatically. Air enters a set of these nozzles at 300 psia and 700°F with a velocity of 80 ft/s and exits at 250 psia and 645°F. Calculate the velocity at the exit of the nozzles.

5–33 The diffuser in a jet engine is designed to decrease the kinetic energy of the air entering the engine compressor without any work or heat interactions. Calculate the velocity at the exit of a diffuser when air at 100 kPa and 20°C enters it with a velocity of 500 m/s and the exit state is 200 kPa and 90°C.

FIGURE P5–33

© Stockbyte/Punchstock.

5–34 Steam enters a nozzle at 400°C and 800 kPa with a velocity of 10 m/s, and leaves at 300°C and 200 kPa while losing heat at a rate of 25 kW. For an inlet area of 800 cm^2, determine the velocity and the volume flow rate of the steam at the nozzle exit. *Answers:* 606 m/s, 2.74 m^3/s

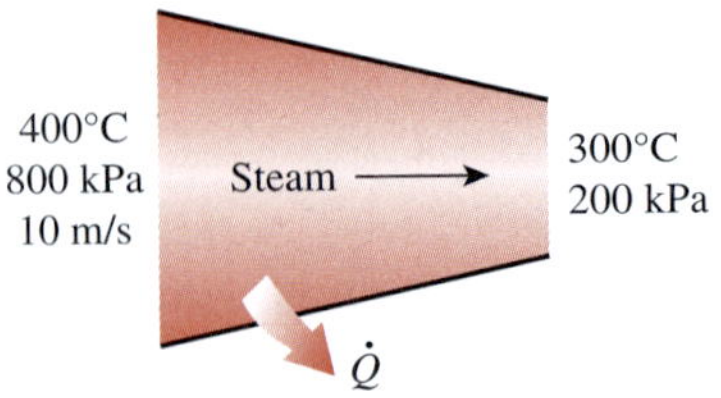

FIGURE P5–34

5–35 Steam at 3 MPa and 400°C enters an adiabatic nozzle steadily with a velocity of 40 m/s and leaves at 2.5 MPa and 300 m/s. Determine (*a*) the exit temperature and (*b*) the ratio of the inlet to exit area A_1/A_2.

5–36E Air at 13 psia and 50°F enters an adiabatic diffuser steadily with a velocity of 600 ft/s and leaves with a low velocity at a pressure of 14.5 psia. The exit area of the diffuser is 4 times the inlet area. Determine (*a*) the exit temperature and (*b*) the exit velocity of the air.

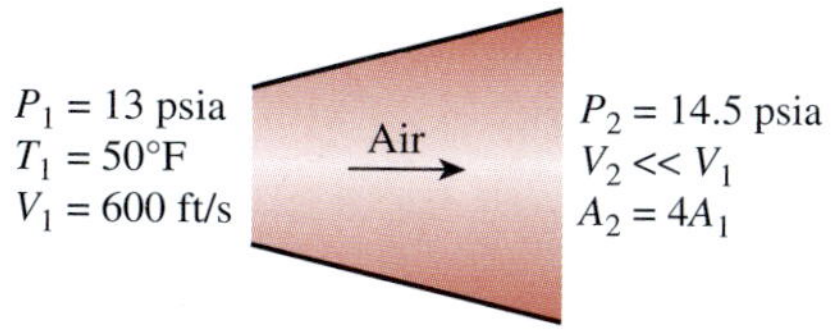

FIGURE P5–36E

5–37 Carbon dioxide enters an adiabatic nozzle steadily at 1 MPa and 500°C with a mass flow rate of 6000 kg/h and leaves at 100 kPa and 450 m/s. The inlet area of the nozzle is 40 cm^2. Determine (*a*) the inlet velocity and (*b*) the exit temperature. *Answers:* (*a*) 60.8 m/s, (*b*) 685.8 K

5–38 Refrigerant-134a at 700 kPa and 120°C enters an adiabatic nozzle steadily with a velocity of 20 m/s and leaves at 400 kPa and 30°C. Determine (*a*) the exit velocity and (*b*) the ratio of the inlet to exit area A_1/A_2.

5–39 Nitrogen gas at 60 kPa and 7°C enters an adiabatic diffuser steadily with a velocity of 275 m/s and leaves at 85 kPa and 27°C. Determine (*a*) the exit velocity of the nitrogen and (*b*) the ratio of the inlet to exit area A_1/A_2.

5–40 Reconsider Prob. 5–39. Using EES (or other) software, investigate the effect of the inlet velocity on the exit velocity and the ratio of the inlet-to-exit area. Let the inlet velocity vary from 210 to 350 m/s. Plot the final results against the inlet velocity, and discuss the results.

5–41 Refrigerant-134a enters a diffuser steadily as saturated vapor at 800 kPa with a velocity of 120 m/s, and it leaves at 900 kPa and 40°C. The refrigerant is gaining heat at a rate of 2 kJ/s as it passes through the diffuser. If the exit area is 80 percent greater than the inlet area, determine (*a*) the exit velocity and (*b*) the mass flow rate of the refrigerant. *Answers:* (*a*) 60.8 m/s, (*b*) 1.308 kg/s

5–42 Steam at 4 MPa and 400°C enters a nozzle steadily with a velocity of 60 m/s, and it leaves at 2 MPa and 300°C. The inlet area of the nozzle is 50 cm^2, and heat is being lost at a rate of 75 kJ/s. Determine (*a*) the mass flow rate of the steam, (*b*) the exit velocity of the steam, and (*c*) the exit area of the nozzle.

Turbines and Compressors

5–43C Consider an adiabatic turbine operating steadily. Does the work output of the turbine have to be equal to the decrease in the energy of the steam flowing through it?

5–44C Consider an air compressor operating steadily. How would you compare the volume flow rates of the air at the compressor inlet and exit?

5–45C Will the temperature of air rise as it is compressed by an adiabatic compressor? Why?

5–46C Somebody proposes the following system to cool a house in the summer: Compress the regular outdoor air, let it cool back to the outdoor temperature, pass it through a turbine, and discharge the cold air leaving the turbine into the house. From a thermodynamic point of view, is the proposed system sound?

5–47 Refrigerant-134a enters a compressor at 100 kPa and –24°C with a flow rate of 1.35 m^3/min and leaves at 800 kPa and 60°C. Determine the mass flow rate of R-134a and the power input to the compressor.

5–48 Refrigerant-134a enters a compressor at 180 kPa as a saturated vapor with a flow rate of 0.35 m^3/min and leaves at 700 kPa. The power supplied to the refrigerant during compression process is 2.35 kW. What is the temperature of R-134a at the exit of the compressor? *Answer:* 48.8°C

5–49 Steam flows steadily through an adiabatic turbine. The inlet conditions of the steam are 6 MPa, 400°C, and 80 m/s, and the exit conditions are 40 kPa, 92 percent quality, and 50 m/s. The mass flow rate of the steam is 20 kg/s. Determine (*a*) the change in kinetic energy, (*b*) the power output, and (*c*) the turbine inlet area. *Answers:* (*a*) −1.95 kJ/kg, (*b*) 14.6 MW, (*c*) 0.0119 m^2

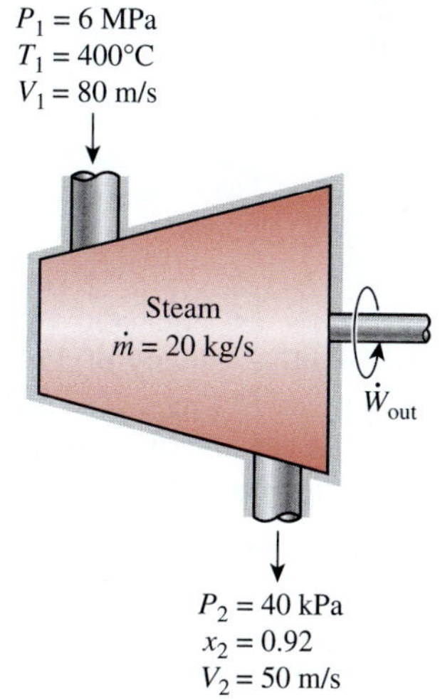

FIGURE P5–49

5–50 Reconsider Prob. 5–49. Using EES (or other) software, investigate the effect of the turbine exit pressure on the power output of the turbine. Let the exit pressure vary from 10 to 200 kPa. Plot the power output against the exit pressure, and discuss the results.

5–51 Steam enters an adiabatic turbine at 10 MPa and 500°C and leaves at 10 kPa with a quality of 90 percent. Neglecting the changes in kinetic and potential energies, determine the mass flow rate required for a power output of 5 MW. *Answer:* 4.852 kg/s

5–52E Steam flows steadily through a turbine at a rate of 45,000 lbm/h, entering at 1000 psia and 900°F and leaving at 5 psia as saturated vapor. If the power generated by the turbine is 4 MW, determine the rate of heat loss from the steam.

5–53 An adiabatic air compressor compresses 10 L/s of air at 120 kPa and 20°C to 1000 kPa and 300°C. Determine (*a*) the work required by the compressor, in kJ/kg, and (*b*) the power required to drive the air compressor, in kW.

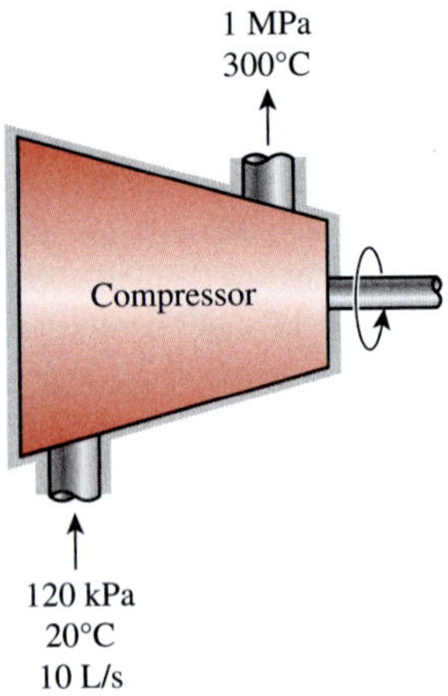

FIGURE P5–53

5–54 Argon gas enters an adiabatic turbine steadily at 1600 kPa and 450°C with a velocity of 55 m/s and leaves at 150 kPa with a velocity of 150 m/s. The inlet area of the turbine is 60 cm². If the power output of the turbine is 190 kW, determine the exit temperature of the argon.

5–55 Helium is to be compressed from 120 kPa and 310 K to 700 kPa and 430 K. A heat loss of 20 kJ/kg occurs during the compression process. Neglecting kinetic energy changes, determine the power input required for a mass flow rate of 90 kg/min.

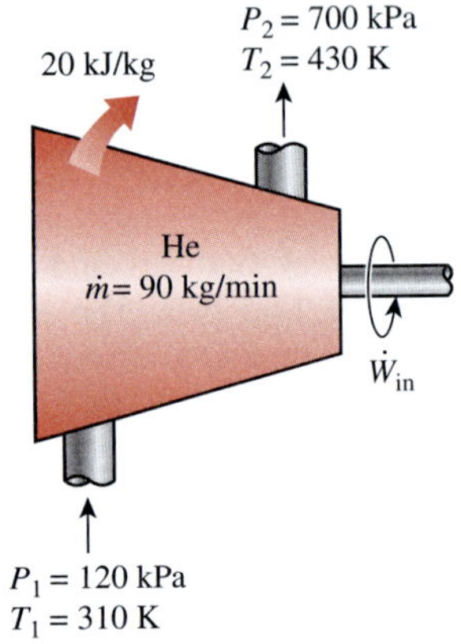

FIGURE P5–55

5–56 Carbon dioxide enters an adiabatic compressor at 100 kPa and 300 K at a rate of 0.5 kg/s and leaves at 600 kPa and 450 K. Neglecting kinetic energy changes, determine (*a*) the volume flow rate of the carbon dioxide at the compressor inlet and (*b*) the power input to the compressor. *Answers:* (*a*) 0.28 m³/s, (*b*) 68.8 kW

5–57 An adiabatic gas turbine expands air at 1300 kPa and 500°C to 100 kPa and 127°C. Air enters the turbine through a 0.2-m² opening with an average velocity of 40 m/s, and exhausts through a 1-m² opening. Determine (*a*) the mass flow rate of air through the turbine and (*b*) the power produced by the turbine. *Answers:* (*a*) 46.9 kg/s, (*b*) 18.3 MW

5–58E Air is expanded by an adiabatic gas turbine from 500 psia and 800°F to 60 psia and 250°F. If the volume flow rate at the exit is 50 ft³/s; the inlet area is 0.6 ft²; and the outlet area is 1.2 ft²; determine the power produced by this turbine.

5–59 Steam enters a steady-flow turbine with a mass flow rate of 20 kg/s at 600°C, 5 MPa, and a negligible velocity. The steam expands in the turbine to a saturated vapor at 500 kPa where 10 percent of the steam is removed for some other use. The remainder of the steam continues to expand to the turbine exit where the pressure is 10 kPa and quality is 85 percent. If the turbine is adiabatic, determine the rate of work done by the steam during this process. *Answer:* 27,790 kW.

5–60 Steam flows steadily into a turbine with a mass flow rate of 26 kg/s and a negligible velocity at 6 MPa and 600°C. The steam leaves the turbine at 0.5 MPa and 200°C

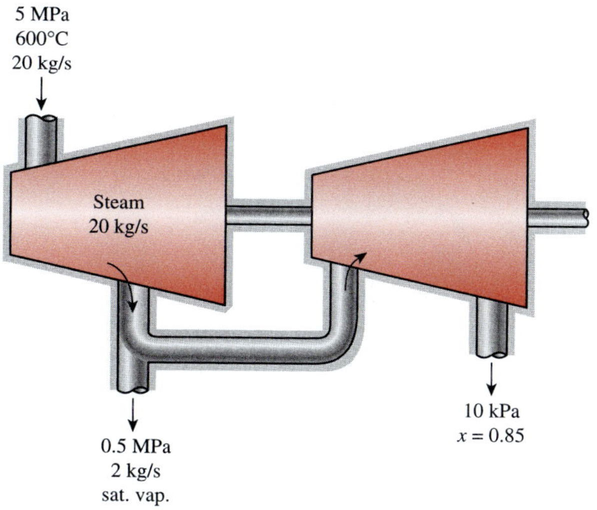

FIGURE P5–59

with a velocity of 180 m/s. The rate of work done by the steam in the turbine is measured to be 20 MW. If the elevation change between the turbine inlet and exit is negligible, determine the rate of heat transfer associated with this process. *Answer:* 455 kW

5–61 Determine the power input for a compressor that compresses helium from 150 kPa and 20°C to 400 kPa and 200°C. Helium enters this compressor through a 0.1-m^2 pipe at a velocity of 15 m/s. What is the increase in the flow power during this process?

Throttling Valves

5–62C Someone claims, based on temperature measurements, that the temperature of a fluid rises during a throttling process in a well-insulated valve with negligible friction. How do you evaluate this claim? Does this process violate any thermodynamic laws?

5–63C Would you expect the temperature of air to drop as it undergoes a steady-flow throttling process? Explain.

5–64C Would you expect the temperature of a liquid to change as it is throttled? Explain.

5–65C During a throttling process, the temperature of a fluid drops from 30 to −20°C. Can this process occur adiabatically?

5–66 An adiabatic capillary tube is used in some refrigeration systems to drop the pressure of the refrigerant from the condenser level to the evaporator level. The R-134a enters the capillary tube as a saturated liquid at 50°C, and leaves at −20°C. Determine the quality of the refrigerant at the inlet of the evaporator.

5–67 Saturated liquid-vapor mixture of water, called wet steam, in a steam line at 2000 kPa is throttled to 100 kPa and 120°C. What is the quality in the steam line? *Answer:* 0.957

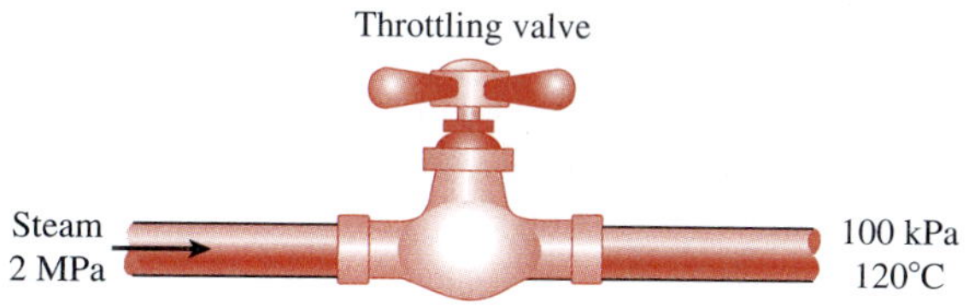

FIGURE P5–67

5–68 Refrigerant-134a at 800 kPa and 25°C is throttled to a temperature of −20°C. Determine the pressure and the internal energy of the refrigerant at the final state. *Answers:* 133 kPa, 80.7 kJ/kg

5–69 A well-insulated valve is used to throttle steam from 8 MPa and 350°C to 2 MPa. Determine the final temperature of the steam. *Answer:* 285°C

5–70 Reconsider Prob. 5–69. Using EES (or other) software, investigate the effect of the exit pressure of steam on the exit temperature after throttling. Let the exit pressure vary from 6 to 1 MPa. Plot the exit temperature of steam against the exit pressure, and discuss the results.

5–71E Refrigerant-134a enters the expansion valve of a refrigeration system at 160 psia as a saturated liquid and leaves at 30 psia. Determine the temperature and internal energy changes across the valve.

Mixing Chambers and Heat Exchangers

5–72C Consider a steady-flow mixing process. Under what conditions will the energy transported into the control volume by the incoming streams be equal to the energy transported out of it by the outgoing stream?

5–73C Consider a steady-flow heat exchanger involving two different fluid streams. Under what conditions will the amount of heat lost by one fluid be equal to the amount of heat gained by the other?

5–74C When two fluid streams are mixed in a mixing chamber, can the mixture temperature be lower than the temperature of both streams? Explain.

5–75 Hot and cold streams of a fluid are mixed in a rigid mixing chamber. The hot fluid flows into the chamber at a mass flow rate of 5 kg/s with an energy in the amount of 150 kJ/kg. The cold fluid flows into the chamber with a mass flow rate of 15 kg/s and carries energy in the amount of 50 kJ/kg. There is heat transfer to the surroundings from the mixing chamber in the amount of 5.5 kW. The mixing chamber operates in a steady-flow manner and does not gain or loose energy or mass with time. Determine the energy carried from the mixing chamber by the fluid mixture per unit mass of fluid, in kJ/kg.

5–76 A hot-water stream at 80°C enters a mixing chamber with a mass flow rate of 0.5 kg/s where it is mixed with a stream of cold water at 20°C. If it is desired that the mixture leave the chamber at 42°C, determine the mass flow rate of the cold-water stream. Assume all the streams are at a pressure of 250 kPa. *Answer:* 0.865 kg/s

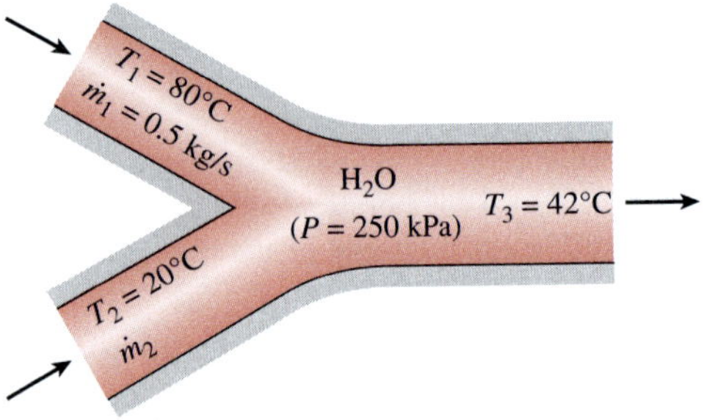

FIGURE P5–76

5–77E Water at 65°F and 20 psia is heated in a chamber by mixing it with saturated water vapor at 20 psia. If both streams enter the mixing chamber at the same mass flow rate, determine the temperature and the quality of the exiting stream. *Answers:* 228°F, 0.415

5–78 A stream of refrigerant-134a at 1 MPa and 20°C is mixed with another stream at 1 MPa and 80°C. If the mass flow rate of the cold stream is twice that of the hot one, determine the temperature and the quality of the exit stream.

5–79 Reconsider Prob. 5–78. Using EES (or other) software, investigate the effect of the mass flow rate of the cold stream of R-134a on the temperature and the quality of the exit stream. Let the ratio of the mass flow rate of the cold stream to that of the hot stream vary from 1 to 4. Plot the mixture temperature and quality against the cold-to-hot mass flow rate ratio, and discuss the results.

5–80E Steam is to be condensed on the shell side of a heat exchanger at 85°F. Cooling water enters the tubes at 60°F at a rate of 138 lbm/s and leaves at 73°F. Assuming the heat exchanger to be well-insulated, determine the rate of heat transfer in the heat exchanger and the rate of condensation of the steam.

5–81 Steam enters the condenser of a steam power plant at 20 kPa and a quality of 95 percent with a mass flow rate of 20,000 kg/h. It is to be cooled by water from a nearby river by circulating the water through the tubes within the condenser. To prevent thermal pollution, the river water is not allowed to experience a temperature rise above 10°C. If the steam is to leave the condenser as saturated liquid at 20 kPa, determine the mass flow rate of the cooling water required. *Answer:* 297.7 kg/s

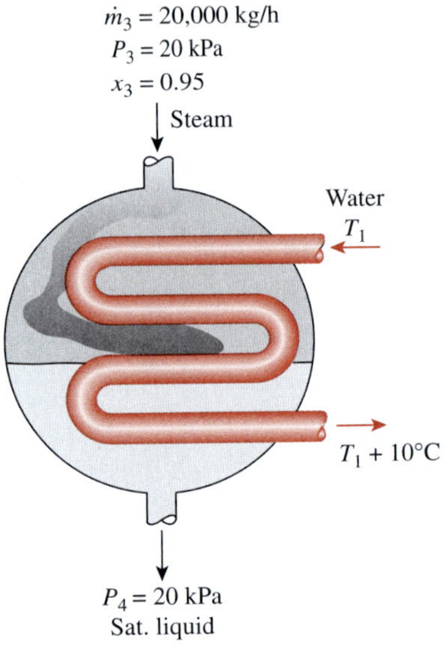

FIGURE P5–81

5–82 A heat exchanger is to cool ethylene glycol (c_p = 2.56 kJ/kg·°C) flowing at a rate of 3.2 kg/s from 80°C to 40°C by water (c_p = 4.18 kJ/kg·°C) that enters at 20°C and leaves at 70°C. Determine (*a*) the rate of heat transfer and (*b*) the mass flow rate of water.

5–83 Reconsider Prob. 5–82. Using EES (or other) software, investigate the effect of the inlet temperature of cooling water on the mass flow rate of water. Let the inlet temperature vary from 10 to 40°C, and assume the exit temperature to remain constant. Plot the mass flow rate of water against the inlet temperature, and discuss the results.

5–84 A thin-walled double-pipe counter-flow heat exchanger is used to cool oil (c_p = 2.20 kJ/kg·°C) from 150 to 40°C at a rate of 2 kg/s by water (c_p = 4.18 kJ/kg·°C) that enters at 22°C at a rate of 1.5 kg/s. Determine the rate of heat transfer in the heat exchanger and the exit temperature of water.

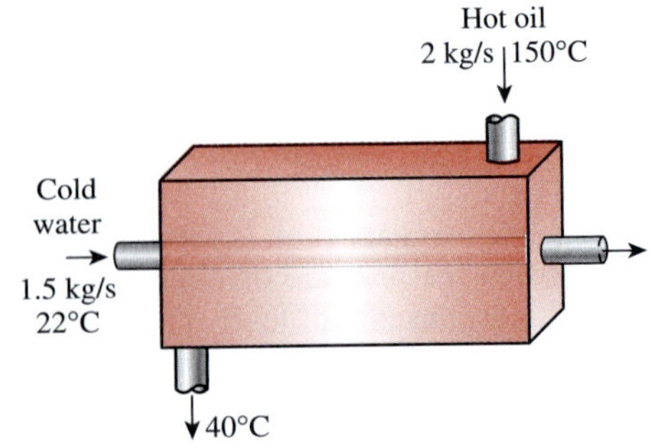

FIGURE P5–84

5–85 Cold water (c_p = 4.18 kJ/kg·°C) leading to a shower enters a thin-walled double-pipe counter-flow heat exchanger at 15°C at a rate of 0.60 kg/s and is heated to 45°C by hot water (c_p = 4.19 kJ/kg·°C) that enters at 100°C at a rate of 3 kg/s. Determine the rate of heat transfer in the heat exchanger and the exit temperature of the hot water.

5–86 Air (c_p = 1.005 kJ/kg·°C) is to be preheated by hot exhaust gases in a cross-flow heat exchanger before it enters the furnace. Air enters the heat exchanger at 95 kPa and 20°C at a rate of 0.6 m^3/s. The combustion gases (c_p = 1.10 kJ/kg·°C) enter at 160°C at a rate of 0.95 kg/s and leave at 95°C. Determine the rate of heat transfer to the air and its outlet temperature.

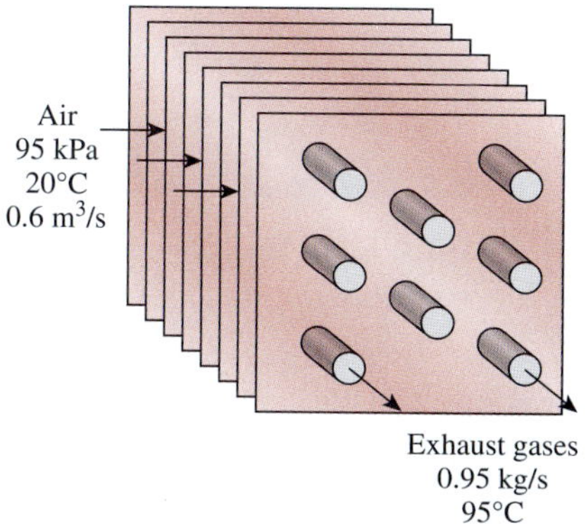

FIGURE P5–86

5–87E An open feedwater heater heats the feedwater by mixing it with hot steam. Consider an electrical power plant with an open feedwater heater that mixes 0.1 lbm/s of steam at 10 psia and 200°F with 2.0 lbm/s of feedwater at 10 psia and 100°F to produce 10 psia and 120°F feedwater at the outlet. The diameter of the outlet pipe is 0.5 ft. Determine the mass flow rate and feedwater velocity at the outlet. Would the outlet flow rate and velocity be significantly different if the temperature at the outlet were 180°F?

5–88E In a steam heating system, air is heated by being passed over some tubes through which steam flows steadily. Steam enters the heat exchanger at 30 psia and 400°F at a rate of 15 lbm/min and leaves at 25 psia and 212°F. Air enters at 14.7 psia and 80°F and leaves at 130°F. Determine the volume flow rate of air at the inlet.

5–89 An air-conditioning system involves the mixing of cold air and warm outdoor air before the mixture is routed to the conditioned room in steady operation. Cold air enters the mixing chamber at 7°C and 105 kPa at a rate of 0.75 m^3/s while warm air enters at 34°C and 105 kPa. The air leaves the room at 24°C. The ratio of the mass flow rates of the hot to cold air streams is 2.2. Using variable specific heats, determine (*a*) the mixture temperature at the inlet of the room and (*b*) the rate of heat gain of the room.

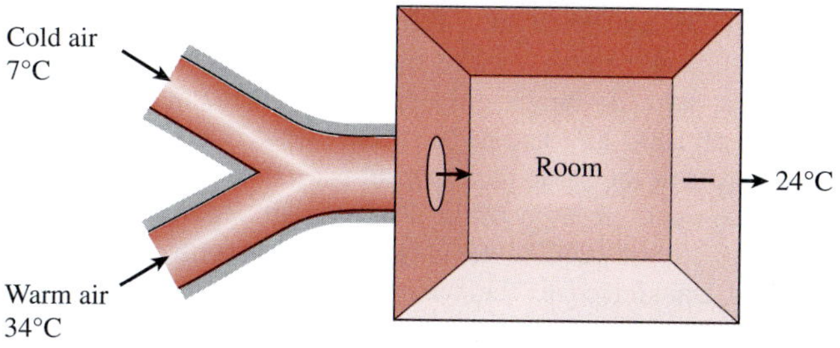

FIGURE P5–89

5–90 Hot exhaust gases of an internal combustion engine are to be used to produce saturated water vapor at 2 MPa pressure. The exhaust gases enter the heat exchanger at 400°C at a rate of 32 kg/min while water enters at 15°C. The heat exchanger is not well insulated, and it is estimated that 10 percent of heat given up by the exhaust gases is lost to the surroundings. If the mass flow rate of the exhaust gases is 15 times that of the water, determine (*a*) the temperature of the exhaust gases at the heat exchanger exit and (*b*) the rate of heat transfer to the water. Use the constant specific heat properties of air for the exhaust gases.

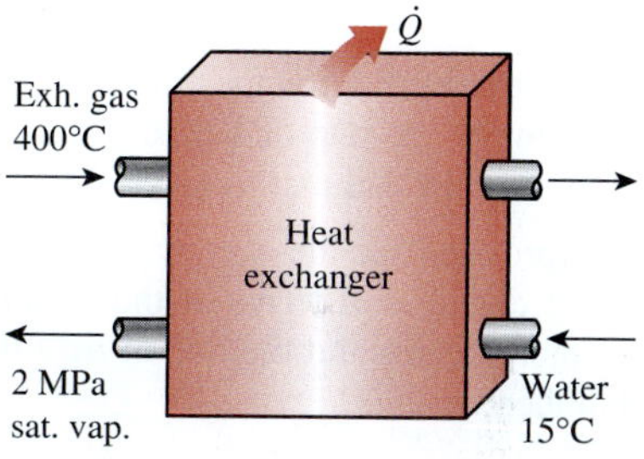

FIGURE P5–90

5–91 A chilled-water heat-exchange unit is designed to cool 5 m^3/s of air at 100 kPa and 30°C to 100 kPa and 18°C by using water at 8°C. Determine the maximum water outlet temperature when the mass flow rate of the water is 2 kg/s. *Answer:* 16.3°C

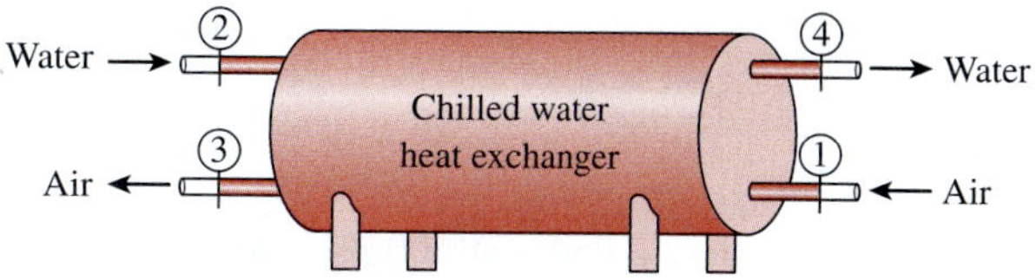

FIGURE P5–91

5–92 The condenser of a refrigeration cycle is basically a heat exchanger in which a refrigerant is condensed by rejecting heat to a cooling fluid. Refrigerant-134a enters a condenser at 1200 kPa and 85°C with a flow rate of 0.042 kg/s and leaves at the same pressure subcooled by 6.3°C. The condensation is done by cooling water which experiences a temperature rise of 12°C in the condenser. Determine (*a*) the rate of heat transferred to the water in the condenser, in kJ/min and (*b*) the mass flow rate of water, in kg/min. *Answers:* (*a*) 525 kJ/min, (*b*) 10.5 kg/min

5–93 The evaporator of a refrigeration cycle is basically a heat exchanger in which a refrigerant is evaporated by absorbing heat from a fluid. Refrigerant-22 enters an evaporator at 200 kPa with a quality of 22 percent and a flow rate of 2.25 L/h. R-22 leaves the evaporator at the same pressure superheated by 5°C. The refrigerant is evaporated by absorbing heat from air whose flow rate is 0.5 kg/s. Determine (*a*) the rate of heat absorbed from the air and (*b*) the temperature change of air. The properties of R-22 at the inlet and exit of the condenser are h_1 = 220.2 kJ/kg, v_1 = 0.0253 m^3/kg, and h_2 = 398.0 kJ/kg.

5–94 Two mass streams of the same ideal gas are mixed in a steady-flow chamber while receiving energy by heat transfer from the surroundings. The mixing process takes place at constant pressure with no work and negligible changes in kinetic and potential energies. Assume the gas has constant specific heats.

(*a*) Determine the expression for the final temperature of the mixture in terms of the rate of heat transfer to the mixing chamber and the inlet and exit mass flow rates.

(*b*) Obtain an expression for the volume flow rate at the exit of the mixing chamber in terms of the volume flow rates of the two inlet streams and the rate of heat transfer to the mixing chamber.

(*c*) For the special case of adiabetic mixing, show that the exit volume flow rate is the sum of the two inlet volume flow rates.

Pipe and Duct Flow

5–95 Argon steadily flows into a constant pressure heater at 300 K and 100 kPa with a mass flow rate of 6.24 kg/s. Heat transfer in the rate of 150 kW is supplied to the argon as it flows through the heater. (*a*) Determine the argon temperature at the heater exit, in °C. (*b*) Determine the argon volume flow rate at the heater exit, in m^3/s.

5–96 Saturated liquid water is heated in a steady-flow steam boiler at a constant pressure of 2 MPa at a rate of 4 kg/s to an outlet temperature of 250°C. Determine the rate of heat transfer in the boiler.

5–97E Water enters a boiler at 500 psia as a saturated liquid and leaves at 600°F at the same pressure. Calculate the heat transfer per unit mass of water.

5–98 A 110-volt electrical heater is used to warm 0.3 m^3/s of air at 100 kPa and 15°C to 100 kPa and 30°C. How much current in amperes must be supplied to this heater?

5–99E The fan on a personal computer draws 0.5 ft^3/s of air at 14.7 psia and 70°F through the box containing the CPU and other components. Air leaves at 14.7 psia and 80°F. Calculate the electrical power, in kW, dissipated by the PC components. *Answer:* 0.0948 kW

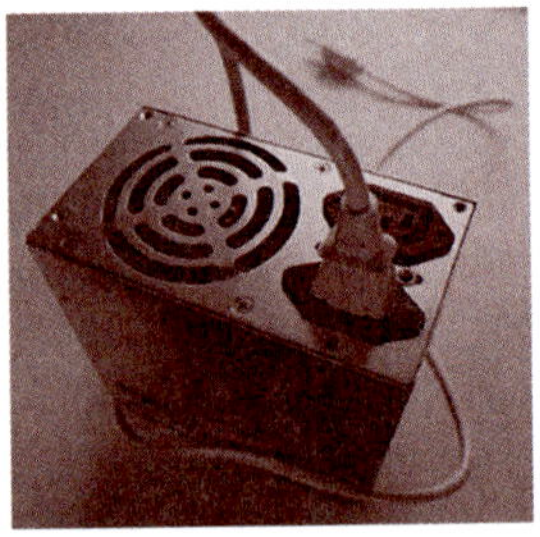

FIGURE P5–99E

5–100 A desktop computer is to be cooled by a fan. The electronic components of the computer consume 60 W of power under full-load conditions. The computer is to operate in environments at temperatures up to 45°C and at elevations up to 3400 m where the average atmospheric pressure is 66.63 kPa. The exit temperature of air is not to exceed 60°C to meet the reliability requirements. Also, the average velocity of air is not to exceed 110 m/min at the exit of the computer case where the fan is installed to keep the noise level down. Determine the flow rate of the fan that needs to be installed and the diameter of the casing of the fan.

5–101 Repeat Prob. 5–100 for a computer that consumes 100 W of power.

5–102E Water enters the tubes of a cold plate at 70°F with an average velocity of 40 ft/min and leaves at 105°F. The diameter of the tubes is 0.25 in. Assuming 15 percent of the heat generated is dissipated from the components to the surroundings by convection and radiation, and the remaining 85 percent is removed by the cooling water, determine the amount of heat generated by the electronic devices mounted on the cold plate. *Answer:* 614 W

5–103 EES The components of an electronic system dissipating 180 W are located in a 1.4-m-long horizontal duct whose cross section is 20 cm × 20 cm. The components in the duct are cooled by forced air that enters the duct at 30°C and 1 atm at a rate of 0.6 m^3/min and leaves at 40°C. Determine the rate of heat transfer from the outer surfaces of the duct to the ambient. *Answer:* 63 W

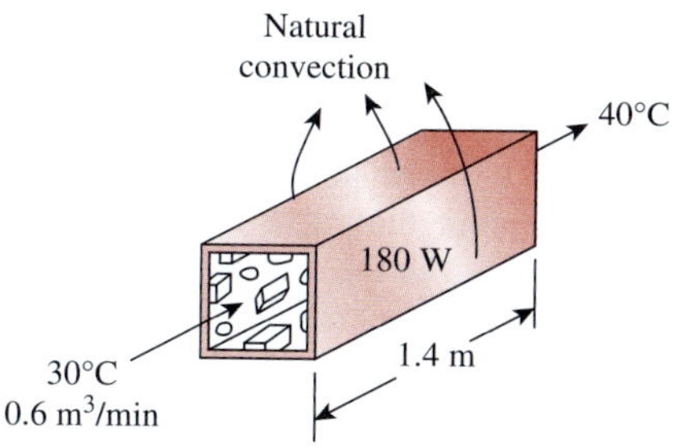

FIGURE P5–103

5–104 Repeat Prob. 5–103 for a circular horizontal duct of diameter 10 cm.

5–105 Consider a hollow-core printed circuit board 9 cm high and 18 cm long, dissipating a total of 15 W. The width of the air gap in the middle of the PCB is 0.25 cm. If the cooling air enters the 12-cm-wide core at 25°C and 1 atm at a rate of 0.8 L/s, determine the average temperature at which the air leaves the hollow core. *Answer:* 46.0°C

5–106 A computer cooled by a fan contains eight PCBs, each dissipating 10 W power. The height of the PCBs is 12 cm and the length is 18 cm. The cooling air is supplied by a 25-W fan mounted at the inlet. If the temperature rise of air as it flows through the case of the computer is not to exceed 10°C, determine (*a*) the flow rate of the air that the fan needs to deliver and (*b*) the fraction of the temperature rise of air that is due to the heat generated by the fan and its motor. *Answers:* (*a*) 0.0104 kg/s, (*b*) 24 percent

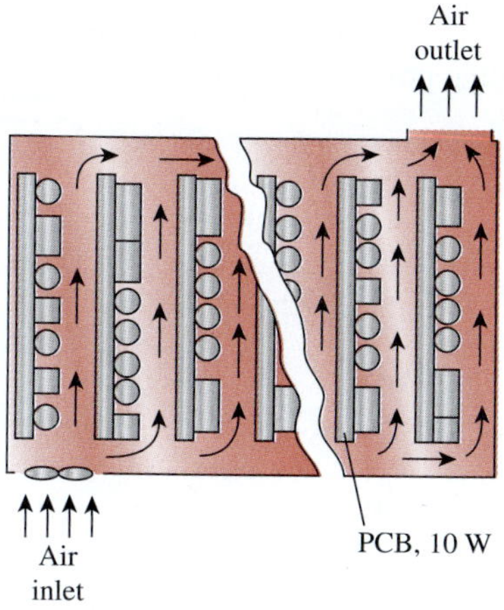

FIGURE P5–106

5–107 Hot water at 90°C enters a 22-m section of a cast iron pipe whose inner diameter is 2.5 cm at an average velocity of 0.6 m/s. The outer surface of the pipe is exposed to the cold air at 10°C in a basement. If water leaves the basement at 88°C, determine the rate of heat loss from the water.

5–108 Reconsider Prob. 5–107. Using EES (or other) software, investigate the effect of the inner pipe diameter on the rate of heat loss. Let the pipe diameter vary from 1.5 to 7.5 cm. Plot the rate of heat loss against the diameter, and discuss the results.

5–109 A 5-m × 6-m × 8-m room is to be heated by an electric resistance heater placed in a short duct in the room. Initially, the room is at 15°C, and the local atmospheric pressure is 98 kPa. The room is losing heat steadily to the outside at a rate of 200 kJ/min. A 200-W fan circulates the air steadily through the duct and the electric heater at an average mass flow rate of 50 kg/min. The duct can be assumed to be adiabatic, and there is no air leaking in or out of the room. If it takes 15 min for the room air to reach an average temperature of 25°C, find (*a*) the power rating of the electric heater and (*b*) the temperature rise that the air experiences each time it passes through the heater.

5–110E The hot-water needs of a household are to be met by heating water at 55°F to 180°F by a parabolic solar collector at a rate of 4 lbm/s. Water flows through a 1.25-in-diameter thin aluminum tube whose outer surface is black-anodized in order to maximize its solar absorption ability. The centerline of the tube coincides with the focal line of the collector, and a glass sleeve is placed outside the tube to minimize the heat losses. If solar energy is transferred to water at a net rate of 400 Btu/h per ft length of the tube, determine the required length of the parabolic collector to meet the hot-water requirements of this house.

5–111 A house has an electric heating system that consists of a 300-W fan and an electric resistance heating element placed in a duct. Air flows steadily through the duct at a rate of 0.6 kg/s and experiences a temperature rise of 7°C. The rate of heat loss from the air in the duct is estimated to be 300 W. Determine the power rating of the electric resistance heating element. *Answer:* 4.22 kW

5–112 Steam enters a long, horizontal pipe with an inlet diameter of $D_1 = 12$ cm at 2 MPa and 300°C with a velocity of 3 m/s. Farther downstream, the conditions are 800 kPa and 250°C, and the diameter is $D_2 = 10$ cm. Determine (*a*) the mass flow rate of the steam and (*b*) the rate of heat transfer. *Answers:* (*a*) 0.270 kg/s, (*b*) 19.9 kJ/s

5–113 Steam enters an insulated pipe at 200 kPa and 200°C and leaves at 150 kPa and 150°C. The inlet-to-outlet diameter ratio for the pipe is $D_1/D_2 = 1.80$. Determine the inlet and exit velocities of the steam.

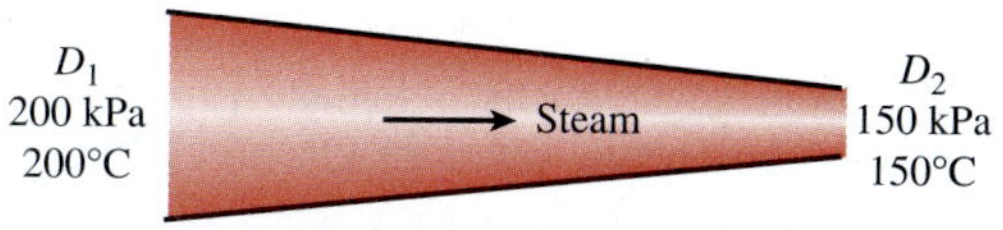

FIGURE P5–113

5–114 Refrigerant-134a enters the condenser of a refrigerator at 900 kPa and 60°C, and leaves as a saturated liquid at the same pressure. Determine the heat transfer from the refrigerant per unit mass.

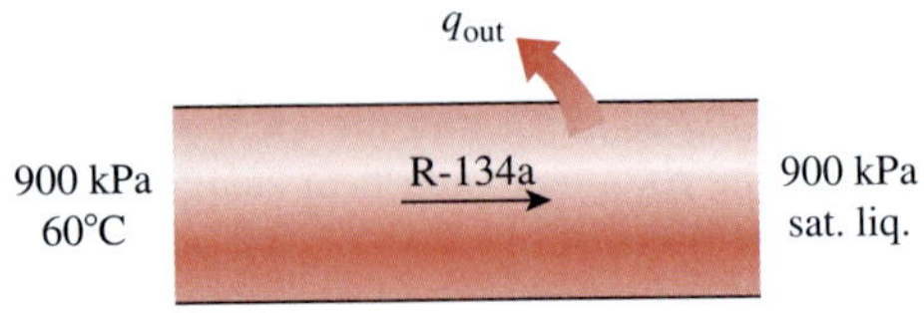

FIGURE P5–114

5–115 Saturated liquid water is heated at constant pressure in a steady-flow device until it is a saturated vapor. Calculate the heat transfer, in kJ/kg, when the vaporization is done at a pressure of 800 kPa.

5–116 Air at 300 K and 100 kPa steadily flows into a hair dryer having electrical work input of 1500 W. Because of the size of the air intake, the inlet velocity of the air is negligible. The air temperature and velocity at the hair dryer exit are 80°C and 21 m/s, respectively. The flow process is both constant pressure and adiabatic. Assume air has constant specific heats evaluated at 300 K. (*a*) Determine the air mass flow rate into the hair dryer, in kg/s. (*b*) Determine the air volume flow rate at the hair dryer exit, in m^3/s. *Answers:* (*a*) 0.0280 kg/s, (*b*) 0.0284 m^3/s

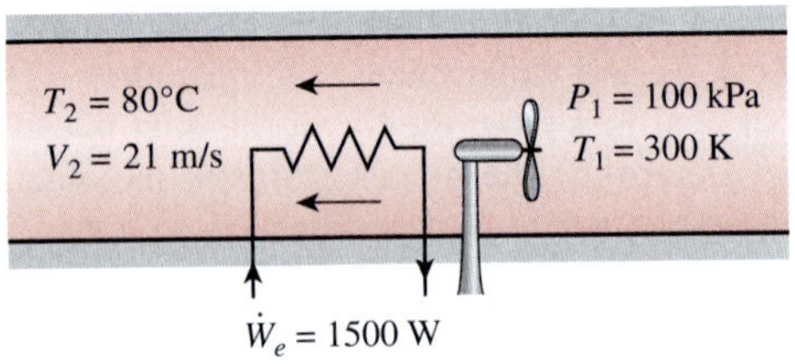

FIGURE P5–116

5–117 Reconsider Prob. 5–116. Using EES (or other) software, investigate the effect of the exit velocity on the mass flow rate and the exit volume flow rate. Let the exit velocity vary from 5 to 25 m/s. Plot the mass flow rate and exit volume flow rate against the exit velocity, and discuss the results.

Charging and Discharging Processes

5–118 An insulated rigid tank is initially evacuated. A valve is opened, and atmospheric air at 95 kPa and 17°C enters the tank until the pressure in the tank reaches 95 kPa, at which point the valve is closed. Determine the final temperature of the air in the tank. Assume constant specific heats. *Answer:* 406 K

5–119 A rigid, insulated tank that is initially evacuated is connected through a valve to a supply line that carries helium at 200 kPa and 120°C. Now the valve is opened, and helium is allowed to flow into the tank until the pressure reaches 200 kPa, at which point the valve is closed. Determine the flow work of the helium in the supply line and the final temperature of the helium in the tank. *Answers:* 816 kJ/kg, 655 K

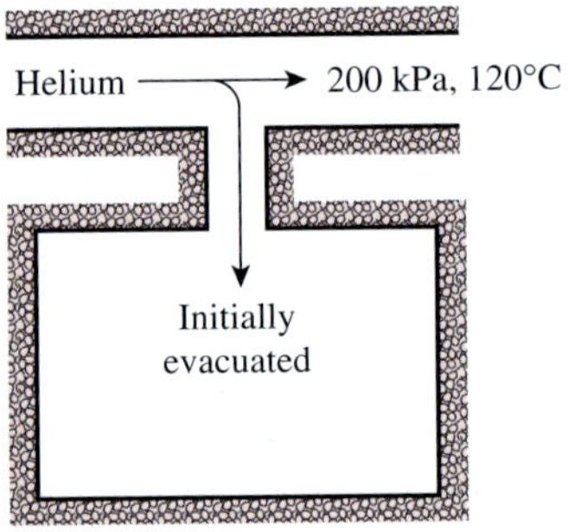

FIGURE P5–119

5–120 Consider a 20-L evacuated rigid bottle that is surrounded by the atmosphere at 100 kPa and 27°C. A valve at the neck of the bottle is now opened and the atmospheric air is allowed to flow into the bottle. The air trapped in the bottle eventually reaches thermal equilibrium with the atmosphere as a result of heat transfer through the wall of the bottle. The valve remains open during the process so that the trapped air also reaches mechanical equilibrium with the atmosphere. Determine the net heat transfer through the wall of the bottle during this filling process. *Answer:* = 2.0 kJ

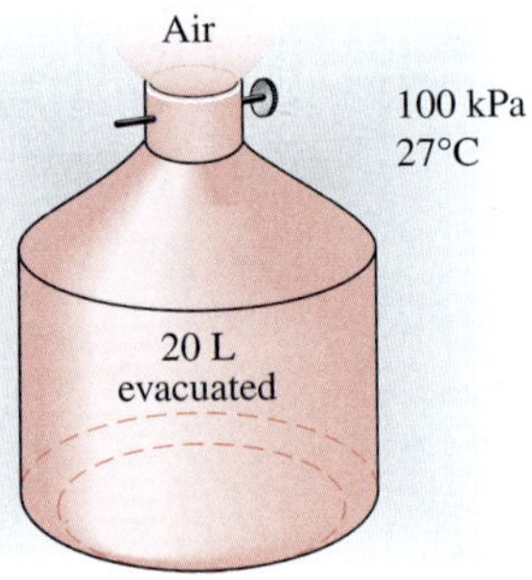

FIGURE P5–120

5–121 A 2-m^3 rigid tank initially contains air at 100 kPa and 22°C. The tank is connected to a supply line through a valve. Air is flowing in the supply line at 600 kPa and 22°C. The valve is opened, and air is allowed to enter the tank until the pressure in the tank reaches the line pressure, at which point the valve is closed. A thermometer placed in the tank indicates that the air temperature at the final state is 77°C. Determine (*a*) the mass of air that has entered the tank and (*b*) the amount of heat transfer. *Answers:* (*a*) 9.58 kg, (*b*) 339 kJ

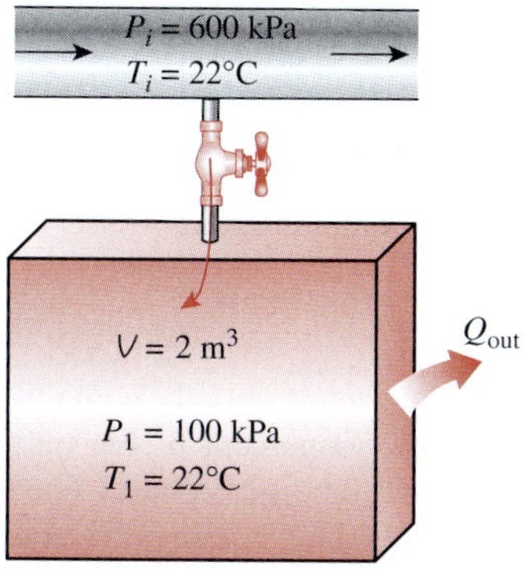

FIGURE P5–121

5–122 A 0.2-m^3 rigid tank equipped with a pressure regulator contains steam at 2 MPa and 300°C. The steam in the tank is now heated. The regulator keeps the steam pressure constant by letting out some steam, but the temperature inside rises. Determine the amount of heat transferred when the steam temperature reaches 500°C.

5–123 An insulated, vertical piston–cylinder device initially contains 10 kg of water, 6 kg of which is in the vapor phase. The mass of the piston is such that it maintains a constant pressure of 200 kPa inside the cylinder. Now steam at 0.5 MPa and 350°C is allowed to enter the cylinder from a supply line until all the liquid in the cylinder has vaporized. Determine (*a*) the final temperature in the cylinder and (*b*) the mass of the steam that has entered. *Answers:* (*a*) 120.2°C, (*b*) 19.07 kg

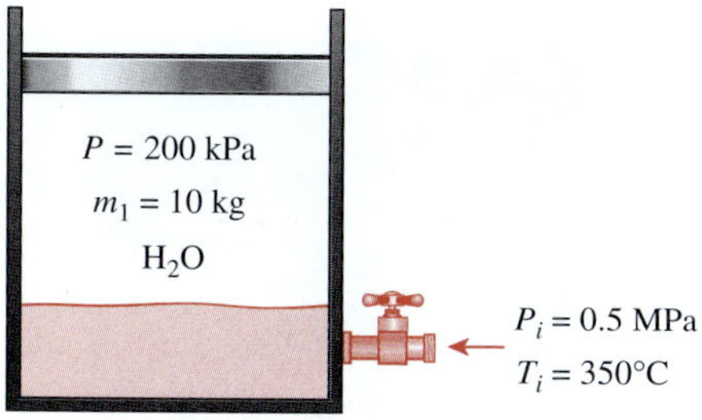

FIGURE P5–123

5–124E A scuba diver's 2-ft^3 air tank is to be filled with air from a compressed air line at 120 psia and 100°F. Initially, the air in this tank is at 20 psia and 70°F. Presuming that the tank is well insulated, determine the temperature and mass in the tank when it is filled to 120 psia.

5–125 An air-conditioning system is to be filled from a rigid container that initially contains 5 kg of liquid R-134a at 24°C. The valve connecting this container to the air-conditioning system is now opened until the mass in the container is 0.25 kg, at which time the valve is closed. During this time, only liquid R-134a flows from the container. Presuming that the process is isothermal while the valve is open, determine the final quality of the R-134a in the container and the total heat transfer. *Answers:* 0.506, 22.6 kJ

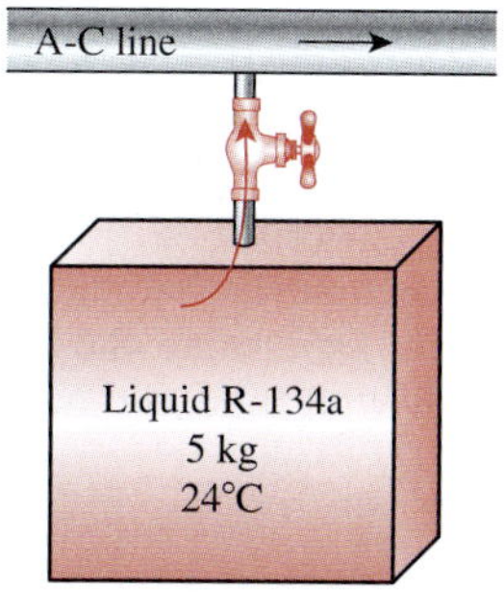

FIGURE P5–125

5–126E Oxygen is supplied to a medical facility from ten 1.5-ft^3 compressed oxygen tanks. Initially, these tanks are at 1500 psia and 80°F. The oxygen is removed from these tanks slowly enough that the temperature in the tanks remains at 80°F. After two weeks, the pressure in the tanks is 300 psia. Determine the mass of oxygen used and the total heat transfer to the tanks.

5–127E The weighted piston of the device shown in Fig. P5–127E maintains the pressure of the piston-cylinder contents at 200 psia. Initially, this system contains no mass. The valve is now opened, and steam from the line flows into the cylinder until the volume is 10 ft^3. This process is adiabatic, and the steam in the line remains at 300 psia and 450°F. Determine the final temperature (and quality if appropriate) of the steam in the cylinder and the total work produced as the device is filled.

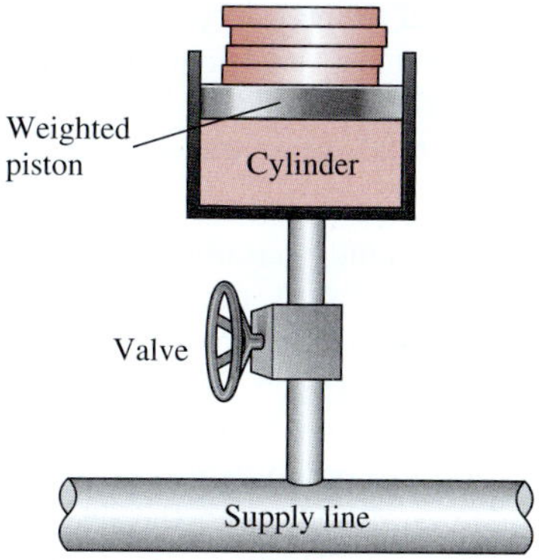

FIGURE P5–127E

5–128E Repeat Prob. 5–127E when the supply line carries oxygen at 300 psia and 450°F. *Answers:* 450°F, 370 Btu

5–129 A 0.03-m^3 rigid tank initially contains refrigerant-134a at 1.4 MPa and 100 percent quality. The tank is connected by a valve to a supply line that carries refrigerant-134a at 1.6 MPa and 36°C. Now the valve is opened, and the refrigerant is allowed to enter the tank. The valve is closed when it is observed that the tank contains saturated liquid at 1.6 MPa. Determine (*a*) the mass of the refrigerant that has entered the tank and (*b*) the amount of heat transfer. *Answers:* (*a*) 29.8 kg, (*b*) 697 kJ

5–130 A 0.3-m^3 rigid tank is filled with saturated liquid water at 200°C. A valve at the bottom of the tank is opened, and liquid is withdrawn from the tank. Heat is transferred to the water such that the temperature in the tank remains constant. Determine the amount of heat that must be transferred by the time one-half of the total mass has been withdrawn.

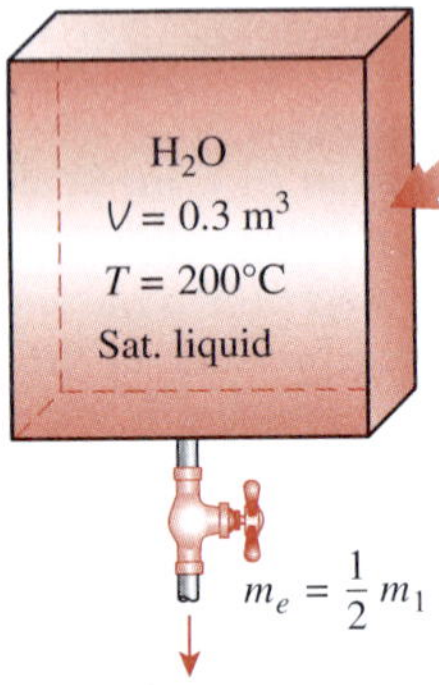

FIGURE P5–130

5–131 A 0.12-m^3 rigid tank contains saturated refrigerant-134a at 800 kPa. Initially, 25 percent of the volume is occupied by liquid and the rest by vapor. A valve at the bottom of the tank is now opened, and liquid is withdrawn from the tank. Heat is transferred to the refrigerant such that the pressure inside the tank remains constant. The valve is closed when no liquid is left in the tank and vapor starts to come out. Determine the total heat transfer for this process. *Answer:* 201.2 kJ

5–132E A 2-ft^3 rigid tank contains saturated refrigerant-134a at 160 psia. Initially, 5 percent of the volume is occupied by liquid and the rest by vapor. A valve at the top of the tank is now opened, and vapor is allowed to escape slowly from the tank. Heat is transferred to the refrigerant such that the pressure inside the tank remains constant. The valve is closed when the last drop of liquid in the tank is vaporized. Determine the total heat transfer for this process.

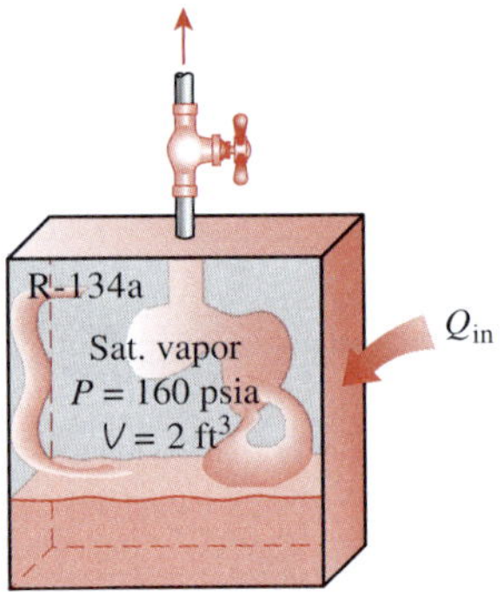

FIGURE P5–132E

5–133 A 0.4-m^3 rigid tank initially contains refrigerant-134a at 14°C. At this state, 70 percent of the mass is in the vapor phase, and the rest is in the liquid phase. The tank is connected by a valve to a supply line where refrigerant at 1 MPa and 100°C flows steadily. Now the valve is opened slightly, and the refrigerant is allowed to enter the tank. When the pressure in the tank reaches 700 kPa, the entire refrigerant in the tank exists in the vapor phase only. At this point the valve is closed. Determine (*a*) the final temperature in the tank, (*b*) the mass of refrigerant that has entered the tank, and (*c*) the heat transfer between the system and the surroundings.

5–134 The air-release flap on a hot-air balloon is used to release hot air from the balloon when appropriate. On one hot-air balloon, the air release opening has an area of 0.5 m^2, and the filling opening has an area of 1 m^2. During a two minute adiabatic flight maneuver, hot air enters the balloon at 100 kPa and 35°C with a velocity of 2 m/s; the air in the balloon remains at 100 kPa and 35°C; and air leaves the balloon through the air-release flap at velocity 1 m/s. At the start of this maneuver, the volume of the balloon is 75 m^3. Determine the final volume of the balloon and work produced by the air inside the balloon as it expands the balloon skin.

FIGURE P5–134

© Vol. /PhotoDisc/Getty RF.

5–135 An insulated 0.15-m^3 tank contains helium at 3 MPa and 130°C. A valve is now opened, allowing some helium to escape. The valve is closed when one-half of the initial mass has escaped. Determine the final temperature and pressure in the tank. *Answers:* 257 K, 956 kPa

5–136E An insulated 60-ft^3 rigid tank contains air at 75 psia and 120°F. A valve connected to the tank is now opened, and air is allowed to escape until the pressure inside drops to 30 psia. The air temperature during this process is maintained constant by an electric resistance heater placed in the tank. Determine the electrical work done during this process.

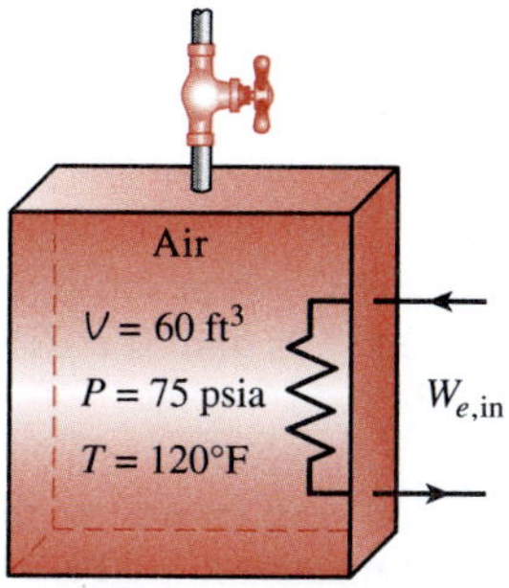

FIGURE P5–136E

5–137 A vertical piston–cylinder device initially contains 0.2 m^3 of air at 20°C. The mass of the piston is such that it maintains a constant pressure of 300 kPa inside. Now a valve connected to the cylinder is opened, and air is allowed to escape until the volume inside the cylinder is decreased by one-half. Heat transfer takes place during the process so that the temperature of the air in the cylinder remains constant. Determine (*a*) the amount of air that has left the cylinder and (*b*) the amount of heat transfer. *Answers:* (*a*) 0.357 kg, (*b*) 0

5–138 A balloon initially contains 40 m^3 of helium gas at atmospheric conditions of 100 kPa and 17°C. The balloon is connected by a valve to a large reservoir that supplies helium gas at 125 kPa and 25°C. Now the valve is opened, and helium is allowed to enter the balloon until pressure equilibrium with the helium at the supply line is reached. The material of the balloon is such that its volume increases linearly with pressure. If no heat transfer takes place during this process, determine the final temperature in the balloon. *Answer:* 315 K

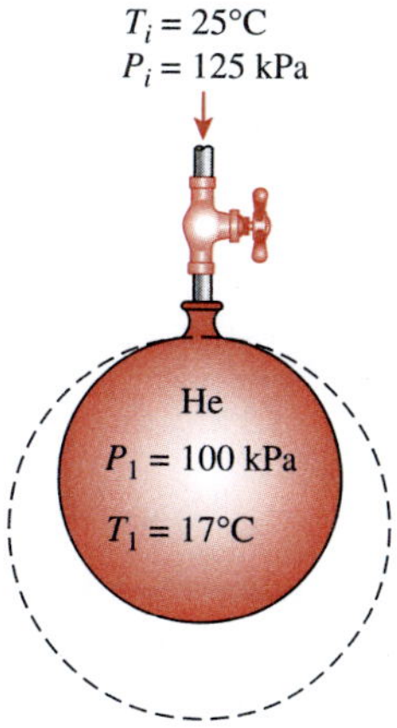

FIGURE P5–138

5–139 The air in an insulated, rigid compressed-air tank whose volume is 0.5 m^3 is initially at 4000 kPa and 20°C. Enough air is now released from the tank to reduce the pressure to 2000 kPa. Following this release, what is the temperature of the remaining air in the tank?

FIGURE P5–139

5–140 An insulated vertical piston–cylinder device initially contains 0.8 m^3 of refrigerant-134a at 1.2 MPa and 120°C. A linear spring at this point applies full force to the piston. A valve connected to the cylinder is now opened, and refrigerant is allowed to escape. The spring unwinds as the piston moves down, and the pressure and volume drop to 0.6 MPa and 0.5 m^3 at the end of the process. Determine (*a*) the amount of refrigerant that has escaped and (*b*) the final temperature of the refrigerant.

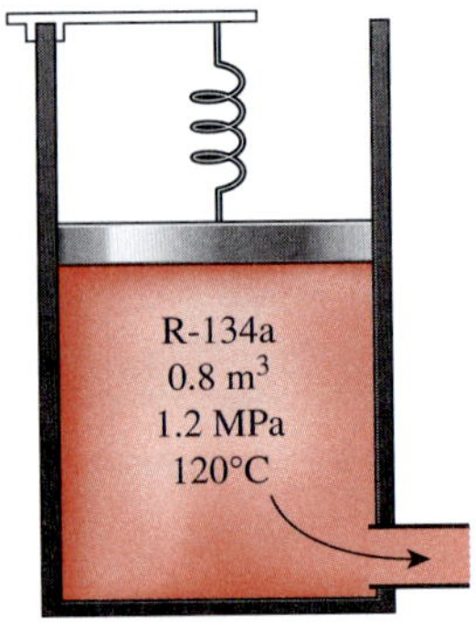

FIGURE P5–140

5–141 A piston–cylinder device initially contains 0.6 kg of steam with a volume of 0.1 m^3. The mass of the piston is such that it maintains a constant pressure of 800 kPa. The cylinder is connected through a valve to a supply line that carries steam at 5 MPa and 500°C. Now the valve is opened and steam is allowed to flow slowly into the cylinder until the volume of the cylinder doubles and the temperature in the cylinder reaches 250°C, at which point the valve is closed. Determine (*a*) the mass of steam that has entered and (*b*) the amount of heat transfer.

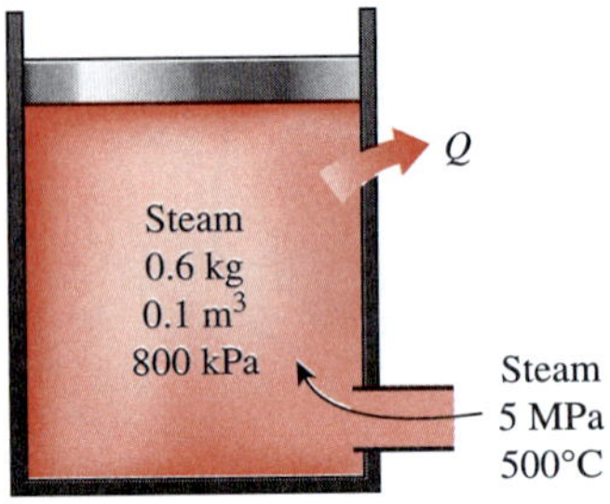

FIGURE P5–141

5–142 An adiabatic piston-cylinder device equipped with a spring maintains the pressure inside at 300 kPa when the volume is 0, and 3000 kPa when the volume is 5 m^3. The device is connected to a steam line maintained at 1500 kPa and 200°C, and initially the volume is 0. Determine the final temperature (and quality if appropriate) when valve is opened and steam in the line is allowed to enter the cylinder until the pressure inside matches that in the line. Also determine the total work produced during this adiabatic filling process.

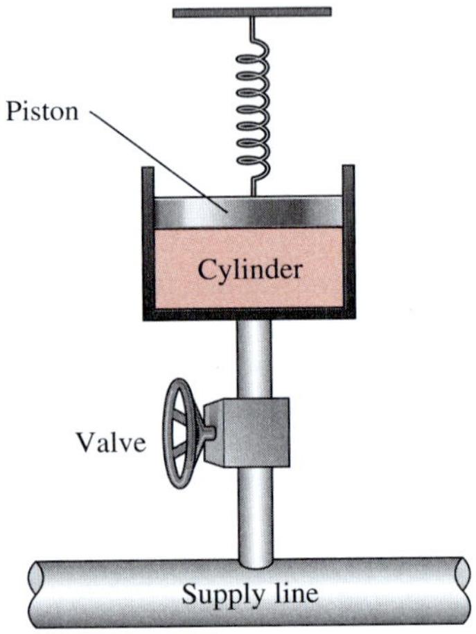

FIGURE P5–142

5–143 Repeat Prob. 5-142 when the line is filled with air that is maintained at 2000 kPa and 327°C.

Review Problems

5–144 Carbon dioxide gas enters a throttling valve at 5 MPa and 100°C and leaves at 100 kPa. Determine the temperature change during this process if CO_2 is assumed to be (*a*) an ideal gas and (*b*) a real gas. Real gas properties of CO_2 may be obtained from EES.

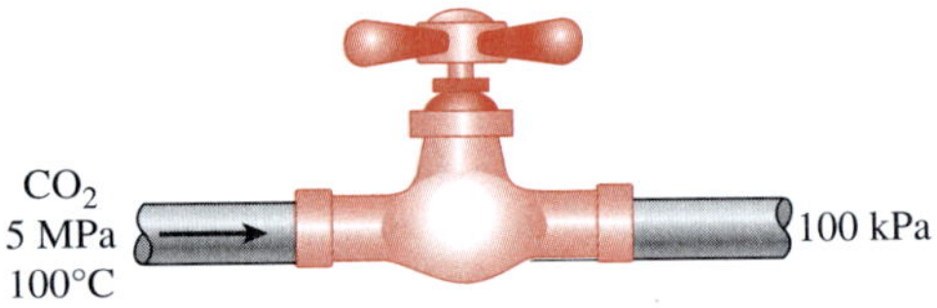

FIGURE P5–144

5–145 Helium steadily enters a pipe with a mass flow rate of 8 kg/s at 427°C and 100 kPa and leaves the pipe at 27°C. The pressure during the process is constant at 100 kPa. (*a*) Determine the heat transfer for the process, in kW. (*b*) Determine the volume flow rate of the helium at the pipe exit, in m^3/s.

5–146 Underground water is being pumped into a pool whose cross section is 6 m × 9 m while water is discharged through a 7-cm-diameter orifice at a constant average velocity of 4 m/s. If the water level in the pool rises at a rate of 2.5 cm/min, determine the rate at which water is supplied to the pool, in m^3/s.

5–147 Air at 4.18 kg/m^3 enters a nozzle that has an inlet-to-exit area ratio of 2:1 with a velocity of 120 m/s and leaves with a velocity of 380 m/s. Determine the density of air at the exit. *Answer:* 2.64 kg/m^3

5–148E Consider a heat exchanger that uses hot air to heat cold water. Air enters this heat exchanger at 20 psia and 200°F at a rate of 100 ft^3/min and leaves at 17 psia and 100°F. Water enters this unit at 20 psia and 50°F at a rate of 0.5 lbm/s and exits at 17 psia and 90°F. Determine the total flow power, in hp, required for this unit and the flow work, in Btu/lbm, for both the air and water streams.

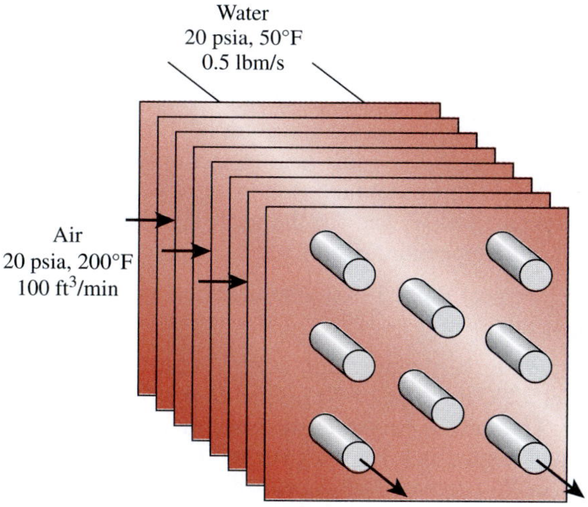

FIGURE P5–148E

5–149 An air compressor compresses 15 L/s of air at 120 kPa and 20°C to 800 kPa and 300°C while consuming 6.2 kW of power. How much of this power is being used to increase the pressure of the air versus the power needed to move the fluid through the compressor? *Answers:* 4.48 kW, 1.72 kW

5–150 A steam turbine operates with 1.6 MPa and 350°C steam at its inlet and saturated vapor at 30°C at its exit. The mass flow rate of the steam is 16 kg/s, and the turbine produces 9000 kW of power. Determine the rate at which heat is lost through the casing of this turbine.

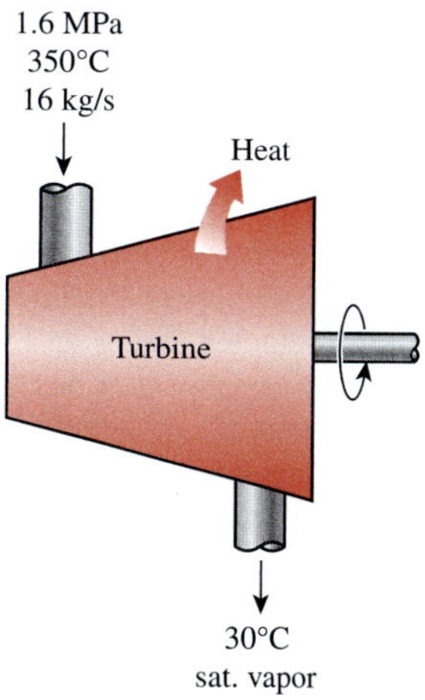

FIGURE P5–150

5–151E Nitrogen gas flows through a long, constant-diameter adiabatic pipe. It enters at 100 psia and 120°F and leaves at 50 psia and 70°F. Calculate the velocity of the nitrogen at the pipe's inlet and outlet.

5–152 A 110-V electric hot-water heater warms 0.1 L/s of water from 18 to 30°C. Calculate the current in amperes that must be supplied to this heater. *Answer:* 45.6 A

5–153 Water is boiled at 100°C electrically by a 3-kW resistance wire. Determine the rate of evaporation of water.

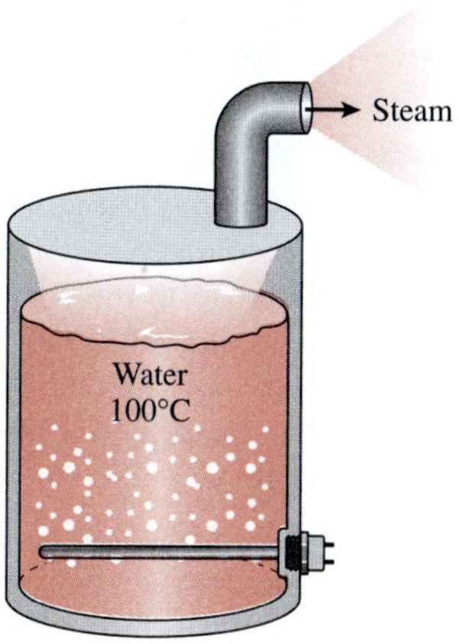

FIGURE P5–153

5–154 Steam enters a long, insulated pipe at 1400 kPa, 350°C, and 10 m/s, and exits at 1000 kPa. The diameter of the pipe is 0.15 m at the inlet, and 0.1 m at the exit. Calculate the mass flow rate of the steam and its speed at the pipe outlet.

5–155 The air flow in a compressed air line is divided into two equal streams by a T-fitting in the line. The compressed air enters this 2.5-cm diameter fitting at 1.6 MPa and 40°C with a velocity of 50 m/s. Each outlet has the same diameter as the inlet, and the air at these outlets has a pressure of 1.4 MPa and a temperature of 36°C. Determine the velocity of the air at the outlets and the rate of change of flow energy (flow power) across the T-fitting.

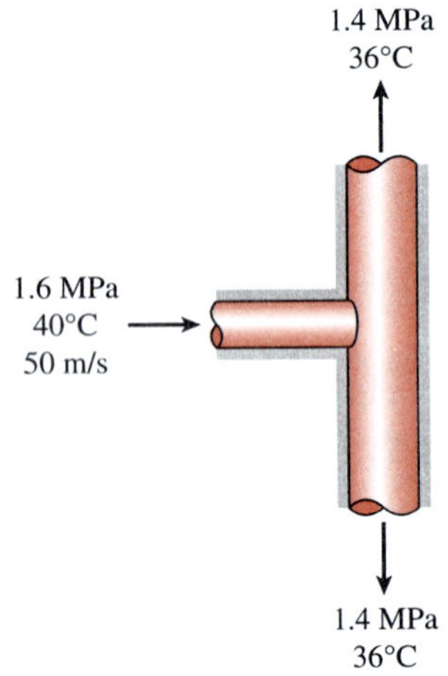

FIGURE P5–155

5–156 Air enters a pipe at 65°C and 200 kPa and leaves at 60°C and 175 kPa. It is estimated that heat is lost from the pipe in the amount of 3.3 kJ per kg of air flowing in the pipe. The diameter ratio for the pipe is $D_1/D_2 = 1.4$. Using constant specific heats for air, determine the inlet and exit velocities of the air. *Answers: 29.9 m/s, 66.1 m/s*

5–157 Steam enters a nozzle with a low velocity at 150°C and 200 kPa, and leaves as a saturated vapor at 75 kPa. There is a heat transfer from the nozzle to the surroundings in the amount of 26 kJ for every kilogram of steam flowing through the nozzle. Determine (*a*) the exit velocity of the steam and (*b*) the mass flow rate of the steam at the nozzle entrance if the nozzle exit area is 0.001 m^2.

5–158 In a gas-fired boiler, water is boiled at 150°C by hot gases flowing through a stainless steel pipe submerged in water. If the rate of heat transfer from the hot gases to water is 74 kJ/s, determine the rate of evaporation of water.

5–159 Cold water enters a steam generator at 20°C and leaves as saturated vapor at 200°C. Determine the fraction of heat used in the steam generator to preheat the liquid water from 20°C to the saturation temperature of 200°C.

5–160 Cold water enters a steam generator at 20°C and leaves as saturated vapor at the boiler pressure. At what pressure will the amount of heat needed to preheat the water to saturation temperature be equal to the heat needed to vaporize the liquid at the boiler pressure?

5–161 An ideal gas expands in an adiabatic turbine from 1200 K and 900 kPa to 700 K. Determine the turbine inlet volume flow rate of the gas, in m^3/s, required to produce turbine work output at the rate of 350 kW. The average values of the specific heats for this gas over the temperature range and the gas constant are c_p = 1.13 kJ/kg·K, c_v = 0.83 kJ/kg·K, and R = 0.30 kJ/kg·K.

5–162 Chickens with an average mass of 2.2 kg and average specific heat of 3.54 kJ/kg·°C are to be cooled by chilled water that enters a continuous-flow-type immersion chiller at 0.5°C. Chickens are dropped into the chiller at a uniform temperature of 15°C at a rate of 500 chickens per hour and are cooled to an average temperature of 3°C before they are taken out. The chiller gains heat from the surroundings at a rate of 200 kJ/h. Determine (*a*) the rate of heat removal from the chickens, in kW, and (*b*) the mass flow rate of water, in kg/s, if the temperature rise of water is not to exceed 2°C.

5–163 Repeat Prob. 5–162 assuming heat gain of the chiller is negligible.

5–164 In a dairy plant, milk at 4°C is pasteurized continuously at 72°C at a rate of 20 L/s for 24 h a day and 365 days a year. The milk is heated to the pasteurizing temperature by hot water heated in a natural-gas-fired boiler that has an efficiency of 90 percent. The pasteurized milk is then cooled by cold water at 18°C before it is finally refrigerated back to 4°C. To save energy and money, the plant installs a regenerator that has an effectiveness of 82 percent. If the cost of natural gas is \$1.10/therm (1 therm = 105,500 kJ), determine how much energy and money the regenerator will save this company per year.

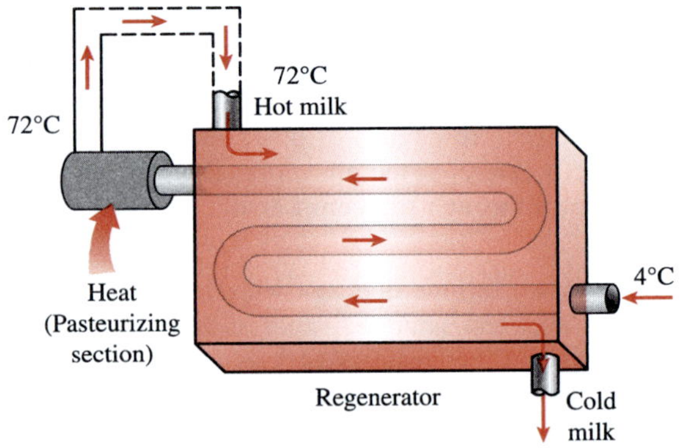

FIGURE P5–164

5–165E A refrigeration system is being designed to cool eggs (ρ = 67.4 lbm/ft^3 and c_p = 0.80 Btu/lbm·°F) with an average mass of 0.14 lbm from an initial temperature of 90°F to a final average temperature of 50°F by air at 34°F at a rate of 10,000 eggs per hour. Determine (*a*) the rate of heat removal from the eggs, in Btu/h and (*b*) the required volume flow rate of air, in ft^3/h, if the temperature rise of air is not to exceed 10°F.

5–166 A glass bottle washing facility uses a well-agitated hot-water bath at 55°C that is placed on the ground. The bottles enter at a rate of 800 per minute at an ambient temperature of 20°C and leave at the water temperature. Each bottle has a mass of 150 g and removes 0.2 g of water as it leaves the bath wet. Make-up water is supplied at 15°C. Disregarding any heat losses from the outer surfaces of the bath, determine the rate at which (*a*) water and (*b*) heat must be supplied to maintain steady operation.

5–167 Long aluminum wires of diameter 5 mm (ρ = 2702 kg/m^3 and c_p = 0.896 kJ/kg·°C) are extruded at a temperature of 350°C and are cooled to 50°C in atmospheric air at 25°C. If the wire is extruded at a velocity of 8 m/min, determine the rate of heat transfer from the wire to the extrusion room.

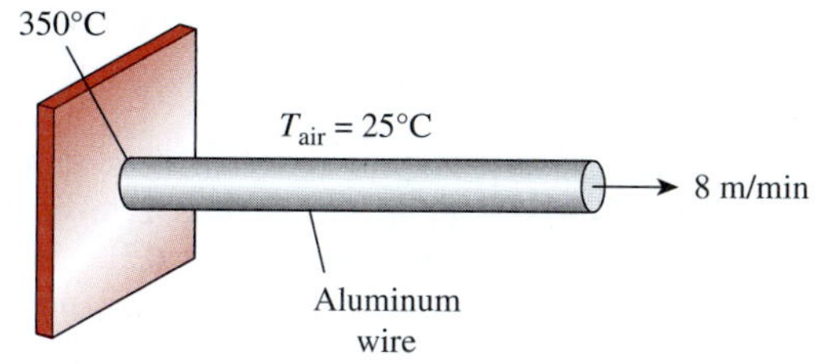

FIGURE P5–167

5–168 Repeat Prob. 5–167 for a copper wire (ρ = 8950 kg/m^3 and c_p = 0.383 kJ/kg·°C).

5–169 Steam at 40°C condenses on the outside of a 5-m-long, 3-cm-diameter thin horizontal copper tube by cooling water that enters the tube at 25°C at an average velocity of 2 m/s and leaves at 35°C. Determine the rate of condensation of steam. *Answer:* 0.0245 kg/s

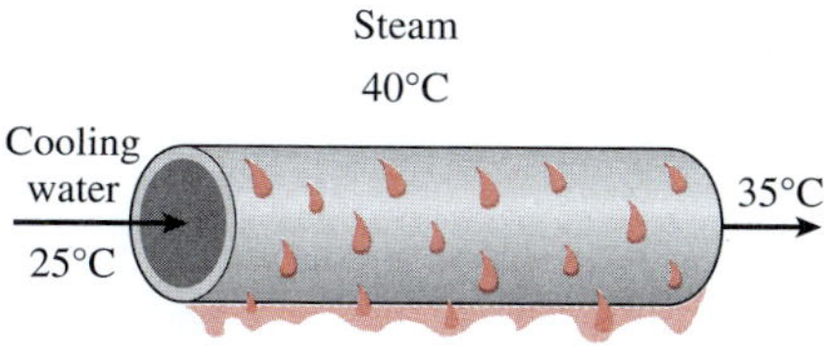

FIGURE P5–169

5–170E Steam at 60 psia and 350°F is mixed with water at 40°F and 60 psia steadily in an adiabatic device. Steam enters the device at a rate of 0.05 lbm/s, while the water enters at 1 lbm/s. Determine the temperature of the mixture leaving this device when the outlet pressure is 60 psia. *Answer:* 97°F

5–171 A constant-pressure R-134a vapor separation unit separates the liquid and vapor portions of a saturated mixture into two separate outlet streams. Determine the flow power needed to pass 6 L/s of R-134a at 320 kPa and 55 percent quality through this unit. What is the mass flow rate, in kg/s, of the two outlet streams?

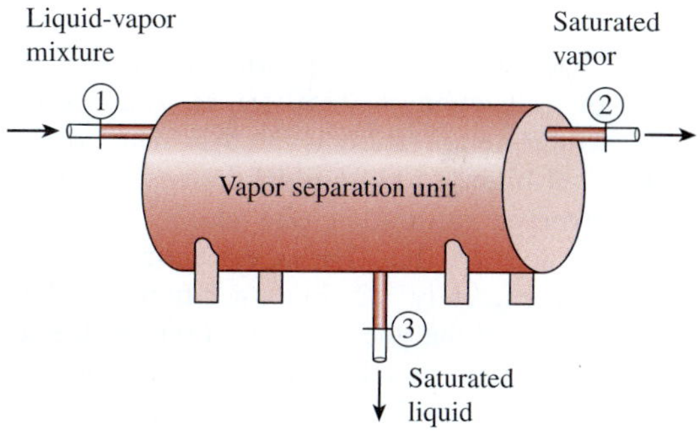

FIGURE P5–171

5–172E A small positioning control rocket in a satellite is driven by a 2-ft^3 container filled with R-134a at −10°F. Upon launch, the container is completely filled with saturated liquid R-134a. The rocket is designed for short bursts of 5-s duration. During each burst, the mass flow rate leaving the rocket is 0.05 lbm/s. How many such bursts can this rocket experience before the quality in the container is 90 percent or more, presuming that the temperature of the container contents is maintained at −10°F? *Answers:* 680

5–173E During the inflation and deflation of a safety airbag in an automobile, the gas enters the airbag with a specific volume of 15 ft^3/lbm and at a mass flow rate that varies with time as illustrated in Fig. P5–173E. The gas leaves this airbag with a specific volume of 13 ft^3/lbm, with a mass flow rate that varies with time, as shown in Fig. P5–173E. Plot the volume of this bag (i.e., airbag size) as a function of time, in ft^3.

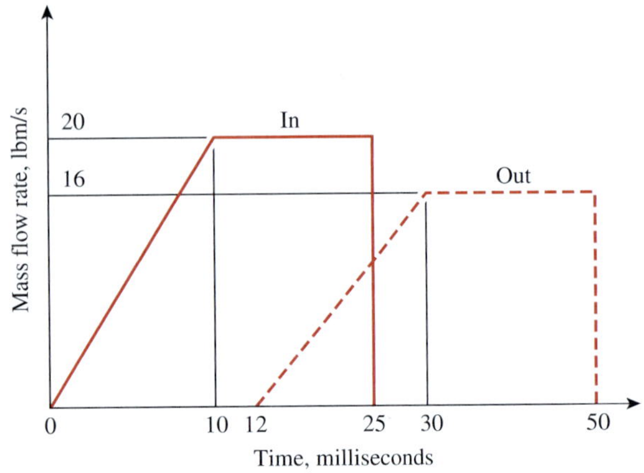

FIGURE P5–173E

5–174E The average atmospheric pressure in Spokane, Washington (elevation = 2350 ft), is 13.5 psia, and the average winter temperature is 36.5°F. The pressurization test of a 9-ft-high, 4500-ft^2 older home revealed that the seasonal average infiltration rate of the house is 2.2 air changes per hour (ACH). That is, the entire air in the house is replaced completely 2.2 times per hour by the outdoor air. It is suggested that the infiltration rate of the house can be reduced by half to 1.1 ACH by winterizing the doors and the windows. If the house is heated by natural gas whose unit cost is $1.24/therm and the heating season can be taken to be six months, determine how much the home owner will save from the heating costs per year by this winterization project. Assume the house is maintained at 72°F at all times and the efficiency of the furnace is 0.92. Also assume the latent heat load during the heating season to be negligible.

5–175 Determine the rate of sensible heat loss from a building due to infiltration if the outdoor air at −5°C and 90 kPa enters the building at a rate of 35 L/s when the indoors is maintained at 20°C.

5–176 Consider a large classroom on a hot summer day with 150 students, each dissipating 60 W of sensible heat. All the lights, with 6.0 kW of rated power, are kept on. The room has no external walls, and thus heat gain through the walls and the roof is negligible. Chilled air is available at 15°C, and the temperature of the return air is not to exceed 25°C. Determine the required flow rate of air, in kg/s, that needs to be supplied to the room to keep the average temperature of the room constant. *Answer:* 1.49 kg/s

5–177 A fan is powered by a 0.5-hp motor and delivers air at a rate of 85 m^3/min. Determine the highest value for the average velocity of air mobilized by the fan. Take the density of air to be 1.18 kg/m^3.

5–178 An air-conditioning system requires airflow at the main supply duct at a rate of 130 m^3/min. The average velocity of air in the circular duct is not to exceed 8 m/s to avoid excessive vibration and pressure drops. Assuming the fan converts 80 percent of the electrical energy it consumes into kinetic energy of air, determine the size of the electric motor needed to drive the fan and the diameter of the main duct. Take the density of air to be 1.20 kg/m^3.

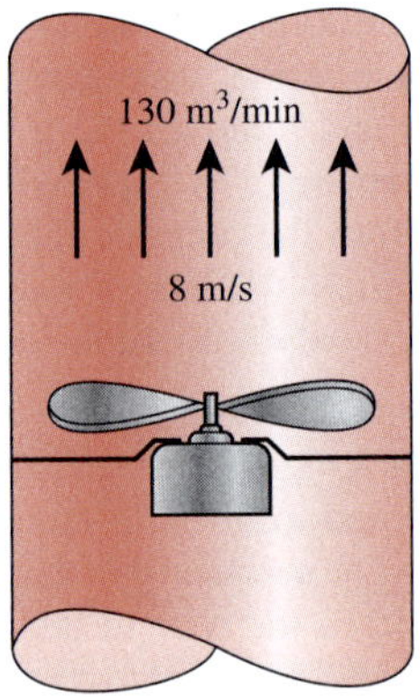

FIGURE P5–178

5–179 An adiabatic air compressor is to be powered by a direct-coupled adiabatic steam turbine that is also driving a generator. Steam enters the turbine at 12.5 MPa and 500°C at a rate of 25 kg/s and exits at 10 kPa and a quality of 0.92. Air enters the compressor at 98 kPa and 295 K at a rate of 10 kg/s and exits at 1 MPa and 620 K. Determine the net power delivered to the generator by the turbine.

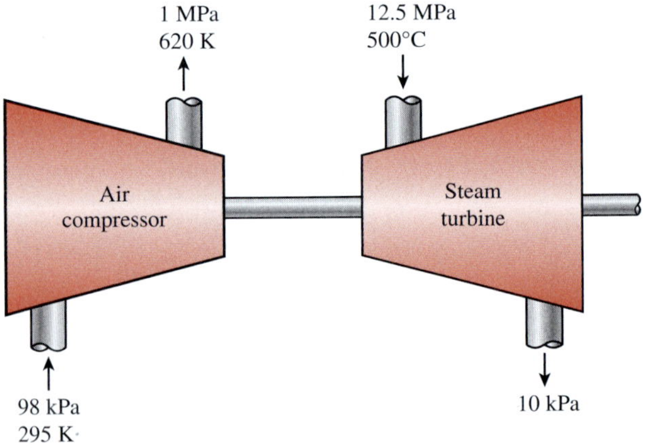

FIGURE P5–179

5–180 Determine the power input for a compressor that compresses helium from 150 kPa and 20°C to 400 kPa and 200°C. Helium enters this compressor through a 0.1-m^2 pipe at a velocity of 15 m/s.

5–181 Refrigerant 134a enters a compressor with a mass flow rate of 5 kg/s and a negligible velocity. The refrigerant enters the compressor as a saturated vapor at 10°C and leaves the compressor at 1400 kPa with an enthalpy of 281.39 kJ/kg and a velocity of 50 m/s. The rate of work done on the refrigerant is measured to be 132.4 kW. If the elevation change between the compressor inlet and exit is negligible, determine the rate of heat transfer associated with this process, in kW.

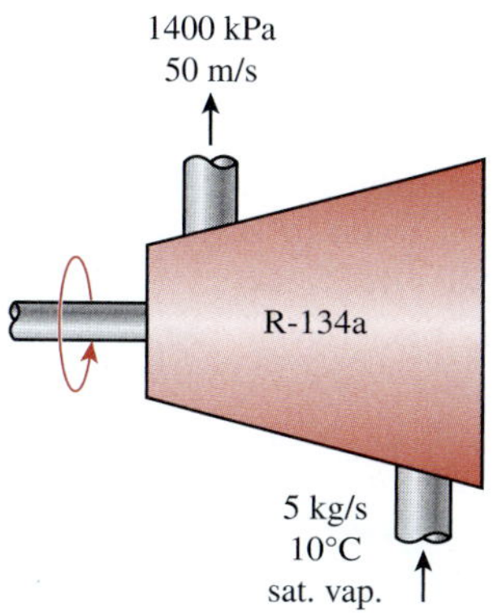

FIGURE P5–181

5–182 Submarines change their depth by adding or removing air from rigid ballast tanks, thereby displacing seawater in the tanks. Consider a submarine that has a 700 m^3 air-ballast tank originally partially filled with 100 m^3 of air at 1500 kPa and 15°C. For the submarine to surface, air at 1500 kPa and 20°C is pumped into the ballast tank, until it is entirely filled with air. The tank is filled so quickly that the process is adiabatic and the seawater leaves the tank at 15°C. Determine the final temperature and mass of the air in the ballast tank.

5–183 In the preceding problem, presume that air is added to the tank in such a way that the temperature and pressure of the air in the tank remain constant. Determine the final mass of the air in the ballast tank under this condition. Also determine the total heat transfer while the tank is being filled in this manner.

5–184 A steel cylinder initially contains nitrogen gas at 200 kPa and 25°C. The cylinder is connected to a supply line containing nitrogen at 800 kPa and 25°C. A valve is opened, allowing nitrogen to flow into the cylinder until the pressure reaches 800 kPa. The cylinder internal volume is 0.1 m^3, its mass (cylinder only) is 50 kg, and its specific heat is 0.43 kJ/kg·K. Find the final mass of nitrogen in the tank and the final temperature of the nitrogen assuming constant specific heats at room temperature for nitrogen, and assuming (*a*) no heat transfer from the nitrogen to the tank, and (*b*) rapid heat transfer between the nitrogen and tank such that the cylinder and nitrogen remain in thermal equilibrium during the process (with negligible heat transfer from the cylinder to its external surroundings.) *Answers:* (*a*) 379 K, 0.711 kg, (*b*) 301 K, 0.896 kg

5–185 A tank with an internal volume of 1 m^3 contains air at 800 kPa and 25°C. A valve on the tank is opened allowing air to escape and the pressure inside quickly drops to 150 kPa, at which point the valve is closed. Assume there is negligible heat transfer from the tank to the air left in the tank.
(*a*) Using the approximation $h_e \approx \text{constant} = h_{e,\text{avg}} = 0.5(h_1 + h_2)$, calculate the mass withdrawn during the process.
(*b*) Consider the same process but broken into two parts. That is, consider an intermediate state at $P_2 = 400$ kPa, calculate the mass removed during the process from $P_1 = 800$ kPa to P_2 and then the mass removed during the process from P_2 to $P_3 = 150$ kPa, using the type of approximation used in part (*a*), and add the two to get the total mass removed.
(*c*) Calculate the mass removed if the variation of h_e is accounted for.

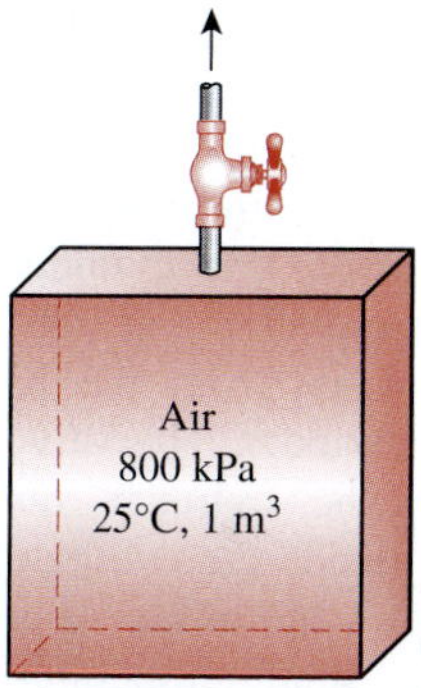

FIGURE P5–185

5–186 A liquid R-134a bottle has an internal volume of 0.001 m^3. Initially it contains 0.4 kg of R-134a (saturated mixture) at 26°C. A valve is opened and R-134a vapor only (no liquid) is allowed to escape slowly such that temperature remains constant until the mass of R-134a remaining is 0.1 kg. Find the heat transfer necessary with the surroundings to maintain the temperature and pressure of the R-134a constant.

5–187 Steam enters a turbine steadily at 7 MPa and 600°C with a velocity of 60 m/s and leaves at 25 kPa with a quality of 95 percent. A heat loss of 20 kJ/kg occurs during the process. The inlet area of the turbine is 150 cm^2, and the exit area is 1400 cm^2. Determine (*a*) the mass flow rate of the steam, (*b*) the exit velocity, and (*c*) the power output.

5–188 Reconsider Prob. 5–187. Using EES (or other) software, investigate the effects of turbine exit area and turbine exit pressure on the exit velocity and power output of the turbine. Let the exit pressure vary from 10 to 50 kPa (with the same quality), and the exit area to vary from 1000 to 3000 cm². Plot the exit velocity and the power outlet against the exit pressure for the exit areas of 1000, 2000, and 3000 cm², and discuss the results.

5–189 In large gas-turbine power plants, air is preheated by the exhaust gases in a heat exchanger called the *regenerator* before it enters the combustion chamber. Air enters the regenerator at 1 MPa and 550 K at a mass flow rate of 800 kg/min. Heat is transferred to the air at a rate of 3200 kJ/s. Exhaust gases enter the regenerator at 140 kPa and 800 K and leave at 130 kPa and 600 K. Treating the exhaust gases as air, determine (*a*) the exit temperature of the air and (*b*) the mass flow rate of exhaust gases. *Answers:* (*a*) 775 K, (*b*) 14.9 kg/s

5–190 It is proposed to have a water heater that consists of an insulated pipe of 5-cm diameter and an electric resistor inside. Cold water at 20°C enters the heating section steadily at a rate of 30 L/min. If water is to be heated to 55°C, determine (*a*) the power rating of the resistance heater and (*b*) the average velocity of the water in the pipe.

5–191 An insulated vertical piston–cylinder device initially contains 0.11 m³ of air at 150 kPa and 22°C. At this state, a linear spring touches the piston but exerts no force on it. The cylinder is connected by a valve to a line that supplies air at 700 kPa and 22°C. The valve is opened, and air from the high-pressure line is allowed to enter the cylinder. The valve is turned off when the pressure inside the cylinder reaches 600 kPa. If the enclosed volume inside the cylinder doubles during this process, determine (*a*) the mass of air that entered the cylinder, and (*b*) the final temperature of the air inside the cylinder.

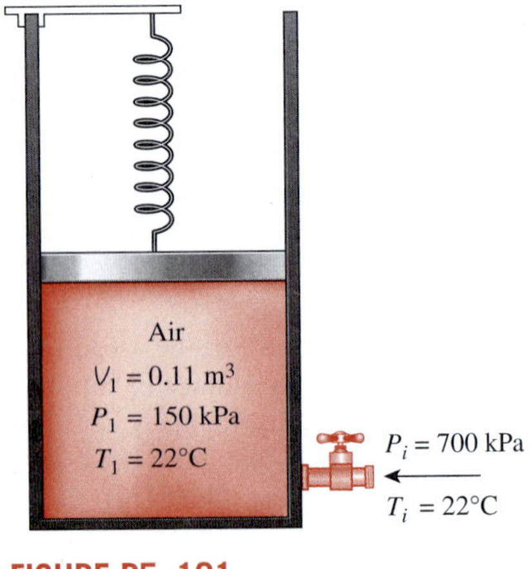

FIGURE P5–191

5–192 A piston–cylinder device initially contains 2 kg of refrigerant-134a at 800 kPa and 80°C. At this state, the piston is touching on a pair of stops at the top. The mass of the piston is such that a 500-kPa pressure is required to move it. A valve at the bottom of the tank is opened, and R-134a is withdrawn from the cylinder. After a while, the piston is observed to move and the valve is closed when half of the refrigerant is withdrawn from the tank and the temperature in the tank drops to 20°C. Determine (*a*) the work done and (*b*) the heat transfer. *Answers:* (*a*) 11.6 kJ, (*b*) 60.7 kJ

5–193 The pump of a water distribution system is powered by a 6-kW electric motor whose efficiency is 95 percent. The water flow rate through the pump is 18 L/s. The diameters of the inlet and outlet pipes are the same, and the elevation difference across the pump is negligible. If the pressures at the inlet and outlet of the pump are measured to be 100 kPa and 300 kPa (absolute), respectively, determine (*a*) the mechanical efficiency of the pump and (*b*) the temperature rise of water as it flows through the pump due to the mechanical inefficiency. *Answers:* (*a*) 63.2 percent, (*b*) 0.028°C

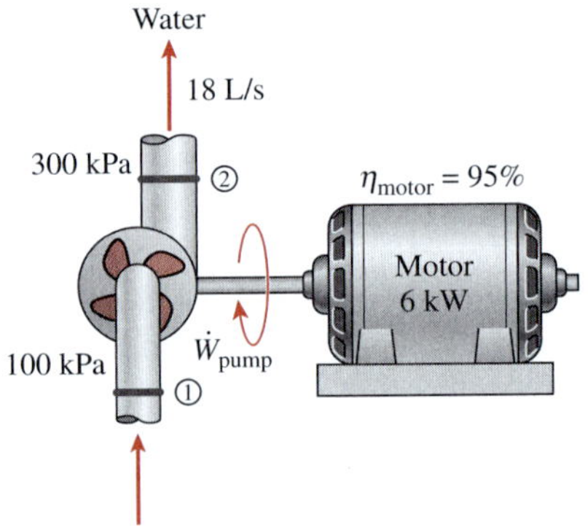

FIGURE P5–193

5–194 The hot-water needs of a household are met by a 60-L electric water heater whose heaters are rated at 1.6 kW. The hot-water tank is initially full with hot water at 80°C. Somebody takes a shower by mixing a constant flow of hot water from the tank with cold water at 20°C at a rate of 0.06 kg/s. After a shower period of 8 min, the water temperature in the tank is measured to drop to 60°C. The heater remained on during the shower and hot water withdrawn from the tank is replaced by cold water at the same flow rate. Determine the mass flow rate of hot water withdrawn from the tank during the shower and the average temperature of mixed water used for the shower.

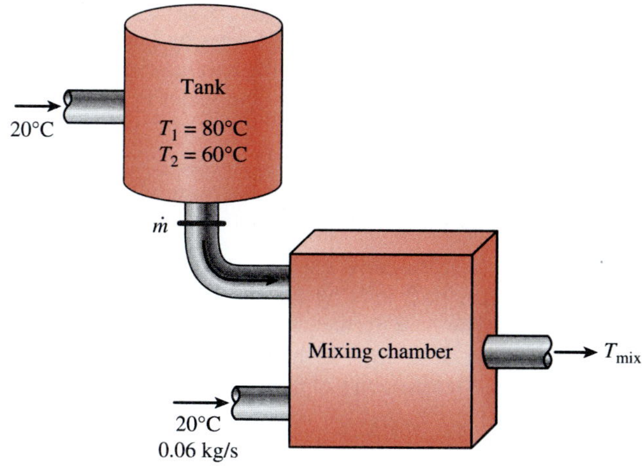

FIGURE P5–194

5–195 The turbocharger of an internal combustion engine consists of a turbine and a compressor. Hot exhaust gases flow through the turbine to produce work and the work output from the turbine is used as the work input to the compressor. The pressure of ambient air is increased as it flows through the compressor before it enters the engine cylinders. Thus, the purpose of a turbocharger is to increase the pressure of air so that more air gets into the cylinder. Consequently, more fuel can be burned and more power can be produced by the engine.

In a turbocharger, exhaust gases enter the turbine at 400°C and 120 kPa at a rate of 0.02 kg/s and leave at 350°C. Air enters the compressor at 50°C and 100 kPa and leaves at 130 kPa at a rate of 0.018 kg/s. The compressor increases the air pressure with a side effect: It also increases the air temperature, which increases the possibility of a gasoline engine to experience an engine knock. To avoid this, an aftercooler is placed after the compressor to cool the warm air by cold ambient air before it enters the engine cylinders. It is estimated that the aftercooler must decrease the air temperature below 80°C if knock is to be avoided. The cold ambient air enters the aftercooler at 30°C and leaves at 40°C. Disregarding any frictional losses in the turbine and the compressor and treating the exhaust gases as air, determine (*a*) the temperature of the air at the compressor outlet and (*b*) the minimum volume flow rate of ambient air required to avoid knock.

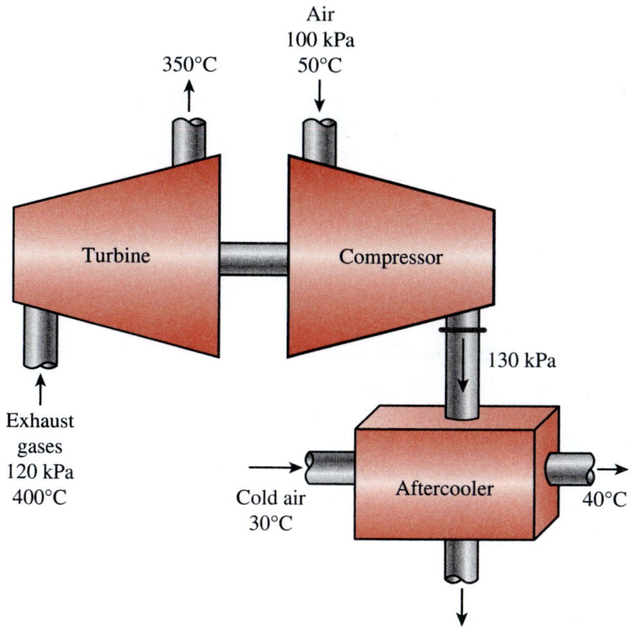

FIGURE P5–195

5–196 A pressure cooker is a pot that cooks food much faster than ordinary pots by maintaining a higher pressure and temperature during cooking. The pressure inside the pot is controlled by a pressure regulator (the petcock) that keeps the pressure at a constant level by periodically allowing some steam to escape, thus preventing any excess pressure buildup. Pressure cookers, in general, maintain a gage pressure of 2 atm (or 3 atm absolute) inside. Therefore, pressure cookers cook at a temperature of about 133°C instead of 100°C, cutting the cooking time by as much as 70 percent while minimizing the loss of nutrients. The newer pressure cookers use a spring valve with several pressure settings rather than a weight on the cover.

A certain pressure cooker has a volume of 6 L and an operating pressure of 75 kPa gage. Initially, it contains 1 kg of water. Heat is supplied to the pressure cooker at a rate of 500 W for 30 min after the operating pressure is reached. Assuming an atmospheric pressure of 100 kPa, determine (*a*) the temperature at which cooking takes place and (*b*) the amount of water left in the pressure cooker at the end of the process. *Answers:* (*a*) 116.04°C, (*b*) 0.6 kg

5–197 A D_0 = 10-m-diameter tank is initially filled with water 2 m above the center of a D = 10-cm-diameter valve near the bottom. The tank surface is open to the atmosphere, and the tank drains through a L = 100-m-long pipe connected to the valve. The friction factor of the pipe is given to be f = 0.015, and the discharge velocity is expressed as $V = \sqrt{\dfrac{2gz}{1.5 + fL/D}}$ where z is the water height above the center of the valve. Determine (*a*) the initial discharge velocity from the tank and (*b*) the time required to empty the tank. The tank can be considered to be empty when the water level drops to the center of the valve.

5–198 The velocity of a liquid flowing in a circular pipe of radius R varies from zero at the wall to a maximum at the pipe center. The velocity distribution in the pipe can be represented as $V(r)$, where r is the radial distance from the pipe center. Based on the definition of mass flow rate $\dot{m}$, obtain a relation for the average velocity in terms of $V(r)$, R, and r.

5–199 Two streams of the same ideal gas having different mass flow rates and temperatures are mixed in a steady-flow, adiabatic mixing device. Assuming constant specific heats, find the simplest expression for the mixture temperature written in the form

$$T_3 = f\left(\frac{\dot{m}_1}{\dot{m}_3}, \frac{\dot{m}_2}{\dot{m}_3}, T_1, T_2\right)$$

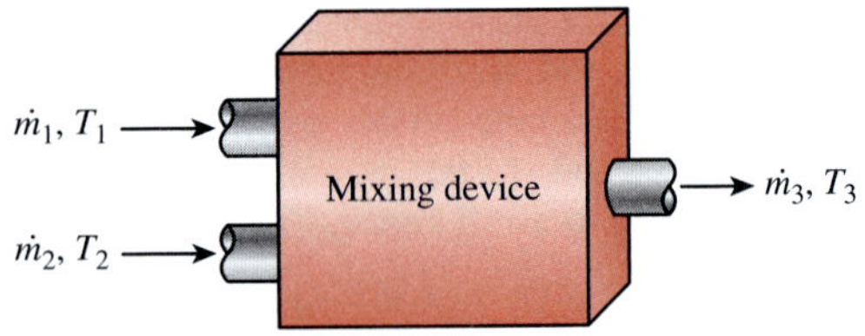

FIGURE P5–199

5–200 A rigid container filled with an ideal gas is heated while the gas is released so that the temperature of the gas remaining in the container stays constant. This container has a single outlet. Derive an expression for the mass flow rate at the outlet as a function of the rate of pressure change in the container.

5–201 Consider an evacuated rigid bottle of volume V that is surrounded by the atmosphere at pressure P_0 and temperature T_0. A valve at the neck of the bottle is now opened and the atmospheric air is allowed to flow into the bottle. The air trapped in the bottle eventually reaches thermal equilibrium with the atmosphere as a result of heat transfer through the wall of the bottle. The valve remains open during the process so that the trapped air also reaches mechanical equilibrium with the atmosphere. Determine the net heat transfer through the wall of the bottle during this filling process in terms of the properties of the system and the surrounding atmosphere.

Fundamentals of Engineering (FE) Exam Problems

5–202 Steam is accelerated by a nozzle steadily from a low velocity to a velocity of 280 m/s at a rate of 2.5 kg/s. If the temperature and pressure of the steam at the nozzle exit are 400°C and 2 MPa, the exit area of the nozzle is

(*a*) 8.4 cm^2 (*b*) 10.7 cm^2
(*c*) 13.5 cm^2 (*d*) 19.6 cm^2
(*e*) 23.0 cm^2

5–203 Steam enters a diffuser steadily at 0.5 MPa, 300°C, and 122 m/s at a rate of 3.5 kg/s. The inlet area of the diffuser is

(*a*) 15 cm^2 (*b*) 50 cm^2
(*c*) 105 cm^2 (*d*) 150 cm^2
(*e*) 190 cm^2

5–204 An adiabatic heat exchanger is used to heat cold water at 15°C entering at a rate of 5 kg/s by hot air at 90°C entering also at a rate of 5 kg/s. If the exit temperature of hot air is 20°C, the exit temperature of cold water is

(*a*) 27°C (*b*) 32°C
(*c*) 52°C (*d*) 85°C
(*e*) 90°C

5–205 A heat exchanger is used to heat cold water at 15°C entering at a rate of 2 kg/s by hot air at 85°C entering at a rate of 3 kg/s. The heat exchanger is not insulated and is losing heat at a rate of 25 kJ/s. If the exit temperature of hot air is 20°C, the exit temperature of cold water is

(*a*) 28°C (*b*) 35°C
(*c*) 38°C (*d*) 41°C
(*e*) 80°C

5–206 An adiabatic heat exchanger is used to heat cold water at 15°C entering at a rate of 5 kg/s by hot water at 90°C entering at a rate of 4 kg/s. If the exit temperature of hot water is 50°C, the exit temperature of cold water is

(*a*) 42°C (*b*) 47°C
(*c*) 55°C (*d*) 78°C
(*e*) 90°C

5–207 In a shower, cold water at 10°C flowing at a rate of 5 kg/min is mixed with hot water at 60°C flowing at a rate of 2 kg/min. The exit temperature of the mixture is

(*a*) 24.3°C (*b*) 35.0°C
(*c*) 40.0°C (*d*) 44.3°C
(*e*) 55.2°C

5–208 In a heating system, cold outdoor air at 7°C flowing at a rate of 4 kg/min is mixed adiabatically with heated air at 70°C flowing at a rate of 3 kg/min. The exit temperature of the mixture is

(*a*) 34°C (*b*) 39°C
(*c*) 45°C (*d*) 63°C
(*e*) 77°C

5–209 Hot combustion gases (assumed to have the properties of air at room temperature) enter a gas turbine at 1 MPa and 1500 K at a rate of 0.1 kg/s, and exit at 0.2 MPa and 900 K. If heat is lost from the turbine to the surroundings at a rate of 15 kJ/s, the power output of the gas turbine is

(*a*) 15 kW (*b*) 30 kW
(*c*) 45 kW (*d*) 60 kW
(*e*) 75 kW

5–210 Steam expands in a turbine from 4 MPa and 500°C to 0.5 MPa and 250°C at a rate of 1350 kg/h. Heat is lost from the turbine at a rate of 25 kJ/s during the process. The power output of the turbine is

(*a*) 157 kW (*b*) 207 kW
(*c*) 182 kW (*d*) 287 kW
(*e*) 246 kW

5–211 Steam is compressed by an adiabatic compressor from 0.2 MPa and 150°C to 0.8 MPa and 350°C at a rate of 1.30 kg/s. The power input to the compressor is

(*a*) 511 kW (*b*) 393 kW
(*c*) 302 kW (*d*) 717 kW
(*e*) 901 kW

5–212 Refrigerant-134a is compressed by a compressor from the saturated vapor state at 0.14 MPa to 0.9 MPa and 60°C at a rate of 0.108 kg/s. The refrigerant is cooled at a rate of 1.10 kJ/s during compression. The power input to the compressor is

(*a*) 4.94 kW (*b*) 6.04 kW
(*c*) 7.14 kW (*d*) 7.50 kW
(*e*) 8.13 kW

5–213 Refrigerant-134a expands in an adiabatic turbine from 1.2 MPa and 100°C to 0.18 MPa and 50°C at a rate of 1.25 kg/s. The power output of the turbine is

(*a*) 44.7 kW (*b*) 66.4 kW
(*c*) 72.7 kW (*d*) 89.2 kW
(*e*) 112.0 kW

5–214 Refrigerant-134a at 1.4 MPa and 90°C is throttled to a pressure of 0.6 MPa. The temperature of the refrigerant after throttling is

(*a*) 22°C (*b*) 56°C
(*c*) 82°C (*d*) 80°C
(*e*) 90°C

5–215 Air at 27°C and 5 atm is throttled by a valve to 1 atm. If the valve is adiabatic and the change in kinetic energy is negligible, the exit temperature of air will be

(*a*) 10°C (*b*) 15°C
(*c*) 20°C (*d*) 23°C
(*e*) 27°C

5–216 Steam at 1 MPa and 300°C is throttled adiabatically to a pressure of 0.4 MPa. If the change in kinetic energy is negligible, the specific volume of the steam after throttling is

(*a*) 0.358 m^3/kg (*b*) 0.233 m^3/kg
(*c*) 0.375 m^3/kg (*d*) 0.646 m^3/kg
(*e*) 0.655 m^3/kg

5–217 Air is to be heated steadily by an 8-kW electric resistance heater as it flows through an insulated duct. If the air enters at 50°C at a rate of 2 kg/s, the exit temperature of air is

(*a*) 46.0°C (*b*) 50.0°C
(*c*) 54.0°C (*d*) 55.4°C
(*e*) 58.0°C

5–218 Saturated water vapor at 40°C is to be condensed as it flows through a tube at a rate of 0.20 kg/s. The condensate leaves the tube as a saturated liquid at 40°C. The rate of heat transfer from the tube is

(*a*) 34 kJ/s (*b*) 481 kJ/s
(*c*) 2406 kJ/s (*d*) 514 kJ/s
(*e*) 548 kJ/s

Design and Essay Problems

5–219 Pneumatic nail drivers used in construction require 0.02 ft^3 of air at 100 psia and 1 Btu of energy to drive a single nail. You have been assigned the task of designing a compressed-air storage tank with enough capacity to drive 500 nails. The pressure in this tank cannot exceed 500 psia, and the temperature cannot exceed that normally found at a construction site. What is the maximum pressure to be used in the tank and what is the tank's volume?

5–220 You have been given the responsibility of picking a steam turbine for an electrical-generation station that is to produce 300 MW of electrical power that will sell for $0.05 per kilowatt-hour. The boiler will produce steam at 700 psia and 700°F, and the condenser is planned to operate at 80°F. The cost of generating and condensing the steam is $0.01 per kilowatt-hour of electricity produced. You have narrowed your selection to the three turbines in the table below. Your criterion for selection is to pay for the equipment as quickly as possible. Which turbine should you choose?

Turbine	Capacity (MW)	η	Cost ($Million)	Operating Cost ($/kWh)
A	50	0.9	5	0.01
B	100	0.92	11	0.01
C	100	0.93	10.5	0.015

5–221E You are to design a small, directional control rocket to operate in space by providing as many as 100 bursts of 5 seconds each with a mass flow rate of 0.5 lbm/s at a velocity of 400 ft/s. Storage tanks that will contain up to 3000 psia are available, and the tanks will be located in an environment whose temperature is 40°F. Your design criterion is to minimize the volume of the storage tank. Should you use a compressed-air or an R-134a system?

5–222 An air cannon uses compressed air to propel a projectile from rest to a final velocity. Consider an air cannon that is to accelerate a 10-gram projectile to a speed of 300 m/s using compressed air, whose temperature cannot exceed 20°C. The volume of the storage tank is not to exceed 0.1 m^3. Select the storage volume size and maximum storage pressure that requires the minimum amount of energy to fill the tank.

5–223 To maintain altitude, the temperature of the air inside a hot-air balloon must remain within a 1°C band, while the volume cannot vary by more than 1 percent. At a 300-m altitude, the air in a 1000 m^3 hot-air balloon needs to maintain a 35°C average temperature. This balloon loses heat at a rate of 3 kW through the fabric. When the burner is activated, it adds 30 kg/s of air at 200°C and 100 kPa to the balloon. When the flap that allows air to escape is opened, air leaves the balloon at a rate of 20 kg/s. Design the burner and exhaust-flap control cycles (on time and off time) necessary to maintain the balloon at a 300-m altitude.

CHAPTER

6

THE SECOND LAW OF THERMODYNAMICS

To this point, we have focused our attention on the first law of thermodynamics, which requires that energy be conserved during a process. In this chapter, we introduce the second law of thermodynamics, which asserts that processes occur in a certain direction and that energy has quality as well as quantity. A process cannot take place unless it satisfies both the first and second laws of thermodynamics. In this chapter, the thermal energy reservoirs, reversible and irreversible processes, heat engines, refrigerators, and heat pumps are introduced first. Various statements of the second law are followed by a discussion of perpetual-motion machines and the thermodynamic temperature scale. The Carnot cycle is introduced next, and the Carnot principles are discussed. Finally, the idealized Carnot heat engines, refrigerators, and heat pumps are examined.

Objectives

The objectives of Chapter 6 are to:

- Introduce the second law of thermodynamics.
- Identify valid processes as those that satisfy both the first and second laws of thermodynamics.
- Discuss thermal energy reservoirs, reversible and irreversible processes, heat engines, refrigerators, and heat pumps.
- Describe the Kelvin–Planck and Clausius statements of the second law of thermodynamics.
- Discuss the concepts of perpetual-motion machines.
- Apply the second law of thermodynamics to cycles and cyclic devices.
- Apply the second law to develop the absolute thermodynamic temperature scale.
- Describe the Carnot cycle.
- Examine the Carnot principles, idealized Carnot heat engines, refrigerators, and heat pumps.
- Determine the expressions for the thermal efficiencies and coefficients of performance for reversible heat engines, heat pumps, and refrigerators.

6–1 ▪ INTRODUCTION TO THE SECOND LAW

FIGURE 6–1
A cup of hot coffee does not get hotter in a cooler room.

In Chaps. 4 and 5, we applied the *first law of thermodynamics*, or the *conservation of energy principle*, to processes involving closed and open systems. As pointed out repeatedly in those chapters, energy is a conserved property, and no process is known to have taken place in violation of the first law of thermodynamics. Therefore, it is reasonable to conclude that a process must satisfy the first law to occur. However, as explained here, satisfying the first law alone does not ensure that the process will actually take place.

It is common experience that a cup of hot coffee left in a cooler room eventually cools off (Fig. 6–1). This process satisfies the first law of thermodynamics since the amount of energy lost by the coffee is equal to the amount gained by the surrounding air. Now let us consider the reverse process—the hot coffee getting even hotter in a cooler room as a result of heat transfer from the room air. We all know that this process never takes place. Yet, doing so would not violate the first law as long as the amount of energy lost by the air is equal to the amount gained by the coffee.

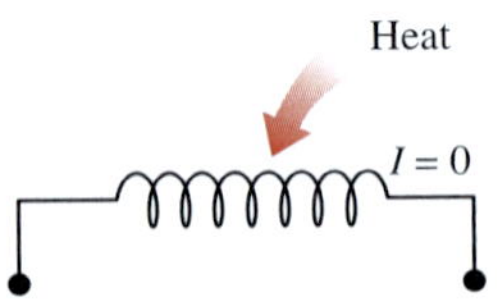

FIGURE 6–2
Transferring heat to a wire will not generate electricity.

As another familiar example, consider the heating of a room by the passage of electric current through a resistor (Fig. 6–2). Again, the first law dictates that the amount of electric energy supplied to the resistance wires be equal to the amount of energy transferred to the room air as heat. Now let us attempt to reverse this process. It will come as no surprise that transferring some heat to the wires does not cause an equivalent amount of electric energy to be generated in the wires.

Finally, consider a paddle-wheel mechanism that is operated by the fall of a mass (Fig. 6–3). The paddle wheel rotates as the mass falls and stirs a fluid within an insulated container. As a result, the potential energy of the mass decreases, and the internal energy of the fluid increases in accordance with the conservation of energy principle. However, the reverse process, raising the mass by transferring heat from the fluid to the paddle wheel, does not occur in nature, although doing so would not violate the first law of thermodynamics.

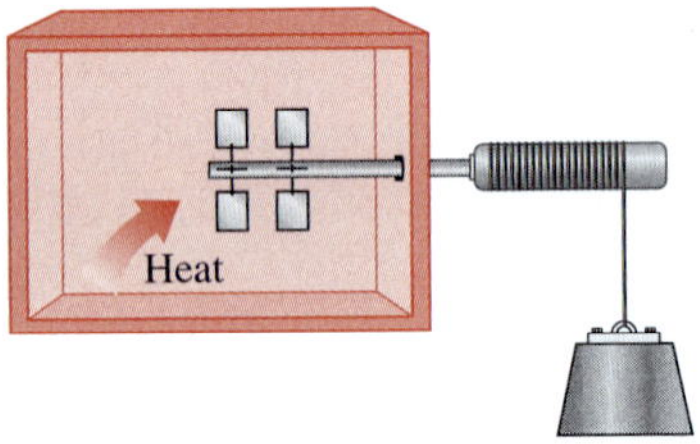

FIGURE 6–3
Transferring heat to a paddle wheel will not cause it to rotate.

It is clear from these arguments that processes proceed in a *certain direction* and not in the reverse direction (Fig. 6–4). The first law places no restriction on the direction of a process, but satisfying the first law does not ensure that the process can actually occur. This inadequacy of the first law to identify whether a process can take place is remedied by introducing another general principle, the *second law of thermodynamics*. We show later in this chapter that the reverse processes discussed above violate the second law of thermodynamics. This violation is easily detected with the help of a property, called *entropy*, defined in Chap. 7. A process cannot occur unless it satisfies both the first and the second laws of thermodynamics (Fig. 6–5).

There are numerous valid statements of the second law of thermodynamics. Two such statements are presented and discussed later in this chapter in relation to some engineering devices that operate on cycles.

FIGURE 6–4
Processes occur in a certain direction, and not in the reverse direction.

The use of the second law of thermodynamics is not limited to identifying the direction of processes. The second law also asserts that energy has *quality* as well as quantity. The first law is concerned with the quantity of energy and the transformations of energy from one form to another with no regard to its quality. Preserving the quality of energy is a major concern to engineers, and the second law provides the necessary means to determine the quality as well as the degree of degradation of energy during a process. As discussed later in this chapter, more of high-temperature energy can be converted to work, and thus it has a higher quality than the same amount of energy at a lower temperature.

The second law of thermodynamics is also used in determining the *theoretical limits* for the performance of commonly used engineering systems, such as heat engines and refrigerators, as well as predicting the *degree of completion* of chemical reactions. The second law is also closely associated with the concept of *perfection*. In fact, the second law *defines* perfection for thermodynamic processes. It can be used to quantify the level of perfection of a process, and point the direction to eliminate imperfections effectively.

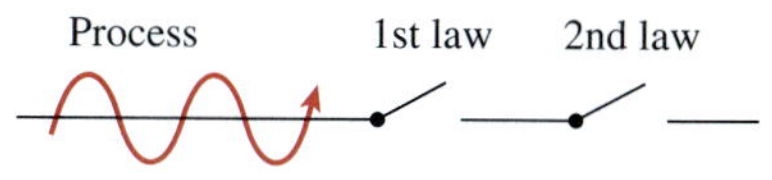

FIGURE 6–5
A process must satisfy both the first and second laws of thermodynamics to proceed.

6–2 ▪ THERMAL ENERGY RESERVOIRS

In the development of the second law of thermodynamics, it is very convenient to have a hypothetical body with a relatively large *thermal energy capacity* (mass × specific heat) that can supply or absorb finite amounts of heat without undergoing any change in temperature. Such a body is called a **thermal energy reservoir**, or just a reservoir. In practice, large bodies of water such as oceans, lakes, and rivers as well as the atmospheric air can be modeled accurately as thermal energy reservoirs because of their large thermal energy storage capabilities or thermal masses (Fig. 6–6). The *atmosphere*, for example, does not warm up as a result of heat losses from residential buildings in winter. Likewise, megajoules of waste energy dumped in large rivers by power plants do not cause any significant change in water temperature.

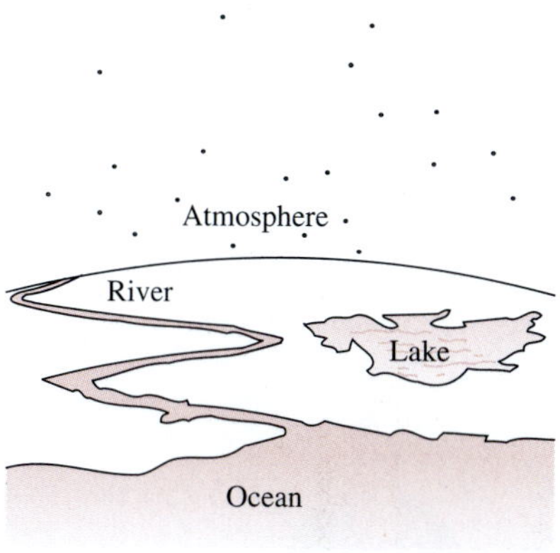

FIGURE 6–6
Bodies with relatively large thermal masses can be modeled as thermal energy reservoirs.

A *two-phase system* can also be modeled as a reservoir since it can absorb and release large quantities of heat while remaining at constant temperature. Another familiar example of a thermal energy reservoir is the *industrial furnace*. The temperatures of most furnaces are carefully controlled, and they are capable of supplying large quantities of thermal energy as heat in an essentially isothermal manner. Therefore, they can be modeled as reservoirs.

A body does not actually have to be very large to be considered a reservoir. Any physical body whose thermal energy capacity is large relative to the amount of energy it supplies or absorbs can be modeled as one. The air in a room, for example, can be treated as a reservoir in the analysis of the heat dissipation from a TV set in the room, since the amount of heat transfer from the TV set to the room air is not large enough to have a noticeable effect on the room air temperature.

A reservoir that supplies energy in the form of heat is called a **source**, and one that absorbs energy in the form of heat is called a **sink** (Fig. 6–7). Thermal energy reservoirs are often referred to as **heat reservoirs** since they supply or absorb energy in the form of heat.

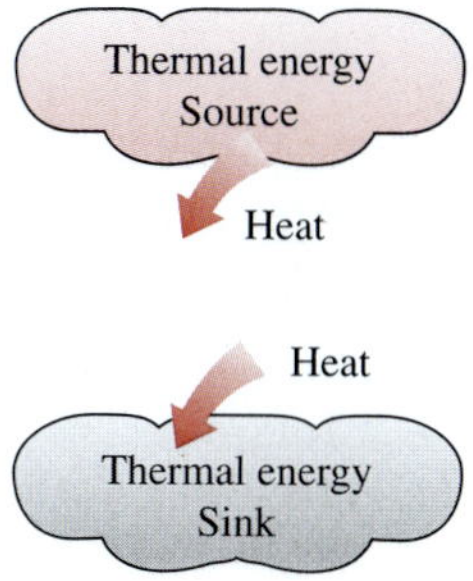

FIGURE 6–7
A source supplies energy in the form of heat, and a sink absorbs it.

Heat transfer from industrial sources to the environment is of major concern to environmentalists as well as to engineers. Irresponsible management of waste energy can significantly increase the temperature of portions of the environment, causing what is called *thermal pollution*. If it is not carefully controlled, thermal pollution can seriously disrupt marine life

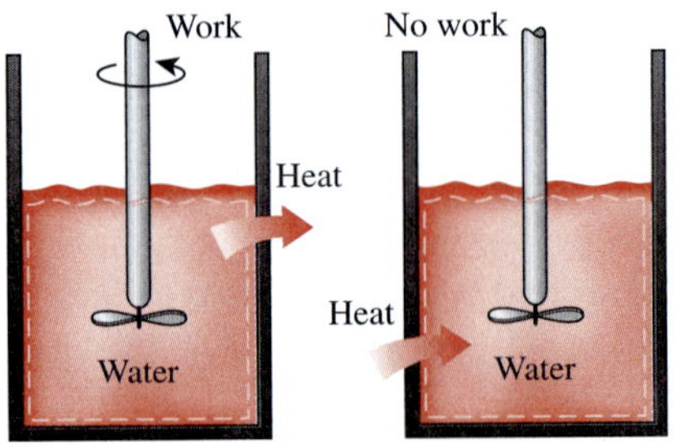

FIGURE 6–8
Work can always be converted to heat directly and completely, but the reverse is not true.

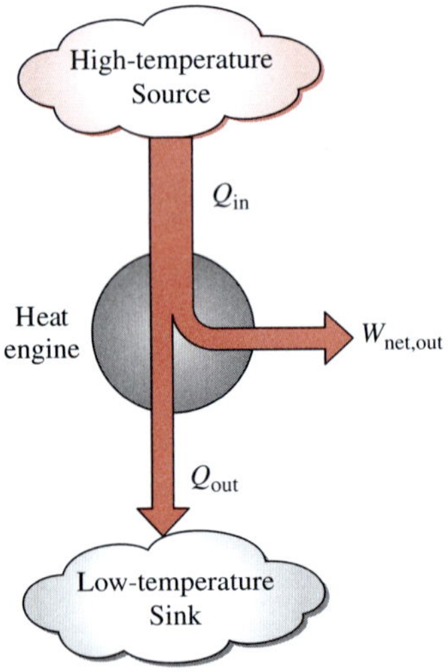

FIGURE 6–9
Part of the heat received by a heat engine is converted to work, while the rest is rejected to a sink.

in lakes and rivers. However, by careful design and management, the waste energy dumped into large bodies of water can be used to improve the quality of marine life by keeping the local temperature increases within safe and desirable levels.

6–3 ▪ HEAT ENGINES

As pointed out earlier, work can easily be converted to other forms of energy, but converting other forms of energy to work is not that easy. The mechanical work done by the shaft shown in Fig. 6–8, for example, is first converted to the internal energy of the water. This energy may then leave the water as heat. We know from experience that any attempt to reverse this process will fail. That is, transferring heat to the water does not cause the shaft to rotate. From this and other observations, we conclude that work can be converted to heat directly and completely, but converting heat to work requires the use of some special devices. These devices are called **heat engines**.

Heat engines differ considerably from one another, but all can be characterized by the following (Fig. 6–9):

1. They receive heat from a high-temperature source (solar energy, oil furnace, nuclear reactor, etc.).
2. They convert part of this heat to work (usually in the form of a rotating shaft).
3. They reject the remaining waste heat to a low-temperature sink (the atmosphere, rivers, etc.).
4. They operate on a cycle.

Heat engines and other cyclic devices usually involve a fluid to and from which heat is transferred while undergoing a cycle. This fluid is called the **working fluid**.

The term *heat engine* is often used in a broader sense to include work-producing devices that do not operate in a thermodynamic cycle. Engines that involve internal combustion such as gas turbines and car engines fall into this category. These devices operate in a mechanical cycle but not in a thermodynamic cycle since the working fluid (the combustion gases) does not undergo a complete cycle. Instead of being cooled to the initial temperature, the exhaust gases are purged and replaced by fresh air-and-fuel mixture at the end of the cycle.

The work-producing device that best fits into the definition of a heat engine is the *steam power plant*, which is an external-combustion engine. That is, combustion takes place outside the engine, and the thermal energy released during this process is transferred to the steam as heat. The schematic of a basic steam power plant is shown in Fig. 6–10. This is a rather simplified diagram, and the discussion of actual steam power plants is given in later chapters. The various quantities shown on this figure are as follows:

Q_{in} = amount of heat supplied to steam in boiler from a high-temperature source (furnace)
Q_{out} = amount of heat rejected from steam in condenser to a low-temperature sink (the atmosphere, a river, etc.)
W_{out} = amount of work delivered by steam as it expands in turbine
W_{in} = amount of work required to compress water to boiler pressure

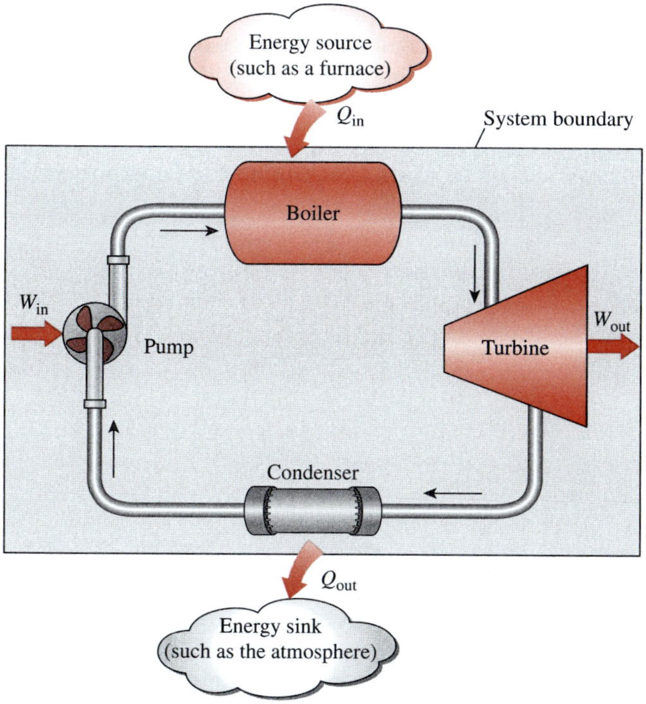

FIGURE 6–10
Schematic of a steam power plant.

Notice that the directions of the heat and work interactions are indicated by the subscripts *in* and *out*. Therefore, all four of the described quantities are always *positive*.

The net work output of this power plant is simply the difference between the total work output of the plant and the total work input (Fig. 6–11):

$$W_{net,out} = W_{out} - W_{in} \qquad \text{(kJ)} \tag{6–1}$$

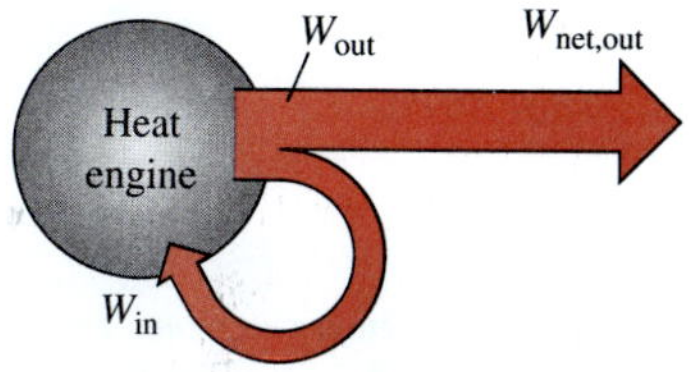

FIGURE 6–11
A portion of the work output of a heat engine is consumed internally to maintain continuous operation.

The net work can also be determined from the heat transfer data alone. The four components of the steam power plant involve mass flow in and out, and therefore should be treated as open systems. These components, together with the connecting pipes, however, always contain the same fluid (not counting the steam that may leak out, of course). No mass enters or leaves this combination system, which is indicated by the shaded area on Fig. 6–10; thus, it can be analyzed as a closed system. Recall that for a closed system undergoing a cycle, the change in internal energy ΔU is zero, and therefore the net work output of the system is also equal to the net heat transfer to the system:

$$W_{net,out} = Q_{in} - Q_{out} \qquad \text{(kJ)} \tag{6–2}$$

Thermal Efficiency

In Eq. 6–2, Q_{out} represents the magnitude of the energy wasted in order to complete the cycle. But Q_{out} is never zero; thus, the net work output of a heat engine is always less than the amount of heat input. That is, only part of the

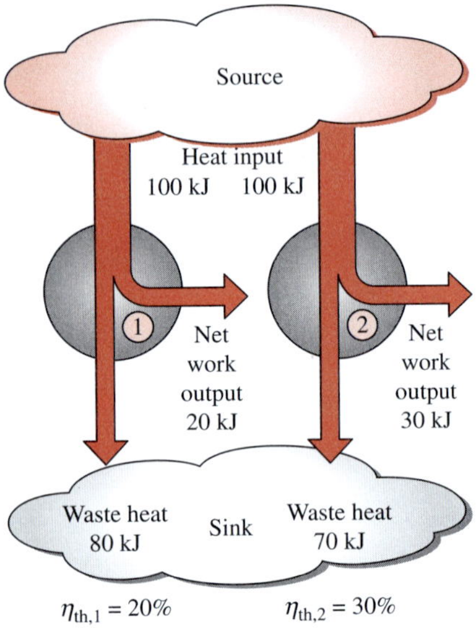

FIGURE 6–12
Some heat engines perform better than others (convert more of the heat they receive to work).

heat transferred to the heat engine is converted to work. The fraction of the heat input that is converted to net work output is a measure of the performance of a heat engine and is called the **thermal efficiency** η_{th} (Fig. 6–12).

For heat engines, the desired output is the net work output, and the required input is the amount of heat supplied to the working fluid. Then the thermal efficiency of a heat engine can be expressed as

$$\text{Thermal efficiency} = \frac{\text{Net work output}}{\text{Total heat input}} \tag{6–3}$$

or

$$\eta_{th} = \frac{W_{net,out}}{Q_{in}} \tag{6–4}$$

It can also be expressed as

$$\eta_{th} = 1 - \frac{Q_{out}}{Q_{in}} \tag{6–5}$$

since $W_{net,out} = Q_{in} - Q_{out}$.

Cyclic devices of practical interest such as heat engines, refrigerators, and heat pumps operate between a high-temperature medium (or reservoir) at temperature T_H and a low-temperature medium (or reservoir) at temperature T_L. To bring uniformity to the treatment of heat engines, refrigerators, and heat pumps, we define these two quantities:

Q_H = magnitude of heat transfer between the cyclic device and the high-temperature medium at temperature T_H
Q_L = magnitude of heat transfer between the cyclic device and the low-temperature medium at temperature T_L

Notice that both Q_L and Q_H are defined as *magnitudes* and therefore are positive quantities. The direction of Q_H and Q_L is easily determined by inspection. Then, the net work output and thermal efficiency relations for any heat engine (shown in Fig. 6–13) can also be expressed as

$$W_{net,out} = Q_H - Q_L$$

and

$$\eta_{th} = \frac{W_{net,out}}{Q_H} \quad \text{or} \quad \eta_{th} = 1 - \frac{Q_L}{Q_H} \tag{6–6}$$

The thermal efficiency of a heat engine is always less than unity since both Q_L and Q_H are defined as positive quantities.

Thermal efficiency is a measure of how efficiently a heat engine converts the heat that it receives to work. Heat engines are built for the purpose of converting heat to work, and engineers are constantly trying to improve the efficiencies of these devices since increased efficiency means less fuel consumption and thus lower fuel bills and less pollution.

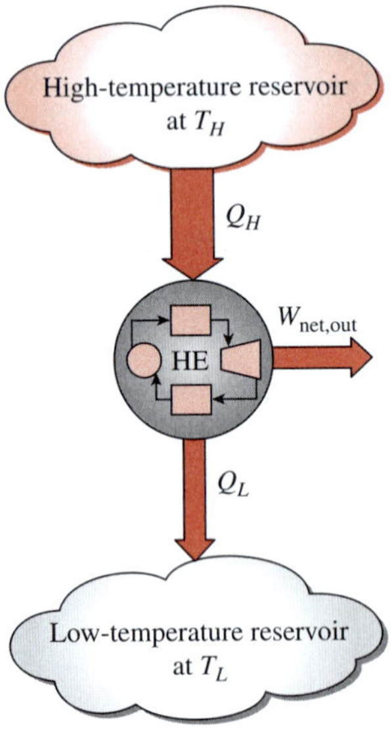

FIGURE 6–13
Schematic of a heat engine.

The thermal efficiencies of work-producing devices are relatively low. Ordinary spark-ignition automobile engines have a thermal efficiency of about 25 percent. That is, an automobile engine converts about 25 percent of the chemical energy of the gasoline to mechanical work. This number is as high as 40 percent for diesel engines and large gas-turbine plants and as high as 60 percent for large combined gas-steam power plants. Thus, even with the most efficient heat engines available today, almost one-half of the energy supplied ends up in the rivers, lakes, or the atmosphere as waste or useless energy (Fig. 6–14).

Can We Save Q_{out}?

In a steam power plant, the condenser is the device where large quantities of waste heat is rejected to rivers, lakes, or the atmosphere. Then one may ask, can we not just take the condenser out of the plant and save all that waste energy? The answer to this question is, unfortunately, a firm *no* for the simple reason that without a heat rejection process in a condenser, the cycle cannot be completed. (Cyclic devices such as steam power plants cannot run continuously unless the cycle is completed.) This is demonstrated next with the help of a simple heat engine.

Consider the simple heat engine shown in Fig. 6–15 that is used to lift weights. It consists of a piston–cylinder device with two sets of stops. The working fluid is the gas contained within the cylinder. Initially, the gas temperature is 30°C. The piston, which is loaded with the weights, is resting on top of the lower stops. Now 100 kJ of heat is transferred to the gas in the cylinder from a source at 100°C, causing it to expand and to raise the loaded piston until the piston reaches the upper stops, as shown in the figure. At this point, the load is removed, and the gas temperature is observed to be 90°C.

The work done on the load during this expansion process is equal to the increase in its potential energy, say 15 kJ. Even under ideal conditions (weightless piston, no friction, no heat losses, and quasi-equilibrium expansion), the amount of heat supplied to the gas is greater than the work done since part of the heat supplied is used to raise the temperature of the gas.

Now let us try to answer this question: *Is it possible to transfer the 85 kJ of excess heat at 90°C back to the reservoir at 100°C for later use?* If it is, then we will have a heat engine that can have a thermal efficiency of 100 percent under ideal conditions. The answer to this question is again *no*, for the very

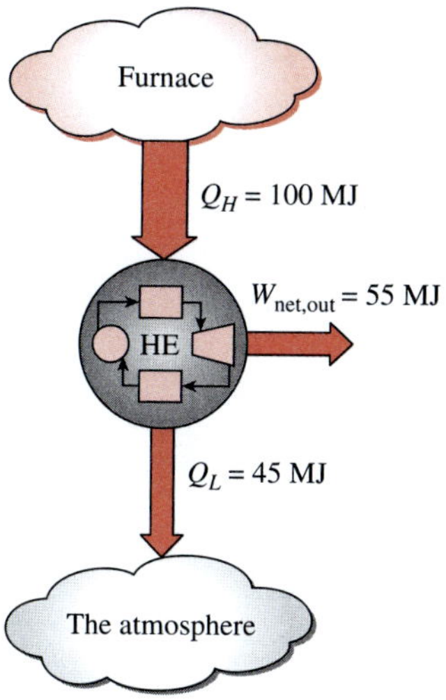

FIGURE 6–14
Even the most efficient heat engines reject almost one-half of the energy they receive as waste heat.

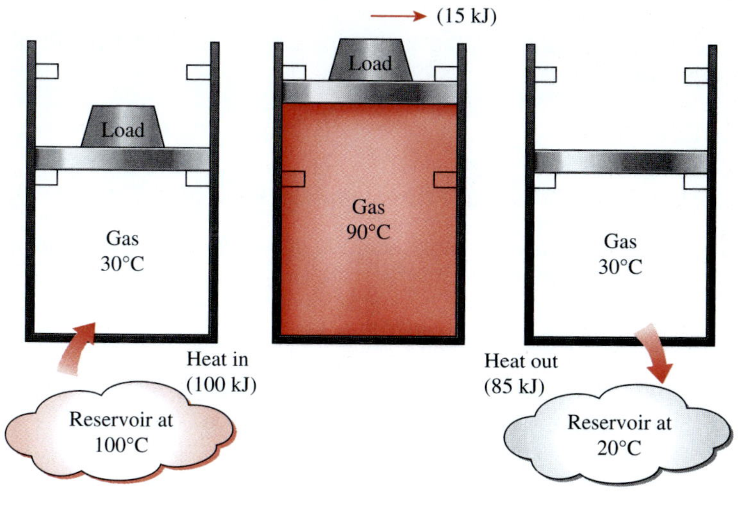

FIGURE 6–15
A heat-engine cycle cannot be completed without rejecting some heat to a low-temperature sink.

simple reason that heat is always transferred from a high-temperature medium to a low-temperature one, and never the other way around. Therefore, we cannot cool this gas from 90 to 30°C by transferring heat to a reservoir at 100°C. Instead, we have to bring the system into contact with a low-temperature reservoir, say at 20°C, so that the gas can return to its initial state by rejecting its 85 kJ of excess energy as heat to this reservoir. This energy cannot be recycled, and it is properly called *waste energy*.

We conclude from this discussion that every heat engine must *waste* some energy by transferring it to a low-temperature reservoir in order to complete the cycle, even under idealized conditions. The requirement that a heat engine exchange heat with at least two reservoirs for continuous operation forms the basis for the Kelvin–Planck expression of the second law of thermodynamics discussed later in this section.

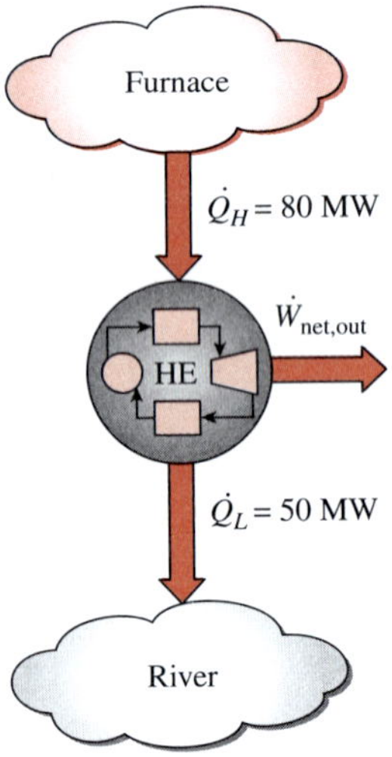

FIGURE 6–16
Schematic for Example 6–1.

EXAMPLE 6–1 Net Power Production of a Heat Engine

Heat is transferred to a heat engine from a furnace at a rate of 80 MW. If the rate of waste heat rejection to a nearby river is 50 MW, determine the net power output and the thermal efficiency for this heat engine.

SOLUTION The rates of heat transfer to and from a heat engine are given. The net power output and the thermal efficiency are to be determined.

Assumptions Heat losses through the pipes and other components are negligible.

Analysis A schematic of the heat engine is given in Fig. 6–16. The furnace serves as the high-temperature reservoir for this heat engine and the river as the low-temperature reservoir. The given quantities can be expressed as

$$\dot{Q}_H = 80 \text{ MW} \quad \text{and} \quad \dot{Q}_L = 50 \text{ MW}$$

The net power output of this heat engine is

$$\dot{W}_{\text{net,out}} = \dot{Q}_H - \dot{Q}_L = (80 - 50) \text{ MW} = \mathbf{30 \text{ MW}}$$

Then the thermal efficiency is easily determined to be

$$\eta_{\text{th}} = \frac{\dot{W}_{\text{net,out}}}{\dot{Q}_H} = \frac{30 \text{ MW}}{80 \text{ MW}} = \mathbf{0.375} \text{ (or 37.5\%)}$$

Discussion Note that the heat engine converts 37.5 percent of the heat it receives to work.

EXAMPLE 6–2 Fuel Consumption Rate of a Car

A car engine with a power output of 65 hp has a thermal efficiency of 24 percent. Determine the fuel consumption rate of this car if the fuel has a heating value of 19,000 Btu/lbm (that is, 19,000 Btu of energy is released for each lbm of fuel burned).

SOLUTION The power output and the efficiency of a car engine are given. The rate of fuel consumption of the car is to be determined.

Assumptions The power output of the car is constant.

Analysis A schematic of the car engine is given in Fig. 6–17. The car engine is powered by converting 24 percent of the chemical energy released during the combustion process to work. The amount of energy input required to produce a power output of 65 hp is determined from the definition of thermal efficiency to be

$$\dot{Q}_H = \frac{\dot{W}_{\text{net,out}}}{\eta_{\text{th}}} = \frac{65 \text{ hp}}{0.24}\left(\frac{2545 \text{ Btu/h}}{1 \text{ hp}}\right) = 689{,}270 \text{ Btu/h}$$

To supply energy at this rate, the engine must burn fuel at a rate of

$$\dot{m} = \frac{689{,}270 \text{ Btu/h}}{19{,}000 \text{ Btu/lbm}} = \mathbf{36.3 \text{ lbm/h}}$$

since 19,000 Btu of thermal energy is released for each lbm of fuel burned.

Discussion Note that if the thermal efficiency of the car could be doubled, the rate of fuel consumption would be reduced by half.

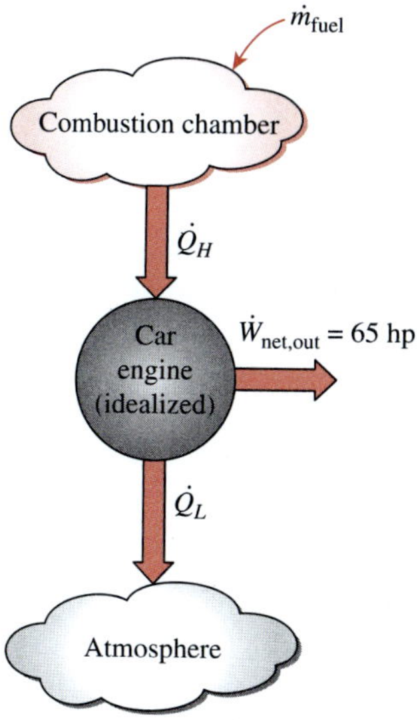

FIGURE 6–17
Schematic for Example 6–2.

The Second Law of Thermodynamics: Kelvin–Planck Statement

We have demonstrated earlier with reference to the heat engine shown in Fig. 6–15 that, even under ideal conditions, a heat engine must reject some heat to a low-temperature reservoir in order to complete the cycle. That is, no heat engine can convert all the heat it receives to useful work. This limitation on the thermal efficiency of heat engines forms the basis for the Kelvin–Planck statement of the second law of thermodynamics, which is expressed as follows:

> It is impossible for any device that operates on a cycle to receive heat from a single reservoir and produce a net amount of work.

That is, a heat engine must exchange heat with a low-temperature sink as well as a high-temperature source to keep operating. The Kelvin–Planck statement can also be expressed as *no heat engine can have a thermal efficiency of 100 percent* (Fig. 6–18), or as *for a power plant to operate, the working fluid must exchange heat with the environment as well as the furnace.*

Note that the impossibility of having a 100 percent efficient heat engine is not due to friction or other dissipative effects. It is a limitation that applies to both the idealized and the actual heat engines. Later in this chapter, we develop a relation for the maximum thermal efficiency of a heat engine. We also demonstrate that this maximum value depends on the reservoir temperatures only.

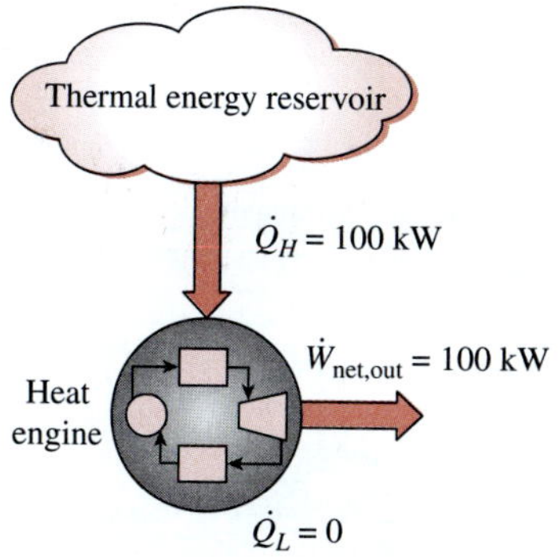

FIGURE 6–18
A heat engine that violates the Kelvin–Planck statement of the second law.

6–4 ▪ REFRIGERATORS AND HEAT PUMPS

We all know from experience that heat is transferred in the direction of decreasing temperature, that is, from high-temperature mediums to low-temperature ones. This heat transfer process occurs in nature without requiring any devices. The reverse process, however, cannot occur by itself.

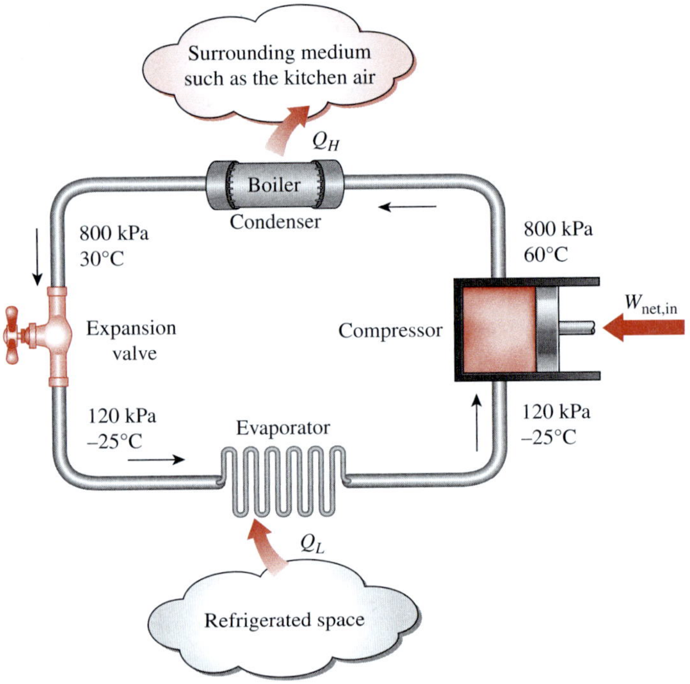

FIGURE 6–19
Basic components of a refrigeration system and typical operating conditions.

The transfer of heat from a low-temperature medium to a high-temperature one requires special devices called **refrigerators**.

Refrigerators, like heat engines, are cyclic devices. The working fluid used in the refrigeration cycle is called a **refrigerant**. The most frequently used refrigeration cycle is the *vapor-compression refrigeration cycle*, which involves four main components: a compressor, a condenser, an expansion valve, and an evaporator, as shown in Fig. 6–19.

The refrigerant enters the compressor as a vapor and is compressed to the condenser pressure. It leaves the compressor at a relatively high temperature and cools down and condenses as it flows through the coils of the condenser by rejecting heat to the surrounding medium. It then enters a capillary tube where its pressure and temperature drop drastically due to the throttling effect. The low-temperature refrigerant then enters the evaporator, where it evaporates by absorbing heat from the refrigerated space. The cycle is completed as the refrigerant leaves the evaporator and reenters the compressor.

In a household refrigerator, the freezer compartment where heat is absorbed by the refrigerant serves as the evaporator, and the coils usually behind the refrigerator where heat is dissipated to the kitchen air serve as the condenser.

A refrigerator is shown schematically in Fig. 6–20. Here Q_L is the magnitude of the heat removed from the refrigerated space at temperature T_L, Q_H is the magnitude of the heat rejected to the warm environment at temperature T_H, and $W_{net,in}$ is the net work input to the refrigerator. As discussed before, Q_L and Q_H represent magnitudes and thus are positive quantities.

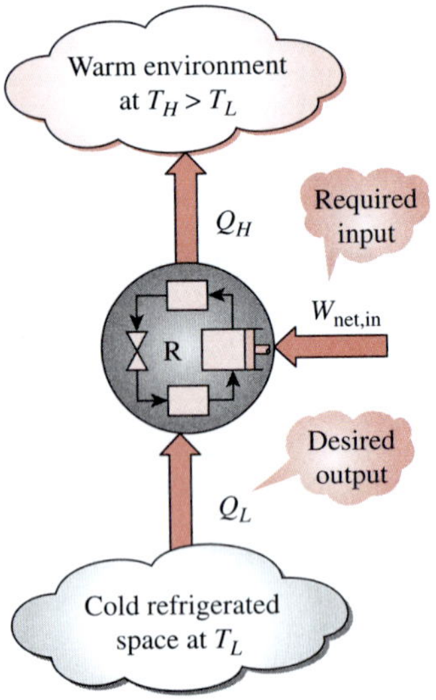

FIGURE 6–20
The objective of a refrigerator is to remove Q_L from the cooled space.

Coefficient of Performance

The *efficiency* of a refrigerator is expressed in terms of the **coefficient of performance** (COP), denoted by COP_R. The objective of a refrigerator is to remove heat (Q_L) from the refrigerated space. To accomplish this objective, it requires a work input of $W_{net,in}$. Then the COP of a refrigerator can be expressed as

$$COP_R = \frac{\text{Desired output}}{\text{Required input}} = \frac{Q_L}{W_{net,in}} \tag{6–7}$$

This relation can also be expressed in rate form by replacing Q_L by $\dot{Q}_L$ and $W_{net,in}$ by $\dot{W}_{net,in}$.

The conservation of energy principle for a cyclic device requires that

$$W_{net,in} = Q_H - Q_L \quad \text{(kJ)} \tag{6–8}$$

Then the COP relation becomes

$$COP_R = \frac{Q_L}{Q_H - Q_L} = \frac{1}{Q_H/Q_L - 1} \tag{6–9}$$

Notice that the value of COP_R can be *greater than unity*. That is, the amount of heat removed from the refrigerated space can be greater than the amount of work input. This is in contrast to the thermal efficiency, which can never be greater than 1. In fact, one reason for expressing the efficiency of a refrigerator by another term—the coefficient of performance—is the desire to avoid the oddity of having efficiencies greater than unity.

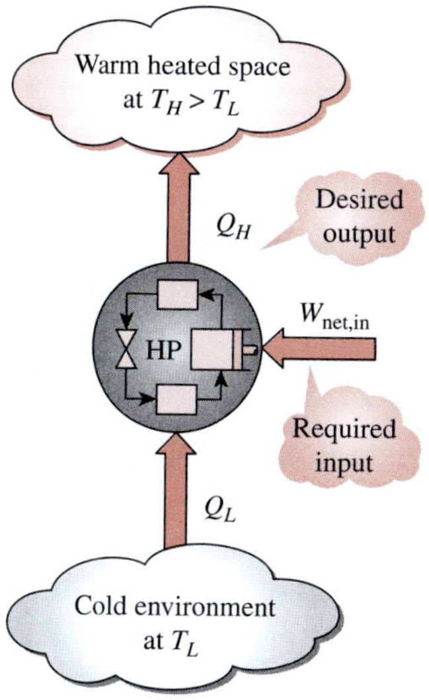

FIGURE 6–21
The objective of a heat pump is to supply heat Q_H into the warmer space.

Heat Pumps

Another device that transfers heat from a low-temperature medium to a high-temperature one is the **heat pump**, shown schematically in Fig. 6–21. Refrigerators and heat pumps operate on the same cycle but differ in their objectives. The objective of a refrigerator is to maintain the refrigerated space at a low temperature by removing heat from it. Discharging this heat to a higher-temperature medium is merely a necessary part of the operation, not the purpose. The objective of a heat pump, however, is to maintain a heated space at a high temperature. This is accomplished by absorbing heat from a low-temperature source, such as well water or cold outside air in winter, and supplying this heat to the high-temperature medium such as a house (Fig. 6–22).

An ordinary refrigerator that is placed in the window of a house with its door open to the cold outside air in winter will function as a heat pump since it will try to cool the outside by absorbing heat from it and rejecting this heat into the house through the coils behind it (Fig. 6–23).

The measure of performance of a heat pump is also expressed in terms of the **coefficient of performance** COP_{HP}, defined as

$$COP_{HP} = \frac{\text{Desired output}}{\text{Required input}} = \frac{Q_H}{W_{net,in}} \tag{6–10}$$

which can also be expressed as

$$COP_{HP} = \frac{Q_H}{Q_H - Q_L} = \frac{1}{1 - Q_L/Q_H} \tag{6–11}$$

A comparison of Eqs. 6–7 and 6–10 reveals that

$$COP_{HP} = COP_R + 1 \tag{6–12}$$

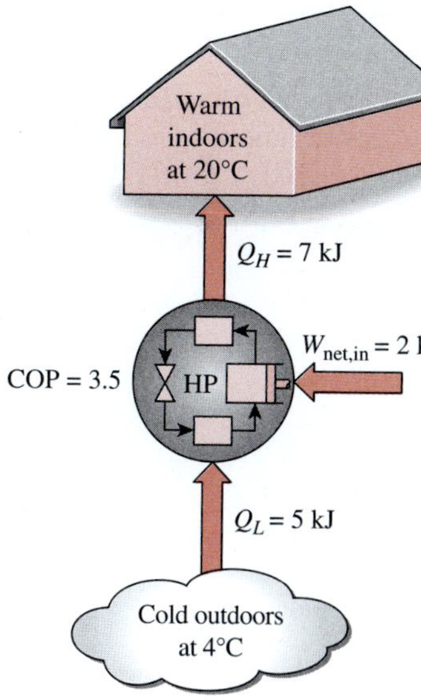

FIGURE 6–22
The work supplied to a heat pump is used to extract energy from the cold outdoors and carry it into the warm indoors.

for fixed values of Q_L and Q_H. This relation implies that the coefficient of performance of a heat pump is always greater than unity since COP_R is a positive quantity. That is, a heat pump will function, at worst, as a resistance heater, supplying as much energy to the house as it consumes. In reality, however, part of Q_H is lost to the outside air through piping and other devices, and COP_{HP} may drop below unity when the outside air temperature is too low. When this happens, the system usually switches to a resistance heating mode. Most heat pumps in operation today have a seasonally averaged COP of 2 to 3.

Most existing heat pumps use the cold outside air as the heat source in winter, and they are referred to as *air-source heat pumps*. The COP of such heat pumps is about 3.0 at design conditions. Air-source heat pumps are not appropriate for cold climates since their efficiency drops considerably when temperatures are below the freezing point. In such cases, geothermal (also called ground-source) heat pumps that use the ground as the heat source can be used. Geothermal heat pumps require the burial of pipes in the ground 1 to 2 m deep. Such heat pumps are more expensive to install, but they are also more efficient (up to 45 percent more efficient than air-source heat pumps). The COP of ground-source heat pumps can be as high as 6 in the cooling mode.

FIGURE 6–23
When installed backward, an air conditioner functions as a heat pump.
BLONDIE © KING FEATURES SYNDICATE.

Air conditioners are basically refrigerators whose refrigerated space is a room or a building instead of the food compartment. A window air-conditioning unit cools a room by absorbing heat from the room air and discharging it to the outside. The same air-conditioning unit can be used as a heat pump in winter by installing it backwards as shown in Fig. 6–23. In this mode, the unit absorbs heat from the cold outside and delivers it to the room. Air-conditioning systems that are equipped with proper controls and a reversing valve operate as air conditioners in summer and as heat pumps in winter.

Performance of Refrigerators, Air-Conditioners, and Heat Pumps

The performance of air conditioners and heat pumps is often expressed in terms of the **energy efficiency ratio** (EER) or **seasonal energy efficiency ratio** (SEER) determined by following certain testing standards. SEER is the ratio the total amount of heat removed by an air conditioner or heat pump during a normal cooling season (in Btu) to the total amount of electricity consumed (in watt-hours, Wh), and it is a measure of seasonal performance of cooling equipment. EER, on the other hand, is a measure of the instantaneous energy efficiency, and is defined as the ratio of the rate of heat removal from the cooled space by the cooling equipment to the rate of electricity consumption in steady operation. Therefore, both EER and SEER have the unit Btu/Wh. Considering that 1 kWh = 3412 Btu and thus 1 Wh = 3.412 Btu, a devicet that removes 1 kWh of heat from the cooled space for each kWh of electricity it consumes (COP = 1) will have an EER of 3.412. Therefore, the relation between EER (or SEER) and COP is

$$EER = 3.412\ COP_R$$

To promote the efficient use of energy, governments worldwide have mandated minimum standards for the performance of energy consuming equipment. Most air conditioners or heat pumps in the market have SEER values from 13 to 21, which correspond to COP values of 3.8 to 6.2. Best performance is achieved using units equipped with variable-speed drives (also called inverters). Variable-speed compressors and fans allow the unit to operate at maximum efficiency for varying heating/cooling needs and weather conditions as determined by a microprocessor. In the air-conditioning mode, for example, they operate at higher speeds on hot days and at lower speeds on cooler days, enhancing both efficiency and comfort.

The EER or COP of a refrigerator decreases with decreasing refrigeration temperature. Therefore, it is not economical to refrigerate to a lower temperature than needed. The COPs of refrigerators are in the range of 2.6–3.0 for cutting and preparation rooms; 2.3–2.6 for meat, deli, dairy, and produce; 1.2–1.5 for frozen foods; and 1.0–1.2 for ice cream units. Note that the COP of freezers is about half of the COP of meat refrigerators, and thus it costs twice as much to cool the meat products with refrigerated air that is cold enough to cool frozen foods. It is good energy conservation practice to use separate refrigeration systems to meet different refrigeration needs.

EXAMPLE 6–3 Heat Rejection by a Refrigerator

The food compartment of a refrigerator, shown in Fig. 6–24, is maintained at 4°C by removing heat from it at a rate of 360 kJ/min. If the required power input to the refrigerator is 2 kW, determine (*a*) the coefficient of performance of the refrigerator and (*b*) the rate of heat rejection to the room that houses the refrigerator.

SOLUTION The power consumption of a refrigerator is given. The COP and the rate of heat rejection are to be determined.

Assumptions Steady operating conditions exist.

Analysis (*a*) The coefficient of performance of the refrigerator is

$$\text{COP}_\text{R} = \frac{\dot{Q}_L}{\dot{W}_{\text{net,in}}} = \frac{360\ \text{kJ/min}}{2\ \text{kW}}\left(\frac{1\ \text{kW}}{60\ \text{kJ/min}}\right) = \mathbf{3}$$

That is, 3 kJ of heat is removed from the refrigerated space for each kJ of work supplied.

(*b*) The rate at which heat is rejected to the room that houses the refrigerator is determined from the conservation of energy relation for cyclic devices,

$$\dot{Q}_H = \dot{Q}_L + \dot{W}_{\text{net,in}} = 360\ \text{kJ/min} + (2\ \text{kW})\left(\frac{60\ \text{kJ/min}}{1\ \text{kW}}\right) = \mathbf{480\ kJ/min}$$

Discussion Notice that both the energy removed from the refrigerated space as heat and the energy supplied to the refrigerator as electrical work eventually show up in the room air and become part of the internal energy of the air. This demonstrates that energy can change from one form to another, can move from one place to another, but is never destroyed during a process.

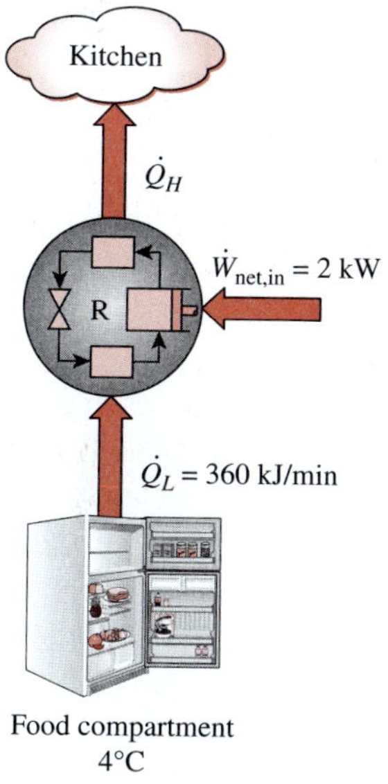

FIGURE 6–24
Schematic for Example 6–3.

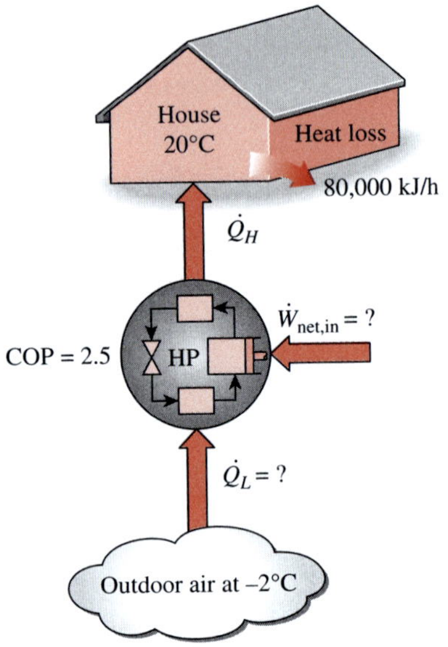

FIGURE 6–25
Schematic for Example 6–4.

EXAMPLE 6–4 Heating a House by a Heat Pump

A heat pump is used to meet the heating requirements of a house and maintain it at 20°C. On a day when the outdoor air temperature drops to −2°C, the house is estimated to lose heat at a rate of 80,000 kJ/h. If the heat pump under these conditions has a COP of 2.5, determine (*a*) the power consumed by the heat pump and (*b*) the rate at which heat is absorbed from the cold outdoor air.

SOLUTION The COP of a heat pump is given. The power consumption and the rate of heat absorption are to be determined.
Assumptions Steady operating conditions exist.
Analysis (*a*) The power consumed by this heat pump, shown in Fig. 6–25, is determined from the definition of the coefficient of performance to be

$$\dot{W}_{\text{net,in}} = \frac{\dot{Q}_H}{\text{COP}_{\text{HP}}} = \frac{80{,}000 \text{ kJ/h}}{2.5} = \mathbf{32{,}000 \text{ kJ/h}} \text{ (or 8.9 kW)}$$

(*b*) The house is losing heat at a rate of 80,000 kJ/h. If the house is to be maintained at a constant temperature of 20°C, the heat pump must deliver heat to the house at the same rate, that is, at a rate of 80,000 kJ/h. Then the rate of heat transfer from the outdoor becomes

$$\dot{Q}_L = \dot{Q}_H - \dot{W}_{\text{net,in}} = (80{,}000 - 32{,}000) \text{ kJ/h} = \mathbf{48{,}000 \text{ kJ/h}}$$

Discussion Note that 48,000 of the 80,000 kJ/h heat delivered to the house is actually extracted from the cold outdoor air. Therefore, we are paying only for the 32,000-kJ/h energy that is supplied as electrical work to the heat pump. If we were to use an electric resistance heater instead, we would have to supply the entire 80,000 kJ/h to the resistance heater as electric energy. This would mean a heating bill that is 2.5 times higher. This explains the popularity of heat pumps as heating systems and why they are preferred to simple electric resistance heaters despite their considerably higher initial cost.

The Second Law of Thermodynamics: Clausius Statement

There are two classical statements of the second law—the Kelvin–Planck statement, which is related to heat engines and discussed in the preceding section, and the Clausius statement, which is related to refrigerators or heat pumps. The Clausius statement is expressed as follows:

> It is impossible to construct a device that operates in a cycle and produces no effect other than the transfer of heat from a lower-temperature body to a higher-temperature body.

It is common knowledge that heat does not, of its own volition, transfer from a cold medium to a warmer one. The Clausius statement does not imply that a cyclic device that transfers heat from a cold medium to a warmer one is impossible to construct. In fact, this is precisely what a common household refrigerator does. It simply states that a refrigerator cannot operate unless its compressor is driven by an external power source, such as an electric motor (Fig. 6–26). This way, the net effect on the surroundings involves the consumption of some energy in the form of work, in addition to the transfer of heat from a colder body to a warmer one. That is, it leaves a trace in the surroundings. Therefore, a household refrigerator is in complete compliance with the Clausius statement of the second law.

Both the Kelvin–Planck and the Clausius statements of the second law are negative statements, and a negative statement cannot be proved. Like any other physical law, the second law of thermodynamics is based on experimental observations. To date, no experiment has been conducted that contradicts the second law, and this should be taken as sufficient proof of its validity.

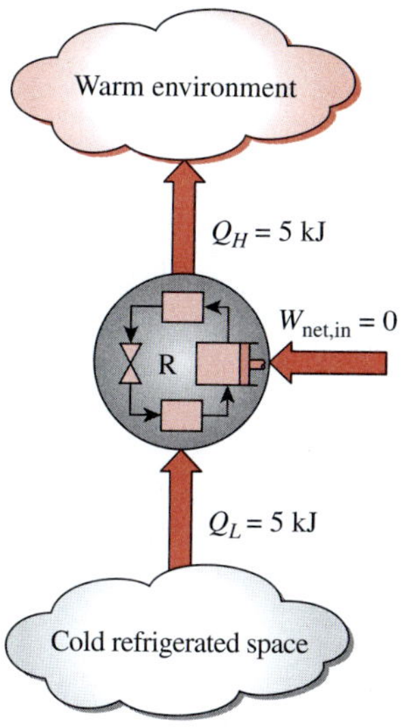

FIGURE 6–26
A refrigerator that violates the Clausius statement of the second law.

Equivalence of the Two Statements

The Kelvin–Planck and the Clausius statements are equivalent in their consequences, and either statement can be used as the expression of the second law of thermodynamics. Any device that violates the Kelvin–Planck statement also violates the Clausius statement, and vice versa. This can be demonstrated as follows.

Consider the heat-engine-refrigerator combination shown in Fig. 6–27*a*, operating between the same two reservoirs. The heat engine is assumed to have, in violation of the Kelvin–Planck statement, a thermal efficiency of 100 percent, and therefore it converts all the heat Q_H it receives to work W. This work is now supplied to a refrigerator that removes heat in the amount of Q_L from the low-temperature reservoir and rejects heat in the amount of $Q_L + Q_H$ to the high-temperature reservoir. During this process, the high-temperature reservoir receives a net amount of heat Q_L (the difference between $Q_L + Q_H$ and Q_H). Thus, the combination of these two devices can be viewed as a refrigerator, as shown in Fig. 6–27*b*, that transfers heat in an

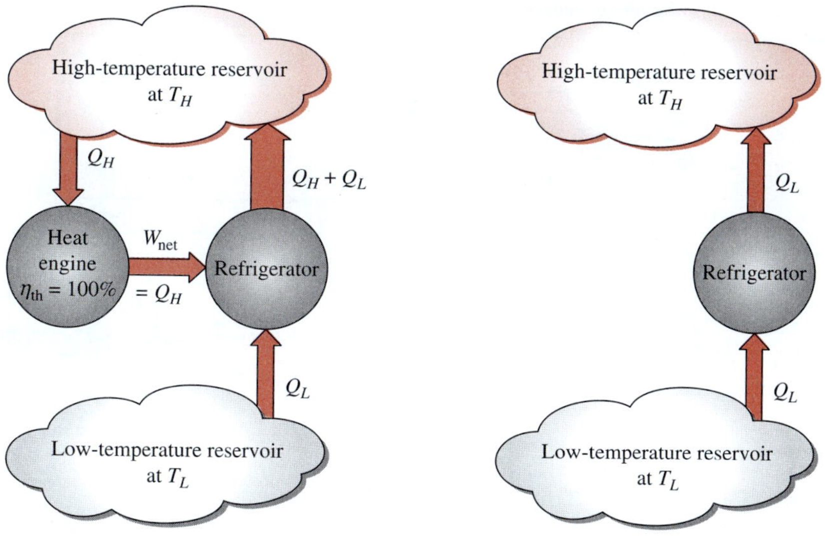

(*a*) A refrigerator that is powered by a 100 percent efficient heat engine

(*b*) The equivalent refrigerator

FIGURE 6–27
Proof that the violation of the Kelvin–Planck statement leads to the violation of the Clausius statement.

amount of Q_L from a cooler body to a warmer one without requiring any input from outside. This is clearly a violation of the Clausius statement. Therefore, a violation of the Kelvin–Planck statement results in the violation of the Clausius statement.

It can also be shown in a similar manner that a violation of the Clausius statement leads to the violation of the Kelvin–Planck statement. Therefore, the Clausius and the Kelvin–Planck statements are two equivalent expressions of the second law of thermodynamics.

6–5 ▪ PERPETUAL-MOTION MACHINES

We have repeatedly stated that a process cannot take place unless it satisfies both the first and second laws of thermodynamics. Any device that violates either law is called a **perpetual-motion machine**, and despite numerous attempts, no perpetual-motion machine is known to have worked. But this has not stopped inventors from trying to create new ones.

A device that violates the first law of thermodynamics (by *creating* energy) is called a **perpetual-motion machine of the first kind** (PMM1), and a device that violates the second law of thermodynamics is called a **perpetual-motion machine of the second kind** (PMM2).

Consider the steam power plant shown in Fig. 6–28. It is proposed to heat the steam by resistance heaters placed inside the boiler, instead of by the energy supplied from fossil or nuclear fuels. Part of the electricity generated by the plant is to be used to power the resistors as well as the pump. The rest of the electric energy is to be supplied to the electric network as the net work output. The inventor claims that once the system is started, this power plant will produce electricity indefinitely without requiring any energy input from the outside.

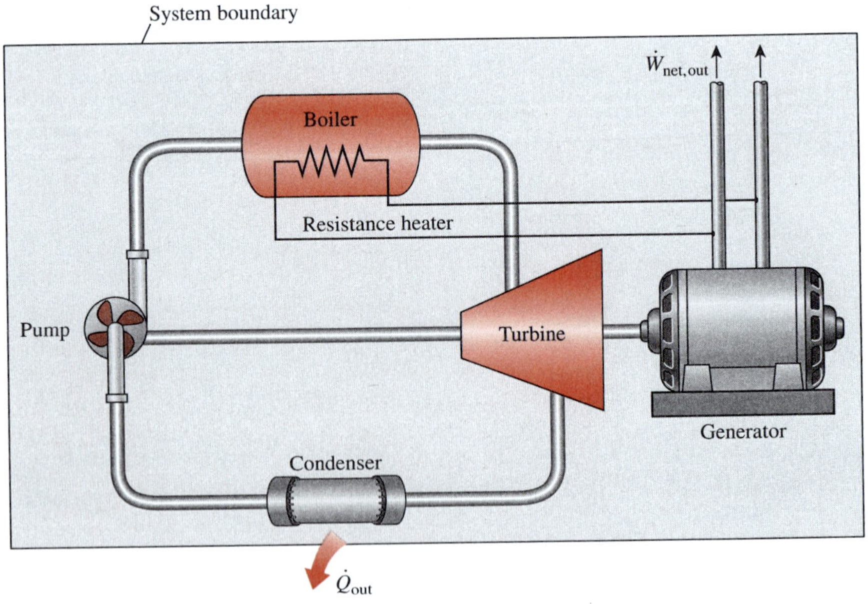

FIGURE 6–28
A perpetual-motion machine that violates the first law of thermodynamics (PMM1).

Well, here is an invention that could solve the world's energy problem—if it works, of course. A careful examination of this invention reveals that the system enclosed by the shaded area is continuously supplying energy to the outside at a rate of $\dot{Q}_{out} + \dot{W}_{net,out}$ without receiving any energy. That is, this system is creating energy at a rate of $\dot{Q}_{out} + \dot{W}_{net,out}$, which is clearly a violation of the first law. Therefore, this wonderful device is nothing more than a PMM1 and does not warrant any further consideration.

Now let us consider another novel idea by the same inventor. Convinced that energy cannot be created, the inventor suggests the following modification that will greatly improve the thermal efficiency of that power plant without violating the first law. Aware that more than one-half of the heat transferred to the steam in the furnace is discarded in the condenser to the environment, the inventor suggests getting rid of this wasteful component and sending the steam to the pump as soon as it leaves the turbine, as shown in Fig. 6–29. This way, all the heat transferred to the steam in the boiler will be converted to work, and thus the power plant will have a theoretical efficiency of 100 percent. The inventor realizes that some heat losses and friction between the moving components are unavoidable and that these effects will hurt the efficiency somewhat, but still expects the efficiency to be no less than 80 percent (as opposed to 40 percent in most actual power plants) for a carefully designed system.

Well, the possibility of doubling the efficiency would certainly be very tempting to plant managers and, if not properly trained, they would probably give this idea a chance, since intuitively they see nothing wrong with it. A student of thermodynamics, however, will immediately label this device as a PMM2, since it works on a cycle and does a net amount of work while exchanging heat with a single reservoir (the furnace) only. It satisfies the first law but violates the second law, and therefore it will not work.

Countless perpetual-motion machines have been proposed throughout history, with many more still being proposed. Some proposers have even gone so far as to patent their inventions, only to find out that what they actually have in their hands is a worthless piece of paper.

Some perpetual-motion machine inventors were very successful in fund-raising. For example, a Philadelphia carpenter named J. W. Kelly collected millions of dollars between 1874 and 1898 from investors in his

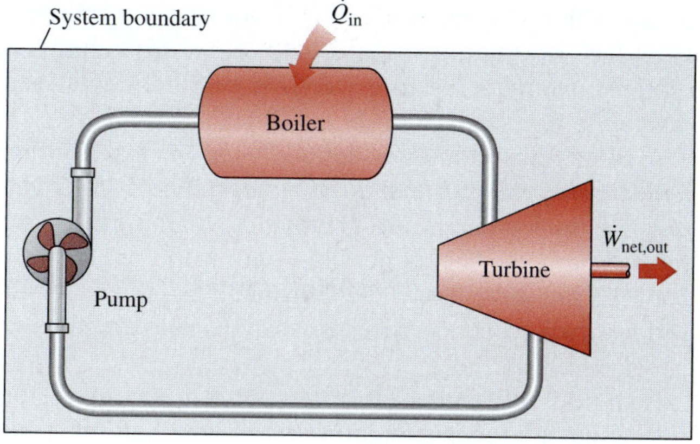

FIGURE 6–29
A perpetual-motion machine that violates the second law of thermodynamics (PMM2).

hydropneumatic-pulsating-vacu-engine, which supposedly could push a railroad train 3000 miles on 1 L of water. Of course, it never did. After his death in 1898, the investigators discovered that the demonstration machine was powered by a hidden motor. Recently, a group of investors was set to invest $2.5 million into a mysterious *energy augmentor*, which multiplied whatever power it took in, but their lawyer wanted an expert opinion first. Confronted by the scientists, the "inventor" fled the scene without even attempting to run his demo machine.

Tired of applications for perpetual-motion machines, the U.S. Patent Office decreed in 1918 that it would no longer consider any perpetual-motion machine applications. However, several such patent applications were still filed, and some made it through the patent office undetected. Some applicants whose patent applications were denied sought legal action. For example, in 1982 the U.S. Patent Office dismissed as just another perpetual-motion machine a huge device that involves several hundred kilograms of rotating magnets and kilometers of copper wire that is supposed to be generating more electricity than it is consuming from a battery pack. However, the inventor challenged the decision, and in 1985 the National Bureau of Standards finally tested the machine just to certify that it is battery-operated. However, it did not convince the inventor that his machine will not work.

The proposers of perpetual-motion machines generally have innovative minds, but they usually lack formal engineering training, which is very unfortunate. No one is immune from being deceived by an innovative perpetual-motion machine. As the saying goes, however, if something sounds too good to be true, it probably is.

6–6 ▪ REVERSIBLE AND IRREVERSIBLE PROCESSES

The second law of thermodynamics states that no heat engine can have an efficiency of 100 percent. Then one may ask, what is the highest efficiency that a heat engine can possibly have? Before we can answer this question, we need to define an idealized process first, which is called the *reversible process*.

The processes that were discussed at the beginning of this chapter occurred in a certain direction. Once having taken place, these processes cannot reverse themselves spontaneously and restore the system to its initial state. For this reason, they are classified as *irreversible processes*. Once a cup of hot coffee cools, it will not heat up by retrieving the heat it lost from the surroundings. If it could, the surroundings, as well as the system (coffee), would be restored to their original condition, and this would be a reversible process.

A **reversible process** is defined as a *process that can be reversed without leaving any trace on the surroundings* (Fig. 6–30). That is, both the system *and* the surroundings are returned to their initial states at the end of the reverse process. This is possible only if the net heat *and* net work exchange between the system and the surroundings is zero for the combined (original and reverse) process. Processes that are not reversible are called **irreversible processes**.

It should be pointed out that a system can be restored to its initial state following a process, regardless of whether the process is reversible or

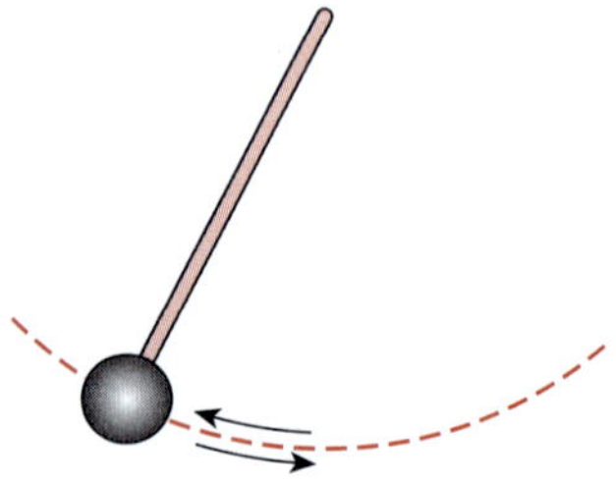

(*a*) Frictionless pendulum

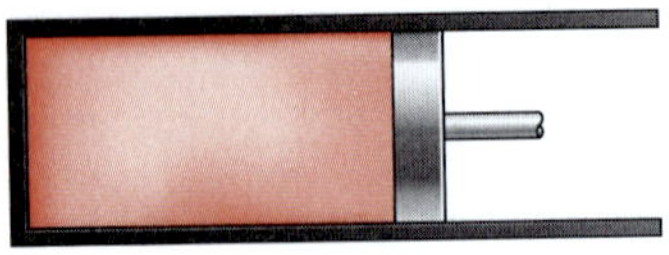

(*b*) Quasi-equilibrium expansion and compression of a gas

FIGURE 6–30
Two familiar reversible processes.

irreversible. But for reversible processes, this restoration is made without leaving any net change on the surroundings, whereas for irreversible processes, the surroundings usually do some work on the system and therefore does not return to their original state.

Reversible processes actually do not occur in nature. They are merely *idealizations* of actual processes. Reversible processes can be approximated by actual devices, but they can never be achieved. That is, all the processes occurring in nature are irreversible. You may be wondering, then, *why* we are bothering with such fictitious processes. There are two reasons. First, they are easy to analyze, since a system passes through a series of equilibrium states during a reversible process. Second, they serve as idealized models to which actual processes can be compared.

In daily life, the concepts of Mr. Right and Ms. Right are also idealizations, just like the concept of a reversible (perfect) process. People who insist on finding Mr. or Ms. Right to settle down are bound to remain Mr. or Ms. Single for the rest of their lives. The possibility of finding the perfect prospective mate is no higher than the possibility of finding a perfect (reversible) process. Likewise, a person who insists on perfection in friends is bound to have no friends.

Engineers are interested in reversible processes because work-producing devices such as car engines and gas or steam turbines *deliver the most work*, and work-consuming devices such as compressors, fans, and pumps *consume the least work* when reversible processes are used instead of irreversible ones (Fig. 6–31).

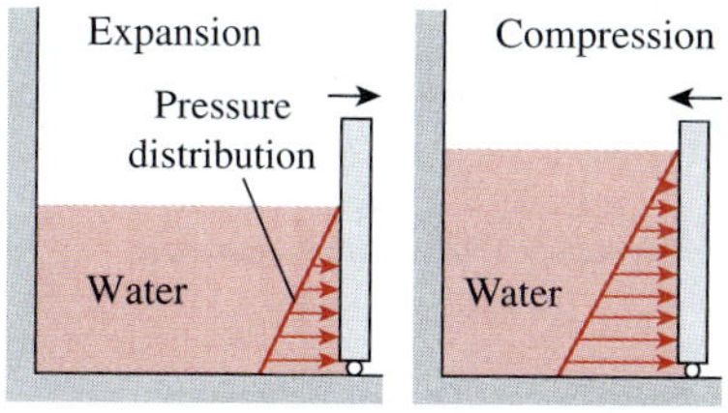

(*a*) Slow (reversible) process

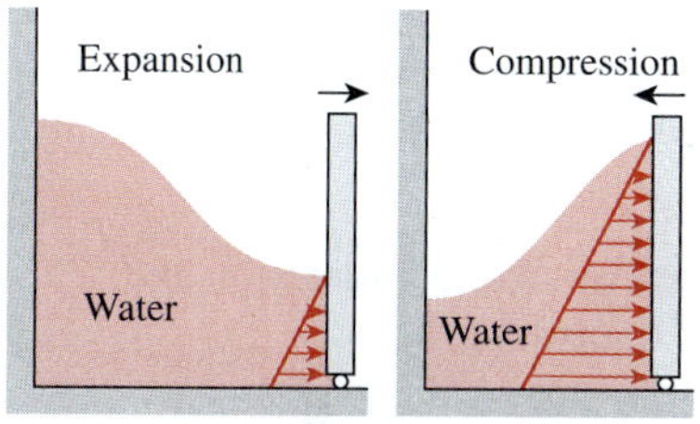

(*b*) Fast (irreversible) process

FIGURE 6–31
Reversible processes deliver the most and consume the least work.

Reversible processes can be viewed as *theoretical limits* for the corresponding irreversible ones. Some processes are more irreversible than others. We may never be able to have a reversible process, but we can certainly approach it. The more closely we approximate a reversible process, the more work delivered by a work-producing device or the less work required by a work-consuming device.

The concept of reversible processes leads to the definition of the **second-law efficiency** for actual processes, which is the degree of approximation to the corresponding reversible processes. This enables us to compare the performance of different devices that are designed to do the same task on the basis of their efficiencies. The better the design, the lower the irreversibilities and the higher the second-law efficiency.

Irreversibilities

The factors that cause a process to be irreversible are called **irreversibilities**. They include friction, unrestrained expansion, mixing of two fluids, heat transfer across a finite temperature difference, electric resistance, inelastic deformation of solids, and chemical reactions. The presence of any of these effects renders a process irreversible. A reversible process involves none of these. Some of the frequently encountered irreversibilities are discussed briefly below.

Friction is a familiar form of irreversibility associated with bodies in motion. When two bodies in contact are forced to move relative to each other (a piston in a cylinder, for example, as shown in Fig. 6–32), a friction force that opposes the motion develops at the interface of these two bodies, and some work is needed to overcome this friction force. The energy supplied as

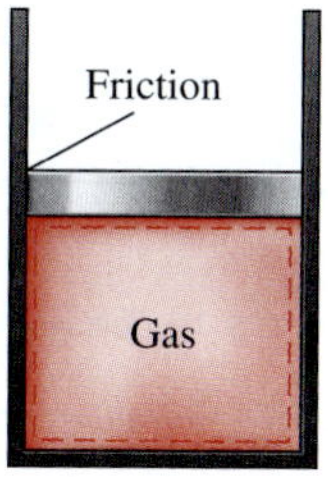

FIGURE 6–32
Friction renders a process irreversible.

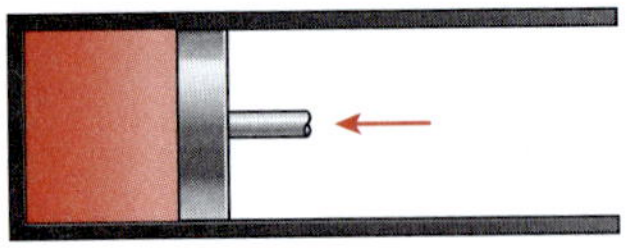

(*a*) Fast compression

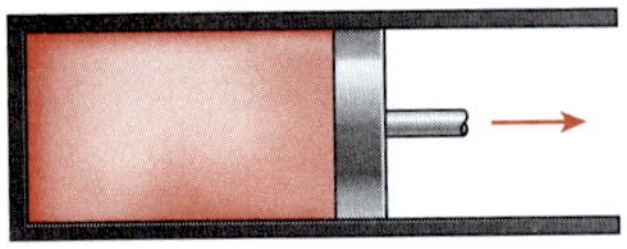

(*b*) Fast expansion

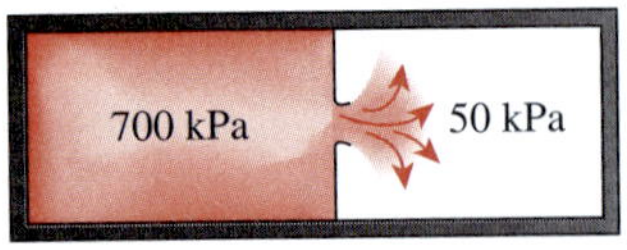

(*c*) Unrestrained expansion

FIGURE 6–33
Irreversible compression and expansion processes.

work is eventually converted to heat during the process and is transferred to the bodies in contact, as evidenced by a temperature rise at the interface. When the direction of the motion is reversed, the bodies are restored to their original position, but the interface does not cool, and heat is not converted back to work. Instead, more of the work is converted to heat while overcoming the friction forces that also oppose the reverse motion. Since the system (the moving bodies) and the surroundings cannot be returned to their original states, this process is irreversible. Therefore, any process that involves friction is irreversible. The larger the friction forces involved, the more irreversible the process is.

Friction does not always involve two solid bodies in contact. It is also encountered between a fluid and solid and even between the layers of a fluid moving at different velocities. A considerable fraction of the power produced by a car engine is used to overcome the friction (the drag force) between the air and the external surfaces of the car, and it eventually becomes part of the internal energy of the air. It is not possible to reverse this process and recover that lost power, even though doing so would not violate the conservation of energy principle.

Another example of irreversibility is the **unrestrained expansion of a gas** separated from a vacuum by a membrane, as shown in Fig. 6–33. When the membrane is ruptured, the gas fills the entire tank. The only way to restore the system to its original state is to compress it to its initial volume, while transferring heat from the gas until it reaches its initial temperature. From the conservation of energy considerations, it can easily be shown that the amount of heat transferred from the gas equals the amount of work done on the gas by the surroundings. The restoration of the surroundings involves conversion of this heat completely to work, which would violate the second law. Therefore, unrestrained expansion of a gas is an irreversible process.

A third form of irreversibility familiar to us all is **heat transfer** through a finite temperature difference. Consider a can of cold soda left in a warm room (Fig. 6–34). Heat is transferred from the warmer room air to the cooler soda. The only way this process can be reversed and the soda restored to its original temperature is to provide refrigeration, which requires some work input. At the end of the reverse process, the soda will be restored to its initial state, but the surroundings will not be. The internal energy of the surroundings will increase by an amount equal in magnitude to the work supplied to the refrigerator. The restoration of the surroundings to the initial state can be done only by converting this excess internal energy completely to work, which is impossible to do without violating the second law. Since only the system, not both the system and the surroundings, can be restored to its initial condition, heat transfer through a finite temperature difference is an irreversible process.

Heat transfer can occur only when there is a temperature difference between a system and its surroundings. Therefore, it is physically impossible to have a reversible heat transfer process. But a heat transfer process becomes less and less irreversible as the temperature difference between the two bodies approaches zero. Then, heat transfer through a differential temperature difference dT can be considered to be reversible. As dT approaches zero, the process can be reversed in direction (at least theoretically) without requiring any refrigeration. Notice that reversible heat transfer is a conceptual process and cannot be duplicated in the real world.

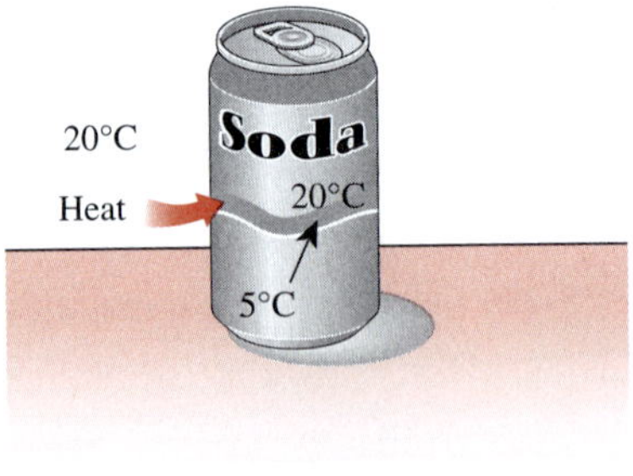

(*a*) An irreversible heat transfer process

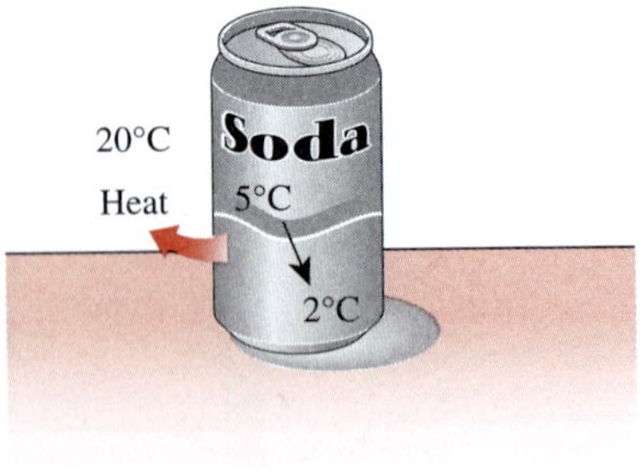

(*b*) An impossible heat transfer process

FIGURE 6–34
(*a*) Heat transfer through a temperature difference is irreversible, and (*b*) the reverse process is impossible.

The smaller the temperature difference between two bodies, the smaller the heat transfer rate will be. Any significant heat transfer through a small temperature difference requires a very large surface area and a very long time. Therefore, even though approaching reversible heat transfer is desirable from a thermodynamic point of view, it is impractical and not economically feasible.

Internally and Externally Reversible Processes

A typical process involves interactions between a system and its surroundings, and a reversible process involves no irreversibilities associated with either of them.

A process is called **internally reversible** if no irreversibilities occur within the boundaries of the system during the process. During an internally reversible process, a system proceeds through a series of equilibrium states, and when the process is reversed, the system passes through exactly the same equilibrium states while returning to its initial state. That is, the paths of the forward and reverse processes coincide for an internally reversible process. The quasi-equilibrium process is an example of an internally reversible process.

A process is called **externally reversible** if no irreversibilities occur outside the system boundaries during the process. Heat transfer between a reservoir and a system is an externally reversible process if the outer surface of the system is at the temperature of the reservoir.

A process is called **totally reversible**, or simply **reversible**, if it involves no irreversibilities within the system or its surroundings (Fig. 6–35). A totally reversible process involves no heat transfer through a finite temperature difference, no nonquasi-equilibrium changes, and no friction or other dissipative effects.

As an example, consider the transfer of heat to two identical systems that are undergoing a constant-pressure (thus constant-temperature) phase-change process, as shown in Fig. 6–36. Both processes are internally reversible, since both take place isothermally and both pass through exactly the same equilibrium states. The first process shown is externally reversible also, since heat transfer for this process takes place through an infinitesimal temperature difference dT. The second process, however, is externally irreversible, since it involves heat transfer through a finite temperature difference ΔT.

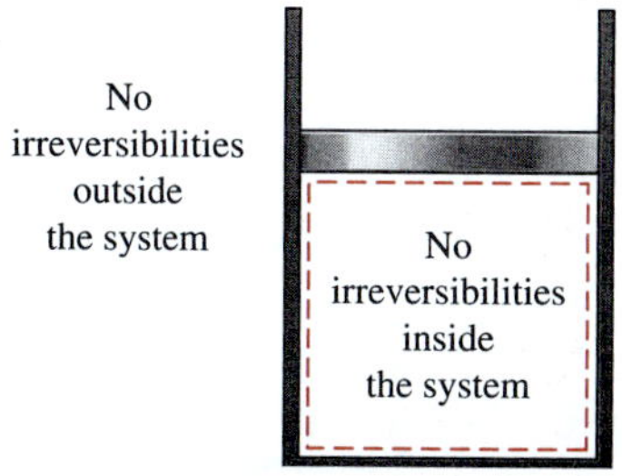

FIGURE 6–35
A reversible process involves no internal and external irreversibilities.

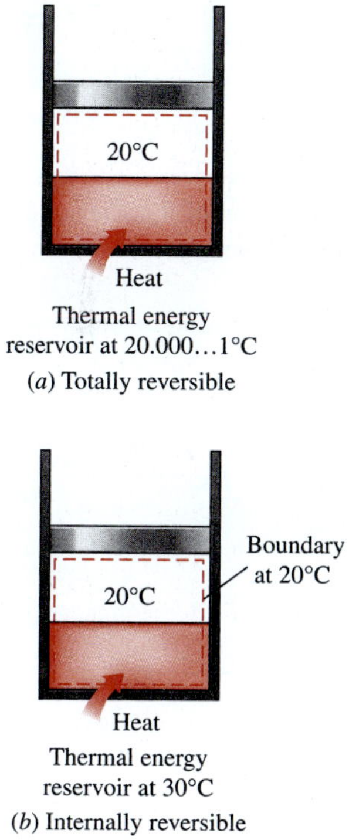

FIGURE 6–36
Totally and internally reversible heat transfer processes.

6–7 ▪ THE CARNOT CYCLE

We mentioned earlier that heat engines are cyclic devices and that the working fluid of a heat engine returns to its initial state at the end of each cycle. Work is done by the working fluid during one part of the cycle and on the working fluid during another part. The difference between these two is the net work delivered by the heat engine. The efficiency of a heat-engine cycle greatly depends on how the individual processes that make up the cycle are executed. The net work, thus the cycle efficiency, can be maximized by using processes that require the least amount of work and deliver the most, that is, by using *reversible processes*. Therefore, it is no surprise that the most efficient cycles are reversible cycles, that is, cycles that consist entirely of reversible processes.

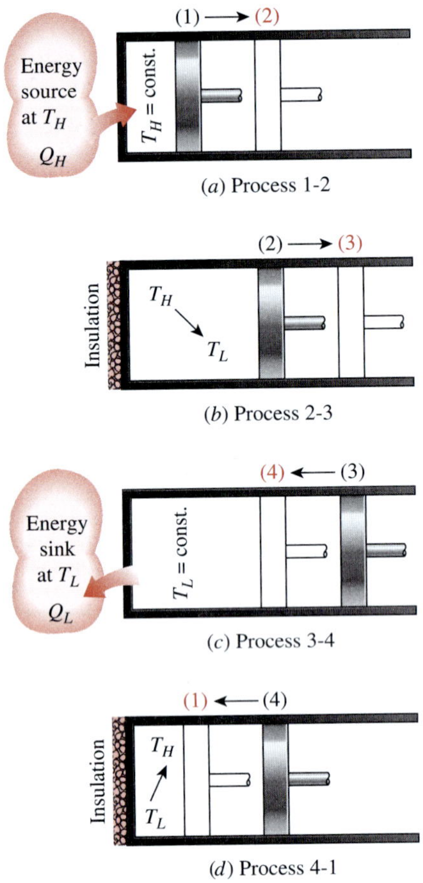

FIGURE 6–37
Execution of the Carnot cycle in a closed system.

Reversible cycles cannot be achieved in practice because the irreversibilities associated with each process cannot be eliminated. However, reversible cycles provide upper limits on the performance of real cycles. Heat engines and refrigerators that work on reversible cycles serve as models to which actual heat engines and refrigerators can be compared. Reversible cycles also serve as starting points in the development of actual cycles and are modified as needed to meet certain requirements.

Probably the best known reversible cycle is the **Carnot cycle**, first proposed in 1824 by French engineer Sadi Carnot. The theoretical heat engine that operates on the Carnot cycle is called the **Carnot heat engine**. The Carnot cycle is composed of four reversible processes—two isothermal and two adiabatic—and it can be executed either in a closed or a steady-flow system.

Consider a closed system that consists of a gas contained in an adiabatic piston–cylinder device, as shown in Fig. 6–37. The insulation of the cylinder head is such that it may be removed to bring the cylinder into contact with reservoirs to provide heat transfer. The four reversible processes that make up the Carnot cycle are as follows:

Reversible Isothermal Expansion (process 1-2, T_H = constant). Initially (state 1), the temperature of the gas is T_H and the cylinder head is in close contact with a source at temperature T_H. The gas is allowed to expand slowly, doing work on the surroundings. As the gas expands, the temperature of the gas tends to decrease. But as soon as the temperature drops by an infinitesimal amount dT, some heat is transferred from the reservoir into the gas, raising the gas temperature to T_H. Thus, the gas temperature is kept constant at T_H. Since the temperature difference between the gas and the reservoir never exceeds a differential amount dT, this is a reversible heat transfer process. It continues until the piston reaches position 2. The amount of total heat transferred to the gas during this process is Q_H.

Reversible Adiabatic Expansion (process 2-3, temperature drops from T_H to T_L). At state 2, the reservoir that was in contact with the cylinder head is removed and replaced by insulation so that the system becomes adiabatic. The gas continues to expand slowly, doing work on the surroundings until its temperature drops from T_H to T_L (state 3). The piston is assumed to be frictionless and the process to be quasi-equilibrium, so the process is reversible as well as adiabatic.

Reversible Isothermal Compression (process 3-4, T_L = constant). At state 3, the insulation at the cylinder head is removed, and the cylinder is brought into contact with a sink at temperature T_L. Now the piston is pushed inward by an external force, doing work on the gas. As the gas is compressed, its temperature tends to rise. But as soon as it rises by an infinitesimal amount dT, heat is transferred from the gas to the sink, causing the gas temperature to drop to T_L. Thus, the gas temperature remains constant at T_L. Since the temperature difference between the gas and the sink never exceeds a differential amount dT, this is a reversible heat transfer process. It continues until the piston reaches state 4. The amount of heat rejected from the gas during this process is Q_L.

Reversible Adiabatic Compression (process 4-1, temperature rises from T_L to T_H). State 4 is such that when the low-temperature reservoir is removed, the insulation is put back on the cylinder head, and the gas is compressed in a reversible manner, the gas returns to its initial state (state 1). The temperature rises from T_L to T_H during this reversible adiabatic compression process, which completes the cycle.

The *P*-*V* diagram of this cycle is shown in Fig. 6–38. Remembering that on a *P*-*V* diagram the area under the process curve represents the boundary work for quasi-equilibrium (internally reversible) processes, we see that the area under curve 1-2-3 is the work done by the gas during the expansion part of the cycle, and the area under curve 3-4-1 is the work done on the gas during the compression part of the cycle. The area enclosed by the path of the cycle (area 1-2-3-4-1) is the difference between these two and represents the net work done during the cycle.

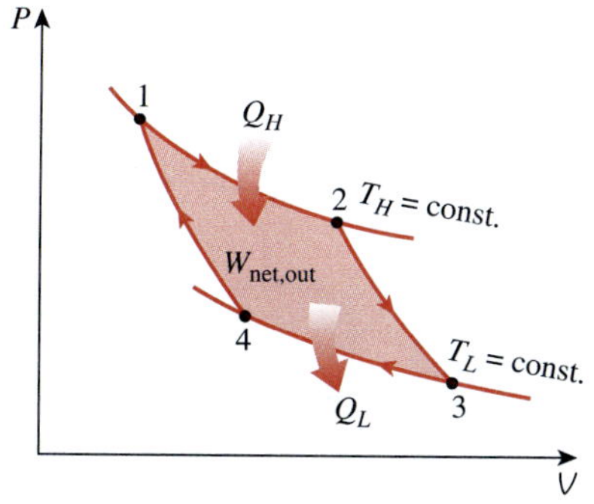

FIGURE 6–38
P-*V* diagram of the Carnot cycle.

Notice that if we acted stingily and compressed the gas at state 3 adiabatically instead of isothermally in an effort *to save* Q_L, we would end up back at state 2, retracing the process path 3-2. By doing so we would save Q_L, but we would not be able to obtain any net work output from this engine. This illustrates once more the necessity of a heat engine exchanging heat with at least two reservoirs at different temperatures to operate in a cycle and produce a net amount of work.

The Carnot cycle can also be executed in a steady-flow system. It is discussed in later chapters in conjunction with other power cycles.

Being a reversible cycle, the Carnot cycle is the most efficient cycle operating between two specified temperature limits. Even though the Carnot cycle cannot be achieved in reality, the efficiency of actual cycles can be improved by attempting to approximate the Carnot cycle more closely.

The Reversed Carnot Cycle

The Carnot heat-engine cycle just described is a totally reversible cycle. Therefore, all the processes that comprise it can be *reversed*, in which case it becomes the **Carnot refrigeration cycle**. This time, the cycle remains exactly the same, except that the directions of any heat and work interactions are reversed: Heat in the amount of Q_L is absorbed from the low-temperature reservoir, heat in the amount of Q_H is rejected to a high-temperature reservoir, and a work input of $W_{\text{net,in}}$ is required to accomplish all this.

The *P*-*V* diagram of the reversed Carnot cycle is the same as the one given for the Carnot cycle, except that the directions of the processes are reversed, as shown in Fig. 6–39.

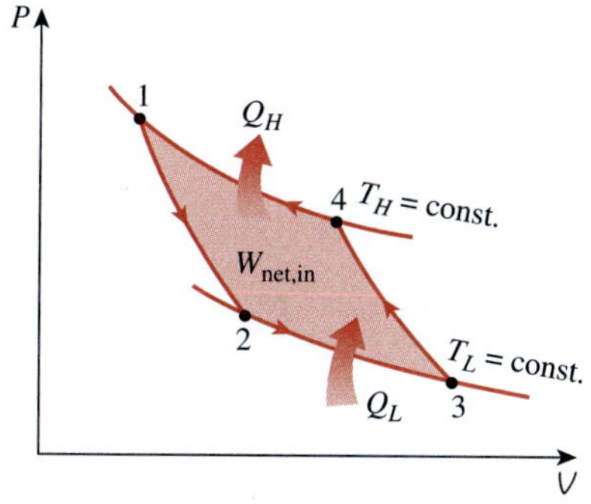

FIGURE 6–39
P-*V* diagram of the reversed Carnot cycle.

6–8 ▪ THE CARNOT PRINCIPLES

The second law of thermodynamics puts limits on the operation of cyclic devices as expressed by the Kelvin–Planck and Clausius statements. A heat engine cannot operate by exchanging heat with a single reservoir, and a refrigerator cannot operate without a net energy input from an external source.

We can draw valuable conclusions from these statements. Two conclusions pertain to the thermal efficiency of reversible and irreversible (i.e., actual) heat

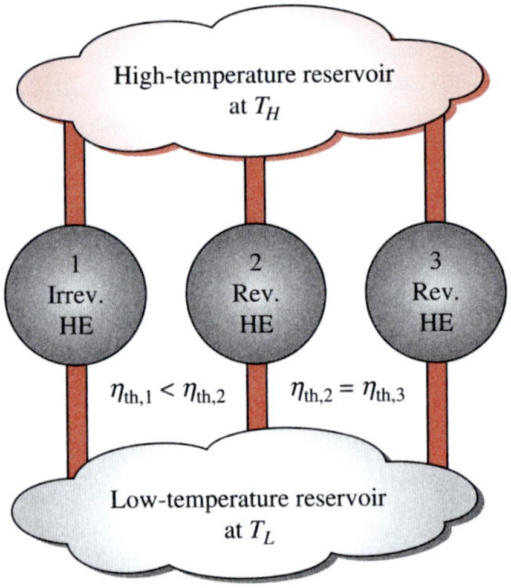

FIGURE 6–40
The Carnot principles.

engines, and they are known as the **Carnot principles** (Fig. 6–40), expressed as follows:

1. The efficiency of an irreversible heat engine is always less than the efficiency of a reversible one operating between the same two reservoirs.
2. The efficiencies of all reversible heat engines operating between the same two reservoirs are the same.

These two statements can be proved by demonstrating that the violation of either statement results in the violation of the second law of thermodynamics.

To prove the first statement, consider two heat engines operating between the same reservoirs, as shown in Fig. 6–41. One engine is reversible and the other is irreversible. Now each engine is supplied with the same amount of heat Q_H. The amount of work produced by the reversible heat engine is W_{rev}, and the amount produced by the irreversible one is W_{irrev}.

In violation of the first Carnot principle, we assume that the irreversible heat engine is more efficient than the reversible one (that is, $\eta_{th,irrev} > \eta_{th,rev}$) and thus delivers more work than the reversible one. Now let the reversible heat engine be reversed and operate as a refrigerator. This refrigerator will receive a work input of W_{rev} and reject heat to the high-temperature reservoir. Since the refrigerator is rejecting heat in the amount of Q_H to the high-temperature reservoir and the irreversible heat engine is receiving the same amount of heat from this reservoir, the net heat exchange for this reservoir is zero. Thus, it could be eliminated by having the refrigerator discharge Q_H directly into the irreversible heat engine.

Now considering the refrigerator and the irreversible engine together, we have an engine that produces a net work in the amount of $W_{irrev} - W_{rev}$ while exchanging heat with a single reservoir—a violation of the Kelvin–Planck

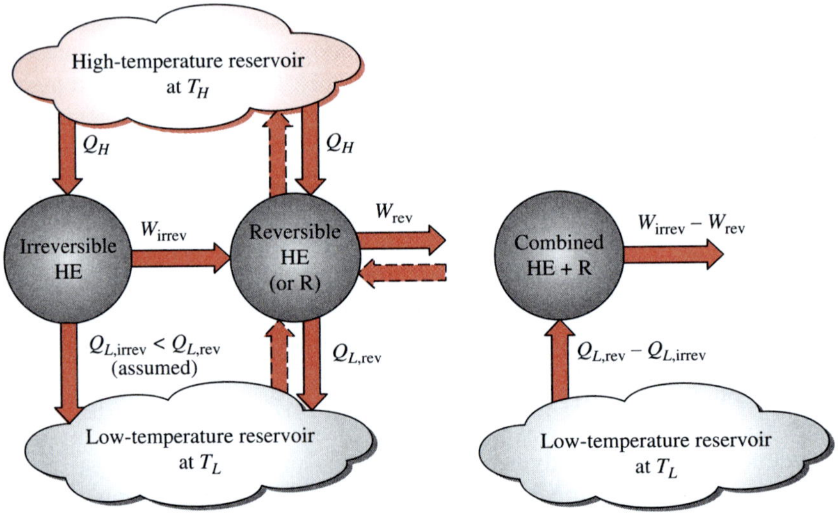

(*a*) A reversible and an irreversible heat engine operating between the same two reservoirs (the reversible heat engine is then reversed to run as a refrigerator)

(*b*) The equivalent combined system

FIGURE 6–41
Proof of the first Carnot principle.

statement of the second law. Therefore, our initial assumption that $\eta_{th,irrev} > \eta_{th,rev}$ is incorrect. Then we conclude that no heat engine can be more efficient than a reversible heat engine operating between the same reservoirs.

The second Carnot principle can also be proved in a similar manner. This time, let us replace the irreversible engine by another reversible engine that is more efficient and thus delivers more work than the first reversible engine. By following through the same reasoning, we end up having an engine that produces a net amount of work while exchanging heat with a single reservoir, which is a violation of the second law. Therefore, we conclude that no reversible heat engine can be more efficient than a reversible one operating between the same two reservoirs, regardless of how the cycle is completed or the kind of working fluid used.

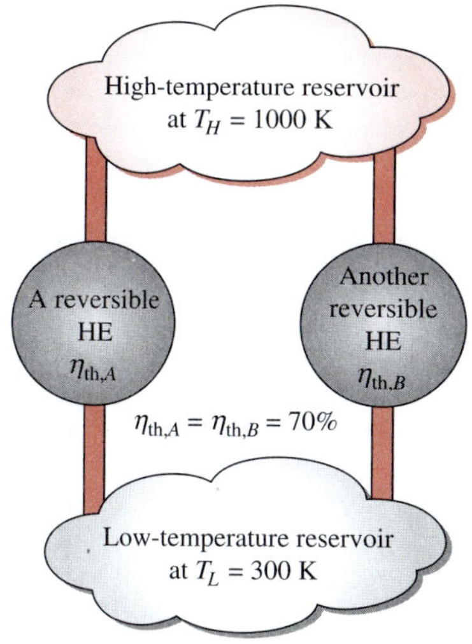

FIGURE 6–42
All reversible heat engines operating between the same two reservoirs have the same efficiency (the second Carnot principle).

6–9 ▪ THE THERMODYNAMIC TEMPERATURE SCALE

A temperature scale that is independent of the properties of the substances that are used to measure temperature is called a **thermodynamic temperature scale**. Such a temperature scale offers great conveniences in thermodynamic calculations, and its derivation is given below using some reversible heat engines.

The second Carnot principle discussed in Section 6–8 states that all reversible heat engines have the same thermal efficiency when operating between the same two reservoirs (Fig. 6–42). That is, the efficiency of a reversible engine is independent of the working fluid employed and its properties, the way the cycle is executed, or the type of reversible engine used. Since energy reservoirs are characterized by their temperatures, the thermal efficiency of reversible heat engines is a function of the reservoir temperatures only. That is,

$$\eta_{th,rev} = g(T_H, T_L)$$

or

$$\frac{Q_H}{Q_L} = f(T_H, T_L) \qquad \textbf{(6–13)}$$

since $\eta_{th} = 1 - Q_L/Q_H$. In these relations T_H and T_L are the temperatures of the high- and low-temperature reservoirs, respectively.

The functional form of $f(T_H, T_L)$ can be developed with the help of the three reversible heat engines shown in Fig. 6–43. Engines A and C are supplied with the same amount of heat Q_1 from the high-temperature reservoir at T_1. Engine C rejects Q_3 to the low-temperature reservoir at T_3. Engine B receives the heat Q_2 rejected by engine A at temperature T_2 and rejects heat in the amount of Q_3 to a reservoir at T_3.

The amounts of heat rejected by engines B and C must be the same since engines A and B can be combined into one reversible engine operating between the same reservoirs as engine C and thus the combined engine will have the same efficiency as engine C. Since the heat input to engine C is the same as the heat input to the combined engines A and B, both systems must reject the same amount of heat.

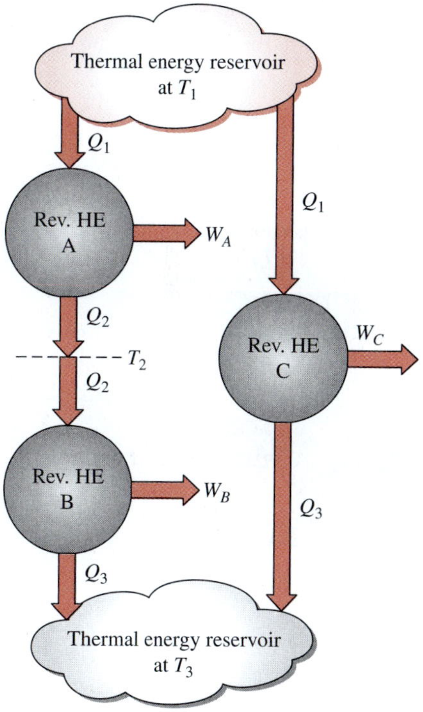

FIGURE 6–43
The arrangement of heat engines used to develop the thermodynamic temperature scale.

Applying Eq. 6–13 to all three engines separately, we obtain

$$\frac{Q_1}{Q_2} = f(T_1, T_2), \quad \frac{Q_2}{Q_3} = f(T_2, T_3), \quad \text{and} \quad \frac{Q_1}{Q_3} = f(T_1, T_3)$$

Now consider the identity

$$\frac{Q_1}{Q_3} = \frac{Q_1}{Q_2}\frac{Q_2}{Q_3}$$

which corresponds to

$$f(T_1, T_3) = f(T_1, T_2) \cdot f(T_2, T_3)$$

A careful examination of this equation reveals that the left-hand side is a function of T_1 and T_3, and therefore the right-hand side must also be a function of T_1 and T_3 only, and not T_2. That is, the value of the product on the right-hand side of this equation is independent of the value of T_2. This condition will be satisfied only if the function f has the following form:

$$f(T_1, T_2) = \frac{\phi(T_1)}{\phi(T_2)} \quad \text{and} \quad f(T_2, T_3) = \frac{\phi(T_2)}{\phi(T_3)}$$

so that $\phi(T_2)$ will cancel from the product of $f(T_1, T_2)$ and $f(T_2, T_3)$, yielding

$$\frac{Q_1}{Q_3} = f(T_1, T_3) = \frac{\phi(T_1)}{\phi(T_3)} \tag{6–14}$$

This relation is much more specific than Eq. 6–13 for the functional form of Q_1/Q_3 in terms of T_1 and T_3.

For a reversible heat engine operating between two reservoirs at temperatures T_H and T_L, Eq. 6–14 can be written as

$$\frac{Q_H}{Q_L} = \frac{\phi(T_H)}{\phi(T_L)} \tag{6–15}$$

This is the only requirement that the second law places on the ratio of heat transfers to and from the reversible heat engines. Several functions $\phi(T)$ satisfy this equation, and the choice is completely arbitrary. Lord Kelvin first proposed taking $\phi(T) = T$ to define a thermodynamic temperature scale as (Fig. 6–44)

$$\left(\frac{Q_H}{Q_L}\right)_{\text{rev}} = \frac{T_H}{T_L} \tag{6–16}$$

This temperature scale is called the **Kelvin scale**, and the temperatures on this scale are called **absolute temperatures**. On the Kelvin scale, the temperature ratios depend on the ratios of heat transfer between a reversible heat engine and the reservoirs and are independent of the physical properties of any substance. On this scale, temperatures vary between zero and infinity.

The thermodynamic temperature scale is not completely defined by Eq. 6–16 since it gives us only a ratio of absolute temperatures. We also need to know the magnitude of a kelvin. At the International Conference on Weights and Measures held in 1954, the triple point of water (the state at which all three phases of water exist in equilibrium) was assigned the value

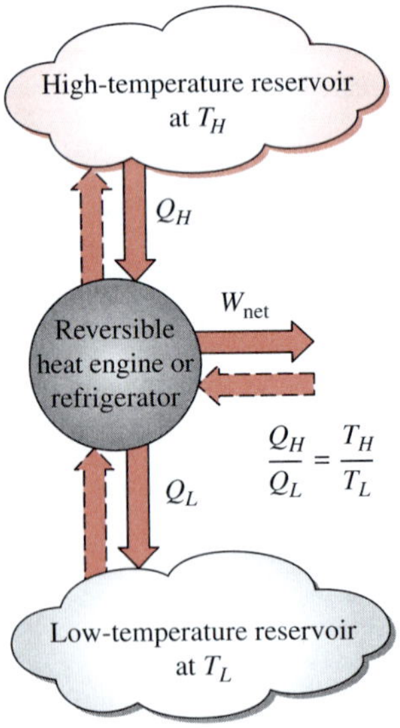

FIGURE 6–44
For reversible cycles, the heat transfer ratio Q_H/Q_L can be replaced by the absolute temperature ratio T_H/T_L.

273.16 K (Fig. 6–45). The *magnitude of a kelvin* is defined as 1/273.16 of the temperature interval between absolute zero and the triple-point temperature of water. The magnitudes of temperature units on the Kelvin and Celsius scales are identical (1 K ≡ 1°C). The temperatures on these two scales differ by a constant 273.15:

$$T(°\text{C}) = T(\text{K}) - 273.15 \quad \text{(6–17)}$$

Even though the thermodynamic temperature scale is defined with the help of the reversible heat engines, it is not possible, nor is it practical, to actually operate such an engine to determine numerical values on the absolute temperature scale. Absolute temperatures can be measured accurately by other means, such as the constant-volume ideal-gas thermometer together with extrapolation techniques as discussed in Chap. 1. The validity of Eq. 6–16 can be demonstrated from physical considerations for a reversible cycle using an ideal gas as the working fluid.

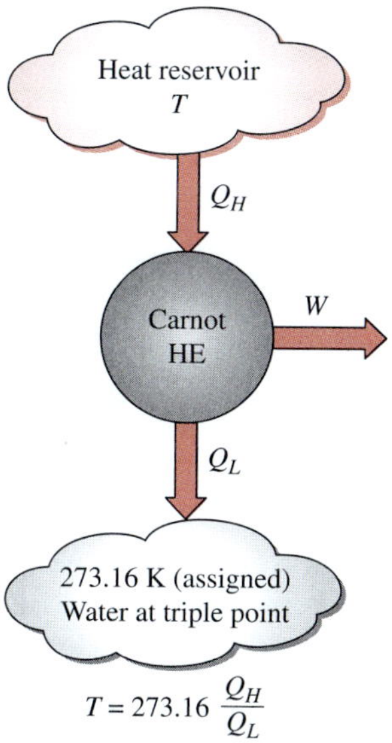

FIGURE 6–45
A conceptual experimental setup to determine thermodynamic temperatures on the Kelvin scale by measuring heat transfers Q_H and Q_L.

6–10 ▪ THE CARNOT HEAT ENGINE

The hypothetical heat engine that operates on the reversible Carnot cycle is called the **Carnot heat engine**. The thermal efficiency of any heat engine, reversible or irreversible, is given by Eq. 6–6 as

$$\eta_{\text{th}} = 1 - \frac{Q_L}{Q_H}$$

where Q_H is heat transferred to the heat engine from a high-temperature reservoir at T_H, and Q_L is heat rejected to a low-temperature reservoir at T_L. For reversible heat engines, the heat transfer ratio in the above relation can be replaced by the ratio of the absolute temperatures of the two reservoirs, as given by Eq. 6–16. Then the efficiency of a Carnot engine, or any reversible heat engine, becomes

$$\eta_{\text{th,rev}} = 1 - \frac{T_L}{T_H} \quad \text{(6–18)}$$

This relation is often referred to as the **Carnot efficiency**, since the Carnot heat engine is the best known reversible engine. *This is the highest efficiency a heat engine operating between the two thermal energy reservoirs at temperatures T_L and T_H can have* (Fig. 6–46). All irreversible (i.e., actual) heat engines operating between these temperature limits (T_L and T_H) have lower efficiencies. An actual heat engine cannot reach this maximum theoretical efficiency value because it is impossible to completely eliminate all the irreversibilities associated with the actual cycle.

Note that T_L and T_H in Eq. 6–18 are *absolute temperatures*. Using °C or °F for temperatures in this relation gives results grossly in error.

The thermal efficiencies of actual and reversible heat engines operating between the same temperature limits compare as follows (Fig. 6–47):

$$\eta_{\text{th}} \begin{cases} < \eta_{\text{th,rev}} & \text{irreversible heat engine} \\ = \eta_{\text{th,rev}} & \text{reversible heat engine} \\ > \eta_{\text{th,rev}} & \text{impossible heat engine} \end{cases} \quad \text{(6–19)}$$

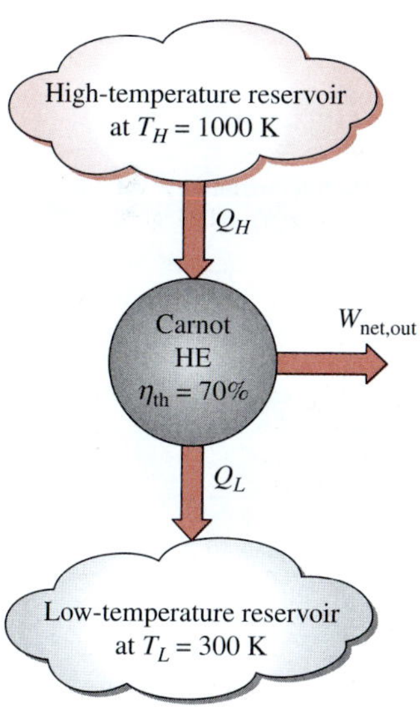

FIGURE 6–46
The Carnot heat engine is the most efficient of all heat engines operating between the same high- and low-temperature reservoirs.

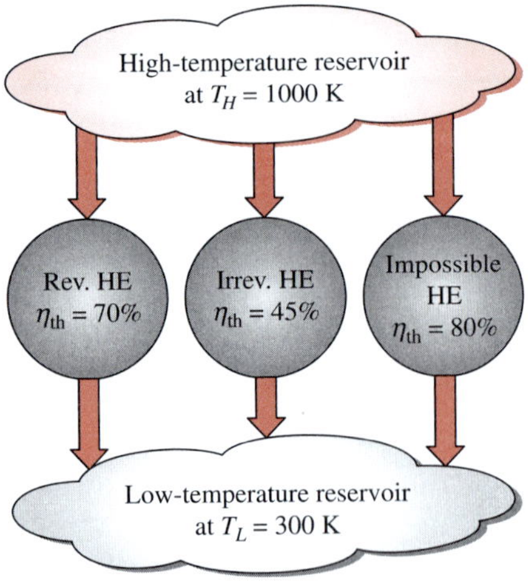

FIGURE 6–47
No heat engine can have a higher efficiency than a reversible heat engine operating between the same high- and low-temperature reservoirs.

Most work-producing devices (heat engines) in operation today have efficiencies under 40 percent, which appear low relative to 100 percent. However, when the performance of actual heat engines is assessed, the efficiencies should not be compared to 100 percent; instead, they should be compared to the efficiency of a reversible heat engine operating between the same temperature limits—because this is the true theoretical upper limit for the efficiency, not 100 percent.

The maximum efficiency of a steam power plant operating between T_H = 1000 K and T_L = 300 K is 70 percent, as determined from Eq. 6–18. Compared with this value, an actual efficiency of 40 percent does not seem so bad, even though there is still plenty of room for improvement.

It is obvious from Eq. 6–18 that the efficiency of a Carnot heat engine increases as T_H is increased, or as T_L is decreased. This is to be expected since as T_L decreases, so does the amount of heat rejected, and as T_L approaches zero, the Carnot efficiency approaches unity. This is also true for actual heat engines. *The thermal efficiency of actual heat engines can be maximized by supplying heat to the engine at the highest possible temperature* (limited by material strength) *and rejecting heat from the engine at the lowest possible temperature* (limited by the temperature of the cooling medium such as rivers, lakes, or the atmosphere).

EXAMPLE 6–5 Analysis of a Carnot Heat Engine

A Carnot heat engine, shown in Fig. 6–48, receives 500 kJ of heat per cycle from a high-temperature source at 652°C and rejects heat to a low-temperature sink at 30°C. Determine (*a*) the thermal efficiency of this Carnot engine and (*b*) the amount of heat rejected to the sink per cycle.

SOLUTION The heat supplied to a Carnot heat engine is given. The thermal efficiency and the heat rejected are to be determined.

Analysis (*a*) The Carnot heat engine is a reversible heat engine, and so its efficiency can be determined from Eq. 6–18 to be

$$\eta_{th,C} = \eta_{th,rev} = 1 - \frac{T_L}{T_H} = 1 - \frac{(30 + 273)\text{ K}}{(652 + 273)\text{ K}} = \mathbf{0.672}$$

That is, this Carnot heat engine converts 67.2 percent of the heat it receives to work.

(*b*) The amount of heat rejected Q_L by this reversible heat engine is easily determined from Eq. 6–16 to be

$$Q_{L,rev} = \frac{T_L}{T_H} Q_{H,rev} = \frac{(30 + 273)\text{ K}}{(652 + 273)\text{ K}} (500\text{ kJ}) = \mathbf{164\ kJ}$$

Discussion Note that this Carnot heat engine rejects to a low-temperature sink 164 kJ of the 500 kJ of heat it receives during each cycle.

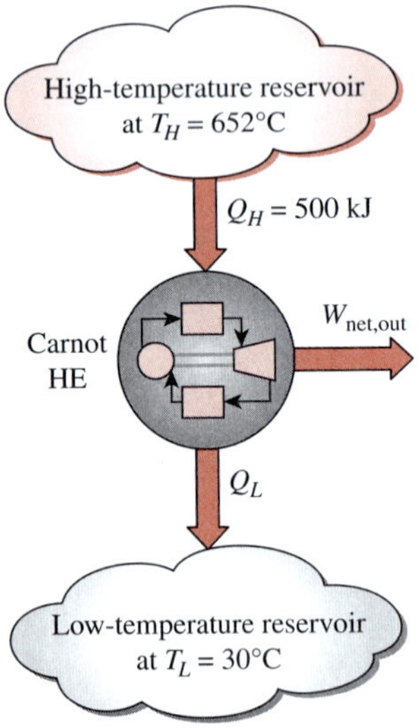

FIGURE 6–48
Schematic for Example 6–5.

The Quality of Energy

The Carnot heat engine in Example 6–5 receives heat from a source at 925 K and converts 67.2 percent of it to work while rejecting the rest (32.8 percent) to a sink at 303 K. Now let us examine how the thermal efficiency varies with the source temperature when the sink temperature is held constant.

The thermal efficiency of a Carnot heat engine that rejects heat to a sink at 303 K is evaluated at various source temperatures using Eq. 6–18 and is listed in Fig. 6–49. Clearly, the thermal efficiency decreases as the source temperature is lowered. When heat is supplied to the heat engine at 500 instead of 925 K, for example, the thermal efficiency drops from 67.2 to 39.4 percent. That is, the fraction of heat that can be converted to work drops to 39.4 percent when the temperature of the source drops to 500 K. When the source temperature is 350 K, this fraction becomes a mere 13.4 percent.

These efficiency values show that energy has **quality** as well as quantity. It is clear from the thermal efficiency values in Fig. 6–49 that *more of the high-temperature thermal energy can be converted to work. Therefore, the higher the temperature, the higher the quality of the energy* (Fig. 6–50).

Large quantities of solar energy, for example, can be stored in large bodies of water called *solar ponds* at about 350 K. This stored energy can then be supplied to a heat engine to produce work (electricity). However, the efficiency of solar pond power plants is very low (under 5 percent) because of the low quality of the energy stored in the source, and the construction and maintenance costs are relatively high. Therefore, they are not competitive even though the energy supply of such plants is free. The temperature (and thus the quality) of the solar energy stored could be raised by utilizing concentrating collectors, but the equipment cost in that case becomes very high.

Work is a more valuable form of energy than heat since 100 percent of work can be converted to heat, but only a fraction of heat can be converted to work. When heat is transferred from a high-temperature body to a lower-temperature one, it is degraded since less of it now can be converted to work. For example, if 100 kJ of heat is transferred from a body at 1000 K to a body at 300 K, at the end we will have 100 kJ of thermal energy stored at 300 K, which has no practical value. But if this conversion is made through a heat engine, up to $1 - 300/1000 = 70$ percent of it could be converted to work, which is a more valuable form of energy. Thus 70 kJ of work potential is wasted as a result of this heat transfer, and energy is degraded.

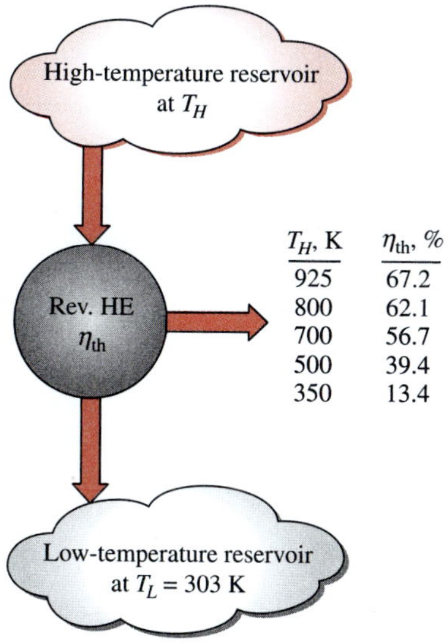

FIGURE 6–49
The fraction of heat that can be converted to work as a function of source temperature (for $T_L = 303$ K).

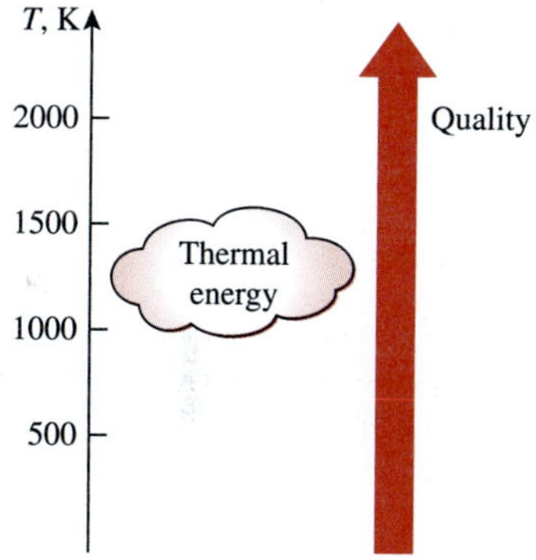

FIGURE 6–50
The higher the temperature of the thermal energy, the higher its quality.

Quantity versus Quality in Daily Life

At times of energy crisis, we are bombarded with speeches and articles on how to "conserve" energy. Yet we all know that the *quantity* of energy is already conserved. What is not conserved is the *quality* of energy, or the work potential of energy. Wasting energy is synonymous to converting it to a less useful form. One unit of high-quality energy can be more valuable than three units of lower-quality energy. For example, a finite amount of thermal energy at high temperature is more attractive to power plant engineers than a vast amount of thermal energy at low temperature, such as the energy stored in the upper layers of the oceans at tropical climates.

As part of our culture, we seem to be fascinated by quantity, and little attention is given to quality. However, quantity alone cannot give the whole picture, and we need to consider quality as well. That is, we need to look at something from both the first- and second-law points of view when evaluating something, even in nontechnical areas. Below we present some ordinary events and show their relevance to the second law of thermodynamics.

Consider two students Andy and Wendy. Andy has 10 friends who never miss his parties and are always around during fun times. However, they seem to be busy when Andy needs their help. Wendy, on the other hand, has five friends. They are never too busy for her, and she can count on them at times of need. Let us now try to answer the question, *Who has more friends?* From the first-law point of view, which considers quantity only, it is obvious that Andy has more friends. However, from the second-law point of view, which considers quality as well, there is no doubt that Wendy is the one with more friends.

Another example with which most people will identify is the multibillion-dollar diet industry, which is primarily based on the first law of thermodynamics. However, considering that 90 percent of the people who lose weight gain it back quickly, with interest, suggests that the first law alone does not give the whole picture. People who seem to be eating whatever they want, whenever they want, without gaining weight are living proof that the calorie-counting technique (the first law) leaves many questions on dieting unanswered. Obviously, more research focused on the second-law effects of dieting is needed before we can fully understand the weight-gain and weight-loss process.

It is tempting to judge things on the basis of their *quantity* instead of their *quality* since assessing quality is much more difficult than assessing quantity. However, assessments made on the basis of quantity only (the first law) may be grossly inadequate and misleading.

6–11 ▪ THE CARNOT REFRIGERATOR AND HEAT PUMP

A refrigerator or a heat pump that operates on the reversed Carnot cycle is called a **Carnot refrigerator**, or a **Carnot heat pump**. The coefficient of performance of any refrigerator or heat pump, reversible or irreversible, is given by Eqs. 6–9 and 6–11 as

$$\text{COP}_\text{R} = \frac{1}{Q_H/Q_L - 1} \quad \text{and} \quad \text{COP}_\text{HP} = \frac{1}{1 - Q_L/Q_H}$$

where Q_L is the amount of heat absorbed from the low-temperature medium and Q_H is the amount of heat rejected to the high-temperature medium. The COPs of all reversible refrigerators or heat pumps can be determined by replacing the heat transfer ratios in the above relations by the ratios of the absolute temperatures of the high- and low-temperature reservoirs, as expressed by Eq. 6–16. Then the COP relations for reversible refrigerators and heat pumps become

$$\text{COP}_{\text{R,rev}} = \frac{1}{T_H/T_L - 1} \tag{6–20}$$

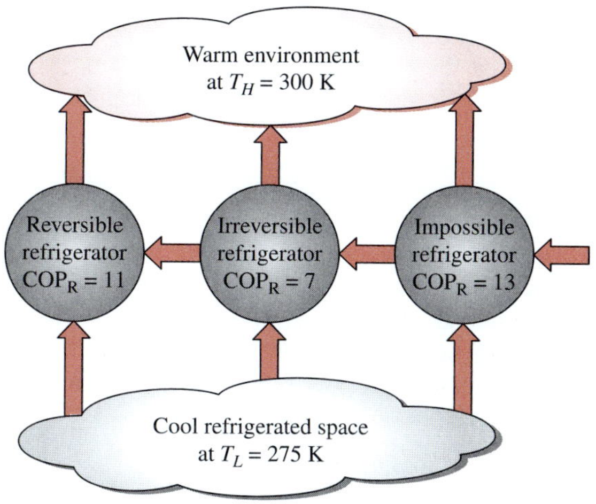

FIGURE 6–51
No refrigerator can have a higher COP than a reversible refrigerator operating between the same temperature limits.

and

$$COP_{HP,rev} = \frac{1}{1 - T_L/T_H} \quad \text{(6–21)}$$

These are the highest coefficients of performance that a refrigerator or a heat pump operating between the temperature limits of T_L and T_H can have. All actual refrigerators or heat pumps operating between these temperature limits (T_L and T_H) have lower coefficients of performance (Fig. 6–51).

The coefficients of performance of actual and reversible refrigerators operating between the same temperature limits can be compared as follows:

$$COP_R \begin{cases} < COP_{R,rev} & \text{irreversible refrigerator} \\ = COP_{R,rev} & \text{reversible refrigerator} \\ > COP_{R,rev} & \text{impossible refrigerator} \end{cases} \quad \text{(6–22)}$$

A similar relation can be obtained for heat pumps by replacing all COP_R's in Eq. 6–22 by COP_{HP}.

The COP of a reversible refrigerator or heat pump is the maximum theoretical value for the specified temperature limits. Actual refrigerators or heat pumps may approach these values as their designs are improved, but they can never reach them.

As a final note, the COPs of both the refrigerators and the heat pumps decrease as T_L decreases. That is, it requires more work to absorb heat from lower-temperature media. As the temperature of the refrigerated space approaches zero, the amount of work required to produce a finite amount of refrigeration approaches infinity and COP_R approaches zero.

EXAMPLE 6–6 A Carnot Refrigeration Cycle Operating in the Saturation Dome

A Carnot refrigeration cycle is executed in a closed system in the saturated liquid–vapor mixture region using 0.8 kg of refrigerant-134a as the working fluid (Fig. 6–52). The maximum and the minimum temperatures in the cycle

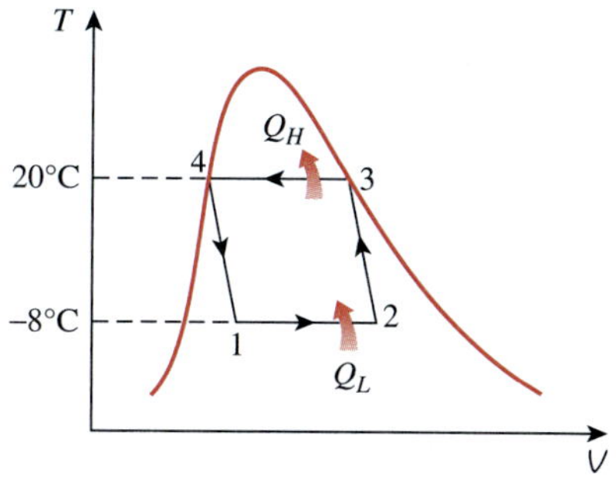

FIGURE 6–52
Schematic for Example 6–6.

are 20 and −8°C, respectively. It is known that the refrigerant is saturated liquid at the end of the heat rejection process, and the net work input to the cycle is 15 kJ. Determine the fraction of the mass of the refrigerant that vaporizes during the heat addition process, and the pressure at the end of the heat rejection process.

SOLUTION A Carnot refrigeration cycle is executed in a closed system. The mass fraction of the refrigerant that vaporizes during the heat addition process and the pressure at the end of the heat rejection process are to be determined.

Assumptions The refrigerator operates on the ideal Carnot cycle.

Analysis Knowing the high and low temperatures, the coefficient of performance of the cycle is

$$\text{COP}_\text{R} = \frac{1}{T_H/T_L - 1} = \frac{1}{(20 + 273\,\text{K})/(-8 + 273\,\text{K}) - 1} = 9.464$$

The amount of cooling is determined from the definition of the coefficient of performance to be

$$Q_L = \text{COP}_\text{R} \times W_\text{in} = (9.464)(15\text{ kJ}) = 142\text{ kJ}$$

The enthalpy of vaporization R-134a at −8°C is h_{fg} = 204.52 kJ/kg (Table A-11). Then the amount of refrigerant that vaporizes during heat absorption becomes

$$Q_L = m_\text{evap}h_{fg@-8°\text{C}} \rightarrow m_\text{evap} = \frac{142\text{ kJ}}{204.52\text{ kJ/kg}} = 0.694\text{ kg}$$

Therefore, the fraction of mass that vaporized during heat addition process to the refrigerant is

$$\text{Mass fraction} = \frac{m_\text{evap}}{m_\text{total}} = \frac{0.694\text{ kg}}{0.8\text{ kg}} = \mathbf{0.868 \text{ or } 86.8\%}$$

The pressure at the end of heat rejection process is simply the saturation pressure at heat rejection temperature,

$$P_4 = P_{\text{sat@20°C}} = \mathbf{572.1\,kPa}$$

Discussion Carnot cycle is an idealized refrigeration cycle, thus it cannot be achieved in practice. Practical refrigeration cycles are analyzed in Chap. 11.

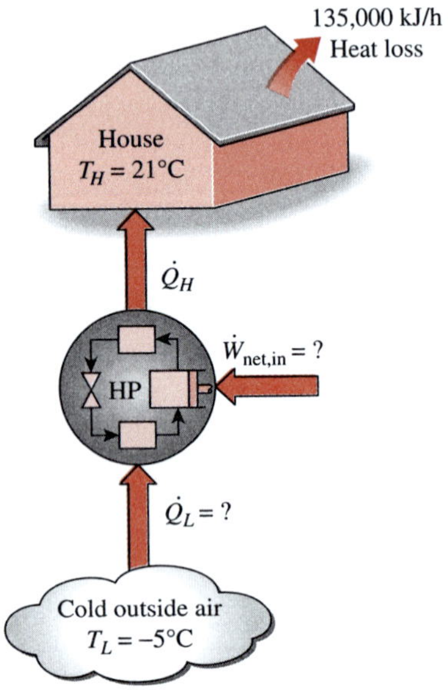

FIGURE 6–53
Schematic for Example 6–7.

EXAMPLE 6–7 Heating a House by a Carnot Heat Pump

A heat pump is to be used to heat a house during the winter, as shown in Fig. 6–53. The house is to be maintained at 21°C at all times. The house is estimated to be losing heat at a rate of 135,000 kJ/h when the outside temperature drops to −5°C. Determine the minimum power required to drive this heat pump.

SOLUTION A heat pump maintains a house at a constant temperature. The required minimum power input to the heat pump is to be determined.

Assumptions Steady operating conditions exist.
Analysis The heat pump must supply heat to the house at a rate of Q_H = 135,000 kJ/h = 37.5 kW. The power requirements are minimum when a reversible heat pump is used to do the job. The COP of a reversible heat pump operating between the house and the outside air is

$$\text{COP}_{\text{HP,rev}} = \frac{1}{1 - T_L/T_H} = \frac{1}{1 - (-5 + 273\text{ K})/(21 + 273\text{ K})} = 11.3$$

Then, the required power input to this reversible heat pump becomes

$$\dot{W}_{\text{net,in}} = \frac{\dot{Q}_H}{\text{COP}_{\text{HP}}} = \frac{37.5\text{ kW}}{11.3} = \mathbf{3.32\text{ kW}}$$

Discussion This reversible heat pump can meet the heating requirements of this house by consuming electric power at a rate of 3.32 kW only. If this house were to be heated by electric resistance heaters instead, the power consumption would jump up 11.3 times to 37.5 kW. This is because in resistance heaters the electric energy is converted to heat at a one-to-one ratio. With a heat pump, however, energy is absorbed from the outside and carried to the inside using a refrigeration cycle that consumes only 3.32 kW. Notice that the heat pump does not create energy. It merely transports it from one medium (the cold outdoors) to another (the warm indoors).

TOPIC OF SPECIAL INTEREST* Household Refrigerators

Refrigerators to preserve perishable foods have long been one of the essential appliances in a household. They have proven to be highly durable and reliable, providing satisfactory service for over 15 years. A typical household refrigerator is actually a combination refrigerator-freezer since it has a freezer compartment to make ice and to store frozen food.

Today's refrigerators use much less energy as a result of using *smaller* and *higher-efficiency* motors and compressors, *better insulation materials, larger coil surface areas*, and *better door seals* (Fig. 6–54). At an average electricity rate of 8.3 cents per kWh, an average refrigerator costs about $72 a year to run, which is half the annual operating cost of a refrigerator 25 years ago. Replacing a 25-year-old, 18-ft^3 refrigerator with a new energy-efficient model will save over 1000 kWh of electricity per year. For the environment, this means a reduction of over 1 ton of CO_2, which causes global climate change, and over 10 kg of SO_2, which causes acid rain.

Despite the improvements made in several areas during the past 100 years in household refrigerators, the basic *vapor-compression refrigeration cycle* has remained unchanged. The alternative *absorption refrigeration* and *thermoelectric refrigeration* systems are currently more expensive and less

*This section can be skipped without a loss in continuity.

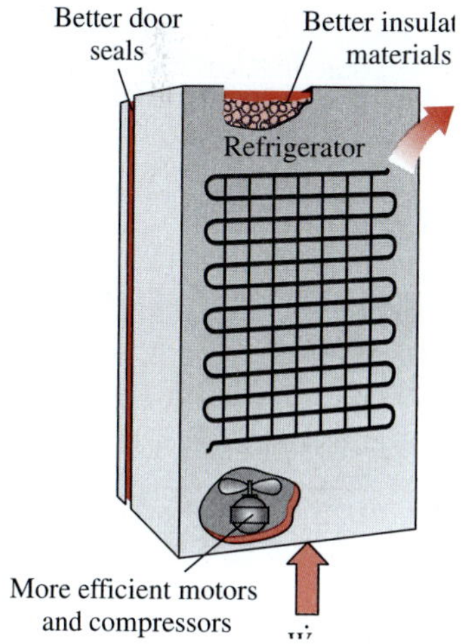

FIGURE 6–54
Today's refrigerators are much more efficient because of the improvements in technology and manufacturing.

TABLE 6–1

Typical operating efficiencies of some refrigeration systems for a freezer temperature of −18°C and ambient temperature of 32°C

Type of refrigeration system	Coefficient of performance
Vapor-compression	1.3
Absorption refrigeration	0.4
Thermoelectric refrigeration	0.1

efficient, and they have found limited use in some specialized applications (Table 6–1).

A household refrigerator is designed to maintain the freezer section at −18°C (0°F) and the refrigerator section at 3°C (37°F). Lower freezer temperatures increase energy consumption without improving the storage life of frozen foods significantly. Different temperatures for the storage of specific foods can be maintained in the refrigerator section by using *special-purpose* compartments.

Practically all full-size refrigerators have a large *air-tight* drawer for leafy vegetables and fresh fruits to seal in moisture and to protect them from the drying effect of cool air circulating in the refrigerator. A covered *egg compartment* in the lid extends the life of eggs by slowing down the moisture loss from the eggs. It is common for refrigerators to have a special warmer compartment for *butter* in the door to maintain butter at spreading temperature. The compartment also isolates butter and prevents it from absorbing *odors* and *tastes* from other food items. Some upscale models have a temperature-controlled *meat compartment* maintained at −0.5°C (31°F), which keeps meat at the lowest safe temperature without freezing it, and thus extending its storage life. The more expensive models come with an automatic *icemaker* located in the freezer section that is connected to the water line, as well as automatic ice and chilled-water dispensers. A typical icemaker can produce 2 to 3 kg of ice per day and store 3 to 5 kg of ice in a removable ice storage container.

Household refrigerators consume from about 90 to 600 W of electrical energy when running and are designed to perform satisfactorily in environments at up to 43°C (110°F). Refrigerators run intermittently, as you may have noticed, running about 30 percent of the time under normal use in a house at 25°C (77°F).

For specified external dimensions, a refrigerator is desired to have *maximum* food storage volume, *minimum* energy consumption, and the *lowest* possible cost to the consumer. The total food storage volume has been increased over the years without an increase in the external dimensions by using thinner but more effective insulation and minimizing the space occupied by the compressor and the condenser. Switching from the fiber-glass insulation (thermal conductivity k = 0.032–0.040 W/m·°C) to expanded-in-place urethane foam insulation (k = 0.019 W/m·°C) made it possible to reduce the wall thickness of the refrigerator by almost half, from about 90 to 48 mm for the freezer section and from about 70 to 40 mm for the refrigerator section. The rigidity and bonding action of the foam also provide additional structural support. However, the entire shell of the refrigerator must be carefully sealed to prevent any water leakage or moisture migration into the insulation since moisture degrades the effectiveness of insulation.

The size of the compressor and the other components of a refrigeration system are determined on the basis of the anticipated heat load (or refrigeration load), which is the rate of heat flow into the refrigerator. The heat load consists of the *predictable part*, such as heat transfer through the walls and door gaskets of the refrigerator, fan motors, and defrost heaters (Fig. 6–55), and the *unpredictable part*, which depends on the user habits such as opening the door, making ice, and loading the refrigerator. The amount of *energy* consumed by

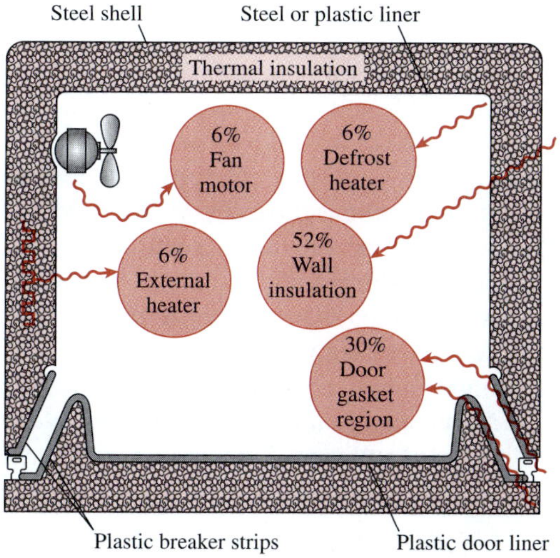

FIGURE 6–55
The cross section of a refrigerator showing the relative magnitudes of various effects that constitute the predictable heat load.
From ASHRAE Handbook of Refrigeration, Chap. 48, Fig. 2.

the refrigerator can be minimized by practicing good *conservation measures* as discussed below.

1. *Open the refrigerator door the fewest times possible* for the shortest duration possible. Each time the refrigerator door is opened, the cool air inside is replaced by the warmer air outside, which needs to be cooled. Keeping the refrigerator or freezer full will save energy by reducing the amount of cold air that can escape each time the door is opened.
2. *Cool the hot foods* to room temperature first before putting them into the refrigerator. Moving a hot pan from the oven directly into the refrigerator not only wastes energy by making the refrigerator work longer, but it also causes the nearby perishable foods to spoil by creating a warm environment in its immediate surroundings (Fig. 6–56).
3. *Clean the condenser coils* located behind or beneath the refrigerator. The dust and grime that collect on the coils act as insulation that slows down heat dissipation through them. Cleaning the coils a couple of times a year with a damp cloth or a vacuum cleaner will improve cooling ability of the refrigerator while cutting down the power consumption by a few percent. Sometimes a fan is used to force-cool the condensers of large or built-in refrigerators, and the strong air motion keeps the coils clean.
4. *Check the door gasket* for air leaks. This can be done by placing a flashlight into the refrigerator, turning off the kitchen lights, and looking for light leaks. Heat transfer through the door gasket region accounts for almost one-third of the regular heat load of the refrigerators, and thus any defective door gaskets must be repaired immediately.
5. *Avoid unnecessarily low temperature settings*. The recommended temperatures for freezers and refrigerators are −18°C (0°F) and 3°C (37°F), respectively. Setting the freezer temperature below −18°C adds significantly to the energy consumption but does not add much to the storage life of frozen foods. Keeping temperatures 6°C (or 10°F) below recommended levels can increase the energy use by as much as 25 percent.

FIGURE 6–56
Putting hot foods into the refrigerator without cooling them first not only wastes energy but also could spoil the foods nearby.

6. *Avoid excessive ice build-up* on the interior surfaces of the evaporator. The ice layer on the surface acts as insulation and slows down heat transfer from the freezer section to the refrigerant. The refrigerator should be defrosted by manually turning off the temperature control switch when the ice thickness exceeds a few millimeters.

   Defrosting is done automatically in no-frost refrigerators by supplying heat to the evaporator by a 300-W to 1000-W resistance heater or by hot refrigerant gas, periodically for short periods. The water is then drained to a pan outside where it is evaporated using the heat dissipated by the condenser. The no-frost evaporators are basically finned tubes subjected to air flow circulated by a fan. Practically all the frost collects on fins, which are the coldest surfaces, leaving the exposed surfaces of the freezer section and the frozen food frost-free.

7. *Use the power-saver switch* that controls the heating coils and prevents condensation on the outside surfaces in humid environments. The low-wattage heaters are used to raise the temperature of the outer surfaces of the refrigerator at critical locations above the dew point in order to avoid water droplets forming on the surfaces and sliding down. Condensation is most likely to occur in summer in hot and humid climates in homes without air-conditioning. The moisture formation on the surfaces is undesirable since it may cause the painted finish of the outer surface to deteriorate and it may wet the kitchen floor. About 10 percent of the total energy consumed by the refrigerator can be saved by turning this heater off and keeping it off unless there is visible condensation on the outer surfaces.

8. *Do not block the air flow passages* to and from the condenser coils of the refrigerator. The heat dissipated by the condenser to the air is carried away by air that enters through the bottom and sides of the refrigerator and leaves through the top. Any blockage of this natural convection air circulation path by large objects such as several cereal boxes on top of the refrigerator will degrade the performance of the condenser and thus the refrigerator (Fig. 6–57).

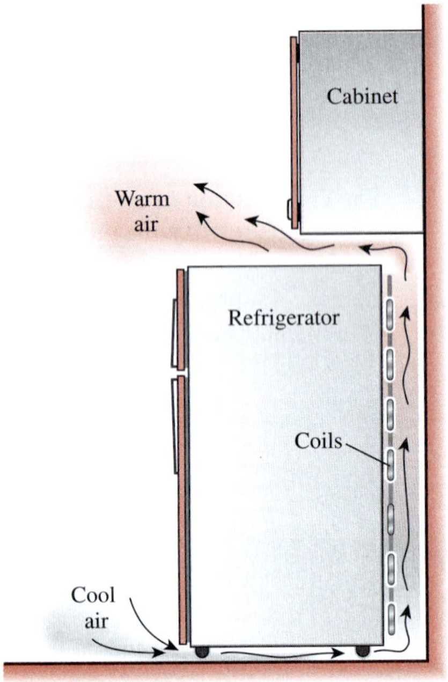

FIGURE 6–57
The condenser coils of a refrigerator must be cleaned periodically, and the airflow passages must not be blocked to maintain high performance.

These and other commonsense conservation measures will result in a reduction in the energy and maintenance costs of a refrigerator as well as an extended trouble-free life of the device.

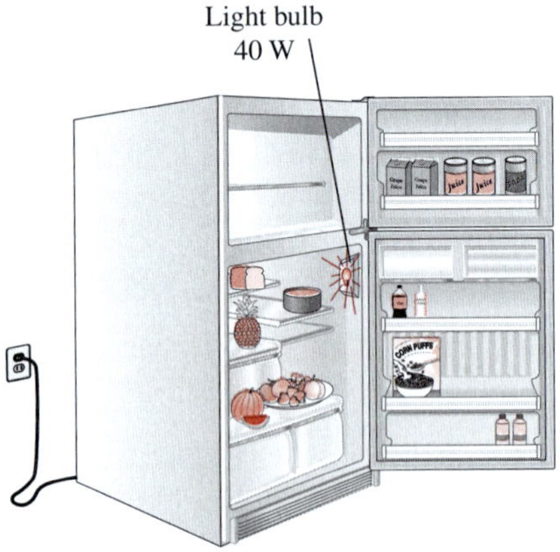

FIGURE 6–58
Schematic for Example 6–8.

EXAMPLE 6–8 Malfunction of a Refrigerator Light Switch

The interior lighting of refrigerators is provided by incandescent lamps whose switches are actuated by the opening of the refrigerator door. Consider a refrigerator whose 40-W lightbulb remains on continuously as a result of a malfunction of the switch (Fig. 6–58). If the refrigerator has a coefficient of performance of 1.3 and the cost of electricity is 8 cents per kWh, determine the increase in the energy consumption of the refrigerator and its cost per year if the switch is not fixed.

SOLUTION The lightbulb of a refrigerator malfunctions and remains on. The increases in the electricity consumption and cost are to be determined.

Assumptions The life of the lightbulb is more than 1 year.

Analysis The lightbulb consumes 40 W of power when it is on, and thus adds 40 W to the heat load of the refrigerator. Noting that the COP of the refrigerator is 1.3, the power consumed by the refrigerator to remove the heat generated by the lightbulb is

$$\dot{W}_{\text{refrig}} = \frac{\dot{Q}_{\text{refrig}}}{\text{COP}_{\text{R}}} = \frac{40\ \text{W}}{1.3} = 30.8\ \text{W}$$

Therefore, the total additional power consumed by the refrigerator is

$$\dot{W}_{\text{total,additional}} = \dot{W}_{\text{light}} + \dot{W}_{\text{refrig}} = 40 + 30.8 = 70.8\ \text{W}$$

The total number of hours in a year is

$$\text{Annual hours} = (365\ \text{days/yr})(24\ \text{h/day}) = 8760\ \text{h/yr}$$

Assuming the refrigerator is opened 20 times a day for an average of 30 s, the light would normally be on for

$$\begin{aligned}\text{Normal operating hours} &= (20\ \text{times/day})(30\ \text{s/time})(1\ \text{h/3600 s})(365\ \text{days/yr}) \\ &= 61\ \text{h/yr}\end{aligned}$$

Then the additional hours the light remains on as a result of the malfunction becomes

$$\begin{aligned}\text{Additional operating hours} &= \text{Annual hours} - \text{Normal operating hours} \\ &= 8760 - 61 = 8699\ \text{h/yr}\end{aligned}$$

Therefore, the additional electric power consumption and its cost per year are

$$\begin{aligned}\text{Additional power consumption} &= \dot{W}_{\text{total,additional}} \times (\text{Additional operating hours}) \\ &= (0.0708\ \text{kW})(8699\ \text{h/yr}) = \mathbf{616\ kWh/yr}\end{aligned}$$

and

$$\begin{aligned}\text{Additional power cost} &= (\text{Additional power consumption})(\text{Unit cost}) \\ &= (616\ \text{kWh/yr})(\$0.08/\text{kWh}) = \mathbf{\$49.3/yr}\end{aligned}$$

Discussion Note that not repairing the switch will cost the homeowner about \$50 a year. This is alarming when we consider that at \$0.08/kWh, a typical refrigerator consumes about \$70 worth of electricity a year.

SUMMARY

The *second law of thermodynamics* states that processes occur in a certain direction, not in any direction. A process does not occur unless it satisfies both the first and the second laws of thermodynamics. Bodies that can absorb or reject finite amounts of heat isothermally are called *thermal energy reservoirs* or *heat reservoirs*.

Work can be converted to heat directly, but heat can be converted to work only by some devices called *heat engines*. The *thermal efficiency* of a heat engine is defined as

$$\eta_{\text{th}} = \frac{W_{\text{net,out}}}{Q_H} = 1 - \frac{Q_L}{Q_H}$$

where $W_{\text{net,out}}$ is the net work output of the heat engine, Q_H is the amount of heat supplied to the engine, and Q_L is the amount of heat rejected by the engine.

Refrigerators and heat pumps are devices that absorb heat from low-temperature media and reject it to higher-temperature ones. The performance of a refrigerator or a heat pump is expressed in terms of the *coefficient of performance*, which is defined as

$$\text{COP}_\text{R} = \frac{Q_L}{W_{\text{net,in}}} = \frac{1}{Q_H/Q_L - 1}$$

$$\text{COP}_\text{HP} = \frac{Q_H}{W_{\text{net,in}}} = \frac{1}{1 - Q_L/Q_H}$$

The *Kelvin–Planck statement* of the second law of thermodynamics states that no heat engine can produce a net amount of work while exchanging heat with a single reservoir only. The *Clausius statement* of the second law states that no device can transfer heat from a cooler body to a warmer one without leaving an effect on the surroundings.

Any device that violates the first or the second law of thermodynamics is called a *perpetual-motion machine*.

A process is said to be *reversible* if both the system and the surroundings can be restored to their original conditions. Any other process is *irreversible*. The effects such as friction, non-quasi-equilibrium expansion or compression, and heat transfer through a finite temperature difference render a process irreversible and are called *irreversibilities*.

The *Carnot cycle* is a reversible cycle that is composed of four reversible processes, two isothermal and two adiabatic. The *Carnot principles* state that the thermal efficiencies of all reversible heat engines operating between the same two reservoirs are the same, and that no heat engine is more efficient than a reversible one operating between the same two reservoirs. These statements form the basis for establishing a *thermodynamic temperature scale* related to the heat transfers between a reversible device and the high- and low-temperature reservoirs by

$$\left(\frac{Q_H}{Q_L}\right)_\text{rev} = \frac{T_H}{T_L}$$

Therefore, the Q_H/Q_L ratio can be replaced by T_H/T_L for reversible devices, where T_H and T_L are the absolute temperatures of the high- and low-temperature reservoirs, respectively.

A heat engine that operates on the reversible Carnot cycle is called a *Carnot heat engine*. The thermal efficiency of a Carnot heat engine, as well as all other reversible heat engines, is given by

$$\eta_{\text{th,rev}} = 1 - \frac{T_L}{T_H}$$

This is the maximum efficiency a heat engine operating between two reservoirs at temperatures T_H and T_L can have.

The COPs of reversible refrigerators and heat pumps are given in a similar manner as

$$\text{COP}_{\text{R,rev}} = \frac{1}{T_H/T_L - 1}$$

and

$$\text{COP}_{\text{HP,rev}} = \frac{1}{1 - T_L/T_H}$$

Again, these are the highest COPs a refrigerator or a heat pump operating between the temperature limits of T_H and T_L can have.

REFERENCES AND SUGGESTED READINGS

1. *ASHRAE Handbook of Refrigeration,* SI version. Atlanta, GA: American Society of Heating, Refrigerating, and Air-Conditioning Engineers, Inc. 1994.
2. D. Stewart. "Wheels Go Round and Round, but Always Run Down." November 1986, *Smithsonian*, pp. 193–208.
3. J. T. Amann, A. Wilson, and K. Ackerly, *Consumer Guide to Home Energy Saving*, 9th ed., American Council for an Energy-Efficient Economy, Washington, D. C., 2007.

PROBLEMS*

Second Law of Thermodynamics and Thermal Energy Reservoirs

6–1C Describe an imaginary process that satisfies the first law but violates the second law of thermodynamics.

6–2C Describe an imaginary process that satisfies the second law but violates the first law of thermodynamics.

6–3C Describe an imaginary process that violates both the first and the second laws of thermodynamics.

* Problems designated by a "C" are concept questions, and students are encouraged to answer them all. Problems designated by an "E" are in English units, and the SI users can ignore them. Problems with the icon are solved using EES, and complete solutions together with parametric studies are included on the enclosed DVD. Problems with the icon are comprehensive in nature, and are intended to be solved with a computer, preferably using the EES software that accompanies this text.

6–4C An experimentalist claims to have raised the temperature of a small amount of water to 150°C by transferring heat from high-pressure steam at 120°C. Is this a reasonable claim? Why? Assume no refrigerator or heat pump is used in the process.

6–5C What is a thermal energy reservoir? Give some examples.

6–6C Consider the process of baking potatoes in a conventional oven. Can the hot air in the oven be treated as a thermal energy reservoir? Explain.

6–7C Consider the energy generated by a TV set. What is a suitable choice for a thermal energy reservoir?

Heat Engines and Thermal Efficiency

6–8C Is it possible for a heat engine to operate without rejecting any waste heat to a low-temperature reservoir? Explain.

6–9C What are the characteristics of all heat engines?

6–10C Baseboard heaters are basically electric resistance heaters and are frequently used in space heating. A home owner claims that her 5-year-old baseboard heaters have a conversion efficiency of 100 percent. Is this claim in violation of any thermodynamic laws? Explain.

6–11C What is the Kelvin–Planck expression of the second law of thermodynamics?

6–12C Does a heat engine that has a thermal efficiency of 100 percent necessarily violate (*a*) the first law and (*b*) the second law of thermodynamics? Explain.

6–13C In the absence of any friction and other irreversibilities, can a heat engine have an efficiency of 100 percent? Explain.

6–14C Are the efficiencies of all the work-producing devices, including the hydroelectric power plants, limited by the Kelvin–Planck statement of the second law? Explain.

6–15C Consider a pan of water being heated (*a*) by placing it on an electric range and (*b*) by placing a heating element in the water. Which method is a more efficient way of heating water? Explain.

6–16E A heat engine has a heat input of 3×10^4 Btu/h and a thermal efficiency of 40 percent. Calculate the power it will produce, in hp.

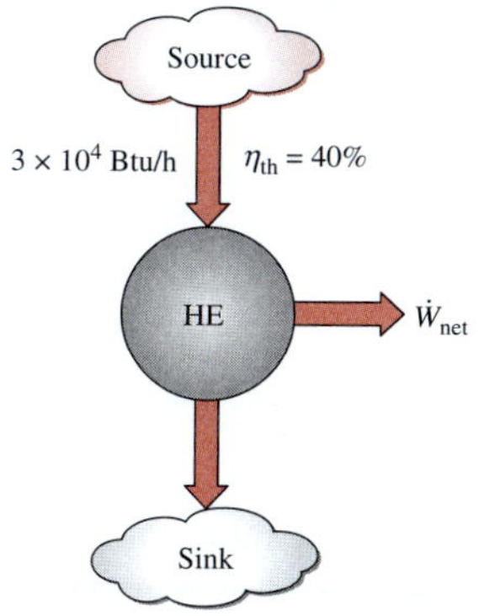

FIGURE P6–16E

6–17 The thermal efficiency of a general heat engine is 40 percent, and it produces 30 hp. At what rate is heat transferred to this engine, in kJ/s?

6–18 A 600-MW steam power plant, which is cooled by a nearby river, has a thermal efficiency of 40 percent. Determine the rate of heat transfer to the river water. Will the actual heat transfer rate be higher or lower than this value? Why?

6–19 A heat engine that pumps water out of an underground mine accepts 700 kJ of heat and produces 250 kJ of work. How much heat does it reject, in kJ?

6–20 A heat engine with a thermal efficiency of 40 percent rejects 1000 kJ/kg of heat. How much heat does it receive? *Answer:* 1667 kJ/kg

6–21 A steam power plant with a power output of 150 MW consumes coal at a rate of 60 tons/h. If the heating value of the coal is 30,000 kJ/kg, determine the overall efficiency of this plant. *Answer:* 30.0 percent

6–22 An automobile engine consumes fuel at a rate of 22 L/h and delivers 55 kW of power to the wheels. If the fuel has a heating value of 44,000 kJ/kg and a density of 0.8 g/cm^3, determine the efficiency of this engine. *Answer:* 25.6 percent

6–23 In 2001, the United States produced 51 percent of its electricity in the amount of 1.878×10^{12} kWh from coal-fired power plants. Taking the average thermal efficiency to be 34 percent, determine the amount of thermal energy rejected by the coal-fired power plants in the United States that year.

6–24 The Department of Energy projects that between the years 1995 and 2010, the United States will need to build new power plants to generate an additional 150,000 MW of electricity to meet the increasing demand for electric power. One possibility is to build coal-fired power plants, which cost \$1300 per kW to construct and have an efficiency of 40 percent. Another possibility is to use the clean-burning Integrated Gasification Combined Cycle (IGCC) plants where the coal is subjected to heat and pressure to gasify it while removing sulfur and particulate matter from it. The gaseous coal is then burned in a gas turbine, and part of the waste heat from the exhaust gases is recovered to generate steam for the steam turbine. Currently the construction of IGCC plants costs about \$1500 per kW, but their efficiency is about 48 percent. The average heating value of the coal is about 28,000,000 kJ per ton (that is, 28,000,000 kJ of heat is released when 1 ton of coal is burned). If the IGCC plant is to recover its cost difference from fuel savings in five years, determine what the price of coal should be in \$ per ton.

6–25 EES Reconsider Prob. 6–24. Using EES (or other) software, investigate the price of coal for varying simple payback periods, plant construction costs, and operating efficiency.

6–26 Repeat Prob. 6–24 for a simple payback period of three years instead of five years.

6–27E Solar energy stored in large bodies of water, called solar pounds, is being used to generate electricity. If such a solar power plant has an efficiency of 4 percent and a net power output of 350 kW, determine the average value of the required solar energy collection rate, in Btu/h.

6–28 A coal-burning steam power plant produces a net power of 300 MW with an overall thermal efficiency of 32 percent. The actual gravimetric air–fuel ratio in the furnace is calculated to be 12 kg air/kg fuel. The heating value of the coal is 28,000 kJ/kg. Determine (*a*) the amount of coal consumed during a 24-hour period and (*b*) the rate of air flowing through the furnace. *Answers:* (*a*) 2.89 × 10^6 kg, (*b*) 402 kg/s

Refrigerators and Heat Pumps

6–29C What is the difference between a refrigerator and a heat pump?

6–30C What is the difference between a refrigerator and an air conditioner?

6–31C In a refrigerator, heat is transferred from a lower-temperature medium (the refrigerated space) to a higher-temperature one (the kitchen air). Is this a violation of the second law of thermodynamics? Explain.

6–32C A heat pump is a device that absorbs energy from the cold outdoor air and transfers it to the warmer indoors. Is this a violation of the second law of thermodynamics? Explain.

6–33C Define the coefficient of performance of a refrigerator in words. Can it be greater than unity?

6–34C Define the coefficient of performance of a heat pump in words. Can it be greater than unity?

6–35C A heat pump that is used to heat a house has a COP of 2.5. That is, the heat pump delivers 2.5 kWh of energy to the house for each 1 kWh of electricity it consumes. Is this a violation of the first law of thermodynamics? Explain.

6–36C A refrigerator has a COP of 1.5. That is, the refrigerator removes 1.5 kWh of energy from the refrigerated space for each 1 kWh of electricity it consumes. Is this a violation of the first law of thermodynamics? Explain.

6–37C What is the Clausius expression of the second law of thermodynamics?

6–38C Show that the Kelvin–Planck and the Clausius expressions of the second law are equivalent.

6–39E A residential heat pump has a coefficient of performance of 2.4. How much heating effect, in Btu/h, will result when 5 hp is supplied to this heat pump?

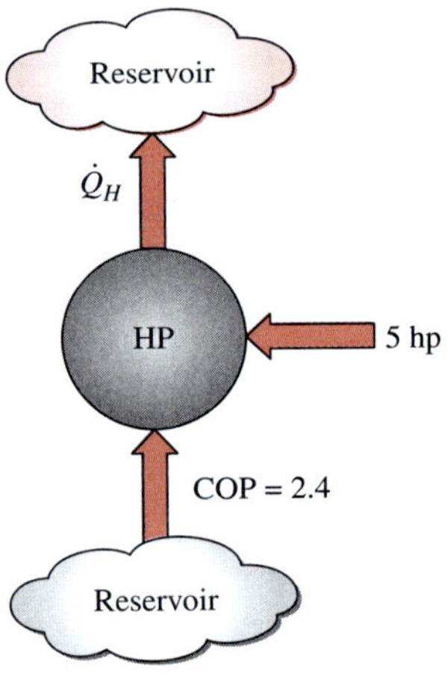

FIGURE P6–39E

6–40 An air conditioner produces a 2-kW cooling effect while rejecting 2.5 kW of heat. What is its COP?

6–41 A refrigerator used to cool a computer requires 3 kW of electrical power and has a COP of 1.4. Calculate the cooling effect of this refrigerator, in kW.

6–42 A food department is kept at −12°C by a refrigerator in an environment at 30°C. The total heat gain to the food department is estimated to be 3300 kJ/h and the heat rejection in the condenser is 4800 kJ/h. Determine the power input to the compressor, in kW and the COP of the refrigerator.

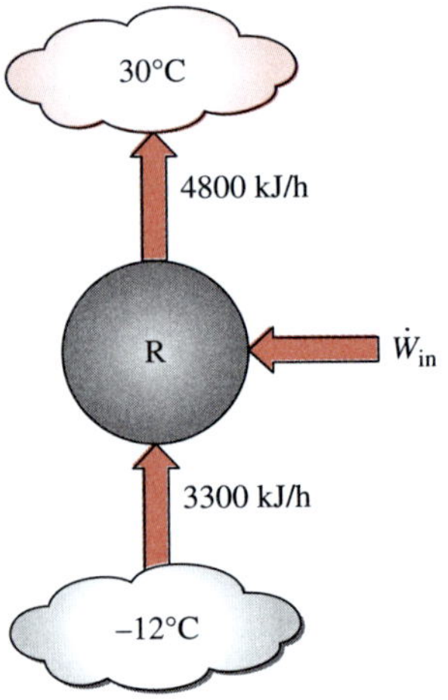

FIGURE P6–42

6–43 A household refrigerator with a COP of 1.2 removes heat from the refrigerated space at a rate of 60 kJ/min. Determine (*a*) the electric power consumed by the refrigerator and (*b*) the rate of heat transfer to the kitchen air. *Answers:* (*a*) 0.83 kW, (*b*) 110 kJ/min

6–44E A commercial heat pump removes 10,000 Btu/h from the source, rejects 15,090 Btu/h to the sink, and requires 2 hp of power. What is this heat pump's coefficient of performance?

6–45 A refrigerator used for cooling food in a grocery store is to produce a 25,000 kJ/h cooling effect, and it has a coefficient of performance of 1.60. How many kilowatts of power will this refrigerator require to operate? *Answer:* 4.34 kW

6–46 A household refrigerator that has a power input of 450 W and a COP of 2.5 is to cool five large watermelons, 10 kg each, to 8°C. If the watermelons are initially at 20°C, determine how long it will take for the refrigerator to cool them. The watermelons can be treated as water whose specific heat is 4.2 kJ/kg·°C. Is your answer realistic or optimistic? Explain. *Answer:* 2240 s

6–47 When a man returns to his well-sealed house on a summer day, he finds that the house is at 35°C. He turns on the air conditioner, which cools the entire house to 20°C in 30 min. If the COP of the air-conditioning system is 2.8, determine the power drawn by the air conditioner. Assume the entire mass within the house is equivalent to 800 kg of air for which c_v = 0.72 kJ/kg·°C and c_p = 1.0 kJ/kg·°C.

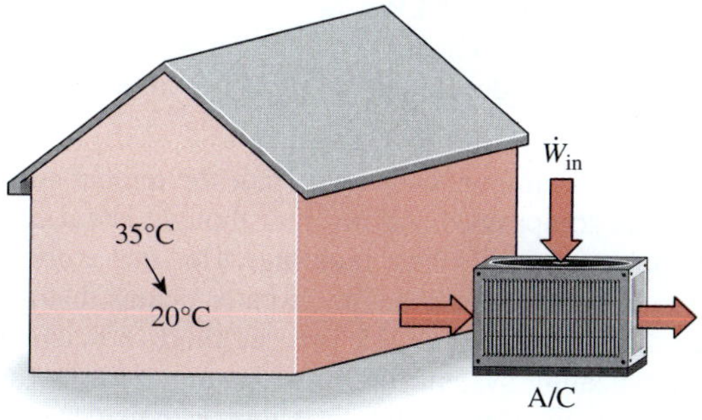

FIGURE P6–47

6–48 Reconsider Prob. 6–47. Using EES (or other) software, determine the power input required by the air conditioner to cool the house as a function for air-conditioner EER ratings in the range 5 to 15. Discuss your results and include representative costs of air-conditioning units in the EER rating range.

6–49 Bananas are to be cooled from 24 to 13°C at a rate of 215 kg/h by a refrigeration system. The power input to the refrigerator is 1.4 kW. Determine the rate of cooling, in kJ/min, and the COP of the refrigerator. The specific heat of banana above freezing is 3.35 kJ/kg·°C.

6–50 A refrigerator is used to cool water from 23 to 5°C in a continuous manner. The heat rejected in the condenser is 570 kJ/min and the power is 2.65 kW. Determine the rate at which water is cooled, in L/min and the COP of the refrigerator. The specific heat of water is 4.18 kJ/kg·°C and its density is 1 kg/L. *Answers:* 5.46 L/min, 2.58

6–51 A heat pump is used to maintain a house at a constant temperature of 23°C. The house is losing heat to the outside air through the walls and the windows at a rate of 60,000 kJ/h while the energy generated within the house from people, lights, and appliances amounts to 4000 kJ/h. For a COP of 2.5, determine the required power input to the heat pump. *Answer:* 6.22 kW

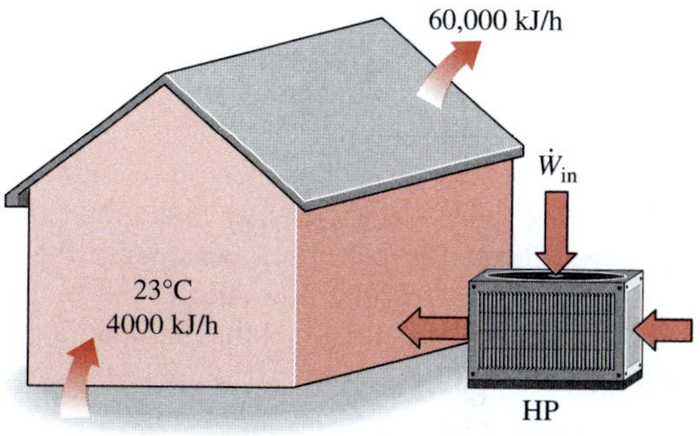

FIGURE P6–51

6–52E Water enters an ice machine at 55°F and leaves as ice at 25°F. If the COP of the ice machine is 2.4 during this operation, determine the required power input for an ice production rate of 28 Ibm/h. (169 Btu of energy needs to be removed from each Ibm of water at 55°F to turn it into ice at 25°F.)

6–53E Consider an office room that is being cooled adequately by a 12,000 Btu/h window air conditioner. Now it is decided to convert this room into a computer room by installing several computers, terminals, and printers with a total rated power of 8.4 kW. The facility has several 7000 Btu/h air conditioners in storage that can be installed to meet the additional cooling requirements. Assuming a usage factor of 0.4 (i.e., only 40 percent of the rated power will be consumed at any given time) and additional occupancy of seven people, each generating heat at a rate of 100 W, determine how many of these air conditioners need to be installed to the room.

6–54 Consider a building whose annual air-conditioning load is estimated to be 120,000 kWh in an area where the unit cost of electricity is $0.10/kWh. Two air conditioners are considered for the building. Air conditioner A has a seasonal average COP of 3.2 and costs $5500 to purchase and install. Air conditioner B has a seasonal average COP of 5.0 and costs $7000 to purchase and install. All else being equal, determine which air conditioner is a better buy.

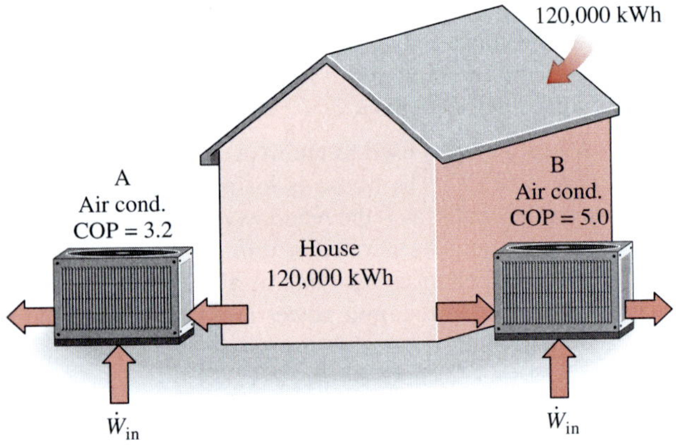

FIGURE P6–54

6–55 A house that was heated by electric resistance heaters consumed 1200 kWh of electric energy in a winter month. If this house were heated instead by a heat pump that has an average COP of 2.4, determine how much money the home owner would have saved that month. Assume a price of 8.5¢/kWh for electricity.

6–56 Refrigerant-134a enters the condenser of a residential heat pump at 800 kPa and 35°C at a rate of 0.018 kg/s and leaves at 800 kPa as a saturated liquid. If the compressor consumes 1.2 kW of power, determine (*a*) the COP of the heat pump and (*b*) the rate of heat absorption from the outside air.

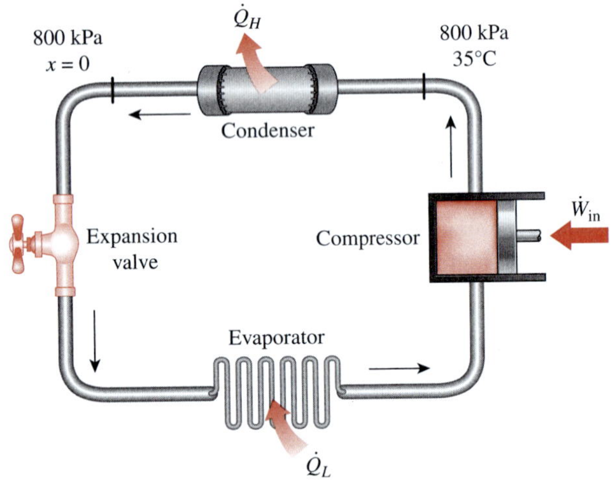

FIGURE P6–56

6–57 Refrigerant-134a enters the evaporator coils placed at the back of the freezer section of a household refrigerator at 100 kPa with a quality of 20 percent and leaves at 100 kPa and −26°C. If the compressor consumes 600 W of power and the COP the refrigerator is 1.2, determine (*a*) the mass flow rate of the refrigerant and (*b*) the rate of heat rejected to the kitchen air. *Answers:* (*a*) 0.00414 kg/s, (*b*) 1320 W

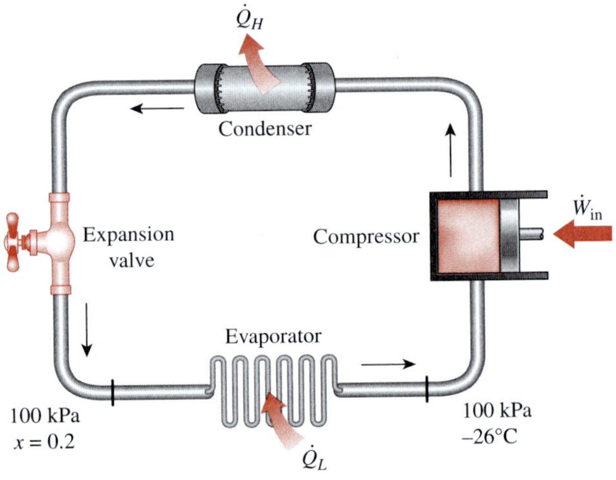

FIGURE P6–57

Perpetual-Motion Machines

6–58C An inventor claims to have developed a resistance heater that supplies 1.2 kWh of energy to a room for each kWh of electricity it consumes. Is this a reasonable claim, or has the inventor developed a perpetual-motion machine? Explain.

6–59C It is common knowledge that the temperature of air rises as it is compressed. An inventor thought about using this high-temperature air to heat buildings. He used a compressor driven by an electric motor. The inventor claims that the compressed hot-air system is 25 percent more efficient than a resistance heating system that provides an equivalent amount of heating. Is this claim valid, or is this just another perpetual-motion machine? Explain.

Reversible and Irreversible Processes

6–60C Show that processes that use work for mixing are irreversible by considering an adiabatic system whose contents are stirred by turning a paddle wheel inside the system (e.g., stirring a cake mix with an electric mixer).

6–61C Show that processes involving rapid chemical reactions are irreversible by considering the combustion of a natural gas (e.g., methane) and air mixture in a rigid container.

6–62C A cold canned drink is left in a warmer room where its temperature rises as a result of heat transfer. Is this a reversible process? Explain.

6–63C A block slides down an inclined plane with friction and no restraining force. Is this process reversible or irreversible? Justify your answer.

6–64C Why are engineers interested in reversible processes even though they can never be achieved?

6–65C Why does a nonquasi-equilibrium compression process require a larger work input than the corresponding quasi-equilibrium one?

6–66C Why does a nonquasi-equilibrium expansion process deliver less work than the corresponding quasi-equilibrium one?

6–67C How do you distinguish between internal and external irreversibilities?

6–68C Is a reversible expansion or compression process necessarily quasi-equilibrium? Is a quasi-equilibrium expansion or compression process necessarily reversible? Explain.

The Carnot Cycle and Carnot Principles

6–69C What are the four processes that make up the Carnot cycle?

6–70C What are the two statements known as the Carnot principles?

6–71C Somebody claims to have developed a new reversible heat-engine cycle that has a higher theoretical efficiency than the Carnot cycle operating between the same temperature limits. How do you evaluate this claim?

6–72C Somebody claims to have developed a new reversible heat-engine cycle that has the same theoretical efficiency as the Carnot cycle operating between the same temperature limits. Is this a reasonable claim?

6–73C Is it possible to develop (*a*) an actual and (*b*) a reversible heat-engine cycle that is more efficient than a Carnot cycle operating between the same temperature limits? Explain.

Carnot Heat Engines

6–74C Is there any way to increase the efficiency of a Carnot heat engine other than by increasing T_H or decreasing T_L?

6–75C Consider two actual power plants operating with solar energy. Energy is supplied to one plant from a solar pond at 80°C and to the other from concentrating collectors that raise the water temperature to 600°C. Which of these power plants will have a higher efficiency? Explain.

6–76E A heat engine operates between two thermal-energy reservoirs at 1260 R and 510 R. What is the maximum work the engine can produce per unit of heat that the engine receives from the source?

6–77 From a work-production perspective, which is more valuable: (*a*) thermal energy reservoirs at 675 K and 325 K or (*b*) thermal energy reservoirs at 625 K and 275 K?

6–78 A heat engine operates between a source at 477°C and a sink at 25°C. If heat is supplied to the heat engine at a steady rate of 65,000 kJ/min, determine the maximum power output of this heat engine.

6–79 Reconsider Prob. 6–78. Using EES (or other) software, study the effects of the temperatures of the heat source and the heat sink on the power produced and the cycle thermal efficiency. Let the source temperature vary from 300 to 1000°C, and the sink temperature to vary from 0 to 50°C. Plot the power produced and the cycle efficiency against the source temperature for sink temperatures of 0°C, 25°C, and 50°C, and discuss the results.

6–80E A heat engine is operating on a Carnot cycle and has a thermal efficiency of 75 percent. The waste heat from this engine is rejected to a nearby lake at 60°F at a rate of 800 Btu/min. Determine (*a*) the power output of the engine and (*b*) the temperature of the source. *Answers:* (*a*) 56.6 hp, (*b*) 2080 R

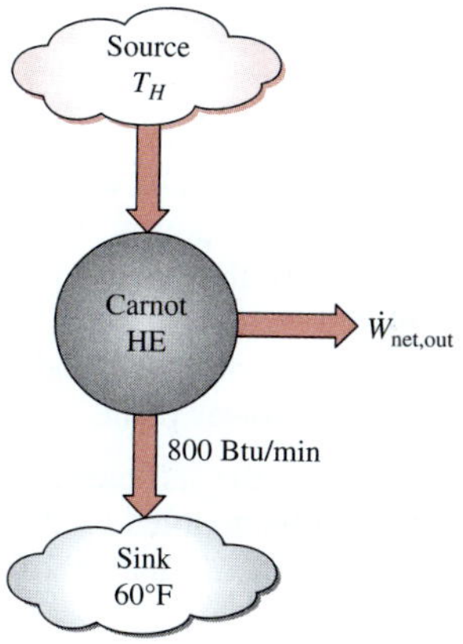

FIGURE P6–80E

6–81E An inventor claims to have devised a cyclical engine for use in space vehicles that operates with a nuclear-fuel-generated energy source whose temperature is 1000 R and a sink at 550 R that radiates waste heat to deep space. He also claims that this engine produces 5 hp while rejecting heat at a rate of 15,000 Btu/h. Is this claim valid?

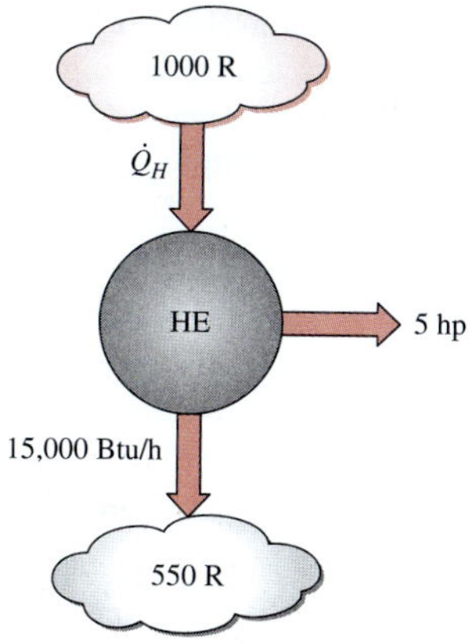

FIGURE P6–81E

6–82 A heat engine receives heat from a heat source at 1200°C and has a thermal efficiency of 40 percent. The heat engine does maximum work equal to 500 kJ. Determine the heat supplied to the heat engine by the heat source, the heat rejected to the heat sink, and the temperature of the heat sink.

6–83 A heat engine operating on the Carnot cycle has a measured work output of 900 kJ and heat rejection of 150 kJ to a heat reservoir at 27°C. Determine the heat supplied to the heat engine by the heat source, in kJ, and the temperature of the heat source, in °C.

6–84 A heat engine is operating on a Carnot cycle and has a thermal efficiency of 75 percent. The waste heat from this engine is rejected to a nearby lake at 15°C at a rate of 14 kW. Determine the power output of the engine and the temperature of the source, in °C. *Answers:* 42 kW, 879°C.

6–85 A geothermal power plant uses geothermal water extracted at 150°C at a rate of 210 kg/s as the heat source and produces 8000 kW of net power. The geothermal water leaves the plant at 90°C. If the environment temperature is 25°C, determine (*a*) the actual thermal efficiency, (*b*) the maximum possible thermal efficiency, and (*c*) the actual rate of heat rejection from this power plant.

6–86 It is claimed that the efficiency of a completely reversible heat engine can be doubled by doubling the temperature of the energy source. Justify the validity of this claim.

Carnot Refrigerators and Heat Pumps

6–87C How can we increase the COP of a Carnot refrigerator?

6–88C A homeowner buys a new refrigerator and a new air conditioner. Which one of these devices would you expect to have a higher COP? Why?

6–89C A homeowner buys a new refrigerator with no freezer compartment and a deep freezer for the new kitchen. Which of these devices would you expect to have a lower COP? Why?

6–90C In an effort to conserve energy in a heat-engine cycle, somebody suggests incorporating a refrigerator that will absorb some of the waste energy Q_L and transfer it to the energy source of the heat engine. Is this a smart idea? Explain.

6–91C It is well established that the thermal efficiency of a heat engine increases as the temperature T_L at which heat is rejected from the heat engine decreases. In an effort to increase the efficiency of a power plant, somebody suggests refrigerating the cooling water before it enters the condenser, where heat rejection takes place. Would you be in favor of this idea? Why?

6–92C It is well known that the thermal efficiency of heat engines increases as the temperature of the energy source increases. In an attempt to improve the efficiency of a power plant, somebody suggests transferring heat from the available energy source to a higher-temperature medium by a heat pump before energy is supplied to the power plant. What do you think of this suggestion? Explain.

6–93 Determine the minimum work per unit of heat transfer from the source that is required to drive a refrigerator with thermal-energy reservoirs at 273 K and 303 K.

6–94 An inventor claims to have developed a heat pump that produces a 200-kW heating effect for a 293 K heated zone while only using 75 kW of power and a heat source at 273 K. Justify the validity of this claim.

6–95 A heat pump operates on a Carnot heat pump cycle with a COP of 8.7. It keeps a space at 26°C by consuming 4.25 kW of power. Determine the temperature of the reservoir from which the heat is absorbed and the heating load provided by the heat pump. *Answers:* 265 K, 37.0 kW

6–96 A refrigerator is to remove heat from the cooled space at a rate of 300 kJ/min to maintain its temperature at −8°C. If the air surrounding the refrigerator is at 25°C, determine the minimum power input required for this refrigerator. *Answer:* 0.623 kW

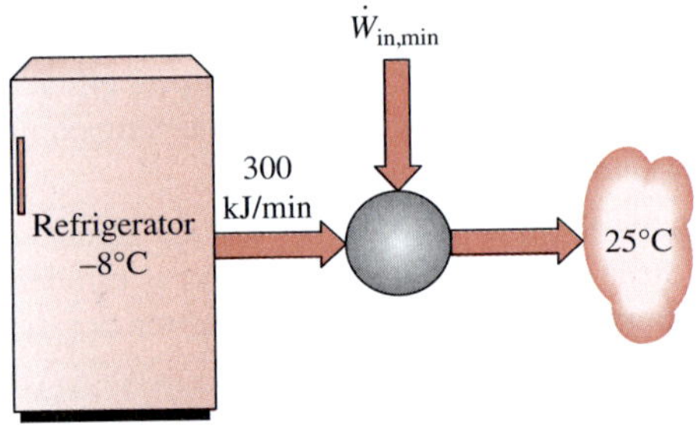

FIGURE P6–96

6–97 An inventor claims to have developed a refrigeration system that removes heat from the closed region at −12°C and transfers it to the surrounding air at 25°C while maintaining a COP of 6.5. Is this claim reasonable? Why?

6–98E An air-conditioning system is used to maintain a house at 70°F when the temperature outside is 100°F. The house is gaining heat through the walls and the windows at a rate of 800 Btu/min, and the heat generation rate within the house from people, lights, and appliances amounts to 100 Btu/min. Determine the minimum power input required for this air-conditioning system. *Answer:* 1.20 hp

6–99 A heat pump is used to maintain a house at 22°C by extracting heat from the outside air on a day when the outside air temperature is 2°C. The house is estimated to lose heat at a rate of 110,000 kJ/h, and the heat pump consumes 5 kW of electric power when running. Is this heat pump powerful enough to do the job?

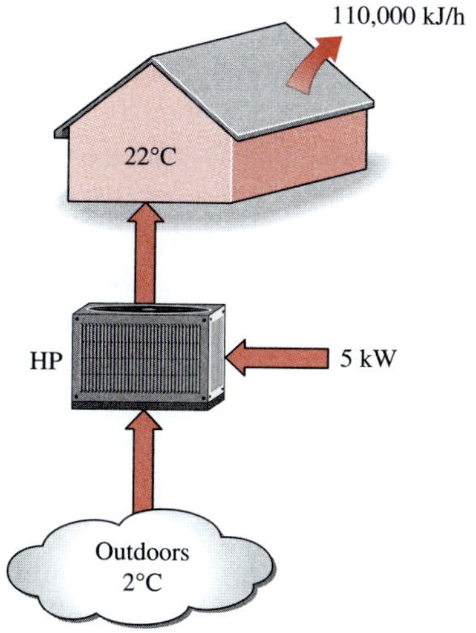

FIGURE P6–99

6–100E A completely reversible refrigerator operates between thermal energy reservoirs at 450 R and 540 R. How many kilowatts of power are required for this device to produce a 15,000-Btu/h cooling effect?

6–101 A refrigerator operating on the reversed Carnot cycle has a measured work input of 200 kW and heat rejection of 2000 kW to a heat reservoir at 27°C. Determine the cooling load supplied to the refrigerator, in kW, and the temperature of the heat source, in °C. *Answers:* 1800 kW, −3°C

6–102 An 18,000 Btu/h split air conditioner running at full load is used to keep a room at 22°C in an environment at 33°C. The power input to the compressor is 3.4 kW. Determine (*a*) the rate of heat rejected in the condenser, in kJ/h, (*b*) the COP of the air conditioner, and (*c*) the rate of cooling, in Btu/h, if the air conditioner operated as a Carnot refrigerator for the same power input.

6–103 A Carnot refrigerator absorbs heat from a space at 15°C at a rate of 16,000 kJ/h and rejects heat to a reservoir at 36°C. Determine the COP of the refrigerator, the power input, in kW, and the rate of heat rejected to high-temperature reservoir, in kJ/h.

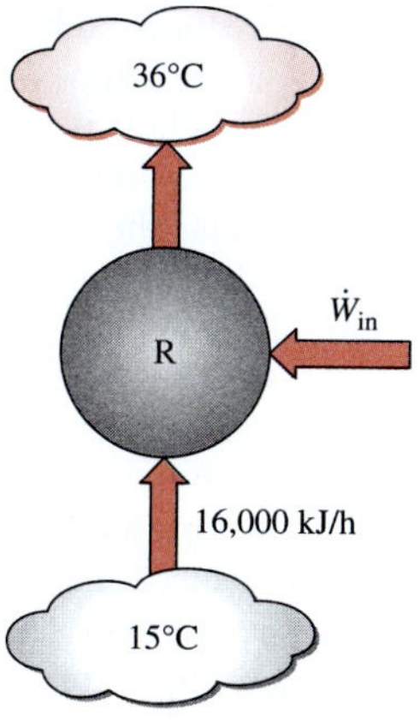

FIGURE P6–103

6–104 The performance of a heat pump degrades (i.e., its COP decreases) as the temperature of the heat source decreases. This makes using heat pumps at locations with severe weather conditions unattractive. Consider a house that is heated and maintained at 20°C by a heat pump during the winter. What is the maximum COP for this heat pump if heat is extracted from the outdoor air at (*a*) 10°C, (*b*) −5°C, and (*c*) −30°C?

6–105E A heat pump is to be used for heating a house in winter. The house is to be maintained at 78°F at all times. When the temperature outdoors drops to 25°F, the heat losses from the house are estimated to be 55,000 Btu/h. Determine the minimum power required to run this heat pump if heat is extracted from (*a*) the outdoor air at 25°F and (*b*) the well water at 50°F.

6–106 A Carnot heat pump is to be used to heat a house and maintain it at 25°C in winter. On a day when the average outdoor temperature remains at about 2°C, the house is estimated to lose heat at a rate of 55,000 kJ/h. If the heat pump consumes 6.6 kW of power while operating, determine (*a*) how long the heat pump ran on that day; (*b*) the total heating costs, assuming an average price of 8.5¢/kWh for electricity; and (*c*) the heating cost for the same day if resistance heating is used instead of a heat pump. *Answers:* (*a*) 4.29 h, (*b*) $2.41, (*c*) $31.2

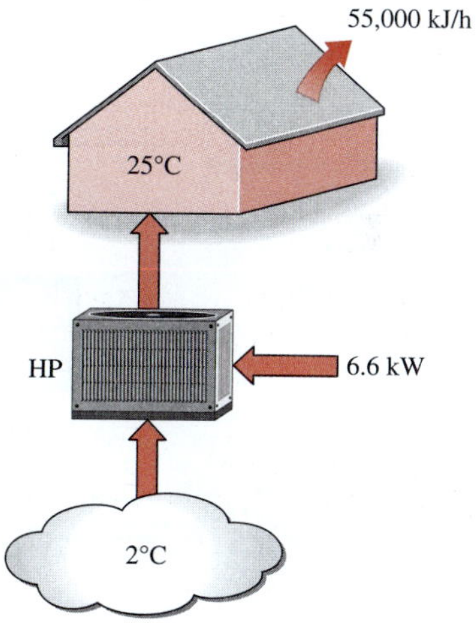

FIGURE P6–106

6–107 A Carnot heat engine receives heat from a reservoir at 900°C at a rate of 800 kJ/min and rejects the waste heat to the ambient air at 27°C. The entire work output of the heat engine is used to drive a refrigerator that removes heat from the refrigerated space at −5°C and transfers it to the same ambient air at 27°C. Determine (*a*) the maximum rate of heat removal from the refrigerated space and (*b*) the total rate of heat rejection to the ambient air. *Answers:* (*a*) 4982 kJ/min, (*b*) 5782 kJ/min

6–108E A Carnot heat engine receives heat from a reservoir at 1700°F at a rate of 700 Btu/min and rejects the waste heat to the ambient air at 80°F. The entire work output of the heat engine is used to drive a refrigerator that removes heat from the refrigerated space at 20°F and transfers it to the same ambient air at 80°F. Determine (*a*) the maximum rate of heat removal from the refrigerated space and (*b*) the total rate of heat rejection to the ambient air. *Answers:* (*a*) 4200 Btu/min, (*b*) 4900 Btu/min

6–109 The structure of a house is such that it loses heat at a rate of 3800 kJ/h per °C difference between the indoors and outdoors. A heat pump that requires a power input of 4 kW is used to maintain this house at 24°C. Determine the lowest outdoor temperature for which the heat pump can meet the heating requirements of this house. *Answer:* −9.5°C

6–110 An air-conditioner with refrigerant-134a as the working fluid is used to keep a room at 23°C by rejecting the waste heat to the outdoor air at 37°C. The room gains heat through the walls and the windows at a rate of 250 kJ/min while the heat generated by the computer, TV, and lights amounts to 900 W. The refrigerant enters the compressor at 400 kPa as a saturated vapor at a rate of 100 L/min and leaves at 1200 kPa and 70°C. Determine (*a*) the actual COP, (*b*) the maximum COP, and (*c*) the minimum volume flow rate of the refrigerant at the compressor inlet for the same compressor inlet and exit conditions. *Answers:* (*a*) 3.45, (*b*) 21.1, (*c*) 16.3 L/min

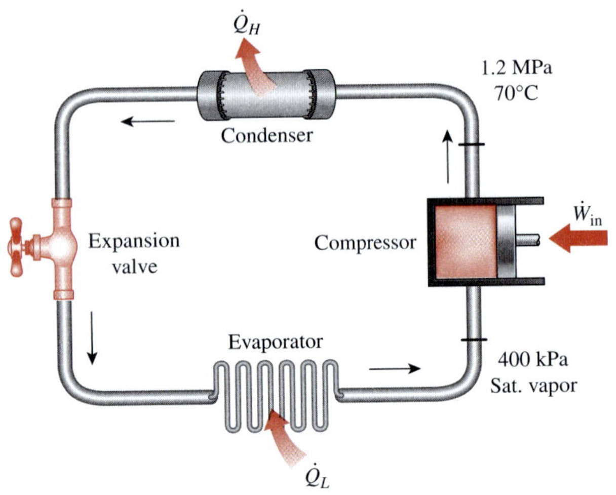

FIGURE P6–110

6–111 Calculate and plot the COP of a completely reversible refrigerator as a function of the temperature of the sink up to 500 K with the temperature of the source fixed at 250 K.

6–112 Derive an expression for the COP of a completely reversible refrigerator in terms of the thermal energy reservoir temperatures, T_L and T_H.

Special Topic: Household Refrigerators

6–113C Explain how you can reduce the energy consumption of your household refrigerator.

6–114C Why is it important to clean the condenser coils of a household refrigerator a few times a year? Also, why is it important not to block airflow through the condenser coils?

6–115C Why are today's refrigerators much more efficient than those built in the past?

6–116C Someone proposes that the refrigeration system of a supermarket be overdesigned so that the entire air-conditioning needs of the store can be met by refrigerated air without installing any air-conditioning system. What do you think of this proposal?

6–117C Someone proposes that the entire refrigerator/freezer requirements of a store be met using a large freezer that supplies sufficient cold air at −20°C instead of installing separate refrigerators and freezers. What do you think of this proposal?

6–118 The "Energy Guide" label of a refrigerator states that the refrigerator will consume $74 worth of electricity per year under normal use if the cost of electricity is $0.07/kWh. If the electricity consumed by the lightbulb is negligible and the refrigerator consumes 300 W when running, determine the fraction of the time the refrigerator will run.

6–119 The interior lighting of refrigerators is usually provided by incandescent lamps whose switches are actuated by the opening of the refrigerator door. Consider a refrigerator whose 40-W lightbulb remains on about 60 h per year. It is proposed to replace the lightbulb by an energy-efficient bulb that consumes only 18 W but costs $25 to purchase and install. If the refrigerator has a coefficient of performance of 1.3 and the cost of electricity is 8 cents per kWh, determine if the energy savings of the proposed lightbulb justify its cost.

6–120 It is commonly recommended that hot foods be cooled first to room temperature by simply waiting a while before they are put into the refrigerator to save energy. Despite this commonsense recommendation, a person keeps cooking a large pan of stew three times a week and putting the pan into the refrigerator while it is still hot, thinking that the money saved is probably too little. But he says he can be convinced if you can show that the money saved is significant. The average mass of the pan and its contents is 5 kg. The average temperature of the kitchen is 23°C, and the average temperature of the food is 95°C when it is taken off the stove. The refrigerated space is maintained at 3°C, and the average specific heat of the food and the pan can be taken to be 3.9 kJ/kg·°C. If the refrigerator has a coefficient of performance of 1.5 and the cost of electricity is 10 cents per kWh, determine how much this person will save a year by waiting for the food to cool to room temperature before putting it into the refrigerator.

FIGURE P6–120

6–121 It is often stated that the refrigerator door should be opened as few times as possible for the shortest duration of time to save energy. Consider a household refrigerator whose interior volume is 0.9 m^3 and average internal temperature is 4°C. At any given time, one-third of the refrigerated space is occupied by food items, and the remaining 0.6 m^3 is filled with air. The average temperature and pressure in the kitchen are 20°C and 95 kPa, respectively. Also, the moisture contents of the air in the kitchen and the refrigerator are 0.010 and 0.004 kg per kg of air, respectively, and thus 0.006 kg of water vapor is condensed and removed for each kg of air that enters. The refrigerator door is opened an average of 8 times a day, and each time half of the air volume in the refrigerator is replaced by the warmer kitchen air. If the refrigerator has a coefficient of performance of 1.4 and the cost of electricity is 7.5 cents per kWh, determine the cost of the energy wasted per year as a result of opening the refrigerator door. What would your answer be if the kitchen air were very dry and thus a negligible amount of water vapor condensed in the refrigerator?

Review Problems

6–122 A heat engine operates with a source at 1280 K and a sink at 290 K. What is the maximum work per unit of heat that the engine can remove from the source?

6–123 A heat engine receives heat from a heat source at 1200°C and rejects heat to a heat sink at 50°C. The heat engine does maximum work equal to 500 kJ. Determine the heat supplied to the heat engine by the heat source and the heat rejected to the heat sink.

6–124E A heat pump creates a heating effect of 32,000 Btu/h for a space maintained at 530 R while using 1.8 kW of electrical power. What is the minimum temperature of the source that satisfies the second law of thermodynamics? *Answer:* 428 R

6–125 A heat pump with a COP of 2.4 is used to heat a house. When running, the heat pump consumes 8 kW of electric power. If the house is losing heat to the outside at an average rate of 40,000 kJ/h and the temperature of the house is 3°C when the heat pump is turned on, determine how long it will take for the temperature in the house to rise to 22°C. Assume the house is well sealed (i.e., no air leaks) and take the entire mass within the house (air, furniture, etc.) to be equivalent to 2000 kg of air.

6–126E A refrigeration system uses water-cooled condenser for rejecting the waste heat. The system absorbs heat from a space at 25°F at a rate of 24,000 Btu/h. Water enters the condenser at 58°F at a rate of 1.45 lbm/s. The COP of the system is estimated to be 1.77. Determine (*a*) the power input to the system, in kW, (*b*) the temperature of the water at the exit of the condenser, in °F and (*c*) the maximum possible COP of the system. The specific heat of water is 1.0 Btu/bm·°F.

6–127 Consider a Carnot heat-engine cycle executed in a closed system using 0.01 kg of refrigerant-134a as the working fluid. The cycle has a thermal efficiency of 15 percent, and the refrigerant-134a changes from saturated liquid to saturated vapor at 50°C during the heat addition process. Determine the net work output of this engine per cycle.

6–128 A heat pump with a COP of 2.8 is used to heat an air-tight house. When running, the heat pump consumes 5 kW of power. If the temperature in the house is 7°C when the heat pump is turned on, how long will it take for the heat pump to raise the temperature of the house to 22°C? Is this answer realistic or optimistic? Explain. Assume the entire mass within the house (air, furniture, etc.) is equivalent to 1500 kg of air. *Answer:* 19.2 min

6–129 A promising method of power generation involves collecting and storing solar energy in large artificial lakes a few meters deep, called solar ponds. Solar energy is absorbed by all parts of the pond, and the water temperature rises everywhere. The top part of the pond, however, loses to the atmosphere much of the heat it absorbs, and as a result, its temperature drops. This cool water serves as insulation for the bottom part of the pond and helps trap the energy there. Usually, salt is planted at the bottom of the pond to prevent the rise of this hot water to the top. A power plant that uses an organic fluid, such as alcohol, as the working fluid can be operated between the top and the bottom portions of the pond. If the water temperature is 35°C near the surface and 80°C near the bottom of the pond, determine the maximum thermal efficiency that this power plant can have. Is it realistic to use 35 and 80°C for temperatures in the calculations? Explain. *Answer:* 12.7 percent

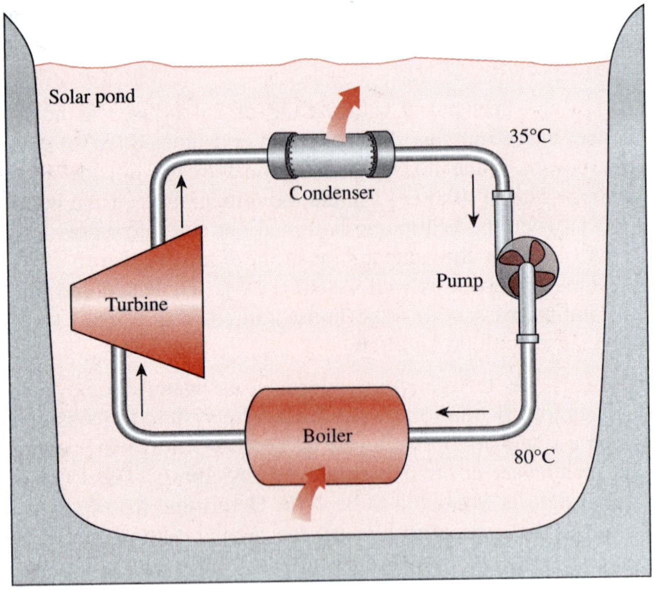

FIGURE P6–129

6–130 Consider a Carnot heat-engine cycle executed in a closed system using 0.025 kg of steam as the working fluid. It is known that the maximum absolute temperature in the cycle is twice the minimum absolute temperature, and the net work output of the cycle is 60 kJ. If the steam changes from saturated vapor to saturated liquid during heat rejection, determine the temperature of the steam during the heat rejection process.

6–131 Reconsider Prob. 6–130. Using EES (or other) software, investigate the effect of the net work output on the required temperature of the steam during the heat rejection process. Let the work output vary from 40 to 60 kJ.

6–132 Consider a Carnot refrigeration cycle executed in a closed system in the saturated liquid–vapor mixture region using 0.96 kg of refrigerant-134a as the working fluid. It is known that the maximum absolute temperature in the cycle is 1.2 times the minimum absolute temperature, and the net work input to the cycle is 22 kJ. If the refrigerant changes from saturated vapor to saturated liquid during the heat rejection process, determine the minimum pressure in the cycle.

6–133 Reconsider Prob. 6–132. Using EES (or other) software, investigate the effect of the net work input on the minimum pressure. Let the work input vary from 10 to 30 kJ. Plot the minimum pressure in the refrigeration cycle as a function of net work input, and discuss the results.

6–134 Consider two Carnot heat engines operating in series. The first engine receives heat from the reservoir at 1800 K and rejects the waste heat to another reservoir at temperature *T*. The second engine receives this energy rejected by the first one, converts some of it to work, and rejects the rest to a reservoir at 300 K. If the thermal efficiencies of both engines are the same, determine the temperature *T*. *Answer:* 735 K

6–135E Calculate and plot the thermal efficiency of a completely reversible heat engine as a function of the source temperature up to 2000 R with the sink temperature fixed at 500 R.

6–136 A Carnot heat engine receives heat at 900 K and rejects the waste heat to the environment at 300 K. The entire work output of the heat engine is used to drive a Carnot refrigerator that removes heat from the cooled space at −15°C at a rate of 250 kJ/min and rejects it to the same environment at 300 K. Determine (*a*) the rate of heat supplied to the heat engine and (*b*) the total rate of heat rejection to the environment.

6–137 Reconsider Prob. 6–136. Using EES (or other) software, investigate the effects of the heat engine source temperature, the environment temperature, and the cooled space temperature on the required heat supply to the heat engine and the total rate of heat rejection to the environment. Let the source temperature vary from 500 to 1000 K, the environment temperature vary from 275 to 325 K, and the cooled space temperature vary from −20 to 0°C. Plot the required heat supply against the source temperature for the cooled space temperature of −15°C and environment temperatures of 275, 300, and 325 K, and discuss the results.

6–138 A heat engine operates between two reservoirs at 800 and 20°C. One-half of the work output of the heat engine is used to drive a Carnot heat pump that removes heat from the cold surroundings at 2°C and transfers it to a house maintained at 22°C. If the house is losing heat at a rate of 62,000 kJ/h, determine the minimum rate of heat supply to the heat engine required to keep the house at 22°C.

6–139E An inventor claims to have developed a refrigerator that maintains the refrigerated space at 35°F while operating in a room where the temperature is 75°F and that has a COP of 13.5. Is this claim reasonable?

6–140 Consider a Carnot heat-pump cycle executed in a steady-flow system in the saturated liquid–vapor mixture region using refrigerant-134a flowing at a rate of 0.22 kg/s as the working fluid. It is known that the maximum absolute temperature in the cycle is 1.2 times the minimum absolute temperature, and the net power input to the cycle is 5 kW. If the refrigerant changes from saturated vapor to saturated liquid during the heat rejection process, determine the ratio of the maximum to minimum pressures in the cycle.

6–141 Replacing incandescent lights with energy-efficient fluorescent lights can reduce the lighting energy consumption to one-fourth of what it was before. The energy consumed by the lamps is eventually converted to heat, and thus switching to energy-efficient lighting also reduces the cooling load in summer but increases the heating load in winter. Consider a building that is heated by a natural gas furnace with an efficiency of 80 percent and cooled by an air conditioner with a COP of 3.5. If electricity costs \$0.08/kWh and natural gas costs \$1.40/therm (1 therm = 105,500 kJ), determine if efficient lighting will increase or decrease the total energy cost of the building (*a*) in summer and (*b*) in winter.

6–142 A heat pump supplies heat energy to a house at the rate of 140,000 kJ/h when the house is maintained at 25°C. Over a period of one month, the heat pump operates for 100 hours to transfer energy from a heat source outside the house to inside the house. Consider a heat pump receiving heat from two different outside energy sources. In one application the heat pump receives heat from the outside air at 0°C. In a second application the heat pump receives heat from a lake having a water temperature of 10°C. If electricity costs $0.085/kWh, determine the maximum money saved by using the lake water rather than the outside air as the outside energy source.

6–143 The cargo space of a refrigerated truck whose inner dimensions are 12 m × 2.3 m × 3.5 m is to be precooled from 25°C to an average temperature of 5°C. The construction of the truck is such that a transmission heat gain occurs at a rate of 80 W/°C. If the ambient temperature is 25°C, determine how long it will take for a system with a refrigeration capacity of 8 kW to precool this truck.

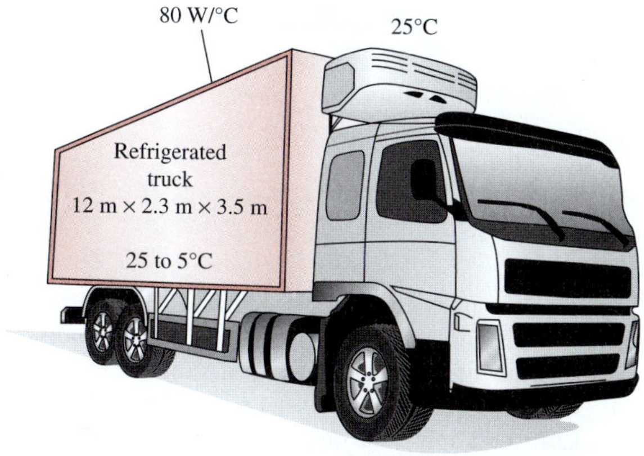

FIGURE P6–143

6–144 A refrigeration system is to cool bread loaves with an average mass of 350 g from 30 to −10°C at a rate of 1200 loaves per hour by refrigerated air at −30°C. Taking the average specific and latent heats of bread to be 2.93 kJ/kg·°C and 109.3 kJ/kg, respectively, determine (*a*) the rate of heat removal from the breads, in kJ/h; (*b*) the required volume flow rate of air, in m³/h, if the temperature rise of air is not to exceed 8°C; and (*c*) the size of the compressor of the refrigeration system, in kW, for a COP of 1.2 for the refrigeration system.

6–145 The drinking water needs of a production facility with 20 employees is to be met by a bubbler type water fountain. The refrigerated water fountain is to cool water from 22 to 8°C and supply cold water at a rate of 0.4 L per hour per person. Heat is transferred to the reservoir from the surroundings at 25°C at a rate of 45 W. If the COP of the refrigeration system is 2.9, determine the size of the compressor, in W, that will be suitable for the refrigeration system of this water cooler.

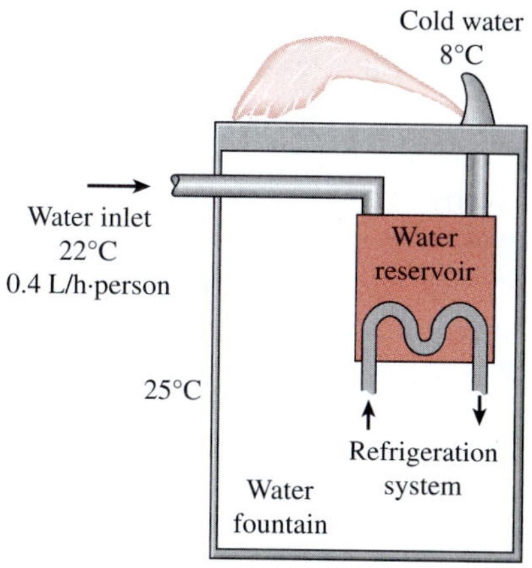

FIGURE P6–145

6–146E The "Energy Guide" label on a washing machine indicates that the washer will use $33 worth of hot water if the water is heated by a gas water heater at a natural gas rate of $1.21/therm. If the water is heated from 60 to 130°F, determine how many gallons of hot water an average family uses per week. Disregard the electricity consumed by the washer, and take the efficiency of the gas water heater to be 58 percent.

6–147 A typical electric water heater has an efficiency of 95 percent and costs $250 a year to operate at a unit cost of electricity of $0.08/kWh. A typical heat pump–powered water heater has a COP of 3.3 but costs about $800 more to install. Determine how many years it will

Water
heater

© The McGraw-Hill Companies, Inc.
Jill Braaten, photographer.

FIGURE P6–147

take for the heat pump water heater to pay for its cost differential from the energy it saves.

6–148 EES Reconsider Prob. 6–147. Using EES (or other) software, investigate the effect of the heat pump COP on the yearly operation costs and the number of years required to break even. Let the COP vary from 2 to 5. Plot the payback period against the COP and discuss the results.

6–149 A homeowner is trying to decide between a high-efficiency natural gas furnace with an efficiency of 97 percent and a ground-source heat pump with a COP of 3.5. The unit costs of electricity and natural gas are \$0.092/kWh and \$1.42/therm (1 therm = 105,500 kJ). Determine which system will have a lower energy cost.

6–150 The kitchen, bath, and other ventilation fans in a house should be used sparingly since these fans can discharge a houseful of warmed or cooled air in just one hour. Consider a 200-m^2 house whose ceiling height is 2.8 m. The house is heated by a 96 percent efficient gas heater and is maintained at 22°C and 92 kPa. If the unit cost of natural gas is \$1.20/therm (1 therm = 105,500 kJ), determine the cost of energy "vented out" by the fans in 1 h. Assume the average outdoor temperature during the heating season to be 5°C.

6–151 Repeat Prob. 6–150 for the air-conditioning cost in a dry climate for an outdoor temperature of 28°C. Assume the COP of the air-conditioning system to be 2.3, and the unit cost of electricity to be \$0.10/kWh.

6–152 A heat pump with refrigerant-134a as the working fluid is used to keep a space at 25°C by absorbing heat from geothermal water that enters the evaporator at 60°C at a rate of 0.065 kg/s and leaves at 40°C. Refrigerant enters the evaporator at 12°C with a quality of 15 percent and leaves at the same pressure as saturated vapor. If the compressor consumes 1.6 kW of power, determine (*a*) the mass flow rate of the refrigerant, (*b*) the rate of heat supply, (*c*) the COP, and (*d*) the minimum power input to the compressor for the same rate of heat supply. *Answers:* (*a*) 0.0338 kg/s, (*b*) 7.04 kW, (*c*) 4.40, (*d*) 0.740 kW

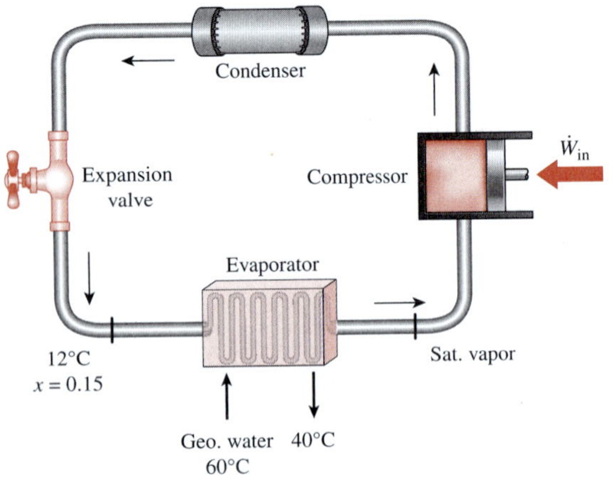

FIGURE P6–152

6–153 Cold water at 10°C enters a water heater at the rate of 0.02 m^3/min and leaves the water heater at 50°C. The water heater receives heat from a heat pump that receives heat from a heat source at 0°C.

(*a*) Assuming the water to be an incompressible liquid that does not change phase during heat addition, determine the rate of heat supplied to the water, in kJ/s.

(*b*) Assuming the water heater acts as a heat sink having an average temperature of 30°C, determine the minimum power supplied to the heat pump, in kW.

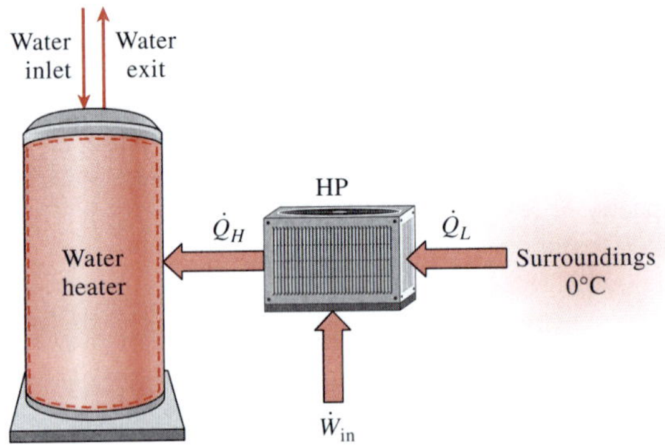

FIGURE P6–153

6–154 A heat pump receives heat from a lake that has an average winter time temperature of 6°C and supplies heat into a house having an average temperature of 27°C.

(*a*) If the house loses heat to the atmosphere at the rate of 64,000 kJ/h, determine the minimum power supplied to the heat pump, in kW.

(*b*) A heat exchanger is used to transfer the energy from the lake water to the heat pump. If the lake water temperature decreases by 5°C as it flows through the lake water-to-heat pump heat exchanger, determine the minimum mass flow rate of lake water, in kg/s. Neglect the effect of the lake water pump.

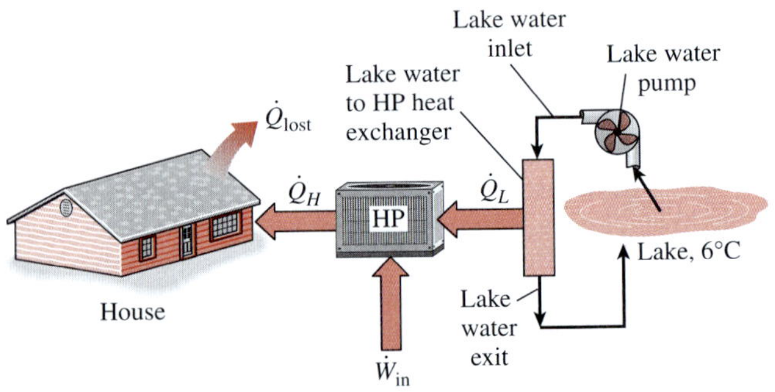

FIGURE P6–154

6–155 Prove that a refrigerator's COP cannot exceed that of a completely reversible refrigerator that shares the same thermal energy reservoirs.

6–156 Prove that the COP of all completely reversible refrigerators must be the same when the reservoir temperatures are the same.

6–157 Derive an expression for the COP of a completely reversible heat pump in terms of the thermal energy reservoir temperatures, T_L and T_H.

6–158 A Carnot heat engine is operating between a source at T_H and a sink at T_L. If it is desired to double the thermal efficiency of this engine, what should the new source temperature be? Assume the sink temperature is held constant.

6–159 When discussing Carnot engines, it is assumed that the engine is in thermal equilibrium with the source and the sink during the heat addition and heat rejection processes, respectively. That is, it is assumed that $T^*_H = T_H$ and $T^*_L = T_L$ so that there is no external irreversibility. In that case, the thermal efficiency of the Carnot engine is $\eta_C = 1 - T_L/T_H$.

In reality, however, we must maintain a reasonable temperature difference between the two heat transfer media in order to have an acceptable heat transfer rate through a finite heat exchanger surface area. The heat transfer rates in that case can be expressed as

$$\dot{Q}_H = (h_A)_H(T_H - T^*_H)$$
$$\dot{Q}_L = (hA)_L(T^*_L - T_L)$$

where h and A are the heat transfer coefficient and heat transfer surface area, respectively. When the values of h, A, T_H, and T_L are fixed, show that the power output will be a maximum when

$$\frac{T^*_L}{T^*_H} = \left(\frac{T_L}{T_H}\right)^{1/2}$$

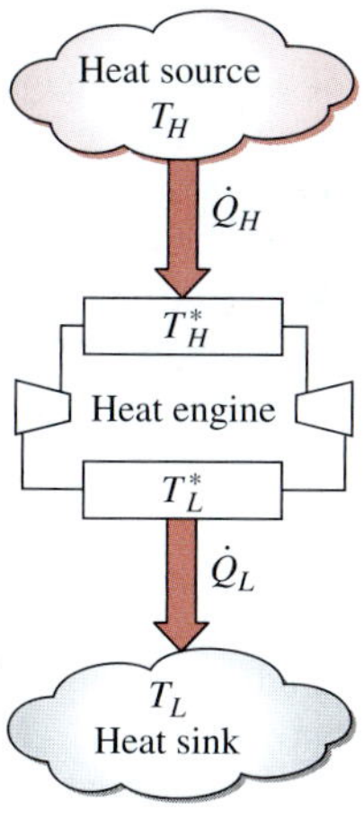

FIGURE P6–159

Also, show that the maximum net power output in this case is

$$\dot{W}_{C,\max} = \frac{(hA)_H T_H}{1 + (hA)_H/(hA)_L}\left[1 - \left(\frac{T_L}{T_H}\right)^{1/2}\right]^2$$

Fundamentals of Engineering (FE) Exam Problems

6–160 The label on a washing machine indicates that the washer will use \$85 worth of hot water if the water is heated by a 90 percent efficient electric heater at an electricity rate of \$0.09/kWh. If the water is heated from 18 to 45°C, the amount of hot water an average family uses per year is

(*a*) 11.6 tons (*b*) 15.8 tons (*c*) 27.1 tons
(*d*) 30.1 tons (*e*) 33.5 tons

6–161 A 2.4-m high 200-m² house is maintained at 22°C by an air-conditioning system whose COP is 3.2. It is estimated that the kitchen, bath, and other ventilating fans of the house discharge a houseful of conditioned air once every hour. If the average outdoor temperature is 32°C, the density of air is 1.20 kg/m³, and the unit cost of electricity is \$0.10/kWh, the amount of money "vented out" by the fans in 10 hours is

(*a*) \$0.50 (*b*) \$1.60 (*c*) \$5.00
(*d*) \$11.00 (*e*) \$16.00

6–162 The drinking water needs of an office are met by cooling tab water in a refrigerated water fountain from 23 to 6°C at an average rate of 10 kg/h. If the COP of this refrigerator is 3.1, the required power input to this refrigerator is

(*a*) 197 W (*b*) 612 W (*c*) 64 W
(*d*) 109 W (*e*) 403 W

6–163 A heat pump is absorbing heat from the cold outdoors at 5°C and supplying heat to a house at 25°C at a rate of 18,000 kJ/h. If the power consumed by the heat pump is 1.9 kW, the coefficient of performance of the heat pump is

(*a*) 1.3 (*b*) 2.6 (*c*) 3.0
(*d*) 3.8 (*e*) 13.9

6–164 A heat engine cycle is executed with steam in the saturation dome. The pressure of steam is 1 MPa during heat addition, and 0.4 MPa during heat rejection. The highest possible efficiency of this heat engine is

(*a*) 8.0% (*b*) 15.6% (*c*) 20.2%
(*d*) 79.8% (*e*) 100%

6–165 A heat engine receives heat from a source at 1000°C and rejects the waste heat to a sink at 50°C. If heat is supplied to this engine at a rate of 100 kJ/s, the maximum power this heat engine can produce is

(*a*) 25.4 kW (*b*) 55.4 kW (*c*) 74.6 kW
(*d*) 95.0 kW (*e*) 100 kW

6–166 A heat pump cycle is executed with R–134a under the saturation dome between the pressure limits of 1.4 and 0.16 MPa. The maximum coefficient of performance of this heat pump is

(*a*) 1.1 (*b*) 3.8 (*c*) 4.8

(*d*) 5.3 (*e*) 2.9

6–167 A refrigeration cycle is executed with R-134a under the saturation dome between the pressure limits of 1.6 and 0.2 MPa. If the power consumption of the refrigerator is 3 kW, the maximum rate of heat removal from the cooled space of this refrigerator is

(*a*) 0.45 kJ/s (*b*) 0.78 kJ/s (*c*) 3.0 kJ/s

(*d*) 11.6 kJ/s (*e*) 14.6 kJ/s

6–168 A heat pump with a COP of 3.2 is used to heat a perfectly sealed house (no air leaks). The entire mass within the house (air, furniture, etc.) is equivalent to 1200 kg of air. When running, the heat pump consumes electric power at a rate of 5 kW. The temperature of the house was 7°C when the heat pump was turned on. If heat transfer through the envelope of the house (walls, roof, etc.) is negligible, the length of time the heat pump must run to raise the temperature of the entire contents of the house to 22°C is

(*a*) 13.5 min (*b*) 43.1 min (*c*) 138 min

(*d*) 18.8 min (*e*) 808 min

6–169 A heat engine cycle is executed with steam in the saturation dome between the pressure limits of 7 and 2 MPa. If heat is supplied to the heat engine at a rate of 150 kJ/s, the maximum power output of this heat engine is

(*a*) 8.1 kW (*b*) 19.7 kW (*c*) 38.6 kW

(*d*) 107 kW (*e*) 130 kW

6–170 An air-conditioning system operating on the reversed Carnot cycle is required to remove heat from the house at a rate of 32 kJ/s to maintain its temperature constant at 20°C. If the temperature of the outdoors is 35°C, the power required to operate this air-conditioning system is

(*a*) 0.58 kW (*b*) 3.20 kW (*c*) 1.56 kW

(*d*) 2.26 kW (*e*) 1.64 kW

6–171 A refrigerator is removing heat from a cold medium at 3°C at a rate of 7200 kJ/h and rejecting the waste heat to a medium at 30°C. If the coefficient of performance of the refrigerator is 2, the power consumed by the refrigerator is

(*a*) 0.1 kW (*b*) 0.5 kW (*c*) 1.0 kW

(*d*) 2.0 kW (*e*) 5.0 kW

6–172 Two Carnot heat engines are operating in series such that the heat sink of the first engine serves as the heat source of the second one. If the source temperature of the first engine is 1300 K and the sink temperature of the second engine is 300 K and the thermal efficiencies of both engines are the same, the temperature of the intermediate reservoir is

(*a*) 625 K (*b*) 800 K (*c*) 860 K

(*d*) 453 K (*e*) 758 K

6–173 Consider a Carnot refrigerator and a Carnot heat pump operating between the same two thermal energy reservoirs. If the COP of the refrigerator is 3.4, the COP of the heat pump is

(*a*) 1.7 (*b*) 2.4 (*c*) 3.4

(*d*) 4.4 (*e*) 5.0

6–174 A typical new household refrigerator consumes about 680 kWh of electricity per year and has a coefficient of performance of 1.4. The amount of heat removed by this refrigerator from the refrigerated space per year is

(*a*) 952 MJ/yr (*b*) 1749 MJ/yr (*c*) 2448 MJ/yr

(*d*) 3427 MJ/yr (*e*) 4048 MJ/yr

6–175 A window air conditioner that consumes 1 kW of electricity when running and has a coefficient of performance of 3 is placed in the middle of a room, and is plugged in. The rate of cooling or heating this air conditioner will provide to the air in the room when running is

(*a*) 3 kJ/s, cooling (*b*) 1 kJ/s, cooling (*c*) 0.33 kJ/s, heating

(*d*) 1 kJ/s, heating (*e*) 3 kJ/s, heating

Design and Essay Problems

6–176 At one time, it was proposed that the thermodynamic temperature function should be ln T rather than T. How would this suggestion have changed the expressions for the performance indices of completely reversible heat engines, refrigerators, and heat pumps?

6–177 Show that the work produced by a reversible process exceeds that produced by an equivalent irreversible process by considering a weight moving down a plane both with and without friction.

6–178 The sun supplies electromagnetic energy to the earth. It appears to have an effective temperature of approximately 5800 K. On a clear summer day in North America, the energy incident on a surface facing the sun is approximately 0.95 kW/m2. The electromagnetic solar energy can be converted into thermal energy by being absorbed on a darkened surface. How might you characterize the work potential of the sun's energy when it is to be used to produce work?

6–179 In the search to reduce thermal pollution and take advantage of renewable energy sources, some people have proposed that we take advantage of such sources as discharges from electrical power plants, geothermal energy, and ocean thermal energy. Although many of these sources contain an enormous amount of energy, the amount of work they are capable of producing is limited. How might you use the work potential to assign an "energy quality" to these proposed sources? Test your proposed "energy quality" measure by applying it to the ocean thermal source, where the temperature 30 m below the surface is perhaps 5°C lower than at the surface. Apply it also to the geothermal water source, where the temperature 2 to 3 km below the surface is perhaps 150°C hotter than at the surface.

6–180 The maximum work that can be produced by a heat engine may be increased by lowering the temperature of its sink. Hence, it seems logical to use refrigerators to cool the sink and reduce its temperature. What is the fallacy in this logic?

6–181 Using a timer (or watch) and a thermometer, conduct the following experiment to determine the rate of heat gain of your refrigerator. First make sure that the door of the refrigerator is not opened for at least a few hours so that steady operating conditions are established. Start the timer when the refrigerator stops running and measure the time Δt_1 it stays off before it kicks in. Then, measure the time Δt_2 it stays on. Noting that the heat removed during Δt_2 is equal to the heat gain of the refrigerator during $\Delta t_1 + \Delta t_2$ and using the power consumed by the refrigerator when it is running, determine the average rate of heat gain for your refrigerator, in W. Take the COP (coefficient of performance) of your refrigerator to be 1.3 if it is not available.

6–182 Design a hydrocooling unit that can cool fruits and vegetables from 30 to 5°C at a rate of 20,000 kg/h under the following conditions:

The unit will be of flood type, which will cool the products as they are conveyed into the channel filled with water. The products will be dropped into the channel filled with water at one end and be picked up at the other end. The channel can be as wide as 3 m and as high as 90 cm. The water is to be circulated and cooled by the evaporator section of a refrigeration system. The refrigerant temperature inside the coils is to be −2°C, and the water temperature is not to drop below 1°C and not to exceed 6°C.

Assuming reasonable values for the average product density, specific heat, and porosity (the fraction of air volume in a box), recommend reasonable values for (*a*) the water velocity through the channel and (*b*) the refrigeration capacity of the refrigeration system.

CHAPTER

7

ENTROPY

In Chap. 6, we introduced the second law of thermodynamics and applied it to cycles and cyclic devices. In this chapter, we apply the second law to processes. The first law of thermodynamics deals with the property *energy* and the conservation of it. The second law leads to the definition of a new property called *entropy*. Entropy is a somewhat abstract property, and it is difficult to give a physical description of it without considering the microscopic state of the system. Entropy is best understood and appreciated by studying its uses in commonly encountered engineering processes, and this is what we intend to do.

This chapter starts with a discussion of the Clausius inequality, which forms the basis for the definition of entropy, and continues with the increase of entropy principle. Unlike energy, entropy is a nonconserved property, and there is no such thing as *conservation of entropy*. Next, the entropy changes that take place during processes for pure substances, incompressible substances, and ideal gases are discussed, and a special class of idealized processes, called *isentropic processes*, is examined. Then, the reversible steady-flow work and the isentropic efficiencies of various engineering devices such as turbines and compressors are considered. Finally, entropy balance is introduced and applied to various systems.

Objectives

The objectives of Chapter 7 are to:

- Apply the second law of thermodynamics to processes.
- Define a new property called *entropy* to quantify the second-law effects.
- Establish the *increase of entropy principle.*
- Calculate the entropy changes that take place during processes for pure substances, incompressible substances, and ideal gases.
- Examine a special class of idealized processes, called *isentropic processes*, and develop the property relations for these processes.
- Derive the reversible steady-flow work relations.
- Develop the isentropic efficiencies for various steady-flow devices.
- Introduce and apply the entropy balance to various systems.

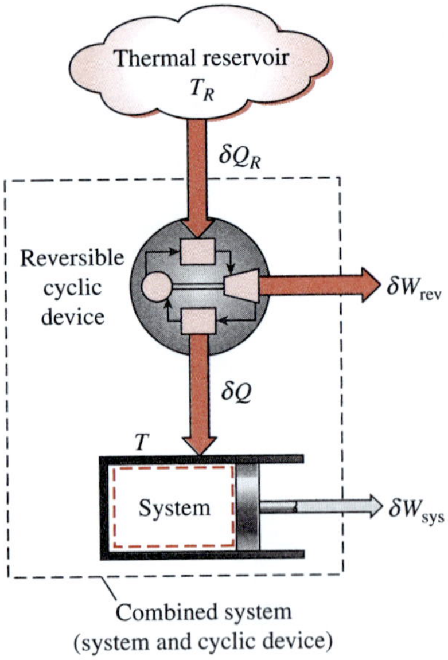

FIGURE 7–1
The system considered in the development of the Clausius inequality.

7–1 ▪ ENTROPY

The second law of thermodynamics often leads to expressions that involve inequalities. An irreversible (i.e., actual) heat engine, for example, is less efficient than a reversible one operating between the same two thermal energy reservoirs. Likewise, an irreversible refrigerator or a heat pump has a lower coefficient of performance (COP) than a reversible one operating between the same temperature limits. Another important inequality that has major consequences in thermodynamics is the **Clausius inequality**. It was first stated by the German physicist R. J. E. Clausius (1822–1888), one of the founders of thermodynamics, and is expressed as

$$\oint \frac{\delta Q}{T} \le 0$$

That is, *the cyclic integral of $\delta Q/T$ is always less than or equal to zero*. This inequality is valid for all cycles, reversible or irreversible. The symbol $\oint$ (integral symbol with a circle in the middle) is used to indicate that the integration is to be performed over the entire cycle. Any heat transfer to or from a system can be considered to consist of differential amounts of heat transfer. Then the cyclic integral of $\delta Q/T$ can be viewed as the sum of all these differential amounts of heat transfer divided by the temperature at the boundary.

To demonstrate the validity of the Clausius inequality, consider a system connected to a thermal energy reservoir at a constant thermodynamic (i.e., absolute) temperature of T_R through a *reversible* cyclic device (Fig. 7–1). The cyclic device receives heat δQ_R from the reservoir and supplies heat δQ to the system whose temperature at that part of the boundary is T (a variable) while producing work δW_{rev}. The system produces work δW_{sys} as a result of this heat transfer. Applying the energy balance to the combined system identified by dashed lines yields

$$\delta W_C = \delta Q_R - dE_C$$

where δW_C is the total work of the combined system ($\delta W_{rev} + \delta W_{sys}$) and dE_C is the change in the total energy of the combined system. Considering that the cyclic device is a *reversible* one, we have

$$\frac{\delta Q_R}{T_R} = \frac{\delta Q}{T}$$

where the sign of δQ is determined with respect to the system (positive if *to* the system and negative if *from* the system) and the sign of δQ_R is determined with respect to the reversible cyclic device. Eliminating δQ_R from the two relations above yields

$$\delta W_C = T_R \frac{\delta Q}{T} - dE_C$$

We now let the system undergo a cycle while the cyclic device undergoes an integral number of cycles. Then the preceding relation becomes

$$W_C = T_R \oint \frac{\delta Q}{T}$$

since the cyclic integral of energy (the net change in the energy, which is a property, during a cycle) is zero. Here W_C is the cyclic integral of δW_C, and it represents the net work for the combined cycle.

It appears that the combined system is exchanging heat with a single thermal energy reservoir while involving (producing or consuming) work W_C during a cycle. On the basis of the Kelvin–Planck statement of the second law, which states that *no system can produce a net amount of work while operating in a cycle and exchanging heat with a single thermal energy reservoir*, we reason that W_C cannot be a work output, and thus it cannot be a positive quantity. Considering that T_R is the thermodynamic temperature and thus a positive quantity, we must have

$$\oint \frac{\delta Q}{T} \leq 0 \qquad \textbf{(7–1)}$$

which is the *Clausius inequality*. This inequality is valid for all thermodynamic cycles, reversible or irreversible, including the refrigeration cycles.

If no irreversibilities occur within the system as well as the reversible cyclic device, then the cycle undergone by the combined system is internally reversible. As such, it can be reversed. In the reversed cycle case, all the quantities have the same magnitude but the opposite sign. Therefore, the work W_C, which could not be a positive quantity in the regular case, cannot be a negative quantity in the reversed case. Then it follows that $W_{C,\text{int rev}} = 0$ since it cannot be a positive or negative quantity, and therefore

$$\oint \left(\frac{\delta Q}{T}\right)_{\text{int rev}} = 0 \qquad \textbf{(7–2)}$$

for internally reversible cycles. Thus, we conclude that *the equality in the Clausius inequality holds for totally or just internally reversible cycles and the inequality for the irreversible ones.*

To develop a relation for the definition of entropy, let us examine Eq. 7–2 more closely. Here we have a quantity whose cyclic integral is zero. Let us think for a moment what kind of quantities can have this characteristic. We know that the cyclic integral of *work* is not zero. (It is a good thing that it is not. Otherwise, heat engines that work on a cycle such as steam power plants would produce zero net work.) Neither is the cyclic integral of heat.

Now consider the volume occupied by a gas in a piston–cylinder device undergoing a cycle, as shown in Fig. 7–2. When the piston returns to its initial position at the end of a cycle, the volume of the gas also returns to its initial value. Thus, the net change in volume during a cycle is zero. This is also expressed as

$$\oint dV = 0 \qquad \textbf{(7–3)}$$

That is, the cyclic integral of volume (or any other property) is zero. Conversely, a quantity whose cyclic integral is zero depends on the *state* only and not the process path, and thus it is a property. Therefore, the quantity $(\delta Q/T)_{\text{int rev}}$ must represent a property in the differential form.

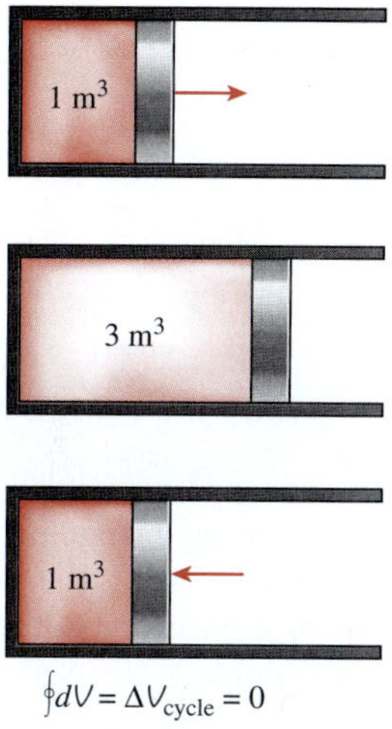

FIGURE 7–2
The net change in volume (a property) during a cycle is always zero.

Clausius realized in 1865 that he had discovered a new thermodynamic property, and he chose to name this property **entropy**. It is designated S and is defined as

$$dS = \left(\frac{\delta Q}{T}\right)_{\text{int rev}} \qquad \text{(kJ/K)} \tag{7–4}$$

Entropy is an extensive property of a system and sometimes is referred to as *total entropy*. Entropy per unit mass, designated s, is an intensive property and has the unit kJ/kg·K. The term *entropy* is generally used to refer to both total entropy and entropy per unit mass since the context usually clarifies which one is meant.

The entropy change of a system during a process can be determined by integrating Eq. 7–4 between the initial and the final states:

$$\Delta S = S_2 - S_1 = \int_1^2 \left(\frac{\delta Q}{T}\right)_{\text{int rev}} \qquad \text{(kJ/K)} \tag{7–5}$$

Notice that we have actually defined the *change* in entropy instead of entropy itself, just as we defined the change in energy instead of the energy itself when we developed the first-law relation. Absolute values of entropy are determined on the basis of the third law of thermodynamics, which is discussed later in this chapter. Engineers are usually concerned with the *changes* in entropy. Therefore, the entropy of a substance can be assigned a zero value at some arbitrarily selected reference state, and the entropy values at other states can be determined from Eq. 7–5 by choosing state 1 to be the reference state ($S = 0$) and state 2 to be the state at which entropy is to be determined.

To perform the integration in Eq. 7–5, one needs to know the relation between Q and T during a process. This relation is often not available, and the integral in Eq. 7–5 can be performed for a few cases only. For the majority of cases we have to rely on tabulated data for entropy.

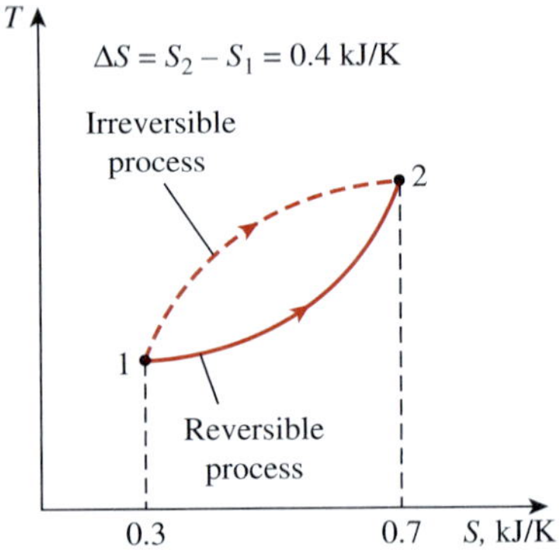

FIGURE 7–3
The entropy change between two specified states is the same whether the process is reversible or irreversible.

Note that entropy is a property, and like all other properties, it has fixed values at fixed states. Therefore, the entropy change ΔS between two specified states is the same no matter what path, reversible or irreversible, is followed during a process (Fig. 7–3).

Also note that the integral of $\delta Q/T$ gives us the value of entropy change *only if* the integration is carried out along an *internally reversible* path between the two states. The integral of $\delta Q/T$ along an irreversible path is not a property, and in general, different values will be obtained when the integration is carried out along different irreversible paths. Therefore, even for irreversible processes, the entropy change should be determined by carrying out this integration along some convenient *imaginary* internally reversible path between the specified states.

A Special Case: Internally Reversible Isothermal Heat Transfer Processes

Recall that isothermal heat transfer processes are internally reversible. Therefore, the entropy change of a system during an internally reversible isothermal heat transfer process can be determined by performing the integration in Eq. 7–5:

$$\Delta S = \int_1^2 \left(\frac{\delta Q}{T}\right)_{\text{int rev}} = \int_1^2 \left(\frac{\delta Q}{T_0}\right)_{\text{int rev}} = \frac{1}{T_0}\int_1^2 (\delta Q)_{\text{int rev}}$$

which reduces to

$$\Delta S = \frac{Q}{T_0} \qquad (\text{kJ/K}) \tag{7–6}$$

where T_0 is the constant temperature of the system and Q is the heat transfer for the internally reversible process. Equation 7–6 is particularly useful for determining the entropy changes of thermal energy reservoirs that can absorb or supply heat indefinitely at a constant temperature.

Notice that the entropy change of a system during an internally reversible isothermal process can be positive or negative, depending on the direction of heat transfer. Heat transfer to a system increases the entropy of a system, whereas heat transfer from a system decreases it. In fact, losing heat is the only way the entropy of a system can be decreased.

EXAMPLE 7–1 Entropy Change during an Isothermal Process

A piston–cylinder device contains a liquid–vapor mixture of water at 300 K. During a constant-pressure process, 750 kJ of heat is transferred to the water. As a result, part of the liquid in the cylinder vaporizes. Determine the entropy change of the water during this process.

SOLUTION Heat is transferred to a liquid–vapor mixture of water in a piston–cylinder device at constant pressure. The entropy change of water is to be determined.

Assumptions No irreversibilities occur within the system boundaries during the process.

Analysis We take the *entire water* (liquid + vapor) in the cylinder as the system (Fig. 7–4). This is a *closed system* since no mass crosses the system boundary during the process. We note that the temperature of the system remains constant at 300 K during this process since the temperature of a pure substance remains constant at the saturation value during a phase-change process at constant pressure.

The system undergoes an internally reversible, isothermal process, and thus its entropy change can be determined directly from Eq. 7–6 to be

$$\Delta S_{\text{sys,isothermal}} = \frac{Q}{T_{\text{sys}}} = \frac{750 \text{ kJ}}{300 \text{ K}} = \mathbf{2.5 \text{ kJ/K}}$$

Discussion Note that the entropy change of the system is positive, as expected, since heat transfer is *to* the system.

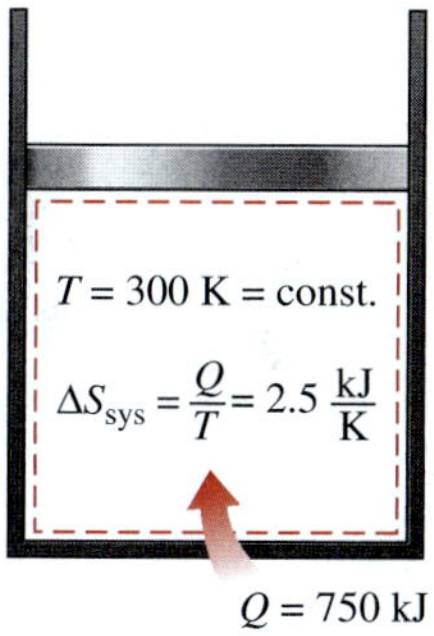

FIGURE 7–4
Schematic for Example 7–1.

7–2 ▪ THE INCREASE OF ENTROPY PRINCIPLE

Consider a cycle that is made up of two processes: process 1-2, which is arbitrary (reversible or irreversible), and process 2-1, which is internally reversible, as shown in Figure 7–5. From the Clausius inequality,

$$\oint \frac{\delta Q}{T} \leq 0$$

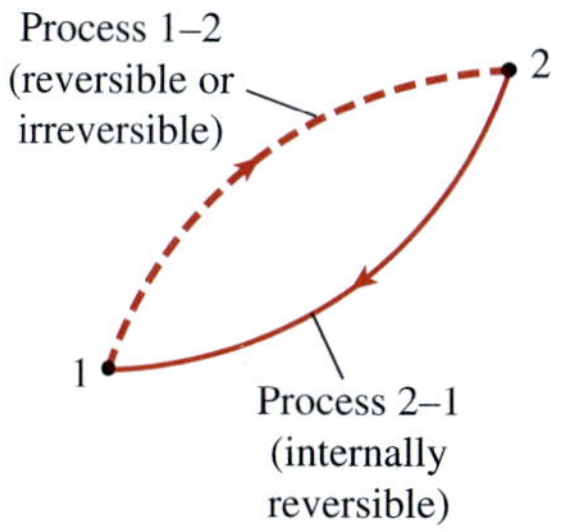

FIGURE 7–5
A cycle composed of a reversible and an irreversible process.

or

$$\int_1^2 \frac{\delta Q}{T} + \int_2^1 \left(\frac{\delta Q}{T}\right)_{\text{int rev}} \leq 0$$

The second integral in the previous relation is recognized as the entropy change $S_1 - S_2$. Therefore,

$$\int_1^2 \frac{\delta Q}{T} + S_1 - S_2 \leq 0$$

which can be rearranged as

$$S_2 - S_1 \geq \int_1^2 \frac{\delta Q}{T} \qquad \textbf{(7–7)}$$

It can also be expressed in differential form as

$$dS \geq \frac{\delta Q}{T} \qquad \textbf{(7–8)}$$

where the equality holds for an internally reversible process and the inequality for an irreversible process. We may conclude from these equations that the entropy change of a closed system during an irreversible process is greater than the integral of $\delta Q/T$ evaluated for that process. In the limiting case of a reversible process, these two quantities become equal. We again emphasize that T in these relations is the *thermodynamic temperature* at the *boundary* where the differential heat δQ is transferred between the system and the surroundings.

The quantity $\Delta S = S_2 - S_1$ represents the *entropy change* of the system. For a reversible process, it becomes equal to $\int_1^2 \delta Q/T$, which represents the *entropy transfer* with heat.

The inequality sign in the preceding relations is a constant reminder that the entropy change of a closed system during an irreversible process is always greater than the entropy transfer. That is, some entropy is *generated* or *created* during an irreversible process, and this generation is due entirely to the presence of irreversibilities. The entropy generated during a process is called **entropy generation** and is denoted by S_{gen}. Noting that the difference between the entropy change of a closed system and the entropy transfer is equal to entropy generation, Eq. 7–7 can be rewritten as an equality as

$$\Delta S_{\text{sys}} = S_2 - S_1 = \int_1^2 \frac{\delta Q}{T} + S_{\text{gen}} \qquad \textbf{(7–9)}$$

Note that the entropy generation S_{gen} is always a *positive* quantity or zero. Its value depends on the process, and thus it is *not* a property of the system. Also, in the absence of any entropy transfer, the entropy change of a system is equal to the entropy generation.

Equation 7–7 has far-reaching implications in thermodynamics. For an isolated system (or simply an adiabatic closed system), the heat transfer is zero, and Eq. 7–7 reduces to

$$\Delta S_{\text{isolated}} \geq 0 \qquad \textbf{(7–10)}$$

This equation can be expressed as *the entropy of an isolated system during a process always increases or, in the limiting case of a reversible process, remains constant*. In other words, it *never* decreases. This is known as the **increase of entropy principle**. Note that in the absence of any heat transfer, entropy change is due to irreversibilities only, and their effect is always to increase entropy.

Entropy is an extensive property, thus the total entropy of a system is equal to the sum of the entropies of the parts of the system. An isolated system may consist of any number of subsystems (Fig. 7–6). A system and its surroundings, for example, constitute an isolated system since both can be enclosed by a sufficiently large arbitrary boundary across which there is no heat, work, or mass transfer (Fig. 7–7). Therefore, a system and its surroundings can be viewed as the two subsystems of an isolated system, and the entropy change of this isolated system during a process is the sum of the entropy changes of the system and its surroundings, which is equal to the entropy generation since an isolated system involves no entropy transfer. That is,

$$S_{gen} = \Delta S_{total} = \Delta S_{sys} + \Delta S_{surr} \geq 0 \quad \textbf{(7-11)}$$

where the equality holds for reversible processes and the inequality for irreversible ones. Note that ΔS_{surr} refers to the change in the entropy of the surroundings as a result of the occurrence of the process under consideration.

Since no actual process is truly reversible, we can conclude that some entropy is generated during a process, and therefore the entropy of the universe, which can be considered to be an isolated system, is continuously increasing. The more irreversible a process, the larger the entropy generated during that process. No entropy is generated during reversible processes ($S_{gen} = 0$).

Entropy increase of the universe is a major concern not only to engineers but also to philosophers, theologians, economists, and environmentalists since entropy is viewed as a measure of the disorder (or "mixed-up-ness") in the universe.

The increase of entropy principle does not imply that the entropy of a system cannot decrease. The entropy change of a system *can* be negative during a process (Fig. 7–8), but entropy generation cannot. The increase of entropy principle can be summarized as follows:

$$S_{gen} \begin{cases} > 0 & \text{Irreversible process} \\ = 0 & \text{Reversible process} \\ < 0 & \text{Impossible process} \end{cases}$$

This relation serves as a criterion in determining whether a process is reversible, irreversible, or impossible.

Things in nature have a tendency to change until they attain a state of equilibrium. The increase of entropy principle dictates that the entropy of an isolated system increases until the entropy of the system reaches a *maximum* value. At that point, the system is said to have reached an equilibrium state since the increase of entropy principle prohibits the system from undergoing any change of state that results in a decrease in entropy.

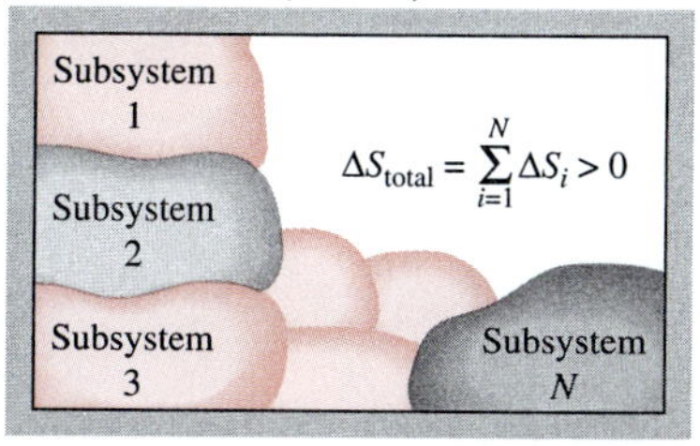

FIGURE 7–6
The entropy change of an isolated system is the sum of the entropy changes of its components, and is never less than zero.

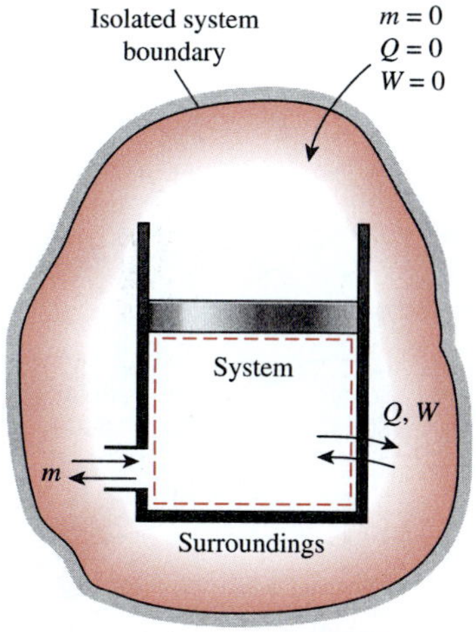

FIGURE 7–7
A system and its surroundings form an isolated system.

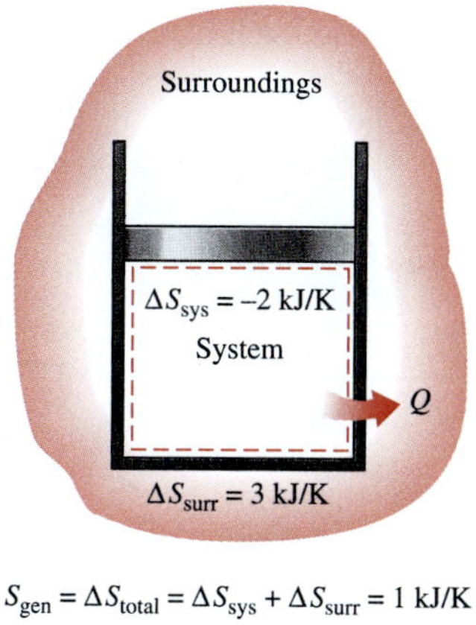

FIGURE 7–8
The entropy change of a system can be negative, but the entropy generation cannot.

Some Remarks about Entropy

In light of the preceding discussions, we draw the following conclusions:

1. Processes can occur in a *certain* direction only, not in *any* direction. A process must proceed in the direction that complies with the increase of entropy principle, that is, $S_{gen} \geq 0$. A process that violates this principle is impossible. This principle often forces chemical reactions to come to a halt before reaching completion.
2. Entropy is a *nonconserved property*, and there is *no* such thing as the *conservation of entropy principle*. Entropy is conserved during the idealized reversible processes only and increases during *all* actual processes.
3. The performance of engineering systems is degraded by the presence of irreversibilities, and *entropy generation* is a measure of the magnitudes of the irreversibilities present during that process. The greater the extent of irreversibilities, the greater the entropy generation. Therefore, entropy generation can be used as a quantitative measure of irreversibilities associated with a process. It is also used to establish criteria for the performance of engineering devices. This point is illustrated further in Example 7–2.

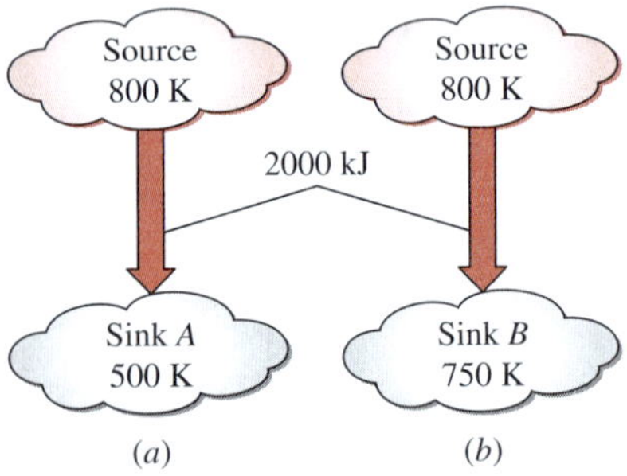

FIGURE 7–9
Schematic for Example 7–2.

EXAMPLE 7–2 **Entropy Generation during Heat Transfer Processes**

A heat source at 800 K loses 2000 kJ of heat to a sink at (*a*) 500 K and (*b*) 750 K. Determine which heat transfer process is more irreversible.

SOLUTION Heat is transferred from a heat source to two heat sinks at different temperatures. The heat transfer process that is more irreversible is to be determined.

Analysis A sketch of the reservoirs is shown in Fig. 7–9. Both cases involve heat transfer through a finite temperature difference, and therefore both are irreversible. The magnitude of the irreversibility associated with each process can be determined by calculating the total entropy change for each case. The total entropy change for a heat transfer process involving two reservoirs (a source and a sink) is the sum of the entropy changes of each reservoir since the two reservoirs form an adiabatic system.

Or do they? The problem statement gives the impression that the two reservoirs are in direct contact during the heat transfer process. But this cannot be the case since the temperature at a point can have only one value, and thus it cannot be 800 K on one side of the point of contact and 500 K on the other side. In other words, the temperature function cannot have a jump discontinuity. Therefore, it is reasonable to assume that the two reservoirs are separated by a partition through which the temperature drops from 800 K on one side to 500 K (or 750 K) on the other. In that case, the entropy change of the partition should also be considered when evaluating the total entropy change for this process. However, considering that entropy is a property and the values of properties depend on the state of a system, we can argue that the entropy change of the partition is zero since the partition appears to have undergone a *steady* process and thus experienced no change in its properties at any point. We base this argument on the fact that the temperature on both sides of the partition and thus throughout remains constant during this process. Therefore, we are justified to assume that $\Delta S_{\text{partition}} = 0$ since the entropy (as well as the energy) content of the partition remains constant during this process.

The entropy change for each reservoir can be determined from Eq. 7–6 since each reservoir undergoes an internally reversible, isothermal process.

(*a*) For the heat transfer process to a sink at 500 K:

$$\Delta S_{\text{source}} = \frac{Q_{\text{source}}}{T_{\text{source}}} = \frac{-2000 \text{ kJ}}{800 \text{ K}} = -2.5 \text{ kJ/K}$$

$$\Delta S_{\text{sink}} = \frac{Q_{\text{sink}}}{T_{\text{sink}}} = \frac{2000 \text{ kJ}}{500 \text{ K}} = +4.0 \text{ kJ/K}$$

and

$$S_{\text{gen}} = \Delta S_{\text{total}} = \Delta S_{\text{source}} + \Delta S_{\text{sink}} = (-2.5 + 4.0) \text{ kJ/K} = \mathbf{1.5 \text{ kJ/K}}$$

Therefore, 1.5 kJ/K of entropy is generated during this process. Noting that both reservoirs have undergone internally reversible processes, the entire entropy generation took place in the partition.

(*b*) Repeating the calculations in part (*a*) for a sink temperature of 750 K, we obtain

$$\Delta S_{\text{source}} = -2.5 \text{ kJ/k}$$

$$\Delta S_{\text{sink}} = +2.7 \text{ kJ/K}$$

and

$$S_{\text{gen}} = \Delta S_{\text{total}} = (-2.5 + 2.7) \text{ kJ/K} = \mathbf{0.2 \text{ kJ/K}}$$

The total entropy change for the process in part (*b*) is smaller, and therefore it is less irreversible. This is expected since the process in (*b*) involves a smaller temperature difference and thus a smaller irreversibility.

Discussion The irreversibilities associated with both processes could be eliminated by operating a Carnot heat engine between the source and the sink. For this case it can be shown that $\Delta S_{\text{total}} = 0$.

7–3 ▪ ENTROPY CHANGE OF PURE SUBSTANCES

Entropy is a property, and thus the value of entropy of a system is fixed once the state of the system is fixed. Specifying two intensive independent properties fixes the state of a simple compressible system, and thus the value of entropy and of other properties at that state. Starting with its defining relation, the entropy change of a substance can be expressed in terms of other properties (see Sec. 7–7). But in general, these relations are too complicated and are not practical to use for hand calculations. Therefore, using a suitable reference state, the entropies of substances are evaluated from measurable property data following rather involved computations, and the results are tabulated in the same manner as the other properties such as v, u, and h (Fig. 7–10).

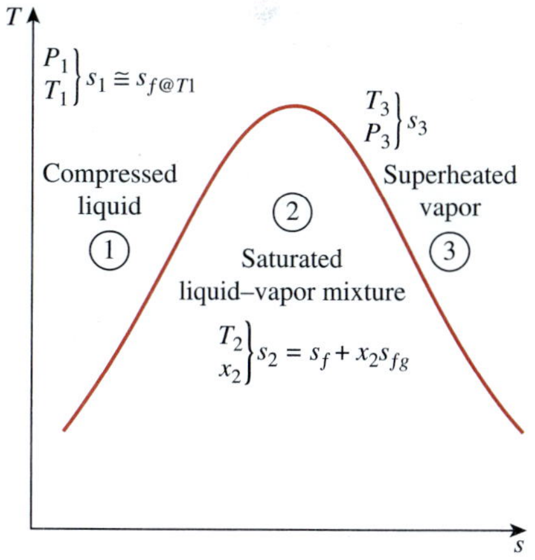

FIGURE 7–10
The entropy of a pure substance is determined from the tables (like other properties).

The entropy values in the property tables are given relative to an arbitrary reference state. In steam tables the entropy of saturated liquid s_f at 0.01°C is assigned the value of zero. For refrigerant-134a, the zero value is assigned to saturated liquid at −40°C. The entropy values become negative at temperatures below the reference value.

The value of entropy at a specified state is determined just like any other property. In the compressed liquid and superheated vapor regions, it can be obtained directly from the tables at the specified state. In the saturated mixture region, it is determined from

$$s = s_f + xs_{fg} \qquad \text{(kJ/kg·K)}$$

where x is the quality and s_f and s_{fg} values are listed in the saturation tables. In the absence of compressed liquid data, the entropy of the compressed liquid can be approximated by the entropy of the saturated liquid at the given temperature:

$$s_{@\,T,P} \cong s_{f@\,T} \qquad \text{(kJ/kg·K)}$$

The entropy change of a specified mass m (a closed system) during a process is simply

$$\Delta S = m\Delta s = m(s_2 - s_1) \qquad \text{(kJ/K)} \tag{7-12}$$

which is the difference between the entropy values at the final and initial states.

When studying the second-law aspects of processes, entropy is commonly used as a coordinate on diagrams such as the *T-s* and *h-s* diagrams. The general characteristics of the *T-s* diagram of pure substances are shown in Fig. 7–11 using data for water. Notice from this diagram that the constant-volume lines are steeper than the constant-pressure lines and the constant-pressure lines are parallel to the constant-temperature lines in the saturated liquid–vapor mixture region. Also, the constant-pressure lines almost coincide with the saturated liquid line in the compressed liquid region.

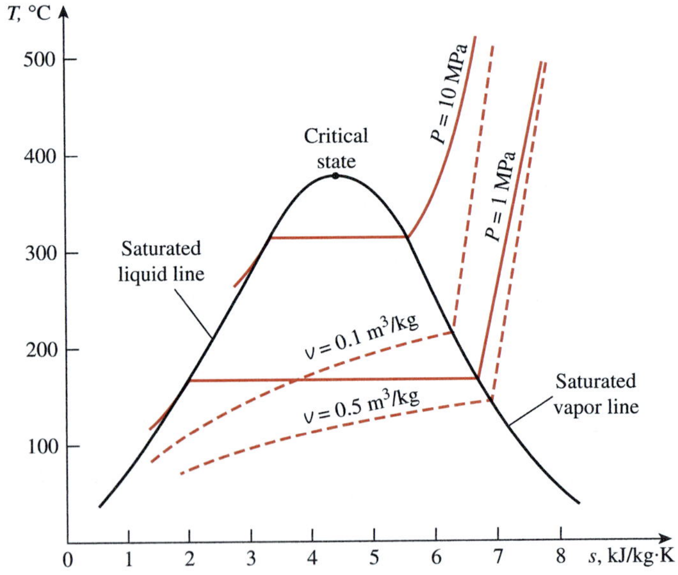

FIGURE 7–11
Schematic of the *T-s* diagram for water.

EXAMPLE 7–3 Entropy Change of a Substance in a Tank

A rigid tank contains 5 kg of refrigerant-134a initially at 20°C and 140 kPa. The refrigerant is now cooled while being stirred until its pressure drops to 100 kPa. Determine the entropy change of the refrigerant during this process.

SOLUTION The refrigerant in a rigid tank is cooled while being stirred. The entropy change of the refrigerant is to be determined.

Assumptions The volume of the tank is constant and thus $v_2 = v_1$.

Analysis We take the refrigerant in the tank as the *system* (Fig. 7–12). This is a *closed system* since no mass crosses the system boundary during the process. We note that the change in entropy of a substance during a process is simply the difference between the entropy values at the final and initial states. The initial state of the refrigerant is completely specified.

Recognizing that the specific volume remains constant during this process, the properties of the refrigerant at both states are

State 1:
$$\left.\begin{aligned} P_1 &= 140 \text{ kPa} \\ T_1 &= 20°\text{C} \end{aligned}\right\} \quad \begin{aligned} s_1 &= 1.0624 \text{ kJ/kg·K} \\ v_1 &= 0.16544 \text{ m}^3/\text{kg} \end{aligned}$$

State 2:
$$\left.\begin{aligned} P_2 &= 100 \text{ kPa} \\ (v_2 &= v_1) \end{aligned}\right\} \quad \begin{aligned} v_f &= 0.0007259 \text{ m}^3/\text{kg} \\ v_g &= 0.19254 \text{ m}^3/\text{kg} \end{aligned}$$

The refrigerant is a saturated liquid–vapor mixture at the final state since $v_f < v_2 < v_g$ at 100 kPa pressure. Therefore, we need to determine the quality first:

$$x_2 = \frac{v_2 - v_f}{v_{fg}} = \frac{0.16544 - 0.0007259}{0.19254 - 0.0007259} = 0.859$$

Thus,

$$s_2 = s_f + x_2 s_{fg} = 0.07188 + (0.859)(0.87995) = 0.8278 \text{ kJ/kg·K}$$

Then, the entropy change of the refrigerant during this process is

$$\Delta S = m(s_2 - s_1) = (5 \text{ kg})(0.8278 - 1.0624) \text{ kJ/kg·K}$$
$$= \mathbf{-1.173 \text{ kJ/K}}$$

Discussion The negative sign indicates that the entropy of the system is decreasing during this process. This is not a violation of the second law, however, since it is the *entropy generation* S_{gen} that cannot be negative.

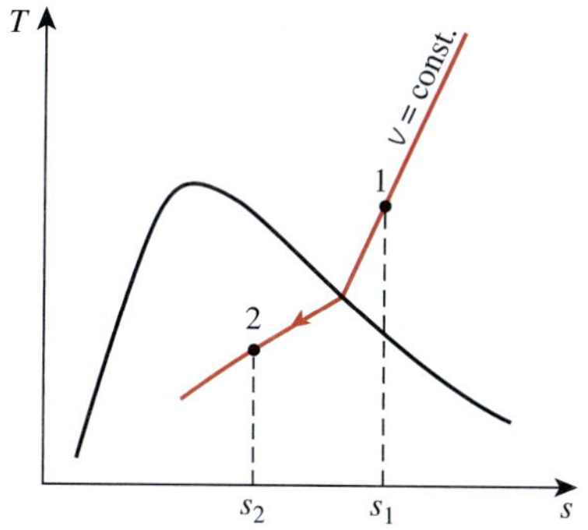

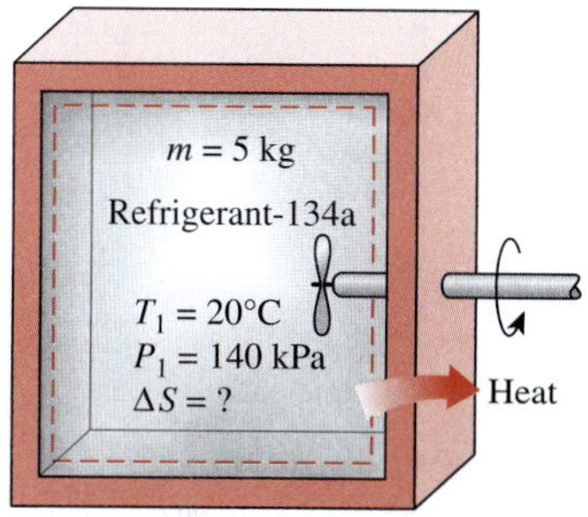

FIGURE 7–12
Schematic and *T-s* diagram for Example 7–3.

EXAMPLE 7–4 Entropy Change during a Constant-Pressure Process

A piston–cylinder device initially contains 3 lbm of liquid water at 20 psia and 70°F. The water is now heated at constant pressure by the addition of 3450 Btu of heat. Determine the entropy change of the water during this process.

SOLUTION Liquid water in a piston–cylinder device is heated at constant pressure. The entropy change of water is to be determined.

Assumptions **1** The tank is stationary and thus the kinetic and potential energy changes are zero, $\Delta KE = \Delta PE = 0$. **2** The process is quasi-equilibrium. **3** The pressure remains constant during the process and thus $P_2 = P_1$.

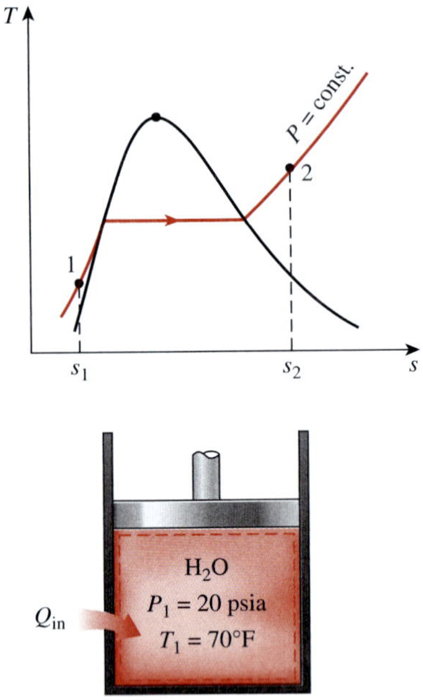

FIGURE 7–13
Schematic and *T-s* diagram for Example 7–4.

Analysis We take the water in the cylinder as the *system* (Fig. 7–13). This is a *closed system* since no mass crosses the system boundary during the process. We note that a piston–cylinder device typically involves a moving boundary and thus boundary work W_b. Also, heat is transferred to the system.

Water exists as a compressed liquid at the initial state since its pressure is greater than the saturation pressure of 0.3632 psia at 70°F. By approximating the compressed liquid as a saturated liquid at the given temperature, the properties at the initial state are

State 1:
$$\left.\begin{array}{l} P_1 = 20 \text{ psia} \\ T_1 = 70°\text{F} \end{array}\right\} \quad \begin{array}{l} s_1 \cong s_{f\,@\,70°\text{F}} = 0.07459 \text{ Btu/lbm·R} \\ h_1 \cong h_{f\,@\,70°\text{F}} = 38.08 \text{ Btu/lbm} \end{array}$$

At the final state, the pressure is still 20 psia, but we need one more property to fix the state. This property is determined from the energy balance,

$$\underbrace{E_{\text{in}} - E_{\text{out}}}_{\substack{\text{Net energy transfer} \\ \text{by heat, work, and mass}}} = \underbrace{\Delta E_{\text{system}}}_{\substack{\text{Change in internal, kinetic,} \\ \text{potential, etc., energies}}}$$

$$Q_{\text{in}} - W_b = \Delta U$$
$$Q_{\text{in}} = \Delta H = m(h_2 - h_1)$$
$$3450 \text{ Btu} = (3 \text{ lbm})(h_2 - 38.08 \text{ Btu/lbm})$$
$$h_2 = 1188.1 \text{ Btu/lbm}$$

since $\Delta U + W_b = \Delta H$ for a constant-pressure quasi-equilibrium process. Then,

State 2:
$$\left.\begin{array}{l} P_2 = 20 \text{ psia} \\ h_2 = 1188.1 \text{ Btu/lbm} \end{array}\right\} \quad \begin{array}{l} s_2 = 1.7761 \text{ Btu/lbm·R} \\ (\text{Table A-6E, interpolation}) \end{array}$$

Therefore, the entropy change of water during this process is

$$\Delta S = m(s_2 - s_1) = (3 \text{ lbm})(1.7761 - 0.07459) \text{ Btu/lbm·R}$$
$$= \mathbf{5.105 \text{ Btu/R}}$$

7–4 ▪ ISENTROPIC PROCESSES

We mentioned earlier that the entropy of a fixed mass can be changed by (1) heat transfer and (2) irreversibilities. Then it follows that the entropy of a fixed mass does not change during a process that is *internally reversible* and *adiabatic* (Fig. 7–14). A process during which the entropy remains constant is called an **isentropic process**. It is characterized by

Isentropic process: $\Delta s = 0$ or $s_2 = s_1$ (kJ/kg·K) **(7–13)**

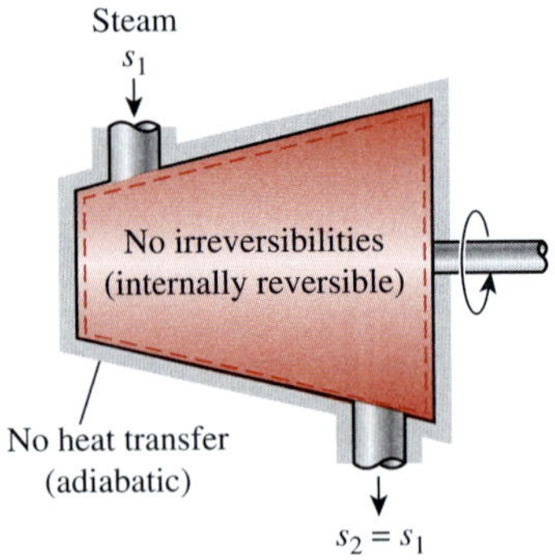

FIGURE 7–14
During an internally reversible, adiabatic (isentropic) process, the entropy remains constant.

That is, a substance will have the same entropy value at the end of the process as it does at the beginning if the process is carried out in an isentropic manner.

Many engineering systems or devices such as pumps, turbines, nozzles, and diffusers are essentially adiabatic in their operation, and they perform best when the irreversibilities, such as the friction associated with the process, are minimized. Therefore, an isentropic process can serve as an appropriate model for actual processes. Also, isentropic processes enable us to define efficiencies for processes to compare the actual performance of these devices to the performance under idealized conditions.

It should be recognized that a *reversible adiabatic* process is necessarily isentropic ($s_2 = s_1$), but an *isentropic* process is not necessarily a reversible adiabatic process. (The entropy increase of a substance during a process as a result of irreversibilities may be offset by a decrease in entropy as a result of heat losses, for example.) However, the term *isentropic process* is customarily used in thermodynamics to imply an *internally reversible, adiabatic process*.

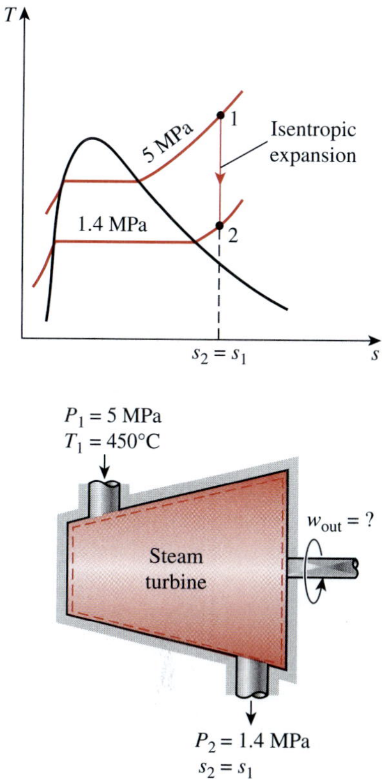

FIGURE 7–15
Schematic and *T-s* diagram for Example 7–5.

EXAMPLE 7–5 Isentropic Expansion of Steam in a Turbine

Steam enters an adiabatic turbine at 5 MPa and 450°C and leaves at a pressure of 1.4 MPa. Determine the work output of the turbine per unit mass of steam if the process is reversible.

SOLUTION Steam is expanded in an adiabatic turbine to a specified pressure in a reversible manner. The work output of the turbine is to be determined.

Assumptions **1** This is a steady-flow process since there is no change with time at any point and thus $\Delta m_{CV} = 0$, $\Delta E_{CV} = 0$, and $\Delta S_{CV} = 0$. **2** The process is reversible. **3** Kinetic and potential energies are negligible. **4** The turbine is adiabatic and thus there is no heat transfer.

Analysis We take the *turbine* as the system (Fig. 7–15). This is a *control volume* since mass crosses the system boundary during the process. We note that there is only one inlet and one exit, and thus $\dot{m}_1 = \dot{m}_2 = \dot{m}$.

The power output of the turbine is determined from the rate form of the energy balance,

$$\underbrace{\dot{E}_{in} - \dot{E}_{out}}_{\substack{\text{Rate of net energy transfer}\\ \text{by heat, work, and mass}}} = \underbrace{dE_{system}/dt}_{\substack{\text{Rate of change in internal, kinetic,}\\ \text{potential, etc., energies}}}^{\nearrow 0\text{ (steady)}} = 0$$

$$\dot{E}_{in} = \dot{E}_{out}$$

$$\dot{m}h_1 = \dot{W}_{out} + \dot{m}h_2 \qquad (\text{since } \dot{Q} = 0,\ \text{ke} \cong \text{pe} \cong 0)$$

$$\dot{W}_{out} = \dot{m}(h_1 - h_2)$$

The inlet state is completely specified since two properties are given. But only one property (pressure) is given at the final state, and we need one more property to fix it. The second property comes from the observation that the process is reversible and adiabatic, and thus isentropic. Therefore, $s_2 = s_1$, and

$$\textit{State 1:} \quad \left.\begin{array}{l} P_1 = 5\text{ MPa} \\ T_1 = 450°\text{C} \end{array}\right\} \quad \begin{array}{l} h_1 = 3317.2\text{ kJ/kg} \\ s_1 = 6.8210\text{ kJ/kg·K} \end{array}$$

$$\textit{State 2:} \quad \left.\begin{array}{l} P_2 = 1.4\text{ MPa} \\ s_2 = s_1 \end{array}\right\} \quad h_2 = 2967.4\text{ kJ/kg}$$

Then, the work output of the turbine per unit mass of the steam becomes

$$w_{out} = h_1 - h_2 = 3317.2 - 2967.4 = \mathbf{349.8\ kJ/kg}$$

7–5 ▪ PROPERTY DIAGRAMS INVOLVING ENTROPY

Property diagrams serve as great visual aids in the thermodynamic analysis of processes. We have used *P-v* and *T-v* diagrams extensively in previous chapters in conjunction with the first law of thermodynamics. In the second-law analysis, it is very helpful to plot the processes on diagrams for which one of the coordinates is entropy. The two diagrams commonly used in the second-law analysis are the *temperature-entropy* and the *enthalpy-entropy* diagrams.

Consider the defining equation of entropy (Eq. 7–4). It can be rearranged as

$$\delta Q_{\text{int rev}} = T\,dS \qquad \text{(kJ)} \tag{7–14}$$

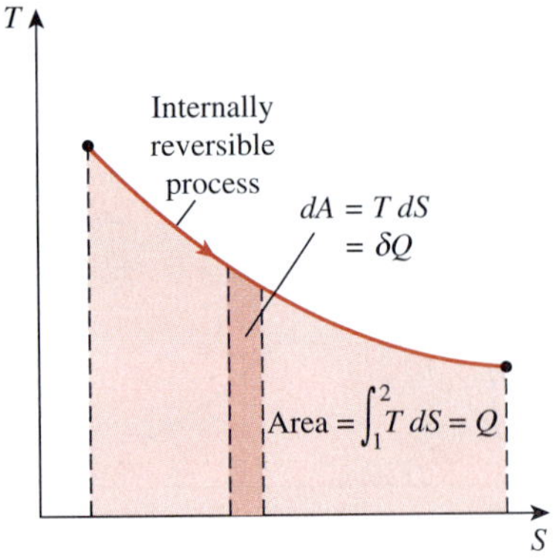

FIGURE 7–16
On a *T-S* diagram, the area under the process curve represents the heat transfer for internally reversible processes.

As shown in Fig. 7–16, $\delta Q_{\text{rev int}}$ corresponds to a differential area on a *T-S* diagram. The total heat transfer during an internally reversible process is determined by integration to be

$$Q_{\text{int rev}} = \int_1^2 T\,dS \qquad \text{(kJ)} \tag{7–15}$$

which corresponds to the area under the process curve on a *T-S* diagram. Therefore, we conclude that *the area under the process curve on a T-S diagram represents heat transfer during an internally reversible process.* This is somewhat analogous to reversible boundary work being represented by the area under the process curve on a *P-V* diagram. Note that the area under the process curve represents heat transfer for processes that are internally (or totally) reversible. The area has no meaning for irreversible processes.

Equations 7–14 and 7–15 can also be expressed on a unit-mass basis as

$$\delta q_{\text{int rev}} = T\,ds \qquad \text{(kJ/kg)} \tag{7–16}$$

and

$$q_{\text{int rev}} = \int_1^2 T\,ds \qquad \text{(kJ/kg)} \tag{7–17}$$

To perform the integrations in Eqs. 7–15 and 7–17, one needs to know the relationship between T and s during a process. One special case for which these integrations can be performed easily is the *internally reversible isothermal process*. It yields

$$Q_{\text{int rev}} = T_0\,\Delta S \qquad \text{(kJ)} \tag{7–18}$$

or

$$q_{\text{int rev}} = T_0\,\Delta s \qquad \text{(kJ/kg)} \tag{7–19}$$

where T_0 is the constant temperature and ΔS is the entropy change of the system during the process.

An isentropic process on a *T-s* diagram is easily recognized as a *vertical-line segment*. This is expected since an isentropic process involves no heat transfer, and therefore the area under the process path must be zero (Fig. 7–17). The *T-s* diagrams serve as valuable tools for visualizing the second-law aspects of processes and cycles, and thus they are frequently used in thermodynamics. The *T-s* diagram of water is given in the appendix in Fig. A–9.

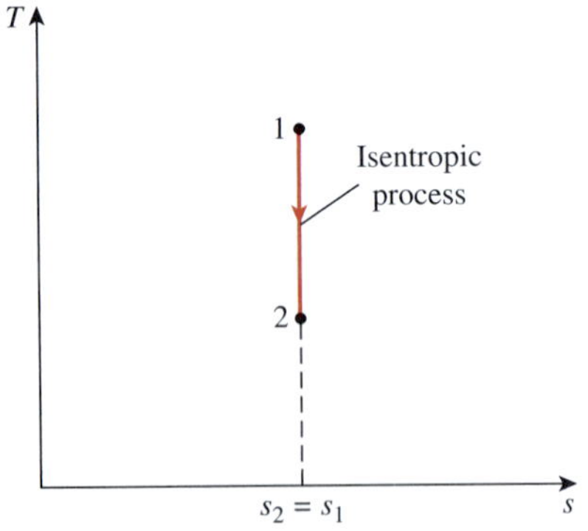

FIGURE 7–17
The isentropic process appears as a *vertical* line segment on a *T-s* diagram.

Another diagram commonly used in engineering is the enthalpy-entropy diagram, which is quite valuable in the analysis of steady-flow devices such as turbines, compressors, and nozzles. The coordinates of an *h-s* diagram represent two properties of major interest: enthalpy, which is a primary property in the first-law analysis of the steady-flow devices, and entropy, which is the property that accounts for irreversibilities during adiabatic processes. In analyzing the steady flow of steam through an adiabatic turbine, for example, the vertical distance between the inlet and the exit states Δh is a measure of the work output of the turbine, and the horizontal distance Δs is a measure of the irreversibilities associated with the process (Fig. 7–18).

The *h-s* diagram is also called a **Mollier diagram** after the German scientist R. Mollier (1863–1935). An *h-s* diagram is given in the appendix for steam in Fig. A–10.

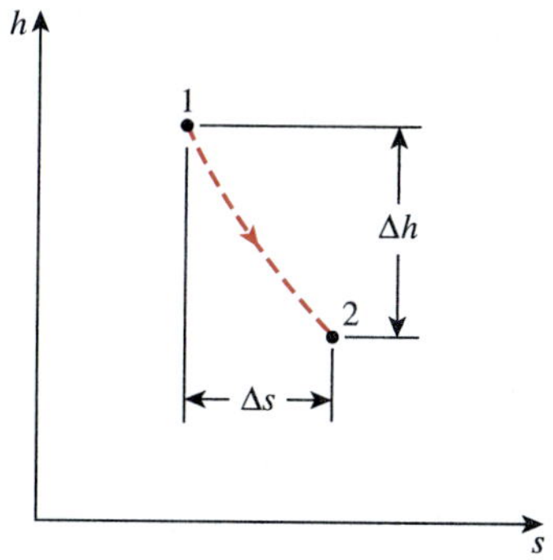

FIGURE 7–18
For adiabatic steady-flow devices, the vertical distance Δh on an *h-s* diagram is a measure of work, and the horizontal distance Δs is a measure of irreversibilities.

EXAMPLE 7–6 The *T-S* Diagram of the Carnot Cycle

Show the Carnot cycle on a *T-S* diagram and indicate the areas that represent the heat supplied Q_H, heat rejected Q_L, and the net work output $W_{net,out}$ on this diagram.

SOLUTION The Carnot cycle is to be shown on a *T-S* diagram, and the areas that represent Q_H, Q_L, and $W_{net,out}$ are to be indicated.

Analysis Recall that the Carnot cycle is made up of two reversible isothermal (T = constant) processes and two isentropic (s = constant) processes. These four processes form a rectangle on a *T-S* diagram, as shown in Fig. 7–19.

On a *T-S* diagram, the area under the process curve represents the heat transfer for that process. Thus the area *A*12*B* represents Q_H, the area *A*43*B* represents Q_L, and the difference between these two (the area in color) represents the net work since

$$W_{net,out} = Q_H - Q_L$$

Therefore, the area enclosed by the path of a cycle (area 1234) on a *T-S* diagram represents the net work. Recall that the area enclosed by the path of a cycle also represents the net work on a *P-V* diagram.

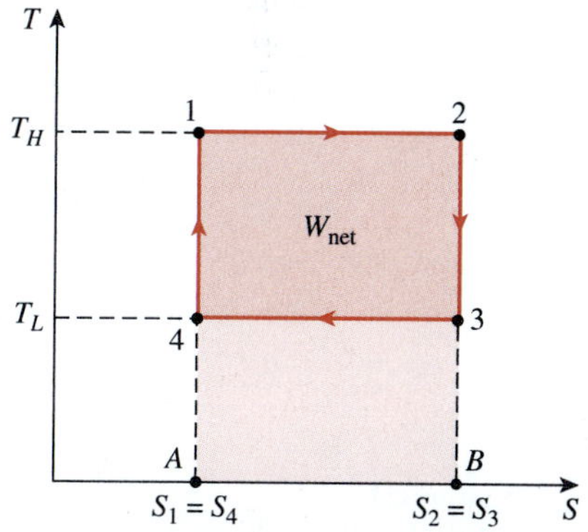

FIGURE 7–19
The *T-S* diagram of a Carnot cycle (Example 7–6).

7–6 ▪ WHAT IS ENTROPY?

It is clear from the previous discussion that entropy is a useful property and serves as a valuable tool in the second-law analysis of engineering devices. But this does not mean that we know and understand entropy well. Because we do not. In fact, we cannot even give an adequate answer to the question, what is entropy? Not being able to describe entropy fully, however, does not take anything away from its usefulness. We could not define *energy* either, but it did not interfere with our understanding of energy transformations and the conservation of energy principle. Granted, entropy is not a household word like energy. But with continued use, our understanding of entropy will deepen, and our appreciation of it will grow. The next discussion should shed some light on the physical meaning of entropy by considering the microscopic nature of matter.

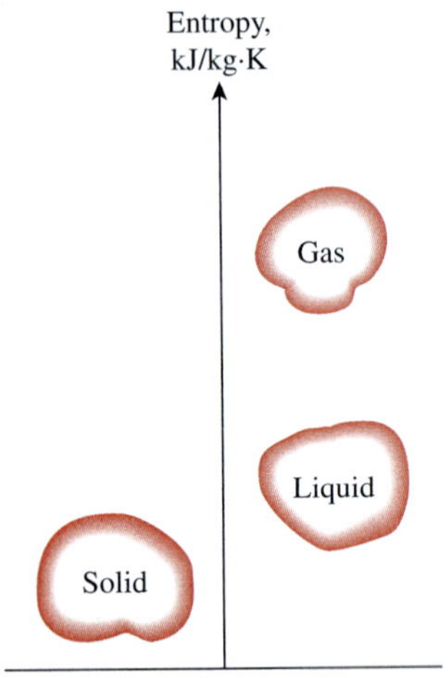

FIGURE 7–20
The level of molecular disorder (entropy) of a substance increases as it melts or evaporates.

Entropy can be viewed as a measure of *molecular disorder*, or *molecular randomness*. As a system becomes more disordered, the positions of the molecules become less predictable and the entropy increases. Thus, it is not surprising that the entropy of a substance is lowest in the solid phase and highest in the gas phase (Fig. 7–20). In the solid phase, the molecules of a substance continually oscillate about their equilibrium positions, but they cannot move relative to each other, and their position at any instant can be predicted with good certainty. In the gas phase, however, the molecules move about at random, collide with each other, and change direction, making it extremely difficult to predict accurately the microscopic state of a system at any instant. Associated with this molecular chaos is a high value of entropy.

When viewed microscopically (from a statistical thermodynamics point of view), an isolated system that appears to be at a state of equilibrium may exhibit a high level of activity because of the continual motion of the molecules. To each state of macroscopic equilibrium there corresponds a large number of possible microscopic states or molecular configurations. The entropy of a system is related to the total number of possible microscopic states of that system, called *thermodynamic probability p*, by the **Boltzmann relation**, expressed as

$$S = k \ln p \tag{7–20}$$

where $k = 1.3806 \times 10^{-23}$ J/K is the **Boltzmann constant**. Therefore, from a microscopic point of view, the entropy of a system increases whenever the molecular randomness or uncertainty (i.e., molecular probability) of a system increases. Thus, entropy is a measure of molecular disorder, and the molecular disorder of an isolated system increases anytime it undergoes a process.

As mentioned earlier, the molecules of a substance in solid phase continually oscillate, creating an uncertainty about their position. These oscillations, however, fade as the temperature is decreased, and the molecules supposedly become motionless at absolute zero. This represents a state of ultimate molecular order (and minimum energy). Therefore, *the entropy of a pure crystalline substance at absolute zero temperature is zero* since there is no uncertainty about the state of the molecules at that instant (Fig. 7–21). This statement is known as the **third law of thermodynamics**. The third law of thermodynamics provides an absolute reference point for the determination of entropy. The entropy determined relative to this point is called **absolute entropy**, and it is extremely useful in the thermodynamic analysis of chemical reactions. Notice that the entropy of a substance that is not pure crystalline (such as a solid solution) is not zero at absolute zero temperature. This is because more than one molecular configuration exists for such substances, which introduces some uncertainty about the microscopic state of the substance.

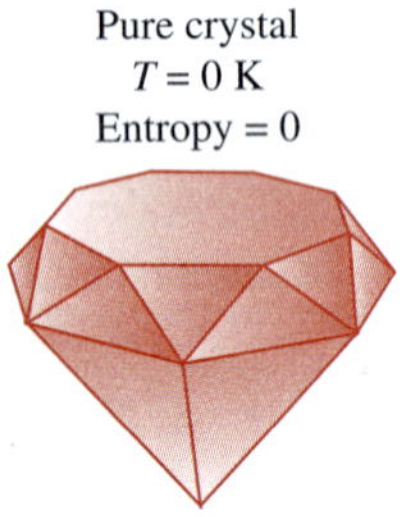

FIGURE 7–21
A pure crystalline substance at absolute zero temperature is in perfect order, and its entropy is zero (the third law of thermodynamics).

Molecules in the gas phase possess a considerable amount of kinetic energy. However, we know that no matter how large their kinetic energies are, the gas molecules do not rotate a paddle wheel inserted into the container and produce work. This is because the gas molecules, and the energy they possess, are disorganized. Probably the number of molecules trying to rotate the wheel in one direction at any instant is equal to the number of molecules that are trying to rotate it in the opposite direction, causing the wheel to remain motionless. Therefore, we cannot extract any useful work directly from disorganized energy (Fig. 7–22).

Now consider a rotating shaft shown in Fig. 7–23. This time the energy of the molecules is completely organized since the molecules of the shaft are rotating in the same direction together. This organized energy can readily be used to perform useful tasks such as raising a weight or generating electricity. Being an organized form of energy, work is free of disorder or randomness and thus free of entropy. *There is no entropy transfer associated with energy transfer as work.* Therefore, in the absence of any friction, the process of raising a weight by a rotating shaft (or a flywheel) does not produce any entropy. Any process that does not produce a net entropy is reversible, and thus the process just described can be reversed by lowering the weight. Therefore, energy is not degraded during this process, and no potential to do work is lost.

Instead of raising a weight, let us operate the paddle wheel in a container filled with a gas, as shown in Fig. 7–24. The paddle-wheel work in this case is converted to the internal energy of the gas, as evidenced by a rise in gas temperature, creating a higher level of molecular disorder in the container. This process is quite different from raising a weight since the organized paddle-wheel energy is now converted to a highly disorganized form of energy, which cannot be converted back to the paddle wheel as the rotational kinetic energy. Only a portion of this energy can be converted to work by partially reorganizing it through the use of a heat engine. Therefore, energy is degraded during this process, the ability to do work is reduced, molecular disorder is produced, and associated with all this is an increase in entropy.

The *quantity* of energy is always preserved during an actual process (the first law), but the *quality* is bound to decrease (the second law). This decrease in quality is always accompanied by an increase in entropy. As an example, consider the transfer of 10 kJ of energy as heat from a hot medium to a cold one. At the end of the process, we still have the 10 kJ of energy, but at a lower temperature and thus at a lower quality.

Heat is, in essence, a form of *disorganized energy*, and some disorganization (entropy) flows with heat (Fig. 7–25). As a result, the entropy and the level of molecular disorder or randomness of the hot body decreases with the entropy and the level of molecular disorder of the cold body increases. The second law requires that the increase in entropy of the cold body be greater than the decrease in entropy of the hot body, and thus the net entropy of the combined system (the cold body and the hot body) increases. That is, the combined system is at a state of greater disorder at the final state. Thus we can conclude that processes can occur only in the direction of increased overall entropy or molecular disorder. That is, the entire universe is getting more and more chaotic every day.

Entropy and Entropy Generation in Daily Life

The concept of entropy can also be applied to other areas. Entropy can be viewed as a measure of disorder or disorganization in a system. Likewise, entropy generation can be viewed as a measure of disorder or disorganization generated during a process. The concept of entropy is not used in daily life nearly as extensively as the concept of energy, even though entropy is

FIGURE 7–22
Disorganized energy does not create much useful effect, no matter how large it is.

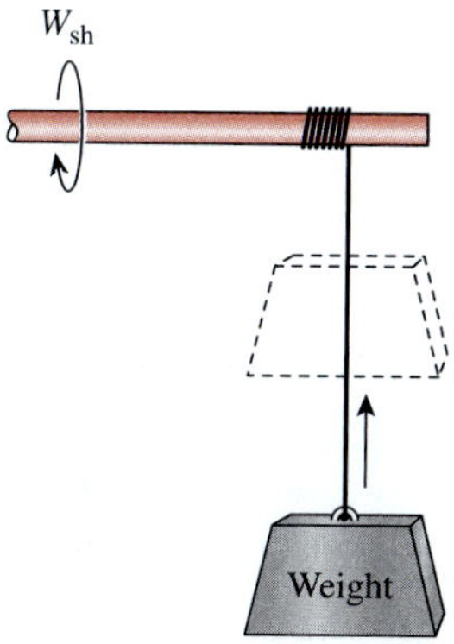

FIGURE 7–23
In the absence of friction, raising a weight by a rotating shaft does not create any disorder (entropy), and thus energy is not degraded during this process.

FIGURE 7–24
The paddle-wheel work done on a gas increases the level of disorder (entropy) of the gas, and thus energy is degraded during this process.

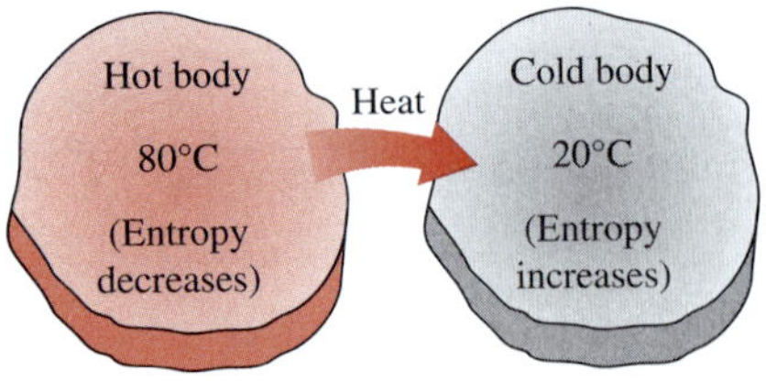

FIGURE 7–25
During a heat transfer process, the net entropy increases. (The increase in the entropy of the cold body more than offsets the decrease in the entropy of the hot body.)

FIGURE 7–26
The use of entropy (disorganization, uncertainty) is not limited to thermodynamics.
BLONDIE © KING FEATURES SYNDICATE.

readily applicable to various aspects of daily life. The extension of the entropy concept to nontechnical fields is not a novel idea. It has been the topic of several articles, and even some books. Next, we present several ordinary events and show their relevance to the concept of entropy and entropy generation.

Efficient people lead low-entropy (highly organized) lives. They have a place for everything (minimum uncertainty), and it takes minimum energy for them to locate something. Inefficient people, on the other hand, are disorganized and lead high-entropy lives. It takes them minutes (if not hours) to find something they need, and they are likely to create a bigger disorder as they are searching since they will probably conduct the search in a disorganized manner (Fig. 7–26). People leading high-entropy lifestyles are always on the run, and never seem to catch up.

You probably noticed (with frustration) that some people seem to *learn* fast and remember well what they learn. We can call this type of learning organized or low-entropy learning. These people make a conscientious effort to file the new information properly by relating it to their existing knowledge base and creating a solid information network in their minds. On the other hand, people who throw the information into their minds as they study, with no effort to secure it, may *think* they are learning. They are bound to discover otherwise when they need to locate the information, for example, during a test. It is not easy to retrieve information from a database that is, in a sense, in the gas phase. Students who have blackouts during tests should reexamine their study habits.

A *library* with a good shelving and indexing system can be viewed as a low-entropy library because of the high level of organization. Likewise, a library with a poor shelving and indexing system can be viewed as a high-entropy library because of the high level of disorganization. A library with no indexing system is like no library, since a book is of no value if it cannot be found.

Consider two identical buildings, each containing one million books. In the first building, the books are *piled* on top of each other, whereas in the second building they are *highly organized, shelved, and indexed* for easy reference. There is no doubt about which building a student will prefer to go to for checking out a certain book. Yet, some may argue from the first-law point of view that these two buildings are equivalent since the mass and knowledge content of the two buildings are identical, despite the high level of disorganization (entropy) in the first building. This example illustrates that any realistic comparisons should involve the second-law point of view.

Two *textbooks* that seem to be identical because both cover basically the same topics and present the same information may actually be *very* different depending on *how* they cover the topics. After all, two seemingly identical cars are not so identical if one goes only half as many miles as the other one on the same amount of fuel. Likewise, two seemingly identical books are not so identical if it takes twice as long to learn a topic from one of them as it does from the other. Thus, comparisons made on the basis of the first law only may be highly misleading.

Having a disorganized (high-entropy) *army* is like having no army at all. It is no coincidence that the command centers of any armed forces are among the primary targets during a war. One army that consists of

10 divisions is 10 times more powerful than 10 armies each consisting of a single division. Likewise, one country that consists of 10 states is more powerful than 10 countries, each consisting of a single state. The *United States* would not be such a powerful country if there were 50 independent countries in its place instead of a single country with 50 states. The European Union has the potential to be a new economic and political superpower. The old cliché "divide and conquer" can be rephrased as "increase the entropy and conquer."

We know that mechanical friction is always accompanied by entropy generation, and thus reduced performance. We can generalize this to daily life: *friction in the workplace* with fellow workers is bound to generate entropy, and thus adversely affect performance (Fig. 7–27). It results in reduced productivity.

We also know that *unrestrained expansion* (or explosion) and uncontrolled electron exchange (chemical reactions) generate entropy and are highly irreversible. Likewise, unrestrained opening of the mouth to scatter angry words is highly irreversible since this generates entropy, and it can cause considerable damage. A person who gets up in anger is bound to sit down at a loss. Hopefully, someday we will be able to come up with some procedures to quantify entropy generated during nontechnical activities, and maybe even pinpoint its primary sources and magnitude.

FIGURE 7–27
As in mechanical systems, friction in the workplace is bound to generate entropy and reduce performance.

7–7 ▪ THE *T ds* RELATIONS

Recall that the quantity $(\delta Q/T)_{\text{int rev}}$ corresponds to a differential change in the property *entropy*. The entropy change for a process, then, can be evaluated by integrating $\delta Q/T$ along some imaginary internally reversible path between the actual end states. For isothermal internally reversible processes, this integration is straightforward. But when the temperature varies during the process, we have to have a relation between δQ and T to perform this integration. Finding such relations is what we intend to do in this section.

The differential form of the conservation of energy equation for a closed stationary system (a fixed mass) containing a simple compressible substance can be expressed for an internally reversible process as

$$\delta Q_{\text{int rev}} - \delta W_{\text{int rev,out}} = dU \tag{7–21}$$

But

$$\delta Q_{\text{int rev}} = T\,dS$$

$$\delta W_{\text{int rev,out}} = P\,dV$$

Thus,

$$T\,dS = dU + P\,dV \qquad \text{(kJ)} \tag{7–22}$$

or

$$T\,ds = du + P\,dv \qquad \text{(kJ/kg)} \tag{7–23}$$

This equation is known as the first *T ds*, or *Gibbs, equation*. Notice that the only type of work interaction a simple compressible system may involve as it undergoes an internally reversible process is the boundary work.

The second $T\,ds$ equation is obtained by eliminating du from Eq. 7–23 by using the definition of enthalpy ($h = u + Pv$):

$$\left.\begin{array}{lcl} h = u + Pv & \longrightarrow & dh = du + P\,dv + v\,dP \\ \text{(Eq. 7–23)} & \longrightarrow & T\,ds = du + P\,dv \end{array}\right\} T\,ds = dh - v\,dP \qquad \textbf{(7–24)}$$

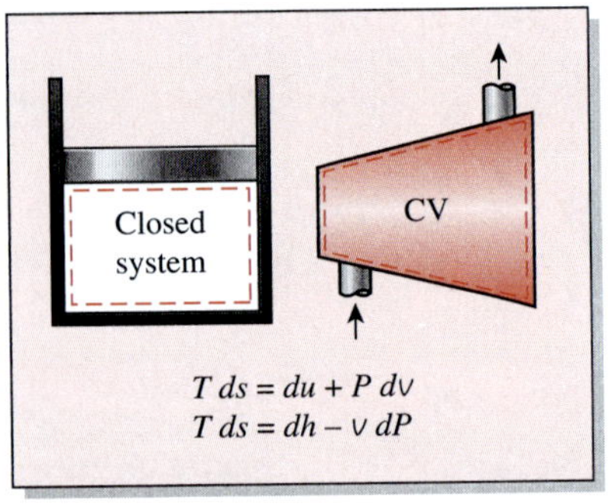

FIGURE 7–28
The $T\,ds$ relations are valid for both reversible and irreversible processes and for both closed and open systems.

Equations 7–23 and 7–24 are extremely valuable since they relate entropy changes of a system to the changes in other properties. Unlike Eq. 7–4, they are property relations and therefore are independent of the type of the processes.

These $T\,ds$ relations are developed with an internally reversible process in mind since the entropy change between two states must be evaluated along a reversible path. However, the results obtained are valid for both reversible and irreversible processes since entropy is a property and the change in a property between two states is independent of the type of process the system undergoes. Equations 7–23 and 7–24 are relations between the properties of a unit mass of a simple compressible system as it undergoes a change of state, and they are applicable whether the change occurs in a closed or an open system (Fig. 7–28).

Explicit relations for differential changes in entropy are obtained by solving for ds in Eqs. 7–23 and 7–24:

$$ds = \frac{du}{T} + \frac{P\,dv}{T} \qquad \textbf{(7–25)}$$

and

$$ds = \frac{dh}{T} - \frac{v\,dP}{T} \qquad \textbf{(7–26)}$$

The entropy change during a process can be determined by integrating either of these equations between the initial and the final states. To perform these integrations, however, we must know the relationship between du or dh and the temperature (such as $du = c_v\,dT$ and $dh = c_p\,dT$ for ideal gases) as well as the equation of state for the substance (such as the ideal-gas equation of state $Pv = RT$). For substances for which such relations exist, the integration of Eq. 7–25 or 7–26 is straightforward. For other substances, we have to rely on tabulated data.

The $T\,ds$ relations for nonsimple systems, that is, systems that involve more than one mode of quasi-equilibrium work, can be obtained in a similar manner by including all the relevant quasi-equilibrium work modes.

7–8 ▪ ENTROPY CHANGE OF LIQUIDS AND SOLIDS

Recall that liquids and solids can be approximated as *incompressible substances* since their specific volumes remain nearly constant during a process. Thus, $dv \cong 0$ for liquids and solids, and Eq. 7–25 for this case reduces to

$$ds = \frac{du}{T} = \frac{c\,dT}{T} \qquad \textbf{(7–27)}$$

since $c_p = c_v = c$ and $du = c\,dT$ for incompressible substances. Then the entropy change during a process is determined by integration to be

Liquids, solids: $$s_2 - s_1 = \int_1^2 c(T)\frac{dT}{T} \cong c_{avg} \ln \frac{T_2}{T_1} \qquad \text{(kJ/kg·K)} \qquad \textbf{(7–28)}$$

where c_{avg} is the *average* specific heat of the substance over the given temperature interval. Note that the entropy change of a truly incompressible substance depends on temperature only and is independent of pressure.

Equation 7–28 can be used to determine the entropy changes of solids and liquids with reasonable accuracy. However, for liquids that expand considerably with temperature, it may be necessary to consider the effects of volume change in calculations. This is especially the case when the temperature change is large.

A relation for isentropic processes of liquids and solids is obtained by setting the entropy change relation above equal to zero. It gives

Isentropic: $$s_2 - s_1 = c_{avg} \ln \frac{T_2}{T_1} = 0 \quad \longrightarrow \quad T_2 = T_1 \qquad \textbf{(7–29)}$$

That is, the temperature of a truly incompressible substance remains constant during an isentropic process. Therefore, the isentropic process of an incompressible substance is also isothermal. This behavior is closely approximated by liquids and solids.

EXAMPLE 7–7 Effect of Density of a Liquid on Entropy

Liquid methane is commonly used in various cryogenic applications. The critical temperature of methane is 191 K (or −82°C), and thus methane must be maintained below 191 K to keep it in liquid phase. The properties of liquid methane at various temperatures and pressures are given in Table 7–1. Determine the entropy change of liquid methane as it undergoes a process from 110 K and 1 MPa to 120 K and 5 MPa (*a*) using tabulated properties and (*b*) approximating liquid methane as an incompressible substance. What is the error involved in the latter case?

TABLE 7–1

Properties of liquid methane

Temp., T, K	Pressure, P, MPa	Density, ρ, kg/m^3	Enthalpy, h, kJ/kg	Entropy, s, kJ/kg·K	Specific heat, c_p, kJ/kg·K
110	0.5	425.3	208.3	4.878	3.476
	1.0	425.8	209.0	4.875	3.471
	2.0	426.6	210.5	4.867	3.460
	5.0	429.1	215.0	4.844	3.432
120	0.5	410.4	243.4	5.185	3.551
	1.0	411.0	244.1	5.180	3.543
	2.0	412.0	245.4	5.171	3.528
	5.0	415.2	249.6	5.145	3.486

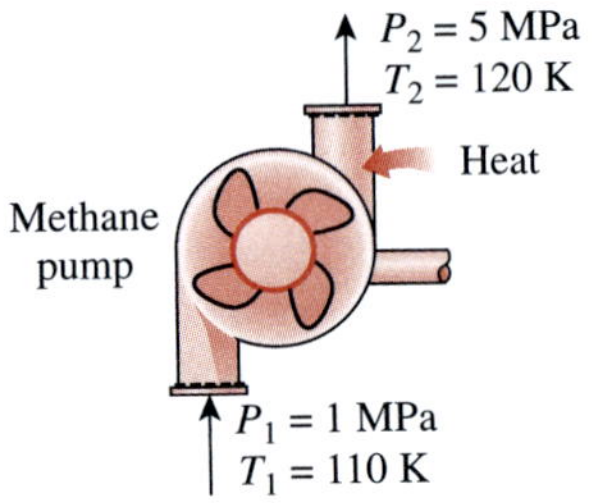

FIGURE 7–29
Schematic for Example 7–7.

SOLUTION Liquid methane undergoes a process between two specified states. The entropy change of methane is to be determined by using actual data and by assuming methane to be incompressible.

Analysis (*a*) We consider a unit mass of liquid methane (Fig. 7–29). The properties of the methane at the initial and final states are

$$\textit{State 1:}\quad \left.\begin{aligned} P_1 &= 1 \text{ MPa} \\ T_1 &= 110 \text{ K} \end{aligned}\right\} \quad \begin{aligned} s_1 &= 4.875 \text{ kJ/kg·K} \\ c_{p1} &= 3.471 \text{ kJ/kg·K} \end{aligned}$$

$$\textit{State 2:}\quad \left.\begin{aligned} P_2 &= 5 \text{ MPa} \\ T_2 &= 120 \text{ K} \end{aligned}\right\} \quad \begin{aligned} s_2 &= 5.145 \text{ kJ/kg·K} \\ c_{p2} &= 3.486 \text{ kJ/kg·K} \end{aligned}$$

Therefore,

$$\Delta s = s_2 - s_1 = 5.145 - 4.875 = \mathbf{0.270\ kJ/kg·K}$$

(*b*) Approximating liquid methane as an incompressible substance, its entropy change is determined to be

$$\Delta s = c_{avg} \ln \frac{T_2}{T_1} = (3.4785 \text{ kJ/kg·K}) \ln \frac{120 \text{ K}}{110 \text{ K}} = \mathbf{0.303\ kJ/kg·K}$$

since

$$c_{avg} = \frac{c_{p1} + c_{p2}}{2} = \frac{3.471 + 3.486}{2} = 3.4785 \text{ kJ/kg·K}$$

Therefore, the error involved in approximating liquid methane as an incompressible substance is

$$\text{Error} = \frac{|\Delta s_{actual} - \Delta s_{ideal}|}{\Delta s_{actual}} = \frac{|0.270 - 0.303|}{0.270} = \mathbf{0.122\ (or\ 12.2\%)}$$

Discussion This result is not surprising since the density of liquid methane changes during this process from 425.8 to 415.2 kg/m^3 (about 3 percent), which makes us question the validity of the incompressible substance assumption. Still, this assumption enables us to obtain reasonably accurate results with less effort, which proves to be very convenient in the absence of compressed liquid data.

EXAMPLE 7–8 Economics of Replacing a Valve by a Turbine

A cryogenic manufacturing facility handles liquid methane at 115 K and 5 MPa at a rate of 0.280 m^3/s . A process requires dropping the pressure of liquid methane to 1 MPa, which is done by throttling the liquid methane by passing it through a flow resistance such as a valve. A recently hired engineer proposes to replace the throttling valve by a turbine in order to produce power while dropping the pressure to 1 MPa. Using data from Table 7–1, determine the maximum amount of power that can be produced by such a turbine. Also, determine how much this turbine will save the facility from electricity usage costs per year if the turbine operates continuously (8760 h/yr) and the facility pays $0.075/kWh for electricity.

SOLUTION Liquid methane is expanded in a turbine to a specified pressure at a specified rate. The maximum power that this turbine can produce and the amount of money it can save per year are to be determined.

Assumptions **1** This is a steady-flow process since there is no change with time at any point and thus $\Delta m_{CV} = 0$, $\Delta E_{CV} = 0$, and $\Delta S_{CV} = 0$. **2** The turbine is adiabatic and thus there is no heat transfer. **3** The process is reversible. **4** Kinetic and potential energies are negligible.

Analysis We take the *turbine* as the system (Fig. 7–30). This is a *control volume* since mass crosses the system boundary during the process. We note that there is only one inlet and one exit and thus $\dot{m}_1 = \dot{m}_2 = \dot{m}$.

The assumptions above are reasonable since a turbine is normally well insulated and it must involve no irreversibilities for best performance and thus *maximum* power production. Therefore, the process through the turbine must be *reversible adiabatic* or *isentropic*. Then, $s_2 = s_1$ and

$$\textit{State 1:} \quad \left.\begin{array}{l} P_1 = 5 \text{ MPa} \\ T_1 = 115 \text{ K} \end{array}\right\} \quad \begin{array}{l} h_1 = 232.3 \text{ kJ/kg} \\ s_1 = 4.9945 \text{ kJ/kg·K} \\ \rho_1 = 422.15 \text{ kg/s} \end{array}$$

$$\textit{State 2:} \quad \left.\begin{array}{l} P_2 = 1 \text{ MPa} \\ s_2 = s_1 \end{array}\right\} \quad h_2 = 222.8 \text{ kJ/kg}$$

Also, the mass flow rate of liquid methane is

$$\dot{m} = \rho_1 \dot{V}_1 = (422.15 \text{ kg/m}^3)(0.280 \text{ m}^3\text{/s}) = 118.2 \text{ kg/s}$$

Then the power output of the turbine is determined from the rate form of the energy balance to be

$$\underbrace{\dot{E}_{in} - \dot{E}_{out}}_{\substack{\text{Rate of net energy transfer} \\ \text{by heat, work, and mass}}} = \underbrace{dE_{system}/dt}_{\substack{\text{Rate of change in internal,} \\ \text{kinetic, potential, etc., energies}}} \overset{\nearrow 0 \text{ (steady)}}{} = 0$$

$$\dot{E}_{in} = \dot{E}_{out}$$

$$\dot{m}h_1 = \dot{W}_{out} + \dot{m}h_2 \qquad (\text{since } \dot{Q} = 0, \text{ ke} \cong \text{ pe} \cong 0)$$

$$\dot{W}_{out} = \dot{m}(h_1 - h_2)$$

$$= (118.2 \text{ kg/s})(232.3 - 222.8) \text{ kJ/kg}$$

$$= \mathbf{1123 \text{ kW}}$$

For continuous operation (365 × 24 = 8760 h), the amount of power produced per year is

$$\text{Annual power production} = \dot{W}_{out} \times \Delta t = (1123 \text{ kW})(8760 \text{ h/yr})$$

$$= 0.9837 \times 10^7 \text{ kWh/yr}$$

At \$0.075/kWh, the amount of money this turbine can save the facility is

$$\text{Annual power savings} = (\text{Annual power production})(\text{Unit cost of power})$$

$$= (0.9837 \times 10^7 \text{ kWh/yr})(\$0.075\text{/kWh})$$

$$= \mathbf{\$737{,}800/yr}$$

That is, this turbine can save the facility \$737,800 a year by simply taking advantage of the potential that is currently being wasted by a throttling valve, and the engineer who made this observation should be rewarded.

FIGURE 7–30
A 1.0-MW liquified natural gas (LNG) turbine with 95-cm turbine runner diameter being installed in a cryogenic test facility.
Courtesy of Ebara International Corporation, Cryodynamics Division, Sparks, Nevada.

Discussion This example shows the importance of the property entropy since it enabled us to quantify the work potential that is being wasted. In practice, the turbine will not be isentropic, and thus the power produced will be less. The analysis above gave us the upper limit. An actual turbine-generator assembly can utilize about 80 percent of the potential and produce more than 900 kW of power while saving the facility more than $600,000 a year.

It can also be shown that the temperature of methane drops to 113.9 K (a drop of 1.1 K) during the isentropic expansion process in the turbine instead of remaining constant at 115 K as would be the case if methane were assumed to be an incompressible substance. The temperature of methane would rise to 116.6 K (a rise of 1.6 K) during the throttling process.

7–9 ▪ THE ENTROPY CHANGE OF IDEAL GASES

An expression for the entropy change of an ideal gas can be obtained from Eq. 7–25 or 7–26 by employing the property relations for ideal gases (Fig. 7–31). By substituting $du = c_v\, dT$ and $P = RT/v$ into Eq. 7–25, the differential entropy change of an ideal gas becomes

$$ds = c_v \frac{dT}{T} + R\frac{dv}{v} \quad \textbf{(7–30)}$$

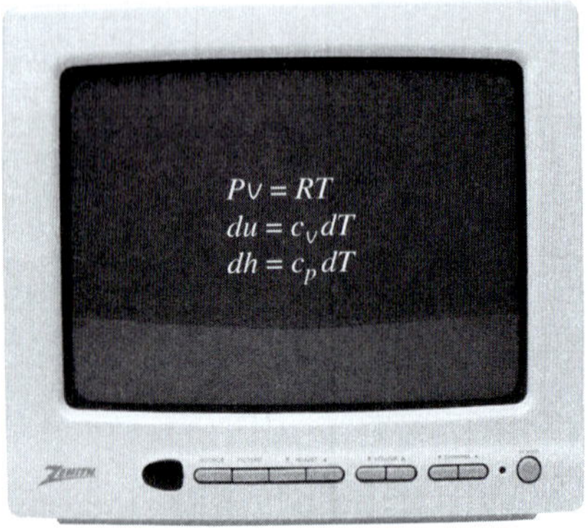

FIGURE 7–31
A broadcast from channel IG.
© Vol. 1/PhotoDisc/Getty RF.

The entropy change for a process is obtained by integrating this relation between the end states:

$$s_2 - s_1 = \int_1^2 c_v(T)\frac{dT}{T} + R \ln\frac{v_2}{v_1} \quad \textbf{(7–31)}$$

A second relation for the entropy change of an ideal gas is obtained in a similar manner by substituting $dh = c_p\, dT$ and $v = RT/P$ into Eq. 7–26 and integrating. The result is

$$s_2 - s_1 = \int_1^2 c_p(T)\,\frac{dT}{T} - R \ln\frac{P_2}{P_1} \quad \textbf{(7–32)}$$

The specific heats of ideal gases, with the exception of monatomic gases, depend on temperature, and the integrals in Eqs. 7–31 and 7–32 cannot be performed unless the dependence of c_v and c_p on temperature is known. Even when the $c_v(T)$ and $c_p(T)$ functions are available, performing long integrations every time entropy change is calculated is not practical. Then two reasonable choices are left: either perform these integrations by simply assuming constant specific heats or evaluate those integrals once and tabulate the results. Both approaches are presented next.

Constant Specific Heats (Approximate Analysis)

Assuming constant specific heats for ideal gases is a common approximation, and we used this assumption before on several occasions. It usually simplifies the analysis greatly, and the price we pay for this convenience is some loss in accuracy. The magnitude of the error introduced by this assumption depends on the situation at hand. For example, for monatomic ideal gases such as helium, the specific heats are independent of temperature, and therefore the constant-specific-heat assumption introduces no error. For ideal gases whose specific heats vary almost linearly in the temperature range of interest, the possible error is minimized by using specific heat values evaluated at the average temperature (Fig. 7–32). The results obtained in this way usually are sufficiently accurate if the temperature range is not greater than a few hundred degrees.

The entropy-change relations for ideal gases under the constant-specific-heat assumption are easily obtained by replacing $c_v(T)$ and $c_p(T)$ in Eqs. 7–31 and 7–32 by $c_{v,\text{avg}}$ and $c_{p,\text{avg}}$, respectively, and performing the integrations. We obtain

$$s_2 - s_1 = c_{v,\text{avg}} \ln \frac{T_2}{T_1} + R \ln \frac{v_2}{v_1} \quad \text{(kJ/kg·K)} \tag{7–33}$$

and

$$s_2 - s_1 = c_{p,\text{avg}} \ln \frac{T_2}{T_1} - R \ln \frac{P_2}{P_1} \quad \text{(kJ/kg·K)} \tag{7–34}$$

Entropy changes can also be expressed on a unit-mole basis by multiplying these relations by molar mass:

$$\bar{s}_2 - \bar{s}_1 = \bar{c}_{v,\text{avg}} \ln \frac{T_2}{T_1} + R_u \ln \frac{v_2}{v_1} \quad \text{(kJ/kmol·K)} \tag{7–35}$$

and

$$\bar{s}_2 - \bar{s}_1 = \bar{c}_{p,\text{avg}} \ln \frac{T_2}{T_1} - R_u \ln \frac{P_2}{P_1} \quad \text{(kJ/kmol·K)} \tag{7–36}$$

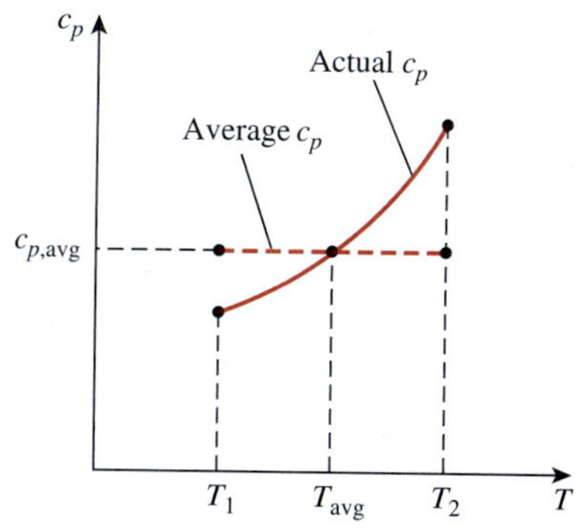

FIGURE 7–32
Under the constant-specific-heat assumption, the specific heat is assumed to be constant at some average value.

Variable Specific Heats (Exact Analysis)

When the temperature change during a process is large and the specific heats of the ideal gas vary nonlinearly within the temperature range, the assumption of constant specific heats may lead to considerable errors in entropy-change calculations. For those cases, the variation of specific heats with temperature should be properly accounted for by utilizing accurate relations for the specific heats as a function of temperature. The entropy change during a process is then determined by substituting these $c_v(T)$ or $c_p(T)$ relations into Eq. 7–31 or 7–32 and performing the integrations.

Instead of performing these laborious integrals each time we have a new process, it is convenient to perform these integrals once and tabulate the results. For this purpose, we choose absolute zero as the reference temperature and define a function $s°$ as

$$s° = \int_0^T c_p(T) \frac{dT}{T} \tag{7–37}$$

Obviously, $s°$ is a function of temperature alone, and its value is zero at absolute zero temperature. The values of $s°$ are calculated at various temperatures, and the results are tabulated in the appendix as a function of temperature for air. Given this definition, the integral in Eq. 7–32 becomes

$$\int_1^2 c_p(T) \frac{dT}{T} = s_2° - s_1° \tag{7–38}$$

T, K	$s°$, kJ/kg·K
⋮	⋮
300	1.70203
310	1.73498
320	1.76690
⋮	⋮

(Table A-17)

FIGURE 7–33
The entropy of an ideal gas depends on both T and P. The function $s°$ represents only the temperature-dependent part of entropy.

where s_2° is the value of $s°$ at T_2 and s_1° is the value at T_1. Thus,

$$s_2 - s_1 = s_2^\circ - s_1^\circ - R \ln \frac{P_2}{P_1} \qquad \text{(kJ/kg·K)} \tag{7–39}$$

It can also be expressed on a unit-mole basis as

$$\bar{s}_2 - \bar{s}_1 = \bar{s}_2^\circ - \bar{s}_1^\circ - R_u \ln \frac{P_2}{P_1} \qquad \text{(kJ/kmol·K)} \tag{7–40}$$

Note that unlike internal energy and enthalpy, the entropy of an ideal gas varies with specific volume or pressure as well as the temperature. Therefore, entropy cannot be tabulated as a function of temperature alone. The $s°$ values in the tables account for the temperature dependence of entropy (Fig. 7–33). The variation of entropy with pressure is accounted for by the last term in Eq. 7–39. Another relation for entropy change can be developed based on Eq. 7–31, but this would require the definition of another function and tabulation of its values, which is not practical.

EXAMPLE 7–9 Entropy Change of an Ideal Gas

Air is compressed from an initial state of 100 kPa and 17°C to a final state of 600 kPa and 57°C. Determine the entropy change of air during this compression process by using (*a*) property values from the air table and (*b*) average specific heats.

SOLUTION Air is compressed between two specified states. The entropy change of air is to be determined by using tabulated property values and also by using average specific heats.
Assumptions Air is an ideal gas since it is at a high temperature and low pressure relative to its critical-point values. Therefore, entropy change relations developed under the ideal-gas assumption are applicable.
Analysis A sketch of the system and the *T-s* diagram for the process are given in Fig. 7–34. We note that both the initial and the final states of air are completely specified.

(*a*) The properties of air are given in the air table (Table A–17). Reading $s°$ values at given temperatures and substituting, we find

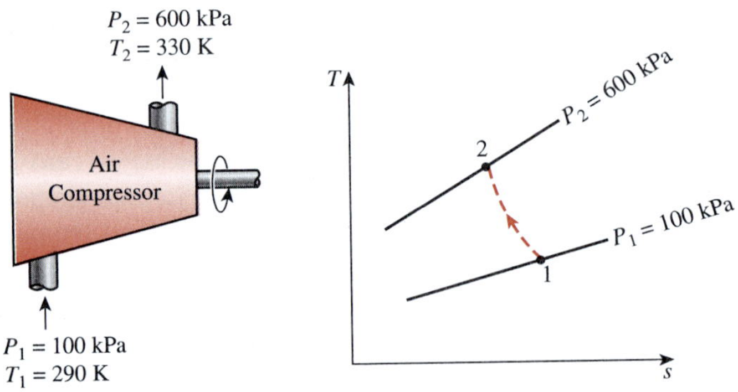

FIGURE 7–34
Schematic and *T-s* diagram for Example 7–9.

$$s_2 - s_1 = s_2^\circ - s_1^\circ - R \ln \frac{P_2}{P_1}$$

$$= [(1.79783 - 1.66802)\ \text{kJ/kg·K}] - (0.287\ \text{kJ/kg·K}) \ln \frac{600\ \text{kPa}}{100\ \text{kPa}}$$

$$= \mathbf{-0.3844\ kJ/kg·K}$$

(*b*) The entropy change of air during this process can also be determined approximately from Eq. 7–34 by using a c_p value at the average temperature of 37°C (Table A–2b) and treating it as a constant:

$$s_2 - s_1 = c_{p,\text{avg}} \ln \frac{T_2}{T_1} - R \ln \frac{P_2}{P_1}$$

$$= (1.006\ \text{kJ/kg·K}) \ln \frac{330\ \text{K}}{290\ \text{K}} - (0.287\ \text{kJ/kg·K}) \ln \frac{600\ \text{kPa}}{100\ \text{kPa}}$$

$$= \mathbf{-0.3842\ kJ/kg·K}$$

Discussion The two results above are almost identical since the change in temperature during this process is relatively small (Fig. 7–35). When the temperature change is large, however, they may differ significantly. For those cases, Eq. 7–39 should be used instead of Eq. 7–34 since it accounts for the variation of specific heats with temperature.

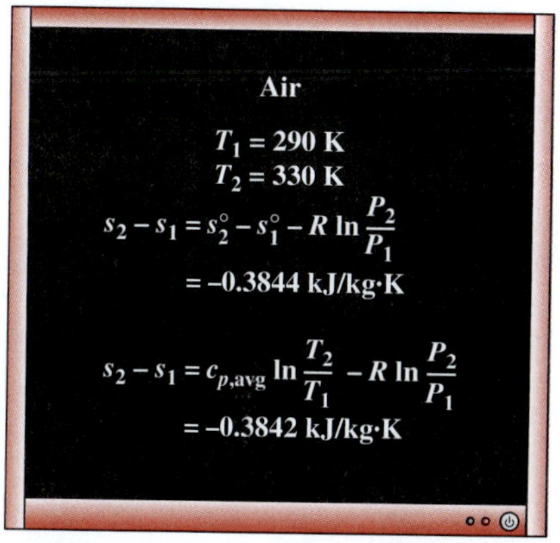

FIGURE 7–35
For small temperature differences, the exact and approximate relations for entropy changes of ideal gases give almost identical results.

Isentropic Processes of Ideal Gases

Several relations for the isentropic processes of ideal gases can be obtained by setting the entropy-change relations developed previously equal to zero. Again, this is done first for the case of constant specific heats and then for the case of variable specific heats.

Constant Specific Heats (Approximate Analysis)

When the constant-specific-heat assumption is valid, the isentropic relations for ideal gases are obtained by setting Eqs. 7–33 and 7–34 equal to zero. From Eq. 7–33,

$$\ln \frac{T_2}{T_1} = -\frac{R}{c_v} \ln \frac{v_2}{v_1}$$

which can be rearranged as

$$\ln \frac{T_2}{T_1} = \ln \left(\frac{v_1}{v_2}\right)^{R/c_v} \qquad \textbf{(7–41)}$$

or

$$\left(\frac{T_2}{T_1}\right)_{s=\text{const.}} = \left(\frac{v_1}{v_2}\right)^{k-1} \qquad \text{(ideal gas)} \qquad \textbf{(7–42)}$$

since $R = c_p - c_v$, $k = c_p/c_v$, and thus $R/c_v = k - 1$.

Equation 7–42 is the *first isentropic relation* for ideal gases under the constant-specific-heat assumption. The *second isentropic relation* is obtained in a similar manner from Eq. 7–34 with the following result:

$$\left(\frac{T_2}{T_1}\right)_{s=\text{const.}} = \left(\frac{P_2}{P_1}\right)^{(k-1)/k} \qquad \text{(ideal gas)} \qquad \textbf{(7–43)}$$

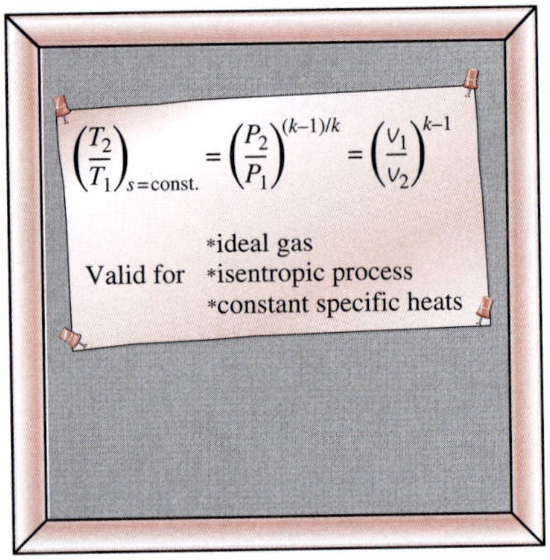

FIGURE 7–36
The isentropic relations of ideal gases are valid for the isentropic processes of ideal gases only.

The *third isentropic relation* is obtained by substituting Eq. 7–43 into Eq. 7–42 and simplifying:

$$\left(\frac{P_2}{P_1}\right)_{s=\text{const.}} = \left(\frac{v_1}{v_2}\right)^k \qquad \text{(ideal gas)} \tag{7–44}$$

Equations 7–42 through 7–44 can also be expressed in a compact form as

$$Tv^{k-1} = \text{constant} \tag{7–45}$$

$$TP^{(1-k)/k} = \text{constant} \qquad \text{(ideal gas)} \tag{7–46}$$

$$Pv^k = \text{constant} \tag{7–47}$$

The specific heat ratio k, in general, varies with temperature, and thus an average k value for the given temperature range should be used.

Note that the ideal-gas isentropic relations above, as the name implies, are strictly valid for isentropic processes only when the constant-specific-heat assumption is appropriate (Fig. 7–36).

Variable Specific Heats (Exact Analysis)

When the constant-specific-heat assumption is not appropriate, the isentropic relations developed previously yields results that are not quite accurate. For such cases, we should use an isentropic relation obtained from Eq. 7–39 that accounts for the variation of specific heats with temperature. Setting this equation equal to zero gives

$$0 = s_2^\circ - s_1^\circ - R \ln \frac{P_2}{P_1}$$

or

$$s_2^\circ = s_1^\circ + R \ln \frac{P_2}{P_1} \tag{7–48}$$

where s_2° is the s° value at the end of the isentropic process.

Relative Pressure and Relative Specific Volume

Equation 7–48 provides an accurate way of evaluating property changes of ideal gases during isentropic processes since it accounts for the variation of specific heats with temperature. However, it involves tedious iterations when the volume ratio is given instead of the pressure ratio. This is quite an inconvenience in optimization studies, which usually require numerous repetitive calculations. To remedy this deficiency, we define two new dimensionless quantities associated with isentropic processes.

The definition of the first is based on Eq. 7–48, which can be rearranged as

$$\frac{P_2}{P_1} = \exp \frac{s_2^\circ - s_1^\circ}{R}$$

or

$$\frac{P_2}{P_1} = \frac{\exp(s_2^\circ/R)}{\exp(s_1^\circ/R)}$$

The quantity $\exp(s^\circ/R)$ is defined as the **relative pressure** P_r. With this definition, the last relation becomes

$$\left(\frac{P_2}{P_1}\right)_{s=\text{const.}} = \frac{P_{r2}}{P_{r1}} \tag{7–49}$$

Note that the relative pressure P_r is a *dimensionless* quantity that is a function of temperature only since s° depends on temperature alone. Therefore, values of P_r can be tabulated against temperature. This is done for air in Table A–17. The use of P_r data is illustrated in Fig. 7–37.

Sometimes specific volume ratios are given instead of pressure ratios. This is particularly the case when automotive engines are analyzed. In such cases, one needs to work with volume ratios. Therefore, we define another quantity related to specific volume ratios for isentropic processes. This is done by utilizing the ideal-gas relation and Eq. 7–49:

$$\frac{P_1 v_1}{T_1} = \frac{P_2 v_2}{T_2} \rightarrow \frac{v_2}{v_1} = \frac{T_2}{T_1}\frac{P_1}{P_2} = \frac{T_2}{T_1}\frac{P_{r1}}{P_{r2}} = \frac{T_2/P_{r2}}{T_1/P_{r1}}$$

The quantity T/P_r is a function of temperature only and is defined as **relative specific volume** v_r. Thus,

$$\left(\frac{v_2}{v_1}\right)_{s=\text{ const.}} = \frac{v_{r2}}{v_{r1}} \tag{7–50}$$

Equations 7–49 and 7–50 are strictly valid for isentropic processes of ideal gases only. They account for the variation of specific heats with temperature and therefore give more accurate results than Eqs. 7–42 through 7–47. The values of P_r and v_r are listed for air in Table A–17.

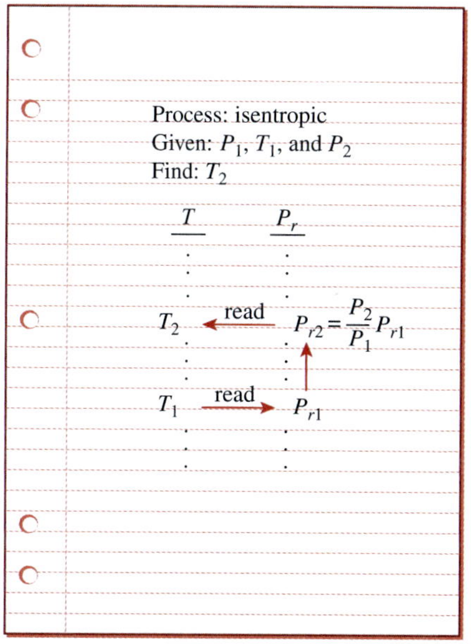

FIGURE 7–37
The use of P_r data for calculating the final temperature during an isentropic process.

EXAMPLE 7–10 Isentropic Compression of Air in a Car Engine

Air is compressed in a car engine from 22°C and 95 kPa in a reversible and adiabatic manner. If the compression ratio V_1/V_2 of this engine is 8, determine the final temperature of the air.

SOLUTION Air is compressed in a car engine isentropically. For a given compression ratio, the final air temperature is to be determined.

Assumptions At specified conditions, air can be treated as an ideal gas. Therefore, the isentropic relations for ideal gases are applicable.

Analysis A sketch of the system and the *T-s* diagram for the process are given in Fig. 7–38.

This process is easily recognized as being isentropic since it is both reversible and adiabatic. The final temperature for this isentropic process can be determined from Eq. 7–50 with the help of relative specific volume data (Table A–17), as illustrated in Fig. 7–39.

For closed systems: $\quad \frac{V_2}{V_1} = \frac{v_2}{v_1}$

At $T_1 = 295$ K: $\quad v_{r1} = 647.9$

From Eq. 7–50: $\quad v_{r2} = v_{r1}\left(\frac{v_2}{v_1}\right) = (647.9)\left(\frac{1}{8}\right) = 80.99 \rightarrow T_2 = \mathbf{662.7\ K}$

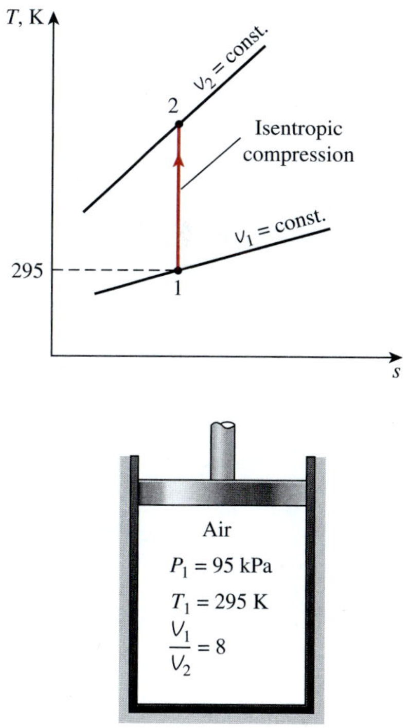

FIGURE 7–38
Schematic and *T-s* diagram for Example 7–10.

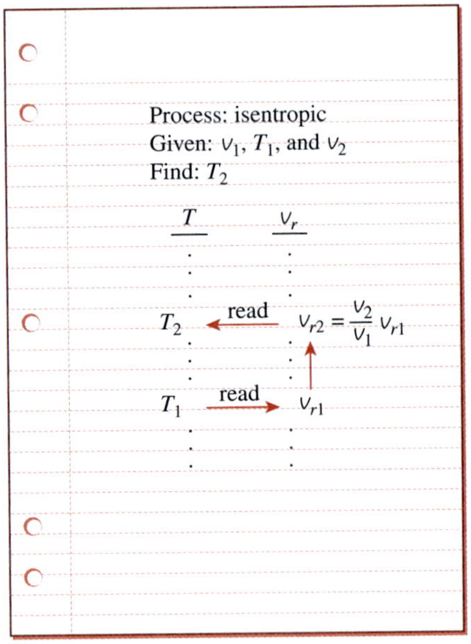

FIGURE 7–39
The use of v_r data for calculating the final temperature during an isentropic process (Example 7–10).

Therefore, the temperature of air will increase by 367.7°C during this process.

ALTERNATIVE SOLUTION The final temperature could also be determined from Eq. 7–42 by assuming constant specific heats for air:

$$\left(\frac{T_2}{T_1}\right)_{s=\text{const.}} = \left(\frac{v_1}{v_2}\right)^{k-1}$$

The specific heat ratio k also varies with temperature, and we need to use the value of k corresponding to the average temperature. However, the final temperature is not given, and so we cannot determine the average temperature in advance. For such cases, calculations can be started with a k value at the initial or the anticipated average temperature. This value could be refined later, if necessary, and the calculations can be repeated. We know that the temperature of the air will rise considerably during this adiabatic compression process, so we *guess* the average temperature to be about 450 K. The k value at this anticipated average temperature is determined from Table A–2*b* to be 1.391. Then, the final temperature of air becomes

$$T_2 = (295\text{ K})(8)^{1.391-1} = 665.2\text{ K}$$

This gives an average temperature value of 480.1 K, which is sufficiently close to the assumed value of 450 K. Therefore, it is not necessary to repeat the calculations by using the k value at this average temperature.

The result obtained by assuming constant specific heats for this case is in error by about 0.4 percent, which is rather small. This is not surprising since the temperature change of air is relatively small (only a few hundred degrees) and the specific heats of air vary almost linearly with temperature in this temperature range.

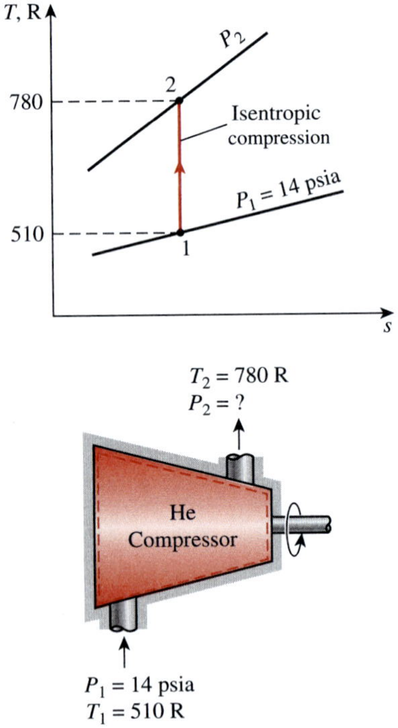

FIGURE 7–40
Schematic and *T-s* diagram for Example 7–11.

EXAMPLE 7–11 Isentropic Compression of an Ideal Gas

Helium gas is compressed by an adiabatic compressor from an initial state of 14 psia and 50°F to a final temperature of 320°F in a reversible manner. Determine the exit pressure of helium.

SOLUTION Helium is compressed from a given state to a specified pressure isentropically. The exit pressure of helium is to be determined.

Assumptions At specified conditions, helium can be treated as an ideal gas. Therefore, the isentropic relations developed earlier for ideal gases are applicable.

Analysis A sketch of the system and the *T-s* diagram for the process are given in Fig. 7–40.

The specific heat ratio k of helium is 1.667 and is independent of temperature in the region where it behaves as an ideal gas. Thus the final pressure of helium can be determined from Eq. 7–43:

$$P_2 = P_1\left(\frac{T_2}{T_1}\right)^{k/(k-1)} = (14\text{ psia})\left(\frac{780\text{ R}}{510\text{ R}}\right)^{1.667/0.667} = \mathbf{40.5\ psia}$$

7–10 ■ REVERSIBLE STEADY-FLOW WORK

The work done during a process depends on the path followed as well as on the properties at the end states. Recall that reversible (quasi-equilibrium) moving boundary work associated with closed systems is expressed in terms of the fluid properties as

$$W_b = \int_1^2 P\, dV$$

We mentioned that the quasi-equilibrium work interactions lead to the maximum work output for work-producing devices and the minimum work input for work-consuming devices.

It would also be very insightful to express the work associated with steady-flow devices in terms of fluid properties.

Taking the positive direction of work to be from the system (work output), the energy balance for a steady-flow device undergoing an internally reversible process can be expressed in differential form as

$$\delta q_{\text{rev}} - \delta w_{\text{rev}} = dh + d\text{ke} + d\text{pe}$$

But

$$\left.\begin{aligned} \delta q_{\text{rev}} &= T\,ds && \text{(Eq. 7–16)} \\ T\,ds &= dh - v\,dP && \text{(Eq. 7–24)} \end{aligned}\right\} \quad \delta q_{\text{rev}} = dh - v\,dP$$

Substituting this into the relation above and canceling dh yield

$$-\delta w_{\text{rev}} = v\,dP + d\text{ke} + d\text{pe}$$

Integrating, we find

$$w_{\text{rev}} = -\int_1^2 v\,dP - \Delta\text{ke} - \Delta\text{pe} \qquad \text{(kJ/kg)} \qquad \textbf{(7–51)}$$

When the changes in kinetic and potential energies are negligible, this equation reduces to

$$w_{\text{rev}} = -\int_1^2 v\,dP \qquad \text{(kJ/kg)} \qquad \textbf{(7–52)}$$

Equations 7–51 and 7–52 are relations for the *reversible work output* associated with an internally reversible process in a steady-flow device. They will give a negative result when work is done on the system. To avoid the negative sign, Eq. 7–51 can be written for work input to steady-flow devices such as compressors and pumps as

$$w_{\text{rev,in}} = \int_1^2 v\,dP + \Delta\text{ke} + \Delta\text{pe} \qquad \textbf{(7–53)}$$

The resemblance between the $v\,dP$ in these relations and $P\,dv$ is striking. They should not be confused with each other, however, since $P\,dv$ is associated with reversible boundary work in closed systems (Fig. 7–41).

Obviously, one needs to know v as a function of P for the given process to perform the integration. When the working fluid is *incompressible*, the

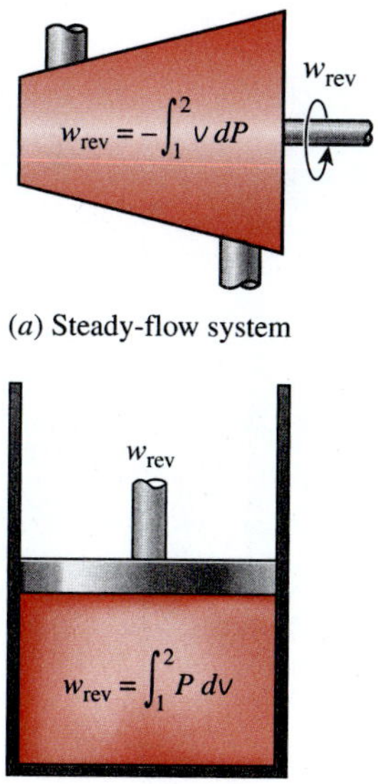

(*a*) Steady-flow system

(*b*) Closed system

FIGURE 7–41
Reversible work relations for steady-flow and closed systems.

specific volume v remains constant during the process and can be taken out of the integration. Then Eq. 7–51 simplifies to

$$w_{rev} = -v(P_2 - P_1) - \Delta ke - \Delta pe \qquad (kJ/kg) \tag{7-54}$$

For the steady flow of a liquid through a device that involves no work interactions (such as a nozzle or a pipe section), the work term is zero, and the equation above can be expressed as

$$v(P_2 - P_1) + \frac{V_2^2 - V_1^2}{2} + g(z_2 - z_1) = 0 \tag{7-55}$$

which is known as the **Bernoulli equation** in fluid mechanics. It is developed for an internally reversible process and thus is applicable to incompressible fluids that involve no irreversibilities such as friction or shock waves. This equation can be modified, however, to incorporate these effects.

Equation 7–52 has far-reaching implications in engineering regarding devices that produce or consume work steadily such as turbines, compressors, and pumps. It is obvious from this equation that the reversible steady-flow work is closely associated with the specific volume of the fluid flowing through the device. *The larger the specific volume, the larger the reversible work produced or consumed by the steady-flow device* (Fig. 7–42). This conclusion is equally valid for actual steady-flow devices. Therefore, every effort should be made to keep the specific volume of a fluid as small as possible during a compression process to minimize the work input and as large as possible during an expansion process to maximize the work output.

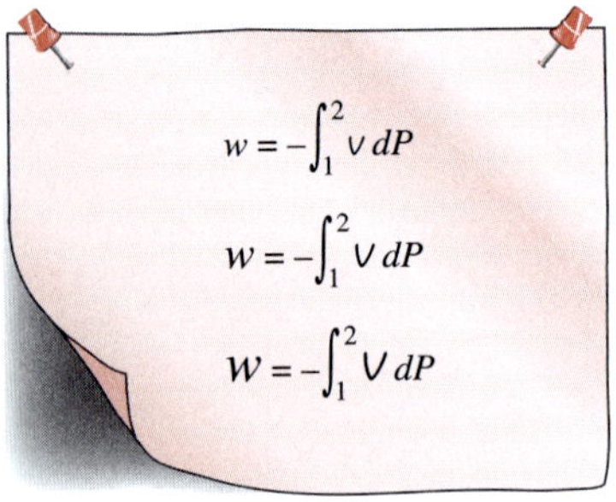

FIGURE 7–42
The larger the specific volume, the greater the work produced (or consumed) by a steady-flow device.

In steam or gas power plants, the pressure rise in the pump or compressor is equal to the pressure drop in the turbine if we disregard the pressure losses in various other components. In steam power plants, the pump handles liquid, which has a very small specific volume, and the turbine handles vapor, whose specific volume is many times larger. Therefore, the work output of the turbine is much larger than the work input to the pump. This is one of the reasons for the wide-spread use of steam power plants in electric power generation.

If we were to compress the steam exiting the turbine back to the turbine inlet pressure before cooling it first in the condenser in order to "save" the heat rejected, we would have to supply all the work produced by the turbine back to the compressor. In reality, the required work input would be even greater than the work output of the turbine because of the irreversibilities present in both processes.

In gas power plants, the working fluid (typically air) is compressed in the gas phase, and a considerable portion of the work output of the turbine is consumed by the compressor. As a result, a gas power plant delivers less net work per unit mass of the working fluid.

EXAMPLE 7–12 **Compressing a Substance in the Liquid versus Gas Phases**

Determine the compressor work input required to compress steam isentropically from 100 kPa to 1 MPa, assuming that the steam exists as (*a*) saturated liquid and (*b*) saturated vapor at the inlet state.

SOLUTION Steam is to be compressed from a given pressure to a specified pressure isentropically. The work input is to be determined for the cases of steam being a saturated liquid and saturated vapor at the inlet.

Assumptions **1** Steady operating conditions exist. **2** Kinetic and potential energy changes are negligible. **3** The process is given to be isentropic.

Analysis We take first the turbine and then the pump as the *system*. Both are *control volumes* since mass crosses the boundary. Sketches of the pump and the turbine together with the *T-s* diagram are given in Fig. 7–43.

(*a*) In this case, steam is a saturated liquid initially, and its specific volume is

$$v_1 = v_{f\,@\,100\text{ kPa}} = 0.001043\text{ m}^3/\text{kg} \qquad \text{(Table A–5)}$$

which remains essentially constant during the process. Thus,

$$w_{\text{rev,in}} = \int_1^2 v\,dP \cong v_1(P_2 - P_1)$$

$$= (0.001043\text{ m}^3/\text{kg})[(1000 - 100)\text{ kPa}]\left(\frac{1\text{ kJ}}{1\text{ kPa}\cdot\text{m}^3}\right)$$

$$= \mathbf{0.94\ kJ/kg}$$

(*b*) This time, steam is a saturated vapor initially and remains a vapor during the entire compression process. Since the specific volume of a gas changes considerably during a compression process, we need to know how v varies with P to perform the integration in Eq. 7–53. This relation, in general, is not readily available. But for an isentropic process, it is easily obtained from the second $T\,ds$ relation by setting $ds = 0$:

$$\left.\begin{array}{ll} T\,ds = dh - v\,dP & \text{(Eq. 7-24)} \\ ds = 0 & \text{(isentropic process)} \end{array}\right\} \quad v\,dP = dh$$

Thus,

$$w_{\text{rev,in}} = \int_1^2 v\,dP = \int_1^2 dh = h_2 - h_1$$

This result could also be obtained from the energy balance relation for an isentropic steady-flow process. Next we determine the enthalpies:

$$\textit{State 1:} \quad \left.\begin{array}{l} P_1 = 100\text{ kPa} \\ \text{(sat. vapor)} \end{array}\right\} \quad \begin{array}{l} h_1 = 2675.0\text{ kJ/kg} \\ s_1 = 7.3589\text{ kJ/kg}\cdot\text{K} \end{array} \qquad \text{(Table A–5)}$$

$$\textit{State 2:} \quad \left.\begin{array}{l} P_2 = 1\text{ MPa} \\ s_2 = s_1 \end{array}\right\} \quad h_2 = 3194.5\text{ kJ/kg} \qquad \text{(Table A–6)}$$

Thus,

$$w_{\text{rev,in}} = (3194.5 - 2675.0)\text{ kJ/kg} = \mathbf{519.5\ kJ/kg}$$

Discussion Note that compressing steam in the vapor form would require over 500 times more work than compressing it in the liquid form between the same pressure limits.

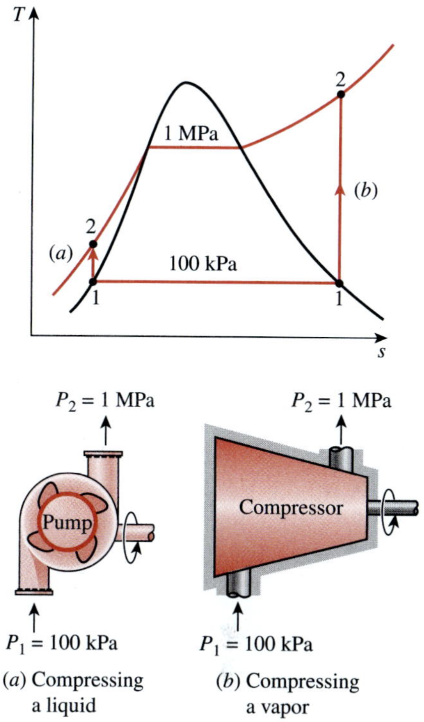

FIGURE 7–43
Schematic and *T-s* diagram for Example 7–12.

Proof that Steady-Flow Devices Deliver the Most and Consume the Least Work When the Process Is Reversible

We have shown in Chap. 6 that cyclic devices (heat engines, refrigerators, and heat pumps) deliver the most work and consume the least when reversible processes are used. Now we demonstrate that this is also the case for individual devices such as turbines and compressors in steady operation.

Consider two steady-flow devices, one reversible and the other irreversible, operating between the same inlet and exit states. Again taking heat transfer to the system and work done by the system to be positive quantities, the energy balance for each of these devices can be expressed in the differential form as

Actual: $$\delta q_{act} - \delta w_{act} = dh + d\text{ke} + d\text{pe}$$

Reversible: $$\delta q_{rev} - \delta w_{rev} = dh + d\text{ke} + d\text{pe}$$

The right-hand sides of these two equations are identical since both devices are operating between the same end states. Thus,

$$\delta q_{act} - \delta w_{act} = \delta q_{rev} - \delta w_{rev}$$

or

$$\delta w_{rev} - \delta w_{act} = \delta q_{rev} - \delta q_{act}$$

However,

$$\delta q_{rev} = T\,ds$$

Substituting this relation into the preceding equation and dividing each term by T, we obtain

$$\frac{\delta w_{rev} - \delta w_{act}}{T} = ds - \frac{\delta q_{act}}{T} \geq 0$$

since

$$ds \geq \frac{\delta q_{act}}{T}$$

Also, T is the absolute temperature, which is always positive. Thus,

$$\partial w_{rev} \geq \partial w_{act}$$

or

$$w_{rev} \geq w_{act}$$

Therefore, work-producing devices such as turbines (w is positive) deliver more work, and work-consuming devices such as pumps and compressors (w is negative) require less work when they operate reversibly (Fig. 7–44).

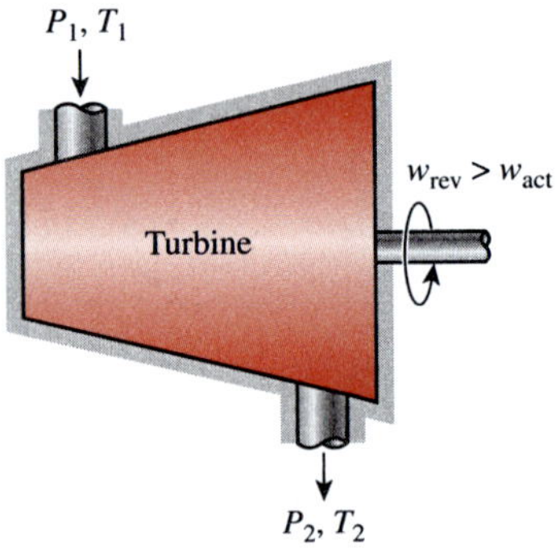

FIGURE 7–44
A reversible turbine delivers more work than an irreversible one if both operate between the same end states.

7–11 ▪ MINIMIZING THE COMPRESSOR WORK

We have just shown that the work input to a compressor is minimized when the compression process is executed in an internally reversible manner. When the changes in kinetic and potential energies are negligible, the compressor work is given by (Eq. 7–53)

$$w_{\text{rev,in}} = \int_1^2 v\, dP \qquad \textbf{(7–56)}$$

Obviously one way of minimizing the compressor work is to approximate an internally reversible process as much as possible by minimizing the irreversibilities such as friction, turbulence, and nonquasi-equilibrium compression. The extent to which this can be accomplished is limited by economic considerations. A second (and more practical) way of reducing the compressor work is to keep the specific volume of the gas as small as possible during the compression process. This is done by maintaining the temperature of the gas as low as possible during compression since the specific volume of a gas is proportional to temperature. Therefore, reducing the work input to a compressor requires that the gas be cooled as it is compressed.

To have a better understanding of the effect of cooling during the compression process, we compare the work input requirements for three kinds of processes: *an isentropic process* (involves no cooling), *a polytropic process* (involves some cooling), and *an isothermal process* (involves maximum cooling). Assuming all three processes are executed between the same pressure levels (P_1 and P_2) in an internally reversible manner and the gas behaves as an ideal gas ($Pv = RT$) with constant specific heats, we see that the compression work is determined by performing the integration in Eq. 7–56 for each case, with the following results:

Isentropic (Pv^k = constant):

$$w_{\text{comp,in}} = \frac{kR(T_2 - T_1)}{k-1} = \frac{kRT_1}{k-1}\left[\left(\frac{P_2}{P_1}\right)^{(k-1)/k} - 1\right] \qquad \textbf{(7–57a)}$$

Polytropic (Pv^n = constant):

$$w_{\text{comp,in}} = \frac{nR(T_2 - T_1)}{n-1} = \frac{nRT_1}{n-1}\left[\left(\frac{P_2}{P_1}\right)^{(n-1)/n} - 1\right] \qquad \textbf{(7–57b)}$$

Isothermal (Pv = constant):

$$w_{\text{comp,in}} = RT \ln \frac{P_2}{P_1} \qquad \textbf{(7–57c)}$$

The three processes are plotted on a *P*-*v* diagram in Fig. 7–45 for the same inlet state and exit pressure. On a *P*-*v* diagram, the area to the left of the process curve is the integral of $v\,dP$. Thus it is a measure of the steady-flow compression work. It is interesting to observe from this diagram that of the three internally reversible cases considered, the adiabatic compression (Pv^k = constant) requires the maximum work and the isothermal compression (T = constant or Pv = constant) requires the minimum. The work input requirement for the polytropic case (Pv^n = constant) is between these two and decreases as the polytropic exponent n is decreased, by increasing the heat rejection during the compression process. If sufficient heat is removed, the value of n approaches unity and the process becomes isothermal. One common way of cooling the gas during compression is to use cooling jackets around the casing of the compressors.

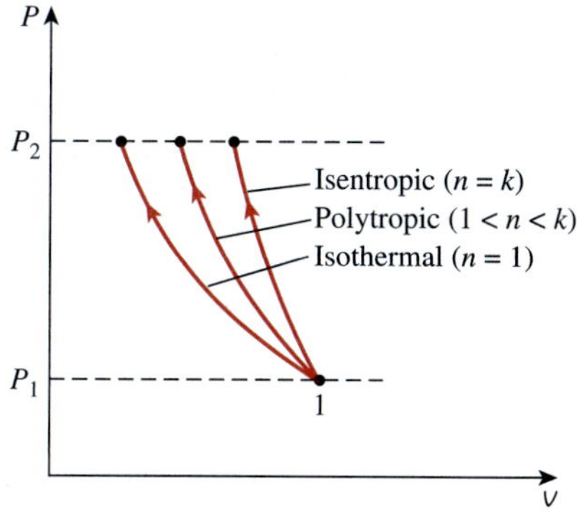

FIGURE 7–45
P-*v* diagrams of isentropic, polytropic, and isothermal compression processes between the same pressure limits.

Multistage Compression with Intercooling

It is clear from these arguments that cooling a gas as it is compressed is desirable since this reduces the required work input to the compressor. However, often it is not possible to have adequate cooling through the casing of the compressor, and it becomes necessary to use other techniques to achieve effective cooling. One such technique is **multistage compression with intercooling**, where the gas is compressed in stages and cooled between each stage by passing it through a heat exchanger called an *intercooler*. Ideally, the cooling process takes place at constant pressure, and the gas is cooled to the initial temperature T_1 at each intercooler. Multistage compression with intercooling is especially attractive when a gas is to be compressed to very high pressures.

The effect of intercooling on compressor work is graphically illustrated on P-v and T-s diagrams in Fig. 7–46 for a two-stage compressor. The gas is compressed in the first stage from P_1 to an intermediate pressure P_x, cooled at constant pressure to the initial temperature T_1, and compressed in the second stage to the final pressure P_2. The compression processes, in general, can be modeled as polytropic (Pv^n = constant) where the value of n varies between k and 1. The colored area on the P-v diagram represents the work saved as a result of two-stage compression with intercooling. The process paths for single-stage isothermal and polytropic processes are also shown for comparison.

The size of the colored area (the saved work input) varies with the value of the intermediate pressure P_x, and it is of practical interest to determine the conditions under which this area is maximized. The total work input for a two-stage compressor is the sum of the work inputs for each stage of compression, as determined from Eq. 7–57*b*:

$$w_{\text{comp,in}} = w_{\text{comp I,in}} + w_{\text{comp II,in}} \tag{7–58}$$

$$= \frac{nRT_1}{n-1}\left[\left(\frac{P_x}{P_1}\right)^{(n-1)/n} - 1\right] + \frac{nRT_1}{n-1}\left[\left(\frac{P_2}{P_x}\right)^{(n-1)/n} - 1\right]$$

The only variable in this equation is P_x. The P_x value that minimizes the total work is determined by differentiating this expression with respect to P_x and setting the resulting expression equal to zero. It yields

$$P_x = (P_1 P_2)^{1/2} \quad \text{or} \quad \frac{P_x}{P_1} = \frac{P_2}{P_x} \tag{7–59}$$

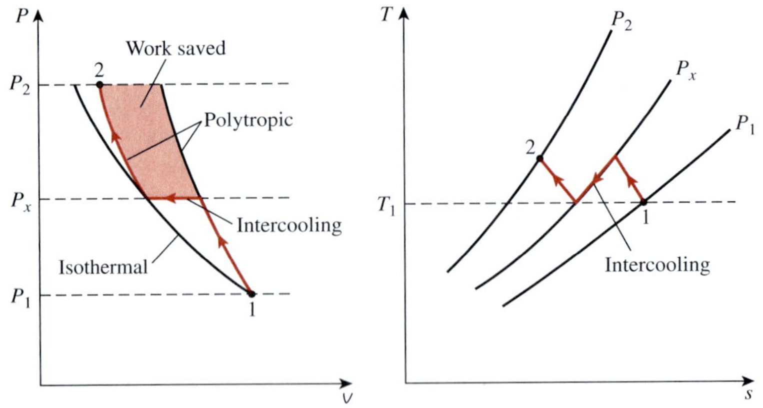

FIGURE 7–46
P-v and T-s diagrams for a two-stage steady-flow compression process.

That is, *to minimize compression work during two-stage compression, the pressure ratio across each stage of the compressor must be the same*. When this condition is satisfied, the compression work at each stage becomes identical, that is, $w_{comp\ I,in} = w_{comp\ II,in}$.

EXAMPLE 7–13 Work Input for Various Compression Processes

Air is compressed steadily by a reversible compressor from an inlet state of 100 kPa and 300 K to an exit pressure of 900 kPa. Determine the compressor work per unit mass for (*a*) isentropic compression with $k = 1.4$, (*b*) polytropic compression with $n = 1.3$, (*c*) isothermal compression, and (*d*) ideal two-stage compression with intercooling with a polytropic exponent of 1.3.

SOLUTION Air is compressed reversibly from a specified state to a specified pressure. The compressor work is to be determined for the cases of isentropic, polytropic, isothermal, and two-stage compression.

Assumptions **1** Steady operating conditions exist. **2** At specified conditions, air can be treated as an ideal gas. **3** Kinetic and potential energy changes are negligible.

Analysis We take the compressor to be the system. This is a control volume since mass crosses the boundary. A sketch of the system and the *T-s* diagram for the process are given in Fig. 7–47.

The steady-flow compression work for all these four cases is determined by using the relations developed earlier in this section:

(*a*) Isentropic compression with $k = 1.4$:

$$w_{comp,in} = \frac{kRT_1}{k-1}\left[\left(\frac{P_2}{P_1}\right)^{(k-1)/k} - 1\right]$$

$$= \frac{(1.4)(0.287\ \text{kJ/kg·K})(300\ \text{K})}{1.4-1}\left[\left(\frac{900\ \text{kPa}}{100\ \text{kPa}}\right)^{(1.4-1)/1.4} - 1\right]$$

$$= \mathbf{263.2\ kJ/kg}$$

(*b*) Polytropic compression with $n = 1.3$:

$$w_{comp,in} = \frac{nRT_1}{n-1}\left[\left(\frac{P_2}{P_1}\right)^{(n-1)/n} - 1\right]$$

$$= \frac{(1.3)(0.287\ \text{kJ/kg·K})(300\ \text{K})}{1.3-1}\left[\left(\frac{900\ \text{kPa}}{100\ \text{kPa}}\right)^{(1.3-1)/1.3} - 1\right]$$

$$= \mathbf{246.4\ kJ/kg}$$

(*c*) Isothermal compression:

$$w_{comp,in} = RT\ln\frac{P_2}{P_1} = (0.287\ \text{kJ/kg·K})(300\ \text{K})\ln\frac{900\ \text{kPa}}{100\ \text{kPa}}$$

$$= \mathbf{189.2\ kJ/kg}$$

(*d*) Ideal two-stage compression with intercooling ($n = 1.3$): In this case, the pressure ratio across each stage is the same, and its value is

$$P_x = (P_1P_2)^{1/2} = [(100\ \text{kPa})(900\ \text{kPa})]^{1/2} = 300\ \text{kPa}$$

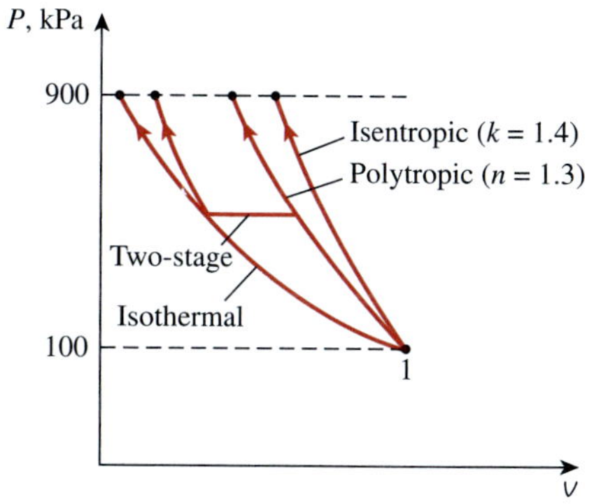

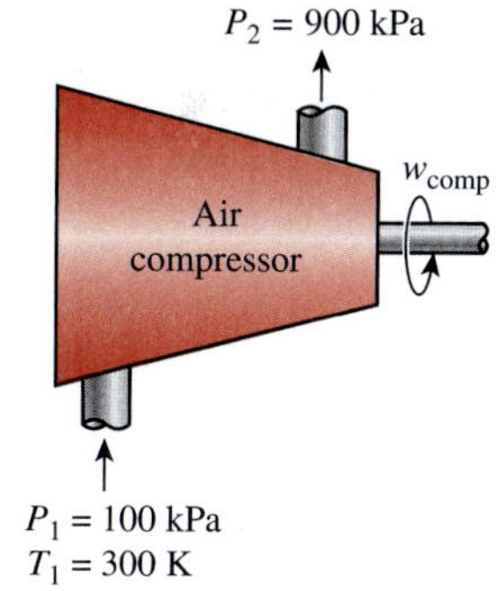

FIGURE 7–47
Schematic and *P-v* diagram for Example 7–13.

The compressor work across each stage is also the same. Thus the total compressor work is twice the compression work for a single stage:

$$w_{\text{comp,in}} = 2w_{\text{comp I,in}} = 2\,\frac{nRT_1}{n-1}\left[\left(\frac{P_x}{P_1}\right)^{(n-1)/n} - 1\right]$$

$$= \frac{2(1.3)(0.287\text{ kJ/kg·K})(300\text{ K})}{1.3-1}\left[\left(\frac{300\text{ kPa}}{100\text{ kPa}}\right)^{(1.3-1)/1.3} - 1\right]$$

$$= \mathbf{215.3\ kJ/kg}$$

Discussion Of all four cases considered, the isothermal compression requires the minimum work and the isentropic compression the maximum. The compressor work is decreased when two stages of polytropic compression are utilized instead of just one. As the number of compressor stages is increased, the compressor work approaches the value obtained for the isothermal case.

7–12 ▪ ISENTROPIC EFFICIENCIES OF STEADY-FLOW DEVICES

We mentioned repeatedly that irreversibilities inherently accompany all actual processes and that their effect is always to downgrade the performance of devices. In engineering analysis, it would be very desirable to have some parameters that would enable us to quantify the degree of degradation of energy in these devices. In the last chapter we did this for cyclic devices, such as heat engines and refrigerators, by comparing the actual cycles to the idealized ones, such as the Carnot cycle. A cycle that was composed entirely of reversible processes served as the *model cycle* to which the actual cycles could be compared. This idealized model cycle enabled us to determine the theoretical limits of performance for cyclic devices under specified conditions and to examine how the performance of actual devices suffered as a result of irreversibilities.

Now we extend the analysis to discrete engineering devices working under steady-flow conditions, such as turbines, compressors, and nozzles, and we examine the degree of degradation of energy in these devices as a result of irreversibilities. However, first we need to define an ideal process that serves as a model for the actual processes.

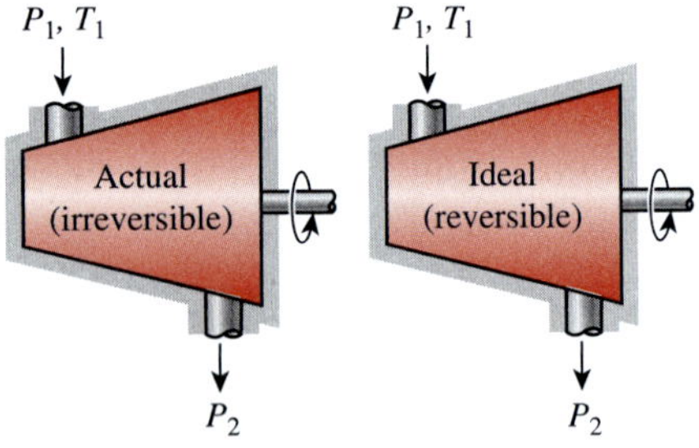

FIGURE 7–48
The isentropic process involves no irreversibilities and serves as the ideal process for adiabatic devices.

Although some heat transfer between these devices and the surrounding medium is unavoidable, many steady-flow devices are intended to operate under adiabatic conditions. Therefore, the model process for these devices should be an adiabatic one. Furthermore, an ideal process should involve no irreversibilities since the effect of irreversibilities is always to downgrade the performance of engineering devices. Thus, the ideal process that can serve as a suitable model for adiabatic steady-flow devices is the *isentropic* process (Fig. 7–48).

The more closely the actual process approximates the idealized isentropic process, the better the device performs. Thus, it would be desirable to have a parameter that expresses quantitatively how efficiently an actual device approximates an idealized one. This parameter is the **isentropic** or **adiabatic efficiency**, which is a measure of the deviation of actual processes from the corresponding idealized ones.

Isentropic efficiencies are defined differently for different devices since each device is set up to perform different tasks. Next, we define the isentropic efficiencies of turbines, compressors, and nozzles by comparing the actual performance of these devices to their performance under isentropic conditions for the same inlet state and exit pressure.

Isentropic Efficiency of Turbines

For a turbine under steady operation, the inlet state of the working fluid and the exhaust pressure are fixed. Therefore, the ideal process for an adiabatic turbine is an isentropic process between the inlet state and the exhaust pressure. The desired output of a turbine is the work produced, and the **isentropic efficiency of a turbine** is defined as *the ratio of the actual work output of the turbine to the work output that would be achieved if the process between the inlet state and the exit pressure were isentropic:*

$$\eta_T = \frac{\text{Actual turbine work}}{\text{Isentropic turbine work}} = \frac{w_a}{w_s} \tag{7-60}$$

Usually the changes in kinetic and potential energies associated with a fluid stream flowing through a turbine are small relative to the change in enthalpy and can be neglected. Then, the work output of an adiabatic turbine simply becomes the change in enthalpy, and Eq. 7–60 becomes

$$\eta_T \cong \frac{h_1 - h_{2a}}{h_1 - h_{2s}} \tag{7-61}$$

where h_{2a} and h_{2s} are the enthalpy values at the exit state for actual and isentropic processes, respectively (Fig. 7–49).

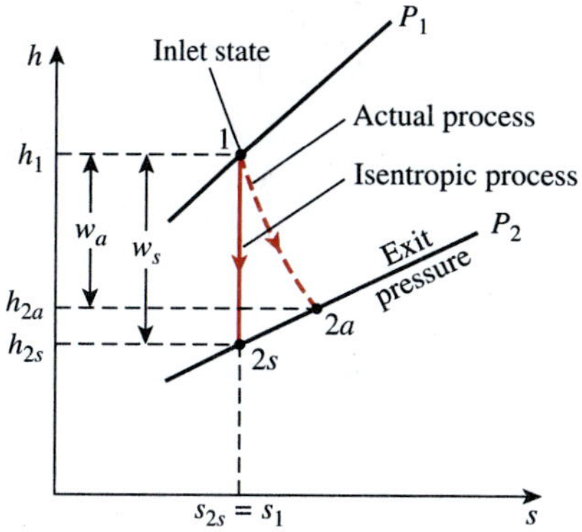

FIGURE 7–49
The *h-s* diagram for the actual and isentropic processes of an adiabatic turbine.

The value of ε_T greatly depends on the design of the individual components that make up the turbine. Well-designed, large turbines have isentropic efficiencies above 90 percent. For small turbines, however, it may drop even below 70 percent. The value of the isentropic efficiency of a turbine is determined by measuring the actual work output of the turbine and by calculating the isentropic work output for the measured inlet conditions and the exit pressure. This value can then be used conveniently in the design of power plants.

EXAMPLE 7–14 **Isentropic Efficiency of a Steam Turbine**

Steam enters an adiabatic turbine steadily at 3 MPa and 400°C and leaves at 50 kPa and 100°C. If the power output of the turbine is 2 MW, determine (*a*) the isentropic efficiency of the turbine and (*b*) the mass flow rate of the steam flowing through the turbine.

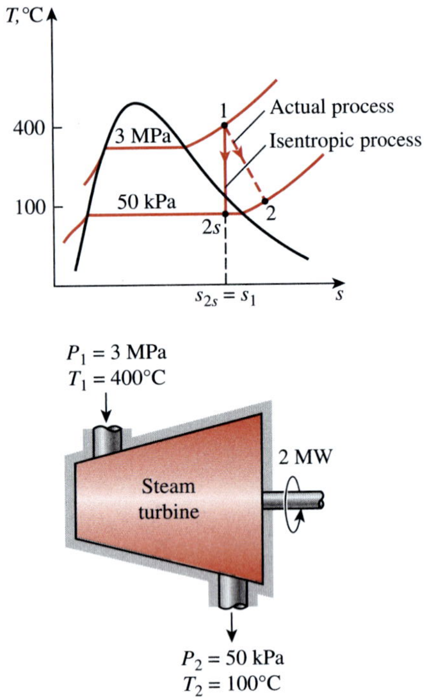

FIGURE 7–50
Schematic and *T-s* diagram for Example 7–14.

SOLUTION Steam flows steadily in a turbine between inlet and exit states. For a specified power output, the isentropic efficiency and the mass flow rate are to be determined.

Assumptions **1** Steady operating conditions exist. **2** The changes in kinetic and potential energies are negligible.

Analysis A sketch of the system and the *T-s* diagram of the process are given in Fig. 7–50.

(*a*) The enthalpies at various states are

State 1: $$\left.\begin{array}{l} P_1 = 3\text{ MPA} \\ T_1 = 400°\text{C} \end{array}\right\} \quad \begin{array}{l} h_1 = 3231.7\text{ kJ/kg} \\ s_1 = 6.9235\text{ kJ/kg·K} \end{array} \qquad \text{(Table A–6)}$$

State 2a: $$\left.\begin{array}{l} P_{2a} = 50\text{ kPa} \\ T_{2a} = 100°\text{C} \end{array}\right\} \quad h_{2a} = 2682.4\text{ kJ/kg} \qquad \text{(Table A–6)}$$

The exit enthalpy of the steam for the isentropic process h_{2s} is determined from the requirement that the entropy of the steam remain constant ($s_{2s} = s_1$):

State 2s: $$\begin{array}{l} P_{2s} = 50\text{ kPa} \\ (s_{2s} = s_1) \end{array} \longrightarrow \begin{array}{l} s_f = 1.0912\text{ kJ/kg·K} \\ s_g = 7.5931\text{ kJ/kg·K} \end{array} \qquad \text{(Table A–5)}$$

Obviously, at the end of the isentropic process steam exists as a saturated mixture since $s_f < s_{2s} < s_g$. Thus, we need to find the quality at state 2*s* first:

$$x_{2s} = \frac{s_{2s} - s_f}{s_{fg}} = \frac{6.9235 - 1.0912}{6.5019} = 0.897$$

and

$$h_{2s} = h_f + x_{2s} h_{fg} = 340.54 + 0.897(2304.7) = 2407.9\text{ kJ/kg}$$

By substituting these enthalpy values into Eq. 7–61, the isentropic efficiency of this turbine is determined to be

$$\eta_T \cong \frac{h_1 - h_{2a}}{h_1 - h_{2s}} = \frac{3231.7 - 2682.4}{3231.7 - 2407.9} = \mathbf{0.667,\ or\ 66.7\%}$$

(*b*) The mass flow rate of steam through this turbine is determined from the energy balance for steady-flow systems:

$$\dot{E}_{\text{in}} = \dot{E}_{\text{out}}$$

$$\dot{m}h_1 = \dot{W}_{a,\text{out}} + \dot{m}h_{2a}$$

$$\dot{W}_{a,\text{out}} = \dot{m}(h_1 - h_{2a})$$

$$2\text{ MW}\left(\frac{1000\text{ kJ/s}}{1\text{ MW}}\right) = \dot{m}(3231.7 - 2682.4)\text{ kJ/kg}$$

$$\dot{m} = \mathbf{3.64\ kg/s}$$

Isentropic Efficiencies of Compressors and Pumps

The **isentropic efficiency of a compressor** is defined as *the ratio of the work input required to raise the pressure of a gas to a specified value in an isentropic manner to the actual work input:*

$$\eta_C = \frac{\text{Isentropic compressor work}}{\text{Actual compressor work}} = \frac{w_s}{w_a} \tag{7-62}$$

Notice that the isentropic compressor efficiency is defined with the *isentropic work input in the numerator* instead of in the denominator. This is because w_s is a smaller quantity than w_a, and this definition prevents η_C from becoming greater than 100 percent, which would falsely imply that the actual compressors performed better than the isentropic ones. Also notice that the inlet conditions and the exit pressure of the gas are the same for both the actual and the isentropic compressor.

When the changes in kinetic and potential energies of the gas being compressed are negligible, the work input to an adiabatic compressor becomes equal to the change in enthalpy, and Eq. 7–62 for this case becomes

$$\eta_C \cong \frac{h_{2s} - h_1}{h_{2a} - h_1} \tag{7-63}$$

where h_{2a} and h_{2s} are the enthalpy values at the exit state for actual and isentropic compression processes, respectively, as illustrated in Fig. 7–51. Again, the value of η_C greatly depends on the design of the compressor. Well-designed compressors have isentropic efficiencies that range from 80 to 90 percent.

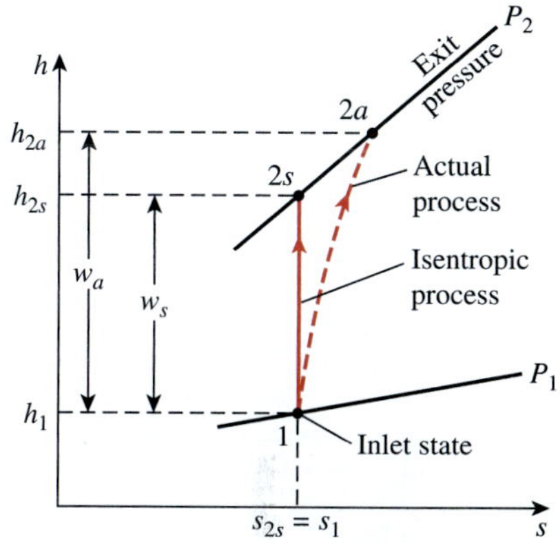

FIGURE 7–51
The *h-s* diagram of the actual and isentropic processes of an adiabatic compressor.

When the changes in potential and kinetic energies of a liquid are negligible, the isentropic efficiency of a pump is defined similarly as

$$\eta_P = \frac{w_s}{w_a} = \frac{v(P_2 - P_1)}{h_{2a} - h_1} \tag{7-64}$$

When no attempt is made to cool the gas as it is compressed, the actual compression process is nearly adiabatic and the reversible adiabatic (i.e., isentropic) process serves well as the ideal process. However, sometimes *compressors are cooled intentionally* by utilizing fins or a water jacket placed around the casing to reduce the work input requirements (Fig. 7–52). In this case, the isentropic process is not suitable as the model process since the device is no longer adiabatic and the isentropic compressor efficiency defined above is meaningless. A realistic model process for compressors that are intentionally cooled during the compression process is the *reversible isothermal process*. Then we can conveniently define an **isothermal efficiency** for such cases by comparing the actual process to a reversible isothermal one:

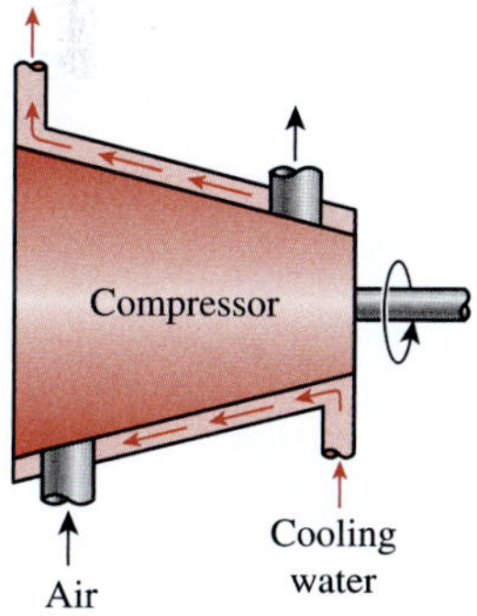

FIGURE 7–52
Compressors are sometimes intentionally cooled to minimize the work input.

$$\eta_C = \frac{w_t}{w_a} \tag{7-65}$$

where w_t and w_a are the required work inputs to the compressor for the reversible isothermal and actual cases, respectively.

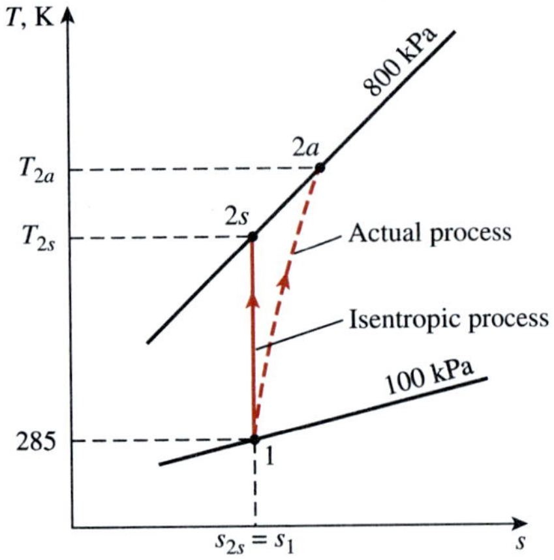

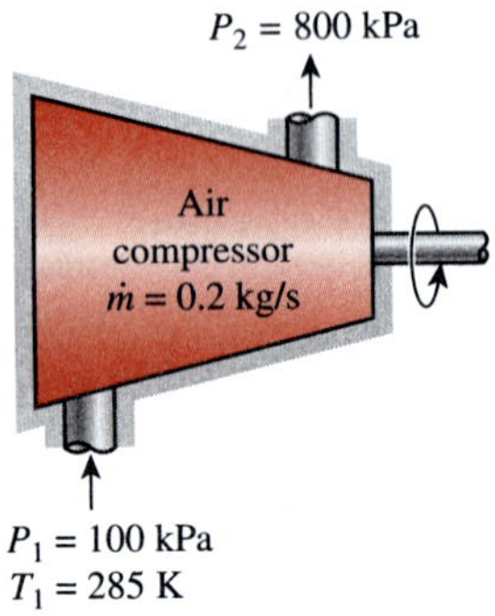

FIGURE 7–53
Schematic and *T-s* diagram for Example 7–15.

EXAMPLE 7–15 Effect of Efficiency on Compressor Power Input

Air is compressed by an adiabatic compressor from 100 kPa and 12°C to a pressure of 800 kPa at a steady rate of 0.2 kg/s. If the isentropic efficiency of the compressor is 80 percent, determine (*a*) the exit temperature of air and (*b*) the required power input to the compressor.

SOLUTION Air is compressed to a specified pressure at a specified rate. For a given isentropic efficiency, the exit temperature and the power input are to be determined.

Assumptions **1** Steady operating conditions exist. **2** Air is an ideal gas. **3** The changes in kinetic and potential energies are negligible.

Analysis A sketch of the system and the *T-s* diagram of the process are given in Fig. 7–53.

(*a*) We know only one property (pressure) at the exit state, and we need to know one more to fix the state and thus determine the exit temperature. The property that can be determined with minimal effort in this case is h_{2a} since the isentropic efficiency of the compressor is given. At the compressor inlet,

$$T_1 = 285 \text{ K} \quad \rightarrow \quad h_1 = 285.14 \text{ kJ/kg} \qquad \text{(Table A–17)}$$
$$(P_{r1} = 1.1584)$$

The enthalpy of the air at the end of the isentropic compression process is determined by using one of the isentropic relations of ideal gases,

$$P_{r2} = P_{r1}\left(\frac{P_2}{P_1}\right) = 1.1584\left(\frac{800 \text{ kPa}}{100 \text{ kPa}}\right) = 9.2672$$

and

$$P_{r2} = 9.2672 \quad \rightarrow \quad h_{2s} = 517.05 \text{ kJ/kg}$$

Substituting the known quantities into the isentropic efficiency relation, we have

$$\eta_C \cong \frac{h_{2s} - h_1}{h_{2a} - h_1} \quad \rightarrow \quad 0.80 = \frac{(517.05 - 285.14) \text{ kJ/kg}}{(h_{2a} - 285.14) \text{ kJ/kg}}$$

Thus,

$$h_{2a} = 575.03 \text{ kJ/kg} \quad \rightarrow \quad T_{2a} = \mathbf{569.5 \text{ K}}$$

(*b*) The required power input to the compressor is determined from the energy balance for steady-flow devices,

$$\dot{E}_{\text{in}} = \dot{E}_{\text{out}}$$
$$\dot{m}h_1 + \dot{W}_{a,\text{in}} = \dot{m}h_{2a}$$
$$\dot{W}_{a,\text{in}} = \dot{m}(h_{2a} - h_1)$$
$$= (0.2 \text{ kg/s})[(575.03 - 285.14) \text{ kJ/kg}]$$
$$= \mathbf{58.0 \text{ kW}}$$

Discussion Notice that in determining the power input to the compressor, we used h_{2a} instead of h_{2s} since h_{2a} is the actual enthalpy of the air as it exits the compressor. The quantity h_{2s} is a hypothetical enthalpy value that the air would have if the process were isentropic.

Isentropic Efficiency of Nozzles

Nozzles are essentially adiabatic devices and are used to accelerate a fluid. Therefore, the isentropic process serves as a suitable model for nozzles. The **isentropic efficiency of a nozzle** is defined as *the ratio of the actual kinetic energy of the fluid at the nozzle exit to the kinetic energy value at the exit of an isentropic nozzle for the same inlet state and exit pressure*. That is,

$$\eta_N = \frac{\text{Actual KE at nozzle exit}}{\text{Isentropic KE at nozzle exit}} = \frac{V_{2a}^2}{V_{2s}^2} \tag{7–66}$$

Note that the exit pressure is the same for both the actual and isentropic processes, but the exit state is different.

Nozzles involve no work interactions, and the fluid experiences little or no change in its potential energy as it flows through the device. If, in addition, the inlet velocity of the fluid is small relative to the exit velocity, the energy balance for this steady-flow device reduces to

$$h_1 = h_{2a} + \frac{V_{2a}^2}{2}$$

Then the isentropic efficiency of the nozzle can be expressed in terms of enthalpies as

$$\eta_N \cong \frac{h_1 - h_{2a}}{h_1 - h_{2s}} \tag{7–67}$$

where h_{2a} and h_{2s} are the enthalpy values at the nozzle exit for the actual and isentropic processes, respectively (Fig. 7–54). Isentropic efficiencies of nozzles are typically above 90 percent, and nozzle efficiencies above 95 percent are not uncommon.

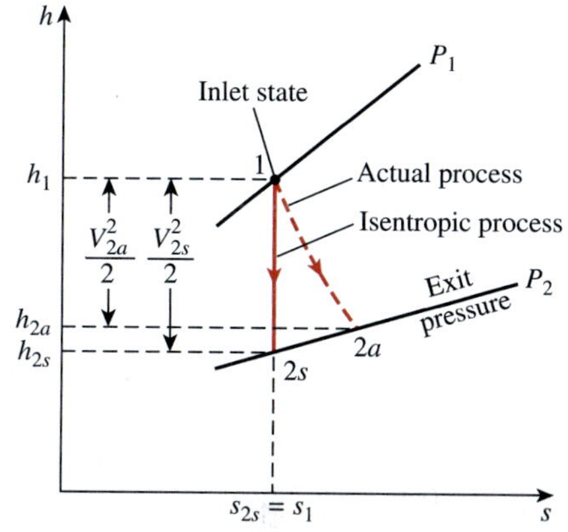

FIGURE 7–54
The *h-s* diagram of the actual and isentropic processes of an adiabatic nozzle.

EXAMPLE 7–16 Effect of Efficiency on Nozzle Exit Velocity

Air at 200 kPa and 950 K enters an adiabatic nozzle at low velocity and is discharged at a pressure of 80 kPa. If the isentropic efficiency of the nozzle is 92 percent, determine (*a*) the maximum possible exit velocity, (*b*) the exit temperature, and (*c*) the actual exit velocity of the air. Assume constant specific heats for air.

SOLUTION The acceleration of air in a nozzle is considered. For specified exit pressure and isentropic efficiency, the maximum and actual exit velocities and the exit temperature are to be determined.

Assumptions **1** Steady operating conditions exist. **2** Air is an ideal gas. **3** The inlet kinetic energy is negligible.

Analysis A sketch of the system and the *T-s* diagram of the process are given in Fig. 7–55.

The temperature of air will drop during this acceleration process because some of its internal energy is converted to kinetic energy. This problem can be solved accurately by using property data from the air table. But we will assume constant specific heats (thus sacrifice some accuracy) to demonstrate their use. Let us guess the average temperature of the air to be about 800 K. Then, the average values of c_p and k at this anticipated average

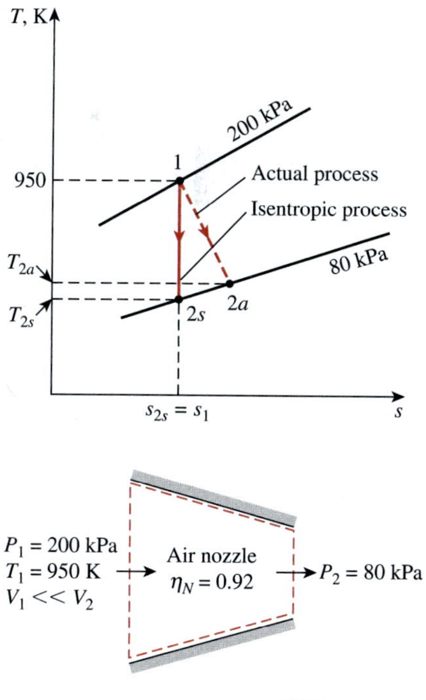

FIGURE 7–55
Schematic and *T-s* diagram for Example 7–16.

temperature are determined from Table A–2*b* to be c_p = 1.099 kJ/kg·K and k = 1.354.

(*a*) The exit velocity of the air will be a maximum when the process in the nozzle involves no irreversibilities. The exit velocity in this case is determined from the steady-flow energy equation. However, first we need to determine the exit temperature. For the isentropic process of an ideal gas we have:

$$\frac{T_{2s}}{T_1} = \left(\frac{P_{2s}}{P_1}\right)^{(k-1)/k}$$

or

$$T_{2s} = T_1\left(\frac{P_{2s}}{P_1}\right)^{(k-1)/k} = (950\text{ K})\left(\frac{80\text{ kPa}}{200\text{ kPa}}\right)^{0.354/1.354} = 748\text{ K}$$

This gives an average temperature of 849 K, which is somewhat higher than the assumed average temperature (800 K). This result could be refined by reevaluating the k value at 749 K and repeating the calculations, however, it is not warranted since the two average temperatures are sufficiently close (doing so would change the temperature by only 1.5 K, which is not significant).

Now we can determine the isentropic exit velocity of the air from the energy balance for this isentropic steady-flow process:

$$e_{\text{in}} = e_{\text{out}}$$

$$h_1 + \frac{V_1^2}{2} = h_{2s} + \frac{V_{2s}^2}{2}$$

or

$$V_{2s} = \sqrt{2(h_1 - h_{2s})} = \sqrt{2c_{p,\text{avg}}(T_1 - T_{2s})}$$

$$= \sqrt{2(1.099\text{ kJ/kg·K})[(950 - 748)\text{ K}]\left(\frac{1000\text{ m}^2/\text{s}^2}{1\text{ kJ/kg}}\right)}$$

$$= \mathbf{666\ m/s}$$

(*b*) The actual exit temperature of the air is higher than the isentropic exit temperature evaluated above and is determined from

$$\eta_N \cong \frac{h_1 - h_{2a}}{h_1 - h_{2s}} = \frac{c_{p,\text{avg}}(T_1 - T_{2a})}{c_{p,\text{avg}}(T_1 - T_{2s})}$$

or

$$0.92 = \frac{950 - T_{2a}}{950 - 748} \quad \rightarrow \quad T_{2a} = \mathbf{764\ K}$$

That is, the temperature is 16 K higher at the exit of the actual nozzle as a result of irreversibilities such as friction. It represents a loss since this rise in temperature comes at the expense of kinetic energy (Fig. 7–56).

(*c*) The actual exit velocity of air can be determined from the definition of isentropic efficiency of a nozzle,

$$\eta_N = \frac{V_{2a}^2}{V_{2s}^2} \quad \rightarrow \quad V_{2a} = \sqrt{\eta_N V_{2s}^2} = \sqrt{0.92(666\text{ m/s})^2} = \mathbf{639\ m/s}$$

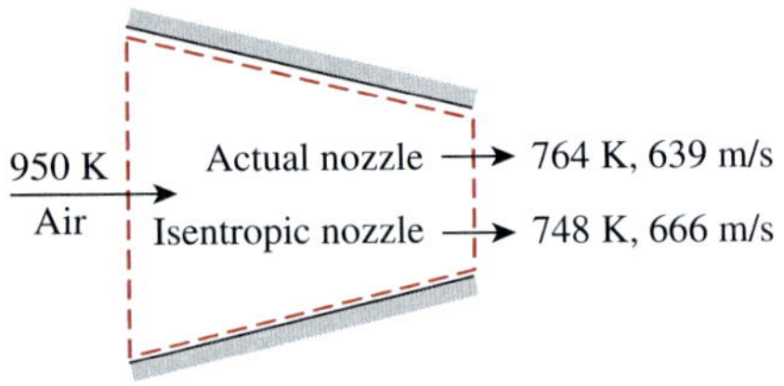

FIGURE 7–56
A substance leaves actual nozzles at a higher temperature (thus a lower velocity) as a result of friction.

7–13 ▪ ENTROPY BALANCE

The property *entropy* is a measure of molecular disorder or randomness of a system, and the second law of thermodynamics states that entropy can be created but it cannot be destroyed. Therefore, the entropy change of a system during a process is greater than the entropy transfer by an amount equal to the entropy generated during the process within the system, and the *increase of entropy principle* for any system is expressed as (Fig. 7–57)

$$\begin{pmatrix}\text{Total}\\ \text{entropy}\\ \text{entering}\end{pmatrix} - \begin{pmatrix}\text{Total}\\ \text{entropy}\\ \text{leaving}\end{pmatrix} + \begin{pmatrix}\text{Total}\\ \text{entropy}\\ \text{generated}\end{pmatrix} = \begin{pmatrix}\text{Change in the}\\ \text{total entropy}\\ \text{of the system}\end{pmatrix}$$

or

$$S_{\text{in}} - S_{\text{out}} + S_{\text{gen}} = \Delta S_{\text{system}} \tag{7–68}$$

which is a verbal statement of Eq. 7–9. This relation is often referred to as the **entropy balance** and is applicable to any system undergoing any process. The entropy balance relation above can be stated as: *the entropy change of a system during a process is equal to the net entropy transfer through the system boundary and the entropy generated within the system.* Next, we discuss the various terms in that relation.

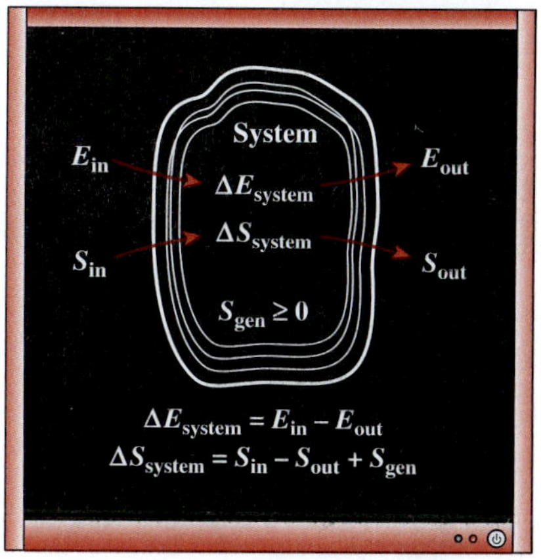

FIGURE 7–57
Energy and entropy balances for a system.

Entropy Change of a System, ΔS_{system}

Despite the reputation of entropy as being vague and abstract and the intimidation associated with it, entropy balance is actually easier to deal with than energy balance since, unlike energy, entropy does not exist in various forms. Therefore, the determination of entropy change of a system during a process involves evaluating entropy of the system at the beginning and at the end of the process and taking their difference. That is,

$$\text{Entropy change} = \text{Entropy at final state} - \text{Entropy at initial state}$$

or

$$\Delta S_{\text{system}} = S_{\text{final}} - S_{\text{initial}} = S_2 - S_1 \tag{7–69}$$

Note that entropy is a property, and the value of a property does not change unless the state of the system changes. Therefore, the entropy change of a system is zero if the state of the system does not change during the process. For example, the entropy change of steady-flow devices such as nozzles, compressors, turbines, pumps, and heat exchangers is zero during steady operation.

When the properties of the system are not uniform, the entropy of the system can be determined by integration from

$$S_{\text{system}} = \int s\,\delta m = \int_V s\rho\, dV \tag{7–70}$$

where V is the volume of the system and ρ is density.

Mechanisms of Entropy Transfer, S_{in} and S_{out}

Entropy can be transferred to or from a system by two mechanisms: *heat transfer* and *mass flow* (in contrast, energy is transferred by work also). Entropy transfer is recognized at the system boundary as it crosses the boundary, and it represents the entropy gained or lost by a system during a process. The only form of entropy interaction associated with a fixed mass or closed system is *heat transfer*, and thus the entropy transfer for an adiabatic closed system is zero.

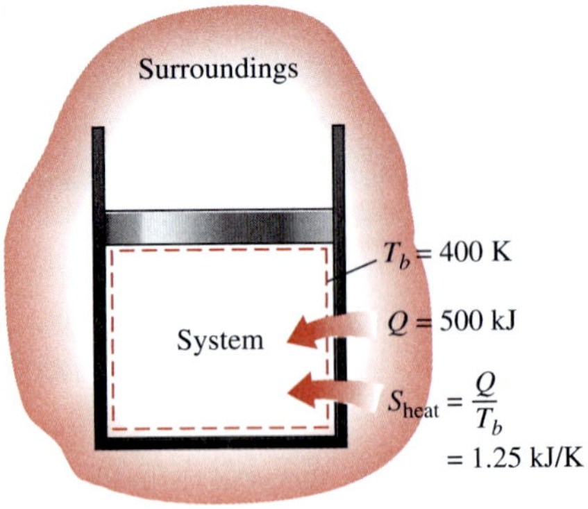

FIGURE 7–58
Heat transfer is always accompanied by entropy transfer in the amount of Q/T, where T is the boundary temperature.

1 Heat Transfer

Heat is, in essence, a form of disorganized energy, and some disorganization (entropy) will flow with heat. Heat transfer to a system increases the entropy of that system and thus the level of molecular disorder or randomness, and heat transfer from a system decreases it. In fact, heat rejection is the only way the entropy of a fixed mass can be decreased. The ratio of the heat transfer Q at a location to the absolute temperature T at that location is called the *entropy flow* or *entropy transfer* and is expressed as (Fig. 7–58)

Entropy transfer by heat transfer:
$$S_{heat} = \frac{Q}{T} \qquad (T = \text{constant}) \tag{7–71}$$

The quantity Q/T represents the entropy transfer accompanied by heat transfer, and the direction of entropy transfer is the same as the direction of heat transfer since thermodynamic temperature T is always a positive quantity.

When the temperature T is not constant, the entropy transfer during a process 1-2 can be determined by integration (or by summation if appropriate) as

$$S_{heat} = \int_1^2 \frac{\delta Q}{T} \cong \sum \frac{Q_k}{T_k} \tag{7–72}$$

where Q_k is the heat transfer through the boundary at temperature T_k at location k.

When two systems are in contact, the entropy transfer from the warmer system is equal to the entropy transfer into the cooler one at the point of contact. That is, no entropy can be created or destroyed at the boundary since the boundary has no thickness and occupies no volume.

Note that **work** is entropy-free, and no entropy is transferred by work. Energy is transferred by both heat and work, whereas entropy is transferred only by heat. That is,

Entropy transfer by work:
$$S_{work} = 0 \tag{7–73}$$

The first law of thermodynamics makes no distinction between heat transfer and work; it considers them as *equals*. The distinction between heat transfer and work is brought out by the second law: *an energy interaction that is accompanied by entropy transfer is heat transfer, and an energy interaction that is not accompanied by entropy transfer is work.* That is, no entropy is exchanged during a work interaction between a system and its surroundings. Thus, only *energy* is exchanged during work interaction whereas both *energy* and *entropy* are exchanged during heat transfer (Fig. 7–59).

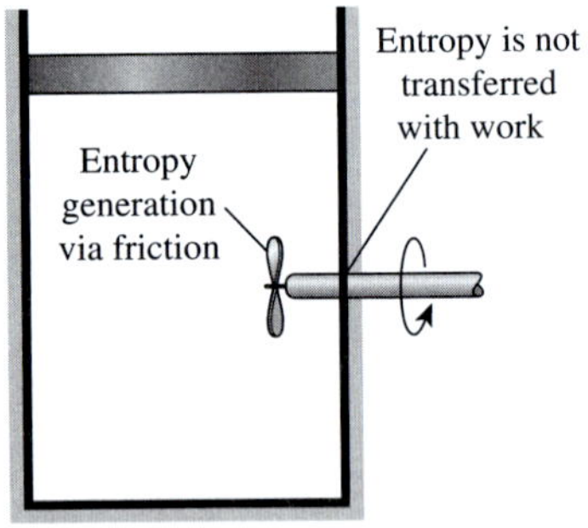

FIGURE 7–59
No entropy accompanies work as it crosses the system boundary. But entropy may be generated within the system as work is dissipated into a less useful form of energy.

2 Mass Flow

Mass contains entropy as well as energy, and the entropy and energy contents of a system are proportional to the mass. (When the mass of a system is doubled, so are the entropy and energy contents of the system.) Both entropy and energy are carried into or out of a system by streams of matter, and the rates of entropy and energy transport into or out of a system are proportional to the mass flow rate. Closed systems do not involve any mass flow and thus any entropy transfer by mass. When a mass in the amount of m enters or leaves a system, entropy in the amount of ms, where s is the specific entropy (entropy per unit mass entering or leaving), accompanies it (Fig. 7–60). That is,

FIGURE 7–60
Mass contains entropy as well as energy, and thus mass flow into or out of system is always accompanied by energy and entropy transfer.

Entropy transfer by mass flow: $$S_{\text{mass}} = ms \quad \textbf{(7–74)}$$

Therefore, the entropy of a system increases by ms when mass in the amount of m enters and decreases by the same amount when the same amount of mass at the same state leaves the system. When the properties of the mass change during the process, the entropy transfer by mass flow can be determined by integration from

$$\dot{S}_{\text{mass}} = \int_{A_c} s\rho V_n\, dA_c \quad \text{and} \quad S_{\text{mass}} = \int s\,\delta m = \int_{\Delta t} \dot{S}_{\text{mass}}\, dt \quad \textbf{(7–75)}$$

where A_c is the cross-sectional area of the flow and V_n is the local velocity normal to dA_c.

Entropy Generation, S_{gen}

Irreversibilities such as friction, mixing, chemical reactions, heat transfer through a finite temperature difference, unrestrained expansion, nonquasi-equilibrium compression, or expansion always cause the entropy of a system to increase, and entropy generation is a measure of the entropy created by such effects during a process.

For a *reversible process* (a process that involves no irreversibilities), the entropy generation is zero and thus the *entropy change* of a system is equal to the *entropy transfer*. Therefore, the entropy balance relation in the reversible case becomes analogous to the energy balance relation, which states that *energy change* of a system during a process is equal to the *energy transfer* during that process. However, note that the energy change of a system equals the energy transfer for *any* process, but the entropy change of a system equals the entropy transfer only for a *reversible* process.

The entropy transfer by heat Q/T is zero for adiabatic systems, and the entropy transfer by mass ms is zero for systems that involve no mass flow across their boundary (i.e., closed systems).

Entropy balance for *any system* undergoing *any process* can be expressed more explicitly as

$$\underbrace{S_{\text{in}} - S_{\text{out}}}_{\substack{\text{Net entropy transfer} \\ \text{by heat and mass}}} + \underbrace{S_{\text{gen}}}_{\substack{\text{Entropy} \\ \text{generation}}} = \underbrace{\Delta S_{\text{system}}}_{\substack{\text{Change} \\ \text{in entropy}}} \quad (\text{kJ/K}) \quad \textbf{(7–76)}$$

or, in the **rate form**, as

$$\underbrace{\dot{S}_{\text{in}} - \dot{S}_{\text{out}}}_{\substack{\text{Rate of net entropy}\\ \text{transfer by heat}\\ \text{and mass}}} + \underbrace{\dot{S}_{\text{gen}}}_{\substack{\text{Rate of entropy}\\ \text{generation}}} = \underbrace{dS_{\text{system}}/dt}_{\substack{\text{Rate of change}\\ \text{in entropy}}} \quad (\text{kW/K}) \tag{7–77}$$

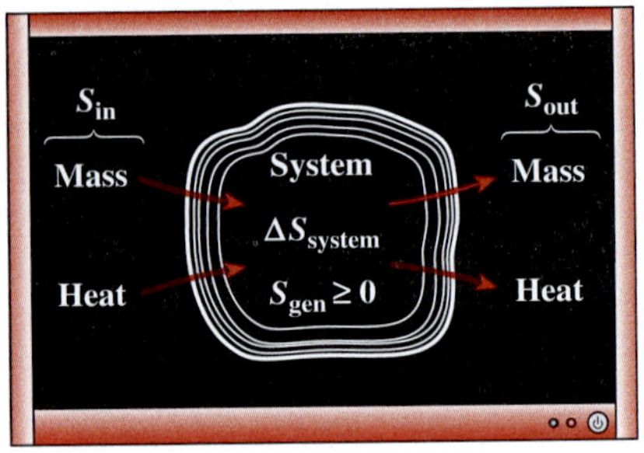

FIGURE 7–61
Mechanisms of entropy transfer for a general system.

where the rates of entropy transfer by heat transferred at a rate of $\dot{Q}$ and mass flowing at a rate of $\dot{m}$ are $\dot{S}_{\text{heat}} = \dot{Q}/T$ and $\dot{S}_{\text{mass}} = \dot{m}s$. The entropy balance can also be expressed on a **unit-mass basis** as

$$(s_{\text{in}} - s_{\text{out}}) + s_{\text{gen}} = \Delta s_{\text{system}} \quad (\text{kJ/kg·K}) \tag{7–78}$$

where all the quantities are expressed per unit mass of the system. Note that for a *reversible process*, the entropy generation term S_{gen} drops out from all of the relations above.

The term S_{gen} represents the entropy generation *within the system boundary* only (Fig. 7–61), and not the entropy generation that may occur outside the system boundary during the process as a result of external irreversibilities. Therefore, a process for which $S_{\text{gen}} = 0$ is *internally reversible*, but not necessarily *totally* reversible. The *total* entropy generated during a process can be determined by applying the entropy balance to an *extended system* that includes the system itself and its immediate surroundings where external irreversibilities might be occurring (Fig. 7–62). Also, the entropy change in this case is equal to the sum of the entropy change of the system and the entropy change of the immediate surroundings. Note that under steady conditions, the state and thus the entropy of the immediate surroundings (let us call it the "buffer zone") at any point does not change during the process, and the entropy change of the buffer zone is zero. The entropy change of the buffer zone, if any, is usually small relative to the entropy change of the system, and thus it is usually disregarded.

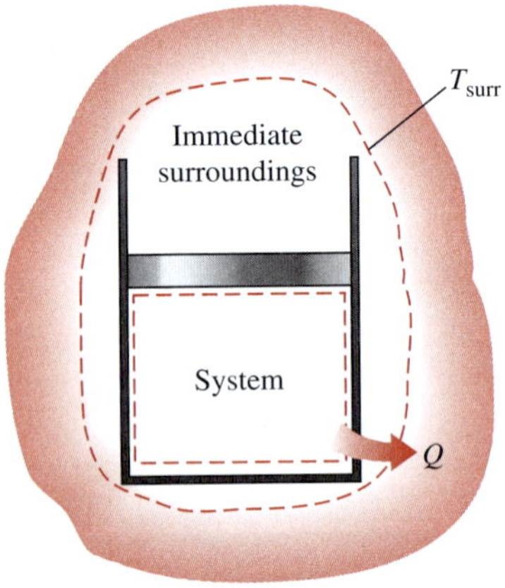

FIGURE 7–62
Entropy generation outside system boundaries can be accounted for by writing an entropy balance on an extended system that includes the system and its immediate surroundings.

When evaluating the entropy transfer between an extended system and the surroundings, the boundary temperature of the extended system is simply taken to be the *environment temperature*.

Closed Systems

A closed system involves *no mass flow* across its boundaries, and its entropy change is simply the difference between the initial and final entropies of the system. The *entropy change* of a closed system is due to the *entropy transfer* accompanying heat transfer and the *entropy generation* within the system boundaries. Taking the positive direction of heat transfer to be *to* the system, the general entropy balance relation (Eq. 7–76) can be expressed for a closed system as

Closed system:
$$\sum \frac{Q_k}{T_k} + S_{\text{gen}} = \Delta S_{\text{system}} = S_2 - S_1 \quad (\text{kJ/K}) \tag{7–79}$$

The entropy balance relation above can be stated as:

> The entropy change of a closed system during a process is equal to the sum of the net entropy transferred through the system boundary by heat transfer and the entropy generated within the system boundaries.

For an *adiabatic process* ($Q = 0$), the entropy transfer term in the above relation drops out and the entropy change of the closed system becomes equal to the entropy generation within the system boundaries. That is,

Adiabatic closed system: $$S_{gen} = \Delta S_{adiabatic\ system} \tag{7–80}$$

Noting that any closed system and its surroundings can be treated as an adiabatic system and the total entropy change of a system is equal to the sum of the entropy changes of its parts, the entropy balance for a closed system and its surroundings can be written as

System + Surroundings: $$S_{gen} = \sum \Delta S = \Delta S_{system} + \Delta S_{surroundings} \tag{7–81}$$

where $\Delta S_{system} = m(s_2 - s_1)$ and the entropy change of the surroundings can be determined from $\Delta S_{surr} = Q_{surr}/T_{surr}$ if its temperature is constant. At initial stages of studying entropy and entropy transfer, it is more instructive to start with the general form of the entropy balance (Eq. 7–76) and to simplify it for the problem under consideration. The specific relations above are convenient to use after a certain degree of intuitive understanding of the material is achieved.

Control Volumes

The entropy balance relations for control volumes differ from those for closed systems in that they involve one more mechanism of entropy exchange: *mass flow across the boundaries*. As mentioned earlier, mass possesses entropy as well as energy, and the amounts of these two extensive properties are proportional to the amount of mass (Fig. 7–63).

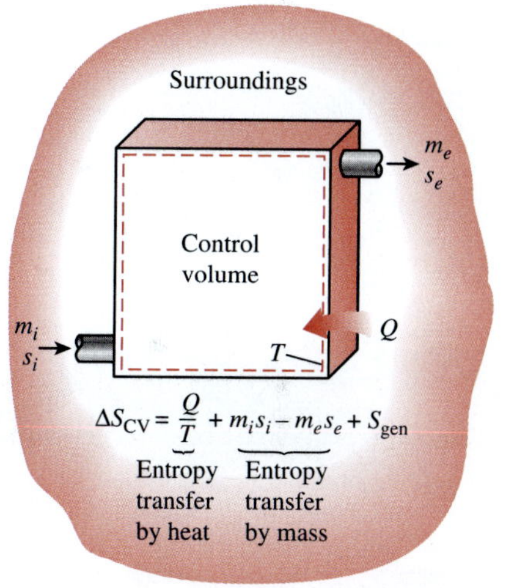

FIGURE 7–63
The entropy of a control volume changes as a result of mass flow as well as heat transfer.

Taking the positive direction of heat transfer to be *to* the system, the general entropy balance relations (Eqs. 7–76 and 7–77) can be expressed for control volumes as

$$\sum \frac{Q_k}{T_k} + \sum m_i s_i - \sum m_e s_e + S_{gen} = (S_2 - S_1)_{CV} \qquad \text{(kJ/K)} \tag{7–82}$$

or, in the rate form, as

$$\sum \frac{\dot{Q}_k}{T_k} + \sum \dot{m}_i s_i - \sum \dot{m}_e s_e + \dot{S}_{gen} = dS_{CV}/dt \qquad \text{(kW/K)} \tag{7–83}$$

This entropy balance relation can be stated as:

> The rate of entropy change within the control volume during a process is equal to the sum of the rate of entropy transfer through the control volume boundary by heat transfer, the net rate of entropy transfer into the control volume by mass flow, and the rate of entropy generation within the boundaries of the control volume as a result of irreversibilities.

Most control volumes encountered in practice such as turbines, compressors, nozzles, diffusers, heat exchangers, pipes, and ducts operate steadily, and thus they experience no change in their entropy. Therefore, the entropy balance relation for a general **steady-flow process** can be obtained from Eq. 7–83 by setting $dS_{CV}/dt = 0$ and rearranging to give

Steady-flow: $$\dot{S}_{gen} = \sum \dot{m}_e s_e - \sum \dot{m}_i s_i - \sum \frac{\dot{Q}_k}{T_k} \tag{7–84}$$

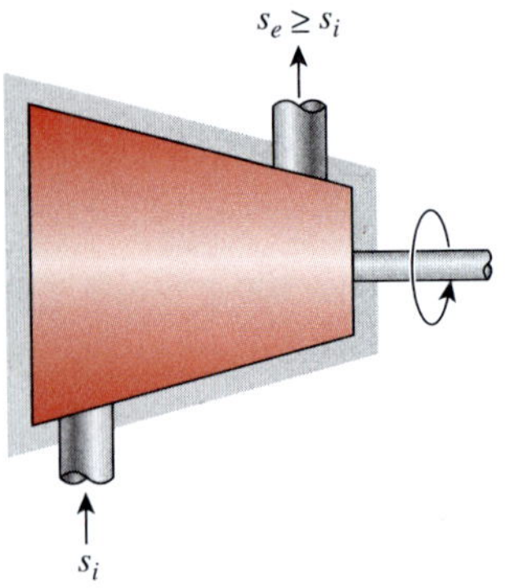

FIGURE 7–64
The entropy of a substance always increases (or remains constant in the case of a reversible process) as it flows through a single-stream, adiabatic, steady-flow device.

For *single-stream* (one inlet and one exit) steady-flow devices, the entropy balance relation simplifies to

Steady-flow, single-stream:
$$\dot{S}_{gen} = \dot{m}(s_e - s_i) - \sum \frac{\dot{Q}_k}{T_k} \quad (7\text{–}85)$$

For the case of an *adiabatic* single-stream device, the entropy balance relation further simplifies to

Steady-flow, single-stream, adiabatic:
$$\dot{S}_{gen} = \dot{m}(s_e - s_i) \quad (7\text{–}86)$$

which indicates that the specific entropy of the fluid must increase as it flows through an adiabatic device since $\dot{S}_{gen} \geq 0$ (Fig. 7–64). If the flow through the device is *reversible* and *adiabatic*, then the entropy remains constant, $s_e = s_i$, regardless of the changes in other properties.

EXAMPLE 7–17 Entropy Generation in a Wall

Consider steady heat transfer through a 5-m × 7-m brick wall of a house of thickness 30 cm. On a day when the temperature of the outdoors is 0°C, the house is maintained at 27°C. The temperatures of the inner and outer surfaces of the brick wall are measured to be 20°C and 5°C, respectively, and the rate of heat transfer through the wall is 1035 W. Determine the rate of entropy generation in the wall, and the rate of total entropy generation associated with this heat transfer process.

SOLUTION Steady heat transfer through a wall is considered. For specified heat transfer rate, wall temperatures, and environment temperatures, the entropy generation rate within the wall and the total entropy generation rate are to be determined.

Assumptions **1** The process is steady, and thus the rate of heat transfer through the wall is constant. **2** Heat transfer through the wall is one-dimensional.

Analysis We first take the *wall* as the system (Fig. 7–65). This is a *closed system* since no mass crosses the system boundary during the process. We note that the entropy change of the wall is zero during this process since the state and thus the entropy of the wall do not change anywhere in the wall. Heat and entropy are entering from one side of the wall and leaving from the other side.

The rate form of the entropy balance for the wall simplifies to

$$\underbrace{\dot{S}_{in} - \dot{S}_{out}}_{\substack{\text{Rate of net entropy}\\ \text{transfer by heat}\\ \text{and mass}}} + \underbrace{\dot{S}_{gen}}_{\substack{\text{Rate of entropy}\\ \text{generation}}} = \underbrace{dS_{system}/dt}_{\substack{\text{Rate of change}\\ \text{in entropy}}}^{\nearrow 0 \text{ (steady)}}$$

$$\left(\frac{\dot{Q}}{T}\right)_{in} - \left(\frac{\dot{Q}}{T}\right)_{out} + \dot{S}_{gen} = 0$$

$$\frac{1035 \text{ W}}{293 \text{ K}} - \frac{1035 \text{ W}}{278 \text{ K}} + \dot{S}_{gen} = 0$$

Therefore, the rate of entropy generation in the wall is

$$\dot{S}_{gen,wall} = \mathbf{0.191\ W/K}$$

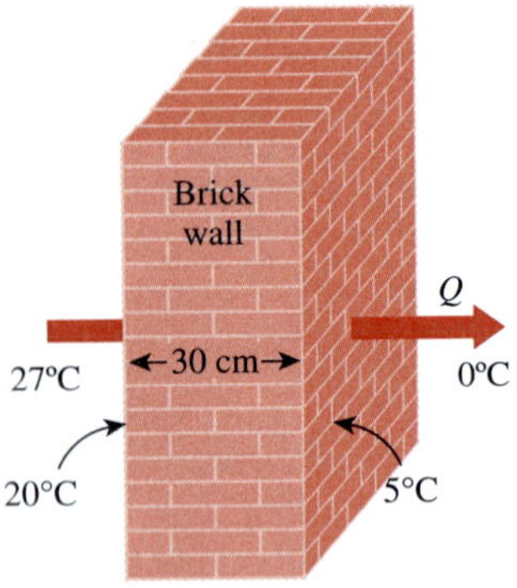

FIGURE 7–65
Schematic for Example 7–17.

Note that entropy transfer by heat at any location is Q/T at that location, and the direction of entropy transfer is the same as the direction of heat transfer.

To determine the rate of total entropy generation during this heat transfer process, we extend the system to include the regions on both sides of the wall that experience a temperature change. Then, one side of the system boundary becomes room temperature while the other side becomes the temperature of the outdoors. The entropy balance for this *extended system* (system + immediate surroundings) is the same as that given above, except the two boundary temperatures are now 300 and 273 K instead of 293 and 278 K, respectively. Then the rate of total entropy generation becomes

$$\frac{1035\ \text{W}}{300\ \text{K}} - \frac{1035\ \text{W}}{273\ \text{K}} + \dot{S}_{\text{gen,total}} = 0 \quad \rightarrow \quad \dot{S}_{\text{gen,total}} = \mathbf{0.341\ W/K}$$

Discussion Note that the entropy change of this extended system is also zero since the state of air does not change at any point during the process. The differences between the two entropy generations is 0.150 W/K, and it represents the entropy generated in the air layers on both sides of the wall. The entropy generation in this case is entirely due to irreversible heat transfer through a finite temperature difference.

EXAMPLE 7–18 Entropy Generation During a Throttling Process

Steam at 7 MPa and 450°C is throttled in a valve to a pressure of 3 MPa during a steady-flow process. Determine the entropy generated during this process and check if the increase of entropy principle is satisfied.

SOLUTION Steam is throttled to a specified pressure. The entropy generated during this process is to be determined, and the validity of the increase of entropy principle is to be verified.

Assumptions **1** This is a steady-flow process since there is no change with time at any point and thus $\Delta m_{CV} = 0$, $\Delta E_{CV} = 0$, and $\Delta S_{CV} = 0$. **2** Heat transfer to or from the valve is negligible. **3** The kinetic and potential energy changes are negligible, $\Delta ke = \Delta pe = 0$.

Analysis We take the throttling valve as the *system* (Fig. 7–66). This is a *control volume* since mass crosses the system boundary during the process. We note that there is only one inlet and one exit and thus $\dot{m}_1 = \dot{m}_2 = \dot{m}$. Also, the enthalpy of a fluid remains nearly constant during a throttling process and thus $h_2 \sum \cong h_1$.

The entropy of the steam at the inlet and the exit states is determined from the steam tables to be

State 1:
$$\left.\begin{array}{l} P_1 = 7\ \text{MPa} \\ T_1 = 450°\text{C} \end{array}\right\} \begin{array}{l} h_1 = 3288.3\ \text{kJ/kg} \\ s_1 = 6.6353\ \text{kJ/kg·K} \end{array}$$

State 2:
$$\left.\begin{array}{l} P_2 = 3\ \text{MPa} \\ h_2 = h_1 \end{array}\right\} \quad s_2 = 7.0046\ \text{kJ/kg·K}$$

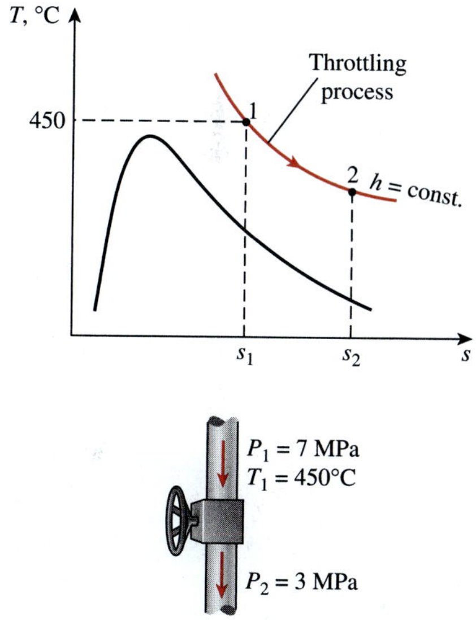

FIGURE 7–66
Schematic and *T-s* diagram for Example 7–18.

Then, the entropy generation per unit mass of the steam is determined from the entropy balance applied to the throttling valve,

$$\underbrace{\dot{S}_{in} - \dot{S}_{out}}_{\substack{\text{Rate of net entropy}\\\text{transfer by heat}\\\text{and mass}}} + \underbrace{\dot{S}_{gen}}_{\substack{\text{Rate of entropy}\\\text{generation}}} = \underbrace{dS_{system}/dt}_{\substack{\text{Rate of change}\\\text{in entropy}}}^{\nearrow 0\text{ (steady)}}$$

$$\dot{m}s_1 - \dot{m}s_2 + \dot{S}_{gen} = 0$$

$$\dot{S}_{gen} = \dot{m}(s_2 - s_1)$$

Dividing by mass flow rate and substituting gives

$$s_{gen} = s_2 - s_1 = 7.0046 - 6.6353 = \mathbf{0.3693\ kJ/kg \cdot K}$$

This is the amount of entropy generated per unit mass of steam as it is throttled from the inlet state to the final pressure, and it is caused by unrestrained expansion. The increase of entropy principle is obviously satisfied during this process since the entropy generation is positive.

EXAMPLE 7–19 Entropy Generated when a Hot Block Is Dropped in a Lake

A 50-kg block of iron casting at 500 K is thrown into a large lake that is at a temperature of 285 K. The iron block eventually reaches thermal equilibrium with the lake water. Assuming an average specific heat of 0.45 kJ/kg·K for the iron, determine (*a*) the entropy change of the iron block, (*b*) the entropy change of the lake water, and (*c*) the entropy generated during this process.

FIGURE 7–67
Schematic for Example 7–19.

SOLUTION A hot iron block is thrown into a lake, and cools to the lake temperature. The entropy changes of the iron and of the lake as well as the entropy generated during this process are to be determined.

Assumptions **1** Both the water and the iron block are incompressible substances. **2** Constant specific heats can be used for the water and the iron. **3** The kinetic and potential energy changes of the iron are negligible, $\Delta KE = \Delta PE = 0$ and thus $\Delta E = \Delta U$.

Properties The specific heat of the iron is 0.45 kJ/kg·K (Table A–3).

Analysis We take the *iron casting* as the system (Fig. 7–67). This is a *closed system* since no mass crosses the system boundary during the process.

To determine the entropy change for the iron block and for the lake, first we need to know the final equilibrium temperature. Given that the thermal energy capacity of the lake is very large relative to that of the iron block, the lake will absorb all the heat rejected by the iron block without experiencing any change in its temperature. Therefore, the iron block will cool to 285 K during this process while the lake temperature remains constant at 285 K.

(*a*) The entropy change of the iron block can be determined from

$$\Delta S_{iron} = m(s_2 - s_1) = mc_{avg} \ln \frac{T_2}{T_1}$$

$$= (50\text{ kg})(0.45\text{ kJ/kg}\cdot\text{K}) \ln \frac{285\text{ K}}{500\text{ K}}$$

$$= \mathbf{-12.65\ kJ/K}$$

(*b*) The temperature of the lake water remains constant during this process at 285 K. Also, the amount of heat transfer from the iron block to the lake is determined from an energy balance on the iron block to be

$$\underbrace{E_{in} - E_{out}}_{\substack{\text{Net energy transfer}\\\text{by heat, work, and mass}}} = \underbrace{\Delta E_{system}}_{\substack{\text{Change in internal, kinetic,}\\\text{potential, etc., energies}}}$$

$$-Q_{out} = \Delta U = mc_{avg}(T_2 - T_1)$$

or

$$Q_{out} = mc_{avg}(T_1 - T_2) = (50\text{ kg})(0.45\text{ kJ/kg·K})(500 - 285)\text{ K} = 4838\text{ kJ}$$

Then, the entropy change of the lake becomes

$$\Delta S_{lake} = \frac{Q_{lake}}{T_{lake}} = \frac{+4838\text{ kJ}}{285\text{ K}} = \mathbf{16.97\ kJ/K}$$

(*c*) The entropy generated during this process can be determined by applying an entropy balance on an *extended system* that includes the iron block and its immediate surroundings so that the boundary temperature of the extended system is at 285 K at all times:

$$\underbrace{S_{in} - S_{out}}_{\substack{\text{Net entropy transfer}\\\text{by heat and mass}}} + \underbrace{S_{gen}}_{\substack{\text{Entropy}\\\text{generation}}} = \underbrace{\Delta S_{system}}_{\substack{\text{Change}\\\text{in entropy}}}$$

$$-\frac{Q_{out}}{T_b} + S_{gen} = \Delta S_{system}$$

or

$$S_{gen} = \frac{Q_{out}}{T_b} + \Delta S_{system} = \frac{4838\text{ kJ}}{285\text{ K}} - (12.65\text{ kJ/K}) = \mathbf{4.32\ kJ/K}$$

Discussion The entropy generated can also be determined by taking the iron block and the entire lake as the system, which is an isolated system, and applying an entropy balance. An isolated system involves no heat or entropy transfer, and thus the entropy generation in this case becomes equal to the total entropy change,

$$S_{gen} = \Delta S_{total} = \Delta S_{system} + \Delta S_{lake} = -12.65 + 16.97 = 4.32\text{ kJ/K}$$

which is the same result obtained above.

EXAMPLE 7-20 Entropy Generation in a Heat Exchanger

Air in a large building is kept warm by heating it with steam in a heat exchanger (Fig. 7-68). Saturated water vapor enters this unit at 35°C at a rate of 10,000 kg/h and leaves as saturated liquid at 32°C. Air at 1-atm pressure enters the unit at 20°C and leaves at 30°C at about the same pressure. Determine the rate of entropy generation associated with this process.

SOLUTION Air is heated by steam in a heat exchanger. The rate of entropy generation associated with this process is to be determined.

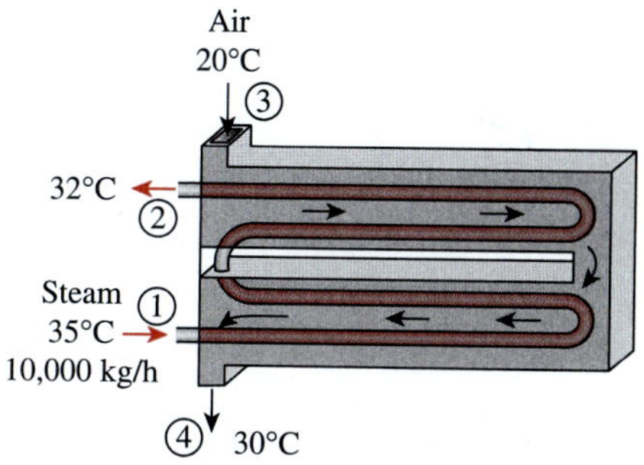

FIGURE 7-68
Schematic for Example 7-20

Assumptions **1** Steady operating conditions exist. **2** The heat exchanger is well-insulated so that heat loss to the surroundings is negligible and thus heat transfer from the hot fluid is equal to the heat transfer to the cold fluid. **3** Changes in the kinetic and potential energies of fluid streams are negligible. **4** Air is an ideal gas with constant specific heats at room temperature. **5** The pressure of air remains constant.

Analysis The rate of entropy generation within the heat exchanger is determined by applying the rate form of the entropy balance on the entire heat exchanger:

$$\underbrace{\dot{S}_{in} - \dot{S}_{out}}_{\substack{\text{Rate of net entropy}\\ \text{transfer by heat}\\ \text{and mass}}} + \underbrace{\dot{S}_{gen}}_{\substack{\text{Rate of entropy}\\ \text{generation}}} = \underbrace{\Delta \dot{S}_{system}^{\nearrow 0\ (\text{steady})}}_{\substack{\text{Rate of change}\\ \text{in entropy}}}$$

$$\dot{m}_{steam} s_1 + \dot{m}_{air} s_3 - \dot{m}_{steam} s_2 - \dot{m}_{air} s_4 + \dot{S}_{gen} = 0$$

$$\dot{S}_{gen} = \dot{m}_{steam}(s_2 - s_1) + \dot{m}_{air}(s_4 - s_3)$$

The specific heat of air at room temperature is $c_p = 1.005$ kJ/kg·°C (Table A-2*a*). The properties of the steam at the inlet and exit states are

$$\left.\begin{aligned} T_1 &= 35°\text{C} \\ x_1 &= 1 \end{aligned}\right\} \quad \begin{aligned} h_1 &= 2564.6 \text{ kJ/kg} \\ s_1 &= 8.3517 \text{ kJ/kg·K} \end{aligned} \qquad \text{(Table A–4)}$$

$$\left.\begin{aligned} T_2 &= 32°\text{C} \\ x_2 &= 0 \end{aligned}\right\} \quad \begin{aligned} h_2 &= 134.10 \text{ kJ/kg} \\ s_2 &= 0.4641 \text{ kJ/kg·K} \end{aligned} \qquad \text{(Table A–4)}$$

From an energy balance the heat transferred from steam is equal to the heat transferred to the air. Then, the mass flow rate of air is determined to be

$$\dot{Q} = \dot{m}_{steam}(h_1 - h_2) = (10{,}000/3600 \text{ kg/s})(2564.6 - 134.10) \text{ kJ/kg} = 6751 \text{ kW}$$

$$\dot{m}_{air} = \frac{\dot{Q}}{c_p(T_4 - T_3)} = \frac{6751 \text{ kW}}{(1.005 \text{ kJ/kg·°C})(30 - 20)°\text{C}} = 671.7 \text{ kg/s}$$

Substituting into the entropy balance relation, the rate of entropy generation becomes

$$\begin{aligned} \dot{S}_{gen} &= \dot{m}_{steam}(s_2 - s_1) + \dot{m}_{air}(s_4 - s_3) \\ &= \dot{m}_{steam}(s_2 - s_1) + \dot{m}_{air} c_p \ln \frac{T_4}{T_3} \\ &= (10{,}000/3600 \text{ kg/s})(0.4641 - 8.3517) \text{ kJ/kg·K} \\ &\quad + (671.7 \text{ kg/s})(1.005 \text{ kJ/kg·K}) \ln \frac{303 \text{ K}}{293 \text{ K}} \\ &= \mathbf{0.745 \text{ kW/K}} \end{aligned}$$

Discussion Note that the pressure of air remains nearly constant as it flows through the heat exchanger, and thus the pressure term is not included in the entropy change expression for air.

EXAMPLE 7–21 **Entropy Generation Associated with Heat Transfer**

A frictionless piston–cylinder device contains a saturated liquid–vapor mixture of water at 100°C. During a constant-pressure process, 600 kJ of heat is transferred to the surrounding air at 25°C. As a result, part of the water vapor contained in the cylinder condenses. Determine (*a*) the entropy change of the water and (*b*) the total entropy generation during this heat transfer process.

SOLUTION Saturated liquid–vapor mixture of water loses heat to its surroundings, and some of the vapor condenses. The entropy change of the water and the total entropy generation are to be determined.
Assumptions **1** There are no irreversibilities involved within the system boundaries, and thus the process is internally reversible. **2** The water temperature remains constant at 100°C everywhere, including the boundaries.
Analysis We first take the *water in the cylinder* as the system (Fig. 7–69). This is a *closed system* since no mass crosses the system boundary during the process. We note that the pressure and thus the temperature of water in the cylinder remain constant during this process. Also, the entropy of the system decreases during the process because of heat loss.

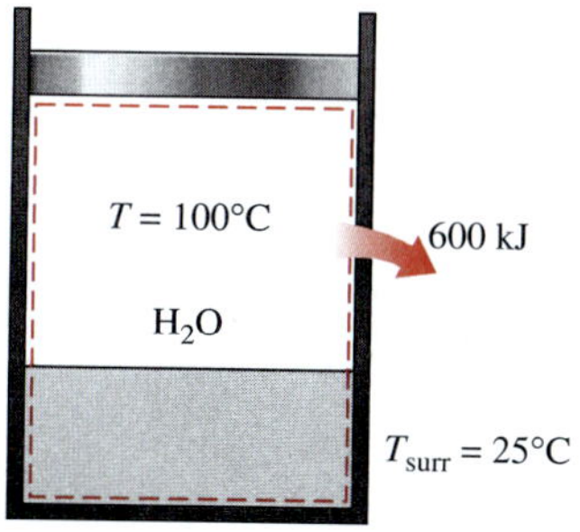

FIGURE 7–69
Schematic for Example 7–21.

(*a*) Noting that water undergoes an internally reversible isothermal process, its entropy change can be determined from

$$\Delta S_{\text{system}} = \frac{Q}{T_{\text{system}}} = \frac{-600 \text{ kJ}}{(100 + 273 \text{ K})} = \mathbf{-1.61 \text{ kJ/K}}$$

(*b*) To determine the total entropy generation during this process, we consider the *extended system*, which includes the water, the piston–cylinder device, and the region immediately outside the system that experiences a temperature change so that the entire boundary of the extended system is at the surrounding temperature of 25°C. The entropy balance for this *extended system* (system + immediate surroundings) yields

$$\underbrace{S_{\text{in}} - S_{\text{out}}}_{\substack{\text{Net entropy transfer} \\ \text{by heat and mass}}} + \underbrace{S_{\text{gen}}}_{\substack{\text{Entropy} \\ \text{generation}}} = \underbrace{\Delta S_{\text{system}}}_{\substack{\text{Change} \\ \text{in entropy}}}$$

$$-\frac{Q_{\text{out}}}{T_b} + S_{\text{gen}} = \Delta S_{\text{sysem}}$$

or

$$S_{\text{gen}} = \frac{Q_{\text{out}}}{T_b} + \Delta S_{\text{system}} = \frac{600 \text{ kJ}}{(25 + 273) \text{ K}} + (-1.61 \text{ kJ/K}) = \mathbf{0.40 \text{ kJ/K}}$$

The entropy generation in this case is entirely due to irreversible heat transfer through a finite temperature difference.

Note that the entropy change of this extended system is equivalent to the entropy change of water since the piston–cylinder device and the immediate surroundings do not experience any change of state at any point, and thus any change in any property, including entropy.
Discussion For the sake of argument, consider the reverse process (i.e., the transfer of 600 kJ of heat from the surrounding air at 25°C to saturated water at 100°C) and see if the increase of entropy principle can detect the impossibility of this process. This time, heat transfer will be to the water (heat gain instead of heat loss), and thus the entropy change of water will be +1.61 kJ/K. Also, the entropy transfer at the boundary of the extended system will have the same magnitude but opposite direction. This will result in an entropy generation of −0.4 kJ/K. The negative sign for the entropy generation indicates that the reverse process is *impossible*.

To complete the discussion, let us consider the case where the surrounding air temperature is a differential amount below 100°C (say 99.999 . . . 9°C) instead of being 25°C. This time, heat transfer from the saturated water to the surrounding air will take place through a differential temperature difference rendering this process *reversible*. It can be shown that $S_{gen} = 0$ for this process.

Remember that reversible processes are idealized processes, and they can be approached but never reached in reality.

Entropy Generation Associated with a Heat Transfer Process

In Example 7–21 it is determined that 0.4 kJ/K of entropy is generated during the heat transfer process, but it is not clear where exactly the entropy generation takes place, and how. To pinpoint the location of entropy generation, we need to be more precise about the description of the system, its surroundings, and the system boundary.

In that example, we assumed both the system and the surrounding air to be isothermal at 100 and 25°C, respectively. This assumption is reasonable if both fluids are well mixed. The inner surface of the wall must also be at 100°C while the outer surface is at 25°C since two bodies in physical contact must have the same temperature at the point of contact. Considering that entropy transfer with heat transfer Q through a surface at constant temperature T is Q/T, the entropy transfer from the water into the wall is $Q/T_{sys} = 1.61$ kJ/K. Likewise, entropy transfer from the outer surface of the wall into the surrounding air is $Q/T_{surr} = 2.01$ kJ/K. Obviously, entropy in the amount of $2.01 - 1.61 = 0.4$ kJ/K is generated in the wall, as illustrated in Fig. 7–70*b*.

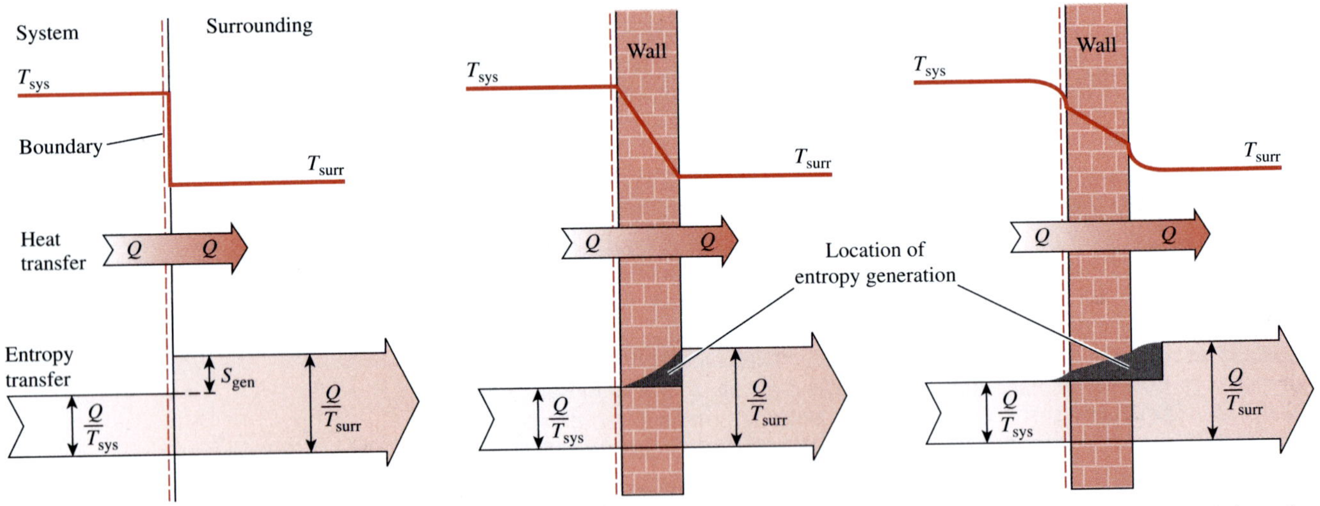

(*a*) The wall is ignored

(*b*) The wall is considered

(*c*) The wall as well as the variations of temperature in the system and the surroundings are considered

FIGURE 7–70
Graphical representation of entropy generation during a heat transfer process through a finite temperature difference.

Identifying the location of entropy generation enables us to determine whether a process is internally reversible. A process is internally reversible if no entropy is generated within the system boundaries. Therefore, the heat transfer process discussed in Example 7–21 is internally reversible if the inner surface of the wall is taken as the system boundary, and thus the system excludes the container wall. If the system boundary is taken to be the outer surface of the container wall, then the process is no longer internally reversible since the wall, which is the site of entropy generation, is now part of the system.

For thin walls, it is very tempting to ignore the mass of the wall and to regard the wall as the boundary between the system and the surroundings. This seemingly harmless choice hides the site of the entropy generation from view and is a source of confusion. The temperature in this case drops suddenly from T_{sys} to T_{surr} at the boundary surface, and confusion arises as to which temperature to use in the relation Q/T for entropy transfer at the boundary.

Note that if the system and the surrounding air are not isothermal as a result of insufficient mixing, then part of the entropy generation will occur in both the system and the surrounding air in the vicinity of the wall, as shown in Fig. 7–70*c*.

TOPIC OF SPECIAL INTEREST* Reducing the Cost of Compressed Air

Compressed air at gage pressures of 550 to 1000 kPa (80 to 150 psig) is commonly used in industrial facilities to perform a wide variety of tasks such as *cleaning, operating pneumatic equipment*, and even *refrigeration*. It is often referred to as the *fourth utility* after electricity, water, and natural gas or oil. In production facilities, there is a widespread waste of energy associated with compressed-air systems and a general lack of awareness about the opportunities to conserve energy. A considerable portion of the energy waste associated with compressed-air systems can be avoided by following some commonsense measures. In this section we discuss the energy losses associated with compressed-air systems and their costs to manufacturers. We also show how to reduce the cost of compressed air in existing facilities by making some modifications with attractive payback periods. With the exception of a few compressors that are driven by natural gas engines, all compressors are driven by electric motors (Fig. 7–71).

FIGURE 7–71
A 1250-hp compressor assembly.
Courtesy of Dresser Rand Company, Painted Post, NY.

Some primitive methods of producing an *air blast* to keep the fire in furnaces alive, such as air-threading bags and the Chinese wind box, date back at least to 2000 BC. The *water trompe* that compresses air by the fall of water in a tube to blow forges (metal heat shops) is believed to have been in use by 150 BC. In 1650, Otto van Guericke made great improvements in both the compressor and vacuum pump. In 1683, Papin proposed using

*This section can be skipped without a loss in continuity.

compressed air to *transmit power* over long distances. In 1829, William Mann received a patent for multistage compression of air. In 1830, Thilorier was recognized for compressing gases to high pressures in *stages*. In 1890, Edward Rix transmitted power with air several miles to operate lifting machines in the North Star mine near Grass Valley, California, by using a compressor driven by Pelton wheels. In 1872, *cooling* was adapted to increase efficiency by spraying water directly into the cylinder through the air inlet valves. This "wet compression" was abandoned later because of the problems it caused. The cooling then was accomplished externally by *water jacketing* the cylinders. The first large-scale compressor used in the United States was a four-cylinder unit built in 1866 for use in the Hoosac tunnel. The cooling was first accomplished by water injection into the cylinder, and later by running a stream of water over the cylinder. Major advances in recent compressor technology are due to Burleigh, Ingersoll, Sergeant, Rand, and Clayton, among others.

The compressors used range from a few horsepower to more than 10,000 hp in size, and they are among the major energy-consuming equipment in most manufacturing facilities. Manufacturers are quick to identify energy (and thus money) losses from *hot surfaces* and to insulate those surfaces. However, somehow they are not so sensitive when it comes to saving *compressed air* since they view air as being free, and the only time the air leaks and dirty air filters get some attention is when the air and pressure losses interfere with the normal operation of the plant. However, paying attention to the compressed-air system and practicing some simple conservation measures can result in considerable energy and cost savings for the plants.

The hissing of *air leaks* can sometimes be heard even in high-noise manufacturing facilities. *Pressure drops* at end-use points in the order of 40 percent of the compressor-discharged pressure are not uncommon. Yet a common response to such a problem is the installation of a larger compressor instead of checking the system and finding out what the problem is. The latter corrective action is usually taken only after the larger compressor also fails to eliminate the problem. The energy wasted in compressed-air systems because of poor installation and maintenance can account for up to 50 percent of the energy consumed by the compressor, and about half of this amount can be saved by simple measures.

Compressor: 125 hp = 93.21 kW
Operating hours: 6000 h/yr
Unit cost of electricity: $0.085/kWh
Motor efficiency: 0.90

Annual energy usage: 621,417 kWh
Annual electricity cost: $52,820/yr

FIGURE 7–72
The cost of electricity to operate a compressor for one year can exceed the purchase price of the compressor.

The cost of electricity to operate a compressor for one year can exceed the purchase price of the compressor. This is especially the case for larger compressors operating two or three shifts. For example, operating a 125-hp compressor powered by a 90-percent efficient electric motor at full load for 6000 hours a year at $0.085/kWh will cost $52,820 a year in electricity cost, which greatly exceeds the purchase and installation cost of a typical unit (Fig. 7–72).

Below we describe some procedures to reduce the cost of compressed air in industrial facilities and quantify the energy and cost savings associated with them. Once the compressor power wasted is determined, the *annual energy* (usually electricity) and *cost savings* can be determined from

$$\text{Energy savings} = (\text{Power saved})(\text{Operating hours})/\eta_{\text{motor}} \quad \textbf{(7–87)}$$

and

$$\text{Cost savings} = (\text{Energy savings})(\text{Unit cost of energy}) \quad \textbf{(7–88)}$$

where η_{motor} is the efficiency of the motor driving the compressor and the unit cost of energy is usually expressed in dollars per kilowatt hour (1 kWh = 3600 kJ).

1 Repairing Air Leaks on Compressed-Air Lines

Air leaks are the greatest single cause of energy loss in manufacturing facilities associated with compressed-air systems. It takes energy to compress the air, and thus the loss of compressed air is a loss of energy for the facility. A compressor must work harder and longer to make up for the lost air and must use more energy in the process. Several studies at plants have revealed that up to 40 percent of the compressed air is lost through leaks. Eliminating the air leaks totally is impractical, and a leakage rate of 10 percent is considered acceptable.

Air leaks, in general, occur at the *joints, flange connections, elbows, reducing bushes, sudden expansions, valve systems, filters, hoses, check valves, relief valves, extensions*, and the *equipment* connected to the compressed-air lines (Fig. 7–73). Expansion and contraction as a result of thermal cycling and vibration are common causes of loosening at the joints, and thus air leaks. Therefore, it is a good practice to *check* the joints for tightness and to *tighten* them periodically. Air leaks also commonly occur at the points of end use or where the compressed-air lines are connected to the equipment that operates on compressed air. Because of the frequent opening and closing of the compressed-air lines at these points, the gaskets wear out quickly, and they need to be replaced periodically.

There are many ways of detecting air leaks in a compressed-air system. Perhaps the simplest way of detecting a large air leak is to listen for it. The high velocity of the air escaping the line produces a *hissing sound* that is difficult not to notice except in environments with a high noise level. Another way of detecting air leaks, especially small ones, is to test the suspected area with *soap water* and to watch for soap bubbles. This method is obviously not practical for a large system with many connections. A modern way of checking for air leaks is to use an acoustic leak detector, which consists of a directional microphone, amplifiers, audio filters, and digital indicators.

A practical way of quantifying the air leaks in a production facility in its entirety is to conduct a *pressure drop test*. The test is conducted by stopping all the operations that use compressed air and by shutting down the compressors and closing the pressure relief valve, which relieves pressure automatically if the compressor is equipped with one. This way, any pressure drop in the compressed-air lines is due to the cumulative effects of air leaks. The drop in pressure in the system with time is observed, and the test is conducted until the pressure drops by an amount that can be measured accurately, usually 0.5 atm. The time it takes for the pressure to drop by this amount is measured, and the decay of pressure as a function of time is recorded. The total volume of the compressed-air system, including the compressed-air tanks, the headers, accumulators, and the primary compressed-air lines, is calculated. Ignoring the small lines will make the job easier and will cause the result to be more conservative. The rate of air leak can be determined using the ideal gas equation of state.

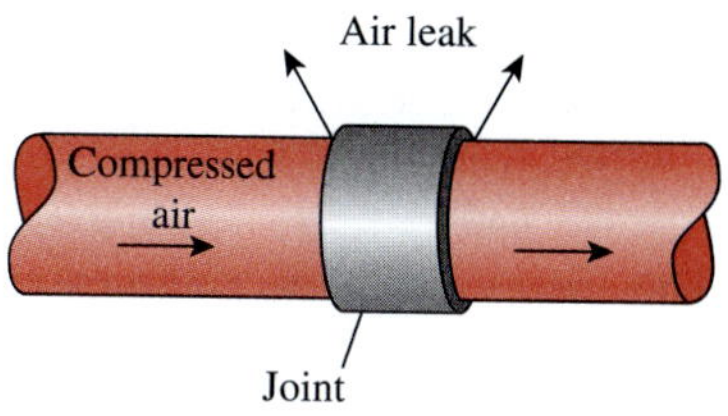

FIGURE 7–73
Air leaks commonly occur at joints and connections.

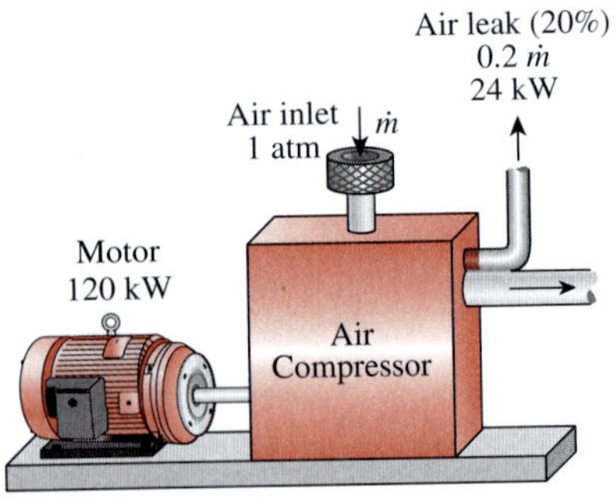

FIGURE 7–74
The energy wasted as compressed air escapes through the leaks is equivalent to the energy it takes to compress it.

The amount of *mechanical energy wasted* as a unit mass of air escapes through the leaks is equivalent to the actual amount of energy it takes to compress it, and is determined from Eq. 7–57, modified as (Fig. 7–74)

$$w_{\text{comp,in}} = \frac{w_{\text{reversible comp,in}}}{\eta_{\text{comp}}} = \frac{nRT_1}{\eta_{\text{comp}}(n-1)}\left[\left(\frac{P_2}{P_1}\right)^{(n-1)/n} - 1\right] \quad \textbf{(7–89)}$$

where n is the polytropic compression exponent ($n = 1.4$ when the compression is isentropic and $1 < n < 1.4$ when there is intercooling) and η_{comp} is the compressor efficiency, whose value usually ranges between 0.7 and 0.9.

Using compressible-flow theory (see Chap. 17), it can be shown that whenever the line pressure is above 2 atm, which is usually the case, the velocity of air at the leak site must be equal to the local *speed of sound*. Then the **mass flow rate of air** through a leak of minimum cross-sectional area A becomes

$$\dot{m}_{\text{air}} = C_{\text{discharge}}\left(\frac{2}{k+1}\right)^{1/(k-1)}\frac{P_{\text{line}}}{RT_{\text{line}}}A\sqrt{kR\left(\frac{2}{k+1}\right)T_{\text{line}}} \quad \textbf{(7–90)}$$

where k is the specific heat ratio ($k = 1.4$ for air) and $C_{\text{discharge}}$ is a discharge (or loss) coefficient that accounts for imperfections in flow at the leak site. Its value ranges from about 0.60 for an orifice with sharp edges to 0.97 for a well-rounded circular hole. The air-leak sites are imperfect in shape, and thus the discharge coefficient can be taken to be 0.65 in the absence of actual data. Also, T_{line} and P_{line} are the temperature and pressure in the compressed-air line, respectively.

Once $\dot{m}_{air}$ and $w_{\text{comp,in}}$ are available, the **power wasted** by the leaking compressed air (or the power saved by repairing the leak) is determined from

$$\text{Power saved} = \text{Power wasted} = \dot{m}_{\text{air}} w_{\text{comp,in}} \quad \textbf{(7–91)}$$

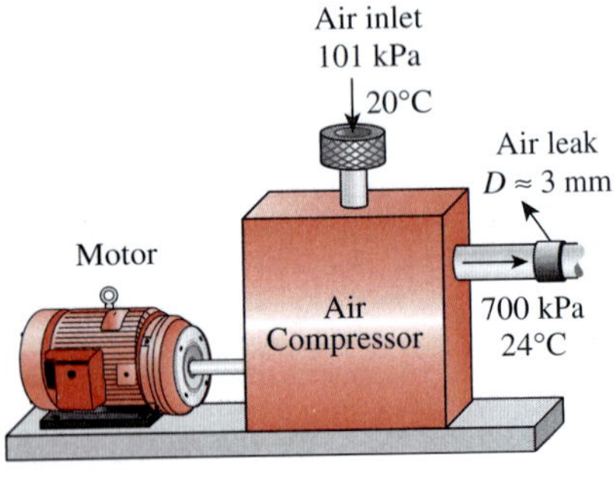

FIGURE 7–75
Schematic for Example 7–22.

EXAMPLE 7–22 Energy and Cost Savings by Fixing Air Leaks

The compressors of a production facility maintain the compressed-air lines at a (gauge) pressure of 700 kPa at sea level where the atmospheric pressure is 101 kPa (Fig. 7–75). The average temperature of air is 20°C at the compressor inlet and 24°C in the compressed-air lines. The facility operates 4200 hours a year, and the average price of electricity is $0.078/kWh. Taking the compressor efficiency to be 0.8, the motor efficiency to be 0.92, and the discharge coefficient to be 0.65, determine the energy and money saved per year by sealing a leak equivalent to a 3-mm-diameter hole on the compressed-air line.

SOLUTION An air leak in the compressed air lines of a facility is considered. The energy and money saved per year by sealing the leak are to be determined.

Assumptions **1** Steady operating conditions exist. **2** Air is an ideal gas. **3** Pressure losses in the compressed air lines are negligible.

Analysis We note that the absolute pressure is the sum of the gauge and atmospheric pressures.

The work needed to compress a unit mass of air at 20°C from the atmospheric pressure of 101 kPa to 700 + 101 = 801 kPa is

$$w_{\text{comp,in}} = \frac{nRT_1}{\eta_{\text{comp}}(n-1)}\left[\left(\frac{P_2}{P_1}\right)^{(n-1)/n} - 1\right]$$

$$= \frac{(1.4)(0.287 \text{ kJ/kg·K})(293 \text{ K})}{(0.8)(1.4-1)}\left[\left(\frac{801 \text{ kPa}}{101 \text{ kPa}}\right)^{0.4/1.4} - 1\right] = 296.9 \text{ kJ/kg}$$

The cross-sectional area of the 3-mm-diameter hole is

$$A = \pi D^2/4 = \pi(3 \times 10^{-3} \text{ m})^2/4 = 7.069 \times 10^{-6} \text{ m}^2$$

Noting that the line conditions are 297 K and 801 kPa, the mass flow rate of the air leaking through the hole is determined to be

$$\dot{m}_{\text{air}} = C_{\text{discharge}}\left(\frac{2}{k+1}\right)^{1/(k-1)}\frac{P_{\text{line}}}{RT_{\text{line}}}A\sqrt{kR\left(\frac{2}{k+1}\right)T_{\text{line}}}$$

$$= (0.65)\left(\frac{2}{1.4+1}\right)^{1/(1.4-1)}\frac{801 \text{ kPa}}{(0.287 \text{ kPa·m}^3/\text{kg·K})(297 \text{ K})}(7.069 \times 10^{-6} \text{ m}^2)$$

$$\times \sqrt{(1.4)(0.287 \text{ kJ/kg·K})\left(\frac{1000 \text{ m}^2/\text{s}^2}{1 \text{ kJ/kg}}\right)\left(\frac{2}{1.4+1}\right)(297 \text{ K})}$$

$$= 0.008632 \text{ kg/s}$$

Then, the power wasted by the leaking compressed air becomes

$$\text{Power wasted} = \dot{m}_{\text{air}}w_{\text{comp,in}}$$

$$= (0.008632 \text{ kg/s})(296.9 \text{ kJ/kg})$$

$$= 2.563 \text{ kW}$$

The compressor operates 4200 h/yr, and the motor efficiency is 0.92. Then the annual energy and cost savings resulting from repairing this leak are determined to be

$$\text{Energy savings} = (\text{Power saved})(\text{Operating hours})/\eta_{\text{motor}}$$

$$= (2.563 \text{ kW})(4200 \text{ h/yr})/0.92$$

$$= \mathbf{11{,}700 \text{ kWh/yr}}$$

$$\text{Cost savings} = (\text{Energy savings})(\text{Unit cost of energy})$$

$$= (11{,}700 \text{ kWh/yr})(\$0.078/\text{kWh})$$

$$= \mathbf{\$913/yr}$$

Discussion Note that the facility will save 11,700 kWh of electricity worth $913 a year when this air leak is fixed. This is a substantial amount for a single leak whose equivalent diameter is 3 mm.

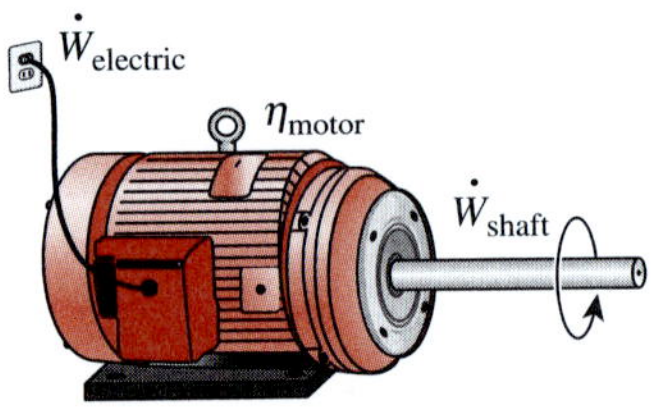

Motor efficiency η_{motor}	Electrical power consumed per kW of mechanical (shaft) power output, $\dot{W}_{electric} = \dot{W}_{shaft}/\eta_{motor}$
100%	1.00 kW
90	1.11
80	1.25
70	1.43
60	1.67
50	2.00
40	2.50
30	3.33
20	5.00
10	10.00

FIGURE 7–76
The electrical energy consumed by a motor is inversely proportional to its efficiency.

2 Installing High-Efficiency Motors

Practically all compressors are powered by electric motors, and the *electrical energy* a motor draws for a specified power output is *inversely proportional* to its efficiency. Electric motors cannot convert the electrical energy they consume into mechanical energy completely, and the ratio of the mechanical power supplied to the electrical power consumed during operation is called the **motor efficiency**, η_{motor}. Therefore, the electric power consumed by the motor and the mechanical (shaft) power supplied to the compressor are related to each other by (Fig. 7–76)

$$\dot{W}_{electric} = \dot{W}_{comp}/\eta_{motor} \quad \textbf{(7–92)}$$

For example, assuming no transmission losses, a motor that is 80 percent efficient will draw 1/0.8 = 1.25 kW of electric power for each kW of shaft power it delivers to the compressor, whereas a motor that is 95 percent efficient will draw only 1/0.95 = 1.05 kW to deliver 1 kW. Therefore, high-efficiency motors cost less to operate than their standard counterparts, but they also usually cost more to purchase. However, the energy savings usually make up for the price differential during the first few years. This is especially true for large compressors that operate more than one regular shift. The *electric power saved* by replacing the existing standard motor of efficiency $\eta_{standard}$ by a high-efficiency one of efficiency $\eta_{efficient}$ is determined from

$$\begin{aligned} \dot{W}_{electric,saved} &= \dot{W}_{electric,standard} - \dot{W}_{electric,efficient} \\ &= \dot{W}_{comp}(1/\eta_{standard} - 1/\eta_{efficient}) \quad \textbf{(7–93)} \\ &= \text{(Rated power)(Load factor)}(1/\eta_{standard} - 1/\eta_{efficient}) \end{aligned}$$

where *rated power* is the nominal power of the motor listed on its label (the power the motor delivers at full load) and the *load factor* is the fraction of the rated power at which the motor normally operates. Then, the annual energy savings as a result of replacing a motor by a high-efficiency motor instead of a comparable standard one is

$$\text{Energy savings} = \dot{W}_{electric,saved} \times \text{Annual operating hours} \quad \textbf{(7–94)}$$

The efficiencies of motors used to power compressors usually range from about 70 percent to over 96 percent. The portion of electric energy not converted to mechanical energy is converted to heat. The amount of heat generated by the motors may reach high levels, especially at part load, and it may cause overheating if not dissipated effectively. It may also cause the air temperature in the compressor room to rise to undesirable levels. For example, a 90-percent-efficient 100-kW motor generates as much heat as a 10-kW resistance heater in the confined space of the compressor room, and it contributes greatly to the heating of the air in the room. If this heated air is not vented properly, and the air into the compressor is drawn from inside the compressor room, the performance of the compressor will also decline, as explained later.

Important considerations in the selection of a motor for a compressor are the operating profile of the compressor (i.e., the variation of the load with time), and the efficiency of the motor at part-load conditions. The part-load efficiency of a motor is as important as the full-load efficiency if the compressor is expected to operate at part load during a significant portion of the total operating time. A typical motor has a nearly flat efficiency curve between half load and full load, and peak efficiency is usually at about 75% load. Efficiency falls off pretty steeply below half load, and thus operation below 50% load should be avoided as much as possible. For example, the efficiency of a motor may drop from 90 percent at full load to 87 percent at half load and 80 percent at quarter load (Fig. 7–77). The efficiency of another motor of similar specifications, on the other hand, may drop from 91 percent at full load to 75 percent at quarter load. The first motor is obviously better suited for a situation in which a compressor is expected to operate at quarter load during a significant portion of the time. The efficiency at part-load conditions can be improved greatly by installing variable voltage controllers if it is economical to do so. Also, *oversizing* a motor just to be on the safe side and to have some excess power just in case is a bad practice since this will cause the motor to operate almost always at *part load* and thus at a *lower efficiency*. Besides, oversized motors have a higher initial cost. However, oversized motors waste little energy as long as they operate at loads above 50% of design.

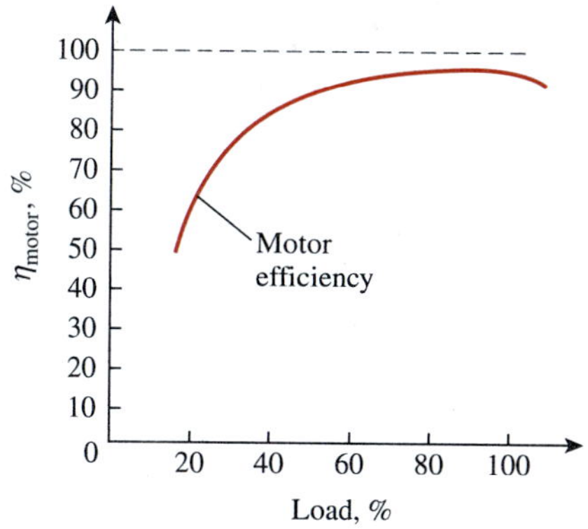

FIGURE 7–77
The efficiency of an electric motor decreases at part load.

Using a Smaller Motor at High Capacity

We tend to purchase *larger equipment* than needed for reasons like having a safety margin or anticipated future expansion, and compressors are no exception. The uncertainties in plant operation are partially responsible for opting for a larger compressor, since it is preferred to have an oversized compressor than an undersized one. Sometimes compressors that have several times the required capacity are purchased with the perception that the extra capacity may be needed some day. The result is a compressor that runs intermittently at full load, or one that runs continuously at part load.

A compressor that operates at part load also causes the motor to operate less efficiently since the efficiency of an electric motor decreases as the point of operation shifts down from its rated power, as explained above. The result is a motor that consumes more electricity per unit power delivered, and thus a more expensive operation. The operating costs can be reduced by switching to a smaller motor that runs at rated power and thus at a higher efficiency.

3 Using Outside Air for Compressor Intake

We have pointed out earlier that the power consumed by a compressor is proportional to the *specific volume*, which is proportional to the *absolute temperature* of the gas at a given pressure. It is also clear from Eq. 7–89 that the compressor work is directly proportional to the *inlet temperature* of air. Therefore, the lower the inlet temperature of the air, the smaller the compressor work. Then the *power reduction factor*, which is the fraction of compressor power reduced as a result of taking intake air from the outside, becomes

$$f_{\text{reduction}} = \frac{W_{\text{comp,inside}} - W_{\text{comp,outside}}}{W_{\text{comp,inside}}} = \frac{T_{\text{inside}} - T_{\text{outside}}}{T_{\text{inside}}} = 1 - \frac{T_{\text{outside}}}{T_{\text{inside}}} \qquad \textbf{(7–95)}$$

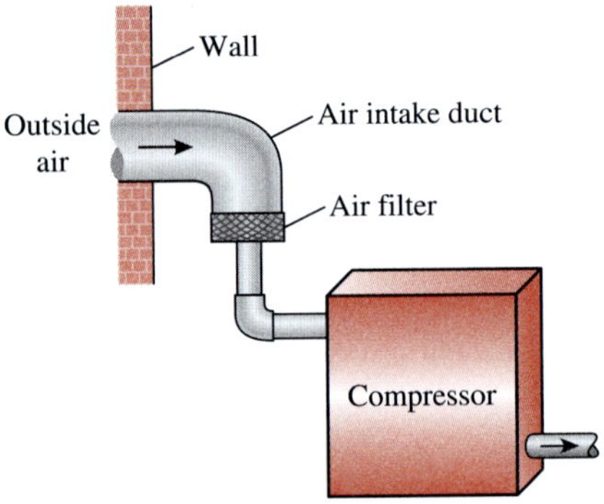

FIGURE 7–78
The power consumption of a compressor can be reduced by taking in air from the outside.

where T_{inside} and T_{outside} are the absolute temperatures (in K or R) of the ambient air inside and outside the facility, respectively. Thus, reducing the absolute inlet temperature by 5 percent, for example, will reduce the compressor power input by 5 percent. As a rule of thumb, for a specified amount of compressed air, the power consumption of the compressor decreases (or, for a fixed power input, the amount of compressed air increases) by 1 percent for each 3°C drop in the temperature of the inlet air to the compressor.

Compressors are usually located inside the production facilities or in adjacent shelters specifically built outside these facilities. The intake air is normally drawn from inside the building or the shelter. However, in many locations the air temperature in the building is higher than the outside air temperature, because of space heaters in the winter and the heat given up by a large number of mechanical and electrical equipment as well as the furnaces year round. The temperature rise in the shelter is also due to the heat dissipation from the compressor and its motor. The outside air is generally *cooler* and thus *denser* than the air in the compressor room even on hot summer days. Therefore, it is advisable to install an *intake duct* to the compressor inlet so that the air is supplied directly from the outside of the building instead of the inside, as shown in Fig. 7–78. This will reduce the energy consumption of the compressor since it takes less energy to compress a specified amount of cool air than the same amount of warm air. Compressing the warm air in a building in winter also wastes the energy used to heat the air.

4 Reducing the Air Pressure Setting

Another source of energy waste in compressed-air systems is compressing the air to a higher pressure than required by the air-driven equipment since it takes more energy to compress air to a higher pressure. In such cases considerable energy savings can be realized by determining the minimum required pressure and then reducing the air pressure control setting on the compressor accordingly. This can be done on both screw-type and reciprocating compressors by simply adjusting the pressure setting to match the needs.

The amount of energy it takes to compress a unit mass of air is determined from Eq. 7–89. We note from that relation that the higher the pressure P_2 at the compressor exit, the larger the work required for compression. Reducing the exit pressure of the compressor to $P_{2,\text{reduced}}$ will reduce the power input requirements of the compressor by a factor of

$$f_{\text{reduction}} = \frac{w_{\text{comp,current}} - w_{\text{comp,reduced}}}{w_{\text{comp,current}}} = 1 - \frac{(P_{2,\text{reduced}}/P_1)^{(n-1)/n} - 1}{(P_2/P_1)^{(n-1)/n} - 1} \qquad \textbf{(7–96)}$$

A power reduction (or savings) factor of $f_{\text{reduction}} = 0.08$, for example, indicates that the power consumption of the compressor is reduced by 8 percent as a result of reducing the pressure setting.

Some applications require slightly compressed air. In such cases, the need can be met by a blower instead of a compressor. Considerable energy can be saved in this manner since a blower requires a small fraction of the power needed by a compressor for a specified mass flow rate.

EXAMPLE 7–23 Reducing the Pressure Setting to Reduce Cost

The compressed-air requirements of a plant located at 1400-m elevation is being met by a 75-hp compressor that takes in air at the local atmospheric pressure of 85.6 kPa and the average temperature of 15°C and compresses it to 900 kPa gauge (Fig. 7–79). The plant is currently paying $12,000 a year in electricity costs to run the compressor. An investigation of the compressed-air system and the equipment using the compressed air reveals that compressing the air to 800 kPa is sufficient for this plant. Determine how much money will be saved as a result of reducing the pressure of the compressed air.

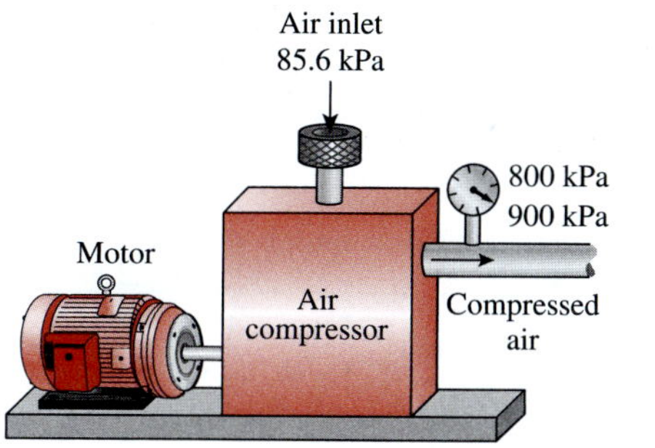

FIGURE 7–79
Schematic for Example 7–23.

Solution It is observed that the compressor of a facility compresses the air to much higher pressures than needed. The cost savings associated with pressure reduction of the compressor are to be determined.

Assumptions **1** Air is an ideal gas. **2** Compression process is isentropic, and thus $n = k = 1.4$.

Analysis The fraction of energy saved as a result of reducing the pressure setting of the compressor is

$$f_{\text{reduction}} = 1 - \frac{(P_{2,\text{reduced}}/P_1)^{(n-1)/n} - 1}{(P_2/P_1)^{(n-1)/n} - 1}$$

$$= 1 - \frac{(885.6/85.6)^{(1.4-1)/1.4} - 1}{(985.6/85.6)^{(1.4-1)/1.4} - 1} = 0.060$$

That is, reducing the pressure setting will reduce the energy consumed by the compressor by about 6 percent. Then,

$$\text{Cost savings} = (\text{Current cost})f_{\text{reduction}} = (\$12{,}000/\text{yr})(0.06) = \mathbf{\$720/yr}$$

Therefore, reducing the pressure setting by 100 kPa will result in annual savings of $720 in this case.

There are also other ways to reduce the cost of compressed air in industrial facilities. An obvious way is *turning the compressor off* during nonproduction periods such as lunch hours, nights, and even weekends. A considerable amount of power can be wasted during this stand-by mode. This is especially the case for screw-type compressors since they consume up to 85 percent of their rated power in this mode. The reciprocating compressors are not immune from this deficiency, however, since they also must cycle on and off because of the air leaks present in the compressed-air lines. The system can be shut down manually during nonproduction periods to save energy, but installing a timer (with manual override) is preferred to do this automatically since it is human nature to put things off when the benefits are not obvious or immediate.

The compressed air is sometimes cooled considerably below its dew point in *refrigerated dryers* in order to condense and remove a large fraction of the water vapor in the air as well as other noncondensable gases such as oil vapors. The temperature of air rises considerably as it is compressed, sometimes exceeding 250°C at compressor exit when compressed adiabatically to just 700 kPa. Therefore, it is desirable to cool air after compression in order

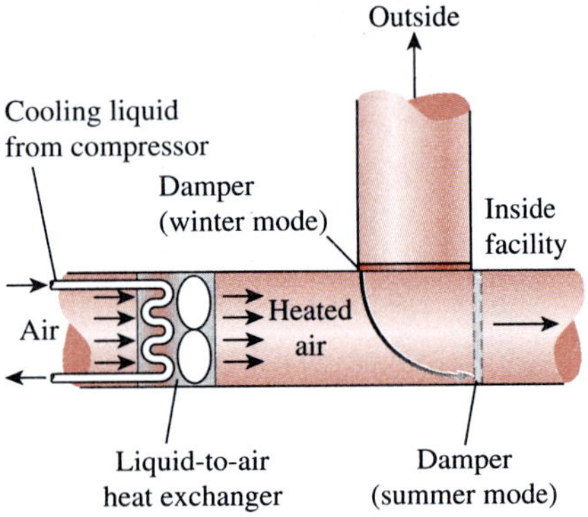

FIGURE 7–80
Waste heat from a compressor can be used to heat a building in winter.

to minimize the amount of power consumed by the refrigeration system, just as it is desirable to let the hot food in a pan cool to the ambient temperature before putting it into the refrigerator. The cooling can be done by either ambient air or water, and the heat picked up by the cooling medium can be used for space heating, feedwater heating, or process-related heating.

Compressors are commonly cooled directly by air or by circulating a liquid such as oil or water through them in order to minimize the power consumption. The heat picked up by the oil or water is usually rejected to the ambient in a liquid-to-air heat exchanger. This *heat rejected* usually amounts to 60 to 90 percent of the power input, and thus it represents a huge amount of energy that can be used for a useful purpose such as *space heating* in winter, *preheating* the air or water in a furnace, or other process-related purposes (Fig. 7–80). For example, assuming 80 percent of the power input is converted to heat, a 150-hp compressor can reject as much heat as a 90-kW electric resistance heater or a 400,000-Btu/h natural gas heater when operating at full load. Thus, the proper utilization of the waste heat from a compressor can result in significant energy and cost savings.

SUMMARY

The second law of thermodynamics leads to the definition of a new property called *entropy*, which is a quantitative measure of microscopic disorder for a system. Any quantity whose cyclic integral is zero is a property, and entropy is defined as

$$dS = \left(\frac{dQ}{T}\right)_{\text{int rev}}$$

For the special case of an internally reversible, isothermal process, it gives

$$\Delta S = \frac{Q}{T_0}$$

The inequality part of the Clausius inequality combined with the definition of entropy yields an inequality known as the *increase of entropy principle*, expressed as

$$S_{\text{gen}} \geq 0$$

where S_{gen} is the *entropy generated* during the process. Entropy change is caused by heat transfer, mass flow, and irreversibilities. Heat transfer to a system increases the entropy, and heat transfer from a system decreases it. The effect of irreversibilities is always to increase the entropy.

The *entropy-change* and *isentropic relations* for a process can be summarized as follows:

1. *Pure substances:*

Any process: $\Delta s = s_2 - s_1$

Isentropic process: $s_2 = s_1$

2. *Incompressible substances:*

Any process: $s_2 - s_1 = c_{\text{avg}} \ln \dfrac{T_2}{T_1}$

Isentropic process: $T_2 = T_1$

3. *Ideal gases:*

a. Constant specific heats (approximate treatment):
Any process:

$$s_2 - s_1 = c_{v,\text{avg}} \ln \frac{T_2}{T_1} + R \ln \frac{v_2}{v_1}$$

$$s_2 - s_1 = c_{p,\text{avg}} \ln \frac{T_2}{T_1} - R \ln \frac{P_2}{P_1}$$

Isentropic process:

$$\left(\frac{T_2}{T_1}\right)_{s=\text{const.}} = \left(\frac{v_1}{v_2}\right)^{k-1}$$

$$\left(\frac{T_2}{T_1}\right)_{s=\text{const.}} = \left(\frac{P_2}{P_1}\right)^{(k-1)/k}$$

$$\left(\frac{P_2}{P_1}\right)_{s=\text{const.}} = \left(\frac{v_1}{v_2}\right)^{k}$$

b. Variable specific heats (exact treatment):
Any process:

$$s_2 - s_1 = s_2^\circ - s_1^\circ - R \ln \frac{P_2}{P_1}$$

Isentropic process:

$$s_2^\circ = s_1^\circ + R \ln \frac{P_2}{P_1}$$

$$\left(\frac{P_2}{P_1}\right)_{s=\text{const.}} = \frac{P_{r2}}{P_{r1}}$$

$$\left(\frac{v_2}{v_1}\right)_{s=\text{const.}} = \frac{v_{r2}}{v_{r1}}$$

where P_r is the *relative pressure* and v_r is the *relative specific volume*. The function s° depends on temperature only.

The *steady-flow work* for a reversible process can be expressed in terms of the fluid properties as

$$w_{\text{rev}} = -\int_1^2 v\,dP - \Delta\text{ke} - \Delta\text{pe}$$

For incompressible substances (v = constant) it simplifies to

$$w_{\text{rev}} = -v(P_2 - P_1) - \Delta\text{ke} - \Delta\text{pe}$$

The work done during a steady-flow process is proportional to the specific volume. Therefore, v should be kept as small as possible during a compression process to minimize the work input and as large as possible during an expansion process to maximize the work output.

The reversible work inputs to a compressor compressing an ideal gas from T_1, P_1 to P_2 in an isentropic (Pv^k = constant), polytropic (Pv^n = constant), or isothermal (Pv = constant) manner, are determined by integration for each case with the following results:

Isentropic: $$w_{\text{comp,in}} = \frac{kR(T_2 - T_1)}{k-1} = \frac{kRT_1}{k-1}\left[\left(\frac{P_2}{P_1}\right)^{(k-1)/k} - 1\right]$$

Polytropic: $$w_{\text{comp,in}} = \frac{nR(T_2 - T_1)}{n-1} = \frac{nRT_1}{n-1}\left[\left(\frac{P_2}{P_1}\right)^{(n-1)/n} - 1\right]$$

Isothermal: $$w_{\text{comp,in}} = RT \ln \frac{P_2}{P_1}$$

The work input to a compressor can be reduced by using multistage compression with intercooling. For maximum savings from the work input, the pressure ratio across each stage of the compressor must be the same.

Most steady-flow devices operate under adiabatic conditions, and the ideal process for these devices is the isentropic process. The parameter that describes how efficiently a device approximates a corresponding isentropic device is called *isentropic* or *adiabatic efficiency*. It is expressed for turbines, compressors, and nozzles as follows:

$$\eta_T = \frac{\text{Actual turbine work}}{\text{Isentropic turbine work}} = \frac{w_a}{w_s} \cong \frac{h_1 - h_{2a}}{h_1 - h_{2s}}$$

$$\eta_C = \frac{\text{Isentropic compressor work}}{\text{Actual compressor work}} = \frac{w_s}{w_a} \cong \frac{h_{2s} - h_1}{h_{2a} - h_1}$$

$$\eta_N = \frac{\text{Actual KE at nozzle exit}}{\text{Isentropic KE at nozzle exit}} = \frac{V_{2a}^2}{V_{2s}^2} \cong \frac{h_1 - h_{2a}}{h_1 - h_{2s}}$$

In the relations above, h_{2a} and h_{2s} are the enthalpy values at the exit state for actual and isentropic *processes*, respectively.

The entropy balance for any system undergoing any process can be expressed in the general form as

$$\underbrace{S_{\text{in}} - S_{\text{out}}}_{\substack{\text{Net entropy transfer} \\ \text{by heat and mass}}} + \underbrace{S_{\text{gen}}}_{\substack{\text{Entropy} \\ \text{generation}}} = \underbrace{\Delta S_{\text{system}}}_{\substack{\text{Change} \\ \text{in entropy}}}$$

or, in the *rate form,* as

$$\underbrace{\dot{S}_{\text{in}} - \dot{S}_{\text{out}}}_{\substack{\text{Rate of net entropy} \\ \text{transfer by} \\ \text{heat and mass}}} + \underbrace{\dot{S}_{\text{gen}}}_{\substack{\text{Rate of entropy} \\ \text{generation}}} = \underbrace{dS_{\text{system}}/dt}_{\substack{\text{Rate of change} \\ \text{in entropy}}}$$

For a general *steady-flow process* it simplifies to

$$\dot{S}_{\text{gen}} = \sum \dot{m}_e s_e - \sum \dot{m}_i s_i - \sum \frac{\dot{Q}_k}{T_k}$$

REFERENCES AND SUGGESTED READINGS

1. A. Bejan. *Advanced Engineering Thermodynamics*. 2nd ed. New York: Wiley Interscience, 1997.
2. A. Bejan. *Entropy Generation through Heat and Fluid Flow*. New York: Wiley Interscience, 1982.
3. Y. A. Çengel and H. Kimmel. "Optimization of Expansion in Natural Gas Liquefaction Processes." *LNG Journal*, U.K., May–June, 1998.
4. Y. Çerci, Y. A. Çengel, and R. H. Turner, "Reducing the Cost of Compressed Air in Industrial Facilities." *International Mechanical Engineering Congress and Exposition*, San Francisco, California, November 12–17, 1995.
5. W. F. E. Feller. *Air Compressors: Their Installation, Operation, and Maintenance*. New York: McGraw-Hill, 1944.
6. D. W. Nutter, A. J. Britton, and W. M. Heffington. "Conserve Energy to Cut Operating Costs." *Chemical Engineering*, September 1993, pp. 127–137.
7. J. Rifkin. *Entropy*. New York: The Viking Press, 1980.

PROBLEMS*

Entropy and the Increase of Entropy Principle

7–1C Does the cyclic integral of work have to be zero (i.e., does a system have to produce as much work as it consumes to complete a cycle)? Explain.

7–2C A system undergoes a process between two fixed states first in a reversible manner and then in an irreversible manner. For which case is the entropy change greater? Why?

7–3C Is the value of the integral $\int_1^2 \delta Q/T$ the same for all processes between states 1 and 2? Explain.

7–4C To determine the entropy change for an irreversible process between states 1 and 2, should the integral $\int_1^2 \delta Q/T$ be performed along the actual process path or an imaginary reversible path? Explain.

7–5C Is an isothermal process necessarily internally reversible? Explain your answer with an example.

7–6C How do the values of the integral $\int_1^2 \delta Q/T$ compare for a reversible and irreversible process between the same end states?

7–7C The entropy of a hot baked potato decreases as it cools. Is this a violation of the increase of entropy principle? Explain.

7–8C Is it possible to create entropy? Is it possible to destroy it?

7–9C When a system is adiabatic, what can be said about the entropy change of the substance in the system?

7–10C Work is entropy free, and sometimes the claim is made that work will not change the entropy of a fluid passing through an adiabatic steady-flow system with a single inlet and outlet. Is this a valid claim?

7–11C A piston–cylinder device contains helium gas. During a reversible, isothermal process, the entropy of the helium will (*never, sometimes, always*) increase.

7–12C A piston–cylinder device contains nitrogen gas. During a reversible, adiabatic process, the entropy of the nitrogen will (*never, sometimes, always*) increase.

* Problems designated by a "C" are concept questions, and students are encouraged to answer them all. Problems designated by an "E" are in English units, and the SI users can ignore them. Problems with the icon are solved using EES, and complete solutions together with parametric studies are included on the enclosed DVD. Problems with the icon are comprehensive in nature, and are intended to be solved with a computer, preferably using the EES software that accompanies this text.

7–13C A piston–cylinder device contains superheated steam. During an actual adiabatic process, the entropy of the steam will (*never, sometimes, always*) increase.

7–14C The entropy of steam will (*increase, decrease, remain the same*) as it flows through an actual adiabatic turbine.

7–15C The entropy of the working fluid of the ideal Carnot cycle (*increases, decreases, remains the same*) during the isothermal heat addition process.

7–16C The entropy of the working fluid of the ideal Carnot cycle (*increases, decreases, remains the same*) during the isothermal heat rejection process.

7–17C During a heat transfer process, the entropy of a system (*always, sometimes, never*) increases.

7–18C Steam is accelerated as it flows through an actual adiabatic nozzle. The entropy of the steam at the nozzle exit will be (*greater than, equal to, less than*) the entropy at the nozzle inlet.

7–19C Is it possible for the entropy change of a closed system to be zero during an irreversible process? Explain.

7–20C What three different mechanisms can cause the entropy of a control volume to change?

7–21E A completely reversible heat engine operates with a source at 1500 R and a sink at 500 R. If the entropy of the sink increases by 10 Btu/R, how much will the entropy of the source decrease? How much heat, in Btu, is transferred from the source?

7–22 An energy-source at 1000 K transfers heat to a completely reversible heat engine. This engine transfers heat to a sink at 300 K. How much heat must be transferred from the energy-source to increase the entropy of the energy-sink by 20 kJ/K?

7–23E A heat engine accepts 200,000 Btu of heat from a source at 1500 R and rejects 100,000 Btu of heat to a sink at 600 R. Calculate the entropy change of all the components of this engine and determine if it is completely reversible. How much total work does it produce?

7–24 Air is compressed by a 30-kW compressor from P_1 to P_2. The air temperature is maintained constant at 25°C during this process as a result of heat transfer to the surrounding medium at 17°C. Determine the rate of entropy change of the air. State the assumptions made in solving this problem. *Answer:* −0.101 kW/K

7–25 Heat in the amount of 100 kJ is transferred directly from a hot reservoir at 1200 K to a cold reservoir at 600 K. Calculate the entropy change of the two reservoirs and determine if the increase of entropy principle is satisfied.

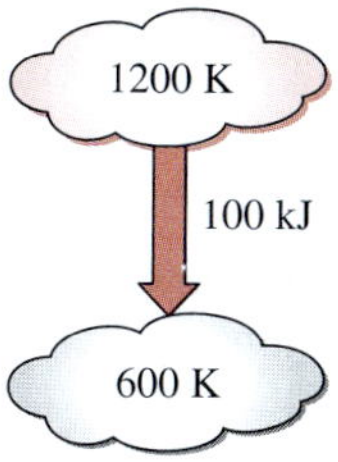

FIGURE P7–25

7–26 In the previous problem, assume that the heat is transferred from the cold reservoir to the hot reservoir contrary to the Clausius statement of the second law. Prove that this violates the increase of entropy principle—as it must according to Clausius.

7–27 A completely reversible heat pump produces heat at a rate of 300 kW to warm a house maintained at 24°C. The exterior air, which is at 7°C, serves as the source. Calculate the rate of entropy change of the two reservoirs and determine if this heat pump satisfies the second law according to the increase of entropy principle.

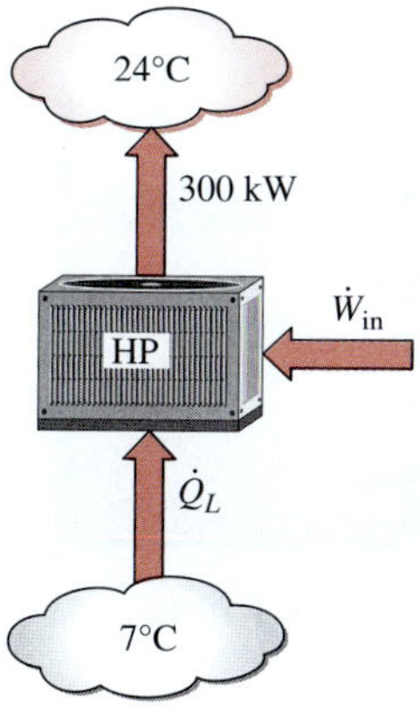

FIGURE P7–27

7–28E During the isothermal heat rejection process of a Carnot cycle, the working fluid experiences an entropy change of −0.7 Btu/R. If the temperature of the heat sink is 95°F, determine (*a*) the amount of heat transfer, (*b*) the entropy change of the sink, and (*c*) the total entropy change for this process. *Answers:* (*a*) 388.5 Btu, (*b*) 0.7 Btu/R, (*c*) 0

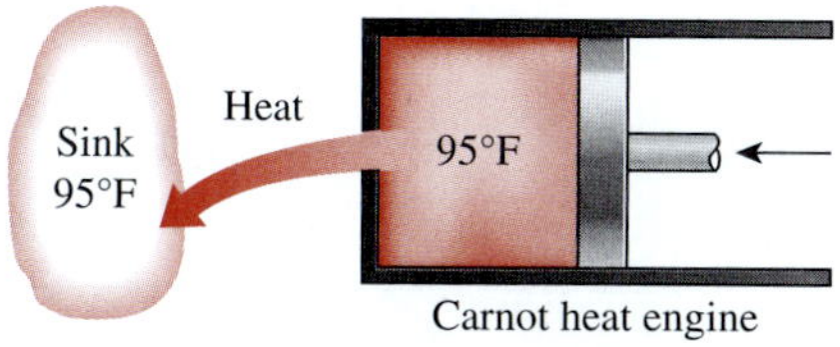

FIGURE P7–28E

7–29 Refrigerant-134a enters the coils of the evaporator of a refrigeration system as a saturated liquid–vapor mixture at a pressure of 160 kPa. The refrigerant absorbs 180 kJ of heat from the cooled space, which is maintained at −5°C, and leaves as saturated vapor at the same pressure. Determine (*a*) the entropy change of the refrigerant, (*b*) the entropy change of the cooled space, and (*c*) the total entropy change for this process.

Entropy Changes of Pure Substances

7–30C Is a process that is internally reversible and adiabatic necessarily isentropic? Explain.

7–31E 2-lbm of water at 300 psia fill a weighted piston-cylinder device whose volume is 2.5 ft^3. The water is then heated at constant pressure until the temperature reaches 500°F. Determine the resulting change in the water's total entropy. *Answer:* 0.474 Btu/R

7–32 A well-insulated rigid tank contains 5 kg of a saturated liquid–vapor mixture of water at 150 kPa. Initially, three-quarters of the mass is in the liquid phase. An electric resistance heater placed in the tank is now turned on and kept on until all the liquid in the tank is vaporized. Determine the entropy change of the steam during this process. *Answer:* 19.2 kJ/K

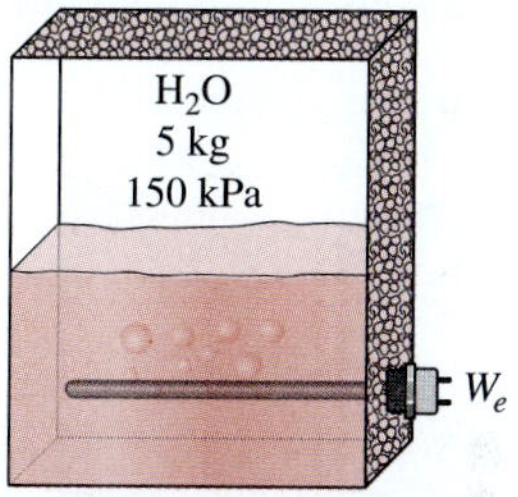

FIGURE P7–32

7–33  A rigid tank is divided into two equal parts by a partition. One part of the tank contains 2.5 kg of compressed liquid water at 400 kPa and 60°C while the other part is evacuated. The partition is now removed, and the water expands to fill the entire tank. Determine the entropy change of water during this process, if the final pressure in the tank is 40 kPa. *Answer:* 0.492 kJ/K

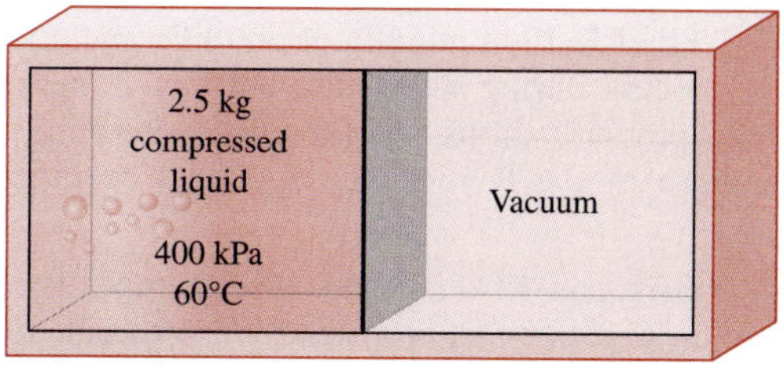

FIGURE P7–33

7–34 Reconsider Prob. 7–33. Using EES (or other) software, evaluate and plot the entropy generated as a function of surrounding temperature, and determine the values of the surrounding temperatures that are valid for this problem. Let the surrounding temperature vary from 0 to 100°C. Discuss your results.

7–35E A piston–cylinder device contains 2 lbm of refrigerant-134a at 120 psia and 100°F. The refrigerant is now cooled at constant pressure until it exists as a liquid at 50°F. Determine the entropy change of the refrigerant during this process.

7–36 An insulated piston–cylinder device contains 5 L of saturated liquid water at a constant pressure of 150 kPa. An electric resistance heater inside the cylinder is now turned on, and 2200 kJ of energy is transferred to the steam. Determine the entropy change of the water during this process. *Answer:* 5.72 kJ/K

7–37 Calculate the change in the specific entropy of water as it is cooled at a constant pressure of 300 kPa from a saturated vapor to a saturated liquid by using a Gibbs equation ($Tds = dh - vdP$). Use the steam tables to verify your result.

7–38E Saturated R-134a vapor enters a compressor at 0°F. At compressor exit, the specific entropy is the same as that at the inlet, and the pressure is 60 psia. Determine the R-134a exit temperature and the change in the enthalpy of R-134a.

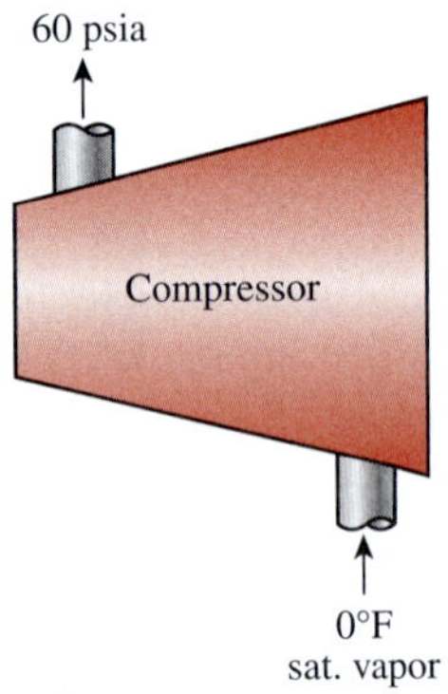

FIGURE P7–38E

7–39 Water vapor enters a turbine at 6 MPa and 400°C, and leaves the turbine at 100 kPa with the same specific entropy as that at the inlet. Calculate the difference between the specific enthalpy of the water at the turbine inlet and exit.

7–40 1-kg of R-134a initially at 600 kPa and 25°C undergoes a process during which the entropy is kept constant until the pressure drops to 100 kPa. Determine the final temperature of the R-134a and the final specific internal energy.

7–41 Refrigerant-134a is expanded isentropically from 800 kPa and 60°C at the inlet of a steady-flow turbine to 100 kPa at the outlet. The outlet area is 1 m^2, and the inlet area is 0.5 m^2. Calculate the inlet and outlet velocities when the mass flow rate is 0.5 kg/s. *Answers:* 0.030 m/s, 0.105 m/s

7–42 A heavily insulated piston–cylinder device contains 0.02 m^3 of steam at 300 kPa and 200°C. Steam is now compressed in a reversible manner to a pressure of 1.2 MPa. Determine the work done on the steam during this process.

7–43 Reconsider Prob. 7–42. Using EES (or other) software, evaluate and plot the work done on the steam as a function of final pressure as the pressure varies from 300 kPa to 1.2 MPa.

7–44 A piston–cylinder device contains 1.2 kg of saturated water vapor at 200°C. Heat is now transferred to steam, and steam expands reversibly and isothermally to a final pressure of 800 kPa. Determine the heat transferred and the work done during this process.

7–45 Reconsider Prob. 7–44. Using EES (or other) software, evaluate and plot the heat transferred to the steam and the work done as a function of final pressure as the pressure varies from the initial value to the final value of 800 kPa.

7–46 Refrigerant-134a at 240 kPa and 20°C undergoes an isothermal process in a closed system until its quality is 20 percent. On per unit mass basis, determine how much work and heat transfer are required. *Answers:* 37.0 kJ/kg, 172 kJ/kg

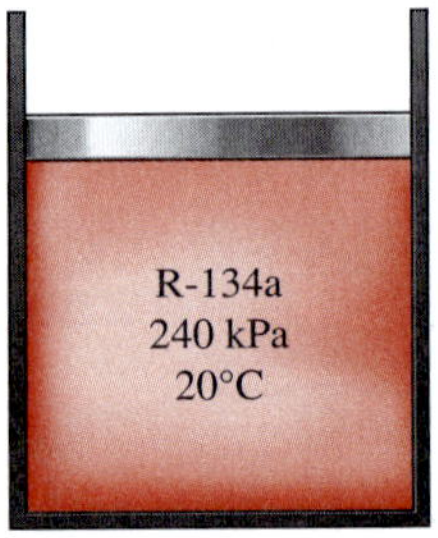

FIGURE P7–46

7–47 Determine the heat transfer, in kJ/kg, for the reversible process 1-3 shown in Fig. P7–47.

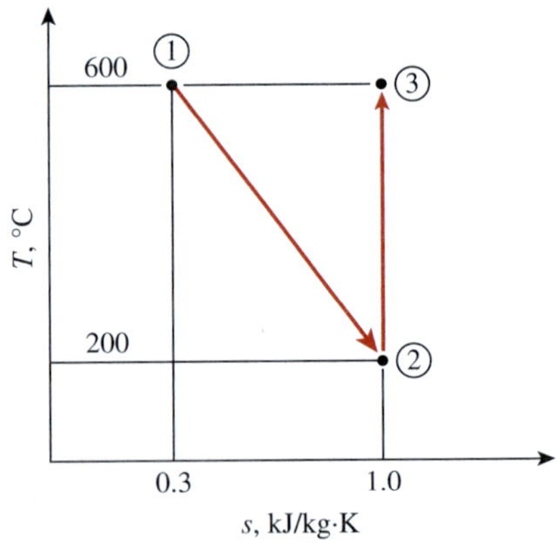

FIGURE P7–47

7–48E Calculate the heat transfer, in Btu/lbm, for the reversible process 1-3 shown in Fig. P7–48E. *Answer:* 515 Btu/lbm

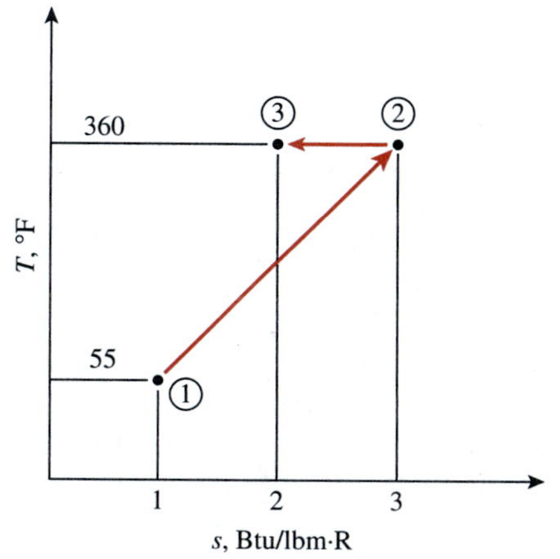

FIGURE P7–48E

7–49 Calculate the heat transfer, in kJ/kg, for the reversible steady-flow process 1-3 shown in Fig. P7–49

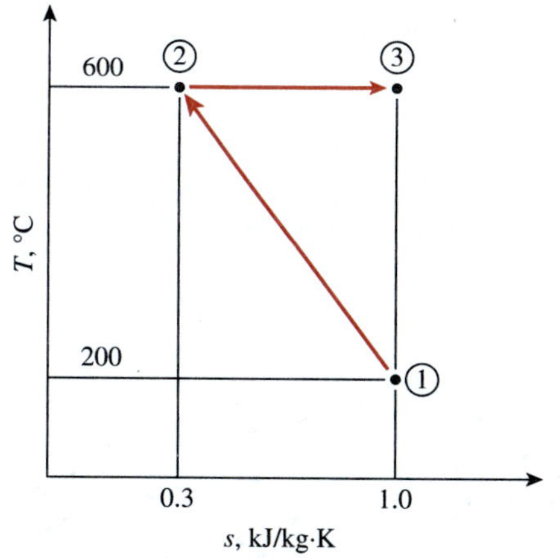

FIGURE P7–49

7–50 Steam is expanded in an isentropic turbine with a single inlet and outlet. At the inlet, the steam is at 2 MPa and 360°C. The steam pressure at the outlet is 100 kPa. Calculate the work produced by this turbine, in kJ/kg.

7–51 An isentropic steam turbine processes 5 kg/s of steam at 4 MPa, which is exhausted at 50 kPa and 100°C. Five percent of this flow is diverted for feedwater heating at 700 kPa. Determine the power produced by this turbine, in kW. *Answer:* 6328 kW

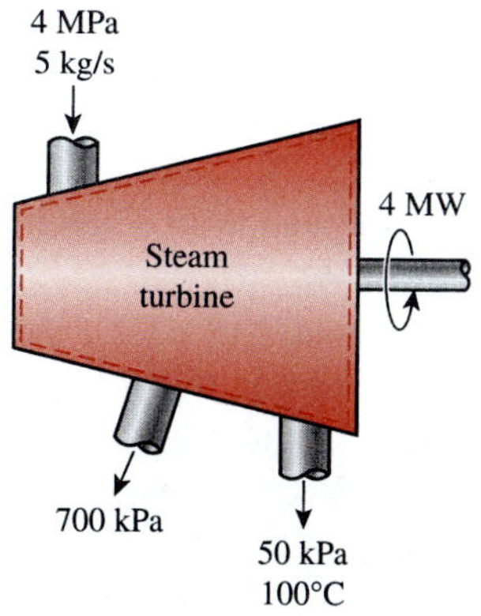

FIGURE P7–51

7–52 Water at 70 kPa and 100°C is compressed isentropically in a closed system to 4 MPa. Determine the final temperature of the water and the work required, in kJ/kg, for this compression.

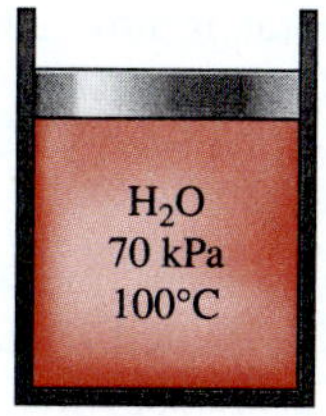

FIGURE P7–52

7–53 0.5-kg of R-134a is expanded isentropically from 600 kPa and 30°C to 140 kPa. Determine the total heat transfer and work production for this expansion.

7–54 Refrigerant-134a enters a steady-flow, adiabatic turbine as a saturated vapor at 1200 kPa and expands to 100 kPa. The power produced by the turbine is found to be 100 kW when the process is also reversible.

(*a*) Sketch the *T-s* diagram with respect to the saturation lines for this process.

(*b*) Determine the volume flow rate of the refrigerant-134a at the turbine exit, in m^3/s.

Answer: 0.376 m^3/s

7–55 Two kilograms of saturated water vapor at 600 kPa are contained in a piston-cylinder device. The water expands adiabatically until the pressure is 100 kPa and is said to produce 700 kJ of work output.

(*a*) Determine the entropy change of the water, in kJ/kg·K.

(*b*) Is this process realistic? Using the *T-s* diagram for the process and the concepts of second law, support your answer.

7–56 Refrigerant-134a enters a steady-flow adiabatic compressor as a saturated vapor at 320 kPa and is compressed to 1200 kPa. The minimum power supplied to the compressor is found to be 100 kW.

(*a*) Sketch the *T-s* diagram with respect to the saturation lines for this process.

(*b*) Determine the volume flow rate of the refrigerant-134a at the compressor inlet, in m^3/s.

7–57 Steam enters a steady-flow adiabatic nozzle with a low inlet velocity as a saturated vapor at 6 MPa and expands to 1.2 MPa.

(*a*) Under the conditions that the exit velocity is to be the maximum possible value, sketch the *T-s* diagram with respect to the saturation lines for this process.

(*b*) Determine the maximum exit velocity of the steam, in m/s.

Answer: 764 m/s

7–58 A rigid, 20-L steam cooker is arranged with a pressure relief valve set to release vapor and maintain the pressure once the pressure inside the cooker reaches 150 kPa. Initially, this cooker is filled with water at 175 kPa with a quality of 10 percent. Heat is now added until the quality inside the cooker is 40 percent. Determine the minimum entropy change of the thermal energy reservoir supplying this heat.

7–59 In the preceding problem, the water is stirred at the same time that it is being heated. Determine the minimum entropy change of the heat-supplying source if 100 kJ of work is done on the water as it is being heated.

7–60 A piston–cylinder device contains 5 kg of steam at 100°C with a quality of 50 percent. This steam undergoes two processes as follows:

1-2 Heat is transferred to the steam in a reversible manner while the temperature is held constant until the steam exists as a saturated vapor.

2-3 The steam expands in an adiabatic, reversible process until the pressure is 15 kPa.

(*a*) Sketch these processes with respect to the saturation lines on a single *T-s* diagram.

(*b*) Determine the heat transferred to the steam in process 1-2, in kJ.

(*c*) Determine the work done by the steam in process 2-3, in kJ.

7–61E A 0.8-ft^3 well-insulated rigid can initially contains refrigerant-134a at 140 psia and 50°F. Now a crack develops in the can, and the refrigerant starts to leak out slowly, Assuming the refrigerant remaining in the can has undergone a reversible, adiabatic process, determine the final mass in the can when the pressure drops to 30 psia.

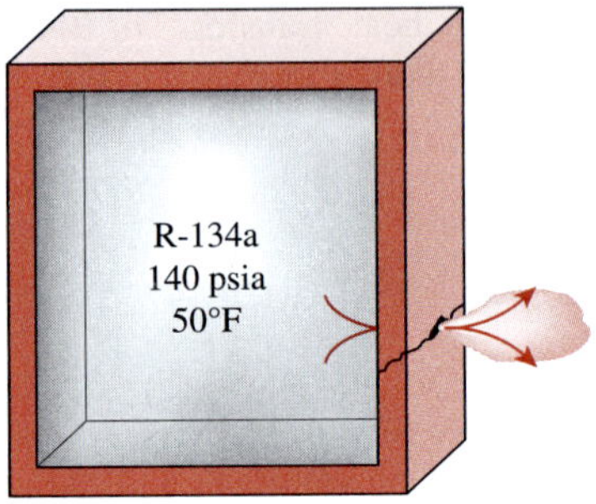

FIGURE P7–61E

7–62E An electric windshield defroster is used to remove 0.25-in of ice from a windshield. The properties of the ice are T_{sat} = 32°F, $u_{if} = h_{if}$ = 144 Btu/lbm, and v = 0.01602 ft^3/lbm. Determine the electrical energy required per square foot of windshield surface area to melt this ice and remove it as liquid water at 32°F. What is the minimum temperature at which the defroster may be operated? Assume that no heat is transferred from the defroster or ice to the surroundings.

Entropy Change of Incompressible Substances

7–63C Consider two solid blocks, one hot and the other cold, brought into contact in an adiabatic container. After a while, thermal equilibrium is established in the container as a result of heat transfer. The first law requires that the amount of energy lost by the hot solid be equal to the amount of energy gained by the cold one. Does the second law require that the decrease in entropy of the hot solid be equal to the increase in entropy of the cold one?

7–64 A 75-kg copper block initially at 110°C is dropped into an insulated tank that contains 160 L of water at 15°C. Determine the final equilibrium temperature and the total entropy change for this process.

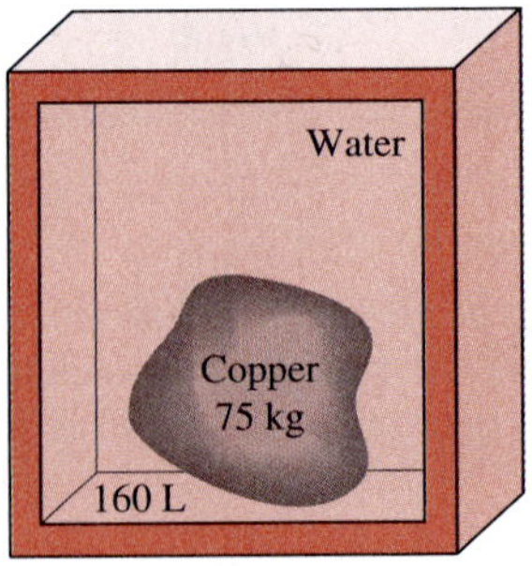

FIGURE P7–64

7–65 Ten grams of computer chips with a specific heat of 0.3 kJ/kg·K are initially at 20°C. These chips are cooled by placement in 5 grams of saturated liquid R-134a at −40°C. Presuming that the pressure remains constant while the chips are being cooled, determine the entropy change of (*a*) the chips, (*b*) the R-134a, and (*c*) the entire system. Is this process possible? Why?

7–66 A 25-kg iron block initially at 350°C is quenched in an insulated tank that contains 100 kg of water at 18°C. Assuming the water that vaporizes during the process condenses back in the tank, determine the total entropy change during this process.

7–67 A 30-kg aluminum block initially at 140°C is brought into contact with a 40-kg block of iron at 60°C in an insulated enclosure. Determine the final equilibrium temperature and the total entropy change for this process. *Answers:* 109°C, 0.251 kJ/K

7–68 Reconsider Prob. 7–67. Using EES (or other) software, study the effect of the mass of the iron block on the final equilibrium temperature and the total entropy change for the process. Let the mass of the iron vary from 10 to 100 kg. Plot the equilibrium temperature and the total entropy change as a function of iron mass, and discuss the results.

7–69 A 50-kg iron block and a 20-kg copper block, both initially at 80°C, are dropped into a large lake at 15°C. Thermal equilibrium is established after a while as a result of heat transfer between the blocks and the lake water. Determine the total entropy change for this process.

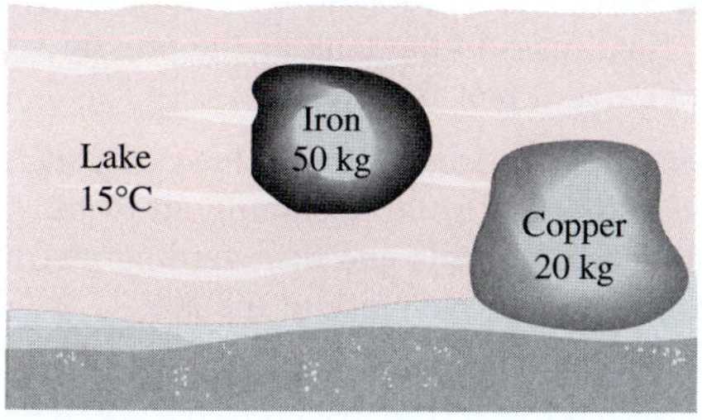

FIGURE P7–69

7–70 An adiabatic pump is to be used to compress saturated liquid water at 10 kPa to a pressure to 15 MPa in a reversible manner. Determine the work input using (*a*) entropy data from the compressed liquid table, (*b*) inlet specific volume and pressure values, (*c*) average specific volume and pressure values. Also, determine the errors involved in parts (*b*) and (*c*).

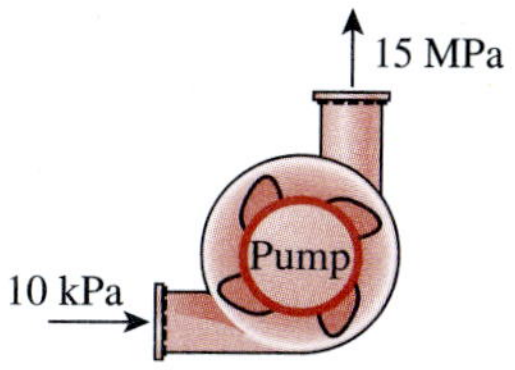

FIGURE P7–70

Entropy Change of Ideal Gases

7–71C Some properties of ideal gases such as internal energy and enthalpy vary with temperature only [that is, $u = u(T)$ and $h = h(T)$]. Is this also the case for entropy?

7–72C Can the entropy of an ideal gas change during an isothermal process?

7–73C An ideal gas undergoes a process between two specified temperatures, first at constant pressure and then at constant volume. For which case will the ideal gas experience a larger entropy change? Explain.

7–74 Prove that the two relations for entropy change of ideal gases under the constant-specific-heat assumption (Eqs. 7–33 and 7–34) are equivalent.

7–75 Starting with the second $T\,ds$ relation (Eq. 7–26), obtain Eq. 7–34 for the entropy change of ideal gases under the constant-specific-heat assumption.

7–76 Starting with Eq. 7–34, obtain Eq. 7–43.

7–77 Which of the two gases—helium or nitrogen—experiences the greatest entropy change as its state is changed from 2000 kPa and 427°C to 200 kPa and 27°C?

7–78 Air is expanded from 2000 kPa and 500°C to 100 kPa and 50°C. Assuming constant specific heats, determine the change in the specific entropy of air.

7–79E What is the difference between the entropies of air at 15 psia and 70°F and air at 40 psia and 250°F per unit mass basis.

7–80 Nitrogen at 900 kPa and 300°C is expanded adiabatically in a closed system to 100 kPa. Determine the minimum nitrogen temperature after the expansion.

7–81E Air at 15 psia and 70°F is compressed adiabatically in a closed system to 200 psia. What is the minimum temperature of the air following this compression?

7–82 An insulated piston–cylinder device initially contains 300 L of air at 120 kPa and 17°C. Air is now heated for 15 min by a 200-W resistance heater placed inside the cylinder. The pressure of air is maintained constant during this process. Determine the entropy change of air, assuming (*a*) constant specific heats and (*b*) variable specific heats.

7–83 A piston–cylinder device contains 0.75 kg of nitrogen gas at 140 kPa and 37°C. The gas is now compressed slowly in a polytropic process during which $PV^{1.3}$ = constant. The process ends when the volume is reduced by one-half. Determine the entropy change of nitrogen during this process. *Answer:* −0.0385 kJ/K

7–84 Reconsider Prob. 7–83. Using EES (or other) software, investigate the effect of varying the polytropic exponent from 1 to 1.4 on the entropy change of the nitrogen. Show the processes on a common *P*-*v* diagram.

7–85E A mass of 15 lbm of helium undergoes a process from an initial state of 50 ft³/lbm and 80°F to a final state of 10 ft³/lbm and 200°F. Determine the entropy change of helium during this process, assuming (*a*) the process is reversible and (*b*) the process is irreversible.

7–86 1-kg of air at 200 kPa and 127°C is contained in a piston-cylinder device. Air is now allowed to expand in a reversible, isothermal process until its pressure is 100 kPa. Determine the amount of heat transferred to the air during this expansion.

7–87 Argon is expanded in an isentropic turbine from 2 MPa and 500°C to 200 kPa. Determine the outlet temperature and the work produced by this turbine per unit mass of argon.

7–88E Air is compressed in an isentropic compressor from 15 psia and 70°F to 200 psia. Determine the outlet temperature and the work consumed by this compressor per unit mass of air. *Answers:* 1095 R, 138 Btu/lbm

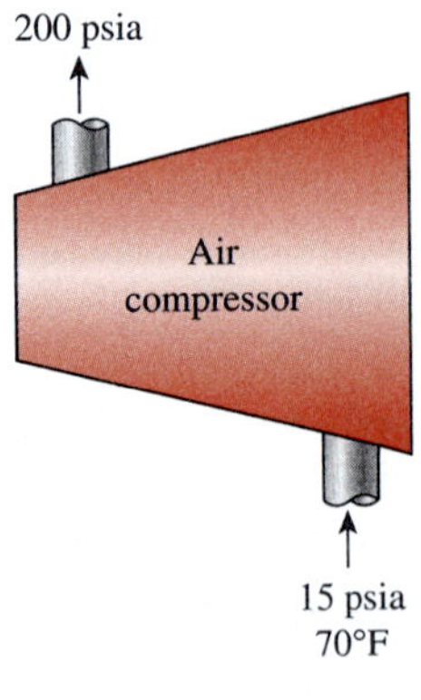

FIGURE P7–88E

7–89 An insulated rigid tank is divided into two equal parts by a partition. Initially, one part contains 5 kmol of an ideal gas at 250 kPa and 40°C, and the other side is evacuated. The partition is now removed, and the gas fills the entire tank. Determine the total entropy change during this process. *Answer:* 28.81 kJ/K

7–90 Air is compressed in a piston–cylinder device from 90 kPa and 22°C to 900 kPa in a reversible, adiabatic process. Determine the final temperature and the work done during this process, assuming (*a*) constant specific heats and (*b*) variable specific heats for air. *Answers:* (*a*) 565 K, 197 kJ/kg

7–91 Reconsider Prob. 7–90. Using EES (or other) software, evaluate and plot the work done and final temperature during the compression process as functions of the final pressure for the two cases as the final pressure varies from 100 to 1200 kPa.

7–92 An insulated rigid tank contains 4 kg of argon gas at 450 kPa and 30°C. A valve is now opened, and argon is allowed to escape until the pressure inside drops to 200 kPa. Assuming the argon remaining inside the tank has undergone a reversible, adiabatic process, determine the final mass in the tank. *Answer:* 2.46 kg

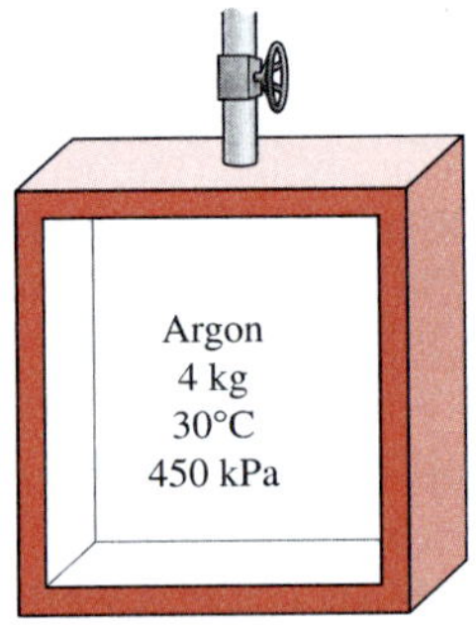

FIGURE P7–92

7–93 Reconsider Prob. 7–92. Using EES (or other) software, investigate the effect of the final pressure on the final mass in the tank as the pressure varies from 450 to 150 kPa, and plot the results.

7–94E Air enters an adiabatic nozzle at 60 psia, 540°F, and 200 ft/s and exits at 12 psia. Assuming air to be an ideal gas with variable specific heats and disregarding any irreversibilities, determine the exit velocity of the air.

7–95 Air is expanded in an adiabatic nozzle during a polytropic process with n = 1.3. It enters the nozzle at 700 kPa and 100°C with a velocity of 30 m/s and exits at a pressure of 200 kPa. Calculate the air temperature and velocity at the nozzle exit.

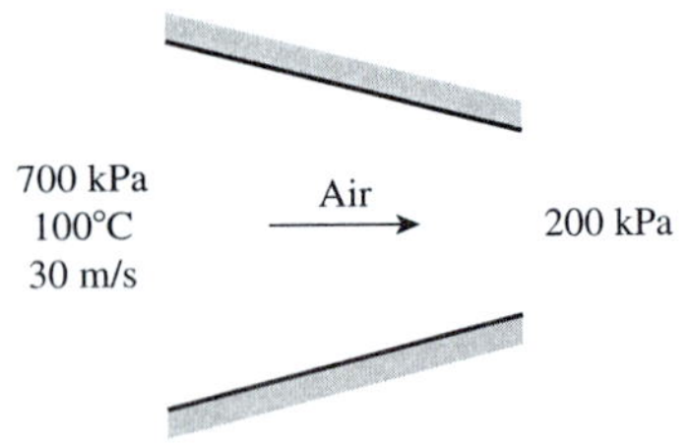

FIGURE P7–95

7–96 Repeat Prob. 7–95 for the polytropic exponent of $n = 1.1$.

7–97 Air at 427°C and 600 kPa is contained in a piston-cylinder device. The air expands adiabatically until the pressure is 100 kPa. Determine the mass of air needed to produce maximum work of 1000 kJ. Assume air has constant specific heats evaluated at 300 K. *Answer:* 4.97 kg

7–98 Helium enters a steady-flow adiabatic compressor at 0.6 kg/s, 100 kPa, and 27°C with a low inlet velocity and is compressed to 600 kPa.

(*a*) Determine the exit temperature to make the work input and kinetic energy at the compressor exit have the minimum values.

(*b*) If the rate of work input into the compressor is measured to be a minimum with a value of 1000 kW, determine the compressor exit velocity, in m/s.

7–99 5-kg of air at 427°C and 600 kPa are contained in a piston-cylinder device. The air expands adiabatically until the pressure is 100 kPa and produces 600 kJ of work output. Assume air has constant specific heats evaluated at 300 K.

(*a*) Determine the entropy change of the air, in kJ/kg·K

(*b*) Since the process is adiabatic, is the process realistic? Using concepts of the second law, support your answer.

7–100 A constant-volume tank contains 5 kg of air at 100 kPa and 327°C. The air is cooled to the surroundings temperature of 27°C. Assume constant specific heats at 300 K. (*a*) Determine the entropy change of the air in the tank during the process, in kJ/K, (*b*) determine the net entropy change of the universe due to this process, in kJ/K, and (*c*) sketch the processes for the air in the tank and the surroundings on a single *T-s* diagram. Be sure to label the initial and final states for both processes.

7–101 A container filled with 45 kg of liquid water at 95°C is placed in a 90-m^3 room that is initially at 12°C. Thermal equilibrium is established after a while as a result of heat transfer between the water and the air in the room. Using constant specific heats, determine (*a*) the final equilibrium temperature, (*b*) the amount of heat transfer between the water and the air in the room, and (*c*) the entropy generation. Assume the room is well sealed and heavily insulated.

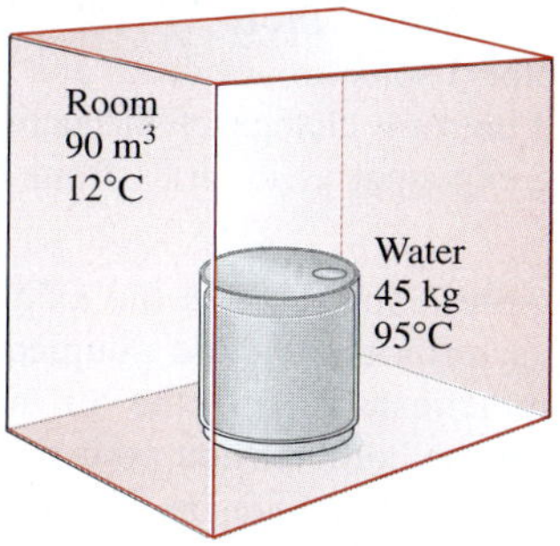

FIGURE P7–101

7–102 An ideal gas at 100 kPa and 15°C enters a steady-flow compressor. The gas is compressed to 600 kPa, and 10 percent of the mass that entered the compressor is removed for some other use. The remaining 90 percent of the inlet gas is compressed to 800 kPa before leaving the compressor. The entire compression process is assumed to be reversible and adiabatic. The power supplied to the compressor is measured to be 32 kW. If the ideal gas has constant specific heats such that $c_v = 0.8$ kJ/kg·K and $c_p = 1.1$ kJ/kg·K, (*a*) sketch the compression process on a *T-s* diagram, (*b*) determine the temperature of the gas at the two compressor exits, in K, and (*c*) determine the mass flow rate of the gas into the compressor, in kg/s.

7–103E The well-insulated container shown in Fig. P7–103E is initially evacuated. The supply line contains air that is maintained at 200 psia and 100°F. The valve is opened until the pressure in the container is the same as the pressure in the supply line. Determine the minimum temperature in the container when the valve is closed.

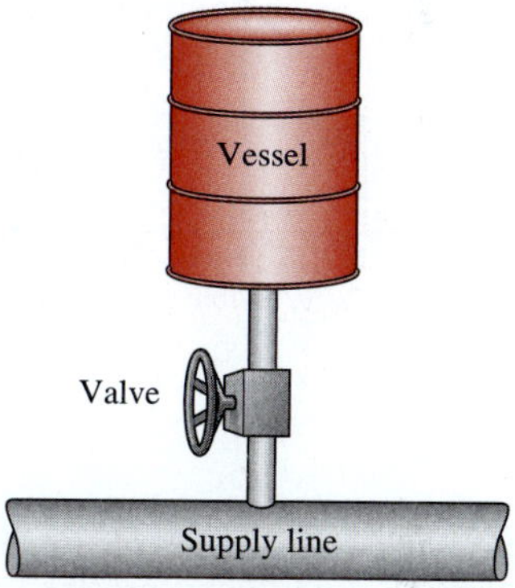

FIGURE P7–103E

Reversible Steady-Flow Work

7–104C In large compressors, the gas is frequently cooled while being compressed to reduce the power consumed by the compressor. Explain how cooling the gas during a compression process reduces the power consumption.

7–105C The turbines in steam power plants operate essentially under adiabatic conditions. A plant engineer suggests to end this practice. She proposes to run cooling water through the outer surface of the casing to cool the steam as it flows through the turbine. This way, she reasons, the entropy of the steam will decrease, the performance of the turbine will improve, and as a result the work output of the turbine will increase. How would you evaluate this proposal?

7–106C It is well known that the power consumed by a compressor can be reduced by cooling the gas during compression.

Inspired by this, somebody proposes to cool the liquid as it flows through a pump, in order to reduce the power consumption of the pump. Would you support this proposal? Explain.

7–107E Air is compressed isothermally from 13 psia and 90°F to 80 psia in a reversible steady-flow device. Calculate the work required, in Btu/lbm, for this compression. *Answer:* 68.5 Btu/lbm

7–108 Saturated water vapor at 150°C is compressed in a reversible steady-flow device to 1000 kPa while its specific volume remains constant. Determine the work required, in kJ/kg.

7–109E Calculate the work produced, in Btu/lbm, for the reversible steady-flow process 1-3 shown in Fig. P7–109E.

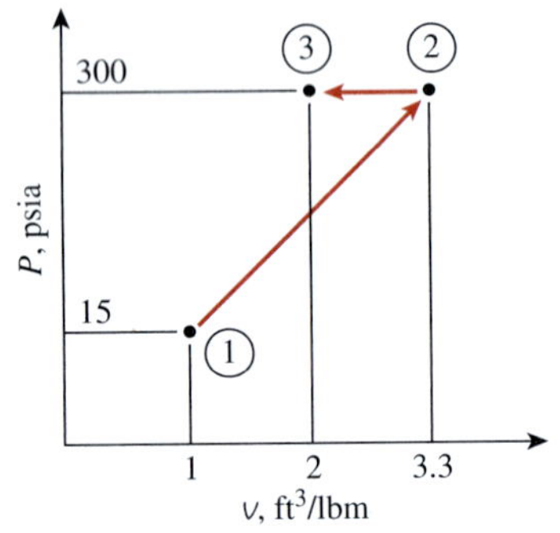

FIGURE P7–109E

7–110 Calculate the work produced, in kJ/kg, for the reversible isothermal, steady-flow process 1-3 shown in Fig. P7–110 when the working fluid is an ideal gas.

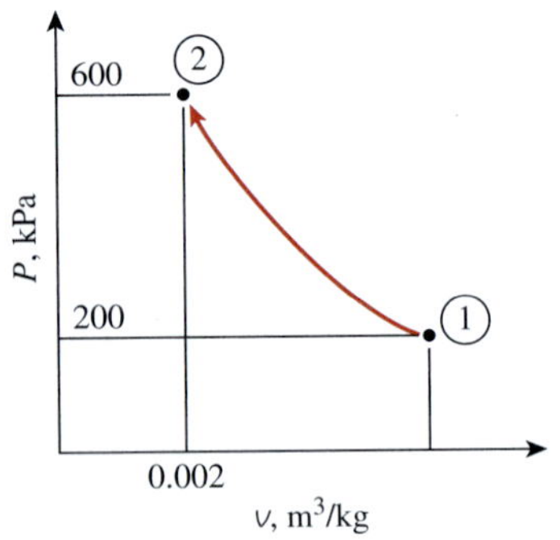

FIGURE P7–110

7–111 Liquid water enters a 25-kW pump at 100-kPa pressure at a rate of 5 kg/s. Determine the highest pressure the liquid water can have at the exit of the pump. Neglect the kinetic and potential energy changes of water, and take the specific volume of water to be 0.001 m^3/kg. *Answer:* 5100 kPa

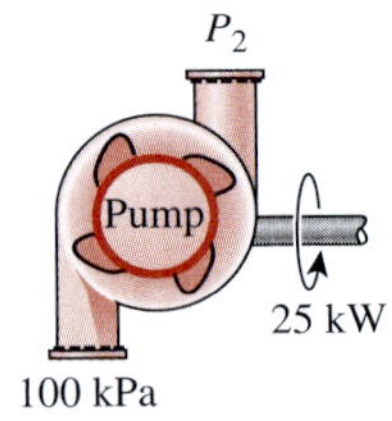

FIGURE P7–111

7–112 Consider a steam power plant that operates between the pressure limits of 5 MPa and 10 kPa. Steam enters the pump as saturated liquid and leaves the turbine as saturated vapor. Determine the ratio of the work delivered by the turbine to the work consumed by the pump. Assume the entire cycle to be reversible and the heat losses from the pump and the turbine to be negligible.

7–113 EES Reconsider Prob. 7–112. Using EES (or other) software, investigate the effect of the quality of the steam at the turbine exit on the net work output. Vary the quality from 0.5 to 1.0, and plot the net work output as a function of this quality.

7–114 Liquid water at 120 kPa enters a 7-kW pump where its pressure is raised to 5 MPa. If the elevation difference between the exit and the inlet levels is 10 m, determine the highest mass flow rate of liquid water this pump can handle. Neglect the kinetic energy change of water, and take the specific volume of water to be 0.001 m^3/kg.

7–115E Helium gas is compressed from 16 psia and 85°F to 120 psia at a rate of 10 ft^3/s. Determine the power input to the compressor, assuming the compression process to be (*a*) isentropic, (*b*) polytropic with $n = 1.2$, (*c*) isothermal, and (*d*) ideal two-stage polytropic with $n = 1.2$.

7–116E EES Reconsider Prob. 7–115E. Using EES (or other) software, evaluate and plot the work of compression and entropy change of the helium as functions of the polytropic exponent as it varies from 1 to 1.667. Discuss your results.

7–117 The compression stages in the axial compressor of the industrial gas turbine are close coupled, making intercooling very impractical. To cool the air in such compressors and to reduce the compression power, it is proposed to spray water mist with drop size on the order of 5 microns into the air stream as it is compressed and to cool the air continuously as the water evaporates. Although the collision of water droplets with turbine blades is a concern, experience with steam turbines indicates that they can cope with water droplet concentrations of up to 14 percent. Assuming air is compressed isentropically at a rate of 2 kg/s from 300 K and 100 kPa to 1200 kPa and the water is injected at a temperature of 20°C at a rate of 0.2 kg/s, determine the reduction in the exit temperature of the compressed air and the compressor power saved. Assume the water vaporizes completely before leaving the compressor, and assume an average mass flow rate of 2.1 kg/s throughout the compressor.

7–118 Reconsider Prob. 7–117. The water-injected compressor is used in a gas turbine power plant. It is claimed that the power output of a gas turbine will increase because of the increase in the mass flow rate of the gas (air + water vapor) through the turbine. Do you agree?

Isentropic Efficiencies of Steady-Flow Devices

7–119C Describe the ideal process for an (*a*) adiabatic turbine, (*b*) adiabatic compressor, and (*c*) adiabatic nozzle, and define the isentropic efficiency for each device.

7–120C Is the isentropic process a suitable model for compressors that are cooled intentionally? Explain.

7–121C On a *T-s* diagram, does the actual exit state (state 2) of an adiabatic turbine have to be on the right-hand side of the isentropic exit state (state 2*s*)? Why?

7–122E Steam at 100 psia and 650°F is expanded adiabatically in a closed system to 10 psia. Determine the work produced, in Btu/lbm, and the final temperature of steam for an isentropic expansion efficiency of 80 percent. *Answers:* 132 Btu/lbm, 275°F

7–123 Steam at 3 MPa and 400°C is expanded to 30 kPa in an adiabatic turbine with an isentropic efficiency of 92 percent. Determine the power produced by this turbine, in kW, when the mass flow rate is 2 kg/s.

7–124 Repeat Prob. 7–123 for a turbine efficiency of 85 percent.

7–125 Steam enters an adiabatic turbine at 7 MPa, 600°C, and 80 m/s and leaves at 50 kPa, 150°C, and 140 m/s. If the power output of the turbine is 6 MW, determine (*a*) the mass flow rate of the steam flowing through the turbine and (*b*) the isentropic efficiency of the turbine. *Answers:* (*a*) 6.95 kg/s, (*b*) 73.4 percent

7–126E Combustion gases enter an adiabatic gas turbine at 1540°F and 120 psia and leave at 60 psia with a low velocity. Treating the combustion gases as air and assuming an isentropic efficiency of 82 percent, determine the work output of the turbine. *Answer:* 71.7 Btu/lbm

7–127 Air at 100 kPa and 20°C is compressed to 700 kPa steadily and adiabatically at a rate of 2 kg/s. Determine the power required to compress this air if the isentropic compression efficiency is 95 percent.

7–128 Steam at 4 MPa and 350°C is expanded in an adiabatic turbine to 120 kPa. What is the isentropic efficiency of this turbine if the steam is exhausted as a saturated vapor?

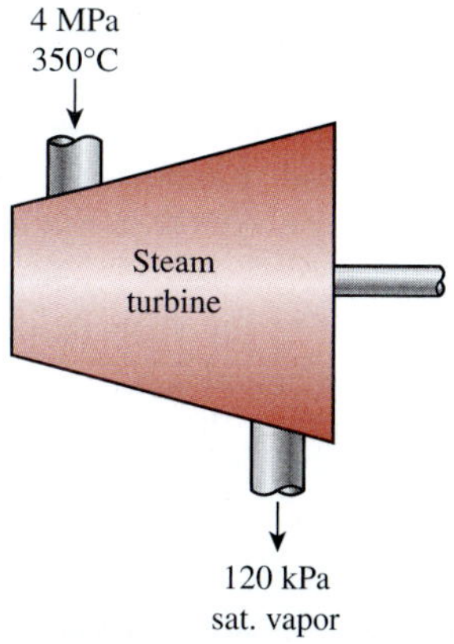

FIGURE P7–128

7–129 Air is expanded from 2 MPa and 327°C to 100 kPa in an adiabatic turbine. Determine the isentropic efficiency of this turbine if the air is exhausted at 0°C.

7–130 A refrigeration unit compresses saturated R-134a vapor at 20°C to 1000 kPa. How much power is required to compress 0.5 kg/s of R-134a with a compressor efficiency of 85 percent? *Answer:* 6.78 kW

7–131 Refrigerant-134a enters an adiabatic compressor as saturated vapor at 100 kPa at a rate of 0.7 m^3/min and exits at 1-MPa pressure. If the isentropic efficiency of the compressor is 87 percent, determine (*a*) the temperature of the refrigerant at the exit of the compressor and (*b*) the power input, in kW. Also, show the process on a *T-s* diagram with respect to saturation lines.

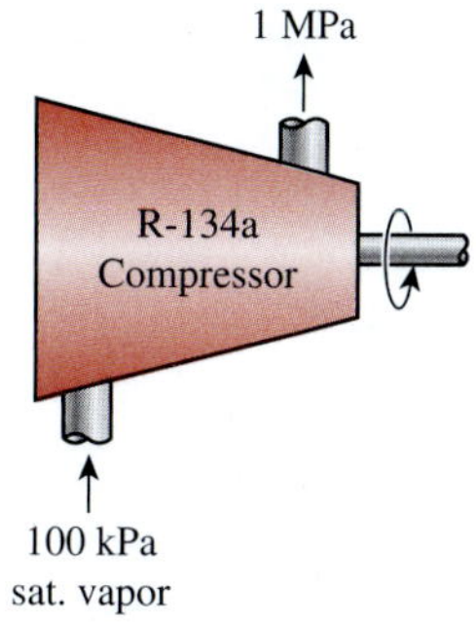

FIGURE P7–131

7–132 Reconsider Prob. 7–131. Using EES (or other) software, redo the problem by including the effects of the kinetic energy of the flow by assuming an inlet-to-exit area ratio of 1.5 for the compressor when the compressor exit pipe inside diameter is 2 cm.

7–133 Air enters an adiabatic compressor at 100 kPa and 17°C at a rate of 2.4 m^3/s, and it exits at 257°C. The compressor has an isentropic efficiency of 84 percent. Neglecting the changes in kinetic and potential energies, determine (*a*) the exit pressure of air and (*b*) the power required to drive the compressor.

7–134 Air is compressed by an adiabatic compressor from 95 kPa and 27°C to 600 kPa and 277°C. Assuming variable specific heats and neglecting the changes in kinetic and potential energies, determine (*a*) the isentropic efficiency of the compressor and (*b*) the exit temperature of air if the process were reversible. *Answers:* (*a*) 81.9 percent, (*b*) 505.5 K

7–135E Argon gas enters an adiabatic compressor at 20 psia and 90°F with a velocity of 60 ft/s, and it exits at 200 psia and 240 ft/s. If the isentropic efficiency of the compressor is 80 percent, determine (*a*) the exit temperature of the argon and (*b*) the work input to the compressor.

7–136E Air enters an adiabatic nozzle at 45 psia and 940°F with low velocity and exits at 650 ft/s. If the isentropic efficiency of the nozzle is 85 percent, determine the exit temperature and pressure of the air.

7–137E Reconsider Prob. 7–136E. Using EES (or other) software, study the effect of varying the nozzle isentropic efficiency from 0.8 to 1.0 on both the exit temperature and pressure of the air, and plot the results.

7–138 The exhaust nozzle of a jet engine expands air at 300 kPa and 180°C adiabatically to 100 kPa. Determine the air velocity at the exit when the inlet velocity is low and the nozzle isentropic efficiency is 96 percent.

7–139E An adiabatic diffuser at the inlet of a jet engine increases the pressure of the air that enters the diffuser at 13 psia and 30°F to 20 psia. What will the air velocity at the diffuser exit be if the diffuser isentropic efficiency defined as the ratio of the actual kinetic energy change to the isentropic kinetic energy change is 82 percent and the diffuser inlet velocity is 1000 ft/s? *Answer:* 606 ft/s

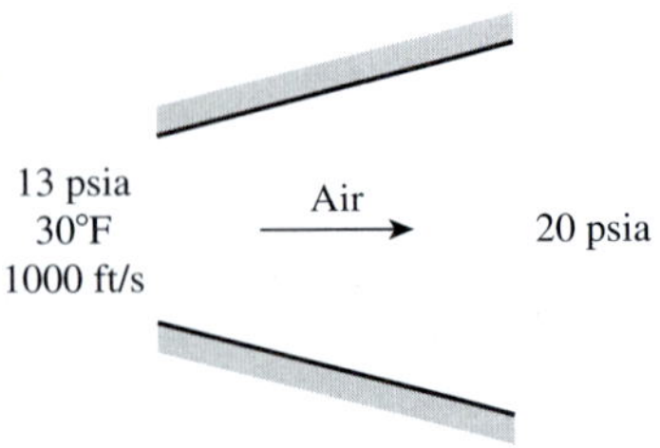

FIGURE P7–139E

Entropy Balance

7–140E Refrigerant-134a is expanded adiabatically from 100 psia and 100°F to a saturated vapor at 10 psia. Determine the entropy generation for this process, in Btu/lbm·R.

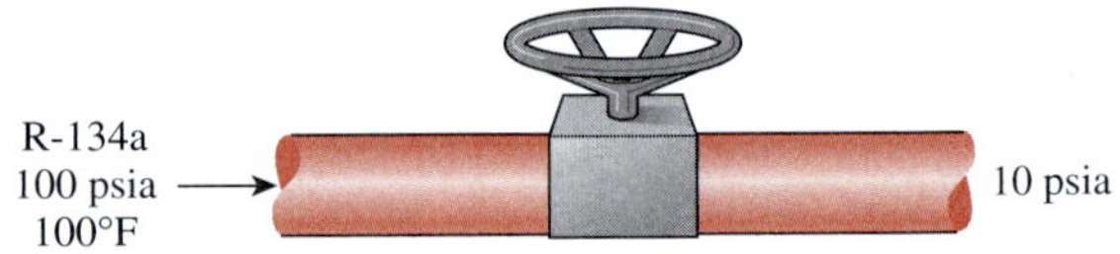

FIGURE P7–140E

7–141 Oxygen enters an insulated 12-cm-diameter pipe with a velocity of 70 m/s. At the pipe entrance, the oxygen is at 240 kPa and 20°C; and, at the exit, it is at 200 kPa and 18°C. Calculate the rate at which entropy is generated in the pipe.

7–142 Nitrogen is compressed by an adiabatic compressor from 100 kPa and 25°C to 800 kPa and 307°C. Calculate the entropy generation for this process, in kJ/kg·K.

7–143 Consider a family of four, with each person taking a 5-min shower every morning. The average flow rate through the shower head is 12 L/min. City water at 15°C is heated to 55°C in an electric water heater and tempered to 42°C by cold water at the T-elbow of the shower before being routed to the shower head. Determine the amount of entropy generated by this family per year as a result of taking daily showers.

7–144 Cold water (c_p = 4.18 kJ/kg·°C) leading to a shower enters a well-insulated, thin-walled, double-pipe, counter-flow heat exchanger at 10°C at a rate of 0.95 kg/s and is heated to 70°C by hot water (c_p = 4.19 kJ/kg·°C) that enters at 85°C at a rate of 1.6 kg/s. Determine (*a*) the rate of heat transfer and (*b*) the rate of entropy generation in the heat exchanger.

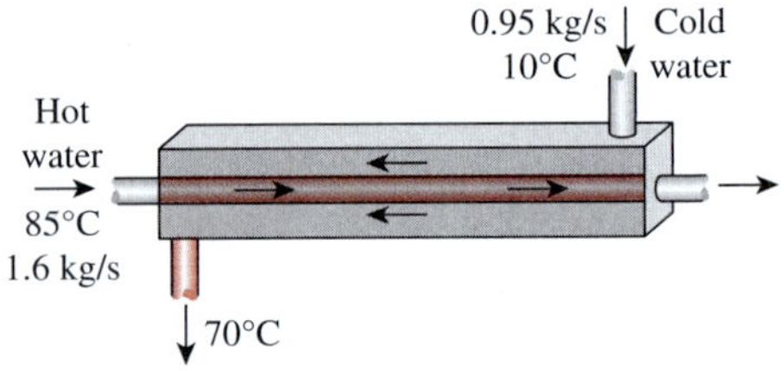

FIGURE P7–144

7–145 Air (c_p = 1.005 kJ/kg·°C) is to be preheated by hot exhaust gases in a cross-flow heat exchanger before it enters the furnace. Air enters the heat exchanger at 95 kPa and 20°C at a rate of 1.6 m^3/s. The combustion gases (c_p = 1.10 kJ/kg·°C)

enter at 180°C at a rate of 2.2 kg/s and leave at 95°C. Determine (*a*) the rate of heat transfer to the air, (*b*) the outlet temperature of the air, and (*c*) the rate of entropy generation.

7–146 A well-insulated, shell-and-tube heat exchanger is used to heat water (c_p = 4.18 kJ/kg·°C) in the tubes from 20 to 70°C at a rate of 4.5 kg/s. Heat is supplied by hot oil (c_p = 2.30 kJ/kg·°C) that enters the shell side at 170°C at a rate of 10 kg/s. Disregarding any heat loss from the heat exchanger, determine (*a*) the exit temperature of the oil and (*b*) the rate of entropy generation in the heat exchanger.

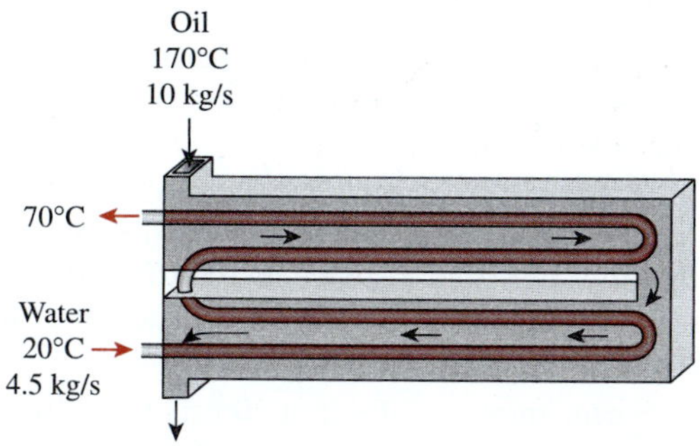

FIGURE P7–146

7–147 Refrigerant-134a is throttled from 1200 kPa and 40°C to 200 kPa. Heat is lost from the refrigerant in the amount of 0.5 kJ/kg to the surroundings at 25°C. Determine (*a*) the exit temperature of the refrigerant and (*b*) the entropy generation during this process.

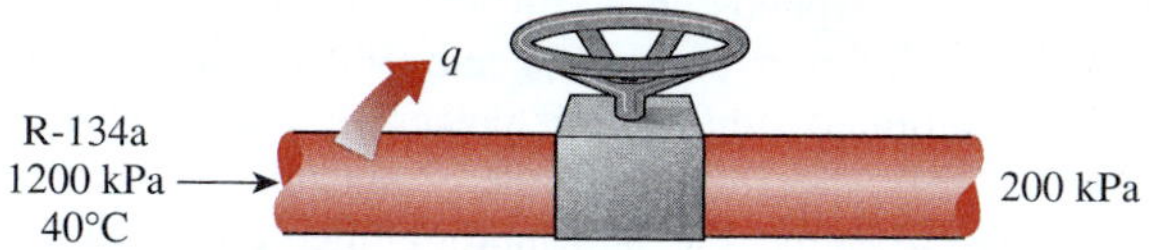

FIGURE P7–147

7–148 In an ice-making plant, water at 0°C is frozen at atmospheric pressure by evaporating saturated R-134a liquid at −16°C. The refrigerant leaves this evaporator as a saturated vapor, and the plant is sized to produce ice at 0°C at a rate of 2500 kg/h. Determine the rate of entropy generation in this plant. *Answer:* 0.0528 kW/K

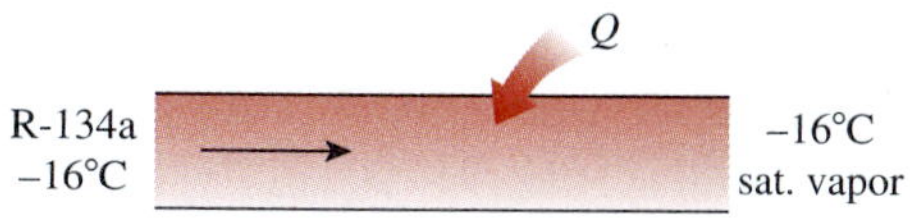

FIGURE P7–148

7–149E Water at 20 psia and 50°F enters a mixing chamber at a rate of 300 lbm/min where it is mixed steadily with steam entering at 20 psia and 240°F. The mixture leaves the chamber at 20 psia and 130°F, and heat is lost to the surrounding air at 70°F at a rate of 180 Btu/min. Neglecting the changes in kinetic and potential energies, determine the rate of entropy generation during this process?

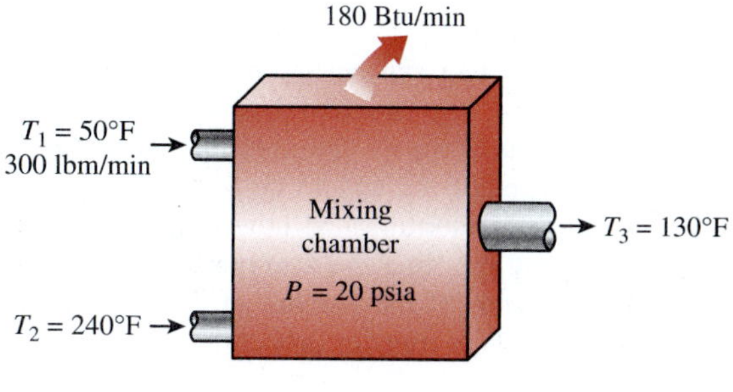

FIGURE P7–149E

7–150E Steam is to be condensed on the shell side of a heat exchanger at 120°F. Cooling water enters the tubes at 60°F at a rate of 92 lbm/s and leaves at 73°F. Assuming the heat exchanger to be well-insulated, determine (*a*) the rate of heat transfer in the heat exchanger and (*b*) the rate of entropy generation in the heat exchanger.

7–151 In a dairy plant, milk at 4°C is pasteurized continuously at 72°C at a rate of 12 L/s for 24 hours a day and 365 days a year. The milk is heated to the pasteurizing temperature by hot water heated in a natural-gas-fired boiler that has an efficiency of 82 percent. The pasteurized milk is then cooled by cold water at 18°C before it is finally refrigerated back to 4°C. To save energy and money, the plant installs a regenerator that has an effectiveness of 82 percent. If the cost of natural gas is $1.04/therm (1 therm = 105,500 kJ), determine how much energy and money the regenerator will save this company per year and the annual reduction in entropy generation.

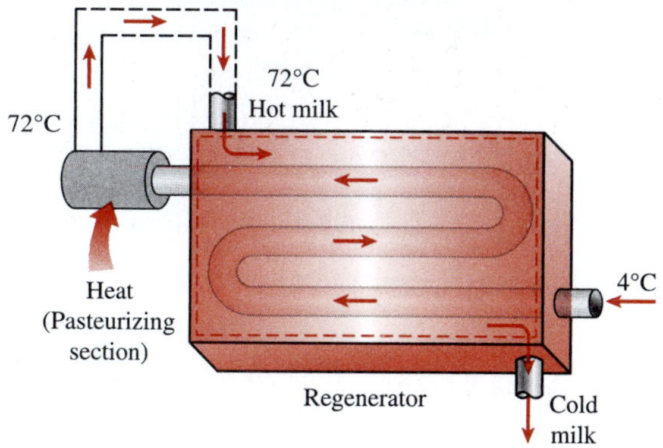

FIGURE P7–151

7–152 Stainless-steel ball bearings (ρ = 8085 kg/m^3 and c_p = 0.480 kJ/kg·°C) having a diameter of 1.8 cm are to be quenched in water at a rate of 1100 per minute. The balls leave the oven at a uniform temperature of 900°C and are exposed to air at 20°C for a while before they are dropped into the water. If the temperature of the balls drops to 850°C prior to quenching, determine (*a*) the rate of heat transfer from the balls to the air and (*b*) the rate of entropy generation due to heat loss from the balls to the air.

7–153 An ordinary egg can be approximated as a 5.5-cm-diameter sphere. The egg is initially at a uniform temperature of 8°C and is dropped into boiling water at 97°C. Taking the properties of the egg to be ρ = 1020 kg/m^3 and c_p = 3.32 kJ/kg·°C, determine (*a*) how much heat is transferred to the egg by the time the average temperature of the egg rises to 70°C and (*b*) the amount of entropy generation associated with this heat transfer process.

FIGURE P7–153

7–154 Long cylindrical steel rods (ρ = 7833 kg/m^3 and c_p = 0.465 kJ/kg·°C) of 10-cm diameter are heat treated by drawing them at a velocity of 3 m/min through a 7-m-long oven maintained at 900°C. If the rods enter the oven at 30°C and leave at 700°C, determine (*a*) the rate of heat transfer to the rods in the oven and (*b*) the rate of entropy generation associated with this heat transfer process.

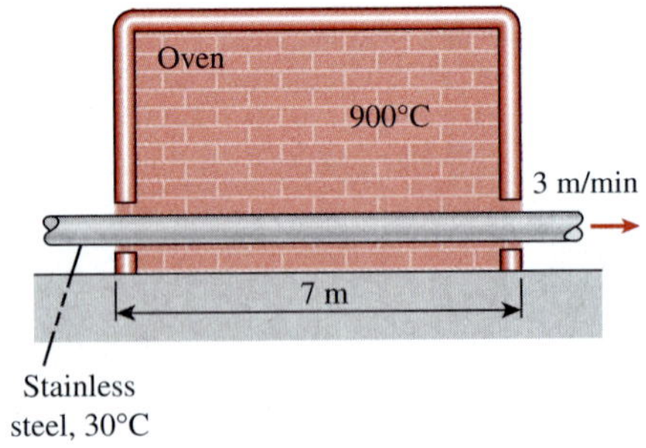

FIGURE P7–154

7–155 The inner and outer surfaces of a 3-m × 8-m brick wall of thickness 20 cm are maintained at temperatures of 20°C and 2°C, respectively. If the rate of heat transfer through the wall is 1550 W, determine the rate of entropy generation within the wall.

7–156 For heat transfer purposes, a standing man can be modeled as a 30-cm-diameter, 170-cm-long vertical cylinder with both the top and bottom surfaces insulated and with the side surface at an average temperature of 34°C. If the rate of heat loss from this man to the environment at 20°C is 336 W, determine the rate of entropy transfer from the body of this person accompanying heat transfer, in W/K.

7–157 A 1000-W iron is left on the ironing board with its base exposed to the air at 20°C. If the surface temperature is 400°C, determine the rate of entropy generation during this process in steady operation. How much of this entropy generation occurs within the iron?

7–158E A frictionless piston–cylinder device contains saturated liquid water at 40-psia pressure. Now 600 Btu of heat is transferred to water from a source at 1000°F, and part of the liquid vaporizes at constant pressure. Determine the total entropy generated during this process, in Btu/R.

7–159E Steam enters a diffuser at 20 psia and 240°F with a velocity of 900 ft/s and exits as saturated vapor at 240°F and 100 ft/s. The exit area of the diffuser is 1 ft^2. Determine (*a*) the mass flow rate of the steam and (*b*) the rate of entropy generation during this process. Assume an ambient temperature of 77°F.

7–160 Steam enters an adiabatic nozzle at 2.5 MPa and 450°C with a velocity of 55 m/s and exits at 1 MPa and 390 m/s. If the nozzle has an inlet area of 6 cm^2, determine (*a*) the exit temperature and (*b*) the rate of entropy generation for this process. *Answers:* (*a*) 406°C, (*b*) 0.0783 kW/K

7–161 Steam expands in a turbine steadily at a rate of 40,000 kg/h, entering at 8 MPa and 500°C and leaving at 40 kPa as saturated vapor. If the power generated by the turbine is 8.2 MW, determine the rate of entropy generation for this process. Assume the surrounding medium is at 25°C. *Answer:* 11.4 kW/K

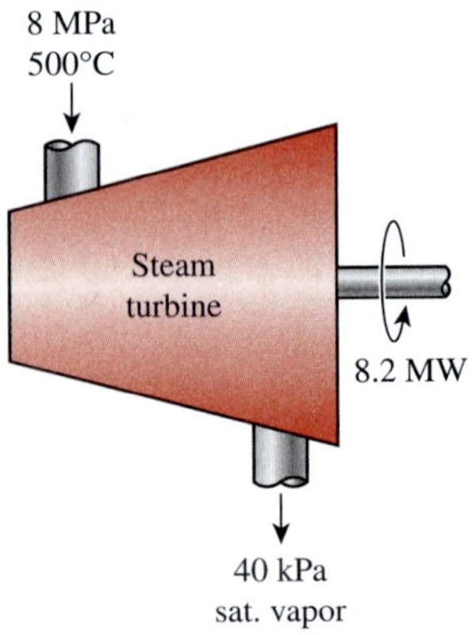

FIGURE P7–161

7–162E Air enters a compressor at ambient conditions of 15 psia and 60°F with a low velocity and exits at 150 psia, 620°F, and 350 ft/s. The compressor is cooled by the ambient air at 60°F at a rate of 1500 Btu/min. The power input to the compressor is 400 hp. Determine (*a*) the mass flow rate of air and (*b*) the rate of entropy generation.

7–163 A hot-water stream at 70°C enters an adiabatic mixing chamber with a mass flow rate of 3.6 kg/s, where it is mixed with a stream of cold water at 20°C. If the mixture leaves the chamber at 42°C, determine (*a*) the mass flow rate of the cold water and (*b*) the rate of entropy generation during this adiabatic mixing process. Assume all the streams are at a pressure of 200 kPa.

7–164 Liquid water at 200 kPa and 20°C is heated in a chamber by mixing it with superheated steam at 200 kPa and 150°C. Liquid water enters the mixing chamber at a rate of 2.5 kg/s, and the chamber is estimated to lose heat to the surrounding air at 25°C at a rate of 1200 kJ/min. If the mixture leaves the mixing chamber at 200 kPa and 60°C, determine (*a*) the mass flow rate of the superheated steam and (*b*) the rate of entropy generation during this mixing process. *Answers:* (*a*) 0.166 kg/s, (*b*) 0.333 kW/K

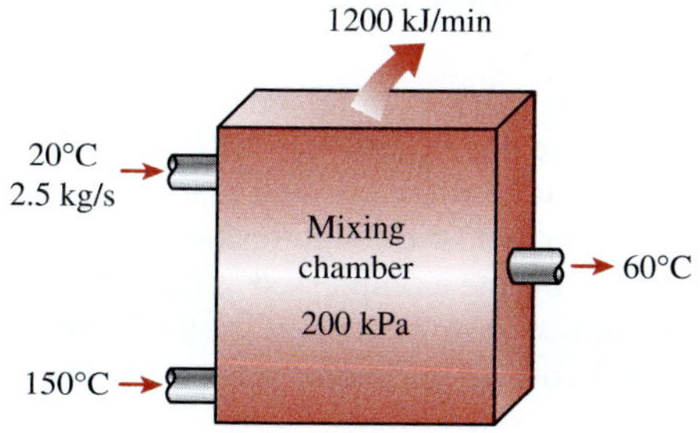

FIGURE P7–164

7–165 A 0.18-m^3 rigid tank is filled with saturated liquid water at 120°C. A valve at the bottom of the tank is now opened, and one-half of the total mass is withdrawn from the tank in the liquid form. Heat is transferred to water from a source at 230°C so that the temperature in the tank remains constant. Determine (*a*) the amount of heat transfer and (*b*) the total entropy generation for this process.

7–166E An iron block of unknown mass at 185°F is dropped into an insulated tank that contains 0.8 ft^3 of water at 70°F. At the same time, a paddle wheel driven by a 200-W motor is activated to stir the water. Thermal equilibrium is established after 10 min with a final temperature of 75°F. Determine (*a*) the mass of the iron block and (*b*) the entropy generated during this process.

7–167 A rigid tank contains 7.5 kg of saturated water mixture at 400 kPa. A valve at the bottom of the tank is now opened, and liquid is withdrawn from the tank. Heat is transferred to the steam such that the pressure inside the tank remains constant. The valve is closed when no liquid is left in the tank. If it is estimated that a total of 5 kJ of heat is transferred to the tank, determine (*a*) the quality of steam in the tank at the initial state, (*b*) the amount of mass that has escaped, and (*c*) the entropy generation during this process if heat is supplied to the tank from a source at 500°C.

Special Topic: Reducing the Cost of Compressed Air

7–168 Compressed air is one of the key utilities in manufacturing facilities, and the total installed power of compressed-air systems in the United States is estimated to be about 20 million horsepower. Assuming the compressors to operate at full load during one-third of the time on average and the average motor efficiency to be 85 percent, determine how much energy and money will be saved per year if the energy consumed by compressors is reduced by 5 percent as a result of implementing some conservation measures. Take the unit cost of electricity to be $0.07/kWh.

7–169 The compressed-air requirements of a plant at sea level are being met by a 90-hp compressor that takes in air at the local atmospheric pressure of 101.3 kPa and the average temperature of 15°C and compresses it to 1100 kPa. An investigation of the compressed-air system and the equipment using the compressed air reveals that compressing the air to 750 kPa is sufficient for this plant. The compressor operates 3500 h/yr at 75 percent of the rated load and is driven by an electric motor that has an efficiency of 94 percent. Taking the price of electricity to be $0.085/kWh, determine the amount of energy and money saved as a result of reducing the pressure of the compressed air.

7–170 A 150-hp compressor in an industrial facility is housed inside the production area where the average temperature during operating hours is 25°C. The average temperature of outdoors during the same hours is 10°C. The compressor operates 4500 h/yr at 85 percent of rated load and is driven by an electric motor that has an efficiency of 90 percent. Taking the price of electricity to be $0.07/kWh, determine the amount of energy and money saved as a result of drawing outside air to the compressor instead of using the inside air.

7–171 The compressed-air requirements of a plant are being met by a 100-hp screw compressor that runs at full load during 40 percent of the time and idles the rest of the time during operating hours. The compressor consumes 35 percent of the rated power when idling and 90 percent of the power when compressing air. The annual operating hours of the facility are 3800 h, and the unit cost of electricity is \$0.075/kWh.

It is determined that the compressed-air requirements of the facility during 60 percent of the time can be met by a 25-hp reciprocating compressor that consumes 95 percent of the rated power when compressing air and no power when not compressing air. It is estimated that the 25-hp compressor runs 85 percent of the time. The efficiencies of the motors of the large and the small compressors at or near full load are 0.90 and 0.88, respectively. The efficiency of the large motor at 35 percent load is 0.82. Determine the amount of energy and money saved as a result of switching to the 25-hp compressor during 60 percent of the time.

7–172 The compressed-air requirements of a plant are being met by a 125-hp screw compressor. The facility stops production for one hour every day, including weekends, for lunch break, but the compressor is kept operating. The compressor consumes 35 percent of the rated power when idling, and the unit cost of electricity is \$0.09/kWh. Determine the amount of energy and money saved per year as a result of turning the compressor off during lunch break. Take the efficiency of the motor at part load to be 84 percent.

7–173 The compressed-air requirements of a plant are met by a 150-hp compressor equipped with an intercooler, an aftercooler, and a refrigerated dryer. The plant operates 6300 h/yr, but the compressor is estimated to be compressing air during only one-third of the operating hours, that is, 2100 hours a year. The compressor is either idling or is shut off the rest of the time. Temperature measurements and calculations indicate that 25 percent of the energy input to the compressor is removed from the compressed air as heat in the aftercooler. The COP of the refrigeration unit is 2.5, and the cost of electricity is \$0.095/kWh. Determine the amount of the energy and money saved per year as a result of cooling the compressed air before it enters the refrigerated dryer.

7–174 The 1800-rpm, 150-hp motor of a compressor is burned out and is to be replaced by either a standard motor that has a full-load efficiency of 93.0 percent and costs \$9031 or a high-efficiency motor that has an efficiency of 96.2 percent and costs \$10,942. The compressor operates 4368 h/yr at full load, and its operation at part load is negligible. If the cost of electricity is \$0.075/kWh, determine the amount of energy and money this facility will save by purchasing the high-efficiency motor instead of the standard motor. Also, determine if the savings from the high-efficiency motor justify the price differential if the expected life of the motor is 10 years. Ignore any possible rebates from the local power company.

7–175 The space heating of a facility is accomplished by natural gas heaters that are 80 percent efficient. The compressed air needs of the facility are met by a large liquid-cooled compressor. The coolant of the compressor is cooled by air in a liquid-to-air heat exchanger whose airflow section is 1.0-m high and 1.0-m wide. During typical operation, the air is heated from 20 to 52°C as it flows through the heat exchanger. The average velocity of air on the inlet side is measured to be 3 m/s. The compressor operates 20 hours a day and 5 days a week throughout the year. Taking the heating season to be 6 months (26 weeks) and the cost of the natural gas to be \$1.00/therm (1 therm = 100,000 Btu = 105,500 kJ), determine how much money will be saved by diverting the compressor waste heat into the facility during the heating season.

7–176 The compressors of a production facility maintain the compressed-air lines at a (gage) pressure of 700 kPa at 1400-m elevation, where the atmospheric pressure is 85.6 kPa. The average temperature of air is 15°C at the compressor inlet and 25°C in the compressed-air lines. The facility operates 4200 h/yr, and the average price of electricity is \$0.07/kWh. Taking the compressor efficiency to be 0.8, the motor efficiency to be 0.93, and the discharge coefficient to be 0.65, determine the energy and money saved per year by sealing a leak equivalent to a 3-mm-diameter hole on the compressed-air line.

7–177 The energy used to compress air in the United States is estimated to exceed one-half quadrillion (0.5×10^{15}) kJ per year. It is also estimated that 10 to 40 percent of the compressed air is lost through leaks. Assuming, on average, 20 percent of the compressed air is lost through air leaks and the unit cost of electricity is \$0.07/kWh, determine the amount and cost of electricity wasted per year due to air leaks.

Review Problems

7–178E A heat engine whose thermal efficiency is 40 percent uses a hot reservoir at 1300 R and a cold reservoir at 500 R. Calculate the entropy change of the two reservoirs when 1 Btu of heat is transferred from the hot reservoir to the engine. Does this engine satisfy the increase of entropy principle? If the thermal efficiency of the heat engine is increased to 70 percent, will the increase of entropy principle still be satisfied?

7–179 A refrigerator with a coefficient of performance of 4 transfers heat from a cold region at −20°C to a hot region at 30°C. Calculate the total entropy change of the regions when 1 kJ of heat is transferred from the cold region. Is the second law satisfied? Will this refrigerator still satisfy the second law if its coefficient of performance is 6?

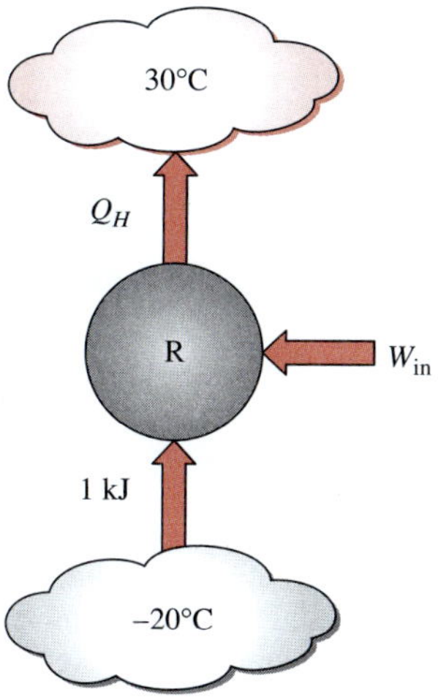

FIGURE P7–179

7–180 Calculate the rate of entropy change of all the components of a refrigerator that uses 10 kW of power, rejects 14 kW of heat, and has a high-temperature energy reservoir at 400 K and a low-temperature energy reservoir at 200 K. What is the rate of cooling produced by this refrigerator? Is this refrigerator completely reversible?

7–181 What is the minimum internal energy that R-134a can achieve as it is compressed adiabatically from 200 kPa and 85 percent quality to 800 kPa in a closed system?

7–182 Is it possible to cool and condense R-134a to a saturated liquid from 1000 kPa and 180°C in a closed system undergoing an isobaric, reversible process while exchanging heat with an isothermal energy reservoir at 100°C?

7–183 It has been suggested that air at 100 kPa and 25°C can be cooled by first compressing it adiabatically in a closed system to 1000 kPa and then expanding it adiabatically back to 100 kPa. Is this possible?

7–184E Can 1 lbm of air in a closed system at 20 psia and 100°F be compressed adiabatically to 120 psia and a volume of 3 ft^3?

7–185E A 100-lbm block of a solid material whose specific heat is 0.5 Btu/lbm·R is at 70°F. It is heated with 10 lbm of saturated water vapor that has a constant pressure of 14.7 psia. Determine the final temperature of the block and water, and the entropy change of (*a*) the block, (*b*) the water, and (*c*) the entire system. Is this process possible? Why?

7–186 One kilogram of air is in a piston–cylinder apparatus that can only exchange heat with a reservoir at 300 K. Initially this air is at 100 kPa and 27°C. Someone claims that the air can be compressed to 250 kPa and 27°C. Determine if this claim is valid by performing a second-law analysis of the claimed process.

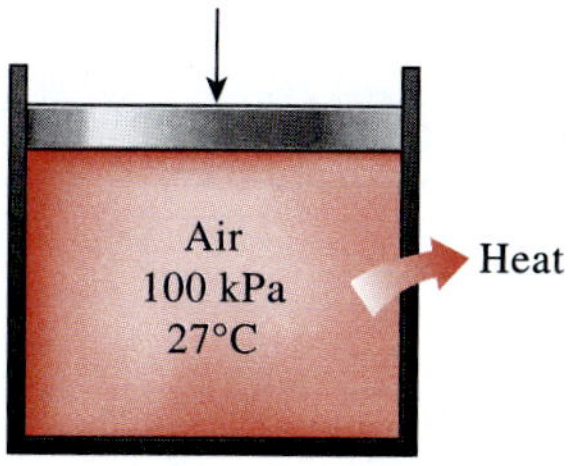

FIGURE P7–186

7–187 A rigid tank contains 3.2 kg of water at 140°C and 400 kPa. Now 48 kJ of shaft work is done on the system and the final temperature in the tank is 80°C. If the entropy change of water is zero and the surroundings are at 15°C, determine (*a*) the final pressure in the tank, (*b*) the amount of heat transfer between the tank and the surroundings, and (*c*) the entropy generation during this process. *Answers:* (*a*) 47.4 kPa, (*b*) 163 kJ, (*c*) 0.565 kJ/K

7–188 A horizontal cylinder is separated into two compartments by an adiabatic, frictionless piston. One side contains 0.2 m^3 of nitrogen and the other side contains 0.1 kg of helium, both initially at 20°C and 95 kPa. The sides of the cylinder and the helium end are insulated. Now heat is added to the nitrogen side from a reservoir at 500°C until the pressure of the helium rises to 120 kPa. Determine (*a*) the final temperature of the helium, (*b*) the final volume of the nitrogen, (*c*) the heat transferred to the nitrogen, and (*d*) the entropy generation during this process.

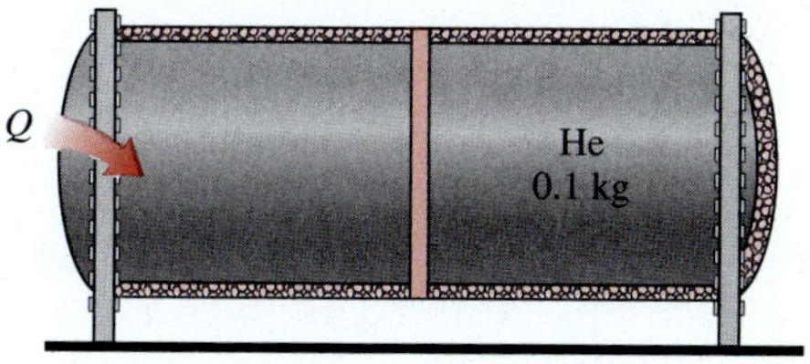

FIGURE P7–188

7–189 A 0.8-m^3 rigid tank contains carbon dioxide (CO_2) gas at 250 K and 100 kPa. A 500-W electric resistance heater placed in the tank is now turned on and kept on for 40 min after which the pressure of CO_2 is measured to be 175 kPa. Assuming the surroundings to be at 300 K and using constant specific heats, determine (*a*) the final temperature of CO_2, (*b*) the net amount of heat transfer from the tank, and (*c*) the entropy generation during this process.

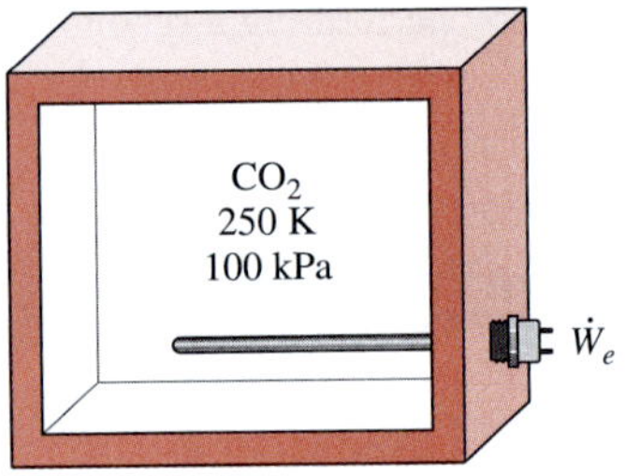

FIGURE P7–189

7–190 Helium gas is throttled steadily from 300 kPa and 50°C. Heat is lost from the helium in the amount of 1.75 kJ/kg to the surroundings at 25°C and 100 kPa. If the entropy of the helium increases by 0.25 kJ/kg·K in the valve, determine (*a*) the exit pressure and temperature and (*b*) the entropy generation during this process. *Answers:* (*a*) 265 kPa, 49.7°C, (*b*) 0.256 kJ/kg·K

7–191 Refrigerant-134a enters a compressor as a saturated vapor at 200 kPa at a rate of 0.03 m^3/s and leaves at 700 kPa. The power input to the compressor is 10 kW. If the surroundings at 20°C experience an entropy increase of 0.008 kW/K, determine (*a*) the rate of heat loss from the compressor, (*b*) the exit temperature of the refrigerant, and (*c*) the rate of entropy generation.

7–192 Air at 500 kPa and 400 K enters an adiabatic nozzle at a velocity of 30 m/s and leaves at 300 kPa and 350 K. Using variable specific heats, determine (*a*) the isentropic efficiency, (*b*) the exit velocity, and (*c*) the entropy generation.

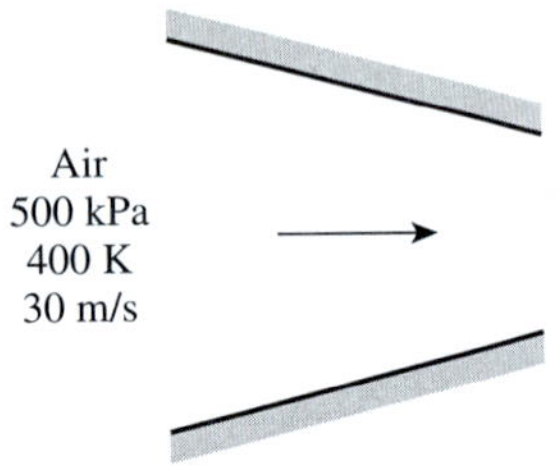

FIGURE P7–192

7–193 An insulated tank containing 0.2 m^3 of saturated water vapor at 350 kPa is connected to an initially evacuated, insulated piston–cylinder device. The mass of the piston is such that a pressure of 200 kPa is required to raise it. Now the valve is opened slightly, and part of the steam flows to the cylinder, raising the piston. This process continues until the pressure in the tank drops to 200 kPa. Assuming the steam that remains in the tank to have undergone a reversible adiabatic process, determine the final temperature (*a*) in the rigid tank and (*b*) in the cylinder.

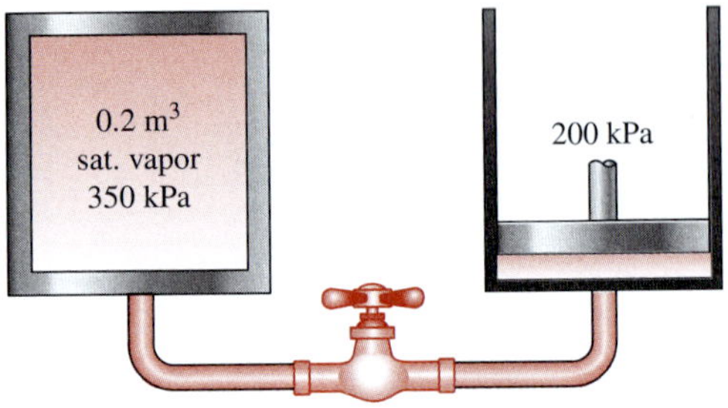

FIGURE P7–193

7–194 3-kg of helium gas at 100 kPa and 27°C are adiabatically compressed to 900 kPa. If the isentropic compression efficiency is 80 percent, determine the required work input and the final temperature of helium.

7–195 Refrigerant-134a at 600 kPa and 100°C undergoes a reversible, isothermal expansion to 200 kPa in a steady-flow device with one inlet and one outlet. Determine the power produced by this device and the rate of heat transfer when the flow rate through the device is 1 kg/s.

7–196 An inventor claims to have invented an adiabatic steady-flow device with a single inlet-outlet that produces 100 kW when expanding 1 kg/s of air from 900 kPa and 300°C to 100 kPa. Is this claim valid?

7–197 You are to expand a gas adiabatically from 3 MPa and 300°C to 80 kPa in a piston-cylinder device. Which of the two choices – air with an isentropic expansion efficiency of 90 percent or neon with an isentropic expansion efficiency of 80 percent – will produce the most work?

7–198 An adiabatic capillary tube is used in some refrigeration systems to drop the pressure of the refrigerant from the condenser level to the evaporator level. R-134a enters the capillary tube as a saturated liquid at 50°C, and leaves at −12°C. Determine the rate of entropy generation in the capillary tube for a mass flow rate of 0.2 kg/s. *Answer:* 0.0077 kW/K

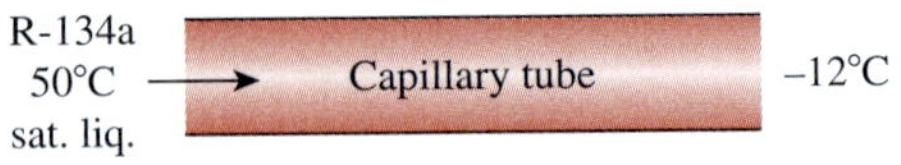

FIGURE P7–198

7–199 Air is compressed steadily by a compressor from 100 kPa and 17°C to 700 kPa at a rate of 5 kg/min. Determine the minimum power input required if the process is (*a*) adiabatic and (*b*) isothermal. Assume air to be an ideal gas with variable specific heats, and neglect the changes in kinetic and potential energies. *Answers:* (*a*) 18.0 kW, (*b*) 13.5 kW

7–200 Air enters a two-stage compressor at 100 kPa and 27°C and is compressed to 625 kPa. The pressure ratio across each stage is the same, and the air is cooled to the initial temperature between the two stages. Assuming the compression process to be isentropic, determine the power input to the compressor for a mass flow rate of 0.15 kg/s. What would your answer be if only one stage of compression were used? *Answers:* 27.1 kW, 31.1 kW

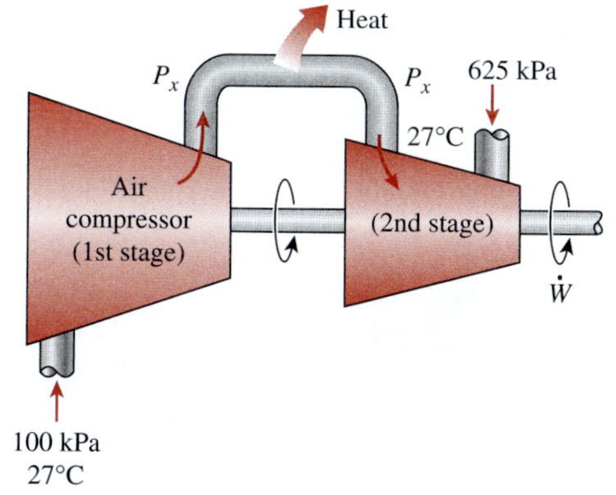

FIGURE P7–200

7–201 Steam at 6 MPa and 500°C enters a two-stage adiabatic turbine at a rate of 15 kg/s. Ten percent of the steam is extracted at the end of the first stage at a pressure of 1.2 MPa for other use. The remainder of the steam is further expanded in the second stage and leaves the turbine at 20 kPa. Determine the power output of the turbine, assuming (*a*) the process is reversible and (*b*) the turbine has an isentropic efficiency of 88 percent. *Answers:* (*a*) 16,291 kW, (*b*) 14,336 kW

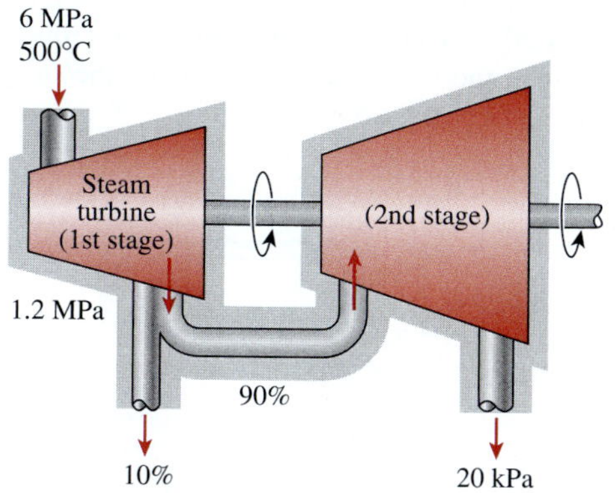

FIGURE P7–201

7–202 Steam enters a two-stage adiabatic turbine at 8 MPa and 550°C. It expands in the first stage to a pressure of 2 MPa. Then steam is reheated at constant pressure to 550°C before it is expanded in a second stage to a pressure of 200 kPa. The power output of the turbine is 80 MW. Assuming an isentropic efficiency of 84 percent for each stage of the turbine, determine the required mass flow rate of steam. Also, show the process on a *T-s* diagram with respect to saturation lines. *Answer:* 85.8 kg/s

7–203 Refrigerant-134a at 100 kPa and −20°C is compressed by an adiabatic 1.3-kW compressor to an exit state of 800 kPa and 60°C. Neglecting the changes in kinetic and potential energies, determine (*a*) the isentropic efficiency of the compressor, (*b*) the volume flow rate of the refrigerant at the compressor inlet, in L/min, and (*c*) the maximum volume flow rate at the inlet conditions that this adiabatic 0.7-kW compressor can handle without violating the second law.

7–204 An adiabatic air compressor is to be powered by a direct-coupled adiabatic steam turbine that is also driving a generator. Steam enters the turbine at 12.5 MPa and 500°C at a rate of 25 kg/s and exits at 10 kPa and a quality of 0.92. Air enters the compressor at 98 kPa and 295 K at a rate of 10 kg/s and exits at 1 MPa and 620 K. Determine (*a*) the net power delivered to the generator by the turbine and (*b*) the rate of entropy generation within the turbine and the compressor during this process.

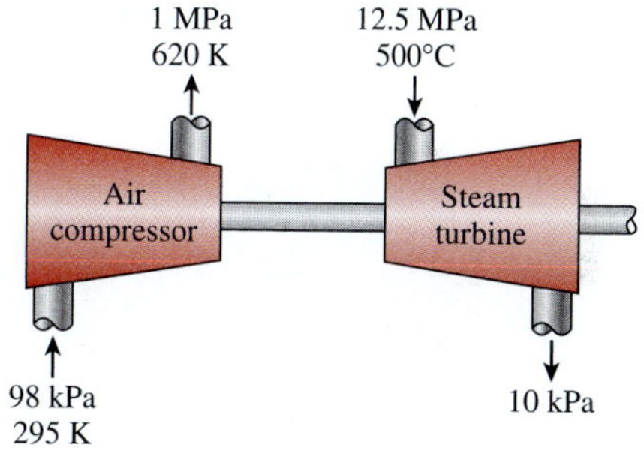

FIGURE P7–204

7–205 Reconsider Prob. 7–204. Using EES (or other) software, determine the isentropic efficiencies for the compressor and turbine. Then use EES to study how varying the compressor efficiency over the range 0.6 to 0.8 and the turbine efficiency over the range 0.7 to 0.95 affect the net work for the cycle and the entropy generated for the process. Plot the net work as a function of the compressor efficiency for turbine efficiencies of 0.7, 0.8, and 0.9, and discuss your results.

7–206 The explosion of a hot water tank in a school in Spencer, Oklahoma, in 1982 killed 7 people while injuring

33 others. Although the number of such explosions has decreased dramatically since the development of the ASME Pressure Vessel Code, which requires the tanks to be designed to withstand four times the normal operating pressures, they still occur as a result of the failure of the pressure relief valves and thermostats. When a tank filled with a high-pressure and high-temperature liquid ruptures, the sudden drop of the pressure of the liquid to the atmospheric level causes part of the liquid to flash into vapor, and thus to experience a huge rise in its volume. The resulting pressure wave that propagates rapidly can cause considerable damage.

Considering that the pressurized liquid in the tank eventually reaches equilibrium with its surroundings shortly after the explosion, the work that a pressurized liquid would do if allowed to expand reversibly and adiabatically to the pressure of the surroundings can be viewed as the *explosive energy* of the pressurized liquid. Because of the very short time period of the explosion and the apparent calm afterward, the explosion process can be considered to be adiabatic with no changes in kinetic and potential energies and no mixing with the air.

Consider a 80-L hot-water tank that has a working pressure of 0.5 MPa. As a result of some malfunction, the pressure in the tank rises to 2 MPa, at which point the tank explodes. Taking the atmospheric pressure to be 100 kPa and assuming the liquid in the tank to be saturated at the time of explosion, determine the total explosion energy of the tank in terms of the TNT equivalence. (The explosion energy of TNT is about 3250 kJ/kg, and 5 kg of TNT can cause total destruction of unreinforced structures within about a 7-m radius.) *Answer:* 1.972 kg TNT

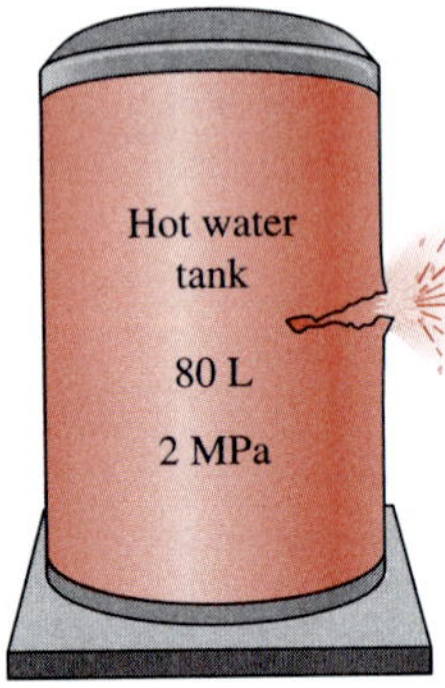

FIGURE P7–206

7–207 Using the arguments in the Prob. 7–206, determine the total explosion energy of a 0.35-L canned drink that explodes at a pressure of 1.2 MPa. To how many kg of TNT is this explosion energy equivalent?

7–208 Air is expanded in an adiabatic turbine of 85 percent isentropic efficiency from an inlet state of 2200 kPa and 300°C to an outlet pressure of 200 kPa. Calculate the outlet temperature of air and the work produced by this turbine per unit mass of air.

7–209 Air is expanded in an adiabatic turbine of 90 percent isentropic efficiency from an inlet state of 2200 kPa and 300°C to an outlet pressure of 200 kPa. Calculate the outlet temperature of air, the work produced by this turbine, and the entropy generation. *Answers:* 319 K, 258 kJ/kg, 0.0944 kJ/kg·K

7–210 To control the power output of an isentropic steam turbine, a throttle valve is placed in the steam line supplying the turbine inlet, as shown in the figure. Steam at 6 MPa and 400°C is supplied to the throttle inlet, and the turbine exhaust pressure is set at 70 kPa. Compare the work produced by this steam turbine, in kJ/kg, when the throttle valve is completely open (so that there is no pressure loss) and when it is partially closed so that the pressure at the turbine inlet is 3 MPa.

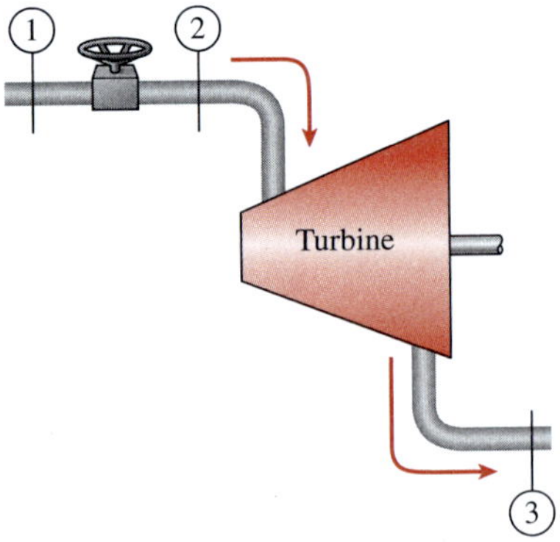

FIGURE P7–210

7–211 Oxygen tanks are filled by compressing the oxygen gas, as shown in the figure. An initially evacuated 1-m^3 tank is to be filled to 13,000 kPa while the temperature of the oxygen in the tank remains constant at 20°C with this system. An isentropic compressor is used. The oxygen enters this compressor with a constant pressure and temperature of 150 kPa and 20°C. Determine the total work required by the compressor and the total heat transfer from the oxygen tank. *Answers:* 80,710 kJ, 93,720 kJ

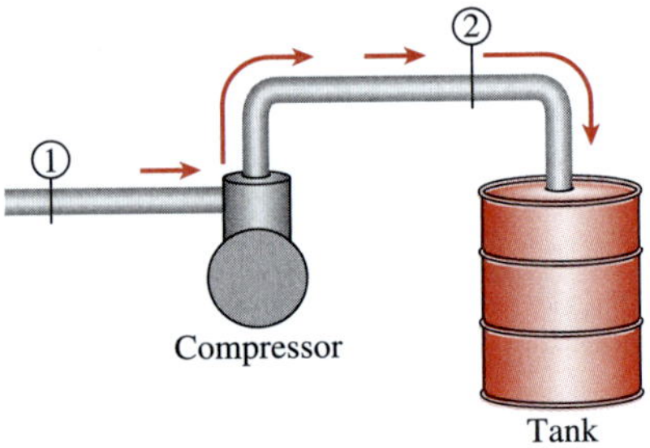

FIGURE P7–211

7–212 Two rigid tanks are connected by a valve. Tank A is insulated and contains 0.3 m³ of steam at 400 kPa and 60 percent quality. Tank B is uninsulated and contains 2 kg of steam at 200 kPa and 250°C. The valve is now opened, and steam flows from tank A to tank B until the pressure in tank A drops to 200 kPa. During this process 300 kJ of heat is transferred from tank B to the surroundings at 17°C. Assuming the steam remaining inside tank A to have undergone a reversible adiabatic process, determine (*a*) the final temperature in each tank and (*b*) the entropy generated during this process. *Answers:* (*a*) 120.2°C, 116.1°C, (*b*) 0.498 kJ/K

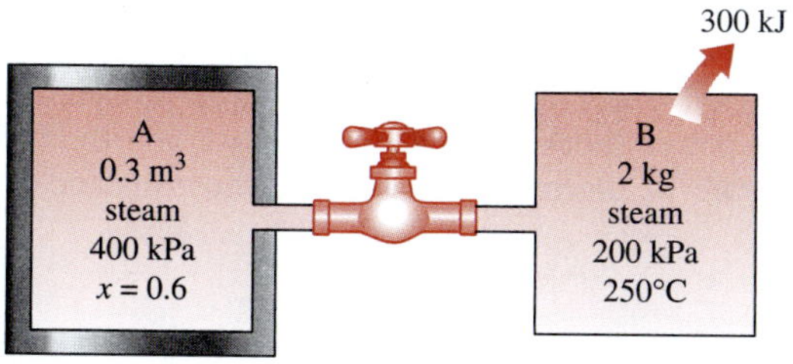

FIGURE P7–212

7–213 Heat is transferred steadily to boiling water in the pan through its flat bottom at a rate of 500 W. If the temperatures of the inner and outer surfaces of the bottom of the tank are 104 and 105°C, respectively, determine the rate of entropy generation within bottom of the pan, in W/K.

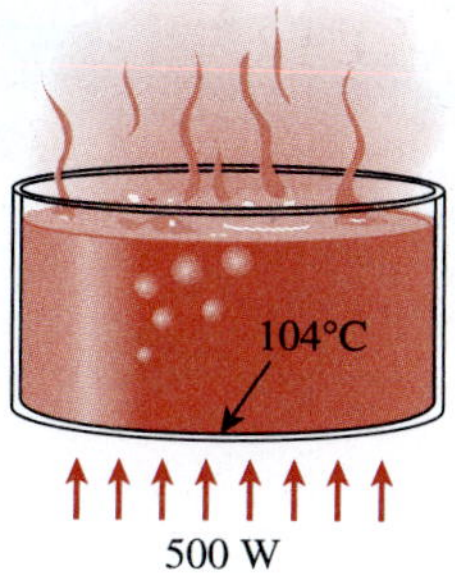

FIGURE P7–213

7–214 A 1200-W electric resistance heating element whose diameter is 0.5 cm is immersed in 40 kg of water initially at 20°C. Assuming the water container is well-insulated, determine how long it will take for this heater to raise the water temperature to 50°C. Also, determine the entropy generated during this process, in kJ/K.

7–215 In large steam power plants, the feedwater is frequently heated in closed feedwater heaters, which are basically heat exchangers, by steam extracted from the turbine at some stage. Steam enters the feedwater heater at 1 MPa and 200°C and leaves as saturated liquid at the same pressure. Feedwater enters the heater at 2.5 MPa and 50°C and leaves 10°C below the exit temperature of the steam. Neglecting any heat losses from the outer surfaces of the heater, determine (*a*) the ratio of the mass flow rates of the extracted steam and the feedwater heater and (*b*) the total entropy change for this process per unit mass of the feedwater.

7–216 Reconsider Prob. 7–215. Using EES (or other) software, investigate the effect of the state of the steam at the inlet of the .feedwater heater. Assume the entropy of the extraction steam is constant at the value for 1 MPa, 200°C and decrease the extraction steam pressure from 1 MPa to 100 kPa. Plot both the ratio of the mass flow rates of the extracted steam and the feedwater heater and the total entropy change for this process per unit mass of the feedwater as functions of the extraction pressure.

7–217E A 5-ft³ rigid tank initially contains refrigerant-134a at 80 psia and 100 percent quality. The tank is connected by a valve to a supply line that carries refrigerant-134a at 160 psia and 80°F. The valve is now opened, allowing the refrigerant to enter the tank, and is closed when it is observed that the tank contains only saturated liquid at 120 psia. Determine (*a*) the mass of the refrigerant that entered the tank, (*b*) the amount of heat transfer with the surroundings at 100°F, and (*c*) the entropy generated during this process.

7–218 A passive solar house that is losing heat to the outdoors at 3°C at an average rate of 50,000 kJ/h is maintained at 22°C at all times during a winter night for 10 h. The house is to be heated by 50 glass containers, each containing 20 L of water that is heated to 80°C during the day by absorbing solar energy. A thermostat controlled 15 kW backup electric resistance heater turns on whenever necessary to keep the house at 22°C. Determine how long the electric heating system was on that night and the amount of entropy generated during the night.

7–219 A 4-m × 5-m × 7-m well-sealed room is to be heated by 1500 kg of liquid water contained in a tank that is placed in the room. The room is losing heat to the outside air at 5°C at an average rate of 10,000 kJ/h. The room is initially at 20°C and 100 kPa and is maintained at a temperature of 20°C at all times. If the hot water is to meet the heating requirements of this room for a 24-h period, determine (*a*) the minimum temperature of the water when it is first brought into the room and (*b*) the entropy generated during a 24-h period. Assume constant specific heats for both air and water at room temperature.

7–220 Consider a well-insulated horizontal rigid cylinder that is divided into two compartments by a piston that is free to move but does not allow either gas to leak into the other side. Initially, one side of the piston contains 2 m³ of N_2 gas at 250 kPa and 100°C while the other side contains 1 m³ of

He gas at 250 kPa and 25°C. Now thermal equilibrium is established in the cylinder as a result of heat transfer through the piston. Using constant specific heats at room temperature, determine (*a*) the final equilibrium temperature in the cylinder and (*b*) the entropy generation during this process. What would your answer be if the piston were not free to move?

7–221 Reconsider Prob. 7–220. Using EES (or other) software, compare the results for constant specific heats to those obtained using variable specific heats built into EES functions.

7–222 Repeat Prob. 7–220 by assuming the piston is made of 5 kg of copper initially at the average temperature of the two gases on both sides.

7–223 An insulated 5-m^3 rigid tank contains air at 500 kPa and 57°C. A valve connected to the tank is now opened, and air is allowed to escape until the pressure inside drops to 200 kPa. The air temperature during this process is maintained constant by an electric resistance heater placed in the tank. Determine (*a*) the electrical energy supplied during this process and (*b*) the total entropy change. *Answers:* (*a*) 1501 kJ, (*b*) 4.40 kJ/K

7–224 An insulated piston–cylinder device initially contains 0.02 m^3 of saturated liquid–vapor mixture of water with a quality of 0.1 at 100°C. Now some ice at −18°C is dropped into the cylinder. If the cylinder contains saturated liquid at 100°C when thermal equilibrium is established, determine (*a*) the amount of ice added and (*b*) the entropy generation during this process. The melting temperature and the heat of fusion of ice at atmospheric pressure are 0°C and 333.7 kJ/kg.

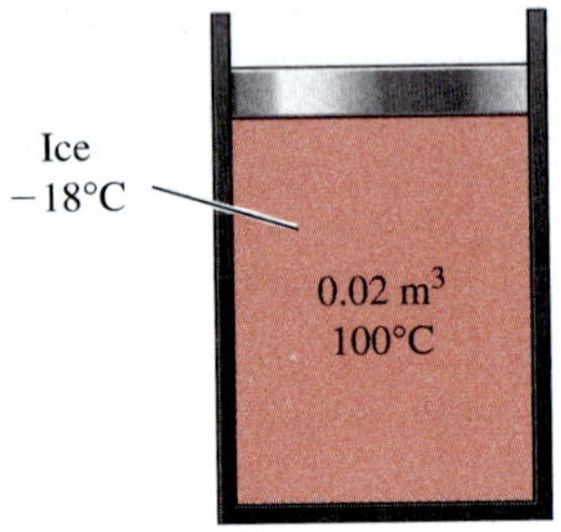

FIGURE P7–224

7–225 Consider a 10-L evacuated rigid bottle that is surrounded by the atmosphere at 90 kPa and 27°C. A valve at the neck of the bottle is now opened and the atmospheric air is allowed to flow into the bottle. The air trapped in the bottle eventually reaches thermal equilibrium with the atmosphere as a result of heat transfer through the wall of the bottle. The valve remains open during the process so that the trapped air also reaches mechanical equilibrium with the atmosphere. Determine the net heat transfer through the wall of the bottle and the entropy generation during this filling process. *Answers:* 0.90 kJ, 0.0030 kJ/K

7–226 (*a*) Water flows through a shower head steadily at a rate of 10 L/min. An electric resistance heater placed in the water pipe heats the water from 16 to 43°C. Taking the density of water to be 1 kg/L, determine the electric power input to the heater, in kW, and the rate of entropy generation during this process, in kW/K.

(*b*) In an effort to conserve energy, it is proposed to pass the drained warm water at a temperature of 39°C through a heat exchanger to preheat the incoming cold water. If the heat exchanger has an effectiveness of 0.50 (that is, it recovers only half of the energy that can possibly be transferred from the drained water to incoming cold water), determine the electric power input required in this case and the reduction in the rate of entropy generation in the resistance heating section.

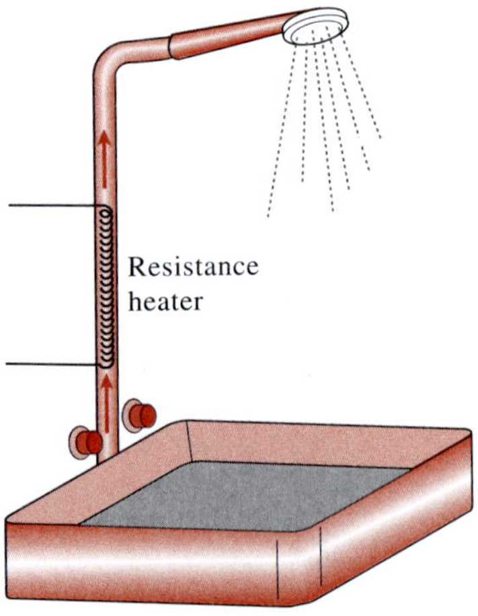

FIGURE P7–226

7–227 Using EES (or other) software, determine the work input to a multistage compressor for a given set of inlet and exit pressures for any number of stages. Assume that the pressure ratio across each stage is identical and the compression process is polytropic. List and plot the compressor work against the number of stages for P_1 = 100 kPa, T_1 = 25°C, P_2 = 1000 kPa, and n = 1.35 for air. Based on this chart, can you justify using compressors with more than three stages?

7–228 Consider the turbocharger of an internal combustion engine. The exhaust gases enter the turbine at 450°C at a rate of 0.02 kg/s and leave at 400°C. Air enters the compressor at

70°C and 95 kPa at a rate of 0.018 kg/s and leaves at 135 kPa. The mechanical efficiency between the turbine and the compressor is 95 percent (5 percent of turbine work is lost during its transmission to the compressor). Using air properties for the exhaust gases, determine (*a*) the air temperature at the compressor exit and (*b*) the isentropic efficiency of the compressor. *Answers:* (*a*) 126.1°C, (*b*) 0.642

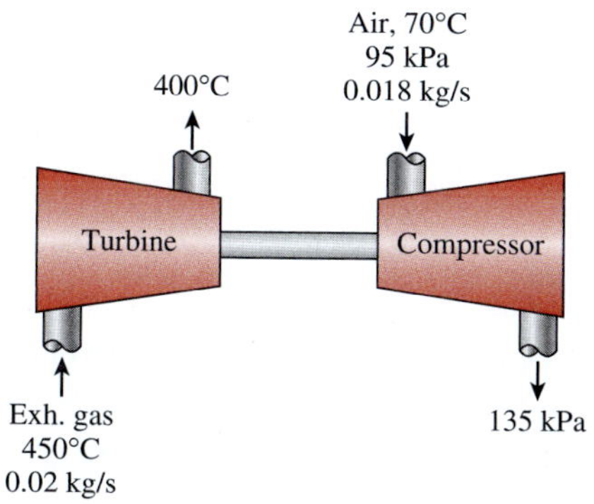

FIGURE P7–228

7–229 A 0.25-m^3 insulated piston–cylinder device initially contains 0.7 kg of air at 20°C. At this state, the piston is free to move. Now air at 500 kPa and 70°C is allowed to enter the cylinder from a supply line until the volume increases by 50 percent. Using constant specific heats at room temperature, determine (*a*) the final temperature, (*b*) the amount of mass that has entered, (*c*) the work done, and (*d*) the entropy generation.

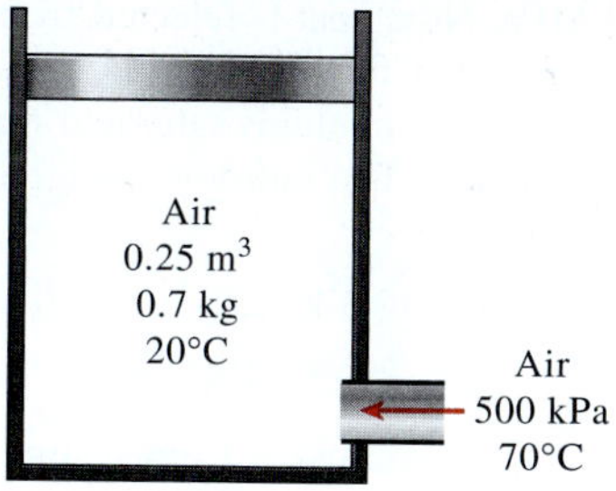

FIGURE P7–229

7–230 When the transportation of natural gas in a pipeline is not feasible for economic reasons, it is first liquefied using nonconventional refrigeration techniques and then transported in super-insulated tanks. In a natural gas liquefaction plant, the liquefied natural gas (LNG) enters a cryogenic turbine at 30 bar and −160°C at a rate of 20 kg/s and leaves at 3 bar. If 115 kW power is produced by the turbine, determine the efficiency of the turbine. Take the density of LNG to be 423.8 kg/m^3. *Answer:* 90.3 percent

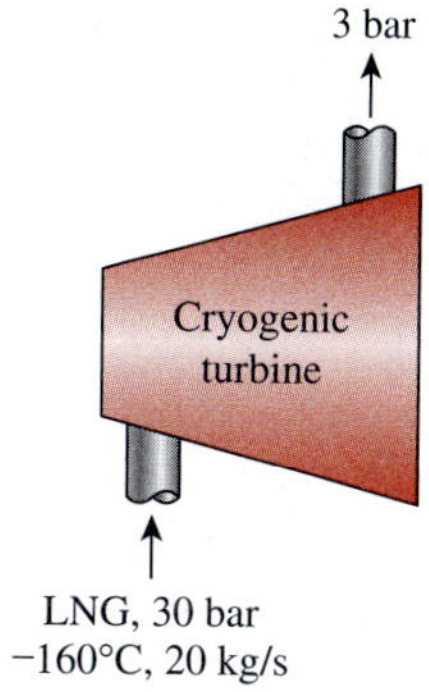

FIGURE P7–230

7–231 A constant volume tank filled with 2 kg of air rejects heat to a heat reservoir at 300 K. During the process the temperature of the air in the tank decreases to the reservoir temperature. Determine the expressions for the entropy changes for the tank and reservoir and the total entropy change or entropy generated of this isolated system. Plot these entropy changes as functions of the initial temperature of the air. Comment on your results. Assume constant specific heats for air at 300 K.

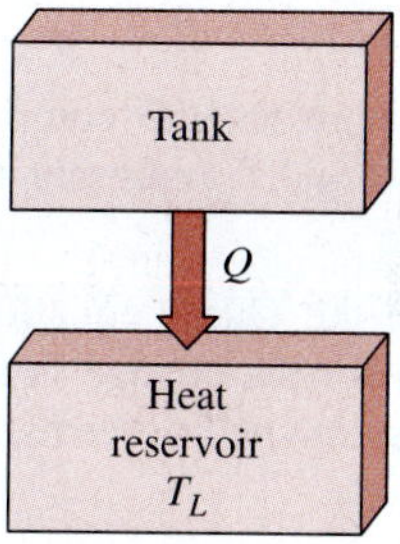

FIGURE P7–231

7–232 A heat engine receives heat from a constant volume tank filled with 2 kg of air. The engine produces work that is stored in a work reservoir and rejects 400 kJ of heat to a heat reservoir at 300 K. During the process the temperature of the air in the tank decreases to 300 K. (*a*) Determine the initial temperature of the air that will maximize the work and the thermal efficiency of the engine. (*b*) Evaluate the total entropy change of this isolated system, the work produced, and the thermal efficiency for the initial air temperature in the tank from part (*a*) and at 100 K above and below the answer to part (*a*). (*c*) Plot the thermal efficiency and the entropy generation as functions of the initial temperature of the air. Comment on your answers. Assume constant specific heats for air at 300 K. *Answers:* (*a*) 759 K, 0.393

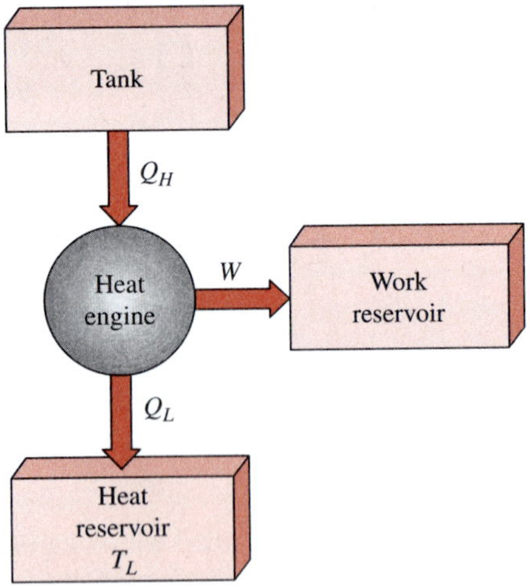

FIGURE P7–232

7–233 For an ideal gas with constant specific heats show that the compressor and turbine isentropic efficiencies may be written as

$$\eta_C = \frac{(P_2/P_1)^{(k-1)/k}}{(T_2/T_1) - 1} \text{ and } \eta_T = \frac{(T_4/T_3) - 1}{(P_4/P_3)^{(k-1)/k} - 1}.$$

The states 1 and 2 represent the compressor inlet and exit states and the states 3 and 4 represent the turbine inlet and exit states.

7–234 Starting with the Gibbs equation $du = Tds - Pdv$, obtain the expression for the change in internal energy of an ideal gas having constant specific heats during the isentropic process Pv^k = constant.

7–235 The temperature of an ideal gas having constant specific heats is given as a function of specific entropy and specific volume as $T(s,v) = Av^{1-k} \exp(s/c_v)$ where A is a constant. Determine the T-v relation for this ideal gas undergoing an isentropic process.

7–236 An ideal gas undergoes a reversible isothermal, steady-flow process. Neglecting the changes in kinetic and potential energies of the flow and assuming constant specific heats, (*a*) obtain the expression for the heat transfer per unit mass flow for the process and (*b*) compare this result with the one obtained from $q_{\text{net}} = \int Tds$ for the process.

7–237 The temperature of an ideal gas having constant specific heats is given as a function of specific entropy and pressure as $T(s,P) = AP^{(k-1)/k} \exp(s/c_p)$ where A is a constant. For a reversible, constant pressure process, find the expression for heat transfer per unit mass as a function of c_p and T using $Q = \int TdS$. Compare this result with that obtained by applying the first law to the closed system undergoing a constant pressure process.

7–238 Consider a two-stage compression with intercooling process when the low-pressure compressor has an isentropic efficiency of $\eta_{C,L}$ and the high-pressure compressor has an isentropic efficiency of $\eta_{C,H}$. Determine the intermediate pressure at which intercooling should take place to minimize the compressor work when the low-pressure compressor exit gases are cooled to the compressor inlet temperature.

Fundamentals of Engineering (FE) Exam Problems

7–239 Steam is condensed at a constant temperature of 30°C as it flows through the condensor of a power plant by rejecting heat at a rate of 55 MW. The rate of entropy change of steam as it flows through the condenser is

(*a*) −1.83 MW/K (*b*) −0.18 MW/K (*c*) 0 MW/K
(*d*) 0.56 MW/K (*e*) 1.22 MW/K

7–240 Steam is compressed from 6 MPa and 300°C to 10 MPa isentropically. The final temperature of the steam is

(*a*) 290°C (*b*) 300°C (*c*) 311°C
(*d*) 371°C (*e*) 422°C

7–241 An apple with an average mass of 0.12 kg and average specific heat of 3.65 kJ/kg·°C is cooled from 25°C to 5°C. The entropy change of the apple is

(*a*) −0.705 kJ/K (*b*) −0.254 kJ/K (*c*) −0.0304 kJ/K
(*d*) 0 kJ/K (*e*) 0.348 kJ/K

7–242 A piston–cylinder device contains 5 kg of saturated water vapor at 3 MPa. Now heat is rejected from the cylinder at constant pressure until the water vapor completely condenses so that the cylinder contains saturated liquid at 3 MPa at the end of the process. The entropy change of the system during this process is

(*a*) 0 kJ/K (*b*) −3.5 kJ/K (*c*) −12.5 kJ/K
(*d*) −17.7 kJ/K (*e*) −19.5 kJ/K

7–243 Helium gas is compressed from 1 atm and 25°C to a pressure of 10 atm adiabatically. The lowest temperature of helium after compression is

(*a*) 25°C (*b*) 63°C (*c*) 250°C
(*d*) 384°C (*e*) 476°C

7–244 Steam expands in an adiabatic turbine from 4 MPa and 500°C to 0.1 MPa at a rate of 2 kg/s. If steam leaves the turbine as saturated vapor, the power output of the turbine is

(*a*) 2058 kW (*b*) 1910 kW (*c*) 1780 kW
(*d*) 1674 kW (*e*) 1542 kW

7–245 Argon gas expands in an adiabatic turbine from 3 MPa and 750°C to 0.2 MPa at a rate of 5 kg/s. The maximum power output of the turbine is

(*a*) 1.06 MW (*b*) 1.29 MW (*c*) 1.43 MW
(*d*) 1.76 MW (*e*) 2.08 MW

7–246 A unit mass of a substance undergoes an irreversible process from state 1 to state 2 while gaining heat from the surroundings at temperature T in the amount of q. If the entropy of the substance is s_1 at state 1, and s_2 at state 2, the entropy change of the substance Δs during this process is

(*a*) $\Delta s < s_2 - s_1$ (*b*) $\Delta s > s_2 - s_1$
(*c*) $\Delta s = s_2 - s_1$ (*d*) $\Delta s = s_2 - s_1 + q/T$
(*e*) $\Delta s > s_2 - s_1 + q/T$

7–247 A unit mass of an ideal gas at temperature T undergoes a reversible isothermal process from pressure P_1 to pressure P_2 while losing heat to the surroundings at temperature T in the amount of q. If the gas constant of the gas is R, the entropy change of the gas Δs during this process is

(*a*) $\Delta s = R\ln(P_2/P_1)$ (*b*) $\Delta s = R\ln(P_2/P_1) - q/T$
(*c*) $\Delta s = R\ln(P_1/P_2)$ (*d*) $\Delta s = R\ln(P_1/P_2) - q/T$
(*e*) $\Delta s = 0$

7–248 Air is compressed from room conditions to a specified pressure in a reversible manner by two compressors: one isothermal and the other adiabatic. If the entropy change of air Δs_{isot} during the reversible isothermal compression, and Δs_{adia} during the reversible adiabatic compression, the correct statement regarding entropy change of air per unit mass is

(*a*) $\Delta s_{isot} = \Delta s_{adia} = 0$ (*b*) $\Delta s_{isot} = \Delta s_{adia} > 0$
(*c*) $\Delta s_{adia} > 0$ (*d*) $\Delta s_{isot} < 0$
(*e*) $\Delta s_{isot} = 0$

7–249 Helium gas is compressed from 27°C and 3.50 m³/kg to 0.775 m³/kg in a reversible and adiabatic manner. The temperature of helium after compression is

(*a*) 74°C (*b*) 122°C (*c*) 547°C
(*d*) 709°C (*e*) 1082°C

7–250 Heat is lost through a plane wall steadily at a rate of 600 W. If the inner and outer surface temperatures of the wall are 20°C and 5°C, respectively, the rate of entropy generation within the wall is

(*a*) 0.11 W/K (*b*) 4.21 W/K (*c*) 2.10 W/K
(*d*) 42.1 W/K (*e*) 90.0 W/K

7–251 Air is compressed steadily and adiabatically from 17°C and 90 kPa to 200°C and 400 kPa. Assuming constant specific heats for air at room temperature, the isentropic efficiency of the compressor is

(*a*) 0.76 (*b*) 0.94 (*c*) 0.86
(*d*) 0.84 (*e*) 1.00

7–252 Argon gas expands in an adiabatic turbine steadily from 600°C and 800 kPa to 80 kPa at a rate of 2.5 kg/s. For isentropic efficiency of 88 percent, the power produced by the turbine is

(*a*) 240 kW (*b*) 361 kW (*c*) 414 kW
(*d*) 602 kW (*e*) 777 kW

7–253 Water enters a pump steadily at 100 kPa at a rate of 35 L/s and leaves at 800 kPa. The flow velocities at the inlet and the exit are the same, but the pump exit where the discharge pressure is measured is 6.1 m above the inlet section. The minimum power input to the pump is

(*a*) 34 kW (*b*) 22 kW (*c*) 27 kW
(*d*) 52 kW (*e*) 44 kW

7–254 Air at 15°C is compressed steadily and isothermally from 100 kPa to 700 kPa at a rate of 0.12 kg/s. The minimum power input to the compressor is

(*a*) 1.0 kW (*b*) 11.2 kW (*c*) 25.8 kW
(*d*) 19.3 kW (*e*) 161 kW

7–255 Air is to be compressed steadily and isentropically from 1 atm to 16 atm by a two-stage compressor. To minimize the total compression work, the intermediate pressure between the two stages must be

(*a*) 3 atm (*b*) 4 atm (*c*) 8.5 atm
(*d*) 9 atm (*e*) 12 atm

7–256 Helium gas enters an adiabatic nozzle steadily at 500°C and 600 kPa with a low velocity, and exits at a pressure of 90 kPa. The highest possible velocity of helium gas at the nozzle exit is

(*a*) 1475 m/s (*b*) 1662 m/s (*c*) 1839 m/s
(*d*) 2066 m/s (*e*) 3040 m/s

7–257 Combustion gases with a specific heat ratio of 1.3 enter an adiabatic nozzle steadily at 800°C and 800 kPa with a low velocity, and exit at a pressure of 85 kPa. The lowest possible temperature of combustion gases at the nozzle exit is

(*a*) 43°C (*b*) 237°C (*c*) 367°C
(*d*) 477°C (*e*) 640°C

7–258 Steam enters an adiabatic turbine steadily at 400°C and 5 MPa, and leaves at 20 kPa. The highest possible percentage of mass of steam that condenses at the turbine exit and leaves the turbine as a liquid is

(*a*) 4% (*b*) 8% (*c*) 12%
(*d*) 18% (*e*) 0%

7–259 Liquid water enters an adiabatic piping system at 15°C at a rate of 8 kg/s. If the water temperature rises by 0.2°C during flow due to friction, the rate of entropy generation in the pipe is

(*a*) 23 W/K (*b*) 55 W/K (*c*) 68 W/K
(*d*) 220 W/K (*e*) 443 W/K

7–260 Liquid water is to be compressed by a pump whose isentropic efficiency is 75 percent from 0.2 MPa to 5 MPa at a rate of 0.15 m^3/min. The required power input to this pump is

(*a*) 4.8 kW (*b*) 6.4 kW (*c*) 9.0 kW
(*d*) 16.0 kW (*e*) 12 kW

7–261 Steam enters an adiabatic turbine at 8 MPa and 500°C at a rate of 18 kg/s, and exits at 0.2 MPa and 300°C. The rate of entropy generation in the turbine is

(*a*) 0 kW/K (*b*) 7.2 kW/K (*c*) 21 kW/K
(*d*) 15 kW/K (*e*) 17 kW/K

7–262 Helium gas is compressed steadily from 90 kPa and 25°C to 800 kPa at a rate of 2 kg/min by an adiabatic compressor. If the compressor consumes 80 kW of power while operating, the isentropic efficiency of this compressor is

(*a*) 54.0% (*b*) 80.5% (*c*) 75.8%
(*d*) 90.1% (*e*) 100%

Design and Essay Problems

7–263 Compare the pumping of water to a higher elevation to isentropic compression of air into a compressed-air tank as means of storing energy for later use.

7–264 It is well-known that the temperature of a gas rises while it is compressed as a result of the energy input in the form of compression work. At high compression ratios, the air temperature may rise above the autoignition temperature of some hydrocarbons, including some lubricating oil. Therefore, the presence of some lubricating oil vapor in high-pressure air raises the possibility of an explosion, creating a fire hazard. The concentration of the oil within the compressor is usually too low to create a real danger. However, the oil that collects on the inner walls of exhaust piping of the compressor may cause an explosion. Such explosions have largely been eliminated by using the proper lubricating oils, carefully designing the equipment, intercooling between compressor stages, and keeping the system clean.

A compressor is to be designed for an industrial application in Los Angeles. If the compressor exit temperature is not to exceed 250°C for safety consideration, determine the maximum allowable compression ratio that is safe for all possible weather conditions for that area.

7–265 Obtain the following information about a power plant that is closest to your town: the net power output; the type and amount of fuel; the power consumed by the pumps, fans, and other auxiliary equipment; stack gas losses; temperatures at several locations; and the rate of heat rejection at the condenser. Using these and other relevant data, determine the rate of entropy generation in that power plant.

7–266 Enthalpy-entropy state diagrams (also known as *Mollier diagrams)* are more useful than temperature-entropy state diagrams for analyzing some systems. Using EES (or other) software and actual property data, plot an enthalpy-entropy diagram for water that includes the saturation lines. Also, sketch isothermal, isobaric, and constant volume processes on this diagram.

7–267 You are given the task of compressing a fixed quantity of gas from P_1, T_1 to P_2, T_1. You wish to do this using a polytropic process coupled with a constant pressure P_2 heat-transfer process. Determine the polytropic process that minimizes the entropy generation.

7–268 When an ideal gas is expanded from P_1 and T_1 to pressure P_2 in a polytropic process Pv^n = constant, heat has to be transferred to the gas for some values of n and rejected for others. Determine the criterion for n when heat has to be transferred to the gas. For those values of n that require a heat addition, this heat is supplied by an energy reservoir maintained at T_1. Which value of n minimizes the entropy generation in this case? When heat must be rejected, it is rejected to an energy reservoir maintained at T_2. Which value of n minimizes the entropy generation in this case?

CHAPTER

8

EXERGY: A MEASURE OF WORK POTENTIAL

The increased awareness that the world's energy resources are limited has caused many countries to reexamine their energy policies and take drastic measures in eliminating waste. It has also sparked interest in the scientific community to take a closer look at the energy conversion devices and to develop new techniques to better utilize the existing limited resources. The first law of thermodynamics deals with the *quantity* of energy and asserts that energy cannot be created or destroyed. This law merely serves as a necessary tool for the bookkeeping of energy during a process and offers no challenges to the engineer. The second law, however, deals with the *quality* of energy. More specifically, it is concerned with the degradation of energy during a process, the entropy generation, and the lost opportunities to do work; and it offers plenty of room for improvement.

The second law of thermodynamics has proved to be a very powerful tool in the optimization of complex thermodynamic systems. In this chapter, we examine the performance of engineering devices in light of the second law of thermodynamics. We start our discussions with the introduction of *exergy* (also called *availability*), which is the maximum useful work that could be obtained from the system at a given state in a specified environment, and we continue with the *reversible work,* which is the maximum useful work that can be obtained as a system undergoes a process between two specified states. Next we discuss the *irreversibility* (also called the *exergy destruction* or *lost work*), which is the wasted work potential during a process as a result of irreversibilities, and we define a *second-law efficiency*. We then develop the *exergy balance* relation and apply it to closed systems and control volumes.

Objectives

The objectives of Chapter 8 are to:

- Examine the performance of engineering devices in light of the second law of thermodynamics.
- Define *exergy*, which is the maximum useful work that could be obtained from the system at a given state in a specified environment.
- Define *reversible work*, which is the maximum useful work that can be obtained as a system undergoes a process between two specified states.
- Define the exergy destruction, which is the wasted work potential during a process as a result of irreversibilities.
- Define the *second-law efficiency*.
- Develop the exergy balance relation.
- Apply exergy balance to closed systems and control volumes.

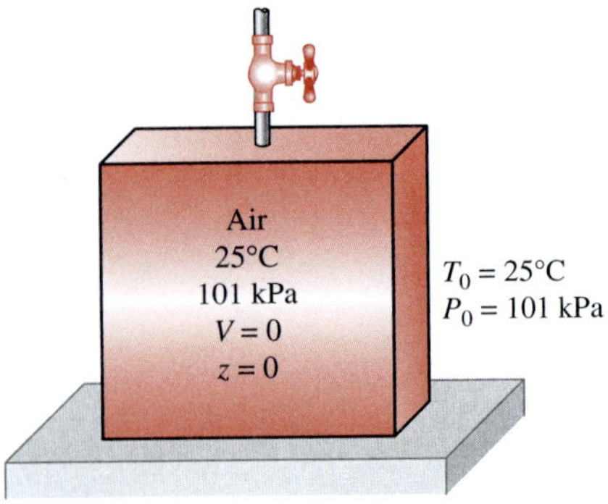

FIGURE 8–1
A system that is in equilibrium with its environment is said to be at the dead state.

FIGURE 8–2
At the dead state, the useful work potential (exergy) of a system is zero.
BLONDIE © KING FEATURES SYNDICATE.

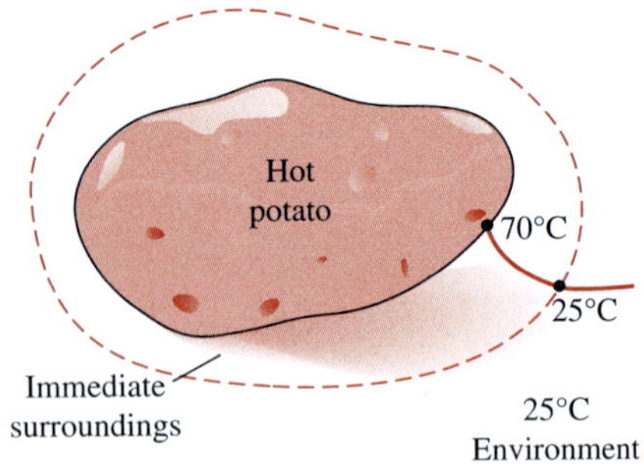

FIGURE 8–3
The immediate surroundings of a hot potato are simply the temperature gradient zone of the air next to the potato.

8–1 ▪ EXERGY: WORK POTENTIAL OF ENERGY

When a new energy source, such as a geothermal well, is discovered, the first thing the explorers do is estimate the amount of energy contained in the source. This information alone, however, is of little value in deciding whether to build a power plant on that site. What we really need to know is the *work potential* of the source—that is, the amount of energy we can extract as useful work. The rest of the energy is eventually discarded as waste energy and is not worthy of our consideration. Thus, it would be very desirable to have a property to enable us to determine the useful work potential of a given amount of energy at some specified state. This property is *exergy,* which is also called the *availability* or *available energy*.

The work potential of the energy contained in a system at a specified state is simply the maximum useful work that can be obtained from the system. You will recall that the work done during a process depends on the initial state, the final state, and the process path. That is,

$$\text{Work} = f(\text{initial state, process path, final state})$$

In an exergy analysis, the *initial state* is specified, and thus it is not a variable. The work output is maximized when the process between two specified states is executed in a *reversible manner,* as shown in Chap. 7. Therefore, all the irreversibilities are disregarded in determining the work potential. Finally, the system must be in the *dead state* at the end of the process to maximize the work output.

A system is said to be in the **dead state** when it is in thermodynamic equilibrium with the environment it is in (Fig. 8–1). At the dead state, a system is at the temperature and pressure of its environment (in thermal and mechanical equilibrium); it has no kinetic or potential energy relative to the environment (zero velocity and zero elevation above a reference level); and it does not react with the environment (chemically inert). Also, there are no unbalanced magnetic, electrical, and surface tension effects between the system and its surroundings, if these are relevant to the situation at hand. The properties of a system at the dead state are denoted by subscript zero, for example, P_0, T_0, h_0, u_0, and s_0. Unless specified otherwise, the dead-state temperature and pressure are taken to be $T_0 = 25°C$ (77°F) and $P_0 = 1$ atm (101.325 kPa or 14.7 psia). A system has zero exergy at the dead state (Fig. 8–2).

Distinction should be made between the *surroundings, immediate surroundings,* and the *environment*. By definition, **surroundings** are everything outside the system boundaries. The **immediate surroundings** refer to the portion of the surroundings that is affected by the process, and **environment** refers to the region beyond the immediate surroundings whose properties are not affected by the process at any point. Therefore, any irreversibilities during a process occur within the system and its immediate surroundings, and the environment is free of any irreversibilities. When analyzing the cooling of a hot baked potato in a room at 25°C, for example, the warm air that surrounds the potato is the immediate surroundings, and the remaining part of the room air at 25°C is the environment. Note that the temperature of the immediate surroundings changes from the temperature of the potato at the boundary to the environment temperature of 25°C (Fig. 8–3).

The notion that a system must go to the dead state at the end of the process to maximize the work output can be explained as follows: If the system temperature at the final state is greater than (or less than) the temperature of the environment it is in, we can always produce additional work by running a heat engine between these two temperature levels. If the final pressure is greater than (or less than) the pressure of the environment, we can still obtain work by letting the system expand to the pressure of the environment. If the final velocity of the system is not zero, we can catch that extra kinetic energy by a turbine and convert it to rotating shaft work, and so on. No work can be produced from a system that is initially at the dead state. The atmosphere around us contains a tremendous amount of energy. However, the atmosphere is in the dead state, and the energy it contains has no work potential (Fig. 8–4).

Therefore, we conclude that a *system delivers the maximum possible work as it undergoes a reversible process from the specified initial state to the state of its environment, that is, the dead state*. This represents the *useful work potential* of the system at the specified state and is called **exergy**. It is important to realize that exergy does not represent the amount of work that a work-producing device will actually deliver upon installation. Rather, it represents the *upper limit on the amount of work a device can deliver without violating any thermodynamic laws*. There will always be a difference, large or small, between exergy and the actual work delivered by a device. This difference represents the room engineers have for improvement.

Note that the exergy of a system at a specified state depends on the conditions of the environment (the dead state) as well as the properties of the system. Therefore, exergy is a property of the *system–environment combination* and not of the system alone. Altering the environment is another way of increasing exergy, but it is definitely not an easy alternative.

The term *availability* was made popular in the United States by the M.I.T. School of Engineering in the 1940s. Today, an equivalent term, *exergy,* introduced in Europe in the 1950s, has found global acceptance partly because it is shorter, it rhymes with energy and entropy, and it can be adapted without requiring translation. In this text the preferred term is *exergy.*

FIGURE 8–4
The atmosphere contains a tremendous amount of energy, but no exergy.

Exergy (Work Potential) Associated with Kinetic and Potential Energy

Kinetic energy is a form of *mechanical energy*, and thus it can be converted to work entirely. Therefore, the *work potential* or *exergy* of the kinetic energy of a system is equal to the kinetic energy itself regardless of the temperature and pressure of the environment. That is,

Exergy of kinetic energy:
$$x_{ke} = ke = \frac{V^2}{2} \quad \text{(kJ/kg)} \qquad \textbf{(8–1)}$$

where V is the velocity of the system relative to the environment.

Potential energy is also a form of *mechanical energy*, and thus it can be converted to work entirely. Therefore, the *exergy* of the potential energy of a

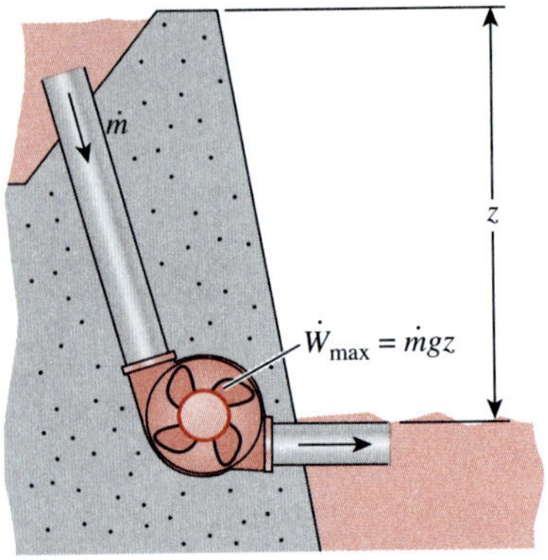

FIGURE 8–5
The *work potential* or *exergy* of potential energy is equal to the potential energy itself.

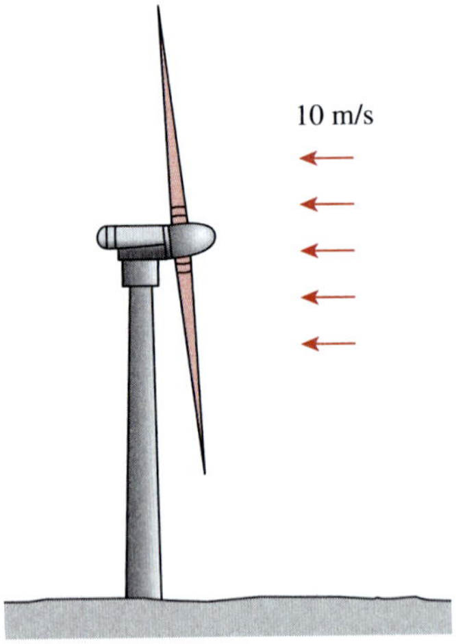

FIGURE 8–6
Schematic for Example 8–1.

system is equal to the potential energy itself regardless of the temperature and pressure of the environment (Fig. 8–5). That is,

Exergy of potential energy $$x_{pe} = pe = gz \qquad (kJ/kg) \tag{8–2}$$

where g is the gravitational acceleration and z is the elevation of the system relative to a reference level in the environment.

Therefore, the exergies of kinetic and potential energies are equal to themselves, and they are entirely available for work. However, the internal energy u and enthalpy h of a system are not entirely available for work, as shown later.

EXAMPLE 8–1 Maximum Power Generation by a Wind Turbine

A wind turbine with a 12-m-diameter rotor, as shown in Fig. 8–6, is to be installed at a location where the wind is blowing steadily at an average velocity of 10 m/s. Determine the maximum power that can be generated by the wind turbine.

SOLUTION A wind turbine is being considered for a specified location. The maximum power that can be generated by the wind turbine is to be determined.

Assumptions Air is at standard conditions of 1 atm and 25°C, and thus its density is 1.18 kg/m^3.

Analysis The air flowing with the wind has the same properties as the stagnant atmospheric air except that it possesses a velocity and thus some kinetic energy. This air will reach the dead state when it is brought to a complete stop. Therefore, the exergy of the blowing air is simply the kinetic energy it possesses:

$$ke = \frac{V^2}{2} = \frac{(10 \text{ m/s})^2}{2}\left(\frac{1 \text{ kJ/kg}}{1000 \text{ m}^2/\text{s}^2}\right) = 0.05 \text{ kJ/kg}$$

That is, every unit mass of air flowing at a velocity of 10 m/s has a work potential of 0.05 kJ/kg. In other words, a perfect wind turbine will bring the air to a complete stop and capture that 0.05 kJ/kg of work potential. To determine the maximum power, we need to know the amount of air passing through the rotor of the wind turbine per unit time, that is, the mass flow rate, which is determined to be

$$\dot{m} = \rho AV = \rho\frac{\pi D^2}{4}V = (1.18 \text{ kg/m}^3)\frac{\pi(12 \text{ m})^2}{4}(10 \text{ m/s}) = 1335 \text{ kg/s}$$

Thus,

$$\text{Maximum power} = \dot{m}(\text{ke}) = (1335 \text{ kg/s})(0.05 \text{ kJ/kg}) = \mathbf{66.8\ kW}$$

This is the maximum power available to the wind turbine. Assuming a conversion efficiency of 30 percent, an actual wind turbine will convert 20.0 kW to electricity. Notice that the work potential for this case is equal to the entire kinetic energy of the air.

Discussion It should be noted that although the entire kinetic energy of the wind is available for power production, Betz's law states that the power output of a wind machine is at maximum when the wind is slowed to one-third of its initial velocity. Therefore, for maximum power (and thus minimum cost per installed power), the highest efficiency of a wind turbine is about 59 percent. In practice, the actual efficiency ranges between 20 and 40 percent and is about 35 percent for many wind turbines.

Wind power is suitable for harvesting when there are steady winds with an average velocity of at least 6 m/s (or 13 mph). Recent improvements in wind turbine design have brought the cost of generating wind power to about 5 cents per kWh, which is competitive with electricity generated from other resources.

EXAMPLE 8–2 Exergy Transfer from a Furnace

Consider a large furnace that can transfer heat at a temperature of 2000 R at a steady rate of 3000 Btu/s. Determine the rate of exergy flow associated with this heat transfer. Assume an environment temperature of 77°F.

SOLUTION Heat is being supplied by a large furnace at a specified temperature. The rate of exergy flow is to be determined.

Analysis The furnace in this example can be modeled as a heat reservoir that supplies heat indefinitely at a constant temperature. The exergy of this heat energy is its useful work potential, that is, the maximum possible amount of work that can be extracted from it. This corresponds to the amount of work that a reversible heat engine operating between the furnace and the environment can produce.

The thermal efficiency of this reversible heat engine is

$$\eta_{\text{th,max}} = \eta_{\text{th,rev}} = 1 - \frac{T_L}{T_H} = 1 - \frac{T_0}{T_H} = 1 - \frac{537\text{ R}}{2000\text{ R}} = 0.732 \text{ (or 73.2\%)}$$

That is, a heat engine can convert, at best, 73.2 percent of the heat received from this furnace to work. Thus, the exergy of this furnace is equivalent to the power produced by the reversible heat engine:

$$\dot{W}_{\text{max}} = \dot{W}_{\text{rev}} = \eta_{\text{th,rev}}\,\dot{Q}_{\text{in}} = (0.732)(3000\text{ Btu/s}) = \mathbf{2196\ Btu/s}$$

Discussion Notice that 26.8 percent of the heat transferred from the furnace is not available for doing work. The portion of energy that cannot be converted to work is called **unavailable energy** (Fig. 8–7). Unavailable energy is simply the difference between the total energy of a system at a specified state and the exergy of that energy.

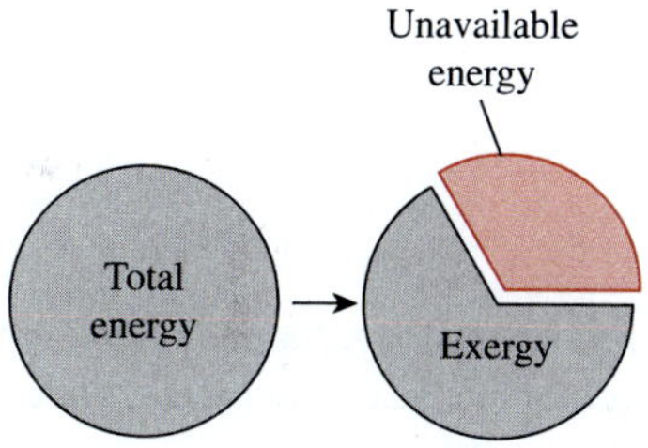

FIGURE 8–7
Unavailable energy is the portion of energy that cannot be converted to work by even a reversible heat engine.

8–2 ▪ REVERSIBLE WORK AND IRREVERSIBILITY

The property exergy serves as a valuable tool in determining the quality of energy and comparing the work potentials of different energy sources or systems. The evaluation of exergy alone, however, is not sufficient for studying engineering devices operating between two fixed states. This is because when evaluating exergy, the final state is always assumed to be the *dead state,* which is hardly ever the case for actual engineering systems. The isentropic efficiencies discussed in Chap. 7 are also of limited use because the exit state

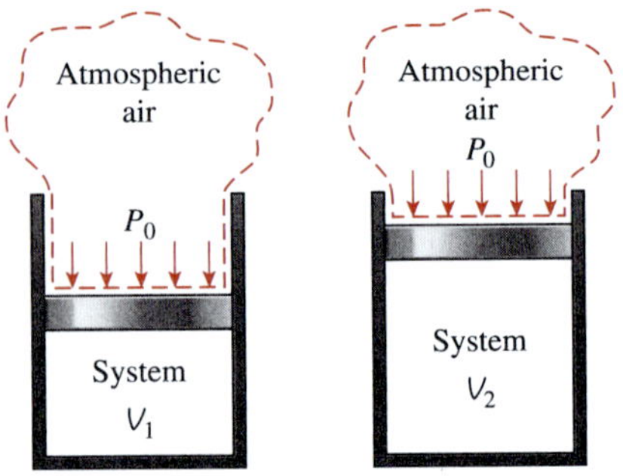

FIGURE 8–8
As a closed system expands, some work needs to be done to push the atmospheric air out of the way (W_{surr}).

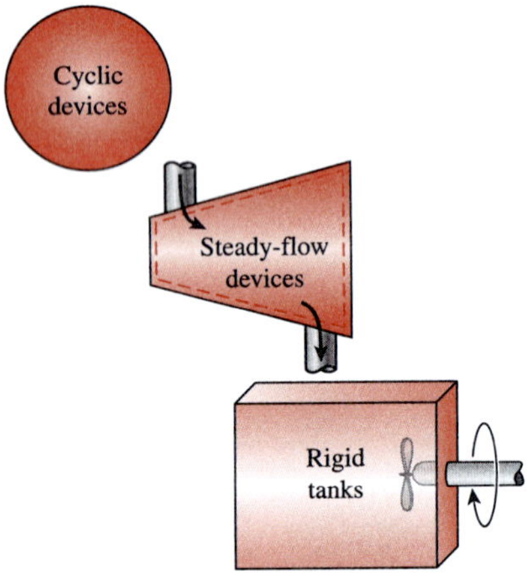

FIGURE 8–9
For constant-volume systems, the total actual and useful works are identical ($W_u = W$).

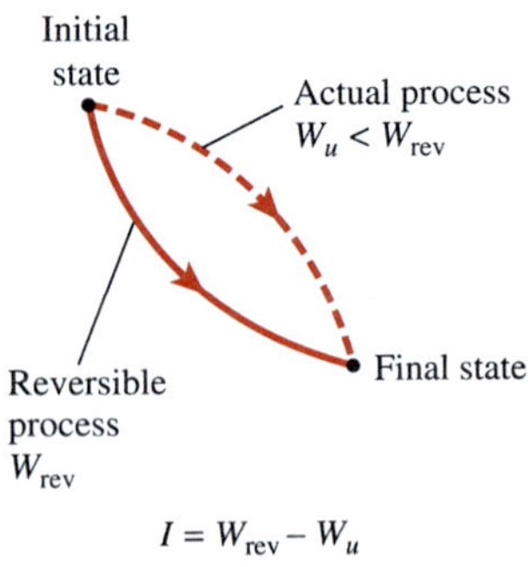

FIGURE 8–10
The difference between reversible work and actual useful work is the irreversibility.

of the model (isentropic) process is not the same as the actual exit state and it is limited to adiabatic processes.

In this section, we describe two quantities that are related to the actual initial and final states of processes and serve as valuable tools in the thermodynamic analysis of components or systems. These two quantities are the *reversible work* and *irreversibility* (or *exergy destruction*). But first we examine the **surroundings work**, which is the work done by or against the surroundings during a process.

The work done by work-producing devices is not always entirely in a usable form. For example, when a gas in a piston–cylinder device expands, part of the work done by the gas is used to push the atmospheric air out of the way of the piston (Fig. 8–8). This work, which cannot be recovered and utilized for any useful purpose, is equal to the atmospheric pressure P_0 times the volume change of the system,

$$W_{surr} = P_0(V_2 - V_1) \tag{8–3}$$

The difference between the actual work W and the surroundings work W_{surr} is called the **useful work** W_u:

$$W_u = W - W_{surr} = W - P_0(V_2 - V_1) \tag{8–4}$$

When a system is expanding and doing work, part of the work done is used to overcome the atmospheric pressure, and thus W_{surr} represents a loss. When a system is compressed, however, the atmospheric pressure helps the compression process, and thus W_{surr} represents a gain.

Note that the work done by or against the atmospheric pressure has significance only for systems whose volume changes during the process (i.e., systems that involve moving boundary work). It has no significance for cyclic devices and systems whose boundaries remain fixed during a process such as rigid tanks and steady-flow devices (turbines, compressors, nozzles, heat exchangers, etc.), as shown in Fig. 8–9.

Reversible work W_{rev} is defined as *the maximum amount of useful work that can be produced (or the minimum work that needs to be supplied) as a system undergoes a process between the specified initial and final states.* This is the useful work output (or input) obtained (or expended) when the process between the initial and final states is executed in a totally reversible manner. When the final state is the dead state, the reversible work equals exergy. For processes that require work, reversible work represents the minimum amount of work necessary to carry out that process. For convenience in presentation, the term *work* is used to denote both work and power throughout this chapter.

Any difference between the reversible work W_{rev} and the useful work W_u is due to the irreversibilities present during the process, and this difference is called **irreversibility** I. It is expressed as (Fig. 8–10)

$$I = W_{rev,out} - W_{u,out} \quad \text{or} \quad I = W_{u,in} - W_{rev,in} \tag{8–5}$$

The irreversibility is equivalent to the *exergy destroyed,* discussed in Sec. 8–4. For a totally reversible process, the actual and reversible work terms are identical, and thus the irreversibility is zero. This is expected since totally reversible processes generate no entropy. Irreversibility is a *positive quantity* for all actual (irreversible) processes since $W_{rev} \geq W_u$ for work-producing devices and $W_{rev} \leq W_u$ for work-consuming devices.

Irreversibility can be viewed as the *wasted work potential* or the *lost opportunity* to do work. It represents the energy that could have been converted to work but was not. The smaller the irreversibility associated with a process, the greater the work that is produced (or the smaller the work that is consumed). The performance of a system can be improved by minimizing the irreversibility associated with it.

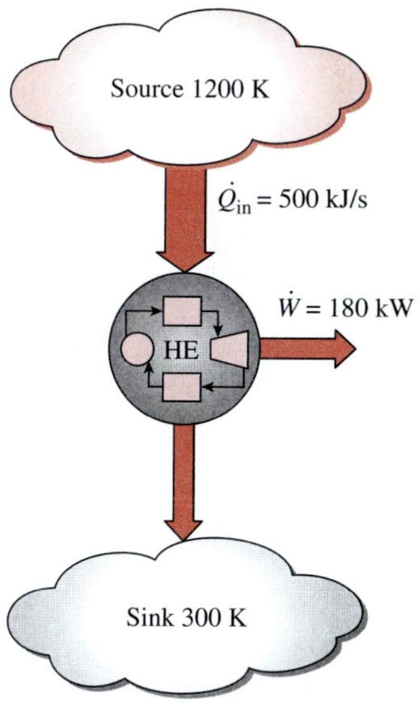

FIGURE 8–11
Schematic for Example 8–3.

EXAMPLE 8–3 The Rate of Irreversibility of a Heat Engine

A heat engine receives heat from a source at 1200 K at a rate of 500 kJ/s and rejects the waste heat to a medium at 300 K (Fig. 8–11). The power output of the heat engine is 180 kW. Determine the reversible power and the irreversibility rate for this process.

SOLUTION The operation of a heat engine is considered. The reversible power and the irreversibility rate associated with this operation are to be determined.

Analysis The reversible power for this process is the amount of power that a reversible heat engine, such as a Carnot heat engine, would produce when operating between the same temperature limits, and is determined to be:

$$\dot{W}_{rev} = \eta_{th,rev}\,\dot{Q}_{in} = \left(1 - \frac{T_{sink}}{T_{source}}\right)\dot{Q}_{in} = \left(1 - \frac{300\text{ K}}{1200\text{ K}}\right)(500\text{ kW}) = \mathbf{375\ kW}$$

This is the maximum power that can be produced by a heat engine operating between the specified temperature limits and receiving heat at the specified rate. This would also represent the *available power* if 300 K were the lowest temperature available for heat rejection.

The irreversibility rate is the difference between the reversible power (maximum power that could have been produced) and the useful power output:

$$\dot{I} = \dot{W}_{rev,out} - \dot{W}_{u,out} = 375 - 180 = \mathbf{195\ kW}$$

Discussion Note that 195 kW of power potential is wasted during this process as a result of irreversibilities. Also, the 500 − 375 = 125 kW of heat rejected to the sink is not available for converting to work and thus is not part of the irreversibility.

EXAMPLE 8–4 Irreversibility during the Cooling of an Iron Block

A 500-kg iron block shown in Fig. 8–12 is initially at 200°C and is allowed to cool to 27°C by transferring heat to the surrounding air at 27°C. Determine the reversible work and the irreversibility for this process.

SOLUTION A hot iron block is allowed to cool in air. The reversible work and irreversibility associated with this process are to be determined.

Assumptions **1** The kinetic and potential energies are negligible. **2** The process involves no work interactions.

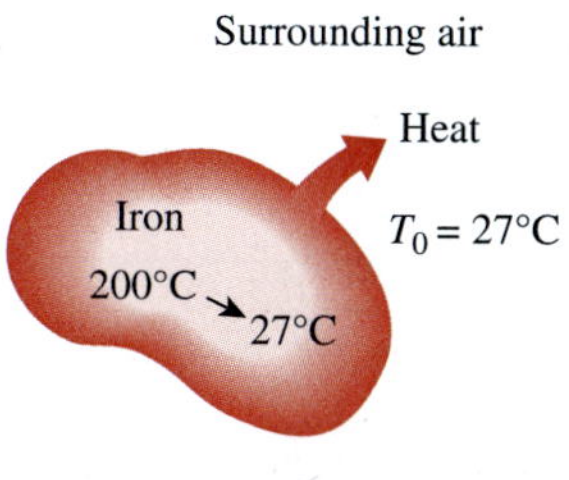

FIGURE 8–12
Schematic for Example 8–4.

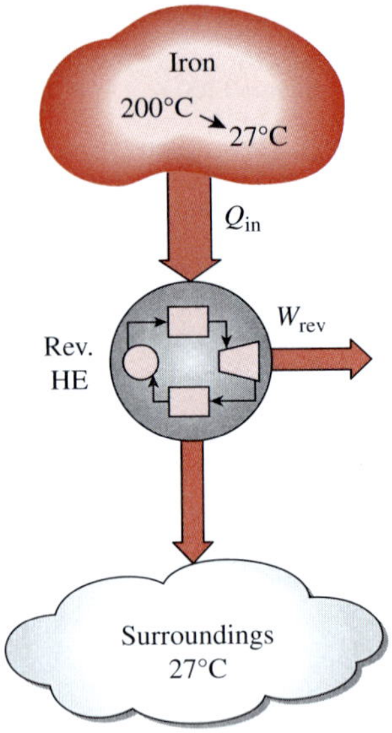

FIGURE 8–13
An irreversible heat transfer process can be made reversible by the use of a reversible heat engine.

Analysis We take the *iron block* as the system. This is a *closed system* since no mass crosses the system boundary. We note that heat is lost from the system.

It probably came as a surprise to you that we are asking to find the "reversible work" for a process that does not involve any work interactions. Well, even if no attempt is made to produce work during this process, the potential to do work still exists, and the reversible work is a quantitative measure of this potential.

The reversible work in this case is determined by considering a series of imaginary reversible heat engines operating between the source (at a variable temperature T) and the sink (at a constant temperature T_0), as shown in Fig. 8–13. Summing their work output:

$$\delta W_{rev} = \eta_{th,rev}\,\delta Q_{in} = \left(1 - \frac{T_{sink}}{T_{source}}\right)\delta Q_{in} = \left(1 - \frac{T_0}{T}\right)\delta Q_{in}$$

and

$$W_{rev} = \int\left(1 - \frac{T_0}{T}\right)\delta Q_{in}$$

The source temperature T changes from $T_1 = 200°C = 473$ K to $T_0 = 27°C = 300$ K during this process. A relation for the differential heat transfer from the iron block can be obtained from the differential form of the energy balance applied on the iron block,

$$\underbrace{\delta E_{in} - \delta E_{out}}_{\substack{\text{Net energy transfer}\\ \text{by heat, work, and mass}}} = \underbrace{dE_{system}}_{\substack{\text{Change in internal, kinetic,}\\ \text{potential, etc., energies}}}$$

$$-\delta Q_{out} = dU = mc_{avg}\,dT$$

Then,

$$\delta Q_{in,heat\ engine} = \delta Q_{out,system} = -mc_{avg}\,dT$$

since heat transfers from the iron and to the heat engine are equal in magnitude and opposite in direction. Substituting and performing the integration, the reversible work is determined to be

$$W_{rev} = \int_{T_1}^{T_0}\left(1 - \frac{T_0}{T}\right)(-mc_{avg}\,dT) = mc_{avg}(T_1 - T_0) - mc_{avg}T_0 \ln\frac{T_1}{T_0}$$

$$= (500\text{ kg})(0.45\text{ kJ/kg}\cdot\text{K})\left[(473 - 300)\text{ K} - (300\text{ K})\ln\frac{473\text{ K}}{300\text{ K}}\right]$$

$$= \mathbf{8191\ kJ}$$

where the specific heat value is obtained from Table A–3. The first term in the above equation [$Q = mc_{avg}(T_1 - T_0) = 38{,}925$ kJ] is the total heat transfer from the iron block to the heat engine. The reversible work for this problem is found to be 8191 kJ, which means that 8191 (21 percent) of the 38,925 kJ of heat transferred from the iron block to the ambient air *could* have been converted to work. If the specified ambient temperature of 27°C is the lowest available environment temperature, the reversible work determined above also represents the exergy, which is the maximum work potential of the sensible energy contained in the iron block.

The irreversibility for this process is determined from its definition,

$$I = W_{rev} - W_u = 8191 - 0 = \mathbf{8191\ kJ}$$

Discussion Notice that the reversible work and irreversibility (the wasted work potential) are the same for this case since the entire work potential is wasted. The source of irreversibility in this process is the heat transfer through a finite temperature difference.

EXAMPLE 8–5 Heating Potential of a Hot Iron Block

The iron block discussed in Example 8–4 is to be used to maintain a house at 27°C when the outdoor temperature is 5°C. Determine the maximum amount of heat that can be supplied to the house as the iron cools to 27°C.

SOLUTION The iron block is now reconsidered for heating a house. The maximum amount of heating this block can provide is to be determined.

Analysis Probably the first thought that comes to mind to make the most use of the energy stored in the iron block is to take it inside and let it cool in the house, as shown in Fig. 8–14, transferring its sensible energy as heat to the indoors air (provided that it meets the approval of the household, of course). The iron block can keep "losing" heat until its temperature drops to the indoor temperature of 27°C, transferring a total of 38,925 kJ of heat. Since we utilized the entire energy of the iron block available for heating without wasting a single kilojoule, it seems like we have a 100-percent-efficient operation, and nothing can beat this, right? Well, not quite.

In Example 8–4 we determined that this process has an irreversibility of 8191 kJ, which implies that things are not as "perfect" as they seem. A "perfect" process is one that involves "zero" irreversibility. The irreversibility in this process is associated with the heat transfer through a finite temperature difference that can be eliminated by running a reversible heat engine between the iron block and the indoor air. This heat engine produces (as determined in Example 8–4) 8191 kJ of work and reject the remaining 38,925 − 8191 = 30,734 kJ of heat to the house. Now we managed to eliminate the irreversibility and ended up with 8191 kJ of work. What can we do with this work? Well, at worst we can convert it to heat by running a paddle wheel, for example, creating an equal amount of irreversibility. Or we can supply this work to a heat pump that transports heat from the outdoors at 5°C to the indoors at 27°C. Such a heat pump, if reversible, has a coefficient of performance of

$$COP_{HP} = \frac{1}{1 - T_L/T_H} = \frac{1}{1 - (278\ K)/(300\ K)} = 13.6$$

That is, this heat pump can supply the house with 13.6 times the energy it consumes as work. In our case, it will consume the 8191 kJ of work and deliver 8191 × 13.6 = 111,398 kJ of heat to the house. Therefore, the hot iron block has the potential to supply

$$(30{,}734 + 111{,}398)\ kJ = 142{,}132\ kJ \cong \mathbf{142\ MJ}$$

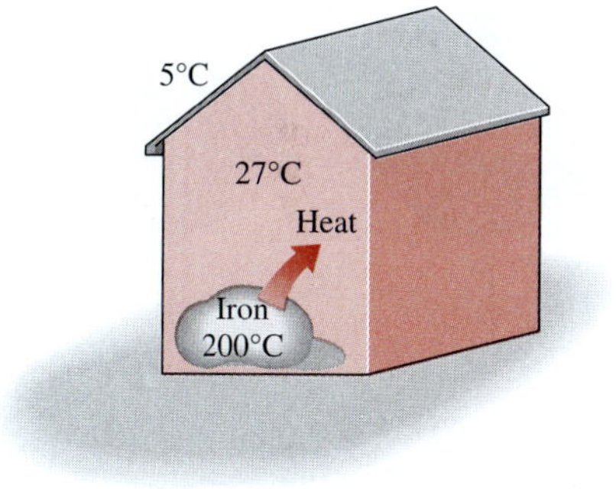

FIGURE 8–14
Schematic for Example 8–5.

of heat to the house. The irreversibility for this process is zero, and this is *the best* we can do under the specified conditions. A similar argument can be given for the electric heating of residential or commercial buildings.

Discussion Now try to answer the following question: What would happen if the heat engine were operated between the iron block and the outside air instead of the house until the temperature of the iron block fell to 27°C? Would the amount of heat supplied to the house still be 142 MJ? Here is a hint: The initial and final states in both cases are the same, and the irreversibility for both cases is zero.

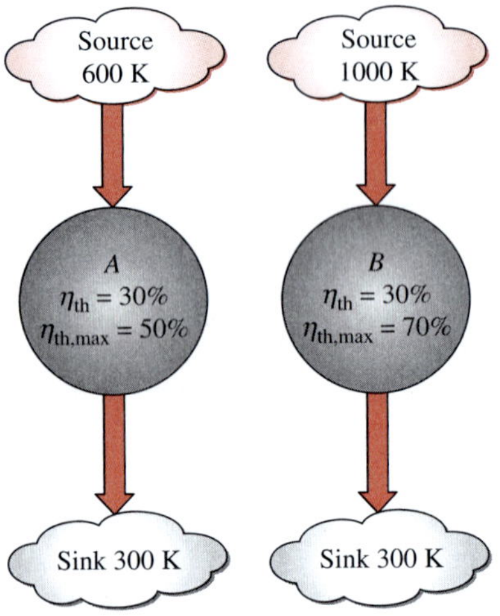

FIGURE 8–15
Two heat engines that have the same thermal efficiency, but different maximum thermal efficiencies.

8–3 ▪ SECOND-LAW EFFICIENCY, η_{II}

In Chap. 6 we defined the *thermal efficiency* and the *coefficient of performance* for devices as a measure of their performance. They are defined on the basis of the first law only, and they are sometimes referred to as the *first-law efficiencies.* The first law efficiency, however, makes no reference to the best possible performance, and thus it may be misleading.

Consider two heat engines, both having a thermal efficiency of 30 percent, as shown in Fig. 8–15. One of the engines (engine A) is supplied with heat from a source at 600 K, and the other one (engine B) from a source at 1000 K. Both engines reject heat to a medium at 300 K. At first glance, both engines seem to convert to work the same fraction of heat that they receive; thus they are performing equally well. When we take a second look at these engines in light of the second law of thermodynamics, however, we see a totally different picture. These engines, at best, can perform as reversible engines, in which case their efficiencies would be

$$\eta_{\text{rev},A} = \left(1 - \frac{T_L}{T_H}\right)_A = 1 - \frac{300\text{ K}}{600\text{ K}} = 50\%$$

$$\eta_{\text{rev},B} = \left(1 - \frac{T_L}{T_H}\right)_B = 1 - \frac{300\text{ K}}{1000\text{ K}} = 70\%$$

Now it is becoming apparent that engine B has a greater work potential available to it (70 percent of the heat supplied as compared to 50 percent for engine A), and thus should do a lot better than engine A. Therefore, we can say that engine B is performing poorly relative to engine A even though both have the same thermal efficiency.

It is obvious from this example that the first-law efficiency alone is not a realistic measure of performance of engineering devices. To overcome this deficiency, we define a **second-law efficiency** η_{II} as the ratio of the actual thermal efficiency to the maximum possible (reversible) thermal efficiency under the same conditions (Fig. 8–16):

$$\eta_{II} = \frac{\eta_{th}}{\eta_{th,rev}} \qquad \text{(heat engines)} \tag{8–6}$$

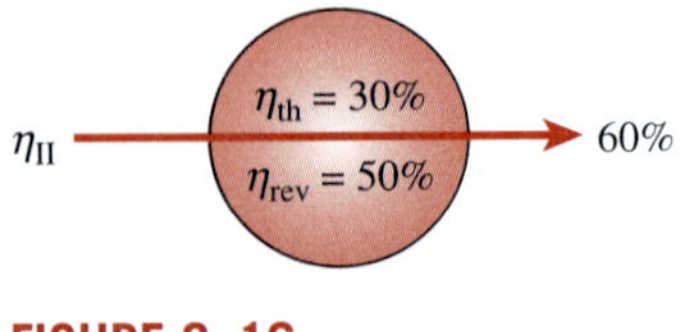

FIGURE 8–16
Second-law efficiency is a measure of the performance of a device relative to its performance under reversible conditions.

Based on this definition, the second-law efficiencies of the two heat engines discussed above are

$$\eta_{II,A} = \frac{0.30}{0.50} = 0.60 \quad \text{and} \quad \eta_{II,B} = \frac{0.30}{0.70} = 0.43$$

That is, engine A is converting 60 percent of the available work potential to useful work. This ratio is only 43 percent for engine B.

The second-law efficiency can also be expressed as the ratio of the useful work output and the maximum possible (reversible) work output:

$$\eta_{II} = \frac{W_u}{W_{rev}} \qquad \text{(work-producing devices)} \tag{8–7}$$

This definition is more general since it can be applied to processes (in turbines, piston–cylinder devices, etc.) as well as to cycles. Note that the second-law efficiency cannot exceed 100 percent (Fig. 8–17).

We can also define a second-law efficiency for work-consuming noncyclic (such as compressors) and cyclic (such as refrigerators) devices as the ratio of the minimum (reversible) work input to the useful work input:

$$\eta_{II} = \frac{W_{rev}}{W_u} \qquad \text{(work-consuming devices)} \tag{8–8}$$

For cyclic devices such as refrigerators and heat pumps, it can also be expressed in terms of the coefficients of performance as

$$\eta_{II} = \frac{\text{COP}}{\text{COP}_{rev}} \qquad \text{(refrigerators and heat pumps)} \tag{8–9}$$

Again, because of the way we defined the second-law efficiency, its value cannot exceed 100 percent. In the above relations, the reversible work W_{rev} should be determined by using the same initial and final states as in the actual process.

The definitions above for the second-law efficiency do not apply to devices that are not intended to produce or consume work. Therefore, we need a more general definition. However, there is some disagreement on a general definition of the second-law efficiency, and thus a person may encounter different definitions for the same device. The second-law efficiency is intended to serve as a measure of approximation to reversible operation, and thus its value should range from zero in the worst case (complete destruction of exergy) to one in the best case (no destruction of exergy). With this in mind, we define the second-law efficiency of a system during a process as (Fig. 8–18)

$$\eta_{II} = \frac{\text{Exergy recovered}}{\text{Exergy expended}} = 1 - \frac{\text{Exergy destroyed}}{\text{Exergy expended}} \tag{8–10}$$

Therefore, when determining the second-law efficiency, the first thing we need to do is determine how much exergy or work potential is expended or consumed during a process. In a reversible operation, we should be able to recover entirely the exergy expended during the process, and the irreversibility in this case should be zero. The second-law efficiency is zero when we recover none of the exergy expended by the system. Note that the exergy can be supplied or recovered at various amounts in various forms such as heat, work, kinetic energy, potential energy, internal energy, and enthalpy. Sometimes there are differing (though valid) opinions on what constitutes expended exergy, and this causes differing definitions for second-law efficiency. At all times, however, the exergy recovered and the exergy destroyed (the irreversibility) must add up to the exergy expended. Also, we need to define the system precisely in order to identify correctly any interactions between the system and its surroundings.

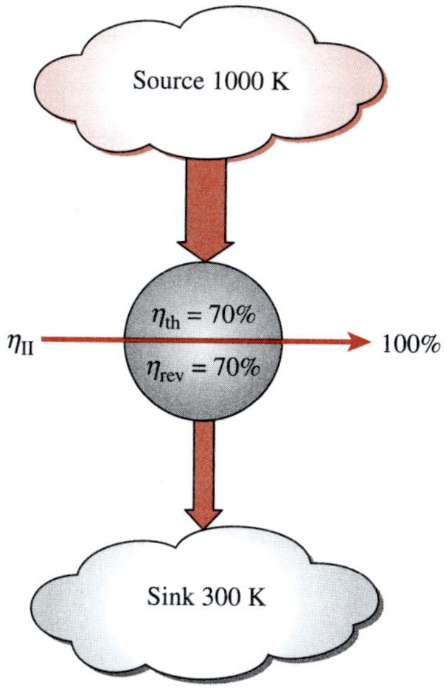

FIGURE 8–17
Second-law efficiency of all reversible devices is 100 percent.

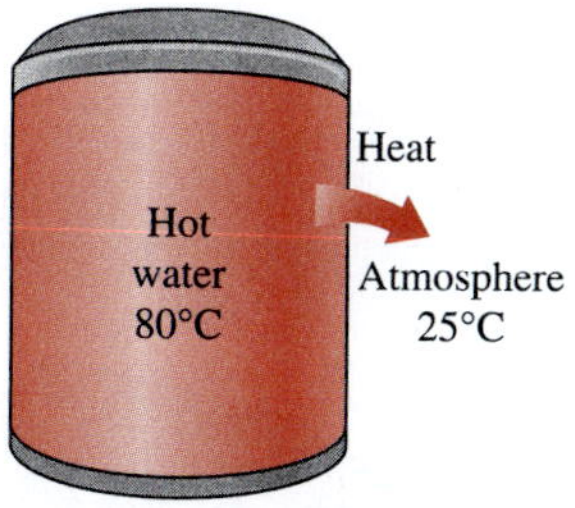

FIGURE 8–18
The second-law efficiency of naturally occurring processes is zero if none of the work potential is recovered.

For a *heat engine,* the exergy expended is the decrease in the exergy of the heat transferred to the engine, which is the difference between the exergy of the heat supplied and the exergy of the heat rejected. (The exergy of the heat rejected at the temperature of the surroundings is zero.) The net work output is the recovered exergy.

For a *refrigerator* or *heat pump,* the exergy expended is the work input since the work supplied to a cyclic device is entirely consumed. The recovered exergy is the exergy of the heat transferred to the high-temperature medium for a heat pump, and the exergy of the heat transferred from the low-temperature medium for a refrigerator.

For a heat exchanger with two unmixed fluid streams, usually the exergy expended is the decrease in the exergy of the higher-temperature fluid stream, and the exergy recovered is the increase in the exergy of the lower-temperature fluid stream. This is discussed further in Sec. 8–8.

In the case of *electric resistance heating*, the exergy expended is the electrical energy the resistance heater consumes from the resource of electric grid. The exergy recovered is the exergy content of the heat supplied to the room, which is the work that can be produced by a Carnot engine receiving this heat. If the heater maintains the heated space at a constant temperature of T_H in an environment at T_0, the second-law efficiency for the electric heater becomes

$$\eta_{\text{II,electric heater}} = \frac{\dot{X}_{\text{recovered}}}{\dot{X}_{\text{expended}}} = \frac{\dot{X}_{\text{heat}}}{\dot{W}_e} = \frac{\dot{Q}_e(1 - T_0/T_H)}{\dot{W}_e} = 1 - \frac{T_0}{T_H} \quad \textbf{(8–11)}$$

since, from the first law considerations, $\dot{Q}_e = \dot{W}_e$. Note that the second-law efficiency of a resistance heater becomes zero when the heater is outdoors (as in a radiant heater) and thus the exergy of the heat supplied to the environment is not recoverable.

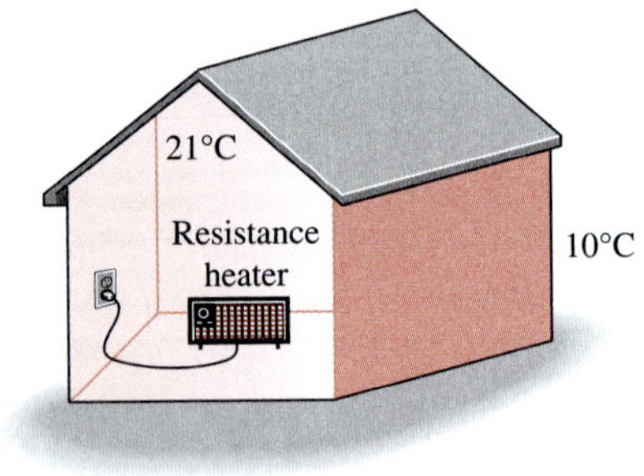

FIGURE 8–19
Schematic for Example 8–6.

EXAMPLE 8–6 **Second-Law Efficiency of Resistance Heaters**

A dealer advertises that he has just received a shipment of electric resistance heaters for residential buildings that have an efficiency of 100 percent (Fig. 8–19). Assuming an indoor temperature of 21°C and outdoor temperature of 10°C, determine the second-law efficiency of these heaters.

SOLUTION Electric resistance heaters are being considered for residential buildings. The second-law efficiency of these heaters is to be determined.

Analysis Obviously the efficiency that the dealer is referring to is the first-law efficiency, meaning that for each unit of electric energy (work) consumed, the heater will supply the house with 1 unit of energy (heat). That is, the advertised heater has a COP of 1.

At the specified conditions, a reversible heat pump would have a coefficient of the performance of

$$\text{COP}_{\text{HP,rev}} = \frac{1}{1 - T_L/T_H} = \frac{1}{1 - (10 + 273\text{ K})/(21 + 273\text{ K})} = 26.7$$

That is, it would supply the house with 26.7 units of heat (extracted from the cold outside air) for each unit of electric energy it consumes.

The second-law efficiency of this resistance heater is

$$\eta_{\mathrm{II}} = \frac{\mathrm{COP}}{\mathrm{COP_{rev}}} = \frac{1.0}{26.7} = \mathbf{0.037 \text{ or } 3.7\%}$$

which does not look so impressive. The dealer will not be happy to see this value. Considering the high price of electricity, a consumer will probably be better off with a "less" efficient gas heater.

Discussion The second-law efficiency of this electric heater can also be determined directly from Eq. 8–11 to be

$$\eta_{\mathrm{II,electric\ heater}} = 1 - \frac{T_0}{T_{\mathrm{H}}} = 1 - \frac{10 + 273\ \mathrm{K}}{21 + 273\ \mathrm{K}} = 0.037 \quad \text{or} \quad 3.7\%$$

Therefore, if we change our mind and decide to convert the heat back to electricity, the best we can do is 3.7 percent. That is, 86.3 percent of the heat can never be converted to electrical energy.

8–4 ▪ EXERGY CHANGE OF A SYSTEM

The property *exergy* is the work potential of a system in a specified environment and represents the maximum amount of useful work that can be obtained as the system is brought to equilibrium with the environment. Unlike energy, the value of exergy depends on the state of the environment as well as the state of the system. Therefore, exergy is a combination property. The exergy of a system that is in equilibrium with its environment is zero. The state of the environment is referred to as the "dead state" since the system is practically "dead" (cannot do any work) from a thermodynamic point of view when it reaches that state.

In this section we limit the discussion to **thermo-mechanical exergy**, and thus disregard any mixing and chemical reactions. Therefore, a system at this "restricted dead state" is at the temperature and pressure of the environment and it has no kinetic or potential energies relative to the environment. However, it may have a different chemical composition than the environment. Exergy associated with different chemical compositions and chemical reactions is discussed in later chapters.

Below we develop relations for the exergies and exergy changes for a fixed mass and a flow stream.

Exergy of a Fixed Mass: Nonflow (or Closed System) Exergy

In general, internal energy consists of *sensible, latent, chemical,* and *nuclear* energies. However, in the absence of any chemical or nuclear reactions, the chemical and nuclear energies can be disregarded and the internal energy can be considered to consist of only sensible and latent energies that can be transferred to or from a system as *heat* whenever there is a temperature difference across the system boundary. The second law of thermodynamics states that heat cannot be converted to work entirely, and thus the work potential of internal energy must be less than the internal energy itself. But how much less?

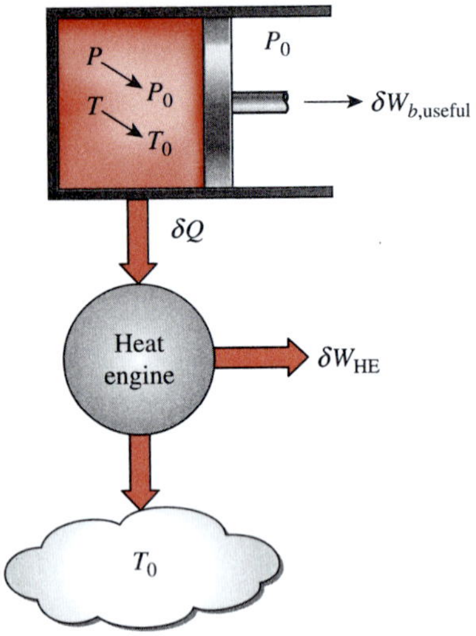

FIGURE 8–20
The *exergy* of a specified mass at a specified state is the useful work that can be produced as the mass undergoes a reversible process to the state of the environment.

To answer that question, we need to consider a stationary closed system at a specified state that undergoes a *reversible* process to the state of the environment (that is, the final temperature and pressure of the system should be T_0 and P_0, respectively). The useful work delivered during this process is the exergy of the system at its initial state (Fig. 8–20).

Consider a piston–cylinder device that contains a fluid of mass m at temperature T and pressure P. The system (the mass inside the cylinder) has a volume V, internal energy U, and entropy S. The system is now allowed to undergo a differential change of state during which the volume changes by a differential amount dV and heat is transferred in the differential amount of δQ. Taking the direction of heat and work transfers to be *from* the system (heat and work outputs), the energy balance for the system during this differential process can be expressed as

$$\underbrace{\delta E_{\text{in}} - \delta E_{\text{out}}}_{\substack{\text{Net energy transfer} \\ \text{by heat, work, and mass}}} = \underbrace{dE_{\text{system}}}_{\substack{\text{Change in internal, kinetic,} \\ \text{potential, etc., energies}}} \tag{8–12}$$

$$-\delta Q - \delta W = dU$$

since the only form of energy the system contains is *internal energy,* and the only forms of energy transfer a fixed mass can involve are heat and work. Also, the only form of work a simple compressible system can involve during a reversible process is the boundary work, which is given to be $\delta W = P\,dV$ when the direction of work is taken to be from the system (otherwise it would be $-P\,dV$). The pressure P in the $P\,dV$ expression is the absolute pressure, which is measured from absolute zero. Any useful work delivered by a piston–cylinder device is due to the pressure above the atmospheric level. Therefore,

$$\delta W = P\,dV = (P - P_0)\,dV + P_0\,dV = \delta W_{b,\text{useful}} + P_0\,dV \tag{8–13}$$

A reversible process cannot involve any heat transfer through a finite temperature difference, and thus any heat transfer between the system at temperature T and its surroundings at T_0 must occur through a reversible heat engine. Noting that $dS = \delta Q/T$ for a reversible process, and the thermal efficiency of a reversible heat engine operating between the temperatures of T and T_0 is $\eta_{\text{th}} = 1 - T_0/T$, the differential work produced by the engine as a result of this heat transfer is

$$\delta W_{\text{HE}} = \left(1 - \frac{T_0}{T}\right)\delta Q = \delta Q - \frac{T_0}{T}\delta Q = \delta Q - (-T_0\,dS) \rightarrow$$

$$\delta Q = \delta W_{\text{HE}} - T_0\,dS \tag{8–14}$$

Substituting the δW and δQ expressions in Eqs. 8–13 and 8–14 into the energy balance relation (Eq. 8–12) gives, after rearranging,

$$\delta W_{\text{total useful}} = \delta W_{\text{HE}} + \delta W_{b,\text{useful}} = -dU - P_0\,dV + T_0\,dS$$

Integrating from the given state (no subscript) to the dead state (0 subscript) we obtain

$$W_{\text{total useful}} = (U - U_0) + P_0(V - V_0) - T_0(S - S_0)$$

where $W_{\text{total useful}}$ is the total useful work delivered as the system undergoes a reversible process from the given state to the dead state, which is *exergy* by definition.

A closed system, in general, may possess kinetic and potential energies, and the total energy of a closed system is equal to the sum of its internal, kinetic, and potential energies. Noting that kinetic and potential energies themselves are forms of exergy, the exergy of a closed system of mass m is

$$X = (U - U_0) + P_0(V - V_0) - T_0(S - S_0) + m\frac{V^2}{2} + mgz \quad \textbf{(8–15)}$$

On a unit mass basis, the **closed system** (or **nonflow**) **exergy** ϕ is expressed as

$$\begin{aligned}\phi &= (u - u_0) + P_0(v - v_0) - T_0(s - s_0) + \frac{V^2}{2} + gz \quad \textbf{(8–16)}\\ &= (e - e_0) + P_0(v - v_0) - T_0(s - s_0)\end{aligned}$$

where u_0, v_0, and s_0 are the properties of the *system* evaluated at the dead state. Note that the exergy of a system is zero at the dead state since $e = e_0$, $v = v_0$, and $s = s_0$ at that state.

The exergy change of a closed system during a process is simply the difference between the final and initial exergies of the system,

$$\begin{aligned}\Delta X = X_2 - X_1 = m(\phi_2 - \phi_1) &= (E_2 - E_1) + P_0(V_2 - V_1) - T_0(S_2 - S_1) \quad \textbf{(8–17)}\\ &= (U_2 - U_1) + P_0(V_2 - V_1) - T_0(S_2 - S_1) + m\frac{V_2^2 - V_1^2}{2} + mg(z_2 - z_1)\end{aligned}$$

or, on a unit mass basis,

$$\begin{aligned}\Delta\phi = \phi_2 - \phi_1 &= (u_2 - u_1) + P_0(v_2 - v_1) - T_0(s_2 - s_1) + \frac{V_2^2 - V_1^2}{2} + g(z_2 - z_1)\\ &= (e_2 - e_1) + P_0(v_2 - v_1) - T_0(s_2 - s_1) \quad \textbf{(8–18)}\end{aligned}$$

For *stationary* closed systems, the kinetic and potential energy terms drop out.

When the properties of a system are not uniform, the exergy of the system can be determined by integration from

$$X_{\text{system}} = \int \phi\, \delta m = \int_V \phi\rho\, dV \quad \textbf{(8–19)}$$

where V is the volume of the system and ρ is density.

Note that exergy is a property, and the value of a property does not change unless the *state* changes. Therefore, the *exergy change* of a system is zero if the state of the system or the environment does not change during the process. For example, the exergy change of steady flow devices such as nozzles, compressors, turbines, pumps, and heat exchangers in a given environment is zero during steady operation.

The exergy of a closed system is either *positive* or *zero*. It is never negative. Even a medium at *low temperature* ($T < T_0$) and/or *low pressure* ($P < P_0$) contains exergy since a cold medium can serve as the heat sink to a heat engine that absorbs heat from the environment at T_0, and an evacuated space makes it possible for the atmospheric pressure to move a piston and do useful work (Fig. 8–21).

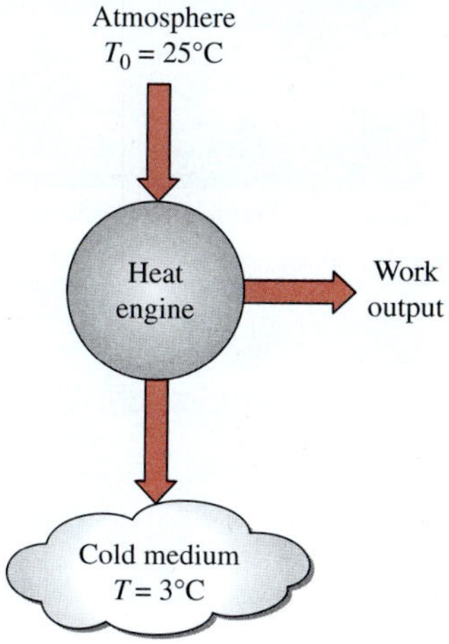

FIGURE 8–21
The *exergy* of a cold medium is also a *positive* quantity since work can be produced by transferring heat to it.

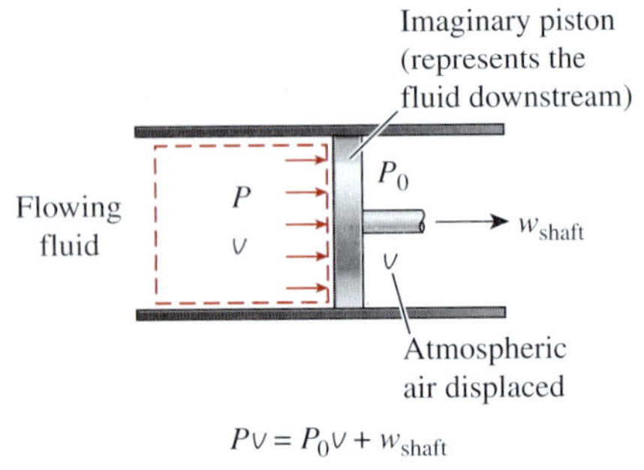

FIGURE 8–22
The *exergy* associated with *flow energy* is the useful work that would be delivered by an imaginary piston in the flow section.

Exergy of a Flow Stream: Flow (or Stream) Exergy

In Chap. 5 it was shown that a flowing fluid has an additional form of energy, called the *flow energy*, which is the energy needed to maintain flow in a pipe or duct, and was expressed as $w_{flow} = Pv$ where v is the specific volume of the fluid, which is equivalent to the *volume change* of a unit mass of the fluid as it is displaced during flow. The flow work is essentially the boundary work done by a fluid on the fluid downstream, and thus the exergy associated with flow work is equivalent to the exergy associated with the boundary work, which is the boundary work in excess of the work done against the atmospheric air at P_0 to displace it by a volume v (Fig. 8–22). Noting that the flow work is Pv and the work done against the atmosphere is P_0v, the *exergy* associated with flow energy can be expressed as

$$x_{flow} = Pv - P_0v = (P - P_0)v \tag{8–20}$$

Therefore, the exergy associated with flow energy is obtained by replacing the pressure P in the flow work relation by the pressure in excess of the atmospheric pressure, $P - P_0$. Then the exergy of a flow stream is determined by simply adding the flow exergy relation above to the exergy relation in Eq. 8–16 for a nonflowing fluid,

$$\begin{aligned} x_{\text{flowing fluid}} &= x_{\text{nonflowing fluid}} + x_{\text{flow}} \qquad (8\text{–}21)\\ &= (u - u_0) + P_0(v - v_0) - T_0(s - s_0) + \frac{V^2}{2} + gz + (P - P_0)v \\ &= (u + Pv) - (u_0 + P_0v_0) - T_0(s - s_0) + \frac{V^2}{2} + gz \\ &= (h - h_0) - T_0(s - s_0) + \frac{V^2}{2} + gz \end{aligned}$$

The final expression is called **flow** (or **stream**) **exergy**, and is denoted by ψ (Fig. 8–23).

Flow exergy:
$$\psi = (h - h_0) - T_0(s - s_0) + \frac{V^2}{2} + gz \tag{8–22}$$

Then the *exergy change* of a fluid stream as it undergoes a process from state 1 to state 2 becomes

$$\Delta\psi = \psi_2 - \psi_1 = (h_2 - h_1) - T_0(s_2 - s_1) + \frac{V_2^2 - V_1^2}{2} + g(z_2 - z_1) \tag{8–23}$$

For fluid streams with negligible kinetic and potential energies, the kinetic and potential energy terms drop out.

Note that the *exergy change* of a closed system or a fluid stream represents the *maximum* amount of useful work that can be done (or the *minimum* amount of useful work that needs to be supplied if it is negative) as the system changes from state 1 to state 2 in a specified environment, and represents the *reversible work* W_{rev}. It is independent of the type of process executed, the kind of system used, and the nature of energy interactions with the surroundings. Also note that the exergy of a closed system cannot be negative, but the exergy of a flow stream can at pressures below the environment pressure P_0.

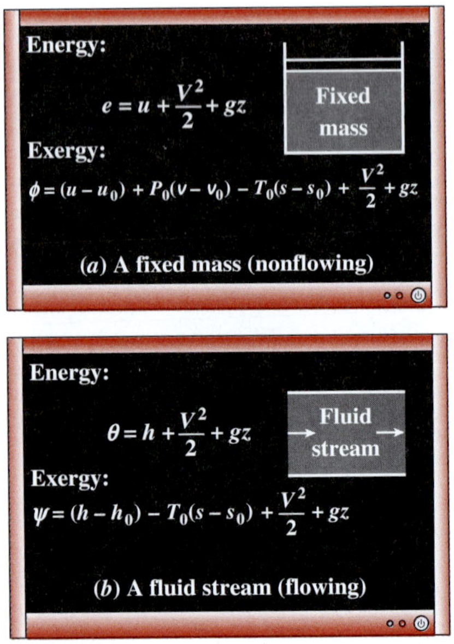

FIGURE 8–23
The *energy* and *exergy* contents of (*a*) a fixed mass and (*b*) a fluid stream.

EXAMPLE 8–7 Work Potential of Compressed Air in a Tank

A 200-m^3 rigid tank contains compressed air at 1 MPa and 300 K. Determine how much work can be obtained from this air if the environment conditions are 100 kPa and 300 K.

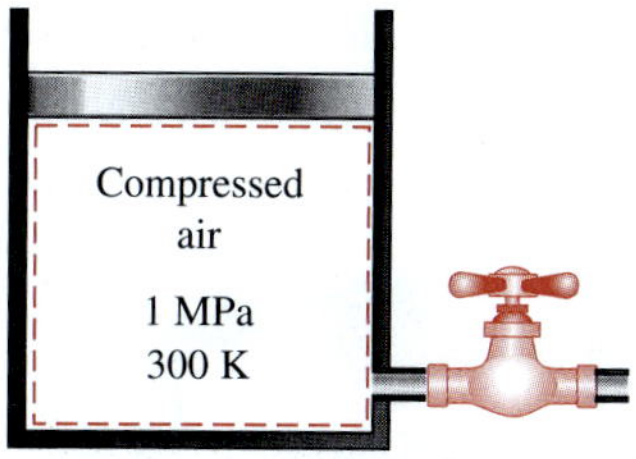

FIGURE 8–24
Schematic for Example 8–7.

SOLUTION Compressed air stored in a large tank is considered. The work potential of this air is to be determined.

Assumptions **1** Air is an ideal gas. **2** The kinetic and potential energies are negligible.

Analysis We take the air in the rigid tank as the system (Fig. 8–24). This is a *closed system* since no mass crosses the system boundary during the process. Here the question is the work potential of a fixed mass, which is the nonflow exergy by definition.

Taking the state of the air in the tank to be state 1 and noting that $T_1 = T_0 = 300$ K, the mass of air in the tank is

$$m_1 = \frac{P_1 V}{RT_1} = \frac{(1000\text{ kPa})(200\text{ m}^3)}{(0.287\text{ kPa·m}^3/\text{kg·K})(300\text{ K})} = 2323\text{ kg}$$

The exergy content of the compressed air can be determined from

$$\begin{aligned} X_1 &= m\phi_1 \\ &= m\left[(u_1 - u_0)^{\nearrow 0} + P_0(v_1 - v_0) - T_0(s_1 - s_0) + \frac{V_1^2}{2}^{\nearrow 0} + gz_1^{\nearrow 0}\right] \\ &= m[P_0(v_1 - v_0) - T_0(s_1 - s_0)] \end{aligned}$$

We note that

$$P_0(v_1 - v_0) = P_0\left(\frac{RT_1}{P_1} - \frac{RT_0}{P_0}\right) = RT_0\left(\frac{P_0}{P_1} - 1\right) \qquad (\text{since } T_1 = T_0)$$

$$T_0(s_1 - s_0) = T_0\left(c_p \ln\frac{T_1}{T_0} - R\ln\frac{P_1}{P_0}\right) = -RT_0 \ln\frac{P_1}{P_0} \qquad (\text{since } T_1 = T_0)$$

Therefore,

$$\begin{aligned} \phi_1 &= RT_0\left(\frac{P_0}{P_1} - 1\right) + RT_0\ln\frac{P_1}{P_0} = RT_0\left(\ln\frac{P_1}{P_0} + \frac{P_0}{P_1} - 1\right) \\ &= (0.287\text{ kJ/kg·K})(300\text{ K})\left(\ln\frac{1000\text{ kPa}}{100\text{ kPa}} + \frac{100\text{ kPa}}{1000\text{ kPa}} - 1\right) \\ &= 120.76\text{ kJ/kg} \end{aligned}$$

and

$$X_1 = m_1\phi_1 = (2323\text{ kg})(120.76\text{ kJ/kg}) = 280{,}525\text{ kJ} \cong \mathbf{281\ MJ}$$

Discussion The work potential of the system is 281 MJ, and thus a maximum of 281 MJ of useful work can be obtained from the compressed air stored in the tank in the specified environment.

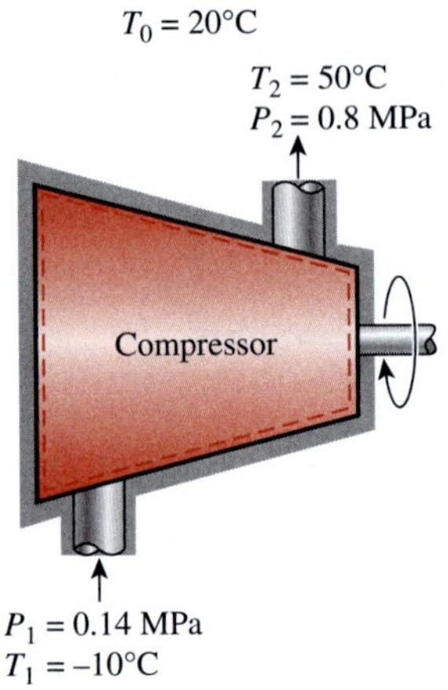

FIGURE 8–25
Schematic for Example 8–8.

EXAMPLE 8–8 Exergy Change During a Compression Process

Refrigerant-134a is to be compressed from 0.14 MPa and −10°C to 0.8 MPa and 50°C steadily by a compressor. Taking the environment conditions to be 20°C and 95 kPa, determine the exergy change of the refrigerant during this process and the minimum work input that needs to be supplied to the compressor per unit mass of the refrigerant.

SOLUTION Refrigerant-134a is being compressed from a specified inlet state to a specified exit state. The exergy change of the refrigerant and the minimum compression work per unit mass are to be determined.

Assumptions **1** Steady operating conditions exist. **2** The kinetic and potential energies are negligible.

Analysis We take the *compressor* as the system (Fig. 8–25). This is a *control volume* since mass crosses the system boundary during the process. Here the question is the exergy change of a fluid stream, which is the change in the flow exergy ψ.

The properties of the refrigerant at the inlet and the exit states are

$$\textit{Inlet state:}\quad \left.\begin{array}{l} P_1 = 0.14\text{ MPa} \\ T_1 = -10°\text{C} \end{array}\right\} \quad \begin{array}{l} h_1 = 246.36\text{ kJ/kg} \\ s_1 = 0.9724\text{ kJ/kg·K} \end{array}$$

$$\textit{Exit state:}\quad \left.\begin{array}{l} P_2 = 0.8\text{ MPa} \\ T_2 = 50°\text{C} \end{array}\right\} \quad \begin{array}{l} h_2 = 286.69\text{ kJ/kg} \\ s_2 = 0.9802\text{ kJ/kg·K} \end{array}$$

The exergy change of the refrigerant during this compression process is determined directly from Eq. 8–23 to be

$$\begin{aligned} \Delta\psi = \psi_2 - \psi_1 &= (h_2 - h_1) - T_0(s_2 - s_1) + \frac{V_2^2 - V_1^2}{2}^{\nearrow 0} + g(z_2 - z_1)^{\nearrow 0} \\ &= (h_2 - h_1) - T_0(s_2 - s_1) \\ &= (286.69 - 246.36)\text{ kJ/kg} - (293\text{ K})[(0.9802 - 0.9724)\text{kJ/kg·K}] \\ &= \mathbf{38.0\ kJ/kg} \end{aligned}$$

Therefore, the exergy of the refrigerant increases during compression by 38.0 kJ/kg.

The exergy change of a system in a specified environment represents the reversible work in that environment, which is the minimum work input required for work-consuming devices such as compressors. Therefore, the increase in exergy of the refrigerant is equal to the minimum work that needs to be supplied to the compressor:

$$w_{\text{in,min}} = \psi_2 - \psi_1 = \mathbf{38.0\ kJ/kg}$$

Discussion Note that if the compressed refrigerant at 0.8 MPa and 50°C were to be expanded to 0.14 MPa and −10°C in a turbine in the same environment in a reversible manner, 38.0 kJ/kg of work would be produced.

8–5 ▪ EXERGY TRANSFER BY HEAT, WORK, AND MASS

Exergy, like energy, can be transferred to or from a system in three forms: *heat, work,* and *mass flow.* Exergy transfer is recognized at the system boundary as exergy crosses it, and it represents the exergy gained or lost by a system during a process. The only two forms of exergy interactions associated with a fixed mass or closed system are *heat transfer* and *work.*

Exergy by Heat Transfer, *Q*

Recall from Chap. 6 that the work potential of the energy transferred from a heat source at temperature T is the maximum work that can be obtained from that energy in an environment at temperature T_0 and is equivalent to the work produced by a Carnot heat engine operating between the source and the environment. Therefore, the Carnot efficiency $\eta_C = 1 - T_0/T$ represents the fraction of energy of a heat source at temperature T that can be converted to work (Fig. 8–26). For example, only 70 percent of the energy transferred from a heat source at $T = 1000$ K can be converted to work in an environment at $T_0 = 300$ K.

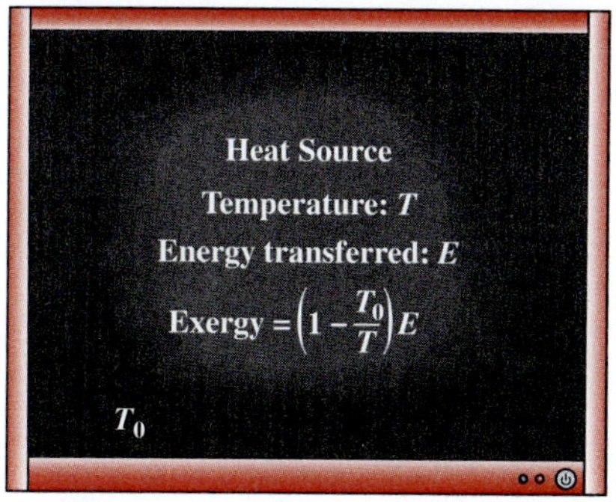

FIGURE 8–26
The Carnot efficiency $\eta_C = 1 - T_0/T$ represents the fraction of the energy transferred from a heat source at temperature T that can be converted to work in an environment at temperature T_0.

Heat is a form of disorganized energy, and thus only a portion of it can be converted to work, which is a form of organized energy (the second law). We can always produce work from heat at a temperature above the environment temperature by transferring it to a heat engine that rejects the waste heat to the environment. Therefore, heat transfer is always accompanied by exergy transfer. Heat transfer Q at a location at thermodynamic temperature T is always accompanied by *exergy transfer* X_{heat} in the amount of

Exergy transfer by heat:
$$X_{\text{heat}} = \left(1 - \frac{T_0}{T}\right)Q \qquad \text{(kJ)} \tag{8–24}$$

This relation gives the exergy transfer accompanying heat transfer Q whether T is greater than or less than T_0. When $T > T_0$, heat transfer to a system increases the exergy of that system and heat transfer from a system decreases it. But the opposite is true when $T < T_0$. In this case, the heat transfer Q is the heat rejected to the cold medium (the waste heat), and it should not be confused with the heat supplied by the environment at T_0. The exergy transferred with heat is zero when $T = T_0$ at the point of transfer.

Perhaps you are wondering what happens when $T < T_0$. That is, what if we have a medium that is at a lower temperature than the environment? In this case it is conceivable that we can run a heat engine between the environment and the "cold" medium, and thus a cold medium offers us an opportunity to produce work. However, this time the environment serves as the heat source and the cold medium as the heat sink. In this case, the relation above gives the negative of the exergy transfer associated with the heat Q transferred to the cold medium. For example, for $T = 100$ K and a heat transfer of $Q = 1$ kJ to the medium, Eq. 8–24 gives $X_{\text{heat}} = (1 - 300/100)(1 \text{ kJ}) = -2$ kJ, which means that the exergy of the cold medium decreases by 2 kJ. It also means that this exergy can be recovered, and the cold medium–environment combination has the potential to produce 2 units of work for each unit of heat rejected to the cold medium at 100 K. That is, a Carnot heat engine operating between $T_0 = 300$ K and $T = 100$ K produces 2 units of work while rejecting 1 unit of heat for each 3 units of heat it receives from the environment.

When $T > T_0$, the exergy and heat transfer are in the same direction. That is, both the exergy and energy content of the medium to which heat is transferred increase. When $T < T_0$ (cold medium), however, the exergy and heat transfer are in opposite directions. That is, the energy of the cold medium increases as a result of heat transfer, but its exergy decreases. The exergy of the cold medium eventually becomes zero when its temperature reaches T_0.

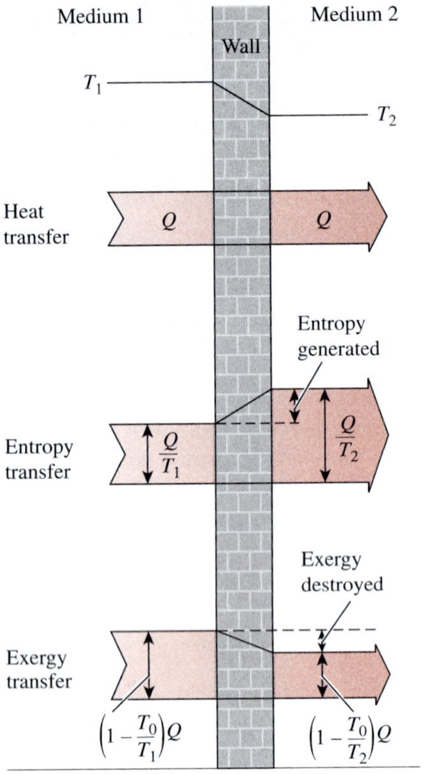

FIGURE 8–27
The transfer and destruction of exergy during a heat transfer process through a finite temperature difference.

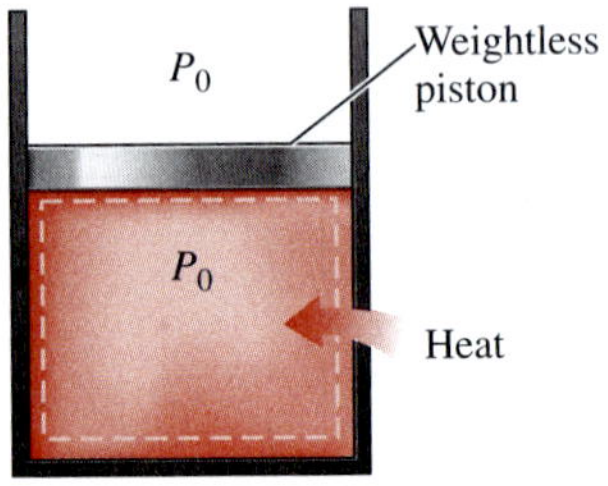

FIGURE 8–28
There is no useful work transfer associated with boundary work when the pressure of the system is maintained constant at atmospheric pressure.

Equation 8–24 can also be viewed as the *exergy associated with thermal energy Q* at temperature *T*.

When the temperature *T* at the location where heat transfer is taking place is not constant, the exergy transfer accompanying heat transfer is determined by integration to be

$$X_{\text{heat}} = \int \left(1 - \frac{T_0}{T}\right) \delta Q \qquad \textbf{(8–25)}$$

Note that heat transfer through a finite temperature difference is irreversible, and some entropy is generated as a result. The entropy generation is always accompanied by exergy destruction, as illustrated in Fig. 8–27. Also note that *heat transfer Q* at a location at temperature *T* is always accompanied by *entropy transfer* in the amount of Q/T and *exergy transfer* in the amount of $(1 - T_0/T)Q$.

Exergy Transfer by Work, *W*

Exergy is the useful work potential, and the exergy transfer by work can simply be expressed as

Exergy transfer by work:
$$X_{\text{work}} = \begin{cases} W - W_{\text{surr}} & \text{(for boundary work)} \\ W & \text{(for other forms of work)} \end{cases} \qquad \textbf{(8–26)}$$

where $W_{\text{surr}} = P_0(V_2 - V_1)$, P_0 is atmospheric pressure, and V_1 and V_2 are the initial and final volumes of the system. Therefore, the exergy transfer with work such as shaft work and electrical work is equal to the work *W* itself. In the case of a system that involves boundary work, such as a piston–cylinder device, the work done to push the atmospheric air out of the way during expansion cannot be transferred, and thus it must be subtracted. Also, during a compression process, part of the work is done by the atmospheric air, and thus we need to supply less useful work from an external source.

To clarify this point further, consider a vertical cylinder fitted with a weightless and frictionless piston (Fig. 8–28). The cylinder is filled with a gas that is maintained at the atmospheric pressure P_0 at all times. Heat is now transferred to the system and the gas in the cylinder expands. As a result, the piston rises and boundary work is done. However, this work cannot be used for any useful purpose since it is just enough to push the atmospheric air aside. (If we connect the piston to an external load to extract some useful work, the pressure in the cylinder will have to rise above P_0 to beat the resistance offered by the load.) When the gas is cooled, the piston moves down, compressing the gas. Again, no work is needed from an external source to accomplish this compression process. Thus we conclude that the work done by or against the atmosphere is not available for any useful purpose, and should be excluded from available work.

Exergy Transfer by Mass, *m*

Mass contains *exergy* as well as energy and entropy, and the exergy, energy, and entropy contents of a system are proportional to mass. Also, the rates of exergy, entropy, and energy transport into or out of a system are proportional to the mass flow rate. Mass flow is a mechanism to transport exergy, entropy, and energy into or out of a system. When mass in the amount of m enters or leaves a system, exergy in the amount of $m\psi$, where $\psi = (h - h_0) - T_0(s - s_0) + V^2/2 + gz$, accompanies it. That is,

Exergy transfer by mass: $$X_{mass} = m\psi \quad \textbf{(8–27)}$$

Therefore, the exergy of a system increases by $m\psi$ when mass in the amount of m enters, and decreases by the same amount when the same amount of mass at the same state leaves the system (Fig. 8–29).

Exergy flow associated with a fluid stream when the fluid properties are variable can be determined by integration from

$$\dot{X}_{mass} = \int_{A_c} \psi\rho V_n \, dA_c \quad \text{and} \quad X_{mass} = \int \psi \, \delta m = \int_{\Delta t} \dot{X}_{mass} \, dt \quad \textbf{(8–28)}$$

where A_c is the cross-sectional area of the flow and V_n is the local velocity normal to dA_c.

Note that exergy transfer by heat X_{heat} is zero for adiabatic systems, and the exergy transfer by mass X_{mass} is zero for systems that involve no mass flow across their boundaries (i.e., closed systems). The total exergy transfer is zero for isolated systems since they involve no heat, work, or mass transfer.

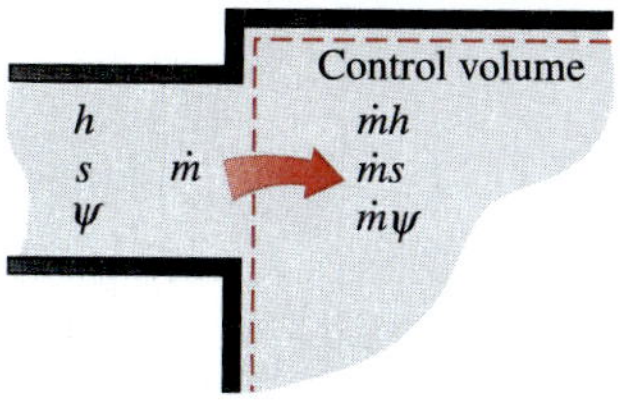

FIGURE 8–29
Mass contains energy, entropy, and exergy, and thus mass flow into or out of a system is accompanied by energy, entropy, and exergy transfer.

8–6 ▪ THE DECREASE OF EXERGY PRINCIPLE AND EXERGY DESTRUCTION

In Chap. 2 we presented the *conservation of energy principle* and indicated that energy cannot be created or destroyed during a process. In Chap. 7 we established the *increase of entropy principle,* which can be regarded as one of the statements of the second law, and indicated that entropy can be created but cannot be destroyed. That is, entropy generation S_{gen} must be positive (actual processes) or zero (reversible processes), but it cannot be negative. Now we are about to establish an alternative statement of the second law of thermodynamics, called the *decrease of exergy principle,* which is the counterpart of the increase of entropy principle.

Consider an *isolated system* shown in Fig. 8–30. By definition, no heat, work, or mass can cross the boundaries of an isolated system, and thus there is no energy and entropy transfer. Then the *energy* and *entropy* balances for an isolated system can be expressed as

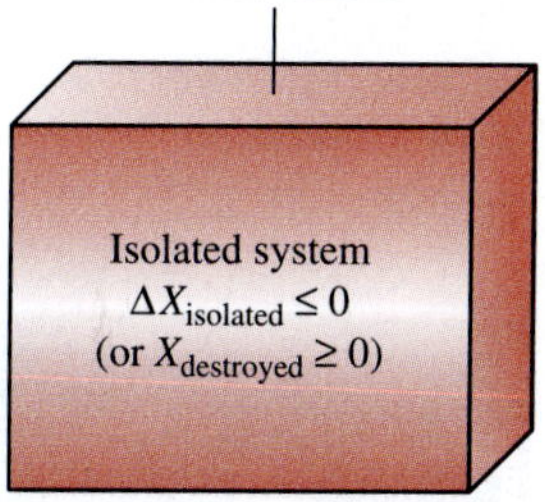

FIGURE 8–30
The isolated system considered in the development of the decrease of exergy principle.

Energy balance: $$E_{in}^{\nearrow 0} - E_{out}^{\nearrow 0} = \Delta E_{system} \rightarrow 0 = E_2 - E_1$$

Entropy balance: $$S_{in}^{\nearrow 0} - S_{out}^{\nearrow 0} + S_{gen} = \Delta S_{system} \rightarrow S_{gen} = S_2 - S_1$$

Multiplying the second relation by T_0 and subtracting it from the first one gives

$$-T_0 S_{gen} = E_2 - E_1 - T_0(S_2 - S_1) \quad \textbf{(8–29)}$$

From Eq. 8–17 we have

$$\begin{aligned} X_2 - X_1 &= (E_2 - E_1) + P_0(V_2 - V_1)^{\nearrow 0} - T_0(S_2 - S_1) \quad \textbf{(8–30)} \\ &= (E_2 - E_1) - T_0(S_2 - S_1) \end{aligned}$$

since $V_2 = V_1$ for an isolated system (it cannot involve any moving boundary and thus any boundary work). Combining Eqs. 8–29 and 8–30 gives

$$-T_0 S_{gen} = X_2 - X_1 \leq 0 \tag{8–31}$$

since T_0 is the thermodynamic temperature of the environment and thus a positive quantity, $S_{gen} \geq 0$, and thus $T_0 S_{gen} \geq 0$. Then we conclude that

$$\Delta X_{isolated} = (X_2 - X_1)_{isolated} \leq 0 \tag{8–32}$$

This equation can be expressed as *the exergy of an isolated system during a process always decreases or, in the limiting case of a reversible process, remains constant.* In other words, it *never* increases and *exergy is destroyed* during an actual process. This is known as the **decrease of exergy principle**. For an isolated system, the decrease in exergy equals exergy destroyed.

Exergy Destruction

Irreversibilities such as friction, mixing, chemical reactions, heat transfer through a finite temperature difference, unrestrained expansion, nonquasi-equilibrium compression or expansion always *generate entropy,* and anything that generates entropy always *destroys exergy.* The **exergy destroyed** is proportional to the entropy generated, as can be seen from Eq. 8–31, and is expressed as

$$X_{destroyed} = T_0 S_{gen} \geq 0 \tag{8–33}$$

Note that exergy destroyed is a *positive quantity* for any actual process and becomes *zero* for a reversible process. Exergy destroyed represents the lost work potential and is also called the *irreversibility* or *lost work.*

Equations 8–32 and 8–33 for the decrease of exergy and the exergy destruction are applicable to *any kind of system* undergoing *any kind of process* since any system and its surroundings can be enclosed by a sufficiently large arbitrary boundary across which there is no heat, work, and mass transfer, and thus any system and its surroundings constitute an *isolated system.*

No actual process is truly reversible, and thus some exergy is destroyed during a process. Therefore, the exergy of the universe, which can be considered to be an isolated system, is continuously decreasing. The more irreversible a process is, the larger the exergy destruction during that process. No exergy is destroyed during a reversible process ($X_{destroyed,rev} = 0$).

The decrease of exergy principle does not imply that the exergy of a system cannot increase. The exergy change of a system *can* be positive or negative during a process (Fig. 8–31), but exergy destroyed cannot be negative. The decrease of exergy principle can be summarized as follows:

$$X_{destroyed} \begin{cases} > 0 & \text{Irreversible process} \\ = 0 & \text{Reversible process} \\ < 0 & \text{Impossible process} \end{cases} \tag{8–34}$$

This relation serves as an alternative criterion to determine whether a process is reversible, irreversible, or impossible.

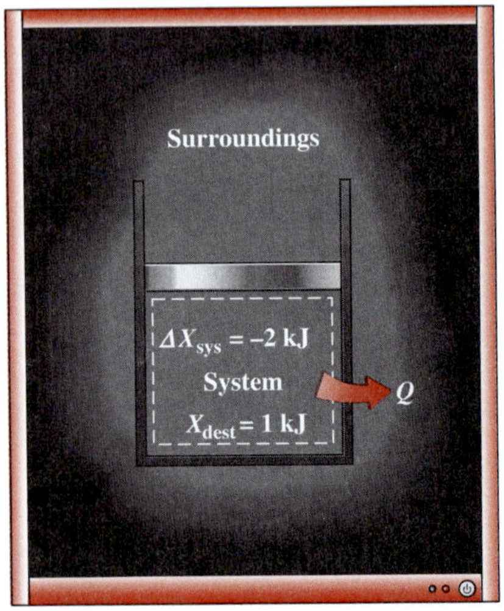

FIGURE 8–31
The exergy change of a system can be negative, but the exergy destruction cannot.

8–7 ▪ EXERGY BALANCE: CLOSED SYSTEMS

The nature of exergy is opposite to that of entropy in that exergy can be *destroyed,* but it cannot be created. Therefore, the *exergy change* of a system during a process is less than the *exergy transfer* by an amount equal to the *exergy destroyed* during the process within the system boundaries. Then the *decrease of exergy principle* can be expressed as (Fig. 8–32)

$$\begin{pmatrix}\text{Total}\\\text{exergy}\\\text{entering}\end{pmatrix} - \begin{pmatrix}\text{Total}\\\text{exergy}\\\text{leaving}\end{pmatrix} - \begin{pmatrix}\text{Total}\\\text{exergy}\\\text{destroyed}\end{pmatrix} = \begin{pmatrix}\text{Change in the}\\\text{total exergy}\\\text{of the system}\end{pmatrix}$$

or

$$X_{\text{in}} - X_{\text{out}} - X_{\text{destroyed}} = \Delta X_{\text{system}} \qquad \textbf{(8–35)}$$

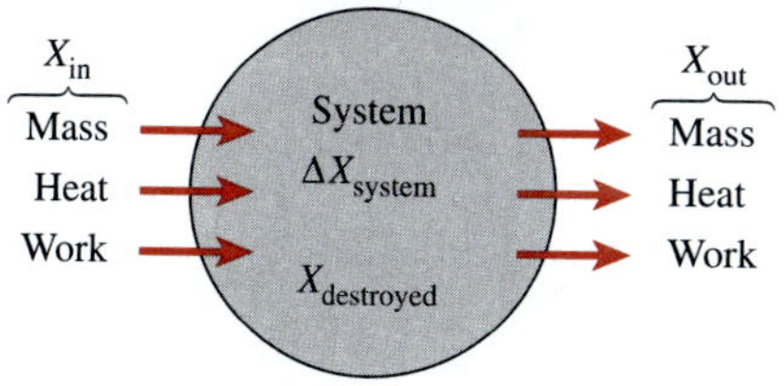

FIGURE 8–32
Mechanisms of exergy transfer.

This relation is referred to as the **exergy balance** and can be stated as *the exergy change of a system during a process is equal to the difference between the net exergy transfer through the system boundary and the exergy destroyed within the system boundaries as a result of irreversibilities.*

We mentioned earlier that exergy can be transferred to or from a system by heat, work, and mass transfer. Then the exergy balance for *any system* undergoing *any process* can be expressed more explicitly as

General:
$$\underbrace{X_{\text{in}} - X_{\text{out}}}_{\substack{\text{Net exergy transfer}\\\text{by heat, work, and mass}}} - \underbrace{X_{\text{destroyed}}}_{\substack{\text{Exergy}\\\text{destruction}}} = \underbrace{\Delta X_{\text{system}}}_{\substack{\text{Change}\\\text{in exergy}}} \quad \text{(kJ)} \qquad \textbf{(8–36)}$$

or, in the **rate form**, as

General, rate form:
$$\underbrace{\dot{X}_{\text{in}} - \dot{X}_{\text{out}}}_{\substack{\text{Rate of net exergy transfer}\\\text{by heat, work, and mass}}} - \underbrace{\dot{X}_{\text{destroyed}}}_{\substack{\text{Rate of exergy}\\\text{destruction}}} = \underbrace{dX_{\text{system}}/dt}_{\substack{\text{Rate of change}\\\text{in exergy}}} \quad \text{(kW)} \qquad \textbf{(8–37)}$$

where the rates of exergy transfer by heat, work, and mass are expressed as $\dot{X}_{\text{heat}} = (1 - T_0/T)\dot{Q}$, $\dot{X}_{\text{work}} = \dot{W}_{\text{useful}}$, and $\dot{X}_{\text{mass}} = \dot{m}\psi$, respectively. The exergy balance can also be expressed per unit mass as

General, unit-mass basis:
$$(x_{\text{in}} - x_{\text{out}}) - x_{\text{destroyed}} = \Delta x_{\text{system}} \quad \text{(kJ/kg)} \qquad \textbf{(8–38)}$$

where all the quantities are expressed per unit mass of the system. Note that for a *reversible process,* the exergy destruction term $X_{\text{destroyed}}$ drops out from all of the relations above. Also, it is usually more convenient to find the entropy generation S_{gen} first, and then to evaluate the exergy destroyed directly from Eq. 8–33. That is,

$$X_{\text{destroyed}} = T_0 S_{\text{gen}} \quad \text{or} \quad \dot{X}_{\text{destroyed}} = T_0 \dot{S}_{\text{gen}} \qquad \textbf{(8–39)}$$

When the environment conditions P_0 and T_0 and the end states of the system are specified, the exergy change of the system $\Delta X_{\text{system}} = X_2 - X_1$ can be determined directly from Eq. 8–17 regardless of how the process is executed. However, the determination of the exergy transfers by heat, work, and mass requires a knowledge of these interactions.

A *closed system* does not involve any mass flow and thus any exergy transfer associated with mass flow. Taking the positive direction of heat transfer to be to the system and the positive direction of work transfer to be

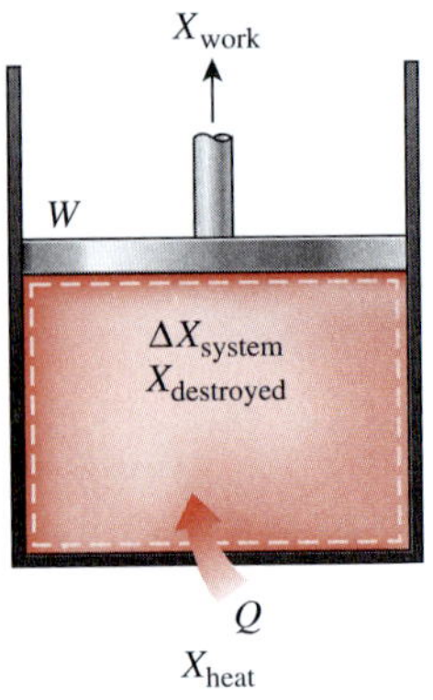

FIGURE 8–33
Exergy balance for a closed system when the direction of heat transfer is taken to be to the system and the direction of work from the system.

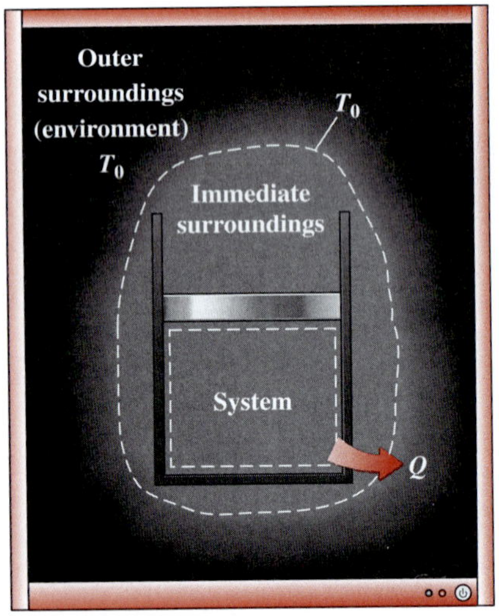

FIGURE 8–34
Exergy destroyed outside system boundaries can be accounted for by writing an exergy balance on the extended system that includes the system and its immediate surroundings.

from the system, the exergy balance for a closed system can be expressed more explicitly as (Fig. 8–33)

$$\text{Closed system:} \qquad X_{heat} - X_{work} - X_{destroyed} = \Delta X_{system} \qquad \textbf{(8–40)}$$

or

$$\text{Closed system:} \sum\left(1 - \frac{T_0}{T_k}\right)Q_k - [W - P_0(V_2 - V_1)] - T_0 S_{gen} = X_2 - X_1 \qquad \textbf{(8–41)}$$

where Q_k is the heat transfer through the boundary at temperature T_k at location k. Dividing the previous equation by the time interval Δt and taking the limit as $\Delta t \to 0$ gives the *rate form* of the exergy balance for a closed system,

$$\text{Rate form:} \sum\left(1 - \frac{T_0}{T_k}\right)\dot{Q}_k - \left(\dot{W} - P_0\frac{dV_{system}}{dt}\right) - T_0\dot{S}_{gen} = \frac{dX_{system}}{dt} \qquad \textbf{(8–42)}$$

Note that the relations above for a closed system are developed by taking the heat transfer to a system and work done by the system to be positive quantities. Therefore, heat transfer from the system and work done on the system should be taken to be negative quantities when using those relations.

The exergy balance relations presented above can be used to determine the *reversible work* W_{rev} by setting the exergy destruction term equal to zero. The work W in that case becomes the reversible work. That is, $W = W_{rev}$ when $X_{destroyed} = T_0 S_{gen} = 0$.

Note that $X_{destroyed}$ represents the exergy destroyed *within the system boundary* only, and not the exergy destruction that may occur outside the system boundary during the process as a result of external irreversibilities. Therefore, a process for which $X_{destroyed} = 0$ is *internally reversible* but not necessarily *totally* reversible. The *total* exergy destroyed during a process can be determined by applying the exergy balance to an *extended system* that includes the system itself and its immediate surroundings where external irreversibilities might be occurring (Fig. 8–34). Also, the exergy change in this case is equal to the sum of the exergy changes of the system and the *exergy change* of the immediate surroundings. Note that under steady conditions, the state and thus the exergy of the immediate surroundings (the "buffer zone") at any point does not change during the process, and thus the exergy change of the immediate surroundings is zero. When evaluating the exergy transfer between an extended system and the environment, the boundary temperature of the extended system is simply taken to be the environment temperature T_0.

For a *reversible process,* the *entropy generation* and thus the *exergy destruction* are *zero,* and the exergy balance relation in this case becomes analogous to the energy balance relation. That is, the exergy change of the system becomes equal to the exergy transfer.

Note that the *energy change* of a system equals the *energy transfer* for *any* process, but the *exergy change* of a system equals the *exergy transfer* only for a *reversible* process. The *quantity* of energy is always preserved during an actual process (the first law), but the *quality* is bound to decrease (the second law). This decrease in quality is always accompanied by an increase in entropy and a decrease in exergy. When 10 kJ of heat is transferred from a hot medium to a cold one, for example, we still have 10 kJ of energy at the end of the process, but at a lower temperature, and thus at a lower quality and at a lower potential to do work.

EXAMPLE 8–9 General Exergy Balance for Closed Systems

Starting with energy and entropy balances, derive the general exergy balance relation for a closed system (Eq. 8–41).

SOLUTION Starting with energy and entropy balance relations, a general relation for exergy balance for a closed system is to be obtained.

Analysis We consider a general closed system (a fixed mass) that is free to exchange heat and work with its surroundings (Fig. 8–35). The system undergoes a process from state 1 to state 2. Taking the positive direction of heat transfer to be *to* the system and the positive direction of work transfer to be *from* the system, the energy and entropy balances for this closed system can be expressed as

Energy balance: $$E_{in} - E_{out} = \Delta E_{system} \rightarrow Q - W = E_2 - E_1$$

Entropy balance: $$S_{in} - S_{out} + S_{gen} = \Delta S_{system} \rightarrow \int_1^2 \left(\frac{\delta Q}{T}\right)_{boundary} + S_{gen} = S_2 - S_1$$

Multiplying the second relation by T_0 and subtracting it from the first one gives

$$Q - T_0 \int_1^2 \left(\frac{\delta Q}{T}\right)_{boundary} - W - T_0 S_{gen} = E_2 - E_1 - T_0(S_2 - S_1)$$

However, the heat transfer for the process 1-2 can be expressed as $Q = \int_1^2 \delta Q$ and the right side of the above equation is, from Eq. 8–17, $(X_2 - X_1) - P_0(V_2 - V_1)$. Thus,

$$\int_1^2 \delta Q - T_0 \int_1^2 \left(\frac{\delta Q}{T}\right)_{boundary} - W - T_0 S_{gen} = X_2 - X_1 - P_0(V_2 - V_1)$$

Letting T_b denote the boundary temperature and rearranging give

$$\int_1^2 \left(1 - \frac{T_0}{T_b}\right) \delta Q - [W - P_0(V_2 - V_1)] - T_0 S_{gen} = X_2 - X_1 \quad \textbf{(8–43)}$$

which is equivalent to Eq. 8–41 for the exergy balance except that the integration is replaced by summation in that equation for convenience. This completes the proof.

Discussion Note that the exergy balance relation above is obtained by adding the energy and entropy balance relations, and thus it is not an independent equation. However, it can be used in place of the entropy balance relation as an alternative second law expression in exergy analysis.

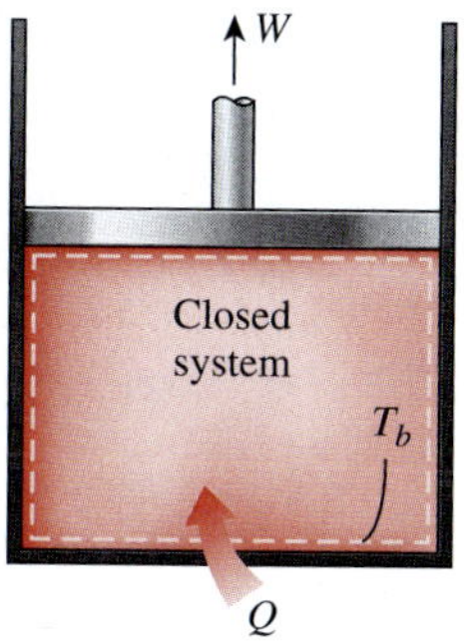

FIGURE 8–35
A general closed system considered in Example 8–9.

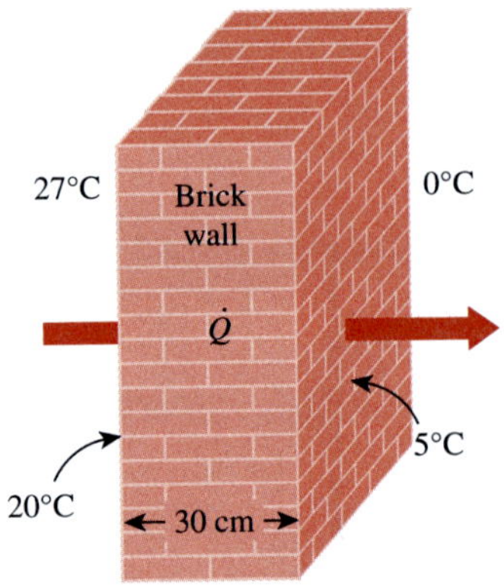

FIGURE 8–36
Schematic for Example 8–10.

EXAMPLE 8–10 Exergy Destruction during Heat Conduction

Consider steady heat transfer through a 5-m × 6-m brick wall of a house of thickness 30 cm. On a day when the temperature of the outdoors is 0°C, the house is maintained at 27°C. The temperatures of the inner and outer surfaces of the brick wall are measured to be 20°C and 5°C, respectively, and the rate of heat transfer through the wall is 1035 W. Determine the rate of exergy destruction in the wall, and the rate of total exergy destruction associated with this heat transfer process.

SOLUTION Steady heat transfer through a wall is considered. For specified heat transfer rate, wall surface temperatures, and environment conditions, the rate of exergy destruction within the wall and the rate of total exergy destruction are to be determined.

Assumptions **1** The process is steady, and thus the rate of heat transfer through the wall is constant. **2** The exergy change of the wall is zero during this process since the state and thus the exergy of the wall do not change anywhere in the wall. **3** Heat transfer through the wall is one-dimensional.

Analysis We first take the *wall* as the system (Fig. 8–36). This is a *closed system* since no mass crosses the system boundary during the process. We note that heat and exergy are entering from one side of the wall and leaving from the other side.

Applying the rate form of the exergy balance to the wall gives

$$\underbrace{\dot{X}_{in} - \dot{X}_{out}}_{\substack{\text{Rate of net exergy transfer}\\ \text{by heat, work, and mass}}} - \underbrace{\dot{X}_{destroyed}}_{\substack{\text{Rate of exergy}\\ \text{destruction}}} = \underbrace{dX_{system}/dt}_{\substack{\text{Rate of change}\\ \text{in exergy}}}^{\nearrow 0 \text{ (steady)}} = 0$$

$$\dot{Q}\left(1 - \frac{T_0}{T}\right)_{in} - \dot{Q}\left(1 - \frac{T_0}{T}\right)_{out} - \dot{X}_{destroyed} = 0$$

$$(1035 \text{ W})\left(1 - \frac{273 \text{ K}}{293 \text{ K}}\right) - (1035 \text{ W})\left(1 - \frac{273 \text{ K}}{278 \text{ K}}\right) - \dot{X}_{destroyed} = 0$$

Solving, the rate of exergy destruction in the wall is determined to be

$$\dot{X}_{destroyed} = \mathbf{52.0 \text{ W}}$$

Note that exergy transfer with heat at any location is $(1 - T_0/T)Q$ at that location, and the direction of exergy transfer is the same as the direction of heat transfer.

To determine the rate of total exergy destruction during this heat transfer process, we extend the system to include the regions on both sides of the wall that experience a temperature change. Then one side of the system boundary becomes room temperature while the other side, the temperature of the outdoors. The exergy balance for this *extended system* (system + immediate surroundings) is the same as that given above, except the two boundary temperatures are 300 and 273 K instead of 293 and 278 K, respectively. Then the rate of total exergy destruction becomes

$$\dot{X}_{destroyed,total} = (1035 \text{ W})\left(1 - \frac{273 \text{ K}}{300 \text{ K}}\right) - (1035 \text{ W})\left(1 - \frac{273 \text{ K}}{273 \text{ K}}\right) = \mathbf{93.2 \text{ W}}$$

The difference between the two exergy destructions is 41.2 W and represents the exergy destroyed in the air layers on both sides of the wall. The exergy destruction in this case is entirely due to irreversible heat transfer through a finite temperature difference.

Discussion This problem was solved in Chap. 7 for entropy generation. We could have determined the exergy destroyed by simply multiplying the entropy generations by the environment temperature of $T_0 = 273$ K.

EXAMPLE 8–11 Exergy Destruction During Expansion of Steam

A piston–cylinder device contains 0.05 kg of steam at 1 MPa and 300°C. Steam now expands to a final state of 200 kPa and 150°C, doing work. Heat losses from the system to the surroundings are estimated to be 2 kJ during this process. Assuming the surroundings to be at $T_0 = 25°C$ and $P_0 = 100$ kPa, determine (*a*) the exergy of the steam at the initial and the final states, (*b*) the exergy change of the steam, (*c*) the exergy destroyed, and (*d*) the second-law efficiency for the process.

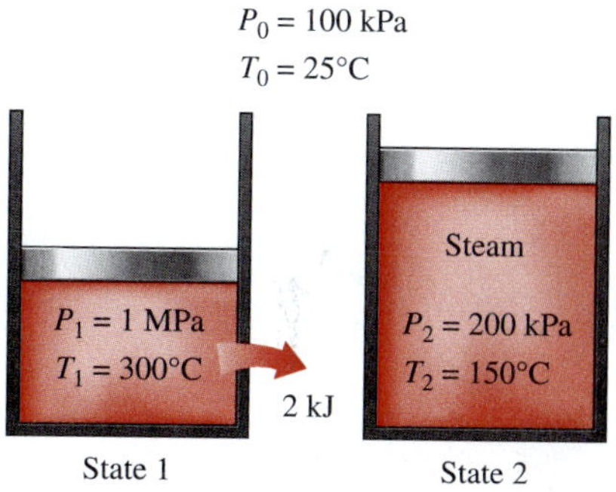

FIGURE 8–37
Schematic for Example 8–11.

SOLUTION Steam in a piston–cylinder device expands to a specified state. The exergies of steam at the initial and final states, the exergy change, the exergy destroyed, and the second-law efficiency for this process are to be determined.

Assumptions The kinetic and potential energies are negligible.

Analysis We take the *steam* contained within the piston–cylinder device as the system (Fig. 8–37). This is a *closed system* since no mass crosses the system boundary during the process. We note that boundary work is done by the system and heat is lost from the system during the process.

(*a*) First we determine the properties of the steam at the initial and final states as well as the state of the surroundings:

State 1: $\left.\begin{array}{l} P_1 = 1 \text{ MPa} \\ T_1 = 300°\text{C} \end{array}\right\} \begin{array}{l} u_1 = 2793.7 \text{ kJ/kg} \\ v_1 = 0.25799 \text{ m}^3/\text{kg} \\ s_1 = 7.1246 \text{ kJ/kg·K} \end{array}$ (Table A–6)

State 2: $\left.\begin{array}{l} P_2 = 200 \text{ kPa} \\ T_2 = 150°\text{C} \end{array}\right\} \begin{array}{l} u_2 = 2577.1 \text{ kJ/kg} \\ v_2 = 0.95986 \text{ m}^3/\text{kg} \\ s_2 = 7.2810 \text{ kJ/kg·K} \end{array}$ (Table A–6)

Dead state: $\left.\begin{array}{l} P_0 = 100 \text{ kPa} \\ T_0 = 25°\text{C} \end{array}\right\} \begin{array}{l} u_0 \cong u_{f @ 25°\text{C}} = 104.83 \text{ kJ/kg} \\ v_0 \cong v_{f @ 25°\text{C}} = 0.00103 \text{ m}^3/\text{kg} \\ s_0 \cong s_{f @ 25°\text{C}} = 0.3672 \text{ kJ/kg·K} \end{array}$ (Table A–4)

The exergies of the system at the initial state X_1 and the final state X_2 are determined from Eq. 8–15 to be

$$\begin{aligned} X_1 &= m[(u_1 - u_0) - T_0(s_1 - s_0) + P_0(v_1 - v_0)] \\ &= (0.05 \text{ kg})\{(2793.7 - 104.83) \text{ kJ/kg} \\ &\quad - (298 \text{ K})[(7.1246 - 0.3672) \text{ kJ/kg·K}] \\ &\quad + (100 \text{ kPa})[(0.25799 - 0.00103) \text{ m}^3/\text{kg}]\}(\text{kJ/kPa·m}^3) \\ &= \mathbf{35.0 \text{ kJ}} \end{aligned}$$

and

$$X_2 = m[(u_2 - u_0) - T_0(s_2 - s_0) + P_0(v_2 - v_0)]$$
$$= (0.05\text{ kg})\{(2577.1 - 104.83)\text{ kJ/kg}$$
$$- (298\text{ K})[(7.2810 - 0.3672)\text{ kJ/kg·K}]$$
$$+ (100\text{ kPa})[(0.95986 - 0.00103)\text{ m}^3\text{/kg}]\}(\text{kJ/kPa·m}^3)$$
$$= \mathbf{25.4\ kJ}$$

That is, steam initially has an exergy content of 35 kJ, which drops to 25.4 kJ at the end of the process. In other words, if the steam were allowed to undergo a reversible process from the initial state to the state of the environment, it would produce 35 kJ of useful work.

(*b*) The exergy change for a process is simply the difference between the exergy at the initial and final states of the process,

$$\Delta X = X_2 - X_1 = 25.4 - 35.0 = \mathbf{-9.6\ kJ}$$

That is, if the process between states 1 and 2 were executed in a reversible manner, the system would deliver 9.6 kJ of useful work.

(*c*) The total exergy destroyed during this process can be determined from the exergy balance applied on the *extended system* (system + immediate surroundings) whose boundary is at the environment temperature of T_0 (so that there is no exergy transfer accompanying heat transfer to or from the environment),

$$\underbrace{X_{in} - X_{out}}_{\substack{\text{Net exergy transfer}\\ \text{by heat, work, and mass}}} - \underbrace{X_{destroyed}}_{\substack{\text{Exergy}\\ \text{destruction}}} = \underbrace{\Delta X_{system}}_{\substack{\text{Change}\\ \text{in exergy}}}$$

$$-X_{work,out} - X_{heat,out}^{\nearrow 0} - X_{destroyed} = X_2 - X_1$$

$$X_{destroyed} = X_1 - X_2 - W_{u,out}$$

where $W_{u,out}$ is the useful boundary work delivered as the system expands. By writing an energy balance on the system, the total boundary work done during the process is determined to be

$$\underbrace{E_{in} - E_{out}}_{\substack{\text{Net energy transfer}\\ \text{by heat, work, and mass}}} = \underbrace{\Delta E_{system}}_{\substack{\text{Change in internal, kinetic,}\\ \text{potential, etc., energies}}}$$

$$-Q_{out} - W_{b,out} = \Delta U$$
$$W_{b,out} = -Q_{out} - \Delta U = -Q_{out} - m(u_2 - u_1)$$
$$= -(2\text{ kJ}) - (0.05\text{ kg})(2577.1 - 2793.7)\text{ kJ/kg}$$
$$= 8.8\text{ kJ}$$

This is the total boundary work done by the system, including the work done against the atmosphere to push the atmospheric air out of the way during the expansion process. The useful work is the difference between the two:

$$W_u = W - W_{surr} = W_{b,out} - P_0(V_2 - V_1) = W_{b,out} - P_0 m(v_2 - v_1)$$
$$= 8.8\text{ kJ} - (100\text{ kPa})(0.05\text{ kg})[(0.9599 - 0.25799)\text{ m}^3\text{/kg}]\left(\frac{1\text{ kJ}}{1\text{ kPa·m}^3}\right)$$
$$= 5.3\text{ kJ}$$

Substituting, the exergy destroyed is determined to be

$$X_{\text{destroyed}} = X_1 - X_2 - W_{u,\text{out}} = 35.0 - 25.4 - 5.3 = \mathbf{4.3\ kJ}$$

That is, 4.3 kJ of work potential is wasted during this process. In other words, an additional 4.3 kJ of energy *could have been* converted to work during this process, but was not.

The exergy destroyed could also be determined from

$$X_{\text{destroyed}} = T_0 S_{\text{gen}} = T_0\left[m(s_2 - s_1) + \frac{Q_{\text{surr}}}{T_0}\right]$$

$$= (298\text{ K})\left\{(0.05\text{ kg})[(7.2810 - 7.1246)\text{ kJ/kg·K}] + \frac{2\text{ kJ}}{298\text{ K}}\right\}$$

$$= 4.3\text{ kJ}$$

which is the same result obtained before.

(*d*) Noting that the decrease in the exergy of the steam is the exergy expended and the useful work output is the exergy recovered, the second-law efficiency for this process can be determined from

$$\eta_{\text{II}} = \frac{\text{Exergy recovered}}{\text{Exergy expended}} = \frac{W_u}{X_1 - X_2} = \frac{5.3}{35.0 - 25.4} = \mathbf{0.552\ or\ 55.2\%}$$

That is, 44.8 percent of the work potential of the steam is wasted during this process.

EXAMPLE 8–12 Exergy Destroyed During Stirring of a Gas

An insulated rigid tank contains 2 lbm of air at 20 psia and 70°F. A paddle wheel inside the tank is now rotated by an external power source until the temperature in the tank rises to 130°F (Fig. 8–38). If the surrounding air is at T_0 = 70°F, determine (*a*) the exergy destroyed and (*b*) the reversible work for this process.

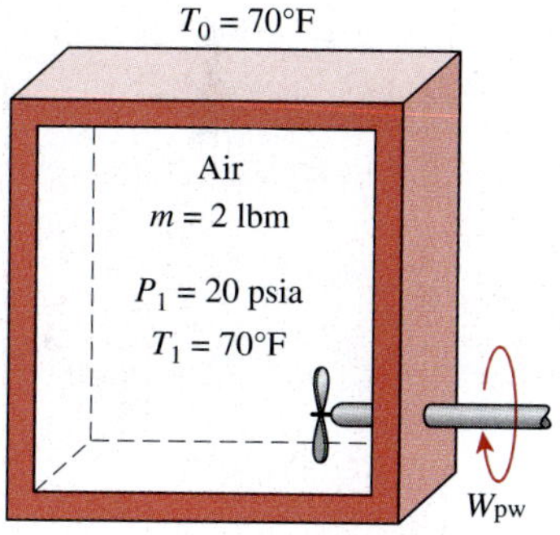

FIGURE 8–38
Schematic for Example 8–12.

SOLUTION The air in an adiabatic rigid tank is heated by stirring it by a paddle wheel. The exergy destroyed and the reversible work for this process are to be determined.

Assumptions **1** Air at about atmospheric conditions can be treated as an ideal gas with constant specific heats at room temperature. **2** The kinetic and potential energies are negligible. **3** The volume of a rigid tank is constant, and thus there is no boundary work. **4** The tank is well insulated and thus there is no heat transfer.

Analysis We take the *air* contained within the tank as the system. This is a *closed system* since no mass crosses the system boundary during the process. We note that shaft work is done on the system.

(*a*) The exergy destroyed during a process can be determined from an exergy balance, or directly from $X_{\text{destroyed}} = T_0 S_{\text{gen}}$. We will use the second approach since it is usually easier. But first we determine the entropy generated from an entropy balance,

$$\underbrace{S_{in} - S_{out}}_{\substack{\text{Net entropy transfer} \\ \text{by heat and mass}}} + \underbrace{S_{gen}}_{\substack{\text{Entropy} \\ \text{generation}}} = \underbrace{\Delta S_{system}}_{\substack{\text{Change} \\ \text{in entropy}}}$$

$$0 + S_{gen} = \Delta S_{system} = m\left(c_v \ln\frac{T_2}{T_1} + R\ln\frac{V_2}{V_1}^{\nearrow 0}\right)$$

$$S_{gen} = mc_v \ln\frac{T_2}{T_1}$$

Taking $c_v = 0.172$ Btu/lbm·°F and substituting, the exergy destroyed becomes

$$X_{destroyed} = T_0 S_{gen} = T_0 mc_v \ln\frac{T_2}{T_1}$$

$$= (530\text{ R})(2\text{ lbm})(0.172\text{ Btu/lbm·°F})\ln\frac{590\text{ R}}{530\text{ R}}$$

$$= \mathbf{19.6\ Btu}$$

(*b*) The reversible work, which represents the minimum work input $W_{rev,in}$ in this case, can be determined from the exergy balance by setting the exergy destruction equal to zero,

$$\underbrace{X_{in} - X_{out}}_{\substack{\text{Net exergy transfer} \\ \text{by heat, work, and mass}}} - \underbrace{X_{destroyed}^{\nearrow 0\text{ (reversible)}}}_{\substack{\text{Exergy} \\ \text{destruction}}} = \underbrace{\Delta X_{system}}_{\substack{\text{Change} \\ \text{in exergy}}}$$

$$W_{rev,in} = X_2 - X_1$$

$$= (E_2 - E_1) + P_0(V_2 - V_1)^{\nearrow 0} - T_0(S_2 - S_1)$$

$$= (U_2 - U_1) - T_0(S_2 - S_1)$$

since $\Delta KE = \Delta PE = 0$ and $V_2 = V_1$. Noting that $T_0(S_2 - S_1) = T_0\,\Delta S_{system} =$ 19.6 Btu, the reversible work becomes

$$W_{rev,in} = mc_v(T_2 - T_1) - T_0(S_2 - S_1)$$

$$= (2\text{ lbm})(0.172\text{ Btu/lbm·°F})(130 - 70)\text{°F} - 19.6\text{ Btu}$$

$$= (20.6 - 19.6)\text{ Btu}$$

$$= \mathbf{1.0\ Btu}$$

Therefore, a work input of just 1.0 Btu would be sufficient to accomplish this process (raise the temperature of air in the tank from 70 to 130°F) if all the irreversibilities were eliminated.

Discussion The solution is complete at this point. However, to gain some physical insight, we will set the stage for a discussion. First, let us determine the actual work (the paddle-wheel work W_{pw}) done during this process. Applying the energy balance on the system,

$$\underbrace{E_{in} - E_{out}}_{\substack{\text{Net energy transfer} \\ \text{by heat, work, and mass}}} = \underbrace{\Delta E_{system}}_{\substack{\text{Change in internal, kinetic,} \\ \text{potential, etc., energies}}}$$

$$W_{pw,in} = \Delta U = 20.6\text{ Btu} \qquad [\text{from part } (b)]$$

since the system is adiabatic ($Q = 0$) and involves no moving boundaries ($W_b = 0$).

To put the information into perspective, 20.6 Btu of work is consumed during the process, 19.6 Btu of exergy is destroyed, and the reversible work input for the process is 1.0 Btu. What does all this mean? It simply means that we could have created the same effect on the closed system (raising its temperature to 130°F at constant volume) by consuming 1.0 Btu of work only instead of 20.6 Btu, and thus saving 19.6 Btu of work from going to waste. This would have been accomplished by a reversible heat pump.

To prove what we have just said, consider a Carnot heat pump that absorbs heat from the surroundings at $T_0 = 530$ R and transfers it to the air in the rigid tank until the air temperature T rises from 530 to 590 R, as shown in Fig. 8–39. The system involves no direct work interactions in this case, and the heat supplied to the system can be expressed in differential form as

$$\delta Q_H = dU = mc_v\, dT$$

The coefficient of performance of a reversible heat pump is given by

$$\text{COP}_{\text{HP}} = \frac{\delta Q_H}{\delta W_{\text{net,in}}} = \frac{1}{1 - T_0/T}$$

Thus,

$$\delta W_{\text{net,in}} = \frac{\delta Q_H}{\text{COP}_{\text{HP}}} = \left(1 - \frac{T_0}{T}\right) mc_v\, dT$$

Integrating, we get

$$\begin{aligned} W_{\text{net,in}} &= \int_1^2 \left(1 - \frac{T_0}{T}\right) mc_v\, dT \\ &= mc_{v,\text{avg}}(T_2 - T_1) - T_0\, mc_{v,\text{avg}} \ln \frac{T_2}{T_1} \\ &= (20.6 - 19.6)\ \text{Btu} = 1.0\ \text{Btu} \end{aligned}$$

The first term on the right-hand side of the final expression above is recognized as ΔU and the second term as the exergy destroyed, whose values were determined earlier. By substituting those values, the total work input to the heat pump is determined to be 1.0 Btu, proving our claim. Notice that the system is still supplied with 20.6 Btu of energy; all we did in the latter case is replace the 19.6 Btu of valuable work by an equal amount of "useless" energy captured from the surroundings.

Discussion It is also worth mentioning that the exergy of the system as a result of 20.6 Btu of paddle-wheel work done on it has increased by 1.0 Btu only, that is, by the amount of the reversible work. In other words, if the system were returned to its initial state, it would produce, at most, 1.0 Btu of work.

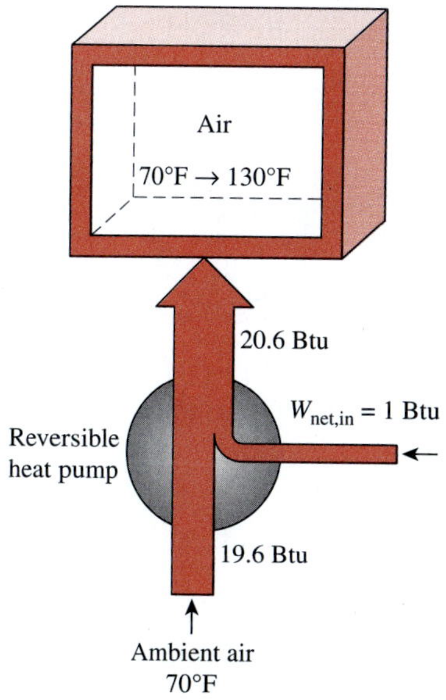

FIGURE 8–39
The same effect on the system can be accomplished by a reversible heat pump that consumes only 1 Btu of work.

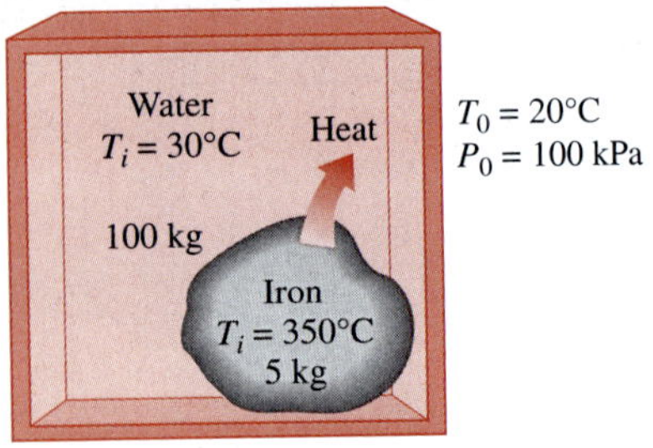

FIGURE 8–40
Schematic for Example 8–13.

EXAMPLE 8–13 Dropping a Hot Iron Block into Water

A 5-kg block initially at 350°C is quenched in an insulated tank that contains 100 kg of water at 30°C (Fig. 8–40). Assuming the water that vaporizes during the process condenses back in the tank and the surroundings are at 20°C and 100 kPa, determine (*a*) the final equilibrium temperature, (*b*) the exergy of the combined system at the initial and the final states, and (*c*) the wasted work potential during this process.

SOLUTION A hot iron block is quenched in an insulated tank by water. The final equilibrium temperature, the initial and final exergies, and the wasted work potential are to be determined.

Assumptions **1** Both water and the iron block are incompressible substances. **2** Constant specific heats at room temperature can be used for both the water and the iron. **3** The system is stationary and thus the kinetic and potential energy changes are zero, ΔKE = ΔPE = 0. **4** There are no electrical, shaft, or other forms of work involved. **5** The system is well-insulated and thus there is no heat transfer.

Analysis We take the entire contents of the tank, *water + iron block*, as the *system*. This is a *closed system* since no mass crosses the system boundary during the process. We note that the volume of a rigid tank is constant, and thus there is no boundary work.

(*a*) Noting that no energy enters or leaves the system during the process, the application of the energy balance gives

$$\underbrace{E_{in} - E_{out}}_{\substack{\text{Net energy transfer}\\ \text{by heat, work, and mass}}} = \underbrace{\Delta E_{system}}_{\substack{\text{Change in internal, kinetic,}\\ \text{potential, etc., energies}}}$$

$$0 = \Delta U$$

$$0 = (\Delta U)_{iron} + (\Delta U)_{water}$$

$$0 = [mc(T_f - T_i)]_{iron} + [mc(T_f - T_i)]_{water}$$

By using the specific-heat values for water and iron at room temperature (from Table A–3), the final equilibrium temperature T_f becomes

$$0 = (5\text{ kg})(0.45\text{ kJ/kg·°C})(T_f - 350°\text{C}) + (100\text{ kg})(4.18\text{ kJ/kg·°C})(T_f - 30°\text{C})$$

which yields

$$T_f = \mathbf{31.7°C}$$

(*b*) Exergy X is an extensive property, and the exergy of a composite system at a specified state is the sum of the exergies of the components of that system at that state. It is determined from Eq. 8–15, which for an incompressible substance reduces to

$$X = (U - U_0) - T_0(S - S_0) + P_0(V - V_0)^{\nearrow 0}$$

$$= mc(T - T_0) - T_0 mc \ln\frac{T}{T_0} + 0$$

$$= mc\left(T - T_0 - T_0 \ln\frac{T}{T_0}\right)$$

where T is the temperature at the specified state and T_0 is the temperature of the surroundings. At the initial state,

$$X_{1,\text{iron}} = (5\text{ kg})(0.45\text{ kJ/kg·K})\left[(623 - 293)\text{ K} - (293\text{ K})\ln\frac{623\text{ K}}{293\text{ K}}\right]$$
$$= 245.2\text{ kJ}$$

$$X_{1,\text{water}} = (100\text{ kg})(4.18\text{ kJ/kg·K})\left[(303 - 293)\text{ K} - (293\text{ K})\ln\frac{303\text{ K}}{293\text{ K}}\right]$$
$$= 69.8\text{ kJ}$$

$$X_{1,\text{total}} = X_{1,\text{iron}} + X_{1,\text{water}} = (245.2 + 69.8)\text{kJ} = \mathbf{315\text{ kJ}}$$

Similarly, the total exergy at the final state is

$$X_{2,\text{iron}} = 0.5\text{ kJ}$$
$$X_{2,\text{water}} = 95.1\text{ kJ}$$
$$X_{2,\text{total}} = X_{2,\text{iron}} + X_{2,\text{water}} = 0.5 + 95.1 = \mathbf{95.6\text{ kJ}}$$

That is, the exergy of the combined system (water + iron) decreased from 315 to 95.6 kJ as a result of this irreversible heat transfer process.

(*c*) The wasted work potential is equivalent to the exergy destroyed, which can be determined from $X_{\text{destroyed}} = T_0S_{\text{gen}}$ or by performing an exergy balance on the system. The second approach is more convenient in this case since the initial and final exergies of the system are already evaluated.

$$\underbrace{X_{\text{in}} - X_{\text{out}}}_{\substack{\text{Net exergy transfer}\\\text{by heat, work, and mass}}} - \underbrace{X_{\text{destroyed}}}_{\substack{\text{Exergy}\\\text{destruction}}} = \underbrace{\Delta X_{\text{system}}}_{\substack{\text{Change}\\\text{in exergy}}}$$

$$0 - X_{\text{destroyed}} = X_2 - X_1$$
$$X_{\text{destroyed}} = X_1 - X_2 = 315 - 95.6 = \mathbf{219.4\text{ kJ}}$$

Discussion Note that 219.4 kJ of work could have been produced as the iron was cooled from 350 to 31.7°C and water was heated from 30 to 31.7°C, but was not.

EXAMPLE 8–14 Work Potential of Heat Transfer Between Two Tanks

Two constant-volume tanks, each filled with 30 kg of air, have temperatures of 900 K and 300 K (Fig. 8–41). A heat engine placed between the two tanks extracts heat from the high temperature tank, produces work, and rejects heat to the low temperature tank. Determine the maximum work that can be produced by the heat engine and the final temperatures of the tanks. Assume constant specific heats at room temperature.

SOLUTION A heat engine operates between two tanks filled with air at different temperatures. The maximum work that can be produced and the final temperature of the tanks are to be determined.
Assumptions Air is an ideal gas with constant specific heats at room temperature.

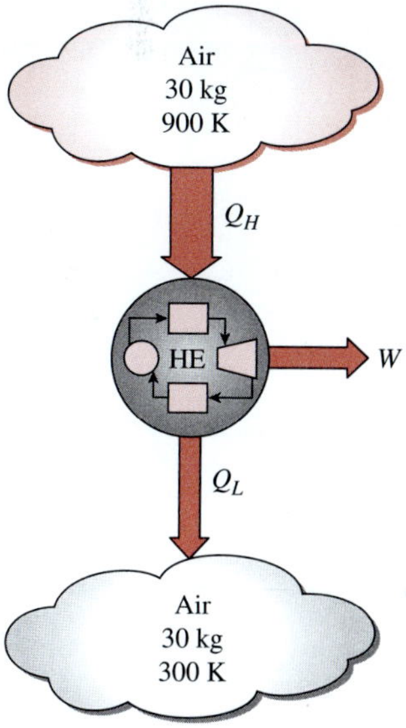

FIGURE 8–41
Schematic for Example 8-14

Properties The gas constant of air is 0.287 kPa·m³/kg·K (Table A-1). The constant volume specific heat of air at room temperature is $c_v = 0.718$ kJ/kg·K (Table A-2*a*)

Analysis For maximum work production, the process must be reversible and thus the entropy generation must be zero. We take the two tanks (the heat source and heat sink) and the heat engine as the system. Noting that the system involves no heat and mass transfer and that the entropy change for cyclic devices is zero, the entropy balance can be expressed as

$$\underbrace{S_{in} - S_{out}}_{\substack{\text{Net entropy transfer}\\ \text{by heat and mass}}} + \underbrace{S_{gen}^{\nearrow 0}}_{\substack{\text{Entropy}\\ \text{generation}}} = \underbrace{\Delta S_{system}}_{\substack{\text{Change}\\ \text{in entropy}}}$$

$$0 + S_{gen}^{\nearrow 0} = \Delta S_{tank,source} + \Delta S_{tank,sink} + \Delta S_{heat\ engine}^{\nearrow 0}$$

$$\Delta S_{tank,source} + \Delta S_{tank,sink} = 0$$

$$\left(mc_v \ln\frac{T_2}{T_1} + mR\ln\frac{V_2}{V_1}^{\nearrow 0}\right)_{source} + \left(mc_v \ln\frac{T_2}{T_1} + mR\ln\frac{V_2}{V_1}^{\nearrow 0}\right)_{sink} = 0$$

$$\ln\frac{T_2}{T_{1,A}}\frac{T_2}{T_{1,B}} = 0 \rightarrow T_2^2 = T_{1,A}T_{1,B}$$

where $T_{1,A}$ and $T_{1,B}$ are the initial temperatures of the source and the sink, respectively, and T_2 is the common final temperature. Therefore, the final temperature of the tanks for maximum power production is

$$T_2 = \sqrt{T_{1,A}T_{1,B}} = \sqrt{(900\text{K})(300\text{K})} = \mathbf{519.6K}$$

The energy balance $E_{in} - E_{out} = \Delta E_{system}$ for the source and sink can be expressed as follows:

Source:

$$-Q_{source,out} = \Delta U = mc_v(T_2 - T_{1A}) \rightarrow Q_{source,out} = mc_v(T_{1,A} - T_2)$$

$$Q_{source,out} = mc_v(T_{1,A} - T_2) = (30\text{ kg})(0.718\text{ kJ/kg·K})(900 - 519.6)\text{K} = 8193\text{ kJ}$$

Sink:

$$Q_{sink,in} = mc_v(T_2 - T_{1,B}) = (30\text{ kg})(0.718\text{ kJ/kg·K})(519.6 - 300)\text{ K} = 4731\text{ kJ}$$

Then the work produced in this case becomes

$$W_{max,out} = Q_H - Q_L = Q_{source,out} - Q_{sink,\,in} = 8193 - 4731 = \mathbf{3463\,kJ}$$

Discussion Note that 3463 kJ of the 8193 kJ heat transferred from the source can be converted to work, and is the best that can be done. This corresponds to a first law efficiency of 3463/8193 = 0.423 or 42.3 percent, but to a second-law efficiency of 100 percent since the process involves no entropy generation and thus no exergy destruction.

8–8 ▪ EXERGY BALANCE: CONTROL VOLUMES

The exergy balance relations for control volumes differ from those for closed systems in that they involve one more mechanism of exergy transfer: *mass flow across the boundaries.* As mentioned earlier, mass possesses exergy as well as energy and entropy, and the amounts of these three extensive properties are proportional to the amount of mass (Fig. 8–42). Again taking the positive direction of heat transfer to be to the system and the positive direction of work transfer to be from the system, the general exergy balance relations (Eqs. 8–36 and 8–37) can be expressed for a control volume more explicitly as

$$X_{heat} - X_{work} + X_{mass,in} - X_{mass,out} - X_{destroyed} = (X_2 - X_1)_{CV} \quad \textbf{(8–44)}$$

or

$$\sum\left(1 - \frac{T_0}{T_k}\right)Q_k - [W - P_0(V_2 - V_1)] + \sum_{in} m\psi - \sum_{out} m\psi - X_{destroyed} = (X_2 - X_1)_{CV} \quad \textbf{(8–45)}$$

It can also be expressed in the **rate form** as

$$\sum\left(1 - \frac{T_0}{T_k}\right)\dot{Q}_k - \left(\dot{W} - P_0\frac{dV_{CV}}{dt}\right) + \sum_{in} \dot{m}\psi - \sum_{out} \dot{m}\psi - \dot{X}_{destroyed} = \frac{dX_{CV}}{dt} \quad \textbf{(8–46)}$$

The exergy balance relation above can be stated as *the rate of exergy change within the control volume during a process is equal to the rate of net exergy transfer through the control volume boundary by heat, work, and mass flow minus the rate of exergy destruction within the boundaries of the control volume.*

When the initial and final states of the control volume are specified, the exergy change of the control volume is $X_2 - X_1 = m_2\phi_2 - m_1\phi_1$.

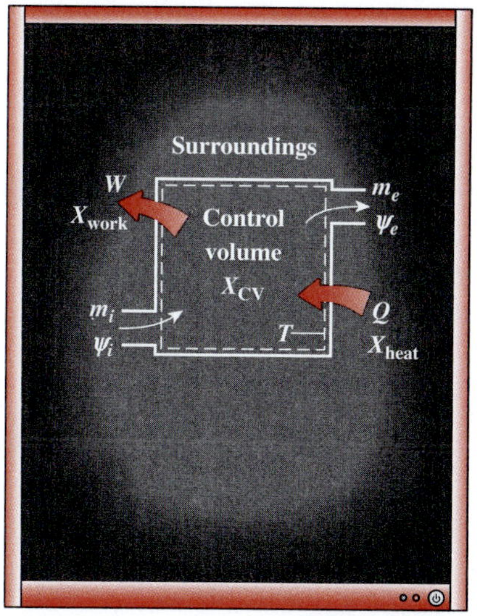

FIGURE 8–42
Exergy is transferred into or out of a control volume by mass as well as heat and work transfer.

Exergy Balance for Steady-Flow Systems

Most control volumes encountered in practice such as turbines, compressors, nozzles, diffusers, heat exchangers, pipes, and ducts operate steadily, and thus they experience no changes in their mass, energy, entropy, and exergy contents as well as their volumes. Therefore, $dV_{CV}/dt = 0$ and $dX_{CV}/dt = 0$ for such systems, and the amount of exergy entering a steady-flow system in all forms (heat, work, mass transfer) must be equal to the amount of exergy leaving plus the exergy destroyed. Then the rate form of the general exergy balance (Eq. 8–46) reduces for a **steady-flow process** to (Fig. 8–43)

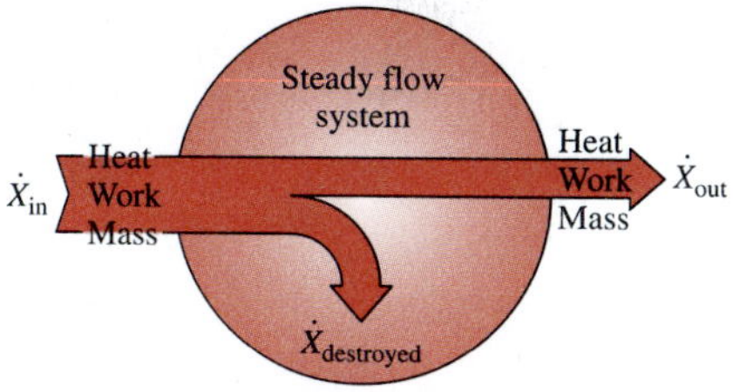

FIGURE 8–43
The exergy transfer to a steady-flow system is equal to the exergy transfer from it plus the exergy destruction within the system.

Steady-flow:
$$\sum\left(1 - \frac{T_0}{T_k}\right)\dot{Q}_k - \dot{W} + \sum_{in} \dot{m}\psi - \sum_{out} \dot{m}\psi - \dot{X}_{destroyed} = 0 \quad \textbf{(8–47)}$$

For a *single-stream* (one-inlet, one-exit) steady-flow device, the relation above further reduces to

Single-stream:
$$\sum\left(1 - \frac{T_0}{T_k}\right)\dot{Q}_k - \dot{W} + \dot{m}(\psi_1 - \psi_2) - \dot{X}_{destroyed} = 0 \quad \textbf{(8–48)}$$

where the subscripts 1 and 2 represent inlet and exit states, $\dot{m}$ is the mass flow rate, and the change in the flow exergy is given by Eq. 8–23 as

$$\psi_1 - \psi_2 = (h_1 - h_2) - T_0(s_1 - s_2) + \frac{V_1^2 - V_2^2}{2} + g(z_1 - z_2)$$

Dividing Eq. 8–48 by $\dot{m}$ gives the exergy balance on a *unit-mass basis* as

Per-unit mass:
$$\sum\left(1-\frac{T_0}{T_k}\right)q_k - w + (\psi_1-\psi_2) - x_{\text{destroyed}} = 0 \qquad \text{(kJ/kg)} \tag{8–49}$$

where q = $\dot{Q}/\dot{m}$ and w = $\dot{W}/\dot{m}$ are the heat transfer and work done per unit mass of the working fluid, respectively.

For the case of an adiabatic single-stream device with no work interactions, the exergy balance relation further simplifies to $\dot{X}_{\text{destroyed}} = \dot{m}(\psi_1 - \psi_2)$, which indicates that the specific exergy of the fluid must decrease as it flows through a work-free adiabatic device or remain the same ($\psi_2 = \psi_1$) in the limiting case of a reversible process regardless of the changes in other properties of the fluid.

Reversible Work, W_{rev}

The exergy balance relations presented above can be used to determine the reversible work W_{rev} by setting the exergy destroyed equal to zero. The work W in that case becomes the reversible work. That is,

General:
$$W = W_{\text{rev}} \qquad \text{when } X_{\text{destroyed}} = 0 \tag{8–50}$$

For example, the reversible power for a single-stream steady-flow device is, from Eq. 8–48,

Single stream:
$$\dot{W}_{\text{rev}} = \dot{m}(\psi_1-\psi_2) + \sum\left(1-\frac{T_0}{T_k}\right)\dot{Q}_k \qquad \text{(kW)} \tag{8–51}$$

which reduces for an adiabatic device to

Adiabatic, single stream:
$$\dot{W}_{\text{rev}} = \dot{m}(\psi_1-\psi_2) \tag{8–52}$$

Note that the exergy destroyed is zero only for a reversible process, and reversible work represents the maximum work output for work-producing devices such as turbines and the minimum work input for work-consuming devices such as compressors.

Second-Law Efficiency of Steady-Flow Devices, η_{II}

The *second-law efficiency* of various steady-flow devices can be determined from its general definition, η_{II} = (Exergy recovered)/(Exergy expended). When the changes in kinetic and potential energies are negligible, the second-law efficiency of an *adiabatic turbine* can be determined from

$$\eta_{\text{II,turb}} = \frac{w_{\text{out}}}{\psi_1-\psi_2} = \frac{h_1-h_2}{\psi_1-\psi_2} = \frac{w_{\text{out}}}{w_{\text{rev,out}}} \quad \text{or} \quad \eta_{\text{II,turb}} = 1 - \frac{T_0 s_{\text{gen}}}{\psi_1-\psi_2} \tag{8–53}$$

where $s_{\text{gen}} = s_2 - s_1$. For an *adiabatic compressor* with negligible kinetic and potential energies, the second-law efficiency becomes

$$\eta_{\text{II,comp}} = \frac{\psi_2-\psi_1}{w_{\text{in}}} = \frac{\psi_2-\psi_1}{h_2-h_1} = \frac{w_{\text{in,rev}}}{w_{\text{in}}} \quad \text{or} \quad \eta_{\text{II,comp}} = 1 - \frac{T_0 s_{\text{gen}}}{h_2-h_1} \tag{8–54}$$

where again $s_{\text{gen}} = s_2 - s_1$. Note that in the case of turbine, the exergy resource utilized is steam, and the expended exergy is simply the decrease in the exergy of the steam. The recovered exergy is the turbine shaft work. In the case of compressor, the exergy resource is mechanical work, and the expended exergy is the work consumed by the compressor. The recovered exergy in this case is the increase in the exergy of the compressed fluid.

For an adiabatic *heat exchanger* with two unmixed fluid streams (Fig. 8–44), the exergy expended is the decrease in the exergy of the hot stream, and the exergy recovered is the increase in the exergy of the cold stream, provided that the cold stream is not at a lower temperature than the surroundings. Then the second-law efficiency of the heat exchanger becomes

$$\eta_{\text{II,HX}} = \frac{\dot{m}_{\text{cold}}(\psi_4 - \psi_3)}{\dot{m}_{\text{hot}}(\psi_1 - \psi_2)} \quad \text{or} \quad \eta_{\text{II,HX}} = 1 - \frac{T_0 \dot{S}_{\text{gen}}}{\dot{m}_{\text{hot}}(\psi_1 - \psi_2)} \tag{8–55}$$

where $\dot{S}_{\text{gen}} = \dot{m}_{\text{hot}}(s_2 - s_1) + \dot{m}_{\text{cold}}(s_4 - s_3)$. Perhaps you are wondering what happens if the heat exchanger is not adiabatic; that is, it is losing some heat to its surroundings at T_0. If the temperature of the boundary (the outer surface of the heat exchanger) T_b is equal T_0, the definition above still holds (except the entropy generation term needs to be modified if the second definition is used). However, if $T_b > T_0$, then the exergy of the lost heat at the boundary should be included in the recovered exergy:

$$\eta_{\text{II,HX}} = \frac{\dot{m}_{\text{cold}}(\psi_4 - \psi_3) + \dot{Q}_{\text{loss}}(1 - T_0/T_b)}{\dot{m}_{\text{hot}}(\psi_1 - \psi_2)} = 1 - \frac{T_0 \dot{S}_{\text{gen}}}{\dot{m}_{\text{hot}}(\psi_1 - \psi_2)} \tag{8–56}$$

where T_b is the temperature of the system boundary through which the lost heat crosses at a rate of $\dot{Q}_{\text{loss}}$. Also, $\dot{S}_{\text{gen}} = \dot{m}_{\text{hot}}\ (s_2 - s_1) + \dot{m}_{\text{cold}}(s_4 - s_3) + \dot{Q}_{\text{loss}}/T_b$ in this case.

Although no attempt is made in practice to utilize this exergy associated with lost heat and it is allowed to be destroyed, the heat exchanger should not be held responsible for this destruction, which occurs outside its boundaries. If we are interested in the exergy destroyed during the process, not just within the boundaries of the device, then it makes sense to consider an *extended system* that includes the immediate surroundings of the device such that the boundaries of the new enlarged system are at T_0. The second-law efficiency of the extended system reflects the effects of the irreversibilities that occur within and just outside the device.

An interesting situation arises when the temperature of the cold stream remains below the temperature of the surroundings at all times. In that case the exergy of the cold stream actually decreases instead of increasing. In such cases it is better to define the second-law efficiency as the ratio of the sum of the exergies of the outgoing streams to the sum of the exergies of the incoming streams.

For an adiabatic *mixing chamber* where a hot steam 1 is mixed with a cold stream 2, forming a mixture 3, the exergy resource is the hot fluid. Then the exergy expanded is the exergy decrease of the hot fluid, and the exergy recovered is the exegy increase of the cold fluid. Nothing that state 3 is the common state of the mixture, the second-law efficiency can be expressed as

$$\eta_{\text{II,mix}} = \frac{\dot{m}_{\text{cold}}(\psi_3 - \psi_2)}{\dot{m}_{\text{hot}}(\psi_1 - \psi_3)} \quad \text{or} \quad \eta_{\text{II,mix}} = 1 - \frac{T_0 \dot{S}_{\text{gen}}}{\dot{m}_{\text{hot}}(\psi_1 - \psi_3)} \tag{8–57}$$

where $\dot{S}_{\text{gen}} = (\dot{m}_{\text{hot}} + \dot{m}_{\text{cold}})s_3 - \dot{m}_{\text{hot}}s_1 - \dot{m}_{\text{cold}}s_2$.

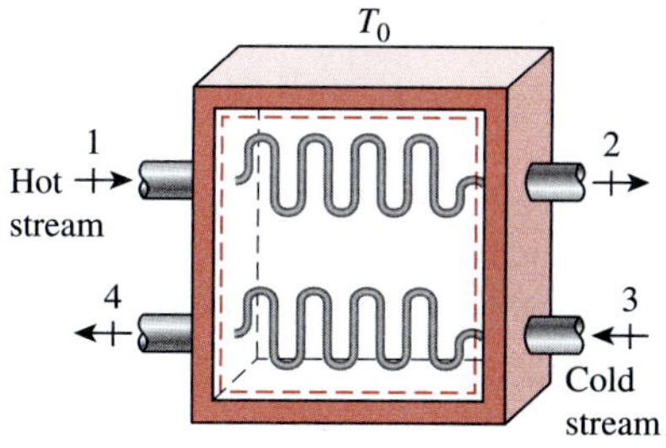

FIGURE 8–44
A heat exchanger with two unmixed fluid streams.

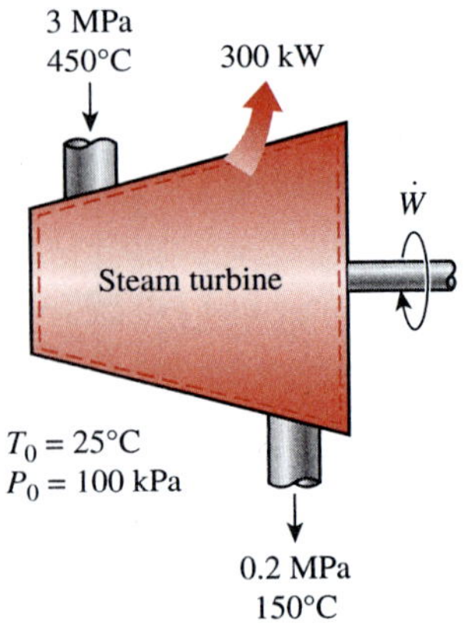

FIGURE 8–45
Schematic for Example 8–15.

EXAMPLE 8–15 Second-Law Analysis of a Steam Turbine

Steam enters a turbine steadily at 3 MPa and 450°C at a rate of 8 kg/s and exits at 0.2 MPa and 150°C, (Fig. 8–45). The steam is losing heat to the surrounding air at 100 kPa and 25°C at a rate of 300 kW, and the kinetic and potential energy changes are negligible. Determine (*a*) the actual power output, (*b*) the maximum possible power output, (*c*) the second-law efficiency, (*d*) the exergy destroyed, and (*e*) the exergy of the steam at the inlet conditions.

SOLUTION A steam turbine operating steadily between specified inlet and exit states is considered. The actual and maximum power outputs, the second-law efficiency, the exergy destroyed, and the inlet exergy are to be determined.

Assumptions **1** This is a steady-flow process since there is no change with time at any point and thus $\Delta m_{CV} = 0$, $\Delta E_{CV} = 0$, and $\Delta X_{CV} = 0$. **2** The kinetic and potential energies are negligible.

Analysis We take the *turbine* as the system. This is a *control volume* since mass crosses the system boundary during the process. We note that there is only one inlet and one exit and thus $\dot{m}_1 = \dot{m}_2 = \dot{m}$. Also, heat is lost to the surrounding air and work is done by the system.

The properties of the steam at the inlet and exit states and the state of the environment are

$$\textit{Inlet state:}\quad \left.\begin{array}{l} P_1 = 3\text{ MPa} \\ T_1 = 450°\text{C} \end{array}\right\} \quad \begin{array}{l} h_1 = 3344.9\text{ kJ/kg} \\ s_1 = 7.0856\text{ kJ/kg·K} \end{array} \qquad \text{(Table A–6)}$$

$$\textit{Exit state:}\quad \left.\begin{array}{l} P_2 = 0.2\text{ MPa} \\ T_2 = 150°\text{C} \end{array}\right\} \quad \begin{array}{l} h_2 = 2769.1\text{ kJ/kg} \\ s_2 = 7.2810\text{ kJ/kg·K} \end{array} \qquad \text{(Table A–6)}$$

$$\textit{Dead state:}\quad \left.\begin{array}{l} P_0 = 100\text{ kPa} \\ T_0 = 25°\text{C} \end{array}\right\} \quad \begin{array}{l} h_0 \cong h_{f\,@\,25°\text{C}} = 104.83\text{ kJ/kg} \\ s_0 \cong s_{f\,@\,25°\text{C}} = 0.3672\text{ kJ/kg·K} \end{array} \qquad \text{(Table A–4)}$$

(*a*) The actual power output of the turbine is determined from the rate form of the energy balance,

$$\underbrace{\dot{E}_{in} - \dot{E}_{out}}_{\substack{\text{Rate of net energy transfer}\\ \text{by heat, work, and mass}}} = \underbrace{dE_{system}/dt}_{\substack{\text{Rate of change in internal, kinetic,}\\ \text{potential, etc., energies}}}^{\nearrow 0\text{ (steady)}} = 0$$

$$\dot{E}_{in} = \dot{E}_{out}$$

$$\dot{m}h_1 = \dot{W}_{out} + \dot{Q}_{out} + \dot{m}h_2 \qquad (\text{since ke} \cong \text{pe} \cong 0)$$

$$\dot{W}_{out} = \dot{m}(h_1 - h_2) - \dot{Q}_{out}$$

$$= (8\text{ kg/s})[(3344.9 - 2769.1)\text{ kJ/kg}] - 300\text{ kW}$$

$$= \mathbf{4306\text{ kW}}$$

(*b*) The maximum power output (reversible power) is determined from the rate form of the exergy balance applied on the *extended system* (system + immediate surroundings), whose boundary is at the environment temperature of T_0, and by setting the exergy destruction term equal to zero,

$$\underbrace{\dot{X}_{in} - \dot{X}_{out}}_{\substack{\text{Rate of net exergy transfer}\\ \text{by heat, work, and mass}}} - \underbrace{\dot{X}_{destroyed}}_{\substack{\text{Rate of exergy}\\ \text{destruction}}}^{\nearrow 0\text{ (reversible)}} = \underbrace{dX_{system}/dt}_{\substack{\text{Rate of change}\\ \text{in exergy}}}^{\nearrow 0\text{ (steady)}} = 0$$

$$\dot{X}_{in} = \dot{X}_{out}$$

$$\dot{m}\psi_1 = \dot{W}_{rev,out} + \dot{X}_{heat}^{\nearrow 0} + \dot{m}\psi_2$$

$$\dot{W}_{rev,out} = \dot{m}(\psi_1 - \psi_2)$$

$$= \dot{m}[(h_1 - h_2) - T_0(s_1 - s_2) - \Delta ke^{\nearrow 0} - \Delta pe^{\nearrow 0}]$$

Note that exergy transfer with heat is zero when the temperature at the point of transfer is the environment temperature T_0. Substituting,

$$\dot{W}_{rev,out} = (8 \text{ kg/s})[(3344.9 - 2769.1) \text{ kJ/kg} - (298 \text{ K})(7.0856 - 7.2810)\text{kJ/kg·K}]$$

$$= \mathbf{5072 \text{ kW}}$$

(*c*) The second-law efficiency of a turbine is the ratio of the actual work delivered to the reversible work,

$$\eta_{II} = \frac{\dot{W}_{out}}{\dot{W}_{rev,out}} = \frac{4306 \text{ kW}}{5072 \text{ kW}} = \mathbf{0.849 \text{ or } 84.9\%}$$

That is, 15.1 percent of the work potential is wasted during this process.

(*d*) The difference between the reversible work and the actual useful work is the exergy destroyed, which is determined to be

$$\dot{X}_{destroyed} = \dot{W}_{rev,out} - \dot{W}_{out} = 5072 - 4306 = \mathbf{776 \text{ kW}}$$

That is, the potential to produce useful work is wasted at a rate of 776 kW during this process. The exergy destroyed could also be determined by first calculating the rate of entropy generation $\dot{S}_{gen}$ during the process.

(*e*) The exergy (maximum work potential) of the steam at the inlet conditions is simply the stream exergy, and is determined from

$$\psi_1 = (h_1 - h_0) - T_0(s_1 - s_0) + \frac{V_1^2}{2}^{\nearrow 0} + gz_1^{\nearrow 0}$$

$$= (h_1 - h_0) - T_0(s_1 - s_0)$$

$$= (3344.9 - 104.83)\text{kJ/kg} - (298 \text{ K})(7.0856 - 0.3672) \text{ kJ/kg·K}$$

$$= \mathbf{1238 \text{ kJ/kg}}$$

That is, not counting the kinetic and potential energies, every kilogram of the steam entering the turbine has a work potential of 1238 kJ. This corresponds to a power potential of (8 kg/s)(1238 kJ/kg) = 9904 kW. Obviously, the turbine is converting 4306/9904 = 43.5 percent of the available work potential of the steam to work.

EXAMPLE 8–16 Exergy Destroyed During Mixing of Fluid Streams

Water at 20 psia and 50°F enters a mixing chamber at a rate of 300 lbm/min, where it is mixed steadily with steam entering at 20 psia and 240°F. The mixture leaves the chamber at 20 psia and 130°F, and heat is being lost to the surrounding air at T_0 = 70°F at a rate of 180 Btu/min (Fig. 8–46).

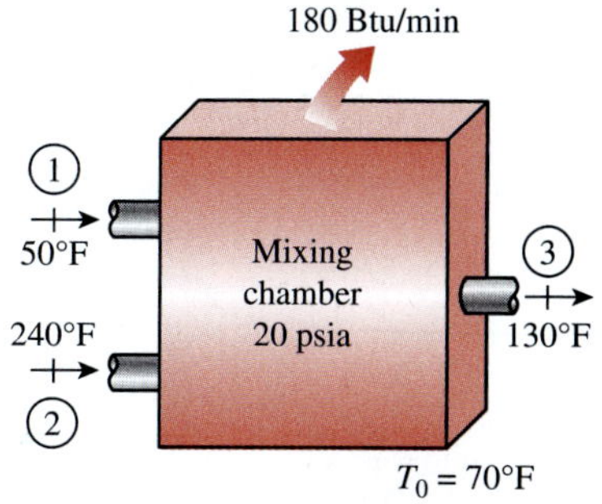

FIGURE 8–46
Schematic for Example 8–16.

Neglecting the changes in kinetic and potential energies, determine the reversible power and the rate of exergy destruction for this process.

SOLUTION Liquid water and steam are mixed in a chamber that is losing heat at a specified rate. The reversible power and the rate of exergy destruction are to be determined.

Assumptions **1** This is a steady-flow process since there is no change with time at any point and thus $\Delta m_{cv} = 0$, $\Delta E_{cv} = 0$, and $\Delta S_{cv} = 0$. **2** There are no work interactions involved. **3** The kinetic and potential energies are negligible, ke $\cong$ pe $\cong$ 0.

Analysis We take the *mixing chamber* as the system (Fig. 8–46). This is a *control volume* since mass crosses the system boundary during the process. We note that there are two inlets and one exit.

Under the stated assumptions and observations, the mass and energy balances for this steady-flow system can be expressed in the rate form as follows:

$$\text{Mass balance:} \quad \dot{m}_{in} - \dot{m}_{out} = dm_{system}/dt \overset{\nearrow 0 \text{ (steady)}}{} = 0 \rightarrow \dot{m}_1 + \dot{m}_2 = \dot{m}_3$$

$$\text{Energy balance:} \quad \underbrace{\dot{E}_{in} - \dot{E}_{out}}_{\substack{\text{Rate of net energy transfer}\\ \text{by heat, work, and mass}}} = \underbrace{dE_{system}/dt \overset{\nearrow 0 \text{ (steady)}}{}}_{\substack{\text{Rate of change in internal, kinetic,}\\ \text{potential, etc., energies}}} = 0$$

$$\dot{E}_{in} = \dot{E}_{out}$$

$$\dot{m}_1 h_1 + \dot{m}_2 h_2 = \dot{m}_3 h_3 + Q_{out} \ (\text{since } \dot{W} = 0 \text{ ke } \cong \text{pe} \cong 0)$$

Combining the mass and energy balances gives

$$\dot{Q}_{out} = \dot{m}_1 h_1 + \dot{m}_2 h_2 - (\dot{m}_1 + \dot{m}_2) h_3$$

The desired properties at the specified states are determined from the steam tables to be

$$\text{State 1:} \quad \left.\begin{array}{l} P_1 = 20 \text{ psia} \\ T_1 = 50°\text{F} \end{array}\right\} \quad \begin{array}{l} h_1 = h_{f\,@\,50°\text{F}} = 18.07 \text{ Btu/lbm} \\ s_1 = s_{f\,@\,50°\text{F}} = 0.03609 \text{ Btu/lbm·R} \end{array}$$

$$\text{State 2:} \quad \left.\begin{array}{l} P_2 = 20 \text{ psia} \\ T_2 = 240°\text{F} \end{array}\right\} \quad \begin{array}{l} h_2 = 1162.3 \text{ Btu/lbm} \\ s_2 = 1.7406 \text{ Btu/lbm·R} \end{array}$$

$$\text{State 3:} \quad \left.\begin{array}{l} P_3 = 20 \text{ psia} \\ T_3 = 130°\text{C} \end{array}\right\} \quad \begin{array}{l} h_3 = h_{f\,@\,130°\text{F}} = 97.99 \text{ Btu/lbm} \\ s_3 = s_{f\,@\,130°\text{F}} = 0.08174 \text{ Btu/lbm·R} \end{array}$$

Substituting,

$$180 \text{ Btu/min} = [300 \times 18.07 + \dot{m}_2 \times 1162.3 - (310 + \dot{m}_2) \times 97.99] \text{ Btu/min}$$

which gives

$$\dot{m}_2 = 22.7 \text{ lbm/min}$$

The maximum power output (reversible power) is determined from the rate form of the exergy balance applied on the *extended system* (system + immediate surroundings), whose boundary is at the environment temperature of T_0, and by setting the exergy destruction term equal to zero,

$$\underbrace{\dot{X}_{in} - \dot{X}_{out}}_{\substack{\text{Rate of net exergy transfer}\\ \text{by heat, work, and mass}}} - \underbrace{\dot{X}_{destroyed} \overset{\nearrow 0 \text{ (reversible)}}{}}_{\substack{\text{Rate of exergy}\\ \text{destruction}}} = \underbrace{dX_{system}/dt \overset{\nearrow 0 \text{ (steady)}}{}}_{\substack{\text{Rate of change}\\ \text{in exergy}}} = 0$$

$$\dot{X}_{in} = \dot{X}_{out}$$

$$\dot{m}_1\psi_1 + \dot{m}_2\psi_2 = \dot{W}_{rev,out} + \dot{X}_{heat}^{\nearrow 0} + \dot{m}_3\psi_3$$

$$\dot{W}_{rev,out} = \dot{m}_1\psi_1 + \dot{m}_2\psi_2 - \dot{m}_3\psi_3$$

Note that exergy transfer by heat is zero when the temperature at the point of transfer is the environment temperature T_0, and the kinetic and potential energies are negligible. Therefore,

$$\begin{aligned}\dot{W}_{rev,out} &= \dot{m}_1(h_1 - T_0s_1) + \dot{m}_2(h_2 - T_0s_2) - \dot{m}_3(h_3 - T_0s_3)\\ &= (300\ \text{lbm/min})[18.07\ \text{Btu/lbm} - (530\ \text{R})(0.03609\ \text{Btu/lbm·R})]\\ &\quad + (22.7\ \text{lbm/min})[1162.3\ \text{Btu/lbm} - (530\ \text{R})(1.7406\ \text{Btu/lbm·R})]\\ &\quad - (322.7\ \text{lbm/min})[97.99\ \text{Btu/lbm} - (530\ \text{R})(0.18174\ \text{Btu/lbm·R})]\\ &= \mathbf{4588\ Btu/min}\end{aligned}$$

That is, we could have produced work at a rate of 4588 Btu/min if we ran a heat engine between the hot and the cold fluid streams instead of allowing them to mix directly.

The exergy destroyed is determined from

$$\dot{X}_{destroyed} = \dot{W}_{rev,out} - \dot{W}_u^{\nearrow 0} = T_0\dot{S}_{gen}$$

Thus,

$$\dot{X}_{destroyed} = \dot{W}_{rev,out} = \mathbf{4588\ Btu/min}$$

since there is no actual work produced during the process (Fig. 8–47).

Discussion The entropy generation rate for this process can be shown to be $\dot{S}_{gen}$ = 8.65 Btu/min·R. Thus the exergy destroyed could also be determined from the second part of the above equation:

$$\dot{X}_{destroyed} = T_0\dot{S}_{gen} = (530\ \text{R})(8.65\ \text{Btu/min·R}) = 4585\ \text{Btu/min}$$

The slight difference between the two results is due to roundoff error.

FIGURE 8–47
For systems that involve no actual work, the reversible work and irreversibility are identical.
BLONDIE © KING FEATURES SYNDICATE.

EXAMPLE 8–17 Charging a Compressed Air Storage System

A 200-m³ rigid tank initially contains atmospheric air at 100 kPa and 300 K and is to be used as a storage vessel for compressed air at 1 MPa and 300 K (Fig. 8–48). Compressed air is to be supplied by a compressor that takes in atmospheric air at P_0 = 100 kPa and T_0 = 300 K. Determine the minimum work requirement for this process.

SOLUTION Air is to be compressed and stored at high pressure in a large tank. The minimum work required is to be determined.

Assumptions **1** Air is an ideal gas. **2** The kinetic and potential energies are negligible. **3** The properties of air at the inlet remain constant during the entire charging process.

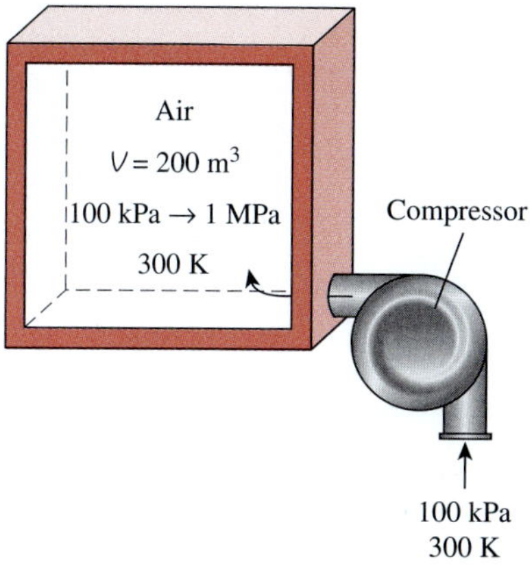

FIGURE 8–48
Schematic for Example 8–17.

Analysis We take the *rigid tank combined with the compressor* as the system. This is a *control volume* since mass crosses the system boundary during the process. We note that this is an unsteady-flow process since the mass content of the system changes as the tank is charged. Also, there is only one inlet and no exit.

The minimum work required for a process is the *reversible work,* which can be determined from the exergy balance applied on the *extended system* (system + immediate surroundings) whose boundary is at the environment temperature of T_0 (so that there is no exergy transfer accompanying heat transfer to or from the environment) and by setting the exergy destruction term equal to zero,

$$\underbrace{X_{\text{in}} - X_{\text{out}}}_{\substack{\text{Net exergy transfer}\\ \text{by heat, work, and mass}}} - \underbrace{X_{\text{destroyed}}^{\nearrow 0 \text{ (reversible)}}}_{\substack{\text{Exergy}\\ \text{destruction}}} = \underbrace{\Delta X_{\text{system}}}_{\substack{\text{Change}\\ \text{in exergy}}}$$

$$X_{\text{in}} - X_{\text{out}} = X_2 - X_1$$

$$W_{\text{rev,in}} + m_1\psi_1^{\nearrow 0} = m_2\phi_2 - m_1\phi_1^{\nearrow 0}$$

$$W_{\text{rev,in}} = m_2\phi_2$$

Note that $\phi_1 = \psi_1 = 0$ since the initial air in the tank and the air entering are at the state of the environment, and the exergy of a substance at the state of the environment is zero. The final mass of air and the exergy of the pressurized air in the tank at the end of the process are

$$m_2 = \frac{P_2 V}{RT_2} = \frac{(1000 \text{ kPa})(200 \text{ m}^3)}{(0.287 \text{ kPa·m}^3/\text{kg·K})(300 \text{ K})} = 2323 \text{ kg}$$

$$\phi_2 = (u_2 - u_0)^{\nearrow 0 (\text{since } T_2 = T_0)} + P_0(v_2 - v_0) - T_0(s_2 - s_0) + \frac{V_2^2}{2}^{\nearrow 0} + gz_2^{\nearrow 0}$$
$$= P_0(v_2 - v_0) - T_0(s_2 - s_0)$$

We note that

$$P_0(v_2 - v_0) = P_0\left(\frac{RT_2}{P_2} - \frac{RT_0}{P_0}\right) = RT_0\left(\frac{P_0}{P_2} - 1\right) \qquad (\text{since } T_2 = T_0)$$

$$T_0(s_2 - s_0) = T_0\left(c_p \ln \frac{T_2}{T_0}^{\nearrow 0} - R \ln \frac{P_2}{P_0}\right) = -RT_0 \ln \frac{P_2}{P_0} \qquad (\text{since } T_2 = T_0)$$

Therefore,

$$\phi_2 = RT_0\left(\frac{P_0}{P_2} - 1\right) + RT_0 \ln \frac{P_2}{P_0} = RT_0\left(\ln \frac{P_2}{P_0} + \frac{P_0}{P_2} - 1\right)$$
$$= (0.287 \text{ kJ/kg·K})(300 \text{ K})\left(\ln \frac{1000 \text{ kPa}}{100 \text{ kPa}} + \frac{100 \text{ kPa}}{1000 \text{ kPa}} - 1\right)$$
$$= 120.76 \text{ kJ/kg}$$

and

$$W_{\text{rev,in}} = m_2\phi_2 = (2323 \text{ kg})(120.76 \text{ kJ/kg}) = 280{,}525 \text{ kJ} \cong \mathbf{281 \text{ MJ}}$$

Discussion Note that a minimum of 281 MJ of work input is required to fill the tank with compressed air at 300 K and 1 MPa. In reality, the required work input will be greater by an amount equal to the exergy destruction during the process. Compare this to the result of Example 8–7. What can you conclude?

TOPIC OF SPECIAL INTEREST* Second-Law Aspects of Daily Life

Thermodynamics is a fundamental natural science that deals with various aspects of energy, and even nontechnical people have a basic understanding of energy and the first law of thermodynamics since there is hardly any aspect of life that does not involve the transfer or transformation of energy in different forms. All the *dieters,* for example, base their lifestyle on the conservation of energy principle. Although the first-law aspects of thermodynamics are readily understood and easily accepted by most people, there is not a public awareness about the second law of thermodynamics, and the second-law aspects are not fully appreciated even by people with technical backgrounds. This causes some students to view the second law as something that is of theoretical interest rather than an important and practical engineering tool. As a result, students show little interest in a detailed study of the second law of thermodynamics. This is unfortunate because the students end up with a one-sided view of thermodynamics and miss the balanced, complete picture.

Many *ordinary events* that go unnoticed can serve as excellent vehicles to convey important concepts of thermodynamics. Below we attempt to demonstrate the relevance of the second-law concepts such as exergy, reversible work, irreversibility, and the second-law efficiency to various aspects of daily life using examples with which even nontechnical people can identify. Hopefully, this will enhance our understanding and appreciation of the second law of thermodynamics and encourage us to use it more often in technical and even nontechnical areas. The critical reader is reminded that the concepts presented below are *soft* and *difficult to quantize,* and that they are offered here to stimulate interest in the study of the second law of thermodynamics and to enhance our understanding and appreciation of it.

The second-law concepts are implicitly used in various aspects of daily life. Many successful people seem to make extensive use of them without even realizing it. There is growing awareness that quality plays as important a role as quantity in even ordinary daily activities. The following appeared in an article in the *Reno Gazette-Journal* on March 3, 1991:

> Dr. Held considers himself a survivor of the tick-tock conspiracy. About four years ago, right around his 40th birthday, he was putting in 21-hour days—working late, working out, taking care of his three children and getting involved in sports. He got about four or five hours of sleep a night. . . . "Now I'm in bed by 9:30 and I'm up by 6," he says. "I get twice as much done as I used to. I don't have to do things twice or read things three times before I understand them."

*This section can be skipped without a loss in continuity.

This statement has a strong relevance to the second-law discussions. It indicates that the problem is not how much time we have (the first law), but, rather, how effectively we use it (the second law). For a person to get *more done in less time* is no different than for a car to go *more miles on less fuel.*

In thermodynamics, *reversible work* for a process is defined as the maximum useful work output (or minimum work input) for that process. It is the useful work that a system would deliver (or consume) during a process between two specified states if that process is executed in a reversible (perfect) manner. The difference between the reversible work and the actual useful work is due to imperfections and is called *irreversibility* (the wasted work potential). For the special case of the final state being the dead state or the state of the surroundings, the reversible work becomes a maximum and is called the *exergy* of the system at the initial state. The irreversibility for a reversible or perfect process is zero.

The *exergy* of a person in daily life can be viewed as the best job that person can do under the most favorable conditions. The *reversible work* in daily life, on the other hand, can be viewed as the best job a person can do under some specified conditions. Then the difference between the reversible work and the actual work done under those conditions can be viewed as the *irreversibility* or the *exergy destroyed.* In engineering systems, we try to identify the major sources of irreversibilities and minimize them in order to maximize performance. In daily life, a person should do just that to maximize his or her performance.

The exergy of a person at a given time and place can be viewed as the maximum amount of work he or she can do at that time and place. Exergy is certainly difficult to quantify because of the interdependence of physical and intellectual capabilities of a person. The ability to perform physical and intellectual tasks simultaneously complicates things even further. *Schooling* and *training* obviously increase the exergy of a person. *Aging* decreases the physical exergy. Unlike most mechanical things, the exergy of human beings is a function of time, and the physical and/or intellectual exergy of a person goes to waste if it is not utilized at the time. A barrel of oil loses nothing from its exergy if left unattended for 40 years. However, a person will lose much of his or her entire exergy during that time period if he or she just sits back.

A hard-working farmer, for example, may make full use of his *physical exergy* but very little use of his *intellectual exergy.* That farmer, for example, could learn a foreign language or a science by listening to some educational CDs at the same time he is doing his physical work. This is also true for people who spend considerable time in the car commuting to work. It is hoped that some day we will be able to do exergy analysis for people and their activities. Such an analysis will point out the way for people to minimize their exergy destruction, and get more done in less time. Computers can perform several tasks at once. Why shouldn't human beings be able to do the same?

Children are born with different levels of *exergies* (talents) in different areas. Giving aptitude tests to children at an early age is simply an attempt to uncover the extent of their "hidden" exergies, or talents. The children are then directed to areas in which they have the greatest exergy. As adults, they are more likely to perform at high levels without stretching the limits if they are naturally fit to be in that area.

We can view the level of *alertness* of a person as his or her *exergy* for intellectual affairs. When a person is well-rested, the degree of alertness, and thus intellectual exergy, is at a maximum and this exergy decreases with time as the person gets tired, as illustrated in Fig. 8–49. Different tasks in daily life require different levels of intellectual exergy, and the difference between available and required alertness can be viewed as the *wasted alertness* or *exergy destruction.* To minimize exergy destruction, there should be a close match between available alertness and required alertness.

Consider a well-rested student who is planning to spend her next 4 h studying and watching a 2-h-long movie. From the *first-law* point of view, it makes no difference in what order these tasks are performed. But from the *second-law* point of view, it makes a lot of difference. Of these two tasks, studying requires more intellectual alertness than watching a movie does, and thus it makes thermodynamic sense to study first when the alertness is high and to watch the movie later when the alertness is lower, as shown in the figure. A student who does it backwards wastes a lot of alertness while watching the movie, as illustrated in Fig. 8–49, and she has to keep going back and forth while studying because of insufficient alertness, thus getting less done in the same time period.

In thermodynamics, *the first-law efficiency* (or thermal efficiency) of a heat engine is defined as the ratio of net work output to total heat input. That is, it is the fraction of the heat supplied that is converted to net work. In general, the first-law efficiency can be viewed as the ratio of the desired output to the required input. The first-law efficiency makes no reference to the *best possible performance,* and thus the first-law efficiency alone is not a realistic measure of performance. To overcome this deficiency, we defined the second-law efficiency, which is a measure of actual performance relative to the best possible performance under the same conditions. For heat engines, the second-law efficiency is defined as the ratio of the actual thermal efficiency to the maximum possible (reversible) thermal efficiency under the same conditions.

In daily life, the *first-law efficiency* or *performance* of a person can be viewed as the accomplishment of that person relative to the effort he or she puts in. The *second-law efficiency* of a person, on the other hand, can be viewed as the performance of that person relative to the best possible performance under the circumstances.

Happiness is closely related to the *second-law efficiency.* Small children are probably the happiest human beings because there is so little they can do, but they do it so well, considering their limited capabilities. That is, children have very high second-law efficiencies in their daily lives. The term "full life" also refers to second-law efficiency. A person is considered to have a full life, and thus a very high second-law efficiency, if he or she has utilized all of his or her abilities to the limit during a lifetime.

Even a person with some disabilities has to put in considerably more effort to accomplish what a physically fit person accomplishes. Yet, despite accomplishing less with more effort, the person with disabilities who gives an

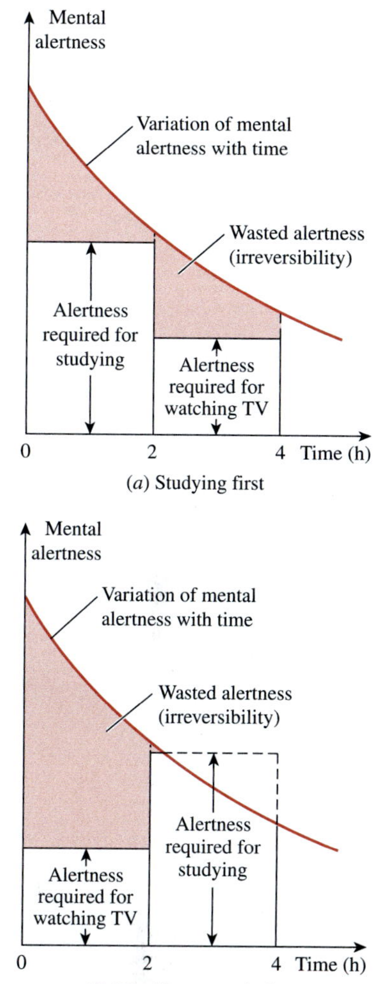

(*a*) Studying first

(*b*) Watching a movie first

FIGURE 8–49
The irreversibility associated with a student studying and watching a movie on television, each for two hours.

FIGURE 8–50
A poetic expression of exergy and exergy destruction.

impressive performance often gets more praise. Thus we can say that this person with disabilities had a low first-law efficiency (accomplishing little with a lot of effort) but a very high second-law efficiency (accomplishing as much as possible under the circumstances).

In daily life, exergy can also be viewed as the *opportunities that we have* and the exergy destruction as the *opportunities wasted.* Time is the biggest asset that we have, and the time wasted is the wasted opportunity to do something useful (Fig. 8–50).

The second law of thermodynamics also has interesting philosophical ramifications. Mass and energy are conserved quantities and are associated with the first law of thermodynamics, while entropy and exergy are non-conserved quantities and are associated with the second law. The universe we perceive through our five senses consists of conserved quantities, and thus we tend to view the non-conserved quantities as being non-real and even out of this universe. The widely accepted big bang theory about the origin of the universe gave rise to the notion that this is an all-material universe, and everything is made of matter (more correctly, mass-energy) only. As conserved quantities, mass and energy fit into the description of truly physical quantities, but entropy and exergy do not since entropy can be created and exergy can be destroyed. Thus entropy and exergy are not truly physical quanities although they are closely related to the physical quantities of mass and energy. Therefore, the second law deals with quantities that are of a different kind of existence—a universe in which things come into existence out of nothing and go out of existence into nothing—and opens up a universe that is beyond the conserved all-material universe we know of.

A similar argument can be given for the laws of nature that rule over matter. There is no question that both the first and the second laws of thermodynamics exist, and these and other laws like Newton's laws of motion govern the physical universe behind the scenes. As Alfred Montapert puts it, "*Nature's laws are the invisible government of the earth.*" Albert Einstein expresses this phenomenon as "*A spirit is manifest in the laws of the universe.*" Yet these laws that constitute the core of sciences cannot be sensed by our five senses and do not have a material existence, and thus they are not subject to the limitations of time and space. As such, the laws that seem to have infused all matter like a spirit rule everywhere, but they are not anywhere. It appears that quantities like entropy and exergy that come into existence out of nothing and go out of existence into nothing together with the laws of nature like the first and the second laws that govern the big-bang universe with an invisible powerful hand, are pointing the way for a broadened definition of existence that is more in line with perceived and observed phenomena.

The arguments presented here are exploratory in nature, and they are hoped to initiate some interesting discussions and research that may lead into better understanding of performance in various aspects of daily life. The second law may eventually be used to determine quantitatively the most effective way to improve the quality of life and performance in daily life, as it is presently used to improve the performance of engineering systems.

SUMMARY

The energy content of the universe is constant, just as its mass content is. Yet at times of crisis we are bombarded with speeches and articles on how to "conserve" energy. As engineers, we know that energy is already conserved. What is not conserved is *exergy,* which is the useful work potential of the energy. Once the exergy is wasted, it can never be recovered. When we use energy (to heat our homes for example), we are not destroying any energy; we are merely converting it to a less useful form, a form of less exergy.

The useful work potential of a system at the specified state is called *exergy.* Exergy is a property and is associated with the state of the system and the environment. A system that is in equilibrium with its surroundings has zero exergy and is said to be at the *dead state.* The exergy of heat supplied by thermal energy reservoirs is equivalent to the work output of a Carnot heat engine operating between the reservoir and the environment.

Reversible work W_{rev} is defined as the maximum amount of useful work that can be produced (or the minimum work that needs to be supplied) as a system undergoes a process between the specified initial and final states. This is the useful work output (or input) obtained when the process between the initial and final states is executed in a totally reversible manner. The difference between the reversible work W_{rev} and the useful work W_u is due to the irreversibilities present during the process and is called the *irreversibility I.* It is equivalent to the *exergy destroyed* and is expressed as

$$I = X_{destroyed} = T_0 S_{gen} = W_{rev,out} - W_{u,out} = W_{u,in} - W_{rev,in}$$

where S_{gen} is the entropy generated during the process. For a totally reversible process, the useful and reversible work terms are identical and thus exergy destruction is zero. Exergy destroyed represents the lost work potential and is also called the *wasted work* or *lost work.*

The *second-law efficiency* is a measure of the performance of a device relative to the performance under reversible conditions for the same end states and is given by

$$\eta_{II} = \frac{\eta_{th}}{\eta_{th,rev}} = \frac{W_u}{W_{rev}}$$

for heat engines and other work-producing devices and

$$\eta_{II} = \frac{COP}{COP_{rev}} = \frac{W_{rev}}{W_u}$$

for refrigerators, heat pumps, and other work-consuming devices. In general, the second-law efficiency is expressed as

$$\eta_{II} = \frac{\text{Exergy recovered}}{\text{Exergy expended}} = 1 - \frac{\text{Exergy destroyed}}{\text{exergy expended}}$$

The exergies of a fixed mass (nonflow exergy) and of a flow stream are expressed as

Nonflow exergy:

$$\phi = (u - u_0) + P_0(v - v_0) - T_0(s - s_0) + \frac{V^2}{2} + gz$$
$$= (e - e_0) + P_0(v - v_0) - T_0(s - s_0)$$

Flow exergy: $$\psi = (h - h_0) - T_0(s - s_0) + \frac{V^2}{2} + gz$$

Then the *exergy change* of a fixed mass or fluid stream as it undergoes a process from state 1 to state 2 is given by

$$\begin{aligned}\Delta X = X_2 - X_1 &= m(\phi_2 - \phi_1)\\ &= (E_2 - E_1) + P_0(V_2 - V_1) - T_0(S_2 - S_1)\\ &= (U_2 - U_1) + P_0(V_2 - V_1) - T_0(S_2 - S_1)\\ &\quad + m\frac{V_2^2 - V_1^2}{2} + mg(z_2 - z_1)\end{aligned}$$

$$\begin{aligned}\Delta\psi = \psi_2 - \psi_1 &= (h_2 - h_1) - T_0(s_2 - s_1)\\ &\quad + \frac{V_2^2 - V_1^2}{2} + g(z_2 - z_1)\end{aligned}$$

Exergy can be transferred by heat, work, and mass flow, and exergy transfer accompanied by heat, work, and mass transfer are given by

Exergy transfer by heat: $$X_{heat} = \left(1 - \frac{T_0}{T}\right)Q$$

Exergy transfer by work: $$X_{work} = \begin{cases} W - W_{surr} & \text{(for boundary work)} \\ W & \text{(for other forms of work)} \end{cases}$$

Exergy transfer by mass: $$X_{mass} = m\psi$$

The exergy of an isolated system during a process always decreases or, in the limiting case of a reversible process, remains constant. This is known as the *decrease of exergy principle* and is expressed as

$$\Delta X_{isolated} = (X_2 - X_1)_{isolated} \leq 0$$

Exergy balance for *any system* undergoing *any process* can be expressed as

General: $$\underbrace{X_{in} - X_{out}}_{\substack{\text{Net exergy transfer}\\\text{by heat, work, and mass}}} - \underbrace{X_{destroyed}}_{\substack{\text{Exergy}\\\text{destruction}}} = \underbrace{\Delta X_{system}}_{\substack{\text{Change}\\\text{in exergy}}}$$

General, rate form: $$\underbrace{\dot{X}_{in} - \dot{X}_{out}}_{\substack{\text{Rate of net exergy transfer}\\\text{by heat, work, and mass}}} - \underbrace{\dot{X}_{destroyed}}_{\substack{\text{Rate of exergy}\\\text{destruction}}} = \underbrace{dX_{system}/dt}_{\substack{\text{Rate of change}\\\text{in exergy}}}$$

General, unit-mass basis: $(x_{in} - x_{out}) - x_{destroyed} = \Delta x_{system}$

where

$$\dot{X}_{heat} = (1 - T_0/T)\dot{Q}$$

$$\dot{X}_{work} = \dot{W}_{useful}$$

$$\dot{X}_{mass} = \dot{m}\psi$$

For a *reversible process,* the exergy destruction term $X_{destroyed}$ drops out. Taking the positive direction of heat transfer to be to the system and the positive direction of work transfer to be from the system, the general exergy balance relations can be expressed more explicitly as

$$\sum\left(1 - \frac{T_0}{T_k}\right)Q_k - [W - P_0(V_2 - V_1)] + \sum_{in} m\psi - \sum_{out} m\psi - X_{destroyed} = X_2 - X_1$$

$$\sum\left(1 - \frac{T_0}{T_k}\right)\dot{Q}_k - \left(\dot{W} - P_0\frac{dV_{CV}}{dt}\right) + \sum_{in} \dot{m}\psi - \sum_{out} \dot{m}\psi - \dot{X}_{destroyed} = \frac{dX_{CV}}{dt}$$

REFERENCES AND SUGGESTED READINGS

1. J. E. Ahern. *The Exergy Method of Energy Systems Analysis.* New York: John Wiley & Sons, 1980.
2. A. Bejan. *Advanced Engineering Thermodynamics.* 2nd ed. New York: Wiley Interscience, 1997.
3. A. Bejan. *Entropy Generation through Heat and Fluid Flow.* New York: John Wiley & Sons, 1982.
4. Y. A. Çengel. "A Unified and Intuitive Approach to Teaching Thermodynamics." ASME International Congress and Exposition, Atlanta, Georgia, November 17–22, 1996.

PROBLEMS*

Exergy, Irreversibility, Reversible Work, and Second-Law Efficiency

8–1C Under what conditions does the reversible work equal irreversibility for a process?

8–2C What final state will maximize the work output of a device?

8–3C Is the exergy of a system different in different environments?

8–4C How does useful work differ from actual work? For what kind of systems are these two identical?

8–5C Consider a process that involves no irreversibilities. Will the actual useful work for that process be equal to the reversible work?

8–6C Consider two geothermal wells whose energy contents are estimated to be the same. Will the exergies of these wells necessarily be the same? Explain.

8–7C Consider two systems that are at the same pressure as the environment. The first system is at the same temperature as the environment, whereas the second system is at a lower temperature than the environment. How would you compare the exergies of these two systems?

8–8C Consider an environment of zero absolute pressure (such as outer space). How will the actual work and the useful work compare in that environment?

8–9C What is the second-law efficiency? How does it differ from the first-law efficiency?

8–10C Does a power plant that has a higher thermal efficiency necessarily have a higher second-law efficiency than one with a lower thermal efficiency? Explain.

8–11C Does a refrigerator that has a higher COP necessarily have a higher second-law efficiency than one with a lower COP? Explain.

8–12C Can a process for which the reversible work is zero be reversible? Can it be irreversible? Explain.

* Problems designated by a "C" are concept questions, and students are encouraged to answer them all. Problems designated by an "E" are in English units, and the SI users can ignore them. Problems with the icon are solved using EES, and complete solutions together with parametric studies are included on the enclosed DVD. Problems with the icon are comprehensive in nature, and are intended to be solved with a computer, preferably using the EES software that accompanies this text.

8–13C Consider a process during which no entropy is generated ($S_{gen} = 0$). Does the exergy destruction for this process have to be zero?

8–14 The electric power needs of a community are to be met by windmills with 20-m-diameter rotors. The windmills are to be located where the wind is blowing steadily at an average velocity of 6 m/s. Determine the minimum number of windmills that need to be installed if the required power output is 900 kW.

8–15E Saturated steam is generated in a boiler by converting a saturated liquid to a saturated vapor at 200 psia. This is done by transferring heat from the combustion gases, which are at 500°F, to the water in the boiler tubes. Calculate the wasted work potential associated with this heat transfer process. How does increasing the temperature of the combustion gases affect the work potential of steam stream? Take T_0 = 80°F and P_0 = 14.7 psia. *Answer:* 66.8 Btu/lbm

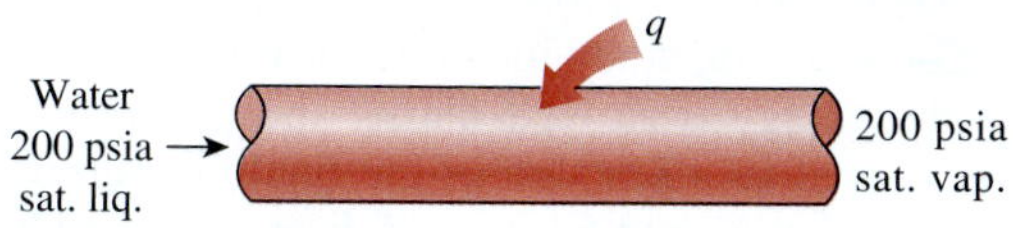

FIGURE P8–15E

8–16 One method of meeting the extra electric power demand at peak periods is to pump some water from a large body of water (such as a lake) to a water reservoir at a higher elevation at times of low demand and to generate electricity at times of high demand by letting this water run down and rotate a turbine (i.e., convert the electric energy to potential energy and then back to electric energy). For an energy storage capacity of 5×10^6 kWh, determine the minimum amount of water that needs to be stored at an average elevation (relative to the ground level) of 75 m. *Answer:* 2.45×10^{10} kg

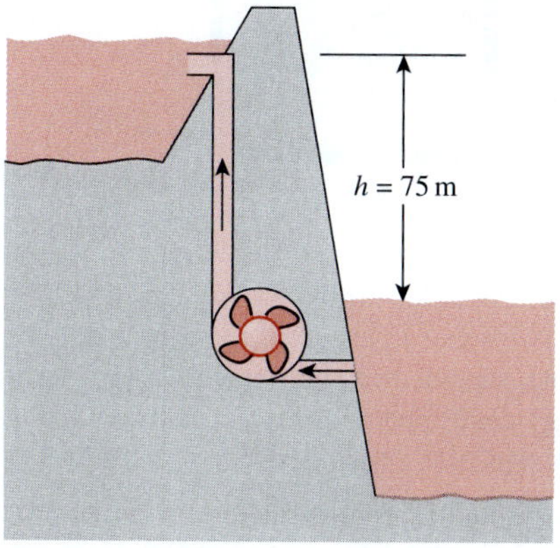

FIGURE P8–16

8–17 How much of the 100 kJ of thermal energy at 800 K can be converted to useful work? Assume the environment to be at 25°C.

8–18 A heat engine that receives heat from a furnace at 1200°C and rejects waste heat to a river at 20°C has a thermal efficiency of 40 percent. Determine the second-law efficiency of this power plant.

8–19 Consider a thermal energy reservoir at 1500 K that can supply heat at a rate of 150,000 kJ/h. Determine the exergy of this supplied energy, assuming an environmental temperature of 25°C.

8–20 A heat engine receives heat from a source at 1100 K at a rate of 400 kJ/s, and it rejects the waste heat to a medium at 320 K. The measured power output of the heat engine is 120 kW, and the environment temperature is 25°C. Determine (*a*) the reversible power, (*b*) the rate of irreversibility, and (*c*) the second-law efficiency of this heat engine. *Answers:* (*a*) 284 kW, (*b*) 164 kW, (*c*) 42.3 percent

8–21 Reconsider Prob. 8–20. Using EES (or other) software, study the effect of reducing the temperature at which the waste heat is rejected on the reversible power, the rate of irreversibility, and the second-law efficiency as the rejection temperature is varied from 500 to 298 K, and plot the results.

8–22E A heat engine that rejects waste heat to a sink at 530 R has a thermal efficiency of 36 percent and a second-law efficiency of 60 percent. Determine the temperature of the source that supplies heat to this engine. *Answer:* 1325 R

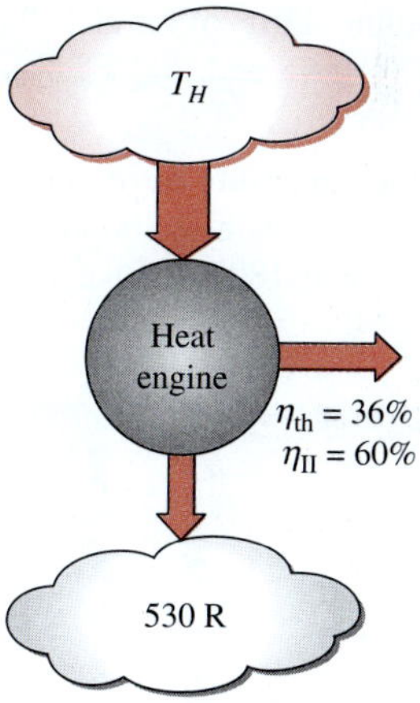

FIGURE P8–22E

8–23 A house that is losing heat at a rate of 50,000 kJ/h when the outside temperature drops to 4°C is to be heated by electric resistance heaters. If the house is to be maintained at 25°C at all times, determine the reversible work input for this process and the irreversibility. *Answers:* 0.978 kW, 12.91 kW

8–24E A freezer is maintained at 20°F by removing heat from it at a rate of 75 Btu/min. The power input to the freezer is 0.70 hp, and the surrounding air is at 75°F. Determine (*a*) the reversible power, (*b*) the irreversibility, and (*c*) the second-law efficiency of this freezer. *Answers:* (*a*) 0.20 hp, (*b*) 0.50 hp, (*c*) 28.9 percent

8–25 A geothermal power plant uses geothermal liquid water at 150°C at a rate of 210 kg/s as the heat source, and produces 5.1 MW of net power in an environment at 25°C. If 7.5 MW of exergy entering the plant with the geothermal water is destructed within the plant, determine (*a*) the exergy of the geothermal water entering the plant, (*b*) the second-law efficiency, and (*c*) the exergy of the heat rejected from the plant.

8–26 Show that the power produced by a wind turbine is proportional to the cube of the wind velocity and to the square of the blade span diameter.

Exergy Analysis of Closed Systems

8–27C Can a system have a higher second-law efficiency than the first-law efficiency during a process? Give examples.

8–28 A mass of 8 kg of helium undergoes a process from an initial state of 3 m^3/kg and 15°C to a final state of 0.5 m^3/kg and 80°C. Assuming the surroundings to be at 25°C and 100 kPa, determine the increase in the useful work potential of the helium during this process.

8–29E Air is expanded in an adiabatic closed system from 150 psia and 100°F to 15 psia with an isentropic expansion efficiency of 95 percent. What is the second-law efficiency of this expansion? Take T_0 = 77°F and P_0 = 14.7 psia.

8–30E Which is a more valuable resource for work production in a closed system—15 ft^3 of air at 100 psia and 250°F or 20 ft^3 of helium at 60 psia and 200°F? Take T_0 = 77°F and P_0 = 14.7 psia.

8–31 Which has the capability to produce the most work in a closed system—1 kg of steam at 800 kPa and 180°C or 1 kg of R-134a at 800 kPa and 180°C? Take T_0 = 25°C and P_0 = 100 kPa. *Answers:* 623 kJ, 5.0 kJ

FIGURE P8–31

8–32 A piston–cylinder device contains 5 kg of refrigerant-134a at 0.7 MPa and 60°C. The refrigerant is now cooled at constant pressure until it exists as a liquid at 24°C. If the surroundings are at 100 kPa and 24°C, determine (*a*) the exergy of the refrigerant at the initial and the final states and (*b*) the exergy destroyed during this process.

8–33E A well-insulated rigid tank contains 6 lbm of saturated liquid–vapor mixture of water at 35 psia. Initially, three-quarters of the mass is in the liquid phase. An electric resistance heater placed in the tank is turned on and kept on until all the liquid in the tank is vaporized. Assuming the surroundings to be at 75°F and 14.7 psia, determine (*a*) the exergy destruction and (*b*) the second-law efficiency for this process.

8–34 A rigid tank is divided into two equal parts by a partition. One part of the tank contains 4 kg of compressed liquid water at 200 kPa and 80°C and the other side is evacuated. Now the partition is removed, and the water expands to fill the entire tank. If the final pressure in the tank is 40 kPa, determine the exergy destroyed during this process. Assume the surroundings to be at 25°C and 100 kPa. *Answer:* 10.3 kJ

8–35 EES Reconsider Prob. 8–34. Using EES (or other) software, study the effect of final pressure in the tank on the exergy destroyed during the process. Plot the exergy destroyed as a function of the final pressure for final pressures between 45 and 5 kPa, and discuss the results.

8–36 An insulated piston–cylinder device contains 0.8 L of saturated liquid water at a constant pressure of 120 kPa. An electric resistance heater inside the cylinder is turned on, and electrical work is done on the water in the amount of 1400 kJ. Assuming the surroundings to be at 25°C and 100 kPa, determine (*a*) the minimum work with which this process could be accomplished and (*b*) the exergy destroyed during this process. *Answers:* (*a*) 278 kJ, (*b*) 1104 kJ

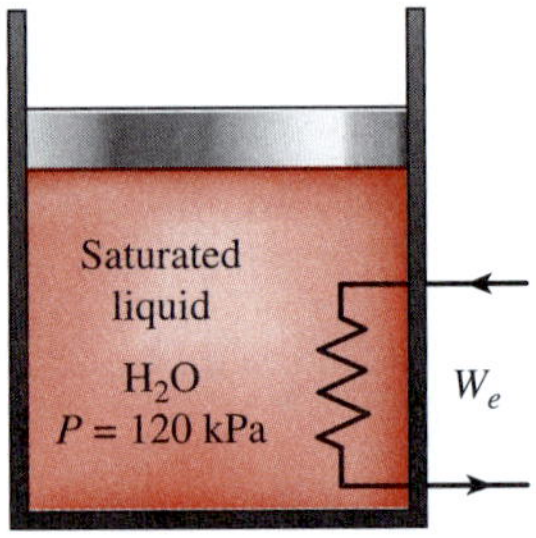

FIGURE P8–36

8–37 EES Reconsider Prob. 8–36. Using EES (or other) software, investigate the effect of the amount of electrical work supplied to the device on the minimum work and the exergy destroyed as the electrical work is varied from 0 to 2000 kJ, and plot your results.

8–38 An insulated piston–cylinder device contains 0.05 m^3 of saturated refrigerant-134a vapor at 0.8 MPa pressure. The refrigerant is now allowed to expand in a reversible manner until the pressure drops to 0.2 MPa. Determine the change in the exergy of the refrigerant during this process and the reversible work. Assume the surroundings to be at 25°C and 100 kPa.

8–39E Oxygen gas is compressed in a piston–cylinder device from an initial state of 12 ft^3/lbm and 75°F to a final state of 1.5 ft^3/lbm and 525°F. Determine the reversible work input and the increase in the exergy of the oxygen during this process. Assume the surroundings to be at 14.7 psia and 75°F. *Answers:* 60.7 Btu/lbm, 60.7 Btu/lbm

8–40 A piston–cylinder device initially contains 2 L of air at 100 kPa and 25°C. Air is now compressed to a final state of 600 kPa and 150°C. The useful work input is 1.2 kJ. Assuming the surroundings are at 100 kPa and 25°C, determine (*a*) the exergy of the air at the initial and the final states, (*b*) the minimum work that must be supplied to accomplish this compression process, and (*c*) the second-law efficiency of this process. *Answers:* (*a*) 0, 0.171 kJ, (*b*) 0.171 kJ, (*c*) 14.3 percent

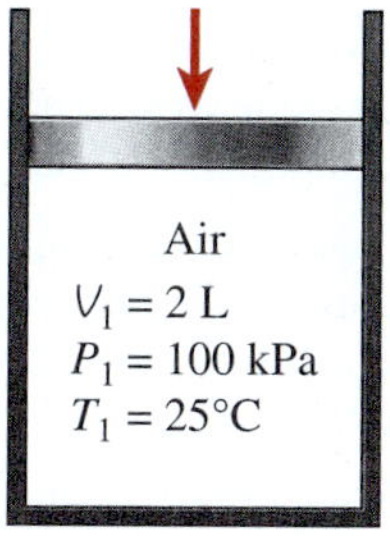

FIGURE P8–40

8–41 A 1.2-m^3 insulated rigid tank contains 2.13 kg of carbon dioxide at 100 kPa. Now paddle-wheel work is done on the system until the pressure in the tank rises to 120 kPa. Determine (*a*) the actual paddle-wheel work done during this process and (*b*) the minimum paddle-wheel work with which this process (between the same end states) could be accomplished. Take T_0 = 298 K. *Answers:* (*a*) 87.0 kJ, (*b*) 7.74 kJ

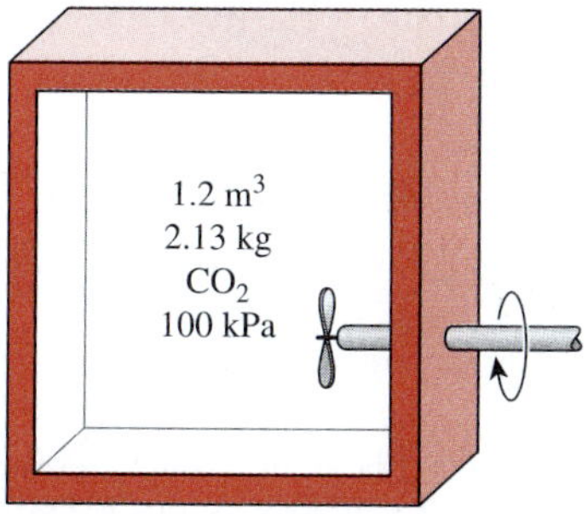

FIGURE P8–41

8–42 An insulated piston–cylinder device initially contains 20 L of air at 140 kPa and 27°C. Air is now heated for 10 min by a 100-W resistance heater placed inside the cylinder. The pressure of air is maintained constant during this process, and the surroundings are at 27°C and 100 kPa. Determine the exergy destroyed during this process. *Answer:* 19.9 kJ

8–43 An insulated rigid tank is divided into two equal parts by a partition. Initially, one part contains 3 kg of argon gas at 300 kPa and 70°C, and the other side is evacuated. The partition is now removed, and the gas fills the entire tank. Assuming the surroundings to be at 25°C, determine the exergy destroyed during this process. *Answer:* 129 kJ

8–44E A 55-lbm copper block initially at 180°F is dropped into an insulated tank that contains 1.2 ft^3 of water at 75°F. Determine (*a*) the final equilibrium temperature and (*b*) the work potential wasted during this process. Assume the surroundings to be at 75°F.

8–45 An iron block of unknown mass at 85°C is dropped into an insulated tank that contains 100 L of water at 20°C. At the same time, a paddle wheel driven by a 200-W motor is activated to stir the water. It is observed that thermal equilibrium is established after 20 min with a final temperature of 24°C. Assuming the surroundings to be at 20°C, determine (*a*) the mass of the iron block and (*b*) the exergy destroyed during this process. *Answers:* (*a*) 52.0 kg, (*b*) 375 kJ

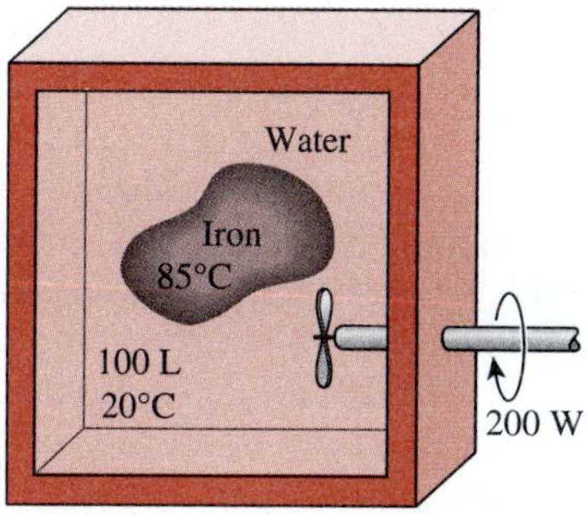

FIGURE P8–45

8–46 A 50-kg iron block and a 20-kg copper block, both initially at 80°C, are dropped into a large lake at 15°C. Thermal equilibrium is established after a while as a result of heat transfer between the blocks and the lake water. Assuming the surroundings to be at 20°C, determine the amount of work that could have been produced if the entire process were executed in a reversible manner.

8–47E A 12-ft^3 rigid tank contains refrigerant-134a at 40 psia and 55 percent quality. Heat is transferred now to the refrigerant from a source at 120°F until the pressure rises to 60 psia. Assuming the surroundings to be at 75°F, determine (*a*) the amount of heat transfer between the source and the refrigerant and (*b*) the exergy destroyed during this process.

8–48 Chickens with an average mass of 1.6 kg and average specific heat of 3.54 kJ/kg·°C are to be cooled by chilled water that enters a continuous-flow-type immersion chiller at 0.5°C and leaves at 2.5°C. Chickens are dropped into the chiller at a uniform temperature of 15°C at a rate of 700 chickens per hour and are cooled to an average temperature of 3°C before they are taken out. The chiller gains heat from the surroundings at a rate of 400 kJ/h. Determine (*a*) the rate of heat removal from the chicken, in kW, and (*b*) the rate of exergy destruction during this chilling process. Take $T_0 = 25°C$.

8–49 Carbon steel balls ($\rho = 7833$ kg/m^3 and $c_p = 0.465$ kJ/kg·°C) 8 mm in diameter are annealed by heating them first to 900°C in a furnace and then allowing them to cool slowly to 100°C in ambient air at 35°C. If 1200 balls are to be annealed per hour, determine (*a*) the rate of heat transfer from the balls to the air and (*b*) the rate of exergy destruction due to heat loss from the balls to the air. *Answers:* (*a*) 260 W, (*b*) 146 W

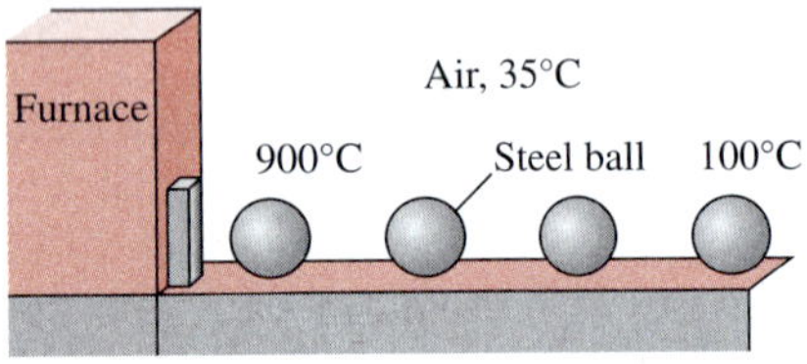

FIGURE P8–49

8–50 A piston–cylinder device initially contains 0.75 kg of refrigerant-134a at 120 kPa and 20°C. Heat is now transferred to the refrigerant from a source at 150°C, and the piston which is resting on a set of stops, starts moving when the pressure inside reaches 140 kPa. Heat transfer continues until the temperature reaches 90°C. Assuming the surroundings to be at 25°C and 100 kPa, determine (*a*) the work done, (*b*) the heat transfer, (*c*) the exergy destroyed, and (*d*) the second-law efficiency of this process. *Answers:* (*a*) 1.53 kJ, (*b*) 44.2 kJ, (*c*) 8.94 kJ, (*d*) 31.6 percent

FIGURE P8–50

8–51 A 0.04-m^3 tank initially contains air at ambient conditions of 100 kPa and 22°C. Now, a 15-liter tank containing liquid water at 85°C is placed into the tank without causing any air to escape. After some heat transfer from the water to the air and the surroundings, both the air and water are measured to be at 44°C. Determine (a) the amount of heat lost to the surroundings and (b) the exergy destruction during this process.

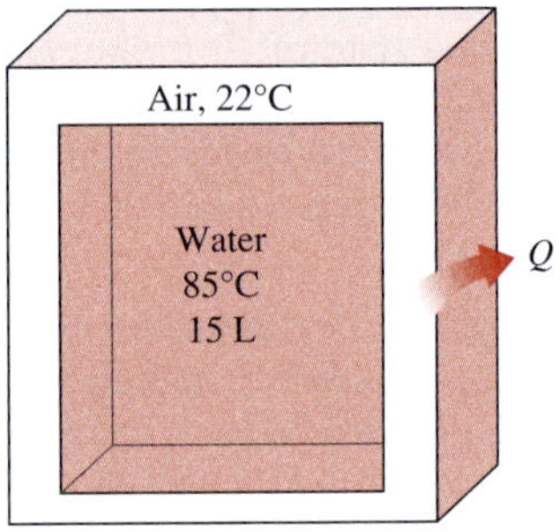

FIGURE P8–51

Exergy Analysis of Control Volumes

8–52 Refrigerant-134a enters an expansion valve at 1200 kPa as a saturated liquid and leaves at 200 kPa. Determine (*a*) the temperature of R-134a at the outlet of the expansion valve and (*b*) the entropy generation and the exergy destruction during this process. Take $T_0 = 25°C$.

8–53 Helium is expanded in a turbine from 1500 kPa and 300°C to 100 kPa and 25°C. Determine the maximum work this turbine can produce, in kJ/kg. Does the maximum work require an adiabatic turbine?

8–54 Air is compressed steadily by an 8-kW compressor from 100 kPa and 17°C to 600 kPa and 167°C at a rate of 2.1 kg/min. Neglecting the changes in kinetic and potential energies, determine (*a*) the increase in the exergy of the air and (*b*) the rate of exergy destroyed during this process. Assume the surroundings to be at 17°C.

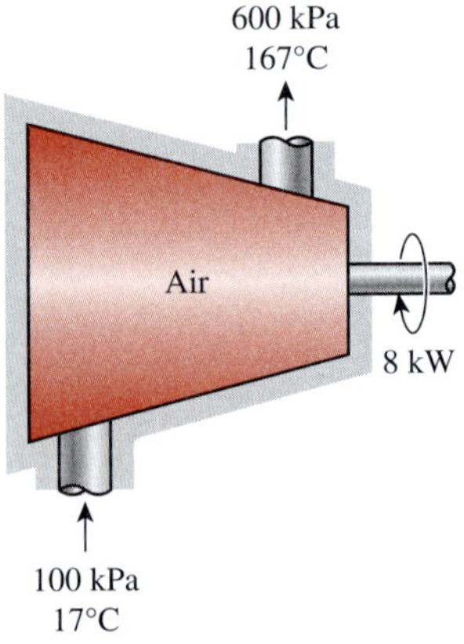

FIGURE P8–54

8–55 Reconsider Prob. 8–54. Using EES (or other) software, solve the problem and in addition determine the actual heat transfer, if any, and its direction, the minimum power input (the reversible power), and the compressor second-law efficiency. Then interpret the results when the outlet temperature is set to, say, 300°C. Explain the values of heat transfer, exergy destroyed, and efficiency when the outlet temperature is set to 209.31°C and mass flow rate to 2.466 kg/min.

8–56 An adiabatic steam diffuser has steam entering at 500 kPa, 200°C, and 30 m/s, and leaving as a saturated vapor at 200 kPa. Calculate the second-law efficiency of the diffuser. Take T_0 = 25°C. *Answer:* 88.4 percent

8–57 Air enters a nozzle steadily at 200 kPa and 65°C with a velocity of 35 m/s and exits at 95 kPa and 240 m/s. The heat loss from the nozzle to the surrounding medium at 17°C is estimated to be 3 kJ/kg. Determine (*a*) the exit temperature and (*b*) the exergy destroyed during this process. *Answers:* (*a*) 34.0°C, (*b*) 36.9 kJ/kg

8–58 EES Reconsider Prob. 8–57. Using EES (or other) software, study the effect of varying the nozzle exit velocity from 100 to 300 m/s on both the exit temperature and exergy destroyed, and plot the results.

8–59 Steam enters a diffuser at 10 kPa and 50°C with a velocity of 300 m/s and exits as saturated vapor at 50°C and 70 m/s. The exit area of the diffuser is 3 m². Determine (*a*) the mass flow rate of the steam and (*b*) the wasted work potential during this process. Assume the surroundings to be at 25°C.

8–60E Air is compressed steadily by a compressor from 14.7 psia and 60°F to 100 psia and 480°F at a rate of 22 lbm/min. Assuming the surroundings to be at 60°F, determine the minimum power input to the compressor. Assume air to be an ideal gas with variable specific heats, and neglect the changes in kinetic and potential energies.

8–61 Steam enters an adiabatic turbine at 6 MPa, 600°C, and 80 m/s and leaves at 50 kPa, 100°C, and 140 m/s. If the power output of the turbine is 5 MW, determine (*a*) the reversible power output and (*b*) the second-law efficiency of the turbine. Assume the surroundings to be at 25°C. *Answers:* (*a*) 5.81 MW, (*b*) 86.1 percent

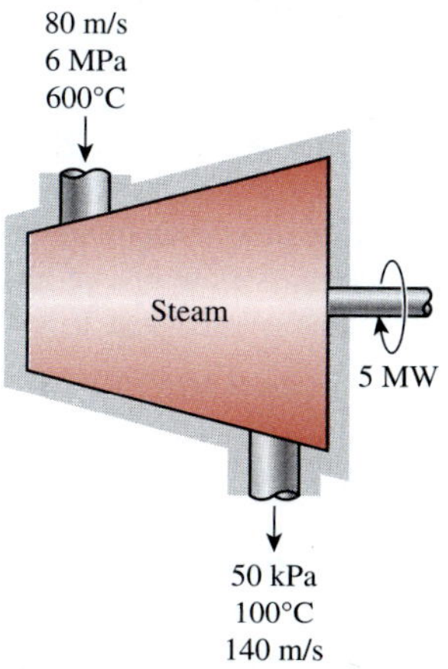

FIGURE P8–61

8–62 Steam is throttled from 6 MPa and 400°C to a pressure of 2 MPa. Determine the decrease in exergy of the steam during this process. Assume the surroundings to be at 25°C. *Answer:* 143 kJ/kg

8–63 Carbon dioxide enters a compressor at 100 kPa and 300 K at a rate of 0.2 kg/s and exits at 600 kPa and 450 K. Determine the power input to the compressor if the process involved no irreversibilities. Assume the surroundings to be at 25°C. *Answer:* 25.5 kW

8–64 Combustion gases enter a gas turbine at 900°C, 800 kPa, and 100 m/s and leave at 650°C, 400 kPa, and 220 m/s. Taking c_p = 1.15 kJ/kg·°C and k = 1.3 for the combustion gases, determine (*a*) the exergy of the combustion gases at the turbine inlet and (*b*) the work output of the turbine under reversible conditions. Assume the surroundings to be at 25°C and 100 kPa. Can this turbine be adiabatic?

8–65E Refrigerant-134a enters an adiabatic compressor as saturated vapor at 30 psia at a rate of 20 ft³/min and exits at 70 psia pressure. If the isentropic efficiency of the compressor is 80 percent, determine (*a*) the actual power input and (*b*) the second-law efficiency of the compressor. Assume the surroundings to be at 75°F. *Answers:* (*a*) 2.85 hp, (*b*) 79.8 percent

8–66 Refrigerant-134a enters an adiabatic compressor at −26°C as a saturated vapor at a rate of 0.45 m³/min and leaves at 800 kPa and 50°C. Determine (*a*) the power input to the compressor, (*b*) the isentropic efficiency of the compressor, and (*c*) the rate of exergy destruction and the second-law efficiency of the compressor. Take T_0 = 27°C. *Answers:* (*a*) 2.06 kW, (*b*) 82.4 percent, (*c*) 0.342 kW, 83.4 percent

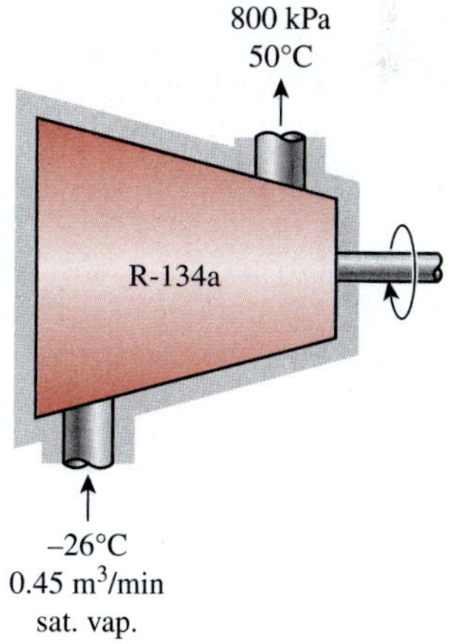

FIGURE P8–66

8–67 Refrigerant-134a is condensed in a refrigeration system by rejecting heat to ambient air at 25°C. R-134a enters the condenser at 700 kPa and 50°C at a rate of 0.05 kg/s and leaves at the same pressure as a saturated liquid. Determine (*a*) the rate of heat rejected in the condenser, (*b*) the

COP of this refrigeration cycle if the cooling load at these conditions is 6 kW, and (*c*) the rate of exergy destruction in the condenser.

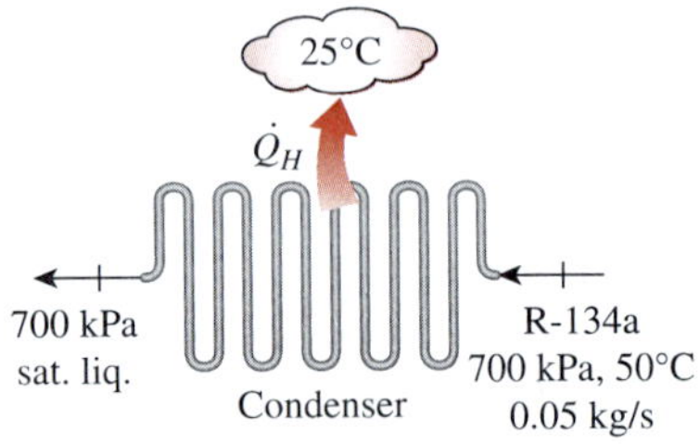

FIGURE P8–67

8–68E Refrigerant-22 absorbs heat from a cooled space at 50°F as it flows through an evaporator of a refrigeration system. R-22 enters the evaporator at 10°F at a rate of 0.08 lbm/s with a quality of 0.3 and leaves as a saturated vapor at the same pressure. Determine (*a*) the rate of cooling provided, in Btu/h, (*b*) the rate of exergy destruction in the evaporator, and (*c*) the second-law efficiency of the evaporator. Take T_0 = 77°F. The properties of R-22 at the inlet and exit of the evaporator are: h_1 = 107.5 Btu/lbm, s_1 = 0.2851 Btu/lbm·R, h_2 = 172.1 Btu/lbm, s_2 = 0.4225 Btu/lbm·R.

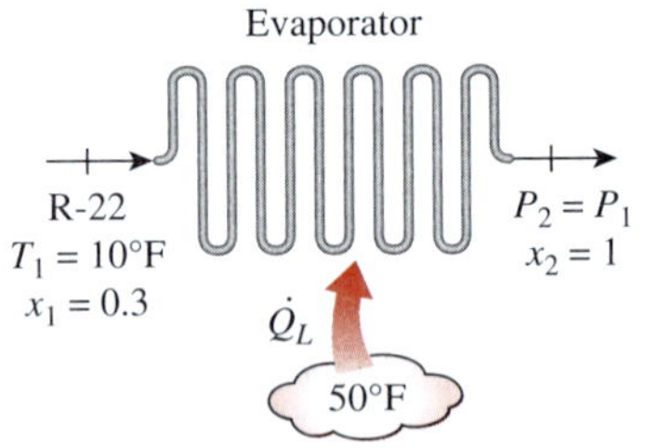

FIGURE P8–68E

8–69 Air is compressed by a compressor from 101 kPa and 27°C to 400 kPa and 220°C at a rate of 0.15 kg/s. Neglecting the changes in kinetic and potential energies and assuming the surroundings to be at 25°C, determine the reversible power input for this process. *Answer:* 24.5 kW

8–70 Reconsider Prob. 8–69. Using EES (or other) software, investigate the effect of compressor exit pressure on reversible power. Vary the compressor exit pressure from 200 to 600 kPa while keeping the exit temperature at 220°C. Plot the reversible power input for this process as a function of the compressor exit pressure.

8–71 How much exergy is lost in a rigid vessel filled with 1 kg of liquid R-134a, whose temperature remains constant at 20°C, as R-134a vapor is released from the vessel? This vessel may exchange heat with the surrounding atmosphere, which is at 100 kPa and 20°C. The vapor is released until the last of the liquid inside the vessel disappears.

8–72E A 40-ft^3 adiabatic container is initially evacuated. The supply line contains air that is maintained at 150 psia and 90°F. The valve is opened until the pressure in the container is the same as the pressure in the supply line. Determine the work potential of the air in this container when it is filled. Take T_0 = 80°F.

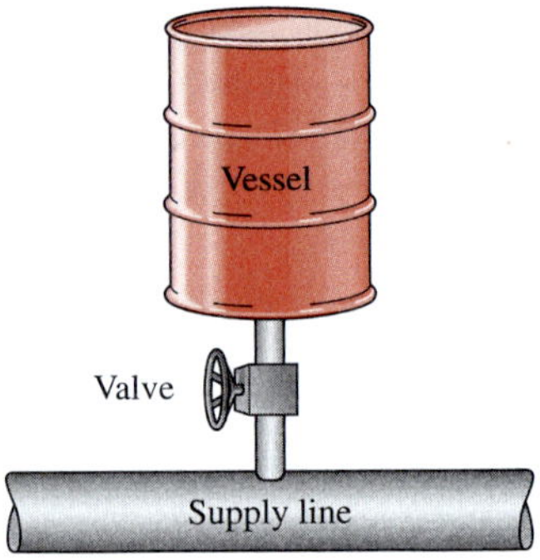

FIGURE P8–72E

8–73E What is the work potential of the air in the filled container of the previous problem if it is filled in such a way that the final pressure and temperature are both the same as in the supply line? The temperature of the surrounding environment is 80°F. Note that the container cannot be adiabatic in this case, and it can exchange heat with the natural environment. *Answer:* 1110 Btu

8–74 Steam expands in a turbine steadily at a rate of 15,000 kg/h, entering at 8 MPa and 450°C and leaving at 50 kPa as saturated vapor. Assuming the surroundings to be at 100 kPa and 25°C, determine (*a*) the power potential of the steam at the inlet conditions and (*b*) the power output of the turbine if there were no irreversibilities present. *Answers:* (*a*) 5515 kW, (*b*) 3902 kW

8–75E Air enters a compressor at ambient conditions of 15 psia and 60°F with a low velocity and exits at 150 psia, 620°F, and 350 ft/s. The compressor is cooled by the ambient air at 60°F at a rate of 1500 Btu/min. The power input to the compressor is 400 hp. Determine (*a*) the mass flow rate of air and (*b*) the portion of the power input that is used just to overcome the irreversibilities.

8–76 Hot combustion gases enter the nozzle of a turbojet engine at 230 kPa, 627°C, and 60 m/s and exit at 70 kPa and 450°C. Assuming the nozzle to be adiabatic and the surroundings to be at 20°C, determine (*a*) the exit velocity and (*b*) the decrease in the exergy of the gases. Take k = 1.3 and c_p = 1.15 kJ/kg·°C for the combustion gases.

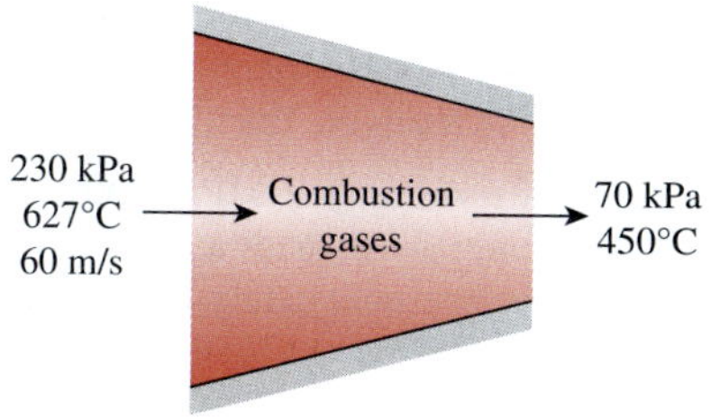

FIGURE P8–76

8–77 Steam is usually accelerated in the nozzle of a turbine before it strikes the turbine blades. Steam enters an adiabatic nozzle at 7 MPa and 500°C with a velocity of 70 m/s and exits at 5 MPa and 450°C. Assuming the surroundings to be at 25°C, determine (*a*) the exit velocity of the steam, (*b*) the isentropic efficiency, and (*c*) the exergy destroyed within the nozzle.

8–78 Ambient air at 100 kPa and 300 K is compressed isentropically in a steady-flow device to 1 MPa. Determine (*a*) the work input to the compressor, (*b*) the exergy of the air at the compressor exit, and (*c*) the exergy of compressed air after it is cooled to 300 K at 1 MPa pressure.

8–79 A 0.1-m^3 rigid tank initially contains refrigerant-134a at 1.2 MPa and 100 percent quality. The tank is connected by a valve to a supply line that carries refrigerant-134a at 1.6 MPa and 30°C. The valve is now opened, allowing the refrigerant to enter the tank, and it is closed when the tank contains only saturated liquid at 1.4 MPa. The refrigerant exchanges heat with its surroundings at 45°C and 100 kPa during this process. Determine (*a*) the mass of the refrigerant that entered the tank and (*b*) the exergy destroyed during this process.

8–80 A 0.6-m^3 rigid tank is filled with saturated liquid water at 170°C. A valve at the bottom of the tank is now opened, and one-half of the total mass is withdrawn from the tank in liquid form. Heat is transferred to water from a source of 210°C so that the temperature in the tank remains constant. Determine (*a*) the amount of heat transfer and (*b*) the reversible work and exergy destruction for this process. Assume the surroundings to be at 25°C and 100 kPa. *Answers:* (*a*) 2545 kJ, (*b*) 141.2 kJ, 141.2 kJ

8–81E An insulated 260-ft^3 rigid tank contains air at 40 psia and 180°F. A valve connected to the tank is opened, and air is allowed to escape until the pressure inside drops to 20 psia. The air temperature during this process is maintained constant by an electric resistance heater placed in the tank. Determine (*a*) the electrical work done during this process and (*b*) the exergy destruction. Assume the surroundings to be at 70°F. *Answers:* (*a*) 962 Btu, (*b*) 782 Btu

8–82 A vertical piston–cylinder device initially contains 0.1 m^3 of helium at 20°C. The mass of the piston is such that it maintains a constant pressure of 300 kPa inside. A valve is now opened, and helium is allowed to escape until the volume inside the cylinder is decreased by one-half. Heat transfer takes place between the helium and its surroundings at 20°C and 95 kPa so that the temperature of helium in the cylinder remains constant. Determine (*a*) the maximum work potential of the helium at the initial state and (*b*) the exergy destroyed during this process.

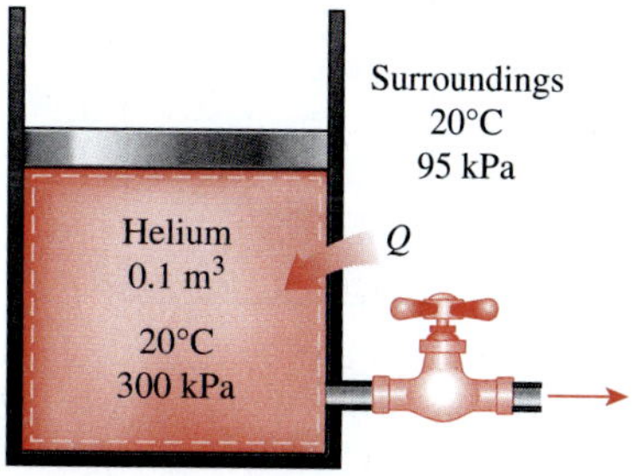

FIGURE P8–82

8–83 A 0.2-m^3 rigid tank initially contains saturated refrigerant-134a vapor at 1 MPa. The tank is connected by a valve to a supply line that carries refrigerant-134a at 1.4 MPa and 60°C. The valve is now opened, and the refrigerant is allowed to enter the tank. The valve is closed when one-half of the volume of the tank is filled with liquid and the rest with vapor at 1.2 MPa. The refrigerant exchanges heat during this process with the surroundings at 25°C. Determine (*a*) the amount of heat transfer and (*b*) the exergy destruction associated with this process.

8–84 An insulated vertical piston–cylinder device initially contains 15 kg of water, 13 kg of which is in the vapor phase. The mass of the piston is such that it maintains a constant pressure of 300 kPa inside the cylinder. Now steam at 2 MPa and 400°C is allowed to enter the cylinder from a supply line until all the liquid in the cylinder is vaporized. Assuming the surroundings to be at 25°C and 100 kPa, determine (*a*) the amount of steam that has entered and (*b*) the exergy destroyed during this process. *Answers:* (*a*) 8.27 kg, (*b*) 2832 kJ

8–85 Consider a family of four, with each person taking a 6-minute shower every morning. The average flow rate through the shower head is 10 L/min. City water at 15°C is heated to 55°C in an electric water heater and tempered to 42°C by cold water at the T-elbow of the shower before being routed to the shower head. Determine the amount of exergy destroyed by this family per year as a result of taking daily showers. Take T_0 = 25°C.

8–86 Liquid water at 200 kPa and 15°C is heated in a chamber by mixing it with superheated steam at 200 kPa and 200°C. Liquid water enters the mixing chamber at a rate of 4 kg/s, and the chamber is estimated to lose heat to the surrounding air at 25°C at a rate of 600 kJ/min. If the mixture leaves the mixing chamber at 200 kPa and 80°C, determine (*a*) the mass flow rate of the superheated steam and (*b*) the wasted work potential during this mixing process.

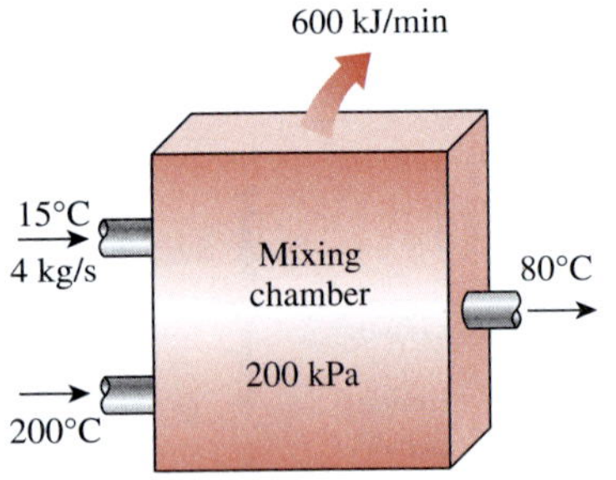

FIGURE P8–86

8–87 Outdoor air (c_p = 1.005 kJ/kg·°C) is to be preheated by hot exhaust gases in a cross-flow heat exchanger before it enters the furnace. Air enters the heat exchanger at 101 kPa and 30°C at a rate of 0.5 m³/s. The combustion gases (c_p = 1.10 kJ/kg·°C) enter at 240°C at a rate of 1.1 kg/s and leave at 190°C. Determine the rate of heat transfer to the air and the rate of exergy destruction in the heat exchanger.

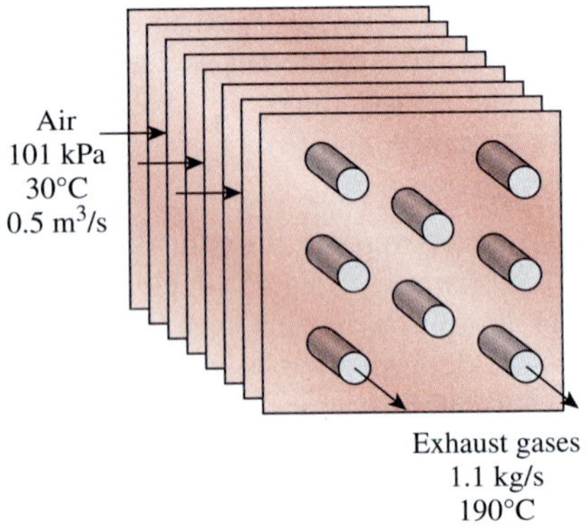

FIGURE P8–87

8–88 A well-insulated shell-and-tube heat exchanger is used to heat water (c_p = 4.18 kJ/kg·°C) in the tubes from 20 to 70°C at a rate of 4.5 kg/s. Heat is supplied by hot oil (c_p = 2.30 kJ/kg·°C) that enters the shell side at 170°C at a rate of 10 kg/s. Disregarding any heat loss from the heat exchanger, determine (*a*) the exit temperature of oil and (*b*) the rate of exergy destruction in the heat exchanger. Take T_0 = 25°C.

8–89E Steam is to be condensed on the shell side of a heat exchanger at 120°F. Cooling water enters the tubes at 60°F at a rate of 115.3 lbm/s and leaves at 73°F. Assuming the heat exchanger to be well insulated, determine (*a*) the rate of heat transfer in the heat exchanger and (*b*) the rate of exergy destruction in the heat exchanger. Take T_0 = 77°F.

8–90 Steam enters a turbine at 9 MPa, 600°C, and 60 m/s and leaves at 20 kPa and 90 m/s with a moisture content of 5 percent. The turbine is not adequately insulated and it estimated that heat is lost from the turbine at a rate of 220 kW. The power output of the turbine is 4.5 MW. Assuming the surroundings to be at 25°C, determine (*a*) the reversible power output of the turbine, (*b*) the exergy destroyed within the turbine, and (*c*) the second-law efficiency of the turbine. (*d*) Also, estimate the possible increase in the power output of the turbine if the turbine were perfectly insulated.

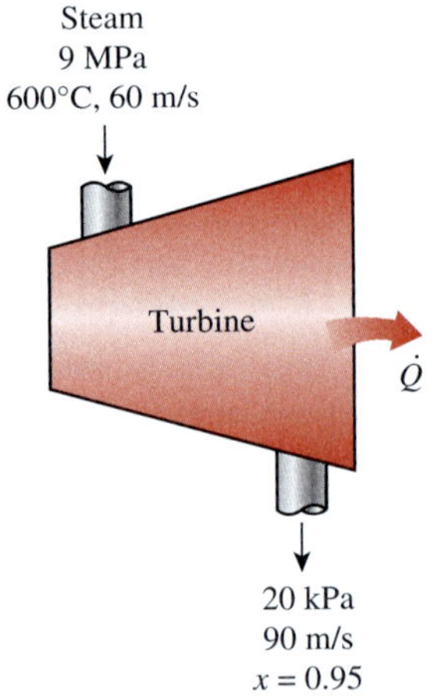

FIGURE P8–90

8–91 Air enters a compressor at ambient conditions of 100 kPa and 20°C at a rate of 4.5 m³/s with a low velocity, and exits at 900 kPa, 60°C, and 80 m/s. The compressor is cooled by cooling water that experiences a temperature rise of 10°C. The isothermal efficiency of the compressor is 70 percent. Determine (*a*) the actual and reversible power inputs, (*b*) the second-law efficiency, and (*c*) the mass flow rate of the cooling water.

8–92 Liquid water at 15°C is heated in a chamber by mixing it with saturated steam. Liquid water enters the chamber at the steam pressure at a rate of 4.6 kg/s and the saturated steam enters at a rate of 0.23 kg/s. The mixture leaves the mixing chamber as a liquid at 45°C. If the surroundings are at 15°C, determine (*a*) the temperature of saturated steam entering the chamber, (*b*) the exergy destruction during this mixing process, and (*c*) the second-law efficiency of the mixing chamber. *Answers: (a) 114.3°C, (b) 114.7 kW, (c) 0.207*

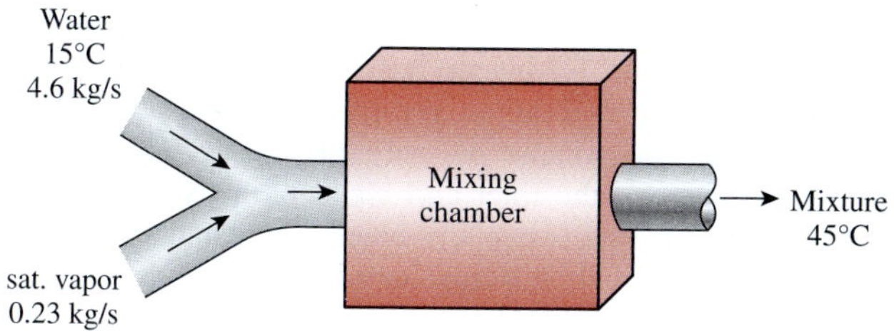

FIGURE P8–92

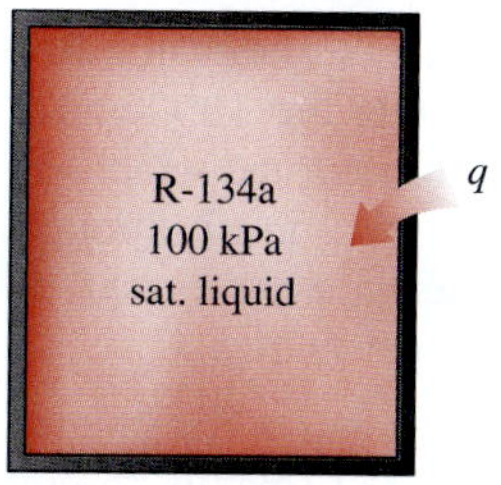

FIGURE P8–99

8–93 Derive an expression for the work potential of the single-phase contents of a rigid adiabatic container when the initially empty container is filled through a single opening from a source of working fluid whose properties remain fixed.

Review Problems

8–94E A refrigerator has a second-law efficiency of 28 percent, and heat is removed from the refrigerated space at a rate of 800 Btu/min. If the space is maintained at 25°F while the surrounding air temperature is 90°F, determine the power input to the refrigerator.

8–95 Refrigerant-134a is expanded adiabatically in an expansion valve from 900 kPa and 30°C to 120 kPa. For environment conditions of 100 kPa and 20°C, determine (*a*) the work potential of R-134a at the inlet, (*b*) the exergy destruction during the process, and (*c*) the second-law efficiency.

8–96 Steam enters an adiabatic nozzle at 3.5 MPa and 300°C with a low velocity and leaves at 1.6 MPa and 250°C at a rate of 0.4 kg/s. If the ambient state is 100 kPa and 18°C, determine (*a*) the exit velocity, (*b*) the rate of exergy destruction, and (*c*) the second-law efficiency.

8–97 Refrigerant-134a at 1600 kPa and 80°C is expanded adiabatically in a closed system to 100 kPa with an isentropic expansion efficiency of 85 percent. Determine the second-law efficiency of this expansion. Take T_0 = 25°C and P_0 = 100 kPa.

8–98 Steam is condensed in a closed system at a constant pressure of 75 kPa from a saturated vapor to a saturated liquid by rejecting heat to a thermal energy reservoir at 37°C. Determine the second-law efficiency of this process. Take T_0 = 25°C and P_0 = 100 kPa.

8–99 Refrigerant-134a is converted from a saturated liquid to a saturated vapor in a closed system using a reversible constant pressure process by transferring heat from a heat reservoir at 0°C. From second-law point of view, is it more effective to do this phase change at 100 kPa or 200 kPa? Take T_0 = 25°C and P_0 = 100 kPa.

8–100 A 50-L electrical radiator containing heating oil is placed in a well-sealed 75-m^3 room. Both the air in the room and the oil in the radiator are initially at the environment temperature of 6°C. Electricity with a rating of 2.4 kW is now turned on. Heat is also lost from the room at an average rate of 0.75 kW. The heater is turned off after some time when the temperatures of the room air and oil are measured to be 20°C and 60°C, respectively. Taking the density and the specific heat of oil to be 950 kg/m^3 and 2.2 kJ/kg·°C, determine (*a*) how long the heater is kept on, (*b*) the exergy destruction, and (*c*) the second-law efficiency for this process. *Answers:* (*a*) 66.6 min, (*b*) 9.09 MJ, (*c*) 5.3 percent

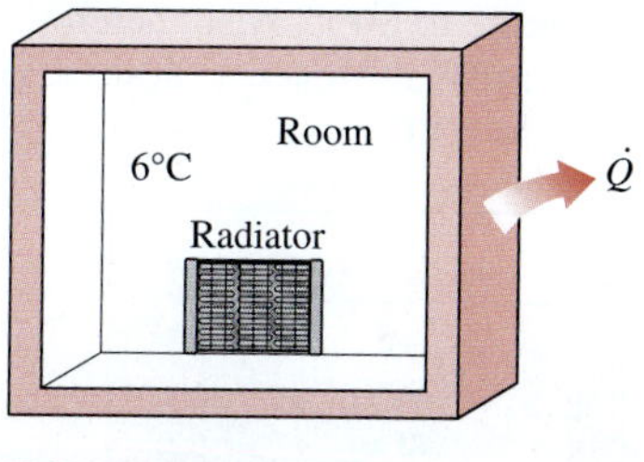

FIGURE P8–100

8–101 Hot exhaust gases leaving an internal combustion engine at 400°C and 150 kPa at a rate of 0.8 kg/s is to be used to produce saturated steam at 200°C in an insulated heat exchanger. Water enters the heat exchanger at the ambient temperature of 20°C, and the exhaust gases leave the heat exchanger at 350°C. Determine (*a*) the rate of steam production, (*b*) the rate of exergy destruction in the heat exchanger, and (*c*) the second-law efficiency of the heat exchanger.

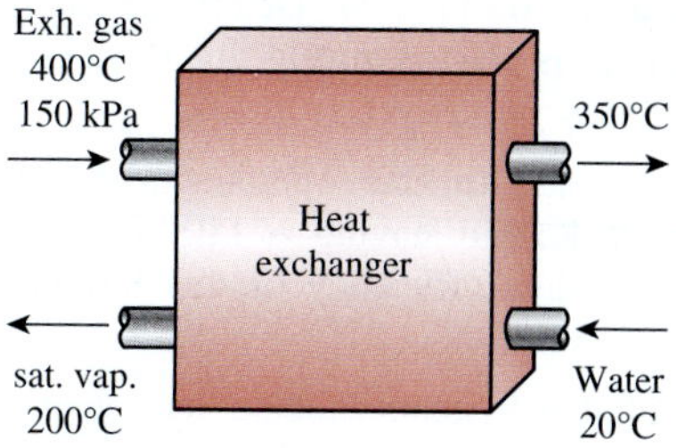

FIGURE P8–101

8–102 The inner and outer surfaces of a 0.5-cm-thick, 2-m × 2-m window glass in winter are 10°C and 3°C, respectively. If the rate of heat loss through the window is 4.4 kJ/s, determine the amount of heat loss, in kJ, through the glass over a period of 5 h. Also, determine the exergy destruction associated with this process. Take T_0 = 5°C.

8–103 An aluminum pan has a flat bottom whose diameter is 30 cm. Heat is transferred steadily to boiling water in the pan through its bottom at a rate of 1100 W. If the temperatures of the inner and outer surfaces of the bottom of the pan are 104°C and 105°C, respectively, determine the rate of exergy destruction within the bottom of the pan during this process, in W. Take T_0 = 25°C.

8–104 A crater lake has a base area of 20,000 m^2, and the water it contains is 12 m deep. The ground surrounding the crater is nearly flat and is 140 m below the base of the lake. Determine the maximum amount of electrical work, in kWh, that can be generated by feeding this water to a hydroelectric power plant. *Answer:* 95,500 kWh

8–105 A 50-cm-long, 800-W electric resistance heating element whose diameter is 0.5 cm is immersed in 40 kg of water initially at 20°C. Assuming the water container is well-insulated, determine how long it will take for this heater to raise the water temperature to 80°C. Also, determine the minimum work input required and exergy destruction for this process, in kJ. Take T_0 = 20°C.

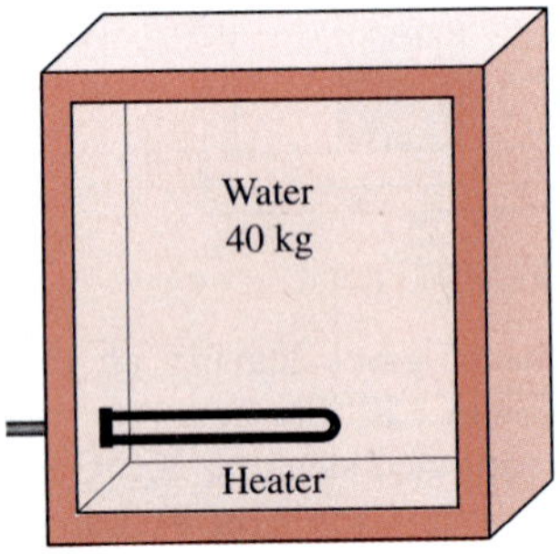

FIGURE P8–105

8–106 A 5-cm-external-diameter, 10-m-long hot water pipe at 80°C is losing heat to the surrounding air at 5°C by natural convection at a rate of 1175 W. Determine the rate at which the work potential is wasted during this process as a result of this heat loss.

8–107 An adiabatic turbine operates with air entering at 550 kPa and 425 K, and leaving at 110 kPa and 325 K. Calculate the second-law efficiency of this turbine. Take T_0 = 25°C. *Answers:* 64.0 percent

8–108 An adiabatic steam nozzle has steam entering at 300 kPa, 150°C, and 45 m/s, and leaving as a saturated vapor at 150 kPa. Calculate the actual and maximum outlet velocity. Take T_0 = 25°C. *Answers:* 372 m/s, 473 m/s

8–109E A steam turbine is equipped to bleed 6 percent of the inlet steam for feedwater heating. It is operated with 500 psia and 600°F steam at the inlet, a bleed pressure of 100 psia, and an exhaust pressure of 5 psia. The turbine efficiency between the inlet and bleed point is 97 percent, and the efficiency between the bleed point and exhaust is 95 percent. Calculate this turbine's second-law efficiency. Take T_0 = 77°F.

8–110 To control an isentropic steam turbine, a throttle valve is placed in the steam line leading to the turbine inlet. Steam at 6 MPa and 700°C is supplied to the throttle inlet, and the turbine exhaust pressure is set at 70 kPa. What is the effect on the stream exergy at the turbine inlet when the throttle valve is partially closed such that the pressure at the turbine inlet is 3 MPa. Compare the second-law efficiency of this system when the valve is partially open to when it is fully open. Take T_0 = 25°C.

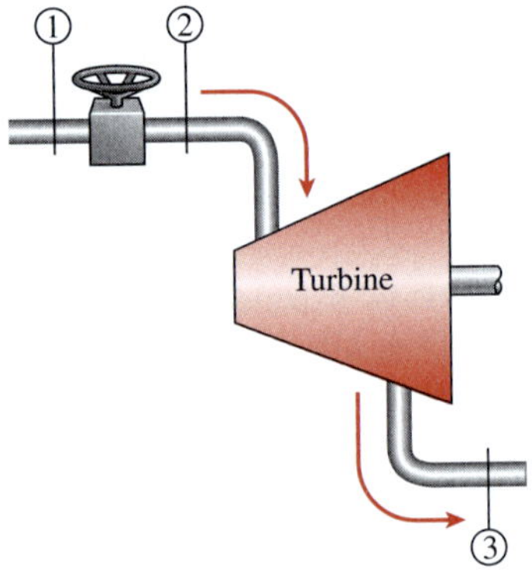

FIGURE P8–110

8–111 Two rigid tanks are connected by a valve. Tank *A* is insulated and contains 0.2 m^3 of steam at 400 kPa and 80 percent quality. Tank *B* is uninsulated and contains 3 kg of steam at 200 kPa and 250°C. The valve is now opened, and steam flows from tank *A* to tank *B* until the pressure in tank *A* drops to 300 kPa. During this process 900 kJ of heat is transferred from tank *B* to the surroundings at 0°C. Assuming the steam remaining inside tank *A* to have undergone a reversible adiabatic process, determine (*a*) the final temperature in each tank and (*b*) the work potential wasted during this process.

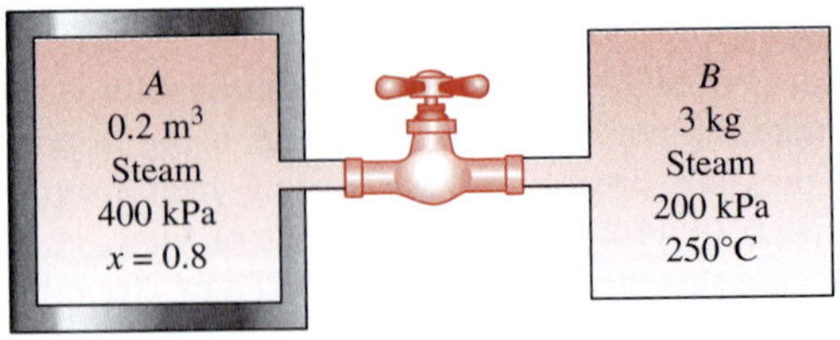

FIGURE P8–111

8–112E A piston–cylinder device initially contains 8 ft^3 of helium gas at 40 psia and 70°F. Helium is now compressed in a polytropic process (PV^n = constant) to 140 psia and 320°F. Assuming the surroundings to be at 14.7 psia and 70°F, determine (*a*) the actual useful work consumed and (*b*) the minimum useful work input needed for this process. *Answers:* (*a*) 50.0 Btu, (*b*) 46.3 Btu

8–113 Steam at 9 MPa and 500°C enters a two-stage adiabatic turbine at a rate of 15 kg/s. Ten percent of the steam is extracted at the end of the first stage at a pressure of 1.4 MPa for other use. The remainder of the steam is further expanded in the second stage and leaves the turbine at 50 kPa. If the turbine has an isentropic efficiency of 88 percent, determine the wasted power potential during this process as a result of irreversibilities. Assume the surroundings to be at 25°C.

8–114 Steam enters a two-stage adiabatic turbine at 8 MPa and 500°C. It expands in the first stage to a state of 2 MPa and 350°C. Steam is then reheated at constant pressure to a temperature of 500°C before it is routed to the second stage, where it exits at 30 kPa and a quality of 97 percent. The work output of the turbine is 5 MW. Assuming the surroundings to be at 25°C, determine the reversible power output and the rate of exergy destruction within this turbine. *Answers:* 5457 kW, 457 kW

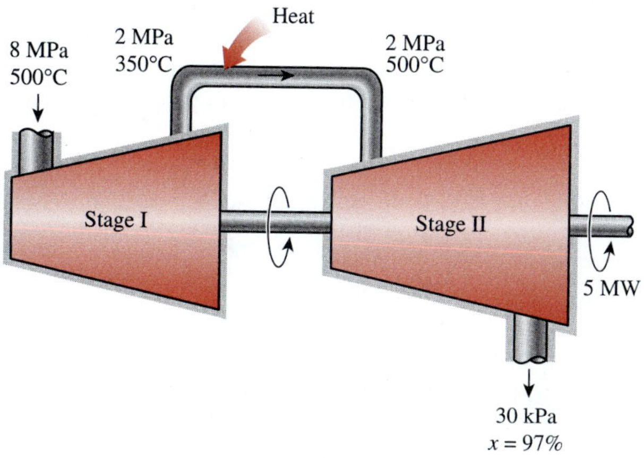

FIGURE P8–114

8–115 Consider a well-insulated horizontal rigid cylinder that is divided into two compartments by a piston that is free to move but does not allow either gas to leak into the other side. Initially, one side of the piston contains 1 m^3 of N_2 gas at 500 kPa and 80°C while the other side contains 1 m^3 of He gas at 500 kPa and 25°C. Now thermal equilibrium is established in the cylinder as a result of heat transfer through the piston. Using constant specific heats at room temperature, determine (*a*) the final equilibrium temperature in the cylinder and (*b*) the wasted work potential during this process. What would your answer be if the piston were not free to move? Take T_0 = 25°C.

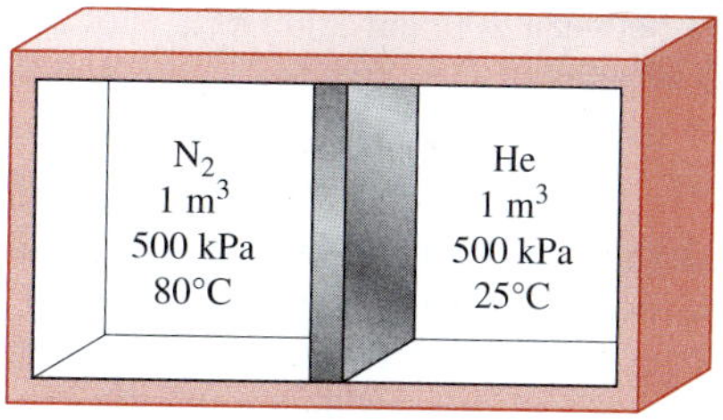

FIGURE P8–115

8–116 Repeat Prob. 8–115 by assuming the piston is made of 5 kg of copper initially at the average temperature of the two gases on both sides.

8–117E Argon gas enters an adiabatic turbine at 1500°F and 200 psia at a rate of 40 lbm/min and exhausts at 30 psia. If the power output of the turbine is 95 hp, determine (*a*) the isentropic efficiency and (*b*) the second-law efficiency of the turbine. Assume the surroundings to be at 77°F.

8–118 In large steam power plants, the feedwater is frequently heated in closed feedwater heaters, which are basically heat exchangers, by steam extracted from the turbine at some stage. Steam enters the feedwater heater at 1.6 MPa and 250°C and leaves as saturated liquid at the same pressure. Feedwater enters the heater at 4 MPa and 30°C and leaves 10°C below the exit temperature of the steam. Neglecting any heat losses from the outer surfaces of the heater, determine (*a*) the ratio of the mass flow rates of the extracted steam and the feedwater heater and (*b*) the reversible work for this process per unit mass of the feedwater. Assume the surroundings to be at 25°C. *Answers:* (*a*) 0.333, (*b*) 110 kJ/kg

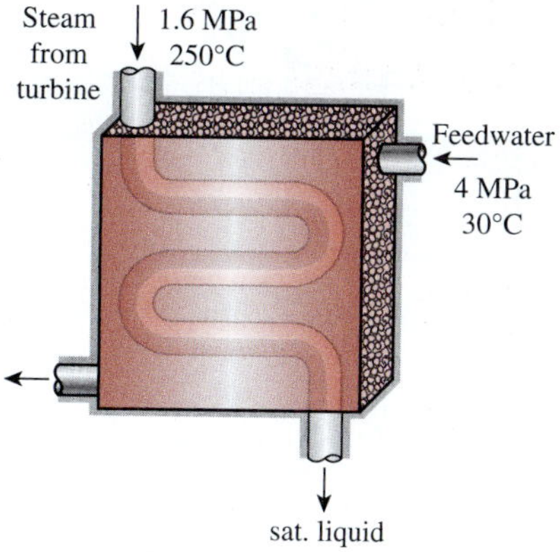

FIGURE P8–118

8–119 Reconsider Prob. 8–118. Using EES (or other) software, investigate the effect of the state of the steam at the inlet of the feedwater heater on the ratio of mass flow rates and the reversible power. Vary the extracted steam pressure between 200 and 2000 kPa. Plot both the ratio of the mass flow rates of the extracted steam and the feedwater heater and the reversible work for this process per unit mass of feedwater as functions of the extraction pressure.

8–120 In order to cool 1 ton of water at 20°C in an insulated tank, a person pours 80 kg of ice at −5°C into the water. Determine (*a*) the final equilibrium temperature in the tank and (*b*) the exergy destroyed during this process. The melting temperature and the heat of fusion of ice at atmospheric pressure are 0°C and 333.7 kJ/kg, respectively. Take T_0 = 20°C.

8–121 One ton of liquid water at 65°C is brought into a well-insulated and well-sealed 3-m × 4-m × 7-m room initially at 16°C and 100 kPa. Assuming constant specific heats for both the air and water at room temperature, determine (*a*) the final equilibrium temperature in the room, (*b*) the exergy destruction, (*c*) the maximum amount of work that can be produced during this process, in kJ. Take T_0 = 10°C.

8–122 Consider a 12-L evacuated rigid bottle that is surrounded by the atmosphere at 100 kPa and 17°C. A valve at the neck of the bottle is now opened and the atmospheric air is allowed to flow into the bottle. The air trapped in the bottle eventually reaches thermal equilibrium with the atmosphere as a result of heat transfer through the wall of the bottle. The valve remains open during the process so that the trapped air also reaches mechanical equilibrium with the atmosphere. Determine the net heat transfer through the wall of the bottle and the exergy destroyed during this filling process.

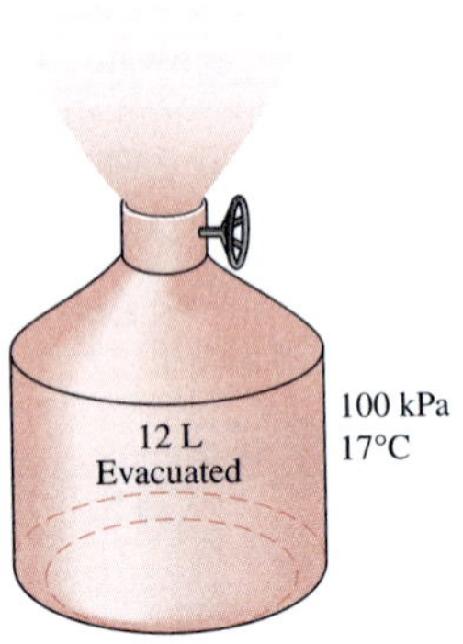

FIGURE P8–122

8–123 A frictionless piston-cylinder device, shown in Fig. P8-123, initially contains 0.01 m³ of argon gas at 400 K and 350 kPa. Heat is now transferred to the argon from a furnace at 1200 K, and the argon expands isothermally until its volume is doubled. No heat transfer takes place between the argon and the surrounding atmospheric air, which is at 300 K and 100 kPa. Determine (*a*) the useful work output, (*b*) the exergy destroyed, and (*c*) the maximum work that can be produced during this process.

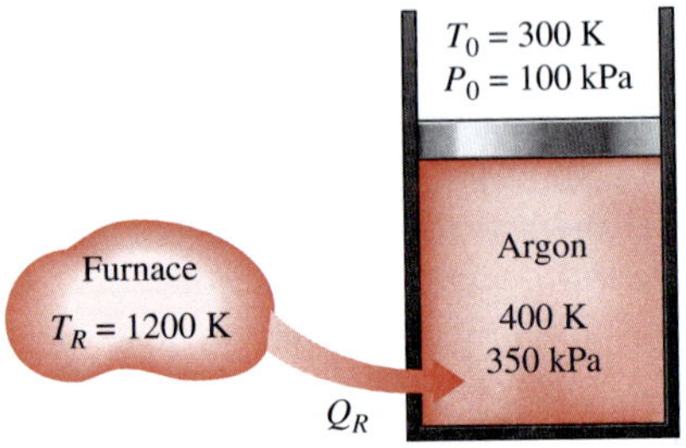

FIGURE P8–123

8–124 Two constant-pressure devices, each filled with 30 kg of air, have temperatures of 900 K and 300 K. A heat engine placed between the two devices extracts heat from the high-temperature device, produces work, and rejects heat to the low-temperature device. Determine the maximum work that can be produced by the heat engine and the final temperatures of the devices. Assume constant specific heats at room temperature.

8–125 A constant-volume tank contains 20 kg of nitrogen at 1000 K, and a constant-pressure device contains 10 kg of argon at 300 K. A heat engine placed between the tank and device extracts heat from the high-temperature tank, produces work, and rejects heat to the low-temperature device. Determine the maximum work that can be produced by the heat engine and the final temperatures of the nitrogen and argon. Assume constant specific heats at room temperature.

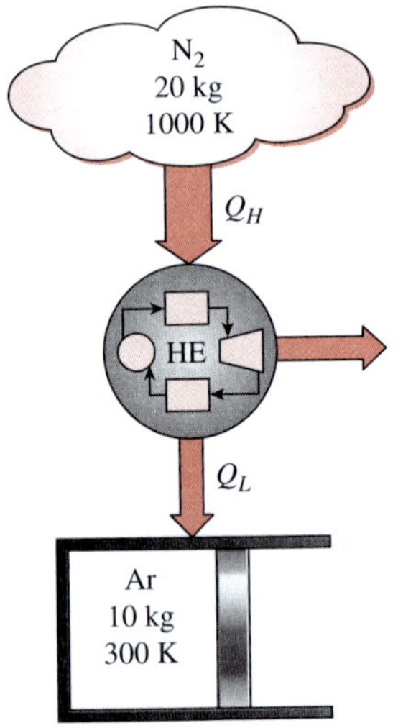

FIGURE P8–125

8–126 A rigid 50-L nitrogen cylinder is equipped with a safety relief valve set at 1200 kPa. Initially, this cylinder contains nitrogen at 1200 kPa and 20°C. Heat is now transferred to the nitrogen from a thermal energy reservoir at 500°C, and nitrogen is allowed to escape until the mass of nitrogen becomes one-half of its initial mass. Determine the change in the nitrogen's work potential as a result of this heating. Take $T_0 = 20°C$.

8–127 A 100-L well-insulated rigid tank is initially filled with nitrogen at 1000 kPa and 20°C. Now a valve is opened and one-half of nitrogen's mass is allowed to escape. Determine the change in the exergy content of the tank.

8–128 Steam is to be condensed in the condenser of a steam power plant at a temperature of 45°C with cooling water from a nearby lake that enters the tubes of the condenser at 12°C at a rate of 330 kg/s and leaves at 20°C. Assuming the condenser to be perfectly insulated, determine (*a*) the rate of condensation of the steam and (*b*) the rate of exergy destruction in the condenser. *Answers: (a) 4.61 kg, (b) 998 kW*

8–129 The compressed-air storage tank shown in Fig. P8–129 has a volume of 500,000 m³, and it initially contains air at 100 kPa and 20°C. The isentropic compressor proceeds to compress air that enters the compressor at 100 kPa and 20°C until the tank is filled at 600 kPa and 20°C. All heat exchanges are with the surrounding air at 20°C. Calculate the change in the work potential of the air stored in the tank. How does this compare to the work required to compress the air as the tank was being filled?

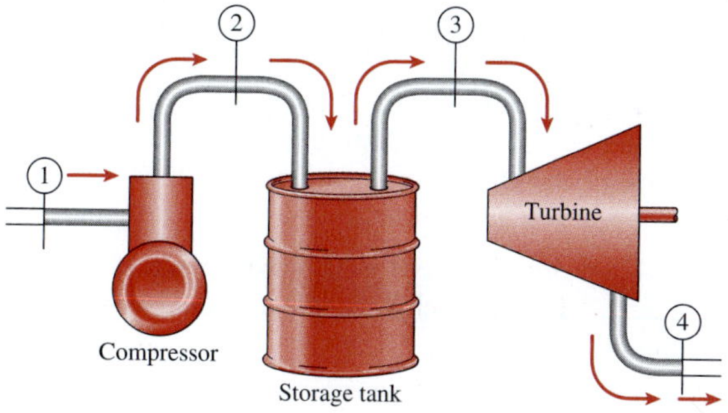

FIGURE P8–129

8–130 The air stored in the tank of the previous problem is now released through the isentropic turbine until the tank contents are at 100 kPa and 20°C. The pressure is always 100 kPa at the turbine outlet, and all heat exchanges are with the surrounding air, which is at 20°C. How does the total work produced by the turbine compare to the change in the work potential of the air in the storage tank?

8–131 A constant-volume tank has a temperature of 600 K and a constant-pressure device has a temperature of 280 K. Both the tank and device are filled with 40 kg of air. A heat engine placed between the tank and device receives heat from the high-temperature tank, produces work, and rejects heat to the low-temperature device. Determine the maximum work that can be produced by the heat engine and the final temperatures of the tank and device. Assume constant specific heats at room temperature.

8–132E In a production facility, 1.2-in-thick, 2-ft × 2-ft square brass plates (ρ = 532.5 lbm/ft³ and c_p = 0.091 Btu/lbm·°F) that are initially at a uniform temperature of 75°F are heated by passing them through an oven at 1300°F at a rate of 300 per minute. If the plates remain in the oven until their average temperature rises to 1000°F, determine the rate of heat transfer to the plates in the furnace and the rate of exergy destruction associated with this heat transfer process.

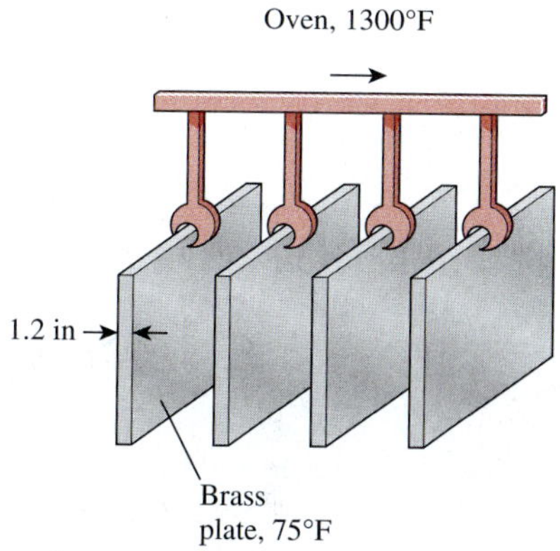

FIGURE P8–132E

8–133 Long cylindrical steel rods (ρ = 7833 kg/m³ and c_p = 0.465 kJ/kg·°C) of 10-cm diameter are heat-treated by drawing them at a velocity of 3 m/min through a 6-m-long oven maintained at 900°C. If the rods enter the oven at 30°C and leave at 700°C, determine (*a*) the rate of heat transfer to the rods in the oven and (*b*) the rate of exergy destruction associated with this heat transfer process. Take T_0 = 25°C.

8–134 A well-insulated heat exchanger is to heat water (c_p = 4.18 kJ/kg·°C) from 25°C to 60°C at a rate of 0.4 kg/s. The heating is to be accomplished by geothermal water (c_p = 4.31 kJ/kg·°C) available at 140°C at a mass flow rate of 0.3 kg/s. The inner tube is thin-walled and has a diameter of 0.6 cm. Determine (*a*) the rate of heat transfer and (*b*) the rate of exergy destruction in the heat exchanger.

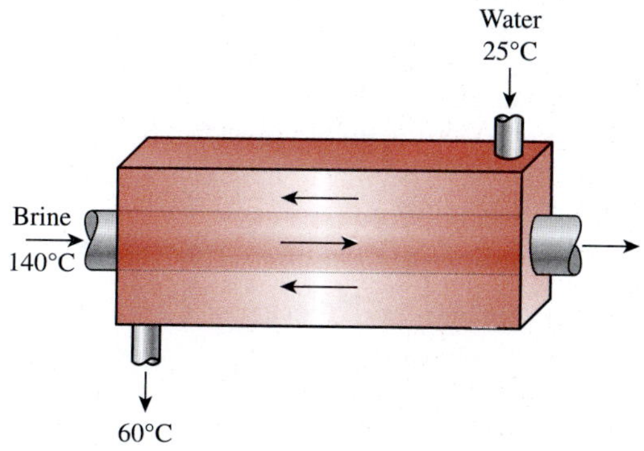

FIGURE P8–134

8–135 In a dairy plant, milk at 4°C is pasteurized continuously at 72°C at a rate of 12 L/s for 24 h/day and 365 days/yr. The milk is heated to the pasteurizing temperature by hot water heated in a natural gas-fired boiler having an efficiency of 82 percent. The pasteurized milk is then cooled by cold water at 18°C before it is finally refrigerated back to 4°C. To save energy and money, the plant installs a regenerator that has an effectiveness of 82 percent. If the cost of natural gas is \$1.04/therm (1 therm = 105,500 kJ), determine how much energy and money the regenerator will save this company per year and the annual reduction in exergy destruction.

8–136 Combustion gases enter a gas turbine at 627°C and 1.2 MPa at a rate of 2.5 kg/s and leave at 527°C and 500 kPa. It is estimated that heat is lost from the turbine at a rate of 20 kW. Using air properties for the combustion gases and assuming the surroundings to be at 25°C and 100 kPa, determine (*a*) the actual and reversible power outputs of the turbine, (*b*) the exergy destroyed within the turbine, and (*c*) the second-law efficiency of the turbine.

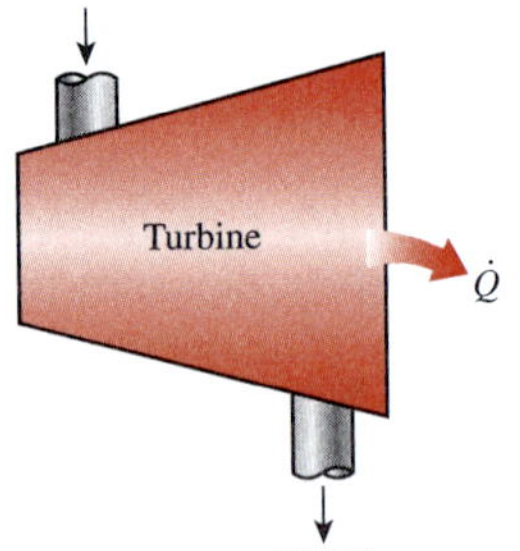

FIGURE P8–136

8–137 Refrigerant-134a enters an adiabatic compressor at 160 kPa superheated by 3°C, and leaves at 1.0 MPa. If the compressor has a second-law efficiency of 80 percent, determine (*a*) the actual work input, (*b*) the isentropic efficiency, and (*c*) the exergy destruction. Take the environment temperature to be 25°C. *Answers:* (*a*) 49.8 kJ/kg, (*b*) 0.78, (*c*) 9.95 kJ/kg

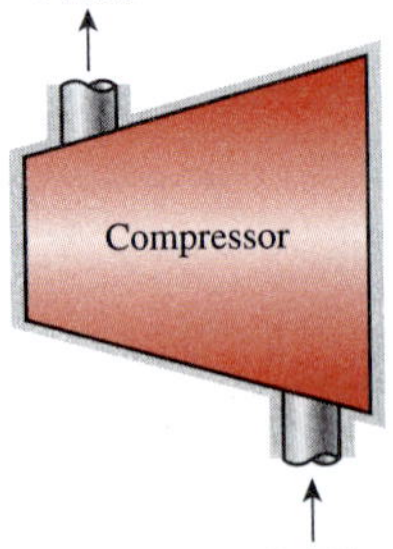

FIGURE P8–137

8–138 Water enters a pump at 100 kPa and 30°C at a rate of 1.35 kg/s, and leaves at 4 MPa. If the pump has an isentropic efficiency of 70 percent, determine (*a*) the actual power input, (*b*) the rate of frictional heating, (*c*) the exergy destruction, and (*d*) the second-law efficiency for an environment temperature of 20°C.

8–139 Argon gas expands from 3.5 MPa and 100°C to 500 kPa in an adiabatic expansion valve. For environment conditions of 100 kPa and 25°C, determine (*a*) the exergy of argon at the inlet, (*b*) the exergy destruction during the process, and (*c*) the second-law efficiency.

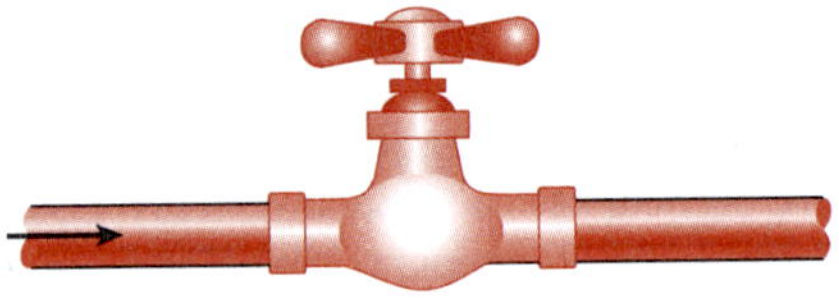

FIGURE P8–139

8–140 Nitrogen gas enters a diffuser at 100 kPa and 110°C with a velocity of 205 m/s, and leaves at 110 kPa and 45 m/s. It is estimated that 2.5 kJ/kg of heat is lost from the diffuser to the surroundings at 100 kPa and 27°C. The exit area of the diffuser is 0.04 m^2. Accounting for the variation of the specific heats with temperature, determine (*a*) the exit temperature, (*b*) the rate of exergy destruction, and (*c*) the second-law efficiency of the diffuser. *Answers:* (*a*) 127°C, (*b*) 12.4 kW, (*c*) 76.1 percent

8–141 Can closed-system exergy be negative? How about flow exergy? Explain using an incompressible substance as an example.

8–142 Obtain a relation for the second-law efficiency of a heat engine that receives heat Q_H from a source at temperature T_H and rejects heat Q_L to a sink at T_L, which is higher than T_0 (the temperature of the surroundings), while producing work in the amount of W.

8–143 Writing the first- and second-law relations and simplifying, obtain the reversible work relation for a closed system that exchanges heat with the surrounding medium at T_0 in the amount of Q_0 as well as a heat reservoir at T_R in the amount of Q_R. (*Hint:* Eliminate Q_0 between the two equations.)

8–144 Writing the first- and second-law relations and simplifying, obtain the reversible work relation for a steady-flow system that exchanges heat with the surrounding medium at T_0 a rate of $\dot{Q}_0$ as well as a thermal reservoir at T_R at a rate of $\dot{Q}_R$. (*Hint:* Eliminate $\dot{Q}_0$ between the two equations.)

8–145 Writing the first- and second-law relations and simplifying, obtain the reversible work relation for a uniform-flow system that exchanges heat with the surrounding medium at T_0 in the amount of Q_0 as well as a heat reservoir at T_R in the amount of Q_R. (*Hint:* Eliminate Q_0 between the two equations.)

Fundamentals of Engineering (FE) Exam Problems

8–146 Heat is lost through a plane wall steadily at a rate of 800 W. If the inner and outer surface temperatures of the wall are 20°C and 5°C, respectively, and the environment temperature is 0°C, the rate of exergy destruction within the wall is

(*a*) 40 W (*b*) 17,500 W (*c*) 765 W
(*d*) 32,800 W (*e*) 0 W

8–147 Liquid water enters an adiabatic piping system at 15°C at a rate of 3 kg/s. It is observed that the water temperature rises by 0.3°C in the pipe due to friction. If the environment temperature is also 15°C, the rate of exergy destruction in the pipe is

(*a*) 3.8 kW (*b*) 24 kW (*c*) 72 kW
(*d*) 98 kW (*e*) 124 kW

8–148 A heat engine receives heat from a source at 1500 K at a rate of 600 kJ/s and rejects the waste heat to a sink at 300 K. If the power output of the engine is 400 kW, the second-law efficiency of this heat engine is

(*a*) 42% (*b*) 53% (*c*) 83%
(*d*) 67% (*e*) 80%

8–149 A water reservoir contains 100 tons of water at an average elevation of 60 m. The maximum amount of electric power that can be generated from this water is

(*a*) 8 kWh (*b*) 16 kWh (*c*) 1630 kWh
(*d*) 16,300 kWh (*e*) 58,800 kWh

8–150 A house is maintained at 21°C in winter by electric resistance heaters. If the outdoor temperature is 9°C, the second-law efficiency of the resistance heaters is

(*a*) 0% (*b*) 4.1% (*c*) 5.7%
(*d*) 25% (*e*) 100%

8–151 A 12-kg solid whose specific heat is 2.8 kJ/kg·°C is at a uniform temperature of −10°C. For an environment temperature of 20°C, the exergy content of this solid is

(*a*) Less than zero (*b*) 0 kJ (*c*) 4.6 kJ
(*d*) 55 kJ (*e*) 1008 kJ

8–152 Keeping the limitations imposed by the second law of thermodynamics in mind, choose the *wrong* statement below:

(*a*) A heat engine cannot have a thermal efficiency of 100%.
(*b*) For all reversible processes, the second-law efficiency is 100%.
(*c*) The second-law efficiency of a heat engine cannot be greater than its thermal efficiency.
(*d*) The second-law efficiency of a process is 100% if no entropy is generated during that process.
(*e*) The coefficient of performance of a refrigerator can be greater than 1.

8–153 A furnace can supply heat steadily at a 1300 K at a rate of 500 kJ/s. The maximum amount of power that can be produced by using the heat supplied by this furnace in an environment at 300 K is

(*a*) 115 kW (*b*) 192 kW (*c*) 385 kW
(*d*) 500 kW (*e*) 650 kW

8–154 Air is throttled from 50°C and 800 kPa to a pressure of 200 kPa at a rate of 0.5 kg/s in an environment at 25°C. The change in kinetic energy is negligible, and no heat transfer occurs during the process. The power potential wasted during this process is

(*a*) 0 (*b*) 0.20 kW (*c*) 47 kW
(*d*) 59 kW (*e*) 119 kW

8–155 Steam enters a turbine steadily at 4 MPa and 400°C and exits at 0.2 MPa and 150°C in an environment at 25°C. The decrease in the exergy of the steam as it flows through the turbine is

(*a*) 58 kJ/kg (*b*) 445 kJ/kg (*c*) 458 kJ/kg
(*d*) 518 kJ/kg (*e*) 597 kJ/kg

Design and Essay Problems

8–156 Human beings are probably the most capable creatures, and they have a high level of physical, intellectual, emotional, and spiritual potentials or exergies. Unfortunately people make little use of their exergies, letting most of their exergies go to waste. Draw four exergy versus time charts, and plot your physical, intellectual, emotional, and spiritual exergies on each of these charts for a 24-h period using your best judgment based on your experience. On these four charts, plot your respective exergies that you have utilized during the last 24 h. Compare the two plots on each chart and determine if you are living a "full" life or if you are wasting your life away. Can you think of any ways to reduce the mismatch between your exergies and your utilization of them?

8–157 The domestic hot-water systems involve a high level of irreversibility and thus they have low second-law efficiencies. The water in these systems is heated from about 15°C to about 60°C, and most of the hot water is mixed with cold water to reduce its temperature to 45°C or even lower before it is used for any useful purpose such as taking a shower or washing clothes at a warm setting. The water is discarded at about the same temperature at which it was used and replaced by fresh cold water at 15°C. Redesign a typical residential hot-water system such that the irreversibility is greatly reduced. Draw a sketch of your proposed design.

8–158 A steam boiler may be thought of as a heat exchanger. The combustion gases may be modeled as a stream of air because their thermodynamic properties are close to those of air. Using this model, consider a boiler that is to convert saturated liquid water at 500 psia to a saturated vapor while maintaining the water pressure constant. Determine the temperature at which the air (i.e., combustion gases) must enter this unit so that the transfer of exergy from the air to the boiling water is done at the minimum loss.

8–159 An adiabatic nozzle is designed to accelerate an ideal gas from nearly 0 m/s, P_1, and T_1 to V m/s. As the efficiency of this nozzle decreases, the pressure at the nozzle exit must also be decreased to maintain the speed at V. Plot the change in the flow exergy as a function of the nozzle efficiency for an ideal gas (say, air).

8–160 The temperature of the air in a building can be maintained at a desirable level during winter by using different methods of heating. Compare heating this air in a heat exchanger unit with condensing steam to heating it with an electric-resistance heater. Perform a second-law analysis to determine the heating method that generates the least entropy and thus causes the least exergy destruction.

CHAPTER

9

GAS POWER CYCLES

Two important areas of application for thermodynamics are power generation and refrigeration. Both are usually accomplished by systems that operate on a thermodynamic cycle. Thermodynamic cycles can be divided into two general categories: *power cycles,* which are discussed in this chapter and Chap. 10, and *refrigeration cycles,* which are discussed in Chap. 11.

The devices or systems used to produce a net power output are often called *engines,* and the thermodynamic cycles they operate on are called *power cycles.* The devices or systems used to produce a refrigeration effect are called *refrigerators, air conditioners,* or *heat pumps,* and the cycles they operate on are called *refrigeration cycles.*

Thermodynamic cycles can also be categorized as *gas cycles* and *vapor cycles,* depending on the *phase* of the working fluid. In gas cycles, the working fluid remains in the gaseous phase throughout the entire cycle, whereas in vapor cycles the working fluid exists in the vapor phase during one part of the cycle and in the liquid phase during another part.

Thermodynamic cycles can be categorized yet another way: *closed* and *open cycles.* In closed cycles, the working fluid is returned to the initial state at the end of the cycle and is recirculated. In open cycles, the working fluid is renewed at the end of each cycle instead of being recirculated. In automobile engines, the combustion gases are exhausted and replaced by fresh air–fuel mixture at the end of each cycle. The engine operates on a mechanical cycle, but the working fluid does not go through a complete thermodynamic cycle.

Heat engines are categorized as *internal combustion* and *external combustion engines,* depending on how the heat is supplied to the working fluid. In external combustion engines (such as steam power plants), heat is supplied to the working fluid from an external source such as a furnace, a geothermal well, a nuclear reactor, or even the sun. In internal combustion engines (such as automobile engines), this is done by burning the fuel within the system boundaries. In this chapter, various gas power cycles are analyzed under some simplifying assumptions.

Objectives

The objectives of Chapter 9 are to:

- Evaluate the performance of gas power cycles for which the working fluid remains a gas throughout the entire cycle.
- Develop simplifying assumptions applicable to gas power cycles.
- Review the operation of reciprocating engines.
- Analyze both closed and open gas power cycles.
- Solve problems based on the Otto, Diesel, Stirling, and Ericsson cycles.
- Solve problems based on the Brayton cycle; the Brayton cycle with regeneration; and the Brayton cycle with intercooling, reheating, and regeneration.
- Analyze jet-propulsion cycles.
- Identify simplifying assumptions for second-law analysis of gas power cycles.
- Perform second-law analysis of gas power cycles.

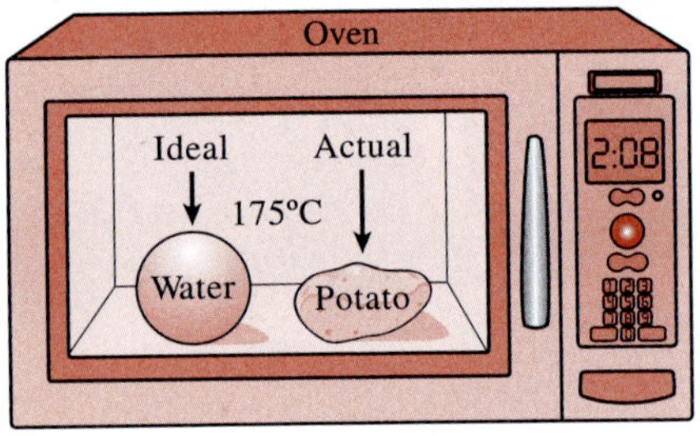

FIGURE 9–1
Modeling is a powerful engineering tool that provides great insight and simplicity at the expense of some loss in accuracy.

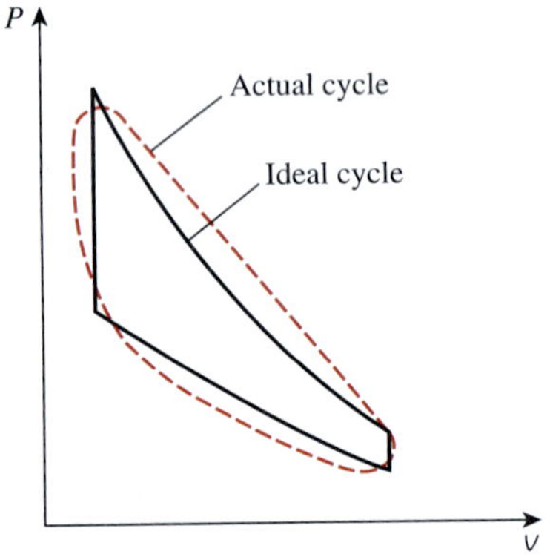

FIGURE 9–2
The analysis of many complex processes can be reduced to a manageable level by utilizing some idealizations.

FIGURE 9–3
Care should be exercised in the interpretation of the results from ideal cycles.

BLONDIE © KING FEATURES SYNDICATE.

9–1 ▪ BASIC CONSIDERATIONS IN THE ANALYSIS OF POWER CYCLES

Most power-producing devices operate on cycles, and the study of power cycles is an exciting and important part of thermodynamics. The cycles encountered in actual devices are difficult to analyze because of the presence of complicating effects, such as friction, and the absence of sufficient time for establishment of the equilibrium conditions during the cycle. To make an analytical study of a cycle feasible, we have to keep the complexities at a manageable level and utilize some idealizations (Fig. 9–1). When the actual cycle is stripped of all the internal irreversibilities and complexities, we end up with a cycle that resembles the actual cycle closely but is made up totally of internally reversible processes. Such a cycle is called an **ideal cycle** (Fig. 9–2).

A simple idealized model enables engineers to study the effects of the major parameters that dominate the cycle without getting bogged down in the details. The cycles discussed in this chapter are somewhat idealized, but they still retain the general characteristics of the actual cycles they represent. The conclusions reached from the analysis of ideal cycles are also applicable to actual cycles. The thermal efficiency of the Otto cycle, the ideal cycle for spark-ignition automobile engines, for example, increases with the compression ratio. This is also the case for actual automobile engines. The numerical values obtained from the analysis of an ideal cycle, however, are not necessarily representative of the actual cycles, and care should be exercised in their interpretation (Fig. 9–3). The simplified analysis presented in this chapter for various power cycles of practical interest may also serve as the starting point for a more in-depth study.

Heat engines are designed for the purpose of converting thermal energy to work, and their performance is expressed in terms of the **thermal efficiency** η_{th}, which is the ratio of the net work produced by the engine to the total heat input:

$$\eta_{th} = \frac{W_{net}}{Q_{in}} \quad \text{or} \quad \eta_{th} = \frac{w_{net}}{q_{in}} \tag{9–1}$$

Recall that heat engines that operate on a totally reversible cycle, such as the Carnot cycle, have the highest thermal efficiency of all heat engines operating between the same temperature levels. That is, nobody can develop a cycle more efficient than the *Carnot cycle.* Then the following question arises naturally: If the Carnot cycle is the best possible cycle, why do we not use it as the model cycle for all the heat engines instead of bothering with several so-called *ideal* cycles? The answer to this question is hardware-related. Most cycles encountered in practice differ significantly from the Carnot cycle, which makes it unsuitable as a realistic model. Each ideal cycle discussed in this chapter is related to a specific work-producing device and is an *idealized* version of the actual cycle.

The ideal cycles are *internally reversible,* but, unlike the Carnot cycle, they are not necessarily externally reversible. That is, they may involve irreversibilities external to the system such as heat transfer through a finite temperature difference. Therefore, the thermal efficiency of an ideal cycle, in general, is less than that of a totally reversible cycle operating between the same

FIGURE 9–4
An automotive engine with the combustion chamber exposed.

Courtesy of General Motors.

temperature limits. However, it is still considerably higher than the thermal efficiency of an actual cycle because of the idealizations utilized (Fig. 9–4).

The idealizations and simplifications commonly employed in the analysis of power cycles can be summarized as follows:

1. The cycle does not involve any *friction.* Therefore, the working fluid does not experience any pressure drop as it flows in pipes or devices such as heat exchangers.
2. All expansion and compression processes take place in a *quasi-equilibrium* manner.
3. The pipes connecting the various components of a system are well insulated, and *heat transfer* through them is negligible.

Neglecting the changes in *kinetic* and *potential energies* of the working fluid is another commonly utilized simplification in the analysis of power cycles. This is a reasonable assumption since in devices that involve shaft work, such as turbines, compressors, and pumps, the kinetic and potential energy terms are usually very small relative to the other terms in the energy equation. Fluid velocities encountered in devices such as condensers, boilers, and mixing chambers are typically low, and the fluid streams experience little change in their velocities, again making kinetic energy changes negligible. The only devices where the changes in kinetic energy are significant are the nozzles and diffusers, which are specifically designed to create large changes in velocity.

In the preceding chapters, *property diagrams* such as the *P*-*v* and *T*-*s* diagrams have served as valuable aids in the analysis of thermodynamic processes. On both the *P*-*v* and *T*-*s* diagrams, the area enclosed by the process curves of a cycle represents the net work produced during the cycle (Fig. 9–5), which is also equivalent to the net heat transfer for that cycle. The *T*-*s* diagram is particularly useful as a visual aid in the analysis of ideal

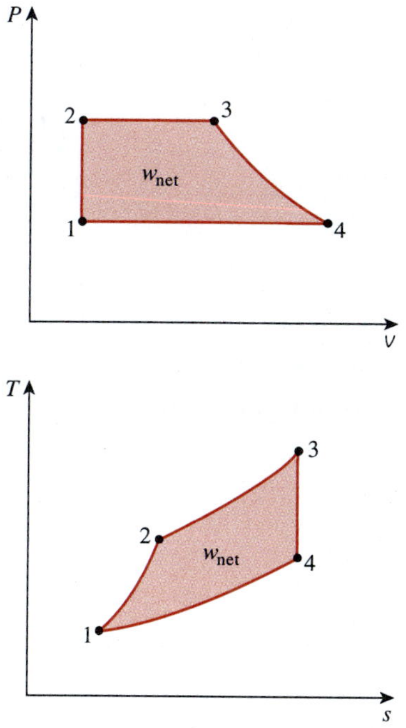

FIGURE 9–5
On both *P*-*v* and *T*-*s* diagrams, the area enclosed by the process curve represents the net work of the cycle.

power cycles. An ideal power cycle does not involve any internal irreversibilities, and so the only effect that can change the entropy of the working fluid during a process is heat transfer.

On a *T-s* diagram, a *heat-addition* process proceeds in the direction of increasing entropy, a *heat-rejection* process proceeds in the direction of decreasing entropy, and an *isentropic* (internally reversible, adiabatic) process proceeds at constant entropy. The area under the process curve on a *T-s* diagram represents the heat transfer for that process. The area under the heat addition process on a *T-s* diagram is a geometric measure of the total heat supplied during the cycle q_{in}, and the area under the heat rejection process is a measure of the total heat rejected q_{out}. The difference between these two (the area enclosed by the cyclic curve) is the net heat transfer, which is also the net work produced during the cycle. Therefore, on a *T-s* diagram, the ratio of the area enclosed by the cyclic curve to the area under the heat-addition process curve represents the thermal efficiency of the cycle. *Any modification that increases the ratio of these two areas will also increase the thermal efficiency of the cycle.*

Although the working fluid in an ideal power cycle operates on a closed loop, the type of individual processes that comprises the cycle depends on the individual devices used to execute the cycle. In the Rankine cycle, which is the ideal cycle for steam power plants, the working fluid flows through a series of steady-flow devices such as the turbine and condenser, whereas in the Otto cycle, which is the ideal cycle for the spark-ignition automobile engine, the working fluid is alternately expanded and compressed in a piston–cylinder device. Therefore, equations pertaining to steady-flow systems should be used in the analysis of the Rankine cycle, and equations pertaining to closed systems should be used in the analysis of the Otto cycle.

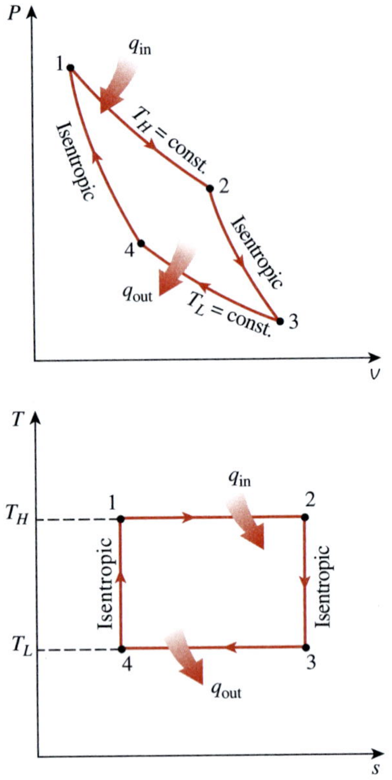

FIGURE 9–6
P-v and *T-s* diagrams of a Carnot cycle.

9–2 ▪ THE CARNOT CYCLE AND ITS VALUE IN ENGINEERING

The Carnot cycle is composed of four totally reversible processes: isothermal heat addition, isentropic expansion, isothermal heat rejection, and isentropic compression. The *P-v* and *T-s* diagrams of a Carnot cycle are replotted in Fig. 9–6. The Carnot cycle can be executed in a closed system (a piston–cylinder device) or a steady-flow system (utilizing two turbines and two compressors, as shown in Fig. 9–7), and either a gas or a vapor can be utilized as the working fluid. The Carnot cycle is the most efficient cycle that can be executed between a heat source at temperature T_H and a sink at temperature T_L, and its thermal efficiency is expressed as

$$\eta_{th,Carnot} = 1 - \frac{T_L}{T_H} \tag{9–2}$$

Reversible isothermal heat transfer is very difficult to achieve in reality because it would require very large heat exchangers and it would take a very long time (a power cycle in a typical engine is completed in a fraction of a second). Therefore, it is not practical to build an engine that would operate on a cycle that closely approximates the Carnot cycle.

The real value of the Carnot cycle comes from its being a standard against which the actual or the ideal cycles can be compared. The thermal

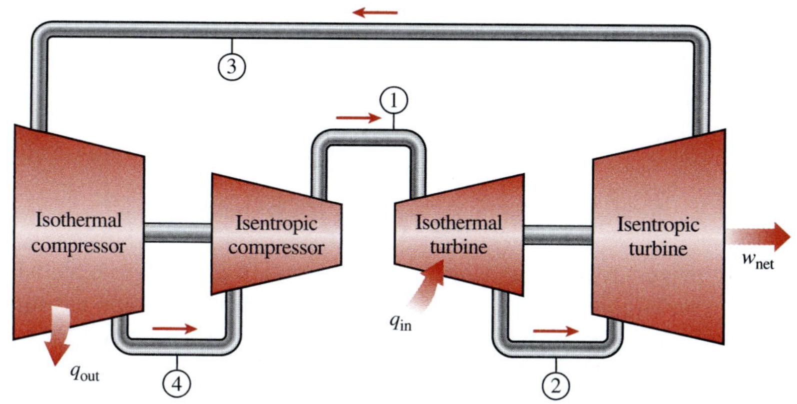

FIGURE 9–7
A steady-flow Carnot engine.

efficiency of the Carnot cycle is a function of the sink and source temperatures only, and the thermal efficiency relation for the Carnot cycle (Eq. 9–2) conveys an important message that is equally applicable to both ideal and actual cycles: *Thermal efficiency increases with an increase in the average temperature at which heat is supplied to the system or with a decrease in the average temperature at which heat is rejected from the system.*

The source and sink temperatures that can be used in practice are not without limits, however. The highest temperature in the cycle is limited by the maximum temperature that the components of the heat engine, such as the piston or the turbine blades, can withstand. The lowest temperature is limited by the temperature of the cooling medium utilized in the cycle such as a lake, a river, or the atmospheric air.

EXAMPLE 9–1 Derivation of the Efficiency of the Carnot Cycle

Show that the thermal efficiency of a Carnot cycle operating between the temperature limits of T_H and T_L is solely a function of these two temperatures and is given by Eq. 9–2.

Solution It is to be shown that the efficiency of a Carnot cycle depends on the source and sink temperatures alone.

Analysis The *T-s* diagram of a Carnot cycle is redrawn in Fig. 9–8. All four processes that comprise the Carnot cycle are reversible, and thus the area under each process curve represents the heat transfer for that process. Heat is transferred to the system during process 1-2 and rejected during process 3-4. Therefore, the amount of heat input and heat output for the cycle can be expressed as

$$q_{in} = T_H(s_2 - s_1) \quad \text{and} \quad q_{out} = T_L(s_3 - s_4) = T_L(s_2 - s_1)$$

since processes 2-3 and 4-1 are isentropic, and thus $s_2 = s_3$ and $s_4 = s_1$. Substituting these into Eq. 9–1, we see that the thermal efficiency of a Carnot cycle is

$$\eta_{th} = \frac{w_{net}}{q_{in}} = 1 - \frac{q_{out}}{q_{in}} = 1 - \frac{T_L(s_2 - s_1)}{T_H(s_2 - s_1)} = 1 - \frac{T_L}{T_H}$$

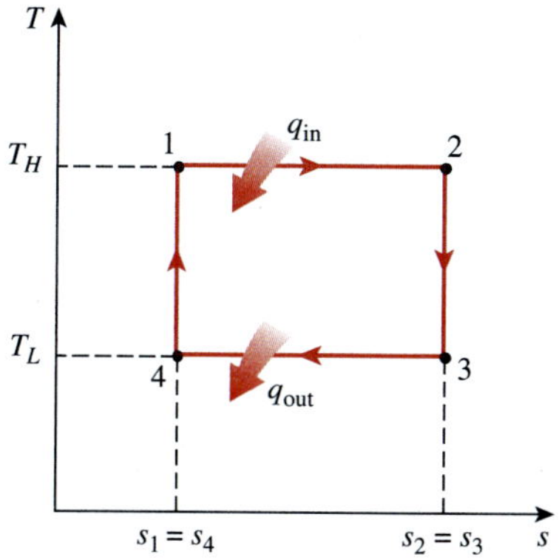

FIGURE 9–8
T-s diagram for Example 9–1.

Discussion Notice that the thermal efficiency of a Carnot cycle is independent of the type of the working fluid used (an ideal gas, steam, etc.) or whether the cycle is executed in a closed or steady-flow system.

9–3 ▪ AIR-STANDARD ASSUMPTIONS

In gas power cycles, the working fluid remains a gas throughout the entire cycle. Spark-ignition engines, diesel engines, and conventional gas turbines are familiar examples of devices that operate on gas cycles. In all these engines, energy is provided by burning a fuel within the system boundaries. That is, they are *internal combustion engines.* Because of this combustion process, the composition of the working fluid changes from air and fuel to combustion products during the course of the cycle. However, considering that air is predominantly nitrogen that undergoes hardly any chemical reactions in the combustion chamber, the working fluid closely resembles air at all times.

Even though internal combustion engines operate on a mechanical cycle (the piston returns to its starting position at the end of each revolution), the working fluid does not undergo a complete thermodynamic cycle. It is thrown out of the engine at some point in the cycle (as exhaust gases) instead of being returned to the initial state. Working on an open cycle is the characteristic of all internal combustion engines.

The actual gas power cycles are rather complex. To reduce the analysis to a manageable level, we utilize the following approximations, commonly known as the **air-standard assumptions**:

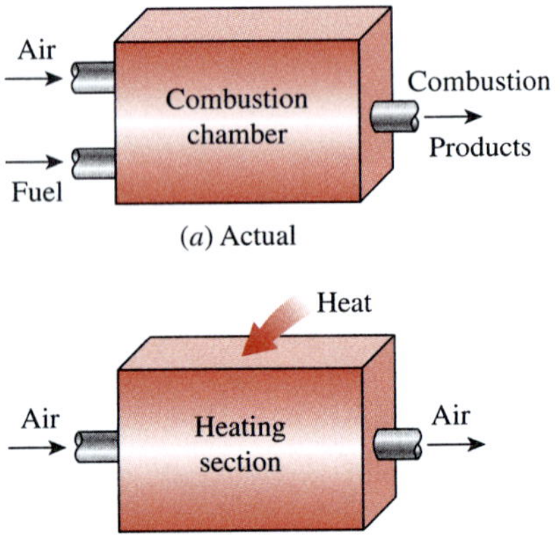

(*b*) Ideal

FIGURE 9–9
The combustion process is replaced by a heat-addition process in ideal cycles.

1. The working fluid is air, which continuously circulates in a closed loop and always behaves as an ideal gas.
2. All the processes that make up the cycle are internally reversible.
3. The combustion process is replaced by a heat-addition process from an external source (Fig. 9–9).
4. The exhaust process is replaced by a heat-rejection process that restores the working fluid to its initial state.

Another assumption that is often utilized to simplify the analysis even more is that air has constant specific heats whose values are determined at *room temperature* (25°C, or 77°F). When this assumption is utilized, the air-standard assumptions are called the **cold-air-standard assumptions**. A cycle for which the air-standard assumptions are applicable is frequently referred to as an **air-standard cycle**.

The air-standard assumptions previously stated provide considerable simplification in the analysis without significantly deviating from the actual cycles. This simplified model enables us to study qualitatively the influence of major parameters on the performance of the actual engines.

9–4 ▪ AN OVERVIEW OF RECIPROCATING ENGINES

Despite its simplicity, the reciprocating engine (basically a piston–cylinder device) is one of the rare inventions that has proved to be very versatile and to have a wide range of applications. It is the powerhouse of the vast majority of automobiles, trucks, light aircraft, ships, and electric power generators, as well as many other devices.

The basic components of a reciprocating engine are shown in Fig. 9–10. The piston reciprocates in the cylinder between two fixed positions called the **top dead center** (TDC)—the position of the piston when it forms the smallest volume in the cylinder—and the **bottom dead center** (BDC)—the position of the piston when it forms the largest volume in the cylinder. The distance between the TDC and the BDC is the largest distance that the piston can travel in one direction, and it is called the **stroke** of the engine. The diameter of the piston is called the **bore**. The air or air–fuel mixture is drawn into the cylinder through the **intake valve**, and the combustion products are expelled from the cylinder through the **exhaust valve**.

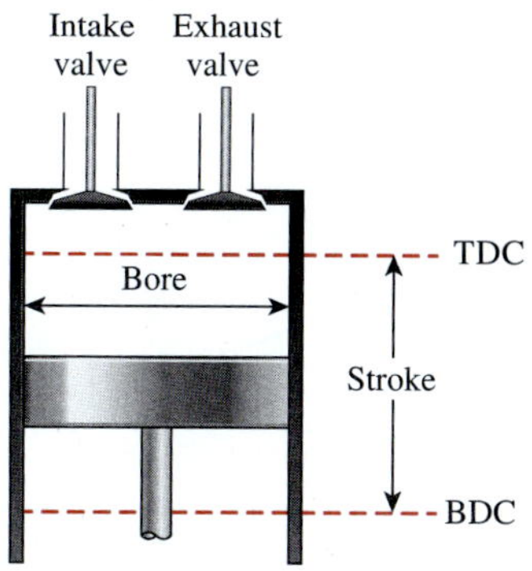

FIGURE 9–10
Nomenclature for reciprocating engines.

The minimum volume formed in the cylinder when the piston is at TDC is called the **clearance volume** (Fig. 9–11). The volume displaced by the piston as it moves between TDC and BDC is called the **displacement volume**. The ratio of the maximum volume formed in the cylinder to the minimum (clearance) volume is called the **compression ratio** r of the engine:

$$r = \frac{V_{max}}{V_{min}} = \frac{V_{BDC}}{V_{TDC}} \qquad \textbf{(9–3)}$$

Notice that the compression ratio is a *volume ratio* and should not be confused with the pressure ratio.

Another term frequently used in conjunction with reciprocating engines is the **mean effective pressure** (MEP). It is a fictitious pressure that, if it acted on the piston during the entire power stroke, would produce the same amount of net work as that produced during the actual cycle (Fig. 9–12). That is,

$$W_{net} = \text{MEP} \times \text{Piston area} \times \text{Stroke} = \text{MEP} \times \text{Displacement volume}$$

or

$$\text{MEP} = \frac{W_{net}}{V_{max} - V_{min}} = \frac{w_{net}}{v_{max} - v_{min}} \quad \text{(kPa)} \qquad \textbf{(9–4)}$$

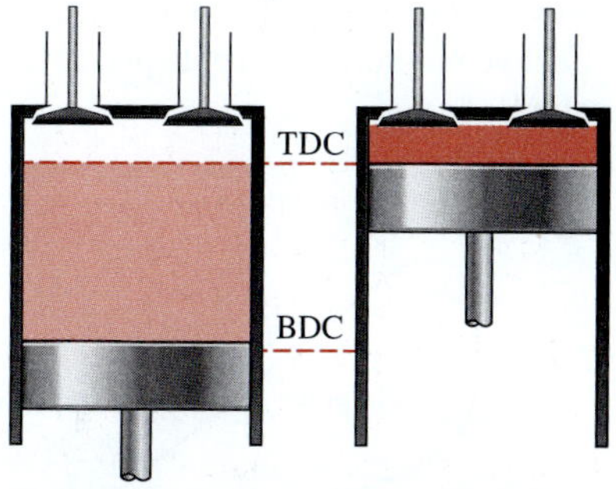

(*a*) Displacement volume

(*b*) Clearance volume

FIGURE 9–11
Displacement and clearance volumes of a reciprocating engine.

The mean effective pressure can be used as a parameter to compare the performances of reciprocating engines of equal size. The engine with a larger value of MEP delivers more net work per cycle and thus performs better.

Reciprocating engines are classified as **spark-ignition (SI) engines** or **compression-ignition (CI) engines**, depending on how the combustion process in the cylinder is initiated. In SI engines, the combustion of the air–fuel mixture is initiated by a spark plug. In CI engines, the air–fuel mixture is self-ignited as a result of compressing the mixture above its self-ignition temperature. In the next two sections, we discuss the *Otto* and *Diesel cycles,* which are the ideal cycles for the SI and CI reciprocating engines, respectively.

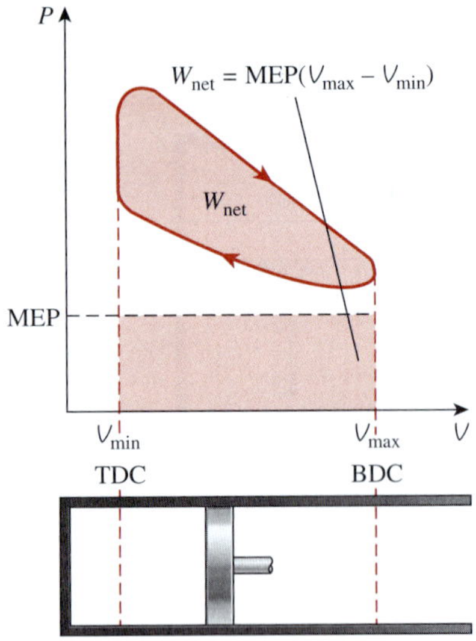

FIGURE 9–12
The net work output of a cycle is equivalent to the product of the mean effective pressure and the displacement volume.

9–5 ▪ OTTO CYCLE: THE IDEAL CYCLE FOR SPARK-IGNITION ENGINES

The Otto cycle is the ideal cycle for spark-ignition reciprocating engines. It is named after Nikolaus A. Otto, who built a successful four-stroke engine in 1876 in Germany using the cycle proposed by Frenchman Beau de Rochas in 1862. In most spark-ignition engines, the piston executes four complete strokes (two mechanical cycles) within the cylinder, and the crankshaft completes two revolutions for each thermodynamic cycle. These engines are called **four-stroke** internal combustion engines. A schematic of each stroke as well as a *P-v* diagram for an actual four-stroke spark-ignition engine is given in Fig. 9–13(*a*).

Initially, both the intake and the exhaust valves are closed, and the piston is at its lowest position (BDC). During the *compression stroke,* the piston moves upward, compressing the air–fuel mixture. Shortly before the piston reaches its highest position (TDC), the spark plug fires and the mixture ignites, increasing the pressure and temperature of the system. The high-pressure gases force the piston down, which in turn forces the crankshaft to rotate, producing a useful work output during the *expansion* or *power stroke.* At the end of this stroke, the piston is at its lowest position (the completion of the first mechanical cycle), and the cylinder is filled with combustion

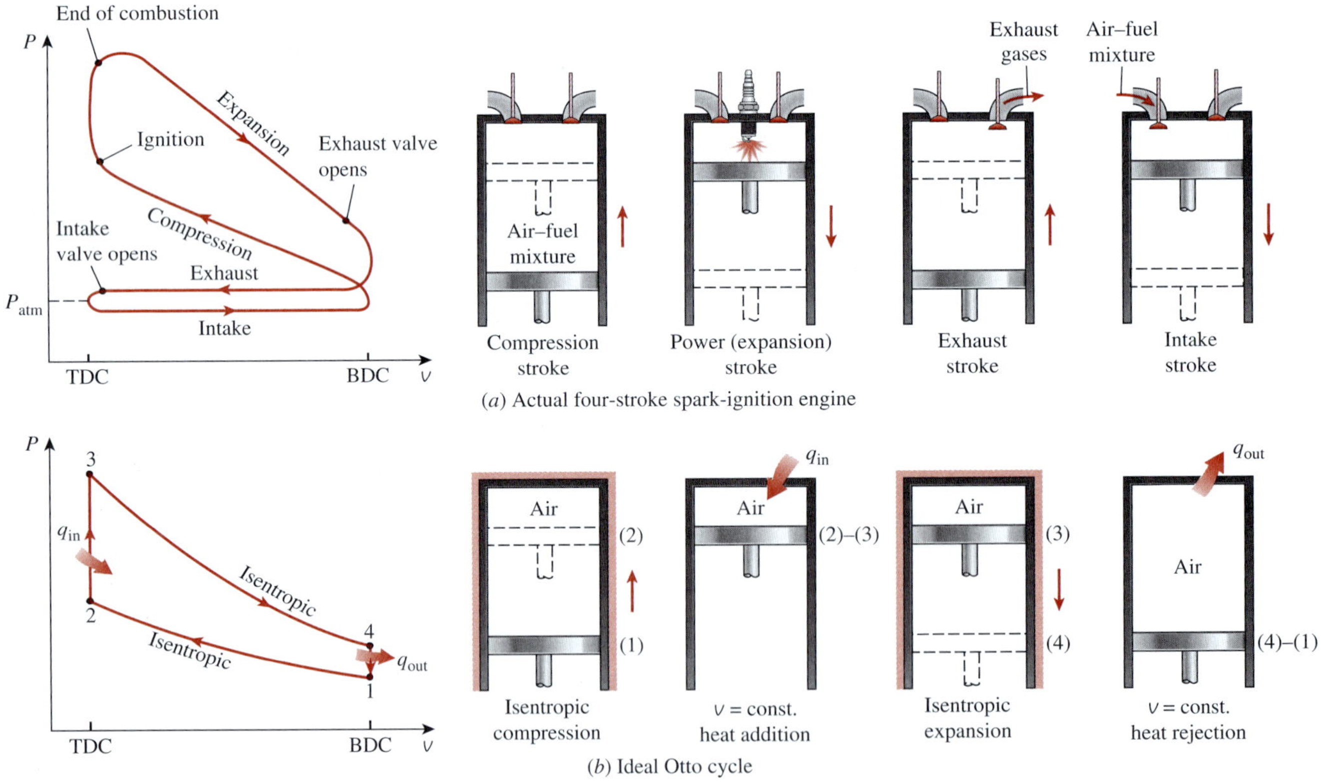

FIGURE 9–13
Actual and ideal cycles in spark-ignition engines and their *P-v* diagrams.

products. Now the piston moves upward one more time, purging the exhaust gases through the exhaust valve (the *exhaust stroke*), and down a second time, drawing in fresh air–fuel mixture through the intake valve (the *intake stroke*). Notice that the pressure in the cylinder is slightly above the atmospheric value during the exhaust stroke and slightly below during the intake stroke.

In **two-stroke engines**, all four functions described above are executed in just two strokes: the power stroke and the compression stroke. In these engines, the crankcase is sealed, and the outward motion of the piston is used to slightly pressurize the air–fuel mixture in the crankcase, as shown in Fig. 9–14. Also, the intake and exhaust valves are replaced by openings in the lower portion of the cylinder wall. During the latter part of the power stroke, the piston uncovers first the exhaust port, allowing the exhaust gases to be partially expelled, and then the intake port, allowing the fresh air–fuel mixture to rush in and drive most of the remaining exhaust gases out of the cylinder. This mixture is then compressed as the piston moves upward during the compression stroke and is subsequently ignited by a spark plug.

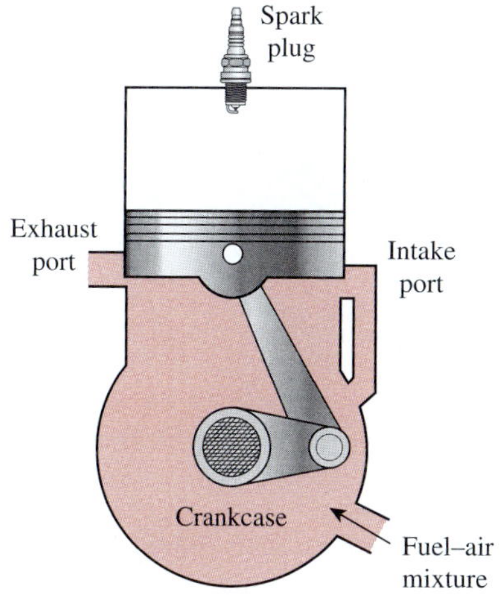

FIGURE 9–14
Schematic of a two-stroke reciprocating engine.

The two-stroke engines are generally less efficient than their four-stroke counterparts because of the incomplete expulsion of the exhaust gases and the partial expulsion of the fresh air–fuel mixture with the exhaust gases. However, they are relatively simple and inexpensive, and they have high power-to-weight and power-to-volume ratios, which make them suitable for applications requiring small size and weight such as for motorcycles, chain saws, and lawn mowers (Fig. 9–15).

Advances in several technologies—such as direct fuel injection, stratified charge combustion, and electronic controls—brought about a renewed interest in two-stroke engines that can offer high performance and fuel economy while satisfying the stringent emission requirements. For a given weight and displacement, a well-designed two-stroke engine can provide significantly more power than its four-stroke counterpart because two-stroke engines produce power on every engine revolution instead of every other one. In the new two-stroke engines, the highly atomized fuel spray that is injected into the combustion chamber toward the end of the compression stroke burns much more completely. The fuel is sprayed after the exhaust valve is closed, which prevents unburned fuel from being ejected into the atmosphere. With stratified combustion, the flame that is initiated by igniting a small amount of the rich fuel–air mixture near the spark plug propagates through the combustion chamber filled with a much leaner mixture, and this results in much cleaner combustion. Also, the advances in electronics have made it possible to ensure the optimum operation under varying engine load and speed conditions. Major car companies have research programs underway on two-stroke engines which are expected to make a comeback in the future.

FIGURE 9–15
Two-stroke engines are commonly used in motorcycles and lawn mowers.

The thermodynamic analysis of the actual four-stroke or two-stroke cycles described is not a simple task. However, the analysis can be simplified significantly if the air-standard assumptions are utilized. The resulting cycle, which closely resembles the actual operating conditions, is the ideal **Otto cycle**. It consists of four internally reversible processes:

1-2 Isentropic compression
2-3 Constant-volume heat addition

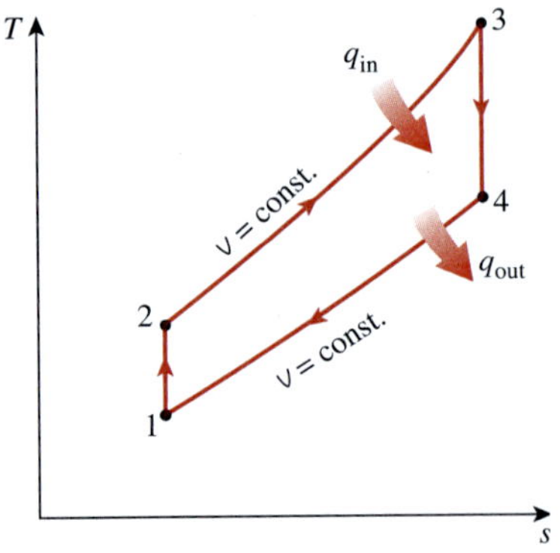

FIGURE 9–16
T-s diagram of the ideal Otto cycle.

3-4 Isentropic expansion
4-1 Constant-volume heat rejection

The execution of the Otto cycle in a piston–cylinder device together with a *P-v* diagram is illustrated in Fig. 9–13*b*. The *T-s* diagram of the Otto cycle is given in Fig. 9–16.

The Otto cycle is executed in a closed system, and disregarding the changes in kinetic and potential energies, the energy balance for any of the processes is expressed, on a unit-mass basis, as

$$(q_{in} - q_{out}) + (w_{in} - w_{out}) = \Delta u \qquad \text{(kJ/kg)} \tag{9–5}$$

No work is involved during the two heat transfer processes since both take place at constant volume. Therefore, heat transfer to and from the working fluid can be expressed as

$$q_{in} = u_3 - u_2 = c_v(T_3 - T_2) \tag{9–6a}$$

and

$$q_{out} = u_4 - u_1 = c_v(T_4 - T_1) \tag{9–6b}$$

Then the thermal efficiency of the ideal Otto cycle under the cold air standard assumptions becomes

$$\eta_{th,Otto} = \frac{w_{net}}{q_{in}} = 1 - \frac{q_{out}}{q_{in}} = 1 - \frac{T_4 - T_1}{T_3 - T_2} = 1 - \frac{T_1(T_4/T_1 - 1)}{T_2(T_3/T_2 - 1)}$$

Processes 1-2 and 3-4 are isentropic, and $v_2 = v_3$ and $v_4 = v_1$. Thus,

$$\frac{T_1}{T_2} = \left(\frac{v_2}{v_1}\right)^{k-1} = \left(\frac{v_3}{v_4}\right)^{k-1} = \frac{T_4}{T_3} \tag{9–7}$$

Substituting these equations into the thermal efficiency relation and simplifying give

$$\eta_{th,Otto} = 1 - \frac{1}{r^{k-1}} \tag{9–8}$$

where

$$r = \frac{V_{max}}{V_{min}} = \frac{V_1}{V_2} = \frac{v_1}{v_2} \tag{9–9}$$

is the **compression ratio** and k is the specific heat ratio c_p/c_v.

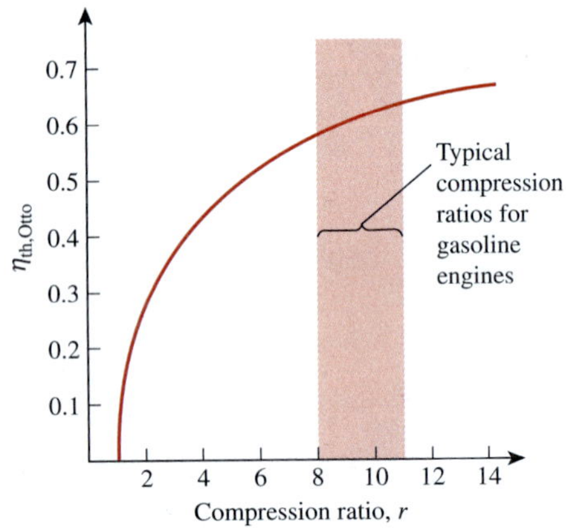

FIGURE 9–17
Thermal efficiency of the ideal Otto cycle as a function of compression ratio ($k = 1.4$).

Equation 9–8 shows that under the cold-air-standard assumptions, the thermal efficiency of an ideal Otto cycle depends on the compression ratio of the engine and the specific heat ratio of the working fluid. The thermal efficiency of the ideal Otto cycle increases with both the compression ratio and the specific heat ratio. This is also true for actual spark-ignition internal combustion engines. A plot of thermal efficiency versus the compression ratio is given in Fig. 9–17 for $k = 1.4$, which is the specific heat ratio value of air at room temperature. For a given compression ratio, the thermal efficiency of an actual spark-ignition engine is less than that of an ideal Otto cycle because of the irreversibilities, such as friction, and other factors such as incomplete combustion.

We can observe from Fig. 9–17 that the thermal efficiency curve is rather steep at low compression ratios but flattens out starting with a compression ratio value of about 8. Therefore, the increase in thermal efficiency with the compression ratio is not as pronounced at high compression ratios. Also, when high compression ratios are used, the temperature of the air–fuel mixture rises above the autoignition temperature of the fuel (the temperature at which the fuel ignites without the help of a spark) during the combustion process, causing an early and rapid burn of the fuel at some point or points ahead of the flame front, followed by almost instantaneous inflammation of the end gas. This premature ignition of the fuel, called **autoignition**, produces an audible noise, which is called **engine knock**. Autoignition in spark-ignition engines cannot be tolerated because it hurts performance and can cause engine damage. The requirement that autoignition not be allowed places an upper limit on the compression ratios that can be used in spark-ignition internal combustion engines.

Improvement of the thermal efficiency of gasoline engines by utilizing higher compression ratios (up to about 12) without facing the autoignition problem has been made possible by using gasoline blends that have good antiknock characteristics, such as gasoline mixed with tetraethyl lead. Tetraethyl lead had been added to gasoline since the 1920s because it is an inexpensive method of raising the *octane rating,* which is a measure of the engine knock resistance of a fuel. Leaded gasoline, however, has a very undesirable side effect: it forms compounds during the combustion process that are hazardous to health and pollute the environment. In an effort to combat air pollution, the government adopted a policy in the mid-1970s that resulted in the eventual phase-out of leaded gasoline. Unable to use lead, the refiners developed other techniques to improve the antiknock characteristics of gasoline. Most cars made since 1975 have been designed to use unleaded gasoline, and the compression ratios had to be lowered to avoid engine knock. The ready availability of high octane fuels made it possible to raise the compression ratios again in recent years. Also, owing to the improvements in other areas (reduction in overall automobile weight, improved aerodynamic design, etc.), today's cars have better fuel economy and consequently get more miles per gallon of fuel. This is an example of how engineering decisions involve compromises, and efficiency is only one of the considerations in final design.

The second parameter affecting the thermal efficiency of an ideal Otto cycle is the specific heat ratio k. For a given compression ratio, an ideal Otto cycle using a monatomic gas (such as argon or helium, $k = 1.667$) as the working fluid will have the highest thermal efficiency. The specific heat ratio k, and thus the thermal efficiency of the ideal Otto cycle, decreases as the molecules of the working fluid get larger (Fig. 9–18). At room temperature it is 1.4 for air, 1.3 for carbon dioxide, and 1.2 for ethane. The working fluid in actual engines contains larger molecules such as carbon dioxide, and the specific heat ratio decreases with temperature, which is one of the reasons that the actual cycles have lower thermal efficiencies than the ideal Otto cycle. The thermal efficiencies of actual spark-ignition engines range from about 25 to 30 percent.

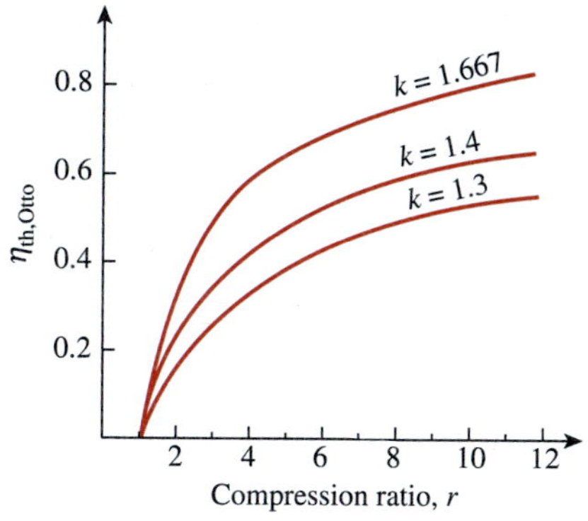

FIGURE 9–18
The thermal efficiency of the Otto cycle increases with the specific heat ratio k of the working fluid.

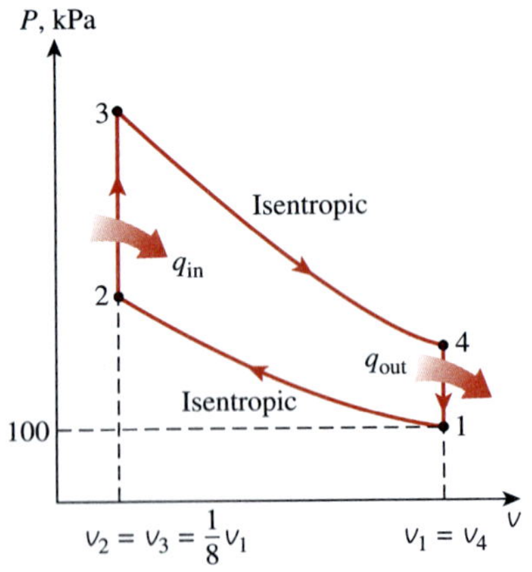

FIGURE 9–19

P-V diagram for the Otto cycle discussed in Example 9–2.

EXAMPLE 9–2 The Ideal Otto Cycle

An ideal Otto cycle has a compression ratio of 8. At the beginning of the compression process, air is at 100 kPa and 17°C, and 800 kJ/kg of heat is transferred to air during the constant-volume heat-addition process. Accounting for the variation of specific heats of air with temperature, determine (*a*) the maximum temperature and pressure that occur during the cycle, (*b*) the net work output, (*c*) the thermal efficiency, and (*d*) the mean effective pressure for the cycle.

Solution An ideal Otto cycle is considered. The maximum temperature and pressure, the net work output, the thermal efficiency, and the mean effective pressure are to be determined.

Assumptions **1** The air-standard assumptions are applicable. **2** Kinetic and potential energy changes are negligible. **3** The variation of specific heats with temperature is to be accounted for.

Analysis The P-V diagram of the ideal Otto cycle described is shown in Fig. 9–19. We note that the air contained in the cylinder forms a closed system.

(*a*) The maximum temperature and pressure in an Otto cycle occur at the end of the constant-volume heat-addition process (state 3). But first we need to determine the temperature and pressure of air at the end of the isentropic compression process (state 2), using data from Table A–17:

$$T_1 = 290 \text{ K} \rightarrow u_1 = 206.91 \text{ kJ/kg}$$
$$V_{r1} = 676.1$$

Process 1-2 (isentropic compression of an ideal gas):

$$\frac{V_{r2}}{V_{r1}} = \frac{V_2}{V_1} = \frac{1}{r} \rightarrow V_{r2} = \frac{V_{r1}}{r} = \frac{676.1}{8} = 84.51 \rightarrow T_2 = 652.4 \text{ K}$$
$$u_2 = 475.11 \text{ kJ/kg}$$

$$\frac{P_2 V_2}{T_2} = \frac{P_1 V_1}{T_1} \rightarrow P_2 = P_1\left(\frac{T_2}{T_1}\right)\left(\frac{V_1}{V_2}\right)$$
$$= (100 \text{ kPa})\left(\frac{652.4 \text{ K}}{290 \text{ K}}\right)(8) = 1799.7 \text{ kPa}$$

Process 2-3 (constant-volume heat addition):

$$q_{in} = u_3 - u_2$$
$$800 \text{ kJ/kg} = u_3 - 475.11 \text{ kJ/kg}$$
$$u_3 = 1275.11 \text{ kJ/kg} \rightarrow T_3 = \mathbf{1575.1 \text{ K}}$$
$$V_{r3} = 6.108$$

$$\frac{P_3 V_3}{T_3} = \frac{P_2 V_2}{T_2} \rightarrow P_3 = P_2\left(\frac{T_3}{T_2}\right)\left(\frac{V_2}{V_3}\right)$$
$$= (1.7997 \text{ MPa})\left(\frac{1575.1 \text{ K}}{652.4 \text{ K}}\right)(1) = \mathbf{4.345 \text{ MPa}}$$

(*b*) The net work output for the cycle is determined either by finding the boundary ($P\,dV$) work involved in each process by integration and adding them or by finding the net heat transfer that is equivalent to the net work done during the cycle. We take the latter approach. However, first we need to find the internal energy of the air at state 4:

Process 3-4 (isentropic expansion of an ideal gas):

$$\frac{v_{r4}}{v_{r3}} = \frac{v_4}{v_3} = r \rightarrow v_{r4} = rv_{r3} = (8)(6.108) = 48.864 \rightarrow T_4 = 795.6\text{ K}$$

$$u_4 = 588.74\text{ kJ/kg}$$

Process 4-1 (constant-volume heat rejection):

$$-q_{\text{out}} = u_1 - u_4 \rightarrow q_{\text{out}} = u_4 - u_1$$

$$q_{\text{out}} = 588.74 - 206.91 = 381.83\text{ kJ/kg}$$

Thus,

$$w_{\text{net}} = q_{\text{net}} = q_{\text{in}} - q_{\text{out}} = 800 - 381.83 = \mathbf{418.17\ kJ/kg}$$

(*c*) The thermal efficiency of the cycle is determined from its definition:

$$\eta_{\text{th}} = \frac{w_{\text{net}}}{q_{\text{in}}} = \frac{418.17\text{ kJ/kg}}{800\text{ kJ/kg}} = \mathbf{0.523\ or\ 52.3\%}$$

Under the cold-air-standard assumptions (constant specific heat values at room temperature), the thermal efficiency would be (Eq. 9–8)

$$\eta_{\text{th,Otto}} = 1 - \frac{1}{r^{k-1}} = 1 - r^{1-k} = 1 - (8)^{1-1.4} = 0.565\text{ or }56.5\%$$

which is considerably different from the value obtained above. Therefore, care should be exercised in utilizing the cold-air-standard assumptions.

(*d*) The mean effective pressure is determined from its definition, Eq. 9–4:

$$\text{MEP} = \frac{w_{\text{net}}}{v_1 - v_2} = \frac{w_{\text{net}}}{v_1 - v_1/r} = \frac{w_{\text{net}}}{v_1(1 - 1/r)}$$

where

$$v_1 = \frac{RT_1}{P_1} = \frac{(0.287\text{ kPa·m}^3/\text{kg·K})(290\text{ K})}{100\text{ kPa}} = 0.832\text{ m}^3/\text{kg}$$

Thus,

$$\text{MEP} = \frac{418.17\text{ kJ/kg}}{(0.832\text{ m}^3/\text{kg})\left(1 - \frac{1}{8}\right)}\left(\frac{1\text{ kPa·m}^3}{1\text{ kJ}}\right) = \mathbf{574\ kPa}$$

Discussion Note that a constant pressure of 574 kPa during the power stroke would produce the same net work output as the entire cycle.

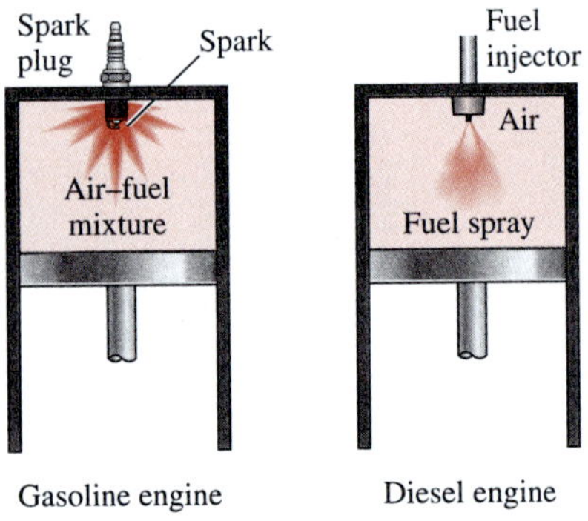

FIGURE 9–20
In diesel engines, the spark plug is replaced by a fuel injector, and only air is compressed during the compression process.

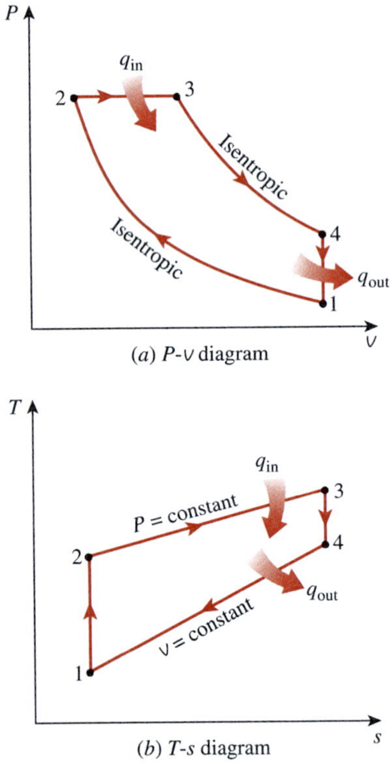

FIGURE 9–21
T-s and *P-v* diagrams for the ideal Diesel cycle.

9–6 ▪ DIESEL CYCLE: THE IDEAL CYCLE FOR COMPRESSION-IGNITION ENGINES

The Diesel cycle is the ideal cycle for CI reciprocating engines. The CI engine, first proposed by Rudolph Diesel in the 1890s, is very similar to the SI engine discussed in the last section, differing mainly in the method of initiating combustion. In spark-ignition engines (also known as *gasoline engines*), the air–fuel mixture is compressed to a temperature that is below the autoignition temperature of the fuel, and the combustion process is initiated by firing a spark plug. In CI engines (also known as *diesel engines*), the air is compressed to a temperature that is above the autoignition temperature of the fuel, and combustion starts on contact as the fuel is injected into this hot air. Therefore, the spark plug and carburetor are replaced by a fuel injector in diesel engines (Fig. 9–20).

In gasoline engines, a mixture of air and fuel is compressed during the compression stroke, and the compression ratios are limited by the onset of autoignition or engine knock. In diesel engines, only air is compressed during the compression stroke, eliminating the possibility of autoignition. Therefore, diesel engines can be designed to operate at much higher compression ratios, typically between 12 and 24. Not having to deal with the problem of autoignition has another benefit: many of the stringent requirements placed on the gasoline can now be removed, and fuels that are less refined (thus less expensive) can be used in diesel engines.

The fuel injection process in diesel engines starts when the piston approaches TDC and continues during the first part of the power stroke. Therefore, the combustion process in these engines takes place over a longer interval. Because of this longer duration, the combustion process in the ideal Diesel cycle is approximated as a constant-pressure heat-addition process. In fact, this is the only process where the Otto and the Diesel cycles differ. The remaining three processes are the same for both ideal cycles. That is, process 1-2 is isentropic compression, 2-3 is constant-pressure heat addition, 3-4 is isentropic expansion, and 4-1 is constant-volume heat rejection. The similarity between the two cycles is also apparent from the *P-v* and *T-s* diagrams of the Diesel cycle, shown in Fig. 9–21.

Noting that the Diesel cycle is executed in a piston–cylinder device, which forms a closed system, the amount of heat transferred to the working fluid at constant pressure and rejected from it at constant volume can be expressed as

$$q_{in} - w_{b,out} = u_3 - u_2 \rightarrow q_{in} = P_2(v_3 - v_2) + (u_3 - u_2)$$
$$= h_3 - h_2 = c_p(T_3 - T_2) \quad \textbf{(9–10a)}$$

and

$$-q_{out} = u_1 - u_4 \rightarrow q_{out} = u_4 - u_1 = c_v(T_4 - T_1) \quad \textbf{(9–10b)}$$

Then the thermal efficiency of the ideal Diesel cycle under the cold-air-standard assumptions becomes

$$\eta_{th,Diesel} = \frac{w_{net}}{q_{in}} = 1 - \frac{q_{out}}{q_{in}} = 1 - \frac{T_4 - T_1}{k(T_3 - T_2)} = 1 - \frac{T_1(T_4/T_1 - 1)}{kT_2(T_3/T_2 - 1)}$$

We now define a new quantity, the **cutoff ratio** r_c, as the ratio of the cylinder volumes after and before the combustion process:

$$r_c = \frac{V_3}{V_2} = \frac{v_3}{v_2} \qquad \textbf{(9–11)}$$

Utilizing this definition and the isentropic ideal-gas relations for processes 1-2 and 3-4, we see that the thermal efficiency relation reduces to

$$\eta_{\text{th,Diesel}} = 1 - \frac{1}{r^{k-1}}\left[\frac{r_c^k - 1}{k(r_c - 1)}\right] \qquad \textbf{(9–12)}$$

where r is the compression ratio defined by Eq. 9–9. Looking at Eq. 9–12 carefully, one would notice that under the cold-air-standard assumptions, the efficiency of a Diesel cycle differs from the efficiency of an Otto cycle by the quantity in the brackets. This quantity is always greater than 1. Therefore,

$$\eta_{\text{th,Otto}} > \eta_{\text{th,Diesel}} \qquad \textbf{(9–13)}$$

when both cycles operate on the same compression ratio. Also, as the cutoff ratio decreases, the efficiency of the Diesel cycle increases (Fig. 9–22). For the limiting case of $r_c = 1$, the quantity in the brackets becomes unity (can you prove it?), and the efficiencies of the Otto and Diesel cycles become identical. Remember, though, that diesel engines operate at much higher compression ratios and thus are usually more efficient than the spark-ignition (gasoline) engines. The diesel engines also burn the fuel more completely since they usually operate at lower revolutions per minute and the air–fuel mass ratio is much higher than spark-ignition engines. Thermal efficiencies of large diesel engines range from about 35 to 40 percent.

The higher efficiency and lower fuel costs of diesel engines make them attractive in applications requiring relatively large amounts of power, such as in locomotive engines, emergency power generation units, large ships, and heavy trucks. As an example of how large a diesel engine can be, a 12-cylinder diesel engine built in 1964 by the Fiat Corporation of Italy had a normal power output of 25,200 hp (18.8 MW) at 122 rpm, a cylinder bore of 90 cm, and a stroke of 91 cm.

In modern high-speed compression ignition engines, fuel is injected into the combustion chamber much sooner compared to the early diesel engines. Fuel starts to ignite late in the compression stroke, and consequently part of the combustion occurs almost at constant volume. Fuel injection continues until the piston reaches the top dead center, and combustion of the fuel keeps the pressure high well into the expansion stroke. Thus, the entire combustion process can better be modeled as the combination of constant-volume and constant-pressure processes. The ideal cycle based on this concept is called the **dual cycle** and P-v diagram for it is given in Fig. 9–23. The relative amounts of heat transferred during each process can be adjusted to approximate the actual cycle more closely. Note that both the Otto and the Diesel cycles can be obtained as special cases of the dual cycle. Dual cycle is a more realistic model than diesel cycle for representing modern, high-speed compression ignition engines.

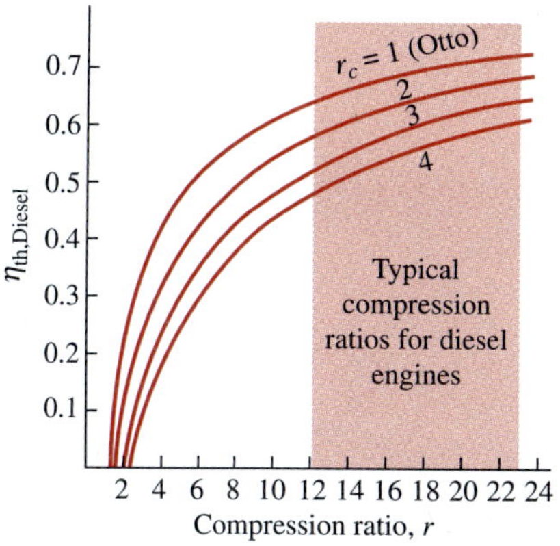

FIGURE 9–22
Thermal efficiency of the ideal Diesel cycle as a function of compression and cutoff ratios ($k = 1.4$).

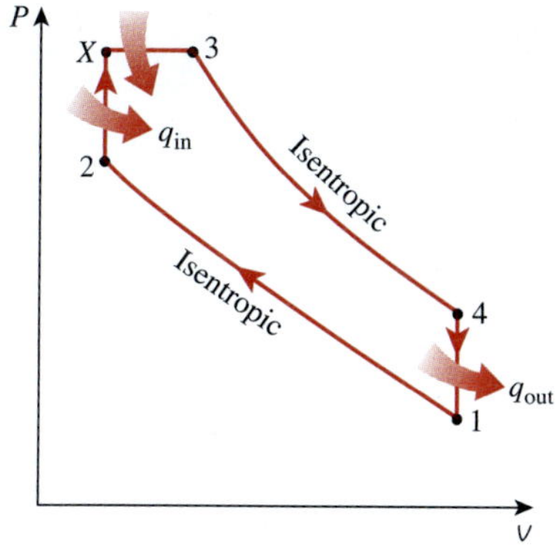

FIGURE 9–23
P-v diagram of an ideal dual cycle.

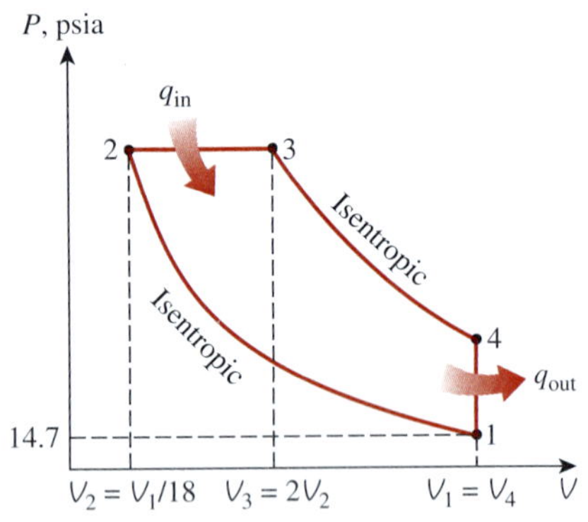

FIGURE 9–24

P-V diagram for the ideal Diesel cycle discussed in Example 9–3.

EXAMPLE 9–3 The Ideal Diesel Cycle

An ideal Diesel cycle with air as the working fluid has a compression ratio of 18 and a cutoff ratio of 2. At the beginning of the compression process, the working fluid is at 14.7 psia, 80°F, and 117 in^3. Utilizing the cold-air-standard assumptions, determine (*a*) the temperature and pressure of air at the end of each process, (*b*) the net work output and the thermal efficiency, and (*c*) the mean effective pressure.

Solution An ideal Diesel cycle is considered. The temperature and pressure at the end of each process, the net work output, the thermal efficiency, and the mean effective pressure are to be determined.

Assumptions **1** The cold-air-standard assumptions are applicable and thus air can be assumed to have constant specific heats at room temperature. **2** Kinetic and potential energy changes are negligible.

Properties The gas constant of air is $R = 0.3704$ psia·ft^3/lbm·R and its other properties at room temperature are $c_p = 0.240$ Btu/lbm·R, $c_v = 0.171$ Btu/lbm·R, and $k = 1.4$ (Table A–2E*a*).

Analysis The *P-V* diagram of the ideal Diesel cycle described is shown in Fig. 9–24. We note that the air contained in the cylinder forms a closed system.

(*a*) The temperature and pressure values at the end of each process can be determined by utilizing the ideal-gas isentropic relations for processes 1-2 and 3-4. But first we determine the volumes at the end of each process from the definitions of the compression ratio and the cutoff ratio:

$$V_2 = \frac{V_1}{r} = \frac{117 \text{ in}^3}{18} = 6.5 \text{ in}^3$$

$$V_3 = r_c V_2 = (2)(6.5 \text{ in}^3) = 13 \text{ in}^3$$

$$V_4 = V_1 = 117 \text{ in}^3$$

Process 1-2 (isentropic compression of an ideal gas, constant specific heats):

$$T_2 = T_1\left(\frac{V_1}{V_2}\right)^{k-1} = (540 \text{ R})(18)^{1.4-1} = \mathbf{1716\ R}$$

$$P_2 = P_1\left(\frac{V_1}{V_2}\right)^{k} = (14.7 \text{ psia})(18)^{1.4} = \mathbf{841\ psia}$$

Process 2-3 (constant-pressure heat addition to an ideal gas):

$$P_3 = P_2 = \mathbf{841\ psia}$$

$$\frac{P_2V_2}{T_2} = \frac{P_3V_3}{T_3} \longrightarrow T_3 = T_2\left(\frac{V_3}{V_2}\right) = (1716 \text{ R})(2) = \mathbf{3432\ R}$$

Process 3-4 (isentropic expansion of an ideal gas, constant specific heats):

$$T_4 = T_3\left(\frac{V_3}{V_4}\right)^{k-1} = (3432 \text{ R})\left(\frac{13 \text{ in}^3}{117 \text{ in}^3}\right)^{1.4-1} = \mathbf{1425\ R}$$

$$P_4 = P_3\left(\frac{V_3}{V_4}\right)^{k} = (841 \text{ psia})\left(\frac{13 \text{ in}^3}{117 \text{ in}^3}\right)^{1.4} = \mathbf{38.8\ psia}$$

(*b*) The net work for a cycle is equivalent to the net heat transfer. But first we find the mass of air:

$$m = \frac{P_1 V_1}{RT_1} = \frac{(14.7\ \text{psia})(117\ \text{in}^3)}{(0.3704\ \text{psia·ft}^3/\text{lbm·R})(540\ \text{R})}\left(\frac{1\ \text{ft}^3}{1728\ \text{in}^3}\right) = 0.00498\ \text{lbm}$$

Process 2-3 is a constant-pressure heat-addition process, for which the boundary work and Δu terms can be combined into Δh. Thus,

$$\begin{aligned} Q_{\text{in}} &= m(h_3 - h_2) = mc_p(T_3 - T_2) \\ &= (0.00498\ \text{lbm})(0.240\ \text{Btu/lbm·R})[(3432 - 1716)\ \text{R}] \\ &= 2.051\ \text{Btu} \end{aligned}$$

Process 4-1 is a constant-volume heat-rejection process (it involves no work interactions), and the amount of heat rejected is

$$\begin{aligned} Q_{\text{out}} &= m(u_4 - u_1) = mc_v(T_4 - T_1) \\ &= (0.00498\ \text{lbm})(0.171\ \text{Btu/lbm·R})[(1425 - 540)\ \text{R}] \\ &= 0.754\ \text{Btu} \end{aligned}$$

Thus,

$$W_{\text{net}} = Q_{\text{in}} - Q_{\text{out}} = 2.051 - 0.754 = \mathbf{1.297\ Btu}$$

Then the thermal efficiency becomes

$$\eta_{\text{th}} = \frac{W_{\text{net}}}{Q_{\text{in}}} = \frac{1.297\ \text{Btu}}{2.051\ \text{Btu}} = \mathbf{0.632\ or\ 63.2\%}$$

The thermal efficiency of this Diesel cycle under the cold-air-standard assumptions could also be determined from Eq. 9–12.

(*c*) The mean effective pressure is determined from its definition, Eq. 9–4:

$$\begin{aligned} \text{MEP} &= \frac{W_{\text{net}}}{V_{\text{max}} - V_{\text{min}}} = \frac{W_{\text{net}}}{V_1 - V_2} = \frac{1.297\ \text{Btu}}{(117 - 6.5)\ \text{in}^3}\left(\frac{778.17\ \text{lbf·ft}}{1\ \text{Btu}}\right)\left(\frac{12\ \text{in.}}{1\ \text{ft}}\right) \\ &= \mathbf{110\ psia} \end{aligned}$$

Discussion Note that a constant pressure of 110 psia during the power stroke would produce the same net work output as the entire Diesel cycle.

9–7 ▪ STIRLING AND ERICSSON CYCLES

The ideal Otto and Diesel cycles discussed in the preceding sections are composed entirely of internally reversible processes and thus are internally reversible cycles. These cycles are not totally reversible, however, since they involve heat transfer through a finite temperature difference during the non-isothermal heat-addition and heat-rejection processes, which are irreversible. Therefore, the thermal efficiency of an Otto or Diesel engine will be less than that of a Carnot engine operating between the same temperature limits.

Consider a heat engine operating between a heat source at T_H and a heat sink at T_L. For the heat-engine cycle to be totally reversible, the

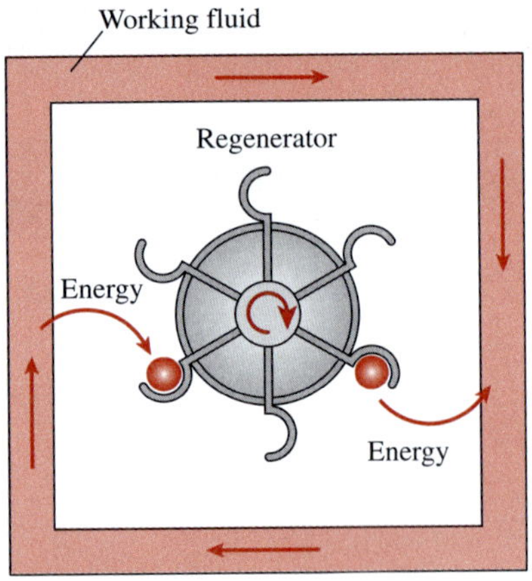

FIGURE 9–25
A regenerator is a device that borrows energy from the working fluid during one part of the cycle and pays it back (without interest) during another part.

temperature difference between the working fluid and the heat source (or sink) should never exceed a differential amount dT during any heat-transfer process. That is, both the heat-addition and heat-rejection processes during the cycle must take place isothermally, one at a temperature of T_H and the other at a temperature of T_L. This is precisely what happens in a Carnot cycle.

There are two other cycles that involve an isothermal heat-addition process at T_H and an isothermal heat-rejection process at T_L: the *Stirling cycle* and the *Ericsson cycle.* They differ from the Carnot cycle in that the two isentropic processes are replaced by two constant-volume regeneration processes in the Stirling cycle and by two constant-pressure regeneration processes in the Ericsson cycle. Both cycles utilize **regeneration**, a process during which heat is transferred to a thermal energy storage device (called a *regenerator*) during one part of the cycle and is transferred back to the working fluid during another part of the cycle (Fig. 9–25).

Figure 9–26(*b*) shows the *T-s* and *P-v* diagrams of the **Stirling cycle**, which is made up of four totally reversible processes:

1-2 $T =$ *constant* expansion (heat addition from the external source)
2-3 $v =$ *constant* regeneration (internal heat transfer from the working fluid to the regenerator)
3-4 $T =$ *constant* compression (heat rejection to the external sink)
4-1 $v =$ *constant* regeneration (internal heat transfer from the regenerator back to the working fluid)

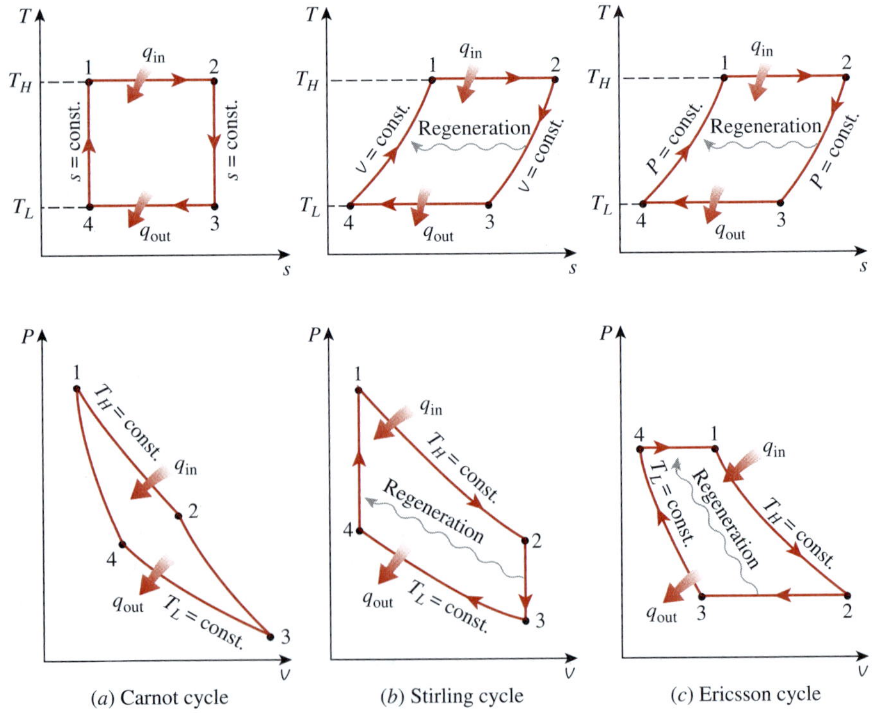

FIGURE 9–26
T-s and *P-v* diagrams of Carnot, Stirling, and Ericsson cycles.

The execution of the Stirling cycle requires rather innovative hardware. The actual Stirling engines, including the original one patented by Robert Stirling, are heavy and complicated. To spare the reader the complexities, the execution of the Stirling cycle in a closed system is explained with the help of the hypothetical engine shown in Fig. 9–27.

This system consists of a cylinder with two pistons on each side and a regenerator in the middle. The regenerator can be a wire or a ceramic mesh or any kind of porous plug with a high thermal mass (mass times specific heat). It is used for the temporary storage of thermal energy. The mass of the working fluid contained within the regenerator at any instant is considered negligible.

Initially, the left chamber houses the entire working fluid (a gas), which is at a high temperature and pressure. During process 1-2, heat is transferred to the gas at T_H from a source at T_H. As the gas expands isothermally, the left piston moves outward, doing work, and the gas pressure drops. During process 2-3, both pistons are moved to the right at the same rate (to keep the volume constant) until the entire gas is forced into the right chamber. As the gas passes through the regenerator, heat is transferred to the regenerator and the gas temperature drops from T_H to T_L. For this heat transfer process to be reversible, the temperature difference between the gas and the regenerator should not exceed a differential amount dT at any point. Thus, the temperature of the regenerator will be T_H at the left end and T_L at the right end of the regenerator when state 3 is reached. During process 3-4, the right piston is moved inward, compressing the gas. Heat is transferred from the gas to a sink at temperature T_L so that the gas temperature remains constant at T_L while the pressure rises. Finally, during process 4-1, both pistons are moved to the left at the same rate (to keep the volume constant), forcing the entire gas into the left chamber. The gas temperature rises from T_L to T_H as it passes through the regenerator and picks up the thermal energy stored there during process 2-3. This completes the cycle.

Notice that the second constant-volume process takes place at a smaller volume than the first one, and the net heat transfer to the regenerator during a cycle is zero. That is, the amount of energy stored in the regenerator during process 2-3 is equal to the amount picked up by the gas during process 4-1.

The *T-s* and *P-v* diagrams of the **Ericsson cycle** are shown in Fig. 9–26*c*. The Ericsson cycle is very much like the Stirling cycle, except that the two constant-volume processes are replaced by two constant-pressure processes.

A steady-flow system operating on an Ericsson cycle is shown in Fig. 9–28. Here the isothermal expansion and compression processes are executed in a compressor and a turbine, respectively, and a counter-flow heat exchanger serves as a regenerator. Hot and cold fluid streams enter the heat exchanger from opposite ends, and heat transfer takes place between the two streams. In the ideal case, the temperature difference between the two fluid streams does not exceed a differential amount at any point, and the cold fluid stream leaves the heat exchanger at the inlet temperature of the hot stream.

Both the Stirling and Ericsson cycles are totally reversible, as is the Carnot cycle, and thus according to the Carnot principle, all three cycles

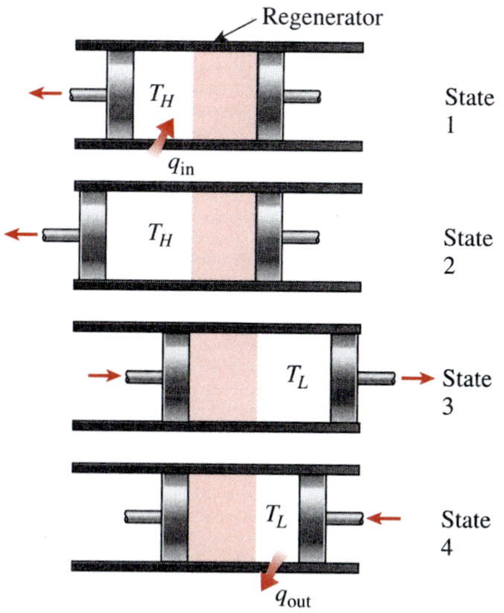

FIGURE 9–27
The execution of the Stirling cycle.

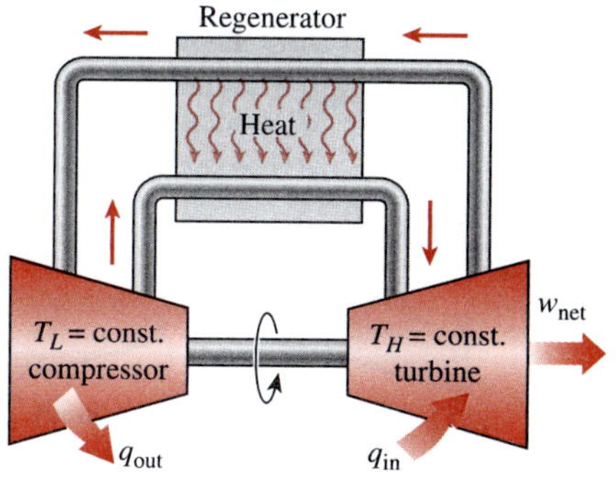

FIGURE 9–28
A steady-flow Ericsson engine.

more than one-half of the turbine work output is used to drive the compressor. The situation is even worse when the isentropic efficiencies of the compressor and the turbine are low. This is quite in contrast to steam power plants, where the back work ratio is only a few percent. This is not surprising, however, since a liquid is compressed in steam power plants instead of a gas, and the steady-flow work is proportional to the specific volume of the working fluid.

A power plant with a high back work ratio requires a larger turbine to provide the additional power requirements of the compressor. Therefore, the turbines used in gas-turbine power plants are larger than those used in steam power plants of the same net power output.

Development of Gas Turbines

The gas turbine has experienced phenomenal progress and growth since its first successful development in the 1930s. The early gas turbines built in the 1940s and even 1950s had simple-cycle efficiencies of about 17 percent because of the low compressor and turbine efficiencies and low turbine inlet temperatures due to metallurgical limitations of those times. Therefore, gas turbines found only limited use despite their versatility and their ability to burn a variety of fuels. The efforts to improve the cycle efficiency concentrated in three areas:

1. Increasing the turbine inlet (or firing) temperatures This has been the primary approach taken to improve gas-turbine efficiency. The turbine inlet temperatures have increased steadily from about 540°C (1000°F) in the 1940s to 1425°C (2600°F) and even higher today. These increases were made possible by the development of new materials and the innovative cooling techniques for the critical components such as coating the turbine blades with ceramic layers and cooling the blades with the discharge air from the compressor. Maintaining high turbine inlet temperatures with an air-cooling technique requires the combustion temperature to be higher to compensate for the cooling effect of the cooling air. However, higher combustion temperatures increase the amount of nitrogen oxides (NO_x), which are responsible for the formation of ozone at ground level and smog. Using steam as the coolant allowed an increase in the turbine inlet temperatures by 200°F without an increase in the combustion temperature. Steam is also a much more effective heat transfer medium than air.

2. Increasing the efficiencies of turbomachinery components The performance of early turbines suffered greatly from the inefficiencies of turbines and compressors. However, the advent of computers and advanced techniques for computer-aided design made it possible to design these components aerodynamically with minimal losses. The increased efficiencies of the turbines and compressors resulted in a significant increase in the cycle efficiency.

3. Adding modifications to the basic cycle The simple-cycle efficiencies of early gas turbines were practically doubled by incorporating intercooling, regeneration (or recuperation), and reheating, discussed in the next two sections. These improvements, of course, come at the expense of increased initial and operation costs, and they cannot be justified unless the decrease in fuel costs offsets the increase in other costs. The relatively low fuel prices, the general desire in the industry to minimize installation costs, and the tremendous increase in the simple-cycle efficiency to about 40 percent left little desire for opting for these modifications.

The first gas turbine for an electric utility was installed in 1949 in Oklahoma as part of a combined-cycle power plant. It was built by General Electric and produced 3.5 MW of power. Gas turbines installed until the mid-1970s suffered from low efficiency and poor reliability. In the past, the base-load electric power generation was dominated by large coal and nuclear power plants. However, there has been an historic shift toward natural gas–fired gas turbines because of their higher efficiencies, lower capital costs, shorter installation times, and better emission characteristics, and the abundance of natural gas supplies, and more and more electric utilities are using gas turbines for base-load power production as well as for peaking. The construction costs for gas-turbine power plants are roughly half that of comparable conventional fossil-fuel steam power plants, which were the primary base-load power plants until the early 1980s. More than half of all power plants to be installed in the foreseeable future are forecast to be gas-turbine or combined gas–steam turbine types.

A gas turbine manufactured by General Electric in the early 1990s had a pressure ratio of 13.5 and generated 135.7 MW of net power at a thermal efficiency of 33 percent in simple-cycle operation. A more recent gas turbine manufactured by General Electric uses a turbine inlet temperature of 1425°C (2600°F) and produces up to 282 MW while achieving a thermal efficiency of 39.5 percent in the simple-cycle mode. A 1.3-ton small-scale gas turbine labeled OP-16, built by the Dutch firm Opra Optimal Radial Turbine, can run on gas or liquid fuel and can replace a 16-ton diesel engine. It has a pressure ratio of 6.5 and produces up to 2 MW of power. Its efficiency is 26 percent in the simple-cycle operation, which rises to 37 percent when equipped with a regenerator.

EXAMPLE 9–5 The Simple Ideal Brayton Cycle

A gas-turbine power plant operating on an ideal Brayton cycle has a pressure ratio of 8. The gas temperature is 300 K at the compressor inlet and 1300 K at the turbine inlet. Utilizing the air-standard assumptions, determine (*a*) the gas temperature at the exits of the compressor and the turbine, (*b*) the back work ratio, and (*c*) the thermal efficiency.

Solution A power plant operating on the ideal Brayton cycle is considered. The compressor and turbine exit temperatures, back work ratio, and the thermal efficiency are to be determined.

Assumptions **1** Steady operating conditions exist. **2** The air-standard assumptions are applicable. **3** Kinetic and potential energy changes are negligible. **4** The variation of specific heats with temperature is to be considered.

Analysis The *T-s* diagram of the ideal Brayton cycle described is shown in Fig. 9–35. We note that the components involved in the Brayton cycle are steady-flow devices.

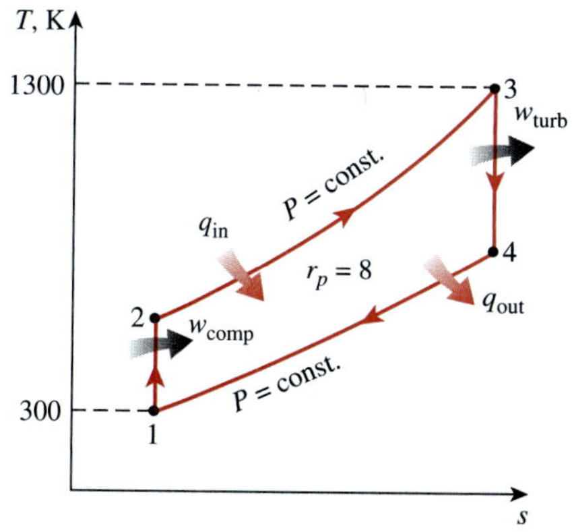

FIGURE 9–35
T-s diagram for the Brayton cycle discussed in Example 9–5.

(*a*) The air temperatures at the compressor and turbine exits are determined from isentropic relations:

Process 1-2 (isentropic compression of an ideal gas):

$$T_1 = 300 \text{ K} \rightarrow h_1 = 300.19 \text{ kJ/kg}$$
$$P_{r1} = 1.386$$

$$P_{r2} = \frac{P_2}{P_1} P_{r1} = (8)(1.386) = 11.09 \rightarrow T_2 = \mathbf{540\ K} \qquad \text{(at compressor exit)}$$
$$h_2 = 544.35 \text{ kJ/kg}$$

Process 3-4 (isentropic expansion of an ideal gas):

$$T_3 = 1300 \text{ K} \rightarrow h_3 = 1395.97 \text{ kJ/kg}$$
$$P_{r3} = 330.9$$

$$P_{r4} = \frac{P_4}{P_3} P_{r3} = \left(\frac{1}{8}\right)(330.9) = 41.36 \rightarrow T_4 = \mathbf{770\ K} \qquad \text{(at turbine exit)}$$
$$h_4 = 789.37 \text{ kJ/kg}$$

(*b*) To find the back work ratio, we need to find the work input to the compressor and the work output of the turbine:

$$w_{\text{comp,in}} = h_2 - h_1 = 544.35 - 300.19 = 244.16 \text{ kJ/kg}$$
$$w_{\text{turb,out}} = h_3 - h_4 = 1395.97 - 789.37 = 606.60 \text{ kJ/kg}$$

Thus,

$$r_{\text{bw}} = \frac{w_{\text{comp,in}}}{w_{\text{turb,out}}} = \frac{244.16 \text{ kJ/kg}}{606.60 \text{ kJ/kg}} = \mathbf{0.403}$$

That is, 40.3 percent of the turbine work output is used just to drive the compressor.

(*c*) The thermal efficiency of the cycle is the ratio of the net power output to the total heat input:

$$q_{\text{in}} = h_3 - h_2 = 1395.97 - 544.35 = 851.62 \text{ kJ/kg}$$
$$w_{\text{net}} = w_{\text{out}} - w_{\text{in}} = 606.60 - 244.16 = 362.4 \text{ kJ/kg}$$

Thus,

$$\eta_{\text{th}} = \frac{w_{\text{net}}}{q_{\text{in}}} = \frac{362.4 \text{ kJ/kg}}{851.62 \text{ kJ/kg}} = \mathbf{0.426\ or\ 42.6\%}$$

The thermal efficiency could also be determined from

$$\eta_{\text{th}} = 1 - \frac{q_{\text{out}}}{q_{\text{in}}}$$

where

$$q_{\text{out}} = h_4 - h_1 = 789.37 - 300.19 = 489.2 \text{ kJ/kg}$$

Discussion Under the cold-air-standard assumptions (constant specific heat values at room temperature), the thermal efficiency would be, from Eq. 9–17,

$$\eta_{\text{th,Brayton}} = 1 - \frac{1}{r_p^{(k-1)/k}} = 1 - \frac{1}{8^{(1.4-1)/1.4}} = 0.448$$

which is sufficiently close to the value obtained by accounting for the variation of specific heats with temperature.

Deviation of Actual Gas-Turbine Cycles from Idealized Ones

The actual gas-turbine cycle differs from the ideal Brayton cycle on several accounts. For one thing, some pressure drop during the heat-addition and heat-rejection processes is inevitable. More importantly, the actual work input to the compressor is more, and the actual work output from the turbine is less because of irreversibilities. The deviation of actual compressor and turbine behavior from the idealized isentropic behavior can be accurately accounted for by utilizing the isentropic efficiencies of the turbine and compressor as

$$\eta_C = \frac{w_s}{w_a} \cong \frac{h_{2s} - h_1}{h_{2a} - h_1} \quad \textbf{(9–19)}$$

and

$$\eta_T = \frac{w_a}{w_s} \cong \frac{h_3 - h_{4a}}{h_3 - h_{4s}} \quad \textbf{(9–20)}$$

where states $2a$ and $4a$ are the actual exit states of the compressor and the turbine, respectively, and $2s$ and $4s$ are the corresponding states for the isentropic case, as illustrated in Fig. 9–36. The effect of the turbine and compressor efficiencies on the thermal efficiency of the gas-turbine engines is illustrated below with an example.

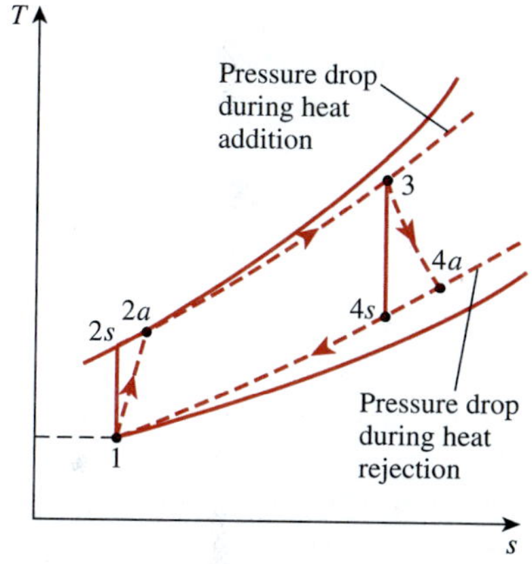

FIGURE 9–36
The deviation of an actual gas-turbine cycle from the ideal Brayton cycle as a result of irreversibilities.

EXAMPLE 9–6 An Actual Gas-Turbine Cycle

Assuming a compressor efficiency of 80 percent and a turbine efficiency of 85 percent, determine (*a*) the back work ratio, (*b*) the thermal efficiency, and (*c*) the turbine exit temperature of the gas-turbine cycle discussed in Example 9–5.

Solution The Brayton cycle discussed in Example 9–5 is reconsidered. For specified turbine and compressor efficiencies, the back work ratio, the thermal efficiency, and the turbine exit temperature are to be determined.

Analysis (*a*) The *T-s* diagram of the cycle is shown in Fig. 9–37. The actual compressor work and turbine work are determined by using the definitions of compressor and turbine efficiencies, Eqs. 9–19 and 9–20:

Compressor:
$$w_{\text{comp,in}} = \frac{w_s}{\eta_C} = \frac{244.16 \text{ kJ/kg}}{0.80} = 305.20 \text{ kJ/kg}$$

Turbine:
$$w_{\text{turb,out}} = \eta_T w_s = (0.85)(606.60 \text{ kJ/kg}) = 515.61 \text{ kJ/kg}$$

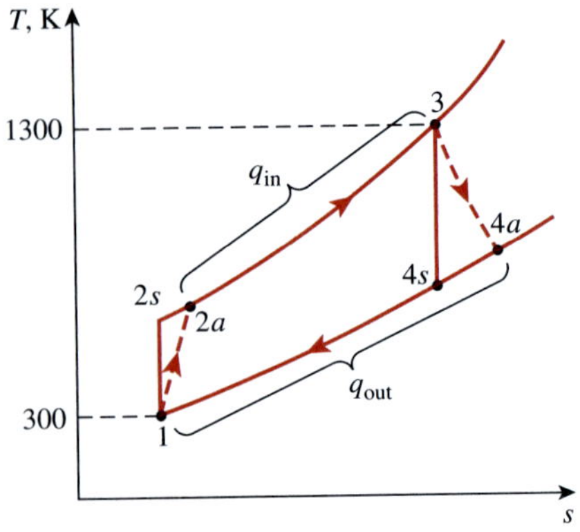

FIGURE 9–37
T-s diagram of the gas-turbine cycle discussed in Example 9–6.

Thus,

$$r_{bw} = \frac{w_{comp,in}}{w_{turb,out}} = \frac{305.20 \text{ kJ/kg}}{515.61 \text{ kJ/kg}} = \mathbf{0.592}$$

That is, the compressor is now consuming 59.2 percent of the work produced by the turbine (up from 40.3 percent). This increase is due to the irreversibilities that occur within the compressor and the turbine.

(*b*) In this case, air leaves the compressor at a higher temperature and enthalpy, which are determined to be

$$w_{comp,in} = h_{2a} - h_1 \rightarrow h_{2a} = h_1 + w_{comp,in}$$
$$= 300.19 + 305.20$$
$$= 605.39 \text{ kJ/kg} \quad (\text{and } T_{2a} = 598 \text{ K})$$

Thus,

$$q_{in} = h_3 - h_{2a} = 1395.97 - 605.39 = 790.58 \text{ kJ/kg}$$
$$w_{net} = w_{out} - w_{in} = 515.61 - 305.20 = 210.41 \text{ kJ/kg}$$

and

$$\eta_{th} = \frac{w_{net}}{q_{in}} = \frac{210.41 \text{ kJ/kg}}{790.58 \text{ kJ/kg}} = \mathbf{0.266 \text{ or } 26.6\%}$$

That is, the irreversibilities occurring within the turbine and compressor caused the thermal efficiency of the gas turbine cycle to drop from 42.6 to 26.6 percent. This example shows how sensitive the performance of a gas-turbine power plant is to the efficiencies of the compressor and the turbine. In fact, gas-turbine efficiencies did not reach competitive values until significant improvements were made in the design of gas turbines and compressors.

(*c*) The air temperature at the turbine exit is determined from an energy balance on the turbine:

$$w_{turb,out} = h_3 - h_{4a} \rightarrow h_{4a} = h_3 - w_{turb,out}$$
$$= 1395.97 - 515.61$$
$$= 880.36 \text{ kJ/kg}$$

Then, from Table A–17,

$$T_{4a} = \mathbf{853 \text{ K}}$$

Discussion The temperature at turbine exit is considerably higher than that at the compressor exit (T_{2a} = 598 K), which suggests the use of regeneration to reduce fuel cost.

9–9 ▪ THE BRAYTON CYCLE WITH REGENERATION

In gas-turbine engines, the temperature of the exhaust gas leaving the turbine is often considerably higher than the temperature of the air leaving the compressor. Therefore, the high-pressure air leaving the compressor can be heated by transferring heat to it from the hot exhaust gases in a counter-flow

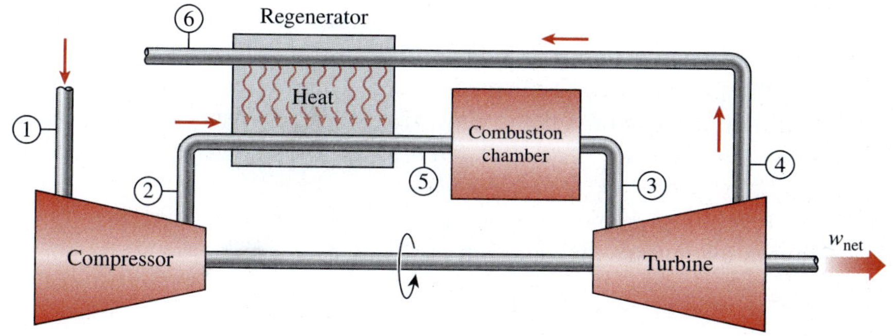

FIGURE 9–38
A gas-turbine engine with regenerator.

heat exchanger, which is also known as a *regenerator* or a *recuperator.* A sketch of the gas-turbine engine utilizing a regenerator and the *T-s* diagram of the new cycle are shown in Figs. 9–38 and 9–39, respectively.

The thermal efficiency of the Brayton cycle increases as a result of regeneration since the portion of energy of the exhaust gases that is normally rejected to the surroundings is now used to preheat the air entering the combustion chamber. This, in turn, decreases the heat input (thus fuel) requirements for the same net work output. Note, however, that the use of a regenerator is recommended only when the turbine exhaust temperature is higher than the compressor exit temperature. Otherwise, heat will flow in the reverse direction (*to* the exhaust gases), decreasing the efficiency. This situation is encountered in gas-turbine engines operating at very high pressure ratios.

The highest temperature occurring within the regenerator is T_4, the temperature of the exhaust gases leaving the turbine and entering the regenerator. Under no conditions can the air be preheated in the regenerator to a temperature above this value. Air normally leaves the regenerator at a lower temperature, T_5. In the limiting (ideal) case, the air exits the regenerator at the inlet temperature of the exhaust gases T_4. Assuming the regenerator to be well insulated and any changes in kinetic and potential energies to be negligible, the actual and maximum heat transfers from the exhaust gases to the air can be expressed as

$$q_{\text{regen,act}} = h_5 - h_2 \tag{9–21}$$

and

$$q_{\text{regen,max}} = h_{5'} - h_2 = h_4 - h_2 \tag{9–22}$$

The extent to which a regenerator approaches an ideal regenerator is called the **effectiveness ϵ** and is defined as

$$\epsilon = \frac{q_{\text{regen,act}}}{q_{\text{regen,max}}} = \frac{h_5 - h_2}{h_4 - h_2} \tag{9–23}$$

When the cold-air-standard assumptions are utilized, it reduces to

$$\epsilon \cong \frac{T_5 - T_2}{T_4 - T_2} \tag{9–24}$$

A regenerator with a higher effectiveness obviously saves a greater amount of fuel since it preheats the air to a higher temperature prior to combustion.

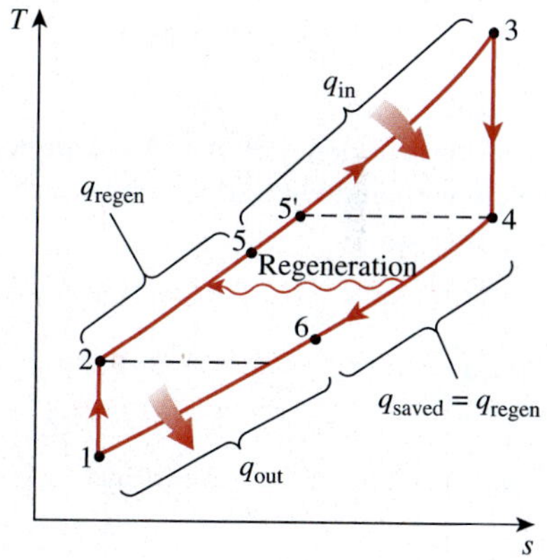

FIGURE 9–39
T-s diagram of a Brayton cycle with regeneration.

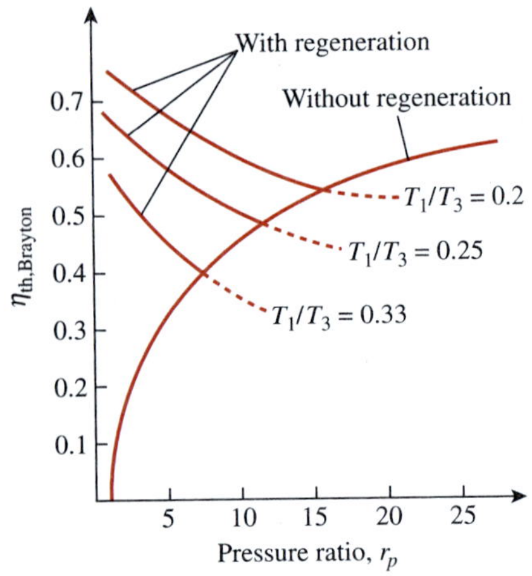

FIGURE 9–40

Thermal efficiency of the ideal Brayton cycle with and without regeneration.

However, achieving a higher effectiveness requires the use of a larger regenerator, which carries a higher price tag and causes a larger pressure drop. Therefore, the use of a regenerator with a very high effectiveness cannot be justified economically unless the savings from the fuel costs exceed the additional expenses involved. The effectiveness of most regenerators used in practice is below 0.85.

Under the cold-air-standard assumptions, the thermal efficiency of an ideal Brayton cycle with regeneration is

$$\eta_{th,regen} = 1 - \left(\frac{T_1}{T_3}\right)(r_p)^{(k-1)/k} \quad \text{(9–25)}$$

Therefore, the thermal efficiency of an ideal Brayton cycle with regeneration depends on the ratio of the minimum to maximum temperatures as well as the pressure ratio. The thermal efficiency is plotted in Fig. 9–40 for various pressure ratios and minimum-to-maximum temperature ratios. This figure shows that regeneration is most effective at lower pressure ratios and low minimum-to-maximum temperature ratios.

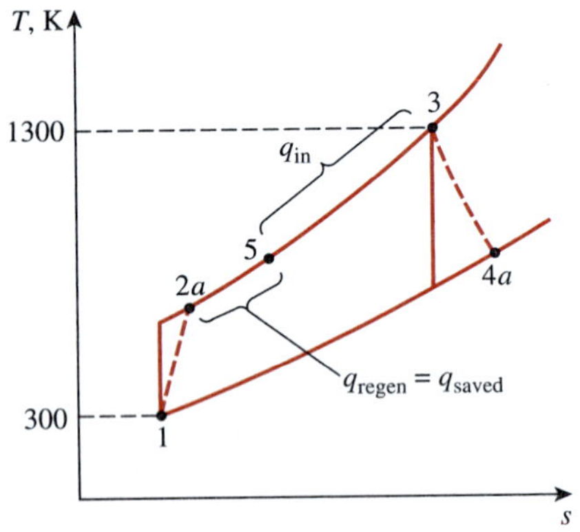

FIGURE 9–41

T-s diagram of the regenerative Brayton cycle described in Example 9–7.

EXAMPLE 9–7 Actual Gas-Turbine Cycle with Regeneration

Determine the thermal efficiency of the gas-turbine described in Example 9–6 if a regenerator having an effectiveness of 80 percent is installed.

Solution The gas-turbine discussed in Example 9–6 is equipped with a regenerator. For a specified effectiveness, the thermal efficiency is to be determined.

Analysis The *T-s* diagram of the cycle is shown in Fig. 9–41. We first determine the enthalpy of the air at the exit of the regenerator, using the definition of effectiveness:

$$\epsilon = \frac{h_5 - h_{2a}}{h_{4a} - h_{2a}}$$

$$0.80 = \frac{(h_5 - 605.39)\text{ kJ/kg}}{(880.36 - 605.39)\text{ kJ/kg}} \rightarrow h_5 = 825.37\text{ kJ/kg}$$

Thus,

$$q_{in} = h_3 - h_5 = (1395.97 - 825.37)\text{ kJ/kg} = 570.60\text{ kJ/kg}$$

This represents a savings of 220.0 kJ/kg from the heat input requirements. The addition of a regenerator (assumed to be frictionless) does not affect the net work output. Thus,

$$\eta_{th} = \frac{w_{net}}{q_{in}} = \frac{210.41\text{ kJ/kg}}{570.60\text{ kJ/kg}} = \mathbf{0.369 \text{ or } 36.9\%}$$

Discussion Note that the thermal efficiency of the gas turbine has gone up from 26.6 to 36.9 percent as a result of installing a regenerator that helps to recuperate some of the thermal energy of the exhaust gases.

9–10 ▪ THE BRAYTON CYCLE WITH INTERCOOLING, REHEATING, AND REGENERATION

The net work of a gas-turbine cycle is the difference between the turbine work output and the compressor work input, and it can be increased by either decreasing the compressor work or increasing the turbine work, or both. It was shown in Chap. 7 that the work required to compress a gas between two specified pressures can be decreased by carrying out the compression process in stages and cooling the gas in between (Fig. 9–42)—that is, using *multistage compression with intercooling.* As the number of stages is increased, the compression process becomes nearly isothermal at the compressor inlet temperature, and the compression work decreases.

Likewise, the work output of a turbine operating between two pressure levels can be increased by expanding the gas in stages and reheating it in between—that is, utilizing *multistage expansion with reheating.* This is accomplished without raising the maximum temperature in the cycle. As the number of stages is increased, the expansion process becomes nearly isothermal. The foregoing argument is based on a simple principle: *The steady-flow compression or expansion work is proportional to the specific volume of the fluid. Therefore, the specific volume of the working fluid should be as low as possible during a compression process and as high as possible during an expansion process.* This is precisely what intercooling and reheating accomplish.

Combustion in gas turbines typically occurs at four times the amount of air needed for complete combustion to avoid excessive temperatures. Therefore, the exhaust gases are rich in oxygen, and reheating can be accomplished by simply spraying additional fuel into the exhaust gases between two expansion states.

The working fluid leaves the compressor at a lower temperature, and the turbine at a higher temperature, when intercooling and reheating are utilized. This makes regeneration more attractive since a greater potential for regeneration exists. Also, the gases leaving the compressor can be heated to a higher temperature before they enter the combustion chamber because of the higher temperature of the turbine exhaust.

A schematic of the physical arrangement and the *T-s* diagram of an ideal two-stage gas-turbine cycle with intercooling, reheating, and regeneration are shown in Figs. 9–43 and 9–44. The gas enters the first stage of the compressor at state 1, is compressed isentropically to an intermediate pressure P_2, is cooled at constant pressure to state 3 ($T_3 = T_1$), and is compressed in the second stage isentropically to the final pressure P_4. At state 4 the gas enters the regenerator, where it is heated to T_5 at constant pressure. In an ideal regenerator, the gas leaves the regenerator at the temperature of the turbine exhaust, that is, $T_5 = T_9$. The primary heat addition (or combustion) process takes place between states 5 and 6. The gas enters the first stage of the turbine at state 6 and expands isentropically to state 7, where it enters the reheater. It is reheated at constant pressure to state 8 ($T_8 = T_6$), where it enters the second stage of the turbine. The gas exits the turbine at state 9 and enters the regenerator, where it is cooled to state 10 at constant pressure. The cycle is completed by cooling the gas to the initial state (or purging the exhaust gases).

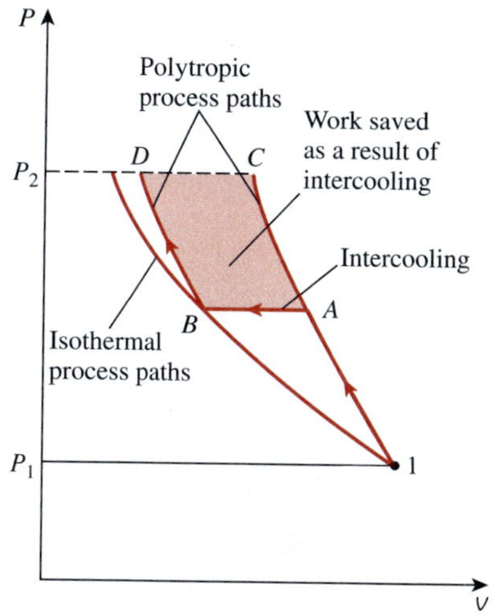

FIGURE 9–42
Comparison of work inputs to a single-stage compressor (1*AC*) and a two-stage compressor with intercooling (1*ABD*).

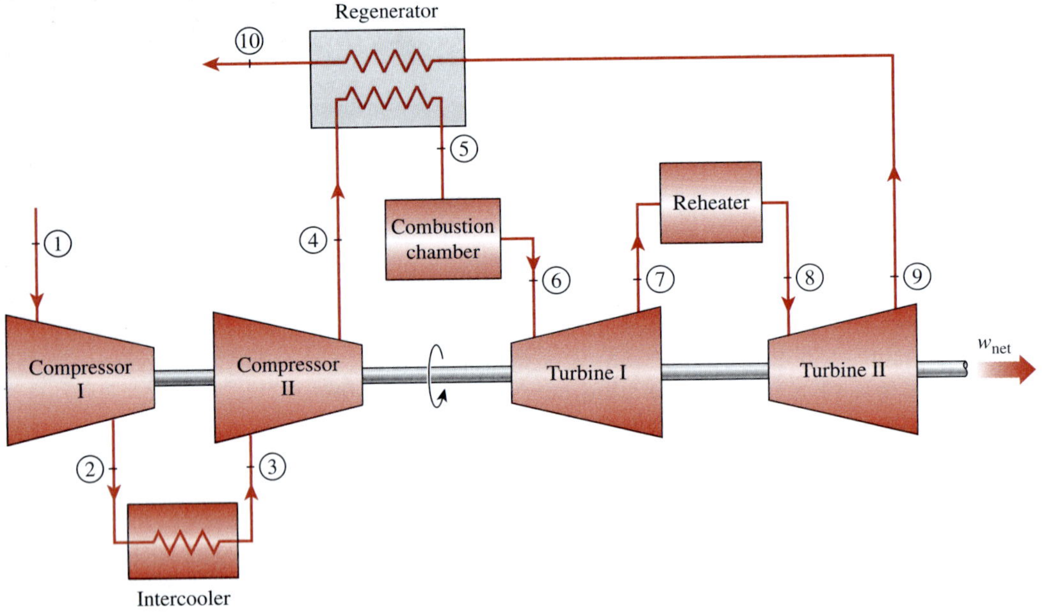

FIGURE 9–43
A gas-turbine engine with two-stage compression with intercooling, two-stage expansion with reheating, and regeneration.

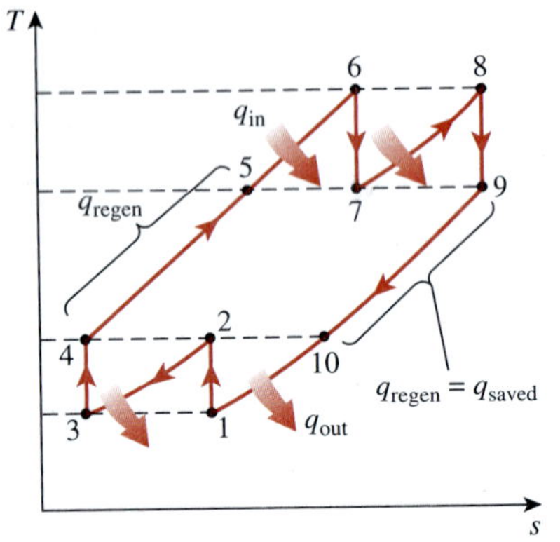

FIGURE 9–44
T-s diagram of an ideal gas-turbine cycle with intercooling, reheating, and regeneration.

It was shown in Chap. 7 that the work input to a two-stage compressor is minimized when equal pressure ratios are maintained across each stage. It can be shown that this procedure also maximizes the turbine work output. Thus, for best performance we have

$$\frac{P_2}{P_1} = \frac{P_4}{P_3} \quad \text{and} \quad \frac{P_6}{P_7} = \frac{P_8}{P_9} \tag{9–26}$$

In the analysis of the actual gas-turbine cycles, the irreversibilities that are present within the compressor, the turbine, and the regenerator as well as the pressure drops in the heat exchangers should be taken into consideration.

The back work ratio of a gas-turbine cycle improves as a result of intercooling and reheating. However, this does not mean that the thermal efficiency also improves. The fact is, intercooling and reheating always decreases the thermal efficiency unless they are accompanied by regeneration. This is because intercooling decreases the average temperature at which heat is added, and reheating increases the average temperature at which heat is rejected. This is also apparent from Fig. 9–44. Therefore, in gas-turbine power plants, intercooling and reheating are always used in conjunction with regeneration.

If the number of compression and expansion stages is increased, the ideal gas-turbine cycle with intercooling, reheating, and regeneration approaches the Ericsson cycle, as illustrated in Fig. 9–45, and the thermal efficiency approaches the theoretical limit (the Carnot efficiency). However, the contribution of each additional stage to the thermal efficiency is less and less, and the use of more than two or three stages cannot be justified economically.

EXAMPLE 9–8 A Gas Turbine with Reheating and Intercooling

An ideal gas-turbine cycle with two stages of compression and two stages of expansion has an overall pressure ratio of 8. Air enters each stage of the compressor at 300 K and each stage of the turbine at 1300 K. Determine the back work ratio and the thermal efficiency of this gas-turbine cycle, assuming (*a*) no regenerators and (*b*) an ideal regenerator with 100 percent effectiveness. Compare the results with those obtained in Example 9–5.

Solution An ideal gas-turbine cycle with two stages of compression and two stages of expansion is considered. The back work ratio and the thermal efficiency of the cycle are to be determined for the cases of no regeneration and maximum regeneration.

Assumptions **1** Steady operating conditions exist. **2** The air-standard assumptions are applicable. **3** Kinetic and potential energy changes are negligible.

Analysis The *T-s* diagram of the ideal gas-turbine cycle described is shown in Fig. 9–46. We note that the cycle involves two stages of expansion, two stages of compression, and regeneration.

For two-stage compression and expansion, the work input is minimized and the work output is maximized when both stages of the compressor and the turbine have the same pressure ratio. Thus,

$$\frac{P_2}{P_1} = \frac{P_4}{P_3} = \sqrt{8} = 2.83 \quad \text{and} \quad \frac{P_6}{P_7} = \frac{P_8}{P_9} = \sqrt{8} = 2.83$$

Air enters each stage of the compressor at the same temperature, and each stage has the same isentropic efficiency (100 percent in this case). Therefore, the temperature (and enthalpy) of the air at the exit of each compression stage will be the same. A similar argument can be given for the turbine. Thus,

At inlets: $T_1 = T_3, \quad h_1 = h_3 \quad \text{and} \quad T_6 = T_8, \quad h_6 = h_8$

At exits: $T_2 = T_4, \quad h_2 = h_4 \quad \text{and} \quad T_7 = T_9, \quad h_7 = h_9$

Under these conditions, the work input to each stage of the compressor will be the same, and so will the work output from each stage of the turbine.

(*a*) In the absence of any regeneration, the back work ratio and the thermal efficiency are determined by using data from Table A–17 as follows:

$$T_1 = 300 \text{ K} \rightarrow h_1 = 300.19 \text{ kJ/kg}$$

$$P_{r1} = 1.386$$

$$P_{r2} = \frac{P_2}{P_1}P_{r1} = \sqrt{8}(1.386) = 3.92 \rightarrow T_2 = 403.3 \text{ K}$$

$$h_2 = 404.31 \text{ kJ/kg}$$

$$T_6 = 1300 \text{ K} \rightarrow h_6 = 1395.97 \text{ kJ/kg}$$

$$P_{r6} = 330.9$$

$$P_{r7} = \frac{P_7}{P_6}P_{r6} = \frac{1}{\sqrt{8}}(330.9) = 117.0 \rightarrow T_7 = 1006.4 \text{ K}$$

$$h_7 = 1053.33 \text{ kJ/kg}$$

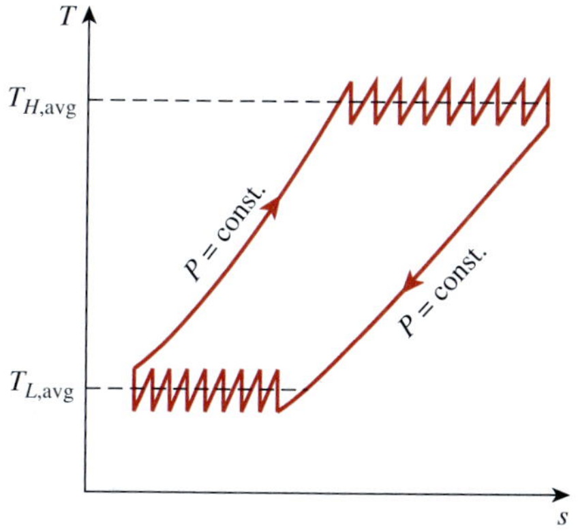

FIGURE 9–45
As the number of compression and expansion stages increases, the gas-turbine cycle with intercooling, reheating, and regeneration approaches the Ericsson cycle.

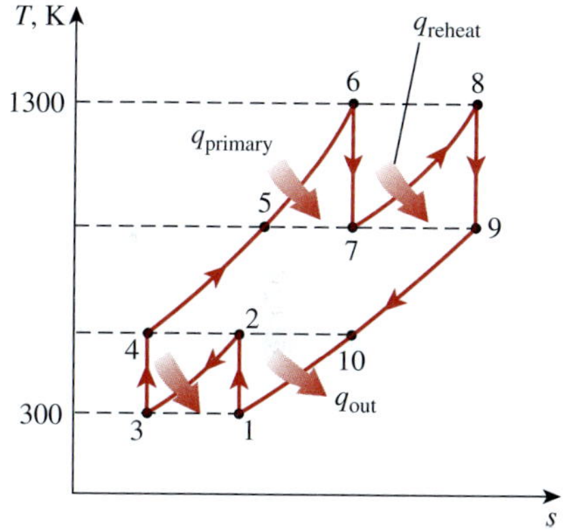

FIGURE 9–46
T-s diagram of the gas-turbine cycle discussed in Example 9–8.

Then

$$w_{comp,in} = 2(w_{comp,in,I}) = 2(h_2 - h_1) = 2(404.31 - 300.19) = 208.24 \text{ kJ/kg}$$

$$w_{turb,out} = 2(w_{turb,out,I}) = 2(h_6 - h_7) = 2(1395.97 - 1053.33) = 685.28 \text{ kJ/kg}$$

$$w_{net} = w_{turb,out} - w_{comp,in} = 685.28 - 208.24 = 477.04 \text{ kJ/kg}$$

$$q_{in} = q_{primary} + q_{reheat} = (h_6 - h_4) + (h_8 - h_7)$$

$$= (1395.97 - 404.31) + (1395.97 - 1053.33) = 1334.30 \text{ kJ/kg}$$

Thus,

$$r_{bw} = \frac{w_{comp,in}}{w_{turb,out}} = \frac{208.24 \text{ kJ/kg}}{685.28 \text{ kJ/kg}} = \mathbf{0.304 \text{ or } 30.4\%}$$

and

$$\eta_{th} = \frac{w_{net}}{q_{in}} = \frac{477.04 \text{ kJ/kg}}{1334.30 \text{ kJ/kg}} = \mathbf{0.358 \text{ or } 35.8\%}$$

A comparison of these results with those obtained in Example 9–5 (single-stage compression and expansion) reveals that multistage compression with intercooling and multistage expansion with reheating improve the back work ratio (it drops from 40.3 to 30.4 percent) but hurt the thermal efficiency (it drops from 42.6 to 35.8 percent). Therefore, intercooling and reheating are not recommended in gas-turbine power plants unless they are accompanied by regeneration.

(*b*) The addition of an ideal regenerator (no pressure drops, 100 percent effectiveness) does not affect the compressor work and the turbine work. Therefore, the net work output and the back work ratio of an ideal gas-turbine cycle are identical whether there is a regenerator or not. A regenerator, however, reduces the heat input requirements by preheating the air leaving the compressor, using the hot exhaust gases. In an ideal regenerator, the compressed air is heated to the turbine exit temperature T_9 before it enters the combustion chamber. Thus, under the air-standard assumptions, $h_5 = h_7 = h_9$.

The heat input and the thermal efficiency in this case are

$$q_{in} = q_{primary} + q_{reheat} = (h_6 - h_5) + (h_8 - h_7)$$

$$= (1395.97 - 1053.33) + (1395.97 - 1053.33) = 685.28 \text{ kJ/kg}$$

and

$$\eta_{th} = \frac{w_{net}}{q_{in}} = \frac{477.04 \text{ kJ/kg}}{685.28 \text{ kJ/kg}} = \mathbf{0.696 \text{ or } 69.6\%}$$

Discussion Note that the thermal efficiency almost doubles as a result of regeneration compared to the no-regeneration case. The overall effect of two-stage compression and expansion with intercooling, reheating, and regeneration on the thermal efficiency is an increase of 63 percent. As the number of compression and expansion stages is increased, the cycle will approach the Ericsson cycle, and the thermal efficiency will approach

$$\eta_{th,Ericsson} = \eta_{th,Carnot} = 1 - \frac{T_L}{T_H} = 1 - \frac{300 \text{ K}}{1300 \text{ K}} = 0.769$$

Adding a second stage increases the thermal efficiency from 42.6 to 69.6 percent, an increase of 27 percentage points. This is a significant increase in efficiency, and usually it is well worth the extra cost associated with the second stage. Adding more stages, however (no matter how many), can increase the efficiency an additional 7.3 percentage points at most, and usually cannot be justified economically.

9–11 ▪ IDEAL JET-PROPULSION CYCLES

Gas-turbine engines are widely used to power aircraft because they are light and compact and have a high power-to-weight ratio. Aircraft gas turbines operate on an open cycle called a **jet-propulsion cycle**. The ideal jet-propulsion cycle differs from the simple ideal Brayton cycle in that the gases are not expanded to the ambient pressure in the turbine. Instead, they are expanded to a pressure such that the power produced by the turbine is just sufficient to drive the compressor and the auxiliary equipment, such as a small generator and hydraulic pumps. That is, the net work output of a jet-propulsion cycle is zero. The gases that exit the turbine at a relatively high pressure are subsequently accelerated in a nozzle to provide the thrust to propel the aircraft (Fig. 9–47). Also, aircraft gas turbines operate at higher pressure ratios (typically between 10 and 25), and the fluid passes through a diffuser first, where it is decelerated and its pressure is increased before it enters the compressor.

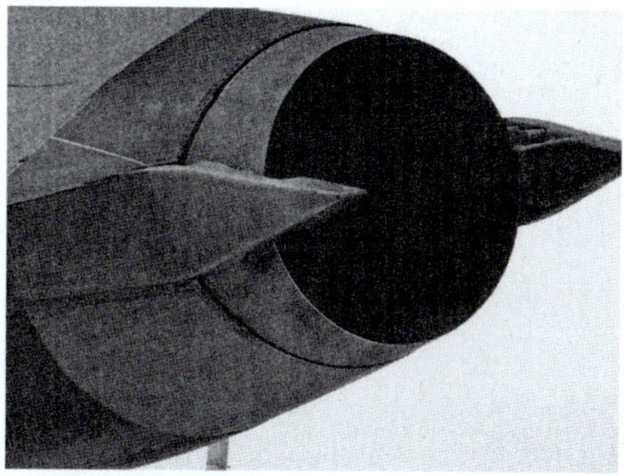

FIGURE 9–47
In jet engines, the high-temperature and high-pressure gases leaving the turbine are accelerated in a nozzle to provide thrust.

Aircraft are propelled by accelerating a fluid in the opposite direction to motion. This is accomplished by either slightly accelerating a large mass of fluid (*propeller-driven engine*) or greatly accelerating a small mass of fluid (*jet* or *turbojet engine*) or both (*turboprop engine*).

A schematic of a turbojet engine and the *T-s* diagram of the ideal turbojet cycle are shown in Fig. 9–48. The pressure of air rises slightly as it is decelerated in the diffuser. Air is compressed by the compressor. It is mixed with fuel in the combustion chamber, where the mixture is burned at constant

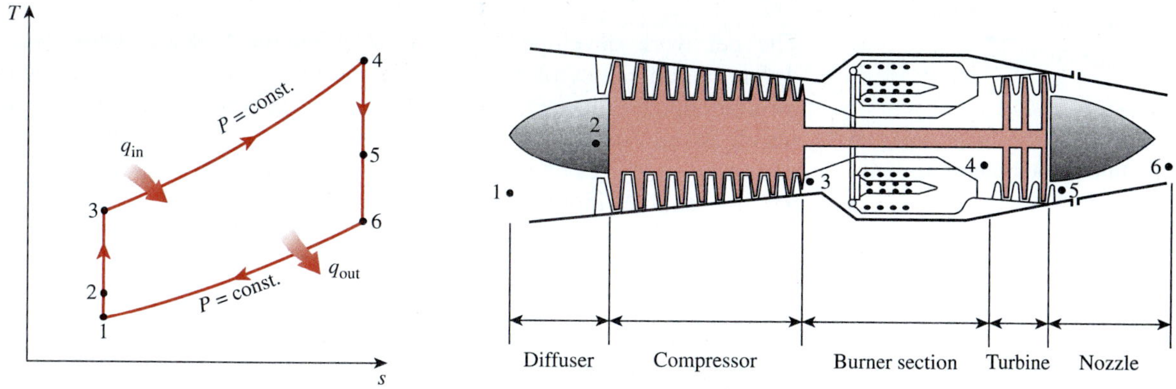

FIGURE 9–48
Basic components of a turbojet engine and the *T-s* diagram for the ideal turbojet cycle.

Source: The Aircraft Gas Turbine Engine and Its Operation. © *United Aircraft Corporation (now United Technologies Corp.), 1951, 1974.*

pressure. The high-pressure and high-temperature combustion gases partially expand in the turbine, producing enough power to drive the compressor and other equipment. Finally, the gases expand in a nozzle to the ambient pressure and leave the engine at a high velocity.

In the ideal case, the turbine work is assumed to equal the compressor work. Also, the processes in the diffuser, the compressor, the turbine, and the nozzle are assumed to be isentropic. In the analysis of actual cycles, however, the irreversibilities associated with these devices should be considered. The effect of the irreversibilities is to reduce the thrust that can be obtained from a turbojet engine.

The **thrust** developed in a turbojet engine is the unbalanced force that is caused by the difference in the momentum of the low-velocity air entering the engine and the high-velocity exhaust gases leaving the engine, and it is determined from Newton's second law. The pressures at the inlet and the exit of a turbojet engine are identical (the ambient pressure); thus, the net thrust developed by the engine is

$$F = (\dot{m}V)_{\text{exit}} - (\dot{m}V)_{\text{inlet}} = \dot{m}(V_{\text{exit}} - V_{\text{inlet}}) \qquad \text{(N)} \tag{9–27}$$

where V_{exit} is the exit velocity of the exhaust gases and V_{inlet} is the inlet velocity of the air, both relative to the aircraft. Thus, for an aircraft cruising in still air, V_{inlet} is the aircraft velocity. In reality, the mass flow rates of the gases at the engine exit and the inlet are different, the difference being equal to the combustion rate of the fuel. However, the air–fuel mass ratio used in jet-propulsion engines is usually very high, making this difference very small. Thus, $\dot{m}$ in Eq. 9–27 is taken as the mass flow rate of air through the engine. For an aircraft cruising at a constant speed, the thrust is used to overcome air drag, and the net force acting on the body of the aircraft is zero. Commercial airplanes save fuel by flying at higher altitudes during long trips since air at higher altitudes is thinner and exerts a smaller drag force on aircraft.

The power developed from the thrust of the engine is called the **propulsive power** $\dot{W}_P$, which is the *propulsive force* (*thrust*) times the *distance* this force acts on the aircraft per unit time, that is, the thrust times the aircraft velocity (Fig. 9–49):

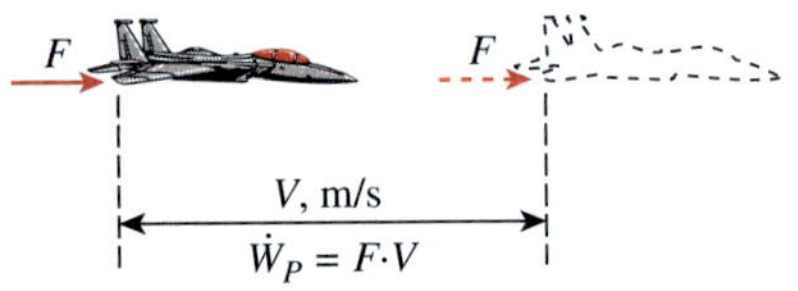

FIGURE 9–49
Propulsive power is the thrust acting on the aircraft through a distance per unit time.

$$\dot{W}_P = FV_{\text{aircraft}} = \dot{m}(V_{\text{exit}} - V_{\text{inlet}})V_{\text{aircraft}} \qquad \text{(kW)} \tag{9–28}$$

The net work developed by a turbojet engine is zero. Thus, we cannot define the efficiency of a turbojet engine in the same way as stationary gas-turbine engines. Instead, we should use the general definition of efficiency, which is the ratio of the desired output to the required input. The desired output in a turbojet engine is the *power produced* to propel the aircraft $\dot{W}_P$, and the required input is the *heating value of the fuel* $\dot{Q}_{\text{in}}$. The ratio of these two quantities is called the **propulsive efficiency** and is given by

$$\eta_P = \frac{\text{Propulsive power}}{\text{Energy input rate}} = \frac{\dot{W}_P}{\dot{Q}_{\text{in}}} \tag{9–29}$$

Propulsive efficiency is a measure of how efficiently the thermal energy released during the combustion process is converted to propulsive energy. The remaining part of the energy released shows up as the kinetic energy of the exhaust gases relative to a fixed point on the ground and as an increase in the enthalpy of the gases leaving the engine.

EXAMPLE 9–9 The Ideal Jet-Propulsion Cycle

A turbojet aircraft flies with a velocity of 850 ft/s at an altitude where the air is at 5 psia and −40°F. The compressor has a pressure ratio of 10, and the temperature of the gases at the turbine inlet is 2000°F. Air enters the compressor at a rate of 100 lbm/s. Utilizing the cold-air-standard assumptions, determine (*a*) the temperature and pressure of the gases at the turbine exit, (*b*) the velocity of the gases at the nozzle exit, and (*c*) the propulsive efficiency of the cycle.

Solution The operating conditions of a turbojet aircraft are specified. The temperature and pressure at the turbine exit, the velocity of gases at the nozzle exit, and the propulsive efficiency are to be determined.

Assumptions **1** Steady operating conditions exist. **2** The cold-air-standard assumptions are applicable and thus air can be assumed to have constant specific heats at room temperature (c_p = 0.240 Btu/lbm·°F and k = 1.4). **3** Kinetic and potential energies are negligible, except at the diffuser inlet and the nozzle exit. **4** The turbine work output is equal to the compressor work input.

Analysis The *T-s* diagram of the ideal jet propulsion cycle described is shown in Fig. 9–50. We note that the components involved in the jet-propulsion cycle are steady-flow devices.

(*a*) Before we can determine the temperature and pressure at the turbine exit, we need to find the temperatures and pressures at other states:

Process 1-2 (isentropic compression of an ideal gas in a diffuser): For convenience, we can assume that the aircraft is stationary and the air is moving toward the aircraft at a velocity of V_1 = 850 ft/s. Ideally, the air exits the diffuser with a negligible velocity ($V_2 \cong 0$):

$$h_2 + \frac{V_2^2 \nearrow^0}{2} = h_1 + \frac{V_1^2}{2}$$

$$0 = c_p(T_2 - T_1) - \frac{V_1^2}{2}$$

$$T_2 = T_1 + \frac{V_1^2}{2c_p}$$

$$= 420\text{ R} + \frac{(850\text{ ft/s})^2}{2(0.240\text{ Btu/lbm·R})}\left(\frac{1\text{ Btu/lbm}}{25{,}037\text{ ft}^2/\text{s}^2}\right)$$

$$= 480\text{ R}$$

$$P_2 = P_1\left(\frac{T_2}{T_1}\right)^{k/(k-1)} = (5\text{ psia})\left(\frac{480\text{ R}}{420\text{ R}}\right)^{1.4/(1.4-1)} = 8.0\text{ psia}$$

Process 2-3 (isentropic compression of an ideal gas in a compressor):

$$P_3 = (r_p)(P_2) = (10)(8.0\text{ psia}) = 80\text{ psia } (= P_4)$$

$$T_3 = T_2\left(\frac{P_3}{P_2}\right)^{(k-1)/k} = (480\text{ R})(10)^{(1.4-1)/1.4} = 927\text{ R}$$

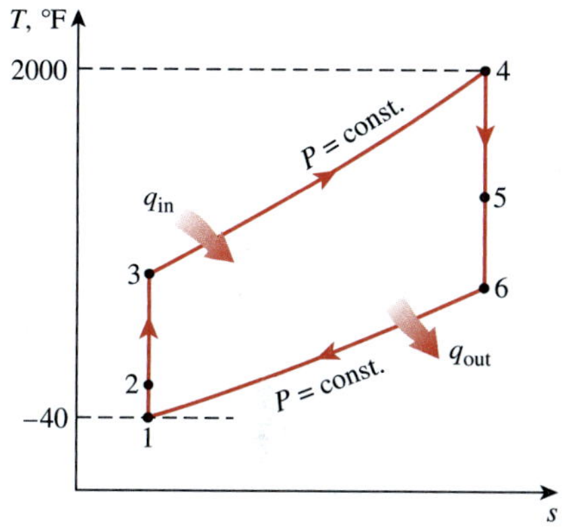

FIGURE 9–50
T-s diagram for the turbojet cycle described in Example 9–9.

Process 4-5 (isentropic expansion of an ideal gas in a turbine): Neglecting the kinetic energy changes across the compressor and the turbine and assuming the turbine work to be equal to the compressor work, we find the temperature and pressure at the turbine exit to be

$$w_{\text{comp,in}} = w_{\text{turb,out}}$$

$$h_3 - h_2 = h_4 - h_5$$

$$c_p(T_3 - T_2) = c_p(T_4 - T_5)$$

$$T_5 = T_4 - T_3 + T_2 = 2460 - 927 + 480 = \mathbf{2013\ R}$$

$$P_5 = P_4\left(\frac{T_5}{T_4}\right)^{k/(k-1)} = (80\text{ psia})\left(\frac{2013\text{ R}}{2460\text{ R}}\right)^{1.4/(1.4-1)} = \mathbf{39.7\ psia}$$

(*b*) To find the air velocity at the nozzle exit, we need to first determine the nozzle exit temperature and then apply the steady-flow energy equation.

Process 5-6 (isentropic expansion of an ideal gas in a nozzle):

$$T_6 = T_5\left(\frac{P_6}{P_5}\right)^{(k-1)/k} = (2013\text{ R})\left(\frac{5\text{ psia}}{39.7\text{ psia}}\right)^{(1.4-1)/1.4} = 1114\text{ R}$$

$$h_6 + \frac{V_6^2}{2} = h_5 + \frac{V_5^2}{2}\nearrow^{0}$$

$$0 = c_p(T_6 - T_5) + \frac{V_6^2}{2}$$

$$V_6 = \sqrt{2c_p(T_5 - T_6)}$$

$$= \sqrt{2(0.240\text{ Btu/lbm·R})[(2013 - 1114)\text{ R}]\left(\frac{25{,}037\text{ ft}^2/\text{s}^2}{1\text{ Btu/lbm}}\right)}$$

$$= \mathbf{3288\ ft/s}$$

(*c*) The propulsive efficiency of a turbojet engine is the ratio of the propulsive power developed $\dot{W}_P$ to the total heat transfer rate to the working fluid:

$$\dot{W}_P = \dot{m}(V_{\text{exit}} - V_{\text{inlet}})V_{\text{aircraft}}$$

$$= (100\text{ lbm/s})[(3288 - 850)\text{ ft/s}](850\text{ ft/s})\left(\frac{1\text{ Btu/lbm}}{25{,}037\text{ ft}^2/\text{s}^2}\right)$$

$$= 8276\text{ Btu/s}\quad(\text{or }11{,}707\text{ hp})$$

$$\dot{Q}_{\text{in}} = \dot{m}(h_4 - h_3) = \dot{m}c_p(T_4 - T_3)$$

$$= (100\text{ lbm/s})(0.240\text{ Btu/lbm·R})[(2460 - 927)\text{ R}]$$

$$= 36{,}794\text{ Btu/s}$$

$$\eta_P = \frac{\dot{W}_P}{\dot{Q}_{\text{in}}} = \frac{8276\text{ Btu/s}}{36{,}794\text{ Btu/s}} = \mathbf{22.5\%}$$

That is, 22.5 percent of the energy input is used to propel the aircraft and to overcome the drag force exerted by the atmospheric air.

Discussion For those who are wondering what happened to the rest of the energy, here is a brief account:

$$\dot{KE}_{out} = \dot{m}\frac{V_g^2}{2} = (100 \text{ lbm/s})\left\{\frac{[(3288 - 850)\text{ft/s}]^2}{2}\right\}\left(\frac{1 \text{ Btu/lbm}}{25{,}037 \text{ ft}^2/\text{s}^2}\right)$$

$$= 11{,}867 \text{ Btu/s} \qquad (32.2\%)$$

$$\dot{Q}_{out} = \dot{m}(h_6 - h_1) = \dot{m}c_p(T_6 - T_1)$$

$$= (100 \text{ lbm/s})(0.24 \text{ Btu/lbm·R})[(1114 - 420) \text{ R}]$$

$$= 16{,}651 \text{ Btu/s} \qquad (45.3\%)$$

Thus, 32.2 percent of the energy shows up as excess kinetic energy (kinetic energy of the gases relative to a fixed point on the ground). Notice that for the highest propulsion efficiency, the velocity of the exhaust gases relative to the ground V_g should be zero. That is, the exhaust gases should leave the nozzle at the velocity of the aircraft. The remaining 45.3 percent of the energy shows up as an increase in enthalpy of the gases leaving the engine. These last two forms of energy eventually become part of the internal energy of the atmospheric air (Fig. 9–51).

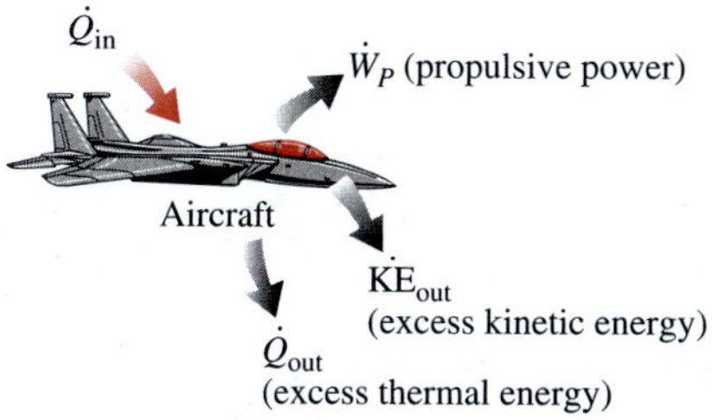

FIGURE 9–51
Energy supplied to an aircraft (from the burning of a fuel) manifests itself in various forms.

Modifications to Turbojet Engines

The first airplanes built were all propeller-driven, with propellers powered by engines essentially identical to automobile engines. The major breakthrough in commercial aviation occurred with the introduction of the turbojet engine in 1952. Both propeller-driven engines and jet-propulsion-driven engines have their own strengths and limitations, and several attempts have been made to combine the desirable characteristics of both in one engine. Two such modifications are the *propjet engine* and the *turbofan engine.*

The most widely used engine in aircraft propulsion is the **turbofan** (or *fanjet*) engine wherein a large fan driven by the turbine forces a considerable amount of air through a duct (cowl) surrounding the engine, as shown in Figs. 9–52 and 9–53. The fan exhaust leaves the duct at a higher velocity, enhancing the total thrust of the engine significantly. A turbofan engine is based on the principle that for the same power, a large volume of slower-moving air

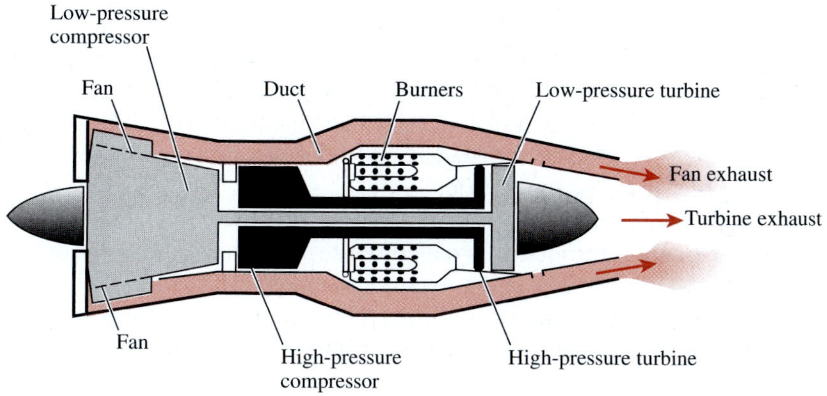

FIGURE 9–52
A turbofan engine.

Source: The Aircraft Gas Turbine and Its Operation. © United Aircraft Corporation (now United Technologies Corp.), 1951, 1974.

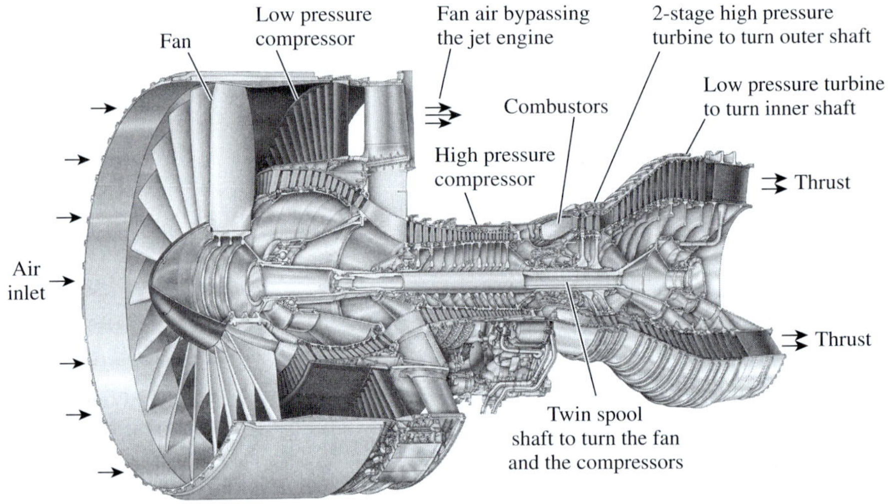

FIGURE 9–53
A modern jet engine used to power Boeing 777 aircraft. This is a Pratt & Whitney PW4084 turbofan capable of producing 84,000 pounds of thrust. It is 4.87 m (192 in.) long, has a 2.84 m (112 in.) diameter fan, and it weighs 6800 kg (15,000 lbm).

Courtesy of Pratt & Whitney Corp.

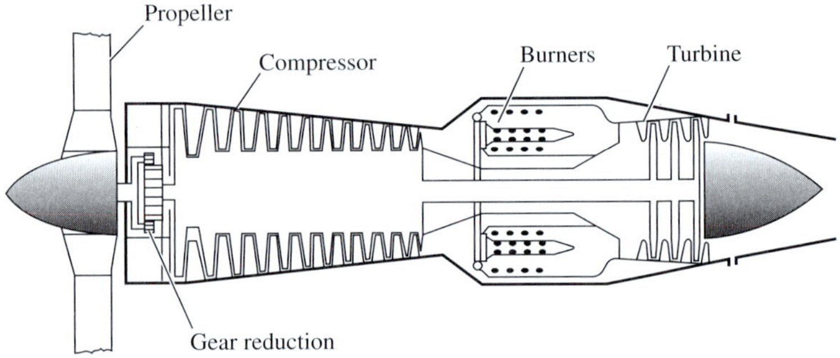

FIGURE 9–54
A turboprop engine.

Source: The Aircraft Gas Turbine Engine and Its Operation. © United Aircraft Corporation (now United Technologies Corp.), 1951, 1974.

produces more thrust than a small volume of fast-moving air. The first commercial turbofan engine was successfully tested in 1955.

The turbofan engine on an airplane can be distinguished from the less efficient turbojet engine by its fat cowling covering the large fan. All the thrust of a turbojet engine is due to the exhaust gases leaving the engine at about twice the speed of sound. In a turbofan engine, the high-speed exhaust gases are mixed with the lower-speed air, which results in a considerable reduction in noise.

New cooling techniques have resulted in considerable increases in efficiencies by allowing gas temperatures at the burner exit to reach over 1500°C, which is more than 100°C above the melting point of the turbine blade materials. Turbofan engines deserve most of the credit for the success of jumbo jets that weigh almost 400,000 kg and are capable of carrying over 400 passengers for up to a distance of 10,000 km at speeds over 950 km/h with less fuel per passenger mile.

The ratio of the mass flow rate of air bypassing the combustion chamber to that of air flowing through it is called the *bypass ratio.* The first commercial high-bypass-ratio engines had a bypass ratio of 5. Increasing the bypass ratio of a turbofan engine increases thrust. Thus, it makes sense to remove the cowl from the fan. The result is a **propjet** engine, as shown in Fig. 9–54. Turbofan and propjet engines differ primarily in their bypass ratios: 5 or 6 for turbofans and as high as 100 for propjets. As a general rule, propellers

FIGURE 9–55
A ramjet engine.

Source: The Aircraft Gas Turbine Engine and Its Operation. © United Aircraft Corporation (now United Technologies Corp.), 1951, 1974.

are more efficient than jet engines, but they are limited to low-speed and low-altitude operation since their efficiency decreases at high speeds and altitudes. The old propjet engines (*turboprops*) were limited to speeds of about Mach 0.62 and to altitudes of around 9100 m. The new propjet engines (*propfans*) are expected to achieve speeds of about Mach 0.82 and altitudes of about 12,200 m. Commercial airplanes of medium size and range propelled by propfans are expected to fly as high and as fast as the planes propelled by turbofans, and to do so on less fuel.

Another modification that is popular in military aircraft is the addition of an **afterburner** section between the turbine and the nozzle. Whenever a need for extra thrust arises, such as for short takeoffs or combat conditions, additional fuel is injected into the oxygen-rich combustion gases leaving the turbine. As a result of this added energy, the exhaust gases leave at a higher velocity, providing a greater thrust.

A **ramjet** engine is a properly shaped duct with no compressor or turbine, as shown in Fig. 9–55, and is sometimes used for high-speed propulsion of missiles and aircraft. The pressure rise in the engine is provided by the ram effect of the incoming high-speed air being rammed against a barrier. Therefore, a ramjet engine needs to be brought to a sufficiently high speed by an external source before it can be fired.

The ramjet performs best in aircraft flying above Mach 2 or 3 (two or three times the speed of sound). In a ramjet, the air is slowed down to about Mach 0.2, fuel is added to the air and burned at this low velocity, and the combustion gases are expended and accelerated in a nozzle.

A **scramjet** engine is essentially a ramjet in which air flows through at supersonic speeds (above the speed of sound). Ramjets that convert to scramjet configurations at speeds above Mach 6 are successfully tested at speeds of about Mach 8.

Finally, a **rocket** is a device where a solid or liquid fuel and an oxidizer react in the combustion chamber. The high-pressure combustion gases are then expanded in a nozzle. The gases leave the rocket at very high velocities, producing the thrust to propel the rocket.

9–12 ▪ SECOND-LAW ANALYSIS OF GAS POWER CYCLES

The ideal Carnot, Ericsson, and Stirling cycles are *totally reversible*; thus they do not involve any irreversibilities. The ideal Otto, Diesel, and Brayton cycles, however, are only *internally reversible,* and they may involve irreversibilities external to the system. A second-law analysis of these cycles reveals where the largest irreversibilities occur and where to start improvements.

Relations for *exergy* and *exergy destruction* for both closed and steady-flow systems are developed in Chap. 8. The exergy destruction for a closed system can be expressed as

$$X_{\text{dest}} = T_0 S_{\text{gen}} = T_0(\Delta S_{\text{sys}} - S_{\text{in}} + S_{\text{out}})$$

$$= T_0\left[(S_2 - S_1)_{\text{sys}} - \frac{Q_{\text{in}}}{T_{b,\text{in}}} + \frac{Q_{\text{out}}}{T_{b,\text{out}}}\right] \quad \text{(kJ)} \qquad \textbf{(9–30)}$$

where $T_{b,\text{in}}$ and $T_{b,\text{out}}$ are the temperatures of the system boundary where heat is transferred into and out of the system, respectively. A similar relation for steady-flow systems can be expressed, in rate form, as

$$\dot{X}_{\text{dest}} = T_0 \dot{S}_{\text{gen}} = T_0(\dot{S}_{\text{out}} - \dot{S}_{\text{in}}) = T_0\left(\sum_{\text{out}} \dot{m}s - \sum_{\text{in}} \dot{m}s - \frac{\dot{Q}_{\text{in}}}{T_{b,\text{in}}} + \frac{\dot{Q}_{\text{out}}}{T_{b,\text{out}}}\right) \quad \text{(kW)}$$

(9–31)

or, on a unit–mass basis for a one-inlet, one-exit steady-flow device, as

$$x_{\text{dest}} = T_0 s_{\text{gen}} = T_0\left(s_e - s_i - \frac{q_{\text{in}}}{T_{b,\text{in}}} + \frac{q_{\text{out}}}{T_{b,\text{out}}}\right) \quad \text{(kJ/kg)} \qquad \textbf{(9–32)}$$

where subscripts i and e denote the inlet and exit states, respectively.

The exergy destruction of a *cycle* is the sum of the exergy destructions of the processes that compose that cycle. The exergy destruction of a cycle can also be determined without tracing the individual processes by considering the entire cycle as a single process and using one of the relations above. Entropy is a property, and its value depends on the state only. For a cycle, reversible or actual, the initial and the final states are identical; thus $s_e = s_i$. Therefore, the exergy destruction of a cycle depends on the magnitude of the heat transfer with the high- and low-temperature reservoirs involved and on their temperatures. It can be expressed on a unit–mass basis as

$$x_{\text{dest}} = T_0\left(\sum \frac{q_{\text{out}}}{T_{b,\text{out}}} - \sum \frac{q_{\text{in}}}{T_{b,\text{in}}}\right) \quad \text{(kJ/kg)} \qquad \textbf{(9–33)}$$

For a cycle that involves heat transfer only with a source at T_H and a sink at T_L, the exergy destruction becomes

$$x_{\text{dest}} = T_0\left(\frac{q_{\text{out}}}{T_L} - \frac{q_{\text{in}}}{T_H}\right) \quad \text{(kJ/kg)} \qquad \textbf{(9–34)}$$

The exergies of a closed system ϕ and a fluid stream ψ at any state can be determined from

$$\phi = (u - u_0) - T_0(s - s_0) + P_0(v - v_0) + \frac{V^2}{2} + gz \quad \text{(kJ/kg)} \qquad \textbf{(9–35)}$$

and

$$\psi = (h - h_0) - T_0(s - s_0) + \frac{V^2}{2} + gz \quad \text{(kJ/kg)} \qquad \textbf{(9–36)}$$

where subscript "0" denotes the state of the surroundings.

EXAMPLE 9–10

Consider an engine operating on the ideal Otto cycle with a compression ratio of 8 (Fig. 9–56). At the beginning of the compression process, air is at 100 kPa and 17°C. During the constant-volume heat-addition process, 800 kJ/kg of heat is transferred to air from a source at 1700 K and waste heat is rejected to the surroundings at 300 K. Accounting for the variation of specific heats of air with temperature, determine (*a*) the exergy destruction associated with each of the four processes and the cycle and (*b*) the second-law efficiency of this cycle.

Solution An engine operating on the ideal Otto cycle is considered. For specific source and sink temperatures, the exergy destruction associated with this cycle and the second-law efficiency are to be determined.

Assumptions **1** Steady operating conditions exist. **2** Kinetic and potential energy changes are negligible.

Analysis We take the engine bordering the heat source at temperature T_H and the environment at temperature T_0 as the system. This cycle was analyzed in Example 9-2, and various quantities were given or determined to be

$$r = 8 \qquad P_2 = 1.7997 \text{ MPa}$$
$$T_0 = 290 \text{ K} \qquad P_3 = 4.345 \text{ MPa}$$
$$T_1 = 290 \text{ K} \qquad q_{in} = 800 \text{ kJ/kg}$$
$$T_2 = 652.4 \text{ K} \qquad q_{out} = 381.83 \text{ kJ/kg}$$
$$T_3 = 1575.1 \text{ K} \qquad w_{net} = 418.17 \text{ kJ/kg}$$

Processes 1-2 and 3-4 are isentropic ($s_1 = s_2$, $s_3 = s_4$) and therefore do not involve any internal or external irreversibilities; that is, $X_{dest,12} = \mathbf{0}$ and $X_{dest,34} = \mathbf{0}$.

Processes 2-3 and 4-1 are constant-volume heat-addition and heat-rejection processes, respectively, and are internally reversible. However, the heat transfer between the working fluid and the source or the sink takes place through a finite temperature difference, rendering both processes irreversible. The exergy destruction associated with each process is determined from Eq. 9-32. However, first we need to determine the entropy change of air during these processes:

$$s_3 - s_2 = s_3^\circ - s_2^\circ - R \ln \frac{P_3}{P_2}$$

$$= (3.5045 - 2.4975) \text{ kJ/kg·K} - (0.287 \text{ kJ/kg·K}) \ln \frac{4.345 \text{ MPa}}{1.7997 \text{ MPa}}$$

$$= 0.7540 \text{ kJ/kg·K}$$

Also,

$$q_{in} = 800 \text{ kJ/kg} \quad \text{and} \quad T_{source} = 1700 \text{ K}$$

Thus

$$x_{dest,23} = T_0\left[(s_3 - s_2)_{sys} - \frac{q_{in}}{T_{source}}\right]$$

$$= (290 \text{ K})\left[0.7540 \text{ kJ/kg·K} - \frac{800 \text{ kJ/kg}}{1700 \text{ K}}\right]$$

$$= \mathbf{82.2 \text{ kJ/kg}}$$

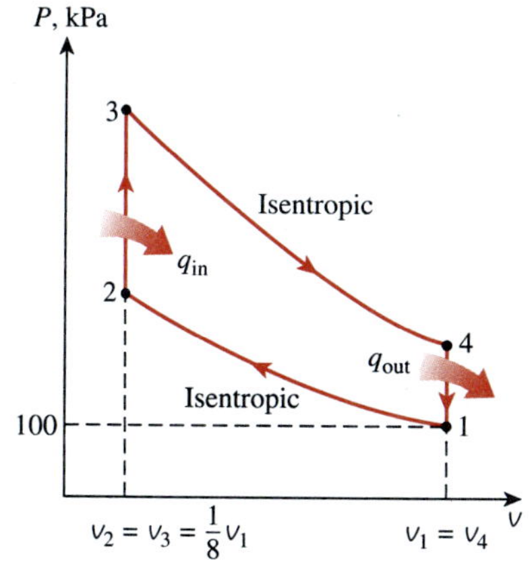

FIGURE 9–56
Schematic for Example 9–10.

For process 4-1, $s_1 - s_4 = s_2 - s_3 = -0.7540$ kJ/kg·K, $q_{R,41} = q_{out} =$ 381.83 kJ/kg, and $T_{sink} = 290$ K. Thus,

$$x_{dest,41} = T_0\left[(s_1 - s_4)_{sys} + \frac{q_{out}}{T_{sink}}\right]$$

$$= (290\text{ K})\left[-0.7540\text{ kJ/kg·K} + \frac{381.83\text{ kJ/kg}}{290\text{ K}}\right]$$

$$= \mathbf{163.2\ kJ/kg}$$

Therefore, the irreversibility of the cycle is

$$x_{dest,cycle} = x_{dest,12} + x_{dest,23} + x_{dest,34} + x_{dest,41}$$

$$= 0 + 82.2\text{ kJ/kg} + 0 + 163.2\text{ kJ/kg}$$

$$= \mathbf{245.4\ kJ/kg}$$

The exergy destruction of the cycle could also be determined from Eq. 9-34. Notice that the largest exergy destruction in the cycle occurs during the heat-rejection process. Therefore, any attempt to reduce the exergy destruction should start with this process.

(*b*) The second-law efficiency is defined as

$$\eta_{II} = \frac{\text{Exergy recovered}}{\text{Exergy expended}} = \frac{x_{recovered}}{x_{expended}} = 1 - \frac{x_{destroyed}}{x_{expended}}$$

Here the expended exergy is the exergy content of the heat supplied to the air in the engine (which is its work potential) and the exergy recovered is the net work output:

$$x_{expended} = x_{heat,\,in} = \left(1 - \frac{T_0}{T_H}\right)q_{in}$$

$$= \left(1 - \frac{290\text{ K}}{1700\text{ K}}\right)(800\text{ kJ/kg}) = 663.5\text{ kJ/kg}$$

$$x_{recovered} = w_{net,\,out} = 418.17\text{ kJ/kg}$$

Substituting, the second-law efficiency of this power plant is determined to be

$$\eta_{II} = \frac{x_{recovered}}{x_{expended}} = \frac{418.17\text{ kJ/kg}}{663.5\text{ kJ/kg}} = 0.630 \text{ or } \mathbf{63.0\%}$$

Discussion The second-law efficiency can also be determined using the exergy destruction data,

$$\eta_{II} = 1 - \frac{x_{destroyed}}{x_{expended}} = 1 - \frac{245.4\text{ kJ/kg}}{663.5\text{ kJ/kg}} = 0.630 \text{ or } 63.0\%$$

Note that the exergy destruction associated with heat transfer involving both the heat source and the environment are accounted for in the results.

TOPIC OF SPECIAL INTEREST* Saving Fuel and Money by Driving Sensibly

Two-thirds of the oil used in the United States is used for transportation. Half of this oil is consumed by passenger cars and light trucks that are used to commute to and from work (38 percent), run a family business (35 percent), and for recreational, social, and religious activities (27 percent). The overall fuel efficiency of the vehicles has increased considerably over the years due to improvements primarily in aerodynamics, materials, and electronic controls. However, the average fuel consumption of new vehicles has not changed much from about 20 miles per gallon (mpg) because of the increasing consumer trend toward purchasing larger and less fuel-efficient cars, trucks, and sport utility vehicles. Motorists also continue to drive more each year: 11,725 miles in 1999 compared to 10,277 miles in 1990. Consequently, the annual gasoline use per vehicle in the United States has increased to 603 gallons in 1999 (worth $1206 at $2.00/gal) from 506 gallons in 1990 (Fig. 9–57).

FIGURE 9–57
The average car in the United States is driven about 12,000 miles a year, uses about 600 gallons of gasoline, worth $1200 at $2.00/gal.

Saving fuel is not limited to good driving habits. It also involves purchasing the right car, using it responsibly, and maintaining it properly. A car does not burn any fuel when it is not running, and thus a sure way to save fuel is not to drive the car at all—but this is not the reason we buy a car. We can reduce driving and thus fuel consumption by considering viable alternatives such as *living close to work and shopping areas, working at home, working longer hours in fewer days, joining a car pool or starting one, using public transportation, combining errands into a single trip and planning ahead, avoiding rush hours and roads with heavy traffic and many traffic lights,* and simply *walking* or *bicycling* instead of driving to nearby places, with the added benefit of good health and physical fitness. Driving only when necessary is the best way to save fuel, money, and the environment too.

Driving efficiently starts before buying a car, just like raising good children starts before getting married. The buying decision made now will affect the fuel consumption for many years. Under average driving conditions, the owner of a 30-mpg vehicle will spend $400 less each year on fuel than the owner of a 20-mpg vehicle (assuming a fuel cost of $2.00 per gallon and 12,000 miles of driving per year). If the vehicle is owned for 5 years, the 30-mpg vehicle will save $2000 during this period (Fig. 9–58). The fuel consumption of a car depends on many factors such as *the type of the vehicle, the weight, the transmission type, the size and efficiency of the engine,* and *the accessories* and *the options installed.* The most fuel-efficient cars are aerodynamically designed compact cars with a small engine, manual transmission, low frontal area (the height times the width of the car), and bare essentials.

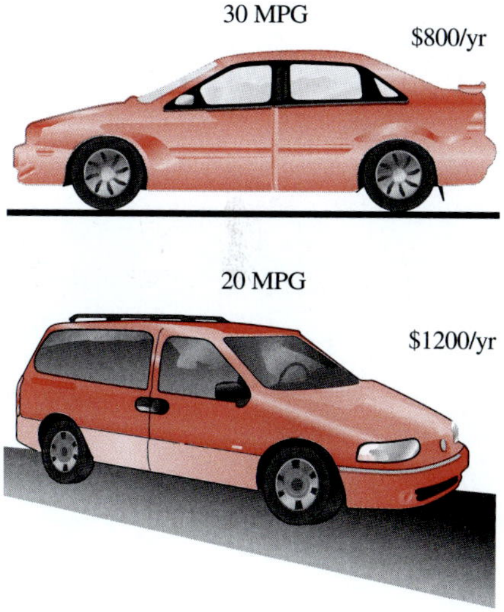

FIGURE 9–58
Under average driving conditions, the owner of a 30-mpg vehicle spends $400 less each year on gasoline than the owner of a 20-mpg vehicle (assuming $2.00/gal and 12,000 miles/yr).

At highway speeds, most fuel is used to overcome aerodynamic drag or air resistance to motion, which is the force needed to move the vehicle through the air. This resistance force is proportional to the drag coefficient and the frontal area. Therefore, for a given frontal area, a sleek-looking aerodynamically designed vehicle with contoured lines that coincide with the streamlines of air flow has a smaller drag coefficient and thus better fuel economy

*This section can be skipped without a loss in continuity. Information in this section is based largely on the publications of the U.S. Department of Energy, Environmental Protection Agency, and the American Automotive Association.

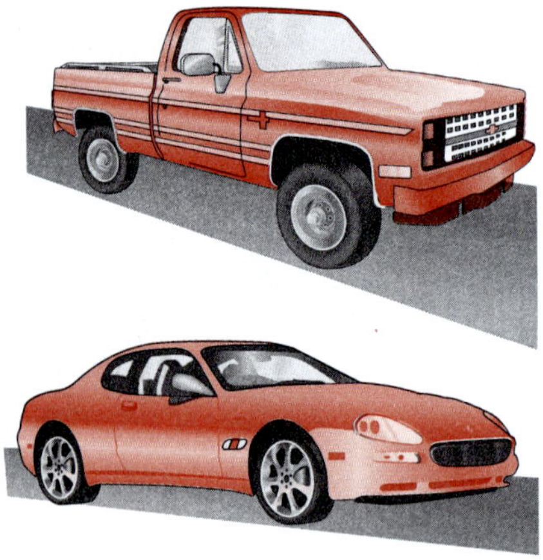

FIGURE 9–59
Aerodynamically designed vehicles have a smaller drag coefficient and thus better fuel economy than boxlike vehicles with sharp corners.

than a boxlike vehicle with sharp corners (Fig. 9–59). For the same overall shape, a compact car has a smaller frontal area and thus better fuel economy compared to a large car.

Moving around the *extra weight* requires more fuel, and thus it hurts fuel economy. Therefore, the lighter the vehicle, the more fuel-efficient it is. Also as a general rule, the larger the engine is, the greater its rate of fuel consumption is. So you can expect a car with a 1.8 L engine to be more fuel efficient than one with a 3.0 L engine. For a given engine size, *diesel engines* operate on much higher compression ratios than the gasoline engines, and thus they are inherently more fuel-efficient. *Manual transmissions* are usually more efficient than the automatic ones, but this is not always the case. A car with automatic transmission generally uses 10 percent more fuel than a car with manual transmission because of the losses associated with the hydraulic connection between the engine and the transmission, and the added weight. Transmissions with an *overdrive gear* (found in four-speed automatic transmissions and five-speed manual transmissions) save fuel and reduce noise and engine wear during highway driving by decreasing the engine rpm while maintaining the same vehicle speed.

Front wheel drive offers better traction (because of the engine weight on top of the front wheels), reduced vehicle weight and thus better fuel economy, with an added benefit of increased space in the passenger compartment. Four-wheel drive mechanisms provide better traction and braking thus safer driving on slippery roads and loose gravel by transmitting torque to all four wheels. However, the added safety comes with increased weight, noise, and cost, and decreased fuel economy. *Radial tires* usually reduce the fuel consumption by 5 to 10 percent by reducing the rolling resistance, but their pressure should be checked regularly since they can look normal and still be underinflated. *Cruise control* saves fuel during long trips on open roads by maintaining steady speed. *Tinted windows* and light interior and exterior colors reduce solar heat gain, and thus the need for air-conditioning.

BEFORE DRIVING

Certain things done before driving can make a significant difference on the fuel cost of the vehicle while driving. Below we discuss some measures such as using the right kind of fuel, minimizing idling, removing extra weight, and keeping the tires properly inflated.

Use Fuel with the Minimum Octane Number Recommended by the Vehicle Manufacturer

Many motorists buy higher-priced premium fuel, thinking that it is better for the engine. Most of today's cars are designed to operate on regular unleaded fuel. If the owner's manual does not call for premium fuel, using anything other than regular gas is simply a waste of money. Octane number is not a measure of the "power" or "quality" of the fuel, it is simply a measure of fuel's resistance to engine knock caused by premature ignition. Despite the implications of flashy names like "premium," "super," or "power plus," a fuel with a higher octane number is not a better fuel; it is simply more expensive because of the extra processing involved to raise the octane number

(Fig. 9–60). Older cars may need to go up one grade level from the recommended new car octane number if they start knocking.

Do Not Overfill the Gas Tank

Topping off the gas tank may cause the fuel to backflow during pumping. In hot weather, an overfilled tank may also cause the fuel to overflow due to thermal expansion. This wastes fuel, pollutes the environment, and may damage the car's paint. Also, fuel tank caps that do not close tightly allow some gasoline to be lost by evaporation. Buying fuel in cool weather such as early in the mornings minimizes evaporative losses. Each gallon of spilled or evaporated fuel emits as much hydrocarbon to the air as 7500 miles of driving.

FIGURE 9–60
Despite the implications of flashy names, a fuel with a higher octane number is not a better fuel; it is simply more expensive.

© Vol. 21/PhotoDisc/Getty RF.

Park in the Garage

The engine of a car parked in a garage overnight is warmer the next morning. This reduces the problems associated with the warming-up period such as starting, excessive fuel consumption, and environmental pollution. In hot weather, a garage blocks the direct sunlight and reduces the need for air conditioning.

Start the Car Properly and Avoid Extended Idling

With today's cars, it is not necessary to prime the engine first by pumping the accelerator pedal repeatedly before starting. This only wastes fuel. Warming up the engine isn't necessary either. Keep in mind that an idling engine wastes fuel and pollutes the environment. Don't race a cold engine to warm it up. An engine warms up faster on the road under a light load, and the catalytic converter begins to function sooner. Start driving as soon as the engine is started, but avoid rapid acceleration and highway driving before the engine and thus the oil fully warms up to prevent engine wear.

In cold weather, the warm-up period is much longer, the fuel consumption during warm-up is much higher, and the exhaust emissions are much larger. At −20°C, for example, a car needs to be driven at least 3 miles to warm up fully. A gasoline engine uses up to 50 percent more fuel during warm-up than it does after it is warmed up. Exhaust emissions from a cold engine during warm-up are much higher since the catalytic converters do not function properly before reaching their normal operating temperature of about 390°C.

Don't Carry Unnecessary Weight in or on the Vehicle

Remove any snow or ice from the vehicle, and avoid carrying unneeded items, especially heavy ones (such as snow chains, old tires, books) in the passenger compartment, trunk, or the cargo area of the vehicle (Fig. 9–61). This wastes fuel since it requires extra fuel to carry around the extra weight. An extra 100 lbm decreases fuel economy of a car by about 1–2 percent.

Some people find it convenient to use a roof rack or carrier for additional cargo space. However, if you must carry some extra items, place them inside the vehicle rather than on roof racks to reduce drag. Any snow that accumulates on a vehicle and distorts its shape must be removed for the same reason.

FIGURE 9–61
A loaded roof rack can increase fuel consumption by up to 5 percent in highway driving.

A loaded roof rack can increase fuel consumption by up to 5 percent in highway driving. Even the most streamlined empty rack increases aerodynamic drag and thus fuel consumption. Therefore, the roof rack should be removed when it is no longer needed.

Keep Tires Inflated to the Recommended Maximum Pressure

Keeping the tires inflated properly is one of the easiest and most important things one can do to improve fuel economy. If a range is recommended by the manufacturer, the higher pressure should be used to maximize fuel efficiency. Tire pressure should be checked when the tire is cold since tire pressure changes with temperature (it increases by 1 psi for every 10°F rise in temperature due to a rise in ambient temperature or just road friction). Underinflated tires run hot and jeopardize safety, cause the tires to wear prematurely, affect the vehicle's handling adversely, and hurt the fuel economy by increasing the rolling resistance. Overinflated tires cause unpleasant bumpy rides, and cause the tires to wear unevenly. Tires lose about 1 psi pressure per month due to air loss caused by the tire hitting holes, bumps, and curbs. Therefore, the tire pressure should be checked at least once a month. Just one tire underinflated by 2 psi results in a 1 percent increase in fuel consumption (Fig. 9–62). Underinflated tires often cause fuel consumption of vehicles to increase by 5 or 6 percent.

It is also important to keep the wheels aligned. Driving a vehicle with the front wheels out of alignment increases rolling resistance and thus fuel consumption while causing handling problems and uneven tire wear. Therefore, the wheels should be aligned properly whenever necessary.

FIGURE 9–62
Underinflated tires often cause fuel consumption of vehicles to increase by 5 or 6 percent.

© The McGraw-Hill Companies/Jill Braaten, photographer

WHILE DRIVING

The driving habits can make a significant difference in the amount of fuel used. Driving sensibly and practicing some fuel-efficient driving techniques such as those discussed below can improve fuel economy easily by more than 10 percent.

Avoid Quick Starts and Sudden Stops

Despite the attention they may get, the abrupt, aggressive "jackrabbit" starts waste fuel, wear the tires, jeopardize safety, and are harder on vehicle components and connectors. The squealing stops wear the brake pads prematurely, and may cause the driver to lose control of the vehicle. Easy starts and stops save fuel, reduce wear and tear, reduce pollution, and are safer and more courteous to other drivers.

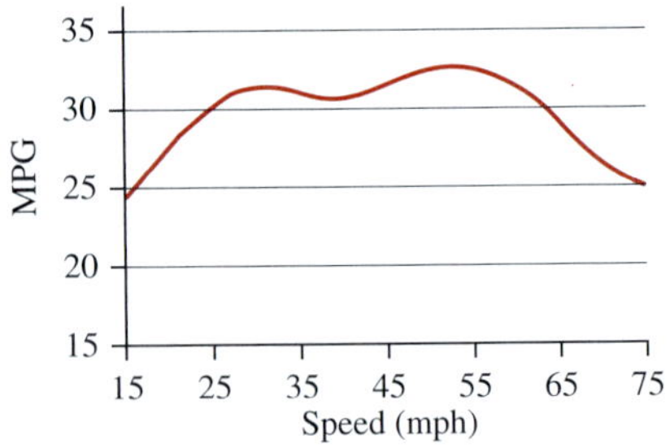

FIGURE 9–63
Aerodynamic drag increases and thus fuel economy decreases rapidly at speeds above 55 mph.

Source: EPA and U.S. Dept. of Energy.

Drive at Moderate Speeds

Avoiding high speeds on open roads results in safer driving and better fuel economy. In highway driving, over 50 percent of the power produced by the engine is used to overcome aerodynamic drag (i.e., to push air out of the way). Aerodynamic drag and thus fuel consumption increase rapidly at speeds above 55 mph, as shown in Fig. 9–63. On average, a car uses about

15 percent more fuel at 65 mph and 25 percent more fuel at 70 mph than it does at 55 mph. (A car uses about 10 percent more fuel at 100 km/h and 20 percent more fuel at 110 km/h than it does at 90 km/h.)

The discussion above should not lead one to conclude that the lower the speed, the better the fuel economy—because it is not. The number of miles that can be driven per gallon of fuel drops sharply at speeds below 30 mph (or 50 km/h), as shown in the chart. Besides, speeds slower than the flow of traffic can create a traffic hazard. Therefore, a car should be driven at moderate speeds for safety and best fuel economy.

Maintain a Constant Speed

The fuel consumption remains at a minimum during steady driving at a moderate speed. Keep in mind that every time the accelerator is hard pressed, more fuel is pumped into the engine. The vehicle should be accelerated gradually and smoothly since extra fuel is squirted into the engine during quick acceleration. Using cruise control on highway trips can help maintain a constant speed and reduce fuel consumption. Steady driving is also safer, easier on the nerves, and better for the heart.

Anticipate Traffic Ahead and Avoid Tailgating

A driver can reduce fuel consumption by up to 10 percent by anticipating traffic conditions ahead and adjusting the speed accordingly, and avoiding tailgating and thus unnecessary braking and acceleration (Fig. 9–64). Accelerations and decelerations waste fuel. Braking and abrupt stops can be minimized, for example, by not following too closely, and slowing down gradually by releasing the gas pedal when approaching a red light, a stop sign, or slow traffic. This relaxed driving style is safer, saves fuel and money, reduces pollution, reduces wear on the tires and brakes, and is appreciated by other drivers. Allowing sufficient time to reach the destination makes it easier to resist the urge to tailgate.

FIGURE 9–64
Fuel consumption can be decreased by up to 10 percent by anticipating traffic conditions ahead and adjusting accordingly.

Avoid Sudden Acceleration and Sudden Braking (Except in Emergencies)

Accelerate gradually and smoothly when passing other vehicles or merging with faster traffic. Pumping or hard pressing the accelerator pedal while driving causes the engine to switch to a "fuel enrichment mode" of operation that wastes fuel. In city driving, nearly half of the engine power is used for acceleration. When accelerating with stick-shifts, the RPM of the engine should be kept to a minimum. Braking wastes the mechanical energy produced by the engine and wears the brake pads.

Avoid Resting Feet on the Clutch or Brake Pedal while Driving

Resting the left foot on the brake pedal increases the temperature of the brake components, and thus reduces their effectiveness and service life while wasting fuel. Similarly, resting the left foot on the clutch pedal lessens the pressure on the clutch pads, causing them to slip and wear prematurely, wasting fuel.

Use Highest Gear (Overdrive) During Highway Driving

Overdrive improves fuel economy during highway driving by decreasing the vehicle's engine speed (or RPM). The lower engine speed reduces fuel consumption per unit time as well as engine wear. Therefore, overdrive (the fifth gear in cars with overdrive manual transmission) should be used as soon as the vehicle's speed is high enough.

FIGURE 9–65
Unnecessary idling during lengthy waits wastes fuel, costs money, and pollutes the air.

Turn the Engine Off Rather Than Letting It Idle

Unnecessary idling during lengthy waits (such as waiting for someone or for service at a drive-up window, being stuck in traffic, etc.) wastes fuel, pollutes the air, and causes engine wear (more wear than driving) (Fig. 9–65). Therefore, the engine should be turned off rather than letting it idle. Idling for more than a minute consumes much more fuel than restarting the engine. Fuel consumption in the lines of drive-up windows and the pollution emitted can be avoided altogether by simply parking the car and going inside.

FIGURE 9–66
Air conditioning increases fuel consumption by 3 to 4 percent during highway driving, and by as much as 10 percent during city driving.

Use the Air Conditioner Sparingly

Air conditioning consumes considerable power and thus increases fuel consumption by 3 to 4 percent during highway driving, and by as much as 10 percent during city driving (Fig. 9–66). The best alternative to air conditioning is to supply fresh outdoor air to the car through the vents by turning on the flow-through ventilation system (usually by running the air conditioner in the "economy" mode) while keeping the windows and the sunroof closed. This measure is adequate to achieve comfort in pleasant weather, and it saves the most fuel since the compressor of the air conditioner is off. In warmer weather, however, ventilation cannot provide adequate cooling effect. In that case we can attempt to achieve comfort by rolling down the windows or opening the sunroof. This is certainly a viable alternative for city driving, but not so on highways since the aerodynamic drag caused by wide-open windows or sunroof at highway speeds consumes more fuel than does the air conditioner. Therefore, at highway speeds, the windows or the sunroof should be closed and the air conditioner should be turned on instead to save fuel. This is especially the case for the newer, aerodynamically designed cars.

Most air conditioners have a "maximum" or "recirculation" setting that reduces the amount of hot outside air that must be cooled, and thus the fuel consumption for air-conditioning. A passive measure to reduce the need for air conditioning is to park the vehicle in the shade, and to leave the windows slightly open to allow for air circulation.

AFTER DRIVING

You cannot be an efficient person and accomplish much unless you take good care of yourself (eating right, maintaining physical fitness, having checkups, etc.), and the cars are no exception. Regular maintenance improves performance, increases gas mileage, reduces pollution, lowers repair costs, and extends engine life. A little time and money saved now may cost a lot later in increased fuel, repair, and replacement costs.

Proper maintenance such as *checking the levels of fluids (engine oil, coolant, transmission, brake, power steering, windshield washer, etc.), the tightness of all belts, and formation of cracks or frays on hoses, belts, and wires, keeping tires properly inflated, lubricating the moving components, and replacing clogged air, fuel, or oil filters* maximizes fuel efficiency (Fig. 9–67). Clogged air filters increase fuel consumption (by up to 10 percent) and pollution by restricting airflow to the engine, and thus they should be replaced. The car should be tuned up regularly unless it has electronic controls and a fuel-injection system. High temperatures (which may be due to a malfunction of the cooling fan) should be avoided as they may cause the break down of the engine oil and thus excessive wear of the engine, and low temperatures (which may be due to a malfunction of the thermostat) may extend the engine's warm-up period and may prevent the engine from reaching the optimum operating conditions. Both effects reduce fuel economy.

Clean oil extends engine life by reducing engine wear caused by friction, removes acids, sludge, and other harmful substances from the engine, improves performance, reduces fuel consumption, and decreases air pollution. Oil also helps to cool the engine, provides a seal between the cylinder walls and the pistons, and prevents the engine from rusting. Therefore, oil and oil filter should be changed as recommended by the vehicle manufacturer. Fuel-efficient oils (indicated by "Energy Efficient API" label) contain certain additives that reduce friction and increase a vehicle's fuel economy by 3 percent or more.

In summary, a person can save fuel, money, and the environment by *purchasing an energy-efficient vehicle, minimizing the amount of driving, being fuel-conscious while driving, and maintaining the car properly.* These measures have the added benefits of enhanced safety, reduced maintenance costs, and extended vehicle life.

FIGURE 9–67
Proper maintenance maximizes fuel efficiency and extends engine life.

SUMMARY

A cycle during which a net amount of work is produced is called a *power cycle,* and a power cycle during which the working fluid remains a gas throughout is called a *gas power cycle.* The most efficient cycle operating between a heat source at temperature T_H and a sink at temperature T_L is the Carnot cycle, and its thermal efficiency is given by

$$\eta_{\text{th,Carnot}} = 1 - \frac{T_L}{T_H}$$

The actual gas cycles are rather complex. The approximations used to simplify the analysis are known as the *air-standard assumptions.* Under these assumptions, all the processes are assumed to be internally reversible; the working fluid is assumed to be air, which behaves as an ideal gas; and the combustion and exhaust processes are replaced by heat-addition and heat-rejection processes, respectively. The air-standard assumptions are called *cold-air-standard assumptions* if air is also assumed to have constant specific heats at room temperature.

In reciprocating engines, the *compression ratio r* and the *mean effective pressure* MEP are defined as

$$r = \frac{V_{\text{max}}}{V_{\text{min}}} = \frac{V_{\text{BDC}}}{V_{\text{TDC}}}$$

$$\text{MEP} = \frac{w_{\text{net}}}{v_{\text{max}} - v_{\text{min}}}$$

The *Otto cycle* is the ideal cycle for the spark-ignition reciprocating engines, and it consists of four internally reversible processes: isentropic compression, constant-volume heat addition, isentropic expansion, and constant-volume heat rejection. Under cold-air-standard assumptions, the thermal efficiency of the ideal Otto cycle is

$$\eta_{\text{th,Otto}} = 1 - \frac{1}{r^{k-1}}$$

where r is the compression ratio and k is the specific heat ratio c_p/c_v.

The *Diesel cycle* is the ideal cycle for the compression-ignition reciprocating engines. It is very similar to the Otto cycle, except that the constant-volume heat-addition process is replaced by a constant-pressure heat-addition process. Its thermal efficiency under cold-air-standard assumptions is

$$\eta_{\text{th,Diesel}} = 1 - \frac{1}{r^{k-1}}\left[\frac{r_c^k - 1}{k(r_c - 1)}\right]$$

where r_c is the *cutoff ratio,* defined as the ratio of the cylinder volumes after and before the combustion process.

Stirling and *Ericsson cycles* are two totally reversible cycles that involve an isothermal heat-addition process at T_H and an isothermal heat-rejection process at T_L. They differ from the Carnot cycle in that the two isentropic processes are replaced by two constant-volume regeneration processes in the Stirling cycle and by two constant-pressure regeneration processes in the Ericsson cycle. Both cycles utilize *regeneration,* a process during which heat is transferred to a thermal energy storage device (called a *regenerator*) during one part of the cycle that is then transferred back to the working fluid during another part of the cycle.

The ideal cycle for modern gas-turbine engines is the *Brayton cycle,* which is made up of four internally reversible processes: isentropic compression, constant-pressure heat addition, isentropic expansion, and constant-pressure heat rejection. Under cold-air-standard assumptions, its thermal efficiency is

$$\eta_{\text{th,Brayton}} = 1 - \frac{1}{r_p^{(k-1)/k}}$$

where $r_p = P_{\text{max}}/P_{\text{min}}$ is the pressure ratio and k is the specific heat ratio. The thermal efficiency of the simple Brayton cycle increases with the pressure ratio.

The deviation of the actual compressor and the turbine from the idealized isentropic ones can be accurately accounted for by utilizing their isentropic efficiencies, defined as

$$\eta_C = \frac{w_s}{w_a} \cong \frac{h_{2s} - h_1}{h_{2a} - h_1}$$

and

$$\eta_T = \frac{w_a}{w_s} \cong \frac{h_3 - h_{4a}}{h_3 - h_{4s}}$$

where states 1 and 3 are the inlet states, $2a$ and $4a$ are the actual exit states, and $2s$ and $4s$ are the isentropic exit states.

In gas-turbine engines, the temperature of the exhaust gas leaving the turbine is often considerably higher than the temperature of the air leaving the compressor. Therefore, the high-pressure air leaving the compressor can be heated by transferring heat to it from the hot exhaust gases in a counter-flow heat exchanger, which is also known as a *regenerator.* The extent to which a regenerator approaches an ideal regenerator is called the *effectiveness* ϵ and is defined as

$$\epsilon = \frac{q_{\text{regen,act}}}{q_{\text{regen,max}}}$$

Under cold-air-standard assumptions, the thermal efficiency of an ideal Brayton cycle with regeneration becomes

$$\eta_{\text{th,regen}} = 1 - \left(\frac{T_1}{T_3}\right)(r_p)^{(k-1)/k}$$

where T_1 and T_3 are the minimum and maximum temperatures, respectively, in the cycle.

The thermal efficiency of the Brayton cycle can also be increased by utilizing *multistage compression with intercooling, regeneration, and multistage expansion with reheating.* The work input to the compressor is minimized when equal pressure ratios are maintained across each stage. This procedure also maximizes the turbine work output.

Gas-turbine engines are widely used to power aircraft because they are light and compact and have a high power-to-weight ratio. The ideal *jet-propulsion cycle* differs from the simple ideal Brayton cycle in that the gases are partially expanded in the turbine. The gases that exit the turbine at a relatively high pressure are subsequently accelerated in a nozzle to provide the thrust needed to propel the aircraft.

The *net thrust* developed by the engine is

$$F = \dot{m}(V_{\text{exit}} - V_{\text{inlet}})$$

where $\dot{m}$ is the mass flow rate of gases, V_{exit} is the exit velocity of the exhaust gases, and V_{inlet} is the inlet velocity of the air, both relative to the aircraft.

The power developed from the thrust of the engine is called the *propulsive power* $\dot{W}_P$, and it is given by

$$\dot{W}_P = \dot{m}(V_{\text{exit}} - V_{\text{inlet}})V_{\text{aircraft}}$$

Propulsive efficiency is a measure of how efficiently the energy released during the combustion process is converted to propulsive energy, and it is defined as

$$\eta_P = \frac{\text{Propulsive power}}{\text{Energy input rate}} = \frac{\dot{W}_P}{\dot{Q}_{\text{in}}}$$

For an ideal cycle that involves heat transfer only with a source at T_H and a sink at T_L, the exergy destruction is

$$x_{\text{dest}} = T_0\left(\frac{q_{\text{out}}}{T_L} - \frac{q_{\text{in}}}{T_H}\right)$$

REFERENCES AND SUGGESTED READINGS

1. V. D. Chase. "Propfans: A New Twist for the Propeller." *Mechanical Engineering,* November 1986, pp. 47–50.
2. C. R. Ferguson and A. T. Kirkpatrick, *Internal Combustion Engines: Applied Thermosciences,* 2nd ed., New York: Wiley, 2000.
3. R. A. Harmon. "The Keys to Cogeneration and Combined Cycles." *Mechanical Engineering,* February 1988, pp. 64–73.
4. J. Heywood, *Internal Combustion Engine Fundamentals,* New York: McGraw-Hill, 1988.
5. L. C. Lichty. *Combustion Engine Processes.* New York: McGraw-Hill, 1967.
6. H. McIntosh. "Jumbo Jet." *10 Outstanding Achievements 1964–1989.* Washington, D.C.: National Academy of Engineering, 1989, pp. 30–33.
7. W. Pulkrabek, *Engineering Fundamentals of the Internal Combustion Engine,* 2nd ed., Upper Saddle River, NJ: Prentice-Hall, 2004.
8. W. Siuru. "Two-stroke Engines: Cleaner and Meaner." *Mechanical Engineering.* June 1990, pp. 66–69.
9. C. F. Taylor. *The Internal Combustion Engine in Theory and Practice.* Cambridge, MA: M.I.T. Press, 1968.

PROBLEMS*

Actual and Ideal Cycles, Carnot Cycle, Air-Standard Assumptions, Reciprocating Engines

9–1C How does the thermal efficiency of an ideal cycle, in general, compare to that of a Carnot cycle operating between the same temperature limits?

9–2C What does the area enclosed by the cycle represent on a *P*-*v* diagram? How about on a *T*-*s* diagram?

9–3C What are the air-standard assumptions?

9–4C What is the difference between air-standard assumptions and the cold-air-standard assumptions?

9–5C What is the difference between the clearance volume and the displacement volume of reciprocating engines?

9–6C Define the compression ratio for reciprocating engines.

9–7C How is the mean effective pressure for reciprocating engines defined?

9–8C Can the mean effective pressure of an automobile engine in operation be less than the atmospheric pressure?

9–9C As a car gets older, will its compression ratio change? How about the mean effective pressure?

* Problems designated by a "C" are concept questions, and students are encouraged to answer them all. Problems designated by an "E" are in English units, and the SI users can ignore them. Problems with the icon are solved using EES, and complete solutions together with parametric studies are included on the enclosed DVD. Problems with the icon are comprehensive in nature, and are intended to be solved with a computer, preferably using the EES software that accompanies this text.

9–10C What is the difference between spark-ignition and compression-ignition engines?

9–11C Define the following terms related to reciprocating engines: stroke, bore, top dead center, and clearance volume.

9–12E What is the maximum possible thermal efficiency of a gas power cycle when using thermal energy reservoirs at 940°F and 40°F?

9–13 An air-standard cycle with constant specific heats is executed in a closed piston-cylinder system and is composed of the following three processes:

1.2 Isentropic compression with a compression ratio $r = 6$, $T_1 = 27°C$, and $P_1 = 100$ kPa

2-3 Constant pressure heat addition

3-1 Constant volume heat rejection

Assume air has constant properties with $c_v = 0.718$ kJ/kg·K, $c_p = 1.005$ kJ/kg·K, $R = 0.287$ kJ/kg·K, and $k = 1.4$.

(*a*) Sketch the *P*-*v* and *T*-*s* diagrams for this cycle.

(*b*) Determine the back work ratio for this cycle.

9–14 An air-standard cycle is executed within a closed piston-cylinder system and consists of three processes as follows:

1-2 V = *constant* heat addition from 100 kPa and 27°C to 700 kPa

2-3 Isothermal expansion until $V_3 = 7V_2$

3-1 P = *constant* heat rejection to the initial state

Assume air has constant properties with $c_v = 0.718$ kJ/kg·K, $c_p = 1.005$ kJ/kg·K, $R = 0.287$ kJ/kg·K, and $k = 1.4$.

(*a*) Sketch the P-v and T-s diagrams for the cycle.

(*b*) Determine the ratio of the compression work to the expansion work (the back work ratio).

(*c*) Determine the cycle thermal efficiency.

Answers: (*b*) 0.440, (c) 26.6 percent

9–15 An ideal gas power cycle is executed within a closed piston-cylinder system and consists of three processes as follows:

1-2 Isentropic compression from an initial temperature of 27°C with a compression ratio of 6.

2-3 Isothermal (constant temperature) expansion to the initial volume.

3-1 Constant volume heat rejection to the initial state.

Assume the gas has constant properties with c_v= 0.6 kJ/kg·K, c_p = 0.9 kJ/kg·K, R = 0.3 kJ/kg·K, and k = 1.5.

(*a*) Sketch the P–v and T-s diagrams for the cycle.

(*b*) Determine the maximum temperature of the cycle, in K.

(*c*) Calculate the work of expansion, in kJ/kg.

(*d*) Calculate the work of compression, in kJ/kg.

(*e*) Calculate the thermal efficiency of the cycle.

9–16 An air-standard cycle with variable specific heats is executed in a closed system and is composed of the following four processes:

1-2 Isentropic compression from 100 kPa and 22°C to 600 kPa

2-3 v = *constant* heat addition to 1500 K

3-4 Isentropic expansion to 100 kPa

4-1 P = *constant* heat rejection to initial state

(*a*) Show the cycle on P-v and T-s diagrams.

(*b*) Calculate the net work output per unit mass.

(*c*) Determine the thermal efficiency.

9–17 Reconsider Prob. 9–16. Using EES (or other) software, study the effect of varying the temperature after the constant-volume heat addition from 1500 K to 2500 K. Plot the net work output and thermal efficiency as a function of the maximum temperature of the cycle. Plot the T-s and P-v diagrams for the cycle when the maximum temperature of the cycle is 1500 K.

9–18 An air-standard cycle is executed in a closed system with 0.004 kg of air and consists of the following three processes:

1-2 Isentropic compression from 100 kPa and 27°C to 1 MPa

2-3 P = *constant* heat addition in the amount of 2.76 kJ

3-1 $P = c_1 v + c_2$ heat rejection to initial state (c_1 and c_2 are constants)

(*a*) Show the cycle on P-v and T-s diagrams.

(*b*) Calculate the heat rejected.

(*c*) Determine the thermal efficiency.

Assume constant specific heats at room temperature.

Answers: (*b*) 1.679 kJ, (*c*) 39.2 percent

9–19E An air-standard cycle with variable specific heats is executed in a closed system and is composed of the following four processes:

1-2 v = *constant* heat addition from 14.7 psia and 80°F in the amount of 300 Btu/lbm

2-3 P = *constant* heat addition to 3200 R

3-4 Isentropic expansion to 14.7 psia

4-1 P = *constant* heat rejection to initial state

(*a*) Show the cycle on P-v and T-s diagrams.

(*b*) Calculate the total heat input per unit mass.

(*c*) Determine the thermal efficiency.

Answers: (*b*) 612.4 Btu/lbm, (*c*) 24.2 percent

9–20E Repeat Prob. 9–19E using constant specific heats at room temperature.

9–21 Consider a Carnot cycle executed in a closed system with 0.6 kg of air. The temperature limits of the cycle are 300 and 1100 K, and the minimum and maximum pressures that occur during the cycle are 20 and 3000 kPa. Assuming constant specific heats, determine the net work output per cycle.

9–22 Consider a Carnot cycle executed in a closed system with air as the working fluid. The maximum pressure in the cycle is 800 kPa while the maximum temperature is 750 K. If the entropy increase during the isothermal heat rejection process is 0.25 kJ/kg·K and the net work output is 100 kJ/kg, determine (a) the minimum pressure in the cycle, (b) the heat rejection from the cycle, and (c) the thermal efficiency of the cycle. (d) If an actual heat engine cycle operates between the same temperature limits and produces 5200 kW of power for an air flow rate of 90 kg/s, determine the second law efficiency of this cycle.

9–23 An ideal gas Carnot cycle uses air as the working fluid, receives heat from a heat reservoir at 1027°C, is repeated 1500 times per minute, and has a compression ratio of 12. The compression ratio is defined as the volume ratio during the compression process. Determine the maximum temperature of the low-temperature heat reservoir, the cycle's thermal efficiency, and the amount of heat that must be supplied per cycle if this device is to produce 500 kW of power.

Answers: 481 K, 63.0 percent, 31.8 kJ

9-24 An ideal gas is contained in a piston-cylinder device and undergoes a power cycle as follows:

1-2 isentropic compression from an initial temperature T_1 = 20°C with a compression ratio r = 5

2-3 constant pressure heat addition

3-1 constant volume heat rejection

The gas has constant specific heats with c_v = **0.7** kJ/kg·K and R = 0.3 kJ/kg·K.

(*a*) Sketch the *P*-*v* and *T*-*s* diagrams for the cycle.

(*b*) Determine the heat and work interactions for each process, in kJ/kg.

(*c*) Determine the cycle thermal efficiency.

(*d*) Obtain the expression for the cycle thermal efficiency as a function of the compression ratio *r* and ratio of specific heats *k*.

Otto Cycle

9–25C How is the rpm (revolutions per minute) of an actual four-stroke gasoline engine related to the number of thermodynamic cycles? What would your answer be for a two-stroke engine?

9–26C How do the efficiencies of the ideal Otto cycle and the Carnot cycle compare for the same temperature limits? Explain.

9–27C What four processes make up the ideal Otto cycle?

9–28C Are the processes that make up the Otto cycle analyzed as closed-system or steady-flow processes? Why?

9–29C How does the thermal efficiency of an ideal Otto cycle change with the compression ratio of the engine and the specific heat ratio of the working fluid?

9–30C Why are high compression ratios not used in spark-ignition engines?

9–31C An ideal Otto cycle with a specified compression ratio is executed using (*a*) air, (*b*) argon, and (*c*) ethane as the working fluid. For which case will the thermal efficiency be the highest? Why?

9–32C What is the difference between fuel-injected gasoline engines and diesel engines?

9–33 An ideal Otto cycle has a compression ratio of 10.5, takes in air at 90 kPa and 40°C, and is repeated 2500 times per minute. Using constant specific heats at room temperature, determine the thermal efficiency of this cycle and the rate of heat input if the cycle is to produce 90 kW of power.

9–34 Repeat Prob. 9–33 for a compression ratio of 8.5.

9-35 Someone has suggested that the air-standard Otto cycle is more accurate if the two isentropic processes are replaced with polytropic processes with a polytropic exponent $n = 1.3$. Consider such a cycle when the compression ratio is 8, $P_1 =$ 95 kPa, $T_1 = 15$°C, and the maximum cycle temperature is 1200°C. Determine the heat transferred to and rejected from this cycle, as well as the cycle's thermal efficiency. Use constant specific heats at room temperature. *Answers:* 835 kJ/kg, 420 kJ/kg, 49.8 percent

9-36 How do the results of Prob. 9-35 change when isentropic processes are used in place of the polytropic processes?

9-37E A six-cylinder, four-stroke, spark-ignition engine operating on the ideal Otto cycle takes in air at 14 psia and 65°F, and is limited to a maximum cycle temperature of 1600°F. Each cylinder has a bore of 3.5 in, and each piston has a stroke of 3.9 in. The minimum enclosed volume is 14 percent of the maximum enclosed volume. How much power will this engine produce when operated at 2500 rpm? Use constant specific heats at room temperature.

9–38E A spark-ignition engine has a compression ratio of 8, an isentropic compression efficiency of 85 percent, and an isentropic expansion efficiency of 95 percent. At the beginning of the compression, the air in the cylinder is at 13 psia and 60°F. The maximum gas temperature is found to be 2300°F by measurement. Determine the heat supplied per unit mass, the thermal efficiency, and the mean effective pressure of this engine when modeled with the Otto cycle. Use constant specific heats at room temperature. *Answers:* 247 Btu/lbm, 47.5 percent, 49.0 psia

9–39 The compression ratio of an air-standard Otto cycle is 9.5. Prior to the isentropic compression process, the air is at 100 kPa, 35°C, and 600 cm^3. The temperature at the end of the isentropic expansion process is 800 K. Using specific heat values at room temperature, determine (*a*) the highest temperature and pressure in the cycle; (*b*) the amount of heat transferred in, in kJ; (*c*) the thermal efficiency; and (*d*) the mean effective pressure. *Answers:* (*a*) 1969 K, 6072 kPa, (*b*) 0.59 kJ, (*c*) 59.4 percent, (*d*) 652 kPa

9–40 Repeat Prob. 9–39, but replace the isentropic expansion process by a polytropic expansion process with the polytropic exponent $n = 1.35$.

9–41E An ideal Otto cycle with air as the working fluid has a compression ratio of 8. The minimum and maximum temperatures in the cycle are 540 and 2400 R. Accounting for the variation of specific heats with temperature, determine (*a*) the amount of heat transferred to the air during the heat-addition process, (*b*) the thermal efficiency, and (*c*) the thermal efficiency of a Carnot cycle operating between the same temperature limits.

9–42E Repeat Prob. 9–41E using argon as the working fluid.

9–43 A four-cylinder, four-stroke, 1.6-L gasoline engine operates on the Otto cycle with a compression ratio of 11. The air is at 100 kPa and 37°C at the beginning of the compression process, and the maximum pressure in the cycle is 8 MPa. The compression and expansion processes may be modeled as polytropic with a polytropic constant of 1.3. Using constant specific heats at 850 K, determine (*a*) the temperature at the end of the expansion process, (*b*) the net work output and the thermal efficiency, (*c*) the mean effective pressure, (*d*) the engine speed for a net power output of 50 kW, and (*e*) the specific fuel consumption, in g/kWh, defined as the ratio of the mass of the fuel consumed to the net work produced. The air–fuel ratio, defined as the amount of air divided by the amount of fuel intake, is 16.

9–44 When we double the compression ratio of an ideal Otto cycle, what happens to the maximum gas temperature and pressure when the state of the air at the beginning of the compression and the amount of heat addition remain the same? Use constant specific heats at room temperature.

9–45 In a spark-ignition engine, some cooling occurs as the gas is expanded. This may be modeled by using a polytropic process in lieu of the isentropic process. Determine if the polytropic exponent used in this model will be greater than or less than the isentropic exponent.

Diesel Cycle

9–46C How does a diesel engine differ from a gasoline engine?

9–47C How does the ideal Diesel cycle differ from the ideal Otto cycle?

9–48C For a specified compression ratio, is a diesel or gasoline engine more efficient?

9–49C Do diesel or gasoline engines operate at higher compression ratios? Why?

9–50C What is the cutoff ratio? How does it affect the thermal efficiency of a Diesel cycle?

9-51 An ideal Diesel cycle has a compression ratio of 20 and a cutoff ratio of 1.3. Determine the maximum temperature of the air and the rate of heat addition to this cycle when it produces 250 kW of power and the state of the air at the beginning of the compression is 90 kPa and 15°C. Use constant specific heats at room temperature.

9–52E An ideal Diesel cycle has a maximum cycle temperature of 2300°F and a cutoff ratio of 1.4. The state of the air at the beginning of the compression is P_1 = 14.4 psia and T_1 = 50°F. This cycle is executed in a four-stroke, eight-cylinder engine with a cylinder bore of 4 in and a piston stroke of 4 in. The minimum volume enclosed in the cylinder is 4.5 percent of the maximum cylinder volume. Determine the power produced by this engine when it is operated at 1800 rpm. Use constant specific heats at room temperature.

9–53 An air-standard dual cycle has a compression ratio of 14 and a cutoff ratio of 1.2. The pressure ratio during the constant-volume heat addition process is 1.5. Determine the thermal efficiency, amount of heat added, the maximum gas pressure and temperature when this cycle is operated at 80 kPa and 20°C at the beginning of the compression. Use constant specific heats at room temperature.

9–54 Repeat Prob. 9–53 when the state of the air at the beginning of the compression is 80 kPa and −20°C.

9–55E An air-standard Diesel cycle has a compression ratio of 18.2. Air is at 80°F and 14.7 psia at the beginning of the compression process and at 3000 R at the end of the heat-addition process. Accounting for the variation of specific heats with temperature, determine (*a*) the cutoff ratio, (*b*) the heat rejection per unit mass, and (*c*) the thermal efficiency.

9–56E Repeat Prob. 9–55E using constant specific heats at room temperature.

9–57 An ideal diesel engine has a compression ratio of 20 and uses air as the working fluid. The state of air at the beginning of the compression process is 95 kPa and 20°C. If the maximum temperature in the cycle is not to exceed 2200 K, determine (*a*) the thermal efficiency and (*b*) the mean effective pressure. Assume constant specific heats for air at room temperature. *Answers:* (*a*) 63.5 percent, (*b*) 933 kPa

9–58 Repeat Prob. 9–57, but replace the isentropic expansion process by polytropic expansion process with the polytropic exponent n = 1.35.

9–59 Reconsider Prob. 9–58. Using EES (or other) software, study the effect of varying the compression ratio from 14 to 24. Plot the net work output, mean effective pressure, and thermal efficiency as a function of the compression ratio. Plot the T-s and P-V diagrams for the cycle when the compression ratio is 20.

9–60 A four-cylinder two-stroke 2.0-L diesel engine that operates on an ideal Diesel cycle has a compression ratio of 22 and a cutoff ratio of 1.8. Air is at 70°C and 97 kPa at the beginning of the compression process. Using the cold-air-standard assumptions, determine how much power the engine will deliver at 2300 rpm.

9–61 Repeat Prob. 9–60 using nitrogen as the working fluid.

9–62 An air-standard dual cycle has a compression ratio of 18 and a cutoff ratio of 1.1. The pressure ratio during constant-volume heat addition process is 1.1. At the beginning of the compression, P_1 = 90 kPa, T_1 = 18°C, and V_1 = 0.003 m^3. How much power will this cycle produce when it is executed 4000 times per minute? Use constant specific heats at room temperature.

9–63 Repeat Prob. 9–62 if the isentropic compression efficiency is 85 percent and the isentropic expansion efficiency is 90 percent. *Answer:* 9.26 kW

9–64E An ideal dual cycle has a compression ratio of 15 and a cutoff ratio of 1.4. The pressure ratio during constant-volume heat addition process is 1.1. The state of the air at the beginning of the compression is P_1 =14.2 psia and T_1 = 75°F. Calculate the cycle's net specific work, specific heat addition, and thermal efficiency. Use constant specific heats at room temperature.

9–65 A six-cylinder, four-stroke, 3.2-L compression-ignition engine operates on the ideal diesel cycle with a compression ratio of 19. The air is at 95 kPa and 67°C at the beginning of the compression process and the engine speed is 1750 rpm. The engine uses light diesel fuel with a heating value of 42,500 kJ/kg, an air–fuel ratio of 28, and a combustion efficiency of 98 percent. Using constant specific heats at 850 K, determine (*a*) the maximum temperature in the cycle and the cutoff ratio (*b*) the net work output per cycle and the thermal efficiency, (*c*) the mean effective pressure, (*d*) the net power output, and (*e*) the specific fuel consumption, in g/kWh, defined as the ratio of the mass of the fuel consumed to the net work produced. *Answers:* (*a*) 2244 K, 2.36, (*b*) 2.71 kJ, 57.4 percent, (*c*) 847 kPa, (d) 39.5 kW, (*e*) 151 g/kWh

9–66 Develop an expression for cutoff ratio r_c which expresses it in terms of $q_{in}/(c_pT_1r^{k-1})$ for an air-standard Diesel cycle.

9–67 Develop an expression for the thermal efficiency of a dual cycle when operated such that $r_c = r_p$ where r_c is the cutoff ratio and r_p is the pressure ratio during the constant-volume heat addition process. What is the thermal efficiency of such engine when the compression ratio is 20 and $r_p = 2$?

9–68 How can one change r_p in Prob. 9–67 so that the same thermal efficiency is maintained when the compression ratio is reduced?

9–69 An air-standard cycle, called the dual cycle, with constant specific heats is executed in a closed piston-cylinder system and is composed of the following five processes:

1-2 Isentropic compression with a compression ratio, $r = V_1/V_2$

2-3 Constant volume heat addition with a pressure ratio, $r_p = P_3P_2$

3-4 Constant pressure heat addition with a volume ratio, $r_c = V_4/V_3$

4-5 Isentropic expansion while work is done until $V_5 = V_1$

5-1 Constant volume heat rejection to the initial state

(*a*) Sketch the *P*-*V* and *T*-*s* diagrams for this cycle.

(*b*) Obtain an expression for the cycle thermal efficiency as a function of *k*, *r*, r_c, and r_p.

(*c*) Evaluate the limit of the efficiency as r_p approaches unity and compare your answer with the expression for the Diesel cycle efficiency.

(*d*) Evaluate the limit of the efficiency as r_c approaches unity and compare your answer with the expression for the Otto cycle efficiency.

Stirling and Ericsson Cycles

9–70C What cycle is composed of two isothermal and two constant-volume processes?

9–71C How does the ideal Ericsson cycle differ from the Carnot cycle?

9–72C Consider the ideal Otto, Stirling, and Carnot cycles operating between the same temperature limits. How would you compare the thermal efficiencies of these three cycles?

9–73C Consider the ideal Diesel, Ericsson, and Carnot cycles operating between the same temperature limits. How would you compare the thermal efficiencies of these three cycles?

9–74 Consider an ideal Ericsson cycle with air as the working fluid executed in a steady-flow system. Air is at 27°C and 120 kPa at the beginning of the isothermal compression process, during which 150 kJ/kg of heat is rejected. Heat transfer to air occurs at 1200 K. Determine (*a*) the maximum pressure in the cycle, (*b*) the net work output per unit mass of air, and (*c*) the thermal efficiency of the cycle. *Answers:* (*a*) 685 kPa, (*b*) 450 kJ/kg, (*c*) 75 percent

9–75 An ideal Stirling cycle operates with 1 kg of air between thermal energy reservoirs at 27°C and 527°C. The maximum cycle pressure is 2000 kPa and the minimum cycle pressure is 100 kPa. Determine the net work produced each time this cycle is executed, and the cycle's thermal efficiency.

9–76 Determine the external rate of heat input and power produced by the Stirling cycle of Prob. 9–75 when it is repeated 1300 times per minute. *Answers:* 10,020 kW, 6264 kW

9–77 An Ericsson cycle operates between thermal energy reservoirs at 627°C and 7°C while producing 500 kW of power. Determine the rate of heat addition to this cycle when it is repeated 2000 times per minute. *Answer:* 726 kW

9–78 If the cycle of Prob. 9–77 is repeated 3000 times per minute while the heat added per cycle remains the same, how much power will the cycle produce?

9–79E An ideal Stirling cycle filled with air uses a 50°F energy reservoir as a sink. The engine is designed so that the maximum air volume is 0.5 ft^3, the minimum air volume is 0.06 ft^3, and the minimum pressure is 10 psia. It is to be operated such that the engine produces 2 Btu of net work when 6 Btu of heat are transferred externally to the engine. Determine the temperature of the energy source, the amount of air contained in the engine, and the maximum air pressure during the cycle.

9–80E Repeat Prob. 9–79E if the engine is to be operated to produce 2.5 Btu of work for the same external heat input?

9–81 An air-standard Stirling cycle operates with a maximum pressure of 3600 kPa and a minimum pressure of 50 kPa. The maximum volume is 12 times the minimum volume, and the low-temperature reservoir is at 20°C. Allowing a 5°C temperature difference between the external reservoirs and the air when appropriate, calculate the specific heat added to the cycle and its net specific work.

9–82 How much heat is stored (and recovered) in the regenerator of Prob. 9-81. Use constant specific heats at room temperature.

Ideal and Actual Gas-Turbine (Brayton) Cycles

9–83C What four processes make up the simple ideal Brayton cycle?

9–84C For fixed maximum and minimum temperatures, what is the effect of the pressure ratio on (*a*) the thermal efficiency and (*b*) the net work output of a simple ideal Brayton cycle?

9–85C What is the back work ratio? What are typical back work ratio values for gas-turbine engines?

9–86C Why are the back work ratios relatively high in gas-turbine engines?

9–87C How do the inefficiencies of the turbine and the compressor affect (*a*) the back work ratio and (*b*) the thermal efficiency of a gas-turbine engine?

9–88E A simple ideal Brayton cycle with air as the working fluid has a pressure ratio of 10. The air enters the compressor at 520 R and the turbine at 2000 R. Accounting for the variation of specific heats with temperature, determine (*a*) the air temperature at the compressor exit, (*b*) the back work ratio, and (*c*) the thermal efficiency.

9–89 A simple Brayton cycle using air as the working fluid has a pressure ratio of 10. The minimum and maximum temperatures in the cycle are 295 and 1240 K. Assuming an isentropic efficiency of 83 percent for the compressor and 87 percent for the turbine, determine (*a*) the air temperature at the turbine exit, (*b*) the net work output, and (*c*) the thermal efficiency.

9–90 Reconsider Prob. 9–89. Using EES (or other) software, allow the mass flow rate, pressure ratio, turbine inlet temperature, and the isentropic efficiencies of the turbine and compressor to vary. Assume the compressor inlet pressure is 100 kPa. Develop a general solution for the problem by taking advantage of the diagram window method for supplying data to EES software.

9–91 Repeat Prob. 9–89 using constant specific heats at room temperature.

9–92E A simple ideal Brayton cycle uses helium as the working fluid; operates with 12 psia and 60°F at the compressor inlet; has a pressure ratio of 14; and a maximum cycle temperature of 1300°F. How much power will this cycle produce when the rate at which the helium is circulated about the cycle is 100 lbm/min? Use constant specific heats at room temperature.

9–93E Repeat Prob. 9-92E when the isentropic efficiency of the compressor is 95 percent.

9–94 Consider a simple Brayton cycle using air as the working fluid; has a pressure ratio of 12; has a maximum cycle temperature of 600°C; and operates the compressor inlet at 90 kPa and 15°C. Which will have the greatest impact on the back-work ratio: a compressor isentropic efficiency of 90 percent or a turbine isentropic efficiency of 90 percent? Use constant specific heats at room temperature.

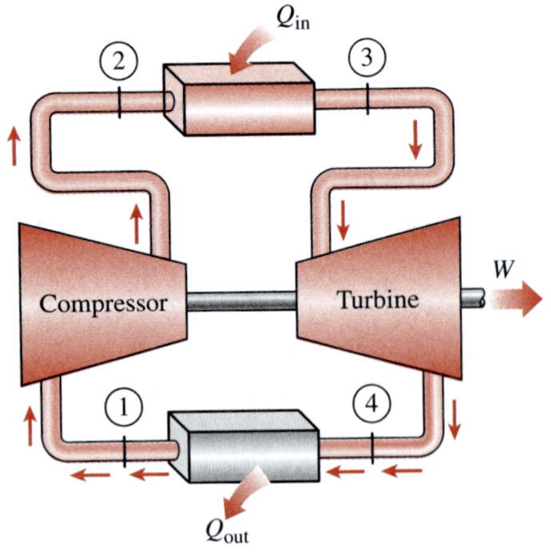

FIGURE P9–94

9–95 Air is used as the working fluid in a simple ideal Brayton cycle that has a pressure ratio of 12, a compressor inlet temperature of 300 K, and a turbine inlet temperature of 1000 K. Determine the required mass flow rate of air for a net power output of 70 MW, assuming both the compressor and the turbine have an isentropic efficiency of (*a*) 100 percent and (*b*) 85 percent. Assume constant specific heats at room temperature. *Answers:* (*a*) 352 kg/s, (*b*) 1037 kg/s

9–96 Air enters the compressor of a gas-turbine engine at 300 K and 100 kPa, where it is compressed to 700 kPa and 580 K. Heat is transferred to air in the amount of 950 kJ/kg before it enters the turbine. For a turbine efficiency of 86 percent, determine (*a*) the fraction of the turbine work output used to drive the compressor and (*b*) the thermal efficiency. Assume variable specific heats for air.

9–97 Repeat Prob. 9–96 using constant specific heats at room temperature.

9–98 An aircraft engine operates on a simple ideal Brayton cycle with a pressure ratio of 10. Heat is added to the cycle at a rate of 500 kW; air passes through the engine at a rate of 1 kg/s; and the air at the beginning of the compression is at 70 kPa and 0°C. Determine the power produced by this engine and its thermal efficiency. Use constant specific heats at room temperature.

9–99 Repeat Prob. 9–98 for a pressure ratio of 15.

9–100 A gas-turbine power plant operates on the simple Brayton cycle between the pressure limits of 100 and 2000 kPa. The working fluid is air, which enters the compressor at 40°C at a rate of 700 m^3/min and leaves the turbine at 650°C. Using variable specific heats for air and assuming a compressor isentropic efficiency of 85 percent and a turbine isentropic efficiency of 88 percent, determine (*a*) the net power output, (*b*) the back work ratio, and (*c*) the thermal efficiency. *Answers:* (*a*) 5404 kW, (*b*) 0.545, (*c*) 39.2 percent

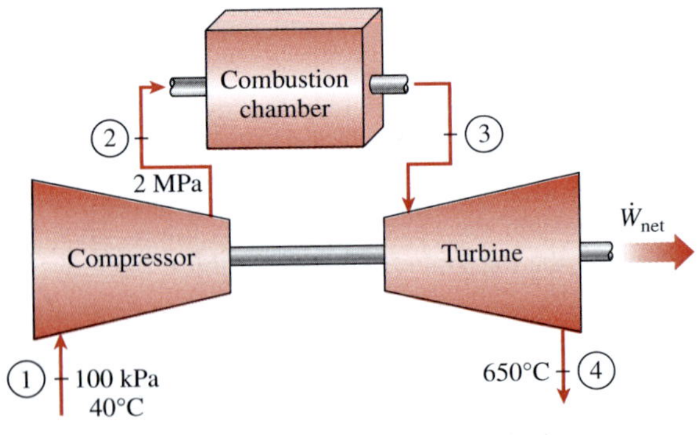

FIGURE P9–100

9–101 A gas-turbine power plant operates on the simple Brayton cycle between the pressure limits of 100 and 800 kPa. Air enters the compressor at 30°C and leaves at 330°C at a mass flow rate of 200 kg/s. The maximum cycle temperature is 1400 K. During operation of the cycle, the net power output is measured experimentally to be 60 MW. Assume constant properties for air at 300 K with c_v = 0.718 kJ/kg·K, c_p = 1.005 kJ/kg·K, R = 0.287 kJ/kg·K, k = 1.4.

(*a*) Sketch the *T-s* diagram for the cycle.

(*b*) Determine the isentropic efficiency of the turbine for these operating conditions.

(*c*) Determine the cycle thermal efficiency.

9–102 A gas-turbine power plant operates on a modified Brayton cycle shown in the figure with an overall pressure ratio of 8. Air enters the compressor at 0°C and 100 kPa. The maximum cycle temperature is 1500 K. The compressor and the turbines are isentropic. The high pressure turbine develops just enough power to run the compressor. Assume constant properties for air at 300 K with c_v = 0.718 kJ/kg·K, c_p = 1.005 kJ/kg·K, R = 0.287 kJ/kg·K, k = 1.4.

(*a*) Sketch the *T-s* diagram for the cycle. Label the data states.

(*b*) Determine the temperature and pressure at state 4, the exit of the high pressure turbine.

(*c*) If the net power output is 200 MW, determine mass flow rate of the air into the compressor, in kg/s.

Answers: (*b*) 1279 K, 457 kPa, (*c*) 442 kg/s

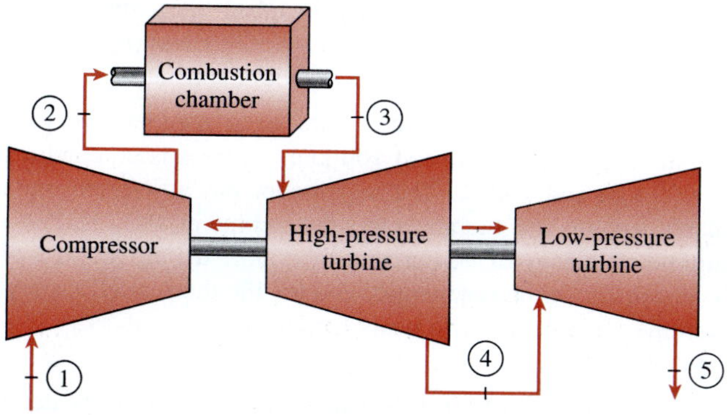

FIGURE P9–102

9–103 A gas-turbine power plant operating on the simple Brayton has a pressure ratio of 7. Air enters the compressor at 0°C and 100 kPa. The maximum cycle temperature is 1500 K. The compressor has an isentropic efficiency of 80 percent, and the turbine has an isentropic efficiency of 90 percent. Assume constant properties for air at 300 K with c_v = 0.718 kJ/kg·K, c_p = 1.005 kJ/kg·K, R = 0.287 kJ/kg·K, k = 1.4.

(*a*) Sketch the *T-s* diagram for the cycle.

(*b*) If the net power output is 150 MW, determine the volume flow rate of the air into the compressor, in m^3/s.

(*c*) For a fixed compressor inlet velocity and flow area, explain the effect of increasing compressor inlet temperature (i.e., summertime operation versus wintertime operation) on the inlet mass flow rate and the net power output with all other parameters of the problem being the same.

Brayton Cycle with Regeneration

9–104C How does regeneration affect the efficiency of a Brayton cycle, and how does it accomplish it?

9–105C Somebody claims that at very high pressure ratios, the use of regeneration actually decreases the thermal efficiency of a gas-turbine engine. Is there any truth in this claim? Explain.

9–106C Define the effectiveness of a regenerator used in gas-turbine cycles.

9–107C In an ideal regenerator, is the air leaving the compressor heated to the temperature at (*a*) turbine inlet, (*b*) turbine exit, (*c*) slightly above turbine exit?

9–108C In 1903, Aegidius Elling of Norway designed and built an 11-hp gas turbine that used steam injection between the combustion chamber and the turbine to cool the combustion gases to a safe temperature for the materials available at the time. Currently there are several gas-turbine power plants that use steam injection to augment power and improve thermal efficiency. For example, the thermal efficiency of the General Electric LM5000 gas turbine is reported to increase from 35.8 percent in simple-cycle operation to 43 percent when steam injection is used. Explain why steam injection increases the power output and the efficiency of gas turbines. Also, explain how you would obtain the steam.

9–109 A gas turbine for an automobile is designed with a regenerator. Air enters the compressor of this engine at 100 kPa and 20°C. The compressor pressure ratio is 8; the maximum cycle temperature is 800°C; and the cold air stream leaves the regenerator 10°C cooler than the hot air stream at the inlet of the regenerator. Assuming both the compressor and the turbine to be isentropic, determine the rates of heat addition and rejection for this cycle when it produces 150 kW. Use constant specific heats at room temperature. *Answers:* 303 kW, 153 kW

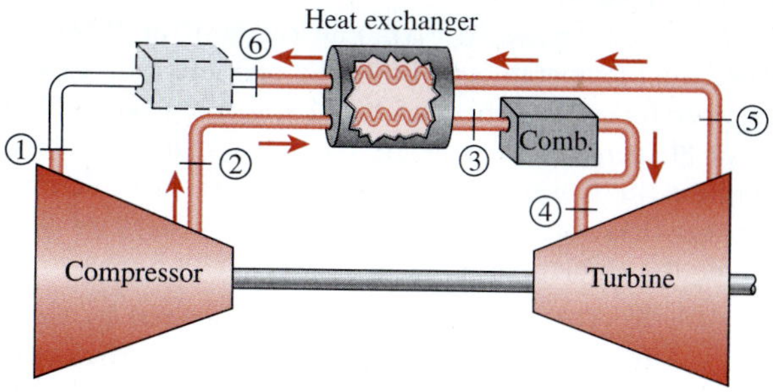

FIGURE P9–109

9–110 Rework Prob. 9–109 when the compressor isentropic efficiency is 87 percent and the turbine isentropic efficiency is 93 percent.

9–111 A gas turbine engine operates on the ideal Brayton cycle with regeneration, as shown in Fig. P9–109. Now the regenerator is rearranged so that the air streams of states 2 and 5 enter at one end of the regenerator and streams 3 and 6 exit at the other end (i.e., parallel flow arrangement of a heat exchanger). Consider such a system when air enters the compressor at 100 kPa and 20°C; the compressor pressure ratio is 7; the maximum cycle temperature is 727°C; and the difference between the hot and cold air stream temperatures is 6°C at the end of the regenerator where the cold stream leaves the regenerator. Is the cycle arrangement shown in the figure more or less efficient than this arrangement? Assume both the compressor and the turbine are isentropic, and use constant specific heats at room temperature.

9–112E An ideal regenerator ($T_3 = T_5$) is added to a simple ideal Brayton cycle (see Fig. P9–109). Air enters the compressor of this cycle at 16 psia and 100°F; the pressure ratio is 11; and the maximum cycle temperature is 1940°F. What is the thermal efficiency of this cycle? Use constant specific heats at room temperature. What would the thermal efficiency of the cycle be without the regenerator?

9–113E The idea of using gas turbines to power automobiles was conceived in the 1930s, and considerable research was done in the 1940s and 1950s to develop automotive gas turbines by major automobile manufacturers such as the Chrysler and Ford corporations in the United States and Rover in the United Kingdom. The world's first gas-turbine-powered automobile, the 200-hp Rover Jet 1, was built in 1950 in the United Kingdom. This was followed by the production of the Plymouth Sport Coupe by Chrysler in 1954 under the leadership of G. J. Huebner. Several hundred gas-turbine-powered Plymouth cars were built in the early 1960s for demonstration purposes and were loaned to a select group of people to gather field experience. The users had no complaints other than slow acceleration. But the cars were never mass-produced because of the high production (especially material) costs and the failure to satisfy the provisions of the 1966 Clean Air Act.

A gas-turbine-powered Plymouth car built in 1960 had a turbine inlet temperature of 1700°F, a pressure ratio of 4, and a regenerator effectiveness of 0.9. Using isentropic efficiencies of 80 percent for both the compressor and the turbine, determine the thermal efficiency of this car. Also, determine the mass flow rate of air for a net power output of 95 hp. Assume the ambient air to be at 540 R and 14.5 psia.

9–114 The 7FA gas turbine manufactured by General Electric is reported to have an efficiency of 35.9 percent in the simple-cycle mode and to produce 159 MW of net power. The pressure ratio is 14.7 and the turbine inlet temperature is 1288°C. The mass flow rate through the turbine is 1,536,000 kg/h. Taking the ambient conditions to be 30°C and 100 kPa, determine the isentropic efficiency of the turbine and the compressor. Also, determine the thermal efficiency of this gas turbine if a regenerator with an effectiveness of 65 percent is added.

9–115 Reconsider Prob. 9–114. Using EES (or other) software, develop a solution that allows different isentropic efficiencies for the compressor and turbine and study the effect of the isentropic efficiencies on net work done and the heat supplied to the cycle. Plot the *T-s* diagram for the cycle.

9–116 A Brayton cycle with regeneration using air as the working fluid has a pressure ratio of 7. The minimum and maximum temperatures in the cycle are 310 and 1150 K. Assuming an isentropic efficiency of 75 percent for the compressor and 82 percent for the turbine and an effectiveness of 65 percent for the regenerator, determine (*a*) the air temperature at the turbine exit, (*b*) the net work output, and (*c*) the thermal efficiency. *Answers:* (*a*) 783 K, (*b*) 108.1 kJ/kg, (*c*) 22.5 percent

9–117 A stationary gas-turbine power plant operates on an ideal regenerative Brayton cycle (ϵ = 100 percent) with air as the working fluid. Air enters the compressor at 95 kPa and 290 K and the turbine at 760 kPa and 1100 K. Heat is transferred to air from an external source at a rate of 75,000 kJ/s. Determine the power delivered by this plant (*a*) assuming constant specific heats for air at room temperature and (*b*) accounting for the variation of specific heats with temperature.

9–118 Air enters the compressor of a regenerative gas-turbine engine at 310 K and 100 kPa, where it is compressed to 900 kPa and 650 K. The regenerator has an effectiveness of 80 percent, and the air enters the turbine at 1400 K. For a turbine efficiency of 90 percent, determine (*a*) the amount of heat transfer in the regenerator and (*b*) the thermal efficiency. Assume variable specific heats for air. *Answers:* (*a*) 193 kJ/kg, (*b*) 40.0 percent

9–119 Repeat Prob. 9–118 using constant specific heats at room temperature.

9–120 Repeat Prob. 9–118 for a regenerator effectiveness of 70 percent.

9–121 Develop an expression for the thermal efficiency of an ideal Brayton cycle with an ideal regenerator of effectiveness 100 percent. Use constant specific heats at room temperature.

Brayton Cycle with Intercooling, Reheating, and Regeneration

9–122C Under what modifications will the ideal simple gas-turbine cycle approach the Ericsson cycle?

9–123C For a specified pressure ratio, why does multistage compression with intercooling decrease the compressor work, and multistage expansion with reheating increase the turbine work?

9–124C The single-stage compression process of an ideal Brayton cycle without regeneration is replaced by a multistage compression process with intercooling between the same pressure limits. As a result of this modification,

(*a*) Does the compressor work increase, decrease, or remain the same?

(*b*) Does the back work ratio increase, decrease, or remain the same?

(*c*) Does the thermal efficiency increase, decrease, or remain the same?

9–125C The single-stage expansion process of an ideal Brayton cycle without regeneration is replaced by a multistage expansion process with reheating between the same pressure limits. As a result of this modification,

(*a*) Does the turbine work increase, decrease, or remain the same?

(*b*) Does the back work ratio increase, decrease, or remain the same?

(*c*) Does the thermal efficiency increase, decrease, or remain the same?

9–126C A simple ideal Brayton cycle without regeneration is modified to incorporate multistage compression with intercooling and multistage expansion with reheating, without changing the pressure or temperature limits of the cycle. As a result of these two modifications,

(*a*) Does the net work output increase, decrease, or remain the same?

(*b*) Does the back work ratio increase, decrease, or remain the same?

(*c*) Does the thermal efficiency increase, decrease, or remain the same?

(*d*) Does the heat rejected increase, decrease, or remain the same?

9–127C A simple ideal Brayton cycle is modified to incorporate multistage compression with intercooling, multistage expansion with reheating, and regeneration without changing the pressure limits of the cycle. As a result of these modifications,

(*a*) Does the net work output increase, decrease, or remain the same?

(*b*) Does the back work ratio increase, decrease, or remain the same?

(*c*) Does the thermal efficiency increase, decrease, or remain the same?

(*d*) Does the heat rejected increase, decrease, or remain the same?

9–128C In an ideal gas-turbine cycle with intercooling, reheating, and regeneration, as the number of compression and expansion stages is increased, the cycle thermal efficiency approaches (*a*) 100 percent, (*b*) the Otto cycle efficiency, or (*c*) the Carnot cycle efficiency.

9–129 Consider an ideal gas-turbine cycle with two stages of compression and two stages of expansion. The pressure ratio across each stage of the compressor and turbine is 3. The air enters each stage of the compressor at 300 K and each stage of the turbine at 1200 K. Determine the back work ratio and the thermal efficiency of the cycle, assuming (*a*) no regenerator is used and (*b*) a regenerator with 75 percent effectiveness is used. Use variable specific heats.

9–130 Repeat Problem 9–129, assuming an efficiency of 84 percent for each compressor stage and an efficiency of 88 percent for each turbine stage.

9-131E A gas turbine has two stages of ideal compression, expansion, intercooling and reheating. Air enters the first compressor at 13 psia and 60°F; the total pressure ratio (across all compressors) is 12; the total rate of heat addition is 500 Btu/s; and the cold air temperature is increased by 50°F in the regenerator. Calculate the power produced by each turbine, power consumed by each compressor, and the rate of heat rejection. Use constant specific heats at room temperature.

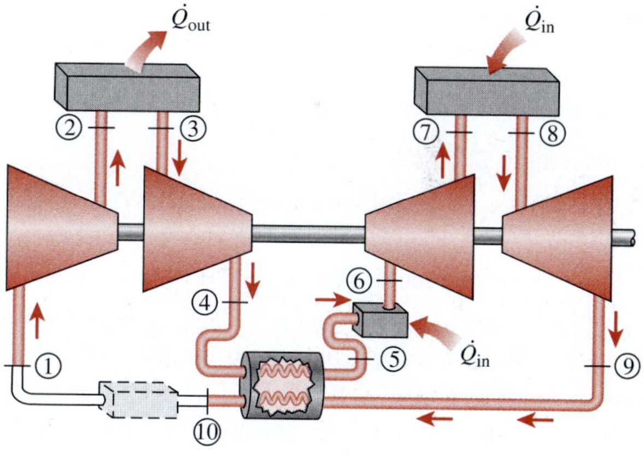

FIGURE P9-131E

9-132E Rework Prob. 9-131E with each compressor having an isentropic efficiency of 85 percent and each turbine having an isentropic efficiency of 90 percent.

9–133 Air enters a gas turbine with two stages of compression and two stages of expansion at 100 kPa and 17°C. This system uses a regenerator as well as reheating and intercooling. The pressure ratio across each compressor is 4; 300 kJ/kg of heat are added to the air in each combustion chamber; and the regenerator operates perfectly while increasing the temperature of the cold air by 20°C. Determine this system's thermal efficiency. Assume isentropic operations for all compressor and the turbine stages and use constant specific heats at room temperature.

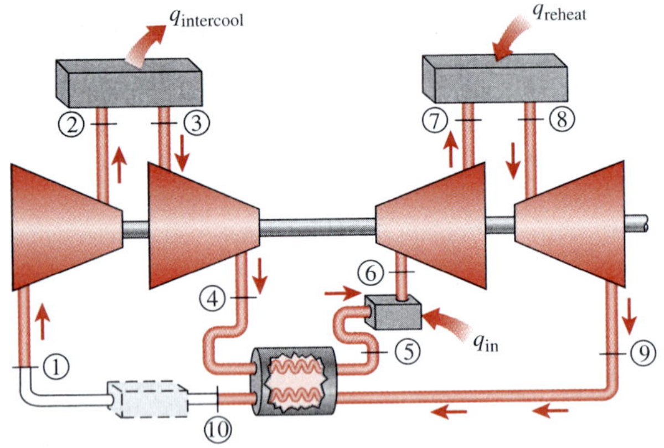

FIGURE P9–133

9–134 Repeat Prob. 9–133 for the case of three stages of compression with intercooling and three stages with expansion with reheating. *Answer:* 40.1 percent

9–135 How much would the thermal efficiency of the cycle in Prob. 9–134 change if the temperature of the cold-air stream leaving the regenerator is 65°C lower than the temperature of the hot-air stream entering the regenerator?

Jet-Propulsion Cycles

9–136C What is propulsive power? How is it related to thrust?

9–137C What is propulsive efficiency? How is it determined?

9–138C Is the effect of turbine and compressor irreversibilities of a turbojet engine to reduce (*a*) the net work, (*b*) the thrust, or (*c*) the fuel consumption rate?

9–139E A turboprop-aircraft propulsion engine operates where the air is at 8 psia and −10°F, on an aircraft flying at a speed of 600 ft/s. The Brayton cycle pressure ratio is 10 and the air temperature at the turbine inlet is 940°F. The propeller diameter is 10 ft and the mass flow rate through the propeller is 20 times that through the compressor. Determine the thrust force generated by this propulsion system. Assume ideal operation for all components and constant specific heats at room temperature.

9–140E How much change would result in the thrust of Prob. 9–139E if the propeller diameter were reduced to 8 ft while maintaining the same mass flow rate through the compressor. Note: the mass flow rate ratio will no longer be 20.

9–141 A turbofan engine operating on an aircraft flying at 200 m/s at an altitude where the air is at 50 kPa and −20°C, is to produce 50,000 N of thrust. The inlet diameter of this engine is 2.5 m; the compressor pressure ratio is 12; and the mass flow rate ratio is 8. Determine the air temperature at the fan outlet needed to produce this thrust. Assume ideal operation for all components and constant specific heats at room temperature. *Answer:* 233 K

9–142 A pure jet engine propels an aircraft at 240 m/s through air at 45 kPa and −13°C. The inlet diameter of this engine is 1.6 m, the compressor pressure ratio is 13, and the temperature at the turbine inlet is 557°C. Determine the velocity at the exit of this engine's nozzle and the thrust produced. Assume ideal operation for all components and constant specific heats at room temperature.

9–143 A turbojet aircraft is flying with a velocity of 320 m/s at an altitude of 9150 m, where the ambient conditions are 32 kPa and −32°C. The pressure ratio across the compressor is 12, and the temperature at the turbine inlet is 1400 K. Air enters the compressor at a rate of 60 kg/s, and the jet fuel has a heating value of 42,700 kJ/kg. Assuming ideal operation for all components and constant specific heats for air at room temperature, determine (*a*) the velocity of the exhaust gases, (*b*) the propulsive power developed, and (*c*) the rate of fuel consumption.

9–144 Repeat Prob. 9–143 using a compressor efficiency of 80 percent and a turbine efficiency of 85 percent.

9–145 Consider an aircraft powered by a turbojet engine that has a pressure ratio of 9. The aircraft is stationary on the ground, held in position by its brakes. The ambient air is at 7°C and 95 kPa and enters the engine at a rate of 20 kg/s. The jet fuel has a heating value of 42,700 kJ/kg, and it is burned completely at a rate of 0.5 kg/s. Neglecting the effect of the diffuser and disregarding the slight increase in mass at the engine exit as well as the inefficiencies of engine components, determine the force that must be applied on the brakes to hold the plane stationary. *Answer:* 19,370 N

9–146 EES Reconsider Prob. 9–145. In the problem statement, replace the inlet mass flow rate by an inlet volume flow rate of 18.1 m^3/s. Using EES (or other) software, investigate the effect of compressor inlet temperature in the range of –20 to 30°C on the force that must be applied to the brakes to hold the plane stationary. Plot this force as a function of compressor inlet temperature.

9–147 Air at 7°C enters a turbojet engine at a rate of 16 kg/s and at a velocity of 300 m/s (relative to the engine). Air is heated in the combustion chamber at a rate 15,000 kJ/s and it leaves the engine at 427°C. Determine the thrust produced by this turbojet engine. (*Hint:* Choose the entire engine as your control volume.)

Second-Law Analysis of Gas Power Cycles

9–148 Which of the Otto cycle processes of Prob. 9–36 loses the greatest amount of work potential? The energy-supply reservoir's temperature is the same as the maximum cycle temperature, and the energy-sink reservoir's temperature is the same as the minimum cycle temperature.

9–149E Determine the exergy destruction associated with the heat rejection process of the Diesel cycle described in Prob. 9–55E, assuming a source temperature of 3500 R and a sink temperature of 540 R. Also, determine the exergy at the end of the isentropic expansion process.

9–150 Calculate the exergy destruction by each process of the dual cycle in Prob. 9–63, Assume standard atmospheric conditions and that the energy-source temperature is same as the maximum cycle temperature and the energy-sink temperature is the same as the minimum cycle temperature.

9–151 Calculate the exergy destruction for each process of Stirling cycle of Prob. 9–81, in kJ/kg.

9–152 Calculate the exergy destruction associated with each of the processes of the Brayton cycle described in Prob. 9–89, assuming a source temperature of 1600 K and a sink temperature of 310 K.

9–153 Repeat Prob. 9–94 using exergy analysis.

9–154 Determine the total exergy destruction associated with the Brayton cycle described in Prob. 9–116, assuming a source temperature of 1500 K and a sink temperature of 290 K. Also, determine the exergy of the exhaust gases at the exit of the regenerator.

9–155 EES Reconsider Prob. 9–154. Using EES (or other) software, investigate the effect of varying the cycle pressure ratio from 6 to 14 on the total exergy destruction for the cycle and the exergy of the exhaust gas leaving the regenerator. Plot these results as functions of pressure ratio. Discuss the results.

9–156 Calculate the lost work potential for each process of Prob. 9–135. The temperature of the hot reservoir is the same as the maximum cycle temperature and the temperature of the cold reservoir is the same as the minimum cycle temperature.

9–157 A gas-turbine power plant operates on the simple Brayton cycle between the pressure limits of 100 and 700 kPa. Air enters the compressor at 30°C at a rate of 12.6 kg/s and leaves at 260°C. A diesel fuel with a heating value of 42,000 kJ/kg is burned in the combustion chamber with an air–fuel ratio of 60 and a combustion efficiency of 97 percent. Combustion gases leave the combustion chamber and enter the turbine whose isentropic efficiency is 85 percent. Treating the combustion gases as air and using constant specific heats at 500°C, determine (*a*) the isentropic efficiency of the compressor, (*b*) the net power output and the back work ratio, (*c*) the thermal efficiency, and (*d*) the second-law efficiency.

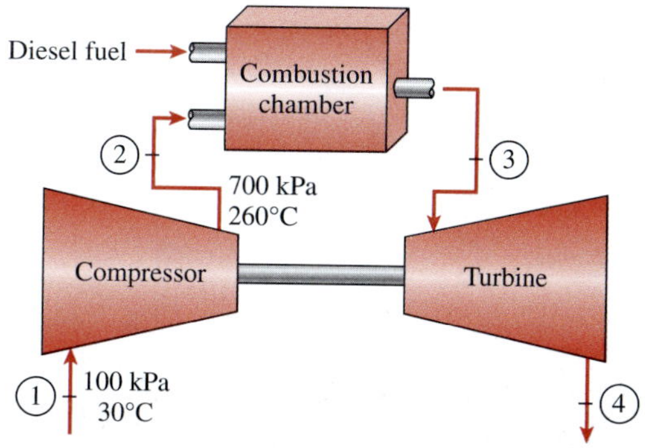

FIGURE P9–157

9–158 A four-cylinder, four-stroke, 1.8-liter modern, high-speed compression-ignition engine operates on the ideal dual cycle with a compression ratio of 16. The air is at 95 kPa and 70°C at the beginning of the compression process and the engine speed is 2200 rpm. Equal amounts of fuel are burned at constant volume and at constant pressure. The maximum allowable pressure in the cycle is 7.5 MPa due to material strength limitations. Using constant specific heats at 1000 K, determine (*a*) the maximum temperature in the cycle, (*b*) the net work output and the thermal efficiency, (*c*) the mean effective pressure, and (*d*) the net power output. Also, determine (*e*) the second-law efficiency of the cycle and the rate of exergy output with the exhaust gases when they are purged. *Answers:* (*a*) 2308 K, (*b*) 836 kJ/kg, 59.5 percent, (*c*) 860 kPa, (*d*) 28.4 kW, (*e*) 68.3 percent, 10.3 kW

Review Problems

9–159 An Otto cycle with a compression ratio of 7 begins its compression at 90 kPa and 15°C. The maximum cycle temperature is 1000°C. Utilizing air-standard assumptions, determine the thermal efficiency of this cycle using (*a*) constant specific heats at room temperature and (*b*) variable specific heats. *Answers:* (*a*) 54.1 percent, (*b*) 51.0 percent

9–160E A Diesel cycle has a compression ratio of 20 and begins its compression at 13 psia and 45°F. The maximum cycle temperature is 1800°F. Utilizing air-standard assumptions, determine the thermal efficiency of this cycle using (*a*) constant specific heats at room temperature and (*b*) variable specific heats.

9–161E A Brayton cycle with a pressure ratio of 12 operates with air entering the compressor at 13 psia and 20°F, and the turbine at 1000°F. Calculate the net specific work produced by this cycle treating the air as an ideal gas with (*a*) constant specific heats at room temperature and (*b*) variable specific heats.

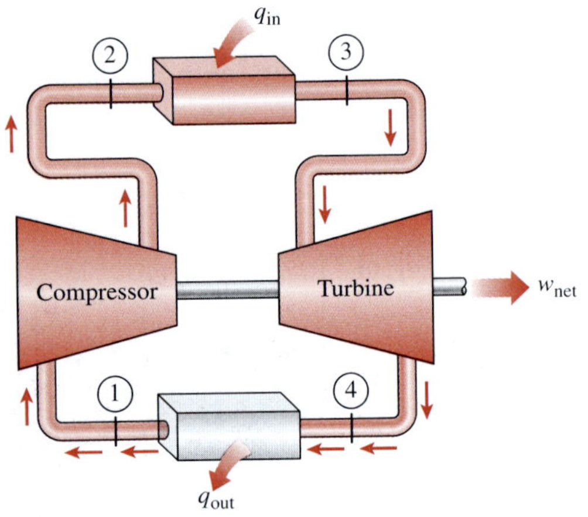

FIGURE P9–161E

9–162 A four-stroke turbocharged V-16 diesel engine built by GE Transportation Systems to power fast trains produces 3500 hp at 1200 rpm. Determine the amount of power produced per cylinder per (*a*) mechanical cycle and (*b*) thermodynamic cycle.

9–163 Consider a simple ideal Brayton cycle operating between the temperature limits of 300 and 1500 K. Using constant specific heats at room temperature, determine the pressure ratio for which the compressor and the turbine exit temperatures of air are equal.

9–164 A four-cylinder spark-ignition engine has a compression ratio of 10.5, and each cylinder has a maximum volume of 0.4 L. At the beginning of the compression process, the air is at 98 kPa and 37°C, and the maximum temperature in the cycle is 2100 K. Assuming the engine to operate on the ideal Otto cycle, determine (*a*) the amount of heat supplied per cylinder, (*b*) the thermal efficiency, and (*c*) the number of revolutions per minute required for a net power output of 45 kW. Assume variable specific heats for air.

9–165 Reconsider Prob. 9–164. Using EES (or other) software, study the effect of varying the compression ratio from 5 to 11 on the net work done and the efficiency of the cycle. Plot the *P-v* and *T-s* diagrams for the cycle, and discuss the results.

9–166 An ideal gas Carnot cycle uses helium as the working fluid and rejects heat to a lake at 15°C. Determine the pressure ratio, compression ratio, and minimum temperature of the heat source for this cycle to have a thermal efficiency of 50 percent. *Answers:* 5.65, 2.83, 576 K

9–167E Repeat Prob. 9–166 when the lake is at 60°F and the Carnot cycle's thermal efficiency is to be 60 percent.

9–168 A typical hydrocarbon fuel produces 43,000 kJ/kg of heat when used in a spark-ignition engine. Determine the compression ratio required for an ideal Otto cycle to use 0.035 grams of fuel to produce 1 kJ of work. Use constant specific heats at room temperature. *Answer:* 15.3

9–169 An ideal Otto cycle has a compression ratio of 9.2 and uses air as the working fluid. At the beginning of the compression process, air is at 98 kPa and 27°C. The pressure is doubled during the constant-volume heat-addition process. Accounting for the variation of specific heats with temperature, determine (*a*) the amount of heat transferred to the air, (*b*) the net work output, (*c*) the thermal efficiency, and (*d*) the mean effective pressure for the cycle.

9–170 Repeat Prob. 9–169 using constant specific heats at room temperature.

9–171E An ideal dual cycle has a compression ratio of 14 and uses air as the working fluid. At the beginning of the compression process, air is at 14.7 psia and 120°F, and occupies a volume of 98 in^3. During the heat-addition process, 0.6 Btu of heat is transferred to air at constant volume and 1.1 Btu at constant pressure. Using constant specific heats evaluated at room temperature, determine the thermal efficiency of the cycle.

9–172 Consider an ideal Stirling cycle using air as the working fluid. Air is at 350 K and 200 kPa at the beginning of the isothermal compression process, and heat is supplied to air from a source at 1800 K in the amount of 900 kJ/kg. Determine (*a*) the maximum pressure in the cycle and (*b*) the net work output per unit mass of air. *Answers:* (*a*) 5873 kPa, (*b*) 725 kJ/kg

9–173 Consider a simple ideal Brayton cycle with air as the working fluid. The pressure ratio of the cycle is 6, and the minimum and maximum temperatures are 300 and 1300 K, respectively. Now the pressure ratio is doubled without changing the minimum and maximum temperatures in the cycle. Determine the change in (*a*) the net work output per unit mass and (*b*) the thermal efficiency of the cycle as a result of this modification. Assume variable specific heats for air. *Answers:* (*a*) 41.5 kJ/kg, (*b*) 10.6 percent

9–174 Repeat Prob. 9–173 using constant specific heats at room temperature.

9–175 A gas-turbine engine with regeneration operates with two stages of compression and two stages of expansion. The pressure ratio across each stage of the compressor and turbine is 4. The air enters each stage of the compressor at 300 K and each stage of the turbine at 1400 K. The compressor and turbine efficiencies are 78 and 86 percent, respectively, and the effectiveness of the regenerator is 75 percent. Determine the back work ratio and the thermal efficiency of the cycle, assuming constant specific heats for air at room temperature. *Answers:* 0.475, 45.1 percent

9–176 Reconsider Prob. 9–175. Using EES (or other) software, study the effects of varying the isentropic efficiencies for the compressor and turbine and regenerator effectiveness on net work done and the heat supplied to the cycle for the variable specific heat case. Let the isentropic efficiencies and the effectiveness vary from 70 percent to 90 percent. Plot the *T-s* diagram for the cycle.

9–177 Repeat Prob. 9–175 using helium as the working fluid.

9–178 Consider an ideal gas-turbine cycle with one stage of compression and two stages of expansion and regeneration. The pressure ratio across each turbine stage is the same. The high-pressure turbine exhaust gas enters the regenerator and then enters the low-pressure turbine for expansion to the compressor inlet pressure. Determine the thermal efficiency of this cycle as a function of the compressor pressure ratio and the high-pressure turbine to compressor inlet temperature ratio. Compare your result with the efficiency of the standard regenerative cycle.

9–179 A gas-turbine plant operates on the regenerative Brayton cycle with two stages of reheating and two-stages of intercooling between the pressure limits of 100 and 1200 kPa. The working fluid is air. The air enters the first and the second stages of the compressor at 300 K and 350 K, respectively, and the first and the second stages of the turbine at 1400 K and 1300 K, respectively. Assuming both the compressor and the turbine have an isentropic efficiency of 80 percent and the regenerator has an effectiveness of 75 percent and using variable specific heats, determine (*a*) the back work ratio and the net work output, (*b*) the thermal efficiency, and (*c*) the second-law efficiency of the cycle. Also determine (*d*) the exergies at the exits of the combustion chamber (state 6) and the regenerator (state 10) (See Fig. 9–43 in the text). *Answers:* (*a*) 0.523, 317 kJ/kg, (*b*) 0.553, (*c*) 0.704, (*d*) 931 kJ/kg, 129 kJ/kg

9–180 Compare the thermal efficiency of a two-stage gas turbine with regeneration, reheating and intercooling to that of a three-stage gas turbine with the same equipment when (*a*) all components operate ideally, (*b*) air enters the first compressor at 100 kPa and 10°C, (*c*) the total pressure ratio across all stages of compression is 16, and (*d*) the maximum cycle temperature is 600°C.

9–181E The specific impulse of an aircraft-propulsion system is the force produced per unit of thrust-producing mass flow rate. Consider a jet engine that operates in an environment at 10 psia and 30°F and propels an aircraft cruising at 1200 ft/s. Determine the specific impulse of this engine when the compressor pressure ratio is 9 and the temperature at the turbine inlet is 700°F. Assume ideal operations for all components and constant specific heats at room temperature.

9–182 Electricity and process heat requirements of a manufacturing facility are to be met by a cogeneration plant consisting of a gas turbine and a heat exchanger for steam production. The plant operates on the simple Brayton cycle between the pressure limits of 100 and 1000 kPa with air as the working fluid. Air enters the compressor at 20°C. Combustion gases leave the turbine and enter the heat exchanger at 450°C, and leave the heat exchanger of 325°C, while the liquid water enters the heat exchanger at 15°C and leaves at 200°C as a saturated vapor. The net power produced by the gas-turbine cycle is 1500 kW. Assuming a compressor isentropic efficiency of 86 percent and a turbine isentropic efficiency of 88 percent and using variable specific heats, determine (*a*) the mass flow rate of air, (*b*) the back work ratio and the thermal efficiency, and (*c*) the rate at which steam is produced in the heat exchanger. Also determine (*d*) the utilization efficiency of the cogeneration plant, defined as the ratio of the total energy utilized to the energy supplied to the plant.

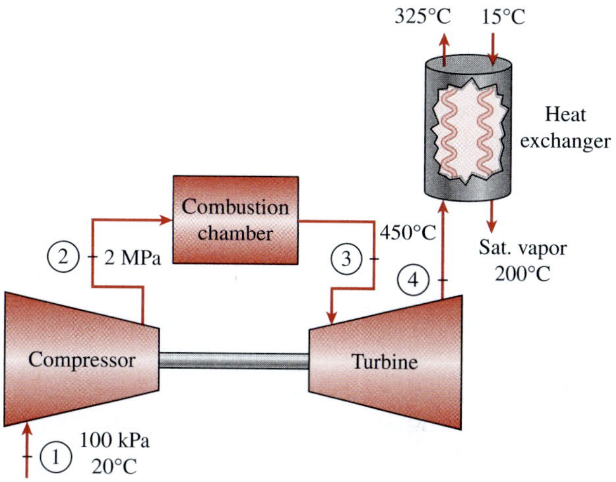

FIGURE P9–182

9–183 A turbojet aircraft flies with a velocity of 900 km/h at an altitude where the air temperature and pressure are −35°C and 40 kPa. Air leaves the diffuser at 50 kPa with a velocity of 15 m/s, and combustion gases enter the turbine at 450 kPa and 950°C. The turbine produces 500 kW of power, all of which is used to drive the compressor. Assuming an isentropic efficiency of 83 percent for the compressor, turbine, and nozzle, and using variable specific heats, determine (*a*) the pressure of combustion gases at the turbine exit, (*b*) the mass flow rate of air through the compressor, (*c*) the velocity of the gases at the nozzle exit, and (*d*) the propul-sive power and the propulsive efficiency for this engine. *Answers:* (*a*) 147 kPa, (*b*) 1.76 kg/s, (*c*) 719 m/s, (*d*) 206 kW, 0.156

9–184 An air standard cycle with constant specific heats is executed in a closed piston-cylinder system and is composed of the following three processes:

1–2 Isentropic compression with a compression ratio $r = V_1 = V_2$

2-3 Constant pressure heat addition

3-1 Constant volume heat rejection

(*a*) Sketch the *P*-*v* and *T*-*s* diagrams for this cycle.

(*b*) Obtain an expression for the back work ratio as a function of *k* and *r*.

(*c*) Obtain an expression for the cycle thermal efficiency as a function of *k* and *r*.

(*d*) Determine the value of the back work ratio and efficiency as *r* goes to unity.

What do your results imply about the net work done by the cycle?

9–185 An air standard cycle with constant specific heats is executed in a closed piston-cylinder system and is composed of the following three processes:

1-2 Constant volume heat addition

2-3 Isentropic expansion with an expansion ratio $r = V_3/V_2$

3-1 Constant pressure heat rejection

(*a*) Sketch the *P-v and T-s* diagrams for this cycle

(*b*) Obtain an expression for the back work ratio as a function of *k* and *r*

(*c*) Obtain an expression for the cycle thermal efficiency as a function of *k* and *r*

(*d*) Determine the value of the back work ratio and efficiency as *r* goes to unity

What do your results imply about the net work done by the cycle?

9–186 An air standard cycle with constant specific heats is executed in a closed piston-cylinder system and is composed of the following four processes:

1–2 Isentropic compression with a compression ratio, $r = V_1/V_2$

2–3 Constant volume heat addition

3–4 Isentropic expansion with an expansion ratio, $r_e = V_4/V_3$

4–1 Constant pressure heat rejection with a volume ratio, $r_p = V_4/V_1$

(*a*) Sketch the *P-v* and *T-s* diagrams for this cycle.

(*b*) Obtain an expression for the cycle thermal efficiency as a function of *k*, *r*, and r_p.

(*c*) Evaluate the limit of the efficiency as r_p approaches unity.

9–187 Reconsider Prob. 9–186. Determine the back work ratio for a fixed minimum-to-maximum temperature ratio T_1/T_3, and determine the limit of the ratio as r_p goes to unity. Compare your result with the expression for the back work ratio for the Otto cycle for same compression ratio and fixed minimum-to-maximum temperature ratio T_1/T_3.

9–188 Using EES (or other) software, determine the effects of compression ratio on the net work output and the thermal efficiency of the Otto cycle for a maximum cycle temperature of 2000 K. Take the working fluid to be air that is at 100 kPa and 300 K at the beginning of the compression process, and assume variable specific heats. Vary the compression ratio from 6 to 15 with an increment of 1. Tabulate and plot your results against the compression ratio.

9–189 Using EES (or other) software, determine the effects of pressure ratio on the net work output and the thermal efficiency of a simple Brayton cycle for a maximum cycle temperature of 1800 K. Take the working fluid to be air that is at 100 kPa and 300 K at the beginning of the compression process, and assume variable specific heats. Vary the pressure ratio from 5 to 24 with an increment of 1. Tabulate and plot your results against the pressure ratio. At what pressure ratio does the net work output become a maximum? At what pressure ratio does the thermal efficiency become a maximum?

9–190 Repeat Problem 9–189 assuming isentropic efficiencies of 80 percent for both the turbine and the compressor.

9–191 Using EES (or other) software, determine the effects of pressure ratio, maximum cycle temperature, and compressor and turbine efficiencies on the net work output per unit mass and the thermal efficiency of a simple Brayton cycle with air as the working fluid. Air is at 100 kPa and 300 K at the compressor inlet. Also, assume constant specific heats for air at room temperature. Determine the net work output and the thermal efficiency for all combinations of the following parameters, and draw conclusions from the results.

Pressure ratio:	5, 8, 14
Maximum cycle temperature:	800, 1200, 1600 K
Compressor isentropic efficiency:	80, 100 percent
Turbine isentropic efficiency:	80, 100 percent

9–192 Repeat Problem 9–191 by considering the variation of specific heats of air with temperature.

9–193 Repeat Problem 9–191 using helium as the working fluid.

9–194 Using EES (or other) software, determine the effect of the number of compression and expansion stages on the thermal efficiency of an ideal regenerative Brayton cycle with multistage compression and expansion. Assume that the overall pressure ratio of the cycle is 18, and the air enters each stage of the compressor at 300 K and each stage of the turbine at 1200 K. Using constant specific heats for air at room temperature, determine the thermal efficiency of the cycle by varying the number of stages from 1 to 22 in increments of 3. Plot the thermal efficiency versus the number of stages. Compare your results to the efficiency of an Ericsson cycle operating between the same temperature limits.

9–195 Repeat Problem 9–194 using helium as the working fluid.

Fundamentals of Engineering (FE) Exam Problems

9–196 An Otto cycle with air as the working fluid has a compression ratio of 10.4. Under cold-air-standard conditions, the thermal efficiency of this cycle is

(*a*) 10 percent (*b*) 39 percent (*c*) 61 percent

(*d*) 79 percent (*e*) 82 percent

9–197 For specified limits for the maximum and minimum temperatures, the ideal cycle with the lowest thermal efficiency is

(*a*) Carnot (*b*) Stirling (*c*) Ericsson
(*d*) Otto (*e*) All are the same

9–198 A Carnot cycle operates between the temperature limits of 300 and 2000 K, and produces 600 kW of net power. The rate of entropy change of the working fluid during the heat addition process is

(*a*) 0 (*b*) 0.300 kW/K (*c*) 0.353 kW/K
(*d*) 0.261 kW/K (*e*) 2.0 kW/K

9–199 Air in an ideal Diesel cycle is compressed from 2 to 0.13 L, and then it expands during the constant pressure heat addition process to 0.30 L. Under cold air standard conditions, the thermal efficiency of this cycle is

(*a*) 41 percent (*b*) 59 percent (*c*) 66 percent
(*d*) 70 percent (*e*) 78 percent

9–200 Helium gas in an ideal Otto cycle is compressed from 20°C and 2.5 to 0.25 L, and its temperature increases by an additional 700°C during the heat addition process. The temperature of helium before the expansion process is

(*a*) 1790°C (*b*) 2060°C (*c*) 1240°C
(*d*) 620°C (*e*) 820°C

9–201 In an ideal Otto cycle, air is compressed from 1.20 kg/m^3 and 2.2 to 0.26 L, and the net work output of the cycle is 440 kJ/kg. The mean effective pressure (MEP) for this cycle is

(*a*) 612 kPa (*b*) 599 kPa (*c*) 528 kPa
(*d*) 416 kPa (*e*) 367 kPa

9–202 In an ideal Brayton cycle, air is compressed from 95 kPa and 25°C to 1100 kPa. Under cold-air-standard conditions, the thermal efficiency of this cycle is

(*a*) 45 percent (*b*) 50 percent (*c*) 62 percent
(*d*) 73 percent (*e*) 86 percent

9–203 Consider an ideal Brayton cycle executed between the pressure limits of 1200 and 100 kPa and temperature limits of 20 and 1000°C with argon as the working fluid. The net work output of the cycle is

(*a*) 68 kJ/kg (*b*) 93 kJ/kg (*c*) 158 kJ/kg
(*d*) 186 kJ/kg (*e*) 310 kJ/kg

9–204 An ideal Brayton cycle has a net work output of 150 kJ/kg and a back work ratio of 0.4. If both the turbine and the compressor had an isentropic efficiency of 85 percent, the net work output of the cycle would be

(*a*) 74 kJ/kg (*b*) 95 kJ/kg (*c*) 109 kJ/kg
(*d*) 128 kJ/kg (*e*) 177 kJ/kg

9–205 In an ideal Brayton cycle, air is compressed from 100 kPa and 25°C to 1 MPa, and then heated to 927°C before entering the turbine. Under cold-air-standard conditions, the air temperature at the turbine exit is

(*a*) 349°C (*b*) 426°C (*c*) 622°C
(*d*) 733°C (*e*) 825°C

9–206 In an ideal Brayton cycle with regeneration, argon gas is compressed from 100 kPa and 25°C to 400 kPa, and then heated to 1200°C before entering the turbine. The highest temperature that argon can be heated in the regenerator is

(*a*) 246°C (*b*) 846°C (*c*) 689°C
(*d*) 368°C (*e*) 573°C

9–207 In an ideal Brayton cycle with regeneration, air is compressed from 80 kPa and 10°C to 400 kPa and 175°C, is heated to 450°C in the regenerator, and then further heated to 1000°C before entering the turbine. Under cold-air-standard conditions, the effectiveness of the regenerator is

(*a*) 33 percent (*b*) 44 percent (*c*) 62 percent
(*d*) 77 percent (*e*) 89 percent

9–208 Consider a gas turbine that has a pressure ratio of 6 and operates on the Brayton cycle with regeneration between the temperature limits of 20 and 900°C. If the specific heat ratio of the working fluid is 1.3, the highest thermal efficiency this gas turbine can have is

(*a*) 38 percent (*b*) 46 percent (*c*) 62 percent
(*d*) 58 percent (*e*) 97 percent

9–209 An ideal gas turbine cycle with many stages of compression and expansion and a regenerator of 100 percent effectiveness has an overall pressure ratio of 10. Air enters every stage of compressor at 290 K, and every stage of turbine at 1200 K. The thermal efficiency of this gas-turbine cycle is

(*a*) 36 percent (*b*) 40 percent (*c*) 52 percent
(*d*) 64 percent (*e*) 76 percent

9–210 Air enters a turbojet engine at 320 m/s at a rate of 30 kg/s, and exits at 650 m/s relative to the aircraft. The thrust developed by the engine is

(*a*) 5 kN (*b*) 10 kN (*c*) 15 kN
(*d*) 20 kN (*e*) 26 kN

Design and Essay Problems

9–211 The amount of fuel introduced into a spark-ignition engine is used in part to control the power produced by the engine. Gasoline produces approximately 42,000 kJ/kg when burned with air in a spark-ignition engine. Develop a schedule for gasoline consumption and maximum cycle temperature versus power production for an Otto cycle with a compression ratio of 8.

9–212 The weight of a diesel engine is directly proportional to the compression ratio ($W = kr$) because extra metal must be used to strengthen the engine for the higher pressures. Examine the net specific work produced by a diesel engine per unit of weight as the pressure ratio is varied and the specific heat input remains fixed. Do this for several heat inputs and proportionality constants k. Are there any optimal combinations of k and specific heat inputs?

9–213 Design a closed-system air-standard gas power cycle composed of three processes and having a minimum thermal efficiency of 20 percent. The processes may be isothermal, isobaric, isochoric, isentropic, polytropic, or pressure as a linear function of volume; however, the Otto, Diesel, Ericsson, and Stirling cycles may not be used. Prepare an engineering report describing your design, showing the system, P-v and T-s diagrams, and sample calculations.

9–214 Write an essay on the most recent developments on the two-stroke engines, and find out when we might be seeing cars powered by two-stroke engines in the market. Why do the major car manufacturers have a renewed interest in two-stroke engines?

9–215 Exhaust gases from the turbine of a simple Brayton cycle are quite hot and may be used for other thermal purposes. One proposed use is generating saturated steam at 110°C from water at 30°C in a boiler. This steam will be distributed to several buildings on a college campus for space heating. A Brayton cycle with a pressure ratio of 6 is to be used for this purpose. Plot the power produced, the flow rate of produced steam, and the maximum cycle temperature as functions of the rate at which heat is added to the cycle. The temperature at the turbine inlet is not to exceed 2000°C.

9–216 A gas turbine operates with a regenerator and two stages of reheating and intercooling. This system is designed so that when air enters the compressor at 100 kPa and 15°C, the pressure ratio for each stage of compression is 3; the air temperature when entering a turbine is 500°C; and the regenerator operates perfectly. At full load, this engine produces 800 kW. For this engine to service a partial load, the heat addition in both combustion chambers is reduced. Develop an optimal schedule of heat addition to the combustion chambers for partial loads ranging from 400 to 800 kW.

9–217 You have been asked to design a power facility for lunar-based laboratory. You have selected a simple Brayton cycle that uses argon as the working fluid and has a pressure ratio of 6. The heat-rejecting heat exchanger maintains the state at the entrance to the compressor at 50 kPa and −20°C. You have elected to use solar collectors to serve as the heat supply. Tests of these collectors give the temperature increase results shown in the figure. Develop a plot of the power that will be produced by this system and its thermal efficiency as a function of the argon mass flow rate. Is there a "best" flow rate at which to operate this power plant?

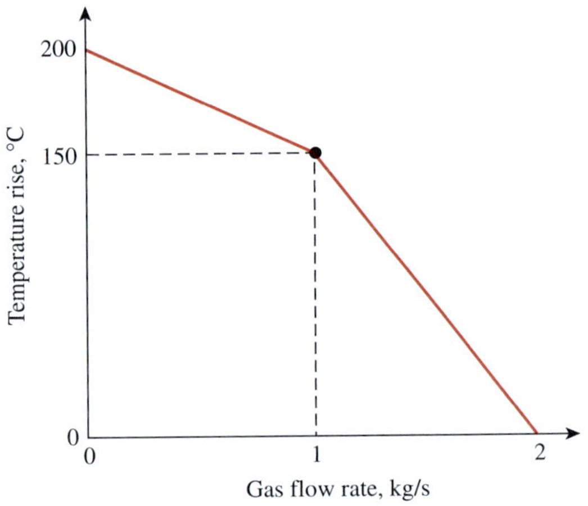

FIGURE P9–217

9–218 If the temperature control at the compressor inlet of Prob. 9–217 is disabled, the temperature at the compressor inlet varies linearly with the gas flow rate. It is −40°C when the flow rate is 0 kg/s, and 0°C when the flow rate is 2 kg/s. Develop a schedule of the power produced and the thermal efficiency as the argon flow rate is varied from 0.01 kg/s to 2 kg/s. Is there an optimal flow rate at which to operate this power plant?

9–219 Since its introduction in 1903 by Aegidius Elling of Norway, steam injection between the combustion chamber and the turbine is used even in some modern gas turbines currently in operation to cool the combustion gases to a metallurgical-safe temperature while increasing the mass flow rate through the turbine. Currently, there are several gas-turbine power plants that use steam injection to augment power and improve thermal efficiency.

Consider a gas-turbine power plant whose pressure ratio is 8. The isentropic efficiencies of the compressor and the turbine are 80 percent, and there is a regenerator with an effectiveness of 70 percent. When the mass flow rate of air through the compressor is 40 kg/s, the turbine inlet temperature becomes 1700 K. But the turbine inlet temperature is limited to 1500 K, and thus steam injection into the combustion gases is being considered. However, to avoid the complexities associated with steam injection, it is proposed to use excess air (that is, to take in much more air than needed for complete combustion) to lower the combustion and thus turbine inlet temperature while increasing the mass flow rate and thus power output of the turbine. Evaluate this proposal, and compare the thermodynamic performance of "high air flow" to that of a "steam-injection" gas-turbine power plant under the following design conditions: the ambient air is at 100 kPa and 25°C, adequate water supply is available at 20°C, and the amount of fuel supplied to the combustion chamber remains constant.

CHAPTER

10

VAPOR AND COMBINED POWER CYCLES

In Chap. 9 we discussed gas power cycles for which the working fluid remains a gas throughout the entire cycle. In this chapter, we consider *vapor power cycles* in which the working fluid is alternatively vaporized and condensed. We also consider power generation coupled with process heating called *cogeneration.*

The continued quest for higher thermal efficiencies has resulted in some innovative modifications to the basic vapor power cycle. Among these, we discuss the *reheat* and *regenerative cycles,* as well as combined gas–vapor power cycles.

Steam is the most common working fluid used in vapor power cycles because of its many desirable characteristics, such as low cost, availability, and high enthalpy of vaporization. Therefore, this chapter is mostly devoted to the discussion of steam power plants. Steam power plants are commonly referred to as *coal plants, nuclear plants,* or *natural gas plants,* depending on the type of fuel used to supply heat to the steam. However, the steam goes through the same basic cycle in all of them. Therefore, all can be analyzed in the same manner.

Objectives

The objectives of Chapter 10 are to:

- Analyze vapor power cycles in which the working fluid is alternately vaporized and condensed.
- Analyze power generation coupled with process heating called *cogeneration.*
- Investigate ways to modify the basic Rankine vapor power cycle to increase the cycle thermal efficiency.
- Analyze the reheat and regenerative vapor power cycles.
- Analyze power cycles that consist of two separate cycles known as combined cycles and binary cycles.

10–1 ▪ THE CARNOT VAPOR CYCLE

We have mentioned repeatedly that the Carnot cycle is the most efficient cycle operating between two specified temperature limits. Thus it is natural to look at the Carnot cycle first as a prospective ideal cycle for vapor power plants. If we could, we would certainly adopt it as the ideal cycle. As explained below, however, the Carnot cycle is not a suitable model for power cycles. Throughout the discussions, we assume *steam* to be the working fluid since it is the working fluid predominantly used in vapor power cycles.

Consider a steady-flow *Carnot cycle* executed within the saturation dome of a pure substance, as shown in Fig. 10–1*a*. The fluid is heated reversibly and isothermally in a boiler (process 1-2), expanded isentropically in a turbine (process 2-3), condensed reversibly and isothermally in a condenser (process 3-4), and compressed isentropically by a compressor to the initial state (process 4-1).

Several impracticalities are associated with this cycle:

1. Isothermal heat transfer to or from a two-phase system is not difficult to achieve in practice since maintaining a constant pressure in the device automatically fixes the temperature at the saturation value. Therefore, processes 1-2 and 3-4 can be approached closely in actual boilers and condensers. Limiting the heat transfer processes to two-phase systems, however, severely limits the maximum temperature that can be used in the cycle (it has to remain under the critical-point value, which is 374°C for water). Limiting the maximum temperature in the cycle also limits the thermal efficiency. Any attempt to raise the maximum temperature in the cycle involves heat transfer to the working fluid in a single phase, which is not easy to accomplish isothermally.

2. The isentropic expansion process (process 2-3) can be approximated closely by a well-designed turbine. However, the quality of the steam decreases during this process, as shown on the *T-s* diagram in Fig. 10–1*a*. Thus the turbine has to handle steam with low quality, that is, steam with a high moisture content. The impingement of liquid droplets on the turbine blades causes erosion and is a major source of wear. Thus steam with qualities less than about 90 percent cannot be tolerated in the operation of power plants. This problem could be eliminated by using a working fluid with a very steep saturated vapor line.

3. The isentropic compression process (process 4-1) involves the compression of a liquid–vapor mixture to a saturated liquid. There are two difficulties associated with this process. First, it is not easy to control the condensation process so precisely as to end up with the desired quality at state 4. Second, it is not practical to design a compressor that handles two phases.

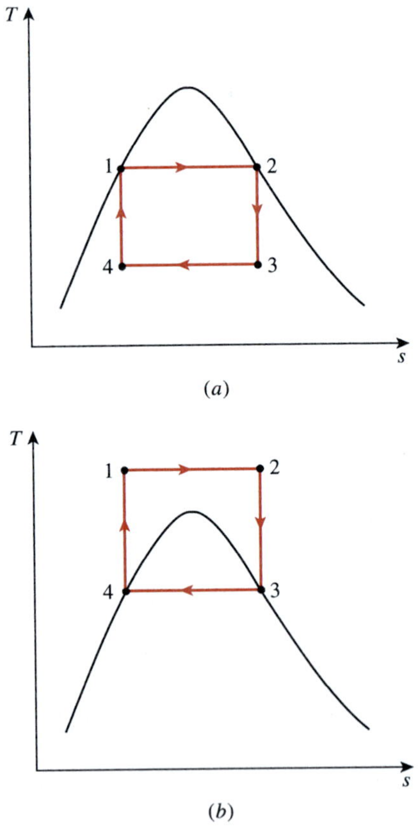

FIGURE 10–1
T-s diagram of two Carnot vapor cycles.

Some of these problems could be eliminated by executing the Carnot cycle in a different way, as shown in Fig. 10–1*b*. This cycle, however, presents other problems such as isentropic compression to extremely high pressures and isothermal heat transfer at variable pressures. Thus we conclude that the Carnot cycle cannot be approximated in actual devices and is not a realistic model for vapor power cycles.

10–2 ▪ RANKINE CYCLE: THE IDEAL CYCLE FOR VAPOR POWER CYCLES

Many of the impracticalities associated with the Carnot cycle can be eliminated by superheating the steam in the boiler and condensing it completely in the condenser, as shown schematically on a *T-s* diagram in Fig. 10–2. The cycle that results is the **Rankine cycle**, which is the ideal cycle for vapor power plants. The ideal Rankine cycle does not involve any internal irreversibilities and consists of the following four processes:

1-2 Isentropic compression in a pump
2-3 Constant pressure heat addition in a boiler
3-4 Isentropic expansion in a turbine
4-1 Constant pressure heat rejection in a condenser

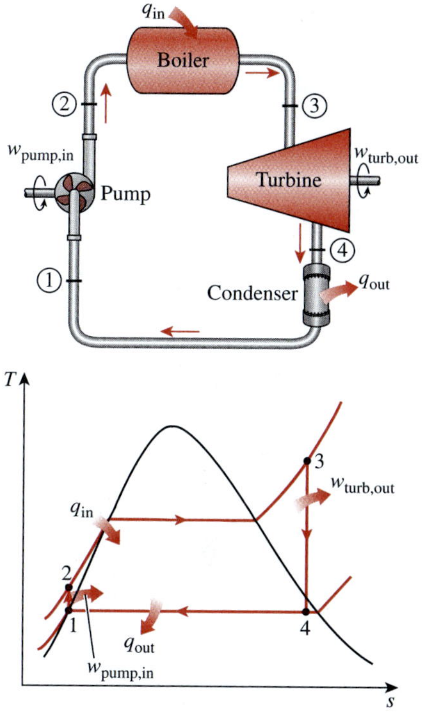

FIGURE 10–2
The simple ideal Rankine cycle.

Water enters the *pump* at state 1 as saturated liquid and is compressed isentropically to the operating pressure of the boiler. The water temperature increases somewhat during this isentropic compression process due to a slight decrease in the specific volume of water. The vertical distance between states 1 and 2 on the *T-s* diagram is greatly exaggerated for clarity. (If water were truly incompressible, would there be a temperature change at all during this process?)

Water enters the *boiler* as a compressed liquid at state 2 and leaves as a superheated vapor at state 3. The boiler is basically a large heat exchanger where the heat originating from combustion gases, nuclear reactors, or other sources is transferred to the water essentially at constant pressure. The boiler, together with the section where the steam is superheated (the superheater), is often called the *steam generator*.

The superheated vapor at state 3 enters the *turbine,* where it expands isentropically and produces work by rotating the shaft connected to an electric generator. The pressure and the temperature of steam drop during this process to the values at state 4, where steam enters the *condenser.* At this state, steam is usually a saturated liquid–vapor mixture with a high quality. Steam is condensed at constant pressure in the condenser, which is basically a large heat exchanger, by rejecting heat to a cooling medium such as a lake, a river, or the atmosphere. Steam leaves the condenser as saturated liquid and enters the pump, completing the cycle. In areas where water is precious, the power plants are cooled by air instead of water. This method of cooling, which is also used in car engines, is called *dry cooling.* Several power plants in the world, including some in the United States, use dry cooling to conserve water.

Remembering that the area under the process curve on a *T-s* diagram represents the heat transfer for internally reversible processes, we see that the area under process curve 2-3 represents the heat transferred to the water in the boiler and the area under the process curve 4-1 represents the heat rejected in the condenser. The difference between these two (the area enclosed by the cycle curve) is the net work produced during the cycle.

Energy Analysis of the Ideal Rankine Cycle

All four components associated with the Rankine cycle (the pump, boiler, turbine, and condenser) are steady-flow devices, and thus all four processes that make up the Rankine cycle can be analyzed as steady-flow processes. The kinetic and potential energy changes of the steam are usually small relative to the work and heat transfer terms and are therefore usually neglected. Then the *steady-flow energy equation* per unit mass of steam reduces to

$$(q_{in} - q_{out}) + (w_{in} - w_{out}) = h_e - h_i \qquad \text{(kJ/kg)} \tag{10–1}$$

The boiler and the condenser do not involve any work, and the pump and the turbine are assumed to be isentropic. Then the conservation of energy relation for each device can be expressed as follows:

Pump ($q = 0$): $$w_{pump,in} = h_2 - h_1 \tag{10–2}$$

or,

$$w_{pump,in} = v(P_2 - P_1) \tag{10–3}$$

where

$$h_1 = h_{f\,@\,P_1} \quad \text{and} \quad v \cong v_1 = v_{f\,@\,P_1} \tag{10–4}$$

Boiler ($w = 0$): $$q_{in} = h_3 - h_2 \tag{10–5}$$

Turbine ($q = 0$): $$w_{turb,out} = h_3 - h_4 \tag{10–6}$$

Condenser ($w = 0$): $$q_{out} = h_4 - h_1 \tag{10–7}$$

The *thermal efficiency* of the Rankine cycle is determined from

$$\eta_{th} = \frac{w_{net}}{q_{in}} = 1 - \frac{q_{out}}{q_{in}} \tag{10–8}$$

where

$$w_{net} = q_{in} - q_{out} = w_{turb,out} - w_{pump,in}$$

The conversion efficiency of power plants in the United States is often expressed in terms of **heat rate**, which is the amount of heat supplied, in Btu's, to generate 1 kWh of electricity. The smaller the heat rate, the greater the efficiency. Considering that 1 kWh = 3412 Btu and disregarding the losses associated with the conversion of shaft power to electric power, the relation between the heat rate and the thermal efficiency can be expressed as

$$\eta_{th} = \frac{3412 \text{ (Btu/kWh)}}{\text{Heat rate (Btu/kWh)}} \tag{10–9}$$

For example, a heat rate of 11,363 Btu/kWh is equivalent to 30 percent efficiency.

The thermal efficiency can also be interpreted as the ratio of the area enclosed by the cycle on a *T-s* diagram to the area under the heat-addition process. The use of these relations is illustrated in the following example.

EXAMPLE 10–1 The Simple Ideal Rankine Cycle

Consider a steam power plant operating on the simple ideal Rankine cycle. Steam enters the turbine at 3 MPa and 350°C and is condensed in the condenser at a pressure of 75 kPa. Determine the thermal efficiency of this cycle.

SOLUTION A steam power plant operating on the simple ideal Rankine cycle is considered. The thermal efficiency of the cycle is to be determined.
Assumptions **1** Steady operating conditions exist. **2** Kinetic and potential energy changes are negligible.
Analysis The schematic of the power plant and the *T-s* diagram of the cycle are shown in Fig. 10–3. We note that the power plant operates on the ideal Rankine cycle. Therefore, the pump and the turbine are isentropic, there are no pressure drops in the boiler and condenser, and steam leaves the condenser and enters the pump as saturated liquid at the condenser pressure.

First we determine the enthalpies at various points in the cycle, using data from steam tables (Tables A–4, A–5, and A–6):

State 1:
$$\left.\begin{array}{l} P_1 = 75 \text{ kPa} \\ \text{Sat. liquid} \end{array}\right\} \quad \begin{array}{l} h_1 = h_{f\,@\,75\text{ kPa}} = 384.44 \text{ kJ/kg} \\ v_1 = v_{f\,@\,75\text{ kPa}} = 0.001037 \text{ m}^3/\text{kg} \end{array}$$

State 2: $P_2 = 3$ MPa
$$s_2 = s_1$$

$$w_{\text{pump,in}} = v_1(P_2 - P_1) = (0.001037 \text{ m}^3/\text{kg})[(3000 - 75) \text{ kPa}]\left(\frac{1 \text{ kJ}}{1 \text{ kPa·m}^3}\right)$$
$$= 3.03 \text{ kJ/kg}$$
$$h_2 = h_1 + w_{\text{pump,in}} = (384.44 + 3.03) \text{ kJ/kg} = 387.47 \text{ kJ/kg}$$

State 3:
$$\left.\begin{array}{l} P_3 = 3 \text{ MPa} \\ T_3 = 350°\text{C} \end{array}\right\} \quad \begin{array}{l} h_3 = 3116.1 \text{ kJ/kg} \\ s_3 = 6.7450 \text{ kJ/kg·K} \end{array}$$

State 4: $P_4 = 75$ kPa (sat. mixture)
$$s_4 = s_3$$
$$x_4 = \frac{s_4 - s_f}{s_{fg}} = \frac{6.7450 - 1.2132}{6.2426} = 0.8861$$
$$h_4 = h_f + x_4 h_{fg} = 384.44 + 0.8861(2278.0) = 2403.0 \text{ kJ/kg}$$

Thus,
$$q_{\text{in}} = h_3 - h_2 = (3116.1 - 387.47) \text{ kJ/kg} = 2728.6 \text{ kJ/kg}$$
$$q_{\text{out}} = h_4 - h_1 = (2403.0 - 384.44) \text{ kJ/kg} = 2018.6 \text{ kJ/kg}$$

and
$$\eta_{\text{th}} = 1 - \frac{q_{\text{out}}}{q_{\text{in}}} = 1 - \frac{2018.6 \text{ kJ/kg}}{2728.6 \text{ kJ/kg}} = \mathbf{0.260 \text{ or } 26.0\%}$$

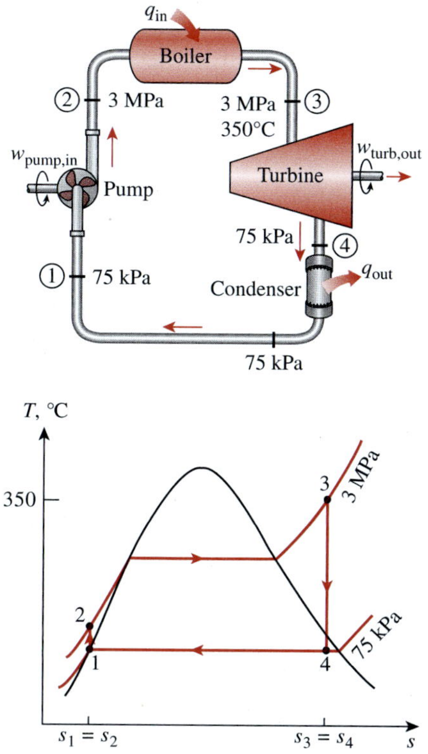

FIGURE 10–3
Schematic and *T-s* diagram for Example 10–1.

The thermal efficiency could also be determined from

$$w_{turb,out} = h_3 - h_4 = (3116.1 - 2403.0)\text{ kJ/kg} = 713.1\text{ kJ/kg}$$

$$w_{net} = w_{turb,out} - w_{pump,in} = (713.1 - 3.03)\text{ kJ/kg} = 710.1\text{ kJ/kg}$$

or

$$w_{net} = q_{in} - q_{out} = (2728.6 - 2018.6)\text{ kJ/kg} = 710.0\text{ kJ/kg}$$

and

$$\eta_{th} = \frac{w_{net}}{q_{in}} = \frac{710.0\text{ kJ/kg}}{2728.6\text{ kJ/kg}} = \mathbf{0.260\ or\ 26.0\%}$$

That is, this power plant converts 26 percent of the heat it receives in the boiler to net work. An actual power plant operating between the same temperature and pressure limits will have a lower efficiency because of the irreversibilities such as friction.

Discussion Notice that the back work ratio ($r_{bw} = w_{in}/w_{out}$) of this power plant is 0.004, and thus only 0.4 percent of the turbine work output is required to operate the pump. Having such low back work ratios is characteristic of vapor power cycles. This is in contrast to the gas power cycles, which typically involve very high back work ratios (about 40 to 80 percent).

It is also interesting to note the thermal efficiency of a Carnot cycle operating between the same temperature limits

$$\eta_{th,Carnot} = 1 - \frac{T_{min}}{T_{max}} = 1 - \frac{(91.76 + 273)\text{ K}}{(350 + 273)\text{ K}} = 0.415$$

The difference between the two efficiencies is due to the large external irreversibility in the Rankine cycle caused by the large temperature difference between steam and combustion gases in the furnace.

10–3 ▪ DEVIATION OF ACTUAL VAPOR POWER CYCLES FROM IDEALIZED ONES

The actual vapor power cycle differs from the ideal Rankine cycle, as illustrated in Fig. 10–4*a*, as a result of irreversibilities in various components. Fluid friction and heat loss to the surroundings are the two common sources of irreversibilities.

Fluid friction causes pressure drops in the boiler, the condenser, and the piping between various components. As a result, steam leaves the boiler at a somewhat lower pressure. Also, the pressure at the turbine inlet is somewhat lower than that at the boiler exit due to the pressure drop in the connecting pipes. The pressure drop in the condenser is usually very small. To compensate for these pressure drops, the water must be pumped to a sufficiently higher pressure than the ideal cycle calls for. This requires a larger pump and larger work input to the pump.

The other major source of irreversibility is the *heat loss* from the steam to the surroundings as the steam flows through various components. To maintain the same level of net work output, more heat needs to be transferred to

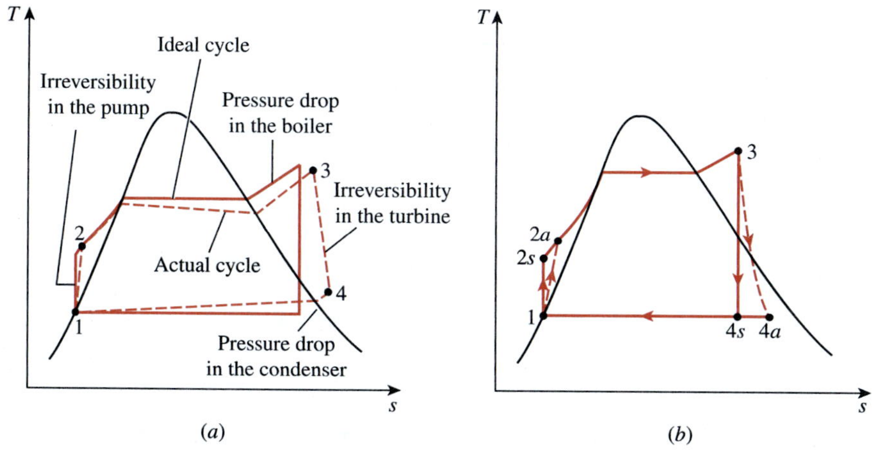

FIGURE 10–4
(*a*) Deviation of actual vapor power cycle from the ideal Rankine cycle. (*b*) The effect of pump and turbine irreversibilities on the ideal Rankine cycle.

the steam in the boiler to compensate for these undesired heat losses. As a result, cycle efficiency decreases.

Of particular importance are the irreversibilities occurring within the pump and the turbine. A pump requires a greater work input, and a turbine produces a smaller work output as a result of irreversibilities. Under ideal conditions, the flow through these devices is isentropic. The deviation of actual pumps and turbines from the isentropic ones can be accounted for by utilizing *isentropic efficiencies,* defined as

$$\eta_P = \frac{w_s}{w_a} = \frac{h_{2s} - h_1}{h_{2a} - h_1} \tag{10–10}$$

and

$$\eta_T = \frac{w_a}{w_s} = \frac{h_3 - h_{4a}}{h_3 - h_{4s}} \tag{10–11}$$

where states $2a$ and $4a$ are the actual exit states of the pump and the turbine, respectively, and $2s$ and $4s$ are the corresponding states for the isentropic case (Fig. 10-4*b*).

Other factors also need to be considered in the analysis of actual vapor power cycles. In actual condensers, for example, the liquid is usually subcooled to prevent the onset of *cavitation,* the rapid vaporization and condensation of the fluid at the low-pressure side of the pump impeller, which may damage it. Additional losses occur at the bearings between the moving parts as a result of friction. Steam that leaks out during the cycle and air that leaks into the condenser represent two other sources of loss. Finally, the power consumed by the auxiliary equipment such as fans that supply air to the furnace should also be considered in evaluating the overall performance of power plants.

The effect of irreversibilities on the thermal efficiency of a steam power cycle is illustrated below with an example.

EXAMPLE 10–2 An Actual Steam Power Cycle

A steam power plant operates on the cycle shown in Fig. 10–5. If the isentropic efficiency of the turbine is 87 percent and the isentropic efficiency of the pump is 85 percent, determine (*a*) the thermal efficiency of the cycle and (*b*) the net power output of the plant for a mass flow rate of 15 kg/s.

SOLUTION A steam power cycle with specified turbine and pump efficiencies is considered. The thermal efficiency and the net power output are to be determined.

Assumptions **1** Steady operating conditions exist. **2** Kinetic and potential energy changes are negligible.

Analysis The schematic of the power plant and the *T-s* diagram of the cycle are shown in Fig. 10–5. The temperatures and pressures of steam at various points are also indicated on the figure. We note that the power plant involves steady-flow components and operates on the Rankine cycle, but the imperfections at various components are accounted for.

(*a*) The thermal efficiency of a cycle is the ratio of the net work output to the heat input, and it is determined as follows:

Pump work input:

$$w_{\text{pump,in}} = \frac{w_{s,\text{pump,in}}}{\eta_P} = \frac{v_1(P_2 - P_1)}{\eta_P}$$

$$= \frac{(0.001009\ \text{m}^3/\text{kg})[(16{,}000 - 9)\ \text{kPa}]}{0.85}\left(\frac{1\ \text{kJ}}{1\ \text{kPa}\cdot\text{m}^3}\right)$$

$$= 19.0\ \text{kJ/kg}$$

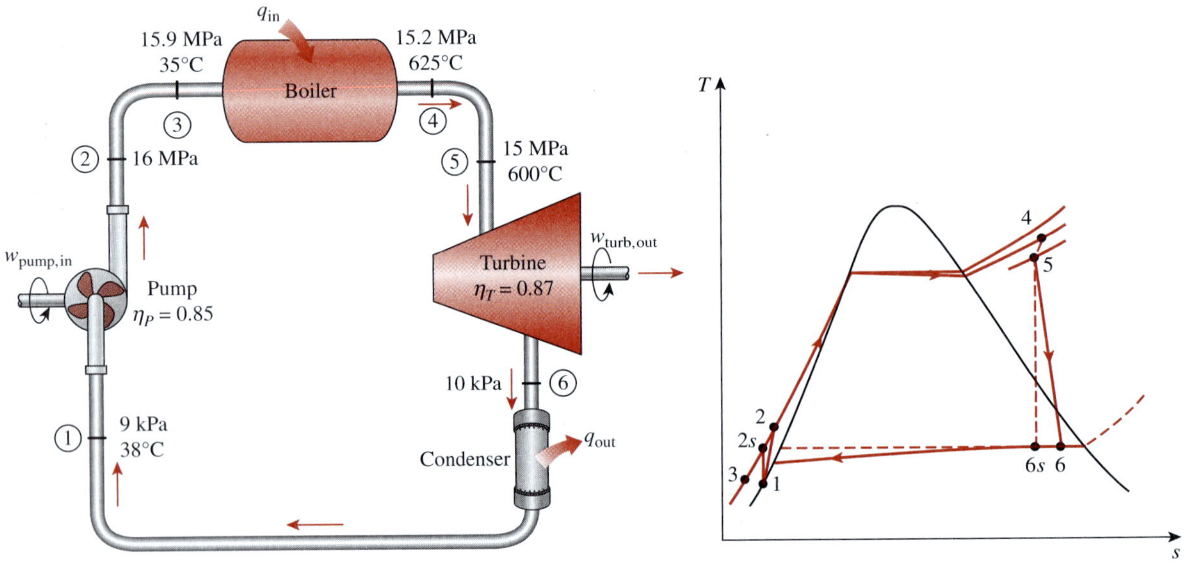

FIGURE 10–5

Schematic and *T-s* diagram for Example 10–2.

Turbine work output:

$$w_{turb,out} = \eta_T w_{s,turb,out}$$

$$= \eta_T(h_5 - h_{6s}) = 0.87(3583.1 - 2115.3)\ kJ/kg$$

$$= 1277.0\ kJ/kg$$

Boiler heat input: $q_{in} = h_4 - h_3 = (3647.6 - 160.1)\ kJ/kg = 3487.5\ kJ/kg$

Thus,

$$w_{net} = w_{turb,out} - w_{pump,in} = (1277.0 - 19.0)\ kJ/kg = 1258.0\ kJ/kg$$

$$\eta_{th} = \frac{w_{net}}{q_{in}} = \frac{1258.0\ kJ/kg}{3487.5\ kJ/kg} = \mathbf{0.361\ or\ 36.1\%}$$

(*b*) The power produced by this power plant is

$$\dot{W}_{net} = \dot{m}(w_{net}) = (15\ kg/s)(1258.0\ kJ/kg) = \mathbf{18.9\ MW}$$

Discussion Without the irreversibilities, the thermal efficiency of this cycle would be 43.0 percent (see Example 10–3*c*).

10–4 ▪ HOW CAN WE INCREASE THE EFFICIENCY OF THE RANKINE CYCLE?

Steam power plants are responsible for the production of most electric power in the world, and even small increases in thermal efficiency can mean large savings from the fuel requirements. Therefore, every effort is made to improve the efficiency of the cycle on which steam power plants operate.

The basic idea behind all the modifications to increase the thermal efficiency of a power cycle is the same: *Increase the average temperature at which heat is transferred to the working fluid in the boiler, or decrease the average temperature at which heat is rejected from the working fluid in the condenser.* That is, the average fluid temperature should be as high as possible during heat addition and as low as possible during heat rejection. Next we discuss three ways of accomplishing this for the simple ideal Rankine cycle.

Lowering the Condenser Pressure (*Lowers* $T_{low,avg}$)

Steam exists as a saturated mixture in the condenser at the saturation temperature corresponding to the pressure inside the condenser. Therefore, lowering the operating pressure of the condenser automatically lowers the temperature of the steam, and thus the temperature at which heat is rejected.

The effect of lowering the condenser pressure on the Rankine cycle efficiency is illustrated on a *T-s* diagram in Fig. 10–6. For comparison purposes, the turbine inlet state is maintained the same. The colored area on this diagram represents the increase in net work output as a result of lowering the condenser pressure from P_4 to P_4'. The heat input requirements also

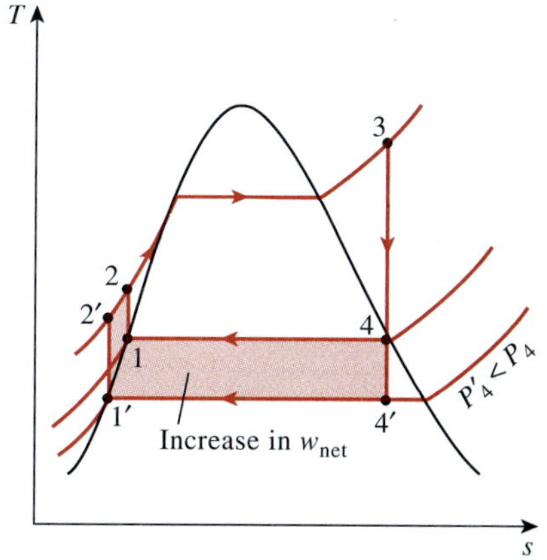

FIGURE 10–6
The effect of lowering the condenser pressure on the ideal Rankine cycle.

increase (represented by the area under curve 2′-2), but this increase is very small. Thus the overall effect of lowering the condenser pressure is an increase in the thermal efficiency of the cycle.

To take advantage of the increased efficiencies at low pressures, the condensers of steam power plants usually operate well below the atmospheric pressure. This does not present a major problem since the vapor power cycles operate in a closed loop. However, there is a lower limit on the condenser pressure that can be used. It cannot be lower than the saturation pressure corresponding to the temperature of the cooling medium. Consider, for example, a condenser that is to be cooled by a nearby river at 15°C. Allowing a temperature difference of 10°C for effective heat transfer, the steam temperature in the condenser must be above 25°C; thus the condenser pressure must be above 3.2 kPa, which is the saturation pressure at 25°C.

Lowering the condenser pressure is not without any side effects, however. For one thing, it creates the possibility of air leakage into the condenser. More importantly, it increases the moisture content of the steam at the final stages of the turbine, as can be seen from Fig. 10–6. The presence of large quantities of moisture is highly undesirable in turbines because it decreases the turbine efficiency and erodes the turbine blades. Fortunately, this problem can be corrected, as discussed next.

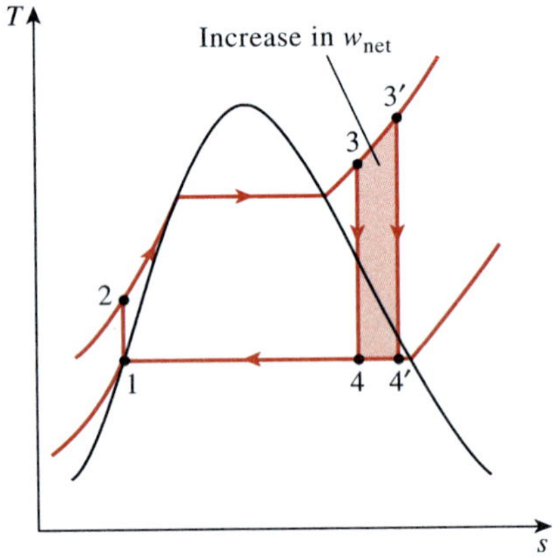

FIGURE 10–7
The effect of superheating the steam to higher temperatures on the ideal Rankine cycle.

Superheating the Steam to High Temperatures (*Increases* $T_{high,avg}$)

The average temperature at which heat is transferred to steam can be increased without increasing the boiler pressure by superheating the steam to high temperatures. The effect of superheating on the performance of vapor power cycles is illustrated on a *T-s* diagram in Fig. 10–7. The colored area on this diagram represents the increase in the net work. The total area under the process curve 3-3′ represents the increase in the heat input. Thus both the net work and heat input increase as a result of superheating the steam to a higher temperature. The overall effect is an increase in thermal efficiency, however, since the average temperature at which heat is added increases.

Superheating the steam to higher temperatures has another very desirable effect: It decreases the moisture content of the steam at the turbine exit, as can be seen from the *T-s* diagram (the quality at state 4′ is higher than that at state 4).

The temperature to which steam can be superheated is limited, however, by metallurgical considerations. Presently the highest steam temperature allowed at the turbine inlet is about 620°C (1150°F). Any increase in this value depends on improving the present materials or finding new ones that can withstand higher temperatures. Ceramics are very promising in this regard.

Increasing the Boiler Pressure (*Increases* $T_{high,avg}$)

Another way of increasing the average temperature during the heat-addition process is to increase the operating pressure of the boiler, which automatically raises the temperature at which boiling takes place. This, in turn, raises the average temperature at which heat is transferred to the steam and thus raises the thermal efficiency of the cycle.

The effect of increasing the boiler pressure on the performance of vapor power cycles is illustrated on a *T-s* diagram in Fig. 10–8. Notice that for a fixed turbine inlet temperature, the cycle shifts to the left and the moisture content of steam at the turbine exit increases. This undesirable side effect can be corrected, however, by reheating the steam, as discussed in the next section.

Operating pressures of boilers have gradually increased over the years from about 2.7 MPa (400 psia) in 1922 to over 30 MPa (4500 psia) today, generating enough steam to produce a net power output of 1000 MW or more in a large power plant. Today many modern steam power plants operate at supercritical pressures ($P > 22.06$ MPa) and have thermal efficiencies of about 40 percent for fossil-fuel plants and 34 percent for nuclear plants. There are over 150 supercritical-pressure steam power plants in operation in the United States. The lower efficiencies of nuclear power plants are due to the lower maximum temperatures used in those plants for safety reasons. The *T-s* diagram of a supercritical Rankine cycle is shown in Fig. 10–9.

The effects of lowering the condenser pressure, superheating to a higher temperature, and increasing the boiler pressure on the thermal efficiency of the Rankine cycle are illustrated below with an example.

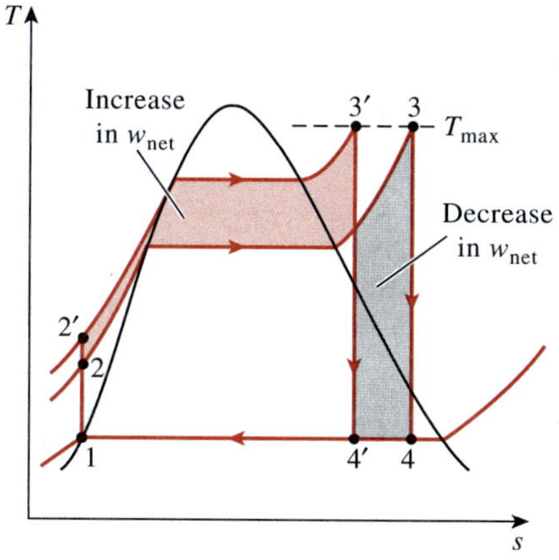

FIGURE 10–8
The effect of increasing the boiler pressure on the ideal Rankine cycle.

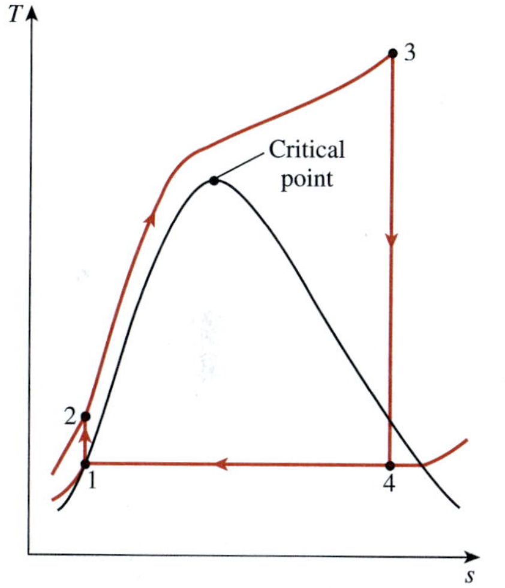

FIGURE 10–9
A supercritical Rankine cycle.

EXAMPLE 10–3 Effect of Boiler Pressure and Temperature on Efficiency

Consider a steam power plant operating on the ideal Rankine cycle. Steam enters the turbine at 3 MPa and 350°C and is condensed in the condenser at a pressure of 10 kPa. Determine (*a*) the thermal efficiency of this power plant, (*b*) the thermal efficiency if steam is superheated to 600°C instead of 350°C, and (*c*) the thermal efficiency if the boiler pressure is raised to 15 MPa while the turbine inlet temperature is maintained at 600°C.

SOLUTION A steam power plant operating on the ideal Rankine cycle is considered. The effects of superheating the steam to a higher temperature and raising the boiler pressure on thermal efficiency are to be investigated.

Analysis The *T-s* diagrams of the cycle for all three cases are given in Fig. 10–10.

(*a*) This is the steam power plant discussed in Example 10–1, except that the condenser pressure is lowered to 10 kPa. The thermal efficiency is determined in a similar manner:

State 1: $\left.\begin{array}{l} P_1 = 10 \text{ kPa} \\ \text{Sat. liquid} \end{array}\right\} \begin{array}{l} h_1 = h_{f\,@\,10\text{ kPa}} = 191.81 \text{ kJ/kg} \\ v_1 = v_{f\,@\,10\text{ kPa}} = 0.00101 \text{ m}^3/\text{kg} \end{array}$

State 2: $P_2 = 3$ MPa

$s_2 = s_1$

$$w_{\text{pump,in}} = v_1(P_2 - P_1) = (0.00101 \text{ m}^3/\text{kg})[(3000 - 10) \text{ kPa}]\left(\frac{1 \text{ kJ}}{1 \text{ kPa}\cdot\text{m}^3}\right)$$

$$= 3.02 \text{ kJ/kg}$$

$$h_2 = h_1 + w_{\text{pump,in}} = (191.81 + 3.02) \text{ kJ/kg} = 194.83 \text{ kJ/kg}$$

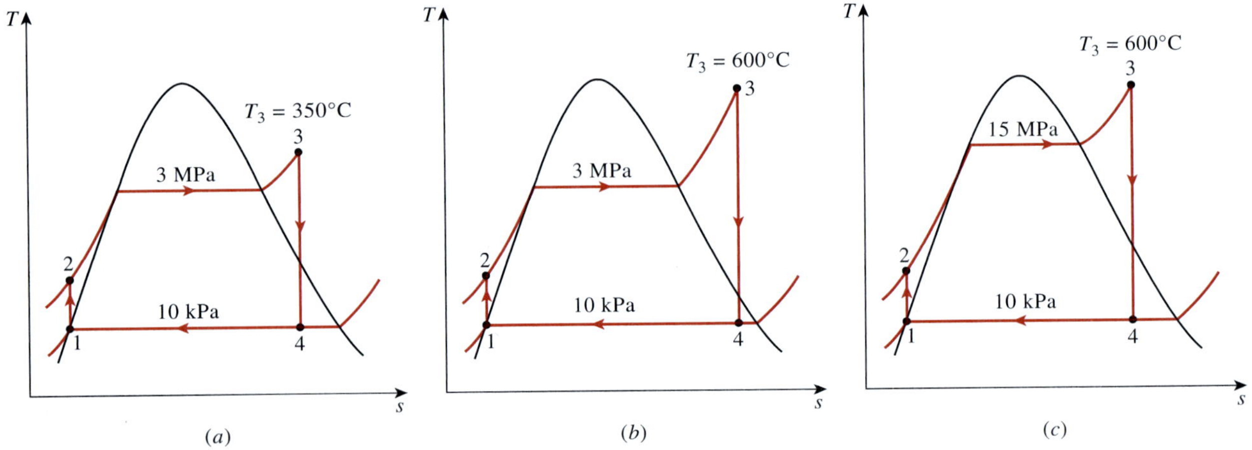

FIGURE 10–10
T-s diagrams of the three cycles discussed in Example 10–3.

State 3: $\left.\begin{array}{l} P_3 = 3 \text{ MPa} \\ T_3 = 350°\text{C} \end{array}\right\}$ $\begin{array}{l} h_3 = 3116.1 \text{ kJ/kg} \\ s_3 = 6.7450 \text{ kJ/kg·K} \end{array}$

State 4: $P_4 = 10 \text{ kPa}$ (sat. mixture)

$$s_4 = s_3$$

$$x_4 = \frac{s_4 - s_f}{s_{fg}} = \frac{6.7450 - 0.6492}{7.4996} = 0.8128$$

Thus,

$$h_4 = h_f + x_4 h_{fg} = 191.81 + 0.8128(2392.1) = 2136.1 \text{ kJ/kg}$$

$$q_{\text{in}} = h_3 - h_2 = (3116.1 - 194.83) \text{ kJ/kg} = 2921.3 \text{ kJ/kg}$$

$$q_{\text{out}} = h_4 - h_1 = (2136.1 - 191.81) \text{ kJ/kg} = 1944.3 \text{ kJ/kg}$$

and

$$\eta_{\text{th}} = 1 - \frac{q_{\text{out}}}{q_{\text{in}}} = 1 - \frac{1944.3 \text{ kJ/kg}}{2921.3 \text{ kJ/kg}} = \mathbf{0.334 \text{ or } 33.4\%}$$

Therefore, the thermal efficiency increases from 26.0 to 33.4 percent as a result of lowering the condenser pressure from 75 to 10 kPa. At the same time, however, the quality of the steam decreases from 88.6 to 81.3 percent (in other words, the moisture content increases from 11.4 to 18.7 percent).

(*b*) States 1 and 2 remain the same in this case, and the enthalpies at state 3 (3 MPa and 600°C) and state 4 (10 kPa and $s_4 = s_3$) are determined to be

$$h_3 = 3682.8 \text{ kJ/kg}$$

$$h_4 = 2380.3 \text{ kJ/kg} \quad (x_4 = 0.915)$$

Thus,

$$q_{\text{in}} = h_3 - h_2 = 3682.8 - 194.83 = 3488.0 \text{ kJ/kg}$$

$$q_{\text{out}} = h_4 - h_1 = 2380.3 - 191.81 = 2188.5 \text{ kJ/kg}$$

and

$$\eta_{th} = 1 - \frac{q_{out}}{q_{in}} = 1 - \frac{2188.5 \text{ kJ/kg}}{3488.0 \text{ kJ/kg}} = \mathbf{0.373 \text{ or } 37.3\%}$$

Therefore, the thermal efficiency increases from 33.4 to 37.3 percent as a result of superheating the steam from 350 to 600°C. At the same time, the quality of the steam increases from 81.3 to 91.5 percent (in other words, the moisture content decreases from 18.7 to 8.5 percent).

(*c*) State 1 remains the same in this case, but the other states change. The enthalpies at state 2 (15 MPa and $s_2 = s_1$), state 3 (15 MPa and 600°C), and state 4 (10 kPa and $s_4 = s_3$) are determined in a similar manner to be

$$h_2 = 206.95 \text{ kJ/kg}$$

$$h_3 = 3583.1 \text{ kJ/kg}$$

$$h_4 = 2115.3 \text{ kJ/kg} \qquad (x_4 = 0.804)$$

Thus,

$$q_{in} = h_3 - h_2 = 3583.1 - 206.95 = 3376.2 \text{ kJ/kg}$$

$$q_{out} = h_4 - h_1 = 2115.3 - 191.81 = 1923.5 \text{ kJ/kg}$$

and

$$\eta_{th} = 1 - \frac{q_{out}}{q_{in}} = 1 - \frac{1923.5 \text{ kJ/kg}}{3376.2 \text{ kJ/kg}} = \mathbf{0.430 \text{ or } 43.0\%}$$

Discussion The thermal efficiency increases from 37.3 to 43.0 percent as a result of raising the boiler pressure from 3 to 15 MPa while maintaining the turbine inlet temperature at 600°C. At the same time, however, the quality of the steam decreases from 91.5 to 80.4 percent (in other words, the moisture content increases from 8.5 to 19.6 percent).

10–5 ▪ THE IDEAL REHEAT RANKINE CYCLE

We noted in the last section that increasing the boiler pressure increases the thermal efficiency of the Rankine cycle, but it also increases the moisture content of the steam to unacceptable levels. Then it is natural to ask the following question:

How can we take advantage of the increased efficiencies at higher boiler pressures without facing the problem of excessive moisture at the final stages of the turbine?

Two possibilities come to mind:

1. Superheat the steam to very high temperatures before it enters the turbine. This would be the desirable solution since the average temperature at which heat is added would also increase, thus increasing the cycle efficiency. This is not a viable solution, however, since it requires raising the steam temperature to metallurgically unsafe levels.

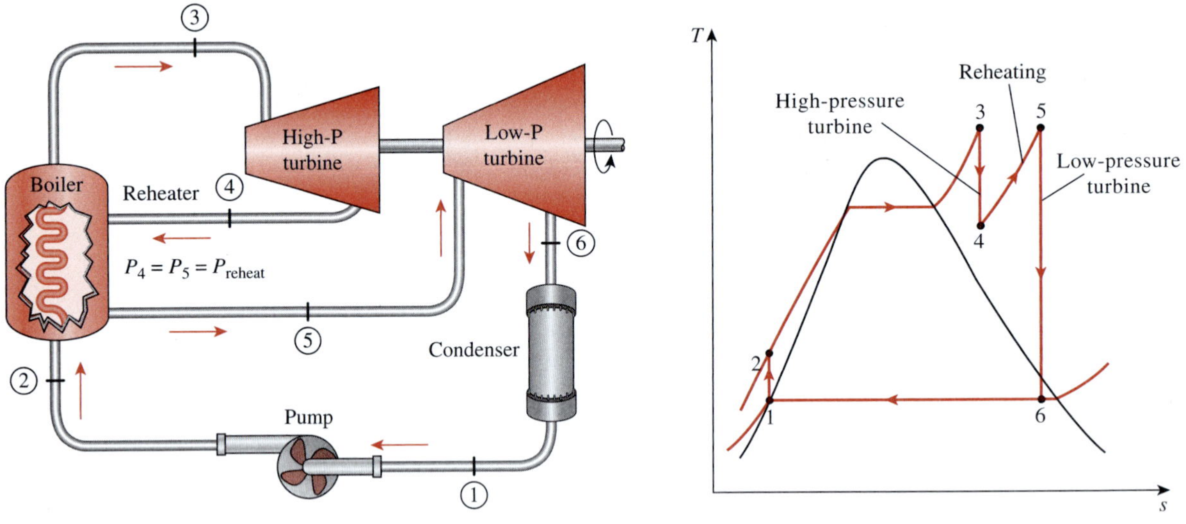

FIGURE 10–11
The ideal reheat Rankine cycle.

2. Expand the steam in the turbine in two stages, and reheat it in between. In other words, modify the simple ideal Rankine cycle with a **reheat** process. Reheating is a practical solution to the excessive moisture problem in turbines, and it is commonly used in modern steam power plants.

The *T-s* diagram of the ideal reheat Rankine cycle and the schematic of the power plant operating on this cycle are shown in Fig. 10–11. The ideal reheat Rankine cycle differs from the simple ideal Rankine cycle in that the expansion process takes place in two stages. In the first stage (the high-pressure turbine), steam is expanded isentropically to an intermediate pressure and sent back to the boiler where it is reheated at constant pressure, usually to the inlet temperature of the first turbine stage. Steam then expands isentropically in the second stage (low-pressure turbine) to the condenser pressure. Thus the total heat input and the total turbine work output for a reheat cycle become

$$q_{in} = q_{primary} + q_{reheat} = (h_3 - h_2) + (h_5 - h_4) \quad \textbf{(10–12)}$$

and

$$w_{turb,out} = w_{turb,I} + w_{turb,II} = (h_3 - h_4) + (h_5 - h_6) \quad \textbf{(10–13)}$$

The incorporation of the single reheat in a modern power plant improves the cycle efficiency by 4 to 5 percent by increasing the average temperature at which heat is transferred to the steam.

The average temperature during the reheat process can be increased by increasing the number of expansion and reheat stages. As the number of stages is increased, the expansion and reheat processes approach an isothermal process at the maximum temperature, as shown in Fig. 10–12. The use of more than two reheat stages, however, is not practical. The theoretical improvement in efficiency from the second reheat is about half of that which results from a single reheat. If the turbine inlet pressure is not high enough, double reheat would result in superheated exhaust. This is undesirable as it would cause the average temperature for heat rejection to increase and thus the cycle efficiency to decrease. Therefore, double reheat is used only on supercritical-pressure

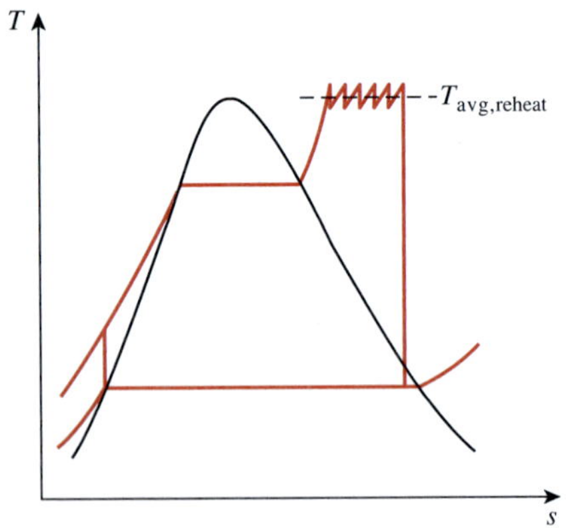

FIGURE 10–12
The average temperature at which heat is transferred during reheating increases as the number of reheat stages is increased.

($P > 22.06$ MPa) power plants. A third reheat stage would increase the cycle efficiency by about half of the improvement attained by the second reheat. This gain is too small to justify the added cost and complexity.

The reheat cycle was introduced in the mid-1920s, but it was abandoned in the 1930s because of the operational difficulties. The steady increase in boiler pressures over the years made it necessary to reintroduce single reheat in the late 1940s and double reheat in the early 1950s.

The reheat temperatures are very close or equal to the turbine inlet temperature. The optimum reheat pressure is about one-fourth of the maximum cycle pressure. For example, the optimum reheat pressure for a cycle with a boiler pressure of 12 MPa is about 3 MPa.

Remember that the sole purpose of the reheat cycle is to reduce the moisture content of the steam at the final stages of the expansion process. If we had materials that could withstand sufficiently high temperatures, there would be no need for the reheat cycle.

EXAMPLE 10–4 The Ideal Reheat Rankine Cycle

Consider a steam power plant operating on the ideal reheat Rankine cycle. Steam enters the high-pressure turbine at 15 MPa and 600°C and is condensed in the condenser at a pressure of 10 kPa. If the moisture content of the steam at the exit of the low-pressure turbine is not to exceed 10.4 percent, determine (*a*) the pressure at which the steam should be reheated and (*b*) the thermal efficiency of the cycle. Assume the steam is reheated to the inlet temperature of the high-pressure turbine.

SOLUTION A steam power plant operating on the ideal reheat Rankine cycle is considered. For a specified moisture content at the turbine exit, the reheat pressure and the thermal efficiency are to be determined.

Assumptions **1** Steady operating conditions exist. **2** Kinetic and potential energy changes are negligible.

Analysis The schematic of the power plant and the *T-s* diagram of the cycle are shown in Fig. 10–13. We note that the power plant operates on the ideal reheat Rankine cycle. Therefore, the pump and the turbines are isentropic, there are no pressure drops in the boiler and condenser, and steam leaves the condenser and enters the pump as saturated liquid at the condenser pressure.

(*a*) The reheat pressure is determined from the requirement that the entropies at states 5 and 6 be the same:

State 6:

$$P_6 = 10 \text{ kPa}$$

$$x_6 = 0.896 \quad \text{(sat. mixture)}$$

$$s_6 = s_f + x_6 s_{fg} = 0.6492 + 0.896(7.4996) = 7.3688 \text{ kJ/kg·K}$$

Also,

$$h_6 = h_f + x_6 h_{fg} = 191.81 + 0.896(2392.1) = 2335.1 \text{ kJ/kg}$$

Thus,

State 5:

$$\left.\begin{aligned} T_5 &= 600°\text{C} \\ s_5 &= s_6 \end{aligned}\right\} \quad \begin{aligned} P_5 &= \mathbf{4.0\ MPa} \\ h_5 &= 3674.9 \text{ kJ/kg} \end{aligned}$$

Therefore, steam should be reheated at a pressure of 4 MPa or lower to prevent a moisture content above 10.4 percent.

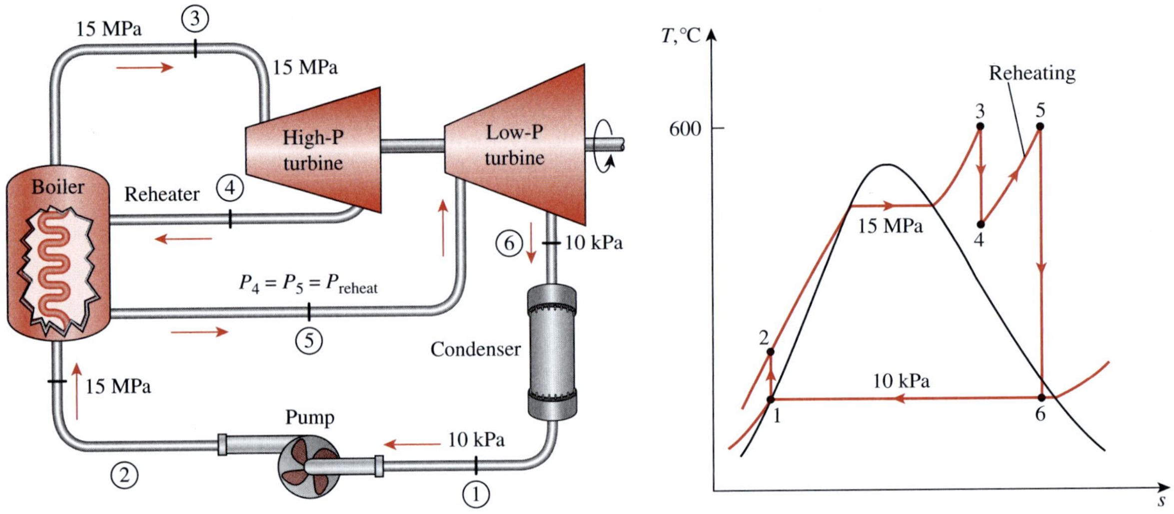

FIGURE 10–13
Schematic and *T-s* diagram for Example 10–4.

(*b*) To determine the thermal efficiency, we need to know the enthalpies at all other states:

State 1: $\left.\begin{array}{l} P_1 = 10 \text{ kPa} \\ \text{Sat. liquid} \end{array}\right\} \begin{array}{l} h_1 = h_{f\,@\,10\text{ kPa}} = 191.81 \text{ kJ/kg} \\ v_1 = v_{f\,@\,10\text{ kPa}} = 0.00101 \text{ m}^3/\text{kg} \end{array}$

State 2: $P_2 = 15$ MPa
$s_2 = s_1$

$$w_{\text{pump,in}} = v_1(P_2 - P_1) = (0.00101 \text{ m}^3/\text{kg}) \times [(15{,}000 - 10)\text{kPa}]\left(\frac{1 \text{ kJ}}{1 \text{ kPa}\cdot\text{m}^3}\right) = 15.14 \text{ kJ/kg}$$

$$h_2 = h_1 + w_{\text{pump,in}} = (191.81 + 15.14) \text{ kJ/kg} = 206.95 \text{ kJ/kg}$$

State 3: $\left.\begin{array}{l} P_3 = 15 \text{ MPa} \\ T_3 = 600°\text{C} \end{array}\right\} \begin{array}{l} h_3 = 3583.1 \text{ kJ/kg} \\ s_3 = 6.6796 \text{ kJ/kg}\cdot\text{K} \end{array}$

State 4: $\left.\begin{array}{l} P_4 = 4 \text{ MPa} \\ s_4 = s_3 \end{array}\right\} \begin{array}{l} h_4 = 3155.0 \text{ kJ/kg} \\ (T_4 = 375.5°\text{C}) \end{array}$

Thus

$$q_{\text{in}} = (h_3 - h_2) + (h_5 - h_4) = (3583.1 - 206.95) \text{ kJ/kg} + (3674.9 - 3155.0) \text{ kJ/kg} = 3896.1 \text{ kJ/kg}$$

$$q_{\text{out}} = h_6 - h_1 = (2335.1 - 191.81) \text{ kJ/kg} = 2143.3 \text{ kJ/kg}$$

and

$$\eta_{th} = 1 - \frac{q_{out}}{q_{in}} = 1 - \frac{2143.3 \text{ kJ/kg}}{3896.1 \text{ kJ/kg}} = \mathbf{0.450 \text{ or } 45.0\%}$$

Discussion This problem was solved in Example 10–3*c* for the same pressure and temperature limits but without the reheat process. A comparison of the two results reveals that reheating reduces the moisture content from 19.6 to 10.4 percent while increasing the thermal efficiency from 43.0 to 45.0 percent.

10–6 ▪ THE IDEAL REGENERATIVE RANKINE CYCLE

A careful examination of the *T-s* diagram of the Rankine cycle redrawn in Fig. 10–14 reveals that heat is transferred to the working fluid during process 2-2′ at a relatively low temperature. This lowers the average heat-addition temperature and thus the cycle efficiency.

To remedy this shortcoming, we look for ways to raise the temperature of the liquid leaving the pump (called the *feedwater*) before it enters the boiler. One such possibility is to transfer heat to the feedwater from the expanding steam in a counterflow heat exchanger built into the turbine, that is, to use **regeneration**. This solution is also impractical because it is difficult to design such a heat exchanger and because it would increase the moisture content of the steam at the final stages of the turbine.

A practical regeneration process in steam power plants is accomplished by extracting, or "bleeding," steam from the turbine at various points. This steam, which could have produced more work by expanding further in the turbine, is used to heat the feedwater instead. The device where the feedwater is heated by regeneration is called a **regenerator**, or a **feedwater heater (FWH)**.

Regeneration not only improves cycle efficiency, but also provides a convenient means of deaerating the feedwater (removing the air that leaks in at the condenser) to prevent corrosion in the boiler. It also helps control the large volume flow rate of the steam at the final stages of the turbine (due to the large specific volumes at low pressures). Therefore, regeneration has been used in all modern steam power plants since its introduction in the early 1920s.

A feedwater heater is basically a heat exchanger where heat is transferred from the steam to the feedwater either by mixing the two fluid streams (open feedwater heaters) or without mixing them (closed feedwater heaters). Regeneration with both types of feedwater heaters is discussed below.

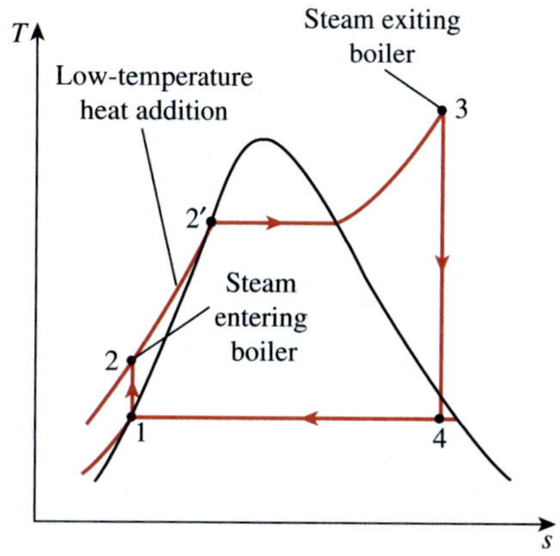

FIGURE 10–14
The first part of the heat-addition process in the boiler takes place at relatively low temperatures.

Open Feedwater Heaters

An **open** (or **direct-contact**) **feedwater heater** is basically a *mixing chamber,* where the steam extracted from the turbine mixes with the feedwater exiting the pump. Ideally, the mixture leaves the heater as a saturated liquid at the heater pressure. The schematic of a steam power plant with one open feedwater heater (also called *single-stage regenerative cycle*) and the *T-s* diagram of the cycle are shown in Fig. 10–15.

In an ideal regenerative Rankine cycle, steam enters the turbine at the boiler pressure (state 5) and expands isentropically to an intermediate pressure (state 6).

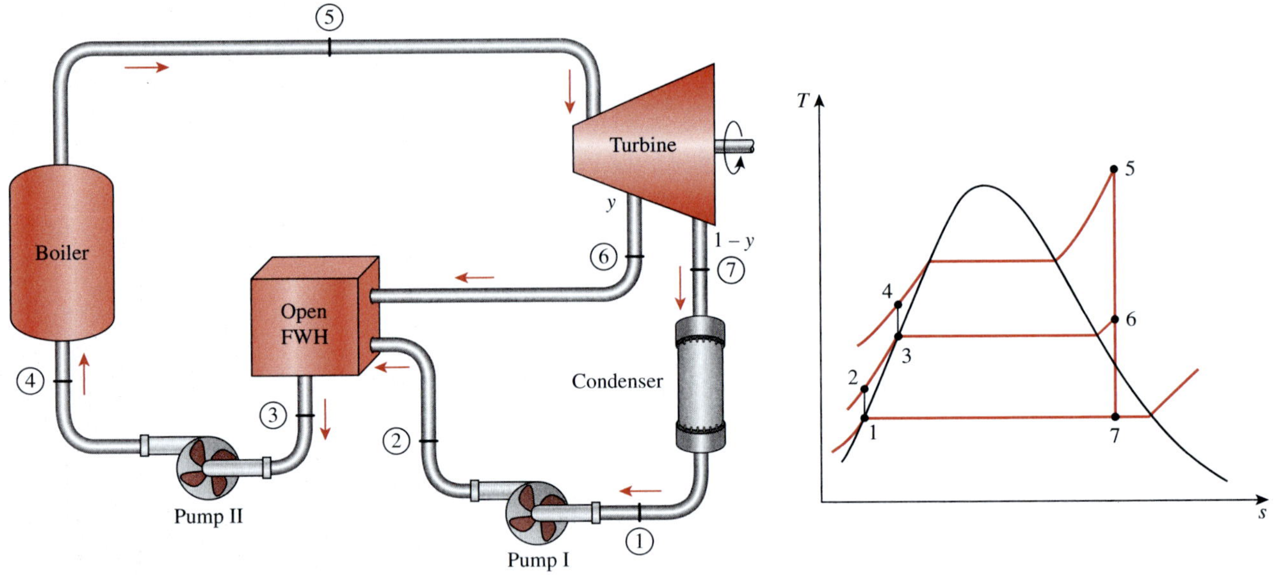

FIGURE 10–15
The ideal regenerative Rankine cycle with an open feedwater heater.

Some steam is extracted at this state and routed to the feedwater heater, while the remaining steam continues to expand isentropically to the condenser pressure (state 7). This steam leaves the condenser as a saturated liquid at the condenser pressure (state 1). The condensed water, which is also called the *feedwater,* then enters an isentropic pump, where it is compressed to the feedwater heater pressure (state 2) and is routed to the feedwater heater, where it mixes with the steam extracted from the turbine. The fraction of the steam extracted is such that the mixture leaves the heater as a saturated liquid at the heater pressure (state 3). A second pump raises the pressure of the water to the boiler pressure (state 4). The cycle is completed by heating the water in the boiler to the turbine inlet state (state 5).

In the analysis of steam power plants, it is more convenient to work with quantities expressed per unit mass of the steam flowing through the boiler. For each 1 kg of steam leaving the boiler, y kg expands partially in the turbine and is extracted at state 6. The remaining $(1 - y)$ kg expands completely to the condenser pressure. Therefore, the mass flow rates are different in different components. If the mass flow rate through the boiler is $\dot{m}$, for example, it is $(1 - y)m\cdot$ through the condenser. This aspect of the regenerative Rankine cycle should be considered in the analysis of the cycle as well as in the interpretation of the areas on the T-s diagram. In light of Fig. 10–15, the heat and work interactions of a regenerative Rankine cycle with one feedwater heater can be expressed per unit mass of steam flowing through the boiler as follows:

$$q_{in} = h_5 - h_4 \tag{10–14}$$

$$q_{out} = (1 - y)(h_7 - h_1) \tag{10–15}$$

$$w_{turb,out} = (h_5 - h_6) + (1 - y)(h_6 - h_7) \tag{10–16}$$

$$w_{pump,in} = (1 - y)w_{pump\ I,in} + w_{pump\ II,in} \tag{10–17}$$

where

$$y = \dot{m}_6/\dot{m}_5 \qquad \text{(fraction of steam extracted)}$$

$$w_{\text{pump I,in}} = v_1(P_2 - P_1)$$

$$w_{\text{pump II,in}} = v_3(P_4 - P_3)$$

The thermal efficiency of the Rankine cycle increases as a result of regeneration. This is because regeneration raises the average temperature at which heat is transferred to the steam in the boiler by raising the temperature of the water before it enters the boiler. The cycle efficiency increases further as the number of feedwater heaters is increased. Many large plants in operation today use as many as eight feedwater heaters. The optimum number of feedwater heaters is determined from economical considerations. The use of an additional feedwater heater cannot be justified unless it saves more from the fuel costs than its own cost.

Closed Feedwater Heaters

Another type of feedwater heater frequently used in steam power plants is the **closed feedwater heater**, in which heat is transferred from the extracted steam to the feedwater without any mixing taking place. The two streams now can be at different pressures, since they do not mix. The schematic of a steam power plant with one closed feedwater heater and the *T-s* diagram of the cycle are shown in Fig. 10–16. In an ideal closed feedwater heater, the feedwater is heated to the exit temperature of the extracted steam, which ideally leaves the heater as a saturated liquid at the extraction pressure. In actual power plants, the feedwater leaves the heater below the exit temperature

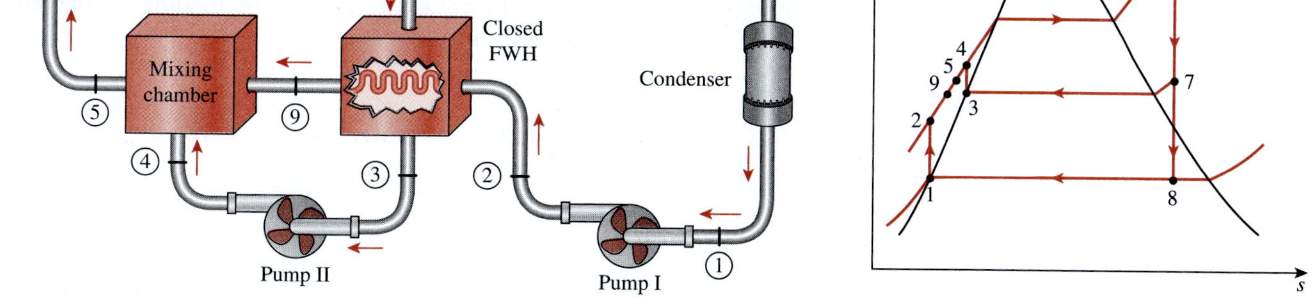

FIGURE 10–16
The ideal regenerative Rankine cycle with a closed feedwater heater.

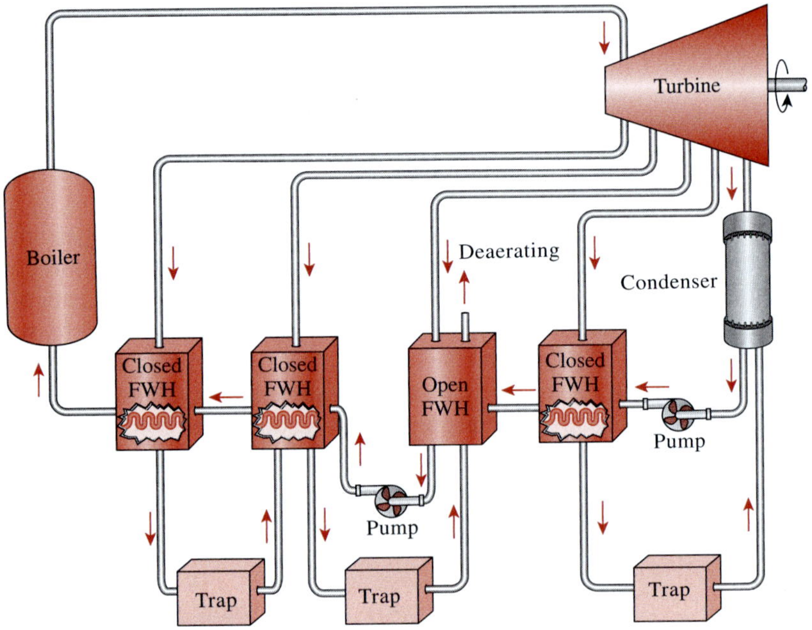

FIGURE 10–17
A steam power plant with one open and three closed feedwater heaters.

of the extracted steam because a temperature difference of at least a few degrees is required for any effective heat transfer to take place.

The condensed steam is then either pumped to the feedwater line or routed to another heater or to the condenser through a device called a **trap**. A trap allows the liquid to be throttled to a lower pressure region but *traps* the vapor. The enthalpy of steam remains constant during this throttling process.

The open and closed feedwater heaters can be compared as follows. Open feedwater heaters are simple and inexpensive and have good heat transfer characteristics. They also bring the feedwater to the saturation state. For each heater, however, a pump is required to handle the feedwater. The closed feedwater heaters are more complex because of the internal tubing network, and thus they are more expensive. Heat transfer in closed feedwater heaters is also less effective since the two streams are not allowed to be in direct contact. However, closed feedwater heaters do not require a separate pump for each heater since the extracted steam and the feedwater can be at different pressures. Most steam power plants use a combination of open and closed feedwater heaters, as shown in Fig. 10–17.

EXAMPLE 10–5 The Ideal Regenerative Rankine Cycle

Consider a steam power plant operating on the ideal regenerative Rankine cycle with one open feedwater heater. Steam enters the turbine at 15 MPa and 600°C and is condensed in the condenser at a pressure of 10 kPa. Some steam leaves the turbine at a pressure of 1.2 MPa and enters the open feedwater heater. Determine the fraction of steam extracted from the turbine and the thermal efficiency of the cycle.

SOLUTION A steam power plant operates on the ideal regenerative Rankine cycle with one open feedwater heater. The fraction of steam extracted from the turbine and the thermal efficiency are to be determined.

Assumptions **1** Steady operating conditions exist. **2** Kinetic and potential energy changes are negligible.

Analysis The schematic of the power plant and the *T-s* diagram of the cycle are shown in Fig. 10–18. We note that the power plant operates on the ideal regenerative Rankine cycle. Therefore, the pumps and the turbines are isentropic; there are no pressure drops in the boiler, condenser, and feedwater heater; and steam leaves the condenser and the feedwater heater as saturated liquid. First, we determine the enthalpies at various states:

State 1: $\left.\begin{array}{l} P_1 = 10 \text{ kPa} \\ \text{Sat. liquid} \end{array}\right\} \begin{array}{l} h_1 = h_{f\,@\,10\text{ kPa}} = 191.81 \text{ kJ/kg} \\ v_1 = v_{f\,@\,10\text{ kPa}} = 0.00101 \text{ m}^3\text{/kg} \end{array}$

State 2: $P_2 = 1.2$ MPa

$s_2 = s_1$

$$w_{\text{pump I,in}} = v_1(P_2 - P_1) = (0.00101 \text{ m}^3\text{/kg})[(1200 - 10) \text{ kPa}]\left(\frac{1 \text{ kJ}}{1 \text{ kPa}\cdot\text{m}^3}\right)$$

$$= 1.20 \text{ kJ/kg}$$

$$h_2 = h_1 + w_{\text{pump I,in}} = (191.81 + 1.20) \text{ kJ/kg} = 193.01 \text{ kJ/kg}$$

State 3: $\left.\begin{array}{l} P_3 = 1.2 \text{ MPa} \\ \text{Sat. liquid} \end{array}\right\} \begin{array}{l} v_3 = v_{f\,@\,1.2\text{ MPa}} = 0.001138 \text{ m}^3\text{/kg} \\ h_3 = h_{f\,@\,1.2\text{ MPa}} = 798.33 \text{ kJ/kg} \end{array}$

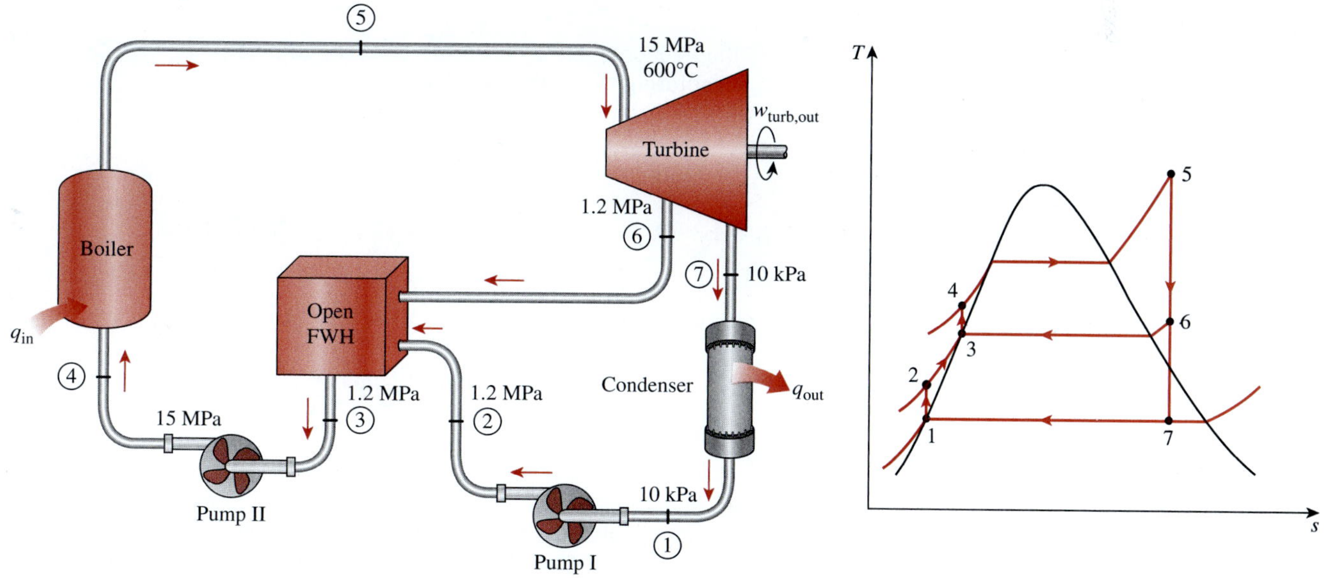

FIGURE 10–18
Schematic and *T-s* diagram for Example 10–5.

State 4: $P_4 = 15 \text{ MPa}$

$$s_4 = s_3$$

$$w_{\text{pump II,in}} = v_3(P_4 - P_3)$$

$$= (0.001138 \text{ m}^3/\text{kg})[(15{,}000 - 1200) \text{ kPa}]\left(\frac{1 \text{ kJ}}{1 \text{ kPa·m}^3}\right)$$

$$= 15.70 \text{ kJ/kg}$$

$$h_4 = h_3 + w_{\text{pump II,in}} = (798.33 + 15.70) \text{ kJ/kg} = 814.03 \text{ kJ/kg}$$

State 5: $\left.\begin{array}{l} P_5 = 15 \text{ MPa} \\ T_5 = 600°\text{C} \end{array}\right\}$ $\begin{array}{l} h_5 = 3583.1 \text{ kJ/kg} \\ s_5 = 6.6796 \text{ kJ/kg·K} \end{array}$

State 6: $\left.\begin{array}{l} P_6 = 1.2 \text{ MPa} \\ s_6 = s_5 \end{array}\right\}$ $\begin{array}{l} h_6 = 2860.2 \text{ kJ/kg} \\ (T_6 = 218.4°\text{C}) \end{array}$

State 7: $P_7 = 10 \text{ kPa}$

$$s_7 = s_5 \quad x_7 = \frac{s_7 - s_f}{s_{fg}} = \frac{6.6796 - 0.6492}{7.4996} = 0.8041$$

$$h_7 = h_f + x_7 h_{fg} = 191.81 + 0.8041(2392.1) = 2115.3 \text{ kJ/kg}$$

The energy analysis of open feedwater heaters is identical to the energy analysis of mixing chambers. The feedwater heaters are generally well insulated ($\dot{Q} = 0$), and they do not involve any work interactions ($\dot{W} = 0$). By neglecting the kinetic and potential energies of the streams, the energy balance reduces for a feedwater heater to

$$\dot{E}_{\text{in}} = \dot{E}_{\text{out}} \longrightarrow \sum_{\text{in}} \dot{m}h = \sum_{\text{out}} \dot{m}h$$

or

$$yh_6 + (1 - y)h_2 = 1(h_3)$$

where y is the fraction of steam extracted from the turbine ($= \dot{m}_6/\dot{m}_5$). Solving for y and substituting the enthalpy values, we find

$$y = \frac{h_3 - h_2}{h_6 - h_2} = \frac{798.33 - 193.01}{2860.2 - 193.01} = \mathbf{0.2270}$$

Thus,

$$q_{\text{in}} = h_5 - h_4 = (3583.1 - 814.03) \text{ kJ/kg} = 2769.1 \text{ kJ/kg}$$

$$q_{\text{out}} = (1 - y)(h_7 - h_1) = (1 - 0.2270)(2115.3 - 191.81) \text{ kJ/kg}$$

$$= 1486.9 \text{ kJ/kg}$$

and

$$\eta_{\text{th}} = 1 - \frac{q_{\text{out}}}{q_{\text{in}}} = 1 - \frac{1486.9 \text{ kJ/kg}}{2769.1 \text{ kJ/kg}} = \mathbf{0.463 \text{ or } 46.3\%}$$

Discussion This problem was worked out in Example 10–3*c* for the same pressure and temperature limits but without the regeneration process. A comparison of the two results reveals that the thermal efficiency of the cycle has increased from 43.0 to 46.3 percent as a result of regeneration. The net work output decreased by 171 kJ/kg, but the heat input decreased by 607 kJ/kg, which results in a net increase in the thermal efficiency.

EXAMPLE 10–6 The Ideal Reheat–Regenerative Rankine Cycle

Consider a steam power plant that operates on an ideal reheat–regenerative Rankine cycle with one open feedwater heater, one closed feedwater heater, and one reheater. Steam enters the turbine at 15 MPa and 600°C and is condensed in the condenser at a pressure of 10 kPa. Some steam is extracted from the turbine at 4 MPa for the closed feedwater heater, and the remaining steam is reheated at the same pressure to 600°C. The extracted steam is completely condensed in the heater and is pumped to 15 MPa before it mixes with the feedwater at the same pressure. Steam for the open feedwater heater is extracted from the low-pressure turbine at a pressure of 0.5 MPa. Determine the fractions of steam extracted from the turbine as well as the thermal efficiency of the cycle.

SOLUTION A steam power plant operates on the ideal reheat–regenerative Rankine cycle with one open feedwater heater, one closed feedwater heater, and one reheater. The fractions of steam extracted from the turbine and the thermal efficiency are to be determined.

Assumptions **1** Steady operating conditions exist. **2** Kinetic and potential energy changes are negligible. **3** In both open and closed feedwater heaters, feedwater is heated to the saturation temperature at the feedwater heater pressure. (Note that this is a conservative assumption since extracted steam enters the closed feedwater heater at 376°C and the saturation temperature at the closed feedwater pressure of 4 MPa is 250°C).

Analysis The schematic of the power plant and the *T-s* diagram of the cycle are shown in Fig. 10–19. The power plant operates on the ideal reheat–regenerative Rankine cycle and thus the pumps and the turbines are isentropic; there are no pressure drops in the boiler, reheater, condenser, and feedwater heaters; and steam leaves the condenser and the feedwater heaters as saturated liquid.

The enthalpies at the various states and the pump work per unit mass of fluid flowing through them are

$$h_1 = 191.81 \text{ kJ/kg} \qquad h_9 = 3155.0 \text{ kJ/kg}$$
$$h_2 = 192.30 \text{ kJ/kg} \qquad h_{10} = 3155.0 \text{ kJ/kg}$$
$$h_3 = 640.09 \text{ kJ/kg} \qquad h_{11} = 3674.9 \text{ kJ/kg}$$
$$h_4 = 643.92 \text{ kJ/kg} \qquad h_{12} = 3014.8 \text{ kJ/kg}$$
$$h_5 = 1087.4 \text{ kJ/kg} \qquad h_{13} = 2335.7 \text{ kJ/kg}$$
$$h_6 = 1087.4 \text{ kJ/kg} \qquad w_{\text{pump I,in}} = 0.49 \text{ kJ/kg}$$
$$h_7 = 1101.2 \text{ kJ/kg} \qquad w_{\text{pump II,in}} = 3.83 \text{ kJ/kg}$$
$$h_8 = 1089.8 \text{ kJ/kg} \qquad w_{\text{pump III,in}} = 13.77 \text{ kJ/kg}$$

The fractions of steam extracted are determined from the mass and energy balances of the feedwater heaters:

Closed feedwater heater:

$$\dot{E}_{\text{in}} = \dot{E}_{\text{out}}$$

$$yh_{10} + (1 - y)h_4 = (1 - y)h_5 + yh_6$$

$$y = \frac{h_5 - h_4}{(h_{10} - h_6) + (h_5 - h_4)} = \frac{1087.4 - 643.92}{(3155.0 - 1087.4) + (1087.4 - 643.92)} = \mathbf{0.1766}$$

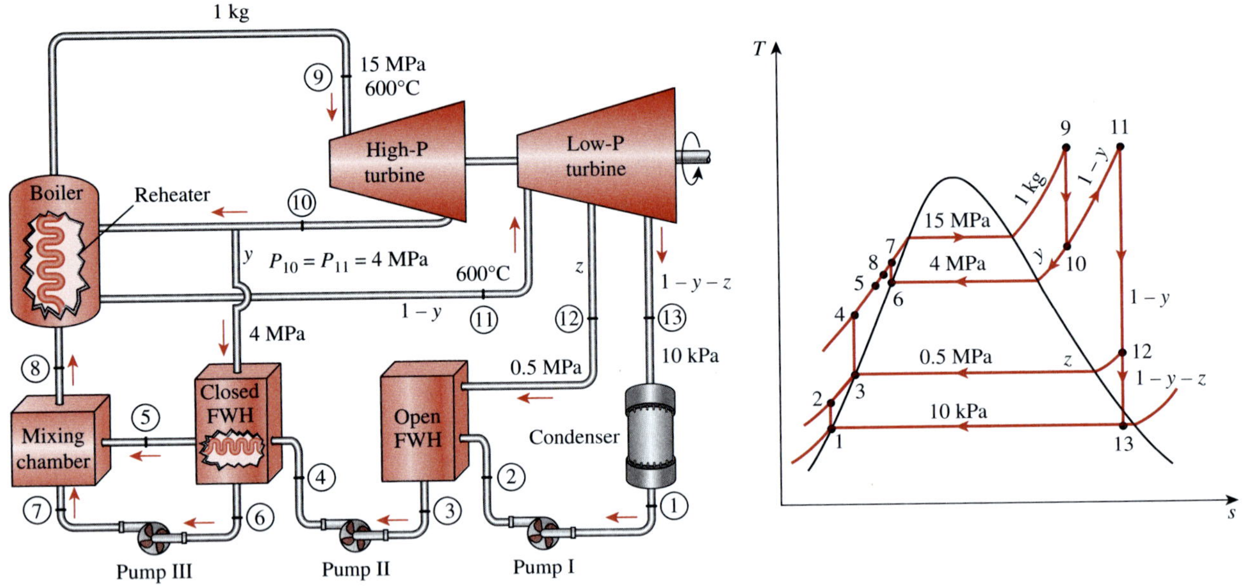

FIGURE 10–19
Schematic and *T-s* diagram for Example 10–6.

Open feedwater heater:

$$\dot{E}_{in} = \dot{E}_{out}$$

$$zh_{12} + (1 - y - z)h_2 = (1 - y)h_3$$

$$z = \frac{(1 - y)(h_3 - h_2)}{h_{12} - h_2} = \frac{(1 - 0.1766)(640.09 - 192.30)}{3014.8 - 192.30} = \mathbf{0.1306}$$

The enthalpy at state 8 is determined by applying the mass and energy equations to the mixing chamber, which is assumed to be insulated:

$$\dot{E}_{in} = \dot{E}_{out}$$

$$(1)h_8 = (1 - y)h_5 + yh_7$$

$$h_8 = (1 - 0.1766)(1087.4)\text{ kJ/kg} + 0.1766(1101.2)\text{ kJ/kg}$$

$$= 1089.8\text{ kJ/kg}$$

Thus,

$$q_{in} = (h_9 - h_8) + (1 - y)(h_{11} - h_{10})$$

$$= (3583.1 - 1089.8)\text{ kJ/kg} + (1 - 0.1766)(3674.9 - 3155.0)\text{ kJ/kg}$$

$$= 2921.4\text{ kJ/kg}$$

$$q_{out} = (1 - y - z)(h_{13} - h_1)$$

$$= (1 - 0.1766 - 0.1306)(2335.7 - 191.81)\text{ kJ/kg}$$

$$= 1485.3\text{ kJ/kg}$$

and

$$\eta_{th} = 1 - \frac{q_{out}}{q_{in}} = 1 - \frac{1485.3 \text{ kJ/kg}}{2921.4 \text{ kJ/kg}} = \mathbf{0.492 \text{ or } 49.2\%}$$

Discussion This problem was worked out in Example 10–4 for the same pressure and temperature limits with reheat but without the regeneration process. A comparison of the two results reveals that the thermal efficiency of the cycle has increased from 45.0 to 49.2 percent as a result of regeneration.

The thermal efficiency of this cycle could also be determined from

$$\eta_{th} = \frac{w_{net}}{q_{in}} = \frac{w_{turb,out} - w_{pump,in}}{q_{in}}$$

where

$$w_{turb,out} = (h_9 - h_{10}) + (1 - y)(h_{11} - h_{12}) + (1 - y - z)(h_{12} - h_{13})$$

$$w_{pump,in} = (1 - y - z)w_{pump\ I,in} + (1 - y)w_{pump\ II,in} + (y)w_{pump\ III,in}$$

Also, if we assume that the feedwater leaves the closed FWH as a saturated liquid at 15 MPa (and thus at T_5 = 342°C and h_5 = 1610.3 kJ/kg), it can be shown that the thermal efficiency would be 50.6.

10–7 ▪ SECOND-LAW ANALYSIS OF VAPOR POWER CYCLES

The ideal Carnot cycle is a *totally reversible cycle,* and thus it does not involve any irreversibilities. The ideal Rankine cycles (simple, reheat, or regenerative), however, are only *internally reversible,* and they may involve irreversibilities external to the system, such as heat transfer through a finite temperature difference. A second-law analysis of these cycles reveals where the largest irreversibilities occur and what their magnitudes are.

Relations for exergy and exergy destruction for steady-flow systems are developed in Chap. 8. The exergy destruction for a steady-flow system can be expressed, in the rate form, as

$$\dot{X}_{dest} = T_0\dot{S}_{gen} = T_0(\dot{S}_{out} - \dot{S}_{in}) = T_0\left(\sum_{out} \dot{m}s + \frac{\dot{Q}_{out}}{T_{b,out}} - \sum_{in} \dot{m}s - \frac{\dot{Q}_{in}}{T_{b,in}}\right) \quad \text{(kW)} \tag{10–18}$$

or on a unit mass basis for a one-inlet, one-exit, steady-flow device as

$$x_{dest} = T_0 s_{gen} = T_0\left(s_e - s_i + \frac{q_{out}}{T_{b,out}} - \frac{q_{in}}{T_{b,in}}\right) \quad \text{(kJ/kg)} \tag{10–19}$$

where $T_{b,in}$ and $T_{b,out}$ are the temperatures of the system boundary where heat is transferred into and out of the system, respectively.

The exergy destruction associated with a *cycle* depends on the magnitude of the heat transfer with the high- and low-temperature reservoirs involved, and their temperatures. It can be expressed on a unit mass basis as

$$x_{dest} = T_0\left(\sum \frac{q_{out}}{T_{b,out}} - \sum \frac{q_{in}}{T_{b,in}}\right) \quad \text{(kJ/kg)} \tag{10–20}$$

For a cycle that involves heat transfer only with a source at T_H and a sink at T_L, the exergy destruction become

$$x_{\text{dest}} = T_0\left(\frac{q_{\text{out}}}{T_L} - \frac{q_{\text{in}}}{T_H}\right) \qquad \text{(kJ/kg)} \tag{10–21}$$

The exergy of a fluid stream ψ at any state can be determined from

$$\psi = (h - h_0) - T_0(s - s_0) + \frac{V^2}{2} + gz \qquad \text{(kJ/kg)} \tag{10–22}$$

where the subscript "0" denotes the state of the surroundings.

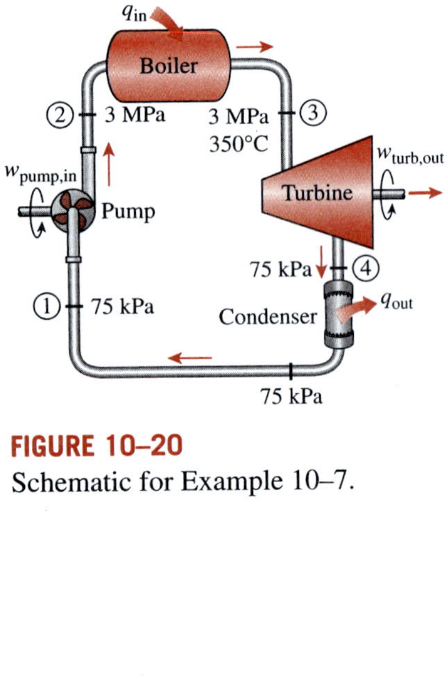

FIGURE 10–20
Schematic for Example 10–7.

EXAMPLE 10–7 Second-Law Analysis of an Ideal Rankine Cycle

Consider a steam power plant operating on the simple ideal Rankine cycle (Fig. 10–20). Steam enters the turbine at 3 MPa and 350°C and is condensed in the condenser at a pressure of 75 kPa. Heat is supplied to the steam in a furnace maintained at 800 K, and waste heat is rejected to the surroundings at 300 K. Determine (*a*) the exergy destruction associated with each of the four processes and the whole cycle and (*b*) the second-law efficiency of this cycle.

SOLUTION A steam power plant operating on the simple ideal Rankine cycle is considered. For specified source and sink temperatures, the exergy destruction associated with this cycle and the second-law efficiency are to be determined.

Assumptions **1** Steady operating conditions exist. **2** Kinetic and potential energy changes are negligible.

Analysis We take the power plant bordering the furnace at temperature T_H and the environment at temperature T_0 as the control volume. This cycle was analyzed in Example 10-1, and various quantities were determined to be q_{in} = 2729 kJ/kg, $w_{\text{pump,in}}$ = 3.0 kJ/kg, $w_{\text{turb,out}}$ = 713 kJ/kg, q_{out} = 2019 kJ/kg, and η_{th} = 26.0 percent.

(*a*) Processes 1-2 and 3-4 are isentropic ($s_1 = s_2$, $s_3 = s_4$) and therefore do not involve any internal or external irreversibilities, that is,

$$x_{\text{dest},12} = \mathbf{0} \quad \text{and} \quad x_{\text{dest},34} = \mathbf{0}$$

Processes 2-3 and 4-1 are constant-pressure heat-addition and heat-rejection processes, respectively, and they are internally reversible. But the heat transfer between the working fluid and the source or the sink takes place through a finite temperature difference, rendering both processes irreversible. The irreversibility associated with each process is determined from Eq. 10–19. The entropy of the steam at each state is determined from the steam tables:

$$s_2 = s_1 = s_{f\,@\,75\text{ kPa}} = 1.2132 \text{ kJ/kg·K}$$

$$s_4 = s_3 = 6.7450 \text{ kJ/kg·K} \qquad \text{(at 3 MPa, 350°C)}$$

Thus,

$$x_{\text{dest},23} = T_0\left(s_3 - s_2 - \frac{q_{\text{in},23}}{T_{\text{source}}}\right)$$

$$= (300 \text{ K})\left[(6.7450 - 1.2132) \text{ kJ/kg·K} - \frac{2729 \text{ kJ/kg}}{800 \text{ K}}\right]$$

$$= \mathbf{636 \text{ kJ/kg}}$$

$$x_{\text{dest},41} = T_0\left(s_1 - s_4 + \frac{q_{\text{out},41}}{T_{\text{sink}}}\right)$$

$$= (300\text{ K})\left[(1.2132 - 6.7450)\text{ kJ/kg·K} + \frac{2019\text{ kJ/kg}}{300\text{ K}}\right]$$

$$= \mathbf{360\ kJ/kg}$$

Therefore, the irreversibility of the cycle is

$$x_{\text{dest,cycle}} = x_{\text{dest},12} + x_{\text{dest},23} + x_{\text{dest},34} + x_{\text{dest},41}$$

$$= 0 + 636\text{ kJ/kg} + 0 + 360\text{ kJ/kg}$$

$$= \mathbf{996\ kJ/kg}$$

The total exergy destroyed during the cycle could also be determined from Eq. 10–21. Notice that the largest exergy destruction in the cycle occurs during the heat-addition process. Therefore, any attempt to reduce the exergy destruction should start with this process. Raising the turbine inlet temperature of the steam, for example, would reduce the temperature difference and thus the exergy destruction.

(*b*) The second-law efficiency is defined as

$$\eta_{\text{II}} = \frac{\text{Exergy recovered}}{\text{Exergy expended}} = \frac{x_{\text{recovered}}}{x_{\text{expended}}} = 1 - \frac{x_{\text{destroyed}}}{x_{\text{expended}}}$$

Here the expended exergy is the exergy content of the heat supplied to steam in boiler (which is its work potential) and the pump input, and the exergy recovered is the work output of the turbine:

$$x_{\text{heat,in}} = \left(1 - \frac{T_0}{T_H}\right)q_{\text{in}} = \left(1 - \frac{300\text{ K}}{800\text{ K}}\right)(2729\text{ kJ/kg}) = 1706\text{ kJ/kg}$$

$$x_{\text{expended}} = x_{\text{heat,in}} + x_{\text{pump,in}} = 1706 + 3.0 = 1709\text{ kJ/kg}$$

$$x_{\text{recovered}} = w_{\text{turbine,out}} = 713\text{ kJ/kg}$$

Substituting, the second-law efficiency of this power plant is determined to be

$$\eta_{\text{II}} = \frac{x_{\text{recovered}}}{x_{\text{expended}}} = \frac{713\text{ kJ/kg}}{1709\text{ kJ/kg}} = 0.417 \quad \text{or} \quad \mathbf{41.7\%}$$

Discussion The second-law efficiency can also be determined using the exergy destruction data,

$$\eta_{\text{II}} = 1 - \frac{x_{\text{destroyed}}}{x_{\text{expended}}} = 1 - \frac{996\text{ kJ/kg}}{1709\text{ kJ/kg}} = 0.417 \text{ or } 41.7\%$$

Also, the system considered contains both the furnace and the condenser, and thus the exergy destruction associated with heat transfer involving both the furnace and the condenser are accounted for.

10–8 ▪ COGENERATION

In all the cycles discussed so far, the sole purpose was to convert a portion of the heat transferred to the working fluid to work, which is the most valuable form of energy. The remaining portion of the heat is rejected to rivers,

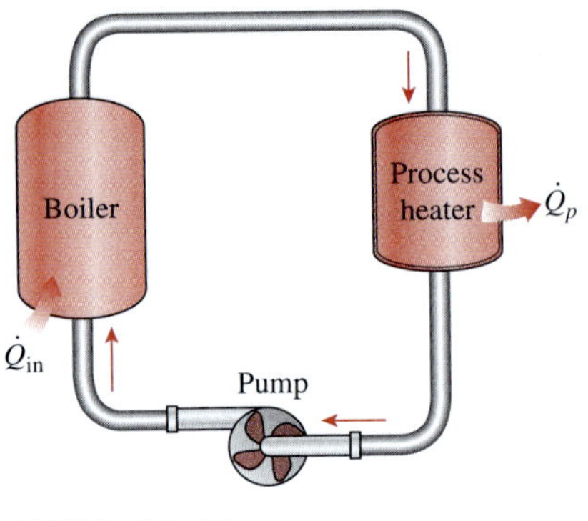

FIGURE 10–21
A simple process-heating plant.

lakes, oceans, or the atmosphere as waste heat, because its quality (or grade) is too low to be of any practical use. Wasting a large amount of heat is a price we have to pay to produce work, because electrical or mechanical work is the only form of energy on which many engineering devices (such as a fan) can operate.

Many systems or devices, however, require energy input in the form of heat, called *process heat.* Some industries that rely heavily on process heat are chemical, pulp and paper, oil production and refining, steel making, food processing, and textile industries. Process heat in these industries is usually supplied by steam at 5 to 7 atm and 150 to 200°C (300 to 400°F). Energy is usually transferred to the steam by burning coal, oil, natural gas, or another fuel in a furnace.

Now let us examine the operation of a process-heating plant closely. Disregarding any heat losses in the piping, all the heat transferred to the steam in the boiler is used in the process-heating units, as shown in Fig. 10–21. Therefore, process heating seems like a perfect operation with practically no waste of energy. From the second-law point of view, however, things do not look so perfect. The temperature in furnaces is typically very high (around 1400°C), and thus the energy in the furnace is of very high quality. This high-quality energy is transferred to water to produce steam at about 200°C or below (a highly irreversible process). Associated with this irreversibility is, of course, a loss in exergy or work potential. It is simply not wise to use high-quality energy to accomplish a task that could be accomplished with low-quality energy.

Industries that use large amounts of process heat also consume a large amount of electric power. Therefore, it makes economical as well as engineering sense to use the already-existing work potential to produce power instead of letting it go to waste. The result is a plant that produces electricity while meeting the process-heat requirements of certain industrial processes. Such a plant is called a *cogeneration plant.* In general, **cogeneration** is *the production of more than one useful form of energy (such as process heat and electric power) from the same energy source.*

Either a steam-turbine (Rankine) cycle or a gas-turbine (Brayton) cycle or even a combined cycle (discussed later) can be used as the power cycle in a cogeneration plant. The schematic of an ideal steam-turbine cogeneration plant is shown in Fig. 10–22. Let us say this plant is to supply process heat $\dot{Q}_p$ at 500 kPa at a rate of 100 kW. To meet this demand, steam is expanded in the turbine to a pressure of 500 kPa, producing power at a rate of, say, 20 kW. The flow rate of the steam can be adjusted such that steam leaves the process-heating section as a saturated liquid at 500 kPa. Steam is then pumped to the boiler pressure and is heated in the boiler to state 3. The pump work is usually very small and can be neglected. Disregarding any heat losses, the rate of heat input in the boiler is determined from an energy balance to be 120 kW.

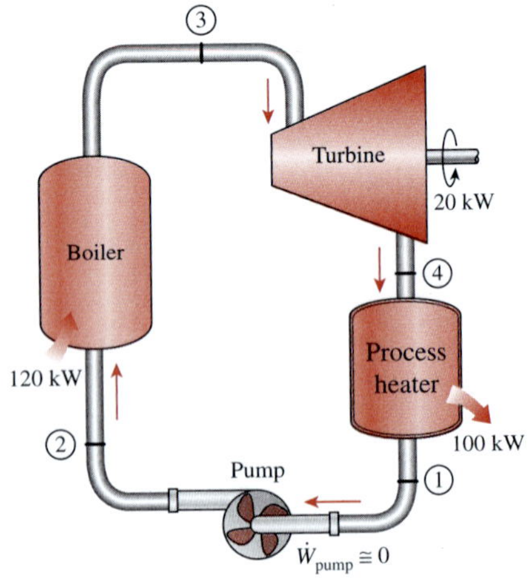

FIGURE 10–22
An ideal cogeneration plant.

Probably the most striking feature of the ideal steam-turbine cogeneration plant shown in Fig. 10–22 is the absence of a condenser. Thus no heat is rejected from this plant as waste heat. In other words, all the energy transferred to the steam in the boiler is utilized as either process heat or electric power. Thus it is appropriate to define a **utilization factor** ε_u for a cogeneration plant as

$$\epsilon_u = \frac{\text{Net work output} + \text{Process heat delivered}}{\text{Total heat input}} = \frac{\dot{W}_{net} + \dot{Q}_p}{\dot{Q}_{in}} \quad \textbf{(10–23)}$$

or

$$\epsilon_u = 1 - \frac{\dot{Q}_{out}}{\dot{Q}_{in}} \quad \textbf{(10–24)}$$

where $\dot{Q}_{out}$ represents the heat rejected in the condenser. Strictly speaking, $\dot{Q}_{out}$ also includes all the undesirable heat losses from the piping and other components, but they are usually small and thus neglected. It also includes combustion inefficiencies such as incomplete combustion and stack losses when the utilization factor is defined on the basis of the heating value of the fuel. The utilization factor of the ideal steam-turbine cogeneration plant is obviously 100 percent. Actual cogeneration plants have utilization factors as high as 80 percent. Some recent cogeneration plants have even higher utilization factors.

Notice that without the turbine, we would need to supply heat to the steam in the boiler at a rate of only 100 kW instead of at 120 kW. The additional 20 kW of heat supplied is converted to work. Therefore, a cogeneration power plant is equivalent to a process-heating plant combined with a power plant that has a thermal efficiency of 100 percent.

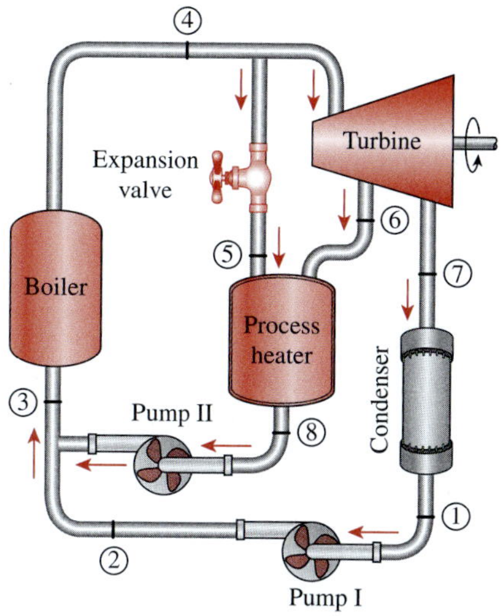

FIGURE 10–23
A cogeneration plant with adjustable loads.

The ideal steam-turbine cogeneration plant described above is not practical because it cannot adjust to the variations in power and process-heat loads. The schematic of a more practical (but more complex) cogeneration plant is shown in Fig. 10–23. Under normal operation, some steam is extracted from the turbine at some predetermined intermediate pressure P_6. The rest of the steam expands to the condenser pressure P_7 and is then cooled at constant pressure. The heat rejected from the condenser represents the waste heat for the cycle.

At times of high demand for process heat, all the steam is routed to the process-heating units and none to the condenser ($\dot{m}_7 = 0$). The waste heat is zero in this mode. If this is not sufficient, some steam leaving the boiler is throttled by an expansion or pressure-reducing valve (PRV) to the extraction pressure P_6 and is directed to the process-heating unit. Maximum process heating is realized when all the steam leaving the boiler passes through the PRV ($\dot{m}_5 = \dot{m}_4$). No power is produced in this mode. When there is no demand for process heat, all the steam passes through the turbine and the condenser ($\dot{m}_5 = \dot{m}_6 = 0$), and the cogeneration plant operates as an ordinary steam power plant. The rates of heat input, heat rejected, and process heat supply as well as the power produced for this cogeneration plant can be expressed as follows:

$$\dot{Q}_{in} = \dot{m}_3(h_4 - h_3) \quad \textbf{(10–25)}$$

$$\dot{Q}_{out} = \dot{m}_7(h_7 - h_1) \quad \textbf{(10–26)}$$

$$\dot{Q}_p = \dot{m}_5 h_5 + \dot{m}_6 h_6 - \dot{m}_8 h_8 \quad \textbf{(10–27)}$$

$$\dot{W}_{turb} = (\dot{m}_4 - \dot{m}_5)(h_4 - h_6) + \dot{m}_7(h_6 - h_7) \quad \textbf{(10–28)}$$

Under optimum conditions, a cogeneration plant simulates the ideal cogeneration plant discussed earlier. That is, all the steam expands in the turbine to the extraction pressure and continues to the process-heating unit. No steam passes through the PRV or the condenser; thus, no waste heat is rejected ($\dot{m}_4 = \dot{m}_6$ and $\dot{m}_5 = \dot{m}_7 = 0$). This condition may be difficult to achieve in practice because of the constant variations in the process-heat and power loads. But the plant should be designed so that the optimum operating conditions are approximated most of the time.

The use of cogeneration dates to the beginning of this century when power plants were integrated to a community to provide district heating, that is, space, hot water, and process heating for residential and commercial buildings. The district heating systems lost their popularity in the 1940s owing to low fuel prices. However, the rapid rise in fuel prices in the 1970s brought about renewed interest in district heating.

Cogeneration plants have proved to be economically very attractive. Consequently, more and more such plants have been installed in recent years, and more are being installed.

EXAMPLE 10–8 An Ideal Cogeneration Plant

Consider the cogeneration plant shown in Fig. 10–24. Steam enters the turbine at 7 MPa and 500°C. Some steam is extracted from the turbine at 500 kPa for process heating. The remaining steam continues to expand to 5 kPa. Steam is then condensed at constant pressure and pumped to the boiler pressure of 7 MPa. At times of high demand for process heat, some steam leaving the boiler is throttled to 500 kPa and is routed to the process heater. The extraction fractions are adjusted so that steam leaves the process heater as a saturated liquid at 500 kPa. It is subsequently pumped to 7 MPa. The mass flow rate of steam through the boiler is 15 kg/s. Disregarding any pressure drops and heat losses in the piping and assuming the turbine and the pump to be isentropic, determine (*a*) the maximum rate at which process heat can be supplied, (*b*) the power produced and the utilization factor when no process heat is supplied, and (*c*) the rate of process heat supply when 10 percent of the steam is extracted before it enters the turbine and 70 percent of the steam is extracted from the turbine at 500 kPa for process heating.

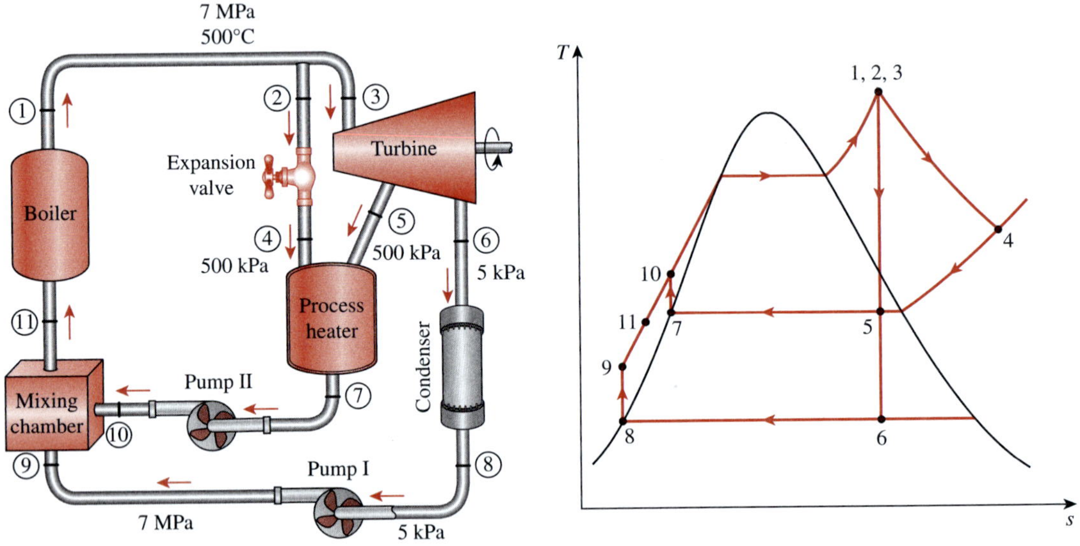

FIGURE 10–24
Schematic and *T-s* diagram for Example 10–8.

SOLUTION A cogeneration plant is considered. The maximum rate of process heat supply, the power produced and the utilization factor when no process heat is supplied, and the rate of process heat supply when steam is extracted from the steam line and turbine at specified ratios are to be determined.

Assumptions **1** Steady operating conditions exist. **2** Pressure drops and heat losses in piping are negligible. **3** Kinetic and potential energy changes are negligible.

Analysis The schematic of the cogeneration plant and the *T-s* diagram of the cycle are shown in Fig. 10–24. The power plant operates on an ideal cycle and thus the pumps and the turbines are isentropic; there are no pressure drops in the boiler, process heater, and condenser; and steam leaves the condenser and the process heater as saturated liquid.

The work inputs to the pumps and the enthalpies at various states are as follows:

$$w_{\text{pump I,in}} = v_8(P_9 - P_8) = (0.001005 \text{ m}^3/\text{kg})[(7000 - 5)\text{kPa}]\left(\frac{1 \text{ kJ}}{1 \text{ kPa}\cdot\text{m}^3}\right)$$
$$= 7.03 \text{ kJ/kg}$$

$$w_{\text{pump II,in}} = v_7(P_{10} - P_7) = (0.001093 \text{ m}^3/\text{kg})[(7000 - 500) \text{ kPa}]\left(\frac{1 \text{ kJ}}{1 \text{ kPa}\cdot\text{m}^3}\right)$$
$$= 7.10 \text{ kJ/kg}$$

$$h_1 = h_2 = h_3 = h_4 = 3411.4 \text{ kJ/kg}$$
$$h_5 = 2739.3 \text{ kJ/kg}$$
$$h_6 = 2073.0 \text{ kJ/kg}$$
$$h_7 = h_{f\,@\,500\text{ kPa}} = 640.09 \text{ kJ/kg}$$
$$h_8 = h_{f\,@\,5\text{ kPa}} = 137.75 \text{ kJ/kg}$$
$$h_9 = h_8 + w_{\text{pump I,in}} = (137.75 + 7.03) \text{ kJ/kg} = 144.78 \text{ kJ/kg}$$
$$h_{10} = h_7 + w_{\text{pump II,in}} = (640.09 + 7.10) \text{ kJ/kg} = 647.19 \text{ kJ/kg}$$

(*a*) The maximum rate of process heat is achieved when all the steam leaving the boiler is throttled and sent to the process heater and none is sent to the turbine (that is, $\dot{m}_4 = \dot{m}_7 = \dot{m}_1 = 15$ kg/s and $\dot{m}_3 = \dot{m}_5 = \dot{m}_6 = 0$). Thus,

$$\dot{Q}_{p,\text{max}} = \dot{m}_1(h_4 - h_7) = (15 \text{ kg/s})[(3411.4 - 640.09) \text{ kJ/kg}] = \mathbf{41{,}570 \text{ kW}}$$

The utilization factor is 100 percent in this case since no heat is rejected in the condenser, heat losses from the piping and other components are assumed to be negligible, and combustion losses are not considered.

(*b*) When no process heat is supplied, all the steam leaving the boiler passes through the turbine and expands to the condenser pressure of 5 kPa (that is, $\dot{m}_3 = \dot{m}_6 = \dot{m}_1 = 15$ kg/s and $\dot{m}_2 = \dot{m}_5 = 0$). Maximum power is produced in this mode, which is determined to be

$$\dot{W}_{\text{turb,out}} = \dot{m}(h_3 - h_6) = (15 \text{ kg/s})[(3411.4 - 2073.0) \text{ kJ/kg}] = 20{,}076 \text{ kW}$$
$$\dot{W}_{\text{pump,in}} = (15 \text{ kg/s})(7.03 \text{ kJ/kg}) = 105 \text{ kW}$$
$$\dot{W}_{\text{net,out}} = \dot{W}_{\text{turb,out}} - \dot{W}_{\text{pump,in}} = (20{,}076 - 105) \text{ kW} = 19{,}971 \text{ kW} \cong \mathbf{20.0 \text{ MW}}$$
$$\dot{Q}_{\text{in}} = \dot{m}_1(h_1 - h_{11}) = (15 \text{ kg/s})[(3411.4 - 144.78) \text{ kJ/kg}] = 48{,}999 \text{ kW}$$

Thus,

$$\epsilon_u = \frac{\dot{W}_{\text{net}} + \dot{Q}_p}{\dot{Q}_{\text{in}}} = \frac{(19{,}971 + 0)\text{ kW}}{48{,}999\text{ kW}} = \mathbf{0.408 \text{ or } 40.8\%}$$

That is, 40.8 percent of the energy is utilized for a useful purpose. Notice that the utilization factor is equivalent to the thermal efficiency in this case.

(*c*) Neglecting any kinetic and potential energy changes, an energy balance on the process heater yields

$$\dot{E}_{\text{in}} = \dot{E}_{\text{out}}$$

$$\dot{m}_4 h_4 + \dot{m}_5 h_5 = \dot{Q}_{p,\text{out}} + \dot{m}_7 h_7$$

or

$$\dot{Q}_{p,\text{out}} = \dot{m}_4 h_4 + \dot{m}_5 h_5 - \dot{m}_7 h_7$$

where

$$\dot{m}_4 = (0.1)(15\text{ kg/s}) = 1.5\text{ kg/s}$$

$$\dot{m}_5 = (0.7)(15\text{ kg/s}) = 10.5\text{ kg/s}$$

$$\dot{m}_7 = \dot{m}_4 + \dot{m}_5 = 1.5 + 10.5 = 12\text{ kg/s}$$

Thus

$$\begin{aligned}\dot{Q}_{p,\text{out}} &= (1.5\text{ kg/s})(3411.4\text{ kJ/kg}) + (10.5\text{ kg/s})(2739.3\text{ kJ/kg}) \\ &\quad - (12\text{ kg/s})(640.09\text{ kJ/kg}) \\ &= \mathbf{26.2\text{ MW}}\end{aligned}$$

Discussion Note that 26.2 MW of the heat transferred will be utilized in the process heater. We could also show that 11.0 MW of power is produced in this case, and the rate of heat input in the boiler is 43.0 MW. Thus the utilization factor is 86.5 percent.

10–9 ▪ COMBINED GAS–VAPOR POWER CYCLES

The continued quest for higher thermal efficiencies has resulted in rather innovative modifications to conventional power plants. The *binary vapor cycle* discussed later is one such modification. A more popular modification involves a gas power cycle topping a vapor power cycle, which is called the **combined gas–vapor cycle**, or just the **combined cycle**. The combined cycle of greatest interest is the gas-turbine (Brayton) cycle topping a steam-turbine (Rankine) cycle, which has a higher thermal efficiency than either of the cycles executed individually.

Gas-turbine cycles typically operate at considerably higher temperatures than steam cycles. The maximum fluid temperature at the turbine inlet is about 620°C (1150°F) for modern steam power plants, but over 1425°C (2600°F) for gas-turbine power plants. It is over 1500°C at the burner exit of turbojet engines. The use of higher temperatures in gas turbines is made possible by recent developments in cooling the turbine blades and coating the blades with high-temperature-resistant materials such as ceramics.

Because of the higher average temperature at which heat is supplied, gas-turbine cycles have a greater potential for higher thermal efficiencies. However, the gas-turbine cycles have one inherent disadvantage: The gas leaves the gas turbine at very high temperatures (usually above 500°C), which erases any potential gains in the thermal efficiency. The situation can be improved somewhat by using regeneration, but the improvement is limited.

It makes engineering sense to take advantage of the very desirable characteristics of the gas-turbine cycle at high temperatures *and* to use the high-temperature exhaust gases as the energy source for the bottoming cycle such as a steam power cycle. The result is a combined gas–steam cycle, as shown in Fig. 10–25. In this cycle, energy is recovered from the exhaust gases by transferring it to the steam in a heat exchanger that serves as the boiler. In general, more than one gas turbine is needed to supply sufficient heat to the steam. Also, the steam cycle may involve regeneration as well as reheating. Energy for the reheating process can be supplied by burning some additional fuel in the oxygen-rich exhaust gases.

Recent developments in gas-turbine technology have made the combined gas–steam cycle economically very attractive. The combined cycle increases the efficiency without increasing the initial cost greatly. Consequently, many new power plants operate on combined cycles, and many more existing steam- or

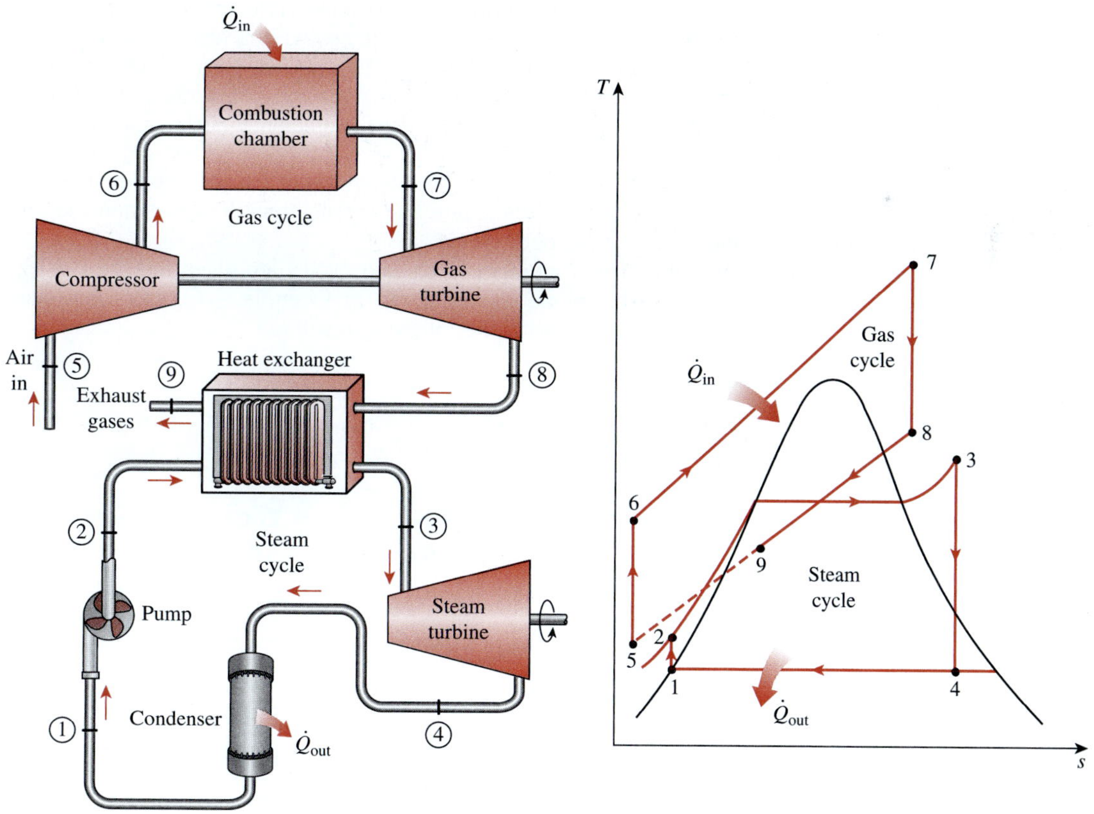

FIGURE 10–25
Combined gas–steam power plant.

gas-turbine plants are being converted to combined-cycle power plants. Thermal efficiencies well over 40 percent are reported as a result of conversion.

A 1090-MW Tohoku combined plant that was put in commercial operation in 1985 in Niigata, Japan, is reported to operate at a thermal efficiency of 44 percent. This plant has two 191-MW steam turbines and six 118-MW gas turbines. Hot combustion gases enter the gas turbines at 1154°C, and steam enters the steam turbines at 500°C. Steam is cooled in the condenser by cooling water at an average temperature of 15°C. The compressors have a pressure ratio of 14, and the mass flow rate of air through the compressors is 443 kg/s.

A 1350-MW combined-cycle power plant built in Ambarli, Turkey, in 1988 by Siemens of Germany is the first commercially operating thermal plant in the world to attain an efficiency level as high as 52.5 percent at design operating conditions. This plant has six 150-MW gas turbines and three 173-MW steam turbines. Some recent combined-cycle power plants have achieved efficiencies above 60 percent.

EXAMPLE 10–8 A Combined Gas–Steam Power Cycle

Consider the combined gas–steam power cycle shown in Fig. 10–26. The topping cycle is a gas-turbine cycle that has a pressure ratio of 8. Air enters the compressor at 300 K and the turbine at 1300 K. The isentropic efficiency of the compressor is 80 percent, and that of the gas turbine is 85 percent. The bottoming cycle is a simple ideal Rankine cycle operating between the pressure limits of 7 MPa and 5 kPa. Steam is heated in a heat exchanger by the exhaust gases to a temperature of 500°C. The exhaust gases leave the heat exchanger at 450 K. Determine (*a*) the ratio of the mass flow rates of the steam and the combustion gases and (*b*) the thermal efficiency of the combined cycle.

SOLUTION A combined gas–steam cycle is considered. The ratio of the mass flow rates of the steam and the combustion gases and the thermal efficiency are to be determined.

Analysis The *T-s* diagrams of both cycles are given in Fig. 10–26. The gas-turbine cycle alone was analyzed in Example 9–6, and the steam cycle in Example 10–8*b*, with the following results:

Gas cycle: $h'_4 = 880.36 \text{ kJ/kg} \qquad (T'_4 = 853 \text{ K})$

$q_{in} = 790.58 \text{ kJ/kg} \qquad w_{net} = 210.41 \text{ kJ/kg} \qquad \eta_{th} = 26.6\%$

$h'_5 = h_{@\,450 \text{ K}} = 451.80 \text{ kJ/kg}$

Steam cycle: $h_2 = 144.78 \text{ kJ/kg} \qquad (T_2 = 33°\text{C})$

$h_3 = 3411.4 \text{ kJ/kg} \qquad (T_3 = 500°\text{C})$

$w_{net} = 1331.4 \text{ kJ/kg} \qquad \eta_{th} = 40.8\%$

(*a*) The ratio of mass flow rates is determined from an energy balance on the heat exchanger:

$$\dot{E}_{in} = \dot{E}_{out}$$

$$\dot{m}_g h'_5 + \dot{m}_s h_3 = \dot{m}_g h'_4 + \dot{m}_s h_2$$

$$\dot{m}_s(h_3 - h_2) = \dot{m}_g(h'_4 - h'_5)$$

$$\dot{m}_s(3411.4 - 144.78) = \dot{m}_g(880.36 - 451.80)$$

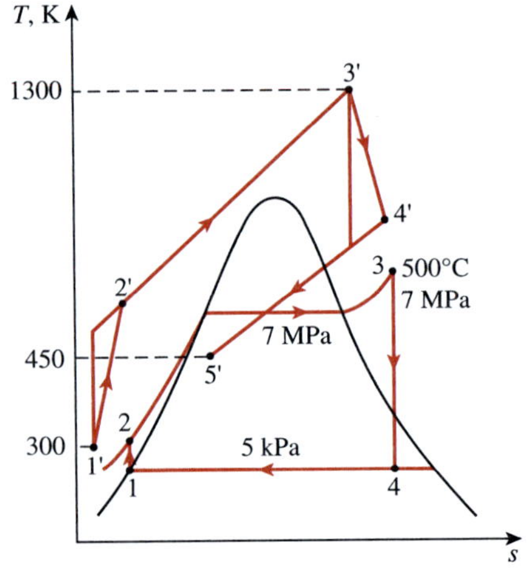

FIGURE 10–26
T-s diagram of the gas–steam combined cycle described in Example 10–9.

Thus,

$$\frac{\dot{m}_s}{\dot{m}_g} = y = \mathbf{0.131}$$

That is, 1 kg of exhaust gases can heat only 0.131 kg of steam from 33 to 500°C as they are cooled from 853 to 450 K. Then the total net work output per kilogram of combustion gases becomes

$$\begin{aligned} w_{\text{net}} &= w_{\text{net,gas}} + y w_{\text{net,steam}} \\ &= (210.41 \text{ kJ/kg gas}) + (0.131 \text{ kg steam/kg gas})(1331.4 \text{ kJ/kg steam}) \\ &= 384.8 \text{ kJ/kg gas} \end{aligned}$$

Therefore, for each kg of combustion gases produced, the combined plant will deliver 384.8 kJ of work. The net power output of the plant is determined by multiplying this value by the mass flow rate of the working fluid in the gas-turbine cycle.

(*b*) The thermal efficiency of the combined cycle is determined from

$$\eta_{\text{th}} = \frac{w_{\text{net}}}{q_{\text{in}}} = \frac{384.8 \text{ kJ/kg gas}}{790.6 \text{ kJ/kg gas}} = \mathbf{0.487 \text{ or } 48.7\%}$$

Discussion Note that this combined cycle converts to useful work 48.7 percent of the energy supplied to the gas in the combustion chamber. This value is considerably higher than the thermal efficiency of the gas-turbine cycle (26.6 percent) or the steam-turbine cycle (40.8 percent) operating alone.

TOPIC OF SPECIAL INTEREST* Binary Vapor Cycles

With the exception of a few specialized applications, the working fluid predominantly used in vapor power cycles is water. Water is the *best* working fluid presently available, but it is far from being the *ideal* one. The binary cycle is an attempt to overcome some of the shortcomings of water and to approach the *ideal* working fluid by using two fluids. Before we discuss the binary cycle, let us list the characteristics of a working fluid most suitable for vapor power cycles:

1. A high critical temperature and a safe maximum pressure. A critical temperature above the metallurgically allowed maximum temperature (about 620°C) makes it possible to transfer a considerable portion of the heat isothermally at the maximum temperature as the fluid changes phase. This makes the cycle approach the Carnot cycle. Very high pressures at the maximum temperature are undesirable because they create material-strength problems.

*This section can be skipped without a loss in continuity.

2. Low triple-point temperature. A triple-point temperature below the temperature of the cooling medium prevents any solidification problems.

3. A condenser pressure that is not too low. Condensers usually operate below atmospheric pressure. Pressures well below the atmospheric pressure create air-leakage problems. Therefore, a substance whose saturation pressure at the ambient temperature is too low is not a good candidate.

4. A high enthalpy of vaporization (h_{fg}) so that heat transfer to the working fluid is nearly isothermal and large mass flow rates are not needed.

5. A saturation dome that resembles an inverted U. This eliminates the formation of excessive moisture in the turbine and the need for reheating.

6. Good heat transfer characteristics (high thermal conductivity).

7. Other properties such as being inert, inexpensive, readily available, and nontoxic.

Not surprisingly, no fluid possesses all these characteristics. Water comes the closest, although it does not fare well with respect to characteristics 1, 3, and 5. We can cope with its subatmospheric condenser pressure by careful sealing, and with the inverted V-shaped saturation dome by reheating, but there is not much we can do about item 1. Water has a low critical temperature (374°C, well below the metallurgical limit) and very high saturation pressures at high temperatures (16.5 MPa at 350°C).

Well, we cannot change the way water behaves during the high-temperature part of the cycle, but we certainly can replace it with a more suitable fluid. The result is a power cycle that is actually a combination of two cycles, one in the high-temperature region and the other in the low-temperature region. Such a cycle is called a **binary vapor cycle**. In binary vapor cycles, the condenser of the high-temperature cycle (also called the *topping cycle*) serves as the boiler of the low-temperature cycle (also called the *bottoming cycle*). That is, the heat output of the high-temperature cycle is used as the heat input to the low-temperature one.

Some working fluids found suitable for the high-temperature cycle are mercury, sodium, potassium, and sodium–potassium mixtures. The schematic and *T-s* diagram for a mercury–water binary vapor cycle are shown in Fig. 10–27. The critical temperature of mercury is 898°C (well above the current metallurgical limit), and its critical pressure is only about 18 MPa. This makes mercury a very suitable working fluid for the topping cycle. Mercury is not suitable as the sole working fluid for the entire cycle, however, since at a condenser temperature of 32°C its saturation pressure is 0.07 Pa. A power plant cannot operate at this vacuum because of air-leakage problems. At an acceptable condenser pressure of 7 kPa, the saturation temperature of mercury is 237°C, which is too high as the minimum temperature in the cycle. Therefore, the use of mercury as a working fluid is limited to the high-temperature cycles. Other disadvantages of mercury

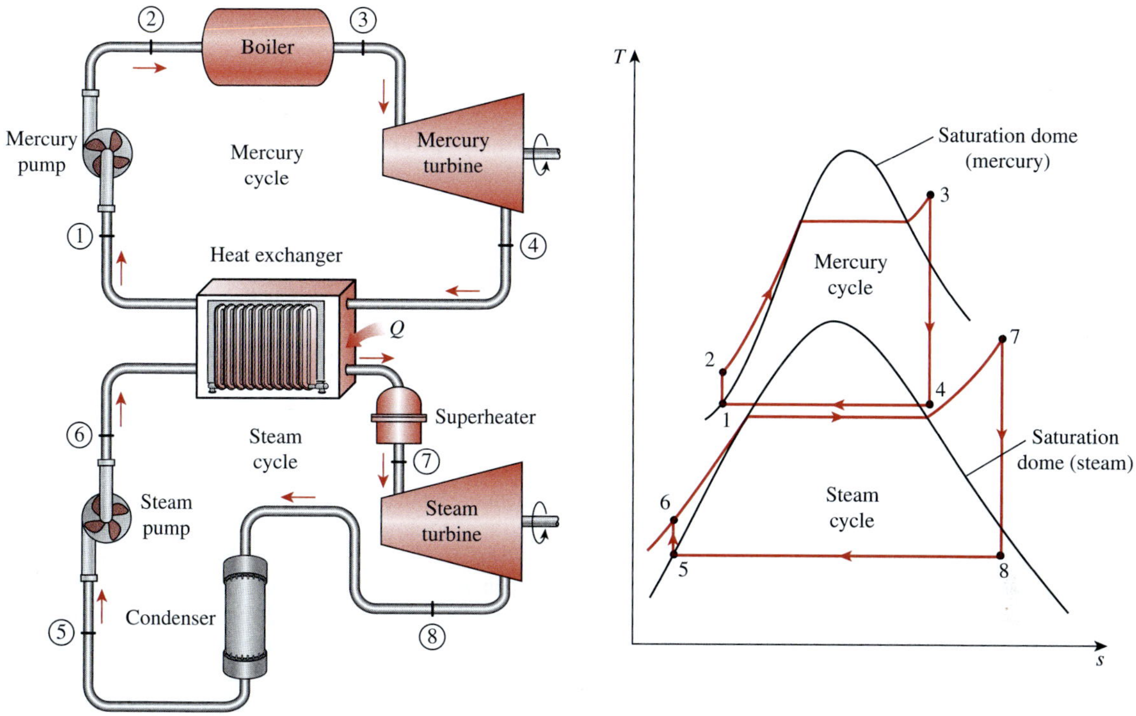

FIGURE 10–27
Mercury–water binary vapor cycle.

are its toxicity and high cost. The mass flow rate of mercury in binary vapor cycles is several times that of water because of its low enthalpy of vaporization.

It is evident from the *T-s* diagram in Fig. 10–27 that the binary vapor cycle approximates the Carnot cycle more closely than the steam cycle for the same temperature limits. Therefore, the thermal efficiency of a power plant can be increased by switching to binary cycles. The use of mercury–water binary cycles in the United States dates back to 1928. Several such plants have been built since then in the New England area, where fuel costs are typically higher. A small (40-MW) mercury–steam power plant that was in service in New Hampshire in 1950 had a higher thermal efficiency than most of the large modern power plants in use at that time.

Studies show that thermal efficiencies of 50 percent or higher are possible with binary vapor cycles. However, binary vapor cycles are not economically attractive because of their high initial cost and the competition offered by the combined gas–steam power plants.

SUMMARY

The *Carnot cycle* is not a suitable model for vapor power cycles because it cannot be approximated in practice. The model cycle for vapor power cycles is the *Rankine cycle,* which is composed of four internally reversible processes: constant-pressure heat addition in a boiler, isentropic expansion in a turbine, constant-pressure heat rejection in a condenser, and isentropic compression in a pump. Steam leaves the condenser as a saturated liquid at the condenser pressure.

The thermal efficiency of the Rankine cycle can be increased by increasing the average temperature at which heat is transferred to the working fluid and/or by decreasing the average temperature at which heat is rejected to the cooling medium. The average temperature during heat rejection can be decreased by lowering the turbine exit pressure. Consequently, the condenser pressure of most vapor power plants is well below the atmospheric pressure. The average temperature during heat addition can be increased by raising the boiler pressure or by superheating the fluid to high temperatures. There is a limit to the degree of superheating, however, since the fluid temperature is not allowed to exceed a metallurgically safe value.

Superheating has the added advantage of decreasing the moisture content of the steam at the turbine exit. Lowering the exhaust pressure or raising the boiler pressure, however, increases the moisture content. To take advantage of the improved efficiencies at higher boiler pressures and lower condenser pressures, steam is usually *reheated* after expanding partially in the high-pressure turbine. This is done by extracting the steam after partial expansion in the high-pressure turbine, sending it back to the boiler where it is reheated at constant pressure, and returning it to the low-pressure turbine for complete expansion to the condenser pressure. The average temperature during the reheat process, and thus the thermal efficiency of the cycle, can be increased by increasing the number of expansion and reheat stages. As the number of stages is increased, the expansion and reheat processes approach an isothermal process at maximum temperature. Reheating also decreases the moisture content at the turbine exit.

Another way of increasing the thermal efficiency of the Rankine cycle is *regeneration.* During a regeneration process, liquid water (feedwater) leaving the pump is heated by steam bled off the turbine at some intermediate pressure in devices called *feedwater heaters.* The two streams are mixed in open feedwater heaters, and the mixture leaves as a saturated liquid at the heater pressure. In closed feedwater heaters, heat is transferred from the steam to the feedwater without mixing.

The production of more than one useful form of energy (such as process heat and electric power) from the same energy source is called *cogeneration.* Cogeneration plants produce electric power while meeting the process heat requirements of certain industrial processes. This way, more of the energy transferred to the fluid in the boiler is utilized for a useful purpose. The fraction of energy that is used for either process heat or power generation is called the *utilization factor* of the cogeneration plant.

The overall thermal efficiency of a power plant can be increased by using a *combined cycle.* The most common combined cycle is the gas–steam combined cycle where a gas-turbine cycle operates at the high-temperature range and a steam-turbine cycle at the low-temperature range. Steam is heated by the high-temperature exhaust gases leaving the gas turbine. Combined cycles have a higher thermal efficiency than the steam- or gas-turbine cycles operating alone.

REFERENCES AND SUGGESTED READINGS

1. R. L. Bannister and G. J. Silvestri. "The Evolution of Central Station Steam Turbines." *Mechanical Engineering*, February 1989, pp. 70–78.
2. R. L. Bannister, G. J. Silvestri, A. Hizume, and T. Fujikawa. "High Temperature Supercritical Steam Turbines." *Mechanical Engineering,* February 1987, pp. 60–65.
3. M. M. El-Wakil. *Powerplant Technology.* New York: McGraw-Hill, 1984.
4. K. W. Li and A. P. Priddy. *Power Plant System Design.* New York: John Wiley & Sons, 1985.
5. H. Sorensen. *Energy Conversion Systems.* New York: John Wiley & Sons, 1983.
6. *Steam, Its Generation and Use.* 39th ed. New York: Babcock and Wilcox Co., 1978.
7. *Turbomachinery* 28, no. 2 (March/April 1987). Norwalk, CT: Business Journals, Inc.
8. J. Weisman and R. Eckart. *Modern Power Plant Engineering.* Englewood Cliffs, NJ: Prentice-Hall, 1985.

PROBLEMS*

Carnot Vapor Cycle

10–1C Why is the Carnot cycle not a realistic model for steam power plants?

10–2E Water enters the boiler of a steady-flow Carnot engine as a saturated liquid at 250 psia and leaves with a quality of 0.95. Steam leaves the turbine at a pressure of 40 psia. Show the cycle on a *T-s* diagram relative to the saturation lines, and determine (*a*) the thermal efficiency, (*b*) the quality at the end of the isothermal heat-rejection process, and (*c*) the net work output. *Answers:* (*a*) 15.5 percent, (*b*) 0.137, (*c*) 122 Btu/lbm

10–3 A steady-flow Carnot cycle uses water as the working fluid. Water changes from saturated liquid to saturated vapor as heat is transferred to it from a source at 250°C. Heat rejection takes place at a pressure of 20 kPa. Show the cycle on a *T-s* diagram relative to the saturation lines, and determine (*a*) the thermal efficiency, (*b*) the amount of heat rejected, and (*c*) the net work output.

10–4 Repeat Prob. 10–3 for a heat rejection pressure of 10 kPa.

10–5 Consider a steady-flow Carnot cycle with water as the working fluid. The maximum and minimum temperatures in the cycle are 350 and 60°C. The quality of water is 0.891 at the beginning of the heat-rejection process and 0.1 at the end. Show the cycle on a *T-s* diagram relative to the saturation lines, and determine (*a*) the thermal efficiency, (*b*) the pressure at the turbine inlet, and (*c*) the net work output. *Answers:* (*a*) 0.465, (*b*) 1.40 MPa, (*c*) 1623 kJ/kg

The Simple Rankine Cycle

10–6C What four processes make up the simple ideal Rankine cycle?

*Problems designated by a "C" are concept questions, and students are encouraged to answer them all. Problems designated by an "E" are in English units, and the SI users can ignore them. Problems with the icon are solved using EES, and complete solutions together with parametric studies are included on the enclosed DVD. Problems with the icon are comprehensive in nature, and are intended to be solved with a computer, preferably using the EES software that accompanies this text.

10–7C Consider a simple ideal Rankine cycle with fixed turbine inlet conditions. What is the effect of lowering the condenser pressure on

Pump work input:	(*a*) increases, (*b*) decreases, (*c*) remains the same
Turbine work output:	(*a*) increases, (*b*) decreases, (*c*) remains the same
Heat supplied:	(*a*) increases, (*b*) decreases, (*c*) remains the same
Heat rejected:	(*a*) increases, (*b*) decreases, (*c*) remains the same
Cycle efficiency:	(*a*) increases, (*b*) decreases, (*c*) remains the same
Moisture content at turbine exit:	(*a*) increases, (*b*) decreases, (*c*) remains the same

10–8C Consider a simple ideal Rankine cycle with fixed turbine inlet temperature and condenser pressure. What is the effect of increasing the boiler pressure on

Pump work input:	(*a*) increases, (*b*) decreases, (*c*) remains the same
Turbine work output:	(*a*) increases, (*b*) decreases, (*c*) remains the same
Heat supplied:	(*a*) increases, (*b*) decreases, (*c*) remains the same
Heat rejected:	(*a*) increases, (*b*) decreases, (*c*) remains the same
Cycle efficiency:	(*a*) increases, (*b*) decreases, (*c*) remains the same
Moisture content at turbine exit:	(*a*) increases, (*b*) decreases, (*c*) remains the same

10–9C Consider a simple ideal Rankine cycle with fixed boiler and condenser pressures. What is the effect of superheating the steam to a higher temperature on

Pump work input:	(*a*) increases, (*b*) decreases, (*c*) remains the same
Turbine work output:	(*a*) increases, (*b*) decreases, (*c*) remains the same
Heat supplied:	(*a*) increases, (*b*) decreases, (*c*) remains the same
Heat rejected:	(*a*) increases, (*b*) decreases, (*c*) remains the same
Cycle efficiency:	(*a*) increases, (*b*) decreases, (*c*) remains the same
Moisture content at turbine exit:	(*a*) increases, (*b*) decreases, (*c*) remains the same

10–10C How do actual vapor power cycles differ from idealized ones?

10–11C Compare the pressures at the inlet and the exit of the boiler for (*a*) actual and (*b*) ideal cycles.

10–12C The entropy of steam increases in actual steam turbines as a result of irreversibilities. In an effort to control entropy increase, it is proposed to cool the steam in the turbine by running cooling water around the turbine casing. It is argued that this will reduce the entropy and the enthalpy of the steam at the turbine exit and thus increase the work output. How would you evaluate this proposal?

10–13C Is it possible to maintain a pressure of 10 kPa in a condenser that is being cooled by river water entering at 20°C?

10–14 An engineer has proposed that a simple ideal Rankine cycle that uses refrigerant-134a be used to produce work with heat from a low-temperature thermal energy reservoir. The boiler operates at 1.6 MPa, the condenser at 0.4 MPa, and the turbine inlet at 80°C. Determine the R-134a mass flow rate needed for this cycle to produce 750 kW of power and the thermal efficiency of the cycle.

10–15 Refrigerant-134a is used as the working fluid in a simple ideal Rankine cycle which operates the boiler at 1400 kPa and the condenser at 10°C. The mixture at the exit of the turbine has a quality of 98 percent. Determine the turbine inlet temperature, the cycle thermal efficiency, and the back-work ratio of this cycle.

10–16 A simple ideal Rankine cycle which uses water as the working fluid operates its condenser at 40°C and its boiler at 300°C. Calculate the work produced by the turbine, the heat supplied in the boiler, and the thermal efficiency of this cycle when the steam enters the turbine without any superheating.

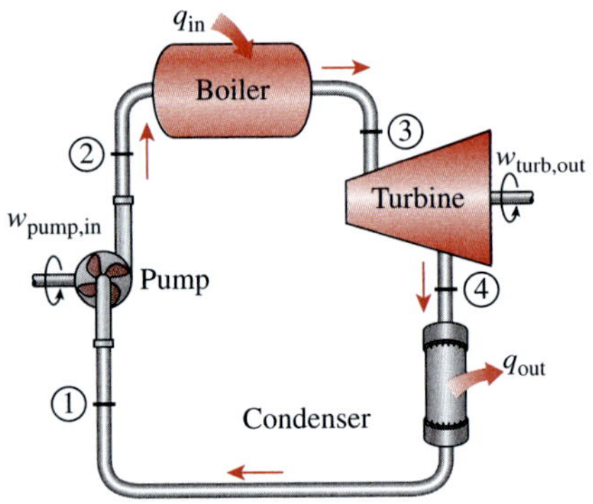

FIGURE P10–16

10–17E The turbine of a steam power plant operating on a simple ideal Rankine cycle produces 1750 kW of power when the boiler is operated at 800 psia, the condenser at 3 psia, and the temperature at the turbine entrance is 900°F. Determine the rate of heat supply in the boiler, the rate of heat rejection in the condenser, and the thermal efficiency of the cycle.

10–18E A simple ideal Rankine cycle with water as the working fluid operates between the pressure limits of 2500 psia in the boiler and 5 psia in the condenser. What is the minimum temperature required at the turbine inlet such that the quality of the steam leaving the turbine is not below 80 percent. When operated at this temperature, what is the thermal efficiency of this cycle?

10–19E A steam Rankine cycle operates between the pressure limits of 2500 psia in the boiler and 1 psia in the condenser. The turbine inlet temperature is 800°F. The turbine isentropic efficiency is 90 percent, the pump losses are negligible, and the cycle is sized to produce 1000 kW of power. Calculate the mass flow rate through the boiler, the power produced by the turbine, the rate of heat supply in the boiler, and the thermal efficiency.

10–20E Reconsider Prob. 10–19E. How much error is caused in the thermal efficiency if the power required by the pump were completely neglected?

10–21 Consider a steam power plant that operates on a simple ideal Rankine cycle and has a net power output of 45 MW. Steam enters the turbine at 7 MPa and 500°C and is cooled in the condenser at a pressure of 10 kPa by running cooling water from a lake through the tubes of the condenser at a rate of 2000 kg/s. Show the cycle on a *T-s* diagram with respect to saturation lines, and determine (*a*) the thermal efficiency of the cycle, (*b*) the mass flow rate of the steam, and (*c*) the temperature rise of the cooling water. *Answers:* (*a*) 38.9 percent, (*b*) 36 kg/s, (*c*) 8.4°C

10–22 Repeat Prob. 10–21 assuming an isentropic efficiency of 87 percent for both the turbine and the pump. *Answers:* (*a*) 33.8 percent, (*b*) 41.4 kg/s, (*c*) 10.5°C

10–23 A simple Rankine cycle uses water as the working fluid. The boiler operates at 6000 kPa and the condenser at 50 kPa. At the entrance to the turbine, the temperature is 450°C. The isentropic efficiency of the turbine is 94 percent, pressure and pump losses are negligible, and the water leaving the condenser is subcooled by 6.3°C. The boiler is sized for a mass flow rate of 20 kg/s. Determine the rate at which heat is added in the boiler, the power required to operate the pumps, the net power produced by the cycle, and the thermal efficiency. *Answers:* 59,660 kW, 122 kW, 18,050 kW, 30.3 percent.

10–24 Using EES (or other) software, determine how much the thermal efficiency of the cycle in Prob. 10–23 would change if there were a 50 kPa pressure drop across the boiler.

10–25 The net work output and the thermal efficiency for the Carnot and the simple ideal Rankine cycles with steam as the working fluid are to be calculated and compared. Steam enters the turbine in both cases at 5 MPa as a saturated vapor, and the condenser pressure is 50 kPa. In the Rankine cycle, the condenser exit state is saturated liquid and in the Carnot cycle, the boiler inlet state is saturated liquid. Draw the *T-s* diagrams for both cycles.

10–26 Consider a coal-fired steam power plant that produces 120 MW of electric power. The power plant operates on a simple ideal Rankine cycle with turbine inlet conditions of 9 MPa and 550°C and a condenser pressure of 15 kPa. The coal has a heating value (energy released when the fuel is burned) of 29,300 kJ/kg. Assuming that 75 percent of this energy is transferred to the steam in the boiler and that the electric generator has an efficiency of 96 percent, determine (*a*) the overall plant efficiency (the ratio of net electric power output to the energy input as fuel) and (*b*) the required rate of coal supply. *Answers:* (*a*) 28.4 percent, (*b*) 51.9 t/h

10–27 The schematic of a single-flash geothermal power plant with state numbers is given in Fig. P10–27. Geothermal resource exists as saturated liquid at 230°C. The geothermal liquid is withdrawn from the production well at a rate of 230 kg/s, and is flashed to a pressure of 500 kPa by an essentially isenthalpic flashing process where the resulting vapor is separated from the liquid in a separator and directed to the turbine. The steam leaves the turbine at 10 kPa with a moisture content of 10 percent and enters the condenser where it is condensed and routed to a reinjection well along with the liquid coming off the separator. Determine (*a*) the mass flow rate of steam through the turbine, (*b*) the isentropic efficiency of the turbine, (*c*) the power output of the turbine, and (*d*) the thermal efficiency of the plant (the ratio of the turbine work output to the energy of the geothermal fluid relative to standard ambient conditions). *Answers:* (*a*) 38.2 kg/s, (*b*) 0.686, (*c*) 15.4 MW, (*d*) 7.6 percent

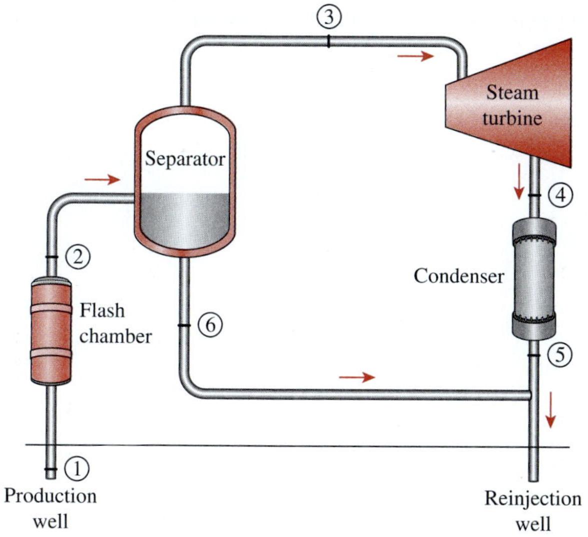

FIGURE P10–27

10–28 Reconsider Prob. 10–27. Now, it is proposed that the liquid water coming out of the separator be routed through another flash chamber maintained at 150 kPa, and the steam produced be directed to a lower stage of the same turbine. Both streams of steam leave the turbine at the same state of 10 kPa and 90 percent quality. Determine (*a*) the temperature of steam at the outlet of the second flash chamber, (*b*) the power produced by the lower stage of the turbine, and (*c*) the thermal efficiency of the plant.

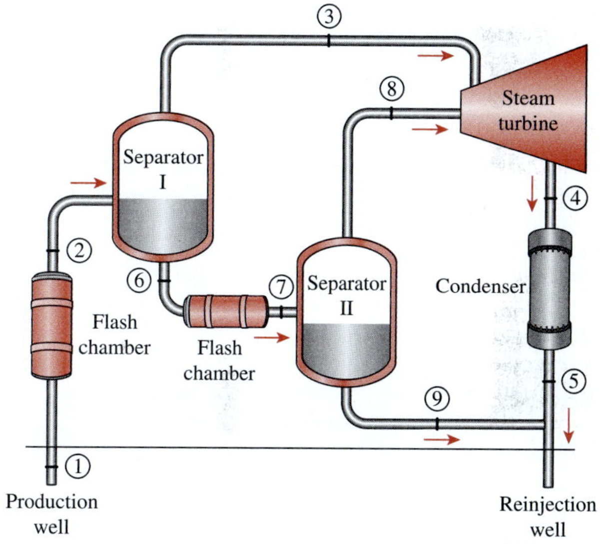

FIGURE P10–28

10–29 Reconsider Prob. 10–27. Now, it is proposed that the liquid water coming out of the separator be used as the heat source in a binary cycle with isobutane as the working fluid. Geothermal liquid water leaves the heat exchanger at 90°C while isobutane enters the turbine at 3.25 MPa and 145°C and leaves at 80°C and 400 kPa. Isobutane is condensed in an air-cooled condenser and then pumped to the heat exchanger pressure. Assuming an isentropic efficiency of 90 percent for the pump, determine (*a*) the mass flow rate of isobutane in the binary cycle, (*b*) the net power outputs of both the flashing and the binary sections of the plant, and (*c*) the thermal efficiencies of the binary cycle and the combined plant. The properties of isobutane may be obtained

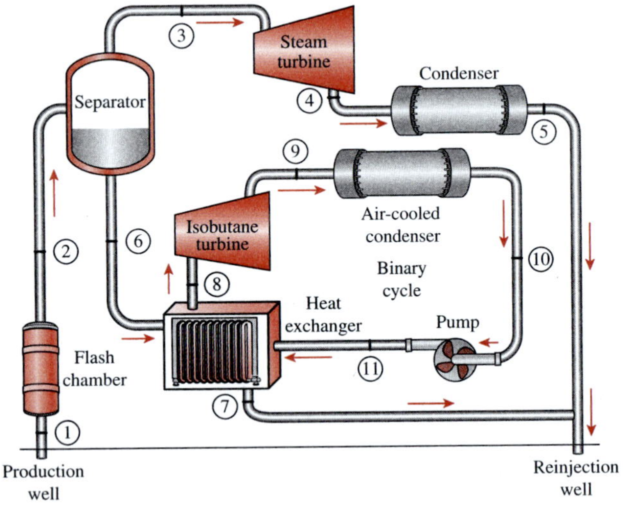

FIGURE P10–29

from EES. *Answers:* (*a*) 105.5 kg/s, (*b*) 15.4 MW, 6.14 MW, (*c*) 12.2 percent, 10.6 percent

The Reheat Rankine Cycle

10–30C How do the following quantities change when a simple ideal Rankine cycle is modified with reheating? Assume the mass flow rate is maintained the same.

Pump work input:	(*a*) increases, (*b*) decreases, (*c*) remains the same
Turbine work output:	(*a*) increases, (*b*) decreases, (*c*) remains the same
Heat supplied:	(*a*) increases, (*b*) decreases, (*c*) remains the same
Heat rejected:	(*a*) increases, (*b*) decreases, (*c*) remains the same
Moisture content at turbine exit:	(*a*) increases, (*b*) decreases, (*c*) remains the same

10–31C Is there an optimal pressure for reheating the steam of a Rankine cycle? Explain.

10–32C Consider a simple ideal Rankine cycle and an ideal Rankine cycle with three reheat stages. Both cycles operate between the same pressure limits. The maximum temperature is 700°C in the simple cycle and 450°C in the reheat cycle. Which cycle do you think will have a higher thermal efficiency?

10–33 An ideal reheat Rankine cycle with water as the working fluid operates the inlet of the high-pressure turbine at 8000 kPa and 450°C; the inlet of the low-pressure turbine at 500 kPa and 500°C; and the condenser at 10 kPa. Determine the mass flow rate through the boiler needed for this system to produce a net 5000 kW of power and the thermal efficiency of the cycle.

10–34 An ideal reheat Rankine cycle with water as the working fluid operates the boiler at 15,000 kPa, the reheater at 2000 kPa, and the condenser at 100 kPa. The temperature is 450°C at the entrance of the high-pressure and low-pressure turbines. The mass flow rate through the cycle is 1.74 kg/s. Determine the power used by pumps, the power produced by the cycle, the rate of heat transfer in the reheater, and the thermal efficiency of this system.

10–35 A steam power plant operates on the ideal reheat Rankine cycle. Steam enters the high-pressure turbine at 6 MPa and 400°C and leaves at 2 MPa. Steam is then reheated at constant pressure to 400°C before it expands to 20 kPa in the low-pressure turbine. Determine the turbine work output, in kJ/kg, and the thermal efficiency of the cycle. Also, show the cycle on a *T-s* diagram with respect to saturation lines.

10–36 Reconsider Prob. 10–35. Using EES (or other) software, solve this problem by the diagram window data entry feature of EES. Include the effects of the turbine and pump efficiencies and also show the effects of reheat on the steam quality at the low-pressure turbine exit. Plot the cycle on a *T-s* diagram with respect to the saturation lines. Discuss the results of your parametric studies.

10–37E Consider a steam power plant that operates on the ideal reheat Rankine cycle. The plant maintains the inlet of the high-pressure turbine at 600 psia and 600°F, the inlet of the low-pressure turbine at 200 psia and 600°F, and the condenser at 10 psia. The net power produced by this plant is 5000 kW. Determine the rate of heat addition and rejection and the thermal efficiency of the cycle.

10–38E In Prob. 10–37E, is there any advantage to operating the reheat section of the boiler at 100 psia rather than 200 psia while maintaining the same low-pressure turbine inlet temperature?

10–39 Consider a steam power plant that operates on the ideal reheat Rankine cycle. The plant maintains the boiler at

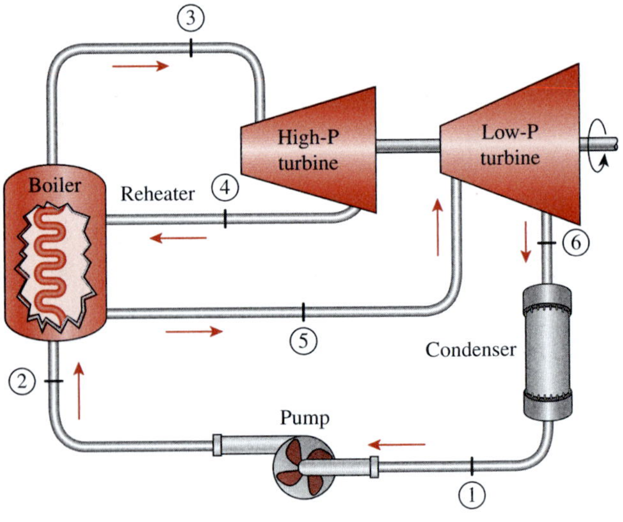

FIGURE P10–39

7000 kPa, the reheat section at 800 kPa, and the condenser at 10 kPa. The mixture quality at the exit of both turbines is 93 percent. Determine the temperature at the inlet of each turbine and the cycle's thermal efficiency. *Answers:* 373°C, 416°C, 37.6 percent

10–40 A steam power plant operates on an ideal reheat Rankine cycle between the pressure limits of 15 MPa and 10 kPa. The mass flow rate of steam through the cycle is 12 kg/s. Steam enters both stages of the turbine at 500°C. If the moisture content of the steam at the exit of the low-pressure turbine is not to exceed 10 percent, determine (*a*) the pressure at which reheating takes place, (*b*) the total rate of heat input in the boiler, and (*c*) the thermal efficiency of the cycle. Also, show the cycle on a *T-s* diagram with respect to saturation lines.

10–41 A steam power plant operates on the reheat Rankine cycle. Steam enters the high-pressure turbine at 12.5 MPa and 550°C at a rate of 7.7 kg/s and leaves at 2 MPa. Steam is then reheated at constant pressure to 450°C before it expands in the low-pressure turbine. The isentropic efficiencies of the turbine and the pump are 85 percent and 90 percent, respectively. Steam leaves the condenser as a saturated liquid. If the moisture content of the steam at the exit of the turbine is not to exceed 5 percent, determine (*a*) the condenser pressure, (*b*) the net power output, and (*c*) the thermal efficiency. *Answers:* (*a*) 9.73 kPa, (*b*) 10.2 MW, (*c*) 36.9 percent

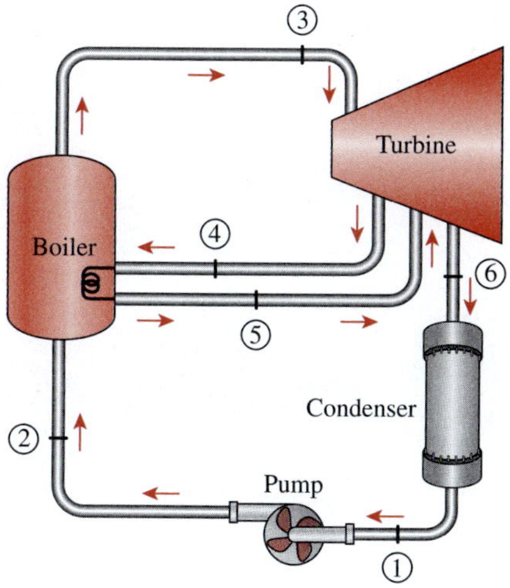

FIGURE P10–41

Regenerative Rankine Cycle

10–42C How do the following quantities change when the simple ideal Rankine cycle is modified with regeneration? Assume the mass flow rate through the boiler is the same.

Turbine work output:	(*a*) increases, (*b*) decreases, (*c*) remains the same
Heat supplied:	(*a*) increases, (*b*) decreases, (*c*) remains the same
Heat rejected:	(*a*) increases, (*b*) decreases, (*c*) remains the same
Moisture content at turbine exit:	(*a*) increases, (*b*) decreases, (*c*) remains the same

10–43C During a regeneration process, some steam is extracted from the turbine and is used to heat the liquid water leaving the pump. This does not seem like a smart thing to do since the extracted steam could produce some more work in the turbine. How do you justify this action?

10–44C How do open feedwater heaters differ from closed feedwater heaters?

10–45C Consider a simple ideal Rankine cycle and an ideal regenerative Rankine cycle with one open feedwater heater. The two cycles are very much alike, except the feedwater in the regenerative cycle is heated by extracting some steam just before it enters the turbine. How would you compare the efficiencies of these two cycles?

10–46C Devise an ideal regenerative Rankine cycle that has the same thermal efficiency as the Carnot cycle. Show the cycle on a *T-s* diagram.

10–47E Turbine bleed steam enters an open feedwater heater of a regenerative Rankine cycle at 20 psia and 250°F while the cold feedwater enters at 110°F. Determine the ratio of the bleed steam mass flow rate to the inlet feedwater mass flow rate required to heat the feedwater to 225°F.

10–48 The closed feedwater heater of a regenerative Rankine cycle is to heat 7000 kPa feedwater from 260°C to a saturated liquid. The turbine supplies bleed steam at 6000 kPa and 325°C to this unit. This steam is condensed to a saturated liquid before entering the pump. Calculate the amount of bleed steam required to heat 1 kg of feedwater in this unit. *Answer:* 0.0779 kg/s

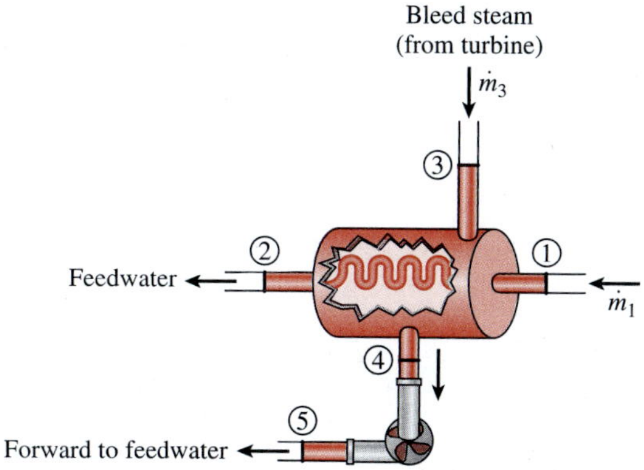

FIGURE P10–48

10–49E The ideal regenerative Rankine cycle with an open feedwater heater uses water as the working fluid. The turbine inlet is operated at 500 psia and 600°F and the condenser at 5 psia. Steam is supplied to the open feedwater heater at 40 psia. Determine the work produced by the turbine, the work consumed by the pumps, the heat rejected in the condenser for this cycle per unit of flow through the boiler.

10–50E Determine the change in the thermal efficiency of the cycle in Prob. 10–49E when the steam supplied to the open feedwater heater is at 60 psia rather than 40 psia.

10–51E Reconsider Prob. 10–49E. Using EES (or other) software, determine the optimum bleed pressure for the open feedwater heater that maximizes the thermal efficiency of the cycle.

10–52 A steam power plant operates on an ideal regenerative Rankine cycle with two open feedwater heaters. Steam enters the turbine at 10 MPa and 600°C and exhausts to the condenser at 5 kPa. Steam is extracted from the turbine at 0.6 and 0.2 MPa. Water leaves both feedwater heaters as a saturated liquid. The mass flow rate of steam through the boiler is 22 kg/s. Show the cycle on a *T-s* diagram, and determine (*a*) the net power output of the power plant and (*b*) the thermal efficiency of the cycle. *Answers:* (*a*) 30.5 MW, (*b*) 47.1 percent

10–53 Consider a steam power plant that operates on the ideal regenerative Rankine cycle with a closed feedwater heater as shown in the figure. The plant maintains the turbine inlet at 3000 kPa and 350°C; and operates the condenser at 20 kPa. Steam is extracted at 1000 kPa to serve the closed feedwater heater, which discharges into the condenser after being throttled to condenser pressure. Calculate the work produced by the turbine, the work consumed by the pump, and the heat supply in the boiler for this cycle per unit of boiler flow rate. *Answers:* 741 kJ/kg, 3.0 kJ/kg, 2353 kJ/kg

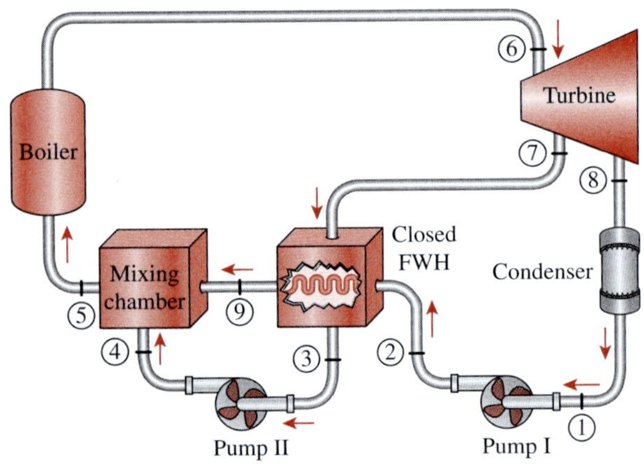

FIGURE P10–53

10–54 Reconsider Prob. 10–53. Using EES (or other) software, determine the optimum bleed pressure for the closed feedwater heater that maximizes the thermal efficiency of the cycle. *Answer:* 220 kPa

10–55 Determine the thermal efficiency of the regenerative Rankine cycle of Prob. 10–53 when the isentropic efficiency of the turbine is 90 percent before and after steam extraction point.

10–56 Determine the thermal efficiency of the regenerative Rankine cycle of Prob. 10–53 when the isentropic efficiency of the turbine before and after steam extraction point is 90 percent and the condenser condensate is subcooled by 10°C.

10–57 Reconsider Prob. 10–53. Using EES (or other) software, determine how much additional heat must be supplied to the boiler when the turbine isentropic efficiency before and after the extraction point is 90 percent and there is a 10 kPa pressure drop across the boiler?

10–58 Consider an ideal steam regenerative Rankine cycle with two feedwater heaters, one closed and one open. Steam enters the turbine at 10 MPa and 600°C and exhausts to the condenser at 10 kPa. Steam is extracted from the turbine at 1.2 MPa for the closed feedwater heater and at 0.6 MPa for the open one. The feedwater is heated to the condensation temperature of the extracted steam in the closed feedwater heater. The extracted steam leaves the closed feedwater heater as a saturated liquid, which is subsequently throttled to the open feedwater heater. Show the cycle on a *T-s* diagram with respect to saturation lines, and determine (*a*) the mass flow rate of steam through the boiler for a net power output of 400 MW and (*b*) the thermal efficiency of the cycle.

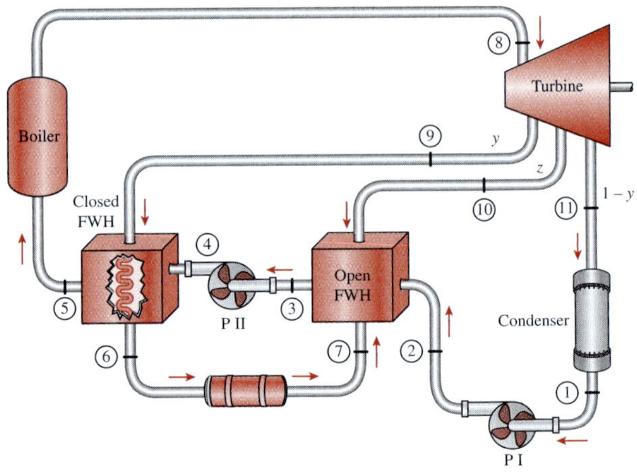

FIGURE P10–58

10–59 Reconsider Prob. 10–58. Using EES (or other) software, investigate the effects of turbine and pump efficiencies as they are varied from 70 percent to 100 percent on the mass flow rate and thermal efficiency. Plot the mass flow rate and the thermal efficiency as a function of turbine efficiency for pump efficiencies of 70, 85, and 100 percent, and discuss the results. Also plot the *T-s* diagram for turbine and pump efficiencies of 85 percent.

10–60 A steam power plant operates on the reheat-regenerative Rankine cycle with a closed feedwater heater. Steam enters the turbine at 8 MPa and 500°C at a rate of 15 kg/s and is condensed in the condenser at a pressure of 20 kPa. Steam is reheated at 3 MPa to 500°C. Some steam is extracted from the low-pressure turbine at 1.0 MPa, is completely condensed in the closed feedwater heater, and pumped to 8 MPa before it mixes with the feedwater at the same pressure. Assuming an isentropic efficiency of 88 percent for both the turbine and the pump, determine (*a*) the temperature of the steam at the inlet of the closed feedwater heater, (*b*) the mass flow rate of the steam extracted from the turbine for the closed feedwater heater, (*c*) the net power output, and (*d*) the thermal efficiency. *Answers:* (*a*) 350°C, (*b*) 2.64 kg/s, (*c*) 16.2 MW, (*d*) 36.7 percent

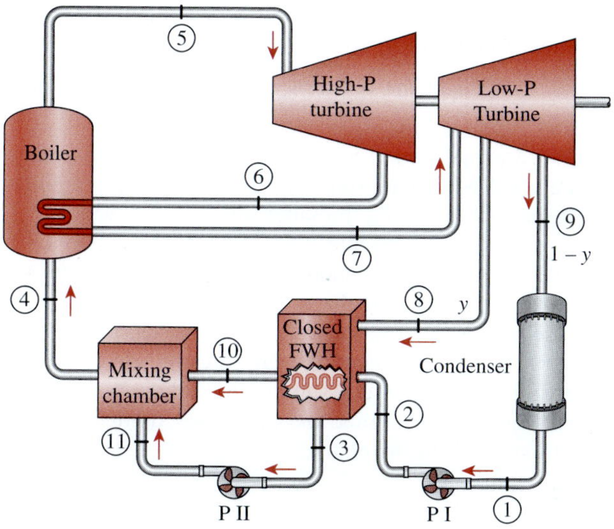

FIGURE P10–60

10–61 An ideal Rankine steam cycle modified with two closed feedwater heaters is shown below. The power cycle receives 50 kg/s of steam at the high pressure inlet to the turbine. The feedwater heater exit states for the boiler feedwater and the condensed steam are the normally assumed ideal states. The fraction of mass entering the high pressure turbine at state 5 that is extracted for the feedwater heater operating at 1400 kPa is $y = 0.1153$. Use the data provided in the tables given below to

(*a*) Sketch the *T-s* diagram for the ideal cycle.
(*b*) Determine the fraction of mass, *z*, that is extracted for the closed feedwater heater operating at the 245 kPa extraction pressure.
(*c*) Determine the required cooling water flow rate, in kg/s, to keep the cooling water temperature rise in the condenser to 10°C. Assume $c_p = 4.18$ kJ/kg·K for cooling water.
(*d*) Determine the net power output and the thermal efficiency of the plant.

Process states and selected data

State	*P*, kPa	*T*, °C	*h*, kJ/kg	*s*, kJ/kg·K
1	20			
2	5000			
3	5000			
4	5000			
5	5000	700	3900	7.512
6	1400		3406	7.512
7	245		2918	7.512
8	20		2477	7.512

Saturation data

P, kPa	v_f, m³/kg	h_f, kJ/kg	s_g, kJ/kg·K
20	0.00102	251	7.907
245		533	7.060
1400		830	6.468
5000	0.00129	1154	5.973

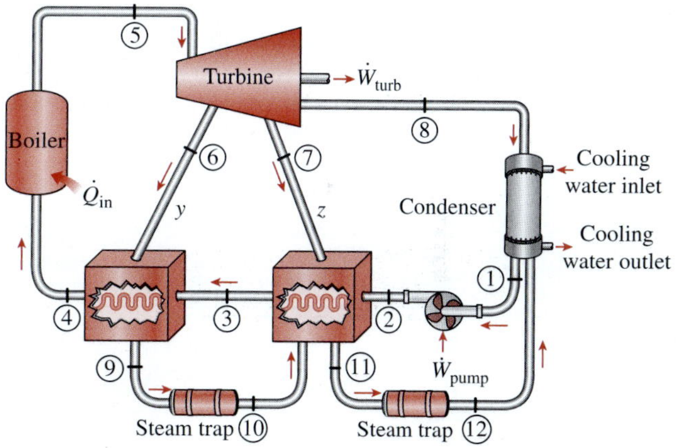

FIGURE P10–61

Second-Law Analysis of Vapor Power Cycles

10–62C How can the second-law efficiency of a simple ideal Rankine cycle be improved?

10–63E Calculate the exergy destruction in each of the components of the simple ideal Rankine cycle of Prob. 10–17E when heat is being rejected to a lake at 40°F, and heat is being supplied by a 1500°F energy reservoir.

10–64 Determine the exergy destruction associated with the heat rejection process in Prob. 10–21. Assume a source temperature of 1500 K and a sink temperature of 290 K. Also, determine the exergy of the steam at the boiler exit. Take $P_0 = 100$ kPa.

10–65 Which component of the ideal reheat Rankine cycle of Prob. 10–33 offers the greatest opportunity to regain lost power potential? The sink is at 10°C and the source is at 600°C.

10–66 Determine the exergy destruction associated with each of the processes of the reheat Rankine cycle described in Prob. 10–35. Assume a source temperature of 1500 K and a sink temperature of 295 K.

10–67 Reconsider Prob. 10–66. Using EES (or other) software, solve this problem by the diagram window data entry feature of EES. Include the effects of the turbine and pump efficiencies to evaluate the irreversibilities associated with each of the processes. Plot the cycle on a *T-s* diagram with respect to the saturation lines. Discuss the results of your parametric studies.

10–68E Which component of the ideal regenerative Rankine cycle of Prob. 10–49E loses the greatest amount of work potential? This cycle rejects heat to a river whose temperature is 60°F, and the source is at 800°F.

10–69 The schematic of a single-flash geothermal power plant with state numbers is given in Fig. P10–69. Geothermal resource exists as saturated liquid at 230°C. The geothermal liquid is withdrawn from the production well at a rate of 230 kg/s and is flashed to a pressure of 500 kPa by an essentially isenthalpic flashing process where the resulting vapor is separated from the liquid in a separator and is directed to the turbine. The steam leaves the turbine at 10 kPa with a moisture content of 5 percent and enters the condenser where it is condensed; it is routed to a reinjection well along with the liquid coming off the separator. Determine (*a*) the power output of the turbine and the thermal efficiency of the plant, (*b*) the exergy of the geothermal liquid at the exit of the flash chamber, and the exergy destructions and the second-law (exergetic) efficiencies for (*c*) the flash chamber, (*d*) the turbine, and (*e*) the entire plant. *Answers:* (*a*) 10.8 MW, 0.053, (*b*) 17.3 MW, (*c*) 5.1 MW, 0.898, (*d*) 10.9 MW, 0.500, (*e*) 39.0 MW, 0.218

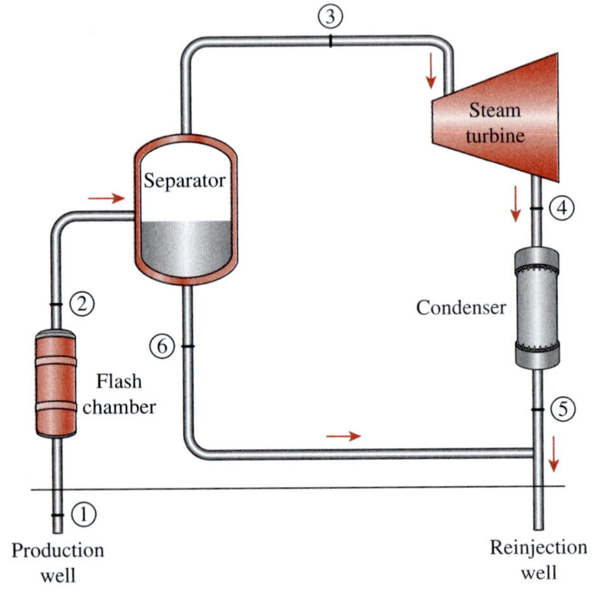

FIGURE P10–69

Cogeneration

10–70C How is the utilization factor ϵ_u for cogeneration plants defined? Could ϵ_u be unity for a cogeneration plant that does not produce any power?

10–71C Consider a cogeneration plant for which the utilization factor is 1. Is the irreversibility associated with this cycle necessarily zero? Explain.

10–72C Consider a cogeneration plant for which the utilization factor is 0.5. Can the exergy destruction associated with this plant be zero? If yes, under what conditions?

10–73C What is the difference between cogeneration and regeneration?

10–74 Steam enters the turbine of a cogeneration plant at 7 MPa and 500°C. One-fourth of the steam is extracted from the turbine at 600-kPa pressure for process heating. The remaining steam continues to expand to 10 kPa. The extracted steam is then condensed and mixed with feedwater at constant pressure and the mixture is pumped to the boiler pressure of 7 MPa. The mass flow rate of steam through the boiler is 30 kg/s. Disregarding any pressure drops and heat losses in the piping, and assuming the turbine and the pump to be isentropic, determine the net power produced and the utilization factor of the plant.

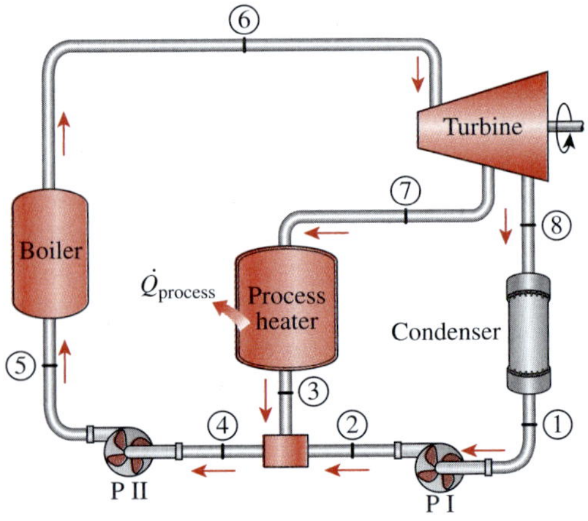

FIGURE P10–74

10–75E A large food-processing plant requires 1.5 lbm/s of saturated or slightly superheated steam at 140 psia, which is extracted from the turbine of a cogeneration plant. The boiler generates steam at 800 psia and 1000°F at a rate of 10 lbm/s, and the condenser pressure is 2 psia. Steam leaves the process heater as a saturated liquid. It is then mixed with the feedwater at the same pressure and this mixture is pumped to the boiler pressure. Assuming both the pumps and the turbine have isentropic efficiencies of 86 percent, determine (*a*) the rate of heat transfer to the boiler and (*b*) the power output of the cogeneration plant. *Answers:* (*a*) 13.810 Btu/s, (*b*) 4440 kW

10–76 Steam is generated in the boiler of a cogeneration plant at 10 MPa and 450°C at a steady rate of 5 kg/s. In normal operation, steam expands in a turbine to a pressure of 0.5 MPa and is then routed to the process heater, where it supplies the process heat. Steam leaves the process heater as a saturated liquid and is pumped to the boiler pressure. In this mode, no steam passes through the condenser, which operates at 20 kPa.

(*a*) Determine the power produced and the rate at which process heat is supplied in this mode.
(*b*) Determine the power produced and the rate of process heat supplied if only 60 percent of the steam is routed to the process heater and the remainder is expanded to the condenser pressure.

10–77 Consider a cogeneration power plant modified with regeneration. Steam enters the turbine at 9 MPa and 400°C and expands to a pressure of 1.6 MPa. At this pressure, 35 percent of the steam is extracted from the turbine, and the remainder expands to 10 kPa. Part of the extracted steam is used to heat the feedwater in an open feedwater heater. The rest of the extracted steam is used for process heating and leaves the process heater as a saturated liquid at 1.6 MPa. It is subsequently mixed with the feedwater leaving the feedwater heater, and the mixture is pumped to the boiler pressure. Assuming the turbines and the pumps to be isentropic, show the cycle on a *T-s* diagram with respect to saturation lines, and determine the mass flow rate of steam through the boiler for a net power output of 25 MW. *Answer:* 29.1 kg/s

10–78 EES Reconsider Prob. 10–77. Using EES (or other) software, investigate the effect of the extraction pressure for removing steam from the turbine to be used for the process heater and open feedwater heater on the required mass flow rate. Plot the mass flow rate through the boiler as a function of the extraction pressure, and discuss the results.

10–79E Steam is generated in the boiler of a cogeneration plant at 600 psia and 800°F at a rate of 18 lbm/s. The plant is to produce power while meeting the process steam requirements for a certain industrial application. One-third of the steam leaving the boiler is throttled to a pressure of 120 psia and is routed to the process heater. The rest of the steam is expanded in an isentropic turbine to a pressure of 120 psia and is also routed to the process heater. Steam leaves the process heater at 240°F. Neglecting the pump work, determine (*a*) the net power produced, (*b*) the rate of process heat supply, and (*c*) the utilization factor of this plant.

10–80 A large city uses a Rankine steam cycle modified with one closed feedwater heater and a process heater shown below to supply nearby buildings with hot water for heating systems and electrical power. The steam flow rate into the turbine is 100 kg/s. Steam entering the turbine is extracted at 2000 kPa, state 5, for the feedwater heater. Steam entering the turbine is extracted at 700 kPa, state 6, for the process heater and leaves

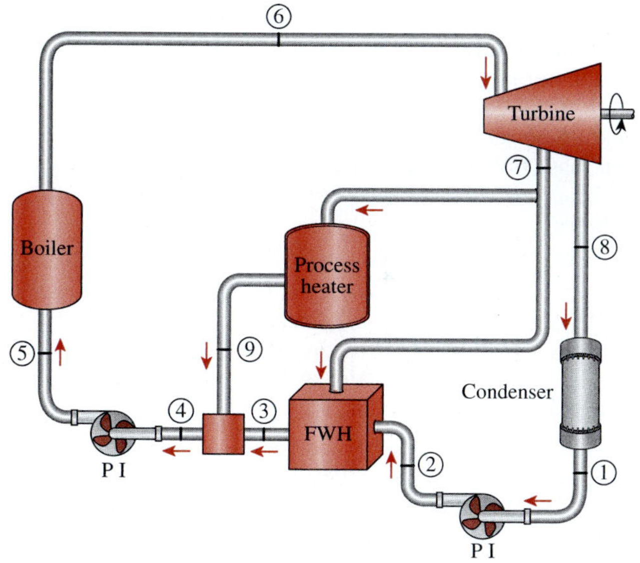

FIGURE P10–77

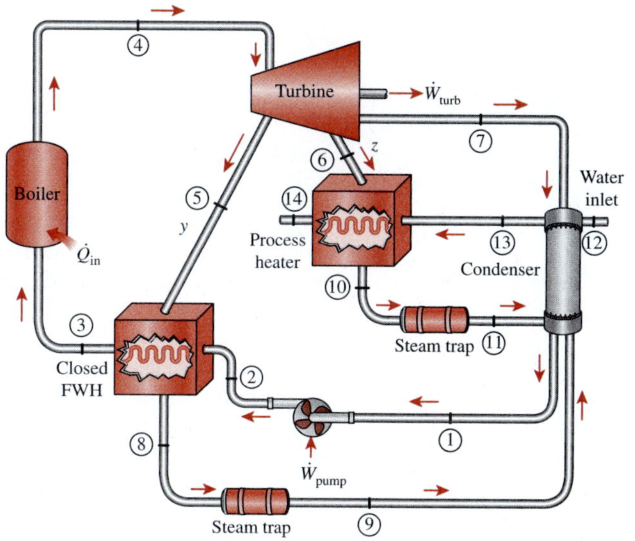

FIGURE P10–80

the process heater as a saturated liquid. The states for the boiler feedwater and the condensed steam leaving the feedwater heater are the normally assumed ideal states. Cold process water serves as the coolant for the condenser and receives the heat transferred from the condensing steam in the condenser. The process water is further heated in the process heater. Use the data provided in the tables given below to determine

(*a*) the *T*-*s* diagram for the ideal cycle,
(*b*) the process water flow rate, in kg/s, when 5 percent of the turbine inlet mass flow rate is extracted for the process heater and the temperature rise of the process water is 40°C, and
(*c*) the utilization efficiency of the plant.

Process states and selected data

State	*P*, kPa	*T*, °C	*h*, kJ/kg	*s*, kJ/kg·K
1	10			
2	10000			
3	10000			
4	10000	500	3374	6.597
5	2000		2930	6.597
6	700		2714	6.597
7	10		2089	6.597

Saturation data

P, kPa	T_{sat}, °C	h_f, kJ/kg	v_f, m^3/kg
10	45.8	191.7	0.00101
700	165	697.3	0.00111
2000	212.4	908.6	0.00118
10000	311	1407.6	0.00145

Combined Gas–Vapor Power Cycles

10–81C In combined gas–steam cycles, what is the energy source for the steam?

10–82C Why is the combined gas–steam cycle more efficient than either of the cycles operated alone?

10–83 Consider a combined gas–steam power plant that has a net power output of 450 MW. The pressure ratio of the gas-turbine cycle is 14. Air enters the compressor at 300 K and the turbine at 1400 K. The combustion gases leaving the gas turbine are used to heat the steam at 8 MPa to 400°C in a heat exchanger. The combustion gases leave the heat exchanger at 460 K. An open feedwater heater incorporated with the steam cycle operates at a pressure of 0.6 MPa. The condenser pressure is 20 kPa. Assuming all the compression and expansion processes to be isentropic, determine (*a*) the mass flow rate ration of air to steam, (*b*) the required rate of heat input in the combustion chamber, and (*c*) thermal efficiency of the combined cycle.

10–84 Reconsider Prob. 10–83. Using EES (or other) software, study the effects of the gas cycle pressure ratio as it is varied from 10 to 20 on the ratio of gas flow rate to steam flow rate and cycle thermal efficiency. Plot your results as functions of gas cycle pressure ratio, and discuss the results.

10–85 A combined gas–steam power cycle uses a simple gas turbine for the topping cycle and simple Rankine cycle for the bottoming cycle. Atmospheric air enters the gas turbine at 101 kPa and 20°C, and the maximum gas cycle temperature is 1100°C. The compressor pressure ratio is 8; the compressor isentropic efficiency is 85 percent; and the gas turbine isentropic efficiency is 90 percent. The gas stream leaves the heat exchanger at the saturation temperature of the steam flowing through the heat exchanger. Steam flows through the heat exchanger with a pressure of 6000 kPa, and leaves at 320°C. The steam-cycle condenser operates at 20 kPa, and the isentropic efficiency of the steam turbine is 90 percent. Determine the mass flow rate of air through the air compressor required for this system to produce 100 MW of power. Use constant specific heats for air at room temperature. *Answer: 279 kg/s*

10–86 An ideal regenerator is added to the gas cycle portion of the combined cycle in Prob. 10–85. How much does this change the efficiency of this combined cycle?

10–87 Determine which components of the combined cycle in Prob. 10–85 are the most wasteful of work potential.

10–88 Consider a combined gas–steam power plant that has a net power output of 280 MW. The pressure ratio of the gas-turbine cycle is 11. Air enters the compressor at 300 K and the turbine at 1100 K. The combustion gases leaving the gas turbine are used to heat the steam at 5 MPa to 350°C in a heat exchanger. The combustion gases leave the heat exchanger at 420 K. An open feedwater heater incorporated with the steam cycle operates at a pressure of 0.8 MPa. The condenser pressure is 10 kPa. Assuming isentropic efficiences of 100 percent for the pump, 82 percent for the compressor, and 86 percent for the gas and steam turbines, determine (*a*) the mass flow rate ratio of air to steam, (*b*) the required rate of heat input in the combustion chamber, and (c) the thermal efficiency of the combined cycle.

10–89 Reconsider Prob. 10–88. Using EES (or other) software, study the effects of the gas cycle pressure ratio as it is varied from 10 to 20 on the ratio of gas flow rate to steam flow rate and cycle thermal efficiency. Plot your results as functions of gas cycle pressure ratio, and discuss the results.

10–90 Consider a combined gas–steam power cycle. The topping cycle is a simple Brayton cycle that has a pressure ratio of 7. Air enters the compressor at 15°C at a rate of 10 kg/s and the gas turbine at 950°C. The bottoming cycle is a reheat Rankine cycle between the pressure limits of 6 MPa and 10 kPa. Steam is heated in a heat exchanger at

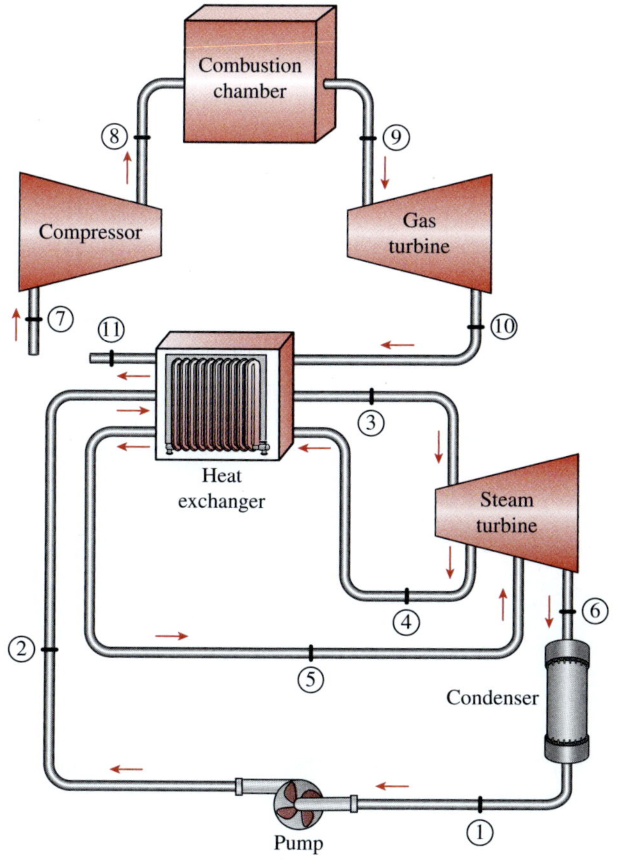

FIGURE P10–90

a rate of 1.15 kg/s by the exhaust gases leaving the gas turbine, and the exhaust gases leave the heat exchanger at 200°C. Steam leaves the high-pressure turbine at 1.0 MPa and is reheated to 400°C in the heat exchanger before it expands in the low-pressure turbine. Assuming 80 percent isentropic efficiency for all pumps and turbines, determine (*a*) the moisture content at the exit of the low-pressure turbine, (*b*) the steam temperature at the inlet of the high-pressure turbine, (*c*) the net power output and the thermal efficiency of the combined plant.

Special Topic: Binary Vapor Cycles

10–91C What is the difference between the binary vapor power cycle and the combined gas–steam power cycle?

10–92C What is a binary power cycle? What is its purpose?

10–93C Why is steam not an ideal working fluid for vapor power cycles?

10–94C Why is mercury a suitable working fluid for the topping portion of a binary vapor cycle but not for the bottoming cycle?

10–95 By writing an energy balance on the heat exchanger of a binary vapor power cycle, obtain a relation for the ratio of mass flow rates of two fluids in terms of their enthalpies.

Review Problems

10–96 A steam power plant operating on a simple ideal Rankine cycle maintains the boiler at 6000 kPa, the turbine inlet at 600°C, and the condenser at 50 kPa. Compare the thermal efficiency of this cycle when it is operated so that the liquid enters the pump as a saturated liquid against that when the liquid enters the pump 11.3°C cooler than a saturated liquid at the condenser pressure.

10–97E The Stillwater geothermal power plant in Nevada, which started full commercial operation in 1986, is designed to operate with seven identical units. Each of these seven units consists of a pair of power cycles, labeled Level I and Level II, operating on the simple Rankine cycle using an organic fluid as the working fluid.

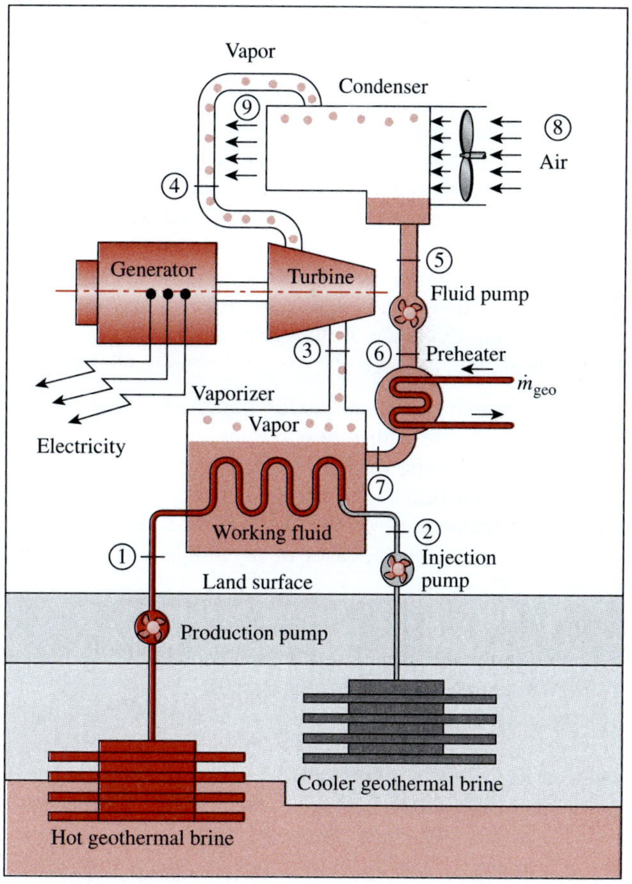

FIGURE P10–97

Schematic of a binary geothermal power plant.

Courtesy of ORMAT Energy Systems, Inc.

The heat source for the plant is geothermal water (brine) entering the vaporizer (boiler) of Level I of each unit at 325°F at a rate of 384,286 lbm/h and delivering 22.79 MBtu/h ("M" stands for "million"). The organic fluid that enters the vaporizer at 202.2°F at a rate of 157,895 lbm/h leaves it at 282.4°F and 225.8 psia as saturated vapor. This saturated vapor expands in the turbine to 95.8°F and 19.0 psia and produces 1271 kW of electric power. About 200 kW of this power is used by the pumps, the auxiliaries, and the six fans of the condenser. Subsequently, the organic working fluid is condensed in an air-cooled condenser by air that enters the condenser at 55°F at a rate of 4,195,100 lbm/h and leaves at 84.5°F. The working fluid is pumped and then preheated in a preheater to 202.2°F by absorbing 11.14 MBtu/h of heat from the geothermal water (coming from the vaporizer of Level II) entering the preheater at 211.8°F and leaving at 154.0°F.

Taking the average specific heat of the geothermal water to be 1.03 Btu/lbm·°F, determine (*a*) the exit temperature of the geothermal water from the vaporizer, (*b*) the rate of heat rejection from the working fluid to the air in the condenser, (*c*) the mass flow rate of the geothermal water at the preheater, and (*d*) the thermal efficiency of the Level I cycle of this geothermal power plant. *Answers:* (*a*) 267.4°F, (*b*) 29.7 MBtu/h, (*c*) 187,120 lbm/h, (*d*) 10.8 percent

10–98 A steam power plant operates on an ideal Rankine cycle with two stages of reheat and has a net power output of 75 MW. Steam enters all three stages of the turbine at 550°C. The maximum pressure in the cycle is 10 MPa, and the minimum pressure is 30 kPa. Steam is reheated at 4 MPa the first time and at 2 MPa the second time. Show the cycle on a *T-s* diagram with respect to saturation lines, and determine (*a*) the thermal efficiency of the cycle, and (*b*) the mass flow rate of the steam. *Answers:* (*a*) 40.5 percent, (*b*) 48.5 kg/s

10–99 Consider a steam power plant operating on the ideal Rankine cycle with reheat between the pressure limits of 25 MPa and 10 kPa with a maximum cycle temperature of 600°C and a moisture content of 8 percent at the turbine exit. For a reheat temperature of 600°C, determine the reheat pressures of the cycle for the cases of (*a*) single and (*b*) double reheat.

10–100 Consider a steam power plant that operates on a regenerative Rankine cycle and has a net power output of 150 MW. Steam enters the turbine at 10 MPa and 500°C and the condenser at 10 kPa. The isentropic efficiency of the turbine is 80 percent, and that of the pumps is 95 percent. Steam is extracted from the turbine at 0.5 MPa to heat the feedwater in an open feedwater heater. Water leaves the feedwater heater as a saturated liquid. Show the cycle on a *T-s* diagram, and determine (*a*) the mass flow rate of steam through the boiler, and (*b*) the thermal efficiency of the cycle. Also, determine the exergy destruction associated with the regeneration process. Assume a source temperature of 1300 K and a sink temperature of 303 K.

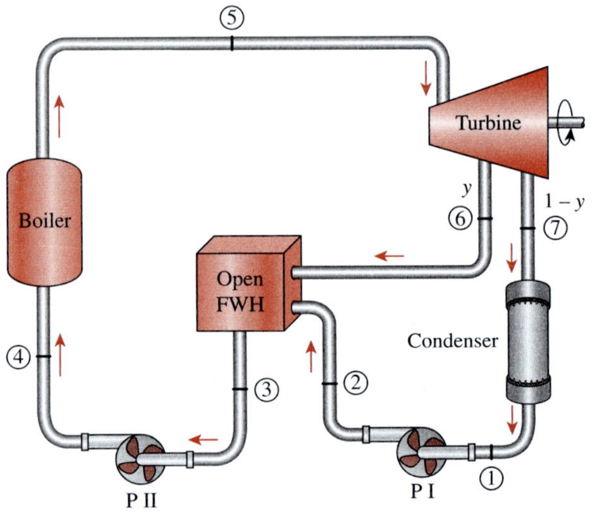

FIGURE P10–100

10–101 Repeat Prob. 10–100 assuming both the pump and the turbine are isentropic.

10–102 Consider an ideal reheat–regenerative Rankine cycle with one open feedwater heater. The boiler pressure is 10 MPa, the condenser pressure is 15 kPa, the reheater pressure is 1 MPa, and the feedwater pressure is 0.6 MPa. Steam enters both the high- and low-pressure turbines at 500°C. Show the cycle on a *T-s* diagram with respect to saturation lines, and determine (*a*) the fraction of steam extracted for regeneration and (*b*) the thermal efficiency of the cycle. *Answers:* (*a*) 0.144, (*b*) 42.1 percent

10–103 Repeat Prob. 10–102 assuming an isentropic efficiency of 84 percent for the turbines and 100 percent for the pumps.

10–104 A steam power plant operating on the ideal reheat-regenerative Rankine cycle with three feedwater heaters as shown in the figure maintains the boiler at 6000 kPa, the condenser at 7.5 kPa, the reheater at 800 kPa, the high-pressure, closed feedwater heater at 3000 kPa, the low-pressure, closed feedwater heater at 1800 kPa, and the open feedwater heater at 100 kPa. The temperature at the inlet of both turbines is 500°C. Determine the following quantities for this system per unit mass flow rate through the boiler.

(*a*) The flow rate required to service the high-pressure, closed feedwater heater.
(*b*) The flow rate required to service the low-pressure, closed feedwater heater.
(*c*) The flow rate required to service the open feedwater heater.
(*d*) The flow through the condenser.
(*e*) The work produced by the high-pressure turbine.

(*f*) The work produced by the low-pressure turbine.
(*g*) The heat supplied in the boiler and reheater.
(*h*) The heat rejected in the condenser.
(*i*) The thermal efficiency.

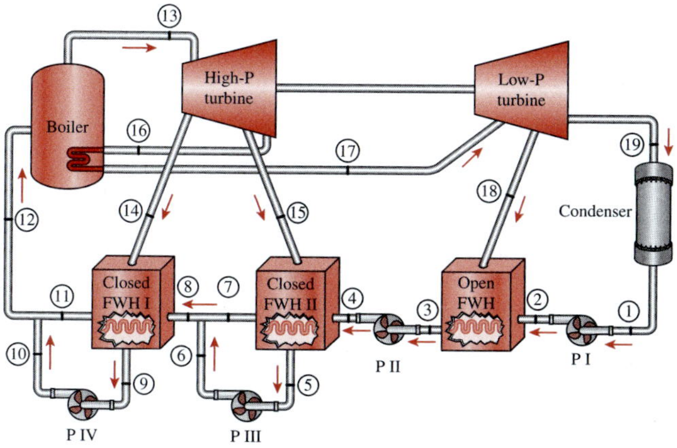

FIGURE P10–104

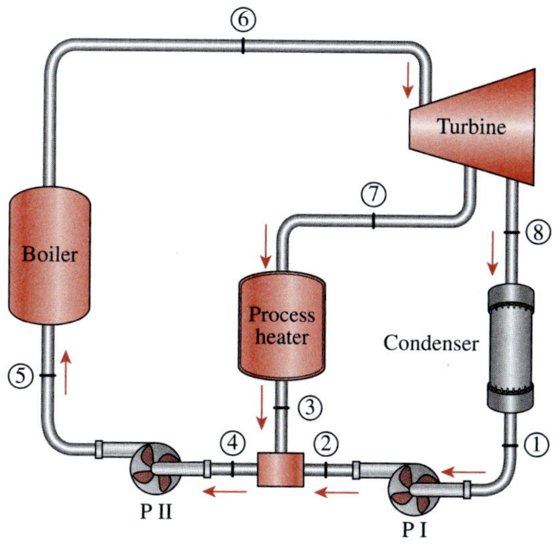

FIGURE P10–107

10–105 Reconsider Prob. 10–104. Using EES (or other) software, determine the optimum bleed pressure for the open feedwater heater that maximizes the thermal efficiency of the cycle.

10–106 Steam is to be supplied from a boiler to a high-pressure turbine whose isentropic efficiency is 75 percent at conditions to be determined. The steam is to leave the high-pressure turbine as a saturated vapor at 1.4 MPa, and the turbine is to produce 1 MW of power. Steam at the turbine exit is extracted at a rate of 1000 kg/min and routed to a process heater while the rest of the steam is supplied to a low-pressure turbine whose isentropic efficiency is 60 percent. The low-pressure turbine allows the steam to expand to 10 kPa pressure and produces 0.8 MW of power. Determine the temperature, pressure, and the flow rate of steam at the inlet of the high-pressure turbine.

10–107 A textile plant requires 4 kg/s of saturated steam at 2 MPa, which is extracted from the turbine of a cogeneration plant. Steam enters the turbine at 8 MPa and 500°C at a rate of 11 kg/s and leaves at 20 kPa. The extracted steam leaves the process heater as a saturated liquid and mixes with the feedwater at constant pressure. The mixture is pumped to the boiler pressure. Assuming an isentropic efficiency of 88 percent for both the turbine and the pumps, determine (*a*) the rate of process heat supply, (*b*) the net power output, and (*c*) the utilization factor of the plant. *Answers:* (*a*) 8.56 MW, (*b*) 8.60 MW, (*c*) 53.8 percent

10–108E Atmospheric air enters the air compressor of a simple combined gas-steam power system at 14.7 psia and 80°F. The air compressor's compression ratio is 10; the gas cycle's maximum temperature is 2100°F; and the air compressor and turbine have an isentropic efficiency of 90 percent. The gas leaves the heat exchanger 50°F hotter than the saturation temperature of the steam in the heat exchanger. The steam pressure in the heat exchanger is 800 psia, and the steam leaves the heat exchanger at 600°F. The steam-condenser pressure is 5 psia and the isentropic efficiency of the steam turbine is 95 percent. Determine the overall thermal efficiency of this combined cycle. For air, use constant specific heats at room temperature. *Answer:* 46.4 percent

10–109E It has been suggested that the steam passing through the condenser of the combined cycle in Prob. 10–108E be routed to buildings during the winter to heat them. When this is done, the pressure in the heating system where the steam is now condensed will have to be increased to 10 psia. How does this change the overall thermal efficiency of the combined cycle?

10–110E During winter, the system of Prob. 10–109E must supply 2×10^6 Btu/h of heat to the buildings. What is the mass flow rate of air through the air compressor and the system's total electrical power production in winter? *Answers:* 27,340 lbm/h, 1286 kW

10–111 The gas-turbine cycle of a combined gas–steam power plant has a pressure ratio of 12. Air enters the compressor at 310 K and the turbine at 1400 K. The combustion gases leaving the gas turbine are used to heat the steam at 12.5 MPa to 500°C in a heat exchanger. The combustion gases leave the heat exchanger at 247°C. Steam expands in a high-pressure turbine to a pressure of 2.5 MPa and is reheated in the combustion chamber to 550°C before it expands in a low-pressure turbine to 10 kPa. The mass flow rate of steam is 12 kg/s. Assuming all the compression and expansion processes to be isentropic, determine (*a*) the mass flow rate of air in the gas-turbine cycle, (*b*) the rate of total heat input, and (*c*) the thermal efficiency of the combined cycle. *Answers:* (*a*) 154 kg/s, (*b*) 1.44×10^5 kJ/s, (*c*) 59.1 percent

10–112 Repeat Prob. 10–111 assuming isentropic efficiencies of 100 percent for the pump, 80 percent for the compressor, and 85 percent for the gas and steam turbines.

10–113 An ideal Rankine steam cycle modified with two closed feedwater heaters and one open feedwater heater is shown below. The power cycle receives 100 kg/s of steam at the high pressure inlet to the turbine. The feedwater heater exit states for the boiler feedwater and the condensed steam are the normally assumed ideal states. Use the data provided in the tables given below to

(*a*) Sketch the *T-s* diagram for the ideal cycle.
(*b*) Determine the fraction of mass *y* extracted for the open feedwater heater.
(*c*) If, in addition to your result from part (*b*), the fraction of mass entering the high pressure turbine at state 7 extracted for the closed feedwater heater operating at 140 kPa is $z = 0.0655$, and at 1910 kPa the extraction fraction is $w = 0.0830$, determine the cooling water temperature rise in the condenser, in °C, when the cooling water flow rate is 4200 kg/s. Assume $c_p = 4.18$ kJ/kg·K for cooling water.
(*d*) Determine the rate of heat rejected in the condenser and the thermal efficiency of the plant.

Process states and selected data

State	*P*, kPa	*T*, °C	*h*, kJ/kg	*s*, kJ/kg·K
1	20			
2	620			
3	620			
4	620			
5	5000			
6	5000			
7	5000	700	3900	7.514
8	1910		3515	7.514
9	620		3154	7.514
10	140		2799	7.514
11	20		2478	7.514

Saturation data

P, kPa	T_{sat}, °C	v_f, m³/kg	h_f, kJ/kg	s_g, kJ/kg·K
20	60.1	0.00102	251	7.907
140	109.3	0.00105	458	7.246
620	160.1	0.00110	676	6.748
1910	210.1	0.00117	898	6.356
5000	263.9	0.00129	1154	5.973

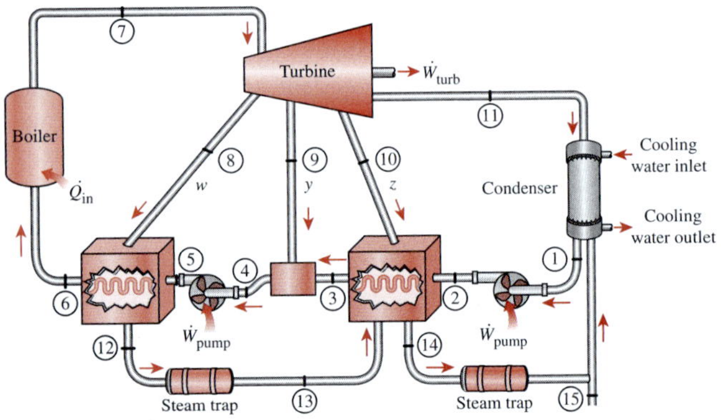

FIGURE P10–113

10–114 An ideal Rankine steam cycle modified for reheat, two closed feedwater heaters and a process heater is shown below. The power cycle produces 300 MW of net power. The feedwater heater exit states for the boiler feedwater and the condensed steam are the normally assumed ideal states. The fraction of mass entering the high pressure turbine at state 5 that is extracted for the feedwater heater operating at 1400 kPa is $y = 0.1160$. 15 percent ($z = 0.15$) of the steam entering the high pressure turbine is extracted from the low pressure turbine at 245 kPa for use in a constant pressure process heater and this steam is sent to the closed feedwater heater at 150 kPa by a steam trap. Use the data provided in the tables gives below to

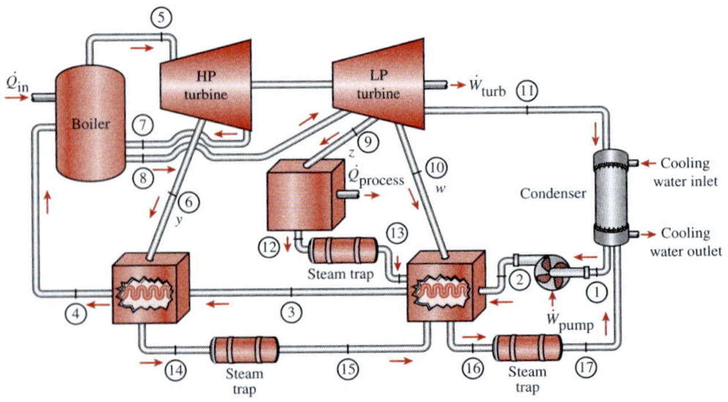

FIGURE 10–114

(*a*) Sketch the *T-s* diagram for the ideal cycle.
(*b*) Determine the fraction of mass, *w*, that is extracted for the closed feedwater heater operating at 150 kPa extraction pressure for the ideal cycle.
(*c*) Determine the mass flow rate through the boiler, the rate of process heat supplied, and the utilization efficiency of this cogeneration plant.

Process states and selected data

State	*P*, kPa	*T*, °C	*h*, kJ/kg	*s*, kJ/kg·K
1	20	60.11		
2	5000			
3	5000			
4	5000			
5	5000	700	3894	7.504
6	1400		3400	7.504
7	1200		3349	7.504
8	1200	600	3692	7.938
9	245		3154	7.938
10	150		3023	7.938
11	20		2620	7.938

Saturation data

P, kPa	v_f, m³/kg	h_f, kJ/kg	s_g, kJ/kg·K
20	0.00102	251.4	7.791
150		467	7.225
245		533	7.061
1200		798	6.519
1400		830	6.465
5000	0.00129	1154	5.979

10–115 Using EES (or other) software, investigate the effect of the condenser pressure on the performance of a simple ideal Rankine cycle. Turbine inlet conditions of steam are maintained constant at 10 MPa and 550°C while the condenser pressure is varied from 5 to 100 kPa. Determine the thermal efficiency of the cycle and plot it against the condenser pressure, and discuss the results.

10–116 Using EES (or other) software, investigate the effect of superheating the steam on the performance of a simple ideal Rankine cycle. Steam enters the turbine at 3 MPa and exits at 10 kPa. The turbine inlet temperature is varied from 250 to 1100°C. Determine the thermal efficiency of the cycle and plot it against the turbine inlet temperature, and discuss the results.

10–117 Using EES (or other) software, investigate the effect of number of reheat stages on the performance of an ideal Rankine cycle. The maximum and minimum pressures in the cycle are 15 MPa and 10 kPa, respectively, and steam enters all stages of the turbine at 500°C. For each case, maintain roughly the same pressure ratio across each turbine stage. Determine the thermal efficiency of the cycle and plot it against the number of reheat stages 1, 2, 4, and 8, and discuss the results.

10–118 Using EES (or other) software, investigate the effect of the number of regeneration stages on the performance of an ideal regenerative Rankine cycle. Steam enters the turbine at 10 MPa and 500°C and the condenser at 10 kPa. For each case, maintain about the same temperature difference between any two regeneration stages. Determine the thermal efficiency of the cycle, and plot it against the number of regeneration stages for 1, 2, 3, 4, 5, 6, 8, and 10 regeneration stages.

10–119 Show that the thermal efficiency of a combined gas–steam power plant η_{cc} can be expressed as

$$\eta_{cc} = \eta_g + \eta_s - \eta_g\eta_s$$

where $\eta_g = W_g/Q_{in}$ and $\eta_s = W_s/Q_{g,out}$ are the thermal efficiencies of the gas and steam cycles, respectively. Using this relation, determine the thermal efficiency of a combined power cycle that consists of a topping gas-turbine cycle with an efficiency of 40 percent and a bottoming steam-turbine cycle with an efficiency of 30 percent.

10–120 It can be shown that the thermal efficiency of a combined gas–steam power plant η_{cc} can be expressed in terms of the thermal efficiencies of the gas- and the steam-turbine cycles as

$$\eta_{cc} = \eta_g + \eta_s - \eta_g\eta_s$$

Prove that the value of η_{cc} is greater than either of η_g or η_s. That is, the combined cycle is more efficient than either of the gas-turbine or steam-turbine cycles alone.

10–121 Starting with Eq. 10–20, show that the exergy destruction associated with a simple ideal Rankine cycle can be expressed as $i = q_{in}(\eta_{th,Carnot} - \eta_{th})$, where η_{th} is efficiency of the Rankine cycle and $\eta_{th,Carnot}$ is the efficiency of the Carnot cycle operating between the same temperature limits.

Fundamentals of Engineering (FE) Exam Problems

10–122 Consider a simple ideal Rankine cycle. If the condenser pressure is lowered while keeping turbine inlet state the same,

(*a*) the turbine work output will decrease.
(*b*) the amount of heat rejected will decrease.
(*c*) the cycle efficiency will decrease.
(*d*) the moisture content at turbine exit will decrease.
(*e*) the pump work input will decrease.

10–123 Consider a simple ideal Rankine cycle with fixed boiler and condenser pressures. If the steam is superheated to a higher temperature,

(*a*) the turbine work output will decrease.
(*b*) the amount of heat rejected will decrease.
(*c*) the cycle efficiency will decrease.
(*d*) the moisture content at turbine exit will decrease.
(*e*) the amount of heat input will decrease.

10–124 Consider a simple ideal Rankine cycle with fixed boiler and condenser pressures. If the cycle is modified with reheating,

(*a*) the turbine work output will decrease.
(*b*) the amount of heat rejected will decrease.
(*c*) the pump work input will decrease.
(*d*) the moisture content at turbine exit will decrease.
(*e*) the amount of heat input will decrease.

10–125 Consider a simple ideal Rankine cycle with fixed boiler and condenser pressures. If the cycle is modified with regeneration that involves one open feedwater heater (select the correct statement per unit mass of steam flowing through the boiler),

(*a*) the turbine work output will decrease.
(*b*) the amount of heat rejected will increase.
(*c*) the cycle thermal efficiency will decrease.
(*d*) the quality of steam at turbine exit will decrease.
(*e*) the amount of heat input will increase.

10–126 Consider a steady-flow Carnot cycle with water as the working fluid executed under the saturation dome between the pressure limits of 3 MPa and 10 kPa. Water changes from saturated liquid to saturated vapor during the heat addition process. The net work output of this cycle is

(*a*) 666 kJ/kg (*b*) 888 kJ/kg (*c*) 1040 kJ/kg
(*d*) 1130 kJ/kg (*e*) 1440 kJ/kg

10–127 A simple ideal Rankine cycle operates between the pressure limits of 10 kPa and 3 MPa, with a turbine inlet temperature of 600°C. Disregarding the pump work, the cycle efficiency is

(*a*) 24 percent (*b*) 37 percent (*c*) 52 percent
(*d*) 63 percent (*e*) 71 percent

10–128 A simple ideal Rankine cycle operates between the pressure limits of 10 kPa and 5 MPa, with a turbine inlet temperature of 600°C. The mass fraction of steam that condenses at the turbine exit is

(*a*) 6 percent (*b*) 9 percent (*c*) 12 percent
(*d*) 15 percent (*e*) 18 percent

10–129 A steam power plant operates on the simple ideal Rankine cycle between the pressure limits of 10 kPa and 5 MPa, with a turbine inlet temperature of 600°C. The rate of heat transfer in the boiler is 300 kJ/s. Disregarding the pump work, the power output of this plant is

(*a*) 93 kW (*b*) 118 kW (*c*) 190 kW
(*d*) 216 kW (*e*) 300 kW

10–130 Consider a combined gas-steam power plant. Water for the steam cycle is heated in a well-insulated heat exchanger by the exhaust gases that enter at 800 K at a rate of 60 kg/s and leave at 400 K. Water enters the heat exchanger at 200°C and 8 MPa and leaves at 350°C and 8 MPa. If the exhaust gases are treated as air with constant specific heats at room temperature, the mass flow rate of water through the heat exchanger becomes

(*a*) 11 kg/s (*b*) 24 kg/s (*c*) 46 kg/s
(*d*) 53 kg/s (*e*) 60 kg/s

10–131 An ideal reheat Rankine cycle operates between the pressure limits of 10 kPa and 8 MPa, with reheat occurring at 4 MPa. The temperature of steam at the inlets of both turbines is 500°C, and the enthalpy of steam is 3185 kJ/kg at the exit of the high-pressure turbine, and 2247 kJ/kg at the exit of the low-pressure turbine. Disregarding the pump work, the cycle efficiency is

(*a*) 29 percent (*b*) 32 percent (*c*) 36 percent
(*d*) 41 percent (*e*) 49 percent

10–132 Pressurized feedwater in a steam power plant is to be heated in an ideal open feedwater heater that operates at a pressure of 2 MPa with steam extracted from the turbine. If the enthalpy of feedwater is 252 kJ/kg and the enthalpy of extracted steam is 2810 kJ/kg, the mass fraction of steam extracted from the turbine is

(*a*) 10 percent (*b*) 14 percent (*c*) 26 percent
(*d*) 36 percent (*e*) 50 percent

10–133 Consider a steam power plant that operates on the regenerative Rankine cycle with one open feedwater heater. The enthalpy of the steam is 3374 kJ/kg at the turbine inlet, 2797 kJ/kg at the location of bleeding, and 2346 kJ/kg at the turbine exit. The net power output of the plant is 120 MW, and the fraction of steam bled off the turbine for regeneration is 0.172. If the pump work is negligible, the mass flow rate of steam at the turbine inlet is

(*a*) 117 kg/s (*b*) 126 kg/s (*c*) 219 kg/s
(*d*) 268 kg/s (*e*) 679 kg/s

10–134 Consider a cogeneration power plant modified with regeneration. Steam enters the turbine at 6 MPa and 450°C at a rate of 20 kg/s and expands to a pressure of 0.4 MPa. At this pressure, 60 percent of the steam is extracted from the turbine, and the remainder expands to a pressure of 10 kPa. Part of the extracted steam is used to heat feedwater in an open feedwater heater. The rest of the extracted steam is used for process heating and leaves the process heater as a saturated liquid at 0.4 MPa. It is subsequently mixed with the feedwater leaving the feedwater heater, and the mixture is pumped to the boiler pressure. The steam in the condenser is cooled and condensed by the cooling water from a nearby river, which enters the adiabatic condenser at a rate of 463 kg/s.

1. The total power output of the turbine is

(*a*) 17.0 MW (*b*) 8.4 MW (*c*) 12.2 MW
(*d*) 20.0 MW (*e*) 3.4 MW

2. The temperature rise of the cooling water from the river in the condenser is

(*a*) 8.0°C (*b*) 5.2°C (*c*) 9.6°C
(*d*) 12.9°C (*e*) 16.2°C

3. The mass flow rate of steam through the process heater is

(*a*) 1.6 kg/s (*b*) 3.8 kg/s (*c*) 5.2 kg/s
(*d*) 7.6 kg/s (*e*) 10.4 kg/s

4. The rate of heat supply from the process heater per unit mass of steam passing through it is

(*a*) 246 kJ/kg (*b*) 893 kJ/kg (*c*) 1344 kJ/kg
(*d*) 1891 kJ/kg (*e*) 2060 kJ/kg

5. The rate of heat transfer to the steam in the boiler is

(*a*) 26.0 MJ/s (*b*) 53.8 MJ/s (*c*) 39.5 MJ/s
(*d*) 62.8 MJ/s (*e*) 125.4 MJ/s

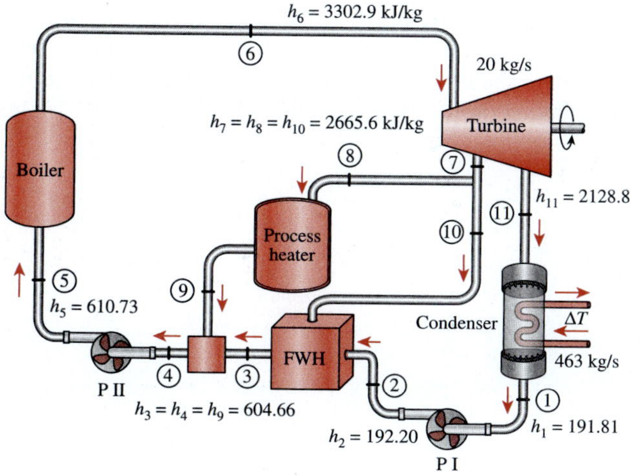

FIGURE P10–134

Design and Essay Problems

10–135 Design a steam power cycle that can achieve a cycle thermal efficiency of at least 40 percent under the conditions that all turbines have isentropic efficiencies of 85 percent and all pumps have isentropic efficiencies of 60 percent. Prepare an engineering report describing your design. Your design report must include, but is not limited to, the following:

(*a*) Discussion of various cycles attempted to meet the goal as well as the positive and negative aspects of your design.
(*b*) System figures and *T-s* diagrams with labeled states and temperature, pressure, enthalpy, and entropy information for your design.
(*c*) Sample calculations.

10–136 A 10-MW geothermal power plant is being considered at a site where geothermal water at 230°C is available. Geothermal water is to be flashed into a chamber to a lower pressure where part of the water evaporates. The liquid is returned to the ground while the vapor is used to drive the steam turbine. The pressures at the turbine inlet and the turbine exit are to remain above 200 kPa and 8 kPa, respectively. High-pressure flash chambers yield a small amount of steam with high exergy whereas lower-pressure flash chambers yield considerably more steam but at a lower exergy. By trying several pressures, determine the optimum pressure of the flash chamber to maximize the power production per unit mass of geothermal water withdrawn. Also, determine the thermal efficiency for each case assuming 10 percent of the power produced is used to drive the pumps and other auxiliary equipment.

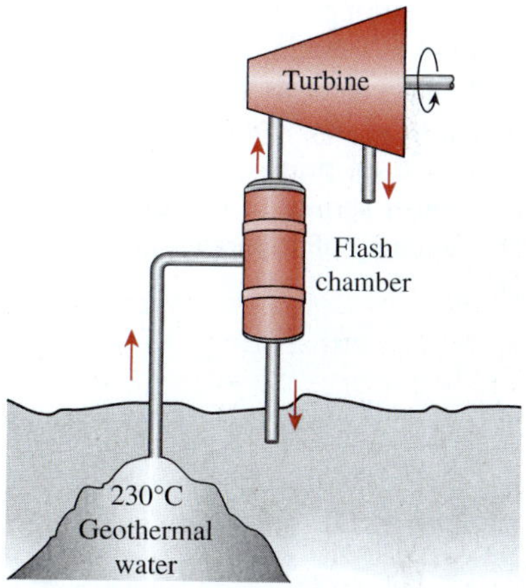

FIGURE P10–136

10–137 A natural gas–fired furnace in a textile plant is used to provide steam at 130°C. At times of high demand, the furnace supplies heat to the steam at a rate of 30 MJ/s. The plant also uses up to 6 MW of electrical power purchased from the local power company. The plant management is considering converting the existing process plant into a cogeneration plant to meet both their process-heat and power requirements. Your job is to come up with some designs. Designs based on a gas turbine or a steam turbine are to be considered. First decide whether a system based on a gas turbine or a steam turbine will best serve the purpose, considering the cost and the complexity. Then propose your design for the cogeneration plant complete with pressures and temperatures and the mass flow rates. Show that the proposed design meets the power and process-heat requirements of the plant.

10–138 Design the condenser of a steam power plant that has a thermal efficiency of 40 percent and generates 10 MW of net electric power. Steam enters the condenser as saturated vapor at 10 kPa, and it is to be condensed outside horizontal tubes through which cooling water from a nearby river flows. The temperature rise of the cooling water is limited to 8°C, and the velocity of the cooling water in the pipes is limited to 6 m/s to keep the pressure drop at an acceptable level. From prior experience, the average heat flux based on the outer surface of the tubes can be taken to be 12,000 W/m^2. Specify the pipe diameter, total pipe length, and the arrangement of the pipes to minimize the condenser volume.

10–139 The technology for power generation using geothermal energy is well established, and numerous geothermal power plants throughout the world are currently generating electricity economically. Binary geothermal plants utilize a volatile secondary fluid such as isobutane, n-pentane, and R-114 in a closed loop. Consider a binary geothermal plant with R-114 as the working fluid that is flowing at a rate of 600 kg/s. The R-114 is vaporized in a boiler at 115°C by the geothermal fluid that enters at 165°C, and is condensed at 30°C outside the tubes by cooling water that enters the tubes at 18°C. Based on prior experience, the average heat flux based on the outer surface of the tubes can be taken to be 4600 W/m^2. The enthalpy of vaporization of R-114 at 30°C is h_{fg} = 121.5 kJ/kg.

Specify (*a*) the length, diameter, and number of tubes and their arrangement in the condenser to minimize overall volume of the condenser; (*b*) the mass flow rate of cooling water; and (*c*) the flow rate of make-up water needed if a cooling tower is used to reject the waste heat from the cooling water. The liquid velocity is to remain under 6 m/s and the length of the tubes is limited to 8 m.

10–140 Stack gases exhausting from electrical power plants are at approximately 150°C. Design a basic Rankine cycle that uses water, refrigerant-134a, or ammonia as the working fluid and that produces the maximum amount of work from this energy source while rejecting heat to the ambient air at 40°C. You are to use a turbine whose efficiency is 92 percent and whose exit quality cannot be less than 85 percent.

10–141 Contact your power company and obtain information on the thermodynamic aspects of their most recently built power plant. If it is a conventional power plant, find out why it is preferred over a highly efficient combined power plant.

10–142 Water-cooled steam condensers are commonly used in steam power plants. Obtain information about water-cooled steam condensers by doing a literature search on the topic and also by contacting some condenser manufacturers. In a report, describe the various types, the way they are designed, the limitation on each type, and the selection criteria.

CHAPTER

11

REFRIGERATION CYCLES

A major application area of thermodynamics is *refrigeration*, which is the transfer of heat from a lower temperature region to a higher temperature one. Devices that produce refrigeration are called *refrigerators,* and the cycles on which they operate are called *refrigeration cycles*. The most frequently used refrigeration cycle is the *vapor-compression refrigeration cycle* in which the refrigerant is vaporized and condensed alternately and is compressed in the vapor phase. Another well-known refrigeration cycle is the *gas refrigeration cycle* in which the refrigerant remains in the gaseous phase throughout. Other refrigeration cycles discussed in this chapter are *cascade refrigeration*, where more than one refrigeration cycle is used; *absorption refrigeration*, where the refrigerant is dissolved in a liquid before it is compressed; and, as a Topic of Special Interest, *thermoelectric refrigeration,* where refrigeration is produced by the passage of electric current through two dissimilar materials.

Objectives

The objectives of Chapter 11 are to:

- Introduce the concepts of refrigerators and heat pumps and the measure of their performance.
- Analyze the ideal vapor-compression refrigeration cycle.
- Analyze the actual vapor-compression refrigeration cycle.
- Review the factors involved in selecting the right refrigerant for an application.
- Discuss the operation of refrigeration and heat pump systems.
- Evaluate the performance of innovative vapor-compression refrigeration systems.
- Analyze gas refrigeration systems.
- Introduce the concepts of absorption-refrigeration systems.
- Review the concepts of thermoelectric power generation and refrigeration.

11–1 ■ REFRIGERATORS AND HEAT PUMPS

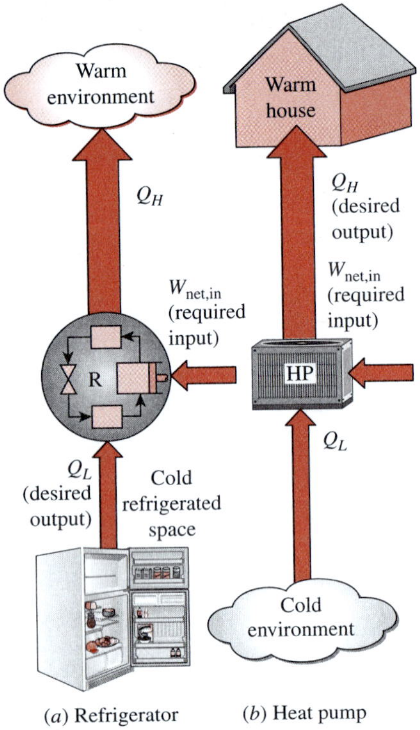

FIGURE 11–1
The objective of a refrigerator is to remove heat (Q_L) from the cold medium; the objective of a heat pump is to supply heat (Q_H) to a warm medium.

We all know from experience that heat flows in the direction of decreasing temperature, that is, from high-temperature regions to low-temperature ones. This heat-transfer process occurs in nature without requiring any devices. The reverse process, however, cannot occur by itself. The transfer of heat from a low-temperature region to a high-temperature one requires special devices called **refrigerators**.

Refrigerators are cyclic devices, and the working fluids used in the refrigeration cycles are called **refrigerants**. A refrigerator is shown schematically in Fig. 11–1*a*. Here Q_L is the magnitude of the heat removed from the refrigerated space at temperature T_L, Q_H is the magnitude of the heat rejected to the warm space at temperature T_H, and $W_{net,in}$ is the net work input to the refrigerator. As discussed in Chap. 6, Q_L and Q_H represent magnitudes and thus are positive quantities.

Another device that transfers heat from a low-temperature medium to a high-temperature one is the **heat pump**. Refrigerators and heat pumps are essentially the same devices; they differ in their objectives only. The objective of a refrigerator is to maintain the refrigerated space at a low temperature by removing heat from it. Discharging this heat to a higher-temperature medium is merely a necessary part of the operation, not the purpose. The objective of a heat pump, however, is to maintain a heated space at a high temperature. This is accomplished by absorbing heat from a low-temperature source, such as well water or cold outside air in winter, and supplying this heat to a warmer medium such as a house (Fig. 11–1*b*).

The performance of refrigerators and heat pumps is expressed in terms of the **coefficient of performance** (COP), defined as

$$\text{COP}_\text{R} = \frac{\text{Desired output}}{\text{Required input}} = \frac{\text{Cooling effect}}{\text{Work input}} = \frac{Q_L}{W_{net,in}} \quad \textbf{(11–1)}$$

$$\text{COP}_\text{HP} = \frac{\text{Desired output}}{\text{Required input}} = \frac{\text{Heating effect}}{\text{Work input}} = \frac{Q_H}{W_{net,in}} \quad \textbf{(11–2)}$$

These relations can also be expressed in the rate form by replacing the quantities Q_L, Q_H, and $W_{net,in}$ by $\dot{Q}_L$, $\dot{Q}_H$, and $\dot{W}_{net,in}$, respectively. Notice that both COP_R and COP_HP can be greater than 1. A comparison of Eqs. 11–1 and 11–2 reveals that

$$\text{COP}_\text{HP} = \text{COP}_\text{R} + 1 \quad \textbf{(11–3)}$$

for fixed values of Q_L and Q_H. This relation implies that $\text{COP}_\text{HP} > 1$ since COP_R is a positive quantity. That is, a heat pump functions, at worst, as a resistance heater, supplying as much energy to the house as it consumes. In reality, however, part of Q_H is lost to the outside air through piping and other devices, and COP_HP may drop below unity when the outside air temperature is too low. When this happens, the system normally switches to the fuel (natural gas, propane, oil, etc.) or resistance-heating mode.

The *cooling capacity* of a refrigeration system—that is, the rate of heat removal from the refrigerated space—is often expressed in terms of **tons of refrigeration**. The capacity of a refrigeration system that can freeze 1 ton

(2000 lbm) of liquid water at 0°C (32°F) into ice at 0°C in 24 h is said to be 1 ton. One ton of refrigeration is equivalent to 211 kJ/min or 200 Btu/min. The cooling load of a typical 200-m^2 residence is in the 3-ton (10-kW) range.

11–2 ▪ THE REVERSED CARNOT CYCLE

Recall from Chap. 6 that the Carnot cycle is a totally reversible cycle that consists of two reversible isothermal and two isentropic processes. It has the maximum thermal efficiency for given temperature limits, and it serves as a standard against which actual power cycles can be compared.

Since it is a reversible cycle, all four processes that comprise the Carnot cycle can be reversed. Reversing the cycle does also reverse the directions of any heat and work interactions. The result is a cycle that operates in the counterclockwise direction on a *T-s* diagram, which is called the **reversed Carnot cycle**. A refrigerator or heat pump that operates on the reversed Carnot cycle is called a **Carnot refrigerator** or a **Carnot heat pump**.

Consider a reversed Carnot cycle executed within the saturation dome of a refrigerant, as shown in Fig. 11–2. The refrigerant absorbs heat isothermally from a low-temperature source at T_L in the amount of Q_L (process 1-2), is compressed isentropically to state 3 (temperature rises to T_H), rejects heat isothermally to a high-temperature sink at T_H in the amount of Q_H (process 3-4), and expands isentropically to state 1 (temperature drops to T_L). The refrigerant changes from a saturated vapor state to a saturated liquid state in the condenser during process 3-4.

The coefficients of performance of Carnot refrigerators and heat pumps are expressed in terms of temperatures as

$$\mathrm{COP}_{\mathrm{R,Carnot}} = \frac{1}{T_H/T_L - 1} \tag{11–4}$$

and

$$\mathrm{COP}_{\mathrm{HP,Carnot}} = \frac{1}{1 - T_L/T_H} \tag{11–5}$$

Notice that both COPs increase as the difference between the two temperatures decreases, that is, as T_L rises or T_H falls.

The reversed Carnot cycle is the *most efficient* refrigeration cycle operating between two specified temperature levels. Therefore, it is natural to look at it first as a prospective ideal cycle for refrigerators and heat pumps. If we could, we certainly would adapt it as the ideal cycle. As explained below, however, the reversed Carnot cycle is not a suitable model for refrigeration cycles.

The two isothermal heat transfer processes are not difficult to achieve in practice since maintaining a constant pressure automatically fixes the temperature of a two-phase mixture at the saturation value. Therefore, processes 1-2 and 3-4 can be approached closely in actual evaporators and condensers. However, processes 2-3 and 4-1 cannot be approximated closely in practice. This is because process 2-3 involves the compression of a liquid–vapor mixture, which requires a compressor that will handle two phases, and process 4-1 involves the expansion of high-moisture-content refrigerant in a turbine.

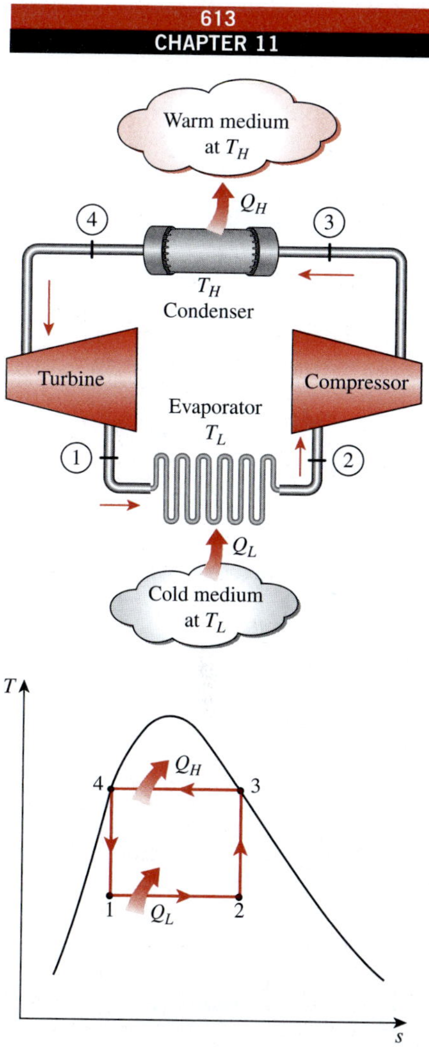

FIGURE 11–2
Schematic of a Carnot refrigerator and *T-s* diagram of the reversed Carnot cycle.

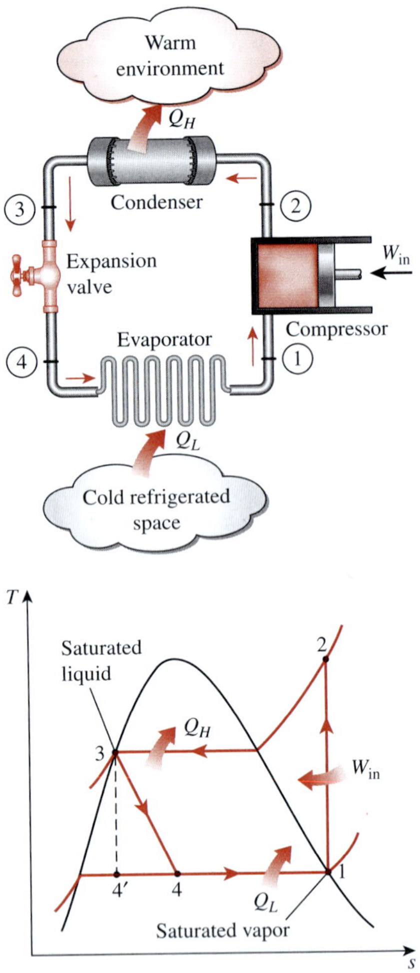

FIGURE 11–3
Schematic and *T-s* diagram for the ideal vapor-compression refrigeration cycle.

It seems as if these problems could be eliminated by executing the reversed Carnot cycle outside the saturation region. But in this case we have difficulty in maintaining isothermal conditions during the heat-absorption and heat-rejection processes. Therefore, we conclude that the reversed Carnot cycle cannot be approximated in actual devices and is not a realistic model for refrigeration cycles. However, the reversed Carnot cycle can serve as a standard against which actual refrigeration cycles are compared.

11–3 ▪ THE IDEAL VAPOR-COMPRESSION REFRIGERATION CYCLE

Many of the impracticalities associated with the reversed Carnot cycle can be eliminated by vaporizing the refrigerant completely before it is compressed and by replacing the turbine with a throttling device, such as an expansion valve or capillary tube. The cycle that results is called the **ideal vapor-compression refrigeration cycle**, and it is shown schematically and on a *T-s* diagram in Fig. 11–3. The vapor-compression refrigeration cycle is the most widely used cycle for refrigerators, air-conditioning systems, and heat pumps. It consists of four processes:

1-2 Isentropic compression in a compressor
2-3 Constant-pressure heat rejection in a condenser
3-4 Throttling in an expansion device
4-1 Constant-pressure heat absorption in an evaporator

In an ideal vapor-compression refrigeration cycle, the refrigerant enters the compressor at state 1 as saturated vapor and is compressed isentropically to the condenser pressure. The temperature of the refrigerant increases during this isentropic compression process to well above the temperature of the surrounding medium. The refrigerant then enters the condenser as superheated vapor at state 2 and leaves as saturated liquid at state 3 as a result of heat rejection to the surroundings. The temperature of the refrigerant at this state is still above the temperature of the surroundings.

The saturated liquid refrigerant at state 3 is throttled to the evaporator pressure by passing it through an expansion valve or capillary tube. The temperature of the refrigerant drops below the temperature of the refrigerated space during this process. The refrigerant enters the evaporator at state 4 as a low-quality saturated mixture, and it completely evaporates by absorbing heat from the refrigerated space. The refrigerant leaves the evaporator as saturated vapor and reenters the compressor, completing the cycle.

In a household refrigerator, the tubes in the freezer compartment where heat is absorbed by the refrigerant serves as the evaporator. The coils behind the refrigerator, where heat is dissipated to the kitchen air, serve as the condenser (Fig. 11–4).

Remember that the area under the process curve on a *T-s* diagram represents the heat transfer for internally reversible processes. The area under the process curve 4-1 represents the heat absorbed by the refrigerant in the evaporator, and the area under the process curve 2-3 represents the heat rejected in the condenser. A rule of thumb is that the *COP improves by 2 to 4 percent for each °C the evaporating temperature is raised or the condensing temperature is lowered.*

Another diagram frequently used in the analysis of vapor-compression refrigeration cycles is the *P-h* diagram, as shown in Fig. 11–5. On this diagram, three of the four processes appear as straight lines, and the heat transfer in the condenser and the evaporator is proportional to the lengths of the corresponding process curves.

Notice that unlike the ideal cycles discussed before, the ideal vapor-compression refrigeration cycle is not an internally reversible cycle since it involves an irreversible (throttling) process. This process is maintained in the cycle to make it a more realistic model for the actual vapor-compression refrigeration cycle. If the throttling device were replaced by an isentropic turbine, the refrigerant would enter the evaporator at state 4′ instead of state 4. As a result, the refrigeration capacity would increase (by the area under process curve 4′-4 in Fig. 11–3) and the net work input would decrease (by the amount of work output of the turbine). Replacing the expansion valve by a turbine is not practical, however, since the added benefits cannot justify the added cost and complexity.

All four components associated with the vapor-compression refrigeration cycle are steady-flow devices, and thus all four processes that make up the cycle can be analyzed as steady-flow processes. The kinetic and potential energy changes of the refrigerant are usually small relative to the work and heat transfer terms, and therefore they can be neglected. Then the steady-flow energy equation on a unit–mass basis reduces to

$$(q_{in} - q_{out}) + (w_{in} - w_{out}) = h_e - h_i \tag{11–6}$$

The condenser and the evaporator do not involve any work, and the compressor can be approximated as adiabatic. Then the COPs of refrigerators and heat pumps operating on the vapor-compression refrigeration cycle can be expressed as

$$\text{COP}_R = \frac{q_L}{w_{net,in}} = \frac{h_1 - h_4}{h_2 - h_1} \tag{11–7}$$

and

$$\text{COP}_{HP} = \frac{q_H}{w_{net,in}} = \frac{h_2 - h_3}{h_2 - h_1} \tag{11–8}$$

where $h_1 = h_{g @ P_1}$ and $h_3 = h_{f @ P_3}$ for the ideal case.

Vapor-compression refrigeration dates back to 1834 when the Englishman Jacob Perkins received a patent for a closed-cycle ice machine using ether or other volatile fluids as refrigerants. A working model of this machine was built, but it was never produced commercially. In 1850, Alexander Twining began to design and build vapor-compression ice machines using ethyl ether, which is a commercially used refrigerant in vapor-compression systems. Initially, vapor-compression refrigeration systems were large and were mainly used for ice making, brewing, and cold storage. They lacked automatic controls and were steam-engine driven. In the 1890s, electric motor-driven smaller machines equipped with automatic controls started to replace the older units, and refrigeration systems began to appear in butcher shops and households. By 1930, the continued improvements made it possible to have vapor-compression refrigeration systems that were relatively efficient, reliable, small, and inexpensive.

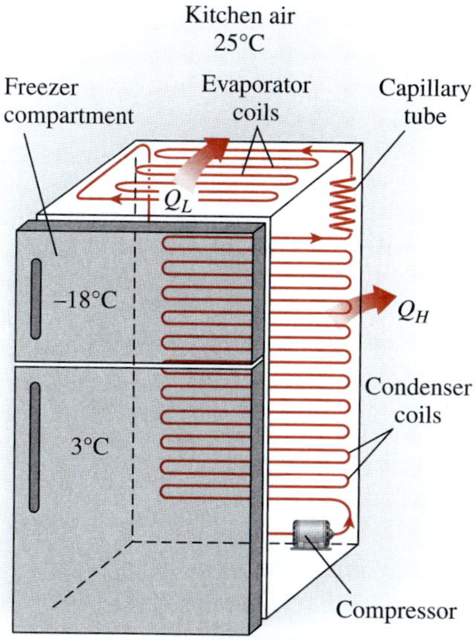

FIGURE 11–4
An ordinary household refrigerator.

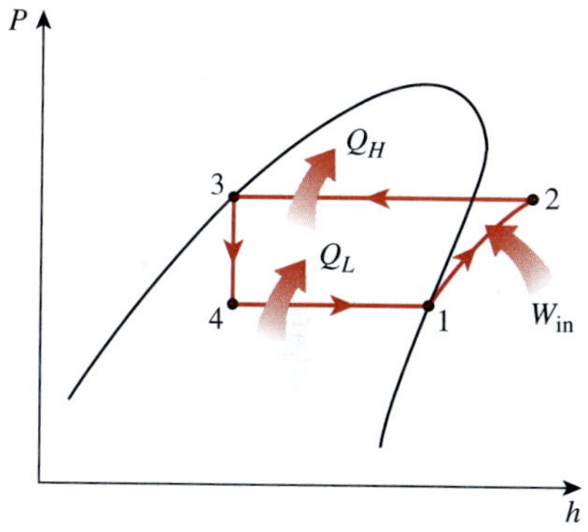

FIGURE 11–5
The *P-h* diagram of an ideal vapor-compression refrigeration cycle.

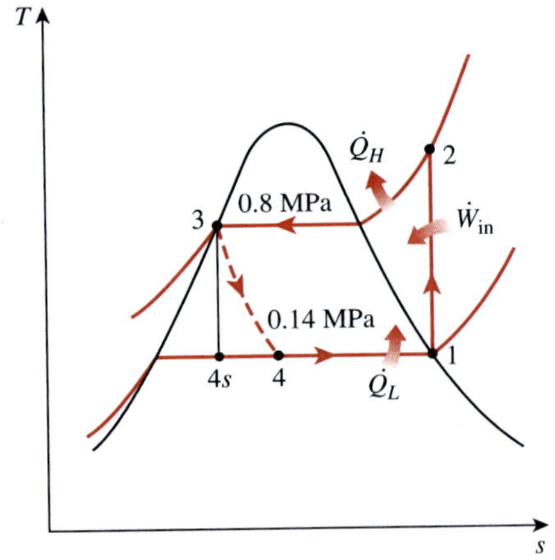

FIGURE 11–6
T-s diagram of the ideal vapor-compression refrigeration cycle described in Example 11–1.

EXAMPLE 11–1 The Ideal Vapor-Compression Refrigeration Cycle

A refrigerator uses refrigerant-134a as the working fluid and operates on an ideal vapor-compression refrigeration cycle between 0.14 and 0.8 MPa. If the mass flow rate of the refrigerant is 0.05 kg/s, determine (*a*) the rate of heat removal from the refrigerated space and the power input to the compressor, (*b*) the rate of heat rejection to the environment, and (*c*) the COP of the refrigerator.

SOLUTION A refrigerator operates on an ideal vapor-compression refrigeration cycle between two specified pressure limits. The rate of refrigeration, the power input, the rate of heat rejection, and the COP are to be determined.

Assumptions **1** Steady operating conditions exist. **2** Kinetic and potential energy changes are negligible.

Analysis The *T-s* diagram of the refrigeration cycle is shown in Fig. 11–6. We note that this is an ideal vapor-compression refrigeration cycle, and thus the compressor is isentropic and the refrigerant leaves the condenser as a saturated liquid and enters the compressor as saturated vapor. From the refrigerant-134a tables, the enthalpies of the refrigerant at all four states are determined as follows:

$$P_1 = 0.14 \text{ MPa} \longrightarrow h_1 = h_{g\,@\,0.14\text{ MPa}} = 239.16 \text{ kJ/kg}$$

$$s_1 = s_{g\,@\,0.14\text{ MPa}} = 0.94456 \text{ kJ/kg·K}$$

$$\left.\begin{array}{l} P_2 = 0.8 \text{ MPa} \\ s_2 = s_1 \end{array}\right\} \quad h_2 = 275.39 \text{ kJ/kg}$$

$$P_3 = 0.8 \text{ MPa} \longrightarrow h_3 = h_{f\,@\,0.8\text{ MPa}} = 95.47 \text{ kJ/kg}$$

$$h_4 \cong h_3 \text{ (throttling)} \longrightarrow h_4 = 95.47 \text{ kJ/kg}$$

(*a*) The rate of heat removal from the refrigerated space and the power input to the compressor are determined from their definitions:

$$\dot{Q}_L = \dot{m}(h_1 - h_4) = (0.05 \text{ kg/s})[(239.16 - 95.47) \text{ kJ/kg}] = \mathbf{7.18 \text{ kW}}$$

and

$$\dot{W}_{in} = \dot{m}(h_2 - h_1) = (0.05 \text{ kg/s})[(275.39 - 239.16) \text{ kJ/kg}] = \mathbf{1.81 \text{ kW}}$$

(*b*) The rate of heat rejection from the refrigerant to the environment is

$$\dot{Q}_H = \dot{m}(h_2 - h_3) = (0.05 \text{ kg/s})[(275.39 - 95.47) \text{ kJ/kg}] = \mathbf{9.0 \text{ kW}}$$

It could also be determined from

$$\dot{Q}_H = \dot{Q}_L + \dot{W}_{in} = 7.18 + 1.81 = 8.99 \text{ kW}$$

(*c*) The coefficient of performance of the refrigerator is

$$\text{COP}_R = \frac{\dot{Q}_L}{\dot{W}_{in}} = \frac{7.18 \text{ kW}}{1.81 \text{ kW}} = \mathbf{3.97}$$

That is, this refrigerator removes about 4 units of thermal energy from the refrigerated space for each unit of electric energy it consumes.

Discussion It would be interesting to see what happens if the throttling valve were replaced by an isentropic turbine. The enthalpy at state 4*s* (the turbine exit with P_{4s} = 0.14 MPa, and $s_{4s} = s_3$ = 0.35404 kJ/kg·K) is 88.94 kJ/kg, and the turbine would produce 0.33 kW of power. This would decrease the power input to the refrigerator from 1.81 to 1.48 kW and increase the rate of heat removal from the refrigerated space from 7.18 to 7.51 kW. As a result, the COP of the refrigerator would increase from 3.97 to 5.07, an increase of 28 percent.

11–4 ▪ ACTUAL VAPOR-COMPRESSION REFRIGERATION CYCLE

An actual vapor-compression refrigeration cycle differs from the ideal one in several ways, owing mostly to the irreversibilities that occur in various components. Two common sources of irreversibilities are fluid friction (causes pressure drops) and heat transfer to or from the surroundings. The *T-s* diagram of an actual vapor-compression refrigeration cycle is shown in Fig. 11–7.

In the ideal cycle, the refrigerant leaves the evaporator and enters the compressor as *saturated vapor*. In practice, however, it may not be possible to control the state of the refrigerant so precisely. Instead, it is easier to design the system so that the refrigerant is slightly superheated at the compressor inlet. This slight overdesign ensures that the refrigerant is completely vaporized when it enters the compressor. Also, the line connecting the evaporator to the compressor is usually very long; thus the pressure drop caused by fluid friction and heat transfer from the surroundings to the refrigerant can be very significant. The result of superheating, heat gain in the connecting line, and pressure drops in the evaporator and the connecting line is an increase in the specific volume, thus an increase in the power input requirements to the compressor since steady-flow work is proportional to the specific volume.

The *compression process* in the ideal cycle is internally reversible and adiabatic, and thus isentropic. The actual compression process, however, involves frictional effects, which increase the entropy, and heat transfer, which may increase or decrease the entropy, depending on the direction. Therefore, the entropy of the refrigerant may increase (process 1-2) or decrease (process 1-2′) during an actual compression process, depending on which effects dominate. The compression process 1-2′ may be even more desirable than the isentropic compression process since the specific volume of the refrigerant and thus the work input requirement are smaller in this case. Therefore, the refrigerant should be cooled during the compression process whenever it is practical and economical to do so.

In the ideal case, the refrigerant is assumed to leave the condenser as *saturated liquid* at the compressor exit pressure. In reality, however, it is unavoidable to have some pressure drop in the condenser as well as in the lines connecting the condenser to the compressor and to the throttling valve. Also, it is not easy to execute the condensation process with such precision

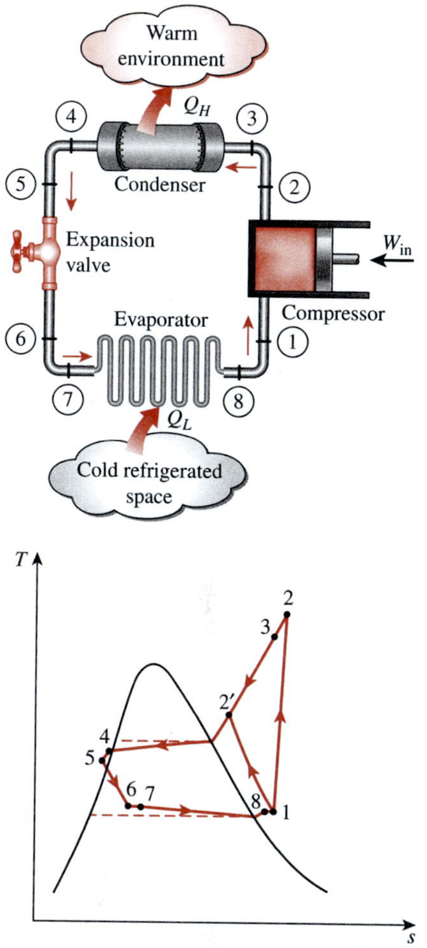

FIGURE 11–7
Schematic and *T-s* diagram for the actual vapor-compression refrigeration cycle.

that the refrigerant is a saturated liquid at the end, and it is undesirable to route the refrigerant to the throttling valve before the refrigerant is completely condensed. Therefore, the refrigerant is subcooled somewhat before it enters the throttling valve. We do not mind this at all, however, since the refrigerant in this case enters the evaporator with a lower enthalpy and thus can absorb more heat from the refrigerated space. The throttling valve and the evaporator are usually located very close to each other, so the pressure drop in the connecting line is small.

EXAMPLE 11–2 The Actual Vapor-Compression Refrigeration Cycle

Refrigerant-134a enters the compressor of a refrigerator as superheated vapor at 0.14 MPa and −10°C at a rate of 0.05 kg/s and leaves at 0.8 MPa and 50°C. The refrigerant is cooled in the condenser to 26°C and 0.72 MPa and is throttled to 0.15 MPa. Disregarding any heat transfer and pressure drops in the connecting lines between the components, determine (*a*) the rate of heat removal from the refrigerated space and the power input to the compressor, (*b*) the isentropic efficiency of the compressor, and (*c*) the coefficient of performance of the refrigerator.

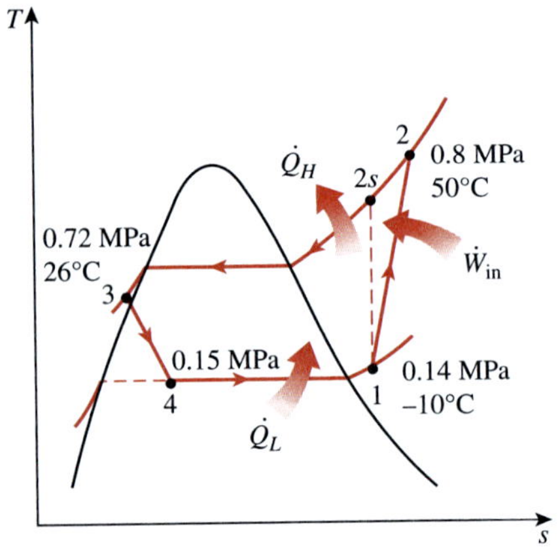

FIGURE 11–8
T-s diagram for Example 11–2.

SOLUTION A refrigerator operating on a vapor-compression cycle is considered. The rate of refrigeration, the power input, the compressor efficiency, and the COP are to be determined.

Assumptions **1** Steady operating conditions exist. **2** Kinetic and potential energy changes are negligible.

Analysis The *T-s* diagram of the refrigeration cycle is shown in Fig. 11–8. We note that the refrigerant leaves the condenser as a compressed liquid and enters the compressor as superheated vapor. The enthalpies of the refrigerant at various states are determined from the refrigerant tables to be

$$\left.\begin{array}{l} P_1 = 0.14\text{ MPa} \\ T_1 = -10°\text{C} \end{array}\right\} \quad h_1 = 246.36\text{ kJ/kg}$$

$$\left.\begin{array}{l} P_2 = 0.8\text{ MPa} \\ T_2 = 50°\text{C} \end{array}\right\} \quad h_2 = 286.69\text{ kJ/kg}$$

$$\left.\begin{array}{l} P_3 = 0.72\text{ MPa} \\ T_3 = 26°\text{C} \end{array}\right\} \quad h_3 \cong h_{f\,@\,26°\text{C}} = 87.83\text{ kJ/kg}$$

$$h_4 \cong h_3\ (\text{throttling}) \longrightarrow h_4 = 87.83\text{ kJ/kg}$$

(*a*) The rate of heat removal from the refrigerated space and the power input to the compressor are determined from their definitions:

$$\dot{Q}_L = \dot{m}(h_1 - h_4) = (0.05\text{ kg/s})[(246.36 - 87.83)\text{ kJ/kg}] = \mathbf{7.93\ kW}$$

and

$$\dot{W}_{\text{in}} = \dot{m}(h_2 - h_1) = (0.05\text{ kg/s})[(286.69 - 246.36)\text{ kJ/kg}] = \mathbf{2.02\ kW}$$

(*b*) The isentropic efficiency of the compressor is determined from

$$\eta_C \cong \frac{h_{2s} - h_1}{h_2 - h_1}$$

where the enthalpy at state 2*s* (P_{2s} = 0.8 MPa and $s_{2s} = s_1$ = 0.9724 kJ/kg·K) is 284.21 kJ/kg. Thus,

$$\eta_C = \frac{284.21 - 246.36}{286.69 - 246.36} = \mathbf{0.939 \text{ or } 93.9\%}$$

(*c*) The coefficient of performance of the refrigerator is

$$\text{COP}_\text{R} = \frac{\dot{Q}_L}{\dot{W}_\text{in}} = \frac{7.93 \text{ kW}}{2.02 \text{ kW}} = \mathbf{3.93}$$

Discussion This problem is identical to the one worked out in Example 11–1, except that the refrigerant is slightly superheated at the compressor inlet and subcooled at the condenser exit. Also, the compressor is not isentropic. As a result, the heat removal rate from the refrigerated space increases (by 10.4 percent), but the power input to the compressor increases even more (by 11.6 percent). Consequently, the COP of the refrigerator decreases from 3.97 to 3.93.

11–5 ▪ SECOND-LAW ANALYSIS OF VAPOR-COMPRESSION REFRIGERATION CYCLE[1]

Consider the vapor-compression refrigeration cycle operating between a low-temperature medium at T_L and a high-temperature medium at T_H as shown in Fig. 11–9. The maximum COP of a refrigeration cycle operating between temperature limits of T_L and T_H was given in Eq. 11–4 as

$$\text{COP}_{\text{R,max}} = \text{COP}_{\text{R,rev}} = \text{COP}_{\text{R,Carnot}} = \frac{T_L}{T_H - T_L} = \frac{1}{T_H/T_L - 1} \qquad \textbf{(11–9)}$$

Actual refrigeration cycles are not as efficient as ideal ones like the Carnot cycle because of the irreversibilities involved. But the conclusion we can draw from Eq. 11–9 that the COP is inversely proportional to the temperature difference $T_H - T_L$ is equally valid for actual refrigeration cycles.

The goal of a second-law or exergy analysis of a refrigeration system is to determine the components that can benefit the most by improvements. This is done identifying the locations of greatest exergy destruction and the components with the lowest exergy or second-law efficiency. Exergy destruction in a component can be determined directly from an exergy balance or indirectly by first calculating the entropy generation and then using the relation

$$\dot{X}_\text{dest} = T_0 \dot{S}_\text{gen} \qquad \textbf{(11–10)}$$

where T_0 is the environment (the dead state) temperature. For a refrigerator, T_0 is usually the temperature of the high-temperature medium T_H (for a heat pump it is T_L). Exergy destructions and exergy or the second-law efficiencies

[1]This section is largely contributed by Professor Mehmet Kanoglu of the University of Gaziantep.

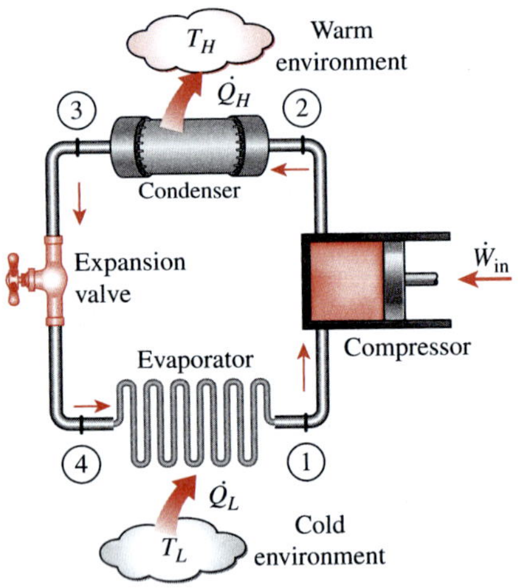

FIGURE 11–9
The vapor-compression refrigeration cycle considered in the second-law analysis.

for major components of a refrigeration system operating on the cycle shown in Fig. 11–9 may be written as follows:

Compressor:

$$\dot{X}_{\text{dest},1\text{-}2} = T_0\dot{S}_{\text{gen},1\text{-}2} = \dot{m}T_0(s_2 - s_1) \tag{11–11}$$

$$\eta_{\text{II,Comp}} = \frac{\dot{X}_{\text{recovered}}}{\dot{X}_{\text{expended}}} = \frac{\dot{W}_{\text{rev}}}{\dot{W}_{\text{act,in}}} = \frac{\dot{m}[h_2 - h_1 - T_0(s_2 - s_1)]}{\dot{m}(h_2 - h_1)} = \frac{\psi_2 - \psi_1}{h_2 - h_1}$$

$$= 1 - \frac{\dot{X}_{\text{dest},1-2}}{\dot{W}_{\text{act,in}}} \tag{11–12}$$

Condenser:

$$\dot{X}_{\text{dest},2\text{-}3} = T_0\dot{S}_{\text{gen},2\text{-}3} = T_0\left[\dot{m}(s_3 - s_2) + \frac{\dot{Q}_H}{T_H}\right] \tag{11–13}$$

$$\eta_{\text{II,Cond}} = \frac{\dot{X}_{\text{recovered}}}{\dot{X}_{\text{expended}}} = \frac{\dot{X}_{\dot{Q}_H}}{\dot{X}_2 - \dot{X}_3} = \frac{\dot{Q}_H(1 - T_0/T_H)}{\dot{X}_2 - \dot{X}_3}$$

$$= \frac{\dot{Q}_H(1 - T_0/T_H)}{\dot{m}[h_2 - h_3 - T_0(s_2 - s_3)]} = 1 - \frac{\dot{X}_{\text{dest, 2-3}}}{\dot{X}_2 - \dot{X}_3} \tag{11–14}$$

Note that when $T_H = T_0$, which is usually the case for refrigerators, $\eta_{\text{II,Cond}} = 0$ since there is no *recoverable* exergy in this case.

Expansion valve:

$$\dot{X}_{\text{dest},3\text{-}4} = T_0\dot{S}_{\text{gen},3\text{-}4} = \dot{m}T_0(s_4 - s_3) \tag{11–15}$$

$$\eta_{\text{II,ExpValve}} = \frac{\dot{X}_{\text{recovered}}}{\dot{X}_{\text{expended}}} = \frac{0}{\dot{X}_3 - \dot{X}_4} = 0 \quad \text{or}$$

$$\eta_{\text{II,ExpValve}} = 1 - \frac{\dot{X}_{\text{dest},3-4}}{\dot{X}_{\text{expended}}} = 1 - \frac{\dot{X}_3 - \dot{X}_4}{\dot{X}_3 - \dot{X}_4} = 0 \tag{11–16}$$

Evaporator:

$$\dot{X}_{\text{dest},4-1} = T_0\dot{S}_{\text{gen},4-1} = T_0\left[\dot{m}(s_1 - s_4) - \frac{\dot{Q}_L}{T_L}\right] \tag{11–17}$$

$$\eta_{\text{II,Evap}} = \frac{\dot{X}_{\text{recovered}}}{\dot{X}_{\text{expended}}} = \frac{\dot{X}_{\dot{Q}_L}}{\dot{X}_4 - \dot{X}_1} = \frac{\dot{Q}_L(T_0 - T_L)/T_L}{\dot{X}_4 - \dot{X}_1}$$

$$= \frac{\dot{Q}_L(T_0 - T_L)/T_L}{\dot{m}[h_4 - h_1 - T_0(s_4 - s_1)]} = 1 - \frac{\dot{X}_{\text{dest},4-1}}{\dot{X}_4 - \dot{X}_1} \tag{11–18}$$

Here $\dot{X}_{\dot{Q}_L}$ represents the positive of the exergy rate associated with the withdrawal of heat from the low-temperature medium at T_L at a rate $\dot{Q}_L$. Note that the directions of heat and exergy transfer become opposite when $T_L < T_0$ (that is, the exergy of the low-temperature medium increases as it loses heat). Also, $\dot{X}_{\dot{Q}_L}$ is equivalent to the power that can be produced by a

Carnot heat engine receiving heat from the environment at T_0 and rejecting heat to the low temperature medium at T_L at a rate of $\dot{Q}_L$, which can be shown to be

$$\dot{X}_{\dot{Q}_L} = \dot{Q}_L \frac{T_0 - T_L}{T_L} \tag{11-19}$$

From the definition of reversibility, this is equivalent to the minimum or reversible power input required to remove heat at a rate of $\dot{Q}_L$ and reject it to the environment at T_0. That is, $\dot{W}_{\text{rev,in}} = \dot{W}_{\text{min,in}} = \dot{X}_{\dot{Q}_L}$.

Note that when $T_L = T_0$, which is often the case for heat pumps, $\eta_{\text{II,Evap}} = 0$ since there is no recoverable exergy in this case.

The total exergy destruction associated with the cycle is the sum of the exergy destructions:

$$\dot{X}_{\text{dest,total}} = \dot{X}_{\text{dest},1-2} + \dot{X}_{\text{dest},2-3} + \dot{X}_{\text{dest},3-4} + \dot{X}_{\text{dest},4-1} \tag{11-20}$$

It can be shown that the total exergy destruction associated with a refrigeration cycle can also be obtained by taking the difference between the exergy supplied (power input) and the exergy recovered (the exergy of the heat withdrawn from the low-temperature medium):

$$\dot{X}_{\text{dest,total}} = \dot{W}_{\text{in}} - \dot{X}_{\dot{Q}_L} \tag{11-21}$$

The second-law or exergy efficiency of the cycle can then be expressed as

$$\eta_{\text{II,cycle}} = \frac{\dot{X}_{\dot{Q}_L}}{\dot{W}_{\text{in}}} = \frac{\dot{W}_{\text{min,in}}}{\dot{W}_{\text{in}}} = 1 - \frac{\dot{X}_{\text{dest,total}}}{\dot{W}_{\text{in}}} \tag{11-22}$$

Substituting $\dot{W}_{\text{in}} = \dfrac{\dot{Q}_L}{\text{COP}_\text{R}}$ and $\dot{X}_{\dot{Q}_L} = \dot{Q}_L \dfrac{T_0 - T_L}{T_L}$ into Eq. 11-22 gives

$$\eta_{\text{II,cycle}} = \frac{\dot{X}_{\dot{Q}_L}}{\dot{W}_{\text{in}}} = \frac{\dot{Q}_L(T_0 - T_L)/T_L}{\dot{Q}_L/\text{COP}_\text{R}} = \frac{\text{COP}_\text{R}}{T_L/(T_H - T_L)} = \frac{\text{COP}_\text{R}}{\text{COP}_\text{R,rev}} \tag{11-23}$$

since $T_0 = T_H$ for a refrigeration cycle. Thus, the second-law efficiency is also equal to the ratio of actual and maximum COPs for the cycle. This second-law efficiency definition accounts for all irreversibilities associated within the refrigerator, including the heat transfers with the refrigerated space and the environment.

EXAMPLE 11–3 Exergy Analysis of Vapor-Compression Refrigeration Cycle

A vapor-compression refrigeration cycle with refrigerant-134a as the working fluid is used to maintain a space at −13°C by rejecting heat to ambient air at 27°C. R-134a enters the compressor at 100 kPa superheated by 6.4°C at a rate of 0.05 kg/s. The isentropic efficiency of the compressor is 85 percent. The refrigerant leaves the condenser at 39.4°C as a saturated liquid. Determine (*a*) the rate of cooling provided and the COP of the system, (*b*) the exergy destruction in each basic component, (*c*) the minimum power input and the second-law efficiency of the cycle, and (*d*) the rate of total exergy destruction.

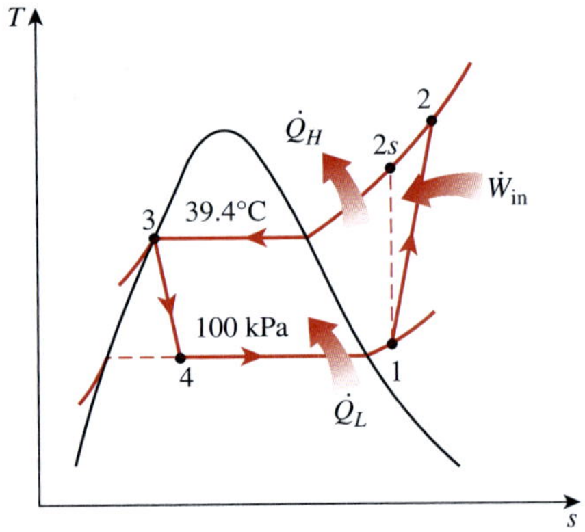

FIGURE 11–10
Temperature-entropy diagram of the vapor-compression refrigeration cycle considered in Example 11–3.

SOLUTION A vapor-compression refrigeration cycle is considered. The cooling rate, the COP, the exergy destructions, the minimum power input, the second-law efficiency, and the total exergy destruction are to be determined.

Assumptions **1** Steady operating conditions exist. **2** Kinetic and potential energy changes are negligible.

Analysis (*a*) The *T-s* diagram of the cycle is given in Fig. 11–10. The properties of R-134a are (Tables A–11 through A-13)

$$\left.\begin{aligned} P_1 &= 100 \text{ kPa} \\ T_1 &= T_{\text{sat@ 100 kPa}} + \Delta T_{\text{superheat}} \\ &= -26.4 + 6.4 = -20°\text{C} \end{aligned}\right\} \begin{aligned} h_1 &= 239.50 \text{ kJ/kg} \\ s_1 &= 0.9721 \text{ kJ/kg·K} \end{aligned}$$

$$P_3 = P_{\text{sat@ 39.4°C}} = 1000 \text{ kPa}$$

$$\left.\begin{aligned} P_2 &= P_3 = 1000 \text{ kPa} \\ s_2 &= s_1 = 0.9446 \text{ kJ/kg·K} \end{aligned}\right\} h_{2s} = 289.11 \text{ kJ/kg}$$

$$\left.\begin{aligned} P_3 &= 1000 \text{ kPa} \\ x_3 &= 0 \end{aligned}\right\} \begin{aligned} h_3 &= 107.32 \text{ kJ/kg} \\ s_3 &= 0.3919 \text{ kJ/kg·K} \end{aligned}$$

$$h_4 = h_3 = 107.32 \text{ kJ/kg}$$

$$\left.\begin{aligned} P_4 &= 100 \text{ kPa} \\ h_4 &= 107.32 \text{ kJ/kg} \end{aligned}\right\} s_4 = 0.4367 \text{ kJ/kg·K}$$

From the definition of isentropic efficiency,

$$\eta_C = \frac{h_{2s} - h_1}{h_2 - h_1}$$

$$0.85 = \frac{289.11 - 239.50}{h_2 - 239.50} \longrightarrow h_2 = 297.86 \text{ kJ/kg}$$

$$\left.\begin{aligned} P_2 &= 1000 \text{ kPa} \\ h_2 &= 297.86 \text{ kJ/kg} \end{aligned}\right\} s_2 = 0.9983 \text{ kJ/kg·K}$$

The refrigeration load, the rate of heat rejected, and the power input are

$$\dot{Q}_L = \dot{m}(h_1 - h_4) = (0.05 \text{ kg/s})(239.50 - 107.32)\text{kJ/kg} = \mathbf{6.609\ kW}$$

$$\dot{Q}_H = \dot{m}(h_2 - h_3) = (0.05 \text{ kg/s})(297.86 - 107.32)\text{kJ/kg} = 9.527 \text{ kW}$$

$$\dot{W}_{\text{in}} = \dot{m}(h_2 - h_1) = (0.05 \text{ kg/s})(297.86 - 239.50)\text{kJ/kg} = 2.918 \text{ kW}$$

Then the COP of the refrigeration cycle becomes

$$\text{COP}_\text{R} = \frac{\dot{Q}_L}{\dot{W}_{\text{in}}} = \frac{6.609 \text{ kW}}{2.918 \text{ kW}} = \mathbf{2.265}$$

(*b*) Noting that the dead-state temperature is $T_0 = T_H = 27 + 273 = 300$ K, the exergy destruction in each component of the cycle is determined as follows:

Compressor:

$$\begin{aligned} \dot{X}_{\text{dest,1-2}} &= T_0\dot{S}_{\text{gen1-2}} = T_0\dot{m}(s_2 - s_1) \\ &= (300 \text{ K})(0.05 \text{ kg/s})(0.9983 - 0.9721)\text{kJ/kg·K} \\ &= \mathbf{0.3940\ kW} \end{aligned}$$

Condenser:

$$\dot{X}_{\text{dest},2\text{-}3} = T_0\dot{S}_{\text{gen},2\text{-}3} = T_0\left[\dot{m}(s_3 - s_2) + \frac{\dot{Q}_H}{T_H}\right]$$

$$= (300\text{ K})\left[(0.05\text{ kg/s})(0.3919 - 0.9983)\text{ kJ/kg·K} + \frac{9.527\text{ kW}}{300\text{ K}}\right]$$

$$= \mathbf{0.4303\ kW}$$

Expansion valve:

$$\dot{X}_{\text{dest},3\text{-}4} = T_0\dot{S}_{\text{gen},3\text{-}4} = T_0\dot{m}(s_4 - s_3)$$

$$= (300\text{ K})(0.05\text{ kg/s})(0.4367 - 0.3919)\text{kJ/kg·K}$$

$$= \mathbf{0.6725\ kW}$$

Evaporator:

$$\dot{X}_{\text{dest},4\text{-}1} = T_0\dot{S}_{\text{gen},4\text{-}1} = T_0\left[\dot{m}(s_1 - s_4) - \frac{\dot{Q}_L}{T_L}\right]$$

$$= (300\text{ K})\left[(0.05\text{ kg/s})(0.9721 - 0.4367)\text{kJ/kg·K} - \frac{6.609\text{ kW}}{260\text{ K}}\right]$$

$$= \mathbf{0.4043\ kW}$$

(*c*) Exergy flow associated with heat transferred from the low-temperature medium is

$$\dot{X}_{\dot{Q}_L} = \dot{Q}_L\frac{T_0 - T_L}{T_L} = (6.609\text{ kW})\frac{300\text{ K} - 260\text{ K}}{260\text{ K}} = 1.017\text{ kW}$$

This is also the minimum or reversible power input for the cycle:

$$\dot{W}_{\text{min,in}} = \dot{X}_{\dot{Q}_L} = \mathbf{1.017\ kW}$$

The second-law efficiency of the cycle is

$$\eta_{\text{II}} = \frac{\dot{X}_{\dot{Q}_L}}{\dot{W}_{\text{in}}} = \frac{1.017\text{ kW}}{2.918\text{ kW}} = 0.3485 = \mathbf{34.9\%}$$

This efficiency may also be determined from $\eta_{\text{II}} = \text{COP}_\text{R}/\text{COP}_{\text{R,rev}}$ where

$$\text{COP}_{\text{R,rev}} = \frac{T_L}{T_H - T_L} = \frac{(-13 + 273)\text{ K}}{[27 - (-13)]\text{K}} = 6.500$$

Substituting,

$$\eta_{\text{II}} = \frac{\text{COP}_\text{R}}{\text{COP}_{\text{R,rev}}} = \frac{2.265}{6.500} = 0.3485 = \mathbf{34.9\%}$$

The results are identical, as expected.

(*d*) The total exergy destruction is the difference between the exergy expended (power input) and the exergy recovered (the exergy of the heat transferred from the low-temperature medium):

$$\dot{X}_{\text{dest,total}} = \dot{W}_{\text{in}} - \dot{X}_{\dot{Q}_L} = 2.918\text{ kW} - 1.017\text{ kW} = \mathbf{1.901\ kW}$$

The total exergy destruction can also be determined by adding exergy destruction in each component:

$$\dot{X}_{\text{dest,total}} = \dot{X}_{\text{dest},1-2} + \dot{X}_{\text{dest},2-3} + \dot{X}_{\text{dest},3-4} + \dot{X}_{\text{dest},4-1}$$
$$= 0.3940 + 0.4303 + 0.6725 + 0.4043$$
$$= 1.901 \text{ kW}$$

The two results are again identical, as expected.

Discussion The exergy input to the cycle is equal to the actual work input, which is 2.92 kW. The same cooling load could have been accomplished by only 34.9 percent of this power (1.02 kW) if a reversible system were used. The difference between the two is the exergy destructed in the cycle (1.90 kW). The expansion valve appears to be the most irreversible component, which accounts for 35.4 percent of the irreversibilities in the cycle. Replacing the expansion valve by a turbine would decrease the irreversibilites while decreasing the net power input. However, this may or may not be practical in an actual system. It can be shown that increasing the evaporating temperature and decreasing the condensing temperature would also decrease the exergy destruction in these components.

11–6 ▪ SELECTING THE RIGHT REFRIGERANT

When designing a refrigeration system, there are several refrigerants from which to choose, such as chlorofluorocarbons (CFCs), ammonia, hydrocarbons (propane, ethane, ethylene, etc.), carbon dioxide, air (in the air-conditioning of aircraft), and even water (in applications above the freezing point). The right choice of refrigerant depends on the situation at hand. Of these, refrigerants such as R-11, R-12, R-22, R-134a, and R-502 account for over 90 percent of the market in the United States.

Ethyl ether was the first commercially used refrigerant in vapor-compression systems in 1850, followed by ammonia, carbon dioxide, methyl chloride, sulphur dioxide, butane, ethane, propane, isobutane, gasoline, and chlorofluorocarbons, among others.

The industrial and heavy-commercial sectors were very satisfied with *ammonia,* and still are, although ammonia is toxic. The advantages of ammonia over other refrigerants are its low cost, higher COPs (and thus lower energy cost), more favorable thermodynamic and transport properties and thus higher heat transfer coefficients (requires smaller and lower-cost heat exchangers), greater detectability in the event of a leak, and no effect on the ozone layer. The major drawback of ammonia is its toxicity, which makes it unsuitable for domestic use. Ammonia is predominantly used in food refrigeration facilities such as the cooling of fresh fruits, vegetables, meat, and fish; refrigeration of beverages and dairy products such as beer, wine, milk, and cheese; freezing of ice cream and other foods; ice production; and low-temperature refrigeration in the pharmaceutical and other process industries.

It is remarkable that the early refrigerants used in the light-commercial and household sectors such as sulfur dioxide, ethyl chloride, and methyl chloride were highly toxic. The widespread publicity of a few instances of leaks that resulted in serious illnesses and death in the 1920s caused a public cry to ban or limit the use of these refrigerants, creating a need for the development of a safe refrigerant for household use. At the request of Frigidaire Corporation, General Motors' research laboratory developed R-21, the first member of the CFC family of refrigerants, within three days in 1928. Of several CFCs developed, the research team settled on R-12 as the refrigerant most suitable for commercial use and gave the CFC family the trade name "Freon." Commercial production of R-11 and R-12 was started in 1931 by a company jointly formed by General Motors and E. I. du Pont de Nemours and Co., Inc. The versatility and low cost of CFCs made them the refrigerants of choice. CFCs were also widely used in aerosols, foam insulations, and the electronic industry as solvents to clean computer chips.

R-11 is used primarily in large-capacity water chillers serving air-conditioning systems in buildings. R-12 is used in domestic refrigerators and freezers, as well as automotive air conditioners. R-22 is used in window air conditioners, heat pumps, air conditioners of commercial buildings, and large industrial refrigeration systems, and offers strong competition to ammonia. R-502 (a blend of R-115 and R-22) is the dominant refrigerant used in commercial refrigeration systems such as those in supermarkets because it allows low temperatures at evaporators while operating at single-stage compression.

The ozone crisis has caused a major stir in the refrigeration and air-conditioning industry and has triggered a critical look at the refrigerants in use. It was realized in the mid-1970s that CFCs allow more ultraviolet radiation into the earth's atmosphere by destroying the protective ozone layer and thus contributing to the greenhouse effect that causes global warming. As a result, the use of some CFCs is banned by international treaties. Fully halogenated CFCs (such as R-11, R-12, and R-115) do the most damage to the ozone layer. The nonfully halogenated refrigerants such as R-22 have about 5 percent of the ozone-depleting capability of R-12. Refrigerants that are friendly to the ozone layer that protects the earth from harmful ultraviolet rays have been developed. The once popular refrigerant R-12 has largely been replaced by the recently developed chlorine-free R-134a.

Two important parameters that need to be considered in the selection of a refrigerant are the temperatures of the two media (the refrigerated space and the environment) with which the refrigerant exchanges heat.

To have heat transfer at a reasonable rate, a temperature difference of 5 to 10°C should be maintained between the refrigerant and the medium with which it is exchanging heat. If a refrigerated space is to be maintained at -10°C, for example, the temperature of the refrigerant should remain at about -20°C while it absorbs heat in the evaporator. The lowest pressure in a refrigeration cycle occurs in the evaporator, and this pressure should be above atmospheric pressure to prevent any air leakage into the refrigeration system. Therefore, a refrigerant should have a saturation pressure of 1 atm or higher at -20°C in this particular case. Ammonia and R-134a are two such substances.

The temperature (and thus the pressure) of the refrigerant on the condenser side depends on the medium to which heat is rejected. Lower temperatures in the condenser (thus higher COPs) can be maintained if the refrigerant is cooled by liquid water instead of air. The use of water cooling cannot be justified economically, however, except in large industrial refrigeration systems. The temperature of the refrigerant in the condenser cannot fall below the temperature of the cooling medium (about 20°C for a household refrigerator), and the saturation pressure of the refrigerant at this temperature should be well below its critical pressure if the heat rejection process is to be approximately isothermal. If no single refrigerant can meet the temperature requirements, then two or more refrigeration cycles with different refrigerants can be used in series. Such a refrigeration system is called a *cascade system* and is discussed later in this chapter.

Other desirable characteristics of a refrigerant include being nontoxic, noncorrosive, nonflammable, and chemically stable; having a high enthalpy of vaporization (minimizes the mass flow rate); and, of course, being available at low cost.

In the case of heat pumps, the minimum temperature (and pressure) for the refrigerant may be considerably higher since heat is usually extracted from media that are well above the temperatures encountered in refrigeration systems.

11–7 ▪ HEAT PUMP SYSTEMS

Heat pumps are generally more expensive to purchase and install than other heating systems, but they save money in the long run in some areas because they lower the heating bills. Despite their relatively higher initial costs, the popularity of heat pumps is increasing. About one-third of all single-family homes built in the United States in the last decade are heated by heat pumps.

The most common energy source for heat pumps is atmospheric air (air-to-air systems), although water and soil are also used. The major problem with air-source systems is *frosting,* which occurs in humid climates when the temperature falls below 2 to 5°C. The frost accumulation on the evaporator coils is highly undesirable since it seriously disrupts heat transfer. The coils can be defrosted, however, by reversing the heat pump cycle (running it as an air conditioner). This results in a reduction in the efficiency of the system. Water-source systems usually use well water from depths of up to 80 m in the temperature range of 5 to 18°C, and they do not have a frosting problem. They typically have higher COPs but are more complex and require easy access to a large body of water such as underground water. Ground-source systems are also rather involved since they require long tubing placed deep in the ground where the soil temperature is relatively constant. The COP of heat pumps usually ranges between 1.5 and 4, depending on the particular system used and the temperature of the source. A new class of recently developed heat pumps that use variable-speed electric motor drives are at least twice as energy efficient as their predecessors.

Both the capacity and the efficiency of a heat pump fall significantly at low temperatures. Therefore, most air-source heat pumps require a supplementary heating system such as electric resistance heaters or an oil or gas furnace. Since water and soil temperatures do not fluctuate much, supplementary heating may not be required for water-source or ground-source systems. However, the heat pump system must be large enough to meet the maximum heating load.

Heat pumps and air conditioners have the same mechanical components. Therefore, it is not economical to have two separate systems to meet the heating and cooling requirements of a building. One system can be used as a heat pump in winter and an air conditioner in summer. This is accomplished by adding a reversing valve to the cycle, as shown in Fig. 11–11.

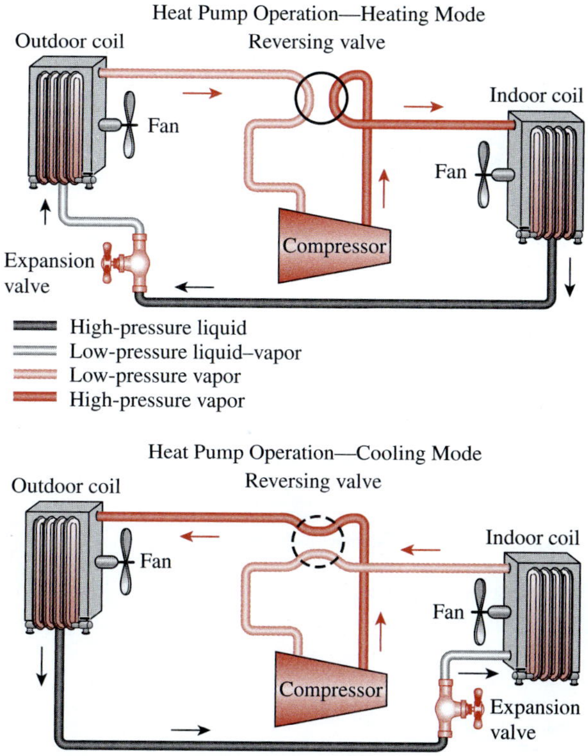

FIGURE 11–11
A heat pump can be used to heat a house in winter and to cool it in summer.

As a result of this modification, the condenser of the heat pump (located indoors) functions as the evaporator of the air conditioner in summer. Also, the evaporator of the heat pump (located outdoors) serves as the condenser of the air conditioner. This feature increases the competitiveness of the heat pump. Such dual-purpose units are commonly used in motels.

Heat pumps are most competitive in areas that have a large cooling load during the cooling season and a relatively small heating load during the heating season, such as in the southern parts of the United States. In these areas, the heat pump can meet the entire cooling and heating needs of residential or commercial buildings. The heat pump is least competitive in areas where the heating load is very large and the cooling load is small, such as in the northern parts of the United States.

11–8 ▪ INNOVATIVE VAPOR-COMPRESSION REFRIGERATION SYSTEMS

The simple vapor-compression refrigeration cycle discussed above is the most widely used refrigeration cycle, and it is adequate for most refrigeration applications. The ordinary vapor-compression refrigeration systems are simple, inexpensive, reliable, and practically maintenance-free (when was the last time you serviced your household refrigerator?). However, for large industrial applications *efficiency,* not simplicity, is the major concern. Also, for some applications the simple vapor-compression refrigeration cycle is inadequate and needs to be modified. We now discuss a few such modifications and refinements.

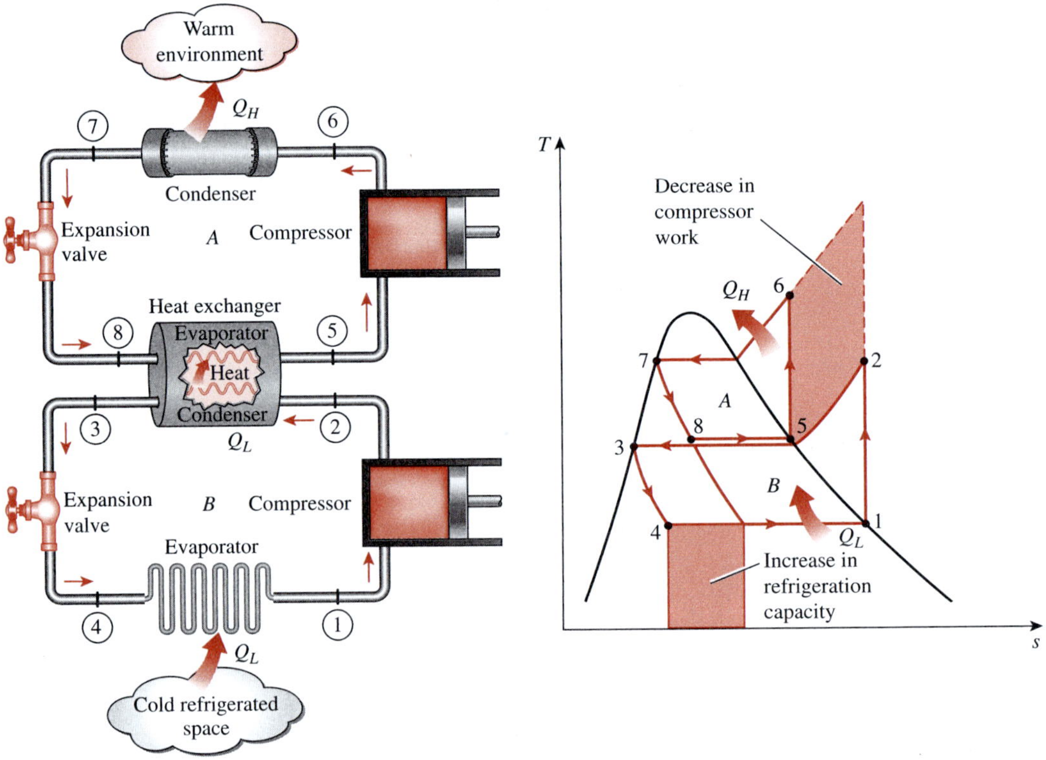

FIGURE 11–12
A two-stage cascade refrigeration system with the same refrigerant in both stages.

Cascade Refrigeration Systems

Some industrial applications require moderately low temperatures, and the temperature range they involve may be too large for a single vapor-compression refrigeration cycle to be practical. A large temperature range also means a large pressure range in the cycle and a poor performance for a reciprocating compressor. One way of dealing with such situations is to perform the refrigeration process in stages, that is, to have two or more refrigeration cycles that operate in series. Such refrigeration cycles are called **cascade refrigeration cycles**.

A two-stage cascade refrigeration cycle is shown in Fig. 11–12. The two cycles are connected through the heat exchanger in the middle, which serves as the evaporator for the topping cycle (cycle *A*) and the condenser for the bottoming cycle (cycle *B*). Assuming the heat exchanger is well insulated and the kinetic and potential energies are negligible, the heat transfer from the fluid in the bottoming cycle should be equal to the heat transfer to the fluid in the topping cycle. Thus, the ratio of mass flow rates through each cycle should be

$$\dot{m}_A(h_5 - h_8) = \dot{m}_B(h_2 - h_3) \quad \longrightarrow \quad \frac{\dot{m}_A}{\dot{m}_B} = \frac{h_2 - h_3}{h_5 - h_8} \tag{11–24}$$

Also,

$$\text{COP}_{\text{R,cascade}} = \frac{\dot{Q}_L}{\dot{W}_{\text{net,in}}} = \frac{\dot{m}_B(h_1 - h_4)}{\dot{m}_A(h_6 - h_5) + \dot{m}_B(h_2 - h_1)} \tag{11-25}$$

In the cascade system shown in the figure, the refrigerants in both cycles are assumed to be the same. This is not necessary, however, since there is no mixing taking place in the heat exchanger. Therefore, refrigerants with more desirable characteristics can be used in each cycle. In this case, there would be a separate saturation dome for each fluid, and the *T-s* diagram for one of the cycles would be different. Also, in actual cascade refrigeration systems, the two cycles would overlap somewhat since a temperature difference between the two fluids is needed for any heat transfer to take place.

It is evident from the *T-s* diagram in Fig. 11–12 that the compressor work decreases and the amount of heat absorbed from the refrigerated space increases as a result of cascading. Therefore, cascading improves the COP of a refrigeration system. Some refrigeration systems use three or four stages of cascading.

EXAMPLE 11–4 A Two-Stage Cascade Refrigeration Cycle

Consider a two-stage cascade refrigeration system operating between the pressure limits of 0.8 and 0.14 MPa. Each stage operates on an ideal vapor-compression refrigeration cycle with refrigerant-134a as the working fluid. Heat rejection from the lower cycle to the upper cycle takes place in an adiabatic counterflow heat exchanger where both streams enter at about 0.32 MPa. (In practice, the working fluid of the lower cycle is at a higher pressure and temperature in the heat exchanger for effective heat transfer.) If the mass flow rate of the refrigerant through the upper cycle is 0.05 kg/s, determine (*a*) the mass flow rate of the refrigerant through the lower cycle, (*b*) the rate of heat removal from the refrigerated space and the power input to the compressor, and (*c*) the coefficient of performance of this cascade refrigerator.

SOLUTION A cascade refrigeration system operating between the specified pressure limits is considered. The mass flow rate of the refrigerant through the lower cycle, the rate of refrigeration, the power input, and the COP are to be determined.

Assumptions **1** Steady operating conditions exist. **2** Kinetic and potential energy changes are negligible. **3** The heat exchanger is adiabatic.

Properties The enthalpies of the refrigerant at all eight states are determined from the refrigerant tables and are indicated on the *T-s* diagram.

Analysis The *T-s* diagram of the refrigeration cycle is shown in Fig. 11–13. The topping cycle is labeled cycle *A* and the bottoming one, cycle *B*. For both cycles, the refrigerant leaves the condenser as a saturated liquid and enters the compressor as saturated vapor.

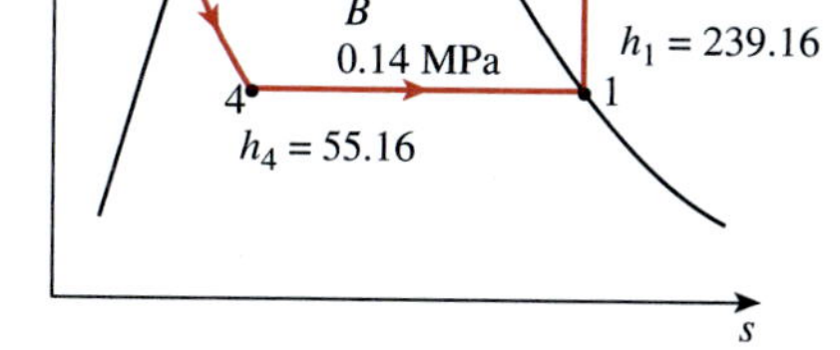

FIGURE 11–13
T-s diagram of the cascade refrigeration cycle described in Example 11–4.

(*a*) The mass flow rate of the refrigerant through the lower cycle is determined from the steady-flow energy balance on the adiabatic heat exchanger,

$$\dot{E}_{\text{out}} = \dot{E}_{\text{in}} \longrightarrow \dot{m}_A h_5 + \dot{m}_B h_3 = \dot{m}_A h_8 + \dot{m}_B h_2$$

$$\dot{m}_A(h_5 - h_8) = \dot{m}_B(h_2 - h_3)$$

$$(0.05 \text{ kg/s})[(251.88 - 95.47) \text{ kJ/kg}] = \dot{m}_B[(255.93 - 55.16) \text{ kJ/kg}]$$

$$\dot{m}_B = \mathbf{0.0390 \text{ kg/s}}$$

(*b*) The rate of heat removal by a cascade cycle is the rate of heat absorption in the evaporator of the lowest stage. The power input to a cascade cycle is the sum of the power inputs to all of the compressors:

$$\dot{Q}_L = \dot{m}_B(h_1 - h_4) = (0.0390 \text{ kg/s})[(239.16 - 55.16) \text{ kJ/kg}] = \mathbf{7.18 \text{ kW}}$$

$$\begin{aligned}\dot{W}_{\text{in}} &= \dot{W}_{\text{comp I,in}} + \dot{W}_{\text{comp II,in}} = \dot{m}_A(h_6 - h_5) + \dot{m}_B(h_2 - h_1)\\ &= (0.05 \text{ kg/s})[(270.92 - 251.88) \text{ kJ/kg}]\\ &\quad + (0.039 \text{ kg/s})[(255.93 - 239.16) \text{ kJ/kg}]\\ &= \mathbf{1.61 \text{ kW}}\end{aligned}$$

(*c*) The COP of a refrigeration system is the ratio of the refrigeration rate to the net power input:

$$\text{COP}_\text{R} = \frac{\dot{Q}_L}{\dot{W}_{\text{net,in}}} = \frac{7.18 \text{ kW}}{1.61 \text{ kW}} = \mathbf{4.46}$$

Discussion This problem was worked out in Example 11–1 for a single-stage refrigeration system. Notice that the COP of the refrigeration system increases from 3.97 to 4.46 as a result of cascading. The COP of the system can be increased even more by increasing the number of cascade stages.

Multistage Compression Refrigeration Systems

When the fluid used throughout the cascade refrigeration system is the same, the heat exchanger between the stages can be replaced by a mixing chamber (called a *flash chamber*) since it has better heat transfer characteristics. Such systems are called **multistage compression refrigeration systems**. A two-stage compression refrigeration system is shown in Fig. 11–14.

In this system, the liquid refrigerant expands in the first expansion valve to the flash chamber pressure, which is the same as the compressor interstage pressure. Part of the liquid vaporizes during this process. This saturated vapor (state 3) is mixed with the superheated vapor from the low-pressure compressor (state 2), and the mixture enters the high-pressure compressor at state 9. This is, in essence, a regeneration process. The saturated liquid (state 7) expands through the second expansion valve into the evaporator, where it picks up heat from the refrigerated space.

The compression process in this system resembles a two-stage compression with intercooling, and the compressor work decreases. Care should be exercised in the interpretations of the areas on the *T-s* diagram in this case since the mass flow rates are different in different parts of the cycle.

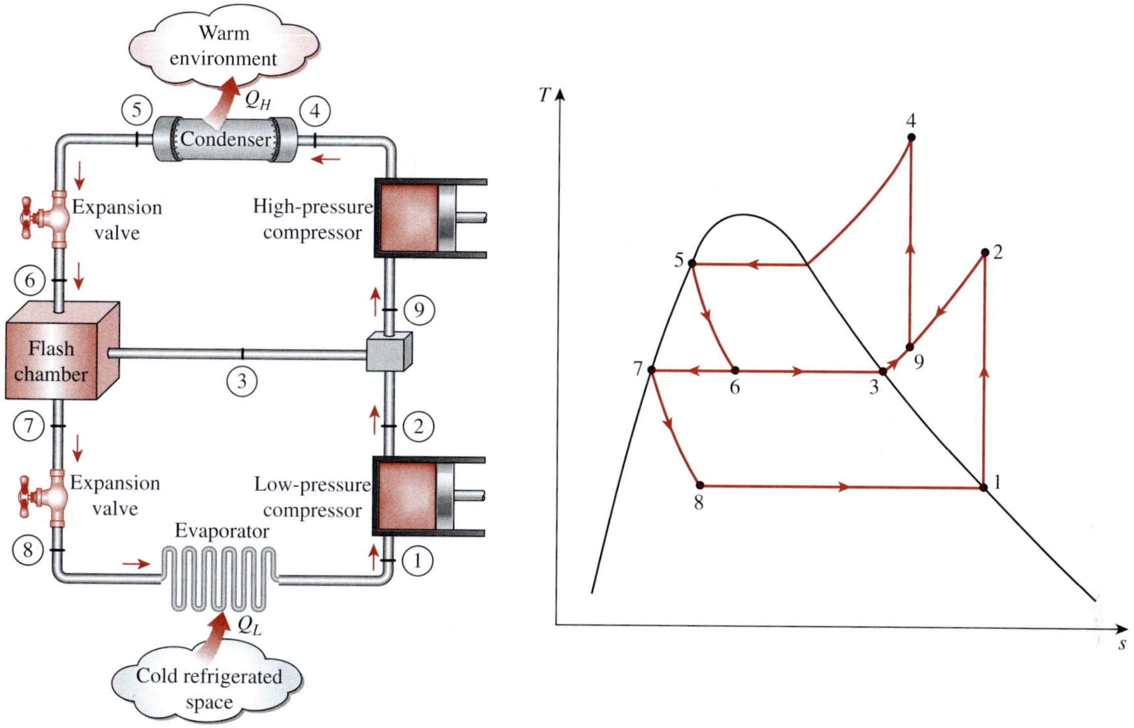

FIGURE 11–14
A two-stage compression refrigeration system with a flash chamber.

EXAMPLE 11–5 A Two-Stage Refrigeration Cycle with a Flash Chamber

Consider a two-stage compression refrigeration system operating between the pressure limits of 0.8 and 0.14 MPa. The working fluid is refrigerant-134a. The refrigerant leaves the condenser as a saturated liquid and is throttled to a flash chamber operating at 0.32 MPa. Part of the refrigerant evaporates during this flashing process, and this vapor is mixed with the refrigerant leaving the low-pressure compressor. The mixture is then compressed to the condenser pressure by the high-pressure compressor. The liquid in the flash chamber is throttled to the evaporator pressure and cools the refrigerated space as it vaporizes in the evaporator. Assuming the refrigerant leaves the evaporator as a saturated vapor and both compressors are isentropic, determine (*a*) the fraction of the refrigerant that evaporates as it is throttled to the flash chamber, (*b*) the amount of heat removed from the refrigerated space and the compressor work per unit mass of refrigerant flowing through the condenser, and (*c*) the coefficient of performance.

SOLUTION A two-stage compression refrigeration system operating between specified pressure limits is considered. The fraction of the refrigerant that evaporates in the flash chamber, the refrigeration and work input per unit mass, and the COP are to be determined.

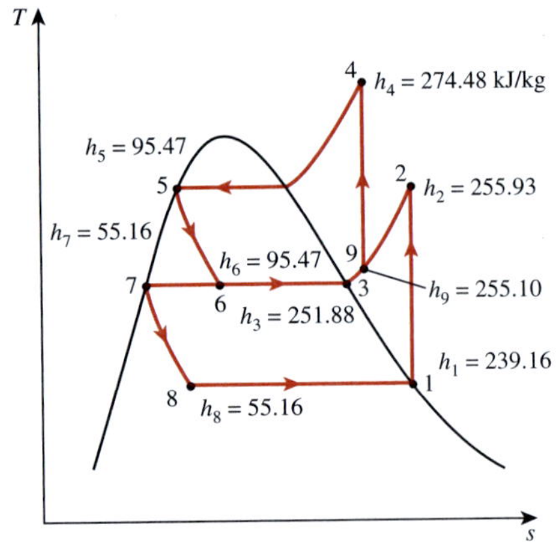

FIGURE 11–15

T-s diagram of the two-stage compression refrigeration cycle described in Example 11–5.

Assumptions **1** Steady operating conditions exist. **2** Kinetic and potential energy changes are negligible. **3** The flash chamber is adiabatic.

Properties The enthalpies of the refrigerant at various states are determined from the refrigerant tables and are indicated on the *T-s* diagram.

Analysis The *T-s* diagram of the refrigeration cycle is shown in Fig. 11–15. We note that the refrigerant leaves the condenser as saturated liquid and enters the low-pressure compressor as saturated vapor.

(*a*) The fraction of the refrigerant that evaporates as it is throttled to the flash chamber is simply the quality at state 6, which is

$$x_6 = \frac{h_6 - h_f}{h_{fg}} = \frac{95.47 - 55.16}{196.71} = \mathbf{0.2049}$$

(*b*) The amount of heat removed from the refrigerated space and the compressor work input per unit mass of refrigerant flowing through the condenser are

$$q_L = (1 - x_6)(h_1 - h_8)$$
$$= (1 - 0.2049)[(239.16 - 55.16)\text{ kJ/kg}] = \mathbf{146.3\ kJ/kg}$$

and

$$w_{in} = w_{comp\ I,in} + w_{comp\ II,in} = (1 - x_6)(h_2 - h_1) + (1)(h_4 - h_9)$$

The enthalpy at state 9 is determined from an energy balance on the mixing chamber,

$$\dot{E}_{out} = \dot{E}_{in}$$
$$(1)h_9 = x_6 h_3 + (1 - x_6)h_2$$
$$h_9 = (0.2049)(251.88) + (1 - 0.2049)(255.93) = 255.10\text{ kJ/kg}$$

Also, $s_9 = 0.9416$ kJ/kg·K. Thus the enthalpy at state 4 (0.8 MPa, $s_4 = s_9$) is $h_4 = 274.48$ kJ/kg. Substituting,

$$w_{in} = (1 - 0.2049)[(255.93 - 239.16)\text{ kJ/kg}] + (274.48 - 255.10)\text{ kJ/kg}$$
$$= \mathbf{32.71\ kJ/kg}$$

(*c*) The coefficient of performance is

$$\text{COP}_R = \frac{q_L}{w_{in}} = \frac{146.3\text{ kJ/kg}}{32.71\text{ kJ/kg}} = \mathbf{4.47}$$

Discussion This problem was worked out in Example 11–1 for a single-stage refrigeration system (COP = 3.97) and in Example 11–4 for a two-stage cascade refrigeration system (COP = 4.46). Notice that the COP of the refrigeration system increased considerably relative to the single-stage compression but did not change much relative to the two-stage cascade compression.

Multipurpose Refrigeration Systems with a Single Compressor

Some applications require refrigeration at more than one temperature. This could be accomplished by using a separate throttling valve and a separate compressor for each evaporator operating at different temperatures. However,

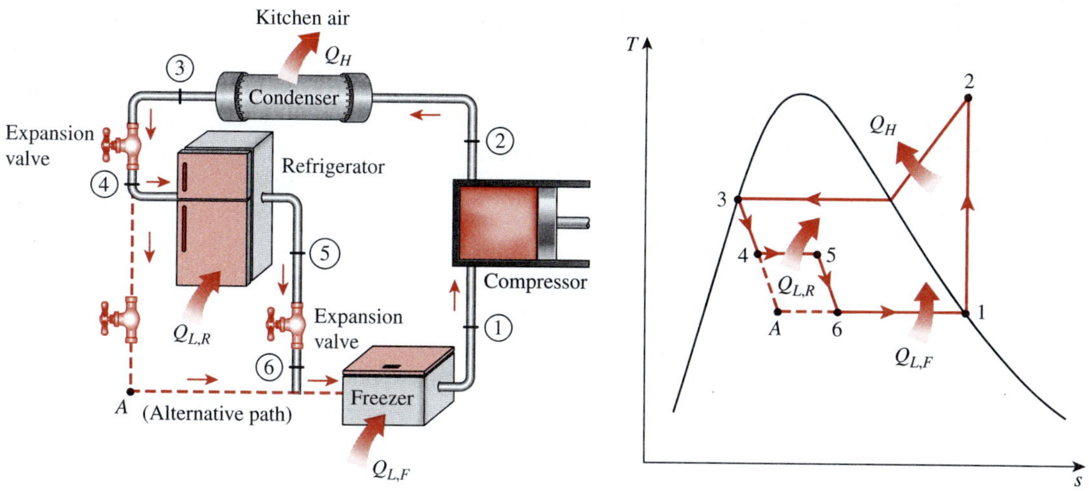

FIGURE 11–16

Schematic and *T-s* diagram for a refrigerator–freezer unit with one compressor.

such a system is bulky and probably uneconomical. A more practical and economical approach would be to route all the exit streams from the evaporators to a single compressor and let it handle the compression process for the entire system.

Consider, for example, an ordinary refrigerator–freezer unit. A simplified schematic of the unit and the *T-s* diagram of the cycle are shown in Fig. 11–16. Most refrigerated goods have a high water content, and the refrigerated space must be maintained above the ice point to prevent freezing. The freezer compartment, however, is maintained at about −18°C. Therefore, the refrigerant should enter the freezer at about −25°C to have heat transfer at a reasonable rate in the freezer. If a single expansion valve and evaporator were used, the refrigerant would have to circulate in both compartments at about −25°C, which would cause ice formation in the neighborhood of the evaporator coils and dehydration of the produce. This problem can be eliminated by throttling the refrigerant to a higher pressure (hence temperature) for use in the refrigerated space and then throttling it to the minimum pressure for use in the freezer. The entire refrigerant leaving the freezer compartment is subsequently compressed by a single compressor to the condenser pressure.

Liquefaction of Gases

The liquefaction of gases has always been an important area of refrigeration since many important scientific and engineering processes at cryogenic temperatures (temperatures below about −100°C) depend on liquefied gases. Some examples of such processes are the separation of oxygen and nitrogen from air, preparation of liquid propellants for rockets, the study of material properties at low temperatures, and the study of some exciting phenomena such as superconductivity.

At temperatures above the critical-point value, a substance exists in the gas phase only. The critical temperatures of helium, hydrogen, and nitrogen

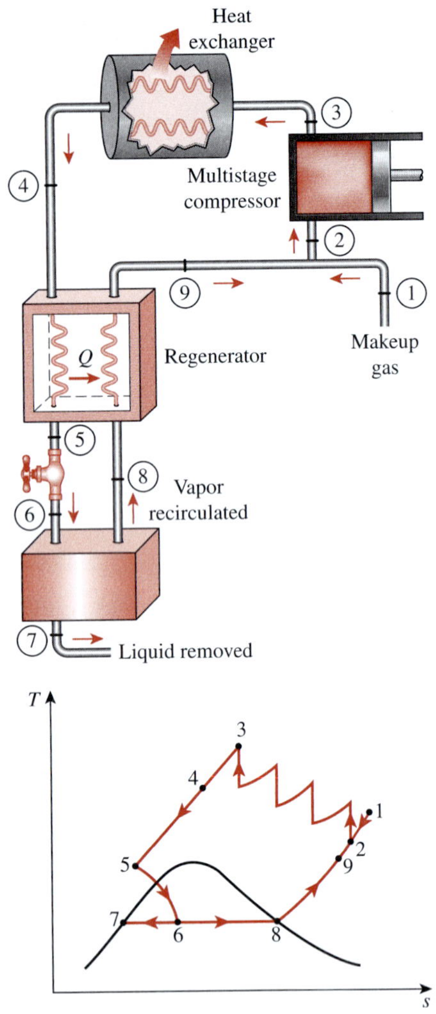

FIGURE 11–17
Linde-Hampson system for liquefying gases.

(three commonly used liquefied gases) are −268, −240, and −147°C, respectively. Therefore, none of these substances exist in liquid form at atmospheric conditions. Furthermore, low temperatures of this magnitude cannot be obtained by ordinary refrigeration techniques. Then the question that needs to be answered in the liquefaction of gases is this: *How can we lower the temperature of a gas below its critical-point value?*

Several cycles, some complex and others simple, are used successfully for the liquefaction of gases. Below we discuss the Linde-Hampson cycle, which is shown schematically and on a *T-s* diagram in Fig. 11–17.

Makeup gas is mixed with the uncondensed portion of the gas from the previous cycle, and the mixture at state 2 is compressed by a multistage compressor to state 3. The compression process approaches an isothermal process due to intercooling. The high-pressure gas is cooled in an aftercooler by a cooling medium or by a separate external refrigeration system to state 4. The gas is further cooled in a regenerative counter-flow heat exchanger by the uncondensed portion of gas from the previous cycle to state 5, and it is throttled to state 6, which is a saturated liquid–vapor mixture state. The liquid (state 7) is collected as the desired product, and the vapor (state 8) is routed through the regenerator to cool the high-pressure gas approaching the throttling valve. Finally, the gas is mixed with fresh makeup gas, and the cycle is repeated.

This and other refrigeration cycles used for the liquefaction of gases can also be used for the solidification of gases.

11–9 ▪ GAS REFRIGERATION CYCLES

As explained in Sec. 11–2, the Carnot cycle (the standard of comparison for power cycles) and the reversed Carnot cycle (the standard of comparison for refrigeration cycles) are identical, except that the reversed Carnot cycle operates in the reverse direction. This suggests that the power cycles discussed in earlier chapters can be used as refrigeration cycles by simply reversing them. In fact, the vapor-compression refrigeration cycle is essentially a modified Rankine cycle operating in reverse. Another example is the reversed Stirling cycle, which is the cycle on which Stirling refrigerators operate. In this section, we discuss the *reversed Brayton cycle,* better known as the **gas refrigeration cycle**.

Consider the gas refrigeration cycle shown in Fig. 11–18. The surroundings are at T_0, and the refrigerated space is to be maintained at T_L. The gas is compressed during process 1-2. The high-pressure, high-temperature gas at state 2 is then cooled at constant pressure to T_0 by rejecting heat to the surroundings. This is followed by an expansion process in a turbine, during which the gas temperature drops to T_4. (Can we achieve the cooling effect by using a throttling valve instead of a turbine?) Finally, the cool gas absorbs heat from the refrigerated space until its temperature rises to T_1.

All the processes described are internally reversible, and the cycle executed is the *ideal* gas refrigeration cycle. In actual gas refrigeration cycles, the compression and expansion processes deviate from the isentropic ones, and T_3 is higher than T_0 unless the heat exchanger is infinitely large.

On a *T-s* diagram, the area under process curve 4-1 represents the heat removed from the refrigerated space, and the enclosed area 1-2-3-4-1 represents the net work input. The ratio of these areas is the COP for the cycle, which may be expressed as

$$\text{COP}_\text{R} = \frac{q_L}{w_{\text{net,in}}} = \frac{q_L}{w_{\text{comp,in}} - w_{\text{turb,out}}} \qquad \textbf{(11–26)}$$

where

$$q_L = h_1 - h_4$$
$$w_{\text{turb,out}} = h_3 - h_4$$
$$w_{\text{comp,in}} = h_2 - h_1$$

The gas refrigeration cycle deviates from the reversed Carnot cycle because the heat transfer processes are not isothermal. In fact, the gas temperature varies considerably during heat transfer processes. Consequently, the gas refrigeration cycles have lower COPs relative to the vapor-compression refrigeration cycles or the reversed Carnot cycle. This is also evident from the *T-s* diagram in Fig. 11–19. The reversed Carnot cycle consumes a fraction of the net work (rectangular area 1*A*3*B*) but produces a greater amount of refrigeration (triangular area under *B*1).

Despite their relatively low COPs, the gas refrigeration cycles have two desirable characteristics: They involve simple, lighter components, which make them suitable for aircraft cooling, and they can incorporate regeneration, which makes them suitable for liquefaction of gases and cryogenic applications. An open-cycle aircraft cooling system is shown in Fig. 11–20. Atmospheric air is compressed by a compressor, cooled by the surrounding air, and expanded in a turbine. The cool air leaving the turbine is then directly routed to the cabin.

The regenerative gas cycle is shown in Fig. 11–21. Regenerative cooling is achieved by inserting a counter-flow heat exchanger into the cycle. Without regeneration, the lowest turbine inlet temperature is T_0, the temperature of the surroundings or any other cooling medium. With regeneration, the high-pressure gas is further cooled to T_4 before expanding in the turbine. Lowering the turbine inlet temperature automatically lowers the turbine exit temperature, which is the minimum temperature in the cycle. Extremely low temperatures can be achieved by repeating this process.

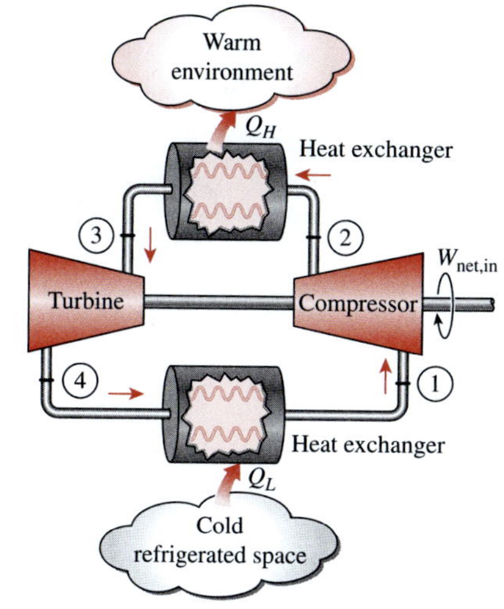

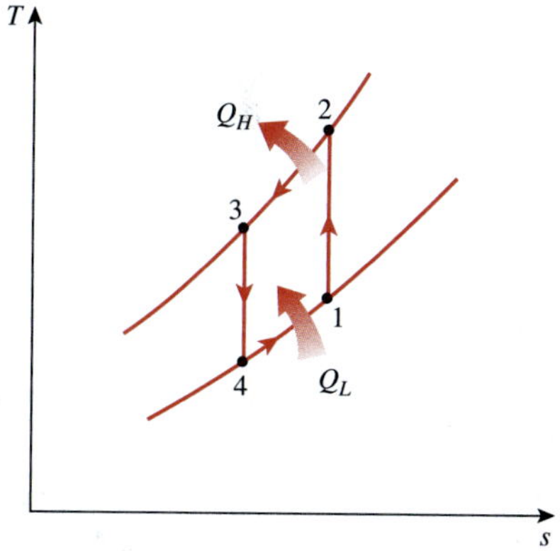

FIGURE 11–18
Simple gas refrigeration cycle.

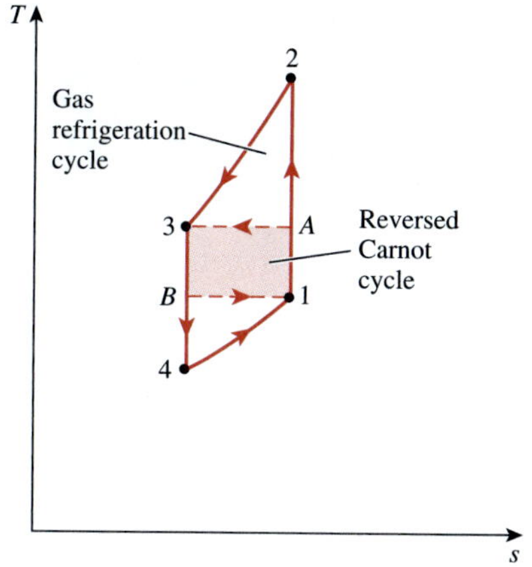

FIGURE 11–19
A reserved Carnot cycle produces more refrigeration (area under *B*1) with less work input (area 1*A*3*B*).

EXAMPLE 11–6 The Simple Ideal Gas Refrigeration Cycle

An ideal gas refrigeration cycle using air as the working medium is to maintain a refrigerated space at 0°F while rejecting heat to the surrounding medium at 80°F. The pressure ratio of the compressor is 4. Determine (*a*) the maximum and minimum temperatures in the cycle, (*b*) the coefficient of performance, and (*c*) the rate of refrigeration for a mass flow rate of 0.1 lbm/s.

SOLUTION An ideal gas refrigeration cycle using air as the working fluid is considered. The maximum and minimum temperatures, the COP, and the rate of refrigeration are to be determined.

Assumptions **1** Steady operating conditions exist. **2** Air is an ideal gas with variable specific heats. **3** Kinetic and potential energy changes are negligible.

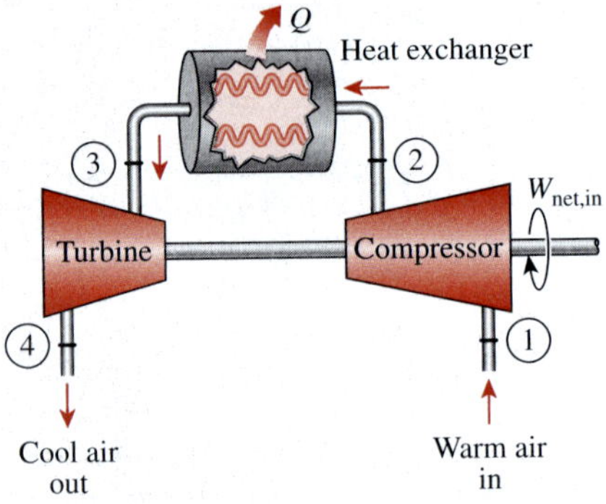

FIGURE 11–20
An open-cycle aircraft cooling system.

Analysis The *T-s* diagram of the gas refrigeration cycle is shown in Fig. 11–22. We note that this is an ideal gas-compression refrigeration cycle, and thus, both the compressor and the turbine are isentropic, and the air is cooled to the environment temperature before it enters the turbine.

(*a*) The maximum and minimum temperatures in the cycle are determined from the isentropic relations of ideal gases for the compression and expansion processes. From Table A–17E,

$$T_1 = 460 \text{ R} \longrightarrow h_1 = 109.90 \text{ Btu/lbm and } P_{r1} = 0.7913$$

$$P_{r2} = \frac{P_2}{P_1} P_{r1} = (4)(0.7913) = 3.165 \longrightarrow \begin{cases} h_2 = 163.5 \text{ Btu/lbm} \\ T_2 = \mathbf{683 \text{ R (or } 223°\text{F)}} \end{cases}$$

$$T_3 = 540 \text{ R} \longrightarrow h_3 = 129.06 \text{ Btu/lbm} \quad \text{and} \quad P_{r3} = 1.3860$$

$$P_{r4} = \frac{P_4}{P_3} P_{r3} = (0.25)(1.386) = 0.3465 \longrightarrow \begin{cases} h_4 = 86.7 \text{ Btu/lbm} \\ T_4 = \mathbf{363 \text{ R (or } -97°\text{F)}} \end{cases}$$

Therefore, the highest and the lowest temperatures in the cycle are 223 and −97°F, respectively.

(*b*) The COP of this ideal gas refrigeration cycle is

$$\text{COP}_R = \frac{q_L}{w_{net,in}} = \frac{q_L}{w_{comp,in} - W_{turb,out}}$$

where

$$q_L = h_1 - h_4 = 109.9 - 86.7 = 23.2 \text{ Btu/lbm}$$

$$W_{turb,out} = h_3 - h_4 = 129.06 - 86.7 = 42.36 \text{ Btu/lbm}$$

$$W_{comp,in} = h_2 - h_1 = 163.5 - 109.9 = 53.6 \text{ Btu/lbm}$$

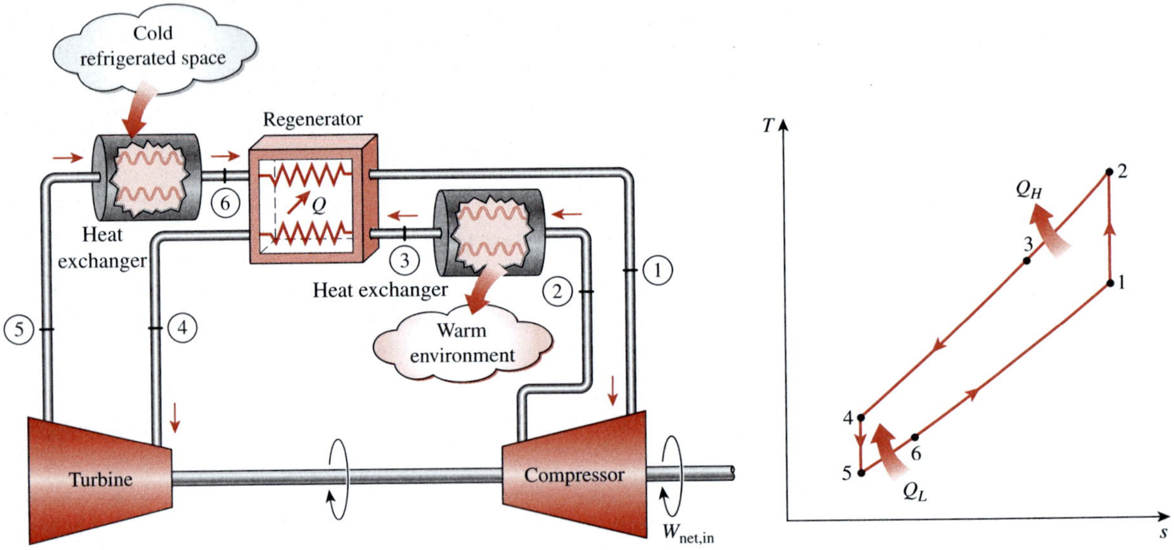

FIGURE 11–21
Gas refrigeration cycle with regeneration.

Thus,

$$\text{COP}_R = \frac{23.2}{53.6 - 42.36} = \mathbf{2.06}$$

(*c*) The rate of refrigeration is

$$\dot{Q}_{refrig} = \dot{m}(q_L) = (0.1 \text{ lbm/s})(23.2 \text{ Btu/lbm}) = \mathbf{2.32 \text{ Btu/s}}$$

Discussion It is worth noting that an ideal vapor-compression cycle working under similar conditions would have a COP greater than 3.

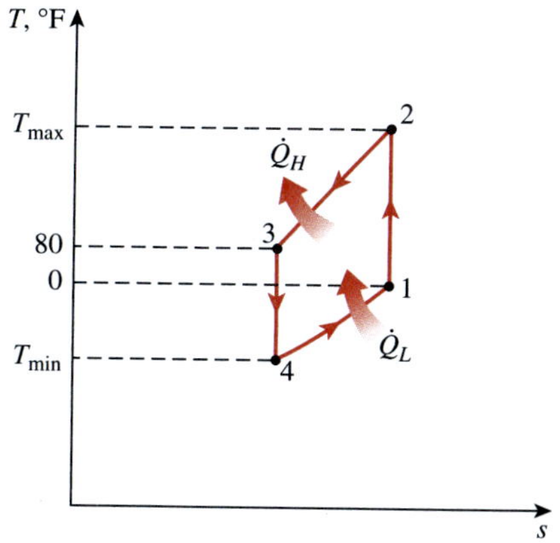

FIGURE 11–22
T-s diagram of the ideal-gas refrigeration cycle described in Example 11–6.

11–10 ▪ ABSORPTION REFRIGERATION SYSTEMS

Another form of refrigeration that becomes economically attractive when there is a source of inexpensive thermal energy at a temperature of 100 to 200°C is **absorption refrigeration**. Some examples of inexpensive thermal energy sources include geothermal energy, solar energy, and waste heat from cogeneration or process steam plants, and even natural gas when it is available at a relatively low price.

As the name implies, absorption refrigeration systems involve the absorption of a *refrigerant* by a *transport medium*. The most widely used absorption refrigeration system is the ammonia–water system, where ammonia (NH_3) serves as the refrigerant and water (H_2O) as the transport medium. Other absorption refrigeration systems include water–lithium bromide and water–lithium chloride systems, where water serves as the refrigerant. The latter two systems are limited to applications such as air-conditioning where the minimum temperature is above the freezing point of water.

To understand the basic principles involved in absorption refrigeration, we examine the NH_3-H_2O system shown in Fig. 11–23. The ammonia–water refrigeration machine was patented by the Frenchman Ferdinand Carre in 1859. Within a few years, the machines based on this principle were being built in the United States primarily to make ice and store food. You will immediately notice from the figure that this system looks very much like the vapor-compression system, except that the compressor has been replaced by a complex absorption mechanism consisting of an absorber, a pump, a generator, a regenerator, a valve, and a rectifier. Once the pressure of NH_3 is raised by the components in the box (this is the only thing they are set up to do), it is cooled and condensed in the condenser by rejecting heat to the surroundings, is throttled to the evaporator pressure, and absorbs heat from the refrigerated space as it flows through the evaporator. So, there is nothing new there. Here is what happens in the box:

Ammonia vapor leaves the evaporator and enters the absorber, where it dissolves and reacts with water to form $NH_3 \cdot H_2O$. This is an exothermic reaction; thus heat is released during this process. The amount of NH_3 that can be dissolved in H_2O is inversely proportional to the temperature. Therefore, it is necessary to cool the absorber to maintain its temperature as low as possible, hence to maximize the amount of NH_3 dissolved in water. The

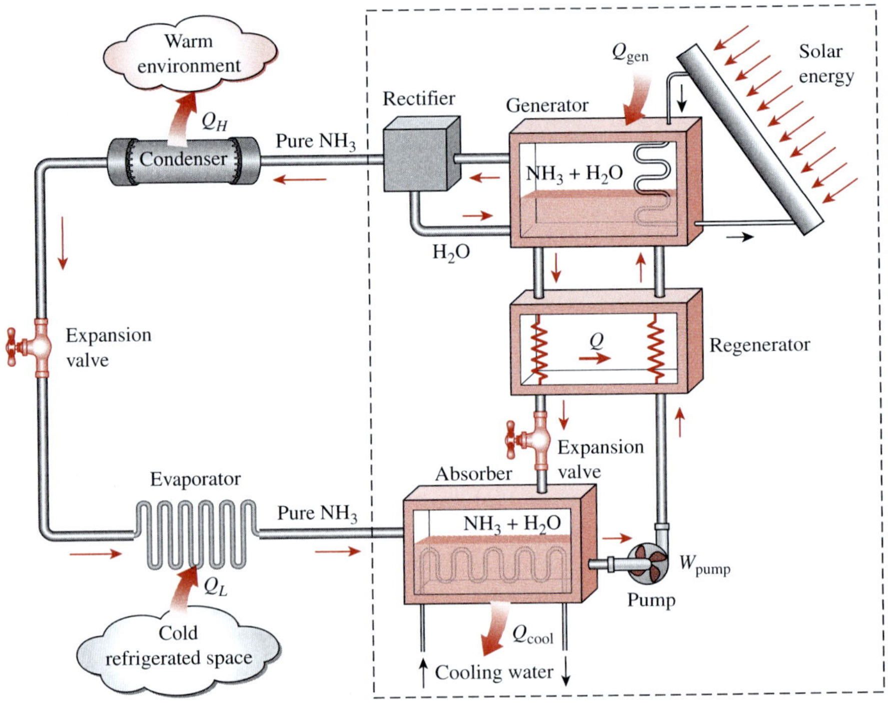

FIGURE 11–23
Ammonia absorption refrigeration cycle.

liquid NH_3 + H_2O solution, which is rich in NH_3, is then pumped to the generator. Heat is transferred to the solution from a source to vaporize some of the solution. The vapor, which is rich in NH_3, passes through a rectifier, which separates the water and returns it to the generator. The high-pressure pure NH_3 vapor then continues its journey through the rest of the cycle. The hot NH_3 + H_2O solution, which is weak in NH_3, then passes through a regenerator, where it transfers some heat to the rich solution leaving the pump, and is throttled to the absorber pressure.

Compared with vapor-compression systems, absorption refrigeration systems have one major advantage: A liquid is compressed instead of a vapor. The steady-flow work is proportional to the specific volume, and thus the work input for absorption refrigeration systems is very small (on the order of one percent of the heat supplied to the generator) and often neglected in the cycle analysis. The operation of these systems is based on heat transfer from an external source. Therefore, absorption refrigeration systems are often classified as *heat-driven systems*.

The absorption refrigeration systems are much more expensive than the vapor-compression refrigeration systems. They are more complex and occupy more space, they are much less efficient thus requiring much larger cooling towers to reject the waste heat, and they are more difficult to service since they are less common. Therefore, absorption refrigeration systems should be considered only when the unit cost of thermal energy is low and is projected to remain low relative to electricity. Absorption refrigeration systems are primarily used in large commercial and industrial installations.

The COP of absorption refrigeration systems is defined as

$$\text{COP}_{\text{absorption}} = \frac{\text{Desired output}}{\text{Required input}} = \frac{Q_L}{Q_{\text{gen}} + W_{\text{pump,in}}} \cong \frac{Q_L}{Q_{\text{gen}}} \tag{11–27}$$

The maximum COP of an absorption refrigeration system is determined by assuming that the entire cycle is totally reversible (i.e., the cycle involves no irreversibilities and any heat transfer is through a differential temperature difference). The refrigeration system would be reversible if the heat from the source (Q_{gen}) were transferred to a Carnot heat engine, and the work output of this heat engine ($W = \varepsilon_{\text{th,rev}}\, Q_{\text{gen}}$) is supplied to a Carnot refrigerator to remove heat from the refrigerated space. Note that $Q_L = W \times \text{COP}_{\text{R,rev}} = \varepsilon_{\text{th,rev}} Q_{\text{gen}} \text{COP}_{\text{R,rev}}$. Then the overall COP of an absorption refrigeration system under reversible conditions becomes (Fig. 11–24)

$$\text{COP}_{\text{rev,absorption}} = \frac{Q_L}{Q_{\text{gen}}} = \eta_{\text{th,rev}} \text{COP}_{\text{R,rev}} = \left(1 - \frac{T_0}{T_s}\right)\left(\frac{T_L}{T_0 - T_L}\right) \tag{11–28}$$

where T_L, T_0, and T_s are the thermodynamic temperatures of the refrigerated space, the environment, and the heat source, respectively. Any absorption refrigeration system that receives heat from a source at T_s and removes heat from the refrigerated space at T_L while operating in an environment at T_0 has a lower COP than the one determined from Eq. 11–28. For example, when the source is at 120°C, the refrigerated space is at −10°C, and the environment is at 25°C, the maximum COP that an absorption refrigeration system can have is 1.8. The COP of actual absorption refrigeration systems is usually less than 1.

Air-conditioning systems based on absorption refrigeration, called *absorption chillers,* perform best when the heat source can supply heat at a high temperature with little temperature drop. The absorption chillers are typically rated at an input temperature of 116°C (240°F). The chillers perform at lower temperatures, but their cooling capacity decreases sharply with decreasing source temperature, about 12.5 percent for each 6°C (10°F) drop in the source temperature. For example, the capacity goes down to 50 percent when the supply water temperature drops to 93°C (200°F). In that case, one needs to double the size (and thus the cost) of the chiller to achieve the same cooling. The COP of the chiller is affected less by the decline of the source temperature. The COP drops by 2.5 percent for each 6°C (10°F) drop in the source temperature. The nominal COP of single-stage absorption chillers at 116°C (240°F) is 0.65 to 0.70. Therefore, for each ton of refrigeration, a heat input of (12,000 Btu/h)/0.65 = 18,460 Btu/h is required. At 88°C (190°F), the COP drops by 12.5 percent and thus the heat input increases by 12.5 percent for the same cooling effect. Therefore, the economic aspects must be evaluated carefully before any absorption refrigeration system is considered, especially when the source temperature is below 93°C (200°F).

Another absorption refrigeration system that is quite popular with campers is a propane-fired system invented by two Swedish undergraduate students. In this system, the pump is replaced by a third fluid (hydrogen), which makes it a truly portable unit.

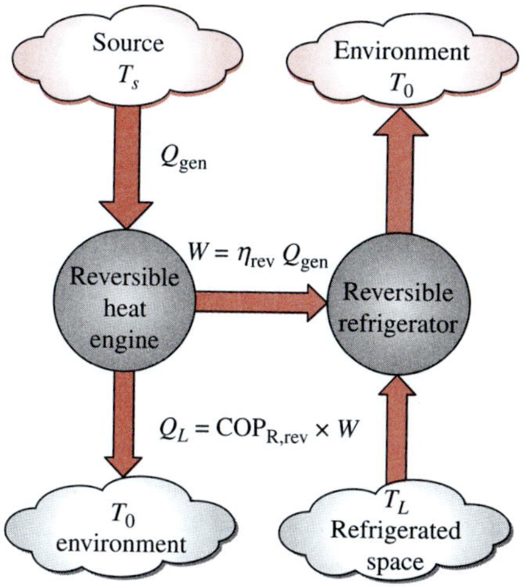

$$W = \eta_{\text{rev}}\, Q_{\text{gen}} = \left(1 - \frac{T_0}{T_s}\right) Q_{\text{gen}}$$

$$Q_L = \text{COP}_{\text{R,rev}} W = \left(\frac{T_L}{T_0 - T_L}\right) W$$

$$\text{COP}_{\text{rev,absorption}} = \frac{Q_L}{Q_{\text{gen}}} = \left(1 - \frac{T_0}{T_s}\right)\left(\frac{T_L}{T_0 - T_L}\right)$$

FIGURE 11–24
Determining the maximum COP of an absorption refrigeration system.

TOPIC OF SPECIAL INTEREST* Thermoelectric Power Generation and Refrigeration Systems

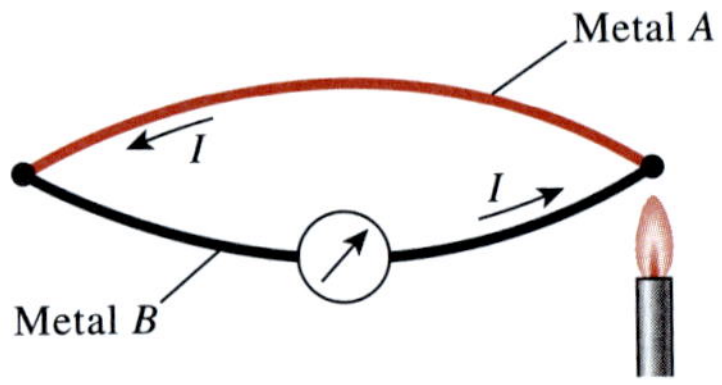

FIGURE 11–25
When one of the junctions of two dissimilar metals is heated, a current I flows through the closed circuit.

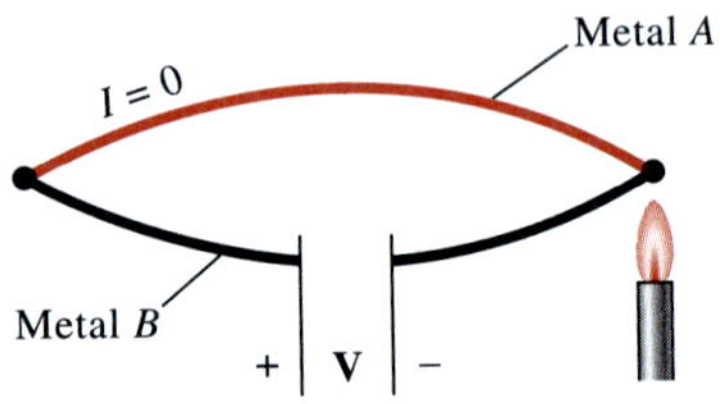

FIGURE 11–26
When a thermoelectric circuit is broken, a potential difference is generated.

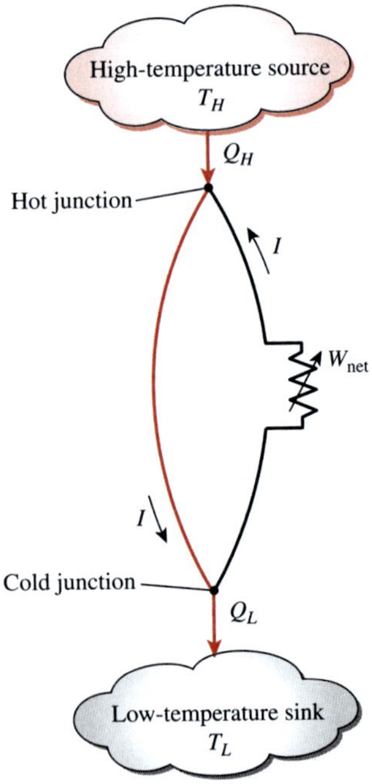

FIGURE 11–27
Schematic of a simple thermoelectric power generator.

All the refrigeration systems discussed above involve many moving parts and bulky, complex components. Then this question comes to mind: Is it really necessary for a refrigeration system to be so complex? Can we not achieve the same effect in a more direct way? The answer to this question is *yes*. It is possible to use electric energy more directly to produce cooling without involving any refrigerants and moving parts. Below we discuss one such system, called *thermoelectric refrigerator*.

Consider two wires made from different metals joined at both ends (junctions), forming a closed circuit. Ordinarily, nothing will happen. However, when one of the ends is heated, something interesting happens: A current flows continuously in the circuit, as shown in Fig. 11–25. This is called the **Seebeck effect**, in honor of Thomas Seebeck, who made this discovery in 1821. The circuit that incorporates both thermal and electrical effects is called a **thermoelectric circuit**, and a device that operates on this circuit is called a **thermoelectric device**.

The Seebeck effect has two major applications: temperature measurement and power generation. When the thermoelectric circuit is broken, as shown in Fig. 11–26, the current ceases to flow, and we can measure the driving force (the electromotive force) or the voltage generated in the circuit by a voltmeter. The voltage generated is a function of the temperature difference and the materials of the two wires used. Therefore, temperature can be measured by simply measuring voltages. The two wires used to measure the temperature in this manner form a *thermocouple,* which is the most versatile and most widely used temperature measurement device. A common T-type thermocouple, for example, consists of copper and constantan wires, and it produces about 40 μV per °C difference.

The Seebeck effect also forms the basis for thermoelectric power generation. The schematic diagram of a **thermoelectric generator** is shown in Fig. 11–27. Heat is transferred from a high-temperature source to the hot junction in the amount of Q_H, and it is rejected to a low-temperature sink from the cold junction in the amount of Q_L. The difference between these two quantities is the net electrical work produced, that is, $W_e = Q_H - Q_L$. It is evident from Fig. 11–27 that the thermoelectric power cycle closely resembles an ordinary heat engine cycle, with electrons serving as the working fluid. Therefore, the thermal efficiency of a thermoelectric generator operating between the temperature limits of T_H and T_L is limited by the efficiency of a Carnot cycle operating between the same temperature limits. Thus, in the absence of any irreversibilities (such as I^2R heating, where R is the total electrical resistance of the wires), the thermoelectric generator will have the Carnot efficiency.

The major drawback of thermoelectric generators is their low efficiency. The future success of these devices depends on finding materials with more desirable characteristics. For example, the voltage output of thermoelectric devices has been increased several times by switching from metal pairs to semiconductors. A practical thermoelectric generator using *n*-type (heavily

*This section can be skipped without a loss in continuity.

doped to create excess electrons) and *p*-type (heavily doped to create a deficiency of electrons) materials connected in series is shown in Fig. 11–28. Despite their low efficiencies, thermoelectric generators have definite weight and reliability advantages and are presently used in rural areas and in space applications. For example, silicon–germanium-based thermoelectric generators have been powering *Voyager* spacecraft since 1980 and are expected to continue generating power for many more years.

If Seebeck had been fluent in thermodynamics, he would probably have tried reversing the direction of flow of electrons in the thermoelectric circuit (by externally applying a potential difference in the reverse direction) to create a refrigeration effect. But this honor belongs to Jean Charles Athanase Peltier, who discovered this phenomenon in 1834. He noticed during his experiments that when a small current was passed through the junction of two dissimilar wires, the junction was cooled, as shown in Fig. 11–29. This is called the **Peltier effect**, and it forms the basis for **thermoelectric refrigeration**. A practical thermoelectric refrigeration circuit using semiconductor materials is shown in Fig. 11–30. Heat is absorbed from the refrigerated space in the amount of Q_L and rejected to the warmer environment in the amount of Q_H. The difference between these two quantities is the net electrical work that needs to be supplied; that is, $W_e = Q_H - Q_L$. Thermoelectric refrigerators presently cannot compete with vapor-compression refrigeration systems because of their low coefficient of performance. They are available in the market, however, and are preferred in some applications because of their small size, simplicity, quietness, and reliability.

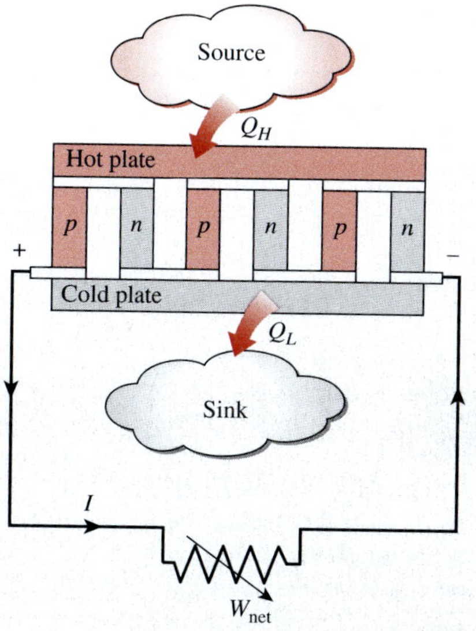

FIGURE 11–28
A thermoelectric power generator.

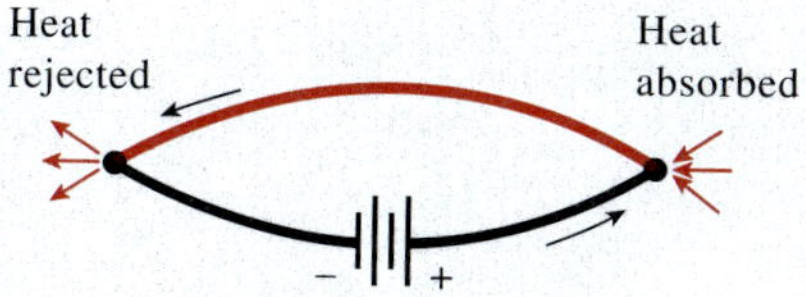

FIGURE 11–29
When a current is passed through the junction of two dissimilar materials, the junction is cooled.

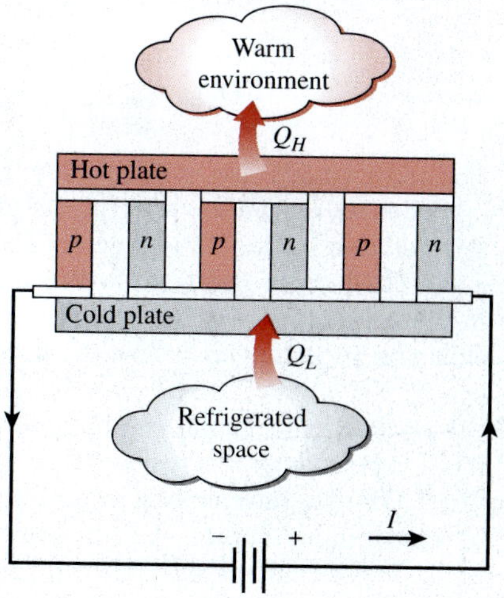

FIGURE 11–30
A thermoelectric refrigerator.

EXAMPLE 11–7 Cooling of a Canned Drink by a Thermoelectric Refrigerator

A thermoelectric refrigerator that resembles a small ice chest is powered by a car battery and has a COP of 0.1. If the refrigerator cools a 0.350-L canned drink from 20 to 4°C in 30 min, determine the average electric power consumed by the thermoelectric refrigerator.

SOLUTION A thermoelectric refrigerator with a specified COP is used to cool canned drinks. The power consumption of the refrigerator is to be determined.

Assumptions Heat transfer through the walls of the refrigerator is negligible during operation.

Properties The properties of canned drinks are the same as those of water at room temperature, $\rho = 1$ kg/L and $c = 4.18$ kJ/kg·°C (Table A–3).

Analysis The cooling rate of the refrigerator is simply the rate of decrease of the energy of the canned drinks,

$$m = \rho V = (1 \text{ kg/L})(0.350 \text{ L}) = 0.350 \text{ kg}$$

$$Q_{\text{cooling}} = mc\,\Delta T = (0.350 \text{ kg})(4.18 \text{ kJ/kg}\cdot\text{°C})(20 - 4)\text{°C} = 23.4 \text{ kJ}$$

$$\dot{Q}_{\text{cooling}} = \frac{Q_{\text{cooling}}}{\Delta t} = \frac{23.4 \text{ kJ}}{30 \times 60 \text{ s}} = 0.0130 \text{ kW} = 13 \text{ W}$$

Then the average power consumed by the refrigerator becomes

$$\dot{W}_{in} = \frac{\dot{Q}_{cooling}}{COP_R} = \frac{13\ W}{0.10} = 130\ W$$

Discussion In reality, the power consumption will be larger because of the heat gain through the walls of the refrigerator.

SUMMARY

The transfer of heat from lower temperature regions to higher temperature ones is called *refrigeration*. Devices that produce refrigeration are called *refrigerators*, and the cycles on which they operate are called *refrigeration cycles*. The working fluids used in refrigerators are called *refrigerants*. Refrigerators used for the purpose of heating a space by transferring heat from a cooler medium are called *heat pumps*.

The performance of refrigerators and heat pumps is expressed in terms of *coefficient of performance* (COP), defined as

$$COP_R = \frac{\text{Desired output}}{\text{Required output}} = \frac{\text{Cooling effect}}{\text{Work input}} = \frac{Q_L}{W_{net,in}}$$

$$COP_{HP} = \frac{\text{Desired output}}{\text{Required input}} = \frac{\text{Heating effect}}{\text{Work input}} = \frac{Q_H}{W_{net,in}}$$

The standard of comparison for refrigeration cycles is the *reversed Carnot cycle*. A refrigerator or heat pump that operates on the reversed Carnot cycle is called a *Carnot refrigerator* or a *Carnot heat pump*, and their COPs are

$$COP_{R,Carnot} = \frac{1}{T_H/T_L - 1}$$

$$COP_{HP,Carnot} = \frac{1}{1 - T_L/T_H}$$

The most widely used refrigeration cycle is the *vapor-compression refrigeration cycle*. In an ideal vapor-compression refrigeration cycle, the refrigerant enters the compressor as a saturated vapor and is cooled to the saturated liquid state in the condenser. It is then throttled to the evaporator pressure and vaporizes as it absorbs heat from the refrigerated space.

Very low temperatures can be achieved by operating two or more vapor-compression systems in series, called *cascading*. The COP of a refrigeration system also increases as a result of cascading. Another way of improving the performance of a vapor-compression refrigeration system is by using *multistage compression with regenerative cooling*. A refrigerator with a single compressor can provide refrigeration at several temperatures by throttling the refrigerant in stages. The vapor-compression refrigeration cycle can also be used to liquefy gases after some modifications.

The power cycles can be used as refrigeration cycles by simply reversing them. Of these, the *reversed Brayton cycle*, which is also known as the *gas refrigeration cycle*, is used to cool aircraft and to obtain very low (cryogenic) temperatures after it is modified with regeneration. The work output of the turbine can be used to reduce the work input requirements to the compressor. Thus the COP of a gas refrigeration cycle is

$$COP_{absorption} = \frac{q_L}{w_{net,in}} = \frac{q_L}{w_{comp,in} - w_{turb,out}}$$

Another form of refrigeration that becomes economically attractive when there is a source of inexpensive thermal energy at a temperature of 100 to 200°C is *absorption refrigeration*, where the refrigerant is absorbed by a transport medium and compressed in liquid form. The most widely used absorption refrigeration system is the ammonia–water system, where ammonia serves as the refrigerant and water as the transport medium. The work input to the pump is usually very small, and the COP of absorption refrigeration systems is defined as

$$COP_{absorption} = \frac{\text{Desired output}}{\text{Required input}} = \frac{Q_L}{Q_{gen} + W_{pump,in}} \cong \frac{Q_L}{Q_{gen}}$$

The maximum COP an absorption refrigeration system can have is determined by assuming totally reversible conditions, which yields

$$COP_{rev,absorption} = \eta_{th,rev}\, COP_{R,rev} = \left(1 - \frac{T_0}{T_s}\right)\left(\frac{T_L}{T_0 - T_L}\right)$$

where T_0, T_L, and T_s are the thermodynamic temperatures of the environment, the refrigerated space, and the heat source, respectively.

REFERENCES AND SUGGESTED READINGS

1. *ASHRAE, Handbook of Fundamentals*. Atlanta: American Society of Heating, Refrigerating, and Air-Conditioning Engineers, 1985.
2. *Heat Pump Systems—A Technology Review*. OECD Report, Paris, 1982.
3. B. Nagengast. "A Historical Look at CFC Refrigerants." *ASHRAE Journal* 30, no. 11 (November 1988), pp. 37–39.
4. W. F. Stoecker. "Growing Opportunities for Ammonia Refrigeration." *Proceedings of the Meeting of the International Institute of Ammonia Refrigeration*, Austin, Texas, 1989.
5. W. F. Stoecker and J. W. Jones. *Refrigeration and Air Conditioning*. 2nd ed. New York: McGraw-Hill, 1982.

PROBLEMS*

The Reversed Carnot Cycle

11–1C Why do we study the reversed Carnot cycle even though it is not a realistic model for refrigeration cycles?

11–2C Why is the reversed Carnot cycle executed within the saturation dome not a realistic model for refrigeration cycles?

11–3 A steady-flow Carnot refrigeration cycle uses refrigerant-134a as the working fluid. The refrigerant changes from saturated vapor to saturated liquid at 40°C in the condenser as it rejects heat. The evaporator pressure is 100 kPa. Show the cycle on a *T-s* diagram relative to saturation lines, and determine (*a*) the coefficient of performance, (*b*) the amount of heat absorbed from the refrigerated space, and (*c*) the net work input. *Answers:* (*a*) 3.72, (*b*) 128 kJ/kg, (*c*) 34.6 kJ/kg

11–4E Refrigerant-134a enters the condenser of a steady-flow Carnot refrigerator as a saturated vapor at 90 psia, and it leaves with a quality of 0.05. The heat absorption from the refrigerated space takes place at a pressure of 30 psia. Show the cycle on a *T-s* diagram relative to saturation lines, and determine (*a*) the coefficient of performance, (*b*) the quality at the beginning of the heat-absorption process, and (*c*) the net work input.

Ideal and Actual Vapor-Compression Refrigeration Cycles

11–5C Does the ideal vapor-compression refrigeration cycle involve any internal irreversibilities?

11–6C Why is the throttling valve not replaced by an isentropic turbine in the ideal vapor-compression refrigeration cycle?

11–7C It is proposed to use water instead of refrigerant-134a as the working fluid in air-conditioning applications where the minimum temperature never falls below the freezing point. Would you support this proposal? Explain.

11–8C In a refrigeration system, would you recommend condensing the refrigerant-134a at a pressure of 0.7 or 1.0 MPa if heat is to be rejected to a cooling medium at 15°C? Why?

11–9C Does the area enclosed by the cycle on a *T-s* diagram represent the net work input for the reversed Carnot cycle? How about for the ideal vapor-compression refrigeration cycle?

11–10C Consider two vapor-compression refrigeration cycles. The refrigerant enters the throttling valve as a saturated liquid at 30°C in one cycle and as subcooled liquid at 30°C in the other one. The evaporator pressure for both cycles is the same. Which cycle do you think will have a higher COP?

11–11C The COP of vapor-compression refrigeration cycles improves when the refrigerant is subcooled before it enters the throttling valve. Can the refrigerant be subcooled indefinitely to maximize this effect, or is there a lower limit? Explain.

11–12E A refrigerator operates on the ideal vapor-compression refrigeration cycle and uses refrigerant-134a as the working fluid. The condenser operates at 300 psia and the evaporator at 20°F. If an adiabatic, reversible expansion device were available and used to expand the liquid leaving the condenser, how much would the COP improve by using this device instead of the throttle device? *Answer:* 16.5 percent

11–13 An ideal vapor-compression refrigeration cycle that uses refrigerant-134a as its working fluid maintains a condenser at 1000 kPa and the evaporator at 4°C. Determine this

*Problems designated by a "C" are concept questions, and students are encouraged to answer them all. Problems designated by an "E" are in English units, and the SI users can ignore them. Problems with the icon are solved using EES, and complete solutions together with parametric studies are included on the enclosed DVD. Problems with the icon are comprehensive in nature, and are intended to be solved with a computer, preferably using the EES software that accompanies this text.

system's COP and the amount of power required to service a 400 kW cooling load. *Answers:* 6.46, 61.9 kW

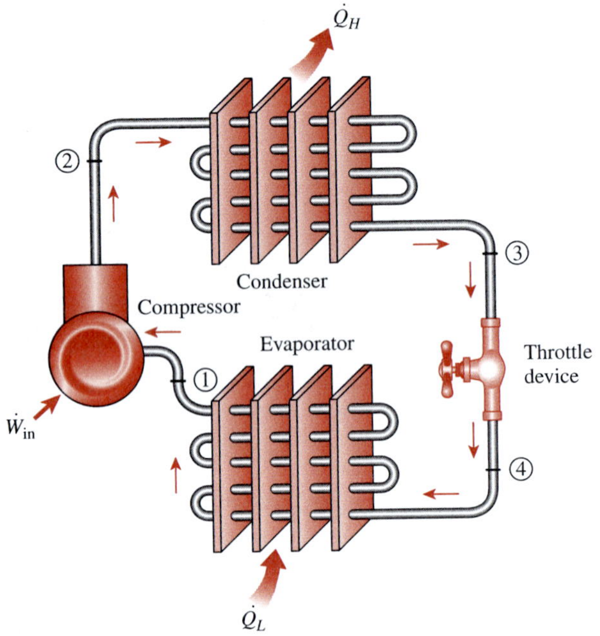

FIGURE P11–13

11–14 A refrigerator uses refrigerant-134a as the working fluid and operates on an ideal vapor-compression refrigeration cycle between 0.12 and 0.7 MPa. The mass flow rate of the refrigerant is 0.05 kg/s. Show the cycle on a *T-s* diagram with respect to saturation lines. Determine (*a*) the rate of heat removal from the refrigerated space and the power input to the compressor, (*b*) the rate of heat rejection to the environment, and (*c*) the coefficient of performance. *Answers:* (*a*) 7.41 kW, 1.83 kW, (*b*) 9.23 kW, (*c*) 4.06

11–15 Repeat Prob. 11–14 for a condenser pressure of 0.9 MPa.

11–16 If the throttling valve in Prob. 11–14 is replaced by an isentropic turbine, determine the percentage increase in the COP and in the rate of heat removal from the refrigerated space. *Answers:* 4.2 percent, 4.2 percent

11–17 Refrigerant-134a enters the compressor of a refrigerator as superheated vapor at 0.20 MPa and −5°C at a rate of 0.07 kg/s, and it leaves at 1.2 MPa and 70°C. The refrigerant is cooled in the condenser to 44°C and 1.15 MPa, and it is throttled to 0.21 MPa. Disregarding any heat transfer and pressure drops in the connecting lines between the components, show the cycle on a *T-s* diagram with respect to saturation lines, and determine (*a*) the rate of heat removal from the refrigerated space and the power input to the compressor, (*b*) the isentropic efficiency of the compressor, and (*c*) the COP of the refrigerator. *Answers:* (a) 9.42 kW, 3.63 kW, (b) 74.1 percent, (c) 2.60

11–18E A refrigerator uses refrigerant-134a as its working fluid and operates on the ideal vapor-compression refrigeration cycle. The refrigerant evaporates at 5°F and condenses at 180 psia. This unit serves a 45,000 Btu/h cooling load. Determine the mass flow rate of the refrigerant and the power that this unit will require.

11–19E EES Using EES (or other) software, repeat Prob. 11–18E if ammonia is used in place of refrigerant-134a.

11–20 A commercial refrigerator with refrigerant-134a as the working fluid is used to keep the refrigerated space at −30°C by rejecting its waste heat to cooling water that enters the condenser at 18°C at a rate of 0.25 kg/s and leaves at 26°C. The refrigerant enters the condenser at 1.2 MPa and 65°C and leaves at 42°C. The inlet state of the compressor is 60 kPa and −34°C and the compressor is estimated to gain a net heat of 450 W from the surroundings. Determine (*a*) the quality of the refrigerant at the evaporator inlet, (*b*) the refrigeration load, (*c*) the COP of the refrigerator, and (*d*) the theoretical maximum refrigeration load for the same power input to the compressor.

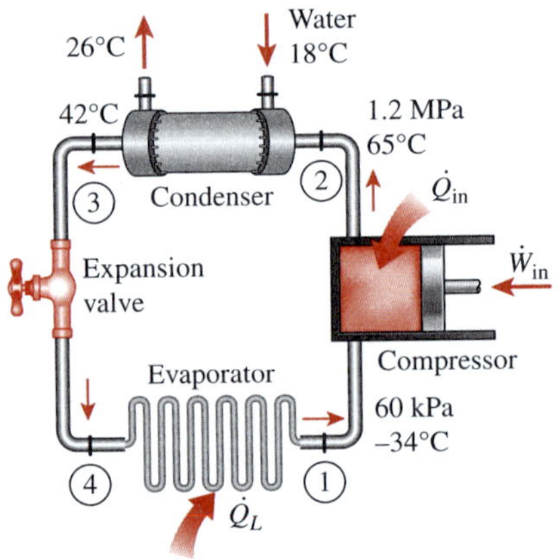

FIGURE P11–20

11–21 Refrigerant-134a enters the compressor of a refrigerator at 100 kPa and −20°C at a rate of 0.5 m³/min and leaves at 0.8 MPa. The isentropic efficiency of the compressor is 78 percent. The refrigerant enters the throttling valve at 0.75 MPa and 26°C and leaves the evaporator as saturated vapor at −26°C. Show the cycle on a *T-s* diagram with respect to saturation lines, and determine (*a*) the power input to the compressor, (*b*) the rate of heat removal from the refrigerated space, and (*c*) the pressure drop and rate of heat gain in the line between the evaporator and the compressor. *Answers:* (*a*) 2.40 kW, (*b*) 6.17 kW, (*c*) 1.73 kPa, 0.203 kW

11–22 Reconsider Prob. 11–21. Using EES (or other) software, investigate the effects of varying the compressor isentropic efficiency over the range 60 to 100 percent and the compressor inlet volume flow rate from 0.1 to 1.0 m^3/min on the power input and the rate of refrigeration. Plot the rate of refrigeration and the power input to the compressor as functions of compressor efficiency for compressor inlet volume flow rates of 0.1, 0.5, and 1.0 m^3/min, and discuss the results.

11–23 A refrigerator uses refrigerant-134a as the working fluid and operates on the ideal vapor-compression refrigeration cycle except for the compression process. The refrigerant enters the evaporator at 120 kPa with a quality of 30 percent and leaves the compressor at 60°C. If the compressor consumes 450 W of power, determine (*a*) the mass flow rate of the refrigerant, (*b*) the condenser pressure, and (*c*) the COP of the refrigerator. *Answers: (a) 0.00727 kg/s, (b) 672 kPa, (c) 2.43*

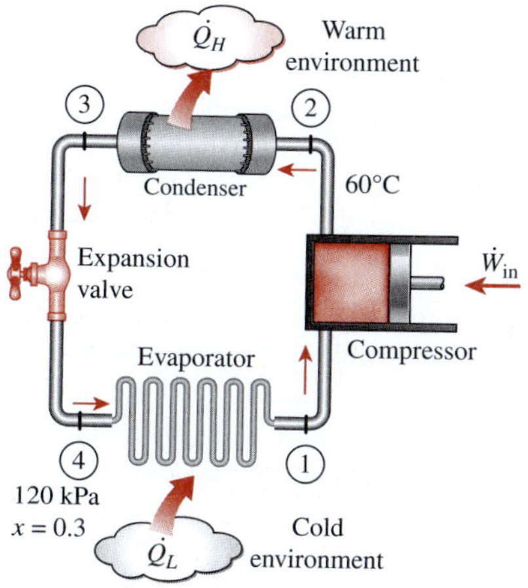

FIGURE P11–23

11–24 The manufacturer of an air conditioner claims a seasonal energy efficiency ratio (SEER) of 16 (Btu/h)/W for one of its units. This unit operates on the normal vapor compression refrigeration cycle and uses refrigerant-22 as the working fluid. This SEER is for the operating conditions when the evaporator saturation temperature is −5°C and the condenser saturation temperature is 45°C. Selected data for refrigerant-22 are provided in the table below.

T, °C	P_{sat}, kPa	h_f, kJ/kg	h_g, kJ/kg	s_g, kJ/kg·K
−5	421.2	38.76	248.1	0.9344
45	1728	101	261.9	0.8682

(*a*) Sketch the hardware and the *T-s* diagram for this air conditioner.

(*b*) Determine the heat absorbed by the refrigerant in the evaporator per unit mass of refrigerant-22, in kJ/kg.

(*c*) Determine the work input to the compressor and the heat rejected in the condenser per unit mass of refrigerant-22, in kJ/kg.

11–25 An actual refrigerator operates on the vapor-compression refrigeration cycle with refrigerant-22 as the working fluid. The refrigerant evaporates at −15°C and condenses at 40°C. The isentropic efficiency of the compressor is 83 percent. The refrigerant is superheated by 5°C at the compressor inlet and subcooled by 5°C at the exit of the condenser. Determine (*a*) the heat removed from the cooled space and the work input, in kJ/kg and the COP of the cycle. Determine (*b*) the same parameters if the cycle operated on the ideal vapor-compression refrigeration cycle between the same evaporating and condensing temperatures.

The properties of R-22 in the case of actual operation are: h_1 = 402.49 kJ/kg, h_2 =454.00 kJ/kg, h_3 = 243.19 kJ/kg

The properties of R-22 in the case of ideal operation are: h_1 = 399.04 kJ/kg, h_2 = 440.71 kJ/kg, h_3, = 249.80 kJ/kg

Note: state 1: compressor inlet, state 2: compressor exit, state 3: condenser exit, state 4: evaporator inlet.

11–26 A refrigerator uses refrigerant-134a as the working fluid and operates on the vapor-compression refrigeration cycle. The evaporator and condenser pressures are 200 kPa and 1400 kPa, respectively. The isentropic efficiency of the compressor is 88 percent. The refrigerant enters the compressor at a rate of 0.025 kg/s superheated by 10.1°C and leaves the condenser subcooled by 4.4°C. Determine (*a*) the rate of cooling provided by the evaporator, the power input, and the COP. Determine (*b*) the same parameters if the cycle operated on the ideal vapor-compression refrigeration cycle between the same pressure limits.

Second-Law Analysis of Vapor-Compression Refrigeration Cycles

11–27C How is the exergy efficiency of a refrigerator operating on the vapor-compression refrigeration cycle defined? Provide two alternative definitions and explain each term.

11–28C How is the exergy efficiency of a heat pump operating on the vapor-compression refrigeration cycle defined? Provide two alternative definitions and show that one can be derived from the other.

11–29C Consider isentropic compressor of a vapor-compression refrigeration cycle. What are the isentropic efficiency and exergy efficiency of this compressor? Justify your answers. Is the exergy efficiency of a compressor necessarily equal to its isentropic efficiency? Explain.

11–30 A space is kept at −23°C by a vapor-compression refrigeration system in an ambient at 25°C. The space gains heat steadily at a rate of 3500 kJ/h and the rate of heat rejection in the condenser is 600 kJ/h. Determine the power input, in kW, the COP of the cycle and the second-law efficiency of the system.

11–31 Bananas are to be cooled from 28°C to 12°C at a rate of 1140 kg/h by a refrigerator that operates on a vapor-compression refrigeration cycle. The power input to the refrigerator is 8.6 kW. Determine (*a*) the rate of heat absorbed from the bananas, in kJ/h, and the COP, (*b*) the minimum power input to the refrigerator, and (*c*) the second-law efficiency and the exergy destruction for the cycle. The specific heat of bananas above freezing is 3.35 kJ/kg·°C. *Answers:* (*a*) 61,100 kJ/h, 1.97, (*b*) 0.463 kW, (*c*) 5.4 percent, 8.14 kW.

11–32 A vapor-compression refrigeration system absorbs heat from a space at 0°C at a rate of 24,000 Btu/h and rejects heat to water in the condenser. The water experiences a temperature rise of 12°C in the condenser. The COP of the system is estimated to be 2.05. Determine (*a*) the power input to the system, in kW, (*b*) the mass flow rate of water through the condenser, and (*c*) the second-law efficiency and the exergy destruction for the refrigerator. Take T_0 = 20°C and $c_{p,\text{water}}$ = 4.18 kJ/kg·°C.

11–33E A refrigerator operating on the vapor-compression refrigeration cycle using refrigerant-134a as the refrigerant is considered. The temperature of the cooled space and the ambient air are at 10°F and 80°F, respectively. R-134a enters the compressor at 20 psia as a saturated vapor and leaves at 140 psia and 160°F. The refrigerant leaves the condenser as a saturated liquid. The rate of cooling provided by the system is 45,000 Btu/h. Determine (*a*) the mass flow rate of R-134a and the COP, (*b*) the exergy destruction in each component of the cycle and the exergy efficiency of the compressor, and (*c*) the second-law efficiency of the cycle and the total exergy destruction in the cycle.

11–34 A room is kept at −12°C by a vapor-compression refrigeration cycle with R-134a as the refrigerant. Heat is rejected to cooling water that enters the condenser at 20°C at a rate of 0.15 kg/s and leaves at 28°C. The refrigerant enters the condenser at 1.2 MPa and 50°C and leave as a saturated liquid. If the compressor consumes 2.2 kW of power, determine (*a*) the refrigeration load, in Btu/h and the COP, (*b*) the second-law efficiency of the refrigerator and the total exergy destruction in the cycle, and (*c*) the exergy destruction in the condenser. Take T_0 = 20°C and $c_{p,\text{water}}$ = 4.18 kJ/kg·°C. *Answers:* (*a*) 9610 Btu/h, 1.28, (*b*) 15.7 percent, 1.85 kW, (*c*) 0.349 kW

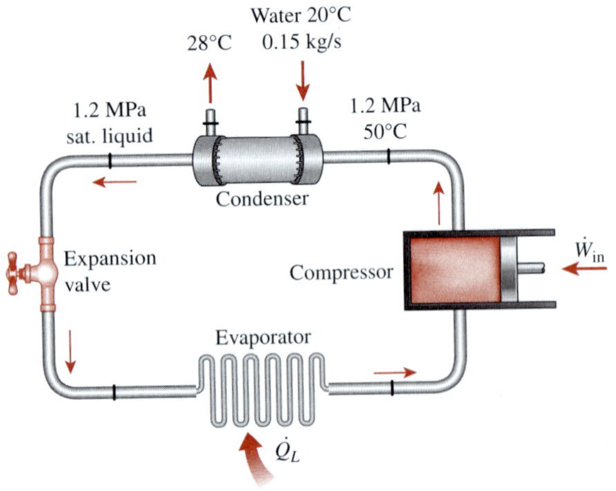

FIGURE P11–34

11–35 A refrigerator operates on the ideal vapor-compression refrigeration cycle with refrigerant-134a as the working fluid. The refrigerant evaporates at −10°C and condenses at 57.9°C. The refrigerant absorbs heat from a space at 5°C and rejects heat to ambient air at 25°C. Determine (*a*) the cooling load, in kJ/kg, and the COP, (*b*) the exergy destruction in each component of the cycle and the total exergy destruction in the cycle, and (*c*) the second-law efficiency of the compressor, evaporator, and the cycle.

11–36 A refrigeration system operates on the ideal vapor-compression refrigeration cycle with ammonia as the refrigerant. The evaporator and condenser pressures are 200 kPa and 2000 kPa, respectively. The temperatures of the low-temperature and high-temperature mediums are −9°C and 27°C, respectively. If the rate of heat rejected in the condenser is 18.0 kW, determine (*a*) the volume flow rate of ammonia at the compressor inlet, in L/s, (*b*) the power input and the COP, and (*c*) the second-law efficiency of the cycle and the total exergy destruction in the cycle. The properties of ammonia at various states are given as follows: h_1 = 1439.3 kJ/kg, s_1= 5.8865 kJ/kg·K, v_1 = 0.5946 m^3/kg, h_2 = 1798.3 kJ/kg, h_3 = 437.4 kJ/kg, s_3 = 1.7892 kJ/kg·K, s_4 = 1.9469 kJ/kg·K. Note: state 1: compressor inlet, state 2: compressor exit, state 3: condenser exit, state 4: evaporator inlet.

11–37 Using EES (or other) software, repeat the previous problem if ammonia, R-134a, and R-22 is used as the refrigerant. Also, for the case of ammonia, investigate the effects of evaporator and condenser pressures on the COP, the second-law efficiency, and the total exergy destruction. Vary the evaporator pressure between 100 and 400 kPa and the condenser pressure between 1000 and 2000 kPa.

Selecting the Right Refrigerant

11–38C When selecting a refrigerant for a certain application, what qualities would you look for in the refrigerant?

11–39C Consider a refrigeration system using refrigerant-134a as the working fluid. If this refrigerator is to operate in an environment at 30°C, what is the minimum pressure to which the refrigerant should be compressed? Why?

11–40C A refrigerant-134a refrigerator is to maintain the refrigerated space at −10°C. Would you recommend an evaporator pressure of 0.12 or 0.14 MPa for this system? Why?

11–41 A refrigerator that operates on the ideal vapor-compression cycle with refrigerant-134a is to maintain the refrigerated space at −10°C while rejecting heat to the environment at 25°C. Select reasonable pressures for the evaporator and the condenser, and explain why you chose those values.

11–42 A heat pump that operates on the ideal vapor-compression cycle with refrigerant-134a is used to heat a house and maintain it at 26°C by using underground water at 14°C as the heat source. Select reasonable pressures for the evaporator and the condenser, and explain why you chose those values.

Heat Pump Systems

11–43C Do you think a heat pump system will be more cost-effective in New York or in Miami? Why?

11–44C What is a water-source heat pump? How does the COP of a water-source heat pump system compare to that of an air-source system?

11–45E A heat pump uses refrigerant-134a as the working fluid and operates on the ideal vapor-compression refrigeration cycle. The condenser pressure is 100 psia while the evaporator temperature is 40°F. What is the COP of this heat pump?

11–46 A heat pump operates on the ideal vapor-compression refrigeration cycle and uses refrigerant-134a as the working fluid. The condenser operates at 1000 kPa and the evaporator at 200 kPa. Determine this system's COP and the rate of heat supplied to the evaporator when the compressor consumes 6 kW.

11–47 A heat pump operates on the ideal vapor-compression cycle with refrigerant-134a as the working fluid. This heat pump is used to keep a space at 25°C by absorbing heat at the rate of 2.7 kW from geothermal water flowing through the evaporator. The evaporator operates at 20°C, and the condenser operates at 1400 kPa. The compressor receives work equal to 20 kJ for each kilogram of refrigerant flowing into it.

(*a*) Sketch the hardware and the *T-s* diagram for this heat pump.

(*b*) Determine the rate of heat transfer to the heated space at 25°C.

(*c*) Determine the COP of the heat pump.

Refrigerant-134a data: T = 20°C: h_f = 79.3 kJ/kg, h_g = 261.6 kJ/kg; P = 1400 kPa: h_f = 127.2 kJ/kg, h_g = 276.2 kJ/kg

11–48 A building requires a 2-ton heat pump for maintaining the interior space at 27°C when the outside temperature is 5°C. The heat pump operates on the normal vapor compression refrigeration cycle and uses refrigerant-134a as the working fluid. The heat pump operating conditions require an evaporator pressure of 240 kPa and a condenser pressure of 1600 kPa while the compressor has an isentropic efficiency of 85 percent. Selected data for refrigerant-134a are provided in the table below.

P, kPa	T_{sat}, °C	h_f, kJ/kg	h_g, kJ/kg	s_g, kJ/kg·K
240	−5.37	43	244	0.9222
1600	57.92	134	275	0.8982

For R-134a at P = 1600 kPa and s = 0.9222 kJ/kg·K, h = 285 kJ/kg. Also, 1 ton = 211 kJ/min.

(*a*) Sketch the hardware and the *T-s* diagram for this heat pump system.

(*b*) Determine the power required to drive the heat pump, in kW and the COP. *Answers:* (*b*) 2.14 kW, 3.29

11–49 A heat pump with refrigerant-134a as the working fluid is used to keep a space at 25°C by absorbing heat from geothermal water that enters the evaporator at 50°C at a rate of 0.065 kg/s and leaves at 40°C. The refrigerant enters the evaporator at 20°C with a quality of 23 percent and leaves at the inlet pressure as saturated vapor. The refrigerant loses 300 W of heat to the surroundings as it flows through the compressor and the refrigerant leaves the compressor at 1.4 MPa at the same entropy as the inlet. Determine (*a*) the degrees of subcooling of the refrigerant in the condenser, (*b*) the mass flow rate of the refrigerant, (*c*) the heating load and the COP of the heat pump, and (*d*) the theoretical minimum power input to the compressor for the same heating load. *Answers:* (*a*) 3.8°C, (*b*) 0.0194 kg/s, (*c*) 3.07 kW, 4.68, (*d*) 0.238 kW

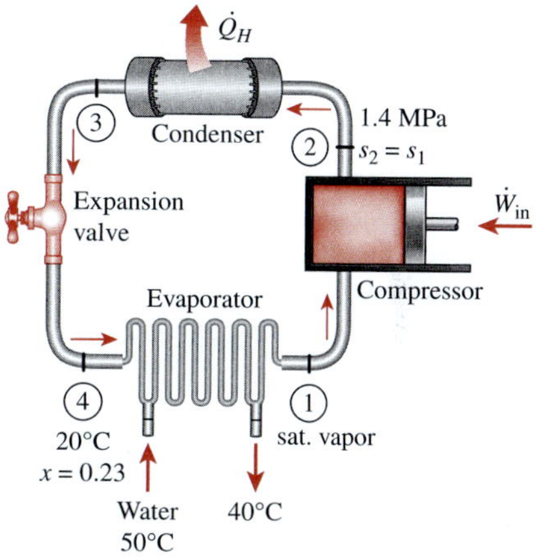

FIGURE P11–49

11–50 Refrigerant-134a enters the condenser of a residential heat pump at 800 kPa and 55°C at a rate of 0.018 kg/s and leaves at 750 kPa subcooled by 3°C. The refrigerant enters the compressor at 200 kPa superheated by 4°C. Determine (*a*) the isentropic efficiency of the compressor, (*b*) the rate of heat supplied to the heated room, and (*c*) the COP of the heat pump. Also, determine (*d*) the COP and the rate of heat supplied to the heated room if this heat pump operated on the ideal vapor-compression cycle between the pressure limits of 200 and 800 kPa.

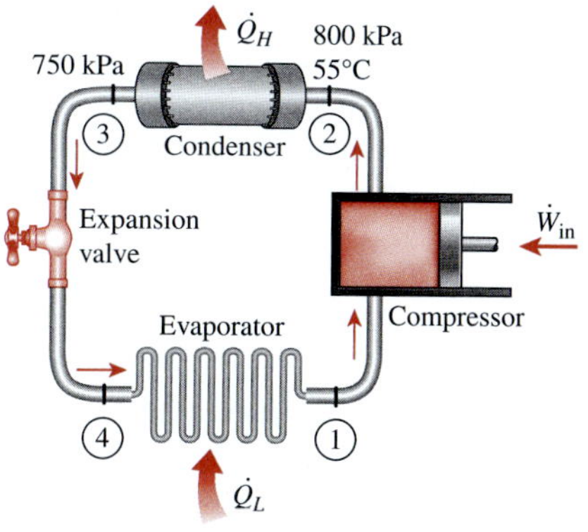

FIGURE P11–50

Innovative Refrigeration Systems

11–51C What is cascade refrigeration? What are the advantages and disadvantages of cascade refrigeration?

11–52C How does the COP of a cascade refrigeration system compare to the COP of a simple vapor-compression cycle operating between the same pressure limits?

11–53C A certain application requires maintaining the refrigerated space at −32°C. Would you recommend a simple refrigeration cycle with refrigerant-134a or a two-stage cascade refrigeration cycle with a different refrigerant at the bottoming cycle? Why?

11–54C Consider a two-stage cascade refrigeration cycle and a two-stage compression refrigeration cycle with a flash chamber. Both cycles operate between the same pressure limits and use the same refrigerant. Which system would you favor? Why?

11–55C Can a vapor-compression refrigeration system with a single compressor handle several evaporators operating at different pressures? How?

11–56C In the liquefaction process, why are gases compressed to very high pressures?

11–57 A two-stage compression refrigeration system operates with refrigerant-134a between the pressure limits of 1.4 and 0.10 MPa. The refrigerant leaves the condenser as a saturated liquid and is throttled to a flash chamber operating at 0.4 MPa. The refrigerant leaving the low-pressure compressor at 0.4 MPa is also routed to the flash chamber. The vapor in the flash chamber is then compressed to the condenser pressure by the high-pressure compressor, and the liquid is throttled to the evaporator pressure. Assuming the refrigerant leaves the evaporator as saturated vapor and both compressors are isentropic, determine (*a*) the fraction of the refrigerant that evaporates as it is throttled to the flash chamber, (*b*) the rate of heat removed from the refrigerated space for a mass flow rate of 0.25 kg/s through the condenser, and (*c*) the coefficient of performance.

11–58 Repeat Prob. 11–57 for a flash chamber pressure of 0.6 MPa.

11–59 Reconsider Prob. 11–57. Using EES (or other) software, investigate the effect of the various refrigerants for compressor efficiencies of 80, 90, and 100 percent. Compare the performance of the refrigeration system with different refrigerants.

11–60 Consider a two-stage cascade refrigeration system operating between the pressure limits of 1.2 MPa and 200 kPa with refrigerant-134a as the working fluid. Heat rejection from the lower cycle to the upper cycle takes place in an adiabatic counterflow heat exchanger where the pressure in the upper and lower cycles are 0.4 and 0.5 MPa, respectively. In both cycles, the refrigerant is a saturated liquid at the condenser exit and a saturated vapor at the compressor inlet, and the isentropic efficiency of the compressor is 80 percent. If the mass flow rate of the refrigerant through the lower cycle is 0.15 kg/s, determine (*a*) the mass flow rate of the refrigerant through the upper cycle, (*b*) the rate of heat removal from the refrigerated space, and (*c*) the COP of this refrigerator. *Answers:* (*a*) 0.212 kg/s, (*b*) 25.7 kW, (*c*) 2.68

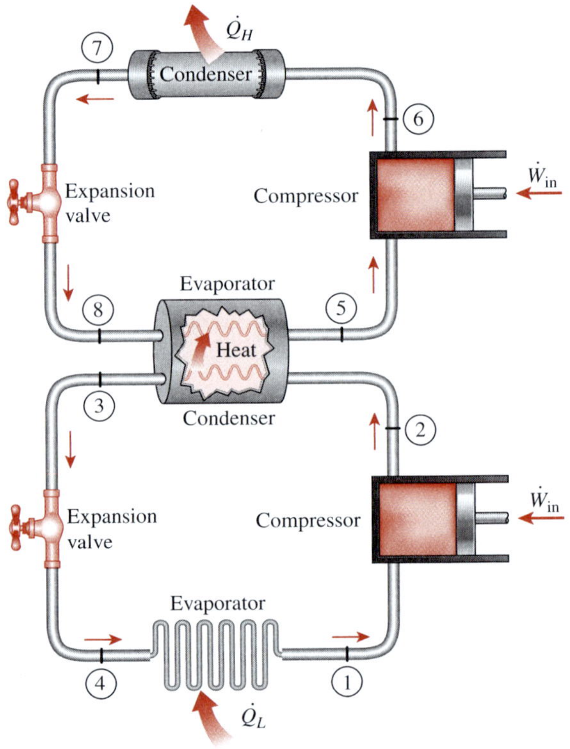

FIGURE P11–60

11–61 A two-evaporator compression refrigeration system as shown in the figure uses refrigerant-134a as the working fluid. The system operates evaporator 1 at 0°C, evaporator 2 at −26.4°C, and the condenser at 800 kPa. The refrigerant is circulated through the compressor at a rate of 0.1 kg/s and the low-temperature evaporator serves a cooling load of 8 kW. Determine the cooling rate of the high-temperature evaporator, the power required by the compressor, and the COP of the system. The refrigerant is saturated liquid at the exit of the condenser and saturated vapor at the exit of each evaporator, and the compressor is isentropic. *Answers:* 6.58 kW, 4.50 kW, 3.24

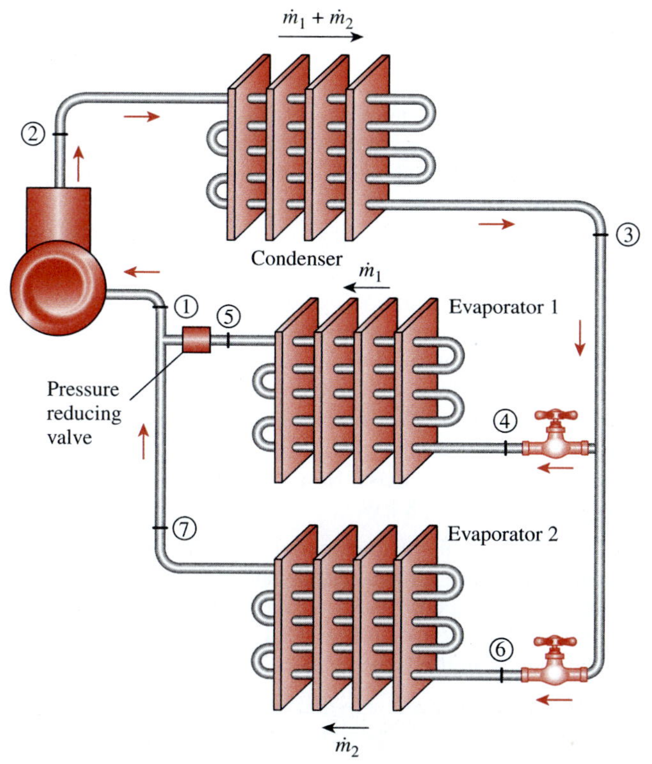

FIGURE P11–61

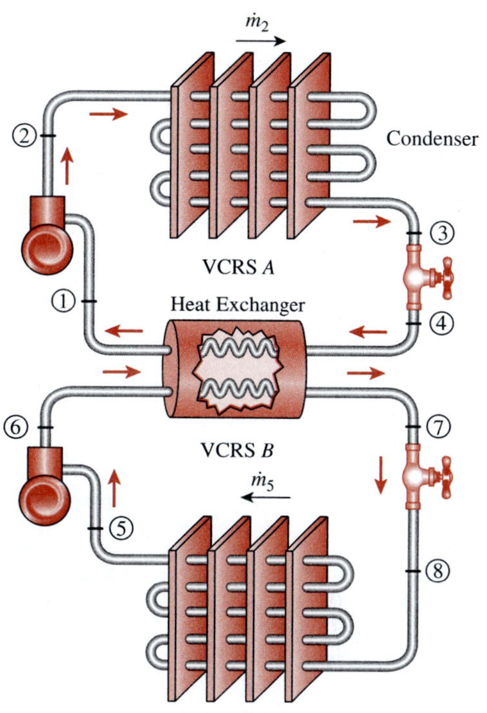

FIGURE P11–64

11–62E A two-evaporator compression refrigeration system like that in Fig. P11–61 uses refrigerant-134a as the working fluid. The system operates evaporator 1 at 30 psia, evaporator 2 at 10 psia, and the condenser at 180 psia. The cooling load for evaporator 1 is 9000 Btu/h and that for evaporator 2 is 24,000 Btu/h. Determine the power required to operate the compressor and the COP of this system. The refrigerant is saturated liquid at the exit of the condenser and saturated vapor at the exit of each evaporator, and the compressor is isentropic.

11–63E Repeat Prob. 11–62E if the 30 psia evaporator is to be replaced with a 60 psia evaporator to serve a 30,000 Btu/h cooling load.

11–64 A two-stage cascade refrigeration system is to provide cooling at −40°C while operating the high-temperature condenser at 1.6 MPa. Each stage operates on the ideal vapor-compression refrigeration cycle. The upper vapor compression refrigeration system (VCRS) uses water as its working fluid and operates its evaporator at 5°C. The lower cycle uses refrigerant-134a as its working fluid and operates its condenser at 400 kPa. This system produces a cooling effect of 20 kJ/s. Determine the mass flow rate of R-134a and water in their respective cycles, and the overall COP of this cascaded system.

11–65 Perform a second-law analysis of the cascaded system of Prob. 11–64 when the low-temperature reservoir is at −30°C and the high-temperature reservoir is at 30°C. Where does the largest exergy destruction occur?

11–66 Consider a two-stage cascade refrigeration cycle with a flash chamber as shown in the figure with refrigerant-134a as the working fluid. The evaporator temperature is −10°C and the condenser pressure is 1600 kPa. The refrigerant leaves the condenser as a saturated liquid and is throttled to a flash chamber operating at 0.45 MPa. Part of the refrigerant evaporates during this flashing process, and this vapor is mixed with the refrigerant leaving the low-pressure compressor. The mixture is then compressed to the condenser pressure by the high-pressure compressor. The liquid in the flash chamber is throttled to the evaporator pressure and cools the refrigerated space as it vaporizes in the evaporator. The mass flow rate of the refrigerant through the low-pressure compressor is 0.11 kg/s. Assuming the refrigerant leaves the evaporator as a saturated vapor and the isentropic efficiency is 86 percent for both compressors, determine (*a*) the mass flow rate of the refrigerant through the high-pressure compressor, (*b*) the rate of refrigeration supplied by the system, and (*c*) the COP of this refrigerator. Also, determine (*d*) the rate of refrigeration and the COP if this refrigerator operated on a single-stage vapor-compression cycle between the same evaporating temperature and condenser pressure with the samc compressor efficiency and the same flow rate as calculated in part (*a*).

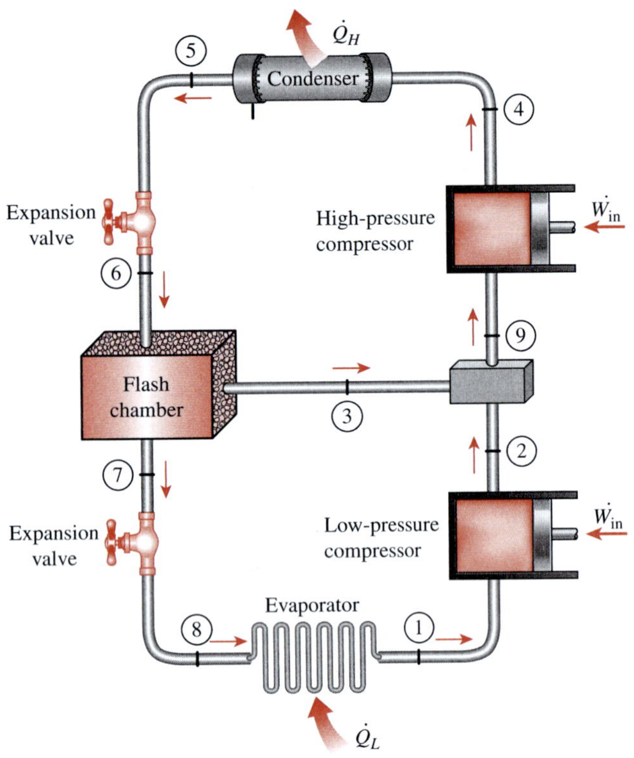

FIGURE P11–66

Gas Refrigeration Cycle

11–67C How does the ideal-gas refrigeration cycle differ from the Brayton cycle?

11–68C How does the ideal-gas refrigeration cycle differ from the Carnot refrigeration cycle?

11–69C Devise a refrigeration cycle that works on the reversed Stirling cycle. Also, determine the COP for this cycle.

11–70C How is the ideal-gas refrigeration cycle modified for aircraft cooling?

11–71C In gas refrigeration cycles, can we replace the turbine by an expansion valve as we did in vapor-compression refrigeration cycles? Why?

11–72C How do we achieve very low temperatures with gas refrigeration cycles?

11–73 Air enters the compressor of an ideal gas refrigeration cycle at 7°C and 35 kPa and the turbine at 37°C and 160 kPa. The mass flow rate of air through the cycle is 0.2 kg/s. Assuming variable specific heats for air, determine (*a*) the rate of refrigeration, (*b*) the net power input, and (*c*) the coefficient of performance. *Answers:* (*a*) 15.9 kW, (*b*) 8.64 kW, (*c*) 1.84

11–74 Repeat Prob. 11–73 for a compressor isentropic efficiency of 80 percent and a turbine isentropic efficiency of 85 percent.

11–75 Reconsider Prob. 11–74. Using EES (or other) software, study the effects of compressor and turbine isentropic efficiencies as they are varied from 70 to 100 percent on the rate of refrigeration, the net power input, and the COP. Plot the *T-s* diagram of the cycle for the isentropic case.

11–76 A gas refrigeration cycle with a pressure ratio of 3 uses helium as the working fluid. The temperature of the helium is −10°C at the compressor inlet and 50°C at the turbine inlet. Assuming isentropic efficiencies of 80 percent for both the turbine and the compressor, determine (*a*) the minimum temperature in the cycle, (*b*) the coefficient of performance, and (*c*) the mass flow rate of the helium for a refrigeration rate of 18 kW.

11–77E An ideal gas refrigeration cycle uses air as the working fluid. The air is at 5 psia and −10°F as it enters the compressor with a compression ratio of 4. The temperature at the turbine entrance is 100°F. Determine this cycle's COP. Use constant specific heats at room temperature.

11–78E Rework Prob. 11–77E when the compressor isentropic efficiency is 87 percent, the turbine isentropic efficiency is 94 percent, and the pressure drop across each heat exchanger is 1 psia. *Answer:* 0.364

11–79 A 36 kJ/kg cooling load at 0°C is to be serviced by an ideal gas refrigeration cycle using air as the working fluid. Waste heat from this cycle is rejected to the surrounding environment at 25°C. At the inlet of the compressor, the air is at 85 kPa and −13°C. Determine the minimum pressure ratio for this system to operate properly. Use constant specific heats at room temperature.

11–80 A gas refrigeration system using air as the working fluid has a pressure ratio of 5. Air enters the compressor at 0°C. The high-pressure air is cooled to 35°C by rejecting heat to the surroundings. The refrigerant leaves the turbine at −80°C and then it absorbs heat from the refrigerated space before entering the regenerator. The mass flow rate of air is 0.4 kg/s. Assuming isentropic efficiencies of 80 percent for the compressor and 85 percent for the turbine and using constant specific heats at room temperature, determine (*a*) the effectiveness of the regenerator, (*b*) the rate of heat removal from the refrigerated space, and (*c*) the COP of the cycle. Also, determine (*d*) the refrigeration load and the COP if this system operated on the simple gas refrigeration cycle. Use the same compressor inlet temperature as given, the same turbine inlet temperature as calculated, and the same compressor and turbine efficiencies. *Answers:* (*a*) 0.434, (*b*) 21.4 kW, (*c*) 0.478, (*d*) 24.7 kW, 0.599

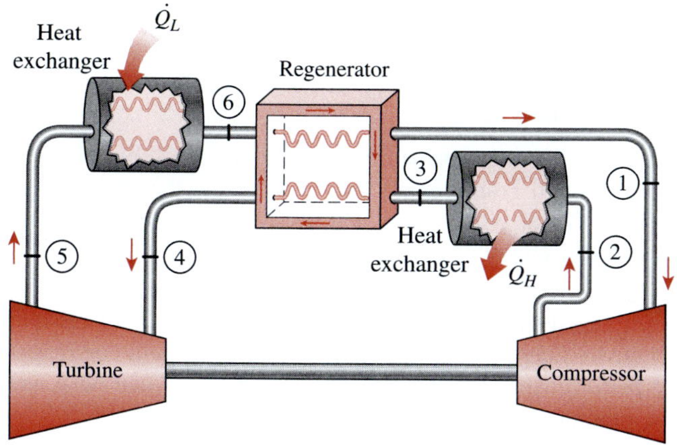

FIGURE P11–80

11–81 An ideal gas refrigeration system with two stages of compression with intercooling as shown in Fig. P11–81 operates with air entering the first compressor at 90 kPa and −18°C. Each compression stage has a pressure ratio of 4 and the two intercoolers can cool the air to 10°C. Calculate the coefficient of performance of this system and the rate at which air must be circulated through this system to service a 75,000 kJ/h cooling load. Use constant specific heats at room temperature. *Answers:* 1.19, 0.163 kg/s

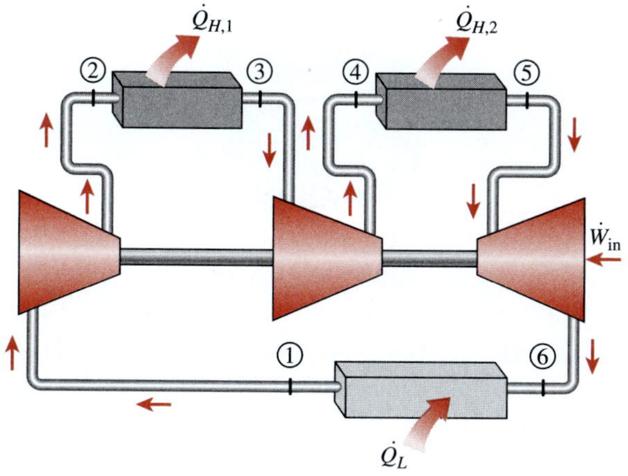

FIGURE P11–81

11–82 How will the answers of Prob. 11–81 change when the isentropic efficiency of each compressor is 85 percent and the isentropic efficiency of the turbine is 95 percent?

11–83 A regenerative gas refrigeration cycle with argon as the working fluid is used to keep a space at −45°C while rejecting heat to the ambient air at 0°C. Argon enters the compressor at −30°C, at a flow rate of 0.08 kg/s. The cycle has a pressure ratio of 5. The temperature of the argon decreases from 15°C to −38°C in the regenerator. The isentropic efficiencies of both the compresor and turbine are 82 percent. Determine (*a*) the refrigeration load and the COP, (*b*) the minimum power input, the second-law efficiency of the cycle, and the total exergy destruction in the cycle. The properties of argon are $c_p = 0.5203$ kJ/kg·K and $k = 1.667$.

Absorption Refrigeration Systems

11–84C Can water be used as a refrigerant in air-conditioning applications? Explain.

11–85C What is absorption refrigeration? How does an absorption refrigeration system differ from a vapor-compression refrigeration system?

11–86C What are the advantages and disadvantages of absorption refrigeration?

11–87C In absorption refrigeration cycles, why is the fluid in the absorber cooled and the fluid in the generator heated?

11–88C How is the coefficient of performance of an absorption refrigeration system defined?

11–89C What are the functions of the rectifier and the regenerator in an absorption refrigeration system?

11–90 An absorption refrigeration system that receives heat from a source at 95°C and maintains the refrigerated space at 0°C is claimed to have a COP of 3.1. If the environmental temperature is 19°C, can this claim be valid? Justify your answer.

11–91 An absorption refrigeration system receives heat from a source at 120°C and maintains the refrigerated space at 0°C. If the temperature of the environment is 25°C, what is the maximum COP this absorption refrigeration system can have?

11–92 Heat is supplied to an absorption refrigeration system from a geothermal well at 130°C at a rate of 5×10^5 kJ/h. The environment is at 25°C, and the refrigerated space is maintained at −30°C. Determine the maximum rate at which this system can remove heat from the refrigerated space. *Answer:* 5.75×10^5 kJ/h

11–93 A reversible absorption refrigerator consists of a reversible heat engine and a reversible refrigerator. The system removes heat from a cooled space at −15°C at a rate of 70 kW. The refrigerator operates in an environment at 25°C. If the heat is supplied to the cycle by condensing saturated steam at 150°C, determine (*a*) the rate at which the steam condenses, and (*b*) the power input to the reversible refrigerator. (*c*) If the COP of an actual absorption chiller at the same temperature limits has a COP of 0.8, determine the second law efficiency of this chiller. *Answers:* (*a*) 0.0174 kg/s, (*b*) 10.9 kW, (*c*) 42.0 percent

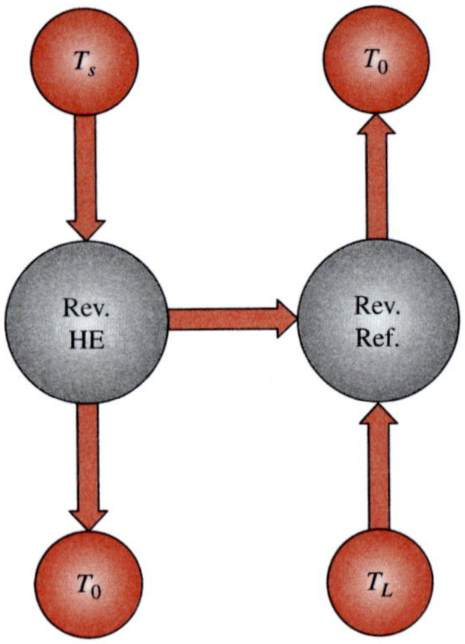

FIGURE P11–93

11–94E An ammonia-water absorption refrigeration cycle is used to keep a space at 25°F when the ambient temperature is 70°F. Pure ammonia enters the condenser at 300 psia and 140°F at a rate of 0.04 lbm/s. Ammonia leaves the condenser as a saturated liquid and is expanded to 30 psia. Ammonia leaves the evaporator as a saturated vapor. Heat is supplied to the generator by geothermal liquid water that enters at 240°F at a rate of 0.55 lbm/s and leaves at 200°F. Determine (*a*) the rate of cooling provided by the system, in Btu/h, the COP, and (*b*) the second-law efficiency of the system. The enthalpies of ammonia at various states of the system are: condenser inlet h_2 = 665.7 Btu/lbm, evaporator inlet h_4 = 190.9 Btu/lbm, evaporator exit h_1 = 619.2 Btu/lbm. Also, take the specific heat of geothermal water to be 1.0 Btu/lbm·°F.

Special Topic: Thermoelectric Power Generation and Refrigeration Systems

11–95C What is a thermoelectric circuit?

11–96C Describe the Seebeck and the Peltier effects.

11–97C Consider a circular copper wire formed by connecting the two ends of a copper wire. The connection point is now heated by a burning candle. Do you expect any current to flow through the wire?

11–98C An iron and a constantan wire are formed into a closed circuit by connecting the ends. Now both junctions are heated and are maintained at the same temperature. Do you expect any electric current to flow through this circuit?

11–99C A copper and a constantan wire are formed into a closed circuit by connecting the ends. Now one junction is heated by a burning candle while the other is maintained at room temperature. Do you expect any electric current to flow through this circuit?

11–100C How does a thermocouple work as a temperature measurement device?

11–101C Why are semiconductor materials preferable to metals in thermoelectric refrigerators?

11–102C Is the efficiency of a thermoelectric generator limited by the Carnot efficiency? Why?

11–103E A thermoelectric generator receives heat from a source at 340°F and rejects the waste heat to the environment at 90°F. What is the maximum thermal efficiency this thermoelectric generator can have? *Answer:* 31.3 percent

11–104 A thermoelectric refrigerator removes heat from a refrigerated space at −5°C at a rate of 130 W and rejects it to an environment at 20°C. Determine the maximum coefficient of performance this thermoelectric refrigerator can have and the minimum required power input. *Answers:* 10.72, 12.1 W

11–105 A thermoelectric cooler has a COP of 0.15 and removes heat from a refrigerated space at a rate of 180 W. Determine the required power input to the thermoelectric cooler, in W.

11–106E A thermoelectric cooler has a COP of 0.18 and the power input to the cooler is 1.8 hp. Determine the rate of heat removed from the refrigerated space, in Btu/min.

11–107 A thermoelectric refrigerator is powered by a 12-V car battery that draws 3 A of current when running. The refrigerator resembles a small ice chest and is claimed to cool nine canned drinks, 0.350-L each, from 25 to 3°C in 12 h. Determine the average COP of this refrigerator.

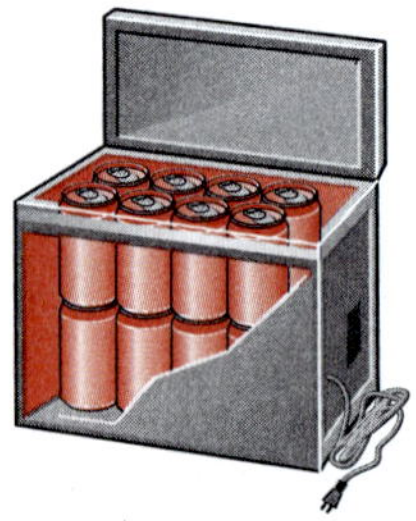

FIGURE P11–107

11–108E Thermoelectric coolers that plug into the cigarette lighter of a car are commonly available. One such cooler is claimed to cool a 12-oz (0.771-lbm) drink from 78 to 38°F or to heat a cup of coffee from 75 to 130°F in about 15 min in a well-insulated cup holder. Assuming an average COP of 0.2 in the cooling mode, determine (*a*) the average rate of heat removal from the drink, (*b*) the average rate of heat supply to the coffee, and (*c*) the electric power drawn from the battery of the car, all in W.

11–109 It is proposed to run a thermoelectric generator in conjunction with a solar pond that can supply heat at a rate of 7×10^6 kJ/h at 90°C. The waste heat is to be rejected to the environment at 22°C. What is the maximum power this thermoelectric generator can produce?

Review Problems

11–110 Consider a steady-flow Carnot refrigeration cycle that uses refrigerant-134a as the working fluid. The maximum and minimum temperatures in the cycle are 30 and −20°C, respectively. The quality of the refrigerant is 0.15 at the beginning of the heat absorption process and 0.80 at the end. Show the cycle on a *T-s* diagram relative to saturation lines, and determine (*a*) the coefficient of performance, (*b*) the condenser and evaporator pressures, and (*c*) the net work input.

11–111 Rooms with floor areas of up to 15-m^2 are cooled adequately by window air conditioners whose cooling capacity is 5000 Btu/h. Assuming the COP of the air conditioner to be 3.5, determine the rate of heat gain of the room, in Btu/h, when the air conditioner is running continuously to maintain a constant room temperature.

11–112 A heat pump water heater (HPWH) heats water by absorbing heat from the ambient air and transferring it to water. The heat pump has a COP of 3.4 and consumes 6 kW of electricity when running. Determine if this heat pump can be used to meet the cooling needs of a room most of the time for "free" by absorbing heat from the air in the room. The rate of heat gain of a room is usually less than 45,000 kJ/h.

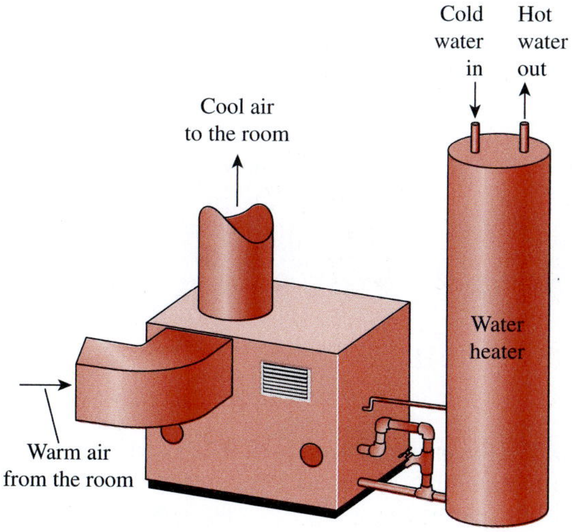

FIGURE P11–112

11–113 A heat pump that operates on the ideal vapor-compression cycle with refrigerant-134a is used to heat a house. The mass flow rate of the refrigerant is 0.32 kg/s. The condenser and evaporator pressures are 900 and 200 kPa, respectively. Show the cycle on a *T-s* diagram with respect to saturation lines, and determine (*a*) the rate of heat supply to the house, (*b*) the volume flow rate of the refrigerant at the compressor inlet, and (*c*) the COP of this heat pump.

11–114 A ground-coupled heat pump operates on the ideal vapor-compression cycle with refrigerant-134a as the working fluid. This heat pump has a COP equal to 6 in this cooling mode and maintains a space at 25°C when absorbing heat from the space at the rate of 18 kW in the evaporator. The evaporator operates at 20°C, and the condenser operates at 1400 kPa. Ground water taken from a well at the rate of 0.32 kg/s absorbs the heat transferred from the refrigerant in the condenser.

(*a*) Sketch the hardware and the *T-s* diagram for this heat pump.
(*b*) If the ground water enters the condenser at 10°C, what will be the temperature of the cooling water as it is returned to the ground? Take c_p = 4.18 kJ/kg·K for cooling water.
(*c*) Determine the COP of the heat pump.

Refrigerant-134a data: T = 20°C: h_f = 79.3 kJ/kg, h_g = 261.6 kJ/kg; P = 1400 kPa: h_f = 127.2 kJ/kg, h_g = 276.2 kJ/kg

11–115 A refrigeration unit operates on the ideal vapor compression refrigeration cycle and uses refrigerant-22 as the working fluid. The operating conditions for this unit are evaporator saturation temperature of −5°C and the condenser saturation temperature of 45°C. Selected data for refrigerant-22 are provided in the table below.

T, °C	P_{sat}, kPa	h_f, kJ/kg	h_g, kJ/kg	s_g, kJ/kg·K
−5	421.2	38.76	248.1	0.9344
45	1728	101	261.9	0.8682

For R-22 at P = 1728 kPa and s = 0.9344 kJ/kg·K, T = 68.15°C and h = 283.7 kJ/kg. Also, take $c_{p,air}$ = 1.005 kJ/kg·K.

(*a*) Sketch the hardware and the *T-s* diagram for this heat pump application.
(*b*) Determine the COP for this refrigeration unit.
(*c*) The evaporator of this unit is located inside the air handler of the building. The air flowing through the air handler enters the air handler at 27°C and is limited to a 20°C temperature drop. Determine the ratio of volume flow rate of air entering the air handler (m^3_{air}/min) to mass flow rate of R-22 ($kg_{R\text{-}22}$/s) through the air handler, in (m^3_{air}/min)/($kg_{R\text{-}22}$/s). Assume the air pressure is 100 kPa.

11–116 An air-conditioner operates on the vapor-compression refrigeration cycle with refrigerant-134a as the refrigerant. The air conditioner is used to keep a space at 21°C while rejecting the waste heat to the ambient air at 37°C. The refrigerant enters the compressor at 180 kPa superheated by 2.7°C at a rate of 0.06 kg/s and leaves the compressor at 1200 kPa and 60°C. R-134a is subcooled by 6.3°C at the exit of the condenser. Determine (*a*) the rate of cooling provided to the space, in Btu/h, and the COP, (*b*) the isentropic efficiency and the exergy efficiency of the compressor, (*c*) the exergy destruction in each component of the cycle and the total exergy destruction in the cycle, and (*d*) the minimum power input and the second-law efficiency of the cycle.

11–117 Consider an ice producing plant that operates on the ideal vapor-compression refrigeration cycle and uses refrigerant-134a as the working fluid. The refrigeration cycle operating conditions require an evaporator pressure of 140 kPa and the condenser pressure of 1200 kPa. Cooling water flows through the water jacket surrounding the condenser and is supplied at the rate of 200 kg/s. The cooling water has a 10°C temperature rise as it flows through the water jacket. To produce ice, potable water is supplied to the chiller section of the refrigeration cycle. For each kg of ice produced, 333 kJ of energy must be removed from the potable water supply.

(*a*) Sketch the hardware for all three working fluids of this refrigerant-ice making system and the *T-s* diagram for refrigeration cycle.
(*b*) Determine the mass flow rate of the refrigerant, in kg/s.
(*c*) Determine the mass flow rate of the potable water supply, in kg/s.

11–118 Rapid freezing of fresh fruit requires air at −34°C. A refrigerator with refrigerant-134a as the working fluid produces this air by operating its evaporator at −37°C and its condenser at 1200 kPa, while rejecting heat to the ambient air at 30°C. If the isentropic efficiency of the compressor is 90 percent and the vapor entering the compressor is superheated by 7°C, determine the process that causes the greatest amount of exergy loss. *Answer:* 22.5 kJ/kg (expansion valve)

11–119 Rework Prob. 11–118 with a 6.3°C subcooling at the exit of the condenser.

11–120 Consider a two-stage compression refrigeration system operating between the pressure limits of 1.4 and 0.18 MPa. The working fluid is refrigerant-134a. The refrigerant leaves the condenser as a saturated liquid and is throttled to a flash chamber operating at 0.6 MPa. Part of the refrigerant evaporates during this flashing process, and this vapor is mixed with the refrigerant leaving the low-pressure compressor. The mixture is then compressed to the condenser pressure by the high-pressure compressor. The liquid in the flash chamber is throttled to the evaporator pressure, and it cools the refrigerated space as it vaporizes in the evaporator. Assuming the refrigerant leaves the evaporator as saturated vapor and both compressors are isentropic, determine (*a*) the fraction of the refrigerant that evaporates as it is throttled to the flash chamber, (*b*) the amount of heat removed from the refrigerated space and the compressor work per unit mass of refrigerant flowing through the condenser, and (*c*) the coefficient of performance. *Answers:* (*a*) 0.253, (*b*) 121 kJ/kg, 36.4 kJ/kg, (*c*) 3.31

11–121E A two-evaporator compression refrigeration system as shown in Fig. P11–121E uses refrigerant-134a as the working fluid. The system operates evaporator 1 at 30°F, evaporator 2 at −29.5°F, and the condenser at 160 psia. The cooling load of evaporator 1 is double that of evaporator 2. Determine the cooling load of both evaporators per unit of flow through the compressor, as well as the COP of this system. The refrigerant is saturated liquid at the exit of the condenser and saturated vapor at the exit of each evaporator, and the compressor is isentropic.

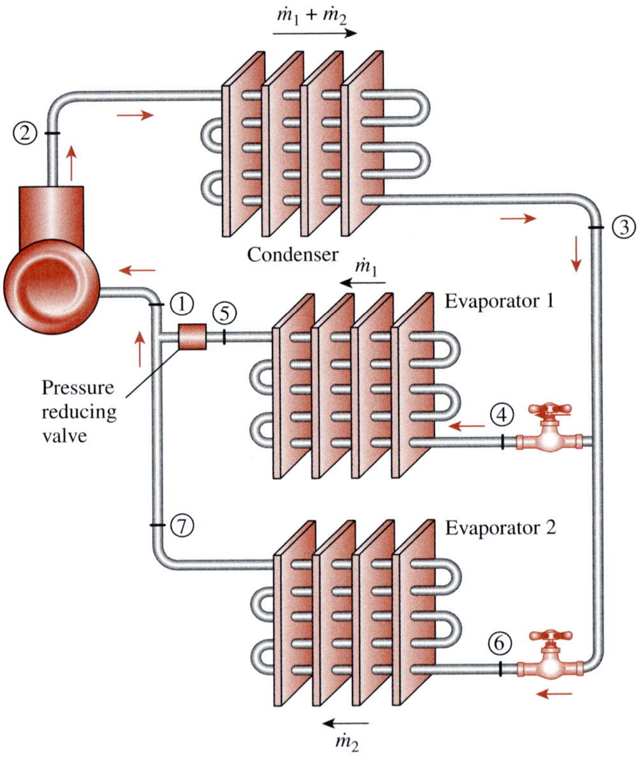

FIGURE P11–121E

11–122E Reconsider Prob. 11–121E. The refrigeration system of that problem cools one reservoir at −15°F and one at 40°F while rejecting heat to a reservoir at 80°F. Which process has the highest exergy destruction?

11–123 A two-stage compression refrigeration system with an adiabatic liquid-vapor separation unit as shown in Fig. P11–123 uses refrigerant-134a as the working fluid. The system operates the evaporator at −32°C, the condenser at 1400 kPa, and the separator at 8.9°C. The refrigerant is circulated through the condenser at a rate of 2 kg/s. Determine the rate of cooling and power requirement for this system. The refrigerant is saturated liquid at the inlet of each expansion valve and saturated vapor at the inlet of each compressor, and the compressors are isentropic.

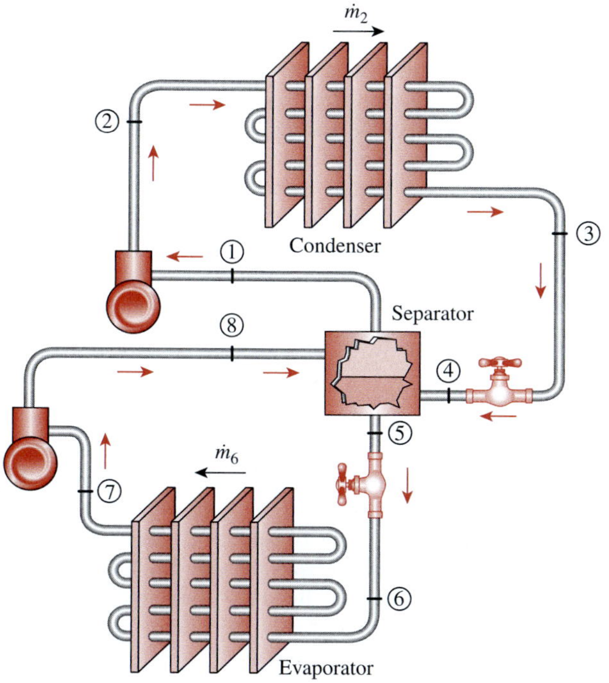

FIGURE P11–123

11–124 Which process of the cycle in Prob. 11–123 has the greatest rate of exergy destruction when the low-temperature reservoir is at −18°C and the high-temperature reservoir is at 25°C? *Answer:* 26.2 kW (condenser)

11–125 Consider a regenerative gas refrigeration cycle using helium as the working fluid. Helium enters the compressor at 100 kPa and −10°C and is compressed to 300 kPa. Helium is then cooled to 20°C by water. It then enters the regenerator where it is cooled further before it enters the turbine. Helium leaves the refrigerated space at −25°C and enters the regenerator. Assuming both the turbine and the compressor to be isentropic, determine (*a*) the temperature of the helium at the turbine inlet, (*b*) the coefficient of performance of the cycle, and (*c*) the net power input required for a mass flow rate of 0.45 kg/s.

11–126 An absorption refrigeration system is to remove heat from the refrigerated space at 2°C at a rate of 28 kW while operating in an environment at 25°C. Heat is to be supplied from a solar pond at 95°C. What is the minimum rate of heat supply required? *Answer:* 12.3 kW

11–127 EES Reconsider Prob. 11–126. Using EES (or other) software, investigate the effect of the source temperature on the minimum rate of heat supply. Let the source temperature vary from 50 to 250°C. Plot the minimum rate of heat supply as a function of source temperature, and discuss the results.

11–128 A gas refrigeration system using air as the working fluid has a pressure ratio of 5. Air enters the compressor at 0°C. The high-pressure air is cooled to 35°C by rejecting heat to the surroundings. The refrigerant leaves the turbine at −80°C and enters the refrigerated space where it absorbs heat before entering the regenerator. The mass flow rate of air is 0.4 kg/s. Assuming isentropic efficiencies of 80 percent for the compressor and 85 percent for the turbine and using variable specific heats, determine (*a*) the effectiveness of the regenerator, (*b*) the rate of heat removal from the refrigerated space, and (*c*) the COP of the cycle. Also, determine (*d*) the refrigeration load and the COP if this system operated on the simple gas refrigeration cycle. Use the same compressor inlet temperature as given, the same turbine inlet temperature as calculated, and the same compressor and turbine efficiencies.

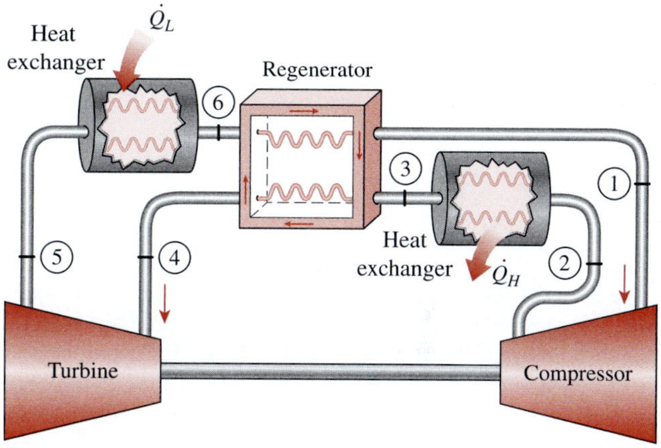

FIGURE P11–128

11–129 The refrigeration system of Fig. P11–129 is another variation of the basic vapor-compression refrigeration system which attempts to reduce the compression work. In this system, a heat exchanger is used to superheat the vapor entering the compressor while subcooling the liquid exiting from the condenser. Consider a system of this type that uses refrigerant-134a as its refrigerant and operates the evaporator at −10.1°C, and the condenser at 800 kPa. Determine the system COP when the heat exchanger provides 11.3°C of subcooling at the throttle valve entrance. Assume the refrigerant leaves the evaporator as a saturated vapor and the compressor is isentropic. *Answer:* 5.23

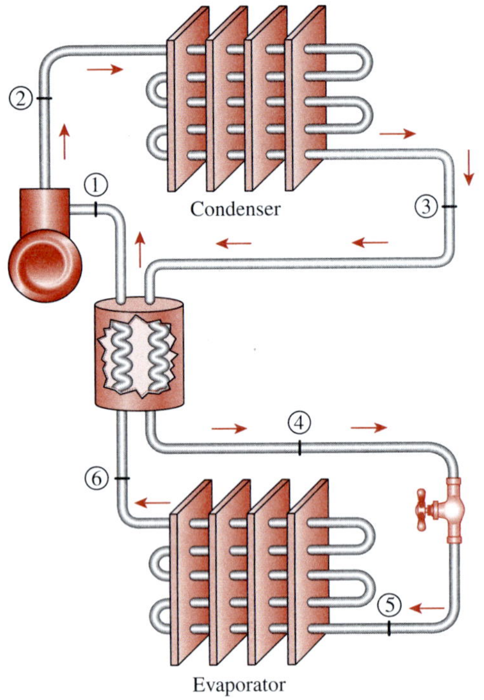

FIGURE P11–129

11–130 Repeat Prob. 11–129 if the heat exchanger provides 21.3°C of subcooling.

11–131 An ideal gas refrigeration system with three stages of compression with intercooling operates with air entering the first compressor at 50 kPa and −30°C. Each compressor in this system has a pressure ratio of 7, and the air temperature at the outlet of all intercoolers is 15°C. Calculate the COP of this system. Use constant specific heats at room temperature.

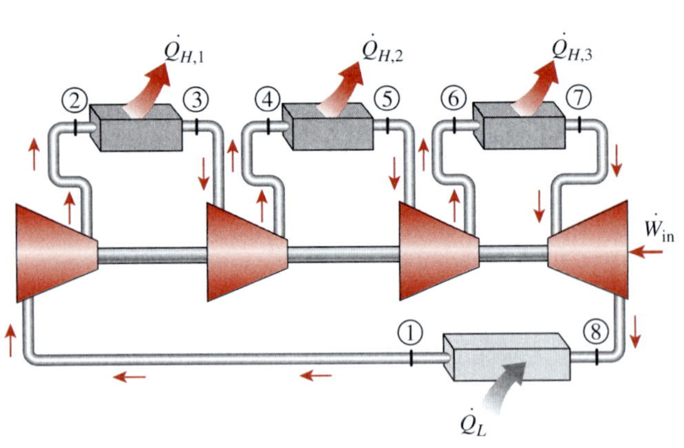

FIGURE P11–131

11–132 The vortex tube (also known as a Ranque or Hirsch tube) is a device that produces a refrigeration effect by expanding pressurized gas such as air in a tube (instead of a turbine as in the reversed Brayton cycle). It was invented and patented by Ranque in 1931 and improved by Hirsch in 1945, and is commercially available in various sizes.

The vortex tube is simply a straight circular tube equipped with a nozzle, as shown in the figure. The compressed gas at temperature T_1 and pressure P_1 is accelerated in the nozzle by expanding it to nearly atmospheric pressure and is introduced into the tube tangentially at a very high (typically supersonic) velocity to produce a swirling motion (vortex) within the tube. The rotating gas is allowed to exit through the full-size tube that extends to the right, and the mass flow rate is controlled by a valve located about 30 diameters downstream. A smaller amount of air at the core region is allowed to escape to the left through a small aperture at the center. It is observed that the gas that is in the core region and escapes through the central aperture is cold while the gas that is in the peripheral region and escapes through the full-size tube is hot. If the temperature and the mass flow rate of the cold stream are T_c and $\dot{m}_c$, respectively, the rate of refrigeration in the vortex tube can be ex-pressed as

$$\dot{Q}_{\text{refrig,vortex tube}} = \dot{m}_c(h_1 - h_c) = \dot{m}_c c_p (T_1 - T_c)$$

where c_p is the specific heat of the gas and $T_1 - T_c$ is the temperature drop of the gas in the vortex tube (the cooling effect). Temperature drops as high as 60°C (or 108°F) are obtained at high pressure ratios of about 10. The coefficient of performance of a vortex tube can be defined as the ratio of the refrigeration rate as given above to the power used to compress the gas. It ranges from about 0.1 to 0.15, which is well below the COPs of ordinary vapor compression refrigerators.

This interesting phenomenon can be explained as follows: the centrifugal force creates a radial pressure gradient in the vortex, and thus the gas at the periphery is pressurized and heated by the gas at the core region, which is cooled as a result. Also, energy is transferred from the inner layers toward the outer layers as the outer layers slow down the inner layers because of fluid viscosity that tends to produce a solid vortex. Both of these effects cause the energy and thus the temperature of the gas in the core region to decline. The conservation of energy requires the energy of the fluid at the outer layers to increase by an equivalent amount.

The vortex tube has no moving parts, and thus it is inherently reliable and durable. The ready availability of the compressed air at pressures up to 10 atm in most industrial facilities makes the vortex tube particularly attractive in such settings. Despite its low efficiency, the vortex tube has found application in small-scale industrial spot-cooling operations such as cooling of soldered parts or critical electronic components, cooling drinking water, and cooling the suits of workers in hot environments.

Consider a vortex tube that receives compressed air at 500 kPa and 300 K and supplies 25 percent of it as cold air at 100 kPa and 278 K. The ambient air is at 300 K and 100 kPa, and the compressor has an isentropic efficiency of 80 percent. The air suffers a pressure drop of 35 kPa in the aftercooler and the compressed air lines between the compressor and the vortex tube.

(*a*) Without performing any calculations, explain how the COP of the vortex tube would compare to the COP of an actual air refrigeration system based on the reversed Brayton cycle for the same pressure ratio. Also, compare the minimum temperatures that can be obtained by the two systems for the same inlet temperature and pressure.
(*b*) Assuming the vortex tube to be adiabatic and using specific heats at room temperature, determine the exit temperature of the hot fluid stream.
(*c*) Show, with calculations, that this process does not violate the second law of thermodynamics.
(*d*) Determine the coefficient of performance of this refrigeration system, and compare it to the COP of a Carnot refrigerator.

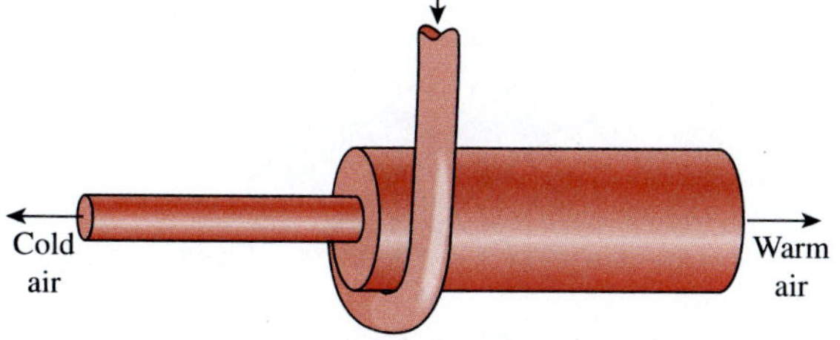

FIGURE P11–132

11–133 Repeat Prob. 11–132 for a pressure of 600 kPa at the vortex tube intake.

11–134 Using EES (or other) software, investigate the effect of the evaporator pressure on the COP of an ideal vapor-compression refrigeration cycle with R-134a as the working fluid. Assume the condenser pressure is kept constant at 1.4 MPa while the evaporator pressure is varied from 100 kPa to 500 kPa. Plot the COP of the refrigeration cycle against the evaporator pressure, and discuss the results.

11–135 Using EES (or other) software, investigate the effect of the condenser pressure on the COP of an ideal vapor-compression refrigeration cycle with R-134a as the working fluid. Assume the evaporator pressure is kept constant at 150 kPa while the condenser pressure is varied from 400 to 1400 kPa. Plot the COP of the refrigeration cycle against the condenser pressure, and discuss the results.

11–136 Derive a relation for the COP of the two-stage refrigeration system with a flash chamber as shown in Fig. 11–14 in terms of the enthalpies and the quality at state 6. Consider a unit mass in the condenser.

Fundamentals of Engineering (FE) Exam Problems

11–137 Consider a heat pump that operates on the reversed Carnot cycle with R-134a as the working fluid executed under the saturation dome between the pressure limits of 140 and 800 kPa. R-134a changes from saturated vapor to saturated liquid during the heat rejection process. The net work input for this cycle is

(*a*) 28 kJ/kg (*b*) 34 kJ/kg (*c*) 49 kJ/kg
(*d*) 144 kJ/kg (*e*) 275 kJ/kg

11–138 A refrigerator removes heat from a refrigerated space at 0°C at a rate of 2.2 kJ/s and rejects it to an environment at 20°C. The minimum required power input is

(*a*) 89 W (*b*) 150 W (*c*) 161 W
(*d*) 557 W (*e*) 2200 W

11–139 A refrigerator operates on the ideal vapor compression refrigeration cycle with R-134a as the working fluid between the pressure limits of 120 and 800 kPa. If the rate of heat removal from the refrigerated space is 32 kJ/s, the mass flow rate of the refrigerant is

(*a*) 0.19 kg/s (*b*) 0.15 kg/s (*c*) 0.23 kg/s
(*d*) 0.28 kg/s (*e*) 0.81 kg/s

11–140 A heat pump operates on the ideal vapor compression refrigeration cycle with R-134a as the working fluid between the pressure limits of 0.32 and 1.2 MPa. If the mass flow rate of the refrigerant is 0.193 kg/s, the rate of heat supply by the heat pump to the heated space is

(*a*) 3.3 kW (*b*) 23 kW (*c*) 26 kW
(*d*) 31 kW (*e*) 45 kW

11–141 An ideal vapor compression refrigeration cycle with R-134a as the working fluid operates between the pressure limits of 120 kPa and 700 kPa. The mass fraction of the refrigerant that is in the liquid phase at the inlet of the evaporator is

(*a*) 0.69 (*b*) 0.63 (*c*) 0.58
(*d*) 0.43 (*e*) 0.35

11–142 Consider a heat pump that operates on the ideal vapor compression refrigeration cycle with R-134a as the working fluid between the pressure limits of 0.32 and 1.2 MPa. The coefficient of performance of this heat pump is

(*a*) 0.17 (*b*) 1.2 (*c*) 3.1
(*d*) 4.9 (*e*) 5.9

11–143 An ideal gas refrigeration cycle using air as the working fluid operates between the pressure limits of 80 and

280 kPa. Air is cooled to 35°C before entering the turbine. The lowest temperature of this cycle is

(*a*) −58°C (*b*) −26°C (*c*) 5°C
(*d*) 11°C (*e*) 24°C

11–144 Consider an ideal gas refrigeration cycle using helium as the working fluid. Helium enters the compressor at 100 kPa and 17°C and compressed to 400 kPa. Helium is then cooled to 20°C before it enters the turbine. For a mass flow rate of 0.2 kg/s, the net power input required is

(*a*) 28.3 kW (*b*) 40.5 kW (*c*) 64.7 kW
(*d*) 93.7 kW (*e*) 113 kW

11–145 An absorption air-conditioning system is to remove heat from the conditioned space at 20°C at a rate of 150 kJ/s while operating in an environment at 35°C. Heat is to be supplied from a geothermal source at 140°C. The minimum rate of heat supply is

(*a*) 86 kJ/s (*b*) 21 kJ/s (*c*) 30 kJ/s
(*d*) 61 kJ/s (*e*) 150 kJ/s

11–146 Consider a refrigerator that operates on the vapor compression refrigeration cycle with R-134a as the working fluid. The refrigerant enters the compressor as saturated vapor at 160 kPa, and exits at 800 kPa and 50°C, and leaves the condenser as saturated liquid at 800 kPa. The coefficient of performance of this refrigerator is

(*a*) 2.6 (*b*) 1.0 (*c*) 4.2
(*d*) 3.2 (*e*) 4.4

Design and Essay Problems

11–147 Write an essay on air-, water-, and soil-based heat pumps. Discuss the advantages and the disadvantages of each system. For each system identify the conditions under which that system is preferable over the other two. In what situations would you not recommend a heat pump heating system?

11–148 Develop and discuss techniques that apply the principle of regeneration to improve the performance of vapor compression refrigeration systems.

11–149 The heat supplied by a heat pump used to maintain a building's temperature is often supplemented by another source of direct heat. The fraction of the total heat required that is supplied by supplemental heat increases as the temperature of the environmental air (which serves as the low-temperature sink) decreases. Develop a supplemental heat schedule as a function of the environmental air temperature that minimizes the total supplemental and heat-pump energy required to service the building.

11–150 Consider a cascaded vapor compression refrigeration system that uses the same working fluid in both systems. Is there a pressure at which to operate the heat exchanger that will optimize the overall COP?

11–151 Consider a solar pond power plant operating on a closed Rankine cycle. Using refrigerant-134a as the working fluid, specify the operating temperatures and pressures in the cycle, and estimate the required mass flow rate of refrigerant-134a for a net power output of 50 kW. Also, estimate the surface area of the pond for this level of continuous power production. Assume that the solar energy is incident on the pond at a rate of 500 W per m^2 of pond area at noontime, and that the pond is capable of storing 15 percent of the incident solar energy in the storage zone.

11–152 It is proposed to use a solar-powered thermoelectric system installed on the roof to cool residential buildings. The system consists of a thermoelectric refrigerator that is powered by a thermoelectric power generator whose top surface is a solar collector. Discuss the feasibility and the cost of such a system, and determine if the proposed system installed on one side of the roof can meet a significant portion of the cooling requirements of a typical house in your area.

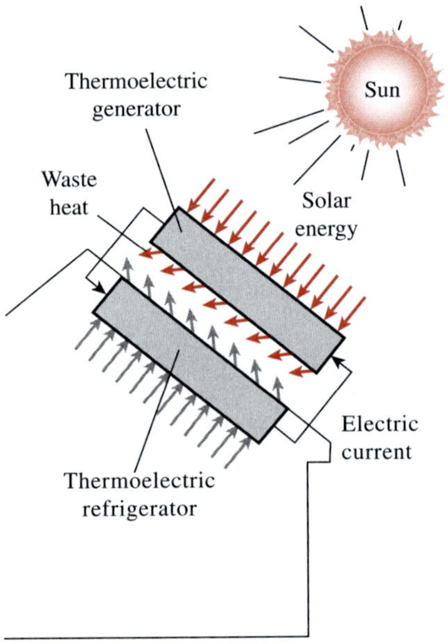

FIGURE P11–152

11–153 A refrigerator using R-12 as the working fluid keeps the refrigerated space at −15°C in an environment at 30°C. You are asked to redesign this refrigerator by replacing R-12 with the ozone-friendly R-134a. What changes in the pressure levels would you suggest in the new system? How do you think the COP of the new system will compare to the COP of the old system?

11–154 Solar or photovoltaic (PV) cells convert sunlight to electricity and are commonly used to power calculators, satellites, remote communication systems, and even pumps. The conversion of light to electricity is called the *photoelectric effect*. It was first discovered in 1839 by Frenchman Edmond Becquerel, and the first PV module, which consisted of several cells connected to each other, was built in 1954 by Bell Laboratories. The PV modules today have conversion efficiencies of about 12 to 15 percent. Noting that the solar energy incident on a normal surface on earth at noontime is about 1000 W/m^2 during a clear day, PV modules on a 1-m^2 surface can provide as much as 150 W of electricity. The annual average daily solar energy incident on a horizontal surface in the United States ranges from about 2 to 6 kWh/m^2.

A PV-powered pump is to be used in Arizona to pump water for wildlife from a depth of 180 m at an average rate of 400 L/day. Assuming a reasonable efficiency for the pumping system, which can be defined as the ratio of the increase in the potential energy of the water to the electrical energy consumed by the pump, and taking the conversion efficiency of the PV cells to be 0.13 to be on the conservative side, determine the size of the PV module that needs to be installed, in m^2.

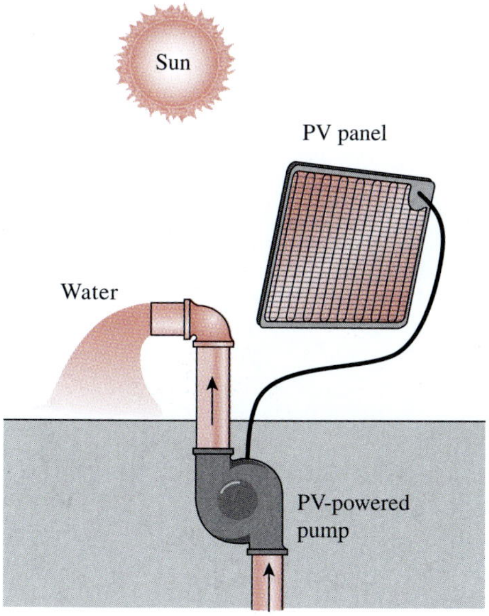

FIGURE P11–154

11–155 A company owns a refrigeration system whose refrigeration capacity is 200 tons (1 ton of refrigeration = 211 kJ/min), and you are to design a forced-air cooling system for fruits whose diameters do not exceed 7 cm under the following conditions: The fruits are to be cooled from 28°C to an average temperature of 8°C. The air temperature is to remain above −2°C and below 10°C at all times, and the velocity of air approaching the fruits must remain under 2 m/s. The cooling section can be as wide as 3.5 m and as high as 2 m.

Assuming reasonable values for the average fruit density, specific heat, and porosity (the fraction of air volume in a box), recommend reasonable values for (*a*) the air velocity approaching the cooling section, (*b*) the product-cooling capacity of the system, in kg·fruit/h, and (*c*) the volume flow rate of air.

APPENDIX

1

PROPERTY TABLES AND CHARTS (SI UNITS)

TABLE A–1

Molar mass, gas constant, and critical-point properties

Substance	Formula	Molar mass, M kg/kmol	Gas constant, R kJ/kg·K*	Critical-point properties: Temperature, K	Pressure, MPa	Volume, m^3/kmol
Air	—	28.97	0.2870	132.5	3.77	0.0883
Ammonia	NH_3	17.03	0.4882	405.5	11.28	0.0724
Argon	Ar	39.948	0.2081	151	4.86	0.0749
Benzene	C_6H_6	78.115	0.1064	562	4.92	0.2603
Bromine	Br_2	159.808	0.0520	584	10.34	0.1355
n-Butane	C_4H_{10}	58.124	0.1430	425.2	3.80	0.2547
Carbon dioxide	CO_2	44.01	0.1889	304.2	7.39	0.0943
Carbon monoxide	CO	28.011	0.2968	133	3.50	0.0930
Carbon tetrachloride	CCl_4	153.82	0.05405	556.4	4.56	0.2759
Chlorine	Cl_2	70.906	0.1173	417	7.71	0.1242
Chloroform	$CHCl_3$	119.38	0.06964	536.6	5.47	0.2403
Dichlorodifluoromethane (R-12)	CCl_2F_2	120.91	0.06876	384.7	4.01	0.2179
Dichlorofluoromethane (R-21)	$CHCl_2F$	102.92	0.08078	451.7	5.17	0.1973
Ethane	C_2H_6	30.070	0.2765	305.5	4.48	0.1480
Ethyl alcohol	C_2H_5OH	46.07	0.1805	516	6.38	0.1673
Ethylene	C_2H_4	28.054	0.2964	282.4	5.12	0.1242
Helium	He	4.003	2.0769	5.3	0.23	0.0578
n-Hexane	C_6H_{14}	86.179	0.09647	507.9	3.03	0.3677
Hydrogen (normal)	H_2	2.016	4.1240	33.3	1.30	0.0649
Krypton	Kr	83.80	0.09921	209.4	5.50	0.0924
Methane	CH_4	16.043	0.5182	191.1	4.64	0.0993
Methyl alcohol	CH_3OH	32.042	0.2595	513.2	7.95	0.1180
Methyl chloride	CH_3Cl	50.488	0.1647	416.3	6.68	0.1430
Neon	Ne	20.183	0.4119	44.5	2.73	0.0417
Nitrogen	N_2	28.013	0.2968	126.2	3.39	0.0899
Nitrous oxide	N_2O	44.013	0.1889	309.7	7.27	0.0961
Oxygen	O_2	31.999	0.2598	154.8	5.08	0.0780
Propane	C_3H_8	44.097	0.1885	370	4.26	0.1998
Propylene	C_3H_6	42.081	0.1976	365	4.62	0.1810
Sulfur dioxide	SO_2	64.063	0.1298	430.7	7.88	0.1217
Tetrafluoroethane (R-134a)	CF_3CH_2F	102.03	0.08149	374.2	4.059	0.1993
Trichlorofluoromethane (R-11)	CCl_3F	137.37	0.06052	471.2	4.38	0.2478
Water	H_2O	18.015	0.4615	647.1	22.06	0.0560
Xenon	Xe	131.30	0.06332	289.8	5.88	0.1186

*The unit kJ/kg·K is equivalent to kPa·m^3/kg·K. The gas constant is calculated from $R = R_u/M$, where R_u = 8.31447 kJ/kmol·K and M is the molar mass.

Source: K. A. Kobe and R. E. Lynn, Jr., *Chemical Review* 52 (1953), pp. 117–236; and ASHRAE, *Handbook of Fundamentals* (Atlanta, GA: American Society of Heating, Refrigerating and Air-Conditioning Engineers, Inc., 1993), pp. 16.4 and 36.1.

TABLE A–2

Ideal-gas specific heats of various common gases

(*a*) At 300 K

Gas	Formula	Gas constant, R kJ/kg·K	c_p kJ/kg·K	c_v kJ/kg·K	k
Air	—	0.2870	1.005	0.718	1.400
Argon	Ar	0.2081	0.5203	0.3122	1.667
Butane	C_4H_{10}	0.1433	1.7164	1.5734	1.091
Carbon dioxide	CO_2	0.1889	0.846	0.657	1.289
Carbon monoxide	CO	0.2968	1.040	0.744	1.400
Ethane	C_2H_6	0.2765	1.7662	1.4897	1.186
Ethylene	C_2H_4	0.2964	1.5482	1.2518	1.237
Helium	He	2.0769	5.1926	3.1156	1.667
Hydrogen	H_2	4.1240	14.307	10.183	1.405
Methane	CH_4	0.5182	2.2537	1.7354	1.299
Neon	Ne	0.4119	1.0299	0.6179	1.667
Nitrogen	N_2	0.2968	1.039	0.743	1.400
Octane	C_8H_{18}	0.0729	1.7113	1.6385	1.044
Oxygen	O_2	0.2598	0.918	0.658	1.395
Propane	C_3H_8	0.1885	1.6794	1.4909	1.126
Steam	H_2O	0.4615	1.8723	1.4108	1.327

Note: The unit kJ/kg·K is equivalent to kJ/kg·°C.

Source: Chemical and Process Thermodynamics 3/E by Kyle, B. G., © 2000. Adapted by permission of Pearson Education, Inc., Upper Saddle River, NJ.

TABLE A–2

Ideal-gas specific heats of various common gases (*Continued*)

(*b*) At various temperatures

Temperature, K	c_p kJ/kg·K	c_v kJ/kg·K	*k*	c_p kJ/kg·K	c_v kJ/kg·K	*k*	c_p kJ/kg·K	c_v kJ/kg·K	*k*
		Air			*Carbon dioxide, CO_2*			*Carbon monoxide, CO*	
250	1.003	0.716	1.401	0.791	0.602	1.314	1.039	0.743	1.400
300	1.005	0.718	1.400	0.846	0.657	1.288	1.040	0.744	1.399
350	1.008	0.721	1.398	0.895	0.706	1.268	1.043	0.746	1.398
400	1.013	0.726	1.395	0.939	0.750	1.252	1.047	0.751	1.395
450	1.020	0.733	1.391	0.978	0.790	1.239	1.054	0.757	1.392
500	1.029	0.742	1.387	1.014	0.825	1.229	1.063	0.767	1.387
550	1.040	0.753	1.381	1.046	0.857	1.220	1.075	0.778	1.382
600	1.051	0.764	1.376	1.075	0.886	1.213	1.087	0.790	1.376
650	1.063	0.776	1.370	1.102	0.913	1.207	1.100	0.803	1.370
700	1.075	0.788	1.364	1.126	0.937	1.202	1.113	0.816	1.364
750	1.087	0.800	1.359	1.148	0.959	1.197	1.126	0.829	1.358
800	1.099	0.812	1.354	1.169	0.980	1.193	1.139	0.842	1.353
900	1.121	0.834	1.344	1.204	1.015	1.186	1.163	0.866	1.343
1000	1.142	0.855	1.336	1.234	1.045	1.181	1.185	0.888	1.335
		Hydrogen, H_2			*Nitrogen, N_2*			*Oxygen, O_2*	
250	14.051	9.927	1.416	1.039	0.742	1.400	0.913	0.653	1.398
300	14.307	10.183	1.405	1.039	0.743	1.400	0.918	0.658	1.395
350	14.427	10.302	1.400	1.041	0.744	1.399	0.928	0.668	1.389
400	14.476	10.352	1.398	1.044	0.747	1.397	0.941	0.681	1.382
450	14.501	10.377	1.398	1.049	0.752	1.395	0.956	0.696	1.373
500	14.513	10.389	1.397	1.056	0.759	1.391	0.972	0.712	1.365
550	14.530	10.405	1.396	1.065	0.768	1.387	0.988	0.728	1.358
600	14.546	10.422	1.396	1.075	0.778	1.382	1.003	0.743	1.350
650	14.571	10.447	1.395	1.086	0.789	1.376	1.017	0.758	1.343
700	14.604	10.480	1.394	1.098	0.801	1.371	1.031	0.771	1.337
750	14.645	10.521	1.392	1.110	0.813	1.365	1.043	0.783	1.332
800	14.695	10.570	1.390	1.121	0.825	1.360	1.054	0.794	1.327
900	14.822	10.698	1.385	1.145	0.849	1.349	1.074	0.814	1.319
1000	14.983	10.859	1.380	1.167	0.870	1.341	1.090	0.830	1.313

Source: Kenneth Wark, *Thermodynamics,* 4th ed. (New York: McGraw-Hill, 1983), p. 783, Table A–4M. Originally published in *Tables of Thermal Properties of Gases,* NBS Circular 564, 1955.

TABLE A–2

Ideal-gas specific heats of various common gases (*Concluded*)

(*c*) As a function of temperature

$$\bar{c}_p = a + bT + cT^2 + dT^3$$
$$(T \text{ in K}, c_p \text{ in kJ/kmol·K})$$

Substance	Formula	*a*	*b*	*c*	*d*	Temperature range, K	% error Max.	% error Avg.
Nitrogen	N_2	28.90	-0.1571×10^{-2}	0.8081×10^{-5}	-2.873×10^{-9}	273–1800	0.59	0.34
Oxygen	O_2	25.48	1.520×10^{-2}	-0.7155×10^{-5}	1.312×10^{-9}	273–1800	1.19	0.28
Air	—	28.11	0.1967×10^{-2}	0.4802×10^{-5}	-1.966×10^{-9}	273–1800	0.72	0.33
Hydrogen	H_2	29.11	-0.1916×10^{-2}	0.4003×10^{-5}	-0.8704×10^{-9}	273–1800	1.01	0.26
Carbon monoxide	CO	28.16	0.1675×10^{-2}	0.5372×10^{-5}	-2.222×10^{-9}	273–1800	0.89	0.37
Carbon dioxide	CO_2	22.26	5.981×10^{-2}	-3.501×10^{-5}	7.469×10^{-9}	273–1800	0.67	0.22
Water vapor	H_2O	32.24	0.1923×10^{-2}	1.055×10^{-5}	-3.595×10^{-9}	273–1800	0.53	0.24
Nitric oxide	NO	29.34	-0.09395×10^{-2}	0.9747×10^{-5}	-4.187×10^{-9}	273–1500	0.97	0.36
Nitrous oxide	N_2O	24.11	5.8632×10^{-2}	-3.562×10^{-5}	10.58×10^{-9}	273–1500	0.59	0.26
Nitrogen dioxide	NO_2	22.9	5.715×10^{-2}	-3.52×10^{-5}	7.87×10^{-9}	273–1500	0.46	0.18
Ammonia	NH_3	27.568	2.5630×10^{-2}	0.99072×10^{-5}	-6.6909×10^{-9}	273–1500	0.91	0.36
Sulfur	S_2	27.21	2.218×10^{-2}	-1.628×10^{-5}	3.986×10^{-9}	273–1800	0.99	0.38
Sulfur dioxide	SO_2	25.78	5.795×10^{-2}	-3.812×10^{-5}	8.612×10^{-9}	273–1800	0.45	0.24
Sulfur trioxide	SO_3	16.40	14.58×10^{-2}	-11.20×10^{-5}	32.42×10^{-9}	273–1300	0.29	0.13
Acetylene	C_2H_2	21.8	9.2143×10^{-2}	-6.527×10^{-5}	18.21×10^{-9}	273–1500	1.46	0.59
Benzene	C_6H_6	−36.22	48.475×10^{-2}	-31.57×10^{-5}	77.62×10^{-9}	273–1500	0.34	0.20
Methanol	CH_4O	19.0	9.152×10^{-2}	-1.22×10^{-5}	-8.039×10^{-9}	273–1000	0.18	0.08
Ethanol	C_2H_6O	19.9	20.96×10^{-2}	-10.38×10^{-5}	20.05×10^{-9}	273–1500	0.40	0.22
Hydrogen chloride	HCl	30.33	-0.7620×10^{-2}	1.327×10^{-5}	-4.338×10^{-9}	273–1500	0.22	0.08
Methane	CH_4	19.89	5.024×10^{-2}	1.269×10^{-5}	-11.01×10^{-9}	273–1500	1.33	0.57
Ethane	C_2H_6	6.900	17.27×10^{-2}	-6.406×10^{-5}	7.285×10^{-9}	273–1500	0.83	0.28
Propane	C_3H_8	−4.04	30.48×10^{-2}	-15.72×10^{-5}	31.74×10^{-9}	273–1500	0.40	0.12
n-Butane	C_4H_{10}	3.96	37.15×10^{-2}	-18.34×10^{-5}	35.00×10^{-9}	273–1500	0.54	0.24
i-Butane	C_4H_{10}	−7.913	41.60×10^{-2}	-23.01×10^{-5}	49.91×10^{-9}	273–1500	0.25	0.13
n-Pentane	C_5H_{12}	6.774	45.43×10^{-2}	-22.46×10^{-5}	42.29×10^{-9}	273–1500	0.56	0.21
n-Hexane	C_6H_{14}	6.938	55.22×10^{-2}	-28.65×10^{-5}	57.69×10^{-9}	273–1500	0.72	0.20
Ethylene	C_2H_4	3.95	15.64×10^{-2}	-8.344×10^{-5}	17.67×10^{-9}	273–1500	0.54	0.13
Propylene	C_3H_6	3.15	23.83×10^{-2}	-12.18×10^{-5}	24.62×10^{-9}	273–1500	0.73	0.17

Source: B. G. Kyle, *Chemical and Process Thermodynamics* (Englewood Cliffs, NJ: Prentice-Hall, 1984). Used with permission.

TABLE A–3

Properties of common liquids, solids, and foods

(*a*) Liquids

	Boiling data at 1 atm		*Freezing data*		*Liquid properties*		
Substance	Normal boiling point, °C	Latent heat of vaporization h_{fg}, kJ/kg	Freezing point, °C	Latent heat of fusion h_{if}, kJ/kg	Temperature, °C	Density ρ, kg/m^3	Specific heat c_p, kJ/kg·K
Ammonia	−33.3	1357	−77.7	322.4	−33.3	682	4.43
					−20	665	4.52
					0	639	4.60
					25	602	4.80
Argon	−185.9	161.6	−189.3	28	−185.6	1394	1.14
Benzene	80.2	394	5.5	126	20	879	1.72
Brine (20% sodium chloride by mass)	103.9	—	−17.4	—	20	1150	3.11
n-Butane	−0.5	385.2	−138.5	80.3	−0.5	601	2.31
Carbon dioxide	−78.4*	230.5 (at 0°C)	−56.6		0	298	0.59
Ethanol	78.2	838.3	−114.2	109	25	783	2.46
Ethyl alcohol	78.6	855	−156	108	20	789	2.84
Ethylene glycol	198.1	800.1	−10.8	181.1	20	1109	2.84
Glycerine	179.9	974	18.9	200.6	20	1261	2.32
Helium	−268.9	22.8	—	—	−268.9	146.2	22.8
Hydrogen	−252.8	445.7	−259.2	59.5	−252.8	70.7	10.0
Isobutane	−11.7	367.1	−160	105.7	−11.7	593.8	2.28
Kerosene	204–293	251	−24.9	—	20	820	2.00
Mercury	356.7	294.7	−38.9	11.4	25	13,560	0.139
Methane	−161.5	510.4	−182.2	58.4	−161.5	423	3.49
					−100	301	5.79
Methanol	64.5	1100	−97.7	99.2	25	787	2.55
Nitrogen	−195.8	198.6	−210	25.3	−195.8	809	2.06
					−160	596	2.97
Octane	124.8	306.3	−57.5	180.7	20	703	2.10
Oil (light)					25	910	1.80
Oxygen	−183	212.7	−218.8	13.7	−183	1141	1.71
Petroleum	—	230–384			20	640	2.0
Propane	−42.1	427.8	−187.7	80.0	−42.1	581	2.25
					0	529	2.53
					50	449	3.13
Refrigerant-134a	−26.1	217.0	−96.6	—	−50	1443	1.23
					−26.1	1374	1.27
					0	1295	1.34
					25	1207	1.43
Water	100	2257	0.0	333.7	0	1000	4.22
					25	997	4.18
					50	988	4.18
					75	975	4.19
					100	958	4.22

* Sublimation temperature. (At pressures below the triple-point pressure of 518 kPa, carbon dioxide exists as a solid or gas. Also, the freezing-point temperature of carbon dioxide is the triple-point temperature of −56.5°C.)

TABLE A–3

Properties of common liquids, solids, and foods (*Concluded*)

(*b*) Solids (values are for room temperature unless indicated otherwise)

Substance	Density, ρ kg/m^3	Specific heat, c_p kJ/kg·K	Substance	Density, ρ kg/m^3	Specific heat, c_p kJ/kg·K
Metals			Nonmetals		
Aluminum			Asphalt	2110	0.920
200 K		0.797	Brick, common	1922	0.79
250 K		0.859	Brick, fireclay (500°C)	2300	0.960
300 K	2,700	0.902	Concrete	2300	0.653
350 K		0.929	Clay	1000	0.920
400 K		0.949	Diamond	2420	0.616
450 K		0.973	Glass, window	2700	0.800
500 K		0.997	Glass, pyrex	2230	0.840
Bronze (76% Cu, 2% Zn, 2% Al)	8,280	0.400	Graphite	2500	0.711
			Granite	2700	1.017
Brass, yellow (65% Cu, 35% Zn)	8,310	0.400	Gypsum or plaster board	800	1.09
			Ice		
Copper			200 K		1.56
−173°C		0.254	220 K		1.71
−100°C		0.342	240 K		1.86
−50°C		0.367	260 K		2.01
0°C		0.381	273 K	921	2.11
27°C	8,900	0.386	Limestone	1650	0.909
100°C		0.393	Marble	2600	0.880
200°C		0.403	Plywood (Douglas Fir)	545	1.21
Iron	7,840	0.45	Rubber (soft)	1100	1.840
Lead	11,310	0.128	Rubber (hard)	1150	2.009
Magnesium	1,730	1.000	Sand	1520	0.800
Nickel	8,890	0.440	Stone	1500	0.800
Silver	10,470	0.235	Woods, hard (maple, oak, etc.)	721	1.26
Steel, mild	7,830	0.500	Woods, soft (fir, pine, etc.)	513	1.38
Tungsten	19,400	0.130			

(*c*) Foods

Food	Water content, % (mass)	Freezing point, °C	Specific heat, kJ/kg·K: Above freezing	Specific heat, kJ/kg·K: Below freezing	Latent heat of fusion, kJ/kg	Food	Water content, % (mass)	Freezing point, °C	Specific heat, kJ/kg·K: Above freezing	Specific heat, kJ/kg·K: Below freezing	Latent heat of fusion, kJ/kg
Apples	84	−1.1	3.65	1.90	281	Lettuce	95	−0.2	4.02	2.04	317
Bananas	75	−0.8	3.35	1.78	251	Milk, whole	88	−0.6	3.79	1.95	294
Beef round	67	—	3.08	1.68	224	Oranges	87	−0.8	3.75	1.94	291
Broccoli	90	−0.6	3.86	1.97	301	Potatoes	78	−0.6	3.45	1.82	261
Butter	16	—	—	1.04	53	Salmon fish	64	−2.2	2.98	1.65	214
Cheese, swiss	39	−10.0	2.15	1.33	130	Shrimp	83	−2.2	3.62	1.89	277
Cherries	80	−1.8	3.52	1.85	267	Spinach	93	−0.3	3.96	2.01	311
Chicken	74	−2.8	3.32	1.77	247	Strawberries	90	−0.8	3.86	1.97	301
Corn, sweet	74	−0.6	3.32	1.77	247	Tomatoes, ripe	94	−0.5	3.99	2.02	314
Eggs, whole	74	−0.6	3.32	1.77	247	Turkey	64	—	2.98	1.65	214
Ice cream	63	−5.6	2.95	1.63	210	Watermelon	93	−0.4	3.96	2.01	311

Source: Values are obtained from various handbooks and other sources or are calculated. Water content and freezing-point data of foods are from *ASHRAE, Handbook of Fundamentals,* SI version (Atlanta, GA: American Society of Heating, Refrigerating and Air-Conditioning Engineers, Inc., 1993), Chapter 30, Table 1. Freezing point is the temperature at which freezing starts for fruits and vegetables, and the average freezing temperature for other foods.

TABLE A–4

Saturated water—Temperature table

		Specific volume, m³/kg		*Internal energy,* kJ/kg			*Enthalpy,* kJ/kg			*Entropy,* kJ/kg·K		
Temp., T °C	Sat. press., P_{sat} kPa	Sat. liquid, v_f	Sat. vapor, v_g	Sat. liquid, u_f	Evap., u_{fg}	Sat. vapor, u_g	Sat. liquid, h_f	Evap., h_{fg}	Sat. vapor, h_g	Sat. liquid, s_f	Evap., s_{fg}	Sat. vapor, s_g
0.01	0.6117	0.001000	206.00	0.000	2374.9	2374.9	0.001	2500.9	2500.9	0.0000	9.1556	9.1556
5	0.8725	0.001000	147.03	21.019	2360.8	2381.8	21.020	2489.1	2510.1	0.0763	8.9487	9.0249
10	1.2281	0.001000	106.32	42.020	2346.6	2388.7	42.022	2477.2	2519.2	0.1511	8.7488	8.8999
15	1.7057	0.001001	77.885	62.980	2332.5	2395.5	62.982	2465.4	2528.3	0.2245	8.5559	8.7803
20	2.3392	0.001002	57.762	83.913	2318.4	2402.3	83.915	2453.5	2537.4	0.2965	8.3696	8.6661
25	3.1698	0.001003	43.340	104.83	2304.3	2409.1	104.83	2441.7	2546.5	0.3672	8.1895	8.5567
30	4.2469	0.001004	32.879	125.73	2290.2	2415.9	125.74	2429.8	2555.6	0.4368	8.0152	8.4520
35	5.6291	0.001006	25.205	146.63	2276.0	2422.7	146.64	2417.9	2564.6	0.5051	7.8466	8.3517
40	7.3851	0.001008	19.515	167.53	2261.9	2429.4	167.53	2406.0	2573.5	0.5724	7.6832	8.2556
45	9.5953	0.001010	15.251	188.43	2247.7	2436.1	188.44	2394.0	2582.4	0.6386	7.5247	8.1633
50	12.352	0.001012	12.026	209.33	2233.4	2442.7	209.34	2382.0	2591.3	0.7038	7.3710	8.0748
55	15.763	0.001015	9.5639	230.24	2219.1	2449.3	230.26	2369.8	2600.1	0.7680	7.2218	7.9898
60	19.947	0.001017	7.6670	251.16	2204.7	2455.9	251.18	2357.7	2608.8	0.8313	7.0769	7.9082
65	25.043	0.001020	6.1935	272.09	2190.3	2462.4	272.12	2345.4	2617.5	0.8937	6.9360	7.8296
70	31.202	0.001023	5.0396	293.04	2175.8	2468.9	293.07	2333.0	2626.1	0.9551	6.7989	7.7540
75	38.597	0.001026	4.1291	313.99	2161.3	2475.3	314.03	2320.6	2634.6	1.0158	6.6655	7.6812
80	47.416	0.001029	3.4053	334.97	2146.6	2481.6	335.02	2308.0	2643.0	1.0756	6.5355	7.6111
85	57.868	0.001032	2.8261	355.96	2131.9	2487.8	356.02	2295.3	2651.4	1.1346	6.4089	7.5435
90	70.183	0.001036	2.3593	376.97	2117.0	2494.0	377.04	2282.5	2659.6	1.1929	6.2853	7.4782
95	84.609	0.001040	1.9808	398.00	2102.0	2500.1	398.09	2269.6	2667.6	1.2504	6.1647	7.4151
100	101.42	0.001043	1.6720	419.06	2087.0	2506.0	419.17	2256.4	2675.6	1.3072	6.0470	7.3542
105	120.90	0.001047	1.4186	440.15	2071.8	2511.9	440.28	2243.1	2683.4	1.3634	5.9319	7.2952
110	143.38	0.001052	1.2094	461.27	2056.4	2517.7	461.42	2229.7	2691.1	1.4188	5.8193	7.2382
115	169.18	0.001056	1.0360	482.42	2040.9	2523.3	482.59	2216.0	2698.6	1.4737	5.7092	7.1829
120	198.67	0.001060	0.89133	503.60	2025.3	2528.9	503.81	2202.1	2706.0	1.5279	5.6013	7.1292
125	232.23	0.001065	0.77012	524.83	2009.5	2534.3	525.07	2188.1	2713.1	1.5816	5.4956	7.0771
130	270.28	0.001070	0.66808	546.10	1993.4	2539.5	546.38	2173.7	2720.1	1.6346	5.3919	7.0265
135	313.22	0.001075	0.58179	567.41	1977.3	2544.7	567.75	2159.1	2726.9	1.6872	5.2901	6.9773
140	361.53	0.001080	0.50850	588.77	1960.9	2549.6	589.16	2144.3	2733.5	1.7392	5.1901	6.9294
145	415.68	0.001085	0.44600	610.19	1944.2	2554.4	610.64	2129.2	2739.8	1.7908	5.0919	6.8827
150	476.16	0.001091	0.39248	631.66	1927.4	2559.1	632.18	2113.8	2745.9	1.8418	4.9953	6.8371
155	543.49	0.001096	0.34648	653.19	1910.3	2563.5	653.79	2098.0	2751.8	1.8924	4.9002	6.7927
160	618.23	0.001102	0.30680	674.79	1893.0	2567.8	675.47	2082.0	2757.5	1.9426	4.8066	6.7492
165	700.93	0.001108	0.27244	696.46	1875.4	2571.9	697.24	2065.6	2762.8	1.9923	4.7143	6.7067
170	792.18	0.001114	0.24260	718.20	1857.5	2575.7	719.08	2048.8	2767.9	2.0417	4.6233	6.6650
175	892.60	0.001121	0.21659	740.02	1839.4	2579.4	741.02	2031.7	2772.7	2.0906	4.5335	6.6242
180	1002.8	0.001127	0.19384	761.92	1820.9	2582.8	763.05	2014.2	2777.2	2.1392	4.4448	6.5841
185	1123.5	0.001134	0.17390	783.91	1802.1	2586.0	785.19	1996.2	2781.4	2.1875	4.3572	6.5447
190	1255.2	0.001141	0.15636	806.00	1783.0	2589.0	807.43	1977.9	2785.3	2.2355	4.2705	6.5059
195	1398.8	0.001149	0.14089	828.18	1763.6	2591.7	829.78	1959.0	2788.8	2.2831	4.1847	6.4678
200	1554.9	0.001157	0.12721	850.46	1743.7	2594.2	852.26	1939.8	2792.0	2.3305	4.0997	6.4302

TABLE A–4

Saturated water—Temperature table (*Concluded*)

		Specific volume, m^3/kg		*Internal energy,* kJ/kg			*Enthalpy,* kJ/kg			*Entropy,* kJ/kg·K		
Temp., T °C	Sat. press., P_{sat} kPa	Sat. liquid, v_f	Sat. vapor, v_g	Sat. liquid, u_f	Evap., u_{fg}	Sat. vapor, u_g	Sat. liquid, h_f	Evap., h_{fg}	Sat. vapor, h_g	Sat. liquid, s_f	Evap., s_{fg}	Sat. vapor, s_g
205	1724.3	0.001164	0.11508	872.86	1723.5	2596.4	874.87	1920.0	2794.8	2.3776	4.0154	6.3930
210	1907.7	0.001173	0.10429	895.38	1702.9	2598.3	897.61	1899.7	2797.3	2.4245	3.9318	6.3563
215	2105.9	0.001181	0.094680	918.02	1681.9	2599.9	920.50	1878.8	2799.3	2.4712	3.8489	6.3200
220	2319.6	0.001190	0.086094	940.79	1660.5	2601.3	943.55	1857.4	2801.0	2.5176	3.7664	6.2840
225	2549.7	0.001199	0.078405	963.70	1638.6	2602.3	966.76	1835.4	2802.2	2.5639	3.6844	6.2483
230	2797.1	0.001209	0.071505	986.76	1616.1	2602.9	990.14	1812.8	2802.9	2.6100	3.6028	6.2128
235	3062.6	0.001219	0.065300	1010.0	1593.2	2603.2	1013.7	1789.5	2803.2	2.6560	3.5216	6.1775
240	3347.0	0.001229	0.059707	1033.4	1569.8	2603.1	1037.5	1765.5	2803.0	2.7018	3.4405	6.1424
245	3651.2	0.001240	0.054656	1056.9	1545.7	2602.7	1061.5	1740.8	2802.2	2.7476	3.3596	6.1072
250	3976.2	0.001252	0.050085	1080.7	1521.1	2601.8	1085.7	1715.3	2801.0	2.7933	3.2788	6.0721
255	4322.9	0.001263	0.045941	1104.7	1495.8	2600.5	1110.1	1689.0	2799.1	2.8390	3.1979	6.0369
260	4692.3	0.001276	0.042175	1128.8	1469.9	2598.7	1134.8	1661.8	2796.6	2.8847	3.1169	6.0017
265	5085.3	0.001289	0.038748	1153.3	1443.2	2596.5	1159.8	1633.7	2793.5	2.9304	3.0358	5.9662
270	5503.0	0.001303	0.035622	1177.9	1415.7	2593.7	1185.1	1604.6	2789.7	2.9762	2.9542	5.9305
275	5946.4	0.001317	0.032767	1202.9	1387.4	2590.3	1210.7	1574.5	2785.2	3.0221	2.8723	5.8944
280	6416.6	0.001333	0.030153	1228.2	1358.2	2586.4	1236.7	1543.2	2779.9	3.0681	2.7898	5.8579
285	6914.6	0.001349	0.027756	1253.7	1328.1	2581.8	1263.1	1510.7	2773.7	3.1144	2.7066	5.8210
290	7441.8	0.001366	0.025554	1279.7	1296.9	2576.5	1289.8	1476.9	2766.7	3.1608	2.6225	5.7834
295	7999.0	0.001384	0.023528	1306.0	1264.5	2570.5	1317.1	1441.6	2758.7	3.2076	2.5374	5.7450
300	8587.9	0.001404	0.021659	1332.7	1230.9	2563.6	1344.8	1404.8	2749.6	3.2548	2.4511	5.7059
305	9209.4	0.001425	0.019932	1360.0	1195.9	2555.8	1373.1	1366.3	2739.4	3.3024	2.3633	5.6657
310	9865.0	0.001447	0.018333	1387.7	1159.3	2547.1	1402.0	1325.9	2727.9	3.3506	2.2737	5.6243
315	10,556	0.001472	0.016849	1416.1	1121.1	2537.2	1431.6	1283.4	2715.0	3.3994	2.1821	5.5816
320	11,284	0.001499	0.015470	1445.1	1080.9	2526.0	1462.0	1238.5	2700.6	3.4491	2.0881	5.5372
325	12,051	0.001528	0.014183	1475.0	1038.5	2513.4	1493.4	1191.0	2684.3	3.4998	1.9911	5.4908
330	12,858	0.001560	0.012979	1505.7	993.5	2499.2	1525.8	1140.3	2666.0	3.5516	1.8906	5.4422
335	13,707	0.001597	0.011848	1537.5	945.5	2483.0	1559.4	1086.0	2645.4	3.6050	1.7857	5.3907
340	14,601	0.001638	0.010783	1570.7	893.8	2464.5	1594.6	1027.4	2622.0	3.6602	1.6756	5.3358
345	15,541	0.001685	0.009772	1605.5	837.7	2443.2	1631.7	963.4	2595.1	3.7179	1.5585	5.2765
350	16,529	0.001741	0.008806	1642.4	775.9	2418.3	1671.2	892.7	2563.9	3.7788	1.4326	5.2114
355	17,570	0.001808	0.007872	1682.2	706.4	2388.6	1714.0	812.9	2526.9	3.8442	1.2942	5.1384
360	18,666	0.001895	0.006950	1726.2	625.7	2351.9	1761.5	720.1	2481.6	3.9165	1.1373	5.0537
365	19,822	0.002015	0.006009	1777.2	526.4	2303.6	1817.2	605.5	2422.7	4.0004	0.9489	4.9493
370	21,044	0.002217	0.004953	1844.5	385.6	2230.1	1891.2	443.1	2334.3	4.1119	0.6890	4.8009
373.95	22,064	0.003106	0.003106	2015.7	0	2015.7	2084.3	0	2084.3	4.4070	0	4.4070

Source: Tables A–4 through A–8 are generated using the Engineering Equation Solver (EES) software developed by S. A. Klein and F. L. Alvarado. The routine used in calculations is the highly accurate Steam_IAPWS, which incorporates the 1995 Formulation for the Thermodynamic Properties of Ordinary Water Substance for General and Scientific Use, issued by The International Association for the Properties of Water and Steam (IAPWS). This formulation replaces the 1984 formulation of Haar, Gallagher, and Kell (NBS/NRC Steam Tables, Hemisphere Publishing Co., 1984), which is also available in EES as the routine STEAM. The new formulation is based on the correlations of Saul and Wagner (J. Phys. Chem. Ref. Data, 16, 893, 1987) with modifications to adjust to the International Temperature Scale of 1990. The modifications are described by Wagner and Pruss (J. Phys. Chem. Ref. Data, 22, 783, 1993). The properties of ice are based on Hyland and Wexler, "Formulations for the Thermodynamic Properties of the Saturated Phases of H_2O from 173.15 K to 473.15 K," *ASHRAE Trans.*, Part 2A, Paper 2793, 1983.

TABLE A–5

Saturated water—Pressure table

		Specific volume, m³/kg		Internal energy, kJ/kg			Enthalpy, kJ/kg			Entropy, kJ/kg·K		
Press., P kPa	Sat. temp., T_{sat} °C	Sat. liquid, v_f	Sat. vapor, v_g	Sat. liquid, u_f	Evap., u_{fg}	Sat. vapor, u_g	Sat. liquid, h_f	Evap., h_{fg}	Sat. vapor, h_g	Sat. liquid, s_f	Evap., s_{fg}	Sat. vapor, s_g
1.0	6.97	0.001000	129.19	29.302	2355.2	2384.5	29.303	2484.4	2513.7	0.1059	8.8690	8.9749
1.5	13.02	0.001001	87.964	54.686	2338.1	2392.8	54.688	2470.1	2524.7	0.1956	8.6314	8.8270
2.0	17.50	0.001001	66.990	73.431	2325.5	2398.9	73.433	2459.5	2532.9	0.2606	8.4621	8.7227
2.5	21.08	0.001002	54.242	88.422	2315.4	2403.8	88.424	2451.0	2539.4	0.3118	8.3302	8.6421
3.0	24.08	0.001003	45.654	100.98	2306.9	2407.9	100.98	2443.9	2544.8	0.3543	8.2222	8.5765
4.0	28.96	0.001004	34.791	121.39	2293.1	2414.5	121.39	2432.3	2553.7	0.4224	8.0510	8.4734
5.0	32.87	0.001005	28.185	137.75	2282.1	2419.8	137.75	2423.0	2560.7	0.4762	7.9176	8.3938
7.5	40.29	0.001008	19.233	168.74	2261.1	2429.8	168.75	2405.3	2574.0	0.5763	7.6738	8.2501
10	45.81	0.001010	14.670	191.79	2245.4	2437.2	191.81	2392.1	2583.9	0.6492	7.4996	8.1488
15	53.97	0.001014	10.020	225.93	2222.1	2448.0	225.94	2372.3	2598.3	0.7549	7.2522	8.0071
20	60.06	0.001017	7.6481	251.40	2204.6	2456.0	251.42	2357.5	2608.9	0.8320	7.0752	7.9073
25	64.96	0.001020	6.2034	271.93	2190.4	2462.4	271.96	2345.5	2617.5	0.8932	6.9370	7.8302
30	69.09	0.001022	5.2287	289.24	2178.5	2467.7	289.27	2335.3	2624.6	0.9441	6.8234	7.7675
40	75.86	0.001026	3.9933	317.58	2158.8	2476.3	317.62	2318.4	2636.1	1.0261	6.6430	7.6691
50	81.32	0.001030	3.2403	340.49	2142.7	2483.2	340.54	2304.7	2645.2	1.0912	6.5019	7.5931
75	91.76	0.001037	2.2172	384.36	2111.8	2496.1	384.44	2278.0	2662.4	1.2132	6.2426	7.4558
100	99.61	0.001043	1.6941	417.40	2088.2	2505.6	417.51	2257.5	2675.0	1.3028	6.0562	7.3589
101.325	99.97	0.001043	1.6734	418.95	2087.0	2506.0	419.06	2256.5	2675.6	1.3069	6.0476	7.3545
125	105.97	0.001048	1.3750	444.23	2068.8	2513.0	444.36	2240.6	2684.9	1.3741	5.9100	7.2841
150	111.35	0.001053	1.1594	466.97	2052.3	2519.2	467.13	2226.0	2693.1	1.4337	5.7894	7.2231
175	116.04	0.001057	1.0037	486.82	2037.7	2524.5	487.01	2213.1	2700.2	1.4850	5.6865	7.1716
200	120.21	0.001061	0.88578	504.50	2024.6	2529.1	504.71	2201.6	2706.3	1.5302	5.5968	7.1270
225	123.97	0.001064	0.79329	520.47	2012.7	2533.2	520.71	2191.0	2711.7	1.5706	5.5171	7.0877
250	127.41	0.001067	0.71873	535.08	2001.8	2536.8	535.35	2181.2	2716.5	1.6072	5.4453	7.0525
275	130.58	0.001070	0.65732	548.57	1991.6	2540.1	548.86	2172.0	2720.9	1.6408	5.3800	7.0207
300	133.52	0.001073	0.60582	561.11	1982.1	2543.2	561.43	2163.5	2724.9	1.6717	5.3200	6.9917
325	136.27	0.001076	0.56199	572.84	1973.1	2545.9	573.19	2155.4	2728.6	1.7005	5.2645	6.9650
350	138.86	0.001079	0.52422	583.89	1964.6	2548.5	584.26	2147.7	2732.0	1.7274	5.2128	6.9402
375	141.30	0.001081	0.49133	594.32	1956.6	2550.9	594.73	2140.4	2735.1	1.7526	5.1645	6.9171
400	143.61	0.001084	0.46242	604.22	1948.9	2553.1	604.66	2133.4	2738.1	1.7765	5.1191	6.8955
450	147.90	0.001088	0.41392	622.65	1934.5	2557.1	623.14	2120.3	2743.4	1.8205	5.0356	6.8561
500	151.83	0.001093	0.37483	639.54	1921.2	2560.7	640.09	2108.0	2748.1	1.8604	4.9603	6.8207
550	155.46	0.001097	0.34261	655.16	1908.8	2563.9	655.77	2096.6	2752.4	1.8970	4.8916	6.7886
600	158.83	0.001101	0.31560	669.72	1897.1	2566.8	670.38	2085.8	2756.2	1.9308	4.8285	6.7593
650	161.98	0.001104	0.29260	683.37	1886.1	2569.4	684.08	2075.5	2759.6	1.9623	4.7699	6.7322
700	164.95	0.001108	0.27278	696.23	1875.6	2571.8	697.00	2065.8	2762.8	1.9918	4.7153	6.7071
750	167.75	0.001111	0.25552	708.40	1865.6	2574.0	709.24	2056.4	2765.7	2.0195	4.6642	6.6837

TABLE A–5

Saturated water—Pressure table (*Concluded*)

		Specific volume, m³/kg		*Internal energy,* kJ/kg			*Enthalpy,* kJ/kg			*Entropy,* kJ/kg·K		
Press., P kPa	Sat. temp., T_{sat} °C	Sat. liquid, v_f	Sat. vapor, v_g	Sat. liquid, u_f	Evap., u_{fg}	Sat. vapor, u_g	Sat. liquid, h_f	Evap., h_{fg}	Sat. vapor, h_g	Sat. liquid, s_f	Evap., s_{fg}	Sat. vapor, s_g
800	170.41	0.001115	0.24035	719.97	1856.1	2576.0	720.87	2047.5	2768.3	2.0457	4.6160	6.6616
850	172.94	0.001118	0.22690	731.00	1846.9	2577.9	731.95	2038.8	2770.8	2.0705	4.5705	6.6409
900	175.35	0.001121	0.21489	741.55	1838.1	2579.6	742.56	2030.5	2773.0	2.0941	4.5273	6.6213
950	177.66	0.001124	0.20411	751.67	1829.6	2581.3	752.74	2022.4	2775.2	2.1166	4.4862	6.6027
1000	179.88	0.001127	0.19436	761.39	1821.4	2582.8	762.51	2014.6	2777.1	2.1381	4.4470	6.5850
1100	184.06	0.001133	0.17745	779.78	1805.7	2585.5	781.03	1999.6	2780.7	2.1785	4.3735	6.5520
1200	187.96	0.001138	0.16326	796.96	1790.9	2587.8	798.33	1985.4	2783.8	2.2159	4.3058	6.5217
1300	191.60	0.001144	0.15119	813.10	1776.8	2589.9	814.59	1971.9	2786.5	2.2508	4.2428	6.4936
1400	195.04	0.001149	0.14078	828.35	1763.4	2591.8	829.96	1958.9	2788.9	2.2835	4.1840	6.4675
1500	198.29	0.001154	0.13171	842.82	1750.6	2593.4	844.55	1946.4	2791.0	2.3143	4.1287	6.4430
1750	205.72	0.001166	0.11344	876.12	1720.6	2596.7	878.16	1917.1	2795.2	2.3844	4.0033	6.3877
2000	212.38	0.001177	0.099587	906.12	1693.0	2599.1	908.47	1889.8	2798.3	2.4467	3.8923	6.3390
2250	218.41	0.001187	0.088717	933.54	1667.3	2600.9	936.21	1864.3	2800.5	2.5029	3.7926	6.2954
2500	223.95	0.001197	0.079952	958.87	1643.2	2602.1	961.87	1840.1	2801.9	2.5542	3.7016	6.2558
3000	233.85	0.001217	0.066667	1004.6	1598.5	2603.2	1008.3	1794.9	2803.2	2.6454	3.5402	6.1856
3500	242.56	0.001235	0.057061	1045.4	1557.6	2603.0	1049.7	1753.0	2802.7	2.7253	3.3991	6.1244
4000	250.35	0.001252	0.049779	1082.4	1519.3	2601.7	1087.4	1713.5	2800.8	2.7966	3.2731	6.0696
5000	263.94	0.001286	0.039448	1148.1	1448.9	2597.0	1154.5	1639.7	2794.2	2.9207	3.0530	5.9737
6000	275.59	0.001319	0.032449	1205.8	1384.1	2589.9	1213.8	1570.9	2784.6	3.0275	2.8627	5.8902
7000	285.83	0.001352	0.027378	1258.0	1323.0	2581.0	1267.5	1505.2	2772.6	3.1220	2.6927	5.8148
8000	295.01	0.001384	0.023525	1306.0	1264.5	2570.5	1317.1	1441.6	2758.7	3.2077	2.5373	5.7450
9000	303.35	0.001418	0.020489	1350.9	1207.6	2558.5	1363.7	1379.3	2742.9	3.2866	2.3925	5.6791
10,000	311.00	0.001452	0.018028	1393.3	1151.8	2545.2	1407.8	1317.6	2725.5	3.3603	2.2556	5.6159
11,000	318.08	0.001488	0.015988	1433.9	1096.6	2530.4	1450.2	1256.1	2706.3	3.4299	2.1245	5.5544
12,000	324.68	0.001526	0.014264	1473.0	1041.3	2514.3	1491.3	1194.1	2685.4	3.4964	1.9975	5.4939
13,000	330.85	0.001566	0.012781	1511.0	985.5	2496.6	1531.4	1131.3	2662.7	3.5606	1.8730	5.4336
14,000	336.67	0.001610	0.011487	1548.4	928.7	2477.1	1571.0	1067.0	2637.9	3.6232	1.7497	5.3728
15,000	342.16	0.001657	0.010341	1585.5	870.3	2455.7	1610.3	1000.5	2610.8	3.6848	1.6261	5.3108
16,000	347.36	0.001710	0.009312	1622.6	809.4	2432.0	1649.9	931.1	2581.0	3.7461	1.5005	5.2466
17,000	352.29	0.001770	0.008374	1660.2	745.1	2405.4	1690.3	857.4	2547.7	3.8082	1.3709	5.1791
18,000	356.99	0.001840	0.007504	1699.1	675.9	2375.0	1732.2	777.8	2510.0	3.8720	1.2343	5.1064
19,000	361.47	0.001926	0.006677	1740.3	598.9	2339.2	1776.8	689.2	2466.0	3.9396	1.0860	5.0256
20,000	365.75	0.002038	0.005862	1785.8	509.0	2294.8	1826.6	585.5	2412.1	4.0146	0.9164	4.9310
21,000	369.83	0.002207	0.004994	1841.6	391.9	2233.5	1888.0	450.4	2338.4	4.1071	0.7005	4.8076
22,000	373.71	0.002703	0.003644	1951.7	140.8	2092.4	2011.1	161.5	2172.6	4.2942	0.2496	4.5439
22,064	373.95	0.003106	0.003106	2015.7	0	2015.7	2084.3	0	2084.3	4.4070	0	4.4070

TABLE A–6

Superheated water

T °C	*v* m³/kg	*u* kJ/kg	*h* kJ/kg	*s* kJ/kg·K	*v* m³/kg	*u* kJ/kg	*h* kJ/kg	*s* kJ/kg·K	*v* m³/kg	*u* kJ/kg	*h* kJ/kg	*s* kJ/kg·K
	P = 0.01 MPa (45.81°C)*				*P* = 0.05 MPa (81.32°C)				*P* = 0.10 MPa (99.61°C)			
Sat.†	14.670	2437.2	2583.9	8.1488	3.2403	2483.2	2645.2	7.5931	1.6941	2505.6	2675.0	7.3589
50	14.867	2443.3	2592.0	8.1741								
100	17.196	2515.5	2687.5	8.4489	3.4187	2511.5	2682.4	7.6953	1.6959	2506.2	2675.8	7.3611
150	19.513	2587.9	2783.0	8.6893	3.8897	2585.7	2780.2	7.9413	1.9367	2582.9	2776.6	7.6148
200	21.826	2661.4	2879.6	8.9049	4.3562	2660.0	2877.8	8.1592	2.1724	2658.2	2875.5	7.8356
250	24.136	2736.1	2977.5	9.1015	4.8206	2735.1	2976.2	8.3568	2.4062	2733.9	2974.5	8.0346
300	26.446	2812.3	3076.7	9.2827	5.2841	2811.6	3075.8	8.5387	2.6389	2810.7	3074.5	8.2172
400	31.063	2969.3	3280.0	9.6094	6.2094	2968.9	3279.3	8.8659	3.1027	2968.3	3278.6	8.5452
500	35.680	3132.9	3489.7	9.8998	7.1338	3132.6	3489.3	9.1566	3.5655	3132.2	3488.7	8.8362
600	40.296	3303.3	3706.3	10.1631	8.0577	3303.1	3706.0	9.4201	4.0279	3302.8	3705.6	9.0999
700	44.911	3480.8	3929.9	10.4056	8.9813	3480.6	3929.7	9.6626	4.4900	3480.4	3929.4	9.3424
800	49.527	3665.4	4160.6	10.6312	9.9047	3665.2	4160.4	9.8883	4.9519	3665.0	4160.2	9.5682
900	54.143	3856.9	4398.3	10.8429	10.8280	3856.8	4398.2	10.1000	5.4137	3856.7	4398.0	9.7800
1000	58.758	4055.3	4642.8	11.0429	11.7513	4055.2	4642.7	10.3000	5.8755	4055.0	4642.6	9.9800
1100	63.373	4260.0	4893.8	11.2326	12.6745	4259.9	4893.7	10.4897	6.3372	4259.8	4893.6	10.1698
1200	67.989	4470.9	5150.8	11.4132	13.5977	4470.8	5150.7	10.6704	6.7988	4470.7	5150.6	10.3504
1300	72.604	4687.4	5413.4	11.5857	14.5209	4687.3	5413.3	10.8429	7.2605	4687.2	5413.3	10.5229
	P = 0.20 MPa (120.21°C)				*P* = 0.30 MPa (133.52°C)				*P* = 0.40 MPa (143.61°C)			
Sat.	0.88578	2529.1	2706.3	7.1270	0.60582	2543.2	2724.9	6.9917	0.46242	2553.1	2738.1	6.8955
150	0.95986	2577.1	2769.1	7.2810	0.63402	2571.0	2761.2	7.0792	0.47088	2564.4	2752.8	6.9306
200	1.08049	2654.6	2870.7	7.5081	0.71643	2651.0	2865.9	7.3132	0.53434	2647.2	2860.9	7.1723
250	1.19890	2731.4	2971.2	7.7100	0.79645	2728.9	2967.9	7.5180	0.59520	2726.4	2964.5	7.3804
300	1.31623	2808.8	3072.1	7.8941	0.87535	2807.0	3069.6	7.7037	0.65489	2805.1	3067.1	7.5677
400	1.54934	2967.2	3277.0	8.2236	1.03155	2966.0	3275.5	8.0347	0.77265	2964.9	3273.9	7.9003
500	1.78142	3131.4	3487.7	8.5153	1.18672	3130.6	3486.6	8.3271	0.88936	3129.8	3485.5	8.1933
600	2.01302	3302.2	3704.8	8.7793	1.34139	3301.6	3704.0	8.5915	1.00558	3301.0	3703.3	8.4580
700	2.24434	3479.9	3928.8	9.0221	1.49580	3479.5	3928.2	8.8345	1.12152	3479.0	3927.6	8.7012
800	2.47550	3664.7	4159.8	9.2479	1.65004	3664.3	4159.3	9.0605	1.23730	3663.9	4158.9	8.9274
900	2.70656	3856.3	4397.7	9.4598	1.80417	3856.0	4397.3	9.2725	1.35298	3855.7	4396.9	9.1394
1000	2.93755	4054.8	4642.3	9.6599	1.95824	4054.5	4642.0	9.4726	1.46859	4054.3	4641.7	9.3396
1100	3.16848	4259.6	4893.3	9.8497	2.11226	4259.4	4893.1	9.6624	1.58414	4259.2	4892.9	9.5295
1200	3.39938	4470.5	5150.4	10.0304	2.26624	4470.3	5150.2	9.8431	1.69966	4470.2	5150.0	9.7102
1300	3.63026	4687.1	5413.1	10.2029	2.42019	4686.9	5413.0	10.0157	1.81516	4686.7	5412.8	9.8828
	P = 0.50 MPa (151.83°C)				*P* = 0.60 MPa (158.83°C)				*P* = 0.80 MPa (170.41°C)			
Sat.	0.37483	2560.7	2748.1	6.8207	0.31560	2566.8	2756.2	6.7593	0.24035	2576.0	2768.3	6.6616
200	0.42503	2643.3	2855.8	7.0610	0.35212	2639.4	2850.6	6.9683	0.26088	2631.1	2839.8	6.8177
250	0.47443	2723.8	2961.0	7.2725	0.39390	2721.2	2957.6	7.1833	0.29321	2715.9	2950.4	7.0402
300	0.52261	2803.3	3064.6	7.4614	0.43442	2801.4	3062.0	7.3740	0.32416	2797.5	3056.9	7.2345
350	0.57015	2883.0	3168.1	7.6346	0.47428	2881.6	3166.1	7.5481	0.35442	2878.6	3162.2	7.4107
400	0.61731	2963.7	3272.4	7.7956	0.51374	2962.5	3270.8	7.7097	0.38429	2960.2	3267.7	7.5735
500	0.71095	3129.0	3484.5	8.0893	0.59200	3128.2	3483.4	8.0041	0.44332	3126.6	3481.3	7.8692
600	0.80409	3300.4	3702.5	8.3544	0.66976	3299.8	3701.7	8.2695	0.50186	3298.7	3700.1	8.1354
700	0.89696	3478.6	3927.0	8.5978	0.74725	3478.1	3926.4	8.5132	0.56011	3477.2	3925.3	8.3794
800	0.98966	3663.6	4158.4	8.8240	0.82457	3663.2	4157.9	8.7395	0.61820	3662.5	4157.0	8.6061
900	1.08227	3855.4	4396.6	9.0362	0.90179	3855.1	4396.2	8.9518	0.67619	3854.5	4395.5	8.8185
1000	1.17480	4054.0	4641.4	9.2364	0.97893	4053.8	4641.1	9.1521	0.73411	4053.3	4640.5	9.0189
1100	1.26728	4259.0	4892.6	9.4263	1.05603	4258.8	4892.4	9.3420	0.79197	4258.3	4891.9	9.2090
1200	1.35972	4470.0	5149.8	9.6071	1.13309	4469.8	5149.6	9.5229	0.84980	4469.4	5149.3	9.3898
1300	1.45214	4686.6	5412.6	9.7797	1.21012	4686.4	5412.5	9.6955	0.90761	4686.1	5412.2	9.5625

*The temperature in parentheses is the saturation temperature at the specified pressure.

† Properties of saturated vapor at the specified pressure.

TABLE A–6

Superheated water (*Concluded*)

T °C	*v* m³/kg	*u* kJ/kg	*h* kJ/kg	*s* kJ/kg·K	*v* m³/kg	*u* kJ/kg	*h* kJ/kg	*s* kJ/kg·K	*v* m³/kg	*u* kJ/kg	*h* kJ/kg	*s* kJ/kg·K
	P = 1.00 MPa (179.88°C)				*P* = 1.20 MPa (187.96°C)				*P* = 1.40 MPa (195.04°C)			
Sat.	0.19437	2582.8	2777.1	6.5850	0.16326	2587.8	2783.8	6.5217	0.14078	2591.8	2788.9	6.4675
200	0.20602	2622.3	2828.3	6.6956	0.16934	2612.9	2816.1	6.5909	0.14303	2602.7	2803.0	6.4975
250	0.23275	2710.4	2943.1	6.9265	0.19241	2704.7	2935.6	6.8313	0.16356	2698.9	2927.9	6.7488
300	0.25799	2793.7	3051.6	7.1246	0.21386	2789.7	3046.3	7.0335	0.18233	2785.7	3040.9	6.9553
350	0.28250	2875.7	3158.2	7.3029	0.23455	2872.7	3154.2	7.2139	0.20029	2869.7	3150.1	7.1379
400	0.30661	2957.9	3264.5	7.4670	0.25482	2955.5	3261.3	7.3793	0.21782	2953.1	3258.1	7.3046
500	0.35411	3125.0	3479.1	7.7642	0.29464	3123.4	3477.0	7.6779	0.25216	3121.8	3474.8	7.6047
600	0.40111	3297.5	3698.6	8.0311	0.33395	3296.3	3697.0	7.9456	0.28597	3295.1	3695.5	7.8730
700	0.44783	3476.3	3924.1	8.2755	0.37297	3475.3	3922.9	8.1904	0.31951	3474.4	3921.7	8.1183
800	0.49438	3661.7	4156.1	8.5024	0.41184	3661.0	4155.2	8.4176	0.35288	3660.3	4154.3	8.3458
900	0.54083	3853.9	4394.8	8.7150	0.45059	3853.3	4394.0	8.6303	0.38614	3852.7	4393.3	8.5587
1000	0.58721	4052.7	4640.0	8.9155	0.48928	4052.2	4639.4	8.8310	0.41933	4051.7	4638.8	8.7595
1100	0.63354	4257.9	4891.4	9.1057	0.52792	4257.5	4891.0	9.0212	0.45247	4257.0	4890.5	8.9497
1200	0.67983	4469.0	5148.9	9.2866	0.56652	4468.7	5148.5	9.2022	0.48558	4468.3	5148.1	9.1308
1300	0.72610	4685.8	5411.9	9.4593	0.60509	4685.5	5411.6	9.3750	0.51866	4685.1	5411.3	9.3036
	P = 1.60 MPa (201.37°C)				*P* = 1.80 MPa (207.11°C)				*P* = 2.00 MPa (212.38°C)			
Sat.	0.12374	2594.8	2792.8	6.4200	0.11037	2597.3	2795.9	6.3775	0.09959	2599.1	2798.3	6.3390
225	0.13293	2645.1	2857.8	6.5537	0.11678	2637.0	2847.2	6.4825	0.10381	2628.5	2836.1	6.4160
250	0.14190	2692.9	2919.9	6.6753	0.12502	2686.7	2911.7	6.6088	0.11150	2680.3	2903.3	6.5475
300	0.15866	2781.6	3035.4	6.8864	0.14025	2777.4	3029.9	6.8246	0.12551	2773.2	3024.2	6.7684
350	0.17459	2866.6	3146.0	7.0713	0.15460	2863.6	3141.9	7.0120	0.13860	2860.5	3137.7	6.9583
400	0.19007	2950.8	3254.9	7.2394	0.16849	2948.3	3251.6	7.1814	0.15122	2945.9	3248.4	7.1292
500	0.22029	3120.1	3472.6	7.5410	0.19551	3118.5	3470.4	7.4845	0.17568	3116.9	3468.3	7.4337
600	0.24999	3293.9	3693.9	7.8101	0.22200	3292.7	3692.3	7.7543	0.19962	3291.5	3690.7	7.7043
700	0.27941	3473.5	3920.5	8.0558	0.24822	3472.6	3919.4	8.0005	0.22326	3471.7	3918.2	7.9509
800	0.30865	3659.5	4153.4	8.2834	0.27426	3658.8	4152.4	8.2284	0.24674	3658.0	4151.5	8.1791
900	0.33780	3852.1	4392.6	8.4965	0.30020	3851.5	4391.9	8.4417	0.27012	3850.9	4391.1	8.3925
1000	0.36687	4051.2	4638.2	8.6974	0.32606	4050.7	4637.6	8.6427	0.29342	4050.2	4637.1	8.5936
1100	0.39589	4256.6	4890.0	8.8878	0.35188	4256.2	4889.6	8.8331	0.31667	4255.7	4889.1	8.7842
1200	0.42488	4467.9	5147.7	9.0689	0.37766	4467.6	5147.3	9.0143	0.33989	4467.2	5147.0	8.9654
1300	0.45383	4684.8	5410.9	9.2418	0.40341	4684.5	5410.6	9.1872	0.36308	4684.2	5410.3	9.1384
	P = 2.50 MPa (223.95°C)				*P* = 3.00 MPa (233.85°C)				*P* = 3.50 MPa (242.56°C)			
Sat.	0.07995	2602.1	2801.9	6.2558	0.06667	2603.2	2803.2	6.1856	0.05706	2603.0	2802.7	6.1244
225	0.08026	2604.8	2805.5	6.2629								
250	0.08705	2663.3	2880.9	6.4107	0.07063	2644.7	2856.5	6.2893	0.05876	2624.0	2829.7	6.1764
300	0.09894	2762.2	3009.6	6.6459	0.08118	2750.8	2994.3	6.5412	0.06845	2738.8	2978.4	6.4484
350	0.10979	2852.5	3127.0	6.8424	0.09056	2844.4	3116.1	6.7450	0.07680	2836.0	3104.9	6.6601
400	0.12012	2939.8	3240.1	7.0170	0.09938	2933.6	3231.7	6.9235	0.08456	2927.2	3223.2	6.8428
450	0.13015	3026.2	3351.6	7.1768	0.10789	3021.2	3344.9	7.0856	0.09198	3016.1	3338.1	7.0074
500	0.13999	3112.8	3462.8	7.3254	0.11620	3108.6	3457.2	7.2359	0.09919	3104.5	3451.7	7.1593
600	0.15931	3288.5	3686.8	7.5979	0.13245	3285.5	3682.8	7.5103	0.11325	3282.5	3678.9	7.4357
700	0.17835	3469.3	3915.2	7.8455	0.14841	3467.0	3912.2	7.7590	0.12702	3464.7	3909.3	7.6855
800	0.19722	3656.2	4149.2	8.0744	0.16420	3654.3	4146.9	7.9885	0.14061	3652.5	4144.6	7.9156
900	0.21597	3849.4	4389.3	8.2882	0.17988	3847.9	4387.5	8.2028	0.15410	3846.4	4385.7	8.1304
1000	0.23466	4049.0	4635.6	8.4897	0.19549	4047.7	4634.2	8.4045	0.16751	4046.4	4632.7	8.3324
1100	0.25330	4254.7	4887.9	8.6804	0.21105	4253.6	4886.7	8.5955	0.18087	4252.5	4885.6	8.5236
1200	0.27190	4466.3	5146.0	8.8618	0.22658	4465.3	5145.1	8.7771	0.19420	4464.4	5144.1	8.7053
1300	0.29048	4683.4	5409.5	9.0349	0.24207	4682.6	5408.8	8.9502	0.20750	4681.8	5408.0	8.8786

TABLE A–6

Superheated water (*Continued*)

T °C	*v* m³/kg	*u* kJ/kg	*h* kJ/kg	*s* kJ/kg·K	*v* m³/kg	*u* kJ/kg	*h* kJ/kg	*s* kJ/kg·K	*v* m³/kg	*u* kJ/kg	*h* kJ/kg	*s* kJ/kg·K
	P = 4.0 MPa (250.35°C)				*P* = 4.5 MPa (257.44°C)				*P* = 5.0 MPa (263.94°C)			
Sat.	0.04978	2601.7	2800.8	6.0696	0.04406	2599.7	2798.0	6.0198	0.03945	2597.0	2794.2	5.9737
275	0.05461	2668.9	2887.3	6.2312	0.04733	2651.4	2864.4	6.1429	0.04144	2632.3	2839.5	6.0571
300	0.05887	2726.2	2961.7	6.3639	0.05138	2713.0	2944.2	6.2854	0.04535	2699.0	2925.7	6.2111
350	0.06647	2827.4	3093.3	6.5843	0.05842	2818.6	3081.5	6.5153	0.05197	2809.5	3069.3	6.4516
400	0.07343	2920.8	3214.5	6.7714	0.06477	2914.2	3205.7	6.7071	0.05784	2907.5	3196.7	6.6483
450	0.08004	3011.0	3331.2	6.9386	0.07076	3005.8	3324.2	6.8770	0.06332	3000.6	3317.2	6.8210
500	0.08644	3100.3	3446.0	7.0922	0.07652	3096.0	3440.4	7.0323	0.06858	3091.8	3434.7	6.9781
600	0.09886	3279.4	3674.9	7.3706	0.08766	3276.4	3670.9	7.3127	0.07870	3273.3	3666.9	7.2605
700	0.11098	3462.4	3906.3	7.6214	0.09850	3460.0	3903.3	7.5647	0.08852	3457.7	3900.3	7.5136
800	0.12292	3650.6	4142.3	7.8523	0.10916	3648.8	4140.0	7.7962	0.09816	3646.9	4137.7	7.7458
900	0.13476	3844.8	4383.9	8.0675	0.11972	3843.3	4382.1	8.0118	0.10769	3841.8	4380.2	7.9619
1000	0.14653	4045.1	4631.2	8.2698	0.13020	4043.9	4629.8	8.2144	0.11715	4042.6	4628.3	8.1648
1100	0.15824	4251.4	4884.4	8.4612	0.14064	4250.4	4883.2	8.4060	0.12655	4249.3	4882.1	8.3566
1200	0.16992	4463.5	5143.2	8.6430	0.15103	4462.6	5142.2	8.5880	0.13592	4461.6	5141.3	8.5388
1300	0.18157	4680.9	5407.2	8.8164	0.16140	4680.1	5406.5	8.7616	0.14527	4679.3	5405.7	8.7124
	P = 6.0 MPa (275.59°C)				*P* = 7.0 MPa (285.83°C)				*P* = 8.0 MPa (295.01°C)			
Sat.	0.03245	2589.9	2784.6	5.8902	0.027378	2581.0	2772.6	5.8148	0.023525	2570.5	2758.7	5.7450
300	0.03619	2668.4	2885.6	6.0703	0.029492	2633.5	2839.9	5.9337	0.024279	2592.3	2786.5	5.7937
350	0.04225	2790.4	3043.9	6.3357	0.035262	2770.1	3016.9	6.2305	0.029975	2748.3	2988.1	6.1321
400	0.04742	2893.7	3178.3	6.5432	0.039958	2879.5	3159.2	6.4502	0.034344	2864.6	3139.4	6.3658
450	0.05217	2989.9	3302.9	6.7219	0.044187	2979.0	3288.3	6.6353	0.038194	2967.8	3273.3	6.5579
500	0.05667	3083.1	3423.1	6.8826	0.048157	3074.3	3411.4	6.8000	0.041767	3065.4	3399.5	6.7266
550	0.06102	3175.2	3541.3	7.0308	0.051966	3167.9	3531.6	6.9507	0.045172	3160.5	3521.8	6.8800
600	0.06527	3267.2	3658.8	7.1693	0.055665	3261.0	3650.6	7.0910	0.048463	3254.7	3642.4	7.0221
700	0.07355	3453.0	3894.3	7.4247	0.062850	3448.3	3888.3	7.3487	0.054829	3443.6	3882.2	7.2822
800	0.08165	3643.2	4133.1	7.6582	0.069856	3639.5	4128.5	7.5836	0.061011	3635.7	4123.8	7.5185
900	0.08964	3838.8	4376.6	7.8751	0.076750	3835.7	4373.0	7.8014	0.067082	3832.7	4369.3	7.7372
1000	0.09756	4040.1	4625.4	8.0786	0.083571	4037.5	4622.5	8.0055	0.073079	4035.0	4619.6	7.9419
1100	0.10543	4247.1	4879.7	8.2709	0.090341	4245.0	4877.4	8.1982	0.079025	4242.8	4875.0	8.1350
1200	0.11326	4459.8	5139.4	8.4534	0.097075	4457.9	5137.4	8.3810	0.084934	4456.1	5135.5	8.3181
1300	0.12107	4677.7	5404.1	8.6273	0.103781	4676.1	5402.6	8.5551	0.090817	4674.5	5401.0	8.4925
	P = 9.0 MPa (303.35°C)				*P* = 10.0 MPa (311.00°C)				*P* = 12.5 MPa (327.81°C)			
Sat.	0.020489	2558.5	2742.9	5.6791	0.018028	2545.2	2725.5	5.6159	0.013496	2505.6	2674.3	5.4638
325	0.023284	2647.6	2857.1	5.8738	0.019877	2611.6	2810.3	5.7596				
350	0.025816	2725.0	2957.3	6.0380	0.022440	2699.6	2924.0	5.9460	0.016138	2624.9	2826.6	5.7130
400	0.029960	2849.2	3118.8	6.2876	0.026436	2833.1	3097.5	6.2141	0.020030	2789.6	3040.0	6.0433
450	0.033524	2956.3	3258.0	6.4872	0.029782	2944.5	3242.4	6.4219	0.023019	2913.7	3201.5	6.2749
500	0.036793	3056.3	3387.4	6.6603	0.032811	3047.0	3375.1	6.5995	0.025630	3023.2	3343.6	6.4651
550	0.039885	3153.0	3512.0	6.8164	0.035655	3145.4	3502.0	6.7585	0.028033	3126.1	3476.5	6.6317
600	0.042861	3248.4	3634.1	6.9605	0.038378	3242.0	3625.8	6.9045	0.030306	3225.8	3604.6	6.7828
650	0.045755	3343.4	3755.2	7.0954	0.041018	3338.0	3748.1	7.0408	0.032491	3324.1	3730.2	6.9227
700	0.048589	3438.8	3876.1	7.2229	0.043597	3434.0	3870.0	7.1693	0.034612	3422.0	3854.6	7.0540
800	0.054132	3632.0	4119.2	7.4606	0.048629	3628.2	4114.5	7.4085	0.038724	3618.8	4102.8	7.2967
900	0.059562	3829.6	4365.7	7.6802	0.053547	3826.5	4362.0	7.6290	0.042720	3818.9	4352.9	7.5195
1000	0.064919	4032.4	4616.7	7.8855	0.058391	4029.9	4613.8	7.8349	0.046641	4023.5	4606.5	7.7269
1100	0.070224	4240.7	4872.7	8.0791	0.063183	4238.5	4870.3	8.0289	0.050510	4233.1	4864.5	7.9220
1200	0.075492	4454.2	5133.6	8.2625	0.067938	4452.4	5131.7	8.2126	0.054342	4447.7	5127.0	8.1065
1300	0.080733	4672.9	5399.5	8.4371	0.072667	4671.3	5398.0	8.3874	0.058147	4667.3	5394.1	8.2819

TABLE A–6

Superheated water (*Concluded*)

T °C	*v* m³/kg	*u* kJ/kg	*h* kJ/kg	*s* kJ/kg·K	*v* m³/kg	*u* kJ/kg	*h* kJ/kg	*s* kJ/kg·K	*v* m³/kg	*u* kJ/kg	*h* kJ/kg	*s* kJ/kg·K
	P = 15.0 MPa (342.16°C)				*P* = 17.5 MPa (354.67°C)				*P* = 20.0 MPa (365.75°C)			
Sat.	0.010341	2455.7	2610.8	5.3108	0.007932	2390.7	2529.5	5.1435	0.005862	2294.8	2412.1	4.9310
350	0.011481	2520.9	2693.1	5.4438								
400	0.015671	2740.6	2975.7	5.8819	0.012463	2684.3	2902.4	5.7211	0.009950	2617.9	2816.9	5.5526
450	0.018477	2880.8	3157.9	6.1434	0.015204	2845.4	3111.4	6.0212	0.012721	2807.3	3061.7	5.9043
500	0.020828	2998.4	3310.8	6.3480	0.017385	2972.4	3276.7	6.2424	0.014793	2945.3	3241.2	6.1446
550	0.022945	3106.2	3450.4	6.5230	0.019305	3085.8	3423.6	6.4266	0.016571	3064.7	3396.2	6.3390
600	0.024921	3209.3	3583.1	6.6796	0.021073	3192.5	3561.3	6.5890	0.018185	3175.3	3539.0	6.5075
650	0.026804	3310.1	3712.1	6.8233	0.022742	3295.8	3693.8	6.7366	0.019695	3281.4	3675.3	6.6593
700	0.028621	3409.8	3839.1	6.9573	0.024342	3397.5	3823.5	6.8735	0.021134	3385.1	3807.8	6.7991
800	0.032121	3609.3	4091.1	7.2037	0.027405	3599.7	4079.3	7.1237	0.023870	3590.1	4067.5	7.0531
900	0.035503	3811.2	4343.7	7.4288	0.030348	3803.5	4334.6	7.3511	0.026484	3795.7	4325.4	7.2829
1000	0.038808	4017.1	4599.2	7.6378	0.033215	4010.7	4592.0	7.5616	0.029020	4004.3	4584.7	7.4950
1100	0.042062	4227.7	4858.6	7.8339	0.036029	4222.3	4852.8	7.7588	0.031504	4216.9	4847.0	7.6933
1200	0.045279	4443.1	5122.3	8.0192	0.038806	4438.5	5117.6	7.9449	0.033952	4433.8	5112.9	7.8802
1300	0.048469	4663.3	5390.3	8.1952	0.041556	4659.2	5386.5	8.1215	0.036371	4655.2	5382.7	8.0574

T °C	*v* m³/kg	*u* kJ/kg	*h* kJ/kg	*s* kJ/kg·K	*v* m³/kg	*u* kJ/kg	*h* kJ/kg	*s* kJ/kg·K	*v* m³/kg	*u* kJ/kg	*h* kJ/kg	*s* kJ/kg·K
	P = 25.0 MPa				*P* = 30.0 MPa				*P* = 35.0 MPa			
375	0.001978	1799.9	1849.4	4.0345	0.001792	1738.1	1791.9	3.9313	0.001701	1702.8	1762.4	3.8724
400	0.006005	2428.5	2578.7	5.1400	0.002798	2068.9	2152.8	4.4758	0.002105	1914.9	1988.6	4.2144
425	0.007886	2607.8	2805.0	5.4708	0.005299	2452.9	2611.8	5.1473	0.003434	2253.3	2373.5	4.7751
450	0.009176	2721.2	2950.6	5.6759	0.006737	2618.9	2821.0	5.4422	0.004957	2497.5	2671.0	5.1946
500	0.011143	2887.3	3165.9	5.9643	0.008691	2824.0	3084.8	5.7956	0.006933	2755.3	2997.9	5.6331
550	0.012736	3020.8	3339.2	6.1816	0.010175	2974.5	3279.7	6.0403	0.008348	2925.8	3218.0	5.9093
600	0.014140	3140.0	3493.5	6.3637	0.011445	3103.4	3446.8	6.2373	0.009523	3065.6	3399.0	6.1229
650	0.015430	3251.9	3637.7	6.5243	0.012590	3221.7	3599.4	6.4074	0.010565	3190.9	3560.7	6.3030
700	0.016643	3359.9	3776.0	6.6702	0.013654	3334.3	3743.9	6.5599	0.011523	3308.3	3711.6	6.4623
800	0.018922	3570.7	4043.8	6.9322	0.015628	3551.2	4020.0	6.8301	0.013278	3531.6	3996.3	6.7409
900	0.021075	3780.2	4307.1	7.1668	0.017473	3764.6	4288.8	7.0695	0.014904	3749.0	4270.6	6.9853
1000	0.023150	3991.5	4570.2	7.3821	0.019240	3978.6	4555.8	7.2880	0.016450	3965.8	4541.5	7.2069
1100	0.025172	4206.1	4835.4	7.5825	0.020954	4195.2	4823.9	7.4906	0.017942	4184.4	4812.4	7.4118
1200	0.027157	4424.6	5103.5	7.7710	0.022630	4415.3	5094.2	7.6807	0.019398	4406.1	5085.0	7.6034
1300	0.029115	4647.2	5375.1	7.9494	0.024279	4639.2	5367.6	7.8602	0.020827	4631.2	5360.2	7.7841

T °C	*v* m³/kg	*u* kJ/kg	*h* kJ/kg	*s* kJ/kg·K	*v* m³/kg	*u* kJ/kg	*h* kJ/kg	*s* kJ/kg·K	*v* m³/kg	*u* kJ/kg	*h* kJ/kg	*s* kJ/kg·K
	P = 40.0 MPa				*P* = 50.0 MPa				*P* = 60.0 MPa			
375	0.001641	1677.0	1742.6	3.8290	0.001560	1638.6	1716.6	3.7642	0.001503	1609.7	1699.9	3.7149
400	0.001911	1855.0	1931.4	4.1145	0.001731	1787.8	1874.4	4.0029	0.001633	1745.2	1843.2	3.9317
425	0.002538	2097.5	2199.0	4.5044	0.002009	1960.3	2060.7	4.2746	0.001816	1892.9	2001.8	4.1630
450	0.003692	2364.2	2511.8	4.9449	0.002487	2160.3	2284.7	4.5896	0.002086	2055.1	2180.2	4.4140
500	0.005623	2681.6	2906.5	5.4744	0.003890	2528.1	2722.6	5.1762	0.002952	2393.2	2570.3	4.9356
550	0.006985	2875.1	3154.4	5.7857	0.005118	2769.5	3025.4	5.5563	0.003955	2664.6	2901.9	5.3517
600	0.008089	3026.8	3350.4	6.0170	0.006108	2947.1	3252.6	5.8245	0.004833	2866.8	3156.8	5.6527
650	0.009053	3159.5	3521.6	6.2078	0.006957	3095.6	3443.5	6.0373	0.005591	3031.3	3366.8	5.8867
700	0.009930	3282.0	3679.2	6.3740	0.007717	3228.7	3614.6	6.2179	0.006265	3175.4	3551.3	6.0814
800	0.011521	3511.8	3972.6	6.6613	0.009073	3472.2	3925.8	6.5225	0.007456	3432.6	3880.0	6.4033
900	0.012980	3733.3	4252.5	6.9107	0.010296	3702.0	4216.8	6.7819	0.008519	3670.9	4182.1	6.6725
1000	0.014360	3952.9	4527.3	7.1355	0.011441	3927.4	4499.4	7.0131	0.009504	3902.0	4472.2	6.9099
1100	0.015686	4173.7	4801.1	7.3425	0.012534	4152.2	4778.9	7.2244	0.010439	4130.9	4757.3	7.1255
1200	0.016976	4396.9	5075.9	7.5357	0.013590	4378.6	5058.1	7.4207	0.011339	4360.5	5040.8	7.3248
1300	0.018239	4623.3	5352.8	7.7175	0.014620	4607.5	5338.5	7.6048	0.012213	4591.8	5324.5	7.5111

TABLE A–7

Compressed liquid water

T °C	*v* m³/kg	*u* kJ/kg	*h* kJ/kg	*s* kJ/kg·K	*v* m³/kg	*u* kJ/kg	*h* kJ/kg	*s* kJ/kg·K	*v* m³/kg	*u* kJ/kg	*h* kJ/kg	*s* kJ/kg·K
	P = 5 MPa (263.94°C)				*P* = 10 MPa (311.00°C)				*P* = 15 MPa (342.16°C)			
Sat.	0.0012862	1148.1	1154.5	2.9207	0.0014522	1393.3	1407.9	3.3603	0.0016572	1585.5	1610.3	3.6848
0	0.0009977	0.04	5.03	0.0001	0.0009952	0.12	10.07	0.0003	0.0009928	0.18	15.07	0.0004
20	0.0009996	83.61	88.61	0.2954	0.0009973	83.31	93.28	0.2943	0.0009951	83.01	97.93	0.2932
40	0.0010057	166.92	171.95	0.5705	0.0010035	166.33	176.37	0.5685	0.0010013	165.75	180.77	0.5666
60	0.0010149	250.29	255.36	0.8287	0.0010127	249.43	259.55	0.8260	0.0010105	248.58	263.74	0.8234
80	0.0010267	333.82	338.96	1.0723	0.0010244	332.69	342.94	1.0691	0.0010221	331.59	346.92	1.0659
100	0.0010410	417.65	422.85	1.3034	0.0010385	416.23	426.62	1.2996	0.0010361	414.85	430.39	1.2958
120	0.0010576	501.91	507.19	1.5236	0.0010549	500.18	510.73	1.5191	0.0010522	498.50	514.28	1.5148
140	0.0010769	586.80	592.18	1.7344	0.0010738	584.72	595.45	1.7293	0.0010708	582.69	598.75	1.7243
160	0.0010988	672.55	678.04	1.9374	0.0010954	670.06	681.01	1.9316	0.0010920	667.63	684.01	1.9259
180	0.0011240	759.47	765.09	2.1338	0.0011200	756.48	767.68	2.1271	0.0011160	753.58	770.32	2.1206
200	0.0011531	847.92	853.68	2.3251	0.0011482	844.32	855.80	2.3174	0.0011435	840.84	858.00	2.3100
220	0.0011868	938.39	944.32	2.5127	0.0011809	934.01	945.82	2.5037	0.0011752	929.81	947.43	2.4951
240	0.0012268	1031.6	1037.7	2.6983	0.0012192	1026.2	1038.3	2.6876	0.0012121	1021.0	1039.2	2.6774
260	0.0012755	1128.5	1134.9	2.8841	0.0012653	1121.6	1134.3	2.8710	0.0012560	1115.1	1134.0	2.8586
280					0.0013226	1221.8	1235.0	3.0565	0.0013096	1213.4	1233.0	3.0410
300					0.0013980	1329.4	1343.3	3.2488	0.0013783	1317.6	1338.3	3.2279
320									0.0014733	1431.9	1454.0	3.4263
340									0.0016311	1567.9	1592.4	3.6555
	P = 20 MPa (365.75°C)				*P* = 30 MPa				*P* = 50 MPa			
Sat.	0.0020378	1785.8	1826.6	4.0146								
0	0.0009904	0.23	20.03	0.0005	0.0009857	0.29	29.86	0.0003	0.0009767	0.29	49.13	−0.0010
20	0.0009929	82.71	102.57	0.2921	0.0009886	82.11	111.77	0.2897	0.0009805	80.93	129.95	0.2845
40	0.0009992	165.17	185.16	0.5646	0.0009951	164.05	193.90	0.5607	0.0009872	161.90	211.25	0.5528
60	0.0010084	247.75	267.92	0.8208	0.0010042	246.14	276.26	0.8156	0.0009962	243.08	292.88	0.8055
80	0.0010199	330.50	350.90	1.0627	0.0010155	328.40	358.86	1.0564	0.0010072	324.42	374.78	1.0442
100	0.0010337	413.50	434.17	1.2920	0.0010290	410.87	441.74	1.2847	0.0010201	405.94	456.94	1.2705
120	0.0010496	496.85	517.84	1.5105	0.0010445	493.66	525.00	1.5020	0.0010349	487.69	539.43	1.4859
140	0.0010679	580.71	602.07	1.7194	0.0010623	576.90	608.76	1.7098	0.0010517	569.77	622.36	1.6916
160	0.0010886	665.28	687.05	1.9203	0.0010823	660.74	693.21	1.9094	0.0010704	652.33	705.85	1.8889
180	0.0011122	750.78	773.02	2.1143	0.0011049	745.40	778.55	2.1020	0.0010914	735.49	790.06	2.0790
200	0.0011390	837.49	860.27	2.3027	0.0011304	831.11	865.02	2.2888	0.0011149	819.45	875.19	2.2628
220	0.0011697	925.77	949.16	2.4867	0.0011595	918.15	952.93	2.4707	0.0011412	904.39	961.45	2.4414
240	0.0012053	1016.1	1040.2	2.6676	0.0011927	1006.9	1042.7	2.6491	0.0011708	990.55	1049.1	2.6156
260	0.0012472	1109.0	1134.0	2.8469	0.0012314	1097.8	1134.7	2.8250	0.0012044	1078.2	1138.4	2.7864
280	0.0012978	1205.6	1231.5	3.0265	0.0012770	1191.5	1229.8	3.0001	0.0012430	1167.7	1229.9	2.9547
300	0.0013611	1307.2	1334.4	3.2091	0.0013322	1288.9	1328.9	3.1761	0.0012879	1259.6	1324.0	3.1218
320	0.0014450	1416.6	1445.5	3.3996	0.0014014	1391.7	1433.7	3.3558	0.0013409	1354.3	1421.4	3.2888
340	0.0015693	1540.2	1571.6	3.6086	0.0014932	1502.4	1547.1	3.5438	0.0014049	1452.9	1523.1	3.4575
360	0.0018248	1703.6	1740.1	3.8787	0.0016276	1626.8	1675.6	3.7499	0.0014848	1556.5	1630.7	3.6301
380					0.0018729	1782.0	1838.2	4.0026	0.0015884	1667.1	1746.5	3.8102

TABLE A–8

Saturated ice–water vapor

		Specific volume, m³/kg		*Internal energy, kJ/kg*			*Enthalpy, kJ/kg*			*Entropy, kJ/kg·K*		
Temp., T °C	Sat. press., P_{sat} kPa	Sat. ice, v_i	Sat. vapor, v_g	Sat. ice, u_i	Subl., u_{ig}	Sat. vapor, u_g	Sat. ice, h_i	Subl., h_{ig}	Sat. vapor, h_g	Sat. ice, s_i	Subl., s_{ig}	Sat. vapor, s_g
0.01	0.61169	0.001091	205.99	−333.40	2707.9	2374.5	−333.40	2833.9	2500.5	−1.2202	10.374	9.154
0	0.61115	0.001091	206.17	−333.43	2707.9	2374.5	−333.43	2833.9	2500.5	−1.2204	10.375	9.154
−2	0.51772	0.001091	241.62	−337.63	2709.4	2371.8	−337.63	2834.5	2496.8	−1.2358	10.453	9.218
−4	0.43748	0.001090	283.84	−341.80	2710.8	2369.0	−341.80	2835.0	2493.2	−1.2513	10.533	9.282
−6	0.36873	0.001090	334.27	−345.94	2712.2	2366.2	−345.93	2835.4	2489.5	−1.2667	10.613	9.347
−8	0.30998	0.001090	394.66	−350.04	2713.5	2363.5	−350.04	2835.8	2485.8	−1.2821	10.695	9.413
−10	0.25990	0.001089	467.17	−354.12	2714.8	2360.7	−354.12	2836.2	2482.1	−1.2976	10.778	9.480
−12	0.21732	0.001089	554.47	−358.17	2716.1	2357.9	−358.17	2836.6	2478.4	−1.3130	10.862	9.549
−14	0.18121	0.001088	659.88	−362.18	2717.3	2355.2	−362.18	2836.9	2474.7	−1.3284	10.947	9.618
−16	0.15068	0.001088	787.51	−366.17	2718.6	2352.4	−366.17	2837.2	2471.0	−1.3439	11.033	9.689
−18	0.12492	0.001088	942.51	−370.13	2719.7	2349.6	−370.13	2837.5	2467.3	−1.3593	11.121	9.761
−20	0.10326	0.001087	1131.3	−374.06	2720.9	2346.8	−374.06	2837.7	2463.6	−1.3748	11.209	9.835
−22	0.08510	0.001087	1362.0	−377.95	2722.0	2344.1	−377.95	2837.9	2459.9	−1.3903	11.300	9.909
−24	0.06991	0.001087	1644.7	−381.82	2723.1	2341.3	−381.82	2838.1	2456.2	−1.4057	11.391	9.985
−26	0.05725	0.001087	1992.2	−385.66	2724.2	2338.5	−385.66	2838.2	2452.5	−1.4212	11.484	10.063
−28	0.04673	0.001086	2421.0	−389.47	2725.2	2335.7	−389.47	2838.3	2448.8	−1.4367	11.578	10.141
−30	0.03802	0.001086	2951.7	−393.25	2726.2	2332.9	−393.25	2838.4	2445.1	−1.4521	11.673	10.221
−32	0.03082	0.001086	3610.9	−397.00	2727.2	2330.2	−397.00	2838.4	2441.4	−1.4676	11.770	10.303
−34	0.02490	0.001085	4432.4	−400.72	2728.1	2327.4	−400.72	2838.5	2437.7	−1.4831	11.869	10.386
−36	0.02004	0.001085	5460.1	−404.40	2729.0	2324.6	−404.40	2838.4	2434.0	−1.4986	11.969	10.470
−38	0.01608	0.001085	6750.5	−408.07	2729.9	2321.8	−408.07	2838.4	2430.3	−1.5141	12.071	10.557
−40	0.01285	0.001084	8376.7	−411.70	2730.7	2319.0	−411.70	2838.3	2426.6	−1.5296	12.174	10.644

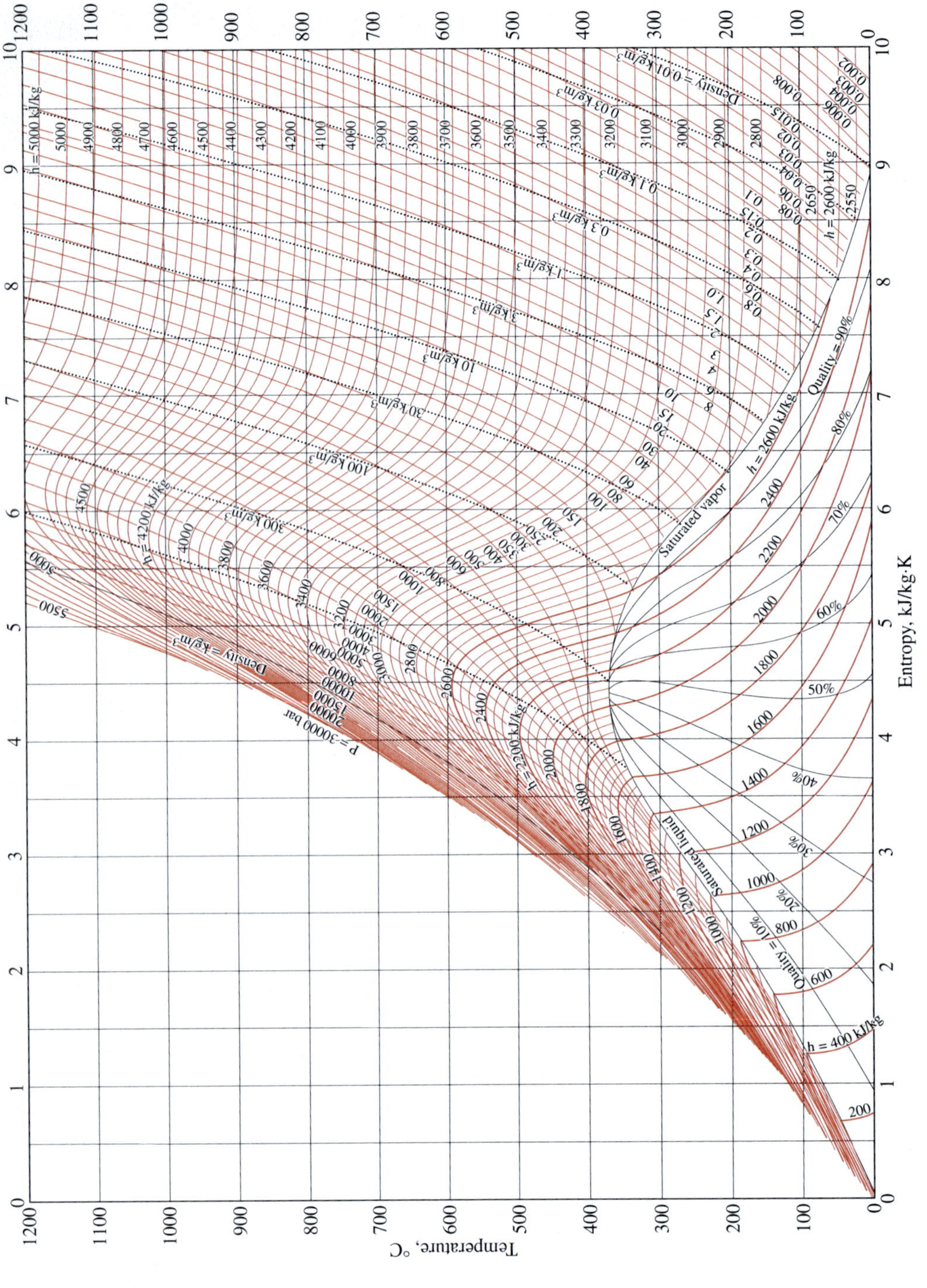

FIGURE A–9

T-s diagram for water.

Copyright © 1984. From NBS/NRC Steam Tables/1 *by Lester Haar, John S. Gallagher, and George S. Kell. Reproduced by permission of Routledge/Taylor & Francis Books, Inc.*

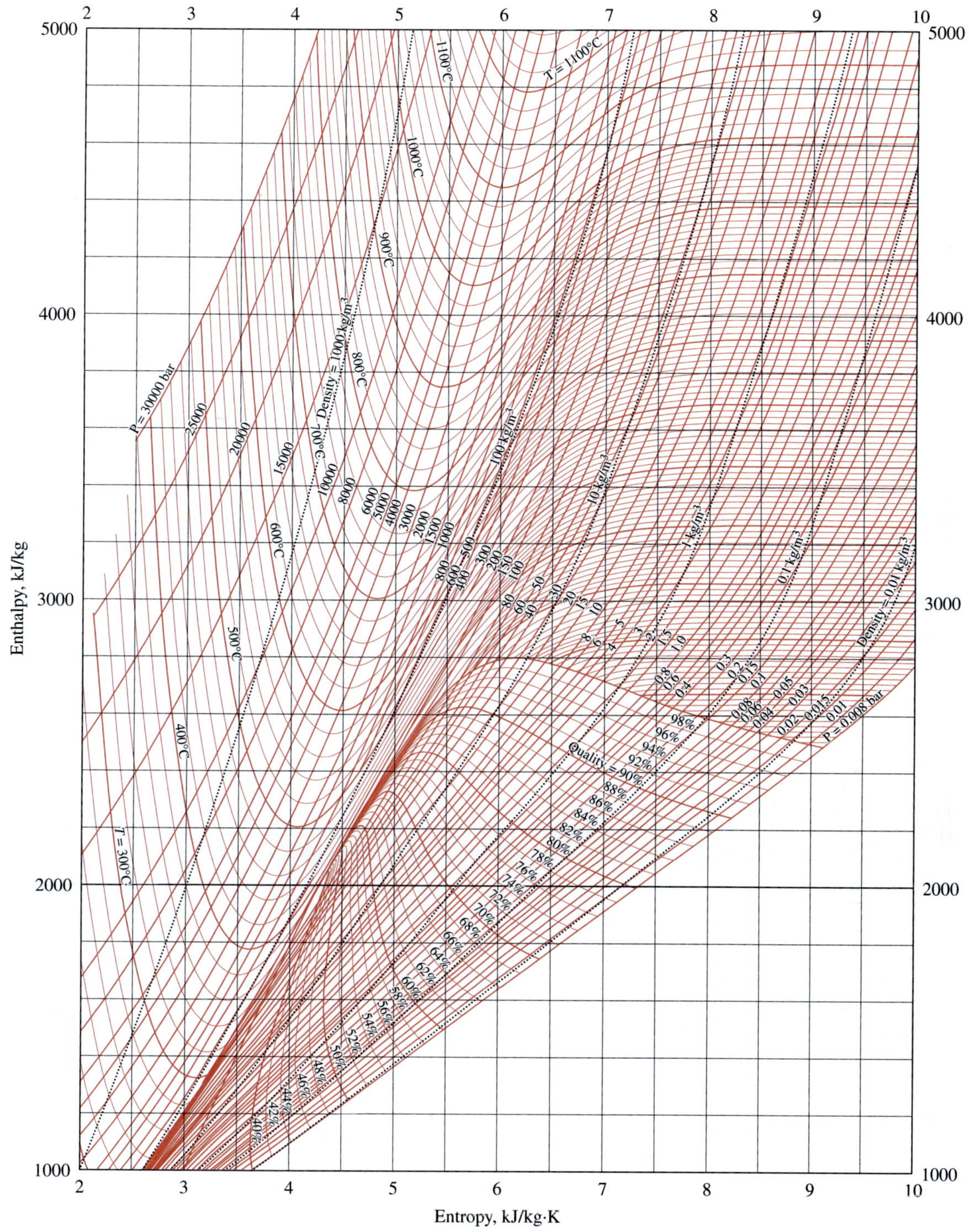

FIGURE A-10

Mollier diagram for water.

TABLE A–11

Saturated refrigerant-134a—Temperature table

		Specific volume, m^3/kg		Internal energy, kJ/kg			Enthalpy, kJ/kg			Entropy, kJ/kg·K		
Temp., T °C	Sat. press., P_{sat} kPa	Sat. liquid, v_f	Sat. vapor, v_g	Sat. liquid, u_f	Evap., u_{fg}	Sat. vapor, u_g	Sat. liquid, h_f	Evap., h_{fg}	Sat. vapor, h_g	Sat. liquid, s_f	Evap., s_{fg}	Sat. vapor, s_g
−40	51.25	0.0007054	0.36081	−0.036	207.40	207.37	0.000	225.86	225.86	0.00000	0.96866	0.96866
−38	56.86	0.0007083	0.32732	2.475	206.04	208.51	2.515	224.61	227.12	0.01072	0.95511	0.96584
−36	62.95	0.0007112	0.29751	4.992	204.67	209.66	5.037	223.35	228.39	0.02138	0.94176	0.96315
−34	69.56	0.0007142	0.27090	7.517	203.29	210.81	7.566	222.09	229.65	0.03199	0.92859	0.96058
−32	76.71	0.0007172	0.24711	10.05	201.91	211.96	10.10	220.81	230.91	0.04253	0.91560	0.95813
−30	84.43	0.0007203	0.22580	12.59	200.52	213.11	12.65	219.52	232.17	0.05301	0.90278	0.95579
−28	92.76	0.0007234	0.20666	15.13	199.12	214.25	15.20	218.22	233.43	0.06344	0.89012	0.95356
−26	101.73	0.0007265	0.18946	17.69	197.72	215.40	17.76	216.92	234.68	0.07382	0.87762	0.95144
−24	111.37	0.0007297	0.17395	20.25	196.30	216.55	20.33	215.59	235.92	0.08414	0.86527	0.94941
−22	121.72	0.0007329	0.15995	22.82	194.88	217.70	22.91	214.26	s237.17	0.09441	0.85307	0.94748
−20	132.82	0.0007362	0.14729	25.39	193.45	218.84	25.49	212.91	238.41	0.10463	0.84101	0.94564
−18	144.69	0.0007396	0.13583	27.98	192.01	219.98	28.09	211.55	239.64	0.11481	0.82908	0.94389
−16	157.38	0.0007430	0.12542	30.57	190.56	221.13	30.69	210.18	240.87	0.12493	0.81729	0.94222
−14	170.93	0.0007464	0.11597	33.17	189.09	222.27	33.30	208.79	242.09	0.13501	0.80561	0.94063
−12	185.37	0.0007499	0.10736	35.78	187.62	223.40	35.92	207.38	243.30	0.14504	0.79406	0.93911
−10	200.74	0.0007535	0.099516	38.40	186.14	224.54	38.55	205.96	244.51	0.15504	0.78263	0.93766
−8	217.08	0.0007571	0.092352	41.03	184.64	225.67	41.19	204.52	245.72	0.16498	0.77130	0.93629
−6	234.44	0.0007608	0.085802	43.66	183.13	226.80	43.84	203.07	246.91	0.17489	0.76008	0.93497
−4	252.85	0.0007646	0.079804	46.31	181.61	227.92	46.50	201.60	248.10	0.18476	0.74896	0.93372
−2	272.36	0.0007684	0.074304	48.96	180.08	229.04	49.17	200.11	249.28	0.19459	0.73794	0.93253
0	293.01	0.0007723	0.069255	51.63	178.53	230.16	51.86	198.60	250.45	0.20439	0.72701	0.93139
2	314.84	0.0007763	0.064612	54.30	176.97	231.27	54.55	197.07	251.61	0.21415	0.71616	0.93031
4	337.90	0.0007804	0.060338	56.99	175.39	232.38	57.25	195.51	252.77	0.22387	0.70540	0.92927
6	362.23	0.0007845	0.056398	59.68	173.80	233.48	59.97	193.94	253.91	0.23356	0.69471	0.92828
8	387.88	0.0007887	0.052762	62.39	172.19	234.58	62.69	192.35	255.04	0.24323	0.68410	0.92733
10	414.89	0.0007930	0.049403	65.10	170.56	235.67	65.43	190.73	256.16	0.25286	0.67356	0.92641
12	443.31	0.0007975	0.046295	67.83	168.92	236.75	68.18	189.09	257.27	0.26246	0.66308	0.92554
14	473.19	0.0008020	0.043417	70.57	167.26	237.83	70.95	187.42	258.37	0.27204	0.65266	0.92470
16	504.58	0.0008066	0.040748	73.32	165.58	238.90	73.73	185.73	259.46	0.28159	0.64230	0.92389
18	537.52	0.0008113	0.038271	76.08	163.88	239.96	76.52	184.01	260.53	0.29112	0.63198	0.92310

TABLE A–11

Saturated refrigerant-134a—Temperature table (*Concluded*)

		Specific volume, m^3/kg		*Internal energy,* kJ/kg			*Enthalpy,* kJ/kg			*Entropy,* kJ/kg·K		
Temp., T °C	Sat. press., P_{sat} kPa	Sat. liquid, v_f	Sat. vapor, v_g	Sat. liquid, u_f	Evap., u_{fg}	Sat. vapor, u_g	Sat. liquid, h_f	Evap., h_{fg}	Sat. vapor, h_g	Sat. liquid, s_f	Evap., s_{fg}	Sat. vapor, s_g
20	572.07	0.0008161	0.035969	78.86	162.16	241.02	79.32	182.27	261.59	0.30063	0.62172	0.92234
22	608.27	0.0008210	0.033828	81.64	160.42	242.06	82.14	180.49	262.64	0.31011	0.61149	0.92160
24	646.18	0.0008261	0.031834	84.44	158.65	243.10	84.98	178.69	263.67	0.31958	0.60130	0.92088
26	685.84	0.0008313	0.029976	87.26	156.87	244.12	87.83	176.85	264.68	0.32903	0.59115	0.92018
28	727.31	0.0008366	0.028242	90.09	155.05	245.14	90.69	174.99	265.68	0.33846	0.58102	0.91948
30	770.64	0.0008421	0.026622	92.93	153.22	246.14	93.58	173.08	266.66	0.34789	0.57091	0.91879
32	815.89	0.0008478	0.025108	95.79	151.35	247.14	96.48	171.14	267.62	0.35730	0.56082	0.91811
34	863.11	0.0008536	0.023691	98.66	149.46	248.12	99.40	169.17	268.57	0.36670	0.55074	0.91743
36	912.35	0.0008595	0.022364	101.55	147.54	249.08	102.33	167.16	269.49	0.37609	0.54066	0.91675
38	963.68	0.0008657	0.021119	104.45	145.58	250.04	105.29	165.10	270.39	0.38548	0.53058	0.91606
40	1017.1	0.0008720	0.019952	107.38	143.60	250.97	108.26	163.00	271.27	0.39486	0.52049	0.91536
42	1072.8	0.0008786	0.018855	110.32	141.58	251.89	111.26	160.86	272.12	0.40425	0.51039	0.91464
44	1130.7	0.0008854	0.017824	113.28	139.52	252.80	114.28	158.67	272.95	0.41363	0.50027	0.91391
46	1191.0	0.0008924	0.016853	116.26	137.42	253.68	117.32	156.43	273.75	0.42302	0.49012	0.91315
48	1253.6	0.0008996	0.015939	119.26	135.29	254.55	120.39	154.14	274.53	0.43242	0.47993	0.91236
52	1386.2	0.0009150	0.014265	125.33	130.88	256.21	126.59	149.39	275.98	0.45126	0.45941	0.91067
56	1529.1	0.0009317	0.012771	131.49	126.28	257.77	132.91	144.38	277.30	0.47018	0.43863	0.90880
60	1682.8	0.0009498	0.011434	137.76	121.46	259.22	139.36	139.10	278.46	0.48920	0.41749	0.90669
65	1891.0	0.0009750	0.009950	145.77	115.05	260.82	147.62	132.02	279.64	0.51320	0.39039	0.90359
70	2118.2	0.0010037	0.008642	154.01	108.14	262.15	156.13	124.32	280.46	0.53755	0.36227	0.89982
75	2365.8	0.0010372	0.007480	162.53	100.60	263.13	164.98	115.85	280.82	0.56241	0.33272	0.89512
80	2635.3	0.0010772	0.006436	171.40	92.23	263.63	174.24	106.35	280.59	0.58800	0.30111	0.88912
85	2928.2	0.0011270	0.005486	180.77	82.67	263.44	184.07	95.44	279.51	0.61473	0.26644	0.88117
90	3246.9	0.0011932	0.004599	190.89	71.29	262.18	194.76	82.35	277.11	0.64336	0.22674	0.87010
95	3594.1	0.0012933	0.003726	202.40	56.47	258.87	207.05	65.21	272.26	0.67578	0.17711	0.85289
100	3975.1	0.0015269	0.002630	218.72	29.19	247.91	224.79	33.58	258.37	0.72217	0.08999	0.81215

Source: Tables A–11 through A–13 are generated using the Engineering Equation Solver (EES) software developed by S. A. Klein and F. L. Alvarado. The routine used in calculations is the R134a, which is based on the fundamental equation of state developed by R. Tillner-Roth and H.D. Baehr, "An International Standard Formulation for the Thermodynamic Properties of 1,1,1,2-Tetrafluoroethane (HFC-134a) for temperatures from 170 K to 455 K and Pressures up to 70 MPa," *J. Phys. Chem, Ref. Data*, Vol. 23, No. 5, 1994. The enthalpy and entropy values of saturated liquid are set to zero at −40°C (and −40°F).

TABLE A–12

Saturated refrigerant-134a—Pressure table

		Specific volume, m³/kg		Internal energy, kJ/kg			Enthalpy, kJ/kg			Entropy, kJ/kg·K		
Press., P kPa	Sat. temp., T_{sat} °C	Sat. liquid, v_f	Sat. vapor, v_g	Sat. liquid, u_f	Evap., u_{fg}	Sat. vapor, u_g	Sat. liquid, h_f	Evap., h_{fg}	Sat. vapor, h_g	Sat. liquid, s_f	Evap., s_{fg}	Sat. vapor, s_g
60	−36.95	0.0007098	0.31121	3.798	205.32	209.12	3.841	223.95	227.79	0.01634	0.94807	0.96441
70	−33.87	0.0007144	0.26929	7.680	203.20	210.88	7.730	222.00	229.73	0.03267	0.92775	0.96042
80	−31.13	0.0007185	0.23753	11.15	201.30	212.46	11.21	220.25	231.46	0.04711	0.90999	0.95710
90	−28.65	0.0007223	0.21263	14.31	199.57	213.88	14.37	218.65	233.02	0.06008	0.89419	0.95427
100	−26.37	0.0007259	0.19254	17.21	197.98	215.19	17.28	217.16	234.44	0.07188	0.87995	0.95183
120	−22.32	0.0007324	0.16212	22.40	195.11	217.51	22.49	214.48	236.97	0.09275	0.85503	0.94779
140	−18.77	0.0007383	0.14014	26.98	192.57	219.54	27.08	212.08	239.16	0.11087	0.83368	0.94456
160	−15.60	0.0007437	0.12348	31.09	190.27	221.35	31.21	209.90	241.11	0.12693	0.81496	0.94190
180	−12.73	0.0007487	0.11041	34.83	188.16	222.99	34.97	207.90	242.86	0.14139	0.79826	0.93965
200	−10.09	0.0007533	0.099867	38.28	186.21	224.48	38.43	206.03	244.46	0.15457	0.78316	0.93773
240	−5.38	0.0007620	0.083897	44.48	182.67	227.14	44.66	202.62	247.28	0.17794	0.75664	0.93458
280	−1.25	0.0007699	0.072352	49.97	179.50	229.46	50.18	199.54	249.72	0.19829	0.73381	0.93210
320	2.46	0.0007772	0.063604	54.92	176.61	231.52	55.16	196.71	251.88	0.21637	0.71369	0.93006
360	5.82	0.0007841	0.056738	59.44	173.94	233.38	59.72	194.08	253.81	0.23270	0.69566	0.92836
400	8.91	0.0007907	0.051201	63.62	171.45	235.07	63.94	191.62	255.55	0.24761	0.67929	0.92691
450	12.46	0.0007985	0.045619	68.45	168.54	237.00	68.81	188.71	257.53	0.26465	0.66069	0.92535
500	15.71	0.0008059	0.041118	72.93	165.82	238.75	73.33	185.98	259.30	0.28023	0.64377	0.92400
550	18.73	0.0008130	0.037408	77.10	163.25	240.35	77.54	183.38	260.92	0.29461	0.62821	0.92282
600	21.55	0.0008199	0.034295	81.02	160.81	241.83	81.51	180.90	262.40	0.30799	0.61378	0.92177
650	24.20	0.0008266	0.031646	84.72	158.48	243.20	85.26	178.51	263.77	0.32051	0.60030	0.92081
700	26.69	0.0008331	0.029361	88.24	156.24	244.48	88.82	176.21	265.03	0.33230	0.58763	0.91994
750	29.06	0.0008395	0.027371	91.59	154.08	245.67	92.22	173.98	266.20	0.34345	0.57567	0.91912
800	31.31	0.0008458	0.025621	94.79	152.00	246.79	95.47	171.82	267.29	0.35404	0.56431	0.91835
850	33.45	0.0008520	0.024069	97.87	149.98	247.85	98.60	169.71	268.31	0.36413	0.55349	0.91762
900	35.51	0.0008580	0.022683	100.83	148.01	248.85	101.61	167.66	269.26	0.37377	0.54315	0.91692
950	37.48	0.0008641	0.021438	103.69	146.10	249.79	104.51	165.64	270.15	0.38301	0.53323	0.91624
1000	39.37	0.0008700	0.020313	106.45	144.23	250.68	107.32	163.67	270.99	0.39189	0.52368	0.91558
1200	46.29	0.0008934	0.016715	116.70	137.11	253.81	117.77	156.10	273.87	0.42441	0.48863	0.91303
1400	52.40	0.0009166	0.014107	125.94	130.43	256.37	127.22	148.90	276.12	0.45315	0.45734	0.91050
1600	57.88	0.0009400	0.012123	134.43	124.04	258.47	135.93	141.93	277.86	0.47911	0.42873	0.90784
1800	62.87	0.0009639	0.010559	142.33	117.83	260.17	144.07	135.11	279.17	0.50294	0.40204	0.90498
2000	67.45	0.0009886	0.009288	149.78	111.73	261.51	151.76	128.33	280.09	0.52509	0.37675	0.90184
2500	77.54	0.0010566	0.006936	166.99	96.47	263.45	169.63	111.16	280.79	0.57531	0.31695	0.89226
3000	86.16	0.0011406	0.005275	183.04	80.22	263.26	186.46	92.63	279.09	0.62118	0.25776	0.87894

TABLE A–13

Superheated refrigerant-134a

T °C	v m³/kg	u kJ/kg	h kJ/kg	s kJ/kg·K	v m³/kg	u kJ/kg	h kJ/kg	s kJ/kg·K	v m³/kg	u kJ/kg	h kJ/kg	s kJ/kg·K
	P = 0.06 MPa (T_{sat} = −36.95°C)				P = 0.10 MPa (T_{sat} = −26.37°C)				P = 0.14 MPa (T_{sat} = −18.77°C)			
Sat.	0.31121	209.12	227.79	0.9644	0.19254	215.19	234.44	0.9518	0.14014	219.54	239.16	0.9446
−20	0.33608	220.60	240.76	1.0174	0.19841	219.66	239.50	0.9721				
−10	0.35048	227.55	248.58	1.0477	0.20743	226.75	247.49	1.0030	0.14605	225.91	246.36	0.9724
0	0.36476	234.66	256.54	1.0774	0.21630	233.95	255.58	1.0332	0.15263	233.23	254.60	1.0031
10	0.37893	241.92	264.66	1.1066	0.22506	241.30	263.81	1.0628	0.15908	240.66	262.93	1.0331
20	0.39302	249.35	272.94	1.1353	0.23373	248.79	272.17	1.0918	0.16544	248.22	271.38	1.0624
30	0.40705	256.95	281.37	1.1636	0.24233	256.44	280.68	1.1203	0.17172	255.93	279.97	1.0912
40	0.42102	264.71	289.97	1.1915	0.25088	264.25	289.34	1.1484	0.17794	263.79	288.70	1.1195
50	0.43495	272.64	298.74	1.2191	0.25937	272.22	298.16	1.1762	0.18412	271.79	297.57	1.1474
60	0.44883	280.73	307.66	1.2463	0.26783	280.35	307.13	1.2035	0.19025	279.96	306.59	1.1749
70	0.46269	288.99	316.75	1.2732	0.27626	288.64	316.26	1.2305	0.19635	288.28	315.77	1.2020
80	0.47651	297.41	326.00	1.2997	0.28465	297.08	325.55	1.2572	0.20242	296.75	325.09	1.2288
90	0.49032	306.00	335.42	1.3260	0.29303	305.69	334.99	1.2836	0.20847	305.38	334.57	1.2553
100	0.50410	314.74	344.99	1.3520	0.30138	314.46	344.60	1.3096	0.21449	314.17	344.20	1.2814
	P = 0.18 MPa (T_{sat} = −12.73°C)				P = 0.20 MPa (T_{sat} = −10.09°C)				P = 0.24 MPa (T_{sat} = −5.38°C)			
Sat.	0.11041	222.99	242.86	0.9397	0.09987	224.48	244.46	0.9377	0.08390	227.14	247.28	0.9346
−10	0.11189	225.02	245.16	0.9484	0.09991	224.55	244.54	0.9380				
0	0.11722	232.48	253.58	0.9798	0.10481	232.09	253.05	0.9698	0.08617	231.29	251.97	0.9519
10	0.12240	240.00	262.04	1.0102	0.10955	239.67	261.58	1.0004	0.09026	238.98	260.65	0.9831
20	0.12748	247.64	270.59	1.0399	0.11418	247.35	270.18	1.0303	0.09423	246.74	269.36	1.0134
30	0.13248	255.41	279.25	1.0690	0.11874	255.14	278.89	1.0595	0.09812	254.61	278.16	1.0429
40	0.13741	263.31	288.05	1.0975	0.12322	263.08	287.72	1.0882	0.10193	262.59	287.06	1.0718
50	0.14230	271.36	296.98	1.1256	0.12766	271.15	296.68	1.1163	0.10570	270.71	296.08	1.1001
60	0.14715	279.56	306.05	1.1532	0.13206	279.37	305.78	1.1441	0.10942	278.97	305.23	1.1280
70	0.15196	287.91	315.27	1.1805	0.13641	287.73	315.01	1.1714	0.11310	287.36	314.51	1.1554
80	0.15673	296.42	324.63	1.2074	0.14074	296.25	324.40	1.1983	0.11675	295.91	323.93	1.1825
90	0.16149	305.07	334.14	1.2339	0.14504	304.92	333.93	1.2249	0.12038	304.60	333.49	1.2092
100	0.16622	313.88	343.80	1.2602	0.14933	313.74	343.60	1.2512	0.12398	313.44	343.20	1.2356
	P = 0.28 MPa (T_{sat} = −1.25°C)				P = 0.32 MPa (T_{sat} = 2.46°C)				P = 0.40 MPa (T_{sat} = 8.91°C)			
Sat.	0.07235	229.46	249.72	0.9321	0.06360	231.52	251.88	0.9301	0.051201	235.07	255.55	0.9269
0	0.07282	230.44	250.83	0.9362								
10	0.07646	238.27	259.68	0.9680	0.06609	237.54	258.69	0.9544	0.051506	235.97	256.58	0.9305
20	0.07997	246.13	268.52	0.9987	0.06925	245.50	267.66	0.9856	0.054213	244.18	265.86	0.9628
30	0.08338	254.06	277.41	1.0285	0.07231	253.50	276.65	1.0157	0.056796	252.36	275.07	0.9937
40	0.08672	262.10	286.38	1.0576	0.07530	261.60	285.70	1.0451	0.059292	260.58	284.30	1.0236
50	0.09000	270.27	295.47	1.0862	0.07823	269.82	294.85	1.0739	0.061724	268.90	293.59	1.0528
60	0.09324	278.56	304.67	1.1142	0.08111	278.15	304.11	1.1021	0.064104	277.32	302.96	1.0814
70	0.09644	286.99	314.00	1.1418	0.08395	286.62	313.48	1.1298	0.066443	285.86	312.44	1.1094
80	0.09961	295.57	323.46	1.1690	0.08675	295.22	322.98	1.1571	0.068747	294.53	322.02	1.1369
90	0.10275	304.29	333.06	1.1958	0.08953	303.97	332.62	1.1840	0.071023	303.32	331.73	1.1640
100	0.10587	313.15	342.80	1.2222	0.09229	312.86	342.39	1.2105	0.073274	312.26	341.57	1.1907
110	0.10897	322.16	352.68	1.2483	0.09503	321.89	352.30	1.2367	0.075504	321.33	351.53	1.2171
120	0.11205	331.32	362.70	1.2742	0.09775	331.07	362.35	1.2626	0.077717	330.55	361.63	1.2431
130	0.11512	340.63	372.87	1.2997	0.10045	340.39	372.54	1.2882	0.079913	339.90	371.87	1.2688
140	0.11818	350.09	383.18	1.3250	0.10314	349.86	382.87	1.3135	0.082096	349.41	382.24	1.2942

TABLE A–13

Superheated refrigerant-134a (*Continued*)

T °C	*v* m³/kg	*u* kJ/kg	*h* kJ/kg	*s* kJ/kg·K	*v* m³/kg	*u* kJ/kg	*h* kJ/kg	*s* kJ/kg·K	*v* m³/kg	*u* kJ/kg	*h* kJ/kg	*s* kJ/kg·K
	P = 0.50 MPa (T_{sat} = 15.71°C)				P = 0.60 MPa (T_{sat} = 21.55°C)				P = 0.70 MPa (T_{sat} = 26.69°C)			
Sat.	0.041118	238.75	259.30	0.9240	0.034295	241.83	262.40	0.9218	0.029361	244.48	265.03	0.9199
20	0.042115	242.40	263.46	0.9383								
30	0.044338	250.84	273.01	0.9703	0.035984	249.22	270.81	0.9499	0.029966	247.48	268.45	0.9313
40	0.046456	259.26	282.48	1.0011	0.037865	257.86	280.58	0.9816	0.031696	256.39	278.57	0.9641
50	0.048499	267.72	291.96	1.0309	0.039659	266.48	290.28	1.0121	0.033322	265.20	288.53	0.9954
60	0.050485	276.25	301.50	1.0599	0.041389	275.15	299.98	1.0417	0.034875	274.01	298.42	1.0256
70	0.052427	284.89	311.10	1.0883	0.043069	283.89	309.73	1.0705	0.036373	282.87	308.33	1.0549
80	0.054331	293.64	320.80	1.1162	0.044710	292.73	319.55	1.0987	0.037829	291.80	318.28	1.0835
90	0.056205	302.51	330.61	1.1436	0.046318	301.67	329.46	1.1264	0.039250	300.82	328.29	1.1114
100	0.058053	311.50	340.53	1.1705	0.047900	310.73	339.47	1.1536	0.040642	309.95	338.40	1.1389
110	0.059880	320.63	350.57	1.1971	0.049458	319.91	349.59	1.1803	0.042010	319.19	348.60	1.1658
120	0.061687	329.89	360.73	1.2233	0.050997	329.23	359.82	1.2067	0.043358	328.55	358.90	1.1924
130	0.063479	339.29	371.03	1.2491	0.052519	338.67	370.18	1.2327	0.044688	338.04	369.32	1.2186
140	0.065256	348.83	381.46	1.2747	0.054027	348.25	380.66	1.2584	0.046004	347.66	379.86	1.2444
150	0.067021	358.51	392.02	1.2999	0.055522	357.96	391.27	1.2838	0.047306	357.41	390.52	1.2699
160	0.068775	368.33	402.72	1.3249	0.057006	367.81	402.01	1.3088	0.048597	367.29	401.31	1.2951
	P = 0.80 MPa (T_{sat} = 31.31°C)				P = 0.90 MPa (T_{sat} = 35.51°C)				P = 1.00 MPa (T_{sat} = 39.37°C)			
Sat.	0.025621	246.79	267.29	0.9183	0.022683	248.85	269.26	0.9169	0.020313	250.68	270.99	0.9156
40	0.027035	254.82	276.45	0.9480	0.023375	253.13	274.17	0.9327	0.020406	251.30	271.71	0.9179
50	0.028547	263.86	286.69	0.9802	0.024809	262.44	284.77	0.9660	0.021796	260.94	282.74	0.9525
60	0.029973	272.83	296.81	1.0110	0.026146	271.60	295.13	0.9976	0.023068	270.32	293.38	0.9850
70	0.031340	281.81	306.88	1.0408	0.027413	280.72	305.39	1.0280	0.024261	279.59	303.85	1.0160
80	0.032659	290.84	316.97	1.0698	0.028630	289.86	315.63	1.0574	0.025398	288.86	314.25	1.0458
90	0.033941	299.95	327.10	1.0981	0.029806	299.06	325.89	1.0860	0.026492	298.15	324.64	1.0748
100	0.035193	309.15	337.30	1.1258	0.030951	308.34	336.19	1.1140	0.027552	307.51	335.06	1.1031
110	0.036420	318.45	347.59	1.1530	0.032068	317.70	346.56	1.1414	0.028584	316.94	345.53	1.1308
120	0.037625	327.87	357.97	1.1798	0.033164	327.18	357.02	1.1684	0.029592	326.47	356.06	1.1580
130	0.038813	337.40	368.45	1.2061	0.034241	336.76	367.58	1.1949	0.030581	336.11	366.69	1.1846
140	0.039985	347.06	379.05	1.2321	0.035302	346.46	378.23	1.2210	0.031554	345.85	377.40	1.2109
150	0.041143	356.85	389.76	1.2577	0.036349	356.28	389.00	1.2467	0.032512	355.71	388.22	1.2368
160	0.042290	366.76	400.59	1.2830	0.037384	366.23	399.88	1.2721	0.033457	365.70	399.15	1.2623
170	0.043427	376.81	411.55	1.3080	0.038408	376.31	410.88	1.2972	0.034392	375.81	410.20	1.2875
180	0.044554	386.99	422.64	1.3327	0.039423	386.52	422.00	1.3221	0.035317	386.04	421.36	1.3124
	P = 1.20 MPa (T_{sat} = 46.29°C)				P = 1.40 MPa (T_{sat} = 52.40°C)				P = 1.60 MPa (T_{sat} = 57.88°C)			
Sat.	0.016715	253.81	273.87	0.9130	0.014107	256.37	276.12	0.9105	0.012123	258.47	277.86	0.9078
50	0.017201	257.63	278.27	0.9267								
60	0.018404	267.56	289.64	0.9614	0.015005	264.46	285.47	0.9389	0.012372	260.89	280.69	0.9163
70	0.019502	277.21	300.61	0.9938	0.016060	274.62	297.10	0.9733	0.013430	271.76	293.25	0.9535
80	0.020529	286.75	311.39	1.0248	0.017023	284.51	308.34	1.0056	0.014362	282.09	305.07	0.9875
90	0.021506	296.26	322.07	1.0546	0.017923	294.28	319.37	1.0364	0.015215	292.17	316.52	1.0194
100	0.022442	305.80	332.73	1.0836	0.018778	304.01	330.30	1.0661	0.016014	302.14	327.76	1.0500
110	0.023348	315.38	343.40	1.1118	0.019597	313.76	341.19	1.0949	0.016773	312.07	338.91	1.0795
120	0.024228	325.03	354.11	1.1394	0.020388	323.55	352.09	1.1230	0.017500	322.02	350.02	1.1081
130	0.025086	334.77	364.88	1.1664	0.021155	333.41	363.02	1.1504	0.018201	332.00	361.12	1.1360
140	0.025927	344.61	375.72	1.1930	0.021904	343.34	374.01	1.1773	0.018882	342.05	372.26	1.1632
150	0.026753	354.56	386.66	1.2192	0.022636	353.37	385.07	1.2038	0.019545	352.17	383.44	1.1900
160	0.027566	364.61	397.69	1.2449	0.023355	363.51	396.20	1.2298	0.020194	362.38	394.69	1.2163
170	0.028367	374.78	408.82	1.2703	0.024061	373.75	407.43	1.2554	0.020830	372.69	406.02	1.2421
180	0.029158	385.08	420.07	1.2954	0.024757	384.10	418.76	1.2807	0.021456	383.11	417.44	1.2676

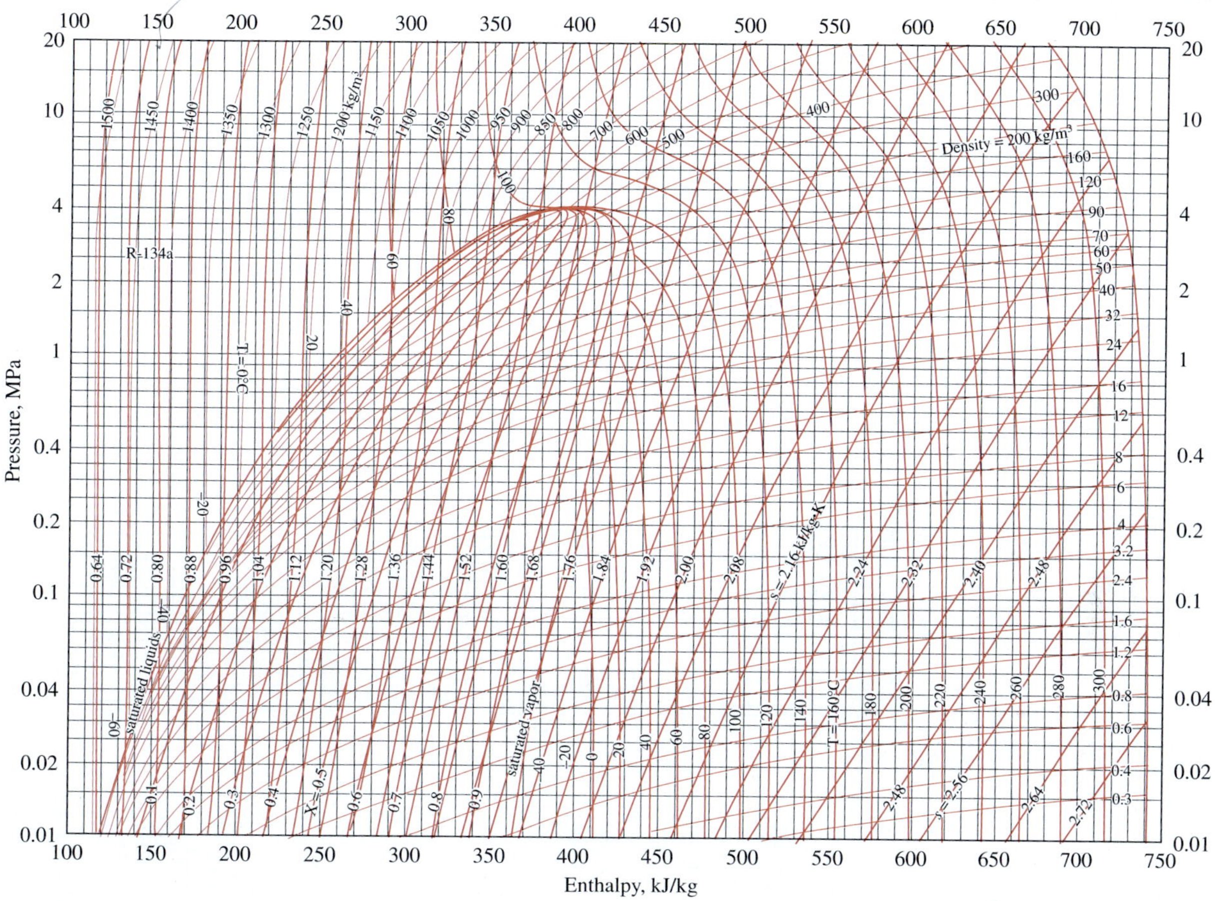

FIGURE A–14

P-h diagram for refrigerant-134a.

Note: The reference point used for the chart is different than that used in the R-134a tables. Therefore, problems should be solved using all property data either from the tables or from the chart, but not from both.

Reprinted by permission of American Society of Heating, Refrigerating, and Air-Conditioning Engineers, Inc., Atlanta, GA.

(*a*) Low pressures, $0 < P_R < 1.0$

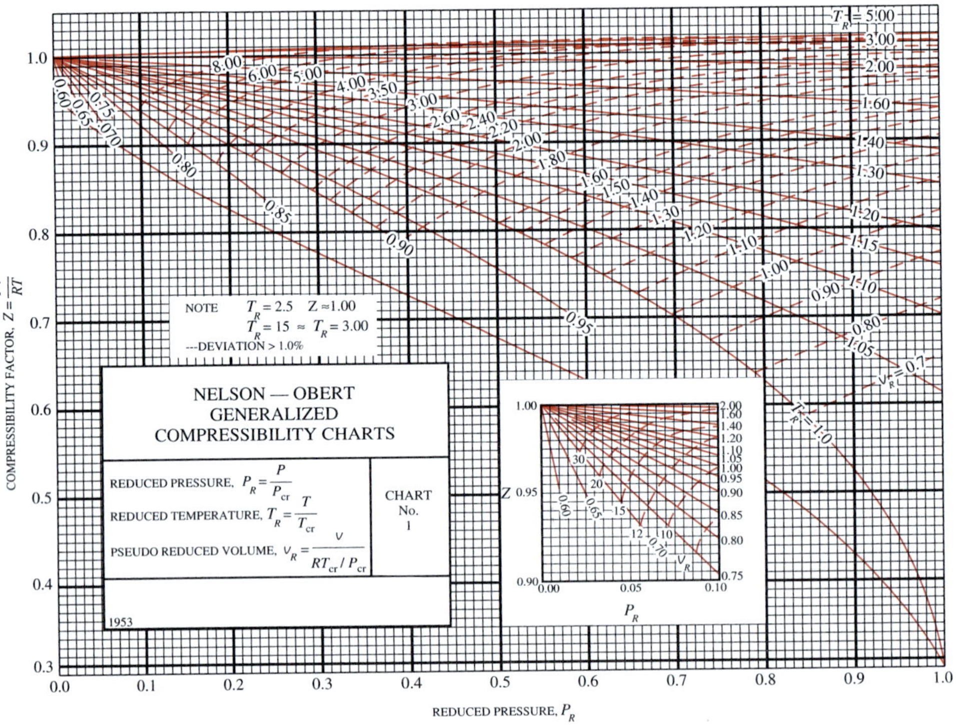

(*b*) Intermediate pressures, $0 < P_R < 7$

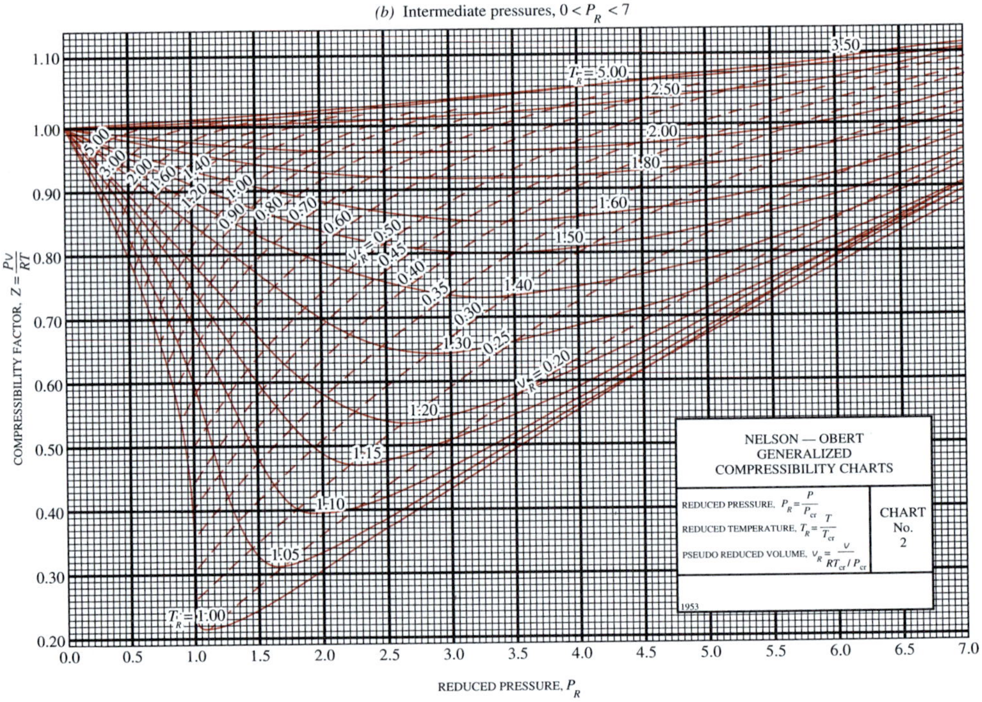

FIGURE A–15

Nelson–Obert generalized compressibility chart.

Used with permission of Dr. Edward E. Obert, University of Wisconsin.

TABLE A–16

Properties of the atmosphere at high altitude

Altitude, m	Temperature, °C	Pressure, kPa	Gravity g, m/s^2	Speed of Sound, m/s	Density, kg/m^3	Viscosity μ, kg/m·s	Thermal Conductivity, W/m·K
0	15.00	101.33	9.807	340.3	1.225	1.789×10^{-5}	0.0253
200	13.70	98.95	9.806	339.5	1.202	1.783×10^{-5}	0.0252
400	12.40	96.61	9.805	338.8	1.179	1.777×10^{-5}	0.0252
600	11.10	94.32	9.805	338.0	1.156	1.771×10^{-5}	0.0251
800	9.80	92.08	9.804	337.2	1.134	1.764×10^{-5}	0.0250
1000	8.50	89.88	9.804	336.4	1.112	1.758×10^{-5}	0.0249
1200	7.20	87.72	9.803	335.7	1.090	1.752×10^{-5}	0.0248
1400	5.90	85.60	9.802	334.9	1.069	1.745×10^{-5}	0.0247
1600	4.60	83.53	9.802	334.1	1.048	1.739×10^{-5}	0.0245
1800	3.30	81.49	9.801	333.3	1.027	1.732×10^{-5}	0.0244
2000	2.00	79.50	9.800	332.5	1.007	1.726×10^{-5}	0.0243
2200	0.70	77.55	9.800	331.7	0.987	1.720×10^{-5}	0.0242
2400	−0.59	75.63	9.799	331.0	0.967	1.713×10^{-5}	0.0241
2600	−1.89	73.76	9.799	330.2	0.947	1.707×10^{-5}	0.0240
2800	−3.19	71.92	9.798	329.4	0.928	1.700×10^{-5}	0.0239
3000	−4.49	70.12	9.797	328.6	0.909	1.694×10^{-5}	0.0238
3200	−5.79	68.36	9.797	327.8	0.891	1.687×10^{-5}	0.0237
3400	−7.09	66.63	9.796	327.0	0.872	1.681×10^{-5}	0.0236
3600	−8.39	64.94	9.796	326.2	0.854	1.674×10^{-5}	0.0235
3800	−9.69	63.28	9.795	325.4	0.837	1.668×10^{-5}	0.0234
4000	−10.98	61.66	9.794	324.6	0.819	1.661×10^{-5}	0.0233
4200	−12.3	60.07	9.794	323.8	0.802	1.655×10^{-5}	0.0232
4400	−13.6	58.52	9.793	323.0	0.785	1.648×10^{-5}	0.0231
4600	−14.9	57.00	9.793	322.2	0.769	1.642×10^{-5}	0.0230
4800	−16.2	55.51	9.792	321.4	0.752	1.635×10^{-5}	0.0229
5000	−17.5	54.05	9.791	320.5	0.736	1.628×10^{-5}	0.0228
5200	−18.8	52.62	9.791	319.7	0.721	1.622×10^{-5}	0.0227
5400	−20.1	51.23	9.790	318.9	0.705	1.615×10^{-5}	0.0226
5600	−21.4	49.86	9.789	318.1	0.690	1.608×10^{-5}	0.0224
5800	−22.7	48.52	9.785	317.3	0.675	1.602×10^{-5}	0.0223
6000	−24.0	47.22	9.788	316.5	0.660	1.595×10^{-5}	0.0222
6200	−25.3	45.94	9.788	315.6	0.646	1.588×10^{-5}	0.0221
6400	−26.6	44.69	9.787	314.8	0.631	1.582×10^{-5}	0.0220
6600	−27.9	43.47	9.786	314.0	0.617	1.575×10^{-5}	0.0219
6800	−29.2	42.27	9.785	313.1	0.604	1.568×10^{-5}	0.0218
7000	−30.5	41.11	9.785	312.3	0.590	1.561×10^{-5}	0.0217
8000	−36.9	35.65	9.782	308.1	0.526	1.527×10^{-5}	0.0212
9000	−43.4	30.80	9.779	303.8	0.467	1.493×10^{-5}	0.0206
10,000	−49.9	26.50	9.776	299.5	0.414	1.458×10^{-5}	0.0201
12,000	−56.5	19.40	9.770	295.1	0.312	1.422×10^{-5}	0.0195
14,000	−56.5	14.17	9.764	295.1	0.228	1.422×10^{-5}	0.0195
16,000	−56.5	10.53	9.758	295.1	0.166	1.422×10^{-5}	0.0195
18,000	−56.5	7.57	9.751	295.1	0.122	1.422×10^{-5}	0.0195

Source: U.S. Standard Atmosphere Supplements, U.S. Government Printing Office, 1966. Based on year-round mean conditions at 45° latitude and varies with the time of the year and the weather patterns. The conditions at sea level ($z = 0$) are taken to be $P = 101.325$ kPa, $T = 15$°C, $\rho = 1.2250$ kg/m^3, $g = 9.80665$ m^2/s.

TABLE A–17

Ideal-gas properties of air

T K	h kJ/kg	P_r	u kJ/kg	v_r	s° kJ/kg·K	T K	h kJ/kg	P_r	u kJ/kg	v_r	s° kJ/kg·K
200	199.97	0.3363	142.56	1707.0	1.29559	580	586.04	14.38	419.55	115.7	2.37348
210	209.97	0.3987	149.69	1512.0	1.34444	590	596.52	15.31	427.15	110.6	2.39140
220	219.97	0.4690	156.82	1346.0	1.39105	600	607.02	16.28	434.78	105.8	2.40902
230	230.02	0.5477	164.00	1205.0	1.43557	610	617.53	17.30	442.42	101.2	2.42644
240	240.02	0.6355	171.13	1084.0	1.47824	620	628.07	18.36	450.09	96.92	2.44356
250	250.05	0.7329	178.28	979.0	1.51917	630	638.63	19.84	457.78	92.84	2.46048
260	260.09	0.8405	185.45	887.8	1.55848	640	649.22	20.64	465.50	88.99	2.47716
270	270.11	0.9590	192.60	808.0	1.59634	650	659.84	21.86	473.25	85.34	2.49364
280	280.13	1.0889	199.75	738.0	1.63279	660	670.47	23.13	481.01	81.89	2.50985
285	285.14	1.1584	203.33	706.1	1.65055	670	681.14	24.46	488.81	78.61	2.52589
290	290.16	1.2311	206.91	676.1	1.66802	680	691.82	25.85	496.62	75.50	2.54175
295	295.17	1.3068	210.49	647.9	1.68515	690	702.52	27.29	504.45	72.56	2.55731
298	298.18	1.3543	212.64	631.9	1.69528	700	713.27	28.80	512.33	69.76	2.57277
300	300.19	1.3860	214.07	621.2	1.70203	710	724.04	30.38	520.23	67.07	2.58810
305	305.22	1.4686	217.67	596.0	1.71865	720	734.82	32.02	528.14	64.53	2.60319
310	310.24	1.5546	221.25	572.3	1.73498	730	745.62	33.72	536.07	62.13	2.61803
315	315.27	1.6442	224.85	549.8	1.75106	740	756.44	35.50	544.02	59.82	2.63280
320	320.29	1.7375	228.42	528.6	1.76690	750	767.29	37.35	551.99	57.63	2.64737
325	325.31	1.8345	232.02	508.4	1.78249	760	778.18	39.27	560.01	55.54	2.66176
330	330.34	1.9352	235.61	489.4	1.79783	780	800.03	43.35	576.12	51.64	2.69013
340	340.42	2.149	242.82	454.1	1.82790	800	821.95	47.75	592.30	48.08	2.71787
350	350.49	2.379	250.02	422.2	1.85708	820	843.98	52.59	608.59	44.84	2.74504
360	360.58	2.626	257.24	393.4	1.88543	840	866.08	57.60	624.95	41.85	2.77170
370	370.67	2.892	264.46	367.2	1.91313	860	888.27	63.09	641.40	39.12	2.79783
380	380.77	3.176	271.69	343.4	1.94001	880	910.56	68.98	657.95	36.61	2.82344
390	390.88	3.481	278.93	321.5	1.96633	900	932.93	75.29	674.58	34.31	2.84856
400	400.98	3.806	286.16	301.6	1.99194	920	955.38	82.05	691.28	32.18	2.87324
410	411.12	4.153	293.43	283.3	2.01699	940	977.92	89.28	708.08	30.22	2.89748
420	421.26	4.522	300.69	266.6	2.04142	960	1000.55	97.00	725.02	28.40	2.92128
430	431.43	4.915	307.99	251.1	2.06533	980	1023.25	105.2	741.98	26.73	2.94468
440	441.61	5.332	315.30	236.8	2.08870	1000	1046.04	114.0	758.94	25.17	2.96770
450	451.80	5.775	322.62	223.6	2.11161	1020	1068.89	123.4	776.10	23.72	2.99034
460	462.02	6.245	329.97	211.4	2.13407	1040	1091.85	133.3	793.36	23.29	3.01260
470	472.24	6.742	337.32	200.1	2.15604	1060	1114.86	143.9	810.62	21.14	3.03449
480	482.49	7.268	344.70	189.5	2.17760	1080	1137.89	155.2	827.88	19.98	3.05608
490	492.74	7.824	352.08	179.7	2.19876	1100	1161.07	167.1	845.33	18.896	3.07732
500	503.02	8.411	359.49	170.6	2.21952	1120	1184.28	179.7	862.79	17.886	3.09825
510	513.32	9.031	366.92	162.1	2.23993	1140	1207.57	193.1	880.35	16.946	3.11883
520	523.63	9.684	374.36	154.1	2.25997	1160	1230.92	207.2	897.91	16.064	3.13916
530	533.98	10.37	381.84	146.7	2.27967	1180	1254.34	222.2	915.57	15.241	3.15916
540	544.35	11.10	389.34	139.7	2.29906	1200	1277.79	238.0	933.33	14.470	3.17888
550	555.74	11.86	396.86	133.1	2.31809	1220	1301.31	254.7	951.09	13.747	3.19834
560	565.17	12.66	404.42	127.0	2.33685	1240	1324.93	272.3	968.95	13.069	3.21751
570	575.59	13.50	411.97	121.2	2.35531						

TABLE A–17

Ideal-gas properties of air (*Concluded*)

T K	h kJ/kg	P_r	u kJ/kg	v_r	$s°$ kJ/kg·K	T K	h kJ/kg	P_r	u kJ/kg	v_r	$s°$ kJ/kg·K
1260	1348.55	290.8	986.90	12.435	3.23638	1600	1757.57	791.2	1298.30	5.804	3.52364
1280	1372.24	310.4	1004.76	11.835	3.25510	1620	1782.00	834.1	1316.96	5.574	3.53879
1300	1395.97	330.9	1022.82	11.275	3.27345	1640	1806.46	878.9	1335.72	5.355	3.55381
1320	1419.76	352.5	1040.88	10.747	3.29160	1660	1830.96	925.6	1354.48	5.147	3.56867
1340	1443.60	375.3	1058.94	10.247	3.30959	1680	1855.50	974.2	1373.24	4.949	3.58335
1360	1467.49	399.1	1077.10	9.780	3.32724	1700	1880.1	1025	1392.7	4.761	3.5979
1380	1491.44	424.2	1095.26	9.337	3.34474	1750	1941.6	1161	1439.8	4.328	3.6336
1400	1515.42	450.5	1113.52	8.919	3.36200	1800	2003.3	1310	1487.2	3.994	3.6684
1420	1539.44	478.0	1131.77	8.526	3.37901	1850	2065.3	1475	1534.9	3.601	3.7023
1440	1563.51	506.9	1150.13	8.153	3.39586	1900	2127.4	1655	1582.6	3.295	3.7354
1460	1587.63	537.1	1168.49	7.801	3.41247	1950	2189.7	1852	1630.6	3.022	3.7677
1480	1611.79	568.8	1186.95	7.468	3.42892	2000	2252.1	2068	1678.7	2.776	3.7994
1500	1635.97	601.9	1205.41	7.152	3.44516	2050	2314.6	2303	1726.8	2.555	3.8303
1520	1660.23	636.5	1223.87	6.854	3.46120	2100	2377.7	2559	1775.3	2.356	3.8605
1540	1684.51	672.8	1242.43	6.569	3.47712	2150	2440.3	2837	1823.8	2.175	3.8901
1560	1708.82	710.5	1260.99	6.301	3.49276	2200	2503.2	3138	1872.4	2.012	3.9191
1580	1733.17	750.0	1279.65	6.046	3.50829	2250	2566.4	3464	1921.3	1.864	3.9474

Note: The properties P_r (relative pressure) and v_r (relative specific volume) are dimensionless quantities used in the analysis of isentropic processes, and should not be confused with the properties pressure and specific volume.

Source: Kenneth Wark, *Thermodynamics,* 4th ed. (New York: McGraw-Hill, 1983), pp. 785–86, table A–5. Originally published in J. H. Keenan and J. Kaye, *Gas Tables* (New York: John Wiley & Sons, 1948).

TABLE A–18

Ideal-gas properties of nitrogen, N_2

T K	$\bar{h}$ kJ/kmol	$\bar{u}$ kJ/kmol	$\bar{s}°$ kJ/kmol·K	T K	$\bar{h}$ kJ/kmol	$\bar{u}$ kJ/kmol	$\bar{s}°$ kJ/kmol·K
0	0	0	0	600	17,563	12,574	212.066
220	6,391	4,562	182.639	610	17,864	12,792	212.564
230	6,683	4,770	183.938	620	18,166	13,011	213.055
240	6,975	4,979	185.180	630	18,468	13,230	213.541
250	7,266	5,188	186.370	640	18,772	13,450	214.018
260	7,558	5,396	187.514	650	19,075	13,671	214.489
270	7,849	5,604	188.614	660	19,380	13,892	214.954
280	8,141	5,813	189.673	670	19,685	14,114	215.413
290	8,432	6,021	190.695	680	19,991	14,337	215.866
298	8,669	6,190	191.502	690	20,297	14,560	216.314
300	8,723	6,229	191.682	700	20,604	14,784	216.756
310	9,014	6,437	192.638	710	20,912	15,008	217.192
320	9,306	6,645	193.562	720	21,220	15,234	217.624
330	9,597	6,853	194.459	730	21,529	15,460	218.059
340	9,888	7,061	195.328	740	21,839	15,686	218.472
350	10,180	7,270	196.173	750	22,149	15,913	218.889
360	10,471	7,478	196.995	760	22,460	16,141	219.301
370	10,763	7,687	197.794	770	22,772	16,370	219.709
380	11,055	7,895	198.572	780	23,085	16,599	220.113
390	11,347	8,104	199.331	790	23,398	16,830	220.512
400	11,640	8,314	200.071	800	23,714	17,061	220.907
410	11,932	8,523	200.794	810	24,027	17,292	221.298
420	12,225	8,733	201.499	820	24,342	17,524	221.684
430	12,518	8,943	202.189	830	24,658	17,757	222.067
440	12,811	9,153	202.863	840	24,974	17,990	222.447
450	13,105	9,363	203.523	850	25,292	18,224	222.822
460	13,399	9,574	204.170	860	25,610	18,459	223.194
470	13,693	9,786	204.803	870	25,928	18,695	223.562
480	13,988	9,997	205.424	880	26,248	18,931	223.927
490	14,285	10,210	206.033	890	26,568	19,168	224.288
500	14,581	10,423	206.630	900	26,890	19,407	224.647
510	14,876	10,635	207.216	910	27,210	19,644	225.002
520	15,172	10,848	207.792	920	27,532	19,883	225.353
530	15,469	11,062	208.358	930	27,854	20,122	225.701
540	15,766	11,277	208.914	940	28,178	20,362	226.047
550	16,064	11,492	209.461	950	28,501	20,603	226.389
560	16,363	11,707	209.999	960	28,826	20,844	226.728
570	16,662	11,923	210.528	970	29,151	21,086	227.064
580	16,962	12,139	211.049	980	29,476	21,328	227.398
590	17,262	12,356	211.562	990	29,803	21,571	227.728

TABLE A–18

Ideal-gas properties of nitrogen, N_2 (*Concluded*)

T K	$\bar{h}$ kJ/kmol	$\bar{u}$ kJ/kmol	$\bar{s}°$ kJ/kmol·K	T K	$\bar{h}$ kJ/kmol	$\bar{u}$ kJ/kmol	$\bar{s}°$ kJ/kmol·K
1000	30,129	21,815	228.057	1760	56,227	41,594	247.396
1020	30,784	22,304	228.706	1780	56,938	42,139	247.798
1040	31,442	22,795	229.344	1800	57,651	42,685	248.195
1060	32,101	23,288	229.973	1820	58,363	43,231	248.589
1080	32,762	23,782	230.591	1840	59,075	43,777	248.979
1100	33,426	24,280	231.199	1860	59,790	44,324	249.365
1120	34,092	24,780	231.799	1880	60,504	44,873	249.748
1140	34,760	25,282	232.391	1900	61,220	45,423	250.128
1160	35,430	25,786	232.973	1920	61,936	45,973	250.502
1180	36,104	26,291	233.549	1940	62,654	46,524	250.874
1200	36,777	26,799	234.115	1960	63,381	47,075	251.242
1220	37,452	27,308	234.673	1980	64,090	47,627	251.607
1240	38,129	27,819	235.223	2000	64,810	48,181	251.969
1260	38,807	28,331	235.766	2050	66,612	49,567	252.858
1280	39,488	28,845	236.302	2100	68,417	50,957	253.726
1300	40,170	29,361	236.831	2150	70,226	52,351	254.578
1320	40,853	29,378	237.353	2200	72,040	53,749	255.412
1340	41,539	30,398	237.867	2250	73,856	55,149	256.227
1360	42,227	30,919	238.376	2300	75,676	56,553	257.027
1380	42,915	31,441	238.878	2350	77,496	57,958	257.810
1400	43,605	31,964	239.375	2400	79,320	59,366	258.580
1420	44,295	32,489	239.865	2450	81,149	60,779	259.332
1440	44,988	33,014	240.350	2500	82,981	62,195	260.073
1460	45,682	33,543	240.827	2550	84,814	63,613	260.799
1480	46,377	34,071	241.301	2600	86,650	65,033	261.512
1500	47,073	34,601	241.768	2650	88,488	66,455	262.213
1520	47,771	35,133	242.228	2700	90,328	67,880	262.902
1540	48,470	35,665	242.685	2750	92,171	69,306	263.577
1560	49,168	36,197	243.137	2800	94,014	70,734	264.241
1580	49,869	36,732	243.585	2850	95,859	72,163	264.895
1600	50,571	37,268	244.028	2900	97,705	73,593	265.538
1620	51,275	37,806	244.464	2950	99,556	75,028	266.170
1640	51,980	38,344	244.896	3000	101,407	76,464	266.793
1660	52,686	38,884	245.324	3050	103,260	77,902	267.404
1680	53,393	39,424	245.747	3100	105,115	79,341	268.007
1700	54,099	39,965	246.166	3150	106,972	80,782	268.601
1720	54,807	40,507	246.580	3200	108,830	82,224	269.186
1740	55,516	41,049	246.990	3250	110,690	83,668	269.763

Source: Tables A–18 through A–25 are adapted from Kenneth Wark, *Thermodynamics,* 4th ed. (New York: McGraw-Hill, 1983), pp. 787–98. Originally published in JANAF, *Thermochemical Tables,* NSRDS-NBS-37, 1971.

TABLE A–19

Ideal-gas properties of oxygen, O_2

T K	$\bar{h}$ kJ/kmol	$\bar{u}$ kJ/kmol	$\bar{s}°$ kJ/kmol·K	T K	$\bar{h}$ kJ/kmol	$\bar{u}$ kJ/kmol	$\bar{s}°$ kJ/kmol·K
0	0	0	0	600	17,929	12,940	226.346
220	6,404	4,575	196.171	610	18,250	13,178	226.877
230	6,694	4,782	197.461	620	18,572	13,417	227.400
240	6,984	4,989	198.696	630	18,895	13,657	227.918
250	7,275	5,197	199.885	640	19,219	13,898	228.429
260	7,566	5,405	201.027	650	19,544	14,140	228.932
270	7,858	5,613	202.128	660	19,870	14,383	229.430
280	8,150	5,822	203.191	670	20,197	14,626	229.920
290	8,443	6,032	204.218	680	20,524	14,871	230.405
298	8,682	6,203	205.033	690	20,854	15,116	230.885
300	8,736	6,242	205.213	700	21,184	15,364	231.358
310	9,030	6,453	206.177	710	21,514	15,611	231.827
320	9,325	6,664	207.112	720	21,845	15,859	232.291
330	9,620	6,877	208.020	730	22,177	16,107	232.748
340	9,916	7,090	208.904	740	22,510	16,357	233.201
350	10,213	7,303	209.765	750	22,844	16,607	233.649
360	10,511	7,518	210.604	760	23,178	16,859	234.091
370	10,809	7,733	211.423	770	23,513	17,111	234.528
380	11,109	7,949	212.222	780	23,850	17,364	234.960
390	11,409	8,166	213.002	790	24,186	17,618	235.387
400	11,711	8,384	213.765	800	24,523	17,872	235.810
410	12,012	8,603	214.510	810	24,861	18,126	236.230
420	12,314	8,822	215.241	820	25,199	18,382	236.644
430	12,618	9,043	215.955	830	25,537	18,637	237.055
440	12,923	9,264	216.656	840	25,877	18,893	237.462
450	13,228	9,487	217.342	850	26,218	19,150	237.864
460	13,525	9,710	218.016	860	26,559	19,408	238.264
470	13,842	9,935	218.676	870	26,899	19,666	238.660
480	14,151	10,160	219.326	880	27,242	19,925	239.051
490	14,460	10,386	219.963	890	27,584	20,185	239.439
500	14,770	10,614	220.589	900	27,928	20,445	239.823
510	15,082	10,842	221.206	910	28,272	20,706	240.203
520	15,395	11,071	221.812	920	28,616	20,967	240.580
530	15,708	11,301	222.409	930	28,960	21,228	240.953
540	16,022	11,533	222.997	940	29,306	21,491	241.323
550	16,338	11,765	223.576	950	29,652	21,754	241.689
560	16,654	11,998	224.146	960	29,999	22,017	242.052
570	16,971	12,232	224.708	970	30,345	22,280	242.411
580	17,290	12,467	225.262	980	30,692	22,544	242.768
590	17,609	12,703	225.808	990	31,041	22,809	242.120

TABLE A–19

Ideal-gas properties of oxygen, O_2 (*Concluded*)

T K	$\bar{h}$ kJ/kmol	$\bar{u}$ kJ/kmol	$\bar{s}°$ kJ/kmol·K	T K	$\bar{h}$ kJ/kmol	$\bar{u}$ kJ/kmol	$\bar{s}°$ kJ/kmol·K
1000	31,389	23,075	243.471	1760	58,880	44,247	263.861
1020	32,088	23,607	244.164	1780	59,624	44,825	264.283
1040	32,789	24,142	244.844	1800	60,371	45,405	264.701
1060	33,490	24,677	245.513	1820	61,118	45,986	265.113
1080	34,194	25,214	246.171	1840	61,866	46,568	265.521
1100	34,899	25,753	246.818	1860	62,616	47,151	265.925
1120	35,606	26,294	247.454	1880	63,365	47,734	266.326
1140	36,314	26,836	248.081	1900	64,116	48,319	266.722
1160	37,023	27,379	248.698	1920	64,868	48,904	267.115
1180	37,734	27,923	249.307	1940	65,620	49,490	267.505
1200	38,447	28,469	249.906	1960	66,374	50,078	267.891
1220	39,162	29,018	250.497	1980	67,127	50,665	268.275
1240	39,877	29,568	251.079	2000	67,881	51,253	268.655
1260	40,594	30,118	251.653	2050	69,772	52,727	269.588
1280	41,312	30,670	252.219	2100	71,668	54,208	270.504
1300	42,033	31,224	252.776	2150	73,573	55,697	271.399
1320	42,753	31,778	253.325	2200	75,484	57,192	272.278
1340	43,475	32,334	253.868	2250	77,397	58,690	273.136
1360	44,198	32,891	254.404	2300	79,316	60,193	273.891
1380	44,923	33,449	254.932	2350	81,243	61,704	274.809
1400	45,648	34,008	255.454	2400	83,174	63,219	275.625
1420	46,374	34,567	255.968	2450	85,112	64,742	276.424
1440	47,102	35,129	256.475	2500	87,057	66,271	277.207
1460	47,831	35,692	256.978	2550	89,004	67,802	277.979
1480	48,561	36,256	257.474	2600	90,956	69,339	278.738
1500	49,292	36,821	257.965	2650	92,916	70,883	279.485
1520	50,024	37,387	258.450	2700	94,881	72,433	280.219
1540	50,756	37,952	258.928	2750	96,852	73,987	280.942
1560	51,490	38,520	259.402	2800	98,826	75,546	281.654
1580	52,224	39,088	259.870	2850	100,808	77,112	282.357
1600	52,961	39,658	260.333	2900	102,793	78,682	283.048
1620	53,696	40,227	260.791	2950	104,785	80,258	283.728
1640	54,434	40,799	261.242	3000	106,780	81,837	284.399
1660	55,172	41,370	261.690	3050	108,778	83,419	285.060
1680	55,912	41,944	262.132	3100	110,784	85,009	285.713
1700	56,652	42,517	262.571	3150	112,795	86,601	286.355
1720	57,394	43,093	263.005	3200	114,809	88,203	286.989
1740	58,136	43,669	263.435	3250	116,827	89,804	287.614

TABLE A–20

Ideal-gas properties of carbon dioxide, CO_2

T K	$\bar{h}$ kJ/kmol	$\bar{u}$ kJ/kmol	$\bar{s}°$ kJ/kmol·K	T K	$\bar{h}$ kJ/kmol	$\bar{u}$ kJ/kmol	$\bar{s}°$ kJ/kmol·K
0	0	0	0	600	22,280	17,291	243.199
220	6,601	4,772	202.966	610	22,754	17,683	243.983
230	6,938	5,026	204.464	620	23,231	18,076	244.758
240	7,280	5,285	205.920	630	23,709	18,471	245.524
250	7,627	5,548	207.337	640	24,190	18,869	246.282
260	7,979	5,817	208.717	650	24,674	19,270	247.032
270	8,335	6,091	210.062	660	25,160	19,672	247.773
280	8,697	6,369	211.376	670	25,648	20,078	248.507
290	9,063	6,651	212.660	680	26,138	20,484	249.233
298	9,364	6,885	213.685	690	26,631	20,894	249.952
300	9,431	6,939	213.915	700	27,125	21,305	250.663
310	9,807	7,230	215.146	710	27,622	21,719	251.368
320	10,186	7,526	216.351	720	28,121	22,134	252.065
330	10,570	7,826	217.534	730	28,622	22,522	252.755
340	10,959	8,131	218.694	740	29,124	22,972	253.439
350	11,351	8,439	219.831	750	29,629	23,393	254.117
360	11,748	8,752	220.948	760	30,135	23,817	254.787
370	12,148	9,068	222.044	770	30,644	24,242	255.452
380	12,552	9,392	223.122	780	31,154	24,669	256.110
390	12,960	9,718	224.182	790	31,665	25,097	256.762
400	13,372	10,046	225.225	800	32,179	25,527	257.408
410	13,787	10,378	226.250	810	32,694	25,959	258.048
420	14,206	10,714	227.258	820	33,212	26,394	258.682
430	14,628	11,053	228.252	830	33,730	26,829	259.311
440	15,054	11,393	229.230	840	34,251	27,267	259.934
450	15,483	11,742	230.194	850	34,773	27,706	260.551
460	15,916	12,091	231.144	860	35,296	28,125	261.164
470	16,351	12,444	232.080	870	35,821	28,588	261.770
480	16,791	12,800	233.004	880	36,347	29,031	262.371
490	17,232	13,158	233.916	890	36,876	29,476	262.968
500	17,678	13,521	234.814	900	37,405	29,922	263.559
510	18,126	13,885	235.700	910	37,935	30,369	264.146
520	18,576	14,253	236.575	920	38,467	30,818	264.728
530	19,029	14,622	237.439	930	39,000	31,268	265.304
540	19,485	14,996	238.292	940	39,535	31,719	265.877
550	19,945	15,372	239.135	950	40,070	32,171	266.444
560	20,407	15,751	239.962	960	40,607	32,625	267.007
570	20,870	16,131	240.789	970	41,145	33,081	267.566
580	21,337	16,515	241.602	980	41,685	33,537	268.119
590	21,807	16,902	242.405	990	42,226	33,995	268.670

TABLE A–20

Ideal-gas properties of carbon dioxide, CO_2 (*Concluded*)

T K	$\bar{h}$ kJ/kmol	$\bar{u}$ kJ/kmol	$\bar{s}°$ kJ/kmol·K	T K	$\bar{h}$ kJ/kmol	$\bar{u}$ kJ/kmol	$\bar{s}°$ kJ/kmol·K
1000	42,769	34,455	269.215	1760	86,420	71,787	301.543
1020	43,859	35,378	270.293	1780	87,612	72,812	302.217
1040	44,953	36,306	271.354	1800	88,806	73,840	302.884
1060	46,051	37,238	272.400	1820	90,000	74,868	303.544
1080	47,153	38,174	273.430	1840	91,196	75,897	304.198
1100	48,258	39,112	274.445	1860	92,394	76,929	304.845
1120	49,369	40,057	275.444	1880	93,593	77,962	305.487
1140	50,484	41,006	276.430	1900	94,793	78,996	306.122
1160	51,602	41,957	277.403	1920	95,995	80,031	306.751
1180	52,724	42,913	278.361	1940	97,197	81,067	307.374
1200	53,848	43,871	297.307	1960	98,401	82,105	307.992
1220	54,977	44,834	280.238	1980	99,606	83,144	308.604
1240	56,108	45,799	281.158	2000	100,804	84,185	309.210
1260	57,244	46,768	282.066	2050	103,835	86,791	310.701
1280	58,381	47,739	282.962	2100	106,864	89,404	312.160
1300	59,522	48,713	283.847	2150	109,898	92,023	313.589
1320	60,666	49,691	284.722	2200	112,939	94,648	314.988
1340	61,813	50,672	285.586	2250	115,984	97,277	316.356
1360	62,963	51,656	286.439	2300	119,035	99,912	317.695
1380	64,116	52,643	287.283	2350	122,091	102,552	319.011
1400	65,271	53,631	288.106	2400	125,152	105,197	320.302
1420	66,427	54,621	288.934	2450	128,219	107,849	321.566
1440	67,586	55,614	289.743	2500	131,290	110,504	322.808
1460	68,748	56,609	290.542	2550	134,368	113,166	324.026
1480	66,911	57,606	291.333	2600	137,449	115,832	325.222
1500	71,078	58,606	292.114	2650	140,533	118,500	326.396
1520	72,246	59,609	292.888	2700	143,620	121,172	327.549
1540	73,417	60,613	292.654	2750	146,713	123,849	328.684
1560	74,590	61,620	294.411	2800	149,808	126,528	329.800
1580	76,767	62,630	295.161	2850	152,908	129,212	330.896
1600	76,944	63,741	295.901	2900	156,009	131,898	331.975
1620	78,123	64,653	296.632	2950	159,117	134,589	333.037
1640	79,303	65,668	297.356	3000	162,226	137,283	334.084
1660	80,486	66,592	298.072	3050	165,341	139,982	335.114
1680	81,670	67,702	298.781	3100	168,456	142,681	336.126
1700	82,856	68,721	299.482	3150	171,576	145,385	337.124
1720	84,043	69,742	300.177	3200	174,695	148,089	338.109
1740	85,231	70,764	300.863	3250	177,822	150,801	339.069

TABLE A–21

Ideal-gas properties of carbon monoxide, CO

T K	$\bar{h}$ kJ/kmol	$\bar{u}$ kJ/kmol	$\bar{s}°$ kJ/kmol·K	T K	$\bar{h}$ kJ/kmol	$\bar{u}$ kJ/kmol	$\bar{s}°$ kJ/kmol·K
0	0	0	0	600	17,611	12,622	218.204
220	6,391	4,562	188.683	610	17,915	12,843	218.708
230	6,683	4,771	189.980	620	18,221	13,066	219.205
240	6,975	4,979	191.221	630	18,527	13,289	219.695
250	7,266	5,188	192.411	640	18,833	13,512	220.179
260	7,558	5,396	193.554	650	19,141	13,736	220.656
270	7,849	5,604	194.654	660	19,449	13,962	221.127
280	8,140	5,812	195.713	670	19,758	14,187	221.592
290	8,432	6,020	196.735	680	20,068	14,414	222.052
298	8,669	6,190	197.543	690	20,378	14,641	222.505
300	8,723	6,229	197.723	700	20,690	14,870	222.953
310	9,014	6,437	198.678	710	21,002	15,099	223.396
320	9,306	6,645	199.603	720	21,315	15,328	223.833
330	9,597	6,854	200.500	730	21,628	15,558	224.265
340	9,889	7,062	201.371	740	21,943	15,789	224.692
350	10,181	7,271	202.217	750	22,258	16,022	225.115
360	10,473	7,480	203.040	760	22,573	16,255	225.533
370	10,765	7,689	203.842	770	22,890	16,488	225.947
380	11,058	7,899	204.622	780	23,208	16,723	226.357
390	11,351	8,108	205.383	790	23,526	16,957	226.762
400	11,644	8,319	206.125	800	23,844	17,193	227.162
410	11,938	8,529	206.850	810	24,164	17,429	227.559
420	12,232	8,740	207.549	820	24,483	17,665	227.952
430	12,526	8,951	208.252	830	24,803	17,902	228.339
440	12,821	9,163	208.929	840	25,124	18,140	228.724
450	13,116	9,375	209.593	850	25,446	18,379	229.106
460	13,412	9,587	210.243	860	25,768	18,617	229.482
470	13,708	9,800	210.880	870	26,091	18,858	229.856
480	14,005	10,014	211.504	880	26,415	19,099	230.227
490	14,302	10,228	212.117	890	26,740	19,341	230.593
500	14,600	10,443	212.719	900	27,066	19,583	230.957
510	14,898	10,658	213.310	910	27,392	19,826	231.317
520	15,197	10,874	213.890	920	27,719	20,070	231.674
530	15,497	11,090	214.460	930	28,046	20,314	232.028
540	15,797	11,307	215.020	940	28,375	20,559	232.379
550	16,097	11,524	215.572	950	28,703	20,805	232.727
560	16,399	11,743	216.115	960	29,033	21,051	233.072
570	16,701	11,961	216.649	970	29,362	21,298	233.413
580	17,003	12,181	217.175	980	29,693	21,545	233.752
590	17,307	12,401	217.693	990	30,024	21,793	234.088

TABLE A–21

Ideal-gas properties of carbon monoxide, CO (*Concluded*)

T K	$\bar{h}$ kJ/kmol	$\bar{u}$ kJ/kmol	$\bar{s}°$ kJ/kmol·K	T K	$\bar{h}$ kJ/kmol	$\bar{u}$ kJ/kmol	$\bar{s}°$ kJ/kmol·K
1000	30,355	22,041	234.421	1760	56,756	42,123	253.991
1020	31,020	22,540	235.079	1780	57,473	42,673	254.398
1040	31,688	23,041	235.728	1800	58,191	43,225	254.797
1060	32,357	23,544	236.364	1820	58,910	43,778	255.194
1080	33,029	24,049	236.992	1840	59,629	44,331	255.587
1100	33,702	24,557	237.609	1860	60,351	44,886	255.976
1120	34,377	25,065	238.217	1880	61,072	45,441	256.361
1140	35,054	25,575	238.817	1900	61,794	45,997	256.743
1160	35,733	26,088	239.407	1920	62,516	46,552	257.122
1180	36,406	26,602	239.989	1940	63,238	47,108	257.497
1200	37,095	27,118	240.663	1960	63,961	47,665	257.868
1220	37,780	27,637	241.128	1980	64,684	48,221	258.236
1240	38,466	28,426	241.686	2000	65,408	48,780	258.600
1260	39,154	28,678	242.236	2050	67,224	50,179	259.494
1280	39,844	29,201	242.780	2100	69,044	51,584	260.370
1300	40,534	29,725	243.316	2150	70,864	52,988	261.226
1320	41,226	30,251	243.844	2200	72,688	54,396	262.065
1340	41,919	30,778	244.366	2250	74,516	55,809	262.887
1360	42,613	31,306	244.880	2300	76,345	57,222	263.692
1380	43,309	31,836	245.388	2350	78,178	58,640	264.480
1400	44,007	32,367	245.889	2400	80,015	60,060	265.253
1420	44,707	32,900	246.385	2450	81,852	61,482	266.012
1440	45,408	33,434	246.876	2500	83,692	62,906	266.755
1460	46,110	33,971	247.360	2550	85,537	64,335	267.485
1480	46,813	34,508	247.839	2600	87,383	65,766	268.202
1500	47,517	35,046	248.312	2650	89,230	67,197	268.905
1520	48,222	35,584	248.778	2700	91,077	68,628	269.596
1540	48,928	36,124	249.240	2750	92,930	70,066	270.285
1560	49,635	36,665	249.695	2800	94,784	71,504	270.943
1580	50,344	37,207	250.147	2850	96,639	72,945	271.602
1600	51,053	37,750	250.592	2900	98,495	74,383	272.249
1620	51,763	38,293	251.033	2950	100,352	75,825	272.884
1640	52,472	38,837	251.470	3000	102,210	77,267	273.508
1660	53,184	39,382	251.901	3050	104,073	78,715	274.123
1680	53,895	39,927	252.329	3100	105,939	80,164	274.730
1700	54,609	40,474	252.751	3150	107,802	81,612	275.326
1720	55,323	41,023	253.169	3200	109,667	83,061	275.914
1740	56,039	41,572	253.582	3250	111,534	84,513	276.494

TABLE A–22

Ideal-gas properties of hydrogen, H_2

T K	$\bar{h}$ kJ/kmol	$\bar{u}$ kJ/kmol	$\bar{s}°$ kJ/kmol·K	T K	$\bar{h}$ kJ/kmol	$\bar{u}$ kJ/kmol	$\bar{s}°$ kJ/kmol·K
0	0	0	0	1440	42,808	30,835	177.410
260	7,370	5,209	126.636	1480	44,091	31,786	178.291
270	7,657	5,412	127.719	1520	45,384	32,746	179.153
280	7,945	5,617	128.765	1560	46,683	33,713	179.995
290	8,233	5,822	129.775	1600	47,990	34,687	180.820
298	8,468	5,989	130.574	1640	49,303	35,668	181.632
300	8,522	6,027	130.754	1680	50,622	36,654	182.428
320	9,100	6,440	132.621	1720	51,947	37,646	183.208
340	9,680	6,853	134.378	1760	53,279	38,645	183.973
360	10,262	7,268	136.039	1800	54,618	39,652	184.724
380	10,843	7,684	137.612	1840	55,962	40,663	185.463
400	11,426	8,100	139.106	1880	57,311	41,680	186.190
420	12,010	8,518	140.529	1920	58,668	42,705	186.904
440	12,594	8,936	141.888	1960	60,031	43,735	187.607
460	13,179	9,355	143.187	2000	61,400	44,771	188.297
480	13,764	9,773	144.432	2050	63,119	46,074	189.148
500	14,350	10,193	145.628	2100	64,847	47,386	189.979
520	14,935	10,611	146.775	2150	66,584	48,708	190.796
560	16,107	11,451	148.945	2200	68,328	50,037	191.598
600	17,280	12,291	150.968	2250	70,080	51,373	192.385
640	18,453	13,133	152.863	2300	71,839	52,716	193.159
680	19,630	13,976	154.645	2350	73,608	54,069	193.921
720	20,807	14,821	156.328	2400	75,383	55,429	194.669
760	21,988	15,669	157.923	2450	77,168	56,798	195.403
800	23,171	16,520	159.440	2500	78,960	58,175	196.125
840	24,359	17,375	160.891	2550	80,755	59,554	196.837
880	25,551	18,235	162.277	2600	82,558	60,941	197.539
920	26,747	19,098	163.607	2650	84,368	62,335	198.229
960	27,948	19,966	164.884	2700	86,186	63,737	198.907
1000	29,154	20,839	166.114	2750	88,008	65,144	199.575
1040	30,364	21,717	167.300	2800	89,838	66,558	200.234
1080	31,580	22,601	168.449	2850	91,671	67,976	200.885
1120	32,802	23,490	169.560	2900	93,512	69,401	201.527
1160	34,028	24,384	170.636	2950	95,358	70,831	202.157
1200	35,262	25,284	171.682	3000	97,211	72,268	202.778
1240	36,502	26,192	172.698	3050	99,065	73,707	203.391
1280	37,749	27,106	173.687	3100	100,926	75,152	203.995
1320	39,002	28,027	174.652	3150	102,793	76,604	204.592
1360	40,263	28,955	175.593	3200	104,667	78,061	205.181
1400	41,530	29,889	176.510	3250	106,545	79,523	205.765

TABLE A–23

Ideal-gas properties of water vapor, H_2O

T K	$\bar{h}$ kJ/kmol	$\bar{u}$ kJ/kmol	$\bar{s}°$ kJ/kmol·K	T K	$\bar{h}$ kJ/kmol	$\bar{u}$ kJ/kmol	$\bar{s}°$ kJ/kmol·K
0	0	0	0	600	20,402	15,413	212.920
220	7,295	5,466	178.576	610	20,765	15,693	213.529
230	7,628	5,715	180.054	620	21,130	15,975	214.122
240	7,961	5,965	181.471	630	21,495	16,257	214.707
250	8,294	6,215	182.831	640	21,862	16,541	215.285
260	8,627	6,466	184.139	650	22,230	16,826	215.856
270	8,961	6,716	185.399	660	22,600	17,112	216.419
280	9,296	6,968	186.616	670	22,970	17,399	216.976
290	9,631	7,219	187.791	680	23,342	17,688	217.527
298	9,904	7,425	188.720	690	23,714	17,978	218.071
300	9,966	7,472	188.928	700	24,088	18,268	218.610
310	10,302	7,725	190.030	710	24,464	18,561	219.142
320	10,639	7,978	191.098	720	24,840	18,854	219.668
330	10,976	8,232	192.136	730	25,218	19,148	220.189
340	11,314	8,487	193.144	740	25,597	19,444	220.707
350	11,652	8,742	194.125	750	25,977	19,741	221.215
360	11,992	8,998	195.081	760	26,358	20,039	221.720
370	12,331	9,255	196.012	770	26,741	20,339	222.221
380	12,672	9,513	196.920	780	27,125	20,639	222.717
390	13,014	9,771	197.807	790	27,510	20,941	223.207
400	13,356	10,030	198.673	800	27,896	21,245	223.693
410	13,699	10,290	199.521	810	28,284	21,549	224.174
420	14,043	10,551	200.350	820	28,672	21,855	224.651
430	14,388	10,813	201.160	830	29,062	22,162	225.123
440	14,734	11,075	201.955	840	29,454	22,470	225.592
450	15,080	11,339	202.734	850	29,846	22,779	226.057
460	15,428	11,603	203.497	860	30,240	23,090	226.517
470	15,777	11,869	204.247	870	30,635	23,402	226.973
480	16,126	12,135	204.982	880	31,032	23,715	227.426
490	16,477	12,403	205.705	890	31,429	24,029	227.875
500	16,828	12,671	206.413	900	31,828	24,345	228.321
510	17,181	12,940	207.112	910	32,228	24,662	228.763
520	17,534	13,211	207.799	920	32,629	24,980	229.202
530	17,889	13,482	208.475	930	33,032	25,300	229.637
540	18,245	13,755	209.139	940	33,436	25,621	230.070
550	18,601	14,028	209.795	950	33,841	25,943	230.499
560	18,959	14,303	210.440	960	34,247	26,265	230.924
570	19,318	14,579	211.075	970	34,653	26,588	231.347
580	19,678	14,856	211.702	980	35,061	26,913	231.767
590	20,039	15,134	212.320	990	35,472	27,240	232.184

TABLE A–23

Ideal-gas properties of water vapor, H_2O (*Continued*)

T K	$\bar{h}$ kJ/kmol	$\bar{u}$ kJ/kmol	$\bar{s}°$ kJ/kmol·K	T K	$\bar{h}$ kJ/kmol	$\bar{u}$ kJ/kmol	$\bar{s}°$ kJ/kmol·K
1000	35,882	27,568	232.597	1760	70,535	55,902	258.151
1020	36,709	28,228	233.415	1780	71,523	56,723	258.708
1040	37,542	28,895	234.223	1800	72,513	57,547	259.262
1060	38,380	29,567	235.020	1820	73,507	58,375	259.811
1080	39,223	30,243	235.806	1840	74,506	59,207	260.357
1100	40,071	30,925	236.584	1860	75,506	60,042	260.898
1120	40,923	31,611	237.352	1880	76,511	60,880	261.436
1140	41,780	32,301	238.110	1900	77,517	61,720	261.969
1160	42,642	32,997	238.859	1920	78,527	62,564	262.497
1180	43,509	33,698	239.600	1940	79,540	63,411	263.022
1200	44,380	34,403	240.333	1960	80,555	64,259	263.542
1220	45,256	35,112	241.057	1980	81,573	65,111	264.059
1240	46,137	35,827	241.773	2000	82,593	65,965	264.571
1260	47,022	36,546	242.482	2050	85,156	68,111	265.838
1280	47,912	37,270	243.183	2100	87,735	70,275	267.081
1300	48,807	38,000	243.877	2150	90,330	72,454	268.301
1320	49,707	38,732	244.564	2200	92,940	74,649	269.500
1340	50,612	39,470	245.243	2250	95,562	76,855	270.679
1360	51,521	40,213	245.915	2300	98,199	79,076	271.839
1380	52,434	40,960	246.582	2350	100,846	81,308	272.978
1400	53,351	41,711	247.241	2400	103,508	83,553	274.098
1420	54,273	42,466	247.895	2450	106,183	85,811	275.201
1440	55,198	43,226	248.543	2500	108,868	88,082	276.286
1460	56,128	43,989	249.185	2550	111,565	90,364	277.354
1480	57,062	44,756	249.820	2600	114,273	92,656	278.407
1500	57,999	45,528	250.450	2650	116,991	94,958	279.441
1520	58,942	46,304	251.074	2700	119,717	97,269	280.462
1540	59,888	47,084	251.693	2750	122,453	99,588	281.464
1560	60,838	47,868	252.305	2800	125,198	101,917	282.453
1580	61,792	48,655	252.912	2850	127,952	104,256	283.429
1600	62,748	49,445	253.513	2900	130,717	106,605	284.390
1620	63,709	50,240	254.111	2950	133,486	108,959	285.338
1640	64,675	51,039	254.703	3000	136,264	111,321	286.273
1660	65,643	51,841	255.290	3050	139,051	113,692	287.194
1680	66,614	52,646	255.873	3100	141,846	116,072	288.102
1700	67,589	53,455	256.450	3150	144,648	118,458	288.999
1720	68,567	54,267	257.022	3200	147,457	120,851	289.884
1740	69,550	55,083	257.589	3250	150,272	123,250	290.756

TABLE A–24

Ideal-gas properties of monatomic oxygen, O

T K	$\bar{h}$ kJ/kmol	$\bar{u}$ kJ/kmol	$\bar{s}°$ kJ/kmol·K	T K	$\bar{h}$ kJ/kmol	$\bar{u}$ kJ/kmol	$\bar{s}°$ kJ/kmol·K
0	0	0	0	2400	50,894	30,940	204.932
298	6,852	4,373	160.944	2450	51,936	31,566	205.362
300	6,892	4,398	161.079	2500	52,979	32,193	205.783
500	11,197	7,040	172.088	2550	54,021	32,820	206.196
1000	21,713	13,398	186.678	2600	55,064	33,447	206.601
1500	32,150	19,679	195.143	2650	56,108	34,075	206.999
1600	34,234	20,931	196.488	2700	57,152	34,703	207.389
1700	36,317	22,183	197.751	2750	58,196	35,332	207.772
1800	38,400	23,434	198.941	2800	59,241	35,961	208.148
1900	40,482	24,685	200.067	2850	60,286	36,590	208.518
2000	42,564	25,935	201.135	2900	61,332	37,220	208.882
2050	43,605	26,560	201.649	2950	62,378	37,851	209.240
2100	44,646	27,186	202.151	3000	63,425	38,482	209.592
2150	45,687	27,811	202.641	3100	65,520	39,746	210.279
2200	46,728	28,436	203.119	3200	67,619	41,013	210.945
2250	47,769	29,062	203.588	3300	69,720	42,283	211.592
2300	48,811	29,688	204.045	3400	71,824	43,556	212.220
2350	49,852	30,314	204.493	3500	73,932	44,832	212.831

TABLE A–25

Ideal-gas properties of hydroxyl, OH

T K	$\bar{h}$ kJ/kmol	$\bar{u}$ kJ/kmol	$\bar{s}°$ kJ/kmol·K	T K	$\bar{h}$ kJ/kmol	$\bar{u}$ kJ/kmol	$\bar{s}°$ kJ/kmol·K
0	0	0	0	2400	77,015	57,061	248.628
298	9,188	6,709	183.594	2450	78,801	58,431	249.364
300	9,244	6,749	183.779	2500	80,592	59,806	250.088
500	15,181	11,024	198.955	2550	82,388	61,186	250.799
1000	30,123	21,809	219.624	2600	84,189	62,572	251.499
1500	46,046	33,575	232.506	2650	85,995	63,962	252.187
1600	49,358	36,055	234.642	2700	87,806	65,358	252.864
1700	52,706	38,571	236.672	2750	89,622	66,757	253.530
1800	56,089	41,123	238.606	2800	91,442	68,162	254.186
1900	59,505	43,708	240.453	2850	93,266	69,570	254.832
2000	62,952	46,323	242.221	2900	95,095	70,983	255.468
2050	64,687	47,642	243.077	2950	96,927	72,400	256.094
2100	66,428	48,968	243.917	3000	98,763	73,820	256.712
2150	68,177	50,301	244.740	3100	102,447	76,673	257.919
2200	69,932	51,641	245.547	3200	106,145	79,539	259.093
2250	71,694	52,987	246.338	3300	109,855	82,418	260.235
2300	73,462	54,339	247.116	3400	113,578	85,309	261.347
2350	75,236	55,697	247.879	3500	117,312	88,212	262.429

TABLE A–26

Enthalpy of formation, Gibbs function of formation, and absolute entropy at 25°C, 1 atm

Substance	Formula	$\bar{h}_f^\circ$ kJ/kmol	$\bar{g}_f^\circ$ kJ/kmol	$\bar{s}^\circ$ kJ/kmol·K
Carbon	$C(s)$	0	0	5.74
Hydrogen	$H_2(g)$	0	0	130.68
Nitrogen	$N_2(g)$	0	0	191.61
Oxygen	$O_2(g)$	0	0	205.04
Carbon monoxide	$CO(g)$	−110,530	−137,150	197.65
Carbon dioxide	$CO_2(g)$	−393,520	−394,360	213.80
Water vapor	$H_2O(g)$	−241,820	−228,590	188.83
Water	$H_2O(\ell)$	−285,830	−237,180	69.92
Hydrogen peroxide	$H_2O_2(g)$	−136,310	−105,600	232.63
Ammonia	$NH_3(g)$	−46,190	−16,590	192.33
Methane	$CH_4(g)$	−74,850	−50,790	186.16
Acetylene	$C_2H_2(g)$	+226,730	+209,170	200.85
Ethylene	$C_2H_4(g)$	+52,280	+68,120	219.83
Ethane	$C_2H_6(g)$	−84,680	−32,890	229.49
Propylene	$C_3H_6(g)$	+20,410	+62,720	266.94
Propane	$C_3H_8(g)$	−103,850	−23,490	269.91
n-Butane	$C_4H_{10}(g)$	−126,150	−15,710	310.12
n-Octane	$C_8H_{18}(g)$	−208,450	+16,530	466.73
n-Octane	$C_8H_{18}(\ell)$	−249,950	+6,610	360.79
n-Dodecane	$C_{12}H_{26}(g)$	−291,010	+50,150	622.83
Benzene	$C_6H_6(g)$	+82,930	+129,660	269.20
Methyl alcohol	$CH_3OH(g)$	−200,670	−162,000	239.70
Methyl alcohol	$CH_3OH(\ell)$	−238,660	−166,360	126.80
Ethyl alcohol	$C_2H_5OH(g)$	−235,310	−168,570	282.59
Ethyl alcohol	$C_2H_5OH(\ell)$	−277,690	−174,890	160.70
Oxygen	$O(g)$	+249,190	+231,770	161.06
Hydrogen	$H(g)$	+218,000	+203,290	114.72
Nitrogen	$N(g)$	+472,650	+455,510	153.30
Hydroxyl	$OH(g)$	+39,460	+34,280	183.70

Source: From JANAF, *Thermochemical Tables* (Midland, MI: Dow Chemical Co., 1971); *Selected Values of Chemical Thermodynamic Properties,* NBS Technical Note 270-3, 1968; and *API Research Project 44* (Carnegie Press, 1953).

TABLE A–27

Properties of some common fuels and hydrocarbons

Fuel (phase)	Formula	Molar mass, kg/kmol	Density,[1] kg/L	Enthalpy of vaporization,[2] kJ/kg	Specific heat,[1] c_p kJ/kg·K	Higher heating value,[3] kJ/kg	Lower heating value,[3] kJ/kg
Carbon (*s*)	C	12.011	2	—	0.708	32,800	32,800
Hydrogen (*g*)	H_2	2.016	—	—	14.4	141,800	120,000
Carbon monoxide (*g*)	CO	28.013	—	—	1.05	10,100	10,100
Methane (*g*)	CH_4	16.043	—	509	2.20	55,530	50,050
Methanol (*ℓ*)	CH_4O	32.042	0.790	1168	2.53	22,660	19,920
Acetylene (*g*)	C_2H_2	26.038	—	—	1.69	49,970	48,280
Ethane (*g*)	C_2H_6	30.070	—	172	1.75	51,900	47,520
Ethanol (*ℓ*)	C_2H_6O	46.069	0.790	919	2.44	29,670	26,810
Propane (*ℓ*)	C_3H_8	44.097	0.500	335	2.77	50,330	46,340
Butane (*ℓ*)	C_4H_{10}	58.123	0.579	362	2.42	49,150	45,370
1-Pentene (*ℓ*)	C_5H_{10}	70.134	0.641	363	2.20	47,760	44,630
Isopentane (*ℓ*)	C_5H_{12}	72.150	0.626	—	2.32	48,570	44,910
Benzene (*ℓ*)	C_6H_6	78.114	0.877	433	1.72	41,800	40,100
Hexene (*ℓ*)	C_6H_{12}	84.161	0.673	392	1.84	47,500	44,400
Hexane (*ℓ*)	C_6H_{14}	86.177	0.660	366	2.27	48,310	44,740
Toluene (*ℓ*)	C_7H_8	92.141	0.867	412	1.71	42,400	40,500
Heptane (*ℓ*)	C_7H_{16}	100.204	0.684	365	2.24	48,100	44,600
Octane (*ℓ*)	C_8H_{18}	114.231	0.703	363	2.23	47,890	44,430
Decane (*ℓ*)	$C_{10}H_{22}$	142.285	0.730	361	2.21	47,640	44,240
Gasoline (*ℓ*)	$C_nH_{1.87n}$	100–110	0.72–0.78	350	2.4	47,300	44,000
Light diesel (*ℓ*)	$C_nH_{1.8n}$	170	0.78–0.84	270	2.2	46,100	43,200
Heavy diesel (*ℓ*)	$C_nH_{1.7n}$	200	0.82–0.88	230	1.9	45,500	42,800
Natural gas (*g*)	$C_nH_{3.8n}N_{0.1n}$	18	—	—	2	50,000	45,000

[1]At 1 atm and 20°C.

[2]At 25°C for liquid fuels, and 1 atm and normal boiling temperature for gaseous fuels.

[3]At 25°C. Multiply by molar mass to obtain heating values in kJ/kmol.

TABLE A–28

Natural logarithms of the equilibrium constant K_p

The equilibrium constant K_p for the reaction $\nu_A A + \nu_B B \rightleftharpoons \nu_C C + \nu_D D$ is defined as $K_p \equiv \frac{P_C^{\nu_C} P_D^{\nu_D}}{P_A^{\nu_A} P_B^{\nu_B}}$

Temp., K	$H_2 \rightleftharpoons 2H$	$O_2 \rightleftharpoons 2O$	$N_2 \rightleftharpoons 2N$	$H_2O \rightleftharpoons H_2 + \frac{1}{2}O_2$	$H_2O \rightleftharpoons \frac{1}{2}H_2 + OH$	$CO_2 \rightleftharpoons CO + \frac{1}{2}O_2$	$\frac{1}{2}N_2 + \frac{1}{2}O_2 \rightleftharpoons NO$
298	−164.005	−186.975	−367.480	−92.208	−106.208	−103.762	−35.052
500	−92.827	−105.630	−213.372	−52.691	−60.281	−57.616	−20.295
1000	−39.803	−45.150	−99.127	−23.163	−26.034	−23.529	−9.388
1200	−30.874	−35.005	−80.011	−18.182	−20.283	−17.871	−7.569
1400	−24.463	−27.742	−66.329	−14.609	−16.099	−13.842	−6.270
1600	−19.637	−22.285	−56.055	−11.921	−13.066	−10.830	−5.294
1800	−15.866	−18.030	−48.051	−9.826	−10.657	−8.497	−4.536
2000	−12.840	−14.622	−41.645	−8.145	−8.728	−6.635	−3.931
2200	−10.353	−11.827	−36.391	−6.768	−7.148	−5.120	−3.433
2400	−8.276	−9.497	−32.011	−5.619	−5.832	−3.860	−3.019
2600	−6.517	−7.521	−28.304	−4.648	−4.719	−2.801	−2.671
2800	−5.002	−5.826	−25.117	−3.812	−3.763	−1.894	−2.372
3000	−3.685	−4.357	−22.359	−3.086	−2.937	−1.111	−2.114
3200	−2.534	−3.072	−19.937	−2.451	−2.212	−0.429	−1.888
3400	−1.516	−1.935	−17.800	−1.891	−1.576	0.169	−1.690
3600	−0.609	−0.926	−15.898	−1.392	−1.088	0.701	−1.513
3800	0.202	−0.019	−14.199	−0.945	−0.501	1.176	−1.356
4000	0.934	0.796	−12.660	−0.542	−0.044	1.599	−1.216
4500	2.486	2.513	−9.414	0.312	0.920	2.490	−0.921
5000	3.725	3.895	−6.807	0.996	1.689	3.197	−0.686
5500	4.743	5.023	−4.666	1.560	2.318	3.771	−0.497
6000	5.590	5.963	−2.865	2.032	2.843	4.245	−0.341

Source: Gordon J. Van Wylen and Richard E. Sonntag, *Fundamentals of Classical Thermodynamics,* English/SI Version, 3rd ed. (New York: John Wiley & Sons, 1986), p. 723, table A.14. Based on thermodynamic data given in JANAF, *Thermochemical Tables* (Midland, MI: Thermal Research Laboratory, The Dow Chemical Company, 1971).

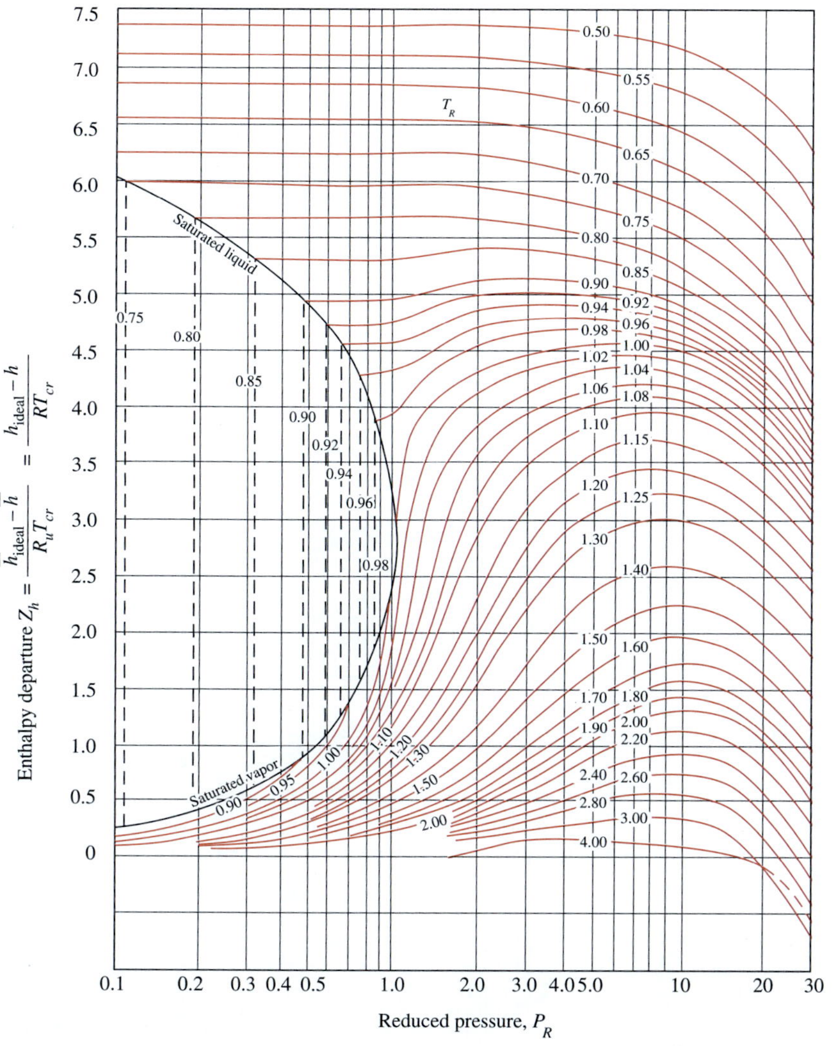

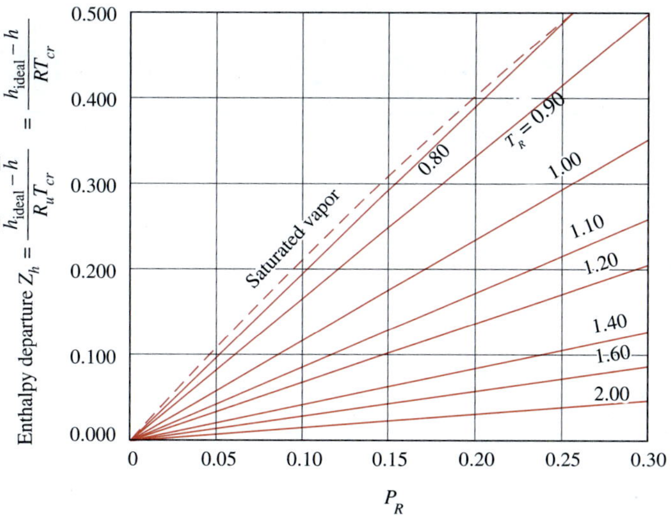

FIGURE A–29

Generalized enthalpy departure chart.

Source: Redrawn from Gordon van Wylen and Richard Sontag, Fundamentals of Classical Thermodynamics, *(SI version), 2d ed., Wiley, New York, 1976. Reprinted by permission of John Wiley and Sons, Inc.*

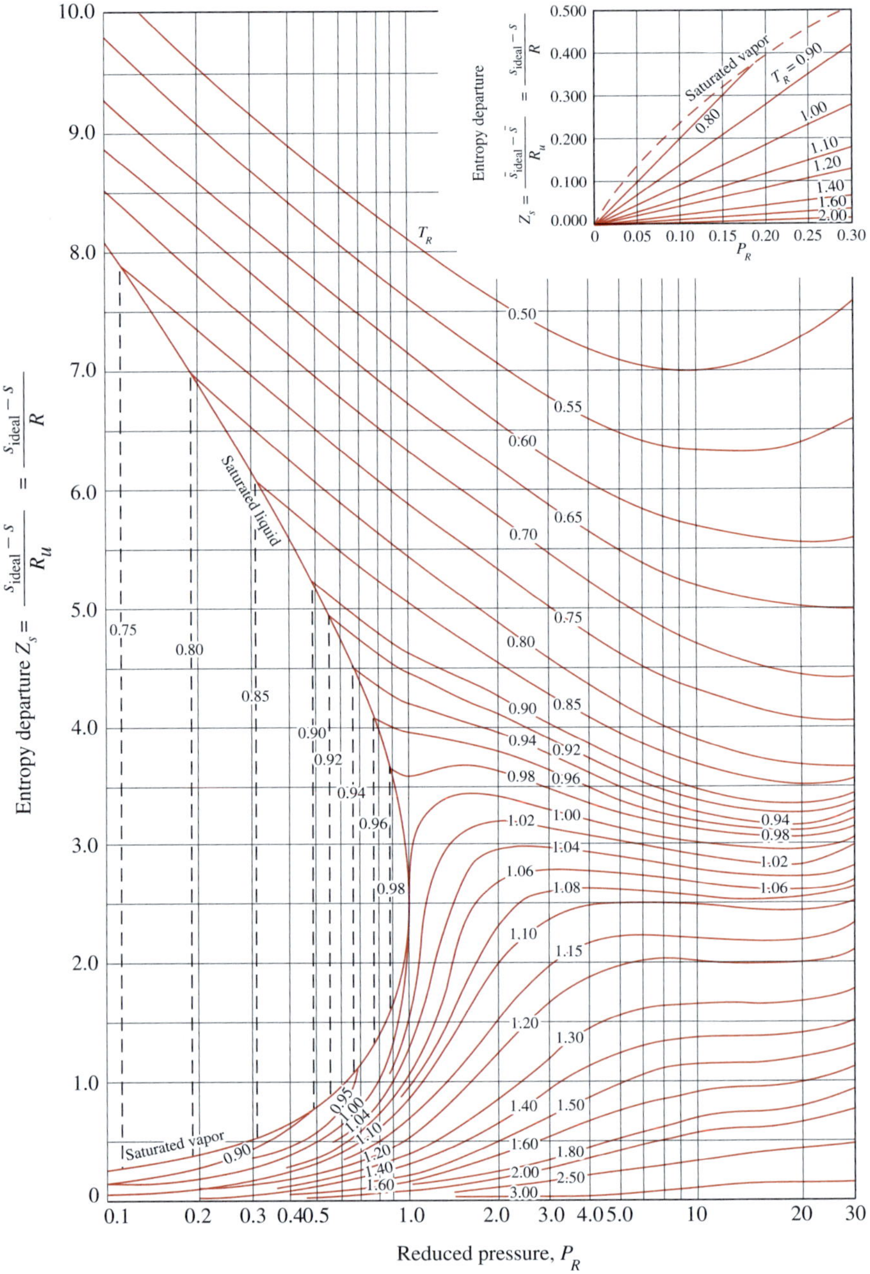

FIGURE A–30

Generalized entropy departure chart.

Source: Redrawn from Gordon van Wylen and Richard Sontag, Fundamentals of Classical Thermodynamics, *(SI version), 2d ed., Wiley, New York, 1976. Reprinted by permission of John Wiley and Sons, Inc.*

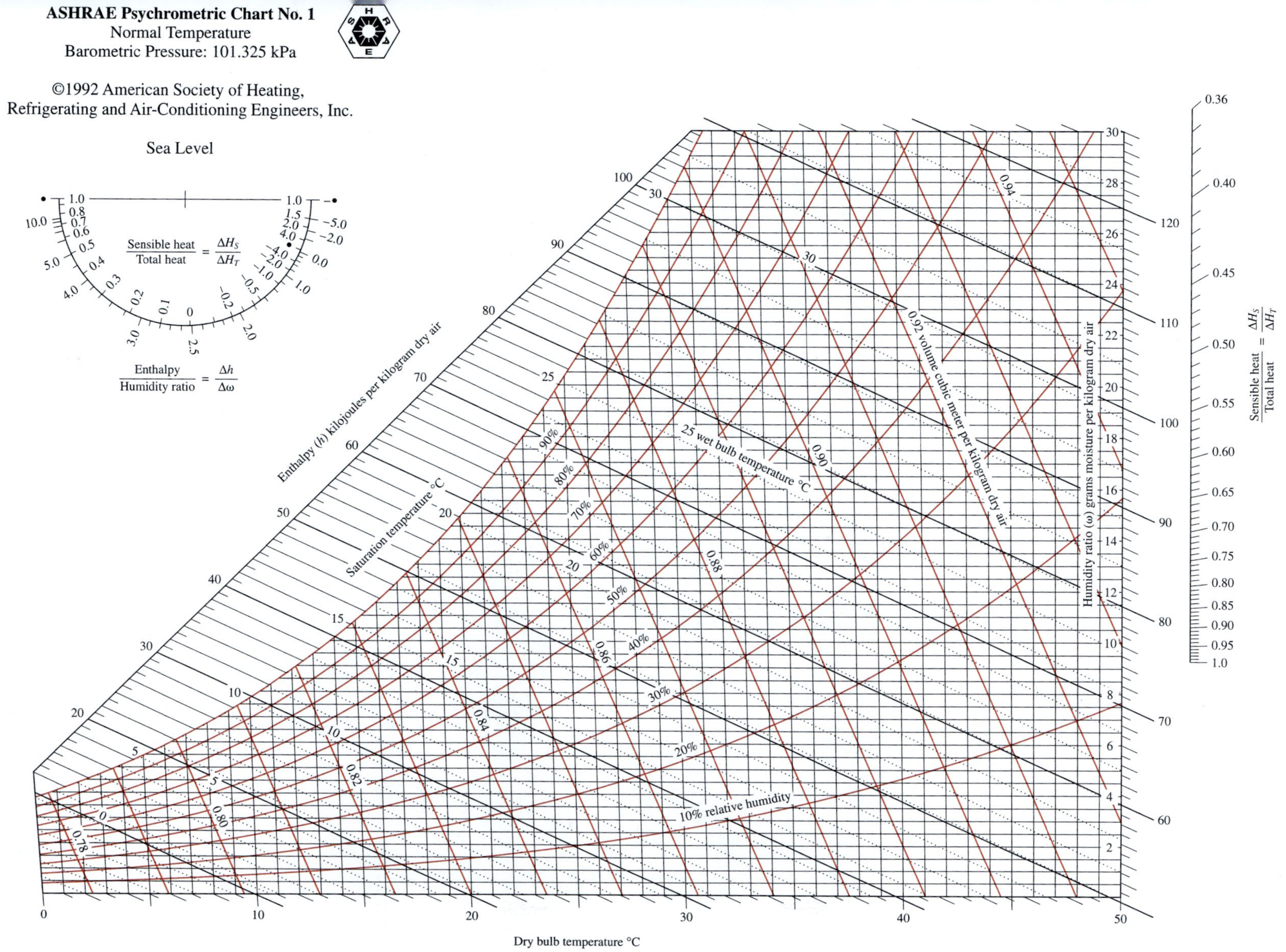

Prepared by Center for Applied Thermodynamic Studies, University of Idaho.

FIGURE A–31

Psychrometric chart at 1 atm total pressure.

Reprinted by permission of the American Society of Heating, Refrigerating and Air-Conditioning Engineers, Inc., Atlanta, GA; used with permission.

$$\mathrm{Ma}^* = \mathrm{Ma}\sqrt{\frac{k+1}{2+(k-1)\mathrm{Ma}^2}}$$

$$\frac{A}{A^*} = \frac{1}{\mathrm{Ma}}\left[\left(\frac{2}{k+1}\right)\left(1+\frac{k-1}{2}\mathrm{Ma}^2\right)\right]^{0.5(k+1)/(k-1)}$$

$$\frac{P}{P_0} = \left(1+\frac{k-1}{2}\mathrm{Ma}^2\right)^{-k/(k-1)}$$

$$\frac{\rho}{\rho_0} = \left(1+\frac{k-1}{2}\mathrm{Ma}^2\right)^{-1/(k-1)}$$

$$\frac{T}{T_0} = \left(1+\frac{k-1}{2}\mathrm{Ma}^2\right)^{-1}$$

TABLE A–32

One-dimensional isentropic compressible-flow functions for an ideal gas with $k = 1.4$

Ma	Ma*	A/A^*	P/P_0	ρ/ρ_0	T/T_0
0	0	∞	1.0000	1.0000	1.0000
0.1	0.1094	5.8218	0.9930	0.9950	0.9980
0.2	0.2182	2.9635	0.9725	0.9803	0.9921
0.3	0.3257	2.0351	0.9395	0.9564	0.9823
0.4	0.4313	1.5901	0.8956	0.9243	0.9690
0.5	0.5345	1.3398	0.8430	0.8852	0.9524
0.6	0.6348	1.1882	0.7840	0.8405	0.9328
0.7	0.7318	1.0944	0.7209	0.7916	0.9107
0.8	0.8251	1.0382	0.6560	0.7400	0.8865
0.9	0.9146	1.0089	0.5913	0.6870	0.8606
1.0	1.0000	1.0000	0.5283	0.6339	0.8333
1.2	1.1583	1.0304	0.4124	0.5311	0.7764
1.4	1.2999	1.1149	0.3142	0.4374	0.7184
1.6	1.4254	1.2502	0.2353	0.3557	0.6614
1.8	1.5360	1.4390	0.1740	0.2868	0.6068
2.0	1.6330	1.6875	0.1278	0.2300	0.5556
2.2	1.7179	2.0050	0.0935	0.1841	0.5081
2.4	1.7922	2.4031	0.0684	0.1472	0.4647
2.6	1.8571	2.8960	0.0501	0.1179	0.4252
2.8	1.9140	3.5001	0.0368	0.0946	0.3894
3.0	1.9640	4.2346	0.0272	0.0760	0.3571
5.0	2.2361	25.000	0.0019	0.0113	0.1667
∝	2.2495	∝	0	0	0

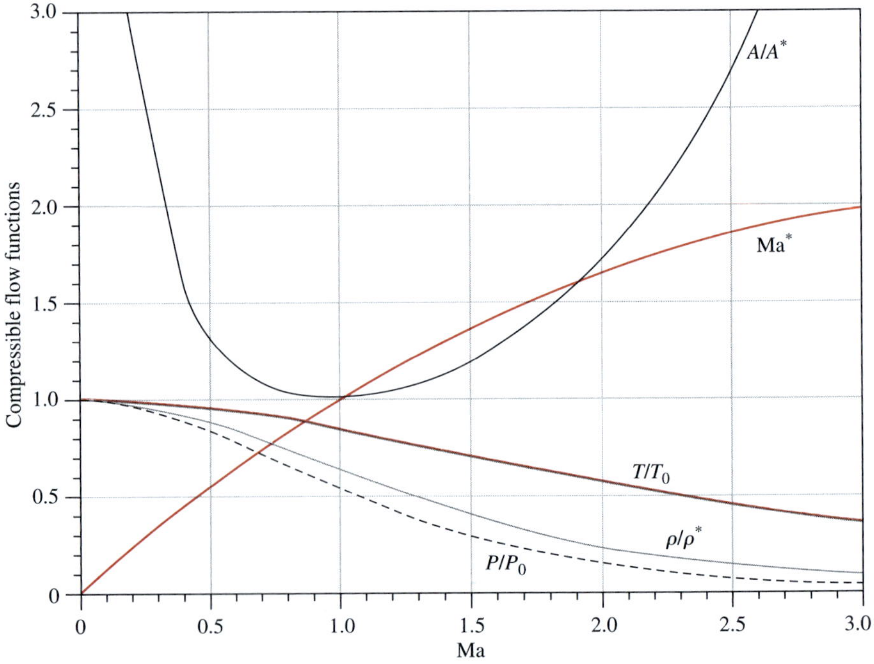

$$T_{01} = T_{02}$$

$$\mathrm{Ma}_2 = \sqrt{\frac{(k-1)\mathrm{Ma}_1^2 + 2}{2k\mathrm{Ma}_1^2 - k + 1}}$$

$$\frac{P_2}{P_1} = \frac{1 + k\mathrm{Ma}_1^2}{1 + k\mathrm{Ma}_2^2} = \frac{2k\mathrm{Ma}_1^2 - k + 1}{k + 1}$$

$$\frac{\rho_2}{\rho_1} = \frac{P_2/P_1}{T_2/T_1} = \frac{(k+1)\mathrm{Ma}_1^2}{2 + (k-1)\mathrm{Ma}_1^2} = \frac{V_1}{V_2}$$

$$\frac{T_2}{T_1} = \frac{2 + \mathrm{Ma}_1^2(k-1)}{2 + \mathrm{Ma}_2^2(k-1)}$$

$$\frac{P_{02}}{P_{01}} = \frac{\mathrm{Ma}_1}{\mathrm{Ma}_2}\left[\frac{1 + \mathrm{Ma}_2^2(k-1)/2}{1 + \mathrm{Ma}_1^2(k-1)/2}\right]^{(k+1)/[2(k-1)]}$$

$$\frac{P_{02}}{P_1} = \frac{(1 + k\mathrm{Ma}_1^2)[1 + \mathrm{Ma}_2^2(k-1)/2]^{k/(k-1)}}{1 + k\mathrm{Ma}_2^2}$$

TABLE A–33

One-dimensional normal-shock functions for an ideal gas with $k = 1.4$

Ma_1	Ma_2	P_2/P_1	ρ_2/ρ_1	T_2/T_1	P_{02}/P_{01}	P_{02}/P_1
1.0	1.0000	1.0000	1.0000	1.0000	1.0000	1.8929
1.1	0.9118	1.2450	1.1691	1.0649	0.9989	2.1328
1.2	0.8422	1.5133	1.3416	1.1280	0.9928	2.4075
1.3	0.7860	1.8050	1.5157	1.1909	0.9794	2.7136
1.4	0.7397	2.1200	1.6897	1.2547	0.9582	3.0492
1.5	0.7011	2.4583	1.8621	1.3202	0.9298	3.4133
1.6	0.6684	2.8200	2.0317	1.3880	0.8952	3.8050
1.7	0.6405	3.2050	2.1977	1.4583	0.8557	4.2238
1.8	0.6165	3.6133	2.3592	1.5316	0.8127	4.6695
1.9	0.5956	4.0450	2.5157	1.6079	0.7674	5.1418
2.0	0.5774	4.5000	2.6667	1.6875	0.7209	5.6404
2.1	0.5613	4.9783	2.8119	1.7705	0.6742	6.1654
2.2	0.5471	5.4800	2.9512	1.8569	0.6281	6.7165
2.3	0.5344	6.0050	3.0845	1.9468	0.5833	7.2937
2.4	0.5231	6.5533	3.2119	2.0403	0.5401	7.8969
2.5	0.5130	7.1250	3.3333	2.1375	0.4990	8.5261
2.6	0.5039	7.7200	3.4490	2.2383	0.4601	9.1813
2.7	0.4956	8.3383	3.5590	2.3429	0.4236	9.8624
2.8	0.4882	8.9800	3.6636	2.4512	0.3895	10.5694
2.9	0.4814	9.6450	3.7629	2.5632	0.3577	11.3022
3.0	0.4752	10.3333	3.8571	2.6790	0.3283	12.0610
4.0	0.4350	18.5000	4.5714	4.0469	0.1388	21.0681
5.0	0.4152	29.000	5.0000	5.8000	0.0617	32.6335
∞	0.3780	∞	6.0000	∞	0	∞

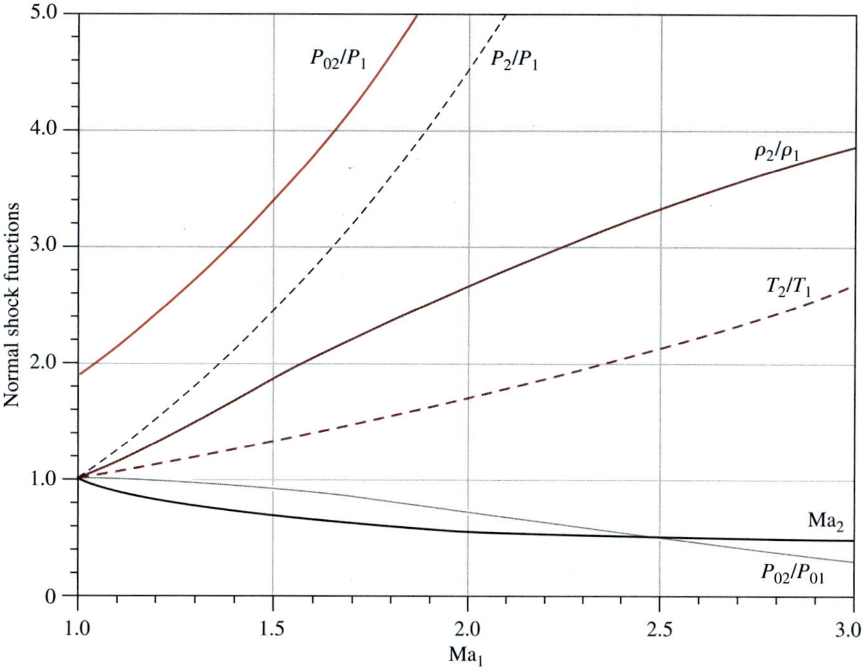

$$\frac{T_0}{T_0^*} = \frac{(k+1)\text{Ma}^2[2+(k-1)\text{Ma}^2]}{(1+k\text{Ma}^2)^2}$$

$$\frac{P_0}{P_0^*} = \frac{k+1}{1+k\text{Ma}^2}\left(\frac{2+(k-1)\text{Ma}^2}{k+1}\right)^{k/(k-1)}$$

$$\frac{T}{T^*} = \left(\frac{\text{Ma}(1+k)}{1+k\text{Ma}^2}\right)^2$$

$$\frac{P}{P^*} = \frac{1+k}{1+k\text{Ma}^2}$$

$$\frac{V}{V^*} = \frac{\rho^*}{\rho} = \frac{(1+k)\text{Ma}^2}{1+k\text{Ma}^2}$$

TABLE A–34

Rayleigh flow functions for an ideal gas with $k = 1.4$

Ma	T_0/T_0^*	P_0/P_0^*	T/T^*	P/P^*	V/V^*
0.0	0.0000	1.2679	0.0000	2.4000	0.0000
0.1	0.0468	1.2591	0.0560	2.3669	0.0237
0.2	0.1736	1.2346	0.2066	2.2727	0.0909
0.3	0.3469	1.1985	0.4089	2.1314	0.1918
0.4	0.5290	1.1566	0.6151	1.9608	0.3137
0.5	0.6914	1.1141	0.7901	1.7778	0.4444
0.6	0.8189	1.0753	0.9167	1.5957	0.5745
0.7	0.9085	1.0431	0.9929	1.4235	0.6975
0.8	0.9639	1.0193	1.0255	1.2658	0.8101
0.9	0.9921	1.0049	1.0245	1.1246	0.9110
1.0	1.0000	1.0000	1.0000	1.0000	1.0000
1.2	0.9787	1.0194	0.9118	0.7958	1.1459
1.4	0.9343	1.0777	0.8054	0.6410	1.2564
1.6	0.8842	1.1756	0.7017	0.5236	1.3403
1.8	0.8363	1.3159	0.6089	0.4335	1.4046
2.0	0.7934	1.5031	0.5289	0.3636	1.4545
2.2	0.7561	1.7434	0.4611	0.3086	1.4938
2.4	0.7242	2.0451	0.4038	0.2648	1.5252
2.6	0.6970	2.4177	0.3556	0.2294	1.5505
2.8	0.6738	2.8731	0.3149	0.2004	1.5711
3.0	0.6540	3.4245	0.2803	0.1765	1.5882

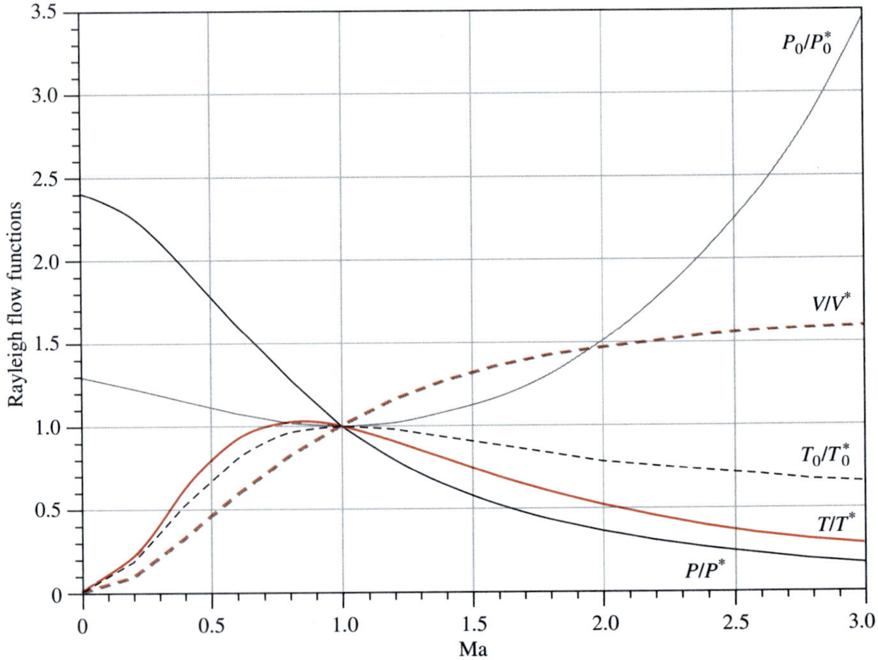

APPENDIX

2

PROPERTY TABLES AND CHARTS (ENGLISH UNITS)

TABLE A–1E

Molar mass, gas constant, and critical-point properties

Substance	Formula	Molar mass, *M* lbm/lbmol	Gas constant, *R* Btu/lbm·R*	Gas constant, *R* psia·ft³/lbm·R*	Critical-point properties Temperature, R	Critical-point properties Pressure, psia	Critical-point properties Volume, ft³/lbmol
Air	—	28.97	0.06855	0.3704	238.5	547	1.41
Ammonia	NH_3	17.03	0.1166	0.6301	729.8	1636	1.16
Argon	Ar	39.948	0.04971	0.2686	272	705	1.20
Benzene	C_6H_6	78.115	0.02542	0.1374	1012	714	4.17
Bromine	Br_2	159.808	0.01243	0.06714	1052	1500	2.17
n-Butane	C_4H_{10}	58.124	0.03417	0.1846	765.2	551	4.08
Carbon dioxide	CO_2	44.01	0.04513	0.2438	547.5	1071	1.51
Carbon monoxide	CO	28.011	0.07090	0.3831	240	507	1.49
Carbon tetrachloride	CCl_4	153.82	0.01291	0.06976	1001.5	661	4.42
Chlorine	Cl_2	70.906	0.02801	0.1517	751	1120	1.99
Chloroform	$CHCl_3$	119.38	0.01664	0.08988	965.8	794	3.85
Dichlorodifluoromethane (R-12)	CCl_2F_2	120.91	0.01643	0.08874	692.4	582	3.49
Dichlorofluoromethane (R-21)	$CHCl_2F$	102.92	0.01930	0.1043	813.0	749	3.16
Ethane	C_2H_6	30.020	0.06616	0.3574	549.8	708	2.37
Ethyl alcohol	C_2H_5OH	46.07	0.04311	0.2329	929.0	926	2.68
Ethylene	C_2H_4	28.054	0.07079	0.3825	508.3	742	1.99
Helium	He	4.003	0.4961	2.6809	9.5	33.2	0.926
n-Hexane	C_6H_{14}	86.178	0.02305	0.1245	914.2	439	5.89
Hydrogen (normal)	H_2	2.016	0.9851	5.3224	59.9	188.1	1.04
Krypton	Kr	83.80	0.02370	0.1280	376.9	798	1.48
Methane	CH_4	16.043	0.1238	0.6688	343.9	673	1.59
Methyl alcohol	CH_3OH	32.042	0.06198	0.3349	923.7	1154	1.89
Methyl chloride	CH_3Cl	50.488	0.03934	0.2125	749.3	968	2.29
Neon	Ne	20.183	0.09840	0.5316	80.1	395	0.668
Nitrogen	N_2	28.013	0.07090	0.3830	227.1	492	1.44
Nitrous oxide	N_2O	44.013	0.04512	0.2438	557.4	1054	1.54
Oxygen	O_2	31.999	0.06206	0.3353	278.6	736	1.25
Propane	C_3H_8	44.097	0.04504	0.2433	665.9	617	3.20
Propylene	C_3H_6	42.081	0.04719	0.2550	656.9	670	2.90
Sulfur dioxide	SO_2	64.063	0.03100	1.1675	775.2	1143	1.95
Tetrafluoroethane (R-134a)	CF_3CH_2F	102.03	0.01946	0.1052	673.6	588.7	3.19
Trichlorofluoromethane (R-11)	CCl_3F	137.37	0.01446	0.07811	848.1	635	3.97
Water	H_2O	18.015	0.1102	0.5956	1164.8	3200	0.90
Xenon	Xe	131.30	0.01513	0.08172	521.55	852	1.90

*Calculated from $R = R_u/M$, where R_u = 1.98588 Btu/lbmol·R = 10.7316 psia·ft³/lbmol·R and *M* is the molar mass.

Source: K. A. Kobe and R. E. Lynn, Jr., *Chemical Review* 52 (1953), pp. 117–236, and ASHRAE, *Handbook of Fundamentals* (Atlanta, GA: American Society of Heating, Refrigerating, and Air-Conditioning Engineers, Inc., 1993), pp. 16.4 and 36.1.

TABLE A–2E

Ideal-gas specific heats of various common gases
(*a*) At 80°F

Gas	Formula	Gas constant, R Btu/lbm·R	c_p Btu/lbm·R	c_v Btu/lbm·R	k
Air	—	0.06855	0.240	0.171	1.400
Argon	Ar	0.04971	0.1253	0.0756	1.667
Butane	C_4H_{10}	0.03424	0.415	0.381	1.09
Carbon dioxide	CO_2	0.04513	0.203	0.158	1.285
Carbon monoxide	CO	0.07090	0.249	0.178	1.399
Ethane	C_2H_6	0.06616	0.427	0.361	1.183
Ethylene	C_2H_4	0.07079	0.411	0.340	1.208
Helium	He	0.4961	1.25	0.753	1.667
Hydrogen	H_2	0.9851	3.43	2.44	1.404
Methane	CH_4	0.1238	0.532	0.403	1.32
Neon	Ne	0.09840	0.246	0.1477	1.667
Nitrogen	N_2	0.07090	0.248	0.177	1.400
Octane	C_8H_{18}	0.01742	0.409	0.392	1.044
Oxygen	O_2	0.06206	0.219	0.157	1.395
Propane	C_3H_8	0.04504	0.407	0.362	1.124
Steam	H_2O	0.1102	0.445	0.335	1.329

Source: Gordon J. Van Wylen and Richard E. Sonntag, *Fundamentals of Classical Thermodynamics*, English/SI Version, 3rd ed. (New York: John Wiley & Sons, 1986), p. 687, Table A–8E.

TABLE A–2E

Ideal-gas specific heats of various common gases (*Continued*)
(*b*) At various temperatures

Temp., °F	c_p Btu/lbm·R	c_v Btu/lbm·R	k	c_p Btu/lbm·R	c_v Btu/lbm·R	k	c_p Btu/lbm·R	c_v Btu/lbm·R	k
	Air			*Carbon dioxide, CO_2*			*Carbon monoxide, CO*		
40	0.240	0.171	1.401	0.195	0.150	1.300	0.248	0.177	1.400
100	0.240	0.172	1.400	0.205	0.160	1.283	0.249	0.178	1.399
200	0.241	0.173	1.397	0.217	0.172	1.262	0.249	0.179	1.397
300	0.243	0.174	1.394	0.229	0.184	1.246	0.251	0.180	1.394
400	0.245	0.176	1.389	0.239	0.193	1.233	0.253	0.182	1.389
500	0.248	0.179	1.383	0.247	0.202	1.223	0.256	0.185	1.384
600	0.250	0.182	1.377	0.255	0.210	1.215	0.259	0.188	1.377
700	0.254	0.185	1.371	0.262	0.217	1.208	0.262	0.191	1.371
800	0.257	0.188	1.365	0.269	0.224	1.202	0.266	0.195	1.364
900	0.259	0.191	1.358	0.275	0.230	1.197	0.269	0.198	1.357
1000	0.263	0.195	1.353	0.280	0.235	1.192	0.273	0.202	1.351
1500	0.276	0.208	1.330	0.298	0.253	1.178	0.287	0.216	1.328
2000	0.286	0.217	1.312	0.312	0.267	1.169	0.297	0.226	1.314
	Hydrogen, H_2			*Nitrogen, N_2*			*Oxygen, O_2*		
40	3.397	2.412	1.409	0.248	0.177	1.400	0.219	0.156	1.397
100	3.426	2.441	1.404	0.248	0.178	1.399	0.220	0.158	1.394
200	3.451	2.466	1.399	0.249	0.178	1.398	0.223	0.161	1.387
300	3.461	2.476	1.398	0.250	0.179	1.396	0.226	0.164	1.378
400	3.466	2.480	1.397	0.251	0.180	1.393	0.230	0.168	1.368
500	3.469	2.484	1.397	0.254	0.183	1.388	0.235	0.173	1.360
600	3.473	2.488	1.396	0.256	0.185	1.383	0.239	0.177	1.352
700	3.477	2.492	1.395	0.260	0.189	1.377	0.242	0.181	1.344
800	3.494	2.509	1.393	0.262	0.191	1.371	0.246	0.184	1.337
900	3.502	2.519	1.392	0.265	0.194	1.364	0.249	0.187	1.331
1000	3.513	2.528	1.390	0.269	0.198	1.359	0.252	0.190	1.326
1500	3.618	2.633	1.374	0.283	0.212	1.334	0.263	0.201	1.309
2000	3.758	2.773	1.355	0.293	0.222	1.319	0.270	0.208	1.298

Note: The unit Btu/lbm·R is equivalent to Btu/lbm·°F.

Source: Kenneth Wark, *Thermodynamics*, 4th ed. (New York: McGraw-Hill, 1983), p. 830, Table A–4. Originally published in *Tables of Properties of Gases*, NBS Circular 564, 1955.

TABLE A–2E

Ideal-gas specific heats of various common gases (*Concluded*)
(*c*) As a function of temperature

$$\bar{c}_p = a + bT + cT^2 + dT^3$$

(T in R, c_p in Btu/lbmol·R)

Substance	Formula	a	b	c	d	Temperature range, R	% error Max.	% error Avg.
Nitrogen	N_2	6.903	-0.02085×10^{-2}	0.05957×10^{-5}	-0.1176×10^{-9}	491-3240	0.59	0.34
Oxygen	O_2	6.085	0.2017×10^{-2}	-0.05275×10^{-5}	0.05372×10^{-9}	491-3240	1.19	0.28
Air	—	6.713	0.02609×10^{-2}	0.03540×10^{-5}	-0.08052×10^{-9}	491-3240	0.72	0.33
Hydrogen	H_2	6.952	-0.02542×10^{-2}	0.02952×10^{-5}	-0.03565×10^{-9}	491-3240	1.02	0.26
Carbon monoxide	CO	6.726	0.02222×10^{-2}	0.03960×10^{-5}	-0.09100×10^{-9}	491-3240	0.89	0.37
Carbon dioxide	CO_2	5.316	0.79361×10^{-2}	-0.2581×10^{-5}	0.3059×10^{-9}	491-3240	0.67	0.22
Water vapor	H_2O	7.700	0.02552×10^{-2}	0.07781×10^{-5}	-0.1472×10^{-9}	491-3240	0.53	0.24
Nitric oxide	NO	7.008	-0.01247×10^{-2}	0.07185×10^{-5}	-0.1715×10^{-9}	491-2700	0.97	0.36
Nitrous oxide	N_2O	5.758	0.7780×10^{-2}	-0.2596×10^{-5}	0.4331×10^{-9}	491-2700	0.59	0.26
Nitrogen dioxide	NO_2	5.48	0.7583×10^{-2}	-0.260×10^{-5}	0.322×10^{-9}	491-2700	0.46	0.18
Ammonia	NH_3	6.5846	0.34028×10^{-2}	0.073034×10^{-5}	-0.27402×10^{-9}	491-2700	0.91	0.36
Sulfur	S_2	6.499	0.2943×10^{-2}	-0.1200×10^{-5}	0.1632×10^{-9}	491-3240	0.99	0.38
Sulfur dioxide	SO_2	6.157	0.7689×10^{-2}	-0.2810×10^{-5}	0.3527×10^{-9}	491-3240	0.45	0.24
Sulfur trioxide	SO_3	3.918	1.935×10^{-2}	-0.8256×10^{-5}	1.328×10^{-9}	491-2340	0.29	0.13
Acetylene	C_2H_2	5.21	1.2227×10^{-2}	-0.4812×10^{-5}	0.7457×10^{-9}	491-2700	1.46	0.59
Benzene	C_6H_6	−8.650	6.4322×10^{-2}	-2.327×10^{-5}	3.179×10^{-9}	491-2700	0.34	0.20
Methanol	CH_4O	4.55	1.214×10^{-2}	-0.0898×10^{-5}	-0.329×10^{-9}	491-1800	0.18	0.08
Ethanol	C_2H_6O	4.75	2.781×10^{-2}	-0.7651×10^{-5}	0.821×10^{-9}	491-2700	0.40	0.22
Hydrogen chloride	HCl	7.244	-0.1011×10^{-2}	0.09783×10^{-5}	-0.1776×10^{-9}	491-2740	0.22	0.08
Methane	CH_4	4.750	0.6666×10^{-2}	0.09352×10^{-5}	-0.4510×10^{-9}	491-2740	1.33	0.57
Ethane	C_2H_6	1.648	2.291×10^{-2}	-0.4722×10^{-5}	0.2984×10^{-9}	491-2740	0.83	0.28
Propane	C_3H_8	−0.966	4.044×10^{-2}	-1.159×10^{-5}	1.300×10^{-9}	491-2740	0.40	0.12
n-Butane	C_4H_{10}	0.945	4.929×10^{-2}	-1.352×10^{-5}	1.433×10^{-9}	491-2740	0.54	0.24
i-Butane	C_4H_{10}	−1.890	5.520×10^{-2}	-1.696×10^{-5}	2.044×10^{-9}	491-2740	0.25	0.13
n-Pentane	C_5H_{12}	1.618	6.028×10^{-2}	-1.656×10^{-5}	1.732×10^{-9}	491-2740	0.56	0.21
n-Hexane	C_6H_{14}	1.657	7.328×10^{-2}	-2.112×10^{-5}	2.363×10^{-9}	491-2740	0.72	0.20
Ethylene	C_2H_4	0.944	2.075×10^{-2}	-0.6151×10^{-5}	0.7326×10^{-9}	491-2740	0.54	0.13
Propylene	C_3H_6	0.753	3.162×10^{-2}	-0.8981×10^{-5}	1.008×10^{-9}	491-2740	0.73	0.17

Source: Chemical and Process Thermodynamics 3/E by Kyle, B. G., © 2000. Adapted by permission of Pearson Education, Inc., Upper Saddle River, NJ.

TABLE A–3E

Properties of common liquids, solids, and foods
(*a*) Liquids

	Boiling data at 1 atm		Freezing data		Liquid properties		
Substance	Normal boiling point, °F	Latent heat of vaporization, h_{fg} Btu/lbm	Freezing point, °F	Latent heat of fusion, h_{if} Btu/lbm	Temperature, °F	Density, ρ lbm/ft³	Specific heat, c_p Btu/lbm·R
Ammonia	−27.9	24.54	−107.9	138.6	−27.9	42.6	1.06
					0	41.3	1.083
					40	39.5	1.103
					80	37.5	1.135
Argon	−302.6	69.5	−308.7	12.0	−302.6	87.0	0.272
Benzene	176.4	169.4	41.9	54.2	68	54.9	0.411
Brine (20% sodium chloride by mass)	219.0	—	0.7	—	68	71.8	0.743
n-Butane	31.1	165.6	−217.3	34.5	31.1	37.5	0.552
Carbon dioxide	−109.2*	99.6 (at 32°F)	−69.8	—	32	57.8	0.583
Ethanol	172.8	360.5	−173.6	46.9	77	48.9	0.588
Ethyl alcohol	173.5	368	−248.8	46.4	68	49.3	0.678
Ethylene glycol	388.6	344.0	12.6	77.9	68	69.2	0.678
Glycerine	355.8	419	66.0	86.3	68	78.7	0.554
Helium	−452.1	9.80	—	—	−452.1	9.13	5.45
Hydrogen	−423.0	191.7	−434.5	25.6	−423.0	4.41	2.39
Isobutane	10.9	157.8	−255.5	45.5	10.9	37.1	0.545
Kerosene	399–559	108	−12.8	—	68	51.2	0.478
Mercury	674.1	126.7	−38.0	4.90	77	847	0.033
Methane	−258.7	219.6	296.0	25.1	−258.7	26.4	0.834
					−160	20.0	1.074
Methanol	148.1	473	−143.9	42.7	77	49.1	0.609
Nitrogen	−320.4	85.4	−346.0	10.9	−320.4	50.5	0.492
					−260	38.2	0.643
Octane	256.6	131.7	−71.5	77.9	68	43.9	0.502
Oil (light)	—	—			77	56.8	0.430
Oxygen	−297.3	91.5	−361.8	5.9	−297.3	71.2	0.408
Petroleum	—	99–165			68	40.0	0.478
Propane	−43.7	184.0	−305.8	34.4	−43.7	36.3	0.538
					32	33.0	0.604
					100	29.4	0.673
Refrigerant-134a	−15.0	93.3	−141.9	—	−40	88.5	0.283
					−15	86.0	0.294
					32	80.9	0.318
					90	73.6	0.348
Water	212	970.1	32	143.5	32	62.4	1.01
					90	62.1	1.00
					150	61.2	1.00
					212	59.8	1.01

*Sublimation temperature. (At pressures below the triple-point pressure of 75.1 psia, carbon dioxide exists as a solid or gas. Also, the freezing-point temperature of carbon dioxide is the triple-point temperature of −69.8°F.)

TABLE A–3E

Properties of common liquids, solids, and foods (*Concluded*)

(*b*) Solids (values are for room temperature unless indicated otherwise)

Substance	Density, ρ lbm/ft^3	Specific heat, c_p Btu/lbm·R	Substance	Density, ρ lbm/ft^3	Specific heat, c_p Btu/lbm·R
Metals			**Nonmetals**		
Aluminum			Asphalt	132	0.220
−100°F		0.192	Brick, common	120	0.189
32°F		0.212	Brick, fireclay (500°C)	144	0.229
100°F	170	0.218	Concrete	144	0.156
200°F		0.224	Clay	62.4	0.220
300°F		0.229	Diamond	151	0.147
400°F		0.235	Glass, window	169	0.191
500°F		0.240	Glass, pyrex	139	0.200
Bronze (76% Cu, 2% Zn, 2% Al)	517	0.0955	Graphite	156	0.170
			Granite	169	0.243
Brass, yellow (65% Cu, 35% Zn)	519	0.0955	Gypsum or plaster board	50	0.260
			Ice		
Copper			−50°F		0.424
−60°F		0.0862	0°F		0.471
0°F		0.0893	20°F		0.491
100°F	555	0.0925	32°F	57.5	0.502
200°F		0.0938	Limestone	103	0.217
390°F		0.0963	Marble	162	0.210
Iron	490	0.107	Plywood (Douglas fir)	34.0	
Lead	705	0.030	Rubber (soft)	68.7	
Magnesium	108	0.239	Rubber (hard)	71.8	
Nickel	555	0.105	Sand	94.9	
Silver	655	0.056	Stone	93.6	
Steel, mild	489	0.119	Woods, hard (maple, oak, etc.)	45.0	
Tungsten	1211	0.031	Woods, soft (fir, pine, etc.)	32.0	

(*c*) Foods

Food	Water content, % (mass)	Freezing point, °F	Specific heat, Btu/lbm·R: Above freezing	Specific heat, Btu/lbm·R: Below freezing	Latent heat of fusion, Btu/lbm	Food	Water content, % (mass)	Freezing point, °F	Specific heat, Btu/lbm·R: Above freezing	Specific heat, Btu/lbm·R: Below freezing	Latent heat of fusion, Btu/lbm
Apples	84	30	0.873	0.453	121	Lettuce	95	32	0.961	0.487	136
Bananas	75	31	0.801	0.426	108	Milk, whole	88	31	0.905	0.465	126
Beef round	67	—	0.737	0.402	96	Oranges	87	31	0.897	0.462	125
Broccoli	90	31	0.921	0.471	129	Potatoes	78	31	0.825	0.435	112
Butter	16	—	—	0.249	23	Salmon fish	64	28	0.713	0.393	92
Cheese, Swiss	39	14	0.513	0.318	56	Shrimp	83	28	0.865	0.450	119
Cherries	80	29	0.841	0.441	115	Spinach	93	31	0.945	0.481	134
Chicken	74	27	0.793	0.423	106	Strawberries	90	31	0.921	0.471	129
Corn, sweet	74	31	0.793	0.423	106	Tomatoes, ripe	94	31	0.953	0.484	135
Eggs, whole	74	31	0.793	0.423	106	Turkey	64	—	0.713	0.393	92
Ice cream	63	22	0.705	0.390	90	Watermelon	93	31	0.945	0.481	134

Source: Values are obtained from various handbooks and other sources or are calculated. Water content and freezing-point data of foods are from ASHRAE, *Handbook of Fundamentals,* I-P version (Atlanta, GA: American Society of Heating, Refrigerating, and Air-Conditioning Engineers, Inc., 1993), Chap. 30, Table 1. Freezing point is the temperature at which freezing starts for fruits and vegetables, and the average freezing temperature for other foods.

TABLE A–4E

Saturated water—Temperature table

		Specific volume, ft³/lbm		*Internal energy, Btu/lbm*			*Enthalpy, Btu/lbm*			*Entropy, Btu/lbm·R*		
Temp., T °F	Sat. press., P_{sat} psia	Sat. liquid, v_f	Sat. vapor, v_g	Sat. liquid, u_f	Evap., u_{fg}	Sat. vapor, u_g	Sat. liquid, h_f	Evap., h_{fg}	Sat. vapor, h_g	Sat. liquid, s_f	Evap., s_{fg}	Sat. vapor, s_g
32.018	0.08871	0.01602	3299.9	0.000	1021.0	1021.0	0.000	1075.2	1075.2	0.00000	2.18672	2.1867
35	0.09998	0.01602	2945.7	3.004	1019.0	1022.0	3.004	1073.5	1076.5	0.00609	2.17011	2.1762
40	0.12173	0.01602	2443.6	8.032	1015.6	1023.7	8.032	1070.7	1078.7	0.01620	2.14271	2.1589
45	0.14756	0.01602	2035.8	13.05	1012.2	1025.3	13.05	1067.8	1080.9	0.02620	2.11587	2.1421
50	0.17812	0.01602	1703.1	18.07	1008.9	1026.9	18.07	1065.0	1083.1	0.03609	2.08956	2.1256
55	0.21413	0.01603	1430.4	23.07	1005.5	1028.6	23.07	1062.2	1085.3	0.04586	2.06377	2.1096
60	0.25638	0.01604	1206.1	28.08	1002.1	1030.2	28.08	1059.4	1087.4	0.05554	2.03847	2.0940
65	0.30578	0.01604	1020.8	33.08	998.76	1031.8	33.08	1056.5	1089.6	0.06511	2.01366	2.0788
70	0.36334	0.01605	867.18	38.08	995.39	1033.5	38.08	1053.7	1091.8	0.07459	1.98931	2.0639
75	0.43016	0.01606	739.27	43.07	992.02	1035.1	43.07	1050.9	1093.9	0.08398	1.96541	2.0494
80	0.50745	0.01607	632.41	48.06	988.65	1036.7	48.07	1048.0	1096.1	0.09328	1.94196	2.0352
85	0.59659	0.01609	542.80	53.06	985.28	1038.3	53.06	1045.2	1098.3	0.10248	1.91892	2.0214
90	0.69904	0.01610	467.40	58.05	981.90	1040.0	58.05	1042.4	1100.4	0.11161	1.89630	2.0079
95	0.81643	0.01612	403.74	63.04	978.52	1041.6	63.04	1039.5	1102.6	0.12065	1.87408	1.9947
100	0.95052	0.01613	349.83	68.03	975.14	1043.2	68.03	1036.7	1104.7	0.12961	1.85225	1.9819
110	1.2767	0.01617	264.96	78.01	968.36	1046.4	78.02	1031.0	1109.0	0.14728	1.80970	1.9570
120	1.6951	0.01620	202.94	88.00	961.56	1049.6	88.00	1025.2	1113.2	0.16466	1.76856	1.9332
130	2.2260	0.01625	157.09	97.99	954.73	1052.7	97.99	1019.4	1117.4	0.18174	1.72877	1.9105
140	2.8931	0.01629	122.81	107.98	947.87	1055.9	107.99	1013.6	1121.6	0.19855	1.69024	1.8888
150	3.7234	0.01634	96.929	117.98	940.98	1059.0	117.99	1007.8	1125.7	0.21508	1.65291	1.8680
160	4.7474	0.01639	77.185	127.98	934.05	1062.0	128.00	1001.8	1129.8	0.23136	1.61670	1.8481
170	5.9999	0.01645	61.982	138.00	927.08	1065.1	138.02	995.88	1133.9	0.24739	1.58155	1.8289
180	7.5197	0.01651	50.172	148.02	920.06	1068.1	148.04	989.85	1137.9	0.26318	1.54741	1.8106
190	9.3497	0.01657	40.920	158.05	912.99	1071.0	158.08	983.76	1141.8	0.27874	1.51421	1.7930
200	11.538	0.01663	33.613	168.10	905.87	1074.0	168.13	977.60	1145.7	0.29409	1.48191	1.7760
210	14.136	0.01670	27.798	178.15	898.68	1076.8	178.20	971.35	1149.5	0.30922	1.45046	1.7597
212	14.709	0.01671	26.782	180.16	897.24	1077.4	180.21	970.09	1150.3	0.31222	1.44427	1.7565
220	17.201	0.01677	23.136	188.22	891.43	1079.6	188.28	965.02	1153.3	0.32414	1.41980	1.7439
230	20.795	0.01684	19.374	198.31	884.10	1082.4	198.37	958.59	1157.0	0.33887	1.38989	1.7288
240	24.985	0.01692	16.316	208.41	876.70	1085.1	208.49	952.06	1160.5	0.35342	1.36069	1.7141
250	29.844	0.01700	13.816	218.54	869.21	1087.7	218.63	945.41	1164.0	0.36779	1.33216	1.6999
260	35.447	0.01708	11.760	228.68	861.62	1090.3	228.79	938.65	1167.4	0.38198	1.30425	1.6862
270	41.877	0.01717	10.059	238.85	853.94	1092.8	238.98	931.76	1170.7	0.39601	1.27694	1.6730
280	49.222	0.01726	8.6439	249.04	846.16	1095.2	249.20	924.74	1173.9	0.40989	1.25018	1.6601
290	57.573	0.01735	7.4607	259.26	838.27	1097.5	259.45	917.57	1177.0	0.42361	1.22393	1.6475
300	67.028	0.01745	6.4663	269.51	830.25	1099.8	269.73	910.24	1180.0	0.43720	1.19818	1.6354
310	77.691	0.01755	5.6266	279.79	822.11	1101.9	280.05	902.75	1182.8	0.45065	1.17289	1.6235
320	89.667	0.01765	4.9144	290.11	813.84	1104.0	290.40	895.09	1185.5	0.46396	1.14802	1.6120
330	103.07	0.01776	4.3076	300.46	805.43	1105.9	300.80	887.25	1188.1	0.47716	1.12355	1.6007
340	118.02	0.01787	3.7885	310.85	796.87	1107.7	311.24	879.22	1190.5	0.49024	1.09945	1.5897
350	134.63	0.01799	3.3425	321.29	788.16	1109.4	321.73	870.98	1192.7	0.50321	1.07570	1.5789
360	153.03	0.01811	2.9580	331.76	779.28	1111.0	332.28	862.53	1194.8	0.51607	1.05227	1.5683
370	173.36	0.01823	2.6252	342.29	770.23	1112.5	342.88	853.86	1196.7	0.52884	1.02914	1.5580
380	195.74	0.01836	2.3361	352.87	761.00	1113.9	353.53	844.96	1198.5	0.54152	1.00628	1.5478
390	220.33	0.01850	2.0842	363.50	751.58	1115.1	364.25	835.81	1200.1	0.55411	0.98366	1.5378

TABLE A–4E

Saturated water—Temperature table (*Concluded*)

		Specific volume, ft^3/lbm		Internal energy, Btu/lbm			Enthalpy, Btu/lbm			Entropy, Btu/lbm·R		
Temp., T °F	Sat. press., P_{sat} psia	Sat. liquid, v_f	Sat. vapor, v_g	Sat. liquid, u_f	Evap., u_{fg}	Sat. vapor, u_g	Sat. liquid, h_f	Evap., h_{fg}	Sat. vapor, h_g	Sat. liquid, s_f	Evap., s_{fg}	Sat. vapor, s_g
400	247.26	0.01864	1.8639	374.19	741.97	1116.2	375.04	826.39	1201.4	0.56663	0.96127	1.5279
410	276.69	0.01878	1.6706	384.94	732.14	1117.1	385.90	816.71	1202.6	0.57907	0.93908	1.5182
420	308.76	0.01894	1.5006	395.76	722.08	1117.8	396.84	806.74	1203.6	0.59145	0.91707	1.5085
430	343.64	0.01910	1.3505	406.65	711.80	1118.4	407.86	796.46	1204.3	0.60377	0.89522	1.4990
440	381.49	0.01926	1.2178	417.61	701.26	1118.9	418.97	785.87	1204.8	0.61603	0.87349	1.4895
450	422.47	0.01944	1.0999	428.66	690.47	1119.1	430.18	774.94	1205.1	0.62826	0.85187	1.4801
460	466.75	0.01962	0.99510	439.79	679.39	1119.2	441.48	763.65	1205.1	0.64044	0.83033	1.4708
470	514.52	0.01981	0.90158	451.01	668.02	1119.0	452.90	751.98	1204.9	0.65260	0.80885	1.4615
480	565.96	0.02001	0.81794	462.34	656.34	1118.7	464.43	739.91	1204.3	0.66474	0.78739	1.4521
490	621.24	0.02022	0.74296	473.77	644.32	1118.1	476.09	727.40	1203.5	0.67686	0.76594	1.4428
500	680.56	0.02044	0.67558	485.32	631.94	1117.3	487.89	714.44	1202.3	0.68899	0.74445	1.4334
510	744.11	0.02067	0.61489	496.99	619.17	1116.2	499.84	700.99	1200.8	0.70112	0.72290	1.4240
520	812.11	0.02092	0.56009	508.80	605.99	1114.8	511.94	687.01	1199.0	0.71327	0.70126	1.4145
530	884.74	0.02118	0.51051	520.76	592.35	1113.1	524.23	672.47	1196.7	0.72546	0.67947	1.4049
540	962.24	0.02146	0.46553	532.88	578.23	1111.1	536.70	657.31	1194.0	0.73770	0.65751	1.3952
550	1044.8	0.02176	0.42465	545.18	563.58	1108.8	549.39	641.47	1190.9	0.75000	0.63532	1.3853
560	1132.7	0.02207	0.38740	557.68	548.33	1106.0	562.31	624.91	1187.2	0.76238	0.61284	1.3752
570	1226.2	0.02242	0.35339	570.40	532.45	1102.8	575.49	607.55	1183.0	0.77486	0.59003	1.3649
580	1325.5	0.02279	0.32225	583.37	515.84	1099.2	588.95	589.29	1178.2	0.78748	0.56679	1.3543
590	1430.8	0.02319	0.29367	596.61	498.43	1095.0	602.75	570.04	1172.8	0.80026	0.54306	1.3433
600	1542.5	0.02362	0.26737	610.18	480.10	1090.3	616.92	549.67	1166.6	0.81323	0.51871	1.3319
610	1660.9	0.02411	0.24309	624.11	460.73	1084.8	631.52	528.03	1159.5	0.82645	0.49363	1.3201
620	1786.2	0.02464	0.22061	638.47	440.14	1078.6	646.62	504.92	1151.5	0.83998	0.46765	1.3076
630	1918.9	0.02524	0.19972	653.35	418.12	1071.5	662.32	480.07	1142.4	0.85389	0.44056	1.2944
640	2059.3	0.02593	0.18019	668.86	394.36	1063.2	678.74	453.14	1131.9	0.86828	0.41206	1.2803
650	2207.8	0.02673	0.16184	685.16	368.44	1053.6	696.08	423.65	1119.7	0.88332	0.38177	1.2651
660	2364.9	0.02767	0.14444	702.48	339.74	1042.2	714.59	390.84	1105.4	0.89922	0.34906	1.2483
670	2531.2	0.02884	0.12774	721.23	307.22	1028.5	734.74	353.54	1088.3	0.91636	0.31296	1.2293
680	2707.3	0.03035	0.11134	742.11	269.00	1011.1	757.32	309.57	1066.9	0.93541	0.27163	1.2070
690	2894.1	0.03255	0.09451	766.81	220.77	987.6	784.24	253.96	1038.2	0.95797	0.22089	1.1789
700	3093.0	0.03670	0.07482	801.75	146.50	948.3	822.76	168.32	991.1	0.99023	0.14514	1.1354
705.10	3200.1	0.04975	0.04975	866.61	0	866.6	896.07	0	896.1	1.05257	0	1.0526

Source: Tables A–4E through A–8E are generated using the Engineering Equation Solver (EES) software developed by S. A. Klein and F. L. Alvarado. The routine used in calculations is the highly accurate Steam_IAPWS, which incorporates the 1995 Formulation for the Thermodynamic Properties of Ordinary Water Substance for General and Scientific Use, issued by The International Association for the Properties of Water and Steam (IAPWS). This formulation replaces the 1984 formulation of Haar, Gallagher, and Kell (NBS/NRC Steam Tables, Hemisphere Publishing Co., 1984), which is also available in EES as the routine STEAM. The new formulation is based on the correlations of Saul and Wagner (J. Phys. Chem. Ref. Data, 16, 893, 1987) with modifications to adjust to the International Temperature Scale of 1990. The modifications are described by Wagner and Pruss (J. Phys. Chem. Ref. Data, 22, 783, 1993). The properties of ice are based on Hyland and Wexler, "Formulations for the Thermodynamic Properties of the Saturated Phases of H_2O from 173.15 K to 473.15 K," *ASHRAE Trans.*, Part 2A, Paper 2793, 1983.

TABLE A–5E

Saturated water—Pressure table

		Specific volume, ft³/lbm		*Internal energy,* Btu/lbm			*Enthalpy,* Btu/lbm			*Entropy,* Btu/lbm·R		
Press., P psia	Sat. temp., T_{sat} °F	Sat. liquid, v_f	Sat. vapor, v_g	Sat. liquid, u_f	Evap., u_{fg}	Sat. vapor, u_g	Sat. liquid, h_f	Evap., h_{fg}	Sat. vapor, h_g	Sat. liquid, s_f	Evap., s_{fg}	Sat. vapor, s_g
1	101.69	0.01614	333.49	69.72	973.99	1043.7	69.72	1035.7	1105.4	0.13262	1.84495	1.9776
2	126.02	0.01623	173.71	94.02	957.45	1051.5	94.02	1021.7	1115.8	0.17499	1.74444	1.9194
3	141.41	0.01630	118.70	109.39	946.90	1056.3	109.40	1012.8	1122.2	0.20090	1.68489	1.8858
4	152.91	0.01636	90.629	120.89	938.97	1059.9	120.90	1006.0	1126.9	0.21985	1.64225	1.8621
5	162.18	0.01641	73.525	130.17	932.53	1062.7	130.18	1000.5	1130.7	0.23488	1.60894	1.8438
6	170.00	0.01645	61.982	138.00	927.08	1065.1	138.02	995.88	1133.9	0.24739	1.58155	1.8289
8	182.81	0.01652	47.347	150.83	918.08	1068.9	150.86	988.15	1139.0	0.26757	1.53800	1.8056
10	193.16	0.01659	38.425	161.22	910.75	1072.0	161.25	981.82	1143.1	0.28362	1.50391	1.7875
14.696	211.95	0.01671	26.805	180.12	897.27	1077.4	180.16	970.12	1150.3	0.31215	1.44441	1.7566
15	212.99	0.01672	26.297	181.16	896.52	1077.7	181.21	969.47	1150.7	0.31370	1.44441	1.7549
20	227.92	0.01683	20.093	196.21	885.63	1081.8	196.27	959.93	1156.2	0.33582	1.39606	1.7319
25	240.03	0.01692	16.307	208.45	876.67	1085.1	208.52	952.03	1160.6	0.35347	1.36060	1.7141
30	250.30	0.01700	13.749	218.84	868.98	1087.8	218.93	945.21	1164.1	0.36821	1.33132	1.6995
35	259.25	0.01708	11.901	227.92	862.19	1090.1	228.03	939.16	1167.2	0.38093	1.30632	1.6872
40	267.22	0.01715	10.501	236.02	856.09	1092.1	236.14	933.69	1169.8	0.39213	1.28448	1.6766
45	274.41	0.01721	9.4028	243.34	850.52	1093.9	243.49	928.68	1172.2	0.40216	1.26506	1.6672
50	280.99	0.01727	8.5175	250.05	845.39	1095.4	250.21	924.03	1174.2	0.41125	1.24756	1.6588
55	287.05	0.01732	7.7882	256.25	840.61	1096.9	256.42	919.70	1176.1	0.41958	1.23162	1.6512
60	292.69	0.01738	7.1766	262.01	836.13	1098.1	262.20	915.61	1177.8	0.42728	1.21697	1.6442
65	297.95	0.01743	6.6560	267.41	831.90	1099.3	267.62	911.75	1179.4	0.43443	1.20341	1.6378
70	302.91	0.01748	6.2075	272.50	827.90	1100.4	272.72	908.08	1180.8	0.44112	1.19078	1.6319
75	307.59	0.01752	5.8167	277.31	824.09	1101.4	277.55	904.58	1182.1	0.44741	1.17895	1.6264
80	312.02	0.01757	5.4733	281.87	820.45	1102.3	282.13	901.22	1183.4	0.45335	1.16783	1.6212
85	316.24	0.01761	5.1689	286.22	816.97	1103.2	286.50	898.00	1184.5	0.45897	1.15732	1.6163
90	320.26	0.01765	4.8972	290.38	813.62	1104.0	290.67	894.89	1185.6	0.46431	1.14737	1.6117
95	324.11	0.01770	4.6532	294.36	810.40	1104.8	294.67	891.89	1186.6	0.46941	1.13791	1.6073
100	327.81	0.01774	4.4327	298.19	807.29	1105.5	298.51	888.99	1187.5	0.47427	1.12888	1.6032
110	334.77	0.01781	4.0410	305.41	801.37	1106.8	305.78	883.44	1189.2	0.48341	1.11201	1.5954
120	341.25	0.01789	3.7289	312.16	795.79	1107.9	312.55	878.20	1190.8	0.49187	1.09646	1.5883
130	347.32	0.01796	3.4557	318.48	790.51	1109.0	318.92	873.21	1192.1	0.49974	1.08204	1.5818
140	353.03	0.01802	3.2202	324.45	785.49	1109.9	324.92	868.45	1193.4	0.50711	1.06858	1.5757
150	358.42	0.01809	3.0150	330.11	780.69	1110.8	330.61	863.88	1194.5	0.51405	1.05595	1.5700
160	363.54	0.01815	2.8347	335.49	776.10	1111.6	336.02	859.49	1195.5	0.52061	1.04405	1.5647
170	368.41	0.01821	2.6749	340.62	771.68	1112.3	341.19	855.25	1196.4	0.52682	1.03279	1.5596
180	373.07	0.01827	2.5322	345.53	767.42	1113.0	346.14	851.16	1197.3	0.53274	1.02210	1.5548
190	377.52	0.01833	2.4040	350.24	763.31	1113.6	350.89	847.19	1198.1	0.53839	1.01191	1.5503
200	381.80	0.01839	2.2882	354.78	759.32	1114.1	355.46	843.33	1198.8	0.54379	1.00219	1.5460
250	400.97	0.01865	1.8440	375.23	741.02	1116.3	376.09	825.47	1201.6	0.56784	0.95912	1.5270
300	417.35	0.01890	1.5435	392.89	724.77	1117.7	393.94	809.41	1203.3	0.58818	0.92289	1.5111
350	431.74	0.01912	1.3263	408.55	709.98	1118.5	409.79	794.65	1204.4	0.60590	0.89143	1.4973
400	444.62	0.01934	1.1617	422.70	696.31	1119.0	424.13	780.87	1205.0	0.62168	0.86350	1.4852
450	456.31	0.01955	1.0324	435.67	683.52	1119.2	437.30	767.86	1205.2	0.63595	0.83828	1.4742
500	467.04	0.01975	0.92819	447.68	671.42	1119.1	449.51	755.48	1205.0	0.64900	0.81521	1.4642
550	476.97	0.01995	0.84228	458.90	659.91	1118.8	460.93	743.60	1204.5	0.66107	0.79388	1.4550
600	486.24	0.02014	0.77020	469.46	648.88	1118.3	471.70	732.15	1203.9	0.67231	0.77400	1.4463

TABLE A–5E

Saturated water—Pressure table (*Concluded*)

		Specific volume, ft^3/lbm		Internal energy, Btu/lbm			Enthalpy, Btu/lbm			Entropy, Btu/lbm·R		
Press., P psia	Sat. temp., T_{sat} °F	Sat. liquid, v_f	Sat. vapor, v_g	Sat. liquid, u_f	Evap., u_{fg}	Sat. vapor, u_g	Sat. liquid, h_f	Evap., h_{fg}	Sat. vapor, h_g	Sat. liquid, s_f	Evap., s_{fg}	Sat. vapor, s_g
700	503.13	0.02051	0.65589	488.96	627.98	1116.9	491.62	710.29	1201.9	0.69279	0.73771	1.4305
800	518.27	0.02087	0.56920	506.74	608.30	1115.0	509.83	689.48	1199.3	0.71117	0.70502	1.4162
900	532.02	0.02124	0.50107	523.19	589.54	1112.7	526.73	669.46	1196.2	0.72793	0.67505	1.4030
1000	544.65	0.02159	0.44604	538.58	571.49	1110.1	542.57	650.03	1192.6	0.74341	0.64722	1.3906
1200	567.26	0.02232	0.36241	566.89	536.87	1103.8	571.85	612.39	1184.2	0.77143	0.59632	1.3677
1400	587.14	0.02307	0.30161	592.79	503.50	1096.3	598.76	575.66	1174.4	0.79658	0.54991	1.3465
1600	604.93	0.02386	0.25516	616.99	470.69	1087.7	624.06	539.18	1163.2	0.81972	0.50645	1.3262
1800	621.07	0.02470	0.21831	640.03	437.86	1077.9	648.26	502.35	1150.6	0.84144	0.46482	1.3063
2000	635.85	0.02563	0.18815	662.33	404.46	1066.8	671.82	464.60	1136.4	0.86224	0.42409	1.2863
2500	668.17	0.02860	0.13076	717.67	313.53	1031.2	730.90	360.79	1091.7	0.91311	0.31988	1.2330
3000	695.41	0.03433	0.08460	783.39	186.41	969.8	802.45	214.32	1016.8	0.97321	0.18554	1.1587
3200.1	705.10	0.04975	0.04975	866.61	0	866.6	896.07	0	896.1	1.05257	0	1.0526

TABLE A–6E

Superheated water

T °F	*v* ft³/lbm	*u* Btu/lbm	*h* Btu/lbm	*s* Btu/lbm·R	*v* ft³/lbm	*u* Btu/lbm	*h* Btu/lbm	*s* Btu/lbm·R	*v* ft³/lbm	*u* Btu/lbm	*h* Btu/lbm	*s* Btu/lbm·R
	P = 1.0 psia (101.69°F)*				*P* = 5.0 psia (162.18°F)				*P* = 10 psia (193.16°F)			
Sat.†	333.49	1043.7	1105.4	1.9776	73.525	1062.7	1130.7	1.8438	38.425	1072.0	1143.1	1.7875
200	392.53	1077.5	1150.1	2.0509	78.153	1076.2	1148.5	1.8716	38.849	1074.5	1146.4	1.7926
240	416.44	1091.2	1168.3	2.0777	83.009	1090.3	1167.1	1.8989	41.326	1089.1	1165.5	1.8207
280	440.33	1105.0	1186.5	2.1030	87.838	1104.3	1185.6	1.9246	43.774	1103.4	1184.4	1.8469
320	464.20	1118.9	1204.8	2.1271	92.650	1118.4	1204.1	1.9490	46.205	1117.6	1203.1	1.8716
360	488.07	1132.9	1223.3	2.1502	97.452	1132.5	1222.6	1.9722	48.624	1131.9	1221.8	1.8950
400	511.92	1147.1	1241.8	2.1722	102.25	1146.7	1241.3	1.9944	51.035	1146.2	1240.6	1.9174
440	535.77	1161.3	1260.4	2.1934	107.03	1160.9	1260.0	2.0156	53.441	1160.5	1259.4	1.9388
500	571.54	1182.8	1288.6	2.2237	114.21	1182.6	1288.2	2.0461	57.041	1182.2	1287.8	1.9693
600	631.14	1219.4	1336.2	2.2709	126.15	1219.2	1335.9	2.0933	63.029	1219.0	1335.6	2.0167
700	690.73	1256.8	1384.6	2.3146	138.09	1256.7	1384.4	2.1371	69.007	1256.5	1384.2	2.0605
800	750.31	1295.1	1433.9	2.3553	150.02	1294.9	1433.7	2.1778	74.980	1294.8	1433.5	2.1013
1000	869.47	1374.2	1535.1	2.4299	173.86	1374.2	1535.0	2.2524	86.913	1374.1	1534.9	2.1760
1200	988.62	1457.1	1640.0	2.4972	197.70	1457.0	1640.0	2.3198	98.840	1457.0	1639.9	2.2433
1400	1107.8	1543.7	1748.7	2.5590	221.54	1543.7	1748.7	2.3816	110.762	1543.6	1748.6	2.3052
	P = 15 psia (212.99°F)				*P* = 20 psia (227.92°F)				*P* = 40 psia (267.22°F)			
Sat.	26.297	1077.7	1150.7	1.7549	20.093	1081.8	1156.2	1.7319	10.501	1092.1	1169.8	1.6766
240	27.429	1087.8	1163.9	1.7742	20.478	1086.5	1162.3	1.7406				
280	29.085	1102.4	1183.2	1.8010	21.739	1101.4	1181.9	1.7679	10.713	1097.3	1176.6	1.6858
320	30.722	1116.9	1202.2	1.8260	22.980	1116.1	1201.2	1.7933	11.363	1112.9	1197.1	1.7128
360	32.348	1131.3	1221.1	1.8496	24.209	1130.7	1220.2	1.8171	11.999	1128.1	1216.9	1.7376
400	33.965	1145.7	1239.9	1.8721	25.429	1145.1	1239.3	1.8398	12.625	1143.1	1236.5	1.7610
440	35.576	1160.1	1258.8	1.8936	26.644	1159.7	1258.3	1.8614	13.244	1157.9	1256.0	1.7831
500	37.986	1181.9	1287.3	1.9243	28.458	1181.6	1286.9	1.8922	14.165	1180.2	1285.0	1.8143
600	41.988	1218.7	1335.3	1.9718	31.467	1218.5	1334.9	1.9398	15.686	1217.5	1333.6	1.8625
700	45.981	1256.3	1383.9	2.0156	34.467	1256.1	1383.7	1.9837	17.197	1255.3	1382.6	1.9067
800	49.967	1294.6	1433.3	2.0565	37.461	1294.5	1433.1	2.0247	18.702	1293.9	1432.3	1.9478
1000	57.930	1374.0	1534.8	2.1312	43.438	1373.8	1534.6	2.0994	21.700	1373.4	1534.1	2.0227
1200	65.885	1456.9	1639.8	2.1986	49.407	1456.8	1639.7	2.1668	24.691	1456.5	1639.3	2.0902
1400	73.836	1543.6	1748.5	2.2604	55.373	1543.5	1748.4	2.2287	27.678	1543.3	1748.1	2.1522
1600	81.784	1634.0	1861.0	2.3178	61.335	1633.9	1860.9	2.2861	30.662	1633.7	1860.7	2.2096
	P = 60 psia (292.69°F)				*P* = 80 psia (312.02°F)				*P* = 100 psia (327.81°F)			
Sat.	7.1766	1098.1	1177.8	1.6442	5.4733	1102.3	1183.4	1.6212	4.4327	1105.5	1187.5	1.6032
320	7.4863	1109.6	1192.7	1.6636	5.5440	1105.9	1187.9	1.6271				
360	7.9259	1125.5	1213.5	1.6897	5.8876	1122.7	1209.9	1.6545	4.6628	1119.8	1206.1	1.6263
400	8.3548	1140.9	1233.7	1.7138	6.2187	1138.7	1230.8	1.6794	4.9359	1136.4	1227.8	1.6521
440	8.7766	1156.1	1253.6	1.7364	6.5420	1154.3	1251.2	1.7026	5.2006	1152.4	1248.7	1.6759
500	9.4005	1178.8	1283.1	1.7682	7.0177	1177.3	1281.2	1.7350	5.5876	1175.9	1279.3	1.7088
600	10.4256	1216.5	1332.2	1.8168	7.7951	1215.4	1330.8	1.7841	6.2167	1214.4	1329.4	1.7586
700	11.4401	1254.5	1381.6	1.8613	8.5616	1253.8	1380.5	1.8289	6.8344	1253.0	1379.5	1.8037
800	12.4484	1293.3	1431.5	1.9026	9.3218	1292.6	1430.6	1.8704	7.4457	1292.0	1429.8	1.8453
1000	14.4543	1373.0	1533.5	1.9777	10.8313	1372.6	1532.9	1.9457	8.6575	1372.2	1532.4	1.9208
1200	16.4525	1456.2	1638.9	2.0454	12.3331	1455.9	1638.5	2.0135	9.8615	1455.6	1638.1	1.9887
1400	18.4464	1543.0	1747.8	2.1073	13.8306	1542.8	1747.5	2.0755	11.0612	1542.6	1747.2	2.0508
1600	20.438	1633.5	1860.5	2.1648	15.3257	1633.3	1860.2	2.1330	12.2584	1633.2	1860.0	2.1083
1800	22.428	1727.6	1976.6	2.2187	16.8192	1727.5	1976.5	2.1869	13.4541	1727.3	1976.3	2.1622
2000	24.417	1825.2	2096.3	2.2694	18.3117	1825.0	2096.1	2.2376	14.6487	1824.9	2096.0	2.2130

*The temperature in parentheses is the saturation temperature at the specified pressure.

† Properties of saturated vapor at the specified pressure.

TABLE A–6E

Superheated water (*Concluded*)

T °F	*v* ft³/lbm	*u* Btu/lbm	*h* Btu/lbm	*s* Btu/lbm·R	*v* ft³/lbm	*u* Btu/lbm	*h* Btu/lbm	*s* Btu/lbm·R	*v* ft³/lbm	*u* Btu/lbm	*h* Btu/lbm	*s* Btu/lbm·R
	P = 120 psia (341.25°F)				*P* = 140 psia (353.03°F)				*P* = 160 psia (363.54°F)			
Sat.	3.7289	1107.9	1190.8	1.5883	3.2202	1109.9	1193.4	1.5757	2.8347	1111.6	1195.5	1.5647
360	3.8446	1116.7	1202.1	1.6023	3.2584	1113.4	1197.8	1.5811				
400	4.0799	1134.0	1224.6	1.6292	3.4676	1131.5	1221.4	1.6092	3.0076	1129.0	1218.0	1.5914
450	4.3613	1154.5	1251.4	1.6594	3.7147	1152.6	1248.9	1.6403	3.2293	1150.7	1246.3	1.6234
500	4.6340	1174.4	1277.3	1.6872	3.9525	1172.9	1275.3	1.6686	3.4412	1171.4	1273.2	1.6522
550	4.9010	1193.9	1302.8	1.7131	4.1845	1192.7	1301.1	1.6948	3.6469	1191.4	1299.4	1.6788
600	5.1642	1213.4	1328.0	1.7375	4.4124	1212.3	1326.6	1.7195	3.8484	1211.3	1325.2	1.7037
700	5.6829	1252.2	1378.4	1.7829	4.8604	1251.4	1377.3	1.7652	4.2434	1250.6	1376.3	1.7498
800	6.1950	1291.4	1429.0	1.8247	5.3017	1290.8	1428.1	1.8072	4.6316	1290.2	1427.3	1.7920
1000	7.2083	1371.7	1531.8	1.9005	6.1732	1371.3	1531.3	1.8832	5.3968	1370.9	1530.7	1.8682
1200	8.2137	1455.3	1637.7	1.9684	7.0367	1455.0	1637.3	1.9512	6.1540	1454.7	1636.9	1.9363
1400	9.2149	1542.3	1746.9	2.0305	7.8961	1542.1	1746.6	2.0134	6.9070	1541.8	1746.3	1.9986
1600	10.2135	1633.0	1859.8	2.0881	8.7529	1632.8	1859.5	2.0711	7.6574	1632.6	1859.3	2.0563
1800	11.2106	1727.2	1976.1	2.1420	9.6082	1727.0	1975.9	2.1250	8.4063	1726.9	1975.7	2.1102
2000	12.2067	1824.8	2095.8	2.1928	10.4624	1824.6	2095.7	2.1758	9.1542	1824.5	2095.5	2.1610
	P = 180 psia (373.07°F)				*P* = 200 psia (381.80°F)				*P* = 225 psia (391.80°F)			
Sat.	2.5322	1113.0	1197.3	1.5548	2.2882	1114.1	1198.8	1.5460	2.0423	1115.3	1200.3	1.5360
400	2.6490	1126.3	1214.5	1.5752	2.3615	1123.5	1210.9	1.5602	2.0728	1119.7	1206.0	1.5427
450	2.8514	1148.7	1243.7	1.6082	2.5488	1146.7	1241.0	1.5943	2.2457	1144.1	1237.6	1.5783
500	3.0433	1169.8	1271.2	1.6376	2.7247	1168.2	1269.0	1.6243	2.4059	1166.2	1266.3	1.6091
550	3.2286	1190.2	1297.7	1.6646	2.8939	1188.9	1296.0	1.6516	2.5590	1187.2	1293.8	1.6370
600	3.4097	1210.2	1323.8	1.6897	3.0586	1209.1	1322.3	1.6771	2.7075	1207.7	1320.5	1.6628
700	3.7635	1249.8	1375.2	1.7361	3.3796	1249.0	1374.1	1.7238	2.9956	1248.0	1372.7	1.7099
800	4.1104	1289.5	1426.5	1.7785	3.6934	1288.9	1425.6	1.7664	3.2765	1288.1	1424.5	1.7528
900	4.4531	1329.7	1478.0	1.8179	4.0031	1329.2	1477.3	1.8059	3.5530	1328.5	1476.5	1.7925
1000	4.7929	1370.5	1530.1	1.8549	4.3099	1370.1	1529.6	1.8430	3.8268	1369.5	1528.9	1.8296
1200	5.4674	1454.3	1636.5	1.9231	4.9182	1454.0	1636.1	1.9113	4.3689	1453.6	1635.6	1.8981
1400	6.1377	1541.6	1746.0	1.9855	5.5222	1541.4	1745.7	1.9737	4.9068	1541.1	1745.4	1.9606
1600	6.8054	1632.4	1859.1	2.0432	6.1238	1632.2	1858.8	2.0315	5.4422	1632.0	1858.6	2.0184
1800	7.4716	1726.7	1975.6	2.0971	6.7238	1726.5	1975.4	2.0855	5.9760	1726.4	1975.2	2.0724
2000	8.1367	1824.4	2095.4	2.1479	7.3227	1824.3	2095.3	2.1363	6.5087	1824.1	2095.1	2.1232
	P = 250 psia (400.97°F)				*P* = 275 psia (409.45°F)				*P* = 300 psia (417.35°F)			
Sat.	1.8440	1116.3	1201.6	1.5270	1.6806	1117.0	1202.6	1.5187	1.5435	1117.7	1203.3	1.5111
450	2.0027	1141.3	1234.0	1.5636	1.8034	1138.5	1230.3	1.5499	1.6369	1135.6	1226.4	1.5369
500	2.1506	1164.1	1263.6	1.5953	1.9415	1162.0	1260.8	1.5825	1.7670	1159.8	1257.9	1.5706
550	2.2910	1185.6	1291.5	1.6237	2.0715	1183.9	1289.3	1.6115	1.8885	1182.1	1287.0	1.6001
600	2.4264	1206.3	1318.6	1.6499	2.1964	1204.9	1316.7	1.6380	2.0046	1203.5	1314.8	1.6270
650	2.5586	1226.8	1345.1	1.6743	2.3179	1225.6	1343.5	1.6627	2.1172	1224.4	1341.9	1.6520
700	2.6883	1247.0	1371.4	1.6974	2.4369	1246.0	1370.0	1.6860	2.2273	1244.9	1368.6	1.6755
800	2.9429	1287.3	1423.5	1.7406	2.6699	1286.5	1422.4	1.7294	2.4424	1285.7	1421.3	1.7192
900	3.1930	1327.9	1475.6	1.7804	2.8984	1327.3	1474.8	1.7694	2.6529	1326.6	1473.9	1.7593
1000	3.4403	1369.0	1528.2	1.8177	3.1241	1368.5	1527.4	1.8068	2.8605	1367.9	1526.7	1.7968
1200	3.9295	1453.3	1635.0	1.8863	3.5700	1452.9	1634.5	1.8755	3.2704	1452.5	1634.0	1.8657
1400	4.4144	1540.8	1745.0	1.9488	4.0116	1540.5	1744.6	1.9381	3.6759	1540.2	1744.2	1.9284
1600	4.8969	1631.7	1858.3	2.0066	4.4507	1631.5	1858.0	1.9960	4.0789	1631.3	1857.7	1.9863
1800	5.3777	1726.2	1974.9	2.0607	4.8882	1726.0	1974.7	2.0501	4.4803	1725.8	1974.5	2.0404
2000	5.8575	1823.9	2094.9	2.1116	5.3247	1823.8	2094.7	2.1010	4.8807	1823.6	2094.6	2.0913

TABLE A–6E

Superheated water (*Continued*)

T °F	*v* ft³/lbm	*u* Btu/lbm	*h* Btu/lbm	*s* Btu/lbm·R	*v* ft³/lbm	*u* Btu/lbm	*h* Btu/lbm	*s* Btu/lbm·R	*v* ft³/lbm	*u* Btu/lbm	*h* Btu/lbm	*s* Btu/lbm·R
	P = 350 psia (431.74°F)				*P* = 400 psia (444.62°F)				*P* = 450 psia (456.31°F)			
Sat.	1.3263	1118.5	1204.4	1.4973	1.1617	1119.0	1205.0	1.4852	1.0324	1119.2	1205.2	1.4742
450	1.3739	1129.3	1218.3	1.5128	1.1747	1122.5	1209.4	1.4901				
500	1.4921	1155.2	1251.9	1.5487	1.2851	1150.4	1245.6	1.5288	1.1233	1145.4	1238.9	1.5103
550	1.6004	1178.6	1282.2	1.5795	1.3840	1174.9	1277.3	1.5610	1.2152	1171.1	1272.3	1.5441
600	1.7030	1200.6	1310.9	1.6073	1.4765	1197.6	1306.9	1.5897	1.3001	1194.6	1302.8	1.5737
650	1.8018	1221.9	1338.6	1.6328	1.5650	1219.4	1335.3	1.6158	1.3807	1216.9	1331.9	1.6005
700	1.8979	1242.8	1365.8	1.6567	1.6507	1240.7	1362.9	1.6401	1.4584	1238.5	1360.0	1.6253
800	2.0848	1284.1	1419.1	1.7009	1.8166	1282.5	1417.0	1.6849	1.6080	1280.8	1414.7	1.6706
900	2.2671	1325.3	1472.2	1.7414	1.9777	1324.0	1470.4	1.7257	1.7526	1322.7	1468.6	1.7117
1000	2.4464	1366.9	1525.3	1.7791	2.1358	1365.8	1523.9	1.7636	1.8942	1364.7	1522.4	1.7499
1200	2.7996	1451.7	1633.0	1.8483	2.4465	1450.9	1632.0	1.8331	2.1718	1450.1	1631.0	1.8196
1400	3.1484	1539.6	1743.5	1.9111	2.7527	1539.0	1742.7	1.8960	2.4450	1538.4	1742.0	1.8827
1600	3.4947	1630.8	1857.1	1.9691	3.0565	1630.3	1856.5	1.9541	2.7157	1629.8	1856.0	1.9409
1800	3.8394	1725.4	1974.0	2.0233	3.3586	1725.0	1973.6	2.0084	2.9847	1724.6	1973.2	1.9952
2000	4.1830	1823.3	2094.2	2.0742	3.6597	1823.0	2093.9	2.0594	3.2527	1822.6	2093.5	2.0462
	P = 500 psia (467.04°F)				*P* = 600 psia (486.24°F)				*P* = 700 psia (503.13°F)			
Sat.	0.92815	1119.1	1205.0	1.4642	0.77020	1118.3	1203.9	1.4463	0.65589	1116.9	1201.9	1.4305
500	0.99304	1140.1	1231.9	1.4928	0.79526	1128.2	1216.5	1.4596				
550	1.07974	1167.1	1267.0	1.5284	0.87542	1158.7	1255.9	1.4996	0.72799	1149.5	1243.8	1.4730
600	1.15876	1191.4	1298.6	1.5590	0.94605	1184.9	1289.9	1.5325	0.79332	1177.9	1280.7	1.5087
650	1.23312	1214.3	1328.4	1.5865	1.01133	1209.0	1321.3	1.5614	0.85242	1203.4	1313.8	1.5393
700	1.30440	1236.4	1357.0	1.6117	1.07316	1231.9	1351.0	1.5877	0.90769	1227.2	1344.8	1.5666
800	1.44097	1279.2	1412.5	1.6576	1.19038	1275.8	1408.0	1.6348	1.01125	1272.4	1403.4	1.6150
900	1.57252	1321.4	1466.9	1.6992	1.30230	1318.7	1463.3	1.6771	1.10921	1316.0	1459.7	1.6581
1000	1.70094	1363.6	1521.0	1.7376	1.41097	1361.4	1518.1	1.7160	1.20381	1359.2	1515.2	1.6974
1100	1.82726	1406.2	1575.3	1.7735	1.51749	1404.4	1572.9	1.7522	1.29621	1402.5	1570.4	1.7341
1200	1.95211	1449.4	1630.0	1.8075	1.62252	1447.8	1627.9	1.7865	1.38709	1446.2	1625.9	1.7685
1400	2.1988	1537.8	1741.2	1.8708	1.82957	1536.6	1739.7	1.8501	1.56580	1535.4	1738.2	1.8324
1600	2.4430	1629.4	1855.4	1.9291	2.0340	1628.4	1854.2	1.9085	1.74192	1627.5	1853.1	1.8911
1800	2.6856	1724.2	1972.7	1.9834	2.2369	1723.4	1971.8	1.9630	1.91643	1722.7	1970.9	1.9457
2000	2.9271	1822.3	2093.1	2.0345	2.4387	1821.7	2092.4	2.0141	2.08987	1821.0	2091.7	1.9969
	P = 800 psia (518.27°F)				*P* = 1000 psia (544.65°F)				*P* = 1250 psia (572.45°F)			
Sat.	0.56920	1115.0	1199.3	1.4162	0.44604	1110.1	1192.6	1.3906	0.34549	1102.0	1181.9	1.3623
550	0.61586	1139.4	1230.5	1.4476	0.45375	1115.2	1199.2	1.3972				
600	0.67799	1170.5	1270.9	1.4866	0.51431	1154.1	1249.3	1.4457	0.37894	1129.5	1217.2	1.3961
650	0.73279	1197.6	1306.0	1.5191	0.56411	1185.1	1289.5	1.4827	0.42703	1167.5	1266.3	1.4414
700	0.78330	1222.4	1338.4	1.5476	0.60844	1212.4	1325.0	1.5140	0.46735	1198.7	1306.8	1.4771
750	0.83102	1246.0	1369.1	1.5735	0.64944	1237.6	1357.8	1.5418	0.50344	1226.4	1342.9	1.5076
800	0.87678	1268.9	1398.7	1.5975	0.68821	1261.7	1389.0	1.5670	0.53687	1252.2	1376.4	1.5347
900	0.96434	1313.3	1456.0	1.6413	0.76136	1307.7	1448.6	1.6126	0.59876	1300.5	1439.0	1.5826
1000	1.04841	1357.0	1512.2	1.6812	0.83078	1352.5	1506.2	1.6535	0.65656	1346.7	1498.6	1.6249
1100	1.13024	1400.7	1568.0	1.7181	0.89783	1396.9	1563.1	1.6911	0.71184	1392.2	1556.8	1.6635
1200	1.21051	1444.6	1623.8	1.7528	0.96327	1441.4	1619.7	1.7263	0.76545	1437.4	1614.5	1.6993
1400	1.36797	1534.2	1736.7	1.8170	1.09101	1531.8	1733.7	1.7911	0.86944	1528.7	1729.8	1.7649
1600	1.52283	1626.5	1851.9	1.8759	1.21610	1624.6	1849.6	1.8504	0.97072	1622.2	1846.7	1.8246
1800	1.67606	1721.9	1970.0	1.9306	1.33956	1720.3	1968.2	1.9053	1.07036	1718.4	1966.0	1.8799
2000	1.82823	1820.4	2091.0	1.9819	1.46194	1819.1	2089.6	1.9568	1.16892	1817.5	2087.9	1.9315

TABLE A–6E

Superheated water (*Concluded*)

T °F	v ft³/lbm	u Btu/lbm	h Btu/lbm	s Btu/lbm·R	v ft³/lbm	u Btu/lbm	h Btu/lbm	s Btu/lbm·R	v ft³/lbm	u Btu/lbm	h Btu/lbm	s Btu/lbm·R
	P = 1500 psia (596.26°F)				*P* = 1750 psia (617.17°F)				*P* = 2000 psia (635.85°F)			
Sat.	0.27695	1092.1	1169.0	1.3362	0.22681	1080.5	1153.9	1.3112	0.18815	1066.8	1136.4	1.2863
600	0.28189	1097.2	1175.4	1.3423								
650	0.33310	1147.2	1239.7	1.4016	0.26292	1122.8	1207.9	1.3607	0.20586	1091.4	1167.6	1.3146
700	0.37198	1183.6	1286.9	1.4433	0.30252	1166.8	1264.7	1.4108	0.24894	1147.6	1239.8	1.3783
750	0.40535	1214.4	1326.9	1.4771	0.33455	1201.5	1309.8	1.4489	0.28074	1187.4	1291.3	1.4218
800	0.43550	1242.2	1363.1	1.5064	0.36266	1231.7	1349.1	1.4807	0.30763	1220.5	1334.3	1.4567
850	0.46356	1268.2	1396.9	1.5328	0.38835	1259.3	1385.1	1.5088	0.33169	1250.0	1372.8	1.4867
900	0.49015	1293.1	1429.2	1.5569	0.41238	1285.4	1419.0	1.5341	0.35390	1277.5	1408.5	1.5134
1000	0.54031	1340.9	1490.8	1.6007	0.45719	1334.9	1482.9	1.5796	0.39479	1328.7	1474.9	1.5606
1100	0.58781	1387.3	1550.5	1.6402	0.49917	1382.4	1544.1	1.6201	0.43266	1377.5	1537.6	1.6021
1200	0.63355	1433.3	1609.2	1.6767	0.53932	1429.2	1603.9	1.6572	0.46864	1425.1	1598.5	1.6400
1400	0.72172	1525.7	1726.0	1.7432	0.61621	1522.6	1722.1	1.7245	0.53708	1519.5	1718.3	1.7081
1600	0.80714	1619.8	1843.8	1.8033	0.69031	1617.4	1840.9	1.7852	0.60269	1615.0	1838.0	1.7693
1800	0.89090	1716.4	1963.7	1.8589	0.76273	1714.5	1961.5	1.8410	0.66660	1712.5	1959.2	1.8255
2000	0.97358	1815.9	2086.1	1.9108	0.83406	1814.2	2084.3	1.8931	0.72942	1812.6	2082.6	1.8778
	P = 2500 psia (668.17°F)				*P* = 3000 psia (695.41°F)				*P* = 3500 psia			
Sat.	0.13076	1031.2	1091.7	1.2330	0.08460	969.8	1016.8	1.1587				
650									0.02492	663.7	679.9	0.8632
700	0.16849	1098.4	1176.3	1.3072	0.09838	1005.3	1059.9	1.1960	0.03065	760.0	779.9	0.9511
750	0.20327	1154.9	1249.0	1.3686	0.14840	1114.1	1196.5	1.3118	0.10460	1057.6	1125.4	1.2434
800	0.22949	1195.9	1302.0	1.4116	0.17601	1167.5	1265.3	1.3676	0.13639	1134.3	1222.6	1.3224
850	0.25174	1230.1	1346.6	1.4463	0.19771	1208.2	1317.9	1.4086	0.15847	1183.8	1286.5	1.3721
900	0.27165	1260.7	1386.4	1.4761	0.21640	1242.8	1362.9	1.4423	0.17659	1223.4	1337.8	1.4106
950	0.29001	1289.1	1423.3	1.5028	0.23321	1273.9	1403.3	1.4716	0.19245	1257.8	1382.4	1.4428
1000	0.30726	1316.1	1458.2	1.5271	0.24876	1302.8	1440.9	1.4978	0.20687	1289.0	1423.0	1.4711
1100	0.33949	1367.3	1524.4	1.5710	0.27732	1356.8	1510.8	1.5441	0.23289	1346.1	1496.9	1.5201
1200	0.36966	1416.6	1587.6	1.6103	0.30367	1408.0	1576.6	1.5850	0.25654	1399.3	1565.4	1.5627
1400	0.42631	1513.3	1710.5	1.6802	0.35249	1507.0	1702.7	1.6567	0.29978	1500.7	1694.8	1.6364
1600	0.48004	1610.1	1832.2	1.7424	0.39830	1605.3	1826.4	1.7199	0.33994	1600.4	1820.5	1.7006
1800	0.53205	1708.6	1954.8	1.7991	0.44237	1704.7	1950.3	1.7773	0.37833	1700.8	1945.8	1.7586
2000	0.58295	1809.4	2079.1	1.8518	0.48532	1806.1	2075.6	1.8304	0.41561	1802.9	2072.1	1.8121
	P = 4000 psia				*P* = 5000 psia				*P* = 6000 psia			
650	0.02448	657.9	676.1	0.8577	0.02379	648.3	670.3	0.8485	0.02325	640.3	666.1	0.8408
700	0.02871	742.3	763.6	0.9347	0.02678	721.8	746.6	0.9156	0.02564	708.1	736.5	0.9028
750	0.06370	962.1	1009.2	1.1410	0.03373	821.8	853.0	1.0054	0.02981	788.7	821.8	0.9747
800	0.10520	1094.2	1172.1	1.2734	0.05937	986.9	1041.8	1.1581	0.03949	897.1	941.0	1.0711
850	0.12848	1156.7	1251.8	1.3355	0.08551	1092.4	1171.5	1.2593	0.05815	1018.6	1083.1	1.1819
900	0.14647	1202.5	1310.9	1.3799	0.10390	1155.9	1252.1	1.3198	0.07584	1103.5	1187.7	1.2603
950	0.16176	1240.7	1360.5	1.4157	0.11863	1203.9	1313.6	1.3643	0.09010	1163.7	1263.7	1.3153
1000	0.17538	1274.6	1404.4	1.4463	0.13128	1244.0	1365.5	1.4004	0.10208	1211.4	1324.7	1.3578
1100	0.19957	1335.1	1482.8	1.4983	0.15298	1312.2	1453.8	1.4590	0.12211	1288.4	1424.0	1.4237
1200	0.22121	1390.3	1554.1	1.5426	0.17185	1372.1	1531.1	1.5070	0.13911	1353.4	1507.8	1.4758
1300	0.24128	1443.0	1621.6	1.5821	0.18902	1427.8	1602.7	1.5490	0.15434	1412.5	1583.8	1.5203
1400	0.26028	1494.3	1687.0	1.6182	0.20508	1481.4	1671.1	1.5868	0.16841	1468.4	1655.4	1.5598
1600	0.29620	1595.5	1814.7	1.6835	0.23505	1585.6	1803.1	1.6542	0.19438	1575.7	1791.5	1.6294
1800	0.33033	1696.8	1941.4	1.7422	0.26320	1689.0	1932.5	1.7142	0.21853	1681.1	1923.7	1.6907
2000	0.36335	1799.7	2068.6	1.7961	0.29023	1793.2	2061.7	1.7689	0.24155	1786.7	2054.9	1.7463

TABLE A–7E

Compressed liquid water

T °F	v ft³/lbm	u Btu/lbm	h Btu/lbm	s Btu/lbm·R	v ft³/lbm	u Btu/lbm	h Btu/lbm	s Btu/lbm·R	v ft³/lbm	u Btu/lbm	h Btu/lbm	s Btu/lbm·R
	P = 500 psia (467.04°F)				P = 1000 psia (544.65°F)				P = 1500 psia (596.26°F)			
Sat.	0.019750	447.68	449.51	0.64900	0.021595	538.58	542.57	0.74341	0.023456	605.07	611.58	0.80836
32	0.015994	0.01	1.49	0.00001	0.015966	0.03	2.99	0.00005	0.015939	0.05	4.48	0.00008
50	0.015998	18.03	19.51	0.03601	0.015972	17.99	20.95	0.03593	0.015946	17.95	22.38	0.03584
100	0.016107	67.86	69.35	0.12930	0.016083	67.69	70.67	0.12899	0.016059	67.53	71.98	0.12869
150	0.016317	117.70	119.21	0.21462	0.016292	117.42	120.43	0.21416	0.016267	117.14	121.66	0.21369
200	0.016607	167.70	169.24	0.29349	0.016580	167.31	170.38	0.29289	0.016553	166.92	171.52	0.29229
250	0.016972	218.04	219.61	0.36708	0.016941	217.51	220.65	0.36634	0.016911	217.00	221.69	0.36560
300	0.017417	268.92	270.53	0.43641	0.017380	268.24	271.46	0.43551	0.017345	267.57	272.39	0.43463
350	0.017954	320.64	322.30	0.50240	0.017910	319.77	323.08	0.50132	0.017866	318.91	323.87	0.50025
400	0.018609	373.61	375.33	0.56595	0.018552	372.48	375.91	0.56463	0.018496	371.37	376.51	0.56333
450	0.019425	428.44	430.24	0.62802	0.019347	426.93	430.51	0.62635	0.019271	425.47	430.82	0.62472
500					0.020368	484.03	487.80	0.68764	0.020258	482.01	487.63	0.68550
550									0.021595	542.50	548.50	0.74731
	P = 2000 psia (635.85°F)				P = 3000 psia (695.41°F)				P = 5000 psia			
Sat.	0.025634	662.33	671.82	0.86224	0.034335	783.39	802.45	0.97321				
32	0.015912	0.07	5.96	0.00010	0.015859	0.10	8.90	0.00011	0.015756	0.13	14.71	0.00002
50	0.015921	17.91	23.80	0.03574	0.015870	17.83	26.64	0.03554	0.015773	17.65	32.25	0.03505
100	0.016035	67.36	73.30	0.12838	0.015988	67.04	75.91	0.12776	0.015897	66.41	81.12	0.12652
200	0.016527	166.54	172.66	0.29170	0.016475	165.79	174.94	0.29053	0.016375	164.36	179.51	0.28824
300	0.017310	266.92	273.33	0.43376	0.017242	265.65	275.22	0.43204	0.017112	263.24	279.07	0.42874
400	0.018442	370.30	377.12	0.56205	0.018338	368.22	378.41	0.55959	0.018145	364.35	381.14	0.55492
450	0.019199	424.06	431.16	0.62314	0.019062	421.36	431.94	0.62010	0.018812	416.40	433.80	0.61445
500	0.020154	480.08	487.54	0.68346	0.019960	476.45	487.53	0.67958	0.019620	469.94	488.10	0.67254
560	0.021739	552.21	560.26	0.75692	0.021405	546.59	558.47	0.75126	0.020862	537.08	556.38	0.74154
600	0.023317	605.77	614.40	0.80898	0.022759	597.42	610.06	0.80086	0.021943	584.42	604.72	0.78803
640					0.024765	654.52	668.27	0.85476	0.023358	634.95	656.56	0.83603
680					0.028821	728.63	744.64	0.92288	0.025366	690.67	714.14	0.88745
700									0.026777	721.78	746.56	0.91564

TABLE A–8E

Saturated ice—water vapor

		Specific volume, ft³/lbm		Internal energy, Btu/lbm			Enthalpy, Btu/lbm			Entropy, Btu/lbm·R		
Temp., T °F	Sat. press., P_{sat} psia	Sat. ice, v_i	Sat. vapor, v_g	Sat. ice, u_i	Subl., u_{ig}	Sat. vapor, u_g	Sat. ice, h_i	Subl., h_{ig}	Sat. vapor, h_g	Sat. ice, s_i	Subl., s_{ig}	Sat. vapor, s_g
32.018	0.08871	0.01747	3299.6	−143.34	1164.2	1020.9	−143.34	1218.3	1075.0	−0.29146	2.4779	2.1864
32	0.08864	0.01747	3302.6	−143.35	1164.2	1020.9	−143.35	1218.4	1075.0	−0.29148	2.4779	2.1865
30	0.08086	0.01747	3605.8	−144.35	1164.6	1020.2	−144.35	1218.5	1074.2	−0.29353	2.4883	2.1948
25	0.06405	0.01746	4505.8	−146.85	1165.4	1018.6	−146.85	1218.8	1072.0	−0.29865	2.5146	2.2160
20	0.05049	0.01746	5657.6	−149.32	1166.2	1016.9	−149.32	1219.1	1069.8	−0.30377	2.5414	2.2376
15	0.03960	0.01745	7138.9	−151.76	1167.0	1015.2	−151.76	1219.3	1067.6	−0.30889	2.5687	2.2598
10	0.03089	0.01744	9054.0	−154.18	1167.8	1013.6	−154.18	1219.5	1065.4	−0.31401	2.5965	2.2825
5	0.02397	0.01743	11,543	−156.57	1168.5	1011.9	−156.57	1219.7	1063.1	−0.31913	2.6248	2.3057
0	0.01850	0.01743	14,797	−158.94	1169.2	1010.3	−158.94	1219.9	1060.9	−0.32426	2.6537	2.3295
−5	0.01420	0.01742	19,075	−161.28	1169.9	1008.6	−161.28	1220.0	1058.7	−0.32938	2.6832	2.3538
−10	0.01083	0.01741	24,731	−163.60	1170.6	1007.0	−163.60	1220.1	1056.5	−0.33451	2.7133	2.3788
−15	0.00821	0.01740	32,257	−165.90	1171.2	1005.3	−165.90	1220.2	1054.3	−0.33964	2.7440	2.4044
−20	0.00619	0.01740	42,335	−168.16	1171.8	1003.6	−168.16	1220.3	1052.1	−0.34478	2.7754	2.4306
−25	0.00463	0.01739	55,917	−170.41	1172.4	1002.0	−170.41	1220.3	1049.9	−0.34991	2.8074	2.4575
−30	0.00344	0.01738	74,345	−172.63	1173.0	1000.3	−172.63	1220.3	1047.7	−0.35505	2.8401	2.4850
−35	0.00254	0.01738	99,526	−174.83	1173.5	998.7	−174.83	1220.3	1045.5	−0.36019	2.8735	2.5133
−40	0.00186	0.01737	134,182	−177.00	1174.0	997.0	−177.00	1220.3	1043.3	−0.36534	2.9076	2.5423

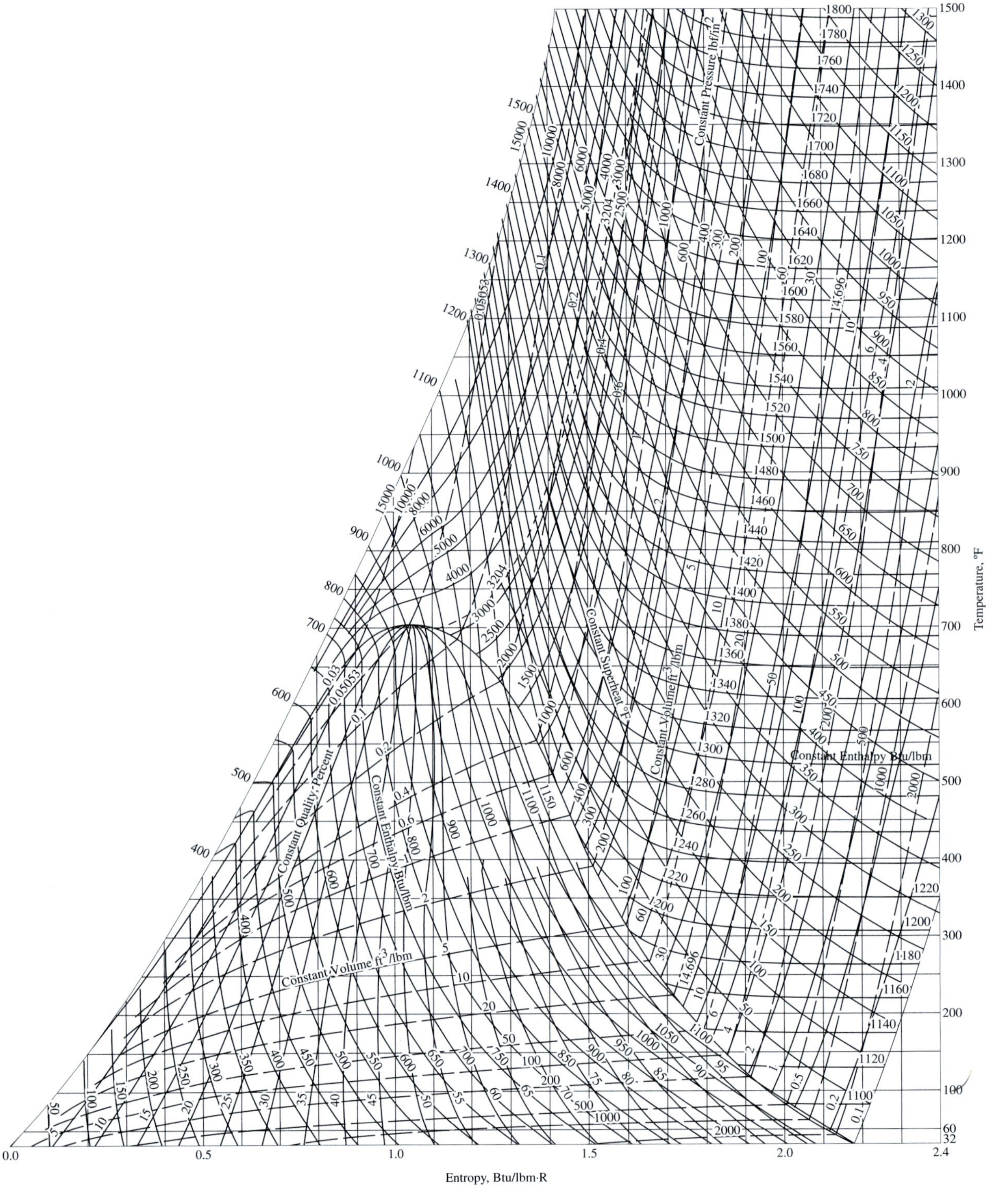

FIGURE A–9E

T-s diagram for water.

Source: Joseph H. Keenan, Frederick G. Keyes, Philip G. Hill, and Joan G. Moore, Steam Tables *(New York: John Wiley & Sons, 1969).*

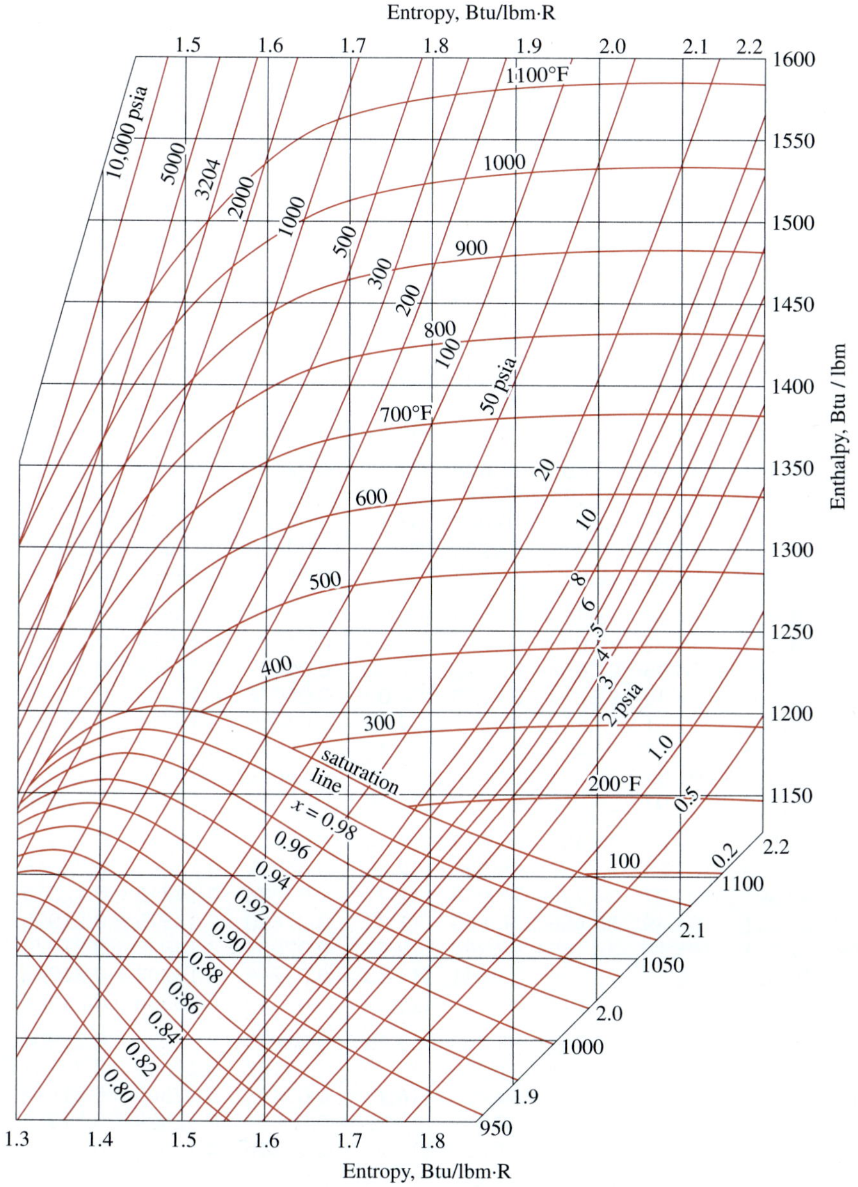

FIGURE A–10E

Mollier diagram for water.

Source: Joseph H. Keenan, Frederick G. Keyes, Philip G. Hill, and Joan G. Moore, Steam Tables *(New York: John Wiley & Sons, 1969).*

TABLE A–11E

Saturated refrigerant-134a—Temperature table

		Specific volume, ft³/lbm		Internal energy, Btu/lbm			Enthalpy, Btu/lbm			Entropy, Btu/lbm·R		
Temp., T °F	Sat. press., P_{sat} psia	Sat. liquid, v_f	Sat. vapor, v_g	Sat. liquid, u_f	Evap., u_{fg}	Sat. vapor, u_g	Sat. liquid, h_f	Evap., h_{fg}	Sat. vapor, h_g	Sat. liquid, s_f	Evap., s_{fg}	Sat. vapor, s_g
−40	7.432	0.01130	5.7796	−0.016	89.167	89.15	0.000	97.100	97.10	0.00000	0.23135	0.23135
−35	8.581	0.01136	5.0509	1.484	88.352	89.84	1.502	96.354	97.86	0.00355	0.22687	0.23043
−30	9.869	0.01143	4.4300	2.990	87.532	90.52	3.011	95.601	98.61	0.00708	0.22248	0.22956
−25	11.306	0.01150	3.8988	4.502	86.706	91.21	4.526	94.839	99.36	0.01058	0.21817	0.22875
−20	12.906	0.01156	3.4426	6.019	85.874	91.89	6.047	94.068	100.12	0.01405	0.21394	0.22798
−15	14.680	0.01163	3.0494	7.543	85.036	92.58	7.574	93.288	100.86	0.01749	0.20978	0.22727
−10	16.642	0.01171	2.7091	9.073	84.191	93.26	9.109	92.498	101.61	0.02092	0.20569	0.22660
−5	18.806	0.01178	2.4137	10.609	83.339	93.95	10.650	91.698	102.35	0.02431	0.20166	0.22598
0	21.185	0.01185	2.1564	12.152	82.479	94.63	12.199	90.886	103.08	0.02769	0.19770	0.22539
5	23.793	0.01193	1.9316	13.702	81.610	95.31	13.755	90.062	103.82	0.03104	0.19380	0.22485
10	26.646	0.01201	1.7345	15.259	80.733	95.99	15.318	89.226	104.54	0.03438	0.18996	0.22434
15	29.759	0.01209	1.5612	16.823	79.846	96.67	16.889	88.377	105.27	0.03769	0.18617	0.22386
20	33.147	0.01217	1.4084	18.394	78.950	97.34	18.469	87.514	105.98	0.04098	0.18243	0.22341
25	36.826	0.01225	1.2732	19.973	78.043	98.02	20.056	86.636	106.69	0.04426	0.17874	0.22300
30	40.813	0.01234	1.1534	21.560	77.124	98.68	21.653	85.742	107.40	0.04752	0.17509	0.22260
35	45.124	0.01242	1.0470	23.154	76.195	99.35	23.258	84.833	108.09	0.05076	0.17148	0.22224
40	49.776	0.01251	0.95205	24.757	75.253	100.01	24.873	83.907	108.78	0.05398	0.16791	0.22189
45	54.787	0.01261	0.86727	26.369	74.298	100.67	26.497	82.963	109.46	0.05720	0.16437	0.22157
50	60.175	0.01270	0.79136	27.990	73.329	101.32	28.131	82.000	110.13	0.06039	0.16087	0.22127
55	65.957	0.01280	0.72323	29.619	72.346	101.97	29.775	81.017	110.79	0.06358	0.15740	0.22098
60	72.152	0.01290	0.66195	31.258	71.347	102.61	31.431	80.013	111.44	0.06675	0.15396	0.22070
65	78.780	0.01301	0.60671	32.908	70.333	103.24	33.097	78.988	112.09	0.06991	0.15053	0.22044
70	85.858	0.01312	0.55681	34.567	69.301	103.87	34.776	77.939	112.71	0.07306	0.14713	0.22019
75	93.408	0.01323	0.51165	36.237	68.251	104.49	36.466	76.866	113.33	0.07620	0.14375	0.21995
80	101.45	0.01334	0.47069	37.919	67.181	105.10	38.169	75.767	113.94	0.07934	0.14038	0.21972
85	110.00	0.01347	0.43348	39.612	66.091	105.70	39.886	74.641	114.53	0.08246	0.13703	0.21949
90	119.08	0.01359	0.39959	41.317	64.979	106.30	41.617	73.485	115.10	0.08559	0.13368	0.21926
95	128.72	0.01372	0.36869	43.036	63.844	106.88	43.363	72.299	115.66	0.08870	0.13033	0.21904
100	138.93	0.01386	0.34045	44.768	62.683	107.45	45.124	71.080	116.20	0.09182	0.12699	0.21881
105	149.73	0.01400	0.31460	46.514	61.496	108.01	46.902	69.825	116.73	0.09493	0.12365	0.21858
110	161.16	0.01415	0.29090	48.276	60.279	108.56	48.698	68.533	117.23	0.09804	0.12029	0.21834
115	173.23	0.01430	0.26913	50.054	59.031	109.08	50.512	67.200	117.71	0.10116	0.11693	0.21809
120	185.96	0.01446	0.24909	51.849	57.749	109.60	52.346	65.823	118.17	0.10428	0.11354	0.21782
130	213.53	0.01482	0.21356	55.495	55.071	110.57	56.080	62.924	119.00	0.11054	0.10670	0.21724
140	244.06	0.01521	0.18315	59.226	52.216	111.44	59.913	59.801	119.71	0.11684	0.09971	0.21655
150	277.79	0.01567	0.15692	63.059	49.144	112.20	63.864	56.405	120.27	0.12321	0.09251	0.21572
160	314.94	0.01619	0.13410	67.014	45.799	112.81	67.958	52.671	120.63	0.12970	0.08499	0.21469
170	355.80	0.01681	0.11405	71.126	42.097	113.22	72.233	48.499	120.73	0.13634	0.07701	0.21335
180	400.66	0.01759	0.09618	75.448	37.899	113.35	76.752	43.726	120.48	0.14323	0.06835	0.21158
190	449.90	0.01860	0.07990	80.082	32.950	113.03	81.631	38.053	119.68	0.15055	0.05857	0.20911
200	504.00	0.02009	0.06441	85.267	26.651	111.92	87.140	30.785	117.93	0.15867	0.04666	0.20533
210	563.76	0.02309	0.04722	91.986	16.498	108.48	94.395	19.015	113.41	0.16922	0.02839	0.19761

Source: Tables A–11E through A–13E are generated using the Engineering Equation Solver (EES) software developed by S. A. Klein and F. L. Alvarado. The routine used in calculations is the R134a, which is based on the fundamental equation of state developed by R. Tillner-Roth and H.D. Baehr, "An International Standard Formulation for the Thermodynamic Properties of 1,1,1,2-Tetrafluoroethane (HFC-134a) for Temperatures from 170 K to 455 K and Pressures up to 70 MPa," *J. Phys. Chem, Ref. Data*, Vol. 23, No. 5, 1994. The enthalpy and entropy values of saturated liquid are set to zero at −40°C (and −40°F).

TABLE A–12E

Saturated refrigerant-134a—Pressure table

		Specific volume, ft³/lbm		Internal energy, Btu/lbm			Enthalpy, Btu/lbm			Entropy, Btu/lbm·R		
Press., P psia	Sat. temp., T °F	Sat. liquid, v_f	Sat. vapor, v_g	Sat. liquid, u_f	Evap., u_{fg}	Sat. vapor, u_g	Sat. liquid, h_f	Evap., h_{fg}	Sat. vapor, h_g	Sat. liquid, s_f	Evap., s_{fg}	Sat. vapor, s_g
5	−53.09	0.01113	8.3785	−3.918	91.280	87.36	−3.907	99.022	95.11	−0.00945	0.24353	0.23408
10	−29.52	0.01144	4.3753	3.135	87.453	90.59	3.156	95.528	98.68	0.00742	0.22206	0.22948
15	−14.15	0.01165	2.9880	7.803	84.893	92.70	7.835	93.155	100.99	0.01808	0.20908	0.22715
20	−2.43	0.01182	2.2772	11.401	82.898	94.30	11.445	91.282	102.73	0.02605	0.19962	0.22567
25	7.17	0.01196	1.8429	14.377	81.231	95.61	14.432	89.701	104.13	0.03249	0.19213	0.22462
30	15.37	0.01209	1.5492	16.939	79.780	96.72	17.006	88.313	105.32	0.03793	0.18589	0.22383
35	22.57	0.01221	1.3369	19.205	78.485	97.69	19.284	87.064	106.35	0.04267	0.18053	0.22319
40	29.01	0.01232	1.1760	21.246	77.307	98.55	21.337	85.920	107.26	0.04688	0.17580	0.22268
45	34.86	0.01242	1.0497	23.110	76.221	99.33	23.214	84.858	108.07	0.05067	0.17158	0.22225
50	40.23	0.01252	0.94791	24.832	75.209	100.04	24.948	83.863	108.81	0.05413	0.16774	0.22188
55	45.20	0.01261	0.86400	26.435	74.258	100.69	26.564	82.924	109.49	0.05733	0.16423	0.22156
60	49.84	0.01270	0.79361	27.939	73.360	101.30	28.080	82.030	110.11	0.06029	0.16098	0.22127
65	54.20	0.01279	0.73370	29.357	72.505	101.86	29.510	81.176	110.69	0.06307	0.15796	0.22102
70	58.30	0.01287	0.68205	30.700	71.688	102.39	30.867	80.357	111.22	0.06567	0.15512	0.22080
75	62.19	0.01295	0.63706	31.979	70.905	102.88	32.159	79.567	111.73	0.06813	0.15245	0.22059
80	65.89	0.01303	0.59750	33.201	70.151	103.35	33.394	78.804	112.20	0.07047	0.14993	0.22040
85	69.41	0.01310	0.56244	34.371	69.424	103.79	34.577	78.064	112.64	0.07269	0.14753	0.22022
90	72.78	0.01318	0.53113	35.495	68.719	104.21	35.715	77.345	113.06	0.07481	0.14525	0.22006
95	76.02	0.01325	0.50301	36.578	68.035	104.61	36.811	76.645	113.46	0.07684	0.14307	0.21991
100	79.12	0.01332	0.47760	37.623	67.371	104.99	37.869	75.962	113.83	0.07879	0.14097	0.21976
110	85.00	0.01347	0.43347	39.612	66.091	105.70	39.886	74.641	114.53	0.08246	0.13703	0.21949
120	90.49	0.01360	0.39644	41.485	64.869	106.35	41.787	73.371	115.16	0.08589	0.13335	0.21924
130	95.64	0.01374	0.36491	43.258	63.696	106.95	43.589	72.144	115.73	0.08911	0.12990	0.21901
140	100.51	0.01387	0.33771	44.945	62.564	107.51	45.304	70.954	116.26	0.09214	0.12665	0.21879
150	105.12	0.01400	0.31401	46.556	61.467	108.02	46.945	69.795	116.74	0.09501	0.12357	0.21857
160	109.50	0.01413	0.29316	48.101	60.401	108.50	48.519	68.662	117.18	0.09774	0.12062	0.21836
170	113.69	0.01426	0.27466	49.586	59.362	108.95	50.035	67.553	117.59	0.10034	0.11781	0.21815
180	117.69	0.01439	0.25813	51.018	58.345	109.36	51.497	66.464	117.96	0.10284	0.11511	0.21795
190	121.53	0.01452	0.24327	52.402	57.349	109.75	52.912	65.392	118.30	0.10524	0.11250	0.21774
200	125.22	0.01464	0.22983	53.743	56.371	110.11	54.285	64.335	118.62	0.10754	0.10998	0.21753
220	132.21	0.01490	0.20645	56.310	54.458	110.77	56.917	62.256	119.17	0.11192	0.10517	0.21710
240	138.73	0.01516	0.18677	58.746	52.591	111.34	59.419	60.213	119.63	0.11603	0.10061	0.21665
260	144.85	0.01543	0.16996	61.071	50.757	111.83	61.813	58.192	120.00	0.11992	0.09625	0.21617
280	150.62	0.01570	0.15541	63.301	48.945	112.25	64.115	56.184	120.30	0.12362	0.09205	0.21567
300	156.09	0.01598	0.14266	65.452	47.143	112.60	66.339	54.176	120.52	0.12715	0.08797	0.21512
350	168.64	0.01672	0.11664	70.554	42.627	113.18	71.638	49.099	120.74	0.13542	0.07814	0.21356
400	179.86	0.01757	0.09642	75.385	37.963	113.35	76.686	43.798	120.48	0.14314	0.06848	0.21161
450	190.02	0.01860	0.07987	80.092	32.939	113.03	81.641	38.041	119.68	0.15056	0.05854	0.20911
500	199.29	0.01995	0.06551	84.871	27.168	112.04	86.718	31.382	118.10	0.15805	0.04762	0.20566

TABLE A–13E

Superheated refrigerant-134a

T °F	v ft³/lbm	u Btu/lbm	h Btu/lbm	s Btu/lbm·R	v ft³/lbm	u Btu/lbm	h Btu/lbm	s Btu/lbm·R	v ft³/lbm	u Btu/lbm	h Btu/lbm	s Btu/lbm·R
	P = 10 psia (T_{sat} = −29.52°F)				P = 15 psia (T_{sat} = −14.15°F)				P = 20 psia (T_{sat} = −2.43°F)			
Sat.	4.3753	90.59	98.68	0.22948	2.9880	92.70	100.99	0.22715	2.2772	94.30	102.73	0.22567
−20	4.4856	92.13	100.43	0.23350								
0	4.7135	95.41	104.14	0.24174	3.1001	95.08	103.68	0.23310	2.2922	94.72	103.20	0.22671
20	4.9380	98.77	107.91	0.24976	3.2551	98.48	107.52	0.24127	2.4130	98.19	107.12	0.23504
40	5.1600	102.20	111.75	0.25761	3.4074	101.95	111.41	0.24922	2.5306	101.70	111.07	0.24311
60	5.3802	105.72	115.67	0.26531	3.5577	105.50	115.38	0.25700	2.6461	105.28	115.07	0.25097
80	5.5989	109.32	119.68	0.27288	3.7064	109.13	119.42	0.26463	2.7600	108.93	119.15	0.25866
100	5.8165	113.01	123.78	0.28033	3.8540	112.84	123.54	0.27212	2.8726	112.66	123.29	0.26621
120	6.0331	116.79	127.96	0.28767	4.0006	116.63	127.74	0.27950	2.9842	116.47	127.52	0.27363
140	6.2490	120.66	132.22	0.29490	4.1464	120.51	132.02	0.28677	3.0950	120.37	131.82	0.28093
160	6.4642	124.61	136.57	0.30203	4.2915	124.48	136.39	0.29393	3.2051	124.35	136.21	0.28812
180	6.6789	128.65	141.01	0.30908	4.4361	128.53	140.84	0.30100	3.3146	128.41	140.67	0.29521
200	6.8930	132.77	145.53	0.31604	4.5802	132.66	145.37	0.30798	3.4237	132.55	145.22	0.30221
220	7.1068	136.98	150.13	0.32292	4.7239	136.88	149.99	0.31487	3.5324	136.78	149.85	0.30912
	P = 30 psia (T_{sat} = 15.37°F)				P = 40 psia (T_{sat} = 29.01°F)				P = 50 psia (T_{sat} = 40.23°F)			
Sat.	1.5492	96.72	105.32	0.22383	1.1760	98.55	107.26	0.22268	0.9479	100.04	108.81	0.22188
20	1.5691	97.56	106.27	0.22581								
40	1.6528	101.17	110.35	0.23414	1.2126	100.61	109.58	0.22738				
60	1.7338	104.82	114.45	0.24219	1.2768	104.34	113.79	0.23565	1.0019	103.84	113.11	0.23031
80	1.8130	108.53	118.59	0.25002	1.3389	108.11	118.02	0.24363	1.0540	107.68	117.43	0.23847
100	1.8908	112.30	122.80	0.25767	1.3995	111.93	122.29	0.25140	1.1043	111.55	121.77	0.24637
120	1.9675	116.15	127.07	0.26517	1.4588	115.82	126.62	0.25900	1.1534	115.48	126.16	0.25406
140	2.0434	120.08	131.42	0.27254	1.5173	119.78	131.01	0.26644	1.2015	119.47	130.59	0.26159
160	2.1185	124.08	135.84	0.27979	1.5750	123.81	135.47	0.27375	1.2488	123.53	135.09	0.26896
180	2.1931	128.16	140.34	0.28693	1.6321	127.91	140.00	0.28095	1.2955	127.66	139.65	0.27621
200	2.2671	132.32	144.91	0.29398	1.6887	132.10	144.60	0.28803	1.3416	131.87	144.28	0.28333
220	2.3408	136.57	149.56	0.30092	1.7449	136.36	149.27	0.29501	1.3873	136.15	148.98	0.29036
240	2.4141	140.89	154.29	0.30778	1.8007	140.70	154.03	0.30190	1.4326	140.50	153.76	0.29728
260	2.4871	145.30	159.10	0.31456	1.8562	145.12	158.86	0.30871	1.4776	144.93	158.60	0.30411
280	2.5598	149.78	163.99	0.32126	1.9114	149.61	163.76	0.31543	1.5223	149.44	163.53	0.31086
	P = 60 psia (T_{sat} = 49.84°F)				P = 70 psia (T_{sat} = 58.30°F)				P = 80 psia (T_{sat} = 65.89°F)			
Sat.	0.7936	101.30	110.11	0.22127	0.6821	102.39	111.22	0.22080	0.59750	103.35	112.20	0.22040
60	0.8179	103.31	112.39	0.22570	0.6857	102.73	111.62	0.22155				
80	0.8636	107.23	116.82	0.23407	0.7271	106.76	116.18	0.23016	0.62430	106.26	115.51	0.22661
100	0.9072	111.16	121.24	0.24211	0.7662	110.76	120.68	0.23836	0.66009	110.34	120.11	0.23499
120	0.9495	115.14	125.68	0.24991	0.8037	114.78	125.19	0.24628	0.69415	114.42	124.69	0.24304
140	0.9908	119.16	130.16	0.25751	0.8401	118.85	129.73	0.25398	0.72698	118.52	129.29	0.25083
160	1.0312	123.25	134.70	0.26496	0.8756	122.97	134.31	0.26149	0.75888	122.68	133.91	0.25841
180	1.0709	127.41	139.30	0.27226	0.9105	127.15	138.94	0.26885	0.79003	126.89	138.58	0.26583
200	1.1101	131.63	143.96	0.27943	0.9447	131.40	143.63	0.27607	0.82059	131.16	143.31	0.27310
220	1.1489	135.93	148.69	0.28649	0.9785	135.71	148.39	0.28317	0.85065	135.49	148.09	0.28024
240	1.1872	140.30	153.48	0.29344	1.0118	140.10	153.21	0.29015	0.88030	139.90	152.93	0.28726
260	1.2252	144.75	158.35	0.30030	1.0449	144.56	158.10	0.29704	0.90961	144.37	157.84	0.29418
280	1.2629	149.27	163.29	0.30707	1.0776	149.10	163.06	0.30384	0.93861	148.92	162.82	0.30100
300	1.3004	153.87	168.31	0.31376	1.1101	153.71	168.09	0.31055	0.96737	153.54	167.86	0.30773
320	1.3377	158.54	173.39	0.32037	1.1424	158.39	173.19	0.31718	0.99590	158.24	172.98	0.31438

TABLE A–13E

Superheated refrigerant-134a (*Concluded*)

T °F	v ft³/lbm	u Btu/lbm	h Btu/lbm	s Btu/lbm·R	v ft³/lbm	u Btu/lbm	h Btu/lbm	s Btu/lbm·R	v ft³/lbm	u Btu/lbm	h Btu/lbm	s Btu/lbm·R
	P = 90 psia (T_{sat} = 72.78°F)				P = 100 psia (T_{sat} = 79.12°F)				P = 120 psia (T_{sat} = 90.49°F)			
Sat.	0.53113	104.21	113.06	0.22006	0.47760	104.99	113.83	0.21976	0.39644	106.35	115.16	0.21924
80	0.54388	105.74	114.80	0.22330	0.47906	105.18	114.05	0.22016				
100	0.57729	109.91	119.52	0.23189	0.51076	109.45	118.90	0.22900	0.41013	108.48	117.59	0.22362
120	0.60874	114.04	124.18	0.24008	0.54022	113.66	123.65	0.23733	0.43692	112.84	122.54	0.23232
140	0.63885	118.19	128.83	0.24797	0.56821	117.86	128.37	0.24534	0.46190	117.15	127.41	0.24058
160	0.66796	122.38	133.51	0.25563	0.59513	122.08	133.09	0.25309	0.48563	121.46	132.25	0.24851
180	0.69629	126.62	138.22	0.26311	0.62122	126.35	137.85	0.26063	0.50844	125.79	137.09	0.25619
200	0.72399	130.92	142.97	0.27043	0.64667	130.67	142.64	0.26801	0.53054	130.17	141.95	0.26368
220	0.75119	135.27	147.78	0.27762	0.67158	135.05	147.47	0.27523	0.55206	134.59	146.85	0.27100
240	0.77796	139.69	152.65	0.28468	0.69605	139.49	152.37	0.28233	0.57312	139.07	151.80	0.27817
260	0.80437	144.19	157.58	0.29162	0.72016	143.99	157.32	0.28931	0.59379	143.61	156.79	0.28521
280	0.83048	148.75	162.58	0.29847	0.74396	148.57	162.34	0.29618	0.61413	148.21	161.85	0.29214
300	0.85633	153.38	167.64	0.30522	0.76749	153.21	167.42	0.30296	0.63420	152.88	166.96	0.29896
320	0.88195	158.08	172.77	0.31189	0.79079	157.93	172.56	0.30964	0.65402	157.62	172.14	0.30569
	P = 140 psia (T_{sat} = 100.50°F)				P = 160 psia (T_{sat} = 109.50°F)				P = 180 psia (T_{sat} = 117.69°F)			
Sat.	0.33771	107.51	116.26	0.21879	0.29316	108.50	117.18	0.21836	0.25813	109.36	117.96	0.21795
120	0.36243	111.96	121.35	0.22773	0.30578	111.01	120.06	0.22337	0.26083	109.94	118.63	0.21910
140	0.38551	116.41	126.40	0.23628	0.32774	115.62	125.32	0.23230	0.28231	114.77	124.17	0.22850
160	0.40711	120.81	131.36	0.24443	0.34790	120.13	130.43	0.24069	0.30154	119.42	129.46	0.23718
180	0.42766	125.22	136.30	0.25227	0.36686	124.62	135.49	0.24871	0.31936	124.00	134.64	0.24540
200	0.44743	129.65	141.24	0.25988	0.38494	129.12	140.52	0.25645	0.33619	128.57	139.77	0.25330
220	0.46657	134.12	146.21	0.26730	0.40234	133.64	145.55	0.26397	0.35228	133.15	144.88	0.26094
240	0.48522	138.64	151.21	0.27455	0.41921	138.20	150.62	0.27131	0.36779	137.76	150.01	0.26837
260	0.50345	143.21	156.26	0.28166	0.43564	142.81	155.71	0.27849	0.38284	142.40	155.16	0.27562
280	0.52134	147.85	161.35	0.28864	0.45171	147.48	160.85	0.28554	0.39751	147.10	160.34	0.28273
300	0.53895	152.54	166.50	0.29551	0.46748	152.20	166.04	0.29246	0.41186	151.85	165.57	0.28970
320	0.55630	157.30	171.71	0.30228	0.48299	156.98	171.28	0.29927	0.42594	156.66	170.85	0.29656
340	0.57345	162.13	176.98	0.30896	0.49828	161.83	176.58	0.30598	0.43980	161.53	176.18	0.30331
360	0.59041	167.02	182.32	0.31555	0.51338	166.74	181.94	0.31260	0.45347	166.46	181.56	0.30996
	P = 200 psia (T_{sat} = 125.22°F)				P = 300 psia (T_{sat} = 156.09°F)				P = 400 psia (T_{sat} = 179.86°F)			
Sat.	0.22983	110.11	118.62	0.21753	0.14266	112.60	120.52	0.21512	0.09642	113.35	120.48	0.21161
140	0.24541	113.85	122.93	0.22481								
160	0.26412	118.66	128.44	0.23384	0.14656	113.82	121.95	0.21745				
180	0.28115	123.35	133.76	0.24229	0.16355	119.52	128.60	0.22802	0.09658	113.41	120.56	0.21173
200	0.29704	128.00	138.99	0.25035	0.17776	124.78	134.65	0.23733	0.11440	120.52	128.99	0.22471
220	0.31212	132.64	144.19	0.25812	0.19044	129.85	140.42	0.24594	0.12746	126.44	135.88	0.23500
240	0.32658	137.30	149.38	0.26565	0.20211	134.83	146.05	0.25410	0.13853	131.95	142.20	0.24418
260	0.34054	141.99	154.59	0.27298	0.21306	139.77	151.59	0.26192	0.14844	137.26	148.25	0.25270
280	0.35410	146.72	159.82	0.28015	0.22347	144.70	157.11	0.26947	0.15756	142.48	154.14	0.26077
300	0.36733	151.50	165.09	0.28718	0.23346	149.65	162.61	0.27681	0.16611	147.65	159.94	0.26851
320	0.38029	156.33	170.40	0.29408	0.24310	154.63	168.12	0.28398	0.17423	152.80	165.70	0.27599
340	0.39300	161.22	175.77	0.30087	0.25246	159.64	173.66	0.29098	0.18201	157.97	171.44	0.28326
360	0.40552	166.17	181.18	0.30756	0.26159	164.70	179.22	0.29786	0.18951	163.15	177.18	0.29035

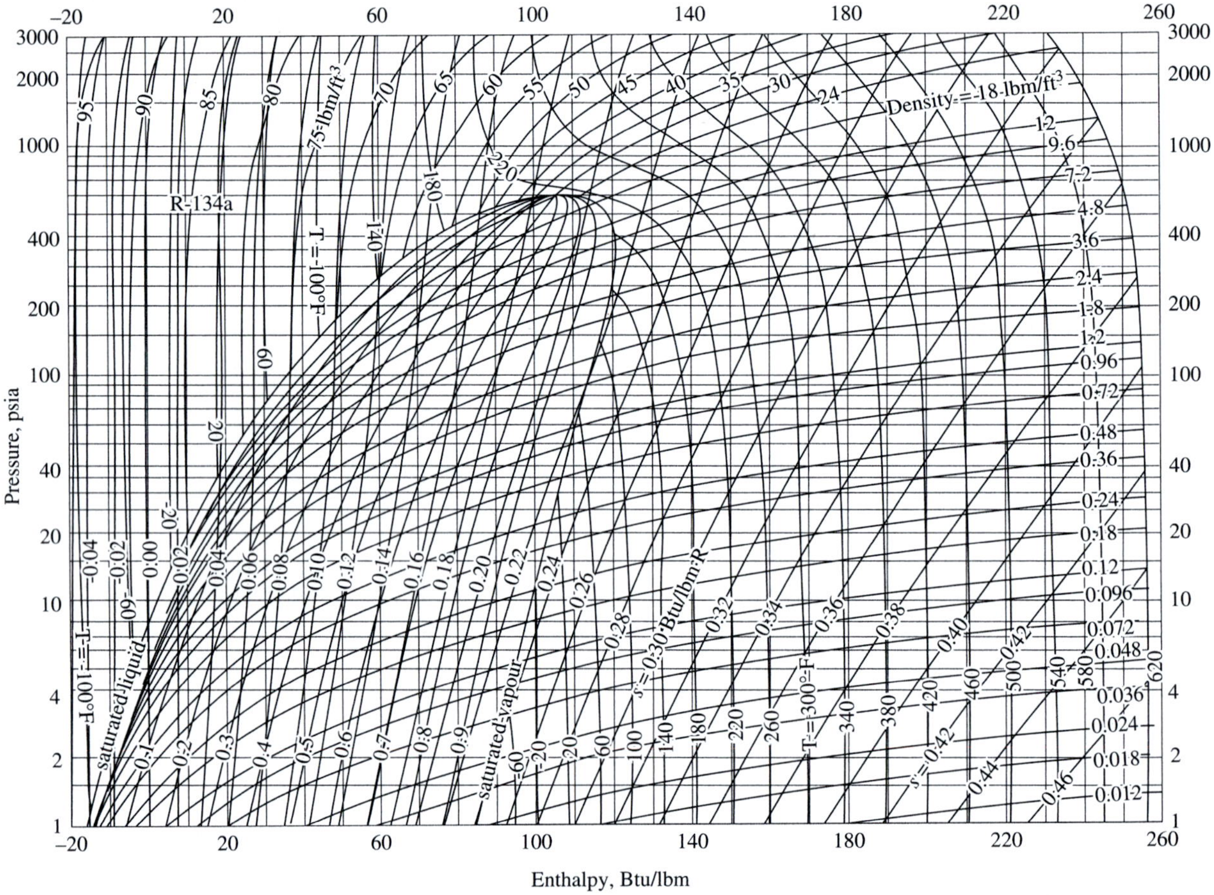

FIGURE A–14E

P-h diagram for refrigerant-134a.

Reprinted by permission of American Society of Heating, Refrigerating, and Air-Conditioning Engineers, Inc., Atlanta, GA.

TABLE A–16E

Properties of the atmosphere at high altitude

Altitude, ft	Temperature, °F	Pressure, psia	Gravity, g, ft/s^2	Speed of sound, ft/s	Density, lbm/ft^3	Viscosity μ, lbm/ft·s	Thermal conductivity, Btu/h·ft·R
0	59.00	14.7	32.174	1116	0.07647	1.202×10^{-5}	0.0146
500	57.22	14.4	32.173	1115	0.07536	1.199×10^{-5}	0.0146
1000	55.43	14.2	32.171	1113	0.07426	1.196×10^{-5}	0.0146
1500	53.65	13.9	32.169	1111	0.07317	1.193×10^{-5}	0.0145
2000	51.87	13.7	32.168	1109	0.07210	1.190×10^{-5}	0.0145
2500	50.09	13.4	32.166	1107	0.07104	1.186×10^{-5}	0.0144
3000	48.30	13.2	32.165	1105	0.06998	1.183×10^{-5}	0.0144
3500	46.52	12.9	32.163	1103	0.06985	1.180×10^{-5}	0.0143
4000	44.74	12.7	32.162	1101	0.06792	1.177×10^{-5}	0.0143
4500	42.96	12.5	32.160	1099	0.06690	1.173×10^{-5}	0.0142
5000	41.17	12.2	32.159	1097	0.06590	1.170×10^{-5}	0.0142
5500	39.39	12.0	32.157	1095	0.06491	1.167×10^{-5}	0.0141
6000	37.61	11.8	32.156	1093	0.06393	1.164×10^{-5}	0.0141
6500	35.83	11.6	32.154	1091	0.06296	1.160×10^{-5}	0.0141
7000	34.05	11.3	32.152	1089	0.06200	1.157×10^{-5}	0.0140
7500	32.26	11.1	32.151	1087	0.06105	1.154×10^{-5}	0.0140
8000	30.48	10.9	32.149	1085	0.06012	1.150×10^{-5}	0.0139
8500	28.70	10.7	32.148	1083	0.05919	1.147×10^{-5}	0.0139
9000	26.92	10.5	32.146	1081	0.05828	1.144×10^{-5}	0.0138
9500	25.14	10.3	32.145	1079	0.05738	1.140×10^{-5}	0.0138
10,000	23.36	10.1	32.145	1077	0.05648	1.137×10^{-5}	0.0137
11,000	19.79	9.72	32.140	1073	0.05473	1.130×10^{-5}	0.0136
12,000	16.23	9.34	32.137	1069	0.05302	1.124×10^{-5}	0.0136
13,000	12.67	8.99	32.134	1065	0.05135	1.117×10^{-5}	0.0135
14,000	9.12	8.63	32.131	1061	0.04973	1.110×10^{-5}	0.0134
15,000	5.55	8.29	32.128	1057	0.04814	1.104×10^{-5}	0.0133
16,000	+1.99	7.97	32.125	1053	0.04659	1.097×10^{-5}	0.0132
17,000	−1.58	7.65	32.122	1049	0.04508	1.090×10^{-5}	0.0132
18,000	−5.14	7.34	32.119	1045	0.04361	1.083×10^{-5}	0.0130
19,000	−8.70	7.05	32.115	1041	0.04217	1.076×10^{-5}	0.0129
20,000	−12.2	6.76	32.112	1037	0.04077	1.070×10^{-5}	0.0128
22,000	−19.4	6.21	32.106	1029	0.03808	1.056×10^{-5}	0.0126
24,000	−26.5	5.70	32.100	1020	0.03553	1.042×10^{-5}	0.0124
26,000	−33.6	5.22	32.094	1012	0.03311	1.028×10^{-5}	0.0122
28,000	−40.7	4.78	32.088	1003	0.03082	1.014×10^{-5}	0.0121
30,000	−47.8	4.37	32.082	995	0.02866	1.000×10^{-5}	0.0119
32,000	−54.9	3.99	32.08	987	0.02661	0.986×10^{-5}	0.0117
34,000	−62.0	3.63	32.07	978	0.02468	0.971×10^{-5}	0.0115
36,000	−69.2	3.30	32.06	969	0.02285	0.956×10^{-5}	0.0113
38,000	−69.7	3.05	32.06	968	0.02079	0.955×10^{-5}	0.0113
40,000	−69.7	2.73	32.05	968	0.01890	0.955×10^{-5}	0.0113
45,000	−69.7	2.148	32.04	968	0.01487	0.955×10^{-5}	0.0113
50,000	−69.7	1.691	32.02	968	0.01171	0.955×10^{-5}	0.0113
55,000	−69.7	1.332	32.00	968	0.00922	0.955×10^{-5}	0.0113
60,000	−69.7	1.048	31.99	968	0.00726	0.955×10^{-5}	0.0113

Source: U.S. Standard Atmosphere Supplements, U.S. Government Printing Office, 1966. Based on year-round mean conditions at 45° latitude and varies with the time of the year and the weather patterns. The conditions at sea level (z = 0) are taken to be P = 14.696 psia, T = 59°F, ρ = 0.076474 lbm/ft^3, g = 32.1741 ft^2/s.

TABLE A–17E

Ideal-gas properties of air

T R	h Btu/lbm	P_r	u Btu/lbm	v_r	s° Btu/lbm·R	T R	h Btu/lbm	P_r	u Btu/lbm	v_r	s° Btu/lbm·R
360	85.97	0.3363	61.29	396.6	0.50369	1600	395.74	71.13	286.06	8.263	0.87130
380	90.75	0.4061	64.70	346.6	0.51663	1650	409.13	80.89	296.03	7.556	0.87954
400	95.53	0.4858	68.11	305.0	0.52890	1700	422.59	90.95	306.06	6.924	0.88758
420	100.32	0.5760	71.52	270.1	0.54058	1750	436.12	101.98	316.16	6.357	0.89542
440	105.11	0.6776	74.93	240.6	0.55172	1800	449.71	114.0	326.32	5.847	0.90308
460	109.90	0.7913	78.36	215.33	0.56235	1850	463.37	127.2	336.55	5.388	0.91056
480	114.69	0.9182	81.77	193.65	0.57255	1900	477.09	141.5	346.85	4.974	0.91788
500	119.48	1.0590	85.20	174.90	0.58233	1950	490.88	157.1	357.20	4.598	0.92504
520	124.27	1.2147	88.62	158.58	0.59173	2000	504.71	174.0	367.61	4.258	0.93205
537	128.10	1.3593	91.53	146.34	0.59945	2050	518.71	192.3	378.08	3.949	0.93891
540	129.06	1.3860	92.04	144.32	0.60078	2100	532.55	212.1	388.60	3.667	0.94564
560	133.86	1.5742	95.47	131.78	0.60950	2150	546.54	223.5	399.17	3.410	0.95222
580	138.66	1.7800	98.90	120.70	0.61793	2200	560.59	256.6	409.78	3.176	0.95919
600	143.47	2.005	102.34	110.88	0.62607	2250	574.69	281.4	420.46	2.961	0.96501
620	148.28	2.249	105.78	102.12	0.63395	2300	588.82	308.1	431.16	2.765	0.97123
640	153.09	2.514	109.21	94.30	0.64159	2350	603.00	336.8	441.91	2.585	0.97732
660	157.92	2.801	112.67	87.27	0.64902	2400	617.22	367.6	452.70	2.419	0.98331
680	162.73	3.111	116.12	80.96	0.65621	2450	631.48	400.5	463.54	2.266	0.98919
700	167.56	3.446	119.58	75.25	0.66321	2500	645.78	435.7	474.40	2.125	0.99497
720	172.39	3.806	123.04	70.07	0.67002	2550	660.12	473.3	485.31	1.996	1.00064
740	177.23	4.193	126.51	65.38	0.67665	2600	674.49	513.5	496.26	1.876	1.00623
760	182.08	4.607	129.99	61.10	0.68312	2650	688.90	556.3	507.25	1.765	1.01172
780	186.94	5.051	133.47	57.20	0.68942	2700	703.35	601.9	518.26	1.662	1.01712
800	191.81	5.526	136.97	53.63	0.69558	2750	717.83	650.4	529.31	1.566	1.02244
820	196.69	6.033	140.47	50.35	0.70160	2800	732.33	702.0	540.40	1.478	1.02767
840	201.56	6.573	143.98	47.34	0.70747	2850	746.88	756.7	551.52	1.395	1.03282
860	206.46	7.149	147.50	44.57	0.71323	2900	761.45	814.8	562.66	1.318	1.03788
880	211.35	7.761	151.02	42.01	0.71886	2950	776.05	876.4	573.84	1.247	1.04288
900	216.26	8.411	154.57	39.64	0.72438	3000	790.68	941.4	585.04	1.180	1.04779
920	221.18	9.102	158.12	37.44	0.72979	3050	805.34	1011	596.28	1.118	1.05264
940	226.11	9.834	161.68	35.41	0.73509	3100	820.03	1083	607.53	1.060	1.05741
960	231.06	10.61	165.26	33.52	0.74030	3150	834.75	1161	618.82	1.006	1.06212
980	236.02	11.43	168.83	31.76	0.74540	3200	849.48	1242	630.12	0.955	1.06676
1000	240.98	12.30	172.43	30.12	0.75042	3250	864.24	1328	641.46	0.907	1.07134
1040	250.95	14.18	179.66	27.17	0.76019	3300	879.02	1418	652.81	0.8621	1.07585
1080	260.97	16.28	186.93	24.58	0.76964	3350	893.83	1513	664.20	0.8202	1.08031
1120	271.03	18.60	194.25	22.30	0.77880	3400	908.66	1613	675.60	0.7807	1.08470
1160	281.14	21.18	201.63	20.29	0.78767	3450	923.52	1719	687.04	0.7436	1.08904
1200	291.30	24.01	209.05	18.51	0.79628	3500	938.40	1829	698.48	0.7087	1.09332
1240	301.52	27.13	216.53	16.93	0.80466	3550	953.30	1946	709.95	0.6759	1.09755
1280	311.79	30.55	224.05	15.52	0.81280	3600	968.21	2068	721.44	0.6449	1.10172
1320	322.11	34.31	231.63	14.25	0.82075	3650	983.15	2196	732.95	0.6157	1.10584
1360	332.48	38.41	239.25	13.12	0.82848	3700	998.11	2330	744.48	0.5882	1.10991
1400	342.90	42.88	246.93	12.10	0.83604	3750	1013.1	2471	756.04	0.5621	1.11393
1440	353.37	47.75	254.66	11.17	0.84341	3800	1028.1	2618	767.60	0.5376	1.11791
1480	363.89	53.04	262.44	10.34	0.85062	3850	1043.1	2773	779.19	0.5143	1.12183
1520	374.47	58.78	270.26	9.578	0.85767	3900	1058.1	2934	790.80	0.4923	1.12571
1560	385.08	65.00	278.13	8.890	0.86456	3950	1073.2	3103	802.43	0.4715	1.12955

TABLE A–17E

Ideal-gas properties of air (*Concluded*)

T R	*h* Btu/lbm	P_r	*u* Btu/lbm	v_r	*s*° Btu/lbm·R	*T* R	*h* Btu/lbm	P_r	*u* Btu/lbm	v_r	*s*° Btu/lbm·R
4000	1088.3	3280	814.06	0.4518	1.13334	4600	1270.4	6089	955.04	0.2799	1.17575
4050	1103.4	3464	825.72	0.4331	1.13709	4700	1300.9	6701	978.73	0.2598	1.18232
4100	1118.5	3656	837.40	0.4154	1.14079	4800	1331.5	7362	1002.5	0.2415	1.18876
4150	1133.6	3858	849.09	0.3985	1.14446	4900	1362.2	8073	1026.3	0.2248	1.19508
4200	1148.7	4067	860.81	0.3826	1.14809	5000	1392.9	8837	1050.1	0.2096	1.20129
4300	1179.0	4513	884.28	0.3529	1.15522	5100	1423.6	9658	1074.0	0.1956	1.20738
4400	1209.4	4997	907.81	0.3262	1.16221	5200	1454.4	10,539	1098.0	0.1828	1.21336
4500	1239.9	5521	931.39	0.3019	1.16905	5300	1485.3	11,481	1122.0	0.1710	1.21923

Note: The properties P_r (relative pressure) and v_r (relative specific volume) are dimensionless quantities used in the analysis of isentropic processes, and should not be confused with the properties pressure and specific volume.

Source: Kenneth Wark, *Thermodynamics*, 4th ed. (New York: McGraw-Hill, 1983), pp. 832–33, Table A–5. Originally published in J. H. Keenan and J. Kaye, *Gas Tables* (New York: John Wiley & Sons, 1948).

TABLE A–18E

Ideal-gas properties of nitrogen, N_2

T R	$\bar{h}$ Btu/lbmol	$\bar{u}$ Btu/lbmol	$\bar{s}°$ Btu/lbmol·R
300	2,082.0	1,486.2	41.695
320	2,221.0	1,585.5	42.143
340	2,360.0	1,684.4	42.564
360	2,498.9	1,784.0	42.962
380	2,638.0	1,883.4	43.337
400	2,777.0	1,982.6	43.694
420	2,916.1	2,082.0	44.034
440	3,055.1	2,181.3	44.357
460	3,194.1	2,280.6	44.665
480	3,333.1	2,379.9	44.962
500	3,472.2	2,479.3	45.246
520	3,611.3	2,578.6	45.519
537	3,729.5	2,663.1	45.743
540	3,750.3	2,678.0	45.781
560	3,889.5	2,777.4	46.034
580	4,028.7	2,876.9	46.278
600	4,167.9	2,976.4	46.514
620	4,307.1	3,075.9	46.742
640	4,446.4	3,175.5	46.964
660	4,585.8	3,275.2	47.178
680	4,725.3	3,374.9	47.386
700	4,864.9	3,474.8	47.588
720	5,004.5	3,574.7	47.785
740	5,144.3	3,674.7	47.977
760	5,284.1	3,774.9	48.164
780	5,424.2	3,875.2	48.345
800	5,564.4	3,975.7	48.522
820	5,704.7	4,076.3	48.696
840	5,845.3	4,177.1	48.865
860	5,985.9	4,278.1	49.031
880	6,126.9	4,379.4	49.193
900	6,268.1	4,480.8	49.352
920	6,409.6	4,582.6	49.507
940	6,551.2	4,684.5	49.659
960	6,693.1	4,786.7	49.808
980	6,835.4	4,889.3	49.955
1000	6,977.9	4,992.0	50.099
1020	7,120.7	5,095.1	50.241
1040	7,263.8	5,198.5	50.380
1060	7,407.2	5,302.2	50.516

T R	$\bar{h}$ Btu/lbmol	$\bar{u}$ Btu/lbmol	$\bar{s}°$ Btu/lbmol·R
1080	7,551.0	5,406.2	50.651
1100	7,695.0	5,510.5	50.783
1120	7,839.3	5,615.2	50.912
1140	7,984.0	5,720.1	51.040
1160	8,129.0	5,825.4	51.167
1180	8,274.4	5,931.0	51.291
1200	8,420.0	6,037.0	51.143
1220	8,566.1	6,143.4	51.534
1240	8,712.6	6,250.1	51.653
1260	8,859.3	6,357.2	51.771
1280	9,006.4	6,464.5	51.887
1300	9,153.9	6,572.3	51.001
1320	9,301.8	6,680.4	52.114
1340	9,450.0	6,788.9	52.225
1360	9,598.6	6,897.8	52.335
1380	9,747.5	7,007.0	52.444
1400	9,896.9	7,116.7	52.551
1420	10,046.6	7,226.7	52.658
1440	10,196.6	7,337.0	52.763
1460	10,347.0	7,447.6	52.867
1480	10,497.8	7,558.7	52.969
1500	10,648.0	7,670.1	53.071
1520	10,800.4	7,781.9	53.171
1540	10,952.2	7,893.9	53.271
1560	11,104.3	8,006.4	53.369
1580	11,256.9	8,119.2	53.465
1600	11,409.7	8,232.3	53.561
1620	11,562.8	8,345.7	53.656
1640	11,716.4	8,459.6	53.751
1660	11,870.2	8,573.6	53.844
1680	12,024.3	8,688.1	53.936
1700	12,178.9	8,802.9	54.028
1720	12,333.7	8,918.0	54.118
1740	12,488.8	9,033.4	54.208
1760	12,644.3	9,149.2	54.297
1780	12,800.2	9,265.3	54.385
1800	12,956.3	9,381.7	54.472
1820	13,112.7	9,498.4	54.559
1840	13,269.5	9,615.5	54.645
1860	13,426.5	9,732.8	54.729

TABLE A–18E

Ideal-gas properties of nitrogen, N_2 (*Concluded*)

T R	$\bar{h}$ Btu/lbmol	$\bar{u}$ Btu/lbmol	$\bar{s}°$ Btu/lbmol·R	T R	$\bar{h}$ Btu/lbmol	$\bar{u}$ Btu/lbmol	$\bar{s}°$ Btu/lbmol·R
1900	13,742	9,968	54.896	3500	27,016	20,065	59.944
1940	14,058	10,205	55.061	3540	27,359	20,329	60.041
1980	14,375	10,443	55.223	3580	27,703	20,593	60.138
2020	14,694	10,682	55.383	3620	28,046	20,858	60.234
2060	15,013	10,923	55.540	3660	28,391	21,122	60.328
2100	15,334	11,164	55.694	3700	28,735	21,387	60.422
2140	15,656	11,406	55.846	3740	29,080	21,653	60.515
2180	15,978	11,649	55.995	3780	29,425	21,919	60.607
2220	16,302	11,893	56.141	3820	29,771	22,185	60.698
2260	16,626	12,138	56.286	3860	30,117	22,451	60.788
2300	16,951	12,384	56.429	3900	30,463	22,718	60.877
2340	17,277	12,630	56.570	3940	30,809	22,985	60.966
2380	17,604	12,878	56.708	3980	31,156	23,252	61.053
2420	17,392	13,126	56.845	4020	31,503	23,520	61.139
2460	18,260	13,375	56.980	4060	31,850	23,788	61.225
2500	18,590	13,625	57.112	4100	32,198	24,056	61.310
2540	18,919	13,875	57.243	4140	32,546	24,324	61.395
2580	19,250	14,127	57.372	4180	32,894	24,593	61.479
2620	19,582	14,379	57.499	4220	33,242	24,862	61.562
2660	19,914	14,631	57.625	4260	33,591	25,131	61.644
2700	20,246	14,885	57.750	4300	33,940	25,401	61.726
2740	20,580	15,139	57.872	4340	34,289	25,670	61.806
2780	20,914	15,393	57.993	4380	34,638	25,940	61.887
2820	21,248	15,648	58.113	4420	34,988	26,210	61.966
2860	21,584	15,905	58.231	4460	35,338	26,481	62.045
2900	21,920	16,161	58.348	4500	35,688	26,751	62.123
2940	22,256	16,417	58.463	4540	36,038	27,022	62.201
2980	22,593	16,675	58.576	4580	36,389	27,293	62.278
3020	22,930	16,933	58.688	4620	36,739	27,565	62.354
3060	23,268	17,192	58.800	4660	37,090	27,836	62.429
3100	23,607	17,451	58.910	4700	37,441	28,108	62.504
3140	23,946	17,710	59.019	4740	37,792	28,379	62.578
3180	24,285	17,970	59.126	4780	38,144	28,651	62.652
3220	24,625	18,231	59.232	4820	38,495	28,924	62.725
3260	24,965	18,491	59.338	4860	38,847	29,196	62.798
3300	25,306	18,753	59.442	4900	39,199	29,468	62.870
3340	25,647	19,014	59.544	5000	40,080	30,151	63.049
3380	25,989	19,277	59.646	5100	40,962	30,834	63.223
3420	26,331	19,539	59.747	5200	41,844	31,518	63.395
3460	26,673	19,802	59.846	5300	42,728	32,203	63.563

Source: Tables A–18E through A–23E are adapted from Kenneth Wark, *Thermodynamics*, 4th ed. (New York: McGraw-Hill, 1983), pp. 834–44. Originally published in J. H. Keenan and J. Kaye, *Gas Tables* (New York: John Wiley & Sons, 1945).

TABLE A–19E

Ideal-gas properties of oxygen, O_2

T R	$\bar{h}$ Btu/lbmol	$\bar{u}$ Btu/lbmol	$\bar{s}°$ Btu/lbmol·R	T R	$\bar{h}$ Btu/lbmol	$\bar{u}$ Btu/lbmol	$\bar{s}°$ Btu/lbmol·R
300	2,073.5	1,477.8	44.927	1080	7,696.8	5,552.1	54.064
320	2,212.6	1,577.1	45.375	1100	7,850.4	5,665.9	54.204
340	2,351.7	1,676.5	45.797	1120	8,004.5	5,780.3	54.343
360	2,490.8	1,775.9	46.195	1140	8,159.1	5,895.2	54.480
380	2,630.0	1,875.3	46.571	1160	8,314.2	6,010.6	54.614
400	2,769.1	1,974.8	46.927	1180	8,469.8	6,126.5	54.748
420	2,908.3	2,074.3	47.267	1200	8,625.8	6,242.8	54.879
440	3,047.5	2,173.8	47.591	1220	8,782.4	6,359.6	55.008
460	3,186.9	2,273.4	47.900	1240	8,939.4	6,476.9	55.136
480	3,326.5	2,373.3	48.198	1260	9,096.7	6,594.5	55.262
500	3,466.2	2,473.2	48.483	1280	9,254.6	6,712.7	55.386
520	3,606.1	2,573.4	48.757	1300	9,412.9	6,831.3	55.508
537	3,725.1	2,658.7	48.982	1320	9,571.9	6,950.2	55.630
540	3,746.2	2,673.8	49.021	1340	9,730.7	7,069.6	55.750
560	3,886.6	2,774.5	49.276	1360	9,890.2	7,189.4	55.867
580	4,027.3	2,875.5	49.522	1380	10,050.1	7,309.6	55.984
600	4,168.3	2,976.8	49.762	1400	10,210.4	7,430.1	56.099
620	4,309.7	3,078.4	49.993	1420	10,371.0	7,551.1	56.213
640	4,451.4	3,180.4	50.218	1440	10,532.0	7,672.4	56.326
660	4,593.5	3,282.9	50.437	1460	10,693.3	7,793.9	56.437
680	4,736.2	3,385.8	50.650	1480	10,855.1	7,916.0	56.547
700	4,879.3	3,489.2	50.858	1500	11,017.1	8,038.3	56.656
720	5,022.9	3,593.1	51.059	1520	11,179.6	8,161.1	56.763
740	5,167.0	3,697.4	51.257	1540	11,342.4	8,284.2	56.869
760	5,311.4	3,802.4	51.450	1560	11,505.4	8,407.4	56.975
780	5,456.4	3,907.5	51.638	1580	11,668.8	8,531.1	57.079
800	5,602.0	4,013.3	51.821	1600	11,832.5	8,655.1	57.182
820	5,748.1	4,119.7	52.002	1620	11,996.6	8,779.5	57.284
840	5,894.8	4,226.6	52.179	1640	12,160.9	8,904.1	57.385
860	6,041.9	4,334.1	52.352	1660	12,325.5	9,029.0	57.484
880	6,189.6	4,442.0	52.522	1680	12,490.4	9,154.1	57.582
900	6,337.9	4,550.6	52.688	1700	12,655.6	9,279.6	57.680
920	6,486.7	4,659.7	52.852	1720	12,821.1	9,405.4	57.777
940	6,636.1	4,769.4	53.012	1740	12,986.9	9,531.5	57.873
960	6,786.0	4,879.5	53.170	1760	13,153.0	9,657.9	57.968
980	6,936.4	4,990.3	53.326	1780	13,319.2	9,784.4	58.062
1000	7,087.5	5,101.6	53.477	1800	13,485.8	9,911.2	58.155
1020	7,238.9	5,213.3	53.628	1820	13,652.5	10,038.2	58.247
1040	7,391.0	5,325.7	53.775	1840	13,819.6	10,165.6	58.339
1060	7,543.6	5,438.6	53.921	1860	13,986.8	10,293.1	58.428

TABLE A–19E

Ideal-gas properties of oxygen, O_2 (*Concluded*)

T R	$\bar{h}$ Btu/lbmol	$\bar{u}$ Btu/lbmol	$\bar{s}°$ Btu/lbmol·R	T R	$\bar{h}$ Btu/lbmol	$\bar{u}$ Btu/lbmol	$\bar{s}°$ Btu/lbmol·R
1900	14,322	10,549	58.607	3500	28,273	21,323	63.914
1940	14,658	10,806	58.782	3540	28,633	21,603	64.016
1980	14,995	11,063	58.954	3580	28,994	21,884	64.114
2020	15,333	11,321	59.123	3620	29,354	22,165	64.217
2060	15,672	11,581	59.289	3660	29,716	22,447	64.316
2100	16,011	11,841	59.451	3700	30,078	22,730	64.415
2140	16,351	12,101	59.612	3740	30,440	23,013	64.512
2180	16,692	12,363	59.770	3780	30,803	23,296	64.609
2220	17,036	12,625	59.926	3820	31,166	23,580	64.704
2260	17,376	12,888	60.077	3860	31,529	23,864	64.800
2300	17,719	13,151	60.228	3900	31,894	24,149	64.893
2340	18,062	13,416	60.376	3940	32,258	24,434	64.986
2380	18,407	13,680	60.522	3980	32,623	24,720	65.078
2420	18,572	13,946	60.666	4020	32,989	25,006	65.169
2460	19,097	14,212	60.808	4060	33,355	25,292	65.260
2500	19,443	14,479	60.946	4100	33,722	25,580	65.350
2540	19,790	14,746	61.084	4140	34,089	25,867	64.439
2580	20,138	15,014	61.220	4180	34,456	26,155	65.527
2620	20,485	15,282	61.354	4220	34,824	26,144	65.615
2660	20,834	15,551	61.486	4260	35,192	26,733	65.702
2700	21,183	15,821	61.616	4300	35,561	27,022	65.788
2740	21,533	16,091	61.744	4340	35,930	27,312	65.873
2780	21,883	16,362	61.871	4380	36,300	27,602	65.958
2820	22,232	16,633	61.996	4420	36,670	27,823	66.042
2860	22,584	16,905	62.120	4460	37,041	28,184	66.125
2900	22,936	17,177	62.242	4500	37,412	28,475	66.208
2940	23,288	17,450	62.363	4540	37,783	28,768	66.290
2980	23,641	17,723	62.483	4580	38,155	29,060	66.372
3020	23,994	17,997	62.599	4620	38,528	29,353	66.453
3060	24,348	18,271	62.716	4660	38,900	29,646	66.533
3100	24,703	18,546	62.831	4700	39,274	29,940	66.613
3140	25,057	18,822	62.945	4740	39,647	30,234	66.691
3180	25,413	19,098	63.057	4780	40,021	30,529	66.770
3220	25,769	19,374	63.169	4820	40,396	30,824	66.848
3260	26,175	19,651	63.279	4860	40,771	31,120	66.925
3300	26,412	19,928	63.386	4900	41,146	31,415	67.003
3340	26,839	20,206	63.494	5000	42,086	32,157	67.193
3380	27,197	20,485	63.601	5100	43,021	32,901	67.380
3420	27,555	20,763	63.706	5200	43,974	33,648	67.562
3460	27,914	21,043	63.811	5300	44,922	34,397	67.743

TABLE A–20E

Ideal-gas properties of carbon dioxide, CO_2

T R	$\bar{h}$ Btu/lbmol	$\bar{u}$ Btu/lbmol	$\bar{s}°$ Btu/lbmol·R	T R	$\bar{h}$ Btu/lbmol	$\bar{u}$ Btu/lbmol	$\bar{s}°$ Btu/lbmol·R
300	2,108.2	1,512.4	46.353	1080	9,575.8	7,431.1	58.072
320	2,256.6	1,621.1	46.832	1100	9,802.6	7,618.1	58.281
340	2,407.3	1,732.1	47.289	1120	10,030.6	7,806.4	58.485
360	2,560.5	1,845.6	47.728	1140	10,260.1	7,996.2	58.689
380	2,716.4	1,961.8	48.148	1160	10,490.6	8,187.0	58.889
400	2,874.7	2,080.4	48.555	1180	10,722.3	8,379.0	59.088
420	3,035.7	2,201.7	48.947	1200	10,955.3	8,572.3	59.283
440	3,199.4	2,325.6	49.329	1220	11,189.4	8,766.6	59.477
460	3,365.7	2,452.2	49.698	1240	11,424.6	8,962.1	59.668
480	3,534.7	2,581.5	50.058	1260	11,661.0	9,158.8	59.858
500	3,706.2	2,713.3	50.408	1280	11,898.4	9,356.5	60.044
520	3,880.3	2,847.7	50.750	1300	12,136.9	9,555.3	60.229
537	4,027.5	2,963.8	51.032	1320	12,376.4	9,755.0	60.412
540	4,056.8	2,984.4	51.082	1340	12,617.0	9,955.9	60.593
560	4,235.8	3,123.7	51.408	1360	12,858.5	10,157.7	60.772
580	4,417.2	3,265.4	51.726	1380	13,101.0	10,360.5	60.949
600	4,600.9	3,409.4	52.038	1400	13,344.7	10,564.5	61.124
620	4,786.6	3,555.6	52.343	1420	13,589.1	10,769.2	61.298
640	4,974.9	3,704.0	52.641	1440	13,834.5	10,974.8	61.469
660	5,165.2	3,854.6	52.934	1460	14,080.8	11,181.4	61.639
680	5,357.6	4,007.2	53.225	1480	14,328.0	11,388.9	61.800
700	5,552.0	4,161.9	53.503	1500	14,576.0	11,597.2	61.974
720	5,748.4	4,318.6	53.780	1520	14,824.9	11,806.4	62.138
740	5,946.8	4,477.3	54.051	1540	15,074.7	12,016.5	62.302
760	6,147.0	4,637.9	54.319	1560	15,325.3	12,227.3	62.464
780	6,349.1	4,800.1	54.582	1580	15,576.7	12,439.0	62.624
800	6,552.9	4,964.2	54.839	1600	15,829.0	12,651.6	62.783
820	6,758.3	5,129.9	55.093	1620	16,081.9	12,864.8	62.939
840	6,965.7	5,297.6	55.343	1640	16,335.7	13,078.9	63.095
860	7,174.7	5,466.9	55.589	1660	16,590.2	13,293.7	63.250
880	7,385.3	5,637.7	55.831	1680	16,845.5	13,509.2	63.403
900	7,597.6	5,810.3	56.070	1700	17,101.4	13,725.4	63.555
920	7,811.4	5,984.4	56.305	1720	17,358.1	13,942.4	63.704
940	8,026.8	6,160.1	56.536	1740	17,615.5	14,160.1	63.853
960	8,243.8	6,337.4	56.765	1760	17,873.5	14,378.4	64.001
980	8,462.2	6,516.1	56.990	1780	18,132.2	14,597.4	64.147
1000	8,682.1	6,696.2	57.212	1800	18,391.5	14,816.9	64.292
1020	8,903.4	6,877.8	57.432	1820	18,651.5	15,037.2	64.435
1040	9,126.2	7,060.9	57.647	1840	18,912.2	15,258.2	64.578
1060	9,350.3	7,245.3	57.861	1860	19,173.4	15,479.7	64.719

TABLE A–20E

Ideal-gas properties of carbon dioxide, CO_2 (*Concluded*)

T R	$\bar{h}$ Btu/lbmol	$\bar{u}$ Btu/lbmol	$\bar{s}°$ Btu/lbmol·R	T R	$\bar{h}$ Btu/lbmol	$\bar{u}$ Btu/lbmol	$\bar{s}°$ Btu/lbmol·R
1900	19,698	15,925	64.999	3500	41,965	35,015	73.462
1940	20,224	16,372	65.272	3540	42,543	35,513	73.627
1980	20,753	16,821	65.543	3580	43,121	36,012	73.789
2020	21,284	17,273	65.809	3620	43,701	36,512	73.951
2060	21,818	17,727	66.069	3660	44,280	37,012	74.110
2100	22,353	18,182	66.327	3700	44,861	37,513	74.267
2140	22,890	18,640	66.581	3740	45,442	38,014	74.423
2180	23,429	19,101	66.830	3780	46,023	38,517	74.578
2220	23,970	19,561	67.076	3820	46,605	39,019	74.732
2260	24,512	20,024	67.319	3860	47,188	39,522	74.884
2300	25,056	20,489	67.557	3900	47,771	40,026	75.033
2340	25,602	20,955	67.792	3940	48,355	40,531	75.182
2380	26,150	21,423	68.025	3980	48,939	41,035	75.330
2420	26,699	21,893	68.253	4020	49,524	41,541	75.477
2460	27,249	22,364	68.479	4060	50,109	42,047	75.622
2500	27,801	22,837	68.702	4100	50,695	42,553	75.765
2540	28,355	23,310	68.921	4140	51,282	43,060	75.907
2580	28,910	23,786	69.138	4180	51,868	43,568	76.048
2620	29,465	24,262	69.352	4220	52,456	44,075	76.188
2660	30,023	24,740	69.563	4260	53,044	44,584	76.327
2700	30,581	25,220	69.771	4300	53,632	45,093	76.464
2740	31,141	25,701	69.977	4340	54,221	45,602	76.601
2780	31,702	26,181	70.181	4380	54,810	46,112	76.736
2820	32,264	26,664	70.382	4420	55,400	46,622	76.870
2860	32,827	27,148	70.580	4460	55,990	47,133	77.003
2900	33,392	27,633	70.776	4500	56,581	47,645	77.135
2940	33,957	28,118	70.970	4540	57,172	48,156	77.266
2980	34,523	28,605	71.160	4580	57,764	48,668	77.395
3020	35,090	29,093	71.350	4620	58,356	49,181	77.581
3060	35,659	29,582	71.537	4660	58,948	49,694	77.652
3100	36,228	30,072	71.722	4700	59,541	50,208	77.779
3140	36,798	30,562	71.904	4740	60,134	50,721	77.905
3180	37,369	31,054	72.085	4780	60,728	51,236	78.029
3220	37,941	31,546	72.264	4820	61,322	51,750	78.153
3260	38,513	32,039	72.441	4860	61,916	52,265	78.276
3300	39,087	32,533	72.616	4900	62,511	52,781	78.398
3340	39,661	33,028	72.788	5000	64,000	54,071	78.698
3380	40,236	33,524	72.960	5100	65,491	55,363	78.994
3420	40,812	34,020	73.129	5200	66,984	56,658	79.284
3460	41,388	34,517	73.297	5300	68,471	57,954	79.569

TABLE A–21E

Ideal-gas properties of carbon monoxide, CO

T R	$\bar{h}$ Btu/lbmol	$\bar{u}$ Btu/lbmol	$\bar{s}°$ Btu/lbmol·R	T R	$\bar{h}$ Btu/lbmol	$\bar{u}$ Btu/lbmol	$\bar{s}°$ Btu/lbmol·R
300	2,081.9	1,486.1	43.223	1080	7,571.1	5,426.4	52.203
320	2,220.9	1,585.4	43.672	1100	7,716.8	5,532.3	52.337
340	2,359.9	1,684.7	44.093	1120	7,862.9	5,638.7	52.468
360	2,498.8	1,783.9	44.490	1140	8,009.2	5,745.4	52.598
380	2,637.9	1,883.3	44.866	1160	8,156.1	5,851.5	52.726
400	2,776.9	1,982.6	45.223	1180	8,303.3	5,960.0	52.852
420	2,916.0	2,081.9	45.563	1200	8,450.8	6,067.8	52.976
440	3,055.0	2,181.2	45.886	1220	8,598.8	6,176.0	53.098
460	3,194.0	2,280.5	46.194	1240	8,747.2	6,284.7	53.218
480	3,333.0	2,379.8	46.491	1260	8,896.0	6,393.8	53.337
500	3,472.1	2,479.2	46.775	1280	9,045.0	6,503.1	53.455
520	3,611.2	2,578.6	47.048	1300	9,194.6	6,613.0	53.571
537	3,725.1	2,663.1	47.272	1320	9,344.6	6,723.2	53.685
540	3,750.3	2,677.9	47.310	1340	9,494.8	6,833.7	53.799
560	3,889.5	2,777.4	47.563	1360	9,645.5	6,944.7	53.910
580	4,028.7	2,876.9	47.807	1380	9,796.6	7,056.1	54.021
600	4,168.0	2,976.5	48.044	1400	9,948.1	7,167.9	54.129
620	4,307.4	3,076.2	48.272	1420	10,100.0	7,280.1	54.237
640	4,446.9	3,175.9	48.494	1440	10,252.2	7,392.6	54.344
660	4,586.6	3,275.8	48.709	1460	10,404.8	7,505.4	54.448
680	4,726.2	3,375.8	48.917	1480	10,557.8	7,618.7	54.522
700	4,886.0	3,475.9	49.120	1500	10,711.1	7,732.3	54.665
720	5,006.1	3,576.3	49.317	1520	10,864.9	7,846.4	54.757
740	5,146.4	3,676.9	49.509	1540	11,019.0	7,960.8	54.858
760	5,286.8	3,777.5	49.697	1560	11,173.4	8,075.4	54.958
780	5,427.4	3,878.4	49.880	1580	11,328.2	8,190.5	55.056
800	5,568.2	3,979.5	50.058	1600	11,483.4	8,306.0	55.154
820	5,709.4	4,081.0	50.232	1620	11,638.9	8,421.8	55.251
840	5,850.7	4,182.6	50.402	1640	11,794.7	8,537.9	55.347
860	5,992.3	4,284.5	50.569	1660	11,950.9	8,654.4	55.411
880	6,134.2	4,386.6	50.732	1680	12,107.5	8,771.2	55.535
900	6,276.4	4,489.1	50.892	1700	12,264.3	8,888.3	55.628
920	6,419.0	4,592.0	51.048	1720	12,421.4	9,005.7	55.720
940	6,561.7	4,695.0	51.202	1740	12,579.0	9,123.6	55.811
960	6,704.9	4,798.5	51.353	1760	12,736.7	9,241.6	55.900
980	6,848.4	4,902.3	51.501	1780	12,894.9	9,360.0	55.990
1000	6,992.2	5,006.3	51.646	1800	13,053.2	9,478.6	56.078
1020	7,136.4	5,110.8	51.788	1820	13,212.0	9,597.7	56.166
1040	7,281.0	5,215.7	51.929	1840	13,371.0	9,717.0	56.253
1060	7,425.9	5,320.9	52.067	1860	13,530.2	9,836.5	56.339

TABLE A–21E

Ideal-gas properties of carbon monoxide, CO (*Concluded*)

T R	$\bar{h}$ Btu/lbmol	$\bar{u}$ Btu/lbmol	$\bar{s}°$ Btu/lbmol·R	T R	$\bar{h}$ Btu/lbmol	$\bar{u}$ Btu/lbmol	$\bar{s}°$ Btu/lbmol·R
1900	13,850	10,077	56.509	3500	27,262	20,311	61.612
1940	14,170	10,318	56.677	3540	27,608	20,576	61.710
1980	14,492	10,560	56.841	3580	27,954	20,844	61.807
2020	14,815	10,803	57.007	3620	28,300	21,111	61.903
2060	15,139	11,048	57.161	3660	28,647	21,378	61.998
2100	15,463	11,293	57.317	3700	28,994	21,646	62.093
2140	15,789	11,539	57.470	3740	29,341	21,914	62.186
2180	16,116	11,787	57.621	3780	29,688	22,182	62.279
2220	16,443	12,035	57.770	3820	30,036	22,450	62.370
2260	16,722	12,284	57.917	3860	30,384	22,719	62.461
2300	17,101	12,534	58.062	3900	30,733	22,988	62.511
2340	17,431	12,784	58.204	3940	31,082	23,257	62.640
2380	17,762	13,035	58.344	3980	31,431	23,527	62.728
2420	18,093	13,287	58.482	4020	31,780	23,797	62.816
2460	18,426	13,541	58.619	4060	32,129	24,067	62.902
2500	18,759	13,794	58.754	4100	32,479	24,337	62.988
2540	19,093	14,048	58.885	4140	32,829	24,608	63.072
2580	19,427	14,303	59.016	4180	33,179	24,878	63.156
2620	19,762	14,559	59.145	4220	33,530	25,149	63.240
2660	20,098	14,815	59.272	4260	33,880	25,421	63.323
2700	20,434	15,072	59.398	4300	34,231	25,692	63.405
2740	20,771	15,330	59.521	4340	34,582	25,934	63.486
2780	21,108	15,588	59.644	4380	34,934	26,235	63.567
2820	21,446	15,846	59.765	4420	35,285	26,508	63.647
2860	21,785	16,105	59.884	4460	35,637	26,780	63.726
2900	22,124	16,365	60.002	4500	35,989	27,052	63.805
2940	22,463	16,225	60.118	4540	36,341	27,325	63.883
2980	22,803	16,885	60.232	4580	36,693	27,598	63.960
3020	23,144	17,146	60.346	4620	37,046	27,871	64.036
3060	23,485	17,408	60.458	4660	37,398	28,144	64.113
3100	23,826	17,670	60.569	4700	37,751	28,417	64.188
3140	24,168	17,932	60.679	4740	38,104	28,691	64.263
3180	24,510	18,195	60.787	4780	38,457	28,965	64.337
3220	24,853	18,458	60.894	4820	38,811	29,239	64.411
3260	25,196	18,722	61.000	4860	39,164	29,513	64.484
3300	25,539	18,986	61.105	4900	39,518	29,787	64.556
3340	25,883	19,250	61.209	5000	40,403	30,473	64.735
3380	26,227	19,515	61.311	5100	41,289	31,161	64.910
3420	26,572	19,780	61.412	5200	42,176	31,849	65.082
3460	26,917	20,045	61.513	5300	43,063	32,538	65.252

TABLE A–22E

Ideal-gas properties of hydrogen, H_2

T R	$\bar{h}$ Btu/lbmol	$\bar{u}$ Btu/lbmol	$\bar{s}°$ Btu/lbmol·R	T R	$\bar{h}$ Btu/lbmol	$\bar{u}$ Btu/lbmol	$\bar{s}°$ Btu/lbmol·R
300	2,063.5	1,467.7	27.337	1400	9,673.8	6,893.6	37.883
320	2,189.4	1,553.9	27.742	1500	10,381.5	7,402.7	38.372
340	2,317.2	1,642.0	28.130	1600	11,092.5	7,915.1	38.830
360	2,446.8	1,731.9	28.501	1700	11,807.4	8,431.4	39.264
380	2,577.8	1,823.2	28.856	1800	12,526.8	8,952.2	39.675
400	2,710.2	1,915.8	29.195	1900	13,250.9	9,477.8	40.067
420	2,843.7	2,009.6	29.520	2000	13,980.1	10,008.4	40.441
440	2,978.1	2,104.3	29.833	2100	14,714.5	10,544.2	40.799
460	3,113.5	2,200.0	30.133	2200	15,454.4	11,085.5	41.143
480	3,249.4	2,296.2	20.424	2300	16,199.8	11,632.3	41.475
500	3,386.1	2,393.2	30.703	2400	16,950.6	12,184.5	41.794
520	3,523.2	2,490.6	30.972	2500	17,707.3	12,742.6	42.104
537	3,640.3	2,573.9	31.194	2600	18,469.7	13,306.4	42.403
540	3,660.9	2,588.5	31.232	2700	19,237.8	13,876.0	42.692
560	3,798.8	2,686.7	31.482	2800	20,011.8	14,451.4	42.973
580	3,937.1	2,785.3	31.724	2900	20,791.5	15,032.5	43.247
600	4,075.6	2,884.1	31.959	3000	21,576.9	15,619.3	43.514
620	4,214.3	2,983.1	32.187	3100	22,367.7	16,211.5	43.773
640	4,353.1	3,082.1	32.407	3200	23,164.1	16,809.3	44.026
660	4,492.1	3,181.4	32.621	3300	23,965.5	17,412.1	44.273
680	4,631.1	3,280.7	32.829	3400	24,771.9	18,019.9	44.513
700	4,770.2	3,380.1	33.031	3500	25,582.9	18,632.4	44.748
720	4,909.5	3,479.6	33.226	3600	26,398.5	19,249.4	44.978
740	5,048.8	3,579.2	33.417	3700	27,218.5	19,870.8	45.203
760	5,188.1	3,678.8	33.603	3800	28,042.8	20,496.5	45.423
780	5,327.6	3,778.6	33.784	3900	28,871.1	21,126.2	45.638
800	5,467.1	3,878.4	33.961	4000	29,703.5	21,760.0	45.849
820	5,606.7	3,978.3	34.134	4100	30,539.8	22,397.7	46.056
840	5,746.3	4,078.2	34.302	4200	31,379.8	23,039.2	46.257
860	5,885.9	4,178.0	34.466	4300	32,223.5	23,684.3	46.456
880	6,025.6	4,278.0	34.627	4400	33,070.9	24,333.1	46.651
900	6,165.3	4,378.0	34.784	4500	33,921.6	24,985.2	46.842
920	6,305.1	4,478.1	34.938	4600	34,775.7	25,640.7	47.030
940	6,444.9	4,578.1	35.087	4700	35,633.0	26,299.4	47.215
960	6,584.7	4,678.3	35.235	4800	36,493.4	26,961.2	47.396
980	6,724.6	4,778.4	35.379	4900	35,356.9	27,626.1	47.574
1000	6,864.5	4,878.6	35.520	5000	38,223.3	28,294.0	47.749
1100	7,564.6	5,380.1	36.188	5100	39,092.8	28,964.9	47.921
1200	8,265.8	5,882.8	36.798	5200	39,965.1	29,638.6	48.090
1300	8,968.7	6,387.1	37.360	5300	40,840.2	30,315.1	48.257

TABLE A–23E

Ideal-gas properties of water vapor, H_2O

T R	$\bar{h}$ Btu/lbmol	$\bar{u}$ Btu/lbmol	$\bar{s}°$ Btu/lbmol·R	T R	$\bar{h}$ Btu/lbmol	$\bar{u}$ Btu/lbmol	$\bar{s}°$ Btu/lbmol·R
300	2,367.6	1,771.8	40.439	1080	8,768.2	6,623.5	50.854
320	2,526.8	1,891.3	40.952	1100	8,942.0	6,757.5	51.013
340	2,686.0	2,010.8	41.435	1120	9,116.4	6,892.2	51.171
360	2,845.1	2,130.2	41.889	1140	9,291.4	7,027.5	51.325
380	3,004.4	2,249.8	42.320	1160	9,467.1	7,163.5	51.478
400	3,163.8	2,369.4	42.728	1180	9,643.4	7,300.1	51.360
420	3,323.2	2,489.1	43.117	1200	9,820.4	7,437.4	51.777
440	3,482.7	2,608.9	43.487	1220	9,998.0	7,575.2	51.925
460	3,642.3	2,728.8	43.841	1240	10,176.1	7,713.6	52.070
480	3,802.0	2,848.8	44.182	1260	10,354.9	7,852.7	52.212
500	3,962.0	2,969.1	44.508	1280	10,534.4	7,992.5	52.354
520	4,122.0	3,089.4	44.821	1300	10,714.5	8,132.9	52.494
537	4,258.0	3,191.9	45.079	1320	10,895.3	8,274.0	52.631
540	4,282.4	3,210.0	45.124	1340	11,076.6	8,415.5	52.768
560	4,442.8	3,330.7	45.415	1360	11,258.7	8,557.9	52.903
580	4,603.7	3,451.9	45.696	1380	11,441.4	8,700.9	53.037
600	4,764.7	3,573.2	45.970	1400	11,624.8	8,844.6	53.168
620	4,926.1	3,694.9	46.235	1420	11,808.8	8,988.9	53.299
640	5,087.8	3,816.8	46.492	1440	11,993.4	9,133.8	53.428
660	5,250.0	3,939.3	46.741	1460	12,178.8	9,279.4	53.556
680	5,412.5	4,062.1	46.984	1480	12,364.8	9,425.7	53.682
700	5,575.4	4,185.3	47.219	1500	12,551.4	9,572.7	53.808
720	5,738.8	4,309.0	47.450	1520	12,738.8	9,720.3	53.932
740	5,902.6	4,433.1	47.673	1540	12,926.8	9,868.6	54.055
760	6,066.9	4,557.6	47.893	1560	13,115.6	10,017.6	54.117
780	6,231.7	4,682.7	48.106	1580	13,305.0	10,167.3	54.298
800	6,396.9	4,808.2	48.316	1600	13,494.4	10,317.6	54.418
820	6,562.6	4,934.2	48.520	1620	13,685.7	10,468.6	54.535
840	6,728.9	5,060.8	48.721	1640	13,877.0	10,620.2	54.653
860	6,895.6	5,187.8	48.916	1660	14,069.2	10,772.7	54.770
880	7,062.9	5,315.3	49.109	1680	14,261.9	10,925.6	54.886
900	7,230.9	5,443.6	49.298	1700	14,455.4	11,079.4	54.999
920	7,399.4	5,572.4	49.483	1720	14,649.5	11,233.8	55.113
940	7,568.4	5,701.7	49.665	1740	14,844.3	11,388.9	55.226
960	7,738.0	5,831.6	49.843	1760	15,039.8	11,544.7	55.339
980	7,908.2	5,962.0	50.019	1780	15,236.1	11,701.2	55.449
1000	8,078.2	6,093.0	50.191	1800	15,433.0	11,858.4	55.559
1020	8,250.4	6,224.8	50.360	1820	15,630.6	12,016.3	55.668
1040	8,422.4	6,357.1	50.528	1840	15,828.7	12,174.7	55.777
1060	8,595.0	6,490.0	50.693	1860	16,027.6	12,333.9	55.884

TABLE A–23E

Ideal-gas properties of water vapor, H_2O (*Cotinued*)

T R	$\bar{h}$ Btu/lbmol	$\bar{u}$ Btu/lbmol	$\bar{s}°$ Btu/lbmol·R	T R	$\bar{h}$ Btu/lbmol	$\bar{u}$ Btu/lbmol	$\bar{s}°$ Btu/lbmol·R
1900	16,428	12,654	56.097	3500	34,324	27,373	62.876
1940	16,830	12,977	56.307	3540	34,809	27,779	63.015
1980	17,235	13,303	56.514	3580	35,296	28,187	63.153
2020	17,643	13,632	56.719	3620	35,785	28,596	63.288
2060	18,054	13,963	56.920	3660	36,274	29,006	63.423
2100	18,467	14,297	57.119	3700	36,765	29,418	63.557
2140	18,883	14,633	57.315	3740	37,258	29,831	63.690
2180	19,301	14,972	57.509	3780	37,752	30,245	63.821
2220	19,722	15,313	57.701	3820	38,247	30,661	63.952
2260	20,145	15,657	57.889	3860	38,743	31,077	64.082
2300	20,571	16,003	58.077	3900	39,240	31,495	64.210
2340	20,999	16,352	58.261	3940	39,739	31,915	64.338
2380	21,429	16,703	58.445	3980	40,239	32,335	64.465
2420	21,862	17,057	58.625	4020	40,740	32,757	64.591
2460	22,298	17,413	58.803	4060	41,242	33,179	64.715
2500	22,735	17,771	58.980	4100	41,745	33,603	64.839
2540	23,175	18,131	59.155	4140	42,250	34,028	64.962
2580	23,618	18,494	59.328	4180	42,755	34,454	65.084
2620	24,062	18,859	59.500	4220	43,267	34,881	65.204
2660	24,508	19,226	59.669	4260	43,769	35,310	65.325
2700	24,957	19,595	59.837	4300	44,278	35,739	65.444
2740	25,408	19,967	60.003	4340	44,788	36,169	65.563
2780	25,861	20,340	60.167	4380	45,298	36,600	65.680
2820	26,316	20,715	60.330	4420	45,810	37,032	65.797
2860	26,773	21,093	60.490	4460	46,322	37,465	65.913
2900	27,231	21,472	60.650	4500	46,836	37,900	66.028
2940	27,692	21,853	60.809	4540	47,350	38,334	66.142
2980	28,154	22,237	60.965	4580	47,866	38,770	66.255
3020	28,619	22,621	61.120	4620	48,382	39,207	66.368
3060	29,085	23,085	61.274	4660	48,899	39,645	66.480
3100	29,553	23,397	61.426	4700	49,417	40,083	66.591
3140	30,023	23,787	61.577	4740	49,936	40,523	66.701
3180	30,494	24,179	61.727	4780	50,455	40,963	66.811
3220	30,967	24,572	61.874	4820	50,976	41,404	66.920
3260	31,442	24,968	62.022	4860	51,497	41,856	67.028
3300	31,918	25,365	62.167	4900	52,019	42,288	67.135
3340	32,396	25,763	62.312	5000	53,327	43,398	67.401
3380	32,876	26,164	62.454	5100	54,640	44,512	67.662
3420	33,357	26,565	62.597	5200	55,957	45,631	67.918
3460	33,839	26,968	62.738	5300	57,279	46,754	68.172

TABLE A–26E

Enthalpy of formation, Gibbs function of formation, and absolute entropy at 77°F, 1 atm

Substance	Formula	$\bar{h}_f^\circ$ Btu/lbmol	$\bar{g}_f^\circ$ Btu/lbmol	$\bar{s}^\circ$ Btu/lbmol·R
Carbon	$C(s)$	0	0	1.36
Hydrogen	$H_2(g)$	0	0	31.21
Nitrogen	$N_2(g)$	0	0	45.77
Oxygen	$O_2(g)$	0	0	49.00
Carbon monoxide	$CO(g)$	−47,540	−59,010	47.21
Carbon dioxide	$CO_2(g)$	−169,300	−169,680	51.07
Water vapor	$H_2O(g)$	−104,040	−98,350	45.11
Water	$H_2O(\ell)$	−122,970	−102,040	16.71
Hydrogen peroxide	$H_2O_2(g)$	−58,640	−45,430	55.60
Ammonia	$NH_3(g)$	−19,750	−7,140	45.97
Methane	$CH_4(g)$	−32,210	−21,860	44.49
Acetylene	$C_2H_2(g)$	+97,540	+87,990	48.00
Ethylene	$C_2H_4(g)$	+22,490	+29,306	52.54
Ethane	$C_2H_6(g)$	−36,420	−14,150	54.85
Propylene	$C_3H_6(g)$	+8,790	+26,980	63.80
Propane	$C_3H_8(g)$	−44,680	−10,105	64.51
n-Butane	$C_4H_{10}(g)$	−54,270	−6,760	74.11
n-Octane	$C_8H_{18}(g)$	−89,680	+7,110	111.55
n-Octane	$C_8H_{18}(\ell)$	−107,530	+2,840	86.23
n-Dodecane	$C_{12}H_{26}(g)$	−125,190	+21,570	148.86
Benzene	$C_6H_6(g)$	+35,680	+55,780	64.34
Methyl alcohol	$CH_3OH(g)$	−86,540	−69,700	57.29
Methyl alcohol	$CH_3OH(\ell)$	−102,670	−71,570	30.30
Ethyl alcohol	$C_2H_5OH(g)$	−101,230	−72,520	67.54
Ethyl alcohol	$C_2H_5OH(\ell)$	−119,470	−75,240	38.40
Oxygen	$O(g)$	+107,210	+99,710	38.47
Hydrogen	$H(g)$	+93,780	+87,460	27.39
Nitrogen	$N(g)$	+203,340	+195,970	36.61
Hydroxyl	$OH(g)$	+16,790	+14,750	43.92

Source: From JANAF, *Thermochemical Tables* (Midland, MI: Dow Chemical Co., 1971), *Selected Values of Chemical Thermodynamic Properties*, NBS Technical Note 270-3, 1968; and *API Research Project 44* (Carnegie Press, 1953).

TABLE A–27E

Properties of some common fuels and hydrocarbons

Fuel (phase)	Formula	Molar mass, lbm/lbmol	Density,[1] lbm/ft³	Enthalpy of vaporization,[2] Btu/lbm	Specific heat,[1] c_p Btu/lbm·°F	Higher heating value,[3] Btu/lbm	Lower heating value,[3] Btu/lbm
Carbon (*s*)	C	12.011	125	—	0.169	14,100	14,100
Hydrogen (*g*)	H_2	2.016	—	—	3.44	60,970	51,600
Carbon monoxide (*g*)	CO	28.013	—	—	0.251	4,340	4,340
Methane (*g*)	CH_4	16.043	—	219	0.525	23,880	21,520
Methanol (ℓ)	CH_4O	32.042	49.3	502	0.604	9,740	8,570
Acetylene (*g*)	C_2H_2	26.038	—	—	0.404	21,490	20,760
Ethane (*g*)	C_2H_6	30.070	—	74	0.418	22,320	20,430
Ethanol (ℓ)	C_2H_6O	46.069	49.3	395	0.583	12,760	11,530
Propane (ℓ)	C_3H_8	44.097	31.2	144	0.662	21,640	19,930
Butane (ℓ)	C_4H_{10}	58.123	36.1	156	0.578	21,130	19,510
1-Pentene (ℓ)	C_5H_{10}	70.134	40.0	156	0.525	20,540	19,190
Isopentane (ℓ)	C_5H_{12}	72.150	39.1	—	0.554	20,890	19,310
Benzene (ℓ)	C_6H_6	78.114	54.7	186	0.411	17,970	17,240
Hexene (ℓ)	C_6H_{12}	84.161	42.0	169	0.439	20,430	19,090
Hexane (ℓ)	C_6H_{14}	86.177	41.2	157	0.542	20,770	19,240
Toluene (ℓ)	C_7H_8	92.141	54.1	177	0.408	18,230	17,420
Heptane (ℓ)	C_7H_{16}	100.204	42.7	157	0.535	20,680	19,180
Octane (ℓ)	C_8H_{18}	114.231	43.9	156	0.533	20,590	19,100
Decane (ℓ)	$C_{10}H_{22}$	142.285	45.6	155	0.528	20,490	19,020
Gasoline (ℓ)	$C_nH_{1.87n}$	100–110	45–49	151	0.57	20,300	18,900
Light diesel (ℓ)	$C_nH_{1.8n}$	170	49–52	116	0.53	19,800	18,600
Heavy diesel (ℓ)	$C_nH_{1.7n}$	200	51–55	99	0.45	19,600	18,400
Natural gas (*g*)	$C_nH_{3.8n}N_{0.1n}$	18	—	—	0.48	21,500	19,400

[1]At 1 atm and 68°F.

[2]At 77°F for liquid fuels, and 1 atm and normal boiling temperature for gaseous fuels.

[3]At 77°F. Multiply by molar mass to obtain heating values in Btu/lbmol.

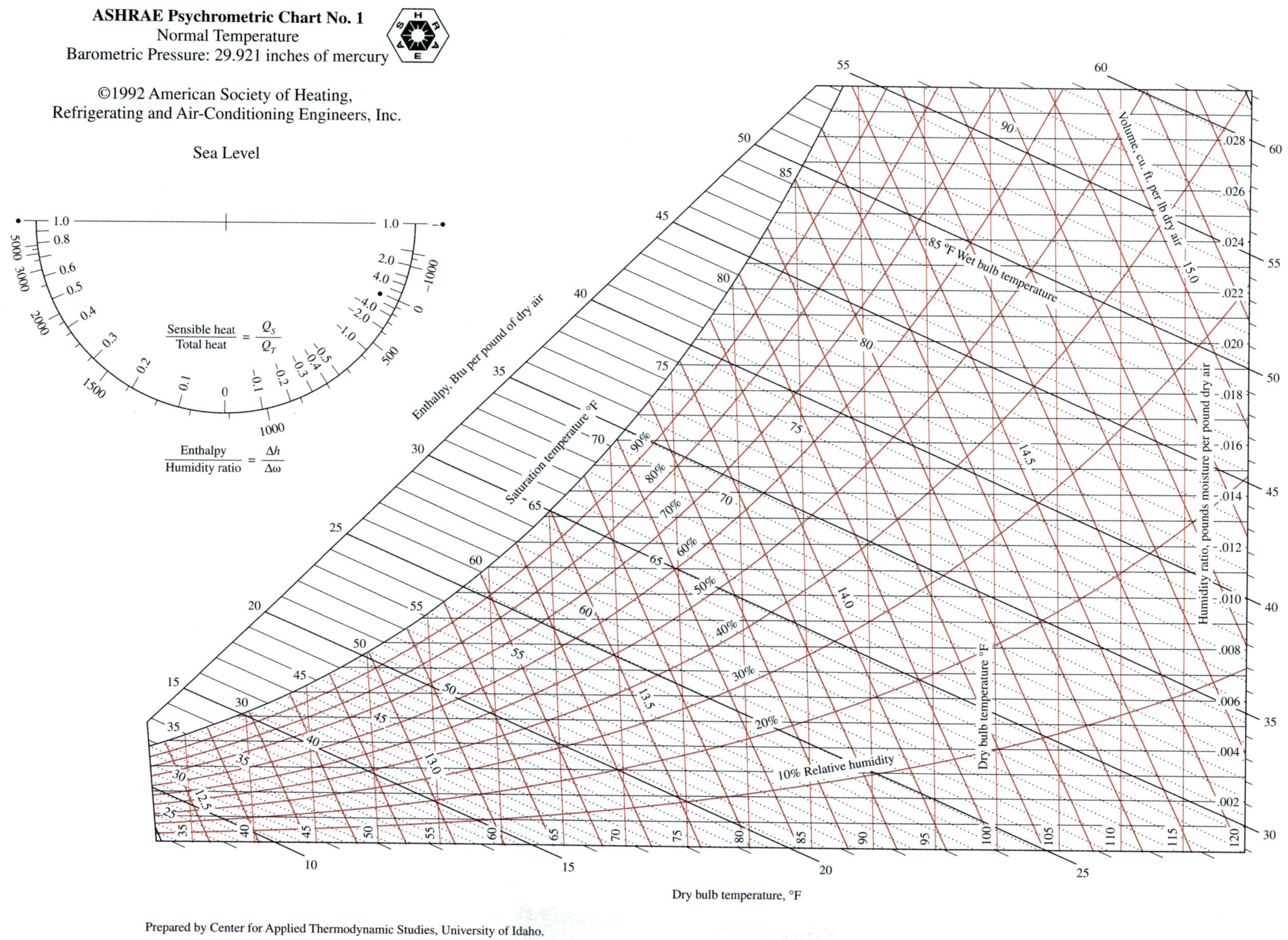

FIGURE A–31E

Psychrometric chart at 1 atm total pressure.

From the American Society of Heating, Refrigerating and Air-Conditioning Engineers, Atlanta, GA: used with permission.

INDEX

A

B

C

D

E

F

G

H

I

N

O

P

Q

R

S

T

U

V

W

X

Z

Conversion Factors

DIMENSION	METRIC	METRIC/ENGLISH
Acceleration	1 m/s^2 = 100 cm/s^2	1 m/s^2 = 3.2808 ft/s^2 1 ft/s^2 = 0.3048* m/s^2
Area	1 m^2 = 10^4 cm^2 = 10^6 mm^2 = 10^{-6} km^2	1 m^2 = 1550 in^2 = 10.764 ft^2 1 ft^2 = 144 in^2 = 0.09290304* m^2
Density	1 g/cm^3 = 1 kg/L = 1000 kg/m^3	1 g/cm^3 = 62.428 lbm/ft^3 = 0.036127 lbm/in^3 1 lbm/in^3 = 1728 lbm/ft^3 1 kg/m^3 = 0.062428 lbm/ft^3
Energy, heat, work, internal energy, enthalpy	1 kJ = 1000 J = 1000 N · m = 1 kPa · m^3 1 kJ/kg = 1000 m^2/s^2 1 kWh = 3600 kJ 1 cal† = 4.184 J 1 IT cal† = 4.1868 J 1 Cal† = 4.1868 kJ	1 kJ = 0.94782 Btu 1 Btu = 1.055056 kJ = 5.40395 psia · ft^3 = 778.169 lbf · ft 1 Btu/lbm = 25,037 ft^2/s^2 = 2.326* kJ/kg 1 kJ/kg = 0.430 Btu/lbm 1 kWh = 3412.14 Btu 1 therm = 10^5 Btu = 1.055 × 10^5 kJ (natural gas)
Force	1 N = 1 kg · m/s^2 = 10^5 dyne 1 kgf = 9.80665 N	1 N = 0.22481 lbf 1 lbf = 32.174 lbm · ft/s^2 = 4.44822 N
Heat flux	1 W/cm^2 = 10^4 W/m^2	1 W/m^2 = 0.3171 Btu/h · ft^2
Heat transfer coefficient	1 W/m^2 · °C = 1 W/m^2 · K	1 W/m^2 · °C = 0.17612 Btu/h · ft^2 · °F
Length	1 m = 100 cm = 1000 mm = 10^6 μm 1 km = 1000 m	1 m = 39.370 in = 3.2808 ft = 1.0926 yd 1 ft = 12 in = 0.3048* m 1 mile = 5280 ft = 1.6093 km 1 in = 2.54* cm
Mass	1 kg = 1000 g 1 metric ton = 1000 kg	1 kg = 2.2046226 lbm 1 lbm = 0.45359237* kg 1 ounce = 28.3495 g 1 slug = 32.174 lbm = 14.5939 kg 1 short ton = 2000 lbm = 907.1847 kg
Power, heat transfer rate	1 W = 1 J/s 1 kW = 1000 W = 1.341 hp 1 hp‡ = 745.7 W	1 kW = 3412.14 Btu/h = 737.56 lbf · ft/s 1 hp = 550 lbf · ft/s = 0.7068 Btu/s = 42.41 Btu/min = 2544.5 Btu/h = 0.74570 kW 1 boiler hp = 33,475 Btu/h 1 Btu/h = 1.055056 kJ/h 1 ton of refrigeration = 200 Btu/min
Pressure	1 Pa = 1 N/m^2 1 kPa = 10^3 Pa = 10^{-3} MPa 1 atm = 101.325 kPa = 1.01325 bars = 760 mm Hg at 0°C = 1.03323 kgf/cm^2 1 mm Hg = 0.1333 kPa	1 Pa = 1.4504 × 10^{-4} psia = 0.020886 lbf/ft^2 1 psi = 144 lbf/ft^2 = 6.894757 kPa 1 atm = 14.696 psia = 29.92 in Hg at 30°F 1 in Hg = 3.387 kPa
Specific heat	1 kJ/kg · °C = 1 kJ/kg · K = 1 J/g · °C	1 Btu/lbm · °F = 4.1868 kJ/kg · °C 1 Btu/lbmol · R = 4.1868 kJ/kmol · K 1 kJ/kg · °C = 0.23885 Btu/lbm · °F = 0.23885 Btu/lbm · R

*Exact conversion factor between metric and English units.

†Calorie is originally defined as the amount of heat needed to raise the temperature of 1 g of water by 1°C, but it varies with temperature. The international steam table (IT) calorie (generally preferred by engineers) is exactly 4.1868 J by definition and corresponds to the specific heat of water at 15°C. The thermochemical calorie (generally preferred by physicists) is exactly 4.184 J by definition and corresponds to the specific heat of water at room temperature. The difference between the two is about 0.06 percent, which is negligible. The capitalized Calorie used by nutritionists is actually a kilocalorie (1000 IT calories).